Algorithms and Computation in Mathematics • Volume 24

Editors

For further volumes:
http://www.springer.com/series/3339

Werner M. Seiler

Involution

The Formal Theory of Differential Equations and its Applications in Computer Algebra

With 18 Figures and 2 Tables

Springer

Werner M. Seiler
Department of Mathematics
University of Kassel
34132 Kassel
Germany
seiler@mathematik.uni-kassel.de

ISSN 1431-1550
ISBN 978-3-642-01286-0 e-ISBN 978-3-642-01287-7
DOI 10.1007/978-3-642-01287-7
Springer Heidelberg Dordrecht London New York

Library of Congress Control Number: 2009932758

Mathematics Subject Classification (2000): 35-02, 35-04, 35A05, 35A30, 35G20, 35N10, 34A09, 58A20, 13-02, 13-04, 13D02, 13N10, 13P10, 68W30

Cover design: deblik, Berlin

Printed on acid-free paper

Springer is part of Springer Science+Business Media (www.springer.com)

To Marion

involution: anything internally complex or intricate

Oxford Advanced Learner's Dictionary of Current English

inuoluere, ML *involvere*, to envelop, roll up, wrap up, whence 'to *involve*'; pp *inuolutus*, ML *involutus*, yields both the tech adj, Geom n, *involute* and the L *inuolutio*, ML o/s *involution* whence E *involution*

Eric Partridge: *Origins: an etymological dictionary of modern English*

Preface

As long as algebra and geometry proceeded along separate paths, their advance was slow and their applications limited. But when these sciences joined company they drew from each other fresh vitality and thenceforward marched on at rapid pace towards perfection

Joseph L. Lagrange

The theory of differential equations is one of the largest fields within mathematics and probably most graduates in mathematics have attended at least one course on differential equations. But differential equations are also of fundamental importance in most applied sciences; whenever a continuous process is modelled mathematically, chances are high that differential equations appear. So it does not surprise that many textbooks exist on both ordinary and partial differential equations. But the huge majority of these books makes an implicit assumption on the structure of the equations: either one deals with scalar equations or with *normal* systems, i. e. with systems in Cauchy–Kovalevskaya form. The main topic of this book is what happens, if this popular assumption is dropped.

This is not just an academic exercise; non-normal systems are ubiquitous in applications. Classical examples include the incompressible Navier–Stokes equations of fluid dynamics, Maxwell's equations of electrodynamics, the Yang–Mills equations of the fundamental gauge theories in modern particle physics or Einstein's equations of general relativity. But also the simulation and control of multibody systems, electrical circuits or chemical reactions lead to non-normal systems of ordinary differential equations, often called differential algebraic equations. In fact, most of the differential equations nowadays encountered by engineers and scientists are probably not normal.

In view of this great importance of non-normal systems, the relative lack of literature on their general theory is all the more surprising. Specific (classes of) systems like the Navier–Stokes equations have been studied in great depth, but the existence of general approaches to non-normal systems seems to be hardly known, although some of them were developed about a century ago! In fact, again and again new attempts have been started for such general theories, in particular for ordinary differential equations where the situation is comparatively straightforward. Classical examples are the Dirac theory of mechanical systems with constraints and the currently fairly popular numerical analysis of differential algebraic equations. However, in both cases researchers have had to learn (sometimes the hard way) that the generalisation to partial differential equations is far from trivial, as new phenomena emerge requiring new techniques.

There are probably many reasons for this rather unfortunate situation. One is surely that the treatment of general systems requires fairly sophisticated tools from differential geometry and algebra which at least practitioners in differential equations are usually not familiar with. Another serious problem is the available literature. Many of the (few) authors in this domain seem to be determined to found a "school." They go to great length to reinvent any piece of mathematics they use and thus effectively create their own language with the result that their works are more or less unreadable for most mathematicians. Furthermore, as in many other fields of science, there is a certain tendency to "religious wars" between the different schools.

While writing this book, I tried to refrain from such ambitions and not to take sides (although experts will easily notice a strong influence by the work of Pommaret which is not surprising, as I made my first steps in this domain under the guidance of his books and lectures). I believe that all approaches contain useful ideas and on closer inspection I have found that the similarities between the various theories are usually much greater than the differences. One of my central goals was to build a coherent framework using only standard notations and terminology in order to make this book as accessible as possible. The general theory of differential equations requires sufficiently much digestion time even without being hidden behind bizarre non-standard constructions.

Another important goal of mine was to strike a reasonable balance between elucidating the underlying theoretical structures and the development of effective algorithms for concrete computations. More applied mathematicians may find some chapters overly abstract. However, over the years I have made the experience that it is highly rewarding and fruitful to study even purely computational problems from a higher theoretical level. One gains sometimes surprising insights opening completely new solution paths.

The main topic of this book is usually called the *formal theory of differential equations*. It combines geometric and algebraic structures into a powerful framework useful not only for theoretical analysis but also for concrete applications. In this context, the adjective "formal" has at least two different meanings. A simple one is that the origin of the theory was the problem of determining formal power series solutions for general systems of differential equations, i. e. of proving the formal integrability. In a broader sense, the word "formal" refers to the fact that we nowhere work with explicit solutions; all analysis is solely based on manipulations of the equations themselves.

Oversimplifying, one could say that most of the works on differential equations up to the early 20th century are of a more formal nature, whereas currently the functional analytic approach dominates the literature. It is often overlooked that in fact formal and functional analytic methods are complementary to each other and that an in-depth treatment of a general system of differential equations will need both. Typically, one starts with a formal analysis, in particular one will first assert the existence of formal solutions, as otherwise there is no point in proceeding any further. Only for normal systems one can skip this step, as it is trivial.

An important part of the formal analysis consists of completing the system to an involutive one. This includes in particular the addition of all hidden integrability conditions. Only after the completion one has a well defined starting point for functional analytic or numerical methods; thus one may consider the formal analysis as a "preconditioning" for the latter. As the title clearly indicates, the notion of involution takes a central place in the book. Indeed, in my opinion *involution is the central principle in the theory of under- or overdetermined systems.*

One may compare the completion to involution of a differential equation with rendering a linear system of equations triangular, say by Gaussian elimination. Most properties of the system, like for example its solvability and the dimension of its solution space, become apparent only after such a transformation. The same is true for differential equations: only after a completion one can decide about the consistency of the system and make statements about the size of the formal solution space.

Purely formal methods will not lead much further; under certain assumptions one can proceed to prove the convergence of the formal solutions and thus obtains an analytic solution theory. However, an applied mathematician will immediately object that such a theory is not nearly sufficient for any practical purposes. Thus the addition of other methods, in particular from functional analysis, is required for further progress. Nevertheless, in this book we will almost exclusively be concerned with the formal analysis of differential equations. One reason is that this alone represents already a fairly demanding task. Another point is the fact that the connection of formal and functional analytic or numerical methods has not been much studied yet. Even many fairly fundamental questions are still unanswered so that aspiring graduate students can find here a wide open field.

As already mentioned, the formal theory relies mainly on a combination of algebraic and geometric techniques. A geometric formalism, the jet bundles, form the foundation. It leads in a natural manner to questions in commutative (not differential!) algebra. Actually, a deeper understanding of involutive systems can be only obtained by taking a closer look at the arising algebraic structures. While the importance of abstract algebra has been evident at the latest since the seminal work of Spencer in 1960s, the formal theory has developed completely independently of commutative algebra. For example, the almost trivial fact that the notion of (the degree of) involution of a differential equation is equivalent to the algebraic notion of Castelnuovo–Mumford regularity of an associated polynomial module has been overlooked until fairly recently.

It is another goal of this book to clarify these relations between differential equations and commutative algebra. For this reason the book contains several chapters dealing exclusively with commutative algebra. These chapters should also be of independent interest to (computer) algebraists, as some of ideas originally developed for the formal analysis of differential equations are quite powerful and considerably simplify many computations with arbitrary polynomial modules.

Unfortunately, the term "involution" is used in different approaches with somewhat different meanings. In fact, even within this book I will use it sometimes in a broader and sometimes in a narrower meaning. In the algebraic theory a fairly general notion of involutive bases of polynomial ideals will be introduced; but later only the special case of Pommaret bases will be used. In the context of differential equations involution is often confused with formal integrability, i. e. with the mere absence of integrability conditions. But while formal integrability is a necessary condition for involution, the latter requires more and I will stress at a number of occasions the importance of this "more".

The book is divided into ten chapters the contents of which I now briefly describe. The first chapter is a short general introduction into the kind of problems treated in the book. It demonstrates the need for treating more general types of differential equations and explains the relationship to some classical algebraic problems.

The second chapter is concerned with the geometry behind the formal theory. It introduces the jet bundle formalism as the geometric foundation of differential equations. In fact, it contains two different introductions of jet bundles. The "pedestrian" one is based on Taylor series and puts more emphasis on local coordinates. The "intrinsic" one uses an abstract construction stressing the affine structure of jet bundles; it does not often appear in the literature but turns out to be rather natural for the formal theory. Then differential equations are intrinsically defined as fibred submanifolds of jet bundles and the basic geometric operations with them, prolongation and projection, are introduced. This leads to an intrinsic picture of integrability conditions and the notion of formal integrability.

While formal integrability is a very natural and intuitive concept, it turns out that for many purposes it does not suffice. The first problem is already to decide effectively whether or not a given differential equation is formally integrable. But it will become evident at several places that it has further shortcomings; for example even in the analytic category an existence *and* uniqueness theorem can be proven only for involutive equations.

In order to resolve these difficulties one must resort to algebraic methods. So the third chapter introduces our main theme—involution—in a purely algebraic framework which, at first sight, is not at all related to differential equations. Involutive bases are introduced for the Abelian monoid $\mathbb{N}_0^n$ and then extended to a fairly general class of rings: the polynomial algebras of solvable type. This approach has the advantage that it does not only allow us to extract the simple combinatorial idea underlying involution in a very clear form, it may also be applied without modifications to many different situations: the classical polynomial ring, rings of differential operators, universal enveloping algebras of Lie algebras, quantum algebras,... Involutive bases are a special form of the familiar Gröbner bases which have become a central algorithmic tool in computer algebra and therefore much of their theory is modelled on the theory of Gröbner bases.

As a kind of interlude, the fourth chapter discusses the computational side of involutive bases. It presents concrete algorithms for their determination. As any involutive basis is simultaneously a Gröbner basis, we obtain here alternatives to the famous Buchberger algorithm and its variants. Benchmarks have shown that the involutive approach is often highly competitive in terms of the computation times. Compared with the classical Gröbner basis theory, the question of termination becomes much more subtle. This is in particular true for the bases of greatest interest to us, the Pommaret bases which will later be used for a constructive definition of involution for differential equations.

This problem of the existence of a Pommaret basis (or more generally of effectively deciding involution) is known under the name δ-regularity and appears in many different disguises in any coordinate based approach to involution. It is often considered as a purely technical nuisance and thus a reason to use other types of involutive bases instead of Pommaret bases. However, later results will show that the "problem" of δ-regularity is in fact a highly useful *feature* of Pommaret bases and, for example, related to Noether normalisation or characteristics in the context of differential equations. Furthermore, a simple and effective solution based on a cheap criterion for δ-singular coordinates exists and can be easily incorporated into completion algorithms.

Instead of an immediate application of the developed algebraic tools to differential equations, the fifth chapter probes deeper into the theory of involutive bases and shows that the fundamental idea behind them consists of the determination of combinatorial decompositions for polynomial modules. Involutive bases unite these decompositions with Gröbner bases. Although quite obvious, this point of view is still fairly new (in commutative algebra, for differential equations it was already exploited by Riquier and Janet at the beginning of the 20th century).

Pommaret bases are not only important for differential equations, but also define a special type of decomposition, a Rees decomposition. The main topic of the fifth chapter is to show that this fact makes them a very powerful tool for computational algebraic geometry. Most of these applications exploit that Pommaret bases possess a highly interesting syzygy theory. For example, they allow for directly reading off the depth, the projective dimension and the Castelnuovo–Mumford regularity of a module and thus for simple constructive proofs of both Hilbert's Syzygy Theorem and the Auslander–Buchsbaum Formula. In addition, they provide a simple realisation of Hironaka's criterion for Cohen-Macaulay modules.

Of course, these results makes one wonder why Pommaret bases are so special compared with other involutive bases. It seems that the answer lies in homological algebra. All invariants that are easy to read off a Pommaret basis are of a homological origin. Hence it is not surprising that Pommaret bases possess a homological interpretation which is the theme of the sixth chapter. It starts with studying the Spencer cohomology and the dual Koszul homology. Then algebraic versions of the classical Cartan test for involution from the theory of exterior differential systems are given. Finally, the relationship between Pommaret bases and these homological constructions is studied in detail.

The seventh chapter returns to differential equations and applies the developed algebraic theory to the analysis of symbols. The (geometric) symbol induces at each point of the differential equation a polynomial (co)module and one may now use either the Spencer cohomology or, in local coordinates, Pommaret bases for defining involutive differential equations. An important structural property of such equations is the existence of a Cartan normal form for local representations. As a first simple application of involution, a rigorous definition of under- and overdetermined equations is given with the help of the principal symbol. This classification makes sense only for involutive equations and it turns out that the classical counting rules comparing the number of equations and unknowns may be misleading.

Rather surprisingly, it seems to be difficult to find a satisfactory treatment of this elementary and natural question in the literature.

We also discuss now the question of rendering an arbitrary differential equation involutive. For ordinary differential equations it is answered by a simple geometric procedure which has been rediscovered many times in different fields; for partial differential equation an answer is provided by the Cartan–Kuranishi completion. In principle, we cover in both cases equations with arbitrary nonlinearities. However, in practice certain steps are hard to perform effectively. In the case of polynomial nonlinearities one can always resort to Gröbner bases techniques, though possibly at the cost of a prohibitively high complexity. Furthermore, in nonlinear equations one must always expect the emergence of singularities, a problem which requires expensive case distinctions and which we will mostly ignore.

The eighth chapter is devoted to some "combinatorial games": abstract measures for the size of the formal solution space like the Hilbert polynomial or the Cartan characters. One may argue about their usefulness, but some classical notions like the number of degrees of freedom of a physical system are in fact based on such "games." The concept of a differential relation between two differential equations generalising Bäcklund transformations is introduced and the induced relation between the respective Hilbert polynomials is studied. For applications in physics, it is often of interest to formally subtract the effect of gauge symmetries. Within the framework of the formal theory, a fairly simple pseudogroup approach to this question exists. Einstein introduced for similar purposes the strength of a differential equation which can be easily related to the formal theory.

A central question in the theory of differential equations is of course the existence and uniqueness of solutions. Since jets may be understood as an intrinsic version of Taylor series, it does not surprise that the simplest results are obtained for analytic solutions. The ninth chapter starts by recalling the Cauchy–Kovalevskaya Theorem as the most general (with respect to the structure of the equation) existence and uniqueness result for analytic solutions of analytic normal systems.

One could try to directly extend its proof to involutive systems. Indeed, this approach is taken in the Janet–Riquier Theory leading to Riquier's theorem on the existence and uniqueness of analytic solutions of involutive[1] systems (here involution is understood in the broader sense of involutive bases). A more flexible approach leads to the Cartan–Kähler Theorem for arbitrary involutive differential equations. Its proof is based on considering a sequence of normal systems to which the Cauchy–Kovalevskaya Theorem is applied; thus power series appear only implicitly in contrast to the proof of Riquier's theorem. The proof also clearly demonstrates why formal integrability is not sufficient for the analysis of differential equations but the crucial tool of the Cartan normal form of an involutive equation is necessary.

Because of the use of the Cauchy–Kovalevskaya Theorem, the Cartan–Kähler Theorem inherits the restriction to the analytic category and this fact seriously limits its practical value. However, if more is known about the structure of the equation, then the presented technique of proof can sometimes be used for obtaining

[1] In the Janet–Riquier Theory one usually speaks of *passive* systems instead of involutive ones.

stronger existence and/or uniqueness results in function spaces of relevance for applied mathematics. This will be the case, if it is possible to substitute the Cauchy–Kovalevskaya Theorem by other results from the extensive literature on existence and uniqueness for normal systems. Two examples of such generalisations concerning linear systems of differential equations will appear in the tenth chapter.

Vessiot developed a dual version of Cartan's theory of exterior differential systems based on a distribution defined over the differential equation. Unfortunately, it has been largely ignored in the literature so far, although it appears to be highly useful for the geometric analysis of differential equations. The remainder of the ninth chapter is devoted to a new presentation of his approach. First of all, it is better embedded into the geometry of jet bundles (in particular the contact structure) which leads to some simplifications. Then it is shown that Vessiot's construction succeeds only for involutive equations (again formal integrability is not sufficient).

The tenth chapter specialises to linear systems of differential equations. It starts with some elementary geometric aspects of the linearity. After studying an extension of the Holmgren Theorem to arbitrary involutive equations and overdetermined elliptic and hyperbolic systems, we consider linear systems from a more algebraic point of view. Firstly, an introduction to basic algebraic analysis is given. Then it is shown how the Cartan–Kuranishi and the involutive completion algorithm may be merged into a new algebraic algorithm that is able to deliver intrinsic geometric results. Compared with straightforward realisations of the Cartan–Kuranishi completion, it is much faster, as due to a clever book-keeping much less prolongations must be computed. Finally, the inverse syzygy problem and the integration of linear systems with constant coefficients of finite type are studied.

As one can see, the book jumps between geometric and algebraic approaches treating quite diverse topics. Somebody reading the whole text will therefore need a certain familiarity with jet bundles and differential geometry on one side and commutative and homological algebra on the other side. Not all readers will have the necessary background or actually be interested in all these topics. This fact has been taken into account in a two-fold way.

Two fairly long appendices collect some basic results from algebra and differential geometry that are used in the main text; in particular, an introduction into the theory of Gröbner bases is given. These appendices try to make the book to a reasonable degree self-contained, but of course they cannot substitute real introductions into the subjects involved. Therefore additional references to standard textbooks are given. I have also included proofs for some of the mentioned results—either because the used techniques also appear elsewhere in the main text or for a comparison with alternative approaches developed in this book.

Many chapters and sections are rather independent of each other. Depending on the interests of a reader many parts may be safely ignored. For example, somebody interested exclusively in the algebraic theory of involutive bases needs only Chapters 3–6. On the other hand, somebody who does not like commutative algebra may ignore precisely these parts, as the most important results needed for the analysis of differential equations reappear in Chapter 7 in an alternative formulation. For the benefit of readers not so familiar with differential geometry most geometric constructions are described both in an intrinsic manner and in local coordinates. Furthermore, many concrete examples are included to illustrate the theoretical results.

Each chapter ends with a brief section called *Notes*. It usually gives some pointers to the literature and to alternative approaches. I have also tried to trace a bit the history of the main ideas. But as I am not a historian of science, the results should be more taken as rough indications than as a definite answer. Furthermore, in several places I rely on other sources and I have not always checked the correctness of the references. A valuable source of information on the state of the art of the theory of differential equations at the end of the 19th century is the article of von Weber [475] in the *Enzyklopädie der Mathematischen Wissenschaften*.

This book evolved out of my Habilitation thesis [406] admitted by Universität Mannheim in 2002. It is, however, not identical with it; in fact, the book has about twice the size of the thesis. Once I had started with some "minor" changes for the publication of the thesis, I could not resist the temptation to include all the things I left out of the thesis for lack of time (on the other hand I excluded the chapter on numerical analysis, as I am not yet satisfied with the status of this part of the theory). In particular, the material on the Vessiot theory and on the homological theory of involutive bases are completely new. In the end, these "minor" changes lead to a delay of several years in the publication and I am very glad that Springer–Verlag, in particular in the person of Ruth Allewelt and Martin Peters, accepted this delay without complaints.

Finally, I would like to thank some people who either helped me to get a deeper understanding of involution and its applications or who read some version of the manuscript and gave me valuable comments. This list includes in particular (in alphabetic order) Ernst Binz, Dirk Fesser, Vladimir P. Gerdt, Ulrich Oberst, Peter J. Olver, Jean-François Pommaret, Alban Quadrat, Julio Rubio, Robin W. Tucker, Jukka Tuomela, Peter J. Vassiliou and Eva Zerz. Furthermore, the editors of the *Algorithms and Computation in Mathematics* series gave me useful feedback. Special thanks are deserved by Marcus Hausdorf who has been my collaborator on most of the topics covered here and who has implemented large parts of the MuPAD code used in concrete computations. Concerning this last point it is also a pleasure to acknowledge the cooperation of Benno Fuchssteiner and the whole MuPAD group in Paderborn over many years.

Over the years, much of my work has been financially supported by Deutsche Forschungsgemeinschaft under various grants. Additional financial support was obtained by two European projects: INTAS grant 99-01222 (*Involutive Systems of Differential and Algebraic Equations*) and NEST-Adventure grant 5006 (*Global Integrability of Field Theories*).

But my deepest gratitude goes to my wife Marion, patiently bearing the life with a scientist. Only her love and support has made it all possible.

Karlsruhe/Kassel, June 2009 *Werner M. Seiler*

Contents

List of Algorithms

Chapter 1
Introduction

One should always generalise.

Carl Jacobi

In this book we will be mainly concerned with the analysis of "general" systems of differential equations. As there are many ways how one might interpret this "generality", we should start by clarifying what we mean by this term. The first differential equation any student of mathematics (or of some natural or engineering sciences) encounters is most probably a scalar ordinary differential equation of the form $u' = \phi(x,u)$ where $u = u(x)$ is the unknown function to be determined.

Entering the theory of such equations, one rapidly notices that the restriction to scalar equations is unnecessary at most places. Almost all proofs remain valid, if we generalise to systems of the form $\mathbf{u}' = \boldsymbol{\phi}(x,\mathbf{u})$ where now $\mathbf{u} = (u^1,\dots,u^m)$ is a vector of unknown functions.[1] A crucial implication of the assumed form is that our system contains as many equations as unknowns. Under rather modest assumptions on the function $\boldsymbol{\phi}$ (e. g. Lipschitz continuity), the classical existence and uniqueness theorems assert that such systems are—at least locally—uniquely solvable for arbitrary choices of the initial data $\mathbf{u}(0) = \mathbf{u}_0$ with a constant vector $\mathbf{u}_0 \in \mathbb{R}^m$. Extending a terminology usually only applied to partial differential equations, we call systems of the above form *normal*.

While the majority of analytical or numerical approaches to differential equations is exclusively concerned with normal systems, many applications lead to a more general form, namely *implicit systems* $\boldsymbol{\Phi}(x,\mathbf{u},\mathbf{u}') = 0$. If we assume that the Jacobian $\partial\boldsymbol{\Phi}/\partial\mathbf{u}'$ is everywhere regular, then the Implicit Function Theorem tells us that any such system is at least locally equivalent to a normal one (whether one can effectively transform it into solved form is a very different question, but theoretically it is always possible). Obviously, a necessary condition for regularity is that the Jacobian is a square matrix, i. e. that there are as many equations as unknown functions contained in the system. After the transformation we can again apply the familiar existence and uniqueness theorems.

If either the Jacobian is no longer a square matrix or its rank is not equal to the number of equations, then our system cannot even theoretically be transformed into

[1] We follow throughout this book the convention in differential geometry that vector components are indexed by a superscript.

W.M. Seiler, *Involution*, Algorithms and Computation in Mathematics 24,
DOI 10.1007/978-3-642-01287-7_1,

a normal one and new phenomena appear. For example, the question of solvability becomes less trivial. In an attempt to move as closely as possible to a normal system, we partition the unknown functions into two subsets $\mathbf{u} = (\mathbf{v}, \mathbf{w})$ such that already the partial Jacobian $\partial \boldsymbol{\Phi} / \partial \mathbf{v}'$ has the same rank as the full one. Applying the Implicit Function Theorem leads now to the *semi-explicit form*

$$\mathbf{v}' = \boldsymbol{\phi}(x, \mathbf{v}, \mathbf{w}, \mathbf{w}') , \tag{1.1a}$$

$$0 = \boldsymbol{\psi}(x, \mathbf{v}, \mathbf{w}) . \tag{1.1b}$$

Here we separated the system into a differential part (1.1a) and a purely algebraic part (1.1b) where no derivatives appear. For this reason, the term *differential algebraic equation* (DAE for short) has become popular for such general systems of ordinary differential equations.

Note that while the differential part (1.1a) obviously consists of as many equations as there are unknowns $\mathbf{v}$, we cannot generally require that similarly the size of the algebraic part (1.1b) is given by the number of unknowns $\mathbf{w}$. If this is the case and additionally the Jacobian $\partial \boldsymbol{\psi} / \partial \mathbf{w}$ is regular, then (again by the Implicit Function Theorem) we may solve (1.1b) for $\mathbf{w}$ and enter the result into (1.1a). This yields a normal system for $\mathbf{v}$ alone.

However, it is also possible that the algebraic part is empty. In this case we are dealing with an underdetermined system, as obviously the functions $\mathbf{w}$ may be chosen arbitrarily in the solution. Alternatively, one may interpret $\mathbf{w}$ now as parameters and considers then again (1.1a) as a normal system for $\mathbf{v}$ alone.

In between these two extremal cases many different possibilities exist. We will see in Section 7.3 that—under some regularity assumptions—*any* system of ordinary differential equations can ultimately be reduced to a combination of these two cases or the system is inconsistent (and thus has no solution at all). The basis for this reduction is a process called *completion to involution*.

The main purpose of the completion of a differential system is to exhibit all hidden *integrability conditions*. Indeed, differentiating (1.1b) yields

$$\frac{\partial \boldsymbol{\psi}}{\partial x} + \frac{\partial \boldsymbol{\psi}}{\partial \mathbf{v}} \mathbf{v}' + \frac{\partial \boldsymbol{\psi}}{\partial \mathbf{w}} \mathbf{w}' = 0 . \tag{1.2}$$

These equations may still be simplified modulo the original ones (1.1), in particular we can eliminate $\mathbf{v}'$ using (1.1a), and it could happen that they vanish identically. However, in general we obtain additional equations. Depending on the properties of the Jacobian $\partial \boldsymbol{\psi} / \partial \mathbf{w}$, these are either algebraic or differential equations. In the former case differentiation of the new algebraic equations may lead to further integrability conditions and so on.

The terminology "integrability conditions" is sometimes misinterpreted: these equations are *not* additional conditions imposed on solutions and do not restrict the solution space, as they are automatically satisfied by any solution of (1.1). Their importance stems from the following fact: before we have constructed all integrability conditions, we cannot make any statements about the existence of solutions, as it is always possible that we obtain during some completion step an equation of the form $1 = 0$ implying that the original system is inconsistent. Furthermore, we need the integrability conditions in order to be able to choose consistent initial data. Suppose that we want to have $\mathbf{v}(0) = \mathbf{v}_0$ and $\mathbf{w}(0) = \mathbf{w}_0$. Then an obvious necessary condition for the existence of a solution is $\boldsymbol{\psi}(0, \mathbf{v}_0, \mathbf{w}_0) = 0$. But this condition is not sufficient, if (algebraic) integrability conditions are hidden in (1.1).

The completion process should be seen in analogy to the transformation of a linear system of equations into triangular form, say by Gaussian elimination. Given some linear system, it is a priori not possible to make any statements about the existence of solutions or the dimension of the solution space. Only after we have given the system a better behaved form by eliminating redundant equations and checking for inconsistencies, we can read off such properties. In the case of differential equations we may not only perform algebraic operations but also differentiate equations. Therefore the situation becomes more complicated, as new equations may show up. But the basic idea that the system must be first brought into a form more amenable to analysis remains the same.

For partial differential equations the situation is similar. Much of the classical theory is concerned with scalar equations, as here the extension to systems is more difficult. Traditionally, one only considers systems where a distinguished independent variable t (the remaining independent variables are collectively denoted by $\mathbf{x}$) exists such that the system may be written in the form

$$\mathbf{u}_t = \boldsymbol{\phi}(t, \mathbf{x}, \mathbf{u}, \mathbf{u}_{\mathbf{x}}) \ . \tag{1.3}$$

As this form is crucial for the famous existence and uniqueness theorem of Cauchy–Kovalevskaya which we will discuss in Section 9.2, one speaks of *systems in Cauchy–Kovalevskaya form* or more briefly of *normal systems*. It is now easy to see why we extended the terminology "normal" to ordinary differential equations in solved form: as the choice of the letter t for the distinguished variable already indicates, one may consider (1.3) as an evolution problem[2] and thus as a kind of abstract ordinary differential equation living on an infinite-dimensional space. This point of view is e. g. the foundation of the semigroup theory of evolution equations.

Of course, the Cauchy–Kovalevskaya form (1.3) does not represent the most general form of a first-order system of partial differential equations: again we must admit implicit systems $\boldsymbol{\Phi}(t, \mathbf{x}, \mathbf{u}, \mathbf{u}_t, \mathbf{u}_{\mathbf{x}}) = 0$ and these are equivalent to a normal one, if and only if—possibly after a transformation of the independent variables $(t, \mathbf{x})$—the Jacobian $\partial\boldsymbol{\Phi}/\partial\mathbf{u}_t$ is regular. Above we saw that (under some regularity assumptions) every system of ordinary differential equations can be transformed into the semi-explicit form (1.1). If we neglect the special case of an empty algebraic part (1.1b), we may say that the characteristic feature of non-normal ordinary differential systems is the presence of equations of differing order. For partial differential

[2] Note that even if a system is in Cauchy–Kovalevskaya form, this fact does not automatically imply that it actually makes sense to treat it as an evolution equation and thus to study its initial value problem. The Laplace equation is as a scalar equation for one unknown function trivially normal, but the initial value problem for it is not well-posed (see Remark 9.3.2). All considerations here are of a purely formal nature.

systems we can similarly deviate from the normal form (1.3) by adding equations of lower order. However, further possibilities exist and are in fact more common.

A straightforward generalisation of the semi-explicit form (1.1) to partial differential equations is a system of the form

$$\mathbf{v}_t = \boldsymbol{\phi}(t, \mathbf{x}, \mathbf{v}, \mathbf{w}, \mathbf{v}_{\mathbf{x}}, \mathbf{w}_{\mathbf{x}}, \mathbf{w}_t) , \tag{1.4a}$$

$$0 = \boldsymbol{\psi}(t, \mathbf{x}, \mathbf{v}, \mathbf{w}, \mathbf{v}_{\mathbf{x}}, \mathbf{w}_{\mathbf{x}}) . \tag{1.4b}$$

Note that in general the equations in (1.4b) are still first-order partial differential equations and not algebraic equations. But they are of lower order with respect to t-derivatives. We will later formalise this idea by saying that the equations in (1.4b) are of lower *class* than those in (1.4a). One sometimes calls such non-normal systems *partial differential algebraic equations* (PDAE), but while such a terminology surely makes sense in the ordinary case (1.1), it seems inappropriate for (1.4), as in general no algebraic equations are present.

If the vector $\mathbf{x}$ contains more than one variable, the game does not end with the form (1.4). We may now distinguish a further independent variable, say y, and denote the remaining ones by $\mathbf{z}$. Then we solve as many equations as possible in (1.4b) for y-derivatives and so on. This process leads to a kind of echelon form of the original system where each "step" consists of equations which are solved for derivatives of one particular independent variables. One speaks of a *Cartan normal form*. A rigorous description of it is rather messy and we defer this topic to Chapter 7.

Above we have seen how for a non-normal system of ordinary differential equations in the form (1.1) the differentiation of the algebraic equations (1.1b) may lead to integrability conditions. Similarly, if a system of partial differential equations contains lower-order equations, then differentiations of these with respect to all independent variables may lead to new equations. However, it is well-known that a second mechanism for the generation of integrability conditions exists: cross-differentiations. Looking at the "semi-explicit" form (1.4), it is obvious that differentiating (1.4b) with respect to t may yield new equations even if the right hand side is of first order. In contrast to the situation by ordinary differential equations, the construction of these new equations takes place in higher order and also the arising integrability conditions can be of higher order than the original system.

In Chapters 4 and 7 we will discuss the explicit construction of integrability conditions in detail. For the moment it suffices to say that for systems of partial differential equations the problem consists not so much of finding *some* integrability condition but of checking whether one has already obtained *all*. It is a priori not even clear that the completion process terminates after a finite number of steps. We will need an algebraic (Noetherian) argument for proving that this is indeed always the case. A system without hidden integrability conditions is called *formally integrable*. As we will see in Section 2.3, this terminology refers to the fact that for such systems it is always possible to construct formal power series solutions in a systematic order-by-order manner (statements about strong solutions are more difficult and will be the topic of Chapter 9).

In order to avoid the explicit use of power series, we will use a geometric approach to differential equations detailed in the next chapter and define them as subspaces in *jet bundles*. At first sight this may appear as an unnecessary abstraction, but it will turn out that this point of view has many advantages. The intrinsic structures present in jet bundles make many constructions very natural and transparent. Furthermore, badly chosen coordinates may lead to artefacts like apparent singularities. One should also note that obviously the properties of any mathematical model of a physical system should be independent of the used coordinates; hence using an intrinsic coordinate-free language for the model is only natural—even if in the end one may resort to local coordinates for concrete computations.

While completion is a comparatively straightforward process for ordinary differential equations requiring only some simple geometric considerations, the situation becomes much more involved for partial differential equations. One basic problem is that integrability conditions may arise in higher order and a priori we do not know a bound (at least not a reasonable one) for the order up to which we have to go. In particular, purely geometric reasoning will not suffice; we must add quite some algebraic theory. In fact, it will be a central theme of this book to explain why the very intuitive notion of formal integrability is often not sufficient and we must introduce the stronger but more abstract concept of *involution*.

Involution is basically a purely algebraic notion. We will see that a natural polynomial structure lies hidden in the inner geometry of the jet bundle. It allows us to associate with any system of differential equation a module, its *symbol module*, over a polynomial ring in as many variables as there are independent variables. The intuitive idea of constructing integrability conditions by taking cross-derivatives can then be formalised via the syzygies of this module. Thus we are automatically lead to classical questions in commutative algebra and algebraic geometry which are of independent interest.

Effective computations with polynomials modules are nowadays almost always done with Gröbner bases (see Appendix B.4 for a brief introduction). However, instead of simply applying such techniques to the symbol module, we will first exploit ideas originating from the Janet–Riquier Theory (one of the oldest approaches to the completion of general systems of partial differential equations) for developing in Chapter 3 a refined version of Gröbner bases: the *involutive bases* possessing additional combinatorial properties. A special kind of them, the *Pommaret bases*, turns out to be particularly useful for many computational tasks in commutative algebra and algebraic geometry; such computations will be the topic of Chapter 5.

One may summarise the main difference between Pommaret bases and arbitrary Gröbner bases that the latter ones take into account only the first syzygy module (most clearly visible in the Schreyer Theorem) whereas the former ones contain information about the full syzygy resolution of the given polynomial module. In fact we will see that while different Gröbner bases for the same module may look very different, the Pommaret basis is to a high degree determined by structural properties of the module.

A simple computational definition of involution based on Pommaret bases suffers from an unpleasant defect. Although we mentioned above, how important it is to work in an intrinsic manner, the construction of the symbol module as a polynomial module requires the introduction of local coordinates and it turns out that in certain "bad" coordinates no finite Pommaret basis exists. This phenomenon is known as the problem of δ-regularity; it will appear repeatedly in this book and we will also discuss several possible solutions for it in concrete computations.

For a fully intrinsic definition of involution we will apply homological algebra in Chapter 6. The symbol of a differential equation induces a restriction of the polynomial de Rham complex which leads in general to a non-trivial cohomology, the *Spencer cohomology* of the differential equation. One can then relate involution to the finiteness of this cohomology. This approach is dual to a very classical construction in commutative algebra, namely the *Koszul homology*. For some reasons, the differential equations community seems to prefer the Spencer cohomology. But at least for computational purposes it is often better and easier to work with the Koszul homology. In particular, this allows us to make contact with results in commutative algebra. As a trivial by-product we will for example see that the degree of involution is nothing but the Castelnuovo–Mumford regularity, a well-known homological invariant in commutative algebra.

Of course, it is legitimate to ask whether one really has to deal with non-normal systems. It emerged already from our superficial discussion so far that we must be prepared to face quite some new phenomena and problems compared with normal systems and we will need a considerable amount of non-trivial algebraic and geometric theory in order to overcome these. Thus the question arises whether non-normal systems are only studied for the sake of generalisation or whether they really appear in applications?

The answer is that in fact the importance of such systems in applications cannot be overestimated. Arguably non-normal systems are even more common than normal ones. Modern approaches to the mathematical modelling of all kinds of engineering problems are usually modular due to the complexity of the systems to be analysed. The interconnection between the different submodels is then realised by *constraints*. These additional equations lead automatically to an overdetermination and thus to non-normal systems.

Mechanical systems are probably one of the largest sources of differential algebraic equations. Traditionally, one tried to explicitly solve the constraints in order to reduce to a normal system (the *state space form*). One reason was the lack of reliable numerical methods for differential algebraic equations. In the last years not only such methods but also ready-to-use software packages have been developed, so that it is now much more common to treat directly the original differential algebraic equations; in particular, because this approach is often more efficient even if one can explicitly solve the constraints, as the arising state space form may contain very complicated expressions.

Similarly, the simulation of electrical networks and many control problems like prescribed path problems lead naturally to differential algebraic equations. Another source is the semi-discretisation of partial differential equations by the method of lines which also often yields differential algebraic equations. The textbook [53] contains many concrete examples (with references).

For partial differential equations the situation is not so easy to survey, as in this field it is still much more common to treat specific equations than to design general theories. Nevertheless, non-normal systems abound here, too. For example, all fundamental interactions in physics are nowadays described by *gauge theories*. As we will see in Chapter 8, due to the existence of a gauge symmetry, the field equations of such a theory are always non-normal.

Because of this great importance of non-normal systems in fundamental physics, theoretical physicists have even developed their own theory of such systems—the (Bergmann-)Dirac theory of constrained Hamiltonian systems. By an unfortunate tradition founded already by Dirac himself, this theory is often developed only for finite-dimensional systems, i. e. ordinary differential equations, but then applied without modifications to field theories. It is not difficult to produce examples where such approaches fail or at the very least run into trouble.

Gauge theories also nicely demonstrate that (although one can find them at many places in the literature) the simple counting rules of linear algebra—a system is well-determined, if it has as many equations as unknowns—are *not* true for differential equations: in a gauge theory one has as many equations as unknown functions, nevertheless the system is underdetermined. This has nothing to do with dependencies between the equations but is directly related to the symmetry which allows that at least one field component can be chosen arbitrarily and thus is not restricted by the field equations. A rigorous definition of under- and overdeterminacy for differential equations will be the topic of Section 7.5.

In continuum mechanics one often considers incompressible media; the incompressibility condition makes the field equations non-normal. The incompressible Navier–Stokes equations describing fluid motion are an example of a system which is not even formally integrable; cross-derivatives lead to a Poisson equation for the pressure. This integrability condition is of great importance for the numerical integration of the Navier–Stokes equations in their standard form: without it neither finite elements nor finite differences (or any other approach) lead to a closed system of equations. In fact, a huge part of the computing time required by a numerical integration is usually spent on the solution of this Poisson equation!

In Section 2.4 we will briefly discuss a number of fundamental differential systems from physics and engineering that are not normal. This list includes in particular Maxwell's equations underlying all electromagnetic processes and the already mentioned incompressible Navier–Stokes equations. These examples will be taken up repeatedly throughout the book in order to demonstrate the concrete application of the developed theory.

Historically, one of the most important sources of overdetermined systems within mathematics itself has been differential geometry. Names like Darboux and Goursat are even today often mentioned in the context of differential equations (in particular of completely integrable systems), but they were actually studying purely geometric problems that in local coordinates could be described by overdetermined systems of partial differential equations.

Thus we may conclude that it is indeed important to be able to handle non-normal systems of differential equations. While huge (and often very sophisticated) theories for specific systems like the Navier–Stokes equations or Maxwell's equations of electromagnetism have been developed, the knowledge about general systems is not so good. More precisely, although a considerable body of literature exists, most people working with differential equations are not aware of it. We hope that this book may help to change this unfortunate situation.

Chapter 2
Formal Geometry of Differential Equations

He who can digest a second or third fluxion, a second or third difference, need not, we think, be squeamish about any point of divinity.

George Berkeley

In this chapter we lay the geometric foundations of the *formal theory of differential equations*. The name "formal theory" stems from the fact that it is, at least indirectly, concerned with the analysis of formal power series solutions, i. e. one ignores the question whether the series actually converge. Another interpretation of the name is that one tries to extract as much information as possible on the solution space by purely formal operations like algebraic manipulations of the equations or their differentiations without actually solving the given equation.

The basic tool for a geometric approach to differential equations is the jet bundle formalism and the first two sections give an introduction to it. We do this first in a more pedestrian way considering jets as a differential geometric approach to (truncated) power series. For most computational purposes this simple point of view is sufficient. In order to obtain a deeper understanding of certain structural properties which will later be of importance, we redevelop the theory in the second section in a more abstract but intrinsic way which does not require power series. In both approaches, special emphasis is put on the contact structure as the key to the geometry of jet bundles. Because of its great importance, we consider different geometric realisations of it, each having its advantages in certain applications.

The third section starts with the geometric definition of (systems of) differential equations as fibred submanifolds of jet bundles and introduces the two basic operations of prolongation to higher order and projection to lower order. They lead almost immediately to the notion of formal integrability. From a more computational point of view, it is related to the problem of constructing order by order formal power series solutions for a given differential equation. For a formally integrable equation such a construction is always possible, as no obstructions in the form of integrability conditions exist. However, the question of effectively deciding whether or not a given equation is formally integrable cannot be answered by purely geometric means but requires the algebraic theory developed in the next chapters.

In the final section we introduce some classical differential equations of mathematical physics like the Einstein or the Navier–Stokes equations. In the subsequent chapters, they will reappear as examples demonstrating that the theory can indeed be effectively used in realistic applications and not only in toy problems.

W.M. Seiler, *Involution*, Algorithms and Computation in Mathematics 24,
DOI 10.1007/978-3-642-01287-7_2,

2.1 A Pedestrian Approach to Jet Bundles

Let $\mathcal{X}$ and $\mathcal{U}$ be two (finite-dimensional) manifolds, for example two open subsets of $\mathbb{R}^n$ and $\mathbb{R}^m$, respectively. Let furthermore $\phi : \mathcal{X} \to \mathcal{U}$ be a smooth, i. e. infinitely differentiable, map between these two manifolds. In order to simplify the notation we always pretend that the map ϕ is globally defined, i. e. on the whole manifold $\mathcal{X}$. Actually, we always study the map around some point $\mathbf{x}_0 \in \mathcal{X}$ and thus it suffices that the map is defined in some neighbourhood $\Omega \subseteq \mathcal{X}$ of this point $\mathbf{x}_0$ which we never mention explicitly.

If we take some charts on the manifolds $\mathcal{X}$ and $\mathcal{U}$ with corresponding local coordinates $\mathbf{x} = (x^1, \dots, x^n)$ and $\mathbf{u} = (u^1, \dots, u^m)$, respectively, then we can locally write the map ϕ in the form $u^\alpha = \phi^\alpha(x^1, \dots, x^n)$ for $1 \leq \alpha \leq m$. In the context of differential equations, the coordinates $\mathbf{x}$ are often called *independent variables*, whereas the coordinates $\mathbf{u}$ represent the *dependent variables*.

According to Taylor's Theorem, we can expand the components ϕ^α of the map ϕ in a neighbourhood of an arbitrary but fixed point $\mathbf{x}_0 \in \mathcal{X}$:

$$\phi^\alpha(\mathbf{x}) = \sum_{0 \leq |\mu| \leq q} \frac{u^\alpha_\mu}{\mu!}(\mathbf{x} - \mathbf{x}_0)^\mu + R^\alpha_q(\mathbf{x}, \mathbf{x}_0) . \tag{2.1}$$

Here $q \geq 0$ is an arbitrary integer, the order of the expansion, and we use the standard multi index notation of multivariate analysis (see Appendix A.1). It is well-known that the Taylor coefficients $u^\alpha_\mu \in \mathbb{R}$ appearing in the expansion (2.1) are the derivatives of the components ϕ^α evaluated at $\mathbf{x}_0$,

$$u^\alpha_\mu = \frac{\partial^{|\mu|}\phi^\alpha}{\partial x^\mu}(\mathbf{x}_0) , \tag{2.2}$$

and that in the Lagrange representation the remainder term takes the form

$$R^\alpha_q(\mathbf{x}, \mathbf{x}_0) = \sum_{|\mu| = q+1} \frac{\partial^{q+1}\phi^\alpha}{\partial x^\mu}\bigl(\mathbf{x}_0 + \theta(\mathbf{x} - \mathbf{x}_0)\bigr)\frac{(\mathbf{x} - \mathbf{x}_0)^\mu}{\mu!} \tag{2.3}$$

for some $\theta \in [0, 1[$. In the sequel the following notations will be convenient: we write $\mathbf{u}_{(q)}$ for all coefficients u^α_μ with $|\mu| = q$ (with the convention $\mathbf{u}_{(0)} = \mathbf{u}$) and $\mathbf{u}^{(q)}$ for $\{\mathbf{u}_{(0)}, \dots, \mathbf{u}_{(q)}\}$, i. e. for all coefficients u^α_μ with $0 \leq |\mu| \leq q$.

We introduce an equivalence relation on the set of smooth maps $\mathcal{X} \to \mathcal{U}$. Two maps ϕ and ψ are equivalent to order q at $\mathbf{x}_0 \in \mathcal{X}$, written $\phi \sim_q \psi$, if their expansions up to order q at $\mathbf{x}_0$ are identical. Geometrically, this condition means that the graphs of the two maps have at $\mathbf{x}_0$ a contact of order q. Thus if ϕ is given by (2.1) and ψ by a similar expansion with coefficients $\bar{u}^\alpha_\mu$, then $\phi \sim_q \psi$ implies that $\mathbf{u}^{(q)} = \bar{\mathbf{u}}^{(q)}$. A *q-jet* at $\mathbf{x}_0$ is defined as an equivalence class

$$[\phi]^{(q)}_{\mathbf{x}_0} = \bigl\{\psi \in \mathcal{C}^\infty(\mathcal{X}, \mathcal{U}) \mid \psi \sim_q \phi \text{ at } \mathbf{x}_0\bigr\} . \tag{2.4}$$

Loosely speaking, the jet bundle of order q is the space of all power series truncated at order q, i. e. a point in it is such an equivalence class $[\phi]^{(q)}_{\mathbf{x}_0}$. As we always consider only a finite part of the series, it does not matter whether the series actually converges and we do not have to restrict to analytic functions.

Definition 2.1.1. The *jet bundle of order* q over the manifolds $\mathcal{X}$ and $\mathcal{U}$ is the space $J_q(\mathcal{X},\mathcal{U})$ of all q-jets $[\phi]^{(q)}_{\mathbf{x}}$ with $\mathbf{x}\in\mathcal{X}$ and ϕ a smooth map $\mathcal{X}\to\mathcal{U}$.

This definition is very similar to the introduction of the tangent bundle $T\mathcal{M}$ of a manifold $\mathcal{M}$ in Appendix C.1. A point in it, i. e. a tangent vector, is also defined as an equivalence class of functions (more precisely, of curves). As for the tangent bundle, it is not difficult to show that $J_q(\mathcal{X},\mathcal{U})$ is a manifold, too. In fact, it is obvious what one may use as local coordinates: given the coordinates $\mathbf{x}$ on $\mathcal{X}$ and $\mathbf{u}$ on $\mathcal{U}$, we take as coordinates for the q-jet $[\phi]^{(q)}_{\mathbf{x}_0}$ the tuple $(\mathbf{x},\mathbf{u}^{(q)})$. The jet bundle of order 0 is the product $\mathcal{X}\times\mathcal{U}$, since a 0-jet tells us only the result $\phi(\mathbf{x}_0)$ of evaluating any map ϕ contained in it at the point $\mathbf{x}_0$. In concrete calculations one often ignores that points in $J_q(\mathcal{X},\mathcal{U})$ are actually equivalence classes of functions. One simply thinks of $J_q(\mathcal{X},\mathcal{U})$ as a space whose coordinates are besides the independent variables $\mathbf{x}$ and dependent variables $\mathbf{u}$ all derivatives $\mathbf{u}_{(p)}$ with $1\leq p\leq q$.

$J_q(\mathcal{X},\mathcal{U})$ is not just any manifold but possesses a rich internal structure. In particular, jet bundles form a natural hierarchy. If $r<q$, we have an obvious projection $\pi^q_r : J_q(\mathcal{X},\mathcal{U})\to J_r(\mathcal{X},\mathcal{U})$ defined by forgetting the higher-order terms of the Taylor expansion. We can project even further by forgetting everything but the expansion point and obtain then maps $\pi^q : J_q(\mathcal{X},\mathcal{U})\to\mathcal{X}$. All these projections are surjective submersions, so that $J_q(\mathcal{X},\mathcal{U})$ may be endowed in many different ways with the structure of a fibred manifold (cf. Appendix C.1). Furthermore, one trivially sees that our coordinates $(\mathbf{x},\mathbf{u}^{(q)})$ are automatically adapted to all these fibrations.

The dimension of a jet bundle is easily calculated by some elementary combinatorics. As $\dim\mathcal{X}=n$, $\dim\mathcal{U}=m$, we get $\dim J_0(\mathcal{X},\mathcal{U})=n+m$. If we proceed from $J_{q-1}(\mathcal{X},\mathcal{U})$ to $J_q(\mathcal{X},\mathcal{U})$, we add one new coordinate for each derivative of order q of m functions of n variables. Thus, we must count the multi indices of length q and obtain by (A.4a) that

$$\dim J_q(\mathcal{X},\mathcal{U})-\dim J_{q-1}(\mathcal{X},\mathcal{U})=m\binom{n+q-1}{q}\,. \tag{2.5}$$

By (A.4c), this result implies that

$$\dim J_q(\mathcal{X},\mathcal{U})=n+m\binom{n+q}{q}\,. \tag{2.6}$$

In practice, one usually considers the fibre dimension over the base space $\mathcal{X}$, i. e. only the number of variables in $\mathbf{u}^{(q)}$, and omits the summand n.

Example 2.1.2. Assume that $\mathcal{X}$ is a two-dimensional manifold with coordinates (x^1,x^2) and $\mathcal{U}$ a one-dimensional manifold with coordinate u^1. The jet bundle

$J_0(\mathcal{X},\mathcal{U}) = \mathcal{X}\times\mathcal{U}$ is then three-dimensional with coordinates $(x^1,x^2;u^1)$. The first-order jet bundle $J_1(\mathcal{X},\mathcal{U})$ is five-dimensional with coordinates $(x^1,x^2;u^1,u^1_{[1,0]},u^1_{[0,1]})$ and for the eight-dimensional second-order jet bundle $J_2(\mathcal{X},\mathcal{U})$ we use the coordinates $(x^1,x^2;u^1,u^1_{[1,0]},u^1_{[0,1]},u^1_{[2,0]},u^1_{[1,1]},u^1_{[0,2]})$ and so on.

Obviously, the multi index notation is rather cumbersome in such concrete examples. For this reason, we restrict its use to theoretical parts and use more natural variable names in examples. Thus we may take variables (x,y) for $\mathcal{X}$ and u for $\mathcal{U}$. The coordinates in $J_2(\mathcal{X},\mathcal{U})$ are then denoted by $(x,y;u;u_x,u_y,u_{xx},u_{xy},u_{yy})$ which brings us back to standard notations in textbooks on differential equations. ◁

Any smooth map $\phi:\mathcal{X}\rightarrow\mathcal{U}$ can be *prolonged* to a smooth map

$$j_q\phi : \begin{cases} \mathcal{X} \longrightarrow J_q(\mathcal{X},\mathcal{U}) \\ \mathbf{x} \longmapsto [\phi]^{(q)}_{\mathbf{x}} = \left(\mathbf{x}, \frac{\partial^{|\mu|}\phi^\alpha}{\partial\mathbf{x}^\mu}(\mathbf{x})\right) \end{cases} \tag{2.7}$$

where μ runs over all multi indices with $0\le|\mu|\le q$. Note that $j_q\phi$ obviously satisfies $\pi^q\circ j_q\phi = \mathrm{id}_{\mathcal{X}}$. Of course, not all smooth maps $\Phi:\mathcal{X}\rightarrow J_q(\mathcal{X},\mathcal{U})$ are of the special form $\Phi = j_q\phi$ for some map $\phi:\mathcal{X}\rightarrow\mathcal{U}$, but prolonged maps are particularly important. In local coordinates, any map Φ with $\pi^q\circ\Phi=\mathrm{id}_{\mathcal{X}}$ has the form $\Phi(\mathbf{x}) = \left(x^i,\phi^\alpha_\mu(\mathbf{x})\right)$, however without any particular relationship between the component functions ϕ^α_μ.

Those maps Φ that come from a prolongation can be characterised with the help of the *contact structure* of the jet bundle $J_q(\mathcal{X},\mathcal{U})$. The name refers to the fact that the equivalence relation underlying our definition of the jet bundle identifies maps whose graphs have a contact of order q. The contact structure is a central property of jet bundles the importance of which cannot be overestimated; it encodes in particular the meaning of the local coordinates $(\mathbf{x},\mathbf{u}^{(q)})$, i. e. that some variables are supposed to be derivatives of other ones. There are many ways to introduce this structure; the simplest one is via certain differential one-forms on $J_q(\mathcal{X},\mathcal{U})$.

Definition 2.1.3. A one-form $\omega\in\Omega^1(J_q(\mathcal{X},\mathcal{U}))$ on the jet bundle $J_q(\mathcal{X},\mathcal{U})$ of order q is a *contact form*, if for any smooth map $\phi:\mathcal{X}\rightarrow\mathcal{U}$ its pull-back on the image of the prolonged map $\mathrm{im}\,(j_q\phi)\subseteq J_q(\mathcal{X},\mathcal{U})$ vanishes. In other words, if we denote by $\iota:\mathrm{im}\,(j_q\phi)\hookrightarrow J_q(\mathcal{X},\mathcal{U})$ the natural inclusion map, then $\iota^*\omega=0$. The contact forms span the *contact codistribution* $\mathcal{C}^0_q\subset T^*\left(J_q(\mathcal{X},\mathcal{U})\right)$.

Proposition 2.1.4. *Let $q\in\mathbb{N}$ be some fixed order. For every $1\le\alpha\le m$ and every multi index μ with $0\le|\mu|<q$ the one-form*[1]

$$\omega^\alpha_\mu = \mathrm{d}u^\alpha_\mu - u^\alpha_{\mu+1_i}\mathrm{d}x^i \tag{2.8}$$

[1] Here we use in the second summand on the right hand side the *summation* or *Einstein convention* that a summation is understood over repeated indices, i. e. in this case over i. The range of the summation should always be clear from the context; here it is obviously $1\le i\le n$. We will follow this convention for most sums in this book.

is a contact form on $J_q(\mathcal{X},\mathcal{U})$. The contact codistribution $\mathcal{C}_q^0$ is locally generated by all these forms and hence $\dim\mathcal{C}_q^0 = m\binom{n+q-1}{q-1}$. A map $\Phi : \mathcal{X} \to J_q(\mathcal{X},\mathcal{U})$ satisfying $\pi^q \circ \Phi = \mathrm{id}_{\mathcal{X}}$ is of the special form $\Phi = j_q\phi$ for some map $\phi : \mathcal{X} \to \mathcal{U}$, if and only if its image $\operatorname{im}\Phi \subseteq J_q(\mathcal{X},\mathcal{U})$ is an integral manifold of $\mathcal{C}_q^0$.

Proof. A general element of $\Omega^1(J_q(\mathcal{X},\mathcal{U}))$ is of the form $\omega = a_i \mathrm{d}x^i + b_\beta^\nu \mathrm{d}u_\nu^\beta$ where the summation in ν is over all multi indices with $0 \leq |\nu| \leq q$. Let $\phi : \mathcal{X} \to \mathcal{U}$ be a smooth map and $\iota : \operatorname{im}(j_q\phi) \hookrightarrow J_q(\mathcal{X},\mathcal{U})$ the corresponding natural inclusion. Then the pull-back of ω on $\operatorname{im}(j_q\phi)$ is of the form

$$\iota^*\omega = \left(a_i + b_\beta^\nu \frac{\partial^{|\nu|+1}\phi^\beta}{\partial \mathbf{x}^{\nu+1_i}}\right) \mathrm{d}x^i \,. \tag{2.9}$$

Now it is straightforward to verify that ω_μ^α is a contact form. For it $a_i = -u_{\mu+1_i}^\alpha$, $b_\alpha^\mu = 1$ and all other coefficients vanish. Since $u_{\mu+1_i}^\alpha = \partial^{|\mu|+1}\phi^\alpha/\partial\mathbf{x}^{\mu+1_i}$ at any point on $\operatorname{im}(j_q\phi)$, we find indeed $\iota^*\omega_\mu^\alpha = 0$.

Let $\rho = [\phi]_{\mathbf{x}_0}^{(q)}$ be an arbitrary point on the submanifold $\operatorname{im}(j_q\phi)$. All maps contained in the equivalence class ρ have identical derivatives in $\mathbf{x}_0$ up to order q, but by (2.9) the form $\iota^*\omega$ also depends on derivatives of order $q+1$. Thus we may choose another representative $\bar\phi \in \rho$ which has in $\mathbf{x}_0$ exactly one different derivative of degree $q+1$, say $c = \partial^{q+1}\phi^\beta/\partial\mathbf{x}^\nu(\mathbf{x}_0) \neq \partial^{q+1}\bar\phi^\beta/\partial\mathbf{x}^\nu(\mathbf{x}_0) = \bar c$ for some multi index ν with $|\nu| = q+1$.

By construction, the two submanifolds $\operatorname{im}(j_q\phi)$ and $\operatorname{im}(j_q\bar\phi)$ intersect in ρ; let $\bar\iota$ denote the natural inclusion map for the latter one. If ω is a contact form, then $\iota^*\omega = \bar\iota^*\omega = 0$. The difference of these two equations evaluated at the point ρ yields $(c - \bar c)b_\beta^{\nu-1_i}(\rho)\mathrm{d}x^i = 0$ where the summation is over all i such that $\nu_i > 0$. Since $c \neq \bar c$ and the forms $\mathrm{d}x^i$ are linearly independent, we must have $b_\beta^{\nu-1_i}(\rho) = 0$. As ρ was an arbitrary point and ν an arbitrary multi index, we can show in this manner that all coefficients b_β^μ with $|\mu| = q$ must vanish for a contact form.

It is easy to see that a non-vanishing one-form $a_i\mathrm{d}x^i$ cannot be a contact form. Hence the subspace of $\Omega^1(J_q(\mathcal{X},\mathcal{U}))$ generated by the forms $\mathrm{d}x^i$ and by the forms $\mathrm{d}u_\mu^\alpha$ with $|\mu| = q$ has zero intersection with $\mathcal{C}_q^0$. Now it follows simply from counting dimensions that the contact codistribution $\mathcal{C}_q^0$ is generated by the forms (2.8). Furthermore, these considerations imply by (A.4c) that $\dim\mathcal{C}_q^0 = \binom{n+q-1}{q-1}$.

For the last assertion let the map $\Phi : \mathcal{X} \to J_q(\mathcal{X},\mathcal{U})$ take in local coordinates the form $\Phi(\mathbf{x}) = \big(x^i, \phi_\mu^\alpha(\mathbf{x})\big)$. If $\operatorname{im}\Phi$ is an integral manifold of the contact codistribution $\mathcal{C}_q^0$, then we must have

$$\phi^*\omega_\mu^\alpha = \left(\frac{\partial\phi_\mu^\alpha}{\partial x^i} - \phi_{\mu+1_i}^\alpha\right)\mathrm{d}x^i = 0 \tag{2.10}$$

for all one-forms (2.8). But this equality immediately implies that $\phi_\mu^\alpha = \partial^{|\mu|}\phi_0^\alpha/\partial\mathbf{x}^\mu$ for all multi indices μ with $0 < |\mu| \leq q$ and hence $\Phi = j_q\phi_0$ is a prolongation. □

Note that the contact codistribution possesses further integral manifolds besides the images of prolonged maps. These are of importance for the description of more general types of solutions exhibiting for example singularities. We will restrict our attention mainly to the regular case; a few examples of singular behaviour will be discussed in Section 9.1. An alternative description of the contact structure uses vector fields instead of differential forms: we simply consider the annihilator of $\mathcal{C}_q^0$.

Definition 2.1.5. The *contact distribution* $\mathcal{C}_q \subset TJ_q(\mathcal{X},\mathcal{U})$ consists of all vectors that are annihilated by the contact forms. A vector field contained in $\mathcal{C}_q$ is called a *contact vector field*.

Proposition 2.1.6. *The contact distribution $\mathcal{C}_q$ is locally spanned by the fields*

$$C_i^{(q)} = \partial_{x^i} + \sum_{\alpha=1}^{m} \sum_{0\leq|\mu|<q} u^\alpha_{\mu+1_i}\partial_{u^\alpha_\mu}\,, \qquad 1\leq i\leq n \tag{2.11a}$$

and

$$C^\mu_\alpha = \partial_{u^\alpha_\mu}\,, \qquad 1\leq\alpha\leq m\,, \quad |\mu| = q\,. \tag{2.11b}$$

Hence $\dim\mathcal{C}_q = n + m\binom{n+q-1}{q}$. *A map $\Phi : \mathcal{X}\to J_q(\mathcal{X},\mathcal{U})$ with $\pi^q\circ\Phi = \mathrm{id}_\mathcal{X}$ is a prolongation $\Phi = j_q\phi$, if and only if $T(\operatorname{im}\Phi)\subset\mathcal{C}_q\big|_{\operatorname{im}\Phi}$, i. e. if and only if its image* $\operatorname{im}\Phi$ *is an integral manifold of the contact distribution $\mathcal{C}_q$.*

Proof. The basis (2.11) of the contact distribution $\mathcal{C}_q$ is trivially obtained from the basis forms (2.8) of the dual contact codistribution $\mathcal{C}_q^0$ and the claim about the dimension follows from (A.4a). The final assertion can be proved in the same manner as the integrable case of Proposition C.3.5. □

Remark 2.1.7. Let us take a quick look at the structure equations of the contact (co)distribution. For $\mathcal{C}_q^0$ this requires to determine the exterior derivative of the contact forms ω^α_μ defined in (2.8). One readily computes

$$\mathrm{d}\omega^\alpha_\mu = -\mathrm{d}u^\alpha_{\mu+1_i}\wedge\mathrm{d}x^i\,. \tag{2.12}$$

If $|\mu| < q-1$, then we may equivalently write $\mathrm{d}\omega^\alpha_\mu = -\omega^\alpha_{\mu+1_i}\wedge\mathrm{d}x^i$ and the exterior derivative vanishes modulo $\mathcal{C}_q^0$. But if $|\mu|$ takes the maximal value $q-1$, such a reformulation is not possible and the exterior derivative does not vanish modulo $\mathcal{C}_q^0$. Hence $\mathcal{C}_q^0$ is not an involutive codistribution for any finite order $q\in\mathbb{N}$.

Dually, we must compute the Lie brackets between the generators of the contact distribution $\mathcal{C}_q$ defined by (2.11). One easily sees that all of them vanish except for

$$\left[C_\alpha^{\nu+1_i}, C_i^{(q)}\right] = \partial_{u^\alpha_\nu} \tag{2.13}$$

where ν is an arbitrary multi index of length $q-1$. As the right hand side is not a contact field, $\mathcal{C}_q$ is not involutive either (this fact follows of course also trivially from Proposition C.3.5). Note that (2.13) together with (2.11a) entails that the derived contact distribution satisfies $T\pi^q_{q-1}(\mathcal{C}_q') = \mathcal{C}_{q-1}$. ◁

For another kind of prolongation we need the *formal* or *total derivatives*. They take as argument a smooth function $\Phi \in \mathcal{F}(J_q(\mathcal{X},\mathcal{U}))$ and return a function in $\mathcal{F}(J_{q+1}(\mathcal{X},\mathcal{U}))$. In local coordinates, the formal derivative D_i with respect to the independent variable x^i is defined as

$$D_i\Phi(\mathbf{x},\mathbf{u}^{(q+1)}) = \frac{\partial \Phi}{\partial x^i}(\mathbf{x},\mathbf{u}^{(q)}) + \sum_{\alpha=1}^{m}\sum_{0\leq|\mu|\leq q}\frac{\partial \Phi}{\partial u^\alpha_\mu}(\mathbf{x},\mathbf{u}^{(q)})\,u^\alpha_{\mu+1_i}\,. \tag{2.14}$$

Note that the function $D_i\Phi$ is always *quasi-linear*, i.e. linear in the derivatives of maximal order: as Φ depends only on $\mathbf{u}^{(q)}$, derivatives $\mathbf{u}_{(q+1)}$ appear in (2.14) only in the terms $u^\alpha_{\mu+1_i}$ with $|\mu| = q$.

In a more intrinsic manner we may introduce the formal derivatives as follows. Let $\Phi \in \mathcal{F}(J_q(\mathcal{X},\mathcal{U}))$ be an arbitrary smooth function on the jet bundle of order q. The concatenation $\Phi \circ \pi^{q+1}_q \in \mathcal{F}(J_{q+1}(\mathcal{X},\mathcal{U}))$ defines a smooth function on the jet bundle of order $q+1$ and thus a natural inclusion $\mathcal{F}(J_q(\mathcal{X},\mathcal{U})) \hookrightarrow \mathcal{F}(J_{q+1}(\mathcal{X},\mathcal{U}))$ exists. Now the formal derivatives can be uniquely defined by the following congruence of one-forms over $J_{q+1}(\mathcal{X},\mathcal{U})$:

$$\mathrm{d}(\Phi \circ \pi^{q+1}_q) \equiv (D_i\Phi)\mathrm{d}x^i \mod \mathcal{C}^0_{q+1}\,. \tag{2.15}$$

Indeed, the left hand side is given by $(\partial\Phi/\partial x^i)\mathrm{d}x^i + (\partial\Phi/\partial u^\alpha_\mu)\mathrm{d}u^\alpha_\mu$. Since modulo the contact codistribution $\mathcal{C}^0_{q+1}$ we may substitute $\mathrm{d}u^\alpha_\mu = u^\alpha_{\mu+1_i}\mathrm{d}x^i$ by (2.8), this expression is equivalent to the right hand side.

Remark 2.1.8. Linear first-order differential operators are closely related to vector fields (see Appendix C.2). However, in the case of the formal derivative it is not possible to consider D_i as a vector field in $\mathfrak{X}(J_q(\mathcal{X},\mathcal{U}))$—for the simple reason that the coefficients are not contained in $\mathcal{F}(J_q(\mathcal{X},\mathcal{U}))$. In the context of Lie symmetry theory, Johnson [240, 241] introduced a generalisation of vector fields as sections of a pull-back bundle (see Appendix C.1 for an explanation of these concepts related to fibred manifolds). The formal derivative may be interpreted as such a generalised vector field: $D_i \in \Gamma_{loc}((\pi^{q+1}_q)^*T(J_q(\mathcal{X},\mathcal{U})))$. ◁

Eq. (2.14) represents the well-known *chain rule* of calculus within the jet bundle formalism. Indeed, suppose that we consider the function $\Phi \in \mathcal{F}(J_q(\mathcal{X},\mathcal{U}))$ restricted to the graph of a map $j_q\phi : \mathcal{X} \to J_q(\mathcal{X},\mathcal{U})$. Then we can substitute the variables u^α_μ by $\partial^{|\mu|}\phi^\alpha/\partial x^\mu$ and obtain the function $\Phi \circ j_q\phi : \mathcal{X} \to \mathbb{R}$. Its partial derivatives $\partial(\Phi \circ j_q\phi)/\partial x^i$ must be computed by the chain rule. Exactly the same result is reached, if we perform the corresponding substitutions in the function $D_i\Phi$. In other words, we have for all maps ϕ the equality

$$\frac{\partial(\Phi \circ j_q\phi)}{\partial x^i} = (D_i\Phi) \circ j_{q+1}\phi\,. \tag{2.16}$$

Example 2.1.9. Let (x,y) be coordinates on $\mathcal{X}$ and u on $\mathcal{U}$. We consider the function $\Phi(x,y,u,u_x,u_y) = (u_y)^2 - xu_x$ defined on the first-order jet bundle $J_1(\mathcal{X},\mathcal{U})$. Its formal derivatives with respect to x and y, respectively, are given by the following two functions defined on the second-order jet bundle $J_2(\mathcal{X},\mathcal{U})$:

$$D_x\Phi(x,y,u,u_x,u_y,u_{xx},u_{xy},u_{yy}) = 2u_yu_{xy} - xu_{xx} - u_x\,, \tag{2.17a}$$

$$D_y\Phi(x,y,u,u_x,u_y,u_{xx},u_{xy},u_{yy}) = 2u_yu_{yy} - xu_{xy}\,. \tag{2.17b}$$

These are indeed quasi-linear functions, i. e. linear in all second-order derivatives, although Φ is quadratic in u_y.

If we substitute the dependent variable u by some function $\psi(x,y)$ in Φ, then we obtain a function $\tilde{\Phi}(x,y) = \Phi\big(x,y,\psi(x,y),\psi_x(x,y),\psi_y(x,y)\big)$. Differentiating this function with respect to x yields according to the chain rule

$$\frac{\partial\tilde{\Phi}}{\partial x}(x,y) = 2\psi_y(x,y)\psi_{xy}(x,y) - x\psi_{xx}(x,y) - \psi_x(x,y) \tag{2.18}$$

and correspondingly for $\partial\tilde{\Phi}/\partial y$. The back substitutions $\psi \to u$, $\psi_x \to u_x$, $\psi_{xx} \to u_{xx}$ etc. lead exactly to the above formal derivatives. ◁

Remark 2.1.10. Note the close relationship between the formal derivative D_i and the contact vector field $C_i^{(q)}$. The only difference is that in the definition of $C_i^{(q)}$ the summation does not include the multi indices of length q. Indeed, the corresponding coefficient would be a derivative of order $q+1$ and thus not defined on the jet bundle $J_q(\mathcal{X},\mathcal{U})$ of order q. If we use the above mentioned natural inclusion $\mathcal{F}\big(J_q(\mathcal{X},\mathcal{U})\big) \hookrightarrow \mathcal{F}\big(J_{q+1}(\mathcal{X},\mathcal{U})\big)$, then we obtain for any function $\Phi \in \mathcal{F}\big(J_q(\mathcal{X},\mathcal{U})\big)$ the equality $D_i\Phi = C_i^{(q+1)}\big(\Phi \circ \pi_q^{q+1}\big)$. ◁

We mentioned above that the jet bundles of different order form a hierarchy via the projections $\pi_r^q : J_q(\mathcal{X},\mathcal{U}) \to J_r(\mathcal{X},\mathcal{U})$ for $q > r$. The case $r = q-1$, where the order is reduced only by one, is of particular interest, as it leads to a special and very important structure.

Proposition 2.1.11. *The jet bundle $J_q(\mathcal{X},\mathcal{U})$ of order q is affine over the jet bundle $J_{q-1}(\mathcal{X},\mathcal{U})$ of order $q-1$.*

Before we prove this proposition, we discuss what it means. Recall that a function is affine, if it is of the form $x \mapsto ax+b$ for constants a, b. Proposition 2.1.11 claims that the highest-order derivatives $\mathbf{u}_{(q)}$ transform under a change of coordinates in $\mathcal{X}$ and $\mathcal{U}$ as affine functions with coefficients a, b depending on the derivatives $\mathbf{u}^{(q-1)}$. A rigorous definition of an affine bundle is given in Appendix C.1.

Proof. We study changes of coordinates $\bar{\mathbf{x}} = \bar{\mathbf{x}}(\mathbf{x})$ and $\bar{\mathbf{u}} = \bar{\mathbf{u}}(\mathbf{u})$ in the manifolds $\mathcal{X}$ and $\mathcal{U}$, respectively. They induce a change of coordinates $\bar{\mathbf{u}}^{(q)} = \bar{\mathbf{u}}^{(q)}\big(\mathbf{x},\mathbf{u}^{(q)}\big)$ in the jet bundle $J_q(\mathcal{X},\mathcal{U})$. It can be determined either by using the chain rule or, equivalently, by requiring that the contact forms (2.8) remain invariant. In either case the result is that in repeated index notation

$$\bar{u}^\alpha_{j_1\cdots j_q} = \left(\frac{\partial\bar{u}^\alpha}{\partial u^\beta}\frac{\partial x^{i_1}}{\partial\bar{x}^{j_1}}\cdots\frac{\partial x^{i_q}}{\partial\bar{x}^{j_q}}\right)u^\beta_{i_1\cdots i_q} + \cdots \tag{2.19}$$

where the dots represent a complicated expression in the variables $\mathbf{u}^{(q-1)}$ and where $\partial\mathbf{x}/\partial\bar{\mathbf{x}}$ denotes the inverse of the Jacobian $\partial\bar{\mathbf{x}}/\partial\mathbf{x}$. But this observation implies that (2.19) is indeed affine in $\mathbf{u}_{(q)}$ as claimed. □

At first sight, Proposition 2.1.11 may appear rather abstract, but it has very important consequences which will be crucial for the application of algebraic methods in the analysis of differential equations. The key is the so-called fundamental identification which we will now discuss. It leads to a natural polynomial structure hidden in the jet bundle hierarchy.

An affine space is always modelled on a vector space: the difference between two points may be interpreted as a vector. In our case it is easy to identify this vector space. Let $[\phi]^{(q)}_{\mathbf{x}}$ and $[\psi]^{(q)}_{\mathbf{x}}$ be two points in $J_q(\mathcal{X},\mathcal{U})$ such that

$$[\phi]^{(q-1)}_{\mathbf{x}} = \pi^q_{q-1}\big([\phi]^{(q)}_{\mathbf{x}}\big) = \pi^q_{q-1}\big([\psi]^{(q)}_{\mathbf{x}}\big) = [\psi]^{(q-1)}_{\mathbf{x}}\,, \tag{2.20}$$

i. e. the two points belong to the same fibre with respect to the fibration given by the projection $\pi^q_{q-1} : J_q(\mathcal{X},\mathcal{U}) \to J_{q-1}(\mathcal{X},\mathcal{U})$. Thus $[\phi]^{(q)}_{\mathbf{x}}$ and $[\psi]^{(q)}_{\mathbf{x}}$ correspond to two Taylor series truncated at degree q which coincide up to degree $q-1$. Obviously, this observation implies that their difference consists of one *homogeneous* polynomial of degree q for each dependent variable u^α.

In a more intrinsic language, we formulate this result as follows. Let $\rho = [\phi]^{(q)}_{\mathbf{x}}$ be a point in $J_q(\mathcal{X},\mathcal{U})$ with $\mathbf{u} = \phi(\mathbf{x})$ and $\bar{\rho} = [\phi]^{(q-1)}_{\mathbf{x}} = \pi^q_{q-1}(\rho)$. Then according to Proposition 2.1.11, the fibre $(\pi^q_{q-1})^{-1}(\bar{\rho})$ is an affine space modelled on the vector space $S_q(T^*_{\mathbf{x}}\mathcal{X}) \otimes T_{\mathbf{u}}\mathcal{U}$ where S_q denotes the q-fold symmetric product. Indeed, this fact follows immediately from our discussion so far. The symmetric algebra is a coordinate-free form of the polynomial ring and thus elements of its components correspond to homogeneous polynomials. Furthermore, one easily verifies that the homogeneous part of (2.19) obtained by dropping the terms represented by the dots describes the behaviour of the coefficients of elements of the vector space $S_q(T^*_{\mathbf{x}}\mathcal{X}) \otimes T_{\mathbf{u}}\mathcal{U}$ under coordinate transformations (note that we must use the *co*tangent space $T^*_{\mathbf{x}}\mathcal{X}$, as the coefficients of tangent vectors would transform with the Jacobian $\partial\bar{\mathbf{x}}/\partial\mathbf{x}$ and not with its inverse).

By Proposition 2.1.11, the qth-order jet bundle $J_q(\mathcal{X},\mathcal{U})$ is an affine bundle over $J_{q-1}(\mathcal{X},\mathcal{U})$. According to Lemma C.4.3, this fact implies (keeping the notations of the discussion above) that the tangent space to the affine space $(\pi^q_{q-1})^{-1}(\bar{\rho})$ at the point $\rho \in J_q(\mathcal{X},\mathcal{U})$ is canonically isomorphic to the corresponding vector space, i. e. to $S_q(T^*_{\mathbf{x}}\mathcal{X}) \otimes T_{\mathbf{u}}\mathcal{U}$. This isomorphism is called the *fundamental identification*.

Lemma C.4.3 provides us only with an abstract definition of this isomorphism; for concrete computations we need a local coordinate expression. On one side we consider the tangent space to the fibre $(\pi^q_{q-1})^{-1}(\bar{\rho})$ at the point ρ, i. e. the *vertical space* $V_\rho\pi^q_{q-1}$ defined as the kernel of the tangent map $T_\rho\pi^q_{q-1}$. Obviously, it is spanned by all the vectors $\partial_{u^\alpha_\mu}$ with $|\mu| = q$. Let us take one of these vectors; it is tangent to the curve $\gamma : t \mapsto \rho(t)$ where $\rho(0) = \rho$ and all coordinates of an arbitrary point $\rho(t)$ coincide with those of ρ except for the one coordinate u^α_μ corresponding to the chosen vector which is increased by t.

On the other side, we may compute the difference quotient $(\rho(t)-\rho)/t \in \mathbb{R}^m$ interpreting the points as above as truncated Taylor series. The αth component of the result is the polynomial $x^\mu/\mu!$. Hence the fundamental identification is the map $\varepsilon_q : V_\rho\pi^q_{q-1} \to S_q(T^*_{\mathbf{x}}\mathcal{X})\otimes T_{\mathbf{u}}\mathcal{U}$ given by

$$\varepsilon_q(\partial_{u^\alpha_\mu}) = \frac{1}{\mu!}\mathrm{d}x^\mu \otimes \partial_{u^\alpha}\ . \tag{2.21}$$

It is important to note here the combinatorial factor $\frac{1}{\mu!}$ having its origin in Taylor's formula (2.1). Later we will use ε_q for the identification of geometric quantities, namely vertical vectors in $V_\rho\pi^q_{q-1}$, with algebraic quantities, namely homogeneous polynomials, which we can study with methods from commutative algebra.

2.2 An Intrinsic Approach to Jet Bundles

In the sequel we will always consider a slightly more general situation: instead of jets of maps $\mathcal{X}\to\mathcal{U}$ between two manifolds $\mathcal{X}$, $\mathcal{U}$ we will take jets of sections of a fibred manifold $\pi:\mathcal{E}\to\mathcal{X}$ (see Appendix C.1 for some basic material on fibred manifolds). Note that this generalisation affects only global properties; locally it makes no real difference. The situation of the last section is recovered, if we take the trivial fibration $\mathrm{pr}_1 : \mathcal{E}=\mathcal{X}\times\mathcal{U}\to\mathcal{X}$ with the canonical projection pr_1 on the first factor (recall from (C.5) that every fibred manifold looks in the neighbourhood of any point $\xi\in\mathcal{E}$ like a trivial one). A section $\sigma:\mathcal{X}\to\mathcal{E}$ is then locally always of the form $\sigma(\mathbf{x}) = \big(\mathbf{x}, s(\mathbf{x})\big)$ with a function $s:\mathcal{X}\to\mathcal{U}$.

For all computational purposes, the introduction of jet bundles via truncated power series presented in the previous section is completely sufficient. For obtaining a deeper understanding of the geometry of jet bundles a more intrinsic approach is of advantage and will be our next topic. Thus this section requires some familiarity with coordinate-free differential geometry. However, readers not so much interested in geometry may safely skip it.

Remark 2.2.1. Before we move on to a completely different way to define jet bundles, we briefly show how the approach of the last section can be formulated in this more general setting without the explicit use of coordinates.

We must now define an equivalence relation on the space $\Gamma_{loc}(\pi)$ of all local sections of the fibred manifold $\pi:\mathcal{E}\to\mathcal{X}$. Let σ_1, σ_2 be two sections defined in the neighbourhood of some point $x\in\mathcal{X}$. We consider them as equivalent to first order at x, again written $\sigma_1\sim_1\sigma_2$, if $\sigma_1(x)=\sigma_2(x)$ and $T_x\sigma_1 = T_x\sigma_2$. If $s_1, s_2:\mathcal{X}\to\mathcal{U}$ are the local functions corresponding to σ_1, σ_2, then the local representation of the tangent map $T_x\sigma_k$ is essentially given by the Jacobian of the function s_k. Hence we trivially find that $\sigma_1\sim_1\sigma_2$, if and only if $s_1\sim_1 s_2$ in the sense of the previous section.

For the definition of higher-order equivalence we use a recursive approach. Any fibred manifold $\pi : \mathcal{E} \rightarrow \mathcal{X}$ trivially induces a fibration of the corresponding tangent bundles, $T\pi : T\mathcal{E} \rightarrow T\mathcal{X}$, and each section $\sigma : \mathcal{X} \rightarrow \mathcal{E}$ induces via its tangent map $T\sigma : T\mathcal{X} \rightarrow T\mathcal{E}$ a section of the new fibration. Fixing a concrete point $x \in \mathcal{X}$, we may also consider the induced restriction $T_{\sigma(x)}\pi : T_{\sigma(x)}\mathcal{E} \rightarrow T_x\mathcal{X}$ as a fibred manifold and $T_x\sigma$ as a section of it. Now we define $\sigma_1 \sim_q \sigma_2$, if and only if $T_x\sigma_1 \sim_{q-1} T_x\sigma_2$. Again it is trivial to verify that $\sigma_1 \sim_q \sigma_2$ holds, if and only if $s_1 \sim_q s_2$, as the iterated tangent maps require higher and higher derivatives of the local functions $s_k : \mathcal{X} \rightarrow \mathcal{U}$ at the point x. Once this equivalence relation is obtained, we proceed as before: a point in the qth order jet bundle over the fibred manifold $\pi : \mathcal{E} \rightarrow \mathcal{X}$ is an equivalence class $[\sigma]_x^{(q)}$ of local sections with respect to the relation $\sim_q$. ◁

We describe now an alternative construction of the first-order jet bundle based on Proposition 2.1.11. It does not seem to be widely known in the literature, but it is very convenient for our purposes. Strictly speaking, we do not construct the jet bundle $J_1\pi$ as introduced in Remark 2.2.1 above but a bundle canonically diffeomorphic to it. However, we will always identify the two bundles.

Definition 2.2.2. Let $\pi : \mathcal{E} \rightarrow \mathcal{X}$ be a fibred manifold. The *first-order jet bundle* over π is the affine bundle $\pi_0^1 : J_1\pi \rightarrow \mathcal{E}$ whose fibre at a point $\xi \in \mathcal{E}$ is the affine space

$$(J_1\pi)_\xi = \mathcal{A}_\xi = \{\gamma \in T_x^*\mathcal{X} \otimes T_\xi\mathcal{E} \mid T_\xi\pi \circ \gamma = \mathrm{id}_{T_x\mathcal{X}}\} \tag{2.22}$$

(where $x = \pi(\xi) \in \mathcal{X}$) modelled on the vector space

$$\mathcal{V}_\xi = T_x^*\mathcal{X} \otimes V_\xi\pi\,. \tag{2.23}$$

Thus any point $\rho \in J_1\pi$ may be considered as a tuple $\rho = (\xi, \gamma_\xi)$ consisting of a point $\xi \in \mathcal{E}$ and a linear map $\gamma_\xi \in \mathcal{A}_\xi$.

If we consider two maps $\gamma_1, \gamma_2 \in \mathcal{A}_\xi$, then their difference $\gamma_1 - \gamma_2$ satisfies

$$T_\xi\pi \circ (\gamma_1 - \gamma_2) = \mathrm{id}_{T_x\mathcal{X}} - \mathrm{id}_{T_x\mathcal{X}} = 0\,. \tag{2.24}$$

Hence $\gamma_1 - \gamma_2$ maps any vector in $T_x\mathcal{X}$ into a vertical vector and $\gamma_1 - \gamma_2 \in \mathcal{V}_\xi^{(1)}$. So $\mathcal{A}_\xi$ is indeed an affine space modelled on the vector space $\mathcal{V}_\xi$. Any element of $\mathcal{A}_\xi$ has the form $\gamma(v) = \big(v, B(v)\big)$ with some linear map B. The whole affine bundle $J_1\pi$ is modelled on the vector bundle $\pi^*(T^*\mathcal{X}) \otimes V\pi \rightarrow \mathcal{E}$. Here the first factor is the pull-back of $\tau_\mathcal{X}^* : T^*\mathcal{X} \rightarrow \mathcal{X}$ over $\pi : \mathcal{E} \rightarrow \mathcal{X}$. As the pull-back notation is rather cumbersome, we will write this bundle in the sequel briefly as $T^*\mathcal{X} \underset{\mathcal{E}}{\otimes} V\pi$. The subscript $\mathcal{E}$ signals that both factors are to be considered as bundles over $\mathcal{E}$ using, if necessary, appropriate pull-backs which are usually obvious from the context.

At first sight, Definition 2.2.2 of the jet bundle seems to be completely unrelated to our previous Definition 2.1.1. So we show next that points in $J_1\pi$ may be interpreted as equivalence classes of sections $\sigma \in \Gamma_{loc}(\pi)$. Noting that the bundle $J_1\pi$ is obviously fibred over $\mathcal{X}$ with the projection $\pi^1 = \pi \circ \pi_0^1 : J_1\pi \rightarrow \mathcal{X}$, we define for such a section its (first) *prolongation* $j_1\sigma \in \Gamma_{loc}(\pi^1)$ by

$$j_1\sigma : \begin{cases} \mathcal{X} & \longrightarrow & J_1\pi \\ x & \longmapsto & (\sigma(x), T_x\sigma) \end{cases} . \tag{2.25}$$

The tangent map $T_x\sigma$ indeed belongs to the space $\mathcal{A}_\xi$ with $\xi = \sigma(x)$. As σ is a map from $\mathcal{X}$ to $\mathcal{E}$, its tangent map $T_x\sigma$ is a linear map $T_x\mathcal{X} \to T_\xi\mathcal{E}$ and thus an element of $T_x^*\mathcal{X} \otimes T_\xi\mathcal{E}$. Furthermore, by definition of a section, σ satisfies $\pi \circ \sigma = \mathrm{id}_\mathcal{X}$ and hence by the chain rule $T_\xi\pi \circ T_x\sigma = \mathrm{id}_{T_x\mathcal{X}}$ (obviously, the condition imposed in (2.22) is motivated by this observation).

Conversely, given an arbitrary element $\gamma \in \mathcal{A}_\xi$, it is trivial to see that many sections $\sigma \in \Gamma_{loc}(\pi)$ exist such that $\sigma(x) = \xi$ and $T_x\sigma = \gamma$, since the last condition refers only to a single point $x \in \mathcal{X}$. The bijection

$$\gamma \in \mathcal{A}_\xi \longleftrightarrow [\sigma]_x^{(1)} = \{\sigma \in \Gamma_{loc}(\pi) \mid \sigma(x) = \xi,\ T_x\sigma = \gamma\} \tag{2.26}$$

provides us thus with a canonical identification between the jet bundles introduced in the last section and those of Definition 2.2.2.

In adapted coordinates $(\mathbf{x}, \mathbf{u})$ on $\mathcal{E}$, we may locally write $\sigma(\mathbf{x}) = \big(\mathbf{x}, \mathbf{s}(\mathbf{x})\big)$. For the tangent map $T_x\sigma$ we obtain then the local expression $(\mathbf{x}, \dot{\mathbf{x}}) \mapsto \big(\sigma(\mathbf{x}), \mathrm{d}\sigma(\mathbf{x}) \cdot \dot{\mathbf{x}}\big)$ with the matrix

$$\mathrm{d}\sigma(\mathbf{x}) = \begin{pmatrix} 1 & 0 \cdots 0 & 0 \\ & \ddots & \\ 0 & 0 \cdots 0 & 1 \\ \frac{\partial s^1}{\partial x^1}(\mathbf{x}) & \cdots & \frac{\partial s^1}{\partial x^n}(\mathbf{x}) \\ \vdots & & \vdots \\ \frac{\partial s^m}{\partial x^1}(\mathbf{x}) & \cdots & \frac{\partial s^m}{\partial x^n}(\mathbf{x}) \end{pmatrix} . \tag{2.27}$$

It consists of the $n \times n$ identity matrix stacked over the Jacobian of the function $\mathbf{s}$ evaluated at the fixed point $x \in \mathcal{X}$. Conversely, the matrix of any linear map contained in $\mathcal{A}_\xi$ has the form (2.27), i.e. the $n \times n$ identity matrix stacked over an arbitrary $m \times n$ matrix. Recall that our local coordinates for $[\sigma]_\mathbf{x}^{(1)} \in J_1\pi$ are $(\mathbf{x}, \mathbf{u}^{(1)})$ with $u_i^\alpha = \frac{\partial s^\alpha}{\partial x^i}(\mathbf{x})$. Hence we may identify the entries of this $m \times n$ matrix with the first-order derivatives u_i^α.

Remark 2.2.3. A point $(\xi, \gamma_\xi) \in J_1\pi$ induces a direct decomposition of the tangent space $T_\xi\mathcal{E}$. By definition, the fibre component γ_ξ represents an element of the affine space A_ξ, i.e. a linear map $T_x\mathcal{X} \to T_\xi\mathcal{E}$ (where again $x = \pi(\xi)$). Because of the condition $T_\xi\pi \circ \gamma_\xi = \mathrm{id}_{T_x\mathcal{X}}$, the intersection $\operatorname{im}\gamma_\xi \cap V_\xi\pi$ contains only the zero vector (the vertical bundle $V_\xi\pi$ is of course just $\mathcal{V}_\xi$) and $T_\xi\mathcal{E} = \operatorname{im}\gamma_\xi \oplus V_\xi\pi$. Hence a global section γ of the bundle $\pi_0^1 : J_1\pi \to \mathcal{E}$ such that $\gamma(\xi) = (\xi, \gamma_\xi)$ may be interpreted as a *connection* on the fibred manifold $\pi : \mathcal{E} \to \mathcal{X}$ (see Appendix C.4) where the horizontal space $H_\xi\pi$ at the point $\xi \in \mathcal{E}$ is given by $\operatorname{im}\gamma_\xi$.[2]

[2] Saunders [393] calls such a section a *jet field*.

In fact, *every* connection on the fibred manifold $\pi : \mathcal{E} \to \mathcal{X}$ can be generated this way. More precisely, there is a bijection between the space of all connections and $\Gamma(\pi_0^1)$, the space of global sections $\mathcal{E} \to J_1\pi$. Given such a section $\gamma \in \Gamma(\pi_0^1)$, we define the corresponding connection via its *horizontal lift* $A[\gamma] : \mathcal{E} \times_{\mathcal{X}} T\mathcal{X} \to T\mathcal{E}$ by setting $A[\gamma](\xi, v_x) = \bigl(\xi, \gamma_\xi(v_x)\bigr)$. Conversely, given the horizontal lift A of an arbitrary connection on the fibred manifold π, we define the associated section $\gamma[A] \in \Gamma(\pi_0^1)$ by setting $(\gamma[A])_\xi(v_x) = A(\xi, v_x)$. ◁

The geometry of the jet bundle is to a large extent determined by its *contact structure*. In the previous section we introduced it in terms of the contact codistribution locally spanned by the one-forms ω_μ^α defined by (2.8). Their definition is obviously not intrinsic, although our characterisation of contact forms via pull-backs is.

Now we take a rather different approach to the contact structure which will simultaneously clarify the relation between the jet and the tangent bundle. It is based on the *contact map*: using Definition 2.2.2 of the first-order jet bundle, it becomes the simple evaluation map:

$$\Gamma_1 : \begin{cases} J_1\pi \times_{\mathcal{X}} T\mathcal{X} & \longrightarrow \quad T\mathcal{E} \\ (\xi, \gamma_\xi, v_x) & \longmapsto \bigl(\xi, \gamma_\xi(v_x)\bigr) \end{cases} . \tag{2.28}$$

If we consider this map not pointwise but along the image of some global section $\gamma : \mathcal{E} \to J_1\pi$, then a comparison with Remark 2.2.3 shows that the contact map is essentially a slight reformulation of the horizontal lift of the connection defined by the section γ, as we may now express the lift as $A[\gamma](\xi, v_x) = \Gamma_1\bigl(\gamma(\xi), v_x\bigr)$.

More abstractly, we may consider the contact map Γ_1 as the unique map such that the diagram

$$\begin{array}{ccc} J_1\pi \times_{\mathcal{X}} T\mathcal{X} & \xrightarrow{\quad\Gamma_1\quad} & T\mathcal{E} \\ & {\scriptstyle ((j_1\sigma)\circ\tau_{\mathcal{X}})\times \mathrm{id}_{T\mathcal{X}}} \nwarrow \quad \nearrow {\scriptstyle T\sigma} & \\ & T\mathcal{X} & \end{array} \tag{2.29}$$

commutes for any section $\sigma \in \Gamma_{loc}(\pi)$.

The *complementary contact map* is defined by

$$\theta_1 : \begin{cases} J_1\pi \times_{\mathcal{E}} T\mathcal{E} & \longrightarrow \quad V\pi \\ (\xi, \gamma_\xi, w_\xi) & \longmapsto w_\xi - \Gamma_1\bigl(\xi, \gamma_\xi, T_\xi\pi(w_\xi)\bigr) \end{cases} . \tag{2.30}$$

Comparing again with Remark 2.2.3, we see that the complementary contact map corresponds to the vertical projector of a connection induced by a global section $\gamma : \mathcal{E} \to J_1\pi$, namely $P^V[\gamma](\xi, w_\xi) = \theta_1\bigl(\gamma(\xi), w_\xi\bigr)$.

As the contact map Γ_1 is obviously linear in its last argument, we may interpret it alternatively as a map $J_1\pi \to T^*\mathcal{X} \otimes_{\mathcal{E}} T\mathcal{E}$. Then in local coordinates we have $\Gamma_1 : (x^i, u^\alpha, u_i^\alpha) \mapsto \bigl(x^i, u^\alpha, dx^i \otimes (\partial_{x^i} + u_i^\alpha \partial_{u^\alpha})\bigr)$. Similarly, we may interpret the complementary contact map θ_1 as a map $J_1\pi \to T^*\mathcal{E} \underset{\mathcal{E}}{\otimes} V\pi$ given in local coordinates by $\theta_1 : (x^i, u^\alpha, u^\alpha_i) \mapsto \bigl(x^i, u^\alpha, (\mathrm{d}u^\alpha - u^\alpha_i \mathrm{d}x^i) \otimes \partial_{u^\alpha}\bigr)$.

A central task of the contact structure is to characterise prolonged sections. If we continue to treat Γ_1 as a map $J_1\pi \to T^*\mathcal{X} \underset{\mathcal{E}}{\otimes} T\mathcal{E}$, then the value $\Gamma_1(\rho)$ of the contact map at a point $\rho \in J_1\pi$ may be considered as a linear map $T_x\mathcal{X} \to T_\xi\mathcal{E}$ where $x = \pi^1(\rho)$ and $\xi = \pi^1_0(\rho)$ and thus we may speak of the image of $\Gamma_1(\rho)$. Using this point of view, we achieve our goal as follows.

Proposition 2.2.4. *A section $\gamma \in \Gamma_{loc}(\pi^1)$ is of the form $\gamma = j_1\sigma$ for some section $\sigma \in \Gamma_{loc}(\pi)$, if and only if $\operatorname{im}\Gamma_1\bigl(\gamma(x)\bigr) = T_{\gamma(x)}\pi^1_0\bigl(T_{\gamma(x)}\operatorname{im}\gamma\bigr)$ for all points $x \in \mathcal{X}$ where the section γ is defined.*

Proof. Any section $\gamma \in \Gamma_{loc}(\pi^1)$ induces a section $\sigma = \pi^1_0 \circ \gamma \in \Gamma_{loc}(\pi)$ and obviously, by the chain rule, $T_{\gamma(x)}\pi^1_0\bigl(T_{\gamma(x)}\operatorname{im}\gamma\bigr) = T_{\sigma(x)}\operatorname{im}\sigma = \operatorname{im}T_x\sigma$. Furthermore, we may write $\gamma(x) = \bigl(\sigma(x), \gamma_{\sigma(x)}\bigr)$ with $\gamma_{\sigma(x)} \in T^*_x\mathcal{X} \otimes T_{\sigma(x)}\mathcal{E}$ and we thus find that $\operatorname{im}\Gamma_1\bigl(\gamma(x)\bigr) = \operatorname{im}\gamma_{\sigma(x)}$. If we now assume that $\gamma = j_1\sigma$, then, by our considerations above about the equivalence of the two approaches to jet bundles, $\gamma_{\sigma(x)} = T_x\sigma$ and thus one direction of our claim is proven.

In the opposite direction our assumption implies that $\operatorname{im}\gamma_{\sigma(x)} = \operatorname{im}T_x\sigma$. Furthermore, a section $\hat\sigma \in \Gamma_{loc}(\pi)$ exists such that at the given point $x \in \mathcal{X}$ both $\hat\sigma(x) = \sigma(x)$ and $T_x\hat\sigma = \gamma_{\sigma(x)}$ (in other words, $\gamma(x) = [\hat\sigma]^{(1)}_x$). Thus we have two sections $\sigma, \hat\sigma \in \Gamma_{loc}(\pi)$ with the same value at x and for which $\operatorname{im}T_x\sigma = \operatorname{im}T_x\hat\sigma$. It follows trivially from the properties of a section that this implies $T_x\sigma = T_x\hat\sigma$ and thus $\gamma(x) = [\sigma]^{(1)}_x$. As the point x was arbitrary, we therefore find $\gamma = j_1\sigma$. □

Thus we find for any section $\sigma \in \Gamma_{loc}(\mathcal{E})$ that $\operatorname{im}\Gamma_1\bigl(j_1\sigma(x)\bigr) = \operatorname{im}T_x\sigma$. Looking at the coordinate expressions of $T_x\sigma$ and Γ_1, respectively, this observation provides another simple proof of the equivalence of our two approaches to the jet bundle $J_1\pi$. Furthermore, as it is easy to see that Γ_1, still considered as a map $J_1\pi \to T^*\mathcal{X} \underset{\mathcal{E}}{\otimes} T\mathcal{E}$, is injective, it follows that we may identify $J_1\pi$ with an affine subbundle of the vector bundle $T^*\mathcal{X} \underset{\mathcal{E}}{\otimes} T\mathcal{E} \to \mathcal{E}$.

Remark 2.2.5. Yet another way to introduce the first-order jet bundle uses *Grassmannians*. It is sometimes useful for comparing constructions based on exterior differential systems with computations within the formal theory of differential equations. Recall that the k-Grassmannian of a manifold $\mathcal{E}$ is the bundle $G_k(\mathcal{E})$ over $\mathcal{E}$ whose fibre at a point $\xi \in \mathcal{E}$ is given by $G_k(\mathcal{E})_\xi = \{\mathcal{V} \subseteq T_\xi\mathcal{E} \mid \dim\mathcal{V} = k\}$, i. e. the space of all k-planes in the tangent space.

If $\pi : \mathcal{E} \to \mathcal{X}$ is a fibred manifold with $\dim\mathcal{X} = n$, then we may consider the distinguished subbundle $G(\pi) = G^{\text{trans}}_n(\mathcal{E}) \subset G_n(\mathcal{E})$ consisting of all transversal n-planes, i. e. $G(\pi)_\xi = \bigl\{\mathcal{V} \in G_n(\mathcal{E})_\xi \mid T_\xi\pi(\mathcal{V}) = T_{\pi(\xi)}\mathcal{X}\bigr\}$. As obviously the elements of $G(\pi)_\xi$ represent all possible horizontal spaces of a connection at the point $\xi \in \mathcal{E}$, it follows from Remark 2.2.3 that we may identify $G(\pi)$ and $J_1\pi$. In fact, as a trivial consequence of these elementary considerations, we see that the contact map Γ_1

induces an explicit isomorphism between $J_1\pi$ and $G(\pi)$: we map a point $\rho \in J_1\pi$ to the point $\bigl(\pi_0^1(\rho), \operatorname{im}\Gamma_1(\rho)\bigr) \in G(\pi)$. ◁

It seems that there is no similarly elegant direct definition of a higher-order jet bundle $J_q\pi$ as an affine bundle over $J_{q-1}\pi$, as the conditions describing the corresponding affine spaces are very cumbersome (this becomes immediately evident, if one starts to write out higher-order tangent maps in components). Thus the simplest approach for an intrinsic introduction of $J_q\pi$ is probably given by the coordinate-free version of the definition from the last section detailed in Remark 2.2.1. However, many aspects of higher-order jet bundles may be discussed in a first-order setting via the following iterative approach.

As each jet bundle $J_q\pi$ is also a fibred manifold over the base space $\mathcal{X}$ via the projection $\pi^q : J_q\pi \to \mathcal{X}$, we may construct iterated jet bundles $J_r\pi^q$ over this fibration. The jet bundle $J_{q+1}\pi$ of order $q+1$ over $\mathcal{E}$ is then identified with a submanifold of the first-order jet bundle $J_1\pi^q$ over the qth order one. The mapping $\iota_{q,1} : J_{q+1}\pi \to J_1\pi^q$ describing this identification is intrinsically defined as the prolongation of the identity map on $J_q\pi$, i.e. as the unique mapping satisfying for any section $\sigma \in \Gamma_{loc}(\pi)$ the condition $\iota_{q,1} \circ j_{q+1}\sigma = j_1(j_q\sigma)$.

If we take on $J_q\pi$ our usual local coordinates $(\mathbf{x}, \mathbf{u}^{(q)})$, then local coordinates on $J_1\pi^q$ are $\bigl(\mathbf{x}, (\mathbf{u}^{(q)})^{(1)}\bigr)$, i.e. all derivatives u^α_μ are considered as dependent variables and all their first-order derivatives $u^\alpha_{\mu,i}$ are introduced. The submanifold of $J_1\pi^q$ with which we identify $J_{q+1}\pi$ is described by the local equations $u^\alpha_{\mu,i} = u^\alpha_{\nu,j}$ whenever $\mu + 1_i = \nu + 1_j$. Obviously, they simply express that partial derivatives commute.

Similarly, we may introduce mappings $\iota_{q,r} : J_{q+r}\pi \to J_r\pi^q$ by the requirement that $\iota_{q,r} \circ j_{q+r}\sigma = j_r(j_q\sigma)$ for any section $\sigma \in \Gamma_{loc}(\pi)$. Local coordinates on $J_r\pi^q$ are of the form $u^\alpha_{\mu,\nu}$ with $0 \leq |\mu| \leq q$ and $0 \leq |\nu| \leq r$. The image of $J_{q+r}\pi$ in $J_r\pi^q$ is then described by the equations $u^\alpha_{\mu,\nu} = u^\alpha_{\bar\mu,\bar\nu}$ whenever $\mu + \nu = \bar\mu + \bar\nu$.

More generally, we define the *prolongation* of any map $\Phi : J_q\pi \to \mathcal{E}'$ where $\pi' : \mathcal{E}' \to \mathcal{X}$ is another fibred manifold over $\mathcal{X}$ and Φ is fibred over the identity $\mathrm{id}_{\mathcal{X}}$. The r-fold prolongation $j_r\Phi$ is a map $J_{q+r}\pi \to J_r\pi'$ such that the diagram

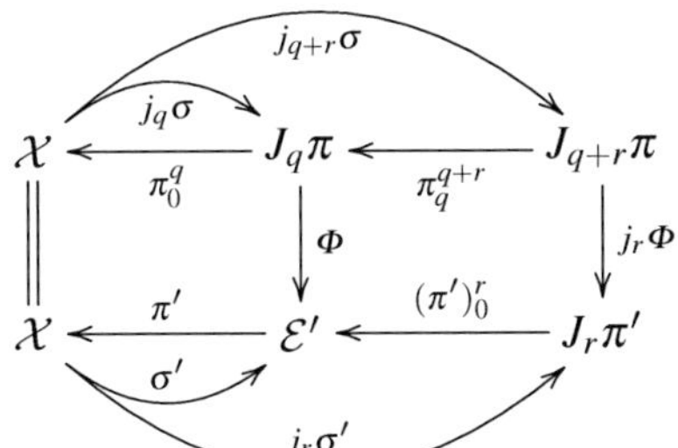

(2.31)

commutes for any section $\sigma \in \Gamma_{loc}(\pi)$, i.e. if we set $\sigma' = \Phi \circ j_q\sigma \in \Gamma_{loc}(\pi')$, then the prolonged map satisfies $(j_r\Phi) \circ (j_{q+r}\sigma) = j_r\sigma'$.

In adapted coordinates $(\mathbf{x}, \mathbf{u})$ on $\mathcal{E}$ and $(\mathbf{x}, \mathbf{v})$ on $\mathcal{E}'$, the map Φ is locally of the form $v^\tau = \Phi^\tau(\mathbf{x}, \mathbf{u}^{(q)})$. The prolongation $j_r\Phi$ has then the local coordinate form $v^\tau_\sigma = D_\sigma\Phi^\tau(\mathbf{x}, \mathbf{u}^{(q+r)})$ with $0 \leq |\sigma| \leq r$. Here $D_\sigma = D_1^{\sigma_1} \cdots D_n^{\sigma_n}$ where D_i is the formal derivative defined by (2.14). Thus the diagram (2.31) is just an abstract representation of the chain rule.

By definition, $(\pi^q)^1_0 : J_1\pi^q \to J_q\pi$ is an affine bundle. It is not difficult to see that that $J_{q+1}\pi \subset J_1\pi^q$ represents an affine subbundle modelled on a vector subbundle of $T^*\mathcal{X} \underset{J_q\pi}{\otimes} V\pi^q$. However, for our purposes this vector subbundle is not so useful. Adapting Proposition 2.1.11 and the subsequent discussion of the affine structure from maps to sections, we immediately obtain the following result.

Proposition 2.2.6. *The jet bundle $J_q\pi$ of order q is an affine bundle over $J_{q-1}\pi$ modelled on the vector bundle $S_q(T^*\mathcal{X}) \underset{J_{q-1}\pi}{\otimes} V\pi$.*

It follows from (2.5) (or (A.4a), respectively) that the fibre dimension of this vector bundle is given by

$$\dim\left(S_q(T^*\mathcal{X}) \underset{J_{q-1}\pi}{\otimes} V\pi\right) = m\binom{n+q-1}{q}\,. \tag{2.32}$$

The *contact structure* of the higher-order jet bundles may be introduced in a manner similar to the first-order case. Generalising (2.29), we define the q-contact map as the uniquely determined map $\Gamma_q : J_q\pi \underset{\mathcal{X}}{\times} T\mathcal{X} \to T(J_{q-1}\pi)$ such that the diagram

$$\begin{array}{ccc}
J_q\pi \underset{\mathcal{X}}{\times} T\mathcal{X} & \xrightarrow{\quad\Gamma_q\quad} & T(J_{q-1}\pi) \\
 & {\scriptstyle ((j_q\sigma)\circ\tau_{\mathcal{X}})\times \mathrm{id}_{T\mathcal{X}}} \nwarrow \quad \nearrow {\scriptstyle T(j_{q-1}\sigma)} & \\
 & T\mathcal{X} &
\end{array} \tag{2.33}$$

commutes for any section $\sigma \in \Gamma_{loc}(\pi)$. For $q > 1$ no straightforward interpretation of Γ_q as an evaluation map exists. However, using iterated jet bundles, we can reduce to the first-order case: consider the contact map $\tilde{\Gamma}_1 : J_1\pi^{q-1} \underset{\mathcal{X}}{\times} T\mathcal{X} \to T(J_{q-1}\pi)$; then $\Gamma_q = \tilde{\Gamma}_1 \circ (\iota_{q-1,1} \times \mathrm{id}_{T\mathcal{X}})$.

It is easy to see that Γ_q is a linear fibred morphism over π^q_{q-1} which can be expressed as the fact that the diagram

$$\begin{array}{ccc}
J_q\pi \underset{\mathcal{X}}{\times} T\mathcal{X} & \xrightarrow{\;\Gamma_q\;} & T(J_{q-1}\pi) \\
{\scriptstyle \mathrm{pr}_1}\big\downarrow & & \big\downarrow{\scriptstyle \tau_{J_{q-1}\pi}} \\
J_q\pi & \xrightarrow[\pi^q_{q-1}]{} & J_{q-1}\pi
\end{array} \tag{2.34}$$

commutes. Here pr_1 is the projection to the first factor and τ the tangent bundle projection. The linearity allows us to consider Γ_q as a map $J_q\pi \to T^*\mathcal{X} \underset{J_{q-1}\pi}{\otimes} T(J_{q-1}\pi)$ given in local coordinates by

$$\Gamma_q : (\mathbf{x}, \mathbf{u}^{(q)}) \mapsto \big(\mathbf{x}, \mathbf{u}^{(q-1)}, \mathrm{d}x^i \otimes (\partial_{x^i} + u^\alpha_{\mu+1_i}\partial_{u^\alpha_\mu})\big) . \tag{2.35}$$

Note that we sum here only over those multi indices μ with $0 \leq |\mu| < q$.

The *complementary contact map* of the qth order jet bundle is now the mapping $\theta_q : J_q\pi \underset{J_{q-1}\pi}{\times} T(J_{q-1}\pi) \to V\pi^{q-1}$ defined by $\theta_q(\rho, w) = w - \Gamma_q\big(\rho, T_{\bar\rho}\pi^{q-1}(w)\big)$ where $\bar\rho = \pi^q_{q-1}\rho$. Interpreting it as a map $J_q\pi \to T^*J_{q-1}\pi \underset{J_{q-1}\pi}{\otimes} V\pi^{q-1}$, it takes in adapted coordinates on the jet bundle $J_q\pi$ the form

$$\theta_q : (\mathbf{x}, \mathbf{u}^{(q)}) \mapsto \big(\mathbf{x}, \mathbf{u}^{(q-1)}, (\mathrm{d}u^\alpha_\mu - u^\alpha_{\mu+1_i}\mathrm{d}x^i) \otimes \partial_{u^\alpha_\mu}\big) \tag{2.36}$$

where again the summation is only over those multi indices μ with $0 \leq |\mu| < q$. For a global section $J_{q-1}\pi \to J_q\pi$ defining a connection on $\pi^{q-1} : J_{q-1}\pi \to \mathcal{X}$, we get the same relation between Γ_q and the horizontal lift on one hand and θ_q and the vertical projector on the other hand as in the first-order case.

Proposition 2.2.7. *A section $\gamma \in \Gamma_{loc}(\pi^q)$ is of the form $\gamma = j_q\sigma$ for some section $\sigma \in \Gamma_{loc}(\pi)$, if and only if $\operatorname{im}\Gamma_q\big(\gamma(x)\big) = T_{\gamma(x)}\pi^q_{q-1}\big(T_{\gamma(x)}\operatorname{im}\gamma\big)$ for all points $x \in \mathcal{X}$ where γ is defined.*

The proof is as in the first-order case (Proposition 2.2.4), though notationally much more messy. Therefore we omit it.

Remark 2.2.8. Again we observe that for any local section $\sigma \in \Gamma_{loc}(\pi)$ the equality $\operatorname{im}\Gamma_{q+1}\big(j_{q+1}\sigma(x)\big) = \operatorname{im}T_x(j_q\sigma)$ holds. Thus we may say that knowing the $(q+1)$-jet $[\sigma]^{(q+1)}_x$ of a section σ at some $x \in \mathcal{X}$ is equivalent to knowing its q-jet $\rho = [\sigma]^{(q)}_x$ at x plus the tangent space $T_\rho \operatorname{im} j_q\sigma$ at this point. ◁

There remains to study the relation between the (complementary) contact map and the contact (co)distribution introduced in the last section. Looking at its expression (2.35) in local coordinates, one already recognises the contact fields $C^{(q)}_i$ introduced in (2.11a). However, one must be careful: in (2.35) one is dealing with vectors in $T(J_{q-1}\pi)$ whereas the fields $C^{(q)}_i$ live in $T(J_q\pi)$. Hence the rigorous statement is that $\Gamma_q(\rho, \partial_{x^i}) = T_\rho\pi^q_{q-1}(C^{(q)}_i|_\rho)$. Since $\ker T\pi^q_{q-1} = V\pi^q_{q-1}$ is spanned by the contact fields C^μ_α defined in (2.11b), we therefore find

$$(\mathcal{C}_q)_\rho = (T_\rho\pi^q_{q-1})^{-1}\big(\operatorname{im}\Gamma_q(\rho)\big) . \tag{2.37}$$

Similarly, in the local form (2.36) of the complementary contact map the contact forms ω^α_μ introduced in (2.8) appear. But again the same problem as above shows up: in (2.36) the forms live on $J_{q-1}\pi$, in (2.8) on $J_q\pi$. This time the solution consists of a simple pull-back. Canonically identifying the vertical bundle $V\pi^{q-1}$ of the fibration $\pi^{q-1} : J_{q-1}\pi \to \mathcal{X}$ with its bidual $V^{**}\pi^{q-1}$ (cf. Remark B.1.7), we may consider the complementary contact map θ_q as a map $J_q\pi \underset{J_{q-1}\pi}{\times} V^*\pi^{q-1} \to T^*(J_{q-1}\pi)$ and find then that $(\pi^q_{q-1})^*(\theta_q(\rho, \mathrm{d}u^\alpha_\mu)) = \omega^\alpha_\mu|_\rho$. In other words,

$$(\mathcal{C}_q^0)_\rho = (\pi^q_{q-1})^*\bigl(\operatorname{im}\theta_q(\rho)\bigr)\,. \tag{2.38}$$

Remark 2.2.9. In the last section, the contact distribution $\mathcal{C}_q$ was generated by two types of vector fields: $C_i^{(q)}$ and C_α^μ defined by (2.11). As these fields depend on the chosen local coordinates, they have of course no intrinsic meaning; however, the splitting into two types has. Obviously, the fields C_α^μ form a basis of the vertical bundle $V\pi^q_{q-1}$. Thus we may decompose $\mathcal{C}_q = \mathcal{H}_q \oplus V\pi^q_{q-1}$ where the n-dimensional complement $\mathcal{H}_q$ is not uniquely defined. It follows from the local form (2.11a) of the contact fields $C_i^{(q)}$ that any such distribution $\mathcal{H}_q$ defines the horizontal bundle of a connection of the fibred manifold $\pi^q : J_q\pi \to \mathcal{X}$.

The local form (2.11a) of the fields $C_i^{(q)}$ also trivially implies the following relation between the contact distributions at different orders:

$$\mathcal{C}_q = T\pi^{q+r}_q(\mathcal{C}_{q+r}) + V\pi^q_{q-1}\,. \tag{2.39}$$

Note that this decomposition is not a direct sum, as the fields $C_i^{(q)}$ are not projectable. Given a global section $\gamma \in \Gamma(\pi^{q+r}_q)$ of the fibration $\pi^{q+r}_q : J_{q+r}\pi \to J_q\pi$, we find for any point $\rho \in J_q\pi$ the direct decomposition

$$\mathcal{C}_q\big|_\rho = T_{\gamma(\rho)}\bigl(\mathcal{C}_{q+r}\big|_{\gamma(\rho)}\bigr) \oplus V\pi^q_{q-1}\,. \tag{2.40}$$

Note, however, that not every possible complement $\mathcal{H}_q$ can be produced that way, as a simple dimensional argument shows. ◁

Remark 2.2.10. As in the first-order case, any global section $J_{q-1}\pi \to J_q\pi$ defines a connection on the fibred manifold $\pi^{q-1} : J_{q-1}\pi \to \mathcal{X}$. We do not show here how the splitting of the tangent bundle into a horizontal and a vertical subbundle is performed, as this is fairly obvious from the relation to iterated jet bundles. ◁

Remark 2.2.11. If we consider—following Remark 2.2.5 on the definition of $J_1\pi$ via Grassmannians—the projection $\pi^1_0 : J_1\pi \to \mathcal{E}$ as a map $G(\pi) \to \mathcal{E}$, then it follows from the above discussion of the relation between the contact map Γ_1 and the contact distribution $\mathcal{C}_1$ on one side and between the complementary contact map θ_1 and the contact codistribution $\mathcal{C}_1^0$ on the other side that at a point $\rho = (\xi, \mathcal{V}) \in G(\pi)$ we have $(\mathcal{C}_1)_\rho = (T_\rho\pi^1_0)^{-1}(\mathcal{V})$ (i. e. $(\mathcal{C}_1)_\rho$ is the preimage of $\mathcal{V}$ under the tangent map of the projection $\pi^1_0 : J_1\pi \to \mathcal{E}$) and $(\mathcal{C}_1^0)_\rho = (\pi^1_0)^*(\mathcal{V}^0)$ (i. e. $(\mathcal{C}_1^0)_\rho$ is the pull-back of the annihilator of the n-plane $\mathcal{V}$). ◁

The contact map provides us with yet another point of view of the fundamental identification. Take some point $\rho \in J_{q-1}\pi$ and choose an arbitrary "reference point" $\hat\rho \in (J_q\pi)_\rho$ in the fibre over it. Then we define on the fibre $(J_q\pi)_\rho$ a map $\chi_q[\hat\rho]$ mapping the point $\bar\rho \in (J_q\pi)_\rho$ to $\Gamma_q(\bar\rho) - \Gamma_q(\hat\rho)$. Here we consider Γ_q again as a map $J_q\pi \to T^*\mathcal{X} \underset{J_{q-1}\pi}{\otimes} T(J_{q-1}\pi)$. By our considerations above, one easily sees that actually $\chi_q[\hat\rho] : (J_q\pi)_\rho \to T_x^*\mathcal{X} \otimes V_\rho\pi^{q-1}_{q-2}$ where as usual $x = \pi^{q-1}(\rho)$: the difference between two images of Γ_q for arguments that lie in the same fibre over $J_{q-1}\pi$ is always vertical with respect to π^{q-1}_{q-2}. Since the tangent space to a vector space may be identified with the vector space itself, the tangent map $T_{\bar\rho}\chi_q[\hat\rho]$ is a homomorphism $T_{\bar\rho}\big((J_q\pi)_\rho\big) = V_{\bar\rho}\pi^q_{q-1} \to T^*_x\mathcal{X}\otimes V_\rho\pi^{q-1}_{q-2}$ which is injective and independent of the chosen "reference point" $\hat\rho$.

In a local chart we get the following picture: let $(\mathbf{x},\hat{\mathbf{u}}^{(q)})$ and $(\mathbf{x},\bar{\mathbf{u}}^{(q)})$ be the coordinates of $\hat\rho$ and $\bar\rho$, respectively; then the value of the map $\chi[\hat\rho]$ at $\bar\rho$ is

$$\chi_q[\hat\rho](\bar\rho) = \sum_{|\mu|=q-1} (\bar u^\alpha_{\mu+1_i} - \hat u^\alpha_{\mu+1_i})\mathrm{d}x^i \otimes \partial_{u^\alpha_\mu} \tag{2.41}$$

and the value of its tangent map is

$$T_{\bar\rho}\chi_q[\hat\rho] = \sum_{|\mu|=q-1} \mathrm{d}u^\alpha_{\mu+1_i} \otimes \mathrm{d}x^i \otimes \partial_{u^\alpha_\mu}\,. \tag{2.42}$$

This local form shows explicitly that the latter one is indeed independent of the "reference point" $\hat\rho$ as remarked above.

For an order $q>1$ we iterate, i.e. we apply on the right hand side the map $\chi_{q-1}[\rho]$ to the vertical component. After $q-1$ iterations we obtain this way a map $V_{\bar\rho}\pi^q_{q-1} \to T^*_x\mathcal{X}\otimes\cdots\otimes T^*_x\mathcal{X}\otimes V_\xi\pi$ where $\xi = \pi^{q-1}_0(\rho)$. It is not difficult to see (for example from the expressions in local coordinates) that its image actually lies in $S_q(T^*_x\mathcal{X})\otimes V_\xi\pi$ and hence it represents the fundamental identification ε_q.

In local coordinates, one easily recovers the expression (2.21) for ε_q by explicitly going through the iteration. The combinatorial factor $\frac{1}{\mu!}$ arises now from the canonical projection $\bigotimes^q T^*_x\mathcal{X} \to S_q(T^*_x\mathcal{X})$.

Addendum: The Contact Structure à la Gardner–Shadwick

Gardner and Shadwick [149] proposed an intrinsic version of the "pedestrian" approach to define the contact codistribution. Let $\rho = [\sigma]^{(q)}_x \in J_q\pi$ be an arbitrary point in the qth-order jet bundle and $\bar\rho = \pi^q_{q-1}(\rho)$ its projection to order $q-1$. Then we may define

$$(\mathcal{C}^0_q)_\rho = \Big((\pi^q_{q-1})^* - (j_{q-1}\sigma\circ\pi^q)^*\Big)T_{\bar\rho}(J_{q-1}\pi)\,. \tag{2.43}$$

As we operate only at a fixed point ρ, this definition is independent of the section σ chosen to represent ρ: the pull-back $(j_{q-1}\sigma)^*$ depends only on the derivatives of σ of order less than or equal to q (the pull-back requires the Jacobian of $j_{q-1}\sigma$); but these possess identical values at x for all sections in the equivalence class ρ.

In local coordinates, it is easy to see that this definition is equivalent to the one given in the last section. We must only determine what the pull-backs do with the basis forms $\mathrm{d}x^i$ and $\mathrm{d}u^\alpha_\mu$. The former ones are annihilated, as all pull-backs are the identity for them. The same happens with the forms $\mathrm{d}u^\alpha_\mu$ with $|\mu|=q$; for them the pull-backs vanish. For the remaining ones we find

$$\Bigl((\pi^q_{q-1})^* - (j_{q-1}\sigma\circ\pi^q)^*\Bigr)\mathrm{d}u^\alpha_\mu = \omega^\alpha_\mu\,. \tag{2.44}$$

Proposition 2.1.4 characterising the contact codistribution may now be reformulated as follows: a section $\gamma\in\Gamma_{loc}(\pi^q)$ satisfies $\gamma^*(\mathcal{C}^0)=0$ if and only if $\gamma=j_q\sigma$ for a section $\sigma\in\Gamma_{loc}(\pi)$. One direction is proved by simple pull-back computations:

$$\begin{aligned}\Bigl((j_q\sigma)^*\mathcal{C}^0_q\Bigr)_x &= \Bigl((\pi^q_{q-1}\circ j_q\sigma)^* - (j_{q-1}\sigma\circ\pi^q\circ j_q\sigma)^*\Bigr)T_{\bar\rho}(J_{q-1}\pi)\\ &= \Bigl((j_{q-1}\sigma)^* - (j_{q-1}\sigma)^*\Bigr)T_{\bar\rho}(J_{q-1}\pi)=0\,.\end{aligned} \tag{2.45}$$

For the converse we use an induction over the order q of the jet bundle. For $q=1$ we set $\sigma=\pi^1_0\circ\gamma\in\Gamma_{loc}(\pi)$ and choose an arbitrary point $x_0\in\mathcal{X}$. Then $\gamma(x)=[\hat\sigma]^{(1)}_{x_0}$ for some section $\hat\sigma\in\Gamma_{loc}(\pi)$ and, by construction, $\sigma(x_0)=\hat\sigma(x_0)$. Now we evaluate our assumption:

$$\begin{aligned}(\gamma^*\mathcal{C}^0_1)_{x_0} &= \Bigl((\pi^1_0\circ\gamma)^* - (\hat\sigma\circ\pi^1\circ\gamma)^*\Bigr)T_{\sigma(x_0)}\mathcal{E}\\ &= \bigl(\sigma^*-\hat\sigma^*\bigr)T_{\sigma(x_0)}\mathcal{E}=0\,.\end{aligned} \tag{2.46}$$

But the last equation can only be satisfied, if $T_{x_0}\sigma=T_{x_0}\hat\sigma$ and hence we must have $\gamma(x_0)=j_1\sigma(x_0)$. As x_0 was arbitrary, this proves our claim for $q=1$.

For the induction step we first note that $(\pi^q_{q-1})^*\mathcal{C}^0_{q-1}\subset\mathcal{C}^0_q$. Indeed, let $\rho\in J_q\pi$, $\bar\rho=\pi^q_{q-1}(\rho)$ and $\bar{\bar\rho}=\pi^q_{q-2}(\rho)$. Then

$$\begin{aligned}\Bigl((\pi^q_{q-1})^*\mathcal{C}^0_{q-1}\Bigr)_\rho &= \Bigl((\pi^{q-1}_{q-2}\circ\pi^q_{q-1})^* - (j_{q-2}\sigma\circ\pi^{q-1}\circ\pi^q_{q-1})^*\Bigr)T_{\bar{\bar\rho}}\mathcal{E}\\ &= \Bigl((\pi^q_{q-1})^* - (j_{q-1}\sigma\circ\pi^q)^*\Bigr)(\pi^{q-1}_{q-2})^*T_{\bar{\bar\rho}}\mathcal{E}\\ &\subset \Bigl((\pi^q_{q-1})^* - (j_{q-1}\sigma\circ\pi^q)^*\Bigr)T_{\bar\rho}\mathcal{E}\,.\end{aligned} \tag{2.47}$$

Now we set again $\sigma=\pi^q_0\circ\gamma$ and consider the point $\rho=\gamma(x_0)=[\hat\sigma]^{(q)}_{x_0}$. Since $\gamma^*\bigl((\pi^q_{q-1})^*\mathcal{C}^0_{q-1}\bigr)=(\pi^q_{q-1}\circ\gamma)^*\mathcal{C}^0_{q-1}=0$, it follows already from our induction hypothesis that $\pi^q_{q-1}\circ\gamma=j_{q-1}\sigma$. Hence in particular $j_{q-1}\sigma(x_0)=j_{q-1}\hat\sigma(x_0)$. Finally, by a similar computation as in the case $q=1$ we obtain that furthermore $(j_{q-1}\sigma)^*=(\pi^q_{q-1}\circ\gamma)^*=(j_{q-1}\hat\sigma)^*$ and hence we may conclude by the same argument as above that $j_q\sigma(x_0)=j_q\hat\sigma(x_0)$. Thus $\gamma=j_q\sigma$ as claimed.

A dual formulation of this approach was proposed already earlier by Goldschmidt and Sternberg [169]. They proved the existence of a unique vector valued one-form $\omega\in\Omega^1(J_q\pi,T(J_{q-1}\pi))$ with the property that $(j_q\sigma)^*\omega=0$ for all local sections $\sigma\in\Gamma_{loc}(\pi)$ and $\omega(v)=T\pi^q_{q-1}(v)$ for all vertical vectors $v\in V\pi^q$. It is not difficult to show that these conditions imply at the point $\rho=[\sigma]^{(q)}_x\in J_q\pi$ that the form ω is defined by

$$\omega(v) = \left(T_\rho \pi^q_{q-1} - T_\rho(j_q\sigma \circ \pi^q)\right)v\,. \tag{2.48}$$

In local coordinates, this leads to the familiar expression

$$\omega = (\mathrm{d}u^\alpha_\mu - u^\alpha_{\mu+1_i}\mathrm{d}x^i) \otimes \partial_{u^\alpha_\mu} \tag{2.49}$$

where the summation is over all multi indices with $0 \leq |\mu| < q$. Thus in the chosen coordinates the components of the intrinsically defined form ω are just the contact forms ω^α_μ spanning the contact codistribution $\mathcal{C}^0_q$.

2.3 Differential Equations

With jet bundles at our hands, we give a geometric definition of (systems of) differential equations. It does not require coordinates or local equations but defines a differential equation as an intrinsic geometric object, namely a fibred submanifold.

Definition 2.3.1. A (non-linear) *differential equation* of order q is a fibred submanifold $\mathcal{R}_q$ of the jet bundle $J_q\pi$.

Since submanifold means for us always *regular submanifold* (see Appendix C.1 for more details), locally, i. e. within some open subset $\mathcal{V}_q \subseteq J_q\pi$, we may always assume that a fibred map

$$\Phi : \begin{cases} \mathcal{V}_q \subseteq J_q\pi \longrightarrow \quad \mathcal{E}' \\ (\mathbf{x}, \mathbf{u}^{(q)}) \longmapsto \Phi(\mathbf{x}, \mathbf{u}^{(q)}) \end{cases} \tag{2.50}$$

exists taking values in a vector bundle $\pi' : \mathcal{E}' \to \mathcal{X}$ over the same base space $\mathcal{X}$ such that $\mathcal{R}_q \cap \mathcal{V}_q$ is the zero set of Φ. Thus locally $\mathcal{R}_q$ can be described by a system of equations $\Phi^\tau(\mathbf{x}, \mathbf{u}^{(q)}) = 0$ with $1 \leq \tau \leq t = \dim \mathcal{E}'$ which brings us back to the traditional picture of a system of differential equations.

Remark 2.3.2. In the sequel we will mostly assume the existence of a *global* morphism $\Phi : J_q\pi \to \mathcal{E}'$ such that the submanifold $\mathcal{R}_q$ is its zero set, i. e. that our equation is defined by the (non-linear) *differential operator* $\Delta[\Phi] : \Gamma_{loc}(\pi) \to \Gamma_{loc}(\pi')$ where $\Delta[\Phi](\sigma) = (\Phi \circ j_q)\sigma$. As we have mainly local applications in mind, this assumption is not a serious restriction but simplifies the notation at some places. Note that in Appendix C.1 we have defined sections as smooth maps, but obviously the notion of a differential operator makes sense in other function spaces, too. ◁

Definition 2.3.1 does not distinguish between scalar equations and systems. We use the following convention. When we speak of a *differential equation*, we always think of it as the geometric object $\mathcal{R}_q$. When we speak of a *system*, we think of a local representation of the submanifold $\mathcal{R}_q$ given by a map of the form (2.50). As such a system consists again of equations $\Phi(\mathbf{x}, \mathbf{u}^{(q)}) = 0$, it becomes of course sometimes a bit difficult to strictly maintain this distinction...

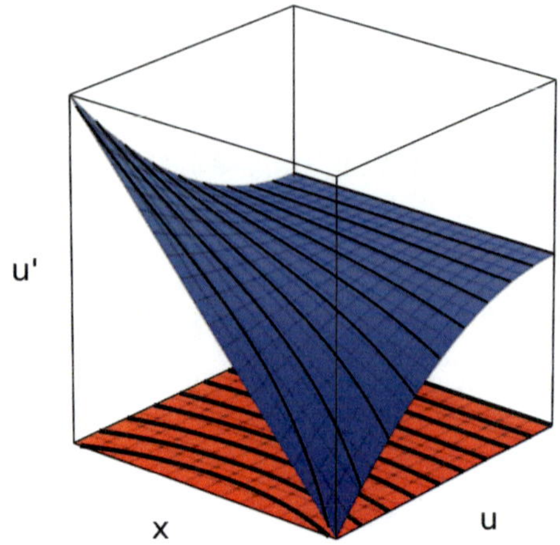

Fig. 2.1 An ordinary differential equation and some of its solutions

Definition 2.3.1 requires $\mathcal{R}_q$ to be a *fibred* submanifold (with respect to the fibration $\pi^q : J_q\pi \to \mathcal{X}$). This condition implies that $\pi^q(\mathcal{R}_q) = \mathcal{X}$. If it were violated, the independent variables might not be really independent, as the differential equation might impose relations between them. More precisely, local representations of $\mathcal{R}_q$ might contain equations $\boldsymbol{\Phi}(\mathbf{x}) = 0$ which obviously should not happen.

We need of course also the notion of a solution. Classically, solutions are functions that entered into a given system let it vanish identically. In our geometric framework, functions are substituted by sections.

Definition 2.3.3. Let $\mathcal{R}_q \subseteq J_q\pi$ be a differential equation of order q. A *(local) solution* is a smooth section $\sigma \in \Gamma_{loc}(\pi)$ such that $\operatorname{im} j_q\sigma \subseteq \mathcal{R}_q$.

In local coordinates, a section $\sigma : \mathcal{X} \to \mathcal{E}$ is written as a map $\mathbf{x} \mapsto \big(\mathbf{x}, \mathbf{s}(\mathbf{x})\big)$. Its prolongation $j_q\sigma : \mathcal{X} \to J_q\pi$ has then the local form $\mathbf{x} \mapsto \big(\mathbf{x}, \partial^{|\mu|}\mathbf{s}(\mathbf{x})/\partial x^\mu\big)$ where μ runs over all multi indices with $0 \le |\mu| \le q$. If the given differential equation $\mathcal{R}_q \subseteq J_q\pi$ is locally represented by the system $\boldsymbol{\Phi}(\mathbf{x}, \mathbf{u}^{(q)}) = 0$, then the condition $\operatorname{im} j_q\sigma \subseteq \mathcal{R}_q$ requires that $\boldsymbol{\Phi}(\mathbf{x}, \partial^{|\mu|}\mathbf{s}(\mathbf{x})/\partial x^\mu) = 0$ for all $\mathbf{x}$ where σ is defined, i. e. the smooth function $\mathbf{s}(\mathbf{x})$ is a local solution in the conventional sense.

Example 2.3.4. As the dimension of the jet bundles $J_q\pi$ is rapidly growing with the order q and with the number n of independent variables, it is difficult to represent differential equations graphically. Only if we restrict to a first-order ordinary differential equation in one unknown function, then $J_1\pi \cong \mathbb{R}^3$ is three-dimensional and we can easily plot the equation and some of its solutions.

Figure 2.1 shows the differential equation $\mathcal{R}_1$ globally defined as the zero set of the function $(x, u, u') \mapsto u' + xu^2$ as a surface in $J_1\pi$. It is not difficult to integrate this separable equation and its solution space consists of the one-parameter family of global sections $\sigma_c : x \mapsto (x, \frac{2c}{2+cx^2})$ with a real parameter c. The red plane in Figure 2.1 represents the x-u plane (for better visibility it is drawn at the bottom of the picture) and the curves on it are the graphs of sections σ_c for some values of the parameter c. As one can see, the graphs of the corresponding 1-jets $j_1\sigma_c$ lie indeed completely on the submanifold $\mathcal{R}_1$. $\lhd$

Remark 2.3.5. It is instructive to compare our definition of a differential equation with the situation in algebraic geometry. There a variety is the zero set of some polynomials, i. e. every point on the variety represents a solution of the algebraic equations defined by the polynomials. Let the differential equation $\mathcal{R}_q \subseteq J_q\pi$ be locally represented by the system $\boldsymbol{\Phi}(\mathbf{x},\mathbf{u}^{(q)}) = 0$. It may be understood as a system of *algebraic* equations on $J_q\pi$ and the points on $\mathcal{R}_q$ are the solutions of these equations. As in algebraic geometry, the individual functions Φ^τ do not really matter but only the ideal in the ring $\mathcal{F}(J_q\pi)$ generated by them.

Recall that a point of the jet bundle $J_q\pi$ is defined as an equivalence class of sections. Let the class $[\sigma]_x^{(q)}$ be a point in $J_q\pi$ and $\mathcal{V} \subseteq \mathcal{X}$ a neighbourhood of x such that σ is defined on $\mathcal{V}$. If the section σ is a solution of $\mathcal{R}_q$ in the sense of Definition 2.3.3, the points $[\sigma]_x^{(q)}$ (for all $x \in \mathcal{V}$) must lie on the submanifold $\mathcal{R}_q$. However, the converse does not necessarily hold and the equivalence classes $[\sigma]_x^{(q)}$ in general contain many sections that are *not* solutions of $\mathcal{R}_q$.

This effect is easy to understand, as by definition the underlying equivalence relation $\sim_q$ takes into account only the values of the derivatives of σ up to order q at a *single* point $x \in \mathcal{X}$. Thus $[\sigma]_x^{(q)} \in \mathcal{R}_q$ only implies that at the considered point x the section σ satisfies the differential equation $\mathcal{R}_q$ up to order q but no statements about higher orders or other points are possible.

As a trivial example consider the ordinary differential equation $\mathcal{R}_2$ defined by $u'' = 0$. An equivalence class $[\sigma]_x^{(2)}$ is contained in $\mathcal{R}_2$ whenever the second derivative of σ vanishes at the point x. Thus for the section σ defined by $\sigma(x) = (x, x^3)$ the class $[\sigma]_0^{(2)}$ lies in $\mathcal{R}_2$, although σ is obviously not a solution. Note, however, that for any other value of x the class $[\sigma]_x^{(2)}$ is not contained in $\mathcal{R}_2$. If $[\sigma]_x^{(2)} \in \mathcal{R}_2$ for *all* points x in some open set $\mathcal{V} \subseteq \mathcal{X}$, then the section σ defines at least on $\mathcal{V}$ a local solution of the differential equation $\mathcal{R}_2$. $\lhd$

Remark 2.3.6. First-order *ordinary* differential equations $\mathcal{R}_1 \subset J_1\pi$ possess a special geometric structure. If $\mathcal{R}_1$ is the image of a global section $\gamma : \mathcal{E} \to J_1\pi$, we may interpret it according to Remark 2.2.3 as a connection on $\pi : \mathcal{E} \to \mathcal{X}$. Locally, such equations correspond to systems of the form $\mathbf{u}' = \boldsymbol{\phi}(x, \mathbf{u})$, i. e. we are dealing with a pure differential equation in solved form without algebraic constraints. If such constraints are present, then they determine a fibred submanifold $\mathcal{R}_0^{(1)} \subset \mathcal{E}$ and we consider equations defined by sections $\gamma : \mathcal{R}_0^{(1)} \to J_1\pi$ corresponding to connections on $\mathcal{R}_0^{(1)}$. This observation extends straightforwardly to an identification of qth-order ordinary differential equations with connections on $\pi^{q-1} : J_{q-1}\pi \to \mathcal{X}$.

In the case of *partial* differential equations, the situation is different; only very special equations correspond to connections. Let $\gamma : \mathcal{E} \to J_1\pi$ be a global section. Then $\mathcal{R}_1 = \operatorname{im}\gamma$ defines an *equation of finite type*, also called a *maximally overdetermined* equation (see Remarks 7.1.2 and 8.2.4 for an explanation of this terminology). Locally, it can be represented by a system of the form $\mathbf{u}_{(1)} = \boldsymbol{\phi}(\mathbf{x},\mathbf{u})$, i. e. the system contains one equation for every first-order derivative u_i^α.

The corresponding connection form is simply $\omega = (\mathrm{d}u^\alpha - \phi_i^\alpha \mathrm{d}x^i) \otimes \partial_{u^\alpha}$, i. e. the horizontal bundle is spanned by the vector fields $X_i = \partial_{x^i} + \phi_i^\alpha \partial_{u^\alpha}$. Indeed, if we interpret as in Section 2.2 the fibre component of each point $(\mathbf{x},\mathbf{u}^{(1)}) \in \mathcal{R}_1$ as a linear map $T_\mathbf{x}\mathcal{X} \to T_{(\mathbf{x},\mathbf{u})}\mathcal{E}$, then in our local coordinates its matrix is given by stacking the $n \times n$ identity matrix over the $m \times n$ matrix defined by ϕ_i^α and the horizontal lift of the vector field $\partial_{x^i} \in \mathfrak{X}(\mathcal{X})$ yields the field $X_i \in \mathfrak{X}(\mathcal{E})$. Assuming that $\pi : \mathcal{E} \to \mathcal{X}$ is a vector bundle and considering the covariant derivative associated with the connection as a map $\nabla : \Gamma_{loc}(\pi) \to \Omega^1(\mathcal{X},\mathcal{E})$, we may identify $\mathcal{R}_1$ with the system $\nabla\sigma = 0$. The inclusion of constraints defining a fibred submanifold $\mathcal{R}_0^{(1)} \subset \mathcal{E}$ is again trivial.

In Example 2.3.17, we will see that a section $\sigma \in \Gamma_{loc}(\pi)$ is a solution of the differential equation $\mathcal{R}_1$, if and only if its prolonged graph $\operatorname{im} j_1\sigma$ defines an integral manifold of the horizontal bundle of the corresponding connection. There we will also discuss when solutions exist. ◁

Two natural geometric operations with differential equations exist: *prolongation* and *projection*. The former one lifts an equation to higher order; the latter one lowers the order. Both operations are of central importance for everything that we will do with differential equations.

Prolonging a differential equation is easy to perform effectively, once a local representation is given. Its intrinsic geometric description is a bit more involved. We need again the embeddings $\iota_{q,r} : J_{q+r}\pi \hookrightarrow J_r\pi^q$. As $\mathcal{R}_q \subseteq J_q\pi$ is also a fibred manifold over $\mathcal{X}$ with the restriction $\hat{\pi}^q$ of π^q to $\mathcal{R}_q$ as projection, we can construct jet bundles over it, for example $J_r\hat{\pi}^q$. It may be considered in a natural way as a submanifold of the iterated jet bundle $J_r\pi^q$. We denote the corresponding embedding by $\hat{\iota}_{q,r}$. Then we define the rth prolongation $\mathcal{R}_{q+r} \subseteq J_{q+r}\pi$ through

$$\mathcal{R}_{q+r} = \iota_{q,r}^{-1}\big(\hat{\iota}_{q,r}(J_r\hat{\pi}^q) \cap \iota_{q,r}(J_{q+r}\pi)\big) \subseteq J_{q+r}\pi \,. \tag{2.51}$$

In words, this somewhat awkward construction has a fairly simple explanation. We consider both $J_r\hat{\pi}^q$ and $J_{q+r}\pi$ as submanifolds of $J_r\pi^q$ using the corresponding embeddings. Then we take their intersection and consider the result via the inverse embedding as a submanifold of $J_{q+r}\pi$. Thus starting with a differential equation of order q, we have obtained an equation of order $q+r$.

In the sequel, we will follow the usual simplification to identify the image of an inclusion map with the source space and to skip the explicit mentioning of all these maps. Hence from now on we will write (2.51) concisely as

$$\mathcal{R}_{q+r} = J_r\hat{\pi}^q \cap J_{q+r}\pi \,. \tag{2.52}$$

If our differential equation $\mathcal{R}_q$ comes from a differential operator, its prolongation can be expressed via the prolongation of the corresponding global map $\boldsymbol{\Phi}$. We get the following commuting diagram

$$\begin{array}{ccccccc} 0 & \longrightarrow & \mathcal{R}_{q+r} & \longrightarrow & J_{q+r}\pi & \xrightarrow{\rho_r\Phi} & J_r\pi' \\ & & \Big\downarrow \hat{\pi}^{q+r}_q & & \Big\downarrow \pi^{q+r}_q & & \Big\downarrow (\pi')^r_0 \\ 0 & \longrightarrow & \mathcal{R}_q & \longrightarrow & J_q\pi & \xrightarrow{\Phi} & \mathcal{E}' \end{array} \tag{2.53}$$

and the prolonged equation $\mathcal{R}_{q+r}$ is the zero set of the prolonged map $\rho_r\boldsymbol{\Phi}$.

In local coordinates, the prolongation is computed with the help of the formal derivative (2.14). If $\boldsymbol{\Phi}(\mathbf{x},\mathbf{u}^{(q)}) = 0$ is a local representation of $\mathcal{R}_q \subseteq J_q\pi$, then its prolongation $\mathcal{R}_{q+r} \subseteq J_{q+r}\pi$ is described locally by the system

$$\mathcal{R}_{q+r}: \begin{cases} \Phi^\tau(\mathbf{x},\mathbf{u}^{(q)}) = 0\,, & 1 \leq \tau \leq t\,, \\ D_j\Phi^\tau(\mathbf{x},\mathbf{u}^{(q+1)}) = 0\,, & 1 \leq j \leq n\,, \\ \vdots & \\ D_\nu\Phi^\tau(\mathbf{x},\mathbf{u}^{(q+r)}) = 0\,, & |\nu| = r\,. \end{cases} \tag{2.54}$$

Indeed, the left hand sides are just the coordinate expressions of $\rho_r\boldsymbol{\Phi}$.

By contrast, projections are easily described geometrically but difficult to perform effectively in local coordinates. Geometrically, we simply apply the natural projections $\pi^q_{q-r}: J_q\pi \rightarrow J_{q-r}\pi$ and define for $0 < r \leq q$

$$\mathcal{R}^{(r)}_{q-r} = \pi^q_{q-r}(\mathcal{R}_q) \subseteq J_{q-r}\pi\,. \tag{2.55}$$

Thus we have obtained a differential equation of order $q-r$ from an equation of order q. In order to construct a local representation of $\mathcal{R}^{(r)}_{q-r}$, one takes a local representation of $\mathcal{R}_q$, say $\boldsymbol{\Phi}(\mathbf{x},\mathbf{u}^{(q)}) = 0$, and tries to eliminate by purely algebraic operations the derivatives of order greater than $q-r$ in as many equations as possible. If we cannot find any equation depending only on lower-order derivatives, then $\mathcal{R}^{(r)}_{q-r} = J_{q-r}\pi$. Otherwise it is a proper submanifold of $J_{q-r}\pi$ locally described by all such equations. We will see later many examples of such computations. Fortunately, we mostly project differential equations generated by prolongations. In this case, the elimination of the higher-order derivatives requires only linear algebra, as the formal derivatives always yield quasi-linear equations.

A slight complication arises from the fact that the prolongation or the projection of a given differential equation $\mathcal{R}_q$ is not necessarily a (fibred) submanifold anymore. Indeed, it is rather obvious that the projection of a manifold leads in general to singularities (some concrete examples with curves may be found in Section 9.1).

Example 2.3.7. It is perhaps less evident that prolongations also do not always yield manifolds. We consider the *non-linear* second-order equation

$$\mathcal{R}_2 : \begin{cases} u_{yy} - \frac{1}{2}(u_{xx})^2 = 0 , \\ u_{xy} - \frac{x}{y}u_{xx} = 0 . \end{cases} \tag{2.56}$$

Adding the formal x- and y-derivative of the second equation and subtracting the formal x-derivative of the first one yields $(u_{xx} - x^2/y^2)u_{xxx} = 0$. Thus $\mathcal{R}_3$ lies in the union of the two hypersurfaces $u_{xx} = x^2/y^2$ and $u_{xxx} = 0$. Such a union is *not* a submanifold, as in the neighbourhood of the intersection it is not homeomorphic to an open subset of some $\mathbb{R}^k$. We must distinguish two cases in which the differential equation may show quite different behaviour; in particular, the solution spaces will generally differ. In our example we have $u_{xxx} = 2x/y^2$ in the first case and $u_{xxx} = 0$ in the second one. ◁

For simplicity, we will make the following blanket assumption (almost) throughout the book. It means that we consider only what is called *regular* equations. Regularity can always be achieved by restricting to some submanifold of $\mathcal{R}_q$, but is difficult to check effectively for non-linear equations.

Blanket Assumption 2.3.8. *All performed projections and prolongations lead again to fibred submanifolds.*

One could be tempted to think that prolongation and projection are some kind of inverse operations: if one first prolongs r times a differential equation $\mathcal{R}_q \subseteq J_q\pi$ in order to obtain $\mathcal{R}_{q+r} \subseteq J_{q+r}\pi$ for some $r > 0$ and subsequently projects back to $J_q\pi$ with $\pi^{q+r}_q : J_{q+r}\pi \to J_q\pi$, then one might naively expect that the obtained equation $\mathcal{R}^{(r)}_q$ is identical with the original one $\mathcal{R}_q$. However, in general this will not be the case, as *integrability conditions* may arise: we only get that always $\mathcal{R}^{(r)}_q \subseteq \mathcal{R}_q$.

Example 2.3.9. A trivial example may explain this phenomenon and also demonstrate explicitly the operations of prolongation and projection. We consider the first-order equation $\mathcal{R}_1$ defined by the linear system

$$\mathcal{R}_1 : \begin{cases} u_z + yu_x = 0 , \\ u_y \qquad\;\; = 0 . \end{cases} \tag{2.57}$$

In order to obtain the system for the prolonged equation $\mathcal{R}_2$, we formally differentiate every equation in (2.57) with respect to all three independent variables x, y, and z. The arising equations are then added to the original system and we get

$$\mathcal{R}_2 : \begin{cases} u_z + yu_x = 0 , & u_y = 0 , \\ u_{xz} + yu_{xx} = 0 , & u_{xy} = 0 , \\ u_{yz} + yu_{xy} + u_x = 0 , & u_{yy} = 0 , \\ u_{zz} + yu_{xz} = 0 , & u_{yz} = 0 . \end{cases} \tag{2.58}$$

In order to compute the projection to $J_1\pi$ of the prolongation $\mathcal{R}_2$ we must try to obtain by algebraic manipulations as many first-order equations as possible. Since we are dealing with a linear system, this is not very difficult. The first two equations are obviously first order, as they stem from the original system. In the first equation of the third row we can eliminate both terms containing second-order derivatives using equations of the second column; no other eliminations are possible.

So we find the *potential* integrability condition $u_x = 0$. In order to verify whether it actually is an integrability condition, we must check its restriction to $\mathcal{R}_1$. If the function u_x vanished on $\mathcal{R}_1$, then we would have $\mathcal{R}_1^{(1)} = \mathcal{R}_1$. However, according to Hadamard's Lemma C.1.5, this would require that u_x is a linear combination of the functions describing $\mathcal{R}_1$ which obviously is not the case. Hence $\mathcal{R}_1^{(1)} \subsetneq \mathcal{R}_1$ and a local representation of it is given by

$$\mathcal{R}_1^{(1)} : \begin{cases} u_z = 0\,, \\ u_y = 0\,, \\ u_x = 0\,. \end{cases} \tag{2.59}$$

It is easy to see that further prolongations and subsequent projections do not yield any additional conditions and thus $\mathcal{R}_1^{(1)} = \mathcal{R}_1^{(2)} = \mathcal{R}_1^{(3)} = \cdots$. ◁

It is important to note that in Example 2.3.9 it is not possible to construct the integrability condition by purely algebraic operations without differentiations. This property may serve as an informal definition of integrability conditions: they are algebraically independent of the original system but can be generated using formally differentiated equations. Obviously, any sufficiently smooth solution of the system *automatically* satisfies all its integrability conditions; they do not represent additional restrictions to the solution space but are equations that have been "hidden" in the structure of the original system.

In Example 2.3.9 we see the classical form of an integrability condition. It arises as a generalised cross-derivative: a linear combination of differentiated equations that yields a lower-order equation. Obviously, cross-derivatives make sense only for partial differential equations. However, a second mechanism for the generation of integrability conditions exists which also works for ordinary differential equations: if the original system contains equations of differing orders, the formal differentiation of a lower-order equation may generate an integrability condition; we call these *integrability conditions of the second kind* in order to distinguish them from generalised cross-derivatives.

Example 2.3.10. We consider the trivial second-order equation $\mathcal{R}_2$ with independent variables x, y and dependent variable u described by the linear system

$$\mathcal{R}_2 : \begin{cases} u_{yy} = 0\,, \\ u_x \;\;= 0\,. \end{cases} \tag{2.60}$$

Prolongation to third order yields a system that contains—among others—the second-order equations $u_{xy} = u_{xx} = 0$ obtained by formally differentiating the second equation in (2.60). Thus the projection back to $\mathcal{R}_2$ is not surjective and we must add these two integrability conditions to the system (2.60) in order to obtain a local representation of the second-order equation $\mathcal{R}_2^{(1)}$.

Classically, this type of integrability conditions is usually ignored, as one considers the equation $u_{xx} = 0$ as an obvious consequence of the equation $u_x = 0$. However, in our geometric approach it is very important to include it explicitly, as otherwise one works with a too large submanifold of $J_2\pi$: $\dim \mathcal{R}_2 = \dim \mathcal{R}_2^{(1)} + 2$. In particular, we will see below that the inclusion of these integrability conditions is crucial for the construction of formal power series solutions. ◁

Geometrically, it makes no sense to speak of individual integrability conditions or to distinguish between cross-derivatives and the prolongation of lower-order equations. Instead both kinds of integrability conditions have the same effect, namely that, given a differential equation $\mathcal{R}_q \subseteq J_q\pi$, for some values of r and s the equation $\mathcal{R}_{q+r}^{(s)} \subseteq J_{q+r}\pi$ obtained by $r+s$ prolongations and s subsequent projections is a proper subset of the equation $\mathcal{R}_{q+r} \subseteq J_{q+r}\pi$ obtained by only r prolongations without any projections. More precisely, $\mathcal{R}_{q+r}^{(s)}$ is a lower-dimensional submanifold of $\mathcal{R}_{q+r}$. In Example 2.3.9 $\mathcal{R}_1$ is two-dimensional[3] ($\dim J_1\pi = 4$ and we have two independent equations), whereas $\dim \mathcal{R}_1^{(1)} = 1$ because of the integrability condition.

Remark 2.3.11. Sometimes it is useful to formulate prolongations in terms of Grassmannians; in fact, in the theory of exterior differential systems this represents the standard approach. Given a differential equation $\mathcal{R}_q \subseteq J_q\pi$, we consider the restriction $\hat{\pi}^q : \mathcal{R}_q \to \mathcal{X}$ of the projection $\pi^q : J_q\pi \to \mathcal{X}$. Then, according to Remark 2.2.5, we may identify $G(\hat{\pi}^q)$ and $J_1\hat{\pi}^q$ and we define

$$\mathcal{R}_{q+1} = \left\{(\rho, V) \in G(\hat{\pi}^q) \mid \exists \sigma \in \rho : V = T_\rho(\operatorname{im} j_q\sigma)\right\} . \tag{2.61}$$

The essence of (2.61) is to require that the point $\rho \in \mathcal{R}_q$, considered as an equivalence class of sections, contains a section $\sigma \in \Gamma_{loc}(\pi)$ such that the tangent space at ρ of the prolonged section $j_q\sigma$ is tangential to the submanifold $\mathcal{R}_q$ (the assumption $(\rho, V) \in G(\hat{\pi}^q)$ trivially implies that $V \subseteq T_\rho\mathcal{R}_q$). The existence of such a section $\sigma \in \Gamma_{loc}(\pi)$ is not automatic, as it implies that actually $\rho \in \mathcal{R}_q^{(1)} \subseteq \mathcal{R}_q$, since then $[\sigma]_x^{(q+1)} \in (\pi_q^{q+1})^{-1}(\rho) \cap \mathcal{R}_{q+1}$ for $x = \pi^q(\rho)$, i.e. the point ρ also satisfies all integrability conditions hidden in the prolongation to $\mathcal{R}_{q+1}$.

The equivalence of this alternative definition of the prolongation $\mathcal{R}_{q+1}$ with the one given above is easily understood via the contact map, as it relies on the observation made in Remark 2.2.8. Let $\hat{\rho} = [\sigma]_x^{(q+1)}$ be an arbitrary point on the prolonged equation $\mathcal{R}_{q+1}$ and $\rho = \pi_q^{q+1}(\hat{\rho})$ the corresponding projected point on the original differential equation $\mathcal{R}_q$. For any representative $\sigma \in \Gamma_{loc}(\pi)$ of the equivalence class $\hat{\rho}$ we find that $\operatorname{im}\Gamma_{q+1}(\hat{\rho}) = T_\rho(\operatorname{im} j_q\sigma)$. Since $\hat{\rho} \in \mathcal{R}_{q+1}$ is equivalent to $\operatorname{im}\Gamma_{q+1}(\hat{\rho}) \subseteq T_\rho\mathcal{R}_q$ (this fact follows for example easily from the local coordinate form of Γ_{q+1}), we may identify $\hat{\rho}$ with the pair $\left(\rho, \operatorname{im}\Gamma_{q+1}(\hat{\rho})\right) \in G(\hat{\pi}^q)$.

---

[3] When we speak about the dimension of a differential equation, we will almost always consider its *fibre dimension*. As differential equations are by definition fibred submanifolds, this is more natural and we will see later that these fibre dimensions are related to the size of the formal solution space.

Conversely, assume $\rho = [\sigma]_x^{(q)} \in \mathcal{R}_q$ and $V = T_\rho(\operatorname{im} j_q\sigma)$. Then on one hand we trivially have $\hat{\rho} = [\sigma]_x^{(q+1)} \in J_{q+1}\pi$ and $\operatorname{im}\Gamma_{q+1}(\hat{\rho}) = V$. On the other hand, $(\rho, V) \in G(\hat{\pi}^q) \cong J_1\hat{\pi}^q$, so that indeed $\hat{\rho} \in J_1\hat{\pi}^q \cap J_{q+1}\pi = \mathcal{R}_{q+1}$ according to our original definition of the prolongation. Thus, although (2.61) defines the prolonged equation $\mathcal{R}_{q+1}$ as a submanifold of $G(\hat{\pi}^q) \cong J_1\hat{\pi}^q$, i. e. of an iterated jet bundle, the condition imposed on the vector space V ensures that any point on it also lies in $J_{q+1}\pi$ and the two definitions are equivalent.

Eq. (2.61) is only of theoretical interest, as it cannot be used for explicit computations. Below we will provide an effective version of it based on the notion of an integral element and ideas from the theory of exterior differential systems. ◁

Example 2.3.12. For first-order differential equations with only one dependent variable u, it is possible to give a closed form expression for the integrability conditions. Let Φ and Ψ be two functions defined on the jet bundle $J_1\pi$ where the fibre dimension m of $\mathcal{E}$ is one. We study the equation $\mathcal{R}_1$ given by $\Phi = \Psi = 0$. For the prolonged equation $\mathcal{R}_2$ we need in addition the formal derivatives

$$D_i\Phi = \frac{\partial\Phi}{\partial x^i} + u_i\frac{\partial\Phi}{\partial u} + u_{ij}\frac{\partial\Phi}{\partial u_j}\,, \tag{2.62a}$$

$$D_i\Psi = \frac{\partial\Psi}{\partial x^i} + u_i\frac{\partial\Psi}{\partial u} + u_{ij}\frac{\partial\Psi}{\partial u_j}\,. \tag{2.62b}$$

It is possible to eliminate all second-order derivatives u_{ij} by taking the following linear combination of these equations:

$$\frac{\partial\Phi}{\partial u_i}D_i\Psi - \frac{\partial\Psi}{\partial u_i}D_i\Psi = \frac{\partial\Phi}{\partial u_i}\left(\frac{\partial\Psi}{\partial x^i} + u_i\frac{\partial\Psi}{\partial u}\right) - \frac{\partial\Psi}{\partial u_i}\left(\frac{\partial\Phi}{\partial x^i} + u_i\frac{\partial\Phi}{\partial u}\right)\,. \tag{2.63}$$

As one can see from the right hand side, it defines a function on $J_1\pi$ which is usually denoted by $[\Phi,\Psi]$. Thus the projected equation $\mathcal{R}_1^{(1)}$ is described by the system $\Phi = \Psi = [\Phi,\Psi] = 0$ and it is identical with $\mathcal{R}_1$, if and only if the function $[\Phi,\Psi]$ vanishes on $\mathcal{R}_1$. By Hadamard's Lemma C.1.5, this vanishing requires the existence of smooth functions $a, b \in \mathcal{F}(J_1\pi)$ such that $[\Phi,\Psi] = a\Phi + b\Psi$.

The extension to more than two equations is straightforward. If the differential equation $\mathcal{R}_1$ is described by the system $\Phi^\tau = 0$ with $1 \leq \tau \leq t$, then we obtain a local representation of $\mathcal{R}_1^{(1)}$ by adding all equations $[\Phi^\sigma, \Phi^\tau] = 0$. $\mathcal{R}_1$ is formally integrable, if and only if all these brackets vanish on $\mathcal{R}_1$.

To some extent we may generalise these considerations to higher-order equations. Let now Φ and Ψ be two functions defined on $J_q\pi$ and $J_r\pi$, respectively. We assume without loss of generality that $q \geq r$. Then the two functions define a differential equation $\mathcal{R}_q \subseteq J_q\pi$. As in Example 2.3.10, we obtain for $q > r$ trivial integrability conditions by prolonging the lower-order equation $\Psi = 0$. More interesting is the question whether or not cross-differentiation yields an additional integrability condition. This cross-derivative takes place at an order less than or equal to $q+r$ depending on the precise form of Φ and Ψ. Generalising (2.63), we consider the linear combination

$$[\Phi,\Psi] = \sum_{|\mu|=q} \frac{\partial\Phi}{\partial u_\mu} D_\mu \Psi - \sum_{|\nu|=r} \frac{\partial\Psi}{\partial u_\nu} D_\nu \Phi\,. \tag{2.64}$$

Again the highest-order derivatives cancel, as in each sum the coefficient of the derivative $u_{\mu+\nu}$ with $|\mu| = q$ and $|\nu| = r$ is the same: $\Phi_{u_\mu}\Psi_{u_\nu}$. Thus $[\Phi,\Psi]$ defines a function on $J_{q+r-1}\pi$ and if it does not vanish on $\mathcal{R}_{q+r-1}$, it represents a non-trivial integrability condition.

However, in contrast to the first-order case, we cannot use here the bracket for deciding the absence of integrability conditions. If the bracket does not vanish, the equation $\mathcal{R}_q$ surely possesses a non-trivial integrability condition. But the converse is not true, as the bracket is not necessarily part of the local representation of suitable projected equations. More precisely, the problem is that (2.64) always computes a cross-derivative in order $q+r$, although it may be possible to form cross-derivatives already at lower orders. Hence it may happen that $[\Phi,\Psi]$ is the formal derivative of some lower order integrability condition and its vanishing is only due to the additional differentiations.

As a concrete example consider the third-order differential equation $\mathcal{R}_3$ in two independent variables x, y defined by $\Phi = u_{xxx} - u$ and $\Psi = u_{xyy}$. Obviously, the "true" integrability condition arises from the cross-derivative $D_{yy}\Phi - D_{xx}\Psi = u_{yy}$, but (2.64) yields $[\Phi,\Psi] = u_{xyy} = -D_x(u_{yy})$ and thus the bracket vanishes on $\mathcal{R}_3$ although the equation has an integrability condition. This effect has its origin in the simple fact that we have an "overlap" in the leading derivatives u_{xxx} and u_{xyy} and therefore an integrability condition arises already at a lower order than the bracket (2.64) considers. In first order this problem cannot occur.

The expression $[\Phi,\Psi]$ is very classical in the first-order case and goes back at least to the 19th century. It appears in the literature under various names like *Mayer bracket*, *Jacobi bracket*, or simply *square bracket* [68, §§57/58] and possesses similar properties as the Poisson bracket in symplectic geometry (cf. Appendix C.6) except that it does not satisfy the Jacobi identity (respectively only a modified form). In fact, restricted to functions Φ, Ψ independent of u, the bracket $[\Phi,\Psi]$ coincides with the Poisson bracket, as one can easily see by comparing (2.63) with (C.48). Restricting even further to functions which are semi-linear in the derivatives, i. e. $\Phi(\mathbf{x},u,\mathbf{u}_{(1)}) = a^i(\mathbf{x})u_i$, $\Psi(\mathbf{x},u,\mathbf{u}_{(1)}) = b^i(\mathbf{x})u_i$, we find that $[\Phi,\Psi]$ may be identified with the Lie bracket of the associated vector fields $X_\Phi = a^i\partial_{x^i}, X_\Psi = b^i\partial_{x^i} \in \mathfrak{X}(\mathcal{X})$: $[\Phi,\Psi] = [X_\Phi, X_\Psi]u$ (this special case will be study in more detail in Example 7.2.12). A modern theory relating the Mayer bracket (2.64) to (co)homological constructions was recently developed by Kruglikov and Lychagin [269, 270]. ◁

We have seen that in general integrability conditions arise, if we prolong and project a differential equation. We now show that such conditions are obstructions for the construction of power series solutions. Obviously, such a computation can only be performed in local coordinates. Thus we assume that we are given a differential equation $\mathcal{R}_q$ described in some neighbourhood of the point $x_0 \in \mathcal{X}$ by the system $\boldsymbol{\Phi}(\mathbf{x},\mathbf{u}^{(q)})=0$ and make the usual ansatz for a power series solution

$$u^{\alpha}(x)=\sum_{|\mu|=0}^{\infty} a_{\mu}^{\alpha}\frac{(\mathbf{x}-\mathbf{x}_0)^{\mu}}{\mu!} \tag{2.65}$$

with real coefficients $a_{\mu}^{\alpha}\in\mathbb{R}$. Note that we are only dealing with formal series, as we do not discuss their convergence (but see Chapter 9).

Of course, without initial or boundary conditions the equation possesses in general many solutions and we cannot determine uniquely all coefficients in (2.65). But we can at least exhibit some relations between them. We enter the ansatz (2.65) into the system describing $\mathcal{R}_q$ and evaluate the arising expressions at the point $x=x_0$. One easily checks that this yields the following *algebraic* equations for the coefficients $\mathbf{a}^{(q)}$ (we use here the same notations as for derivatives):

$$\Phi^{\tau}(\mathbf{x}_0,\mathbf{a}^{(q)})=0\,,\qquad 1\leq\tau\leq t\,, \tag{2.66}$$

i. e. in the functions Φ^{τ} we simply substitute every derivative u_{μ}^{α} by the corresponding Taylor coefficient a_{μ}^{α} of the ansatz (2.65).

Remark 2.3.13. In general, (2.66) represent a non-linear system of equations. It is well-known that the solution space of such a system might have a very complicated structure; for example, it could consist of disjoint parts of different dimensions. From a theoretical point of view we may now argue that we defined a differential equation as a *submanifold* of the jet bundle $J_q\pi$ and as in fact $\mathcal{R}_q$ is nothing but the solution space of the algebraic system (2.66), this solution space must be well-behaved by definition. In many practical examples, the situation is unfortunately less nice: the differential equation is given as a system of equations and this system does not describe a submanifold but some space with singular points (e. g. a variety, if the non-linearity is of a polynomial nature). In such a case we must restrict to some smooth subset which is a submanifold. Thus in general we have to make case distinctions, as different smooth subsets with possibly different dimensions may be possible, and in the different cases the differential equation may show very different behaviour. In the sequel we will ignore this problem and follow the theoretical point of view by assuming that $\mathcal{R}_q$ is indeed a submanifold in the strict differential geometric meaning of this term. ◁

Now we turn our attention to the prolonged equation $\mathcal{R}_{q+1}$. According to (2.54), it is described in a neighbourhood of the expansion point $\mathbf{x}_0$ by the original equations $\boldsymbol{\Phi}(\mathbf{x},\mathbf{u}^{(q)})=0$ and in addition by the equations $D_i\boldsymbol{\Phi}(\mathbf{x},\mathbf{u}^{(q+1)})=0$ for $1\leq i\leq n$. Again we enter our ansatz into this system and evaluate at $x=x_0$. The first part of the arising algebraic system for the coefficients $\mathbf{a}^{(q+1)}$ has already been handled in the previous step. The only new equations are

$$D_i\Phi^{\tau}(\mathbf{x}_0,\mathbf{a}^{(q+1)})=0\,,\qquad \begin{cases}1\leq\tau\leq t\,,\\ 1\leq i\leq n\,.\end{cases} \tag{2.67}$$

If we assume that all relations between the coefficients up to order q have already been taken into account with (2.66), then (2.67) represents an inhomogeneous *linear* system for the coefficients $\mathbf{a}_{(q+1)}$ of order $q+1$. Its matrix and right hand side depends on the coefficients $\mathbf{a}^{(q)}$ of lower order. The linearity of the system stems from the quasi-linearity of the formal derivative (2.14).

In this manner we proceed order by order. At order $q+r$ we get from the local representation of $\mathcal{R}_{q+r}$ as only new equations

$$D_\nu \Phi^\tau(\mathbf{x}_0, \mathbf{a}^{(q+r)}) = 0\,, \qquad \begin{cases} 1\leq\tau\leq t\,, \\ |\nu|=r\,. \end{cases} \tag{2.68}$$

Assuming again that all relations between the Taylor coefficients $\mathbf{a}^{(q+r-1)}$ of order less than $q+r$ have already been taken into account in the previous steps, we may say that these new equations form an inhomogeneous linear system for the coefficients $\mathbf{a}_{(q+r)}$ of order $q+r$ whose matrix and right hand side depends on the lower-order coefficients $\mathbf{a}^{(q+r-1)}$.

In the end we obtain a system of infinitely many equations for the infinitely many coefficients of the ansatz (2.65). Each solution of this system defines a formal power series solution. Except (2.66) all subsystems are linear and their solution spaces have a very simple structure. But if we choose different solutions of (2.66), the solution spaces of the linear systems (2.68) may change their dimension, as their matrices depend on the coefficients of lower order. In line with our above made blanket regularity assumption, we will ignore this possibility (recall also Remark 2.3.13).

Eliminating dependent equations, each of the remaining (infinitely many) equations can be solved for a different coefficient a^α_μ of our ansatz. These coefficients are called *principal*, all the other ones are the *parametric coefficients*. Of course, this distinction is not uniquely defined; in general, there are many different ways to solve the equations for some coefficients. The parametric coefficients may be chosen completely arbitrarily; they parametrise the formal solution space of the differential equation $\mathcal{R}_q$. Once they are fixed, the principal coefficients are determined by the equation $\mathcal{R}_q$ and its prolongations $\mathcal{R}_{q+r}$.

In a concrete computation we can perform only a finite number of prolongations. Thus we must stop the above construction at some order $r\geq q$. Of course, we expect that with the help of the so far found equations for the coefficients $\mathbf{a}^{(r)}$ we have at least obtained a correct truncation of the power series solutions. However, this hope is only satisfied, if no integrability conditions of order less than or equal to r are hidden in the system describing $\mathcal{R}_r$, as otherwise some further conditions on the Taylor coefficients $\mathbf{a}^{(r)}$ exist which we have not taken into account, i. e. if the assumptions we made above are justified.

Example 2.3.14. Recall that in Example 2.3.9 we found that $\mathcal{R}_1^{(1)} \neq \mathcal{R}_1$. Entering a power series ansatz into the local representation (2.57) of $\mathcal{R}_1$ and evaluating at a point (x_0, y_0, z_0) yields for the first-order Taylor coefficients the homogeneous linear system $a_{[0,0,1]} + y_0 a_{[1,0,0]} = a_{[0,1,0]} = 0$ possessing a one-dimensional solution. Using the local representation (2.59) of $\mathcal{R}_1^{(1)}$ we find that in fact also $a_{[0,0,1]} = a_{[1,0,0]} = 0$ for any power series solution of our differential equation.

Alternatively, this additional condition may be derived using the local representation (2.58) of the prolonged equation $\mathcal{R}_2$. Entering again a power series ansatz and evaluating at (x_0, y_0, z_0) yields besides the already familiar equations $a_{[0,0,1]} + y_0 a_{[1,0,0]} = a_{[0,1,0]} = 0$ for the first-order Taylor coefficients the following inhomogeneous linear system for the second-order coefficients:

$$\begin{aligned} a_{[1,0,1]} + y_0 a_{[2,0,0]} &= 0\,, & a_{[1,1,0]} &= 0\,,\\ a_{[0,1,1]} + y_0 a_{[1,1,0]} &= -a_{[1,0,0]}\,, & a_{[0,2,0]} &= 0\,,\\ a_{[2,0,0]} + y_0 a_{[1,0,1]} &= 0\,, & a_{[0,1,1]} &= 0\,. \end{aligned} \tag{2.69}$$

Performing standard Gaussian elimination leads to the equation $0 = a_{[1,0,0]}$. Thus a necessary condition for the solvability of our linear system for the *second-order* coefficients is that this equation is satisfied by our chosen solution of the *first-order* coefficients. However, we would not have seen this equation, if we had stopped our calculations after the first step, and hence would have worked with an incorrect truncation of the true power series solution possessing too many degrees of freedom.

The same argument explains why it was so important to include in Example 2.3.10 the integrability conditions obtained by differentiating the first-order equation: otherwise we would ignore two conditions on the second-order coefficients and the arising power series would generally not be a solution of the given differential equation. ◁

As this example demonstrates, one should always remember that integrability conditions are not additional restrictions of the solution space; any solution of the original system *automatically* satisfies them. They represent conditions implicitly contained in the original system and which can be made visible by performing a suitable sequence of prolongations and projections. As we will see throughout the book, it is for many purposes important to exhibit explicitly all hidden integrability conditions. The simplest point is to verify whether the given differential equation is *consistent*, i. e. whether it possesses at all any formal solution. This will only be the case, if no integrability condition of the form $1 = 0$ arises, as such a condition trivially implies that some equation $\mathcal{R}_{q+r}^{(s)}$ is actually the empty set.

Definition 2.3.15. The differential equation $\mathcal{R}_q$ is called *formally integrable*, if for all $r \geq 0$ the equality $\mathcal{R}_{q+r}^{(1)} = \mathcal{R}_{q+r}$ holds.

Thus a differential equation is formally integrable, if at no order of prolongation integrability conditions arise. For such equations the above made assumptions in the construction of formal power series solutions are always satisfied and thus such solutions indeed exist which explains the name "formal integrability". Furthermore, we can be sure that we will not overlook any hidden conditions on the lower-order coefficients, if we build only a truncated series. In Example 2.3.9, the equation $\mathcal{R}_1$ is obviously not formally integrable, but one easily shows that $\mathcal{R}_1^{(1)}$ is.

In practice, Definition 2.3.15 is rather problematic, as it consists of infinitely many conditions. In order to be able to check effectively whether or not a given differential equation is formally integrable, we need a finite criterion. Such criteria exist (and will be the topic of Section 7.2); however, they are not of a geometric but an algebraic nature and we must first develop the necessary algebraic concepts.

Our discussion above leads straightforwardly to Algorithm 2.1 for the construction of the formal power series solution of a formally integrable equation $\mathcal{R}_q$ up to a given order $r \geq q$. In Line /1/ it performs the non-linear first step of computing the Taylor coefficients up to order q. Note again that, in line with the discussion above, the formulation of the algorithm ignores the possibility that the solution space of the arising non-linear system may may have a non-trivial structure.

Algorithm 2.1 Power series solution of formally integrable differential equation

---

Input: local representation $\boldsymbol{\Phi}(\mathbf{x},\mathbf{u}^{(q)})$ of formally integrable differential equation $\mathcal{R}_q$, expansion point $x_0 \in \mathcal{X}$, truncation order $r \geq q$

Output: power series solution of $\mathcal{R}_q$ up to order r

1: compute parametric representation of solution of algebraic system $\boldsymbol{\Phi}(x_0,\mathbf{a}^{(q)}) = 0$ for coefficients of order up to q

2: **for** s **from** q **to** $r-1$ **do**

3: determine $t = \operatorname{rank}(\partial\boldsymbol{\Phi}/\partial u^\alpha_\mu)$ with $|\mu| = s$

4: choose equations $\Phi_1 = \cdots = \Phi_t = 0$ such that corresponding minor has rank t

5: add equations $D_i\Phi_j = 0$ for $1 \leq i \leq n$ and $1 \leq j \leq t$

6: compute parametric solution of linear system of equations for coefficients of order $s+1$
$\sum_{\alpha=1}^m \sum_{|\mu|=q} C^\mu_\alpha(\Phi_j)(x_0,\mathbf{a}^{(s)})a^\alpha_{\mu+1_i} = -C^{(q)}_i(\Phi_j)(x_0,\mathbf{a}^{(s)})$

7: **end for**

8: **return** $\mathbf{a}^{(r)}$

The Lines /3-5/ determine a local representation of the prolonged equation $\mathcal{R}_{s+1}$. Here we explicitly exploit the assumption that the original equation $\mathcal{R}_q$ (and thus trivially also any of its prolongations) is formally integrable. As in such a system no integrability conditions are hidden, it suffices to choose a maximal set of algebraically independent equations of order s and to differentiate formally each of them with respect to all independent variables.[4] In general, the arising equations will not all be algebraically independent, so that the linear system in Line /6/ will contain some spurious equations. We will see later in Section 7.2 how one can avoid this inefficency, if one is actually dealing with an involutive and not merely a formally integrable equation.

In the formulation of the linear system in Line /6/ we exploited the fact that the formal derivative may be expressed with the help of the contact vector fields introduced in (2.11): if $\Phi \in \mathcal{F}(J_q\pi)$ lives on the jet bundle of order q, then its formal derivative with respect to x^i is given by

---

[4] Note that here it is also quite important that a local representation of a formally integrable equation always comprises all integrability conditions of the second kind, as these are necessary for computing all principal coefficients.

$$D_i\Phi = C_i^{(q)}(\Phi) + \sum_{\alpha=1}^{m} \sum_{|\mu|=q} u^\alpha_{\mu+1_i} C^\mu_\alpha(\Phi) \,. \tag{2.70}$$

This formulation allows us to rewrite (2.67) explicitly as an inhomogeneous system for the coefficients of highest order.

Remark 2.3.16. Closely related to formal integrability is the notion of *local solvability*. Given a differential equation $\mathcal{R}_q \subseteq J_q\pi$, we call it locally solvable, if for every point $\rho \in \mathcal{R}_q$ a local solution σ exists such that $\rho \in \operatorname{im} j_q\sigma$. Note that no uniqueness of this solution is required. Indeed, as we will see later in much more detail, for partial differential equations the data provided by a single point ρ is by far not sufficient for obtaining a unique solution.

Strictly speaking, this notion depends on the considered function space, i. e. what kind of regularity assumptions are made for solutions. If we restrict to formal solutions, then the considerations above imply that local solvability and formal integrability are equivalent. In other function spaces, local solvability may be a stricter condition: in the case of smooth solutions local solvability trivially implies formal integrability but not vice versa (cf. Example 9.2.2 and Corollary 9.4.3). ◁

Example 2.3.17. In Remark 2.3.6 we introduced differential equations of finite type and saw that we may interpret them as connections. Now we exploit this observation for the integration of such equations. For simplicity of notation, we restrict to first-order equations $\mathcal{R}_1 \subset J_1\pi$ which are the image of a section $\gamma: \mathcal{E} \to J_1\pi$. Locally, we choose a representation of the form

$$\mathcal{R}_1 \, : \, \left\{ \, \mathbf{u}_{(1)} = \boldsymbol{\phi}(\mathbf{x},\mathbf{u}) \,. \right. \tag{2.71}$$

A natural question is when such an equation is formally integrable. As (2.71) contains no equations of lower order, the only possibility for integrability conditions are cross-derivatives

$$0 = u^\alpha_{ij} - u^\alpha_{ji} = D_i\phi^\alpha_j - D_j\phi^\alpha_i = \Delta^\alpha_{ij} \,. \tag{2.72}$$

Evaluation of these on the submanifold $\mathcal{R}_1$ yields the integrability conditions

$$\Delta^\alpha_{ij}\big|_{\mathcal{R}_1} = \frac{\partial \phi^\alpha_j}{\partial x^i} + \frac{\partial \phi^\alpha_j}{\partial u^\beta}\phi^\beta_i - \frac{\partial \phi^\alpha_i}{\partial x^j} - \frac{\partial \phi^\alpha_i}{\partial u^\beta}\phi^\beta_j = 0 \,. \tag{2.73}$$

Obviously, the satisfaction of these equations is a necessary condition for the formal integrability of $\mathcal{R}_1$. In our particular case of a maximally overdetermined equation, it is not difficult to see that this is also a sufficient condition: any cross-derivative in a prolonged equation $\mathcal{R}_{1+r}$ is a prolongation of one of the conditions $\Delta^\alpha_{ij}|_{\mathcal{R}_1} = 0$. We do not demonstrate this fact here in detail, as we will prove a much more general result in Section 7.2.

A more geometric way to view these integrability conditions starts with the introduction of the vector fields $X_i \in \mathfrak{X}(\mathcal{E})$ locally defined by

$$X_i\big|_{(\mathbf{x},\mathbf{u})} = \partial_{x^i} + \phi^\alpha_i(\mathbf{x},\mathbf{u})\partial_{u^\alpha} \,, \qquad 1 \leq i \leq n \,. \tag{2.74}$$

As discussed in Remark 2.3.6, they span the horizontal bundle of the connection on the fibred manifold $\pi : \mathcal{E} \to \mathcal{X}$ defined by the section γ.

The integrability conditions (2.73) are equivalent to the vanishing of all Lie brackets $[X_i, X_j]$ between these fields. In other words, the differential equation $\mathcal{R}_1$ is formally integrable, if and only if the section γ defines a *flat* connection where the horizontal bundle is involutive. By the Frobenius Theorem C.3.3, any involutive distribution is integrable and thus the horizontal distribution $\mathcal{H} = \langle X_1, \dots, X_n \rangle \subset T\mathcal{E}$ possesses n-dimensional integral manifolds $\mathcal{N} \subset \mathcal{E}$.

We claim that these integral manifolds are images of solutions of $\mathcal{R}_1$. Indeed, let $\sigma : \mathcal{V} \subseteq \mathcal{X} \to \mathcal{E}$ be a (local) solution defined in some open set $\mathcal{V}$. In our case this implies that $\gamma(\sigma(x)) = j_1\sigma(x)$ for all $x \in \mathcal{V}$. According to our intrinsic approach to jet bundles, we can write $j_1\sigma(x) = \big(\sigma(x), T_x\sigma\big)$ and obtain $\operatorname{im} T_x\sigma = \mathcal{H}_{\sigma(x)}$. This equality entails that $\operatorname{im}\sigma$ is an integral manifold of the horizontal distribution $\mathcal{H}$.

Conversely, let $\mathcal{N} \subset \mathcal{E}$ be an n-dimensional integral manifold of the horizontal distribution $\mathcal{H}$. Obviously, it is transversal to the fibration $\pi : \mathcal{E} \to \mathcal{X}$ and hence the image of a (local) section $\sigma : \mathcal{V} \subseteq \mathcal{X} \to \mathcal{E}$. By definition of an integral manifold, this implies that $\operatorname{im} T_x\sigma = \mathcal{H}_{\sigma(x)}$ or, equivalently, $\gamma(\sigma(x)) = j_1\sigma(x)$ for all $x \in \mathcal{V}$. Thus $\operatorname{im} j_1\sigma \subset \mathcal{R}_1$ and σ is a solution.

Here we are speaking about *smooth* solutions, as the Frobenius Theorem is concerned with the existence of smooth integral manifold. Hence we have obtained an existence theorem for smooth solutions of differential equations of finite type (the extension to a uniqueness theorem is trivial, as each integral manifold $\mathcal{N}$ is uniquely determined by the choice of one point $\rho \in \mathcal{R}_1$). As we will see later in Chapter 9, for more general differential equations we are only able to deduce the existence of analytic solutions.

This result also has practical implications: the concrete determination of integral manifolds of an involutive distribution requires only the solution of *ordinary* differential equations, although we are dealing with a partial differential equation. This was already shown by Lie [292] interpreting the vector fields X_i as generators of Lie point symmetries of $\mathcal{R}_1$. Indeed, let an n-dimensional integral manifold $\mathcal{N}$ of $\mathcal{H}$ be locally described by the m equations $\psi^\alpha(\mathbf{x}, \mathbf{u}) = 0$. As the vector fields X_i span the tangent space of $\mathcal{N}$, we must have $\mathrm{d}\psi^\alpha(X_i) = X_i\psi^\alpha = 0$ (cf. Appendix C.2). Thus it suffices to determine all solutions of the overdetermined linear first-order system $X_i v = 0$ for one unknown $v(\mathbf{x}, \mathbf{u})$. Such systems are studied in Example 7.2.12 where it will be explicitly shown that their resolution requires only the integration of ordinary differential equations. ◁

The basic idea of the construction in the last example was to reduce the problem of solving a differential equation of finite type to an application of the Frobenius Theorem. As we will see in Sections 9.5 and 9.6, this idea can be generalised to arbitrary involutive equations. However, then we will have to consider infinitely many distributions instead of only one as above.

The key for this reduction is to consider *integral elements* or *infinitesimal solutions*. As we will see below, their construction requires essentially only linear algebra. Informally, an integral element at a point $\rho \in \mathcal{R}_q$ is a subspace $\mathcal{U}_\rho \subseteq T_\rho\mathcal{R}_q$

which is potentially part of the tangent space of a prolonged solution $j_q\sigma$. The traditional way to define them is based on a treatment of the differential equation as an exterior differential system; we propose here an alternative direct approach using the contact map.

Definition 2.3.18. Let $\mathcal{R}_q \subseteq J_q\pi$ be a differential equation. We call a linear subspace $\mathcal{U}_\rho \subseteq T_\rho\mathcal{R}_q$ an *integral element* at the point $\rho \in \mathcal{R}_q$, if a point $\hat{\rho} \in \mathcal{R}_{q+1}$ exists such that $\pi_q^{q+1}(\hat{\rho}) = \rho$ and $\mathcal{U}_\rho \subseteq \operatorname{im}\Gamma_{q+1}(\hat{\rho})$.

The motivation for this definition is the following observation. Recall from Remark 2.2.8 that if $\sigma \in \Gamma_{loc}(\pi)$ is an arbitrary section with $\rho = j_q\sigma(x)$, then $T_\rho(\operatorname{im} j_q\sigma) = \operatorname{im}\Gamma_{q+1}(\hat{\rho})$ for $\hat{\rho} = j_{q+1}\sigma(x)$. Now assume that σ is a solution of $\mathcal{R}_q$ and thus $\operatorname{im} j_q\sigma \subseteq \mathcal{R}_q$. Obviously, this implies $T_\rho(\operatorname{im} j_q\sigma) \subseteq T_\rho\mathcal{R}_q$ and hence this subspace is an n-dimensional integral element. This observation also explains the alternative name "infinitesimal solution". Note that the converse is *not* true: given an integral element $\mathcal{U}_\rho \subseteq T_\rho\mathcal{R}_q$, there does not necessarily exist a solution $\sigma \in \Gamma_{loc}(\pi)$ such that $\mathcal{U}_\rho \subseteq T_\rho(\operatorname{im} j_q\sigma)$.

It follows immediately from Definition 2.3.18 that integral elements only occur at points $\rho \in \mathcal{R}_q^{(1)}$, as otherwise no $\hat{\rho} \in \mathcal{R}_{q+1}$ with $\pi_q^{q+1}(\hat{\rho}) = \rho$ exists. Another obvious consequence of the definition is that any integral element lies transversal to the fibration π^q and is thus at most n-dimensional where again $n = \dim\mathcal{X}$, as any image of the contact map Γ_q has this property.

Our next result demonstrates that our definition of an integral element is equivalent to the traditional one. Let $\iota : \mathcal{R}_q \hookrightarrow J_q\pi$ be the inclusion map. Then we may consider the pull-back $\iota^*\mathcal{C}_q^0$ of the contact codistribution or more precisely the differential ideal $\mathcal{I}[\mathcal{R}_q] = \langle \iota^*\mathcal{C}_q^0\rangle_{\text{diff}} \subseteq \Omega(\mathcal{R}_q)$ generated by it (recall that algebraically $\mathcal{I}[\mathcal{R}_q]$ is thus spanned by a basis of $\iota^*\mathcal{C}_q^0$ and its exterior derivatives).

So far we have always implicitly identified the tangent space $T_\rho\mathcal{R}_q$ with the subspace $T_\rho\iota(T_\rho\mathcal{R}_q) \subseteq T_\rho(J_q\pi)$—as it is costumary in differential geometry. For a few calculations in the proof of the following proposition it is important to explicitly write the tangent map, so that we will now be a bit more pedantic with our notations. Note that $T\iota$ is trivially injective, since ι itself is injective.[5] Thus a vector $v \in T_\rho\mathcal{R}_q$ is uniquely defined, as soon as $T_\rho\iota(v)$ is given.

Proposition 2.3.19. *Let $\mathcal{R}_q$ be a differential equation such that $\mathcal{R}_q^{(1)} = \mathcal{R}_q$. A linear subspace $\mathcal{U}_\rho \subseteq T_\rho\mathcal{R}_q$ is an* integral element *at the point $\rho \in \mathcal{R}_q$, if and only if $T_\rho\iota(\mathcal{U}_\rho)$ lies transversal to the fibration π^q and every differential form $\omega \in \mathcal{I}[\mathcal{R}_q]$ vanishes on $\mathcal{U}_\rho$.*

Proof. Assume first that the subspace $\mathcal{U}_\rho$ is an integral element. Thus there exists a point $\hat{\rho} \in \mathcal{R}_{q+1}$ such that $\pi_q^{q+1}(\hat{\rho}) = \rho$ and $T_\rho\iota(\mathcal{U}_\rho) \subseteq \operatorname{im}\Gamma_{q+1}(\hat{\rho})$. This implies firstly that $T_\rho\iota(\mathcal{U}_\rho)$ is transversal to π^q and secondly that every one-form $\omega \in \iota^*\mathcal{C}_q^0$

[5] Recall from Remark C.2.2 that by a slight abuse of notation we call the tangent map $T\iota$ of an injective map ι push-forward ι_* when applied to a vector field.

vanishes on $\mathcal{U}_\rho$, as $\operatorname{im}\Gamma_{q+1}(\hat{\rho}) \subset (\mathcal{C}_q)_\rho$. Thus there only remains to show that the same is true for every two-form $\omega \in \iota^*(\mathrm{d}\mathcal{C}_q^0)$.

Choose a (local) section $\gamma: \mathcal{R}_q \to \mathcal{R}_{q+1}$ with $\gamma(\rho) = \hat{\rho}$ (such a section always exists, as by assumption the restricted projection $\hat{\pi}_q^{q+1}: \mathcal{R}_{q+1} \to \mathcal{R}_q$ is surjective and the imposed condition refers only to a single point). We introduce on $\mathcal{R}_q$ a distribution $\mathcal{D}$ of rank n by requiring that $T_{\tilde{\rho}}\iota(\mathcal{D}_{\tilde{\rho}}) = \operatorname{im}\Gamma_{q+1}\big(\gamma(\tilde{\rho})\big)$ for any point $\tilde{\rho} \in \mathcal{R}_q$. Obviously, by construction $\mathcal{U}_\rho \subseteq \mathcal{D}_\rho$. It follows from the coordinate form (2.35) of the contact map that in local coordinates the distribution $\mathcal{D}$ is spanned by n vector fields X_i such that

$$\iota_* X_i = C_i^{(q)} + \sum_{\alpha=1}^{m} \sum_{|\mu|=q} \gamma^\alpha_{\mu+1_i} C^\mu_\alpha \tag{2.75}$$

where $C_i^{(q)}$ and C^μ_α are the contact fields defined by (2.11) and the coefficients γ^α_ν the highest-order components of the section γ. Using (C.9), the commutator of two such vector fields evaluates to

$$\begin{aligned} \iota_*\big([X_i,X_j]\big) &= \big(C_i^{(q)}(\gamma^\alpha_{\mu+1_j}) - C_j^{(q)}(\gamma^\alpha_{\mu+1_i})\big)C^\mu_\alpha \\ &\quad + \gamma^\alpha_{\mu+1_j}[C_i^{(q)}, C^\mu_\alpha] - \gamma^\alpha_{\mu+1_i}[C_j^{(q)}, C^\mu_\alpha]\,. \end{aligned} \tag{2.76}$$

By the commutation relation (2.13), the commutators in the second line vanish whenever $\mu_i = 0$ or $\mu_j = 0$, respectively. Otherwise we obtain $-\partial_{u^\alpha_{\mu-1_i}}$ and $-\partial_{u^\alpha_{\mu-1_j}}$, respectively. But this implies that the two sums in the second line cancel each other and we find that $\iota_*\big([X_i,X_j]\big) \in \mathcal{C}_q$. Applying formula (C.22) yields for any contact form $\omega \in \mathcal{C}_q^0$ that

$$\begin{aligned} \iota^*(\mathrm{d}\omega)(X_i,X_j) &= \mathrm{d}\omega(\iota_* X_i, \iota_* X_j) \\ &= \iota_* X_i\big(\omega(\iota_* X_j)\big) - \iota_* X_j\big(\omega(\iota_* X_i)\big) + \omega\big(\iota_*([X_i,X_j])\big)\,. \end{aligned} \tag{2.77}$$

Each summand in the last expression vanishes, as all appearing fields are contact fields. Hence we conclude that any form $\omega \in \iota^*(\mathrm{d}\mathcal{C}_q^0)$ vanishes on $\mathcal{D}$ and hence in particular on $\mathcal{U}_\rho \subseteq \mathcal{D}_\rho$.

For the converse note that any subspace $\mathcal{U}_\rho \subseteq T_\rho\mathcal{R}_q$ satisfying the imposed conditions is spanned by some linear combinations of vectors v_i such that $T_\rho\iota(v_i) = C_i^{(q)}|_\rho + \gamma^\alpha_{\mu,i} C^\mu_\alpha|_\rho$ where $\gamma^\alpha_{\mu,i}$ are real coefficients and consider a contact form ω^α_ν defined by (2.8) with $|\nu| = q-1$. Then $\mathrm{d}\omega^\alpha_\nu = \mathrm{d}x^i \wedge \mathrm{d}u^\alpha_{\nu+1_i}$. Evaluating the condition $\iota^*(\mathrm{d}\omega^\alpha_\nu)|_\rho(v_i,v_j) = \mathrm{d}\omega^\alpha_\nu\big(T_\rho\iota(v_i), T_\rho\iota(v_j)\big) = 0$ yields the equation $\gamma^\alpha_{\nu+1_i,j} = \gamma^\alpha_{\nu+1_j,i}$. Hence the coefficients are actually of the form $\gamma^\alpha_{\mu,i} = \gamma^\alpha_{\mu+1_i}$ and a section $\sigma \in \Gamma_{loc}(\pi)$ exists such that $\rho = [\sigma]_x^{(q)}$ and $T_\rho(\operatorname{im} j_q\sigma)$ is spanned by the vectors $T_\rho\iota(v_i)$. But this observation implies that $\mathcal{U}_\rho$ is an integral element. □

It is important for this result that we take the full *differential* ideal generated by the codistribution $\iota^*\mathcal{C}_q^0$, i.e. that we also consider the two-forms $\iota_*(\mathrm{d}\mathcal{C}_q^0)$. As one can see in the proof, only the additional conditions imposed by them ensure that $\mathcal{U}_\rho$ can indeed be a part of the tangent space to a prolonged section.

Based on this proposition, it is now possible to determine effectively integral elements at a given point ρ on a differential equation $\mathcal{R}_q$. One first computes the pull-back $\iota^*\omega_\mu^\alpha$ of the basic contact forms (2.8) and their exterior derivatives. For the construction of a k-dimensional integral element one makes then the ansatz $T_\rho\iota(v_j) = \alpha_j^i C_i^{(q)}|_\rho + \gamma_{\mu,j}^\alpha C_\alpha^\mu|_\rho$ with $1 \leq j \leq k$. The yet unknown coefficients α_j^i, $\gamma_{\mu,j}^\alpha$ must satisfy the linear system of equations arising from the conditions $\iota^*\omega_\mu^\alpha|_\rho(v_j) = 0$ and $\iota^*(\mathrm{d}\omega_\mu^\alpha)|_\rho(v_i,v_j) = 0$ for all $1 \leq i,j \leq k$. Thus once the pull-back has been performed explicitly (which is easily possible only, if the differential equation is given in solved form), the computation reduces to elementary linear algebra. An alternative ansatz avoiding the need to compute a pull-back will be discussed extensively in Section 9.5.

Example 2.3.20. We demonstrate Proposition 2.3.19 for a first-order evolution equation in $1+1$ dimensions:

$$\mathcal{R}_1 : \{u_t = \phi(x,t,u,u_x)\ . \tag{2.78}$$

Let $\rho = (x,t,u,u_x,u_t) \in \mathcal{R}_1$ be a point on the equation (i. e. u_t has the appropriate value) and $\hat{\rho} = (x,t,u,u_x,u_y,u_{xx},u_{xt},u_{tt}) \in \mathcal{R}_2$ a point in the fibre over ρ. All such points are characterised by their u_{xx}-value, as this is the only parametric derivative. Indeed, a local representation of $\mathcal{R}_2$ is obtained by adding the equations

$$\begin{aligned} u_{xt} &= D_x\phi = C_1^{(1)}(\phi) + u_{xx}\phi_{u_x}\ , \\ u_{tt} &= D_t\phi = C_2^{(1)}(\phi) + u_{xt}\phi_{u_x} = C_2^{(1)}(\phi) + \phi_{u_x}\bigl(C_1^{(1)}(\phi) + u_{xx}\phi_{u_x}\bigr)\ . \end{aligned} \tag{2.79}$$

Thus the values of the principle derivatives u_{tt}, u_{xt} are determined, as soon as a value for u_{xx} has been chosen.

According to Definition 2.3.18, any two-dimensional integral element $\mathcal{U}_\rho$ of the differential equation $\mathcal{R}_q$ at the point ρ is generated by vectors

$$\begin{aligned} v_1 &= \partial_x + u_x\partial_u + u_{xx}\partial_{u_x} + u_{xt}\partial_{u_t}\ , \\ v_2 &= \partial_t + u_t\partial_u + u_{xt}\partial_{u_x} + u_{tt}\partial_{u_t}\ , \end{aligned} \tag{2.80}$$

as these vectors will span $\operatorname{im}\Gamma_2(\hat{\rho})$, if we substitute the principal derivatives u_{xt} and u_{tt} by the expressions (2.79). Thus we find a one-parameter family of integral elements at ρ parametrised by the value of u_{xx}.

Now we use the alternative approach via the contact codistribution. In the first-order jet bundle, every contact form is a multiple of $\omega = \mathrm{d}u - u_x\mathrm{d}x - u_t\mathrm{d}t$. Hence the differential ideal $\mathcal{I}[\mathcal{R}_1] \subseteq \Omega(\mathcal{R}_1)$ is generated by the two forms

$$\begin{aligned} \omega_1 &= \mathrm{d}u - u_x\mathrm{d}x - \phi\mathrm{d}t\ , \\ \omega_2 &= \mathrm{d}u_x \wedge \mathrm{d}x - \phi_x\mathrm{d}x \wedge \mathrm{d}t - \phi_u\mathrm{d}u \wedge \mathrm{d}t - \phi_{u_x}\mathrm{d}u_x \wedge \mathrm{d}t\ . \end{aligned} \tag{2.81}$$

Here we use (x,t,u,u_x) as coordinates on the manifold $\mathcal{R}_1$. Any two-dimensional transversal subspace $\mathcal{U}_\rho \subseteq T_\rho\mathcal{R}_1$ satisfying $\omega_1|_\rho(\mathcal{U}_\rho) = 0$ is generated by two contact vectors w_1, w_2 of the form

$$\begin{aligned} T_\rho\iota(w_1) &= C_1^{(1)} + a_1 C^{[1,0]} + b_1 C^{[0,1]} = \partial_x + u_x\partial_u + a_1\partial_{u_x} + b_1\partial_{u_t}\,, \\ T_\rho\iota(w_2) &= C_2^{(1)} + a_2 C^{[1,0]} + b_2 C^{[0,1]} = \partial_t + u_t\partial_u + a_2\partial_{u_x} + b_2\partial_{u_t}\,. \end{aligned} \tag{2.82}$$

We must now first impose the tangency condition that the one-form $\mathrm{d}(u_t - \phi)$ vanishes on these vectors and then the condition $\omega_2|_\rho(w_1,w_2) = 0$. Together they lead to the following inhomogeneous linear system of equations for the unknown coefficients a_1, a_2, b_1, b_2:

$$\begin{aligned} b_1 - \phi_{u_x}(\rho)a_1 &= -C_1^{(1)}(\phi)(\rho)\,, \\ b_2 - \phi_{u_x}(\rho)a_2 &= -C_2^{(1)}(\phi)(\rho)\,, \\ b_1 - a_2 &= 0\,. \end{aligned} \tag{2.83}$$

Obviously, the solution space of this system can be parametrised by a single parameter, say a_1. Taking (2.79) into account and identifying $a_1 = u_{xx}$, we find that $v_1 = T_\rho\iota(w_1)$ and $v_2 = T_\rho\iota(w_2)$ as predicted by Proposition 2.3.19. Note how the last line of (2.83) stemming from the vanishing of the two-form $\omega_2 = \iota^*(\mathrm{d}\omega)$ enforces that the u_t-component of $T_\rho\iota(w_1)$ and the u_x-component of $T_\rho\iota(w_2)$ coincide, as it is automatically the case for the vectors v_1, v_2 because of the commutativity of partial derivatives. ◁

2.4 Some Examples

In this section we briefly discuss some classical differential equations which will be used as examples at many places later on. They stem from physics and are of great importance in applications. Most of them are already formally integrable, however, we have not yet the necessary tools to prove this fact.

Example 2.4.1. Maxwell's equations form the foundations of electrodynamics. They describe the evolution of the electric and magnetic fields and fluxes in a four-dimensional space-time. Their most general (differential) form is

$$\mathbf{D}_t = \nabla\times\mathbf{H} - \mathbf{J}_e\,, \qquad \mathbf{B}_t = -\nabla\times\mathbf{E} - \mathbf{J}_m\,, \tag{2.84a}$$

$$0 = \nabla\cdot\mathbf{D} - \rho_e\,, \qquad 0 = \nabla\cdot\mathbf{B} - \rho_m\,. \tag{2.84b}$$

Here ∇ denotes the gradient with respect to the spatial variables only; $\cdot$ and $\times$ represent the scalar and the vector product, respectively. $\mathbf{E}$ and $\mathbf{H}$ are the electric and magnetic field, respectively, $\mathbf{D}$ and $\mathbf{B}$ the corresponding fluxes, $\mathbf{J}_e$ and $\mathbf{J}_m$ are currents and finally, ρ_e and ρ_m charge densities. In physics, $\mathbf{J}_m$ and ρ_m are usually set to zero, as nobody has ever observed experimentally magnetic charges or currents. However, there is no mathematical reason for not adding such terms (at the level of (2.84); in gauge theory the situation changes). In particular, in the context of perturbation calculations it makes sense to analyse the effect of these "source" terms.

The system (2.84) is yet underdetermined. In a simple medium, we have the constitutive relations $\mathbf{D} = \varepsilon\mathbf{E}$, $\mathbf{B} = \mu\mathbf{H}$ and $\mathbf{J}_e = \sigma\mathbf{E}$ where ε, μ and σ are the permittivity, the permeability and the conductivity, respectively. If the medium is isotropic and homogeneous, these quantities are constant scalars. In general, they are tensor valued functions of space-time. We will mainly study Maxwell's equations in vacuum. There $\sigma = 0$ and $\varepsilon\mu = 1/c^2$ where c is the speed of light. In natural units $c = 1$ and we obtain the following system:

$$\mathbf{E}_t = \nabla\times\mathbf{B}\,, \qquad \mathbf{B}_t = -\nabla\times\mathbf{E}\,, \tag{2.85a}$$

$$0 = \nabla\cdot\mathbf{E}\,, \qquad 0 = \nabla\cdot\mathbf{B}\,. \tag{2.85b}$$

The equations in (2.85a) describe the time evolution of the fields. The equations in (2.85b) are often called the *Gauss* or *divergence laws*; they render (2.85) overdetermined, as we have eight equations for six field components. Because of the well-known identity $\nabla\cdot(\nabla\times\mathbf{X}) = 0$ for any vector field $\mathbf{X}$, we obtain no integrability conditions, if we first prolong (2.85) and then project back. Indeed,

$$\partial_t(\nabla\cdot\mathbf{E}) = \nabla\cdot\mathbf{E}_t = \nabla\cdot(\nabla\times\mathbf{B}) = 0 \tag{2.86}$$

and similarly for $\mathbf{B}$. Note that the same is true for the general form (2.84), if we ignore the currents and charges. We will see later that these considerations suffice to prove that (2.85) is a formally integrable system.

This identity also implies another interesting property. Let us take only the evolutionary part (2.85a); it forms a hyperbolic (in t-direction) system in Cauchy–Kovalevskaya form. So we may consider the initial value problem for it. Assume that we prescribe for $t = 0$ fields $\mathbf{E}_0(\mathbf{x})$ and $\mathbf{B}_0(\mathbf{x})$ satisfying the divergence laws (2.85b). Then it follows from (2.86) that the solution $\mathbf{E}(\mathbf{x},t)$, $\mathbf{B}(\mathbf{x},t)$ of (2.85a) for these initial data satisfy (2.85b) for all times t, i. e. it is a solution of the full Maxwell equations (2.85). In Section 9.4 we will see that this is not accidental but in fact a characteristic and very important property of all *involutive* equations (it will turn out that this is exactly the decisive property that distinguishes merely formally integrable equations from involutive ones).

Sometimes, the Gauss laws are considered "redundant" because of this observation in order to avoid the discussion of an overdetermined problem (which is of course nonsense). As a consequence in the numerical integration of Maxwell's equations the constraints (2.85b) are often ignored. However, the identity (2.86) holds only for true solutions and not necessarily for approximations and indeed numerical solutions often significantly violate the Gauss laws (see Example 10.4.14 for a slightly more extensive discussion).

If we include the currents $\mathbf{J}_e$ and $\mathbf{J}_m$ and the charge densities ρ_e and ρ_m, we see that they must satisfy the *continuity equations*

$$(\rho_e)_t + \nabla \cdot \mathbf{J}_e = 0\,, \qquad (\rho_m)_t + \nabla \cdot \mathbf{J}_m = 0\,. \tag{2.87}$$

Otherwise the equations become inconsistent and do not admit any solutions. This is a prototypical example of a *compatibility condition*, a concept which we will discuss in more details in Remark 7.1.7 and Section 10.5. ◁

Example 2.4.2. The fundamental equations governing the flow of fluids and gas are the *Navier–Stokes equations*. We are mainly interested in *incompressible flows*. The flow lives in some domain $\Omega \subseteq \mathbb{R}^3$; for simplicity we assume it to be open. The base manifold $\mathcal{X}$ is given by the product $\Omega \times \mathbb{R}$ where the factor $\mathbb{R}$ represents time. The dependent variables are the velocity of the flow and the pressure. Thus we use as total space $\mathcal{E} = T\Omega \times \mathbb{R}^2$ with local coordinates $(\mathbf{x}, t, \mathbf{u}, p)$: $\mathbf{x}$ represents a point in Ω, t the time, $\mathbf{u}$ are coordinates of a tangent vector in $T_{\mathbf{x}}\Omega$ and p is the pressure.

The incompressible Navier–Stokes equations are

$$\mathbf{u}_t + (\mathbf{u} \cdot \nabla)\mathbf{u} = \nu \Delta \mathbf{u} - \nabla p\,, \tag{2.88a}$$

$$\nabla \cdot \mathbf{u} = 0\,. \tag{2.88b}$$

Here we use again notations from vector analysis: ∇ is the gradient with respect to the spatial variables $\mathbf{x}$ and thus the scalar product $\nabla \cdot \mathbf{u}$ corresponds to the divergence of the velocity field $\mathbf{u}$. The left hand side of (2.88a) represents the convective derivative of $\mathbf{u}$ expressing that our coordinate system moves with the flow; in components it is given by $\partial u^i/\partial t + u^j \partial u^i/\partial x^j$. Physically speaking, (2.88a) represents a momentum balance, while (2.88b) stems from the conservation of mass and expresses the incompressibility of the flow.

The coefficient ν describes the *viscosity* of the flow. If it is zero, we obtain *Euler's equations* of inviscid flow. They can be described in a much more geometric way as a geodesic flow on the group of volume preserving diffeomorphisms [26]. While analytically quite different, the Euler and the Navier–Stokes equations behave fairly similar with respect to their formal properties.

Eq. (2.88) describes a submanifold $\mathcal{R}_2 \subset J_2\pi$. As the incompressibility constraint (2.88b) is of first order, $\mathcal{R}_2$ is *not* formally integrable; trivial integrability conditions are given by the prolongations of (2.88b) to second order. Thus $\mathcal{R}_2^{(1)}$ is described by (2.88) plus the equations

$$\nabla \cdot \mathbf{u}_t = 0\,, \qquad \nabla(\nabla \cdot \mathbf{u}) = 0\,. \tag{2.89}$$

The equation $\mathcal{R}_2^{(1)}$ is still not formally integrable, as we find in the next prolongation a non-trivial integrability condition: if we take the divergence of (2.88a), we can eliminate the third-order term $\Delta(\nabla \cdot \mathbf{u})$ by the divergence of the second equation in (2.89). There remains the second-order equation

$$\Delta p = -\nabla \cdot \big((\mathbf{u} \cdot \nabla)\mathbf{u}\big)\,, \tag{2.90}$$

i. e. a Poisson equation for the pressure. One can show that the equation $\mathcal{R}_2^{(2)}$ defined by (2.88), (2.89), and (2.90) is formally integrable.

The pressure p is at the origin of many of the problems appearing in the theoretical or numerical analysis of the incompressible Navier–Stokes equations. To some extent, it may be considered as a kind of Lagrange multiplier; without it the incompressibility constraint (2.88b) would not be consistent with the evolution equation (2.88a). Via (2.90) the pressure automatically adapts its value in such a manner that the flow remains incompressible. In the numerical integration of (2.88), often most of the computation time is spent in solving (2.90). One fundamental problem consists of selecting appropriate boundary conditions for this equation, as there seems to be no natural choice.

Note that although the original system (2.88) comprises as many equations as unknown functions, we are actually dealing with an overdetermined system. This is due to the fact that (2.88) is not in Cauchy–Kovalevskaya form. As a consequence besides the less interesting equations (2.89) the non-trivial integrability condition (2.90) arises. This example demonstrates that the classification into under-, well- or overdetermined equations makes sense only after all integrability conditions have been added. We will discuss a rigorous definition of these terms in Section 7.5. ◁

The following two examples are actually of a very geometric nature. Nevertheless, we describe them mainly in local coordinates. There are two reasons for this. First of all, we do not want to introduce all the necessary differential geometry. Secondly, we will take at many places a more applied point of view: in order to be able to demonstrate our computational techniques with the help of these examples, we must write them down in local coordinates.

Example 2.4.3. In modern elementary particle physics three of the four fundamental interactions are described by *Yang–Mills theories*. The proper geometric framework for their formulation are principal fibre bundles, but here we take a simplified approach. Underlying the theory is the space-time, an n-dimensional manifold $\mathcal{M}$ equipped with a Lorentzian metric. Again for simplicity we restrict here to the case of a flat space-time with the Minkowski metric η_{ij} defined by $\eta_{ii} = 1$ for $1 \leq i < n$, $\eta_{nn} = -1$ and $\eta_{ij} = 0$ in all other cases.

Assume we are given a d-dimensional Lie group $\mathcal{G}$. The basic field of the corresponding Yang–Mills theory, the *vector potential*, is a one-form $A \in \Omega^1(\mathcal{M}, \mathfrak{g})$ over space-time taking values in the Lie algebra $\mathfrak{g}$ of $\mathcal{G}$ (from a geometric point of view A is considered as the connection form of a connection on a $\mathcal{G}$-principal fibre bundle). If a basis of $\mathfrak{g}$ is given by the elements T_a with $1 \leq a \leq d$, such a one-form has the local form $A = A^a_i T_a \mathrm{d}x^i$. Thus we are dealing with a fibre bundle over space-time whose fibre dimension is dn.

The Yang–Mills *field strength* is the two-form $F = \mathrm{d}A + \frac{1}{2}A \wedge A = F^a_{ij} T_a \mathrm{d}x^i \wedge \mathrm{d}x^j$ (according to the Maurer–Cartan formula (C.37), it represents just the curvature of the connection defined by A). If the structure constants of the Lie algebra $\mathfrak{g}$ are C^c_{ab} (i. e. if $[T_a, T_b] = C^c_{ab} T_c$), then the coefficients F^a_{ij} are given by

$$F^a_{ij} = \partial_{x^i} A^a_j - \partial_{x^j} A^a_i + \frac{1}{2} C^a_{bc} A^b_i A^c_j \,. \tag{2.91}$$

This geometric language allows us to write the Yang–Mills field equations in a very concise form:

$$\mathrm{D}F = 0\,, \qquad \mathrm{D}{*}F = 0\,. \tag{2.92}$$

where D denotes the exterior covariant derivative for the connection A defined by $\mathrm{D}F = \mathrm{d}F + A \wedge F$ and $*$ the Hodge operator associated with the metric on space-time. The first equation is called *Bianchi identity* and follows immediately from the definition of the field strength (using the Jacobi identity for the structure constants). The second equation are the true field equations; in local coordinates they read

$$\Phi^a_\ell\left(A^b_i, \partial_{x^j}A^b_i, \partial_{x^j x^k}A^b_i\right) = \eta^{ij}\left(\partial_{x^i}F^a_{j\ell} + C^a_{bc}A^b_i F^c_{j\ell}\right) = 0 \tag{2.93}$$

where η^{ij} denote the components of the inverse matrix to (η_{ij}) (i. e. $\eta^{ij}\eta_{jk} = \delta^i_k$).

Substituting (2.91) for the field strength, (2.93) is a semi-linear system of dn second-order equations for the components A^a_i of the vector potential. The non-linearity is introduced by the last term in the definition of the field strength which is quadratic in the vector potential. It disappears for an Abelian Lie algebra $\mathfrak{g}$ where all structure constants vanish. Thus in this case one obtains linear field equations.

In a rather lengthy computation one can show that the differential equation $\mathcal{R}_2$ defined locally by (2.93) is formally integrable. The key is the *Noether identity*

$$\eta^{ij}\left(D_i\Phi^a_j + C^a_{bc}A^b_i\Phi^c_j\right) \equiv 0 \tag{2.94}$$

stemming from the properties of the structure constants C^a_{bc} due to the Jacobi identity. It implies the equality $\mathcal{R}_2^{(1)} = \mathcal{R}_2$ and, as we will see later, this observation suffices to prove formal integrability.

Again it is tempting to classify (2.93) as a well-determined system, as it consists of dn equations $\Phi^a_i = 0$ for the dn unknowns A^a_i. However, we will see in Section 7.5 that we are in fact dealing with an underdetermined system. From a physical point of view this was to be expected, as Yang–Mills theories are the prototype of *gauge theories*: the underdeterminacy is a simple consequence of the presence of a gauge symmetry. We will discuss this aspect in more detail in Section 8.3.

The Yang–Mills equations for the Lie group $U(1)$ are closely related to Maxwell's equations. As we are dealing with a one-dimensional group, we skip the upper index on the coefficients of the potential A and the field strength F. The coefficients of the latter define thus in four dimensions a 4×4 antisymmetric matrix related to the electric and magnetic field:

$$F = \begin{pmatrix} 0 & B^{(z)} & -B^{(y)} & -E^{(x)} \\ -B^{(z)} & 0 & B^{(x)} & -E^{(y)} \\ B^{(y)} & -B^{(x)} & 0 & -E^{(z)} \\ E^{(x)} & E^{(y)} & E^{(z)} & 0 \end{pmatrix}. \tag{2.95}$$

In other words, if we build the three-dimensional vector $\mathbf{A}$ consisting of the first three coefficients of A, then $\mathbf{B} = \nabla \times \mathbf{A}$ and $\mathbf{E} = -\partial_t\mathbf{A} - \nabla \cdot A_4$. Note that these relations automatically imply half of the equations in (2.85), namely $\nabla \cdot \mathbf{B} = 0$ and $\mathbf{B}_t = -\nabla \times \mathbf{E}$ (the Bianchi identity $\mathrm{d}F = 0$). The remaining four equations are equivalent to the four Yang–Mills equations (2.93).

In this case the above mentioned gauge symmetry takes a particularly simple form: the field strength $F = \mathrm{d}A$ (and hence our field equations) remains invariant, if we add to A a term of the form $\mathrm{d}\Lambda$ with a smooth function Λ. This observation implies that one component of the vector potential A may be chosen arbitrarily.

This geometric view of Maxwell's equations also "explains" why there should appear no magnetic charges: only if $\mathrm{d}F = 0$, i.e. if $\nabla \cdot \mathbf{B} = 0$ and $\mathbf{B}_t = -\nabla \times \mathbf{E}$ without additional charges or currents, a vector potential A with $F = \mathrm{d}A$ can exist by Poincaré's Lemma (cf. Remark C.2.9). ◁

Example 2.4.4. General relativity describes the fourth fundamental interaction, gravity; *Einstein's equations* are a cornerstone of the theory. The basic idea is that gravity manifests itself in the geometry of space-time. We consider here only the vacuum case without any matter fields but in arbitrary dimensions.[6] Space-time is modelled again as an n-dimensional Lorentzian manifold, i.e. as a manifold $\mathcal{M}$ equipped with a metric g_{ij} that has the same signature as the Minkowski metric η_{ij}. This manifold $\mathcal{M}$ is our base space $\mathcal{X}$. The dependent variables are the components g_{ij} of the metric g. Thus $\pi : \mathcal{E} \to \mathcal{X}$ is here the vector bundle of symmetric bilinear forms; its fibre dimension is $m = n(n+1)/2$.

Geometrically, the vacuum Einstein equations express the fact that the Ricci tensor R_{ij} of the Levi–Civita connection for the metric g vanishes:

$$R_{ij}[g] = 0\,. \tag{2.96}$$

As the Ricci tensor is symmetric and depends on the second-order derivatives of the metric, (2.96) is a system of $n(n+1)/2$ equations defining a submanifold $\mathcal{R}_2 \subset J_2\pi$. Note that we have again as many equations as unknown functions.

Einstein's equations describe the evolution of the Lorentzian structure of $\mathcal{M}$ and thus how space-time is "curved." If matter is present, it leads to source terms on the right hand side of (2.96) and thus changes the metric. The motion of a particle is always on a geodesic and its interaction with the gravitational field is encoded in these changes of g.

Writing out the components R_{ij} of the Ricci tensor in terms of the metric g and its derivatives leads to rather complicated non-linear expressions which can be found in any textbook on either Riemannian geometry or general relativity. We omit them here and use instead a trick for simplifying the coordinate form of (2.96): we take *locally geodesic coordinates*. Such coordinates exist for every point $x_0 \in \mathcal{M}$ and are characterised by two properties: (i) the metric becomes the Minkowski metric at the point x_0, $g_{ij}(x_0) = \eta_{ij}$, and (ii) the Christoffel symbols vanish at x_0. At x_0, (2.96) takes in these coordinates the following simple linear form

[6] Strictly speaking, we assume that the dimension is at least four; for topological reasons two and three dimensions are special cases where different equations have to be used [105].

$$R_{ij}(x_0) = \frac{1}{2}\eta^{k\ell}\Big[\partial_{x^ix^k}g_{j\ell}(x_0) + \partial_{x^jx^\ell}g_{ik}(x_0) - \partial_{x^ix^j}g_{k\ell}(x_0) - \partial_{x^kx^\ell}g_{ij}(x_0)\Big] = 0\,. \tag{2.97}$$

Note that the defining properties of a locally geodesic coordinate system hold only *at the single point* x_0 and generally not even in a small neighbourhood of it. Thus (2.97) should not be considered as a differential equation but only as a relation between the fibre coordinates of $J_2\pi$ in the fibre over the point $x_0 \in \mathcal{M}$. In particular, we cannot expect that local representations of prolongations $\mathcal{R}_{2+r} \subset J_{2+r}\pi$ are obtained by formally differentiating (2.97). However, this holds for the first prolongation of $\mathcal{R}_2$, as $\partial_{x^k}g_{ij}(x_0) = 0$ because of the metricity of the Levi–Civita connection. In further prolongations, lower-order terms appear. Fortunately, it turns out that $\mathcal{R}_3$ suffices for the formal analysis of the Einstein equations.

Because of the *contracted Bianchi identity* (a formal analogue to the Noether identity (2.94) of the Yang–Mills equations)

$$g^{ki}\partial_{x^k}R_{ij} - \partial_{x^j}(g^{ki}R_{ki}) = \Gamma^k_{ij}g^{i\ell}R_{k\ell} - \Gamma^\ell_{\ell k}g^{ik}R_{ij} \tag{2.98}$$

(here g^{ij} denote the components of the inverse g^{-1} of the metric tensor g), we find that $\mathcal{R}_2^{(1)} = \mathcal{R}_2$. We will see later that the equality (2.98) suffices to prove the formal integrability of Einstein's equations. ◁

Example 2.4.5. With the exception of the Navier–Stokes equations all these examples are of a variational origin. Let as usual $\pi : \mathcal{E} \to \mathcal{X}$ be a fibred manifold with an n-dimensional basis $\mathcal{X}$ and m-dimensional fibres. Assume that we are given an n-form $\lambda \in \Omega^n(J_q\pi)$ such that $\iota_X\lambda = 0$ for all vector fields $X \in \mathfrak{X}(J_q\pi)$ which are vertical for the fibration $\pi^q : J_q\pi \to \mathcal{X}$. This form is called the *Lagrangian* and we would like to determine sections $\sigma : \Omega \to \mathcal{E}$ such that the *action*

$$I[\sigma] = \int_\Omega (j_q\sigma)^*\lambda \tag{2.99}$$

becomes minimal. Here $\Omega \subseteq \mathcal{X}$ is assumed to be an open connected submanifold with a smooth boundary $\partial\Omega$ and the integrand $(j_q\sigma)^*\lambda$ is the pull-back of the Lagrangian n-form λ along the prolonged section $j_q\sigma : \Omega \to J_q\pi$. Usually, the variation is subject to some boundary conditions, e. g. that the values of the section σ on $\partial\Omega$ are prescribed.

In local coordinates, such a Lagrangian is of the form $\lambda = L(\mathbf{x}, \mathbf{u}^{(q)})d\mathbf{x}$ with a *Lagrangian density* $L \in \mathcal{F}(J_q\pi)$. If $\sigma(x) = \big(\mathbf{x}, \mathbf{s}(\mathbf{x})\big)$, then (2.99) becomes

$$I[\mathbf{s}] = \int_\Omega L\Big(\mathbf{x}, \mathbf{s}(\mathbf{x}), \frac{\partial\mathbf{s}(\mathbf{x})}{\partial\mathbf{x}}, \dots\Big)d\mathbf{x}\,. \tag{2.100}$$

A necessary condition for this integral to be minimal for σ is that the variational derivative δL of the Lagrangian density vanishes. Recall that δL is defined by the condition that for all sections $\eta(x) = \big(\mathbf{x}, \mathbf{h}(\mathbf{x})\big)$ with compact support (so that σ and $\sigma + \varepsilon\eta$ for $\varepsilon \in \mathbb{R}$ satisfy the same boundary conditions)

$$\left.\frac{\mathrm{d}}{\mathrm{d}\varepsilon}\right|_{\varepsilon=0} I[\mathbf{s}+\varepsilon\mathbf{h}] = \int_{\Omega} \delta L\Big(\mathbf{x},\mathbf{s}(\mathbf{x}),\frac{\partial\mathbf{s}(\mathbf{x})}{\partial\mathbf{x}},\dots\Big)\cdot\mathbf{h}(\mathbf{x})\,\mathrm{d}\mathbf{x} = 0\,. \tag{2.101}$$

By a straightforward computation involving only some partial integrations (see e. g. [342, Sect. 4.1]), one finds for the components of the variational derivative

$$(\delta L)^{\alpha} = \sum_{0\leq|\mu|\leq q} (-1)^{|\mu|} D_{\mu}\frac{\partial L}{\partial u^{\alpha}_{\mu}}\,. \tag{2.102}$$

The condition $\delta L = 0$ is known as the *Euler–Lagrange equation*. It represents a system of m equations of order $2q$.

If we consider the special case $n = q = 1$ and interpret the independent variable as time t, then $L = L(t, u^{\alpha}, \dot{u}^{\alpha})$. Now the Euler–Lagrange equations are

$$\frac{\mathrm{d}}{\mathrm{d}t}\Big(\frac{\partial L}{\partial \dot{u}^{\alpha}}\Big) - \frac{\partial L}{\partial u^{\alpha}} = \frac{\partial^2 L}{\partial \dot{u}^{\alpha}\partial \dot{u}^{\beta}}\ddot{u}^{\beta} + \frac{\partial^2 L}{\partial \dot{u}^{\alpha}\partial u^{\beta}}\dot{u}^{\beta} + \frac{\partial^2 L}{\partial \dot{u}^{\alpha}\partial t} - \frac{\partial L}{\partial u^{\alpha}} = 0\,. \tag{2.103}$$

In physics the most common situation is that the Lagrangian is of the special form $L = \frac{1}{2}m_{\alpha\beta}(\mathbf{u})\dot{u}^{\alpha}\dot{u}^{\beta} - V(\mathbf{u})$ where the coefficients $m_{\alpha\beta}$ form the positive definite *mass matrix* M and V is a *potential*. If M is constant, then we obtain as Euler–Lagrange equations the classical Newtonian equations

$$M\ddot{\mathbf{u}} = -\nabla V \tag{2.104}$$

or in words: mass times acceleration equals force. In the general case, additional velocity depending terms appear. As we assume that the mass matrix M is positive definite, it is in particular invertible and we may always bring the Euler–Lagrange equations into the normal form $\ddot{\mathbf{u}} = \mathbf{f}(\mathbf{u},\dot{\mathbf{u}})$.

It is easy to see that Euler–Lagrange equations are always linear in the highest-order derivatives. If we assume for simplicity that $q = 1$ (which is almost always the case in physics), then the Euler–Lagrange equations can be brought into Cauchy–Kovalevskaya form, if and only if (possibly after a coordinate transformation) an independent variable x^k exists (preferably the time t) such that the Hessian $\partial^2 L/\partial u^{\alpha}_k \partial u^{\beta}_k$ is invertible.

Otherwise one speaks of a *singular Lagrangian*. In this case one must generally expect that integrability conditions are hidden in the Euler–Lagrange equations. If for example in (2.103) the Hessian $\partial^2 L/\partial \dot{u}^{\alpha}\partial \dot{u}^{\beta}$ is singular, then contraction of (2.103) with any zero eigenvector of it yields an equation of lower order. Such equations are usually called *constraints* and their prolongation may lead to an integrability condition (and potentially to further constraints). In Section 7.3 we will study extensively how to proceed in such a situation. ◁

Finally, we consider a class of ordinary differential equations, namely *Hamiltonian systems*. They are ubiquitous in physics but also appear in other fields of science like economics. Here they serve us mainly as an illustration how natural the more abstract geometric approach to jet bundles presented in Section 2.2 is.

Therefore we only consider the regular case without constraints (the general case will be treated in an Addendum to Section 7.3).

Example 2.4.6. Hamiltonian systems are a cornerstone of classical mechanics. In the last decades, a geometric approach to mechanics has dominated the mathematical research in this field so much that one now often speaks of geometric mechanics. It is based on symplectic geometry and its generalisations (see Appendix C.6).

The classical picture starts with a symplectic manifold $(\mathcal{M},\omega)$, the *phase space*, and a Hamiltonian $H \in \mathcal{F}(\mathcal{M})$. In concrete applications, $\mathcal{M}$ is usually a cotangent bundle $T^*\mathcal{Q}$ where $\mathcal{Q}$ is the *configuration space* of the system. If we take adapted coordinates $(\mathbf{q},\mathbf{p})$ on $T^*\mathcal{Q}$, then $\mathbf{q}$ represents the *position* and $\mathbf{p}$ the *momentum* variables. The symplectic two-form takes now the simple form $\omega = \mathrm{d}\mathbf{q}\wedge\mathrm{d}\mathbf{p}$.

The Hamiltonian H describes the total energy of the system. The evolution of the system is governed by the corresponding Hamiltonian vector field intrinsically defined through the equation

$$\iota_{X_H}\omega = \mathrm{d}H \tag{2.105}$$

(cf. Definition C.6.3). Although (2.105) is often called the Hamiltonian equation of motion, this is not really correct. Equations of motion are differential equations, whereas (2.105) represents a set of algebraic equations on $T\mathcal{M}$.

The proper equations of motion are the ordinary differential equation corresponding to the vector field X_H. Using canonical or Darboux coordinates on $\mathcal{M}$, they take the familiar form (C.42)

$$\dot{\mathbf{q}} = \frac{\partial H}{\partial \mathbf{p}}\,, \qquad \dot{\mathbf{p}} = -\frac{\partial H}{\partial \mathbf{q}} \tag{2.106}$$

where the dot denotes the differentiation with respect to a parameter t which is usually identified with the time. In our context an obvious question is now how these equations can be naturally interpreted as a fibred submanifold $\mathcal{R}_1$ of some jet bundle, i. e. as a differential equation in the sense of Definition 2.3.1? It seems that this problem of relating the intrinsic definition (2.105) of a Hamiltonian vector field with an intrinsic formulation of the Hamiltonian equations of motion (2.106) has not attracted much interest so far and it was solved only recently [134].

It is probably impossible to find a reasonable solution within the framework of symplectic geometry. In the jet bundle formalism the independent variable t plays a more distinguished role than just that of a mere curve parameter and must be included from the very beginning. Therefore we need an approach which does not distinguish between autonomous and explicitly time-dependent systems.

Such an approach is naturally provided by *cosymplectic geometry*. We assume now that $(\mathcal{E},\omega,\eta)$ is a cosymplectic manifold (cf. Definition C.6.6) which is simultaneously a fibred manifold $\pi: \mathcal{E}\rightarrow\mathcal{X}$. Furthermore, we assume that the Reeb vector field on $\mathcal{E}$ is transversal to the fibration, i. e. at no point $\xi\in\mathcal{E}$ the projection $T_\xi\pi(R_\xi)\in T_{\pi(\xi)}\mathcal{X}$ vanishes. The manifold $\mathcal{E}$ is often called the *extended phase space*. In the simplest cases, it is of the form $\mathcal{E} = \mathcal{M}\times\mathcal{X}$ where $(\mathcal{M},\bar{\omega})$ is a symplectic manifold and $\mathcal{X} = \mathbb{R}$ models the time t. A cosymplectic structure on $\mathcal{E}$ is then given by $\omega = \mathrm{pr}_1^*\bar{\omega}$ and $\eta = \mathrm{pr}_2^*(\mathrm{d}t)$ (cf. Example C.6.7).

If the Reeb vector field R is projectable, it provides us with an intrinsic clock, namely the vector field $T\pi(R)$. Given a Hamiltonian $H \in \mathcal{F}(\mathcal{E})$, it serves as reference for the evolution of the system determined by the evolution vector field E_H. Trajectories of the system correspond thus to integral curves of E_H. Recall from Appendix C.6 that in Darboux coordinates $R = \partial_t$ and $E_H = \partial_t + H_{p_i}\partial_{q^i} - H_{q^i}\partial_{p_i}$. Thus we obtain as local coordinate form of the equations of motion

$$\dot{\mathbf{q}} = \frac{\partial H}{\partial \mathbf{p}}\,, \qquad \dot{\mathbf{p}} = -\frac{\partial H}{\partial \mathbf{q}}\,, \qquad \dot{t} = 1 \tag{2.107}$$

(note that in contrast to the classical equations (2.106) the Hamiltonian H may now depend explicitly on the time t).

It is straightforward to give (2.107) an intrinsic description as submanifold of the jet bundle $J_1\pi$. By assumption, the Reeb vector field R and thus also E_H for arbitrary Hamiltonians H is transversal. Since the base space $\mathcal{X}$ is one-dimensional, such a vector field spans the horizontal bundle of a connection on $\mathcal{E}$. According to Remark 2.2.3, any such connection corresponds to a section $\gamma : \mathcal{E} \to J_1\pi$ and (2.107) is just a local representation of $\mathcal{R}_1 = \operatorname{im}\gamma$.

We may also study the inverse problem: decide whether a given differential equation $\mathcal{R}_1 \subseteq J_1\pi$ is Hamiltonian and if yes, construct the corresponding evolution vector field E_H. For simplicity, we assume that we are dealing with an unconstrained system, i. e. $\mathcal{R}_1 = \operatorname{im}\gamma$ for some section $\gamma : \mathcal{E} \to J_1\pi$. The extension to systems subject to constraints will be discussed in an Addendum to Section 7.3.

The basic idea is to compare the connections defined by the section γ and the Reeb vector field R, respectively. One way to do this is to pick vectors on the base manifold $\mathcal{X}$ and to lift them with the respective horizontal lifts. The difference between the two thus obtained vectors is obviously vertical. Recall that in our situation the vertical bundle $V\pi$ is spanned by the Hamiltonian vector fields. We call the given differential equation $\mathcal{R}_1$ Hamiltonian, if and only if the difference is of the form X_H for some function $H \in \mathcal{F}(\mathcal{E})$.

An obvious choice for vectors at $x \in \mathcal{X}$ are the projections $T_\xi\pi(R_\xi)$ for some points $\xi \in \pi^{-1}(x)$. Thus we may formalise the considerations above by introducing the vector field $X = -P^V[\gamma](R)$ where $P^V[\gamma]$ denotes the vertical projector of the connection defined by γ. Obviously, if we lift $T_\xi\pi(R_\xi)$ with the connection associated to the Reeb vector field R, then we obtain just R_ξ. By applying $P^V[\gamma]$ to it, we effectively compute the difference between the vectors obtained by lifting $T_\xi\pi(R_\xi)$ with the connection defined by γ and R.

With the help of the contact map Γ_1 on $J_1\pi$, we may alternatively describe the above constructed vector field X at a given point $\xi \in \mathcal{E}$ as follows

$$X_\xi = P^V[R]\bigl(\Gamma_1\bigl(\gamma(\xi), T_\xi\pi(R_\xi)\bigr)\bigr)\,. \tag{2.108}$$

Indeed, the argument of $P^V[R]$ is nothing but an explicit expression for the horizontal lift of $T\pi(R)$ for the connection associated with γ. Independent of which description of X we use, the differential equation $\mathcal{R}_1$ is Hamiltonian, if and only if a function $H \in \mathcal{F}(\mathcal{E})$ exists such that $X = X_H$.

Finally, we compare the two different notions of a solution. As already mentioned above, in the traditional approach one considers the integral curves of the evolution vector field E_H; whereas for differential equations we use Definition 2.3.3. But again it is straightforward to show the equivalence of these two points of view.

Indeed, it follows from (2.108) that $(E_H)_\xi = \Gamma_1\bigl(\gamma(\xi), T_\xi\pi(R_\xi)\bigr)$. If the section σ is a solution, then $\gamma \circ \sigma = j_1\sigma$. Thus, by definition of the contact map, we get at the point $\xi = \sigma(x)$ that $(E_H)_\xi = T_x\sigma \cdot T_\xi\pi(R_\xi) \in T_\xi(\operatorname{im}\sigma)$ and $\operatorname{im}\sigma$ is an integral curve of E_H. Conversely, it follows from the transversality of E_H that every integral curve is (at least locally) the image of a section $\sigma : \mathcal{X} \to \mathcal{E}$. Applying the same argument backwards, we see that $\gamma \circ \sigma = j_1\sigma$. $\triangleleft$

2.5 Notes

Implicitly, jet bundles have been used for a long time. They were formally introduced by Ehresmann in a series of papers [120, 121] (see also [122]). They have become a standard tool for the geometric analysis of differential equations. By now, many approaches to jet bundles exist with various degrees of abstraction and sophistication. The book by Saunders [394] serves as a standard reference; useful introductions are also contained in [170, 342]. In particular, the intrinsic formulation of our "pedestrian" approach to jet bundles (Remark 2.2.1) can be found in [170].

Another approach to jet bundles is discussed by Palais [347, Ch. IV]. It requires that $\mathcal{U}$ is a real vector space, which is the case in most applications. We define for any point $x_0 \in \mathcal{X}$ the ideal $\mathcal{I}_{x_0}(\mathcal{X}) \subset \mathcal{F}(\mathcal{X})$ of all smooth functions $\phi : \mathcal{X} \to \mathbb{R}$ satisfying $\phi(x_0) = 0$. Then we introduce the spaces

$$\mathcal{Z}^{(q)}_{x_0}(\mathcal{X},\mathcal{U}) = \bigl[\mathcal{I}_{x_0}(\mathcal{X})\bigr]^q \cdot \mathcal{C}^\infty(\mathcal{X},\mathcal{U})\,. \tag{2.109}$$

All functions in $\mathcal{Z}^{(q)}_{x_0}(\mathcal{X},\mathcal{U})$ possess at x_0 an at least q-fold zero and thus all their derivatives up to order $q-1$ vanish at this point. Finally, we identify q-jets at x_0 with the elements of the quotient space $\mathcal{C}^\infty(\mathcal{X},\mathcal{U})/\mathcal{Z}^{(q+1)}_{x_0}(\mathcal{X},\mathcal{U})$.

Some of the more geometric aspects of jet bundles have been described following the presentation by Lychagin [296]. This concerns in particular the decomposition of the contact distribution into the vertical bundle and the horizontal space of a connection discussed in Remark 2.2.9. The term *fundamental identification* seems to have been coined by Kuranishi [275]. Our intrinsic description of it in Section 2.2 follows ideas of Ruiz [389], although he does not explicitly use the contact map.

Throughout this book, we will always assume that we start with a fibred manifold $\pi : \mathcal{E} \to \mathcal{X}$, i. e. that from the beginning we have a clear distinction between independent and dependent variables. This is a natural assumption for most problems in physics (and many other fields of science using differential equations as models) where $\mathcal{X}$ is usually space-time and $\mathcal{E}$ some fibre bundle. However, for many problems in differential geometry this assumption is somewhat artificial, as here one often deals with manifolds $\mathcal{E}$ without any natural fibration. For such problems an approach via exterior differential systems (see below) is often better suited.

There also exist generalisations of jet bundles applicable in such situations. The basic idea in our approach is that each section $\sigma \in \Gamma_{loc}(\mathcal{E})$ defines an n-dimensional submanifold of $\mathcal{E}$, namely $\operatorname{im}\sigma$ (here as usual $n = \dim \mathcal{X}$). However, not every n-dimensional submanifold of $\mathcal{E}$ is of this particular form but only those that are transversal to the fibration $\pi : \mathcal{E} \to \mathcal{X}$. If $\mathcal{E}$ is not a fibred manifold, this restriction makes no sense and we simply drop it. Thus one considers equivalence classes of arbitrary n-dimensional submanifolds[7] and builds jet bundles in a similar manner, as we have done it; for more information see e. g. [163, Sect. 5.3] and references therein. Basically the same construction is described by Olver [342, Sect. 3.5] as "extended" jet bundle and used for a rigorous formulation of symmetry reductions.

Even if one starts with a fibred manifold, there are situations where our concept that a solution is a section and thus transversal is too restrictive. It is well-known that solutions of hyperbolic equations can be highly irregular. For example, it is not uncommon that a shock forms where some derivatives become unbounded; sometimes one even has multivalued solutions (just think of a breaking wave at the beach). In the sense of Definition 2.3.3 these are not solutions. However, if one drops the transversality condition, such phenomena can still be described geometrically with jets, as the solutions define submanifolds (for ordinary differential equations such a generalised notion of solution is briefly discussed in Section 9.1).

Our "pedestrian" treatment of the contact structure is inspired by Johnson [239]. The intrinsic approach to first-order jet bundles $J_1\pi$ presented in Section 2.2 does not seem to be very popular. It is rarely found in the literature, although it possesses a number of advantages for studying the geometry of jet bundles. Recently, it has gained some importance in the context of multisymplectic geometry, as the affine dual of $J_1\pi$ carries a natural multisymplectic structure [174]. The corresponding intrinsic description of the contact structure on $J_1\pi$ via the contact map is taken from Modugno [325]; a similar discussion is contained in [163, Sect. 5.1].

We assume throughout this book that we are always dealing with real manifolds. In some examples this assumption leads to problems with Definition 2.3.1 of a differential equation. Take for example the equation $(u')^2 + u^2 + x^2 = 1$ describing the unit sphere in $J_1\pi$ where we take $\mathcal{E} = \mathbb{R}^2$ and $\mathcal{X} = \mathbb{R}$. Obviously, the unit sphere is *not* a fibred submanifold over $\mathcal{X}$, as the projection to $\mathcal{X}$ is not surjective.[8] This problem disappears, if one considers the problem over the complex numbers $\mathbb{C}$, as now for any value $x \in \mathbb{C}$ a point on the unit sphere exists ($\mathbb{C}$ is algebraically closed).

For a satisfactory complex theory it is important to restrict to analytic manifolds and to consider only holomorphic sections. Only under these assumptions our

---

[7] Using the approach via Grassmannians described in Remark 2.2.5, this generalisation is equivalent to consider the full Grassmannian $G_n(\mathcal{E})$ (which is independent of any fibration) instead of its transversal subspace $G(\pi) = G_n^{\text{trans}}(\mathcal{E})$.

[8] This is not the only problem: at the two points $(\pm 1, 0, 0)$ the projection is not a submersion—a fact which is independent of the used ground field.

treatment of the contact structure can be kept without modification considering now all variables as complex valued. Otherwise, we would have to consider not only derivatives $\partial/\partial x^i$, say, but also derivatives $\partial/\partial \bar{x}^i$ with respect to the complex conjugate variables, as only for holomorphic sections the latter ones always vanish.

One of the earliest geometric approach to differential equations, the *Cartan–Kähler Theory* [58, 74, 231, 248, 326], is not based on jet bundles but represents the equations with differential forms. More precisely, instead of differential equations one considers here a differential ideal in the algebra of differential forms over some manifold $\mathcal{M}$. Solutions are represented by integral manifolds of this ideal. We will study in Section 9.5 a dual version of this approach due to Vessiot in more detail.

Given a differential equation $\mathcal{R}_q \subseteq J_q\pi$, there exist in general many ways to produce a corresponding differential ideal. The simplest—and in particular always applicable—method consists of using the ideal $\mathcal{I}[\mathcal{R}_q]$ generated by the pull-back of the contact codistribution $\mathcal{C}_q^0$ to the submanifold $\mathcal{R}_q$ (we used this ideal already for the characterisation of integral elements in Proposition 2.3.19). In order to recover our notion of a solution, we must in addition impose *independence forms* corresponding to our transversality conditions naturally induced by the underlying fibration $\pi : \mathcal{E} \to \mathcal{X}$: if the forms $\theta_1, \dots, \theta_n$ span the cotangent bundle $T^*\mathcal{X}$, then we consider only those integral manifolds such that the pull-backs of the forms $(\pi^q)^*(\theta_i)$ on them do not vanish.

We have seen that there are some analogies between our geometric view of differential equations and the situation in algebraic geometry. However, one should note that there are also some apparent differences. In particular, not every section contained in an equivalence class $[\sigma]_x^{(q)} \in \mathcal{R}_q$ is a solution of the differential equation $\mathcal{R}_q$. If one wants such a one-to-one relation between points and solutions, then one must take *all* prolongations of $\mathcal{R}_q$ into account. This point of view leads naturally to the introduction of the jet bundle $J_\infty\pi$ of infinite order a point of which defines a unique formal power series.

The proper definition of $J_\infty\pi$ as a manifold requires some mathematical care, as it is an infinite-dimensional space and it is not obvious how to equip it with the necessary topological and differentiable structure. One can show that this goal can be rigorously achieved using the notions of projective and injective limits in directed systems of sets [42]. As the jet bundles of different orders are related by the natural projections π_q^r, we may define

$$J_\infty\pi = \operatorname*{proj\,lim}_{r\to\infty} J_r\pi\,. \tag{2.110}$$

In order to introduce smooth functions on the manifold $J_\infty\pi$, we exploit the natural inclusions $\mathcal{F}(J_q\pi) \hookrightarrow \mathcal{F}(J_r\pi)$ for $r > q$ and set

$$\mathcal{F}(J_\infty\pi) = \operatorname*{inj\,lim}_{r\to\infty} \mathcal{F}(J_r\pi)\,. \tag{2.111}$$

Note that the definition of an injective limit implies that any function in $\mathcal{F}(J_\infty\pi)$ is an element of at least one space $\mathcal{F}(J_q\pi)$. Thus although these functions are defined on an infinite-dimensional manifold, they always depend only on a finite number of variables. In a similar, though slightly more complicated manner, one constructs the tangent and the cotangent bundle of $J_\infty\pi$ [42].

Then we introduce the infinite prolongation $\mathcal{R}_\infty \subseteq J_\infty\pi$ of a differential equation $\mathcal{R}_q \subseteq J_q\pi$ by simply replacing in (2.110) $J_r\pi$ by its submanifold $\mathcal{R}_r$. As each point on the submanifold $\mathcal{R}_\infty$ corresponds to one formal power series solution of the given equation $\mathcal{R}_q$ and vice versa, $\mathcal{R}_\infty$ may indeed be identified with the formal solution space of $\mathcal{R}_q$. Thus, at least at the formal level, we have now arrived in the same situation as in algebraic geometry. This point of view is much used by the school of Vinogradov (see e.g. [12, 84, 263, 461]) who calls such a submanifold $\mathcal{R}_\infty \subseteq J_\infty\pi$ a *differential variety* or *diffiety* for short.

An interesting aspect of the infinite-order jet bundle $J_\infty\pi$ is that its contact codistribution $\mathcal{C}^0_\infty$ is involutive. As discussed in Remark 2.1.7, this is never the case in a jet bundle $J_q\pi$ of finite order q due to the obstructions obtained by taking the exterior derivative of a contact form ω^α_μ with a multi index μ of maximal length $|\mu| = q-1$. In $\mathcal{C}^0_\infty$ no maximal value for $|\mu|$ exists and thus it is involutive.

Dually, we obtain the same picture for the contact distribution. Whereas $\mathcal{C}_q$ cannot be involutive for any finite order q, since the commutators between the transversal vector fields $C^{(q)}_i$ and the vertical fields C^μ_α do not lie in $\mathcal{C}_q$ according to (2.13), the contact distribution $\mathcal{C}_\infty$ does not contain any vertical fields. Note furthermore that the n vector fields $C^{(\infty)}_i$ forming a basis of it may be identified with the formal derivatives D_i. As the formal derivatives commute like ordinary partial derivatives, we find trivially $[C^{(\infty)}_i, C^{(\infty)}_j] = 0$ for all $1 \leq i, j \leq n$ and thus the contact distribution $\mathcal{C}_\infty$ is indeed involutive.

The material on Hamiltonian systems in Example 2.4.6 first appeared in [134]. Rather surprisingly, we have not been able to find an intrinsic definition of a Hamiltonian differential equation in the literature. Lately, jet bundles have become fairly popular in geometric mechanics, not only for the treatment of time-dependent systems but also as a basis of the multisymplectic formalism for field theories [67, 174]. But the mechanics is studied in the conventional way via vector fields and forms living on the jet bundle. Thus one may say that these works study mechanics *on* jet bundles, whereas our approach lives *in* jet bundles.

The main motivation for our formulation lies in the theoretical unification of differential equations theory and geometric mechanics. For the finite-dimensional systems we have studied here, this can, of course, also be achieved in the classical approach, since dynamical systems theory can be intrinsically formulated via vector fields. However, we believe that our approach provides a better basis for the extension to field theories. This aspect becomes even more evident for systems with constraints, where the traditional vector field approach runs into trouble for infinite-dimensional systems (see [134] for concrete examples).

Chapter 3
Involution I: Algebraic Theory

Algebra is generous: she often gives more than is asked for.
Jean d'Alembert

The first chapter was mainly concerned with the geometry behind the formal theory. For many problems a purely geometric approach is not sufficient (a concrete example is the question of proving the formal integrability of a differential equation) and additional algebraic ingredients are needed. These lead us to the title concept of this book: involution. Its algebraic theory will be the topic of the next four chapters. We will start in the current chapter with a more combinatorial approach leading to a special kind of Gröbner bases, the involutive bases; its algorithmic realisation is the topic of Chapter 4. In Chapter 5 we will show that the structure analysis of polynomial modules becomes significantly easier with the help of such bases. Finally, Chapter 6 will provide us with a homological explanation of the remarkable properties of (some) involutive bases. In Chapter 7 we will then return to differential equations and see how these algebraic ingredients appear there naturally in the analysis of the symbol of a differential equation.

In the first section we introduce the notion of an involutive division, a restriction of the usual divisibility relation of power products. As we are interested in applications beyond the classical commutative case, i. e. also in situations where the variables do not commute with each others, we introduce the theory for the monoid of multi indices and not for terms. Here we will already meet most of the key ideas about involution, as the later extension to polynomials will be fairly straightforward.

The next section introduces a rather general class of rings: the polynomials of solvable type. It comprises many classical algebras which are important for applications like rings of linear differential or difference operators or universal enveloping algebras of Lie algebras. Obviously, most of these rings are not commutative. The defining property of solvable polynomial rings is the existence of an associated term order such that the leading terms behave under multiplication like in the familiar commutative case. As most of the theory of Gröbner bases relies solely on an analysis of leading terms and some normal form considerations, this class of rings represents probably the most natural setting for the definition of involutive bases.

Although we are mainly interested in polynomials with coefficients in a field, we define polynomials of solvable type over rings. In this case one of the most important results for a constructive theory, Hilbert's Basis Theorem, becomes non-trivial

W.M. Seiler, *Involution*, Algorithms and Computation in Mathematics 24,
DOI 10.1007/978-3-642-01287-7_3, © Springer-Verlag Berlin Heidelberg 2010

and the whole Section 3.3 is devoted to this question. We will see four different proofs, each making different assumptions on the polynomials and each using different techniques. Two of them will simultaneously give us the existence of Gröbner bases for ideals in solvable polynomial rings.

In the last section we finally define involutive bases in arbitrary polynomials algebras of solvable type; it turns out that they represent a special kind of Gröbner bases with additional combinatorial properties. Our introduction of involutive bases is closely modelled on a classical approach to Gröbner bases and thus assumes some familiarity with their theory. A brief review of the most important points in the form we need is contained in Appendix B.

The question of the existence and the effective construction of involutive bases is postponed to Chapter 4; here we are only concerned with some elementary consequences of the definition. In particular, we will show that, in contrast to ordinary Gröbner bases, involutive bases lead to unique standard representations, a property that will later be the key to their application in commutative algebra in Chapter 5.

3.1 Involutive Divisions

As the concept of involution appears in different contexts, we need a fairly general and flexible approach for its introduction. For this reason we do not define it in terms of power products or differential operators or something similar but in terms of multi indices (introduced in Appendix A.1). Thus we work now in the Abelian monoid $(\mathbb{N}_0^n, +)$ with the addition defined componentwise. The multi indices may be identified in a natural way with the vertices of an n-dimensional integer lattice, so that we can easily visualise subsets of $\mathbb{N}_0^n$ (at least for $n \leq 3$). Choosing an arbitrary field $\Bbbk$, we may of course also identify any multi index $\mu \in \mathbb{N}_0^n$ with the corresponding monomial $x^\mu \in \mathcal{P} = \Bbbk[x^1, \dots, x^n]$.

For any multi index $\nu \in \mathbb{N}_0^n$ we introduce its *cone* $\mathcal{C}(\nu) = \nu + \mathbb{N}_0^n$, i. e. the set of all multi indices that can be reached from ν by adding another multi index. We say that ν *divides* μ, written $\nu \mid \mu$, if $\mu \in \mathcal{C}(\nu)$. It might appear strange to speak here of a "division", since the operation in the monoid $\mathbb{N}_0^n$ is an addition, but if we think of the power products $x^\nu, x^\mu \in \mathcal{P}$, then $\mu \in \mathcal{C}(\nu)$ is equivalent to $x^\nu \mid x^\mu$.

It is obvious that cones are the simplest examples of monoid ideals (i. e. subsets $\mathcal{I} \subseteq \mathbb{N}_0^n$ which are "multiplicatively" closed: $\mathcal{I} + \mathbb{N}_0^n = \mathcal{I}$) and that any monoid ideal may be considered as the union of cones (by Dickson's Lemma A.1.2 a finite number of cones always suffices). Given a finite subset $\mathcal{B} \subset \mathbb{N}_0^n$, we define its *span* as the monoid ideal generated by $\mathcal{B}$:

$$\langle \mathcal{B} \rangle = \bigcup_{\nu \in \mathcal{B}} \mathcal{C}(\nu) . \tag{3.1}$$

Again this definition is easy to interpret for power products: the set $\{x^\nu \mid \nu \in \mathcal{B}\}$ generates a monomial ideal $\mathcal{J} \subset \mathcal{P}$ and we have $\mu \in \langle \mathcal{B} \rangle$, if and only if $x^\mu \in \mathcal{J}$.

The basic idea underlying the definition of an involutive division consists of introducing a restriction of the cone of a multi index, the involutive cone, where it is only allowed to add multi indices certain entries of which vanish. Alternatively, we may speak of a restriction of the above defined divisibility relation. The final goal will be to obtain a *disjoint* union in (3.1) by using only these involutive cones on the right hand side. This idea will naturally lead to the combinatorial decompositions of algebras $\mathcal{A} = \mathcal{P}/\mathcal{J}$ discussed in Section 5.1.

Definition 3.1.1. An *involutive division* L is defined on the monoid $(\mathbb{N}_0^n, +)$, if for any finite subset $\mathcal{B} \subset \mathbb{N}_0^n$ a set $N_{L,\mathcal{B}}(\nu) \subseteq \{1,\dots,n\}$ of *multiplicative indices*, and consequently a submonoid $L(\nu,\mathcal{B}) = \{\mu \in \mathbb{N}_0^n \mid \forall j \notin N_{L,\mathcal{B}}(\nu) : \mu_j = 0\}$, is associated to every multi index $\nu \in \mathcal{B}$ such that the following two conditions on the *involutive cones* $\mathcal{C}_{L,\mathcal{B}}(\nu) = \nu + L(\nu,\mathcal{B}) \subseteq \mathbb{N}_0^n$ are satisfied.

(i) If there exist two elements $\mu,\nu \in \mathcal{B}$ with $\mathcal{C}_{L,\mathcal{B}}(\mu) \cap \mathcal{C}_{L,\mathcal{B}}(\nu) \neq \emptyset$, either $\mathcal{C}_{L,\mathcal{B}}(\mu) \subseteq \mathcal{C}_{L,\mathcal{B}}(\nu)$ or $\mathcal{C}_{L,\mathcal{B}}(\nu) \subseteq \mathcal{C}_{L,\mathcal{B}}(\mu)$ holds.
(ii) If $\mathcal{B}' \subset \mathcal{B}$, then $N_{L,\mathcal{B}}(\nu) \subseteq N_{L,\mathcal{B}'}(\nu)$ for all $\nu \in \mathcal{B}'$.

An arbitrary multi index $\mu \in \mathbb{N}_0^n$ is *involutively divisible* by $\nu \in \mathcal{B}$, written $\nu \,|_{L,\mathcal{B}}\, \mu$, if $\mu \in \mathcal{C}_{L,\mathcal{B}}(\nu)$. In this case ν is called an *involutive divisor* of μ.

Before discussing the precise meaning of this definition and in particular of the two conditions contained in it, we should stress the following important point: as clearly indicated by the notation, involutive divisibility is defined with respect to both an involutive division L *and* a fixed finite set $\mathcal{B} \subset \mathbb{N}_0^n$; only elements of $\mathcal{B}$ can appear as involutive divisors. Obviously, involutive divisibility $\nu \,|_{L,\mathcal{B}}\, \mu$ implies ordinary divisibility $\nu \mid \mu$.

The involutive cone $\mathcal{C}_{L,\mathcal{B}}(\nu)$ of an element $\nu \in \mathcal{B}$ is generally only a subset of the full cone $\mathcal{C}(\nu)$: we are not allowed to add arbitrary multi indices to ν but may increase only certain entries of ν determined by the multiplicative indices. The first condition in the above definition says that involutive cones can intersect only trivially: if two intersect, then one must be a subset of the other.

The *non-multiplicative indices* form the complement of $N_{L,\mathcal{B}}(\nu)$ in N and are denoted by $\overline{N}_{L,\mathcal{B}}(\nu)$. If we remove some elements from the set $\mathcal{B}$ and determine the multiplicative indices of the remaining elements with respect to the subset $\mathcal{B}'$, then we will obtain in general a different result than before. The second condition for an involutive division (sometimes called the *filter axiom*) says now that while it may happen that a non-multiplicative index becomes multiplicative for some $\nu \in \mathcal{B}'$, the converse cannot happen.

Remark 3.1.2. Alternatively, we may define axiomatically the notion of involutive divisibility as follows. For every finite subset $\mathcal{B} \subset \mathbb{N}_0^n$ a relation $|_{L,\mathcal{B}}$ on $\mathcal{B} \times \mathbb{N}_0^n$ is given such that for all multi indices $\mu,\nu \in \mathcal{B}$ and $\rho,\sigma \in \mathbb{N}_0^n$ the following six conditions are satisfied:

1. $\nu \,|_{L,\mathcal{B}}\, \rho$ implies that $\nu \mid \rho$.
2. $\nu \,|_{L,\mathcal{B}}\, \nu$.
3. $\nu \,|_{L,\mathcal{B}}\, (\nu+\rho)$ and $\nu \,|_{L,\mathcal{B}}\, (\nu+\sigma)$, if and only if $\nu \,|_{L,\mathcal{B}}\, (\nu+\rho+\sigma)$.

4. If $\nu\,|_{L,\mathcal{B}}\,\rho$ and $\mu\,|_{L,\mathcal{B}}\,\rho$, then either $\nu\,|_{L,\mathcal{B}}\,\mu$ or $\mu\,|_{L,\mathcal{B}}\,\nu$.
5. If $\nu\,|_{L,\mathcal{B}}\,\mu$ and $\mu\,|_{L,\mathcal{B}}\,\rho$, then also $\nu\,|_{L,\mathcal{B}}\,\rho$ holds.
6. If $\mathcal{B}' \subset \mathcal{B}$ and $\nu\,|_{L,\mathcal{B}}\,\mu$ for some multi indices $\nu \in \mathcal{B}'$ and $\mu \in \mathbb{N}_0^n$, then also $\nu\,|_{L,\mathcal{B}'}\,\mu$.

The equivalence of this approach to Definition 3.1.1 is easily seen. If we define $\mathcal{C}_{L,\mathcal{B}}(\nu) = \{\mu \in \mathbb{N}_0^n \mid \nu\,|_{L,\mathcal{B}}\,\mu\}$, then the first three conditions ensure firstly that $\mathcal{C}_{L,\mathcal{B}}(\nu) \subseteq \mathcal{C}(\nu)$ and secondly that $\mathcal{C}_{L,\mathcal{B}}(\nu)$ is of the form $\nu + L(\nu,B)$ for a submonoid $L(\nu,\mathcal{B}) \subseteq \mathbb{N}_0^n$ which is defined by some set of multiplicative indices (note the "if and only if" in the third condition). The next two conditions require that the intersection of two involutive cones is trivial and the last one is a verbatim translation of the filter axiom. ◁

Example 3.1.3. A simple example of an involutive division is the *Thomas division* T. It is defined as follows: let $\mathcal{B} \subset \mathbb{N}_0^n$ be a finite set and $\mu \in \mathcal{B}$ an arbitrary element; then $i \in N_{T,\mathcal{B}}(\mu)$, if and only if $\mu_i = \max_{\nu\in\mathcal{B}} \nu_i$. One easily verifies that both conditions in Definition 3.1.1 are satisfied. ◁

Example 3.1.4. The *Janet division* J may be considered as a kind of refinement of the Thomas division. In order to define it, we must introduce certain subsets of the given set $\mathcal{B} \subset \mathbb{N}_0^n$:

$$(d_k,\dots,d_n) = \left\{\, \nu \in \mathcal{B} \mid \nu_i = d_i\,,\ k \le i \le n \,\right\}. \tag{3.2}$$

Thus all multi indices in $(d_k,\dots,d_n)$ share the same "k-tail". The index n is multiplicative for $\nu \in \mathcal{B}$, if $\nu_n = \max_{\mu\in\mathcal{B}}\{\mu_n\}$, and $k < n$ is multiplicative for ν, if $\nu_k = \max_{\mu\in(\nu_{k+1},\dots,\nu_n)}\{\mu_k\}$. Obviously, in contrast to the Thomas division, this definition depends on the order in which we process the entries of the multi indices (i. e. on the ordering of the variables $x^1,\dots,x^n$ in polynomial computations) and we obtain simple variants, if we first apply an arbitrary but fixed permutation $\pi \in S_n$ to each multi index.

Consider for $n = 2$ the set $\mathcal{B} = \{\nu^{(1)} = [2,0], \nu^{(2)} = [0,2]\}$. As $\nu_2^{(2)} > \nu_2^{(1)}$, the index 2 is multiplicative only for $\nu^{(2)}$. In order to decide whether 1 is multiplicative, we must look at the subsets (0) and (2), respectively. As each of them consists of only one element, the index 1 is multiplicative for both multi indices and we obtain that $N_{J,\mathcal{B}}(\nu^{(1)}) = \{1\}$ and $N_{J,\mathcal{B}}(\nu^{(2)}) = \{1,2\}$. Thus the involutive cone $\mathcal{C}_{T,\mathcal{B}}(\nu^{(1)}) = \{[k,0] \mid k \ge 2\}$ is a proper subset of the full cone $\mathcal{C}(\nu^{(1)})$. If we consider the subset $\mathcal{B}' \subset \mathcal{B}$ consisting only of $\nu^{(1)}$, then 2 becomes multiplicative for the multi index $\nu^{(1)}$, too, so that now $\mathcal{C}_{T,\mathcal{B}'}(\nu^{(1)}) = \mathcal{C}(\nu^{(1)})$. ◁

Lemma 3.1.5. *The Janet division is an involutive division.*

Proof. The second condition of Definition 3.1.1 is easy to verify. If k is a multiplicative index for $\nu \in \mathcal{B}$, then ν_k is maximal in a certain subset $(d_{k+1},\dots,d_n) \subseteq \mathcal{B}$. If we remove some elements of the set $\mathcal{B}$ in order to obtain a subset $\mathcal{B}' \subset \mathcal{B}$, then this maximality is unchanged, as the corresponding subset $(d_{k+1},\dots,d_n) \subseteq \mathcal{B}'$ may only have become smaller. Thus all multiplicative indices remain multiplicative. It may of course happen that some non-multiplicative indices become multiplicative as in Example 3.1.4 above, but this is allowed for an involutive division.

Concerning the first condition in Definition 3.1.1, we will prove an even stronger result: if $\mathcal{C}_{J,\mathcal{B}}(\mu) \cap \mathcal{C}_{J,\mathcal{B}}(\nu) \neq \emptyset$ for two multi indices $\mu, \nu \in \mathcal{B}$, then μ and ν must in fact be identical. So assume there exists a multi index $\rho \in \mathcal{C}_{J,\mathcal{B}}(\mu) \cap \mathcal{C}_{J,\mathcal{B}}(\nu)$. Then there must exist two multi indices $\bar{\mu} \in \mathbb{N}^n_{N_{J,\mathcal{B}}(\mu)}$ and $\bar{\nu} \in \mathbb{N}^n_{N_{J,\mathcal{B}}(\nu)}$ such that $\rho = \mu + \bar{\mu} = \nu + \bar{\nu}$. We look first at ρ_n. If $\bar{\mu}_n > 0$ and $\bar{\nu}_n > 0$, n is multiplicative for both μ and ν. But by the definition of the Janet division this requires $\mu_n = \nu_n$. It is not possible that $\bar{\mu}_n > 0$ and $\bar{\nu}_n = 0$, as then $\mu_n < \nu_n$ and nevertheless $n \in N_{J,\mathcal{B}}(\mu)$ contradicting the definition of the Janet division. The same holds for $\bar{\mu}_n = 0$ and $\bar{\nu}_n > 0$. Thus we may conclude that $\mu_n = \nu_n$.

We proceed now to ρ_{n-1}. By the considerations above we know already that $\mu, \nu \in (\mu_n)$. But this observation implies that we can repeat the same reasoning for the next entry. For $n-1$ to be a multiplicative index for μ, we must have $\mu_{n-1} \geq \nu_{n-1}$ and vice versa. Hence $\mu_{n-1} = \nu_{n-1}$ and $\mu, \nu \in (\mu_{n-1}, \mu_n)$. Iterating this argument, we get that $\mu = \nu$. □

We have stressed above that involutive divisibility is always defined with respect to a given set $\mathcal{B}$. However, involutive divisions exist where the assignment of the multiplicative variables is independent of $\mathcal{B}$. Such divisions are very convenient in concrete computations and have a special name.

Definition 3.1.6. An involutive division L is *globally defined*, if the assignment of the multiplicative indices to a multi index $\nu \in \mathcal{B}$ is independent of the set $\mathcal{B}$. In this case we simply write $N_L(\nu)$.

Example 3.1.7. The most important division for our purposes is the *Pommaret division P*. It is a global division and assigns the multiplicative indices according to a simple rule: if $k = \operatorname{cls} \nu$ (see Appendix A.1 for the definition of the class of a multi index), then we set $N_P(\nu) = \{1, \dots, k\}$. Thus, like for the Janet division, the position of the entries of the multi indices are relevant and one can introduce simple variants by first applying a permutation to every multi index. ◁

Lemma 3.1.8. *The Pommaret division is an involutive division.*

Proof. The second condition of Definition 3.1.1 obviously holds for *any* globally defined division, as the set $\mathcal{B}$ does not influence the multiplicative indices. Concerning the first condition, assume that a multi index $\rho \in \mathcal{C}_P(\mu) \cap \mathcal{C}_P(\nu)$ exists. As in the proof of Lemma 3.1.5, we write $\rho = \mu + \bar{\mu} = \nu + \bar{\nu}$ with multi indices $\bar{\mu} \in \mathbb{N}^n_{N_P(\mu)}$ and $\bar{\nu} \in \mathbb{N}^n_{N_P(\nu)}$. Without loss of generality, we may assume that $\operatorname{cls} \mu \leq \operatorname{cls} \nu = k$. By definition of the Pommaret division, this assumption implies $\rho_i = \mu_i = \nu_i$ for all $i > k$. We must distinguish two cases. If $\operatorname{cls} \mu = k$, too, then either $\nu_k \leq \mu_k$ and consequently $\mu \in \mathcal{C}_P(\nu)$ or $\nu_k > \mu_k$ and $\nu \in \mathcal{C}_P(\mu)$. If $\operatorname{cls} \mu < k$ and $\nu_k \leq \mu_k$, then again $\mu \in \mathcal{C}_P(\nu)$. This observation finishes the proof, as it is not possible that $\operatorname{cls} \mu < k$ and $\nu_k > \mu_k$: because of $\rho_k \geq \nu_k$, the index k would then have to be multiplicative for μ contradicting $\operatorname{cls} \mu < k$. □

Above we introduced the span of a set $\mathcal{B} \subset \mathbb{N}_0^n$ as the union of the cones of its elements. Given an involutive division it appears natural to consider also the union of the involutive cones. Obviously, this construction yields in general only a subset (without any algebraic structure) of the monoid ideal $\langle \mathcal{B} \rangle$.

Definition 3.1.9. The *involutive span* of a finite set $\mathcal{B} \subset \mathbb{N}_0^n$ is

$$\langle \mathcal{B} \rangle_L = \bigcup_{\nu \in \mathcal{B}} \mathcal{C}_{L,\mathcal{B}}(\nu) \,. \tag{3.3}$$

The set $\mathcal{B}$ is called *weakly involutive* for the division L or a *weak involutive basis* of the monoid ideal $\langle \mathcal{B} \rangle$, if $\langle \mathcal{B} \rangle_L = \langle \mathcal{B} \rangle$. The set $\mathcal{B}$ is a *strong involutive basis* or for short an *involutive basis*, if the union in (3.3) is disjoint, i. e. the intersections of the involutive cones are empty. We call any finite set $\mathcal{B} \subseteq \bar{\mathcal{B}} \subset \mathbb{N}_0^n$ such that $\langle \bar{\mathcal{B}} \rangle_L = \langle \mathcal{B} \rangle$ a *(weak) involutive completion* of $\mathcal{B}$. An *obstruction to involution* for the set $\mathcal{B}$ is a multi index $\nu \in \langle \mathcal{B} \rangle \setminus \langle \mathcal{B} \rangle_L$.

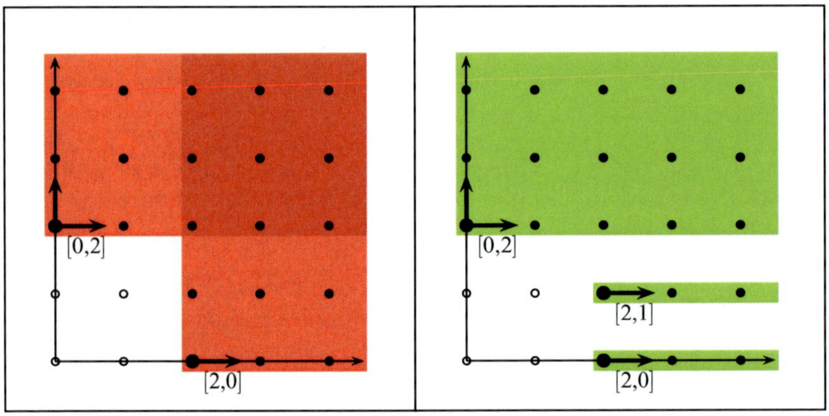

Fig. 3.1 *Left:* intersecting cones. *Right:* involutive cones.

Example 3.1.10. Figure 3.1 demonstrates the geometric interpretation of involutive divisions for $n = 2$. In both diagrams one can see the monoid ideal generated by the set $\mathcal{B} = \{[0,2],[2,0]\}$; the vertices belonging to it are marked by dark points. The arrows represent the multiplicative indices, i. e. the "allowed directions", for both the Janet and the Pommaret division, as they coincide for the set $\mathcal{B}$. The left diagram shows that the full cones of the two elements of $\mathcal{B}$ intersect in the darkly shaded area and that $\mathcal{B}$ is not (weakly) involutive, as the multi indices $[k,1]$ with $k \geq 2$ are obstructions to involution. The right diagram shows a strong involutive basis of $\langle \mathcal{B} \rangle$ for both the Janet and the Pommaret division. We must add to $\mathcal{B}$ the multi index $[2,1]$ and both for it and for $[2,0]$ only the index 1 is multiplicative. One clearly sees how the span $\langle \mathcal{B} \rangle$ is decomposed into three disjoint involutive cones: one of dimension 2, two of dimension 1. ◁

Example 3.1.11. As the monoid $\mathbb{N}_0^n$ possesses a natural grading by the length $|\mu|$ of the multi indices, we may consider for any monoid ideal $\mathcal{I} \subseteq \mathbb{N}_0^n$ the homogeneous component $\mathcal{I}_q$ of all multi indices in $\mathcal{I}$ of length q. The ideal $\mathcal{I}$ is called *(reverse) lexicographic*, if for each length q a number $r_q \geq 0$ exists such that $\mathcal{I}_q$ consists of the r_q greatest multi indices of length q with respect to the (reverse) lexicographic order. Any basis of such an ideal is a weak Pommaret basis, although we have not yet the necessary tools for proving this fact (see Proposition 5.5.6 below). But we may already note the crucial step in the proof.

Let us assume for simplicity that our ideal possesses a basis $\mathcal{B}$ where all generators are of the same length. If $\nu \in \mathcal{B}$ is a generator of class k, then a necessary condition for $\mathcal{B}$ being a Pommaret basis is that each multi index $\nu + 1_j$ where $k < j \leq n$ (i. e. where j is non-multiplicative for ν) is contained in some involutive cone. By our definition of the (reverse) lexicographic order (cf. Example A.1.7), we have $\nu \prec_{\text{lex}} \mu = \nu + 1_j - 1_k$ (and $\nu \prec_{\text{revlex}} \mu$). Thus the multi index μ must also be an element of the basis and, since obviously $\operatorname{cls}\mu \geq \operatorname{cls}\nu$, the index k is multiplicative for it with respect to the Pommaret division implying that $\nu + 1_j \in \mathcal{C}_P(\mu)$. ◁

We are particularly interested in *strong* involutive bases, as only these lead to disjoint decompositions. The following result shows that in the "monomial" case any weak involutive basis can be reduced to a strong one by simply eliminating some redundant elements. Its proof provides a nice motivation for the two conditions imposed in Definition 3.1.1 of an involutive division.

Proposition 3.1.12. *If $\mathcal{B}$ is a weakly involutive set, then a subset $\mathcal{B}' \subseteq \mathcal{B}$ exists such that $\mathcal{B}'$ is a strong involutive basis of $\langle\mathcal{B}\rangle$.*

Proof. If $\mathcal{B}$ is not yet a strong involutive basis, then the union in (3.3) cannot be disjoint. Thus intersection involutive cones must exist. By the first condition, this fact implies that some cones are contained in other ones; no other form of intersection is possible. If we eliminate the corresponding elements of $\mathcal{B}$, we get a subset $\mathcal{B}' \subset \mathcal{B}$ which has by the second condition the same involutive span, as the remaining elements may only gain additional multiplicative indices. Thus after a finite number of such eliminations of elements of $\mathcal{B}$ we arrive at a strong involutive basis. □

Remark 3.1.13. Let $\mathcal{I}_1, \mathcal{I}_2$ be two monoid ideals in $\mathbb{N}_0^n$ and $\mathcal{B}_1, \mathcal{B}_2$ (weak) involutive bases of them for some division L. In general, we cannot expect that $\mathcal{B}_1 \cup \mathcal{B}_2$ is again a weak involutive basis of the ideal $\mathcal{I}_1 + \mathcal{I}_2$, as the involutive cones of the generators may shrink when taken with respect to the larger set $\mathcal{B}_1 \cup \mathcal{B}_2$. Only for a global division we always obtain at least a weak involutive basis (which may then be reduced to a strong basis according to Proposition 3.1.12). ◁

An obvious necessary condition for a set $\mathcal{B}$ to be strongly involutive for the division L is that no element of it involutively divides another one. Sets with this property have a special name.

Definition 3.1.14. A finite subset $\mathcal{B} \subset \mathbb{N}_0^n$ which does not contain two distinct multi indices $\mu \neq \nu$ such that $\mu \,|_{L,\mathcal{B}}\, \nu$ is called *involutively autoreduced* with respect to the involutive division L.

Remark 3.1.15. In the proof of Lemma 3.1.5 we showed that with respect to the Janet division two involutive cones can only intersect, if they are identical. Thus for this division any set $\mathcal{B}$ is involutively autoreduced and any weakly involutive set is automatically strongly involutive. This property is surely convenient for computations, but it has the disadvantage that in general it leads to unnecessary large bases. As a simple example consider for $n = 1$ the set $\mathcal{B} = \{[1],[2]\}$. For most involutive divisions $\mathcal{B}$ would not be involutively autoreduced and the only strong involutive basis of $\langle\mathcal{B}\rangle$ would be $\{[1]\}$. Not so for the Janet division where the presence of the multi index [2] entails that no index is multiplicative for [1]. $\lhd$

The explicit algorithmic construction of (strong) involutive completions for a given set $\mathcal{B} \subset \mathbb{N}_0^n$ will be discussed in detail in Section 4.2. For the moment we only note that we cannot expect that for an arbitrary set $\mathcal{B}$ and an arbitrary involutive division L an involutive basis $\mathcal{B}'$ of $\langle\mathcal{B}\rangle$ exists.

Example 3.1.16. We take the Pommaret division and the set $\mathcal{B} = \{[1,1]\}$. As $\operatorname{cls}[1,1] = 1$, we get $N_P([1,1]) = \{1\}$. So $\mathcal{C}_P([1,1]) \subsetneq \mathcal{C}([1,1])$ and the set $\mathcal{B}$ is not involutive. But any multi index contained in $\langle\mathcal{B}\rangle$ has also class 1 and hence it is easy to see that even by adding finitely many of them we cannot obtain an involutive basis of $\langle\mathcal{B}\rangle$ for the Pommaret division. We can generate $\langle\mathcal{B}\rangle$ involutively only with the *infinite* set $\{[1,k] \mid k \in \mathbb{N}\}$. $\lhd$

Remark 3.1.17. Since by definition an involutive basis is always finite, it does not really make sense to say that an infinite set involutively generates some monoid ideal. Ignoring this inconsistency for a moment, we now show that even if a monoid ideal does not possess a finite Pommaret basis, it has at least an infinite Pommaret basis with so much structure that it admits a simple finite description generalising the one found in the example above.

In order to see this fact, we consider first the case of an *irreducible* monoid ideal $\mathcal{I} \subset \mathbb{N}_0^n$. It is well-known that any such $\mathcal{I}$ has a minimal basis of the form $\{(\ell_1)_{i_1},\dots,(\ell_r)_{i_r}\}$ with $1 \leq r \leq n$, $\ell_j > 0$ and $1 \leq i_1 < \cdots < i_r \leq n$. Here $(\ell_j)_{i_j}$ denotes the multi index where all entries are zero except of the i_jth one which takes the value ℓ_j (see (A.2)). Such an irreducible ideal possesses a Pommaret basis, if and only if there are no "gaps" in the sequence $i_1 < \cdots < i_r \leq n$, i. e. $i_r = n$ and $i_1 = n - r + 1$. Indeed, in order to generate involutively $\mathcal{I}$, we need all multi indices of the form $[0,\dots,0,\ell_{i_j},\mu_{i_j+1},\dots,\mu_n]$ where $1 \leq j \leq r$ and where the possibles values of μ_k are bounded by $\ell_{i_{j'}}$, if and only if $i_{j'} = k$ for some $j' > j$; otherwise we must admit all $\mu_k \geq 0$. Hence only if no gaps appear, the arising set is finite.

For a general monoid ideal $\mathcal{I}$, we exploit that any monoid ideal in $\mathbb{N}_0^n$ possesses a unique irreducible decomposition [322, Thm. 5.27], i. e. we can always express $\mathcal{I}$ as the intersection of finitely many irreducible ideals. In Remark 4.1.6 we will show how a Pommaret basis of the intersection of two (monoid) ideals can be obtained from Pommaret bases of the ideals by simply taking least common multiples.

As a simple corollary of these considerations we find that any Artinian (or zero-dimensional) monoid ideal $\mathcal{I}$ has a Pommaret basis. Indeed it is well-known that $\mathcal{I}$ is Artinian, if and only if it contains an irreducible ideal $\mathcal{J}$ with a minimal basis

$\{(\ell_1)_1,\dots,(\ell_n)_n\}$. As no gaps appear, $\mathcal{J}$ possesses a finite Pommaret basis $\mathcal{B}'$. Now the finite set $\mathcal{B} = \mathcal{B}' \cup (\mathcal{I} \setminus \mathcal{J})$ is trivially a weak Pommaret basis of $\mathcal{I}$. ◁

Some involutive divisions have the property that with respect to them any monoid ideal $\mathcal{I} \subseteq \mathbb{N}_0^n$ possesses an involutive basis. Of course, this property plays an important role in algorithmic considerations and we provide a special name for it.

Definition 3.1.18. An involutive division L is called *Noetherian*, if any finite subset $\mathcal{B} \subset \mathbb{N}_0^n$ possesses a finite involutive completion with respect to L.

It follows trivially from Example 3.1.16 that the Pommaret division is not Noetherian. For the Janet division the situation is different.

Lemma 3.1.19. *The Janet division is Noetherian.*

Proof. Let $\mathcal{B} \subset \mathbb{N}_0^n$ be an arbitrary finite set of multi indices. We will explicitly construct a Janet basis for $\langle\mathcal{B}\rangle$. Let μ be the multi index defined by $\mu_i = \max_{\nu\in\mathcal{B}} \nu_i$ (i. e. $x^\mu = \operatorname{lcm}\mathcal{B}$). We claim that the set

$$\bar{\mathcal{B}} = \{\bar{\nu} \in \mathbb{N}_0^n \cap \langle\mathcal{B}\rangle \mid \mu \in \mathcal{C}(\bar{\nu})\} \tag{3.4}$$

is an involutive completion of $\mathcal{B}$ with respect to the Janet division. Obviously, we have $\mathcal{B} \subseteq \bar{\mathcal{B}}$ and $\bar{\mathcal{B}} \subset \langle\mathcal{B}\rangle$. Let $\rho \in \langle\mathcal{B}\rangle$ be an arbitrary element. If $\rho \in \bar{\mathcal{B}}$, then trivially $\rho \in \langle\bar{\mathcal{B}}\rangle_J$. Otherwise we set $I = \{i \mid \rho_i > \mu_i\}$ and consider the multi index $\bar{\rho}$ defined by $\bar{\rho}_k = \min\{\rho_k, \mu_k\}$ for $1 \leq k \leq n$. By the definition of the set $\bar{\mathcal{B}}$ and of the multi index μ, we have that $\bar{\rho} \in \bar{\mathcal{B}}$ and that $I \subseteq N_{J,\bar{\mathcal{B}}}(\bar{\rho})$. This fact implies that $\rho \in \mathcal{C}_{J,\bar{\mathcal{B}}}(\bar{\rho})$ and thus $\bar{\mathcal{B}}$ is a finite Janet basis for $\langle\mathcal{B}\rangle$. □

Recall that a basis $\mathcal{B}$ of a monoid ideal is called minimal, if it is not possible to remove an element of $\mathcal{B}$ without losing the property that we have a basis. A similar notion can naturally be introduced for involutive bases.

Definition 3.1.20. Let $\mathcal{I} \subseteq \mathbb{N}_0^n$ be a monoid ideal and L an involutive division. An involutive basis $\mathcal{B}$ of $\mathcal{I}$ with respect to L is called *minimal*, if any other involutive basis $\mathcal{B}'$ of $\mathcal{I}$ with respect to L satisfies $\mathcal{B} \subseteq \mathcal{B}'$.

Any monomial ideal has a *unique* minimal monomial basis: take an arbitrary monomial basis and eliminate all terms having a divisor in the basis. Obviously, these eliminations do not change the span and the result is a minimal basis. We will see later in Section 4.2 that the minimal involutive basis of any monoid ideal is also unique (if it exists). For globally defined divisions, any involutive basis is unique.

Proposition 3.1.21. *Let L be a globally defined involutive division and $\mathcal{I} \subseteq \mathbb{N}_0^n$ a monoid ideal. If $\mathcal{I}$ possesses an involutive basis for the division L, then it is unique and thus minimal.*

Proof. Let $\mathcal{B}_1$, $\mathcal{B}_2$ be two distinct involutive bases of the monoid ideal $\mathcal{I}$. Then both $\mathcal{B}_1 \setminus \mathcal{B}_2$ and $\mathcal{B}_2 \setminus \mathcal{B}_1$ must be non-empty, as otherwise one basis was contained in the other one and thus the larger basis could not be involutively autoreduced with respect to the global division L. Consider an arbitrary multi index $\nu \in \mathcal{B}_1 \setminus \mathcal{B}_2$. The basis $\mathcal{B}_2$ must contain a unique multi index μ with $\mu \mid_L \nu$. It cannot be an element of $\mathcal{B}_1$, as $\mathcal{B}_1$ is involutively autoreduced. Thus a unique multi index $\lambda \in \mathcal{B}_1$ must exist with $\lambda \mid_L \mu$. As L is globally defined, this implies that $\lambda \mid_L \nu$, a contradiction. □

Addendum: Some Algorithmic Considerations

In any concrete computation with involutive bases two operations appear repeatedly: the determination of the multiplicative indices for an element of a given subset $\mathcal{B} \subset \mathbb{N}_0^n$ and the search for an involutive divisor $\nu \in \mathcal{B}$ of an arbitrary multi index μ. In this Addendum we briefly discuss how these operations may be efficiently performed for the special cases of the Janet and the Pommaret division, respectively, as these are the most important divisions in applications. Readers only interested in the theory may safely omit this material.

We begin with the Janet division. Its definition is based on the subsets $(d_k, \dots, d_n)$ introduced in (3.2). Via their inclusion relations these sets induce in a natural manner a tree structure for every finite set $\mathcal{B} \subset \mathbb{N}_0^n$: the root is $() = \mathcal{B}$ and the children of a node $(d_k, \dots, d_n)$ are the non-empty subsets $(d_{k-1}, d_k, \dots, d_n)$. Obviously, the leaves correspond to the individual multi indices contained in $\mathcal{B}$. We call this tree the *Janet tree* $\mathcal{J}(\mathcal{B})$ of the set $\mathcal{B}$.

For any set $\mathcal{B} \subset \mathbb{N}_0^n$ the depth of $\mathcal{J}(\mathcal{B})$ is always $n+1$ and the number of children of the node $(d_k, \dots, d_n)$ is bounded by the maximal value of ν_{k-1} for the multi indices $\nu \in (d_k, \dots, d_n)$. If we sort the children of each node such that smaller values of d_{k-1} appear to the left of higher values, then the leaves appear in lexicographic order (according to our definition in Appendix A.1).

As a first application of this idea we discuss the construction of the sets $N_{J,\mathcal{B}}(\nu)$ for all multi indices ν in a given finite set $\mathcal{B} \subset \mathbb{N}_0^n$. In Algorithm 3.1 the use of the Janet tree $\mathcal{J}(\mathcal{B})$ may not become apparent at first sight, but effectively the algorithm traverses the tree from the right to the left. It runs two pointers p_1 and p_2 over the multi indices: in the kth iteration p_1 marks the first entry from the right where the multi indices $\nu^{(k-2)}$ and $\nu^{(k-1)}$ differ and similarly p_2 the first entry from the right where $\nu^{(k-1)}$ and $\nu^{(k)}$ differ. Counting from the leaves, these pointers correspond to the currently considered level in the tree; when their values increase then we climb up until we reach the next node where a branching occurs.

The correctness of Algorithm 3.1 is implied by the following considerations. After the sorting, the multi index $\nu^{(1)}$ is maximal with respect to the lexicographic order hence and thus all indices are multiplicative for it with respect to the Janet division. By the definition of the pointers p_1 and p_2, both multi indices $\nu^{(k-1)}$ and $\nu^{(k)}$ are contained in the subset $\bigl(\nu^{(k)}_{p_2+1}, \dots, \nu^{(k)}_n\bigr)$.

Because of the lexicographic ordering, we have $\nu^{(k)}_{p_2} < \nu^{(k-1)}_{p_2}$ and hence p_2 is not multiplicative for $\nu^{(k)}$. Concerning the multiplicity of the indices $i > p_2$,

Algorithm 3.1 Janet multiplicative indices

Input: finite list $\mathcal{B} = [\nu^{(1)}, \dots, \nu^{(r)}]$ of pairwise different multi indices from $\mathbb{N}_0^n$
Output: list $\mathcal{M} = [N_{J,\mathcal{B}}(\nu^{(1)}), \dots, N_{J,\mathcal{B}}(\nu^{(k)})]$ with multiplicative variables

$\mathcal{B} \leftarrow \mathtt{sort}(\mathcal{B}, \prec_{\text{lex}})$; $\quad \nu \leftarrow \mathcal{B}[1]$
$p_1 \leftarrow n$; $\quad I \leftarrow \{1, \dots, n\}$; $\quad \mathcal{M}[1] \leftarrow I$
for k **from** 2 **to** r **do**
  $p_2 \leftarrow \mathtt{max}\,\{i \mid (\nu - \mathcal{B}[k])_i \neq 0\}$; $\quad I \leftarrow I \setminus \{p_2\}$
  if $p_1 < p_2$ **then**
    $I \leftarrow I \cup \{p_1, \dots, p_2 - 1\}$
  end if
  $\mathcal{M}[k] \leftarrow I$; $\quad \nu \leftarrow \mathcal{B}[k]$; $\quad p_1 \leftarrow p_2$
end for
return $\mathcal{M}$

we simply keep the previous values, as they are identical for all multi indices in $(\nu^{(k)}_{p_2+1}, \dots, \nu^{(k)}_n)$. Note that all indices $i < \min\{p_1, p_2\}$ are always included in the current set I.

If the pointers coincide, $p_1 = p_2$, then we have completed one branch starting from the node $(\nu^{(k)}_{p_2+1}, \dots, \nu^{(k)}_n)$ and continue with the next one. In this case obviously nothing happens except that p_2 cannot be multiplicative as explained above. If $p_1 > p_2$, we follow one branch further down until the next branching takes place. Again this only affects the multiplicity of p_2.

Finally, if $p_1 < p_2$, then we are finished with the analysis of all children of $(\nu^{(k-1)}_{p_1+1}, \dots, \nu^{(k-1)}_n)$ and we climb up in the tree until we reach the next node where a branching occur: this is $(\nu^{(k)}_{p_2+1}, \dots, \nu^{(k)}_n)$. As we have now a new value at the position p_2, all considerations about the indices between p_1 and p_2 are obsolete and within the child $(\nu^{(k)}_{p_2}, \dots, \nu^{(k)}_n)$ the multi index $\nu^{(k)}$ has the highest entry at the position p_2. Hence the indices $p_1, \dots, p_2 - 1$ become multiplicative for it.

Example 3.1.22. We demonstrate Algorithm 3.1 for the (already lexicographically sorted) set

$$\begin{aligned} \mathcal{B} = \{&[2,3,3],\ [1,3,3],\ [0,3,3],\ [1,0,3],\ [1,2,1], \\ &[4,2,0],\ [2,2,0],\ [0,2,0]\}\,. \end{aligned} \tag{3.5}$$

Figure 3.2 shows the corresponding Janet tree.

The greatest multi index is $[2,3,3]$ and hence all indices are multiplicative for it. At the beginning we have $p_1 = 3$ and comparing $[2,3,3]$ and $[1,3,3]$ yields $p_2 = 1$. Hence for $[1,3,3]$ only 2 and 3 are multiplicative. Now p_1 is reset to 1 and the next comparison yields again $p_2 = 1$. Thus $[0,3,3]$ has the same multiplicative indices as $[1,3,3]$. In the next iteration we find $p_2 = 2 > p_1 = 1$ and the new multiplicative indices are 1 and 3. So we encounter already in the first three iterations all cases for the relation between p_1 and p_2. $\lhd$

Our next task consists of searching in a given set $\mathcal{B}$ for a Janet divisor of an arbitrary multi index $\mu \in \mathbb{N}_0^n$, i. e. to check whether $\mathcal{B}$ contains a multi index ν such

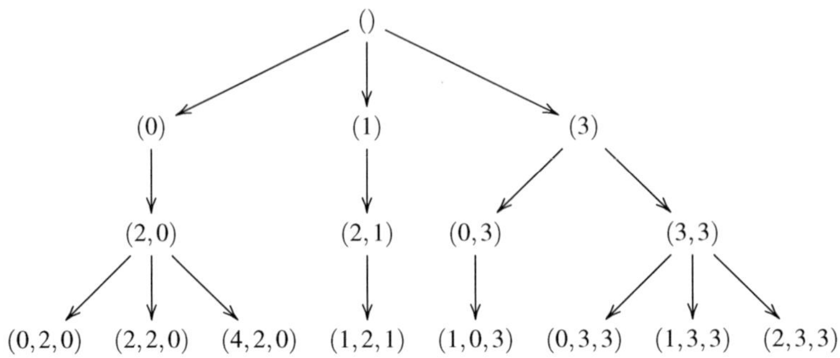

Fig. 3.2 Janet tree for Example 3.1.22

that $\nu\,|_{J,\mathcal{B}}\,\mu$. Note that ν is uniquely determined, if it exists, as by Remark 3.1.15 any set $\mathcal{B}$ is involutively autoreduced for the Janet division. Given the Janet tree $\mathcal{J}(\mathcal{B})$ our task may be solved very efficiently with Algorithm 3.2.

We start at the root of $\mathcal{J}(\mathcal{B})$ and descend level by level. At step i we check whether the current node $\mathcal{S} = (d_{i+1},\dots,d_n)$ either has a child $(d_i,\dots,d_n)$ such that $d_i = \mu_i$ or whether μ_i is greater than the maximal value d_i of any child. If this condition is satisfied, we may continue, as in the latter case i is multiplicative for all multi indices contained in $d_i\,.\,\mathcal{S} = (d_i,\dots,d_n)$. Otherwise the set $\mathcal{B}$ contains no Janet divisor of μ. If the algorithm arrives in Line /12/, then the subset $\mathcal{S}$ consists of precisely one multi index $\nu \in \mathcal{B}$ which is the Janet divisor of μ. Indeed, at each position i we have either $\nu_i = \mu_i$ or $\nu_i < \mu_i$ and the index i is multiplicative for ν.

Algorithm 3.2 Janet divisor

Input: Janet tree $\mathcal{J}(\mathcal{B})$ and multi index $\mu \in \mathbb{N}_0^n$
Output: Element $\nu \in \mathcal{B}$ with $\nu\,|_{J,\mathcal{B}}\,\mu$ or `FAIL`
1: $\mathcal{S} \leftarrow ()$; $i \leftarrow n$
2: **while** $i > 0$ **do**
3:     **if** $\mu_i \geq d_i = \max_{\nu\in\mathcal{S}} \nu_i$ **then**
4:         $\mathcal{S} \leftarrow d_i\,.\,\mathcal{S}$
5:     **else if** $\mathcal{S}$ has child $\mu_i\,.\,\mathcal{S}$ **then**
6:         $\mathcal{S} \leftarrow \mu_i\,.\,\mathcal{S}$
7:     **else**
8:         **return** `FAIL`
9:     **end if**
10:     $i \leftarrow i - 1$
11: **end while**
12: **return** $\mathcal{S}$

In Chapter 4 we will discuss algorithms for the construction of involutive completions. Thus we do not always work with a fixed set $\mathcal{B}$ but multi indices may be added to or removed from $\mathcal{B}$. Of course it would be inefficient to rebuild the complete Janet tree $\mathcal{J}(\mathcal{B})$ each time this happened. However, it is trivial to efficiently insert or delete leaves from $\mathcal{J}(\mathcal{B})$. The corresponding algorithms are so simple that we omit their pseudo code.

For the insertion of a new leaf μ we descend the Janet tree, as long as nodes $(\mu_i,\dots,\mu_n)$ already exist, then we add new nodes for the not yet existing subsets. For the deletion of a leave $\nu \in \mathcal{B}$ we follow the branch ending in ν from the bottom upwards until we find a parent node possessing more than one children; all nodes below this one are eliminated. Thus in a concrete implementation the deletion algorithm requires additional pointers to all leaves and from each child to its parent.

For the Pommaret division the determination of the multiplicative indices is of course trivial. For the determination of a Pommaret divisor we may follow a similar approach as for the Janet division. In fact, it is convenient to express (a part of) the search for a Pommaret divisor in the set $\mathcal{B}$ with the help of the Janet tree $\mathcal{J}(\mathcal{B})$. The relation $\nu \mid_P \mu$ holds, if and only if $\nu_i = \mu_i$ for all $i > k = \operatorname{cls}\nu$ and $\nu_k \leq \mu_k$. Note that a Pommaret divisor ν is uniquely determined, only if the set $\mathcal{B}$ is involutively autoreduced. If several divisors exist, Algorithm 3.3 returns the one with the minimal value of the entry ν_k.

Algorithm 3.3 Pommaret divisor

Input: Janet tree $\mathcal{J}(\mathcal{B})$ and multi index $\mu \in \mathbb{N}_0^n$
Output: Element $\nu \in \mathcal{B}$ with $\nu \mid_P \mu$ or FAIL

$\mathcal{S} \leftarrow ()$; $\quad i \leftarrow n$
while ($\mathcal{S}$ has child $\mu_i.\mathcal{S}$) **and** $(i > 0)$ **do**
  $\mathcal{S} \leftarrow \mu_i.\mathcal{S}$; $\quad i \leftarrow i-1$
end while
if $i = 0$ **then**
  return μ {We even have $\mu \in \mathcal{B}$.}
end if
for d **from** 0 **to** $\mu_i - 1$ **do**
  if $\nu = [0,\dots,0,d,\mu_{i+1},\dots,\mu_n] \in \mathcal{B}$ **then**
    return ν
  end if
end for
return FAIL

The algorithm descends the Janet tree $\mathcal{J}(\mathcal{B})$, as long as nodes corresponding to subsets $(\mu_i,\dots,\mu_n)$ exist. Then it simply checks whether $\mathcal{B}$ contains a multi index of the form $[0,\dots,0,d,\mu_{i+1},\dots,\mu_n]$ with $d < \mu_i$. If this is not the case, then the given set $\mathcal{B}$ does not contain a Pommaret divisor of μ.

The complexity of all these algorithms depends on how a concrete implementation realises the Janet tree, as typically we must search whether a node has a certain child. A simple approach would consist of having in each node a sorted list of the pointers to its children. Then Algorithm 3.2 will require at a node $\mathcal{S}$ on level i at most $\max_{\nu\in\mathcal{S}} \nu_i$ comparisons and the total number of comparisons is bounded by $d = \max_{\nu\in\mathcal{B}} |\nu|$. Thus all our algorithms can easily be realised in $O(n+d)$.

A straightforward optimisation would be to replace the list in node $\mathcal{S}$ by an array of length $\max_{\nu\in\mathcal{S}} \nu_i$. The dth entry of the array would either be a pointer to the child $d\,.\,\mathcal{S}$, if it exists, or an empty pointer otherwise. Assuming that the access to an array element is possible in constant time, this even leads to a complexity $O(n)$ for our algorithms. However, in the case of the insertion and deletion algorithm some additional overhead occurs, as it now may happen that an array must be replaced by a larger or a smaller one.

3.2 Polynomial Algebras of Solvable Type

We introduced in the last section involutive bases in $\mathbb{N}_0^n$. We could now interpret the multi indices as monomials and proceed to define involutive bases for polynomial ideals. But we are also interested in other rings containing, say, linear differential or difference operators. As the basic ideas remain unchanged in many different situations, we need a class of rings within which all interesting cases can be handled in a unified manner: the *polynomial algebras of solvable type*.

Let $\mathcal{P} = \mathcal{R}[x^1,\dots,x^n]$ be a polynomial ring over a ring $\mathcal{R}$. If $\mathcal{R}$ is commutative, then $\mathcal{P}$ is a commutative ring with respect to the usual multiplication. We want to equip the $\mathcal{R}$-module $\mathcal{P}$ with alternative multiplications, in particular with non-commutative ones, where it is allowed that both the variables x^i do not commute any more and that they operate on the coefficients. We will denote in the sequel the usual multiplication either by a dot $\cdot$ or by no symbol at all and alternative multiplications $\mathcal{P}\times\mathcal{P}\to\mathcal{P}$ by $f\star g$.

Like Gröbner bases, involutive bases are defined with respect to a *term order* $\prec$ (cf. Definition A.1.1). It selects in each polynomial $f\in\mathcal{P}$ a *leading term* $\operatorname{lt}_\prec f = x^\mu$ with *leading exponent* $\operatorname{le}_\prec f = \mu$; the coefficient $r\in\mathcal{R}$ of x^μ in f is the *leading coefficient* $\operatorname{lc}_\prec f$ and the product rx^μ is the *leading monomial* $\operatorname{lm}_\prec f$ (see also Appendix B.4). Based on the leading exponents, we can associate to each finite set $\mathcal{F}\subset\mathcal{P}$ a finite set $\operatorname{le}_\prec\mathcal{F}\subset\mathbb{N}_0^n$ to which we may apply the theory developed in the last section. But in order for this approach to make sense, we need a kind of compatibility between the multiplication $\star$ and the chosen term order $\prec$. These considerations naturally lead to the following definition.

Definition 3.2.1. The triple $\bigl(\mathcal{P} = \mathcal{R}[x^1,\dots,x^n],\star,\prec\bigr)$ is a *polynomial algebra of solvable type* over the coefficient ring $\mathcal{R}$ for the term order $\prec$, if the multiplication $\star:\mathcal{P}\times\mathcal{P}\to\mathcal{P}$ satisfies the following three axioms.

(i) $(\mathcal{P},\star)$ is a ring with neutral element 1.
(ii) $\forall r\in\mathcal{R}, f\in\mathcal{P}:\ r\star f = rf$.
(iii) $\forall\mu,\nu\in\mathbb{N}_0^n, r\in\mathcal{R}\setminus\{0\}:\ \operatorname{le}_\prec(x^\mu\star x^\nu) = \mu+\nu\ \wedge\ \operatorname{le}_\prec(x^\mu\star r) = \mu$.

Condition (i) ensures that the arithmetics in $(\mathcal{P},\star)$ obeys the usual associative and distributive laws. Because of Condition (ii), the algebra $(\mathcal{P},\star)$ may be considered as a left $\mathcal{R}$-module. We do not assume that it is also a right $\mathcal{R}$-module, as this requirement would exclude the possibility that the variables x^i operate non-trivially on $\mathcal{R}$. Condition (iii) ensures the compatibility of the new multiplication $\star$

and the term order $\prec$; we say that $\star$ is an *order respecting multiplication*. This property implies the existence of injective maps $\rho_\mu : \mathcal{R} \to \mathcal{R}$, maps $h_\mu : \mathcal{R} \to \mathcal{P}$ with $\mathrm{le}_\prec\bigl(h_\mu(r)\bigr) \prec \mu$ for all $r \in \mathcal{R}$, coefficients $r_{\mu\nu} \in \mathcal{R} \setminus \{0\}$ and polynomials $h_{\mu\nu} \in \mathcal{P}$ with $\mathrm{le}_\prec(h_{\mu\nu}) \prec \mu + \nu$ such that

$$x^\mu \star r = \rho_\mu(r)x^\mu + h_\mu(r)\,, \tag{3.6a}$$

$$x^\mu \star x^\nu = r_{\mu\nu}x^{\mu+\nu} + h_{\mu\nu}\,. \tag{3.6b}$$

Lemma 3.2.2. *The maps ρ_μ and the coefficients $r_{\mu\nu}$ satisfy for arbitrary multi indices $\mu, \nu, \lambda \in \mathbb{N}_0^n$ and for arbitrary ring elements $r \in \mathcal{R}$*

$$\rho_\mu\bigl(\rho_\nu(r)\bigr)r_{\mu\nu} = r_{\mu\nu}\rho_{\mu+\nu}(r)\,, \tag{3.7a}$$

$$\rho_\mu(r_{\nu\lambda})r_{\mu,\nu+\lambda} = r_{\mu\nu}r_{\mu+\nu,\lambda}\,. \tag{3.7b}$$

Furthermore, all maps ρ_μ are ring endomorphisms.

Proof. The first assertion is a trivial consequence of the associativity of the multiplication $\star$. The equations correspond to the leading coefficients of the equalities $x^\mu \star (x^\nu \star r) = (x^\mu \star x^\nu) \star r$ and $x^\mu \star (x^\nu \star x^\lambda) = (x^\mu \star x^\nu) \star x^\lambda$, respectively. The second assertion follows mainly from Condition (i). □

If the coefficient ring $\mathcal{R}$ is a (skew) field, then for arbitrary polynomials $f, g \in \mathcal{P}$ an element $r \in \mathcal{R} \setminus \{0\}$ and a polynomial $h \in \mathcal{P}$ satisfying $\mathrm{le}_\prec(h) \prec \mathrm{le}_\prec(f \cdot g)$ exist such that the new and the usual multiplication on $\mathcal{P}$ are related by

$$f \star g = r(f \cdot g) + h\,. \tag{3.8}$$

Indeed, if $\mathrm{lm}_\prec f = ax^\mu$ and $\mathrm{lm}_\prec g = bx^\nu$, then a simple computation yields that r is the (unique) solution of the equation $a\rho_\mu(b)r_{\mu\nu} = rab$ in $\mathcal{R}$ and h is the difference $f \star g - r(f \cdot g)$. Thus under this assumption on the coefficient ring $\mathcal{R}$ we may reformulate Condition (iii) in Definition 3.2.1 more concisely as

(iii)' $\forall f, g \in \mathcal{P} : \mathrm{le}_\prec(f \star g) = \mathrm{le}_\prec(f \cdot g) = \mathrm{le}_\prec f + \mathrm{le}_\prec g.$

Proposition 3.2.3. *The product $\star$ is fixed, as soon as the following data are given: constants $r_{ij} \in \mathcal{R} \setminus \{0\}$, polynomials $h_{ij} \in \mathcal{P}$ and maps $\rho_i : \mathcal{R} \to \mathcal{R}$, $h_i : \mathcal{R} \to \mathcal{P}$ such that for $1 \le i \le n$*

$$x^i \star r = \rho_i(r)x^i + h_i(r)\,, \quad \forall r \in \mathcal{R}\,, \tag{3.9a}$$

$$x^i \star x^j = r_{ij}x^j \star x^i + h_{ij}\,, \quad \forall 1 \le j < i\,. \tag{3.9b}$$

Proof. The set of all "terms" $x^{i_1} \star x^{i_2} \star \cdots \star x^{i_q}$ with $i_1 \le i_2 \le \cdots \le i_q$ forms a basis of $\mathcal{P}$, as because of Condition (iii) the map $x^{i_1} \star x^{i_2} \star \cdots \star x^{i_q} \mapsto x^{i_1} \cdot x^{i_2} \cdots x^{i_q}$ is an $\mathcal{R}$-module automorphism mapping the new basis into the standard basis. Obviously, it is possible to evaluate any product $f \star g$ by repeated applications of the rewrite rules (3.9) provided f and g are expressed in the new basis. □

Note that this proof is non-constructive in the sense that we are not able to determine the multiplication in terms of the standard basis, as we do not know explicitly the transformation between the new and the standard basis. The advantage of this proof is that it is valid for arbitrary coefficient rings $\mathcal{R}$. Making some assumptions on the ring $\mathcal{R}$ (the simplest possibility is to require that it is a field), one could use Lemma 3.2.2 to express the coefficients $r_{\mu\nu}$ and ρ_μ in (3.6) by the data in the commutation relations (3.9). This would yield a constructive proof.

Of course, the data in Proposition 3.2.3 cannot be chosen arbitrarily. Besides the obvious restrictions for the leading exponents of the polynomials h_{ij} and $h_i(r)$ imposed by Condition (iii), each map ρ_i must be an injective $\mathcal{R}$-endomorphism and each map h_i must satisfy $h_i(r+s) = h_i(r) + h_i(s)$ and a kind of pseudo-Leibniz rule $h_i(rs) = \rho_i(r)h_i(s) + h_i(r) * s$. The associativity of $\star$ imposes further rather complicated conditions. For the case of a multiplication defined by commutation rules (i. e. when Proposition 3.2.3 can be made effective) they have been explicitly determined by Levandovskyy [285, 286] who called them *non-degeneracy conditions* (see also the extensive discussion by Kredel [264, Sect. 3.3]).

Before we continue to study some elementary properties of polynomial algebras of solvable type, we consider a number of examples in order to demonstrate that this concept appears in fact in many different applications. Thus although the motivation for the conditions in Definition 3.2.1 is mainly of an algorithmic nature (see the discussion after Definition 3.3.12 below), it turns out that they are fairly natural in many contexts.

Example 3.2.4. Obviously, the most trivial example of an algebra of solvable type is the usual polynomial ring $\mathcal{P}$, i. e. the case $f \star g = fg$. Its multiplication respects any term order, as—by definition—a term order is compatible with the monoid structure of the set $\mathbb{T}$ of terms.

A non-commutative example is provided by rings of *linear differential operators* (with variable coefficients). We take for the coefficients some ring $\mathbb{F}$ of functions in $x^1,\dots,x^n$, say the rational functions $\mathbb{Q}(x^1,\dots,x^n)$, and consider the polynomial ring $\mathcal{D} = \mathbb{F}[\partial_1,\dots,\partial_n]$ where $\partial_i = \partial/\partial x^i$. More generally, we may take any differential ring $\mathbb{F}$ with commuting derivations ∂_i. According to the Leibniz rule, the product of two monomial operators $a\partial^\mu$ and $b\partial^\nu$ with $a, b \in \mathbb{F}$ is

$$a\partial^\mu \star b\partial^\nu = \sum_{\sigma+\rho=\mu} \binom{\mu}{\sigma} a \frac{\partial^{|\sigma|} b}{\partial x^\sigma} \partial^{\rho+\nu} \tag{3.10}$$

where $\binom{\mu}{\sigma}$ is shorthand for $\prod_{i=1}^n \binom{\mu_i}{\sigma_i}$. By the properties of a term order

$$\mathrm{le}_\prec\left(a\partial^\mu \star b\partial^\nu\right) = \mu + \nu = \mathrm{le}_\prec\left(a\partial^\mu\right) + \mathrm{le}_\prec\left(b\partial^\nu\right), \tag{3.11}$$

as any term $\partial^{\rho+\nu}$ appearing on the right hand side of (3.10) divides $\partial^{\mu+\nu}$ and thus $\partial^{\rho+\nu} \preceq \partial^{\mu+\nu}$ for any term order $\prec$. Hence $\mathcal{D}$ is a polynomial algebra of solvable type for all term orders. ◁

This example can be considerably extended to a fairly large class of non-commutative polynomial algebras originally introduced by Noether and Schmeidler [336] and later systematically studied by Ore [344]; our exposition follows [54].

Example 3.2.5. Let $\mathcal{R}$ be an arbitrary commutative ring and $\sigma : \mathcal{R} \to \mathcal{R}$ an injective endomorphism. A *pseudo-derivation* with respect to σ is a map $\delta : \mathcal{R} \to \mathcal{R}$ such that (i) $\delta(f+g) = \delta(f) + \delta(g)$ and (ii) $\delta(f \cdot g) = \sigma(f) \cdot \delta(g) + \delta(f) \cdot g$ for all $f, g \in \mathcal{R}$. If $\sigma = \mathrm{id}_{\mathcal{R}}$, the identity map, (ii) is the standard Leibniz rule for derivations. Ore [344] called $\sigma(f)$ the *conjugate* and $\delta(f)$ the *derivative* of f.

If the coefficient ring $\mathcal{R}$ is even a field, then the conjugate and the derivative are actually closely related or trivial. Applying the generalised Leibniz rule (ii) to the obvious identity $\delta(f \cdot g) = \delta(g \cdot f)$ yields the equation

$$\big(\sigma(f) - f\big) \cdot \delta(g) = \big(\sigma(g) - g\big) \cdot \delta(f) \,. \tag{3.12}$$

Assume that $\sigma \neq \mathrm{id}_{\mathcal{R}}$. Then there exists an element $f \in \mathcal{R}$ with $\sigma(f) \neq f$ and we set $\tilde{f} = \delta(f)/\big(\sigma(f) - f\big)$. Now (3.12) implies that $\delta(g) = \tilde{f}\big(\sigma(g) - g\big)$ for any $g \in \mathcal{R}$ and hence $\delta = \tilde{f} \cdot (\sigma - \mathrm{id}_{\mathcal{R}})$. Conversely assume that $\delta \neq 0$. Then there exists an $f \in \mathcal{R}$ with $\delta(f) \neq 0$ and we set $\tilde{f} = \big(\sigma(f) - f\big)/\delta(f)$. This time (3.12) implies that $\sigma(g) = \tilde{f} \cdot \delta(g) + g$ for any $g \in \mathcal{R}$ and hence $\sigma = \tilde{f} \cdot \delta + \mathrm{id}_{\mathcal{R}}$.

Given the maps σ and δ, the ring $\mathcal{R}[\partial; \sigma, \delta]$ of univariate *Ore polynomials* consists of all formal polynomials in ∂ with coefficients in $\mathcal{R}$, i. e. of expressions of the form $\theta = \sum_{i=0}^{q} f_i \partial^i$ with $f_i \in \mathcal{R}$ and $q \in \mathbb{N}_0$. The addition is defined as usual. The "variable" ∂ operates on an element $f \in \mathcal{R}$ according to the rule

$$\partial \star f = \sigma(f)\partial + \delta(f) \tag{3.13}$$

which is extended associatively and distributively to define the multiplication in the ring $\mathcal{R}[\partial; \sigma, \delta]$: given two elements $\theta_1, \theta_2 \in \mathcal{R}[\partial; \sigma, \delta]$, we can transform the product $\theta_1 \star \theta_2$ to the above normal form by repeatedly applying (3.13). The product of two linear polynomials evaluates then to

$$\begin{aligned}(f_1 + f_2\partial) \star (g_1 + g_2\partial) = f_1 g_1 + f_2 \delta(g_1) + \\ \big[f_1 g_2 + f_2 \sigma(g_1) + f_2 \delta(g_2)\big]\partial + f_2 \sigma(g_2)\partial^2 \,.\end{aligned} \tag{3.14}$$

The condition that σ is injective ensures that $\deg\big(\theta_1 \star \theta_2\big) = \deg\theta_1 + \deg\theta_2$. We call $\mathcal{R}[\partial; \sigma, \delta]$ the *Ore algebra* generated by σ and δ.

A simple familiar example is given by $\mathcal{R} = \mathbb{Q}(x)$, $\delta = \frac{d}{dx}$ and $\sigma = \mathrm{id}_{\mathcal{R}}$ yielding linear ordinary differential operators with rational functions as coefficients. Similarly, we can obtain *recurrence* and *difference operators*. We set $\mathcal{R} = \mathbb{C}(n)$, the field of sequences $(s_n)_{n \in \mathbb{Z}}$ with complex elements $s_n \in \mathbb{C}$, and take for σ the shift operator, i. e. the automorphism mapping s_n to s_{n+1}. Then $\Delta = \sigma - \mathrm{id}_{\mathcal{R}}$ is a pseudo-derivation. $\mathcal{R}[E; \sigma, 0]$ consists of linear ordinary recurrence operators, $\mathcal{R}[E; \sigma, \Delta]$ of linear ordinary difference operators.

Note that so far it is not yet clear why we call the elements of these classical examples "operators." In fact, the elements of any Ore algebra $\mathcal{R}[\partial; \sigma, \delta]$ may be interpreted as operators on $\mathcal{R}$-modules. Let $\mathcal{V}$ be an $\mathcal{R}$-module and A an *$\mathcal{R}$-pseudo-linear map* $A : \mathcal{V} \to \mathcal{V}$, i. e. $A(u+v) = A(u) + A(v)$ and $A(fu) = \sigma(f)A(u) + \delta(f)u$ for all $f \in \mathcal{R}$ and $u, v \in \mathcal{V}$. Then we introduce an action $\alpha : \mathcal{R} \times \mathcal{V} \to \mathcal{V}$ mapping $(\theta = \sum f_i \partial^i, u)$ to $A_\theta u = \sum f_i A^i(u)$. For $\mathcal{V} = \mathcal{R}^m$ a natural choice for A is $A(v_1, \dots, v_m) = \bigl(\delta(v_1), \dots, \delta(v_m)\bigr)$. If $\delta = 0$ as for the recurrence operators, then we may also take $A(v_1, \dots, v_m) = \bigl(\sigma(v_1), \dots, \sigma(v_m)\bigr)$. With these definitions the above mentioned Ore algebras contain linear operators in the familiar sense.

For multivariate Ore polynomials we take a set $\Sigma = \{\sigma_1, \dots, \sigma_n\}$ of injective $\mathcal{R}$-endomorphisms and a set $\Delta = \{\delta_1, \dots, \delta_n\}$ where each δ_i is a pseudo-derivation with respect to σ_i. For each pair (σ_i, δ_i) we introduce a "variable" ∂_i satisfying a commutation rule of the form (3.13). If we require that all the maps σ_i, δ_j commute with each other, i. e. $\sigma_i \circ \sigma_j = \sigma_j \circ \sigma_i$, $\delta_i \circ \delta_j = \delta_j \circ \delta_i$ and $\sigma_i \circ \delta_j = \delta_j \circ \sigma_i$ for all $i \neq j$, one easily checks that $\partial_i \star \partial_j = \partial_j \star \partial_i$, i. e. the "variables" ∂_i commute.

Setting $\mathcal{D} = \{\partial_1, \dots, \partial_n\}$, we denote by $\mathcal{R}[\mathcal{D}; \Sigma, \Delta]$ the ring of multivariate Ore polynomials. Because of the commutativity of the variables ∂_i, we may write the terms as ∂^μ with multi indices $\mu \in \mathbb{N}_0^n$, so that it indeed makes sense to speak of a polynomial ring. The proof that $\bigl(\mathcal{R}[\mathcal{D}; \Sigma, \Delta], \star\bigr)$ is an algebra of solvable type for any term order goes completely analogously to the case of linear differential operators treated explicitly in Example 3.2.4 because of the great similarity of (3.13) and (3.10) (for $\mu = 1_j$ and $\nu = 0$). ◁

Ore algebras represent the special case where in Proposition 3.2.3 the maps h_i take their values in $\mathcal{R}$ and not in $\mathcal{P}$ and where the variables commute. We consider now an example with non-commutative variables. By the same construction principle, factorisation of a free tensor algebra by a suitable quadratic ideal, one can also generate many other interesting algebras like Clifford algebras.

Example 3.2.6. Let $\mathfrak{A}$ be a finite-dimensional *Lie algebra* over a field $\Bbbk$. As usual, we denote the product in $\mathfrak{A}$ by brackets: $[X, Y]$. Readers unfamiliar with Lie algebras should think of a subalgebra of $\mathfrak{gl}(m, \Bbbk)$, the algebra of all $m \times m$ matrices with entries from $\Bbbk$. The bracket is then the well-known commutator $[X, Y] = XY - YX$. Note that here the commutator is considered as the multiplication in the algebra and not the ordinary matrix product!

As an algebra over a field $\Bbbk$, $\mathfrak{A}$ is in particular a vector space and thus we can choose a basis $\{X_1, \dots, X_n\}$. The structure of the Lie algebra is defined by the commutation relations of the generators X_i:

$$[X_i, X_j] = C_{ij}^k X_k \tag{3.15}$$

where the $C_{ij}^k \in \Bbbk$ are the structure constants of the algebra $\mathfrak{A}$.

We construct now the *universal enveloping algebra* of $\mathfrak{A}$, a very important tool in the representation theory of Lie algebras (see e. g. [467, Chapt. 3]). We start with the tensor algebra $\mathfrak{T}(\mathfrak{A})$ over $\mathfrak{A}$; a basis of it is given by all formal tensor products $X_{i_1} \otimes X_{i_2} \otimes \cdots \otimes X_{i_q}$ with $q \in \mathbb{N}_0$. As $X_i \otimes X_j \neq X_j \otimes X_i$, we cannot identify such "terms" with multi indices in $\mathbb{N}_0^n$. We must first factor by the two-sided ideal generated by all quadratic elements $X_i \otimes X_j - X_j \otimes X_i - [X_i, X_j]$ in order to obtain the universal enveloping algebra $\mathfrak{U}(\mathfrak{A})$.

One can show that $\mathfrak{U}(\mathfrak{A})$ is an associative algebra and has the following universal property. Let $\pi : \mathfrak{A} \to \mathfrak{U}(\mathfrak{A})$ be the canonical linear map defined by $X_i \mapsto [X_i]$, i. e. each generator X_i is mapped on its equivalence class in the factor algebra $\mathfrak{U}(\mathfrak{A})$. The map π satisfies $\pi([X,Y]) = [\pi(X), \pi(Y)]$ for all $X, Y \in \mathfrak{A}$. Let $\mathfrak{B}$ be another associative algebra with a linear map $\phi : \mathfrak{A} \to \mathfrak{B}$ such that $\phi([X,Y]) = [\phi(X), \phi(Y)]$ for all $X, Y \in \mathfrak{A}$. Then there exists a unique algebra homomorphism $\tilde{\phi} : \mathfrak{U}(\mathfrak{A}) \to \mathfrak{B}$ such that $\phi = \tilde{\phi} \circ \pi$.

In each equivalence class there is exactly one representative containing only terms of the form $X_{i_1} \otimes X_{i_2} \otimes \cdots \otimes X_{i_q}$ where the indices form an ascending sequence $i_1 \leq i_2 \leq \cdots \leq i_q$. Given an arbitrary element $t \in \mathfrak{T}(\mathfrak{A})$, this representative can be constructed by substituting each product $X_i \otimes X_j$ with $i > j$ by $X_j \otimes X_i + [X_i, X_j]$ until no more such products appear. This leads to a normal form which can be identified with an element of $\Bbbk[x^1, \dots, x^n]$ where we have written x^i for the equivalence class $[X_i]$ in order to return to our usual notation. Thus as vector spaces $\mathfrak{U}(\mathfrak{A})$ and $\Bbbk[x^1, \dots, x^n]$ are isomorphic (this is the *Poincaré–Birkhoff–Witt Theorem* [467]).

As the right hand side of (3.15) is of lower degree than the left hand side, $\mathfrak{U}(\mathfrak{A})$ is a polynomial algebra of solvable type for any degree compatible term order. Kandry-Rody and Weispfenning [249] showed that if $\mathfrak{A}$ is a *solvable* Lie algebra, one can find bases $\{X_1, \dots, X_n\}$ such that $\mathfrak{U}(\mathfrak{A})$ is also of solvable type with respect to the lexicographic order.

As a concrete example we study the three-dimensional Lie algebra $\mathfrak{so}(3)$. Its commutation relations are $[X_i, X_j] = \varepsilon_{ijk} X_k$ where ε_{ijk} is the totally antisymmetric tensor and thus the multiplication in $\mathfrak{U}\bigl(\mathfrak{so}(3)\bigr)$ is defined by the equations

$$\begin{aligned} x^1 \star x^2 &= x^1 x^2 \,, & x^2 \star x^1 &= x^1 x^2 - x^3 \,, \\ x^1 \star x^3 &= x^1 x^3 \,, & x^3 \star x^1 &= x^1 x^3 + x^2 \,, \\ x^2 \star x^3 &= x^2 x^3 \,, & x^3 \star x^2 &= x^2 x^3 - x^1 \,. \end{aligned} \tag{3.16}$$

It is obvious that while the multiplication respects any degree compatible term order, it does not respect the lexicographic order.

Bell and Goodearl [38] introduced the concept of a *Poincaré–Birkhoff–Witt extension*[1] (*PBW extension* for short) of a ring $\mathcal{R}$ as a ring extension $\mathcal{P} \supseteq \mathcal{R}$ containing a finite number of elements $x^1, \dots, x^n \in \mathcal{P}$ such that (i) $\mathcal{P}$ is freely generated as a left $\mathcal{R}$-module by the monomials x^μ with $\mu \in \mathbb{N}_0^n$, (ii) $x^i \star r - r \star x^i \in \mathcal{R}$ for all $r \in \mathcal{R}$ and (iii) $x^i \star x^j - x^j \star x^i \in \mathcal{R} + \mathcal{R}x^1 + \cdots \mathcal{R}x^n$. Obviously, any such extension is a polynomial algebra of solvable type in the sense of Definition 3.2.1 for any degree compatible term order. Other term orders generally do not respect the multiplication in $\mathcal{P}$. Concrete examples of such extensions are a generalisation of the universal enveloping algebras, namely the *skew enveloping algebras* $\mathcal{R}\#\mathfrak{U}(\mathfrak{A})$ where $\mathcal{R}$ is a $\Bbbk$-algebra on which the elements of $\mathfrak{A}$ act as derivations [319, Sect. 1.7.10]. ◁

[1] One also finds the terminology *almost normalising extension* [319, Sect. 1.6.10].

Example 3.2.7. Non-commuting variables also occur in the *Weyl algebra* consisting of linear differential operators with polynomial coefficients. We may consider it as a polynomial ring $\mathcal{W}_n = \mathbb{k}[x^1,\dots,x^n,\partial_1,\dots,\partial_n]$ where the unknowns satisfy the commutation rules

$$[x^i,x^j]=0\,,\quad [\partial_i,\partial_j]=0\,,\quad [x^i,\partial_j]=\delta^i_j\,. \tag{3.17}$$

$\mathcal{W}_n$ is not an enveloping algebra, as it is simple (see Proposition 3.2.12 below) and one can show that an enveloping algebra is never simple [319, Cor. 1.7.5]. However, $\mathcal{W}_n$ is the quotient of an enveloping algebra. Indeed, adding a further variable z commuting which all other ones and writing the right hand side of the last commutator in (3.17) as $\delta^i_j z$, we recognise that (3.17) defines a $(2n+1)$-dimensional Lie algebra, the *Heisenberg algebra* $\mathfrak{h}$ being the core of much of quantum mechanics. Now we have the isomorphism $\mathcal{W}_n \cong \mathfrak{U}(\mathfrak{h})/\langle z-1\rangle$.

Modules over the Weyl algebra, often called $\mathcal{D}$*-modules*, are used in many different fields, in particular in singularity theory and in the representation theory of algebraic groups. An introduction into their theory may be found in [97]; more advanced material is contained in [46, 50]. Gröbner bases in the Weyl algebra are for example discussed in [391].

With respect to solvability the Weyl algebra behaves like a ring of differential operators with polynomial coefficients. By the same argument as in Example 3.2.4 one can show that it is an algebra of solvable type for any term order. The difference to the there studied rings of differential operators lies in the different definition of terms. A term in the Weyl algebra has the form $x^\mu\partial^\nu$ where $\mu,\nu\in\mathbb{N}_0^n$. Thus the leading exponent of an element of $\mathcal{W}_n$ is a multi index with $2n$ entries. Terms in $\mathcal{D}$ are of the form ∂^μ with $\mu\in\mathbb{N}_0^n$ so that exponents have only n entries. ◁

Example 3.2.8. The quantisation of a mechanical system represents a classical problem in theoretical physics for which many approaches have already been suggested. One of them, the *deformation quantisation* [110, 472], is based on the idea to deform the classical commutative product into a new one, the *Moyal star product* $\star$. Generally, a *Gerstenhaber deformation* of a $\mathbb{k}$-algebra $\mathcal{A}$ is defined as a $\mathbb{k}[[\lambda]]$-algebra $\tilde{\mathcal{A}}$ such that $\tilde{\mathcal{A}}/\langle\lambda\rangle\cong\mathcal{A}$.

We consider here only the simplest case, namely polynomial observables over the symplectic manifold $T^*\mathbb{R}\cong\mathbb{R}^2$. Thus we have the polynomial ring $\mathcal{P}=\mathbb{R}[q,p]$. Via the identification $p=\lambda\partial/\partial q$ with a parameter λ, we may map $\mathcal{P}$ to the Weyl algebra $\mathcal{W}_1$ (in physics $\lambda=-i\hbar$ with the Planck constant $\hbar$). However, in order to obtain a well-defined isomorphism $\rho:\mathcal{P}\to\mathcal{W}_1$, we must specify an ordering, as obviously the terms qp and pq are mapped into different elements of $\mathcal{W}_1$ (this is the famous ordering problem of quantum mechanics). Given such an isomorphism ρ, we may introduce as Moyal star product

$$f\star g=\rho^{-1}\big(\rho(f)\rho(g)\big)\,. \tag{3.18}$$

Two simple orderings are the standard ordering S where q is always written to the left of p and the Weyl ordering W using a total symmetrisation. The corresponding

Moyal star products are

$$f \star_S g = \sum_{r=0}^{\infty} \frac{\lambda^r}{r!} \frac{\partial^r f}{\partial p^r} \frac{\partial^r g}{\partial q^r} , \tag{3.19}$$

$$f \star_W g = \sum_{r=0}^{\infty} \frac{\lambda^r}{2^r r!} \sum_{s=0}^{r} \binom{r}{s} (-1)^{r-s} \frac{\partial^r f}{\partial q^s \partial p^{r-s}} \frac{\partial^r g}{\partial q^{r-s} \partial p^s} . \tag{3.20}$$

In both cases it is obvious that we obtain a solvable algebra with respect to any degree compatible term order. ◁

Example 3.2.9. In all these examples, the coefficients $r_{\mu\nu}$ appearing in (3.6) are one; thus (3.9b) are classical commutation relations. This is no longer true in the *quantised enveloping algebras* $\mathfrak{U}_h(\mathfrak{A})$ introduced by Drinfeld [113] and Jimbo [237]. For these algebra it is non-trivial that a Poincaré–Birkhoff–Witt Theorem holds; it was shown for general Lie algebras $\mathfrak{A}$ by Lusztig [295]. Berger [39] generalised this result later to a larger class of associative algebras, the *q-algebras*. They are characterised by the fact that the polynomials h_{ij} in (3.9b) are at most quadratic with the additional restriction that h_{ij} may contain only those quadratic terms $x^k x^\ell$ that satisfy $i < k \leq \ell < j$ and $k - i = j - \ell$. Thus any such algebra is a polynomial algebra of solvable type for any degree compatible term order.

A simple concrete example is the *q-Heisenberg algebra* for a real $q > 0$ (and $q \neq 1$). Let f be a function of a real variable x lying in some appropriate function space. Then we introduce the operators

$$\delta_q f(x) = \frac{f(x) - f(qx)}{(1-q)x} , \quad \tau_q f(x) = f(qx) , \quad \hat{x} f(x) = x f(x) . \tag{3.21}$$

It is straightforward to verify that these three operators satisfy the following q-deformed form of the classical Heisenberg commutation rules

$$\delta_q \star \hat{x} = \hat{x} \star \delta_q + \tau_q , \quad \delta_q \star \tau_q = q \tau_q \star \delta_q , \quad \tau_q \star \hat{x} = q \hat{x} \star \tau_q . \tag{3.22}$$

Hence the algebra $\mathbb{k}[\delta_q, \tau_q, \hat{x}]$ is a polynomial algebra of solvable type for any degree compatible term order (but also for any lexicographic order where the variables are ordered such that $\tau_q \prec \delta_q$ and $\tau_q \prec \hat{x}$). ◁

Provided that the coefficient ring $\mathcal{R}$ of the polynomial algebra $\mathcal{P}$ is a left Ore domain (so that it possesses a left quotient skew field—cf. Appendix B.1), we may also introduce "left rational functions", i. e. left fractions of the elements of $\mathcal{P}$.

Proposition 3.2.10. *If the coefficient ring $\mathcal{R}$ is a domain, then any polynomial algebra $(\mathcal{P}, \star, \prec)$ of solvable type over it is a domain, too. If $\mathcal{R}$ is even a left Ore domain, then so is $\mathcal{P}$.*

Proof. The first assertion is a trivial consequence of (3.8): if $\mathcal{R}$ has no zero divisors, then $f \cdot g \neq 0$ implies $f \star g \neq 0$ and $\mathcal{P}$ cannot contain zero divisors.

For the second one we must verify the left Ore condition (see Appendix B.1): we must show that one can find for any two polynomials $f,g \in \mathcal{P}$ with $f \star g \neq 0$ two further polynomials $\phi, \psi \in \mathcal{P} \setminus \{0\}$ such that $\phi \star f = \psi \star g$. We describe now a concrete algorithmic solution of this problem.

We set $\mathcal{F}_0 = \{f,g\}$ and choose coefficients $r_0, s_0 \in \mathcal{R}$ such that in the difference $r_0 g \star f - s_0 f \star g = \bar{h}_1$ the leading terms of the two summands cancel (this is always possible, since $\mathcal{R}$ itself satisfies the left Ore condition). Then we perform a (left) pseudo-reduction of $\bar{h}_1$ with respect to $\mathcal{F}_0$. It leads with an appropriately chosen coefficient $t_0 \in \mathcal{R}$ to an equation of the form

$$t_0 \bar{h}_1 = \phi_0 \star f + \psi_0 \star g + h_1 \tag{3.23}$$

where the remainder h_1 satisfies $\operatorname{le}_\prec h_1 \notin \langle \operatorname{le}_\prec \mathcal{F}_0 \rangle$. If $h_1 = 0$, we are done and the polynomials $\phi = t_0 r_0 g - \phi_0$ and $\psi = t_0 s_0 f + \psi_0$ form a solution of our problem. By Part (iii) of Definition 3.2.1 we have $\operatorname{le}_\prec \bar{h}_1 \prec \operatorname{le}_\prec f + \operatorname{le}_\prec g$. This estimate implies by the monotonicity of term orders that $\operatorname{le}_\prec \phi_0 \prec \operatorname{le}_\prec g$ and $\operatorname{le}_\prec \psi_0 \prec \operatorname{le}_\prec f$. Thus we have found a non-trivial solution: $\phi, \psi \neq 0$.

Otherwise we set $\mathcal{F}_1 = \mathcal{F}_0 \cup \{h_1\}$ and choose coefficients $r_1, s_1 \in \mathcal{R}$ such that in the difference $r_1 f \star h_1 - s_1 h_1 \star f = \bar{h}_2$ the leading terms of the two summands cancel. Then we perform a (left) pseudo-reduction of $\bar{h}_2$ with respect to $\mathcal{F}_1$. This computation yields a coefficient $t_1 \in \mathcal{R}$ and polynomials $\phi_1, \psi_1, \rho_1 \in \mathcal{P}$ such that

$$t_1 \bar{h}_2 = \phi_1 \star f + \psi_1 \star g + \rho_1 \star h_1 + h_2 \tag{3.24}$$

where the remainder h_2 satisfies $\operatorname{le}_\prec h_2 \notin \langle \operatorname{le}_\prec \mathcal{F}_1 \rangle$. If $h_2 = 0$, then we are done, as we can substitute h_1 from (3.23) and obtain thus for our problem the solution

$$\begin{aligned} \phi &= (t_1 r_1 f - \rho_1) \star (t_0 r_0 g - \phi_0) - t_1 s_1 h_1 - \phi_1 \,, \\ \psi &= (t_1 r_1 f - \rho_1) \star (t_0 s_0 f + \psi_0) + \psi_1 \,. \end{aligned} \tag{3.25}$$

By the same reasoning on the leading exponents as above, it is a non-trivial one.

Otherwise we iterate: we set $\mathcal{F}_2 = \mathcal{F}_1 \cup \{h_2\}$, choose coefficients $r_2, s_2 \in \mathcal{R}$ such that in the difference $r_2 f \star h_2 - s_2 h_2 \star f = \bar{h}_3$ the leading terms of the two summands cancel, compute the remainder h_3 of a (left) pseudo-reduction of $\bar{h}_3$ with respect to the set $\mathcal{F}_2$ and so on. If the iteration stops, i. e. if the remainder h_N vanishes for some value $N \in \mathbb{N}$, then we can construct non-zero polynomials ϕ, ψ with $\phi \star f = \psi \star g$ by substituting all remainders h_i by their defining equations. The iteration terminates by a simple Noetherian argument: $\langle \operatorname{le}_\prec \mathcal{F}_0 \rangle \subset \langle \operatorname{le}_\prec \mathcal{F}_1 \rangle \subset \langle \operatorname{le}_\prec \mathcal{F}_2 \rangle \subset \cdots$ is a strictly ascending chain of monoid ideals in $\mathbb{N}_0^n$ and thus cannot be infinite. □

We gave here a direct and in particular constructive proof that $\mathcal{P}$ satisfies the left Ore condition. One can show [319, Theorem 2.1.15] that any left/right Noetherian domain is also a left/right Ore domain. However, as we will see in the next section, proving that a polynomial algebra of solvable type over a coefficient ring is left or right Noetherian can be quite subtle.

Expressed in a more formal language, the construction in the proof above leads to Algorithm 3.4. In Line /5/ the subalgorithm PseudoReduce determines for the input $\bar{h}_i$ and $\mathcal{H} = \{h_1, \dots, h_{i-1}\}$ a coefficient $t \in \mathcal{R}$, polynomials $\boldsymbol{\rho} = (\rho_1, \dots, \rho_{i-1})$ and a remainder $h_i \in \mathcal{P}$ such that $t\bar{h}_i = \sum_{j=1}^{i-1} \rho_j \star h_j + h_i$ and $\mathrm{le}_{\prec} h_i$ is not divisible by $\mathrm{le}_{\prec} h_j$ for any $1 \leq j < i$. In the Lines /6/ and /7/ we compute from these data polynomials $\phi_i, \psi_i \in \mathcal{P}$ such that $h_i = \phi_i \star f + \psi_i \star g$. Thus once the remainder vanishes, i. e. $h_i = 0$, Line /12/ returns correct left Ore multipliers.

The left Ore multipliers ϕ, ψ are not uniquely determined, as we could equally well multiply at each step with g or alternate between f and g etc. In Algorithm 3.4 this observation concerns mainly the Lines /3/ and /4/ where we could use h_2 instead of h_1 (but then we must also move the term tsh_{i-1} from Line /6/ to Line /7/ and replace in both lines h_1 by h_2). After some obvious modifications, a similar algorithm allows for the computation of right Ore multipliers.

Algorithm 3.4 Left Ore multipliers

Input: polynomials $f, g \in \mathcal{P}$
Output: polynomials $\phi, \psi \in \mathcal{P}$ such that $\phi \star f = \psi \star g$
1: $h_1 \leftarrow f$; $h_2 \leftarrow g$; $\mathcal{H} \leftarrow \{f, g\}$; $i \leftarrow 3$; $\phi_1 \leftarrow 1$; $\psi_1 \leftarrow 0$; $\phi_2 \leftarrow 0$; $\psi_2 \leftarrow 1$
2: **repeat**
3: choose $r, s \in \mathcal{R}$ such that $\mathrm{le}_{\prec}(rh_1 \star h_{i-1} - sh_{i-1} \star h_1) \prec \mathrm{le}_{\prec}(h_1 \star h_{i-1})$
4: $\bar{h}_i \leftarrow rh_1 \star h_{i-1} - sh_{i-1} \star h_1$
5: $(h_i, t, \boldsymbol{\rho}) \leftarrow$ PseudoReduce$(\bar{h}_i, \mathcal{H})$
6: $\phi_i \leftarrow trh_1 \star \phi_{i-1} - \sum_{j=1}^{i-1} \rho_j \star \phi_j - tsh_{i-1}$
7: $\psi_i \leftarrow trh_1 \star \psi_{i-1} - \sum_{j=1}^{i-1} \rho_j \star \psi_j$
8: **if** $h_i \neq 0$ **then**
9: $\mathcal{H} \leftarrow \mathcal{H} \cup \{h_i\}$; $i \leftarrow i + 1$
10: **end if**
11: **until** $h_i = 0$
12: **return** $(\phi_i, -\psi_i)$

Example 3.2.11. In the commutative polynomial ring one has always the trivial solution $\phi = g$ and $\psi = f$. One might expect that in the non-commutative case one only has to add some lower terms to it. However, this is not the case. Consider the universal enveloping algebra of the Lie algebra $\mathfrak{so}(3)$ introduced above. Its multiplication (3.16) obviously respects any degree compatible term order but not the lexicographic order. Choosing $f = x^1$ and $g = x^2$, possible solutions for $\phi \star f = \psi \star g$ are $\phi = (x^2)^2 - 1$ and $\psi = x^1x^2 - 2x^3$ or $\phi = x^1x^2 + x^3$ and $\psi = (x^1)^2 - 1$. They are easily constructed using Algorithm 3.4 once in the given form and once with h_1 replaced by h_2 in Line /4/. Here we must use polynomials of degree 2; it is not possible to find a solution of degree 1. $\triangleleft$

In most what we will be doing in the sequel there is no real difference between the usual polynomial ring and arbitrary polynomial algebras of solvable type. The non-commutativity has hardly any effect on the theory of Gröbner and involutive bases, respectively, in these algebras. In order to show that this similarity rapidly ends, if one proceeds beyond such purely computational aspects, we close this section by listing three results for the Weyl algebra which are in marked contrast to properties of the commutative polynomial ring. These results are not used later and only mentioned for demonstration purposes; more details can be found in [45].

Proposition 3.2.12. *The Weyl algebra $\mathcal{W}_n$ is simple, i. e. the only two-sided ideals in $\mathcal{W}_n$ are the trivial ones: $\{0\}$ and $\mathcal{W}_n$.*

Proof. We proceed by induction. The case $n = 0$ is trivial, as $\Bbbk$ is a field. Now assume that we know already that $\mathcal{W}_{n-1}$ is simple and let $\mathcal{I} \subseteq \mathcal{W}_n$ be a two-sided ideal containing an element $f \neq 0$. In order to prove that $\mathcal{I} = \mathcal{W}_n$, it suffices to show that $\mathcal{I} \cap \mathcal{W}_{n-1} \neq \{0\}$, as then by assumption $1 \in \mathcal{I}$.

We write $f = d_0 + d_1\partial_n + \cdots + d_q\partial_n^q$ with $d_i \in \mathcal{W}_{n-1}[x^n]$ and $d_q \neq 0$. If $q = 0$, let $f_0 = f = d_0$. Otherwise one easily computes that

$$f_1 = f \star x^n - x^n \star f = d_1 + 2d_2\partial_n + \cdots + qd_q\partial_n^{q-1}\,. \tag{3.26}$$

Since $f_1 \in \mathcal{I}$, we can iterate until we find $f_q = q!d_q \in \mathcal{I} \cap \mathcal{W}_{n-1}[x^n]$. Note that by assumption $f_q \neq 0$.

Now we write $f_q = g_0 + g_1x^n + \cdots + g_r(x^n)^r$ with $g_i \in \mathcal{W}_{n-1}$ and $g_r \neq 0$. For $r = 0$ we are done. Otherwise we compute

$$f_{q,1} = \partial_n \star f_q - f_q \star \partial_n = g_1 + 2g_2x^n + \cdots + rg_r(x^n)^{r-1}\,. \tag{3.27}$$

Again we can iterate until we find $f_{q,r} = r!g_r \in \mathcal{I} \cap \mathcal{W}_{n-1}$. As by assumption this element does not vanish, we are done. □

Example 3.2.13. By the Quillen-Suslin Theorem B.2.13 every projective polynomial module is free. Stafford [429] showed that in the Weyl algebra $\mathcal{W}_1$ the ideal $\mathcal{I} = \langle \partial^2, x\partial - 1\rangle$ is projective but not free. Thus we cannot expect to extend this theorem to arbitrary polynomial algebras of solvable type. ◁

The following rather surprising result is also due to Stafford [430]. Its proof is too long and too technical to be presented here; it can be found in [45, Chap. 1, §7]. Leykin [289] provided an alternative constructive proof permitting the explicit construction of the generators. However, typically they are far too lengthy to be of any practical value for computations.

Theorem 3.2.14. *Every left ideal in $\mathcal{W}_n$ may be generated by two elements.*

3.3 Hilbert's Basis Theorem and Gröbner Bases

A classical property of the ordinary polynomial ring $\mathcal{P} = \mathcal{R}[x^1,\dots,x^n]$, which is crucial in the theory of Gröbner bases, is Hilbert's Basis Theorem B.1.13 stating that $\mathcal{P}$ is a Noetherian ring, if its coefficient ring $\mathcal{R}$ is. For our more general class of polynomial algebras, this property remains true only under additional assumptions. As we are generally dealing with a non-commutative ring, we must furthermore distinguish left, right and two-sided ideals and thus also study separately whether $\mathcal{P}$ is left or right Noetherian.

With the exception of an Addendum to Section 4.2, we will exclusively work with left ideals and thus do not introduce special notations for distinguishing the different types of ideal. The restriction to left ideals is not just for convenience but stems from the fundamental left-right asymmetry of Definition 3.2.1 of a polynomial algebra of solvable type where products $r \star x^\mu$ and $x^\mu \star r$ are treated completely differently. For this reason we will concentrate on the question when $\mathcal{P}$ is left Noetherian where it is possible to provide comparatively comprehensive answers (see also the remarks in Example 3.3.14 below).

Most standard proofs of Hilbert's Basis Theorem consider only the univariate case and then extend inductively to an arbitrary (but finite) number of variables. However, this inductive approach is not possible for arbitrary polynomial algebras of solvable type, as the multiplication $\star$ does not necessarily restrict to a subalgebra with fewer variables. A simple counterexample is provided by the universal enveloping algebra $\mathfrak{U}(\mathfrak{so}(3))$ introduced in Example 3.2.6 where $\star$ cannot be restricted to the subspace $\Bbbk[x^1, x^2]$ since $x^2 \star x^1 = x^1x^2 - x^3$.

In this section we will study four different situations where it is possible to prove a basis theorem for left ideals. The first one deals with iterated polynomial algebras where the inductive reduction to the univariate case is still feasible. For univariate polynomials we then adapt a simple classical proof by contradiction. As second case we consider a class of polynomial algebras where the commutation relations (3.6) are assumed to be of a special form (which will reappear in Section 4.6). The made assumptions permit us to generalise the previous proof for univariate polynomials directly to multivariate polynomial algebras. In the third proof we assume the existence of a filtration Σ on the ring $\mathcal{P}$. A very general argument (which does not even require $\mathcal{P}$ to be a polynomial algebra) shows then that $\mathcal{P}$ is left Noetherian, if its associated graded ring $\mathrm{gr}_\Sigma\mathcal{P}$ is left Noetherian. Finally, we treat the by far simplest situation that the coefficient ring $\mathcal{R}$ is a (skew) field. Here, one can show by a standard argument not only that $\mathcal{P}$ is left Noetherian but simultaneously the existence of Gröbner bases for left ideals.

Iterated Polynomial Algebras of Solvable Type

The key to the inductive approach to proving Hilbert's Basis Theorem is the natural iterative structure of the ordinary polynomial ring. It requires in particular that the multiplication $\star$ restricts to subrings in a lower number of variables; iterated Ore algebras provide a standard example where such a restriction is possible. Within our fairly general class of non-commutative polynomial rings we must furthermore impose conditions on the coefficients appearing in the commutation relations (3.6).

Definition 3.3.1. The polynomial algebra of solvable type $(\mathcal{P},\star,\prec)$ is called *iterated*, if it satisfies the following three conditions.

(i) $\mathcal{P}$ can be written in the form $\mathcal{P}=\mathcal{R}[x^1][x^2]\cdots[x^n]$ where each intermediate ring $\mathcal{P}_{(k)}=\mathcal{R}[x^1][x^2]\cdots[x^k]$ is again solvable for the corresponding restrictions of the multiplication $\star$ and the term order $\prec$.
(ii) The equality $x^k\star\mathcal{P}_{(k-1)}+\mathcal{P}_{(k-1)}=\mathcal{P}_{(k-1)}\star x^k+\mathcal{P}_{(k-1)}$ holds for $1\leq k\leq n$.
(iii) In (3.6b) the coefficients $r_{\mu\nu}$ are units whenever the multi indices are of the form $\mu=\ell_i$, $\nu=m_i$ for $1\leq i\leq n$ and arbitrary values $\ell,m\in\mathbb{N}$.

The first condition just encodes the iterative structure of $\mathcal{P}$. The other two, more technical conditions are needed in the next proof. Assume that we are given a "univariate" polynomial $f=\sum_{\ell=1}^{q}a_\ell(x^k)^\ell\in\mathcal{P}_{(k)}=\mathcal{P}_{(k-1)}[x^k]$ of degree q; Condition (ii) allows us to rewrite it in the "reverse" form $f=\sum_{\ell=1}^{q}(x^k)^\ell\star\bar{a}_\ell$ with uniquely determined coefficients $\bar{a}_\ell\in\mathcal{P}_{(k-1)}$. Assume furthermore that we multiply in this reverse form with the power $(x^k)^m$ from the left. Then the result contains terms of the form $r_{\ell_k,m_k}(x^k)^{\ell+m}\star a_\ell$, i. e. with coefficients on both sides of the power. According to Condition (iii), the coefficients on the left side are units and we can avoid such a coefficient at least for the leading term in the result by multiplying with the monomial $r_{q_k,m_k}^{-1}(x^k)^m$ instead of the pure power product $(x^k)^m$.

These two observations allow us now to adapt a classical proof of Hilbert's Basis Theorem in the commutative case. The basic idea consists of moving to the reverse form of univariate polynomials so that when we multiply with powers from the left (we are interested only in left ideals!), we can at the relevant places ignore the operation of the variables on the coefficients.

Theorem 3.3.2. *If $(\mathcal{P},\star,\prec)$ is an iterated polynomial algebra of solvable type over a left Noetherian ring $\mathcal{R}$, then $\mathcal{P}$ is a left Noetherian ring, too.*

Proof. By the considerations above, it suffices to consider only the univariate case. Thus let $\mathcal{P}=\mathcal{R}[x]$. Let furthermore $\mathcal{I}^{(0)}\subseteq\mathcal{I}^{(1)}\subseteq\mathcal{I}^{(2)}\subseteq\cdots\subseteq\mathcal{P}$ be an ascending chain of left ideals. As discussed above, Condition (ii)—which now takes the form $x\star\mathcal{R}+\mathcal{R}=\mathcal{R}\star x+\mathcal{R}$—allows us to rewrite any polynomial in the reverse form $f=\sum_\ell x^\ell\star r_\ell$ and from now on we assume that all polynomials are written in this form. We introduce the following sets:

$$\mathcal{J}^{(\ell,q)}=\left\{r\in\mathcal{R}\mid\exists f\in\mathcal{I}^{(\ell)}:\deg f\leq q\wedge\operatorname{lc}_\prec f=r\right\}. \tag{3.28}$$

The condition $\operatorname{lc}_\prec f=r$ refers here to the leading coefficient in the reverse representation of f. Conditions (ii) and (iii) imply that these sets are left(!) ideals in $\mathcal{R}$. Indeed, because of Condition (ii) we can find for any $r\in\mathcal{R}$ coefficients $s,t\in\mathcal{R}$ such that $s\star x=x\star r+t$. Hence we can produce at the expense of some additional lower-order terms arbitrary coefficients to the right of x—without leaving the given ideal $\mathcal{I}^{(\ell)}$—and by Condition (iii) this property extends to higher powers of x.

If the equality $\mathcal{J}^{(\ell,q)}=\mathcal{J}^{(\ell',q)}$ holds for all $q\geq 0$, then we have $\mathcal{I}^{(\ell)}=\mathcal{I}^{(\ell')}$. This fact can be seen as follows. Assume that $\ell<\ell'$ and $\mathcal{I}^{(\ell)}\subsetneq\mathcal{I}^{(\ell')}$. We consider a polynomial $f'\in\mathcal{I}^{(\ell')}\setminus\mathcal{I}^{(\ell)}$ of minimal degree, say $\deg f'=d'$. Thus f' is of the form $f' = x^{d'} \star r + \cdots$ where the dots represents terms of lower order. Because of $\mathcal{J}^{(\ell,d')} = \mathcal{J}^{(\ell',d')}$, the ideal $\mathcal{I}^{(\ell)}$ contains a polynomial $f = x^d \star r + \cdots$ with $d \leq d'$. Since we assume that the coefficient $r_{d'-d,d}$ is a unit, we can invert it and obtain that $f' - r_{d'-d,d}^{-1} x^{d'-d} \star f \in \mathcal{I}^{(\ell')}$ is a polynomial of degree less than d. Hence it must in fact be contained in $\mathcal{I}^{(\ell)}$. But this immediately implies $f' \in \mathcal{I}^{(\ell)}$, a contradiction.

The definition of these $\mathcal{R}$-ideals trivially implies that $\mathcal{J}^{(\ell,q)} \subseteq \mathcal{J}^{(\ell',q')}$ whenever $\ell \leq \ell'$ and $q \leq q'$. Thus we have constructed an ascending chain of left ideals: $\mathcal{J}^{(0,0)} \subseteq \mathcal{J}^{(1,1)} \subseteq \mathcal{J}^{(2,2)} \subseteq \cdots \subseteq \mathcal{R}$. Since by assumption $\mathcal{R}$ is left Noetherian, the chain becomes stationary at some index ℓ_∞. Next we consider for $0 \leq q < \ell_\infty$ the ascending chain of left ideals $\mathcal{J}^{(0,q)} \subseteq \mathcal{J}^{(1,q)} \subseteq \mathcal{J}^{(2,q)} \subseteq \cdots$. Again the chain becomes stationary at some index ℓ_q, since $\mathcal{R}$ is left Noetherian. We set $\ell = \max\{\ell_\infty, \ell_0, \ell_1, \ldots, \ell_{\ell_\infty - 1}\}$. Then by construction we have $\mathcal{J}^{(k,q)} = \mathcal{J}^{(\ell,q)}$ for all $k \geq \ell$ and all $q \geq 0$ implying that $\mathcal{I}^{(k)} = \mathcal{I}^{(\ell)}$ for all $k \geq \ell$. Thus the chain becomes stationary and $\mathcal{P}$ is left Noetherian. □

Example 3.3.3. The additional conditions in Definition 3.3.1 cannot be omitted. If they are violated, $\mathcal{P}$ may be non-Noetherian, as the following concrete counterexample demonstrates. Let $\mathcal{R} = \Bbbk[t]$ be the ordinary polynomial ring and consider $\mathcal{P} = \mathcal{R}[x]$ with the multiplication defined by $x \star x = x^2$ and $x \star t = t^2 x$. Note that we do not consider $\mathcal{P}$ as a bivariate polynomial ring; as such it is obviously not solvable. $\mathcal{P}$ is here a univariate polynomial algebra of solvable type over the ring $\mathcal{R}$.

It is easy to see that the second condition in Definition 3.3.1 is violated here, as it is not possible to express $tx \in \mathcal{R} \star x$ as an element of $x \star \mathcal{R} + \mathcal{R}$. Now we consider the left ideals $\mathcal{I}_k = \langle tx, tx^2, \ldots, tx^k \rangle$. Since $x \star tx^k = t^2 x^{k+1}$ etc, $tx^{k+1} \notin \mathcal{I}_k$ and we have found an infinite strictly ascending sequence of ideals $\mathcal{I}_1 \subsetneq \mathcal{I}_2 \subsetneq \cdots$, so that $\mathcal{P}$ is not left Noetherian also $\mathcal{R}$ obviously is. ◁

Polynomial Algebras with Centred Commutation Relations

With some complications the central (univariate) arguments in the proof of Theorem 3.3.2 can be directly generalised to multivariate polynomial rings. However, this approach requires again certain assumptions on the commutation relations (3.6) in order to ensure that all necessary computations are possible.

Definition 3.3.4. The polynomial algebra of solvable type $(\mathcal{P}, \star, \prec)$ has *centred commutation relations*, if the following conditions are satisfied.

(i) There exists a subfield $\Bbbk \subseteq \mathcal{R}$ lying in the centre of the coefficient ring $\mathcal{R}$.
(ii) The functions ρ_μ appearing in (3.6a) are of the form $\rho_\mu(r) = \bar{\rho}_\mu(r) r$ with functions $\bar{\rho}_\mu : \mathcal{R} \to \Bbbk$.
(iii) We have $r_{\mu\nu} \in \Bbbk$ in (3.6b).

As in the proof of Theorem 3.3.2, we proceed by deriving certain ideals in the coefficient ring $\mathcal{R}$ from a given polynomial ideal $\mathcal{I} \subseteq \mathcal{P}$. Then we will show with the help of König's Lemma A.1.3 that if $\mathcal{P}$ is not left Noetherian, then $\mathcal{R}$ cannot be left Noetherian either. The key is the following observation the proof of which partially explains the second condition in Definition 3.3.4.

Lemma 3.3.5. *Let $\mathcal{I} \subseteq \mathcal{P}$ be a left ideal in a solvable algebra $(\mathcal{P}, \star, \prec)$ with centred commutation relations and $\mu \in \mathbb{N}_0^n$ an arbitrary multi index. Then*

$$\mathcal{J}_{\mathcal{I},\mu} = \{\mathrm{lc}_\prec f \mid f \in \mathcal{I} \wedge \mathrm{le}_\prec f = \mu\} \cup \{0\} \tag{3.29}$$

is a left ideal in $\mathcal{R}$. Furthermore, for all multi indices $\mu, \nu \in \mathbb{N}_0^n$ with $\mu \mid \nu$ the inclusion $\mathcal{J}_{\mathcal{I},\mu} \subseteq \mathcal{J}_{\mathcal{I},\nu}$ holds.

Proof. Let $0 \neq s,t \in \mathcal{J}_{\mathcal{I},\mu}$ be arbitrary elements and choose polynomials $f,g \in \mathcal{I}$ with $\mathrm{lm}_\prec f = sx^\mu$ and $\mathrm{lm}_\prec g = tx^\mu$. For $s = t$, we have trivially $s - t = 0 \in \mathcal{J}_{\mathcal{I},\mu}$. Otherwise, $f \neq g$ and their difference $h = f - g \in \mathcal{I}$ satisfies $\mathrm{le}_\prec h = \mu$. This entails $s - t = \mathrm{lc}_\prec h \in \mathcal{J}_{\mathcal{I},\mu}$. For any ring element $r \in \mathcal{R}$, we have either $rs = 0 \in \mathcal{J}_{\mathcal{I},\mu}$ or $0 \neq rs = \mathrm{lc}_\prec(rf)$. Since in the second case $\mathrm{le}_\prec(rf) = \mu$, we find again $rs \in \mathcal{J}_{\mathcal{I},\mu}$ and $\mathcal{J}_{\mathcal{I},\mu}$ is indeed an ideal.

If $\mu \mid \nu$ and $\mu \neq \nu$, then we set $\lambda = \nu - \mu$. By our assumptions, the ring element $\bar{\rho}_\lambda(s) r_{\lambda\mu}$ is invertible and $s r_{\lambda\mu} = r_{\lambda\mu} s$. An easy computation yields then that $\mathrm{lm}_\prec\big((\bar{\rho}_\lambda(s) r_{\lambda\mu})^{-1} x^\lambda \star f\big) = sx^\nu$ and thus any $s \in \mathcal{J}_{\mathcal{I},\mu}$ is contained in $\mathcal{J}_{\mathcal{I},\nu}$. □

Lemma 3.3.6. *Let $\mathcal{I} \subseteq \mathcal{P}$ be a left ideal. Then $\mathcal{S}_\mathcal{I} = \{\mathcal{J}_{\mathcal{I},\mu} \mid \mu \in \mathrm{le}_\prec \mathcal{I}\}$ has the König property for the partial order induced by set inclusion.*

Proof. By Dickson's Lemma A.1.2, the monoid $\mathrm{le}_\prec \mathcal{I} \subseteq \mathbb{N}_0^n$ possesses a finite Dickson basis $\mathcal{B}_\mathcal{I}$. It follows from Lemma 3.3.5 that any minimal element of $\mathcal{S}_\mathcal{I}$ must be a set $\mathcal{J}_{\mathcal{I},\nu}$ for a multi index $\nu \in \mathcal{B}_\mathcal{I}$. Hence there can be only finitely many minimal elements in the set $\mathcal{S}_\mathcal{I}$ itself.

Each $\mathcal{J}_{\mathcal{I},\nu} \in \mathcal{S}_\mathcal{I}$ has an upper set $\mathcal{S}_{\mathcal{I},\nu} = \{\mathcal{J}_{\mathcal{I},\mu} \in \mathcal{S}_\mathcal{I} \mid \mathcal{J}_{\mathcal{I},\nu} \subset \mathcal{J}_{\mathcal{I},\mu}\}$. Introducing the corresponding subset $\mathcal{N}_{\mathcal{I},\nu} \subseteq \mathrm{le}_\prec \mathcal{I}$ and applying the same argument as above shows that each set $\mathcal{S}_{\mathcal{I},\nu}$ has only finitely many minimal elements, too. Thus $\mathcal{S}_\mathcal{I}$ has indeed the König property. □

Recall from Remark A.1.4 that if $\mathcal{R}$ is a (left) Noetherian ring and $\mathcal{S}$ a set of (left) ideals of $\mathcal{R}$, then, by König's Lemma A.1.3, $\mathcal{S}$ can possess the König property only, if it is finite. Since we assume that the coefficient ring $\mathcal{R}$ of our solvable algebra $\mathcal{P}$ is left Noetherian, Lemma 3.3.6 entails that for each ideal $\mathcal{I} \subseteq \mathcal{P}$ the set $\mathcal{S}_\mathcal{I}$ defined in Lemma 3.3.6 is finite.

Theorem 3.3.7. *Let $(\mathcal{P}, \star, \prec)$ be a polynomial algebra of solvable type with centred commutation relations over a left Noetherian ring $\mathcal{R}$. Then $\mathcal{P}$ is left Noetherian.*

Proof. Consider for a left ideal $\mathcal{I} \subseteq \mathcal{P}$ the set $\mathcal{S}_\mathcal{I}$ defined in Lemma 3.3.6. We may then introduce for each ideal $\mathcal{J} \in \mathcal{S}_\mathcal{I}$ contained in it the following set of multi indices: $\mathcal{N}_\mathcal{J} = \{\nu \in \mathrm{le}_\prec \mathcal{I} \mid \mathcal{J}_{\mathcal{I},\nu} = J\}$. By Dickson's Lemma A.1.2, each set $\mathcal{N}_\mathcal{J}$ has a finite Dickson basis $\mathcal{B}_\mathcal{J}$ and we define $\mathcal{B}_\mathcal{I} = \bigcup_{\mathcal{J} \in \mathcal{S}_\mathcal{I}} \mathcal{B}_\mathcal{J}$. According to our considerations above, the set $\mathcal{S}_\mathcal{I}$ is finite implying that $\mathcal{B}_\mathcal{I}$ is finite, too.

Since the ring $\mathcal{R}$ is assumed to be left Noetherian, we can choose for each multi index $\nu \in \mathcal{B}_{\mathcal{I}}$ a finite number of ideal elements $g_{\nu,1},\dots,g_{\nu,k_\nu} \in \mathcal{I}$ such that $\mathrm{le}_\prec g_{\nu,j} = \nu$ for $1 \le j \le k_\nu$ and $\mathcal{J}_{\mathcal{I},\nu} = \langle \mathrm{lc}_\prec g_{\nu,1},\dots,\mathrm{lc}_\prec g_{\nu,k_\nu}\rangle$. We claim that the finite set $\mathcal{G}_{\mathcal{I}} = \bigcup_{\nu\in\mathcal{B}_{\mathcal{I}}}\{g_{\nu,1},\dots,g_{\nu,k_\nu}\}$ generates the ideal $\mathcal{I}$.

Obviously, $\langle\mathcal{G}_{\mathcal{I}}\rangle \subseteq \mathcal{I}$. Assume now that $\langle\mathcal{G}_{\mathcal{I}}\rangle \subsetneq \mathcal{I}$ and choose some polynomial $f \in \mathcal{I}\setminus\langle\mathcal{G}_{\mathcal{I}}\rangle$ with minimal leading exponent $\mu = \mathrm{le}_\prec f$. Let $r = \mathrm{lc}_\prec f$. By construction, a multi index $\nu \in \mathcal{B}_{\mathcal{I}}$ exists such that $\nu \mid \mu$ and $\mathcal{J}_{\mathcal{I},\mu} = \mathcal{J}_{\mathcal{I},\nu}$. If we write $r_j = \mathrm{lc}_\prec g_{\nu,j}$ for $1 \le j \le k_\nu$, then r has a representation $r = \sum_{j=1}^{k_\nu} s_j r_j$ with some coefficients $s_j \in \mathcal{R}$. Let $\lambda = \mu - \nu$. Exploiting that $\bar\rho_\lambda : \mathcal{R} \to \Bbbk$ and $r_{\lambda\nu} \in \Bbbk$, we consider the polynomial $f' = f - \sum_{j=1}^{k_\nu} s_j\bigl(\bar\rho_\lambda(r_j)r_{\lambda\nu}\bigr)^{-1} x^\lambda \star g_{\nu,j}$. One easily verifies that $\mathrm{le}_\prec f' \prec \mathrm{le}_\prec f$ and thus $f' \in \langle\mathcal{G}_{\mathcal{I}}\rangle$ by our choice of f. But this implies $f \in \langle\mathcal{G}_{\mathcal{I}}\rangle$, a contradiction. Hence $\langle\mathcal{G}_{\mathcal{I}}\rangle = \mathcal{I}$. □

Filtered Algebras

Our third approach to prove that $\mathcal{P}$ is left Noetherian assumes that $\mathcal{P}$ possesses a *filtration* Σ and thus is a filtered ring (see Definition B.1.20). This is in particular always the case, if $\mathcal{P}$ is solvable with respect to a degree compatible term order $\prec$, as then we may use the total degree filtration defined by $\Sigma_k = \bigoplus_{i=0}^{k}\mathcal{P}_i$. Note that if $\prec$ is not degree compatible, it may well happen that $\deg(f\star g) > \deg(f\cdot g)$ and then Σ does not define a filtration.

It is straightforward to verify that for the total degree filtration the *associated graded ring* $\mathrm{gr}_\Sigma\mathcal{P}$ is again a polynomial algebra of solvable type. If in the commutation relation (3.9) $\deg h_i(r) = 0$, $\deg h_{ij} \le 1$, $\rho_i = \mathrm{id}_{\mathcal{R}}$ and $r_{ij} = 1$ (which is for example the case for all Poincaré–Birkhoff–Witt extensions), then in fact $\mathrm{gr}_\Sigma\mathcal{P} = (\mathcal{P},\cdot)$, the commutative polynomial ring. Such algebras are sometimes called *almost commutative* [319, Sect. 8.4.2].

Our goal is now to show that $\mathcal{P}$ is left Noetherian, if $\mathrm{gr}_\Sigma\mathcal{P}$ is. As preparation we formulate some results on filtered $\mathcal{P}$-modules $\mathcal{M}$, although we will later apply them only to ideals in $\mathcal{P}$. In fact, this whole approach remains valid for arbitrary filtered rings $\mathcal{P}$, as we make nowhere use of the polynomial nature of $\mathcal{P}$. The key is the following lemma relating finite bases in $\mathcal{M}$ and $\mathrm{gr}_\Gamma\mathcal{M}$ via the Γ-symbol (cf. the discussion after Definition B.1.20).

Lemma 3.3.8. *Let $\mathcal{M}$ be a left $\mathcal{P}$-module with a filtration Γ and let furthermore $\{m_1,\dots,m_r\} \subset \mathcal{M}$ be a finite subset such that $\bigl\{\sigma_\Gamma(m_1),\dots,\sigma_\Gamma(m_r)\bigr\}$ generates $\mathrm{gr}_\Gamma\mathcal{M}$ as a left $\mathrm{gr}_\Sigma\mathcal{P}$-module. Then $\{m_1,\dots,m_r\}$ generates $\mathcal{M}$.*

Proof. We show by an induction on i that each component Γ_i is contained in the span $\langle m_1,\dots,m_r\rangle$. For $i = 0$ this is trivial, as by definition $\bar\Gamma_0 = \Gamma_0$ and hence $\sigma_\Gamma(m) = m$ for all elements $m \in \Gamma_0$.

Assume that $\Gamma_{i-1} \subset \langle m_1,\dots,m_r\rangle$ and consider an arbitrary element $m \in \Gamma_i\setminus\Gamma_{i-1}$. Since $\bigl\{\sigma_\Gamma(m_1),\dots,\sigma_\Gamma(m_r)\bigr\}$ generates $\mathrm{gr}_\Gamma\mathcal{M}$, there exist elements $\bar f_k \in \mathrm{gr}_\Sigma\mathcal{P}$ such that $\sigma_\Gamma(m) = \sum_{k=1}^{r} \bar{f}_k \star \sigma_\Gamma(m_k)$. More precisely, $\bar{f}_k \in \bar{\Sigma}_{\ell_k}$ with $\ell_k + \deg_\Gamma m_k = i$ (obviously, $\bar{f}_k = 0$ for $\deg_\Gamma m_k > i$). If we now choose representatives $f_k \in \Sigma_{\ell_k}$ of the equivalence classes $\bar{f}_k$, then by construction $m - \sum_{k=1}^{r} f_k \star m_k \in \Gamma_{i-1}$ and thus $m \in \langle m_1, \dots, m_r \rangle$. □

Definition 3.3.9. A filtration Γ on a left $\mathcal{P}$-module $\mathcal{M}$ is *good*, if $\mathrm{gr}_\Gamma \mathcal{M}$ is a finitely generated left $\mathrm{gr}_\Sigma \mathcal{P}$-module.

Lemma 3.3.10. *The left $\mathcal{P}$-module $\mathcal{M}$ is finitely generated, if and only if it possesses a good filtration.*

Proof. Let $\{m_1, \dots, m_r\}$ be a finite generating set for the module $\mathcal{M}$. Then we define $\Gamma_i = \sum_{k=1}^{r} \Sigma_i \star m_k$. It is trivial to verify that this yields a filtration. Since $\{\sigma_\Gamma(m_1), \dots, \sigma_\Gamma(m_r)\} \subset \Gamma_0$ generates $\mathrm{gr}_\Gamma \mathcal{M}$, we even have a good filtration.

For the converse, let Γ be a good filtration and let $\{\bar{f}_1, \dots, \bar{f}_r\}$ generate the associated graded module $\mathrm{gr}_\Gamma \mathcal{M}$. We decompose each generator into its Γ-homogeneous parts, i. e. we write $\bar{f}_i = \sum_{j=1}^{k_i} \bar{f}_{ij}$ with $\bar{f}_{ij} \in \bar{\Gamma}_j$. Then we choose elements $m_{ij} \in \mathcal{M}$ such that $\sigma_\Gamma(m_{ij}) = \bar{f}_{ij}$. Obviously, these elements $\sigma_\Gamma(m_{ij})$ generate $\mathrm{gr}_\Gamma \mathcal{M}$ and hence, by Lemma 3.3.8, the elements m_{ij} form a finite generating set of $\mathcal{M}$. □

Theorem 3.3.11. *Let Σ be a filtration on the ring $\mathcal{P}$. If the associated graded ring $\mathrm{gr}_\Sigma \mathcal{P}$ is left Noetherian, then $\mathcal{P}$ is left Noetherian, too.*

Proof. Let $\mathcal{I}$ be an arbitrary left ideal in $\mathcal{P}$. This implies trivially that $\mathcal{I}$ is a left $\mathcal{P}$-module and Σ induces a filtration Γ on it via $\Gamma_i = \mathcal{I} \cap \Sigma_i$. Furthermore, $\mathrm{gr}_\Gamma \mathcal{I}$ is a left ideal in $\mathrm{gr}_\Sigma \mathcal{P}$. Since by assumption the latter ring is left Noetherian, $\mathrm{gr}_\Gamma \mathcal{I}$ is finitely generated and Γ is a good filtration. By Lemma 3.3.10, the ideal $\mathcal{I}$ is thus finitely generated. □

Polynomial Algebras over Fields

Because of Condition (iii) in Definition 3.2.1 of a polynomial algebra of solvable type, we can define Gröbner bases for ideals in such algebras. If the coefficient ring $\mathcal{R}$ is a (skew) field $\Bbbk$, this is straightforward. In this book we will almost exclusively restrict to this case. A more general situation will be discussed only in Section 4.6.

Definition 3.3.12. Let $(\mathcal{P}, \star, \prec)$ be a polynomial algebra of solvable type over a (skew) field $\Bbbk$ and $\mathcal{I} \subseteq \mathcal{P}$ a left ideal. A finite set $\mathcal{G} \subset \mathcal{P}$ is a *Gröbner basis* of the ideal $\mathcal{I}$ (for the term order $\prec$), if $\langle \mathrm{le}_\prec \mathcal{G} \rangle = \mathrm{le}_\prec \mathcal{I}$.

For the ordinary multiplication this definition reduces to the classical Definition B.4.1 of Gröbner bases. The decisive point, explaining the conditions imposed in Definition 3.2.1, is that *normal forms* with respect to a finite set $\mathcal{F} \subset \mathcal{P}$ can be computed in algebras of solvable type in precisely the same way as in the ordinary polynomial ring. Assume we are given a polynomial $f \in \mathcal{P}$ such that $\mathrm{le}_\prec g \mid \mathrm{le}_\prec f$ for some $g \in \mathcal{G}$ and set $\mu = \mathrm{le}_\prec f - \mathrm{le}_\prec g$. If we consider $g_\mu = x^\mu \star g$, then by Condition (iii) $\mathrm{le}_\prec g_\mu = \mathrm{le}_\prec f$. Setting $d = (\mathrm{lc}_\prec f)(\mathrm{lc}_\prec g_\mu)^{-1}$, we find by Condition (ii) that $\mathrm{le}_\prec(f - dg_\mu) \prec \mathrm{le}_\prec f$. Hence we may use the simple Algorithm B.1 for computing normal forms; in particular, it always terminates by the same argument as in the ordinary case. Note that in general $d \neq (\mathrm{lc}_\prec f)(\mathrm{lc}_\prec g)^{-1}$, if $r \neq 1$ in (3.8), and that normal form computations are typically more expensive due to the appearance of the additional polynomial h in (3.8).

The Gröbner basis theory of Appendix B.4 can be straightforwardly extended to arbitrary polynomial algebras of solvable type over (skew) fields, as most proofs are based on the computation of normal forms. The remaining arguments mostly take place in $\mathbb{N}_0^n$ and thus can be applied without changes. As one typical example we present yet another proof of Hilbert's Basis Theorem that gives us as additional benefit the existence of Gröbner bases.

Theorem 3.3.13. *Let $(\mathcal{P}, \star, \prec)$ be a polynomial algebra of solvable type over a (skew) field $\mathcal{R}$. Then $\mathcal{P}$ is a left Noetherian ring and every left ideal $\mathcal{I} \subseteq \mathcal{P}$ possesses a Gröbner basis with respect to $\prec$.*

Proof. By Dickson's Lemma A.1.2, the monoid ideal $\mathrm{le}_\prec \mathcal{I}$ has a finite basis $\mathcal{N}$. Let $\mathcal{G} \subset \mathcal{I}$ be a finite set with $\mathrm{le}_\prec \mathcal{G} = \mathcal{N}$. By definition, such a set $\mathcal{G}$ is a Gröbner basis of $\mathcal{I}$. The ring $\mathcal{P}$ is left Noetherian, if we can show that $\mathcal{G}$ is also a generating set of the ideal $\mathcal{I}$. This is done by a simple normal form argument.

As we are working over a (skew) field, we may assume without loss of generality that $\mathcal{G}$ is a monic basis, i. e. that $\mathrm{lc}_\prec g = 1$ for all $g \in \mathcal{G}$. Let $f \in \mathcal{I}$ be an arbitrary element of the given ideal with $\mathrm{lm}_\prec f = rx^\mu$. Then the Gröbner basis $\mathcal{G}$ contains a generator g such that $\nu = \mathrm{le}_\prec g$ divides μ. Consider the polynomial $\bar{f} = f - rr_{\mu-\nu,\nu}^{-1} x^{\mu-\nu} \star g$. Obviously, $\mathrm{le}_\prec \bar{f} \prec \mathrm{le}_\prec f$ by construction and, since $\bar{f} \in \mathcal{I}$, we may iterate. As $\prec$ is a well-order, we must reach 0 after a finite number of such reduction steps and hence any element of $\mathcal{I}$ is expressible as a linear combination of the elements in $\mathcal{G}$.

This proves that $\mathcal{I} \subseteq \langle \mathcal{G} \rangle$. Since $\mathcal{G} \subset \mathcal{I}$, it is not possible that $\mathcal{G}$ generates a larger ideal than $\mathcal{I}$ and hence $\mathcal{G}$ is indeed a basis of $\mathcal{I}$. □

Example 3.3.14. Even in the case of a coefficient field we cannot generally expect $\mathcal{P}$ to be a *right* Noetherian ring, as the following extension of Example 3.3.3 demonstrates. We take this time $\mathcal{R} = \Bbbk(t)$, the field of rational functions in t, and consider again $\mathcal{P} = \mathcal{R}[x]$ with $x \star x = x^2$ and $x \star r(t) = r(t^2)x$ for any $r \in \mathcal{R}$. If $r = p/q \in \mathcal{R}$ with two polynomials $p, q \in \Bbbk[t]$, then we set $\deg r = \deg p - \deg q$.

If we write any polynomial f contained in $x \star \mathcal{P}$ in the usual form $f = \sum_{i=1}^n a_i x^i$, then all coefficients a_i are of even degree in t. This simple observation implies that $(tx \star \mathcal{P}) \cap (x \star \mathcal{P}) = 0$. Consider the *right* ideals $I_k = \langle tx, x \star tx, \dots, x^k \star tx \rangle$. We claim that they form an infinite strictly ascending sequence, so that $\mathcal{P}$ is not right Noetherian. Assume to the contrary that there were integers $m < n$ and polynomials $f_i \in \mathcal{P}$ such that $x^m \star tx \star f_m + \cdots + x^n \star tx \star f_n = 0$ with $f_m, f_n \neq 0$. This implies that $tx \star f_m = -(x \star tx \star f_{m+1} + \cdots + x^{n-m} \star tx \star f_n) \in tx \star \mathcal{P} \cap x \star \mathcal{P} = 0$ and thus $f_m = 0$ contradicting our assumptions.

In the proof of Theorem 3.3.13 we essentially use that in normal form computations we multiply with elements of $\mathcal{P}$ from the left. Because of the already mentioned left-right asymmetry of Definition 3.2.1, right ideals show in general a completely different behaviour. In order to obtain right Noetherian rings we must either adapt correspondingly our definition of a solvable algebra or impose additional conditions on the commutation relations (3.6).

The simplest possibility is to require that all the maps ρ_μ in (3.6) are automorphisms (by Proposition 3.2.3 it suffices, if the maps ρ_i in (3.9a) satisfy this condition). In this case we have $\mathbb{k}\star x^i + \mathbb{k} = x^i \star \mathbb{k} + \mathbb{k}$ for all variables x^i implying that we can rewrite any polynomial $f = \sum_\mu c_\mu x^\mu$ in the reverse form $f = \sum_\mu x^\mu \star \tilde{c}_\mu$. Now a straightforward adaption of the above proof shows that $\mathcal{P}$ is also right Noetherian. Obviously, our example does not satisfy this condition, as the map $r(t) \mapsto r(t^2)$ is of course not an automorphism. $\lhd$

We stop here and do not give more details on extending the Gröbner basis theory to polynomial algebras of solvable type, as we will use in the next section a completely different approach leading to involutive bases.

3.4 Involutive Bases

We proceed to define involutive bases for left ideals in polynomial algebras of solvable type. Our approach closely follows the introduction of Gröbner bases in Appendix B.4. We could at once consider submodules of free modules over such an algebra. But this only complicates the notation. So we start with the ideal case and the extension to submodules goes as for Gröbner bases.

Definition 3.4.1. Let $(\mathcal{P},\star,\prec)$ be a polynomial algebra of solvable type over a coefficient field $\mathbb{k}$ and $\mathcal{I} \subseteq \mathcal{P}$ a non-zero left ideal. A finite set $\mathcal{H} \subset \mathcal{I}$ is a *weak involutive basis* of $\mathcal{I}$ for an involutive division L on $\mathbb{N}_0^n$, if $\mathrm{le}_\prec \mathcal{H}$ is a weak involutive basis of the monoid ideal $\mathrm{le}_\prec \mathcal{I}$. The set $\mathcal{H}$ is a *(strong) involutive basis* of $\mathcal{I}$, if $\mathrm{le}_\prec \mathcal{H}$ is a strong involutive basis of $\mathrm{le}_\prec \mathcal{I}$ and two distinct elements of $\mathcal{H}$ never possess the same leading exponents.

This definition represents a natural extension of our Definition 3.3.12 of a Gröbner basis in $\mathcal{P}$. In particular, it implies immediately that any weak involutive basis is a Gröbner basis. As in Section 3.1, we call any finite set $\mathcal{F} \subset \mathcal{P}$ *(weakly) involutive*, if it is a (weak) involutive basis of the ideal $\langle \mathcal{F} \rangle$ generated by it.

Definition 3.4.2. Let $\mathcal{F} \subset \mathcal{P} \setminus \{0\}$ be a finite set of polynomials and L an involutive division on $\mathbb{N}_0^n$. We assign to each element $f \in \mathcal{F}$ a set of *multiplicative variables*

$$X_{L,\mathcal{F},\prec}(f) = \left\{x^i \mid i \in N_{L,\mathrm{le}_\prec \mathcal{F}}(\mathrm{le}_\prec f)\right\} . \tag{3.30}$$

The *involutive span* of $\mathcal{F}$ is then the set

$$\langle \mathcal{F} \rangle_{L,\prec} = \sum_{f \in \mathcal{F}} \Bbbk\big[X_{L,\mathcal{F},\prec}(f)\big] \star f \subseteq \langle \mathcal{F} \rangle \,. \tag{3.31}$$

Remark 3.4.3. The involutive span of a set $\mathcal{F}$ depends decisively on both the involutive division L and the term order $\prec$. Choosing a different term order will generally lead to different leading terms in $\mathrm{le}_\prec \mathcal{F}$ and thus we may obtain very different multiplicative variables for each generator $f \in \mathcal{F}$. In particular, the total number of multiplicative variables for $\mathcal{F}$, which we may consider as a rough indicator of the "size" of the involutive span $\langle \mathcal{F} \rangle_{L,\prec}$, may change drastically.

For example, the Pommaret division has a special relationship to class respecting term orders (recall that according to Lemma A.1.8 any class respecting order coincides on terms of the same degree with the degree reverse lexicographic order). For homogeneous polynomials such orders always lead to maximal sets of multiplicative indices and thus generally to smaller bases. We will see in Chapter 5 that also from a theoretical point of view Pommaret bases with respect to such an order are particularly useful. ◁

An important aspect of Gröbner bases is the existence of standard representations for any ideal element (Proposition B.4.8). For (weak) involutive bases a similar characterisation exists and in the case of strong bases we even obtain *unique* representations, a crucial improvement to the classical theory.

Theorem 3.4.4. *Let $\mathcal{I} \subseteq \mathcal{P}$ be a non-zero ideal, $\mathcal{H} \subset \mathcal{I} \setminus \{0\}$ a finite set and L an involutive division on $\mathbb{N}_0^n$. Then the following two statements are equivalent.*

(i) *The set $\mathcal{H} \subset \mathcal{I}$ is a weak involutive basis of $\mathcal{I}$ with respect to L and $\prec$.*
(ii) *Every polynomial $f \in \mathcal{I}$ can be written in the form*

$$f = \sum_{h \in \mathcal{H}} P_h \star h \tag{3.32}$$

with coefficients $P_h \in \Bbbk[X_{L,\mathcal{H},\prec}(h)]$ satisfying $\mathrm{lt}_\prec(P_h \star h) \preceq \mathrm{lt}_\prec f$ for all polynomials $h \in \mathcal{H}$ such that $P_h \neq 0$.

$\mathcal{H}$ is a strong involutive basis, if and only if the representation (3.32) *is unique.*

Proof. Let us first assume that $\mathcal{H}$ is a weak involutive basis and consider an arbitrary polynomial $f \in \mathcal{I}$. According to Definition 3.4.1, its leading exponent $\mathrm{le}_\prec f$ lies in the involutive cone $\mathcal{C}_{L,\mathrm{le}_\prec \mathcal{H}}(\mathrm{le}_\prec h)$ of the leading exponent of at least one generator $h \in \mathcal{H}$. Let $\mu = \mathrm{le}_\prec f - \mathrm{le}_\prec h$ and set $f_1 = f - cx^\mu \star h$ where the coefficient $c \in \Bbbk$ is chosen such that the leading terms cancel. Obviously, $f_1 \in \mathcal{I}$ and $\mathrm{lt}_\prec f_1 \prec \mathrm{lt}_\prec f$. Iteration yields a sequence of polynomials $f_i \in \mathcal{I}$. After a finite number of steps we must reach $f_N = 0$, as the leading terms are always decreasing and by assumption the leading exponent of *any* polynomial in $\mathcal{I}$ possesses an involutive divisor in $\mathrm{le}_\prec \mathcal{H}$. But this observation implies the existence of a representation of the form (3.32).

Now assume that $\mathcal{H}$ is even a strong involutive basis and take an involutive standard representation (3.32). By definition of a strong basis, there exists one and only one generator $h \in \mathcal{H}$ such that $\mathrm{lt}_\prec(P_h \star h) = \mathrm{lt}_\prec f$. This fact determines uniquely

$\mathrm{lt}_\prec P_h$. Applying the same argument to $f - (\mathrm{lt}_\prec P_h) \star h$ shows by recursion that the representation (3.32) is indeed unique.

For the converse note that (ii) trivially implies that $\mathrm{le}_\prec f \in \langle \mathrm{le}_\prec \mathcal{H} \rangle_{L,\prec}$ for any polynomial $f \in \mathcal{I}$. Thus $\mathrm{le}_\prec \mathcal{I} \subseteq \langle \mathrm{le}_\prec \mathcal{H} \rangle_{L,\prec}$. As the converse inclusion is obvious, we have in fact an equality and $\mathcal{H}$ is a weak involutive basis.

Now let us assume that the set $\mathcal{H}$ is only a weak but not a strong involutive basis of $\mathcal{I}$. This implies the existence of two generators $h_1, h_2 \in \mathcal{H}$ such that $\mathcal{C}_{L,\mathrm{le}_\prec \mathcal{H}}(\mathrm{le}_\prec h_2) \subseteq \mathcal{C}_{L,\mathrm{le}_\prec \mathcal{H}}(\mathrm{le}_\prec h_1)$. Hence we have $\mathrm{lm}_\prec h_2 = \mathrm{lm}_\prec (cx^\mu \star h_1)$ for suitably chosen $c \in \Bbbk$ and $\mu \in \mathbb{N}_0^n$. Consider the polynomial $h_2 - cx^\mu \star h_1 \in \mathcal{I}$. If it vanishes, we have found a non-trivial involutive standard representation of 0. Otherwise an involutive standard representation $h_2 - cx^\mu \star h_1 = \sum_{h\in\mathcal{H}} P_h \star h$ with $P_h \in \Bbbk[X_{L,\mathcal{H},\prec}(h)]$ exists. Setting now $P'_h = P_h$ for all generators $h \neq h_1, h_2$ and $P'_{h_1} = P_{h_1} + cx^\mu$, $P'_{h_2} = P_{h_2} - 1$ yields again a non-trivial involutive standard representation $0 = \sum_{h\in\mathcal{H}} P'_h \star h$. The existence of such a non-trivial representation of 0 immediately implies that (3.32) cannot be unique. Thus only for a strong involutive basis the involutive standard representation is always unique. □

Corollary 3.4.5. *Let $\mathcal{H}$ be a weak involutive basis of the left ideal $\mathcal{I} \subseteq \mathcal{P}$. Then $\langle \mathcal{H} \rangle_{L,\prec} = \mathcal{I}$. If $\mathcal{H}$ is even a strong involutive basis of $\mathcal{I}$, then $\mathcal{I}$ considered as a $\Bbbk$-linear space possesses a direct sum decomposition $\mathcal{I} = \bigoplus_{h\in\mathcal{H}} \Bbbk[X_{L,\mathcal{H},\prec}(h)] \star h$.*

Proof. It follows immediately from Theorem 3.4.4 that $\mathcal{I} \subseteq \langle \mathcal{H} \rangle_{L,\prec}$. But as $\mathcal{H}$ is also a Gröbner basis of $\mathcal{I}$, we have in fact equality. The direct sum decomposition for a strong involutive basis is a trivial consequence of the uniqueness of the involutive standard representation in this case. □

Example 3.4.6. It is *not* true that any finite set $\mathcal{F}$ with $\langle \mathcal{F} \rangle_{L,\prec} = \mathcal{I}$ is a weak involutive basis of the ideal $\mathcal{I}$. Consider in the ordinary polynomial ring $\Bbbk[x,y]$ the ideal $\mathcal{I}$ generated by the two polynomials $f_1 = y^2$ and $f_2 = y^2 + x^2$. If we order the variables as $x^1 = x$ and $x^2 = y$, then the set $\mathcal{F} = \{f_1, f_2\}$ trivially satisfies $\langle \mathcal{F} \rangle_{J,\prec} = \mathcal{I}$, as with respect to the Janet division all variables are multiplicative for each generator. However, $\mathrm{le}_\prec \mathcal{F} = \{[0,2]\}$ does *not* generate $\mathrm{le}_\prec \mathcal{I}$, as obviously $[2,0] \in \mathrm{le}_\prec \mathcal{I} \setminus \langle \{[0,2]\} \rangle$. Thus $\mathcal{F}$ is not a weak Janet basis (neither is the autoreduced set $\mathcal{F}' = \{y^2, x^2\}$, as $x^2 y \notin \langle \mathcal{F}' \rangle_{J,\prec}$). ◁

By a generalisation of Proposition 3.1.12, any weak involutive basis $\mathcal{H}$ contains a strong involutive basis as subset. We will see later why it is nevertheless useful to introduce weak bases for some applications.

Proposition 3.4.7. *Let $\mathcal{I} \subseteq \mathcal{P}$ be an ideal and $\mathcal{H} \subset \mathcal{I}$ a weak involutive basis of it for the involutive division L. Then there exists a subset $\mathcal{H}' \subseteq \mathcal{H}$ which is a strong involutive basis of $\mathcal{I}$.*

Proof. If the set $\mathrm{le}_\prec \mathcal{H}$ is already a strong involutive basis of $\mathrm{le}_\prec \mathcal{I}$, we are done. Otherwise $\mathcal{H}$ contains polynomials h_1, h_2 such that $\mathrm{le}_\prec h_1 \mid_{L,\mathrm{le}_\prec \mathcal{H}} \mathrm{le}_\prec h_2$. Consider the subset $\mathcal{H}' = \mathcal{H} \setminus \{h_2\}$. As in the proof of Proposition 3.1.12, one easily shows that $\mathrm{le}_\prec \mathcal{H}' = \mathrm{le}_\prec \mathcal{H} \setminus \{\mathrm{le}_\prec h_2\}$ is still a weak involutive basis of $\mathrm{le}_\prec \mathcal{I}$ and thus $\mathcal{H}'$

is still a weak involutive basis of $\mathcal{I}$. After a finite number of such eliminations we must reach a strong involutive basis. □

In order to proceed, we must adapt some further notions familiar from the theory of Gröbner bases to involutive bases (see Appendix B.4 for the classical versions).

Definition 3.4.8. Let $\mathcal{F} \subset \mathcal{P}$ be a finite set and L an involutive division. A polynomial $g \in \mathcal{P}$ is *involutively reducible* with respect to $\mathcal{F}$, if it contains a term $x^\mu \in \operatorname{supp} g$ such that $\operatorname{le}_\prec f \,|_{L,\operatorname{le}_\prec \mathcal{F}}\, \mu$ for some $f \in \mathcal{F}$. The polynomial g is in *involutive normal form* with respect to $\mathcal{F}$, if it is not involutively reducible. The set $\mathcal{F}$ is *involutively autoreduced*, if no polynomial $f \in \mathcal{F}$ contains a term $x^\mu \in \operatorname{supp} f$ such that another polynomial $f' \in \mathcal{F} \setminus \{f\}$ exists with $\operatorname{le}_\prec f' \,|_{L,\operatorname{le}_\prec \mathcal{F}}\, \mu$.

Remark 3.4.9. The definition of an involutively autoreduced set *cannot* be formulated more concisely by saying that each $f \in \mathcal{F}$ is in involutive normal form with respect to $\mathcal{F} \setminus \{f\}$. If we are not dealing with a global division, the removal of f from $\mathcal{F}$ will generally change the assignment of the multiplicative indices and thus affect the involutive divisibility. ◁

An *obstruction to involution* is a polynomial $g \in \langle \mathcal{F} \rangle \setminus \langle \mathcal{F} \rangle_{L,\prec}$ possessing an ordinary standard representation (B.73) with respect to $\mathcal{F}$ but not an involutive one; i. e. we can write $g = \sum_{f \in \mathcal{F}} P_f \star f$ with coefficients $P_f \in \mathcal{P}$ satisfying $\operatorname{le}_\prec (P_f \star f) \preceq \operatorname{le}_\prec g$ whenever $P_f \neq 0$ but at least one coefficient P_f depends on a variable $x^j \in \overline{X}_{L,\mathcal{F},\prec}(f)$ which is non-multiplicative for the associated generator $f \in \mathcal{F}$. We will later see that these elements make the difference between an involutive and a Gröbner basis.

Remark 3.4.10. If $\mathcal{P} = \mathbb{F}[\partial_1, \dots, \partial_n]$ is actually a ring of linear differential operators over a differential field $\mathbb{F}$ and $\prec$ a degree compatible term order, then the question whether or not an element $f \in \langle \mathcal{F} \rangle$ possesses a standard representation is related to the existence of integrability conditions. Indeed, if in any representation of f in the form of (3.32) the leading terms on the right hand side are of higher degree (order) than f, then we are dealing with an integrability condition in the sense of the informal definition used in Section 2.3 (in the form of a generalised cross-derivative). Note, however, that $\operatorname{lt}_\prec (P_h \star h) \succ \operatorname{lt}_\prec f$ is possible even if both terms are of the same degree. Thus while a Gröbner basis of $\langle \mathcal{F} \rangle$ always induces a formally integrable system,[2] the converse is not necessarily true. ◁

Example 3.4.11. Consider the set $\mathcal{F} = \{f_1, f_2, f_3\} \subset \Bbbk[x,y,z]$ with the polynomials $f_1 = z^2 - xy$, $f_2 = yz - x$ and $f_3 = y^2 - z$. For any degree compatible term order, the leading terms of f_2 and f_3 are unique. For f_1 we have two possibilities: if we use the degree lexicographic order (i. e. for $x \prec y \prec z$), it is z^2, for the degree inverse lexicographic order (i. e. for $x \succ y \succ z$) the leading term is xy. In the first case, we

[2] Generally, a Gröbner basis becomes a formally integrable system in the sense of Section 2.3 only after all integrability conditions of the second kind have been taken into account, too. But this is easily done: if q is the maximal degree of a generator in $\mathcal{F}$, then we must augment $\mathcal{F}$ by all operators of the form $\partial^\mu f$ with $f \in \mathcal{F}$ and $|\mu| + \deg f = q$ where ∂^μ contains only derivatives multiplicative for f.

find for the Janet division that $\langle\mathcal{F}\rangle_{J,\prec_{\text{deglex}}} = \langle\mathcal{F}\rangle$, so that for this term order $\mathcal{F}$ is a Janet basis, i. e. an involutive basis with respect to the Janet division, although we do not have yet the necessary tools to prove this fact. Thus, upon the identification $x \to \partial_x$ etc, the system $\mathcal{F}$ defines a formally integrable differential equation.

In the second case, $f_4 = z^3 - x^2 = zf_1 + xf_2$ does not possess a standard representation although it is not an "integrability condition." Adding this polynomial to $\mathcal{F}$ yields a Gröbner basis $\mathcal{G}$ of $\langle\mathcal{F}\rangle$, as one can easily check using Theorem B.4.14. But now $f_5 = zf_2$ is an obstruction to involution of $\mathcal{G}$, as it is not involutively reducible with respect to the Janet division. Thus $\mathcal{G}$ is not yet a Janet basis, but one can show that adding f_5 yields one. ◁

Remark 3.4.12. If $\mathcal{G}$ is a Gröbner basis of the ideal $\mathcal{I}$, then any element of $\mathcal{I}$ has a standard representation by Proposition B.4.8. But this does not imply that for a given division L the ideal $\mathcal{I}$ is free of obstructions to involution. In order to obtain at least a weak involutive basis, we must add further elements of $\mathcal{I}$ to $\mathcal{G}$ until $\langle \text{le}_\prec \mathcal{G}\rangle_L = \text{le}_\prec \mathcal{I}$. Obviously, this observation allows us to reduce the construction of a polynomial involutive basis to a Gröbner basis computation plus a monomial completion. But we will see in Section 4.2 that better possibilities exist.

It follows immediately from these considerations that in general involutive bases are not reduced Gröbner bases, as we already observed in Example 3.4.11. For the term order $\prec_{\text{deglex}}$ the set $\mathcal{F}$ was simultaneously a Janet basis and a reduced Gröbner basis. But for the order $\prec_{\text{deginvlex}}$ the reduced Gröbner basis is $\mathcal{F}\cup\{f_4\}$, whereas a Janet basis requires in addition the polynomial f_5. We will see later in Chapter 5 that this "redundancy" in involutive bases is the key for their use in the structure analysis of polynomial ideals and modules. ◁

It often suffices, if one does not consider all terms in $\operatorname{supp} g$ but only the leading term $\text{lt}_\prec g$: the polynomial g is *involutively head reducible*, if $\text{le}_\prec f \,|_{L,\text{le}_\prec \mathcal{F}}\, \text{le}_\prec g$ for some $f \in \mathcal{F}$. Similarly, the set $\mathcal{F}$ is *involutively head autoreduced*, if no leading exponent of an element $f \in \mathcal{F}$ is involutively divisible by the leading exponent of another element $f' \in \mathcal{F}\setminus\{f\}$. Note that the definition of a strong involutive basis immediately implies that it is involutively head autoreduced.

Example 3.4.13. In the case of the Janet division, a set $\mathcal{F}$ is involutively head autoreduced as soon as all elements have different leading terms. Indeed, we have seen in the proof of Lemma 3.1.5 that with respect to the Janet division two involutive cones only intersect, if they are identical. ◁

As involutive reducibility is a restriction of ordinary reducibility, involutive normal forms can be determined with trivial adaptions of the familiar algorithms like for example Algorithm B.1. The termination follows by the same argument as usual, namely that $\prec$ is a well-order. If g' is an involutive normal form of the polynomial $g \in \mathcal{P}$ with respect to the set $\mathcal{F}$ for the division L, then we write $g' = \text{NF}_{\mathcal{F},L,\prec}(g)$, although involutive normal forms are in general not unique (like ordinary normal forms—see Example B.4.6). Depending on the order in which reductions are applied different results are generally obtained.

The ordinary normal form is unique, if and only if it is computed with respect to a Gröbner basis (Proposition B.4.8). In fact, this property is often used to define Gröbner bases. For the involutive normal form, the situation is somewhat different.

Lemma 3.4.14. *The sum in the definition* (3.31) *of the involutive span* $\langle\mathcal{F}\rangle_{L,\prec}$ *is direct, if and only if the finite set* $\mathcal{F}\subset\mathcal{P}\setminus\{0\}$ *is involutively head autoreduced with respect to the involutive division L.*

Proof. One direction is obvious. For the converse, let f_1, f_2 be two distinct elements of $\mathcal{F}$ and $X_i = X_{L,\mathcal{F},\prec}(f_i)$ their respective sets of multiplicative variables for L. Assume that two polynomials $P_i \in \Bbbk[X_i]$ exist with $P_1 \star f_1 = P_2 \star f_2$ and hence $\mathrm{le}_\prec(P_1 \star f_1) = \mathrm{le}_\prec(P_2 \star f_2)$. As the multiplication $\star$ respects the order $\prec$, this fact implies that $\mathcal{C}_{L,\mathrm{le}_\prec \mathcal{F}}(\mathrm{le}_\prec f_1) \cap \mathcal{C}_{L,\mathrm{le}_\prec \mathcal{F}}(\mathrm{le}_\prec f_2) \neq \emptyset$. Thus one of the involutive cones is completely contained in the other one and either $\mathrm{le}_\prec f_1 \,|_{L,\mathrm{le}_\prec \mathcal{F}}\, \mathrm{le}_\prec f_2$ or $\mathrm{le}_\prec f_2 \,|_{L,\mathrm{le}_\prec \mathcal{F}}\, \mathrm{le}_\prec f_1$ contradicting that $\mathcal{F}$ is involutively head autoreduced. □

Proposition 3.4.15. *If the finite set* $\mathcal{F}\subset\mathcal{P}\setminus\{0\}$ *is involutively head autoreduced, each polynomial* $g\in\mathcal{P}$ *has a unique involutive normal form* $\mathrm{NF}_{\mathcal{F},L,\prec}(g)$.

Proof. If 0 is an involutive normal form of g, then obviously $g \in \langle\mathcal{F}\rangle_{L,\prec}$. Conversely, assume that $g \in \langle\mathcal{F}\rangle_{L,\prec}$, i. e. the polynomial g can be written in the form $g = \sum_{f\in\mathcal{F}} P_f \star f$ with $P_f \in \Bbbk[X_{L,\mathcal{F},\prec}(f)]$. As $\mathcal{F}$ is involutively head autoreduced, the leading terms of the summands can never cancel (see the proof of Lemma 3.4.14). Thus $\mathrm{lt}_\prec g = \mathrm{lt}_\prec(P_f \star f)$ for some $f\in\mathcal{F}$ and any polynomial $g \in \langle\mathcal{F}\rangle_{L,\prec}$ is involutively head reducible with respect to $\mathcal{F}$. Each reduction step in an involutive normal form algorithm leads to a new polynomial $g' \in \langle\mathcal{F}\rangle_{L,\prec}$ with $\mathrm{lt}_\prec g' \preceq \mathrm{lt}_\prec g$. If the leading term is reduced, we even get $\mathrm{lt}_\prec g' \prec \mathrm{lt}_\prec g$. As each terminating normal form algorithm must sooner or later reduce the leading term, we eventually obtain 0 as unique involutive normal form of any $g \in \langle\mathcal{F}\rangle_{L,\prec}$.

Let g_1 and g_2 be two involutive normal forms of the polynomial g. Obviously, $g_1 - g_2 \in \langle\mathcal{F}\rangle_{L,\prec}$. By definition of a normal form both $\mathrm{supp}\, g_1$ and $\mathrm{supp}\, g_2$ do not contain any term involutively reducible with respect to $\mathcal{F}$ and the same holds for $\mathrm{supp}\,(g_1 - g_2) \subseteq \mathrm{supp}\, g_1 \cup \mathrm{supp}\, g_2$. Hence the difference $g_1 - g_2$ is also in involutive normal form and by our considerations above we must have $g_1 - g_2 = 0$. □

Proposition 3.4.16. *The ordinary and the involutive normal form of any polynomial* $g\in\mathcal{P}$ *with respect to a weakly involutive set* $\mathcal{F}\subset\mathcal{P}\setminus\{0\}$ *are identical.*

Proof. Recalling the proof of the previous proposition, we see that we used the assumption that $\mathcal{F}$ was involutively head autoreduced only for proving the existence of a generator $f\in\mathcal{F}$ such that $\mathrm{lt}_\prec f \,|_{L,\mathrm{le}_\prec \mathcal{F}}\, \mathrm{lt}_\prec g$ for every polynomial $g \in \langle\mathcal{F}\rangle_{L,\prec}$. But it follows immediately from Theorem 3.4.4 that this property also holds for any weak involutive basis. Thus, by the same argument as above, we may conclude that the involutive normal form with respect to a weakly involutive set is unique. For Gröbner bases the uniqueness of the ordinary normal form is a classical property and any weak involutive basis is also a Gröbner basis. As a polynomial in ordinary normal form with respect to $\mathcal{F}$ is trivially in involutive normal form with respect to $\mathcal{F}$, too, the two normal forms must coincide. □

Finally, we extend the notion of a minimal involutive basis from $\mathbb{N}_0^n$ to $\mathcal{P}$. This is done in close analogy to Definition B.4.11 of a minimal Gröbner basis.

Definition 3.4.17. Let $\mathcal{I} \subseteq \mathcal{P}$ be a non-zero ideal and L an involutive division. An involutive basis $\mathcal{H}$ of $\mathcal{I}$ with respect to L is *minimal*, if $\operatorname{le}_\prec \mathcal{H}$ is the minimal involutive basis of the monoid ideal $\operatorname{le}_\prec \mathcal{I}$ for the division L.

By Proposition 3.1.21, any involutive basis for a globally defined division like the Pommaret division is minimal. Uniqueness requires two further assumptions. First of all, we obviously need a full involutive autoreduction instead of only a head autoreduction. Secondly, we must normalise the leading coefficients to one, i. e. we must take a *monic* basis.

Proposition 3.4.18. *Let $\mathcal{I} \subseteq \mathcal{P}$ be a non-zero ideal and L an involutive division. Then $\mathcal{I}$ possesses at most one monic, involutively autoreduced, minimal involutive basis for the division L.*

Proof. Assume that $\mathcal{H}_1$ and $\mathcal{H}_2$ are two different monic, involutively autoreduced, minimal involutive bases of $\mathcal{I}$ with respect to L and $\prec$. By definition of a minimal involutive bases, this implies that $\operatorname{lt}_\prec \mathcal{H}_1 = \operatorname{lt}_\prec \mathcal{H}_2$. As $\mathcal{H}_1$ and $\mathcal{H}_2$ are not identical, we must have two polynomials $h_1 \in \mathcal{H}_1$ and $h_2 \in \mathcal{H}_2$ such that $\operatorname{lt}_\prec h_1 = \operatorname{lt}_\prec h_2$ but $h_1 \neq h_2$. Now consider the polynomial $h = h_1 - h_2 \in \mathcal{I}$. Its leading exponent must lie in the involutive span of the set $\operatorname{le}_\prec \mathcal{H}_1 = \operatorname{le}_\prec \mathcal{H}_2$. On the other hand, the term $\operatorname{lt}_\prec h$ must be contained in either $\operatorname{supp} h_1$ or $\operatorname{supp} h_2$. But this observation implies that either $\mathcal{H}_1$ or $\mathcal{H}_2$ is not involutively autoreduced. □

3.5 Notes

Involutive divisions were introduced by Gerdt and Blinkov [156] generalising ideas used already by Janet [235] (extending earlier works by Riquier [381]) in the context of differential equations. All divisions introduced so far are also much older than the theory of involutive bases; their modern names were coined by Gerdt and Blinkov [156]. One should note a difference in conventions: in the case of a division depending on the ordering of the variables, we always analyse multi indices from the right, i. e. we start with the last entry μ_n and not with the first entry μ_1. Thus for the Janet division we first check whether n is multiplicative and then continue with the lower values;[3] for the Pommaret division multi indices of the form $[0,\dots,0,\mu_n]$ possess the most multiplicative indices. In [156], it is the other way round: they start the analysis of a multi index with μ_1 and not μ_n; for the Janet division they first check whether 1 is multiplicative and for the Pommaret division multi indices $[\mu_1,0,\dots,0]$ possess the most multiplicative indices. Obviously, the two conventions become identical, if one simply inverts the ordering of the variables.

[3] In fact, this is the way the division was introduced by Janet [235, pp. 16–17].

The terminology *Pommaret division* is historically not correct, as this division was already introduced by Janet. He used it mainly in [234, pp. 30–31], probably influenced by Cartan's work on exterior differential systems, but also in the earlier article [233, p. 30]. As we will see in Section 4.3, both divisions are in fact more closely related than their apparently very different definitions indicate. The Thomas division appeared first in [454].

A weaker notion of an involutive division was introduced by Apel [20]. One could say that he uses a "local" approach whereas ours is "global." He defines an involutive division only with respect to a *single* set $\mathcal{N} \subseteq \mathbb{N}_0^n$ and hence does not need the second condition in Definition 3.1.1, the filter axiom relating the set of multiplicative indices for subsets. We will discuss the effect of this modification in more detail in the Notes of Chapter 4 in the context of completion algorithms.

Janet trees were introduced by Gerdt et al [159], though not in the form presented here.[4] We use the natural tree defined by the inclusion relations between the subsets $(d_k,\dots,d_n)$ and thus obtain a tree with fixed depth $n+1$ where each node may possess several children. As already mentioned, as simple concrete realisation would consist of using a list of pointers in each node. The trees of Gerdt et al [159] may be interpreted as a special realisation of our trees. Instead of using such a list, each node is resolved into a binary tree.

This choice does not lead automatically to a particularly good (nor to a particularly bad) realisation. From a conceptual point of view, however, it is much better to use our simpler trees. One has a clear separation between the underlying ideas and details of implementation (compare the simple nature of our algorithms with the algorithms presented in [159]). One also has more freedom in the implementation where one may exploit special features of the used programming environment.[5]

The notion of an involutive basis is due to Gerdt and Blinkov [156]; the special case of a Pommaret basis was earlier introduced by Zharkov and Blinkov [489]. As already mentioned, such bases are implicitly contained in many articles, even in rather old ones. The main credit of these three authors is that, inspired by the work of Janet, they provided a rigorous algebraic formulation in terms of the modern theory of Gröbner bases. In particular, they introduced involutive reductions and normal forms, two important concepts missing in the classical Janet–Riquier Theory.

It should be pointed out that already several years before these works Wu [483] showed how ideas from the Janet–Riquier theory may be exploited for the construction of a special type of Gröbner bases. In our terminology, his "well-behaved bases" are *Thomas bases*, i. e. involutive bases with respect to the Thomas division. As already mentioned, the Janet division may be considered as a refinement or optimisation of the Thomas division, as the involutive cones of the latter one are always contained in those of the former one. Hence for computational purposes the Thomas division is of no interest, but it is sometimes useful in theoretical considerations.

---

[4] Blinkov [47] discusses similar tree structures also for other divisions.

[5] For example *MuPAD* has no pointer types but a list type that behaves like an array with access to the elements in constant time independent of the length of the list. In such a language the use of a binary trees offers no advantages but will actually lead to a higher complexity.

Pommaret bases (for ideals in power series rings) appeared implicitly in the famous work of Hironaka [218] on the resolution of singularities. Much later, Amasaki [14, 15] followed up this idea and explicitly introduced them (for polynomial ideals) under the name *Weierstraß bases* because of their connection to the Weierstraß Preparation Theorem. In his study of their properties, Amasaki obtained results similar to the ones we will present in Chapter 5, however in a fairly different way.

Our Definition 3.4.1 of an involutive basis is not the same as the one given by Gerdt and Blinkov [156]. First of all, they considered only strong bases; the notation of a weak basis was introduced in [409]. Secondly, their approach is based on the involutive normal form (thus mimicking the classical definition of Gröbner bases by Buchberger) and their definition requires always a full involutive autoreduction of the basis. Ignoring such minor differences, the basic equivalence of their and our approach is ensured by Corollary 3.4.5.

One may wonder why we have introduced weak involutive bases. Only strong involutive bases possess the interesting additional combinatorial properties that will be discussed in Chapter 5. One (rather weak) answer is that they arise naturally in our approach to the definition of involutive bases and that is sometimes convenient in proofs to show first that some basis is weakly involutive and then afterwards that it is in fact strongly involutive.

The real answer will become apparent in Sections 4.5 and 4.6 where we will discuss two generalisations of the theory, namely the extension to semigroup orders and solvable algebras over rings. In both cases, generally no strong involutive bases exist but it is still possible to construct weak bases and thus Gröbner bases. From a theoretical point of view that makes involutive bases somewhat uninteresting in such situations. However, the completion algorithms of Chapter 4 may still be used as an alternative to the classical Buchberger algorithm for the construction of Gröbner bases. As discussed there in more detail, from a computational point of view this new algorithm is highly attractive, as it is often faster.

The term "algebra of solvable type" was coined by Kandry-Rody and Weispfenning [249], when they studied Gröbner bases for non-commutative rings. Their definition is more restrictive than ours, as it does not allow that the terms operate on the coefficients and requires a stronger form of compatibility between the multiplication $\star$ and the term order $\prec$. It automatically implies that $\star$ respects $\prec$. For our purposes this is the decisive property and thus we have used it for Definition 3.2.1 instead of the more technical axioms in [249].

Kredel [264] generalised the work of Kandry-Rody and Weispfenning [249] and considered essentially the same class of algebras as here. Various variants of it have appeared under different names in the literature. Popular is in particular the approach to consider the algebras as the quotient of a free tensor algebra by an appropriate quadratic ideal [18, 286]; one speaks then of *G-algebras*. In most cases the authors restrict to the case of a (skew) coefficient field and do not allow that the variables operate on the coefficients. The corresponding theory of Gröbner bases has been treated at many places in the literature; besides the already cited works we mention in particular [61, 62, 63, 165] where the terminology *PBW algebra* is used (PBW is here an acronym for Poincaré–Birkhoff–Witt, as in the case of universal enveloping algebras the theorem with this name ensures that we may consider the algebra as a polynomial ring). Somewhat surprisingly, a complexity theory of Gröbner bases in polynomial algebras of solvable type was developed only very recently: Aschenbrenner and Leykin [29] establish degree bounds similar to the commutative case. Solvable algebras over rings are less frequently studied; our main source was here the thesis of Kredel [264].

The main motivation for our use of solvable algebras was simply that they provide a very natural framework for discussing Gröbner or involutive bases and allow us to give a unified treatment of polynomials and linear differential operators: the key Condition (iii) in Definition 3.2.1 ensures that we can work with the leading ideal just as in the commutative case and then apply simple normal form arguments. Nevertheless, the examples given in Section 3.2 demonstrate that this class of algebras is also of great practical interest. Many further examples and references can be found in the recent thesis of Levandovskyy [286].

From the point of view of the theory of non-commutative rings polynomial algebras of solvable type are fairly close to the commutative theory. There exists a theory of Gröbner bases also for more general types of non-commutative rings, see for example [40, 180, 290, 329, 330, 465]. An extension to involutive bases was recently presented by Evans [129]. However, here it is much less clear how one should define an involutive division and what are useful divisions. The divisions used in [129] are rather different from the ones appearing in the commutative theory.

Our proof of Hilbert's Basis Theorem for iterated polynomial algebras of solvable type (Theorem 3.3.2) is a straightforward adaption of a similar proof in [319, Theorem 1.2.9] for Ore algebras where the conjugate σ is an automorphism. The only novel aspect consists of the third condition in Definition 3.3.1 taking care of the coefficients potentially appearing when multiplying two powers of the same variables (these coefficients are always 1 in the case of an Ore algebra). The basic underlying idea is rather simple but even in the case of the ordinary commutative polynomial ring it was discovered only comparatively recently by Sarges [392] (it has since become the standard approach to demonstrating the basis theorem in textbooks on commutative algebra). The extension of this proof to solvable algebra with centred commutation relations (Theorem 3.3.7) is due to [264, Sect. 3.5] ("algebras satisfying the extended axioms" in his terminology).

The treatment of filtered algebras via the associated graded algebra (Theorem 3.3.11) is taken from [45, Chapt. I, §2]. There only the special case of the Weyl algebra is treated, but the adaption to more general solvable algebras is trivial (see also [319, Theorem 1.6.9]). A more comprehensive discussion of this technique in the context of Gröbner bases is contained in the book of Li [290].

The counterexamples of non-Noetherian rings stem from [319, Example 1.2.11]. [264, Sect. 3.5] considered also the question when a polynomial algebra $\mathcal{P}$ of solvable type is right Noetherian. An essential condition is here that the maps ρ_μ appearing in the commutation relations (3.6a) are automorphisms, i.e. invertible. In the case that $\mathcal{P}$ has centred commutation relations, this additional assumption suffices already to show (with minor additions to our proof of Theorem 3.3.7) that the algebra $\mathcal{P}$ is both left and right Noetherian.

The basic idea of the proof in the case of a coefficient field (Theorem 3.3.13) goes back to Gordan [172]; it was his answer to Hilbert's original non-constructive proof in his two famous papers on invariant theory [213, 214] which also contained the proofs of the Nullstellensatz, of the rationality of the Hilbert series and of the Syzygy Theorem. As additional benefit, we obtain for free the probably simplest proof of the existence of Gröbner bases for arbitrary (left) ideals in $\mathcal{P}$. The original proof via the termination of the Buchberger Algorithm B.3 (cf. Theorem B.4.15) is much longer (and much more technical), as it requires first a proof of the Buchberger criterion for a Gröbner basis (Theorem B.4.14).

Chapter 4
Completion to Involution

I wish to God these calculations had been executed by steam.
Charles Babbage

In the previous chapter we only defined the notion of an involutive basis but we did not consider the question of the existence of such bases. Recall from Section 3.3 that (in the case of a coefficient field) the existence proof for Gröbner bases is straightforward. For involutive bases the situation is considerably more complicated. Indeed, we have already seen in Example 3.1.16 an (even monomial) ideal not possessing a finite Pommaret basis. Thus we surely cannot expect that an arbitrary polynomial ideal has for every involutive division a finite involutive basis.

In Section 4.1 we introduce a special class of involutive divisions, the constructive divisions, which is naturally related to an algorithm for computing involutive bases (contained in Section 4.2). If such a division is in addition Noetherian, then the algorithm will always terminate with an involutive basis and thus provides us with a proof of the existence of such bases for many divisions (including in particular the Janet division). Unfortunately, both the definition of constructive divisions and the termination proof are highly technical and not very intuitive.

As a reward the underlying algorithm turns out to be surprisingly efficient despite its simplicity. However, in general, it does not produce a minimal basis and furthermore still contains some redundancies. In Section 4.4 we show how it can be modified such that the output is always minimal and simultaneously introduce a number of optimisations. This optimised algorithm underlies most implementations of involutive bases in computer algebra systems.

Section 4.3 considers in detail the special case of the Pommaret division. As it is not Noetherian, our completion algorithm does not necessarily always terminate for it. However, it turns out that it does so generically; more precisely, the whole problem (known under the name of δ-regularity) is only a matter of the chosen variables and in a generic coordinate system the algorithm terminates. Because of its great importance, both from a theoretical point of view and for concrete computations, this topic will reappear at a number of places in later chapters. We will prove that every ideal possesses in suitably chosen coordinates a Pommaret basis and also show how such coordinates may be constructed effectively using a simple and cheap criterion for "bad" coordinate systems based on a comparison of the multiplicative variables obtained by the Pommaret and the Janet division, respectively.

W.M. Seiler, *Involution*, Algorithms and Computation in Mathematics 24,
DOI 10.1007/978-3-642-01287-7_4,

The next chapter will show that such coordinate questions are common in commutative algebra and thus effective methods for determining "good" coordinates are of considerable interest. Many statements take a particularly simple form in an appropriately chosen coordinate system. "Good" coordinates are often generic. However, in general it is firstly not easy to verify whether the currently used variables are generic in the corresponding sense and secondly non-trivial to construct systematically "good" coordinates. Usually, one resorts to probabilistic methods which are disadvantageous for computations, as they destroy all the sparsity usually present in polynomial (or differential) systems appearing in practice. For all topics that we will investigate in the next chapter, δ-regular coordinates represent a "good" choice. Furthermore, our approach for their construction is deterministic and tries to preserve as much sparsity as possible.

The final two sections discuss two extensions of the theory of involutive bases: bases with respect to semigroup orders—a wider class of orders which is important for local computations (as needed e. g. in singularity theory) but algorithmically more demanding—and bases in solvable algebras with coefficients from a ring instead of a field. In both cases, the concept of weak involutive bases becomes important, as in general strong bases do not exist any more.

4.1 Constructive Divisions

In Chapter 3 we introduced involutive bases. But we postponed the question how they can actually be constructed. Unfortunately, for arbitrary involutive division no satisfying solution is known so far. One could follow a brute force approach, namely performing a breadth first search through the tree of all possible completions. Obviously, it terminates only, if a finite basis exists. But for divisions satisfying some additional properties one can design a fairly efficient completion algorithm.

The first problem one faces in the search for an involutive completion of a finite subset $\mathcal{B} \subset \mathbb{N}_0^n$ for some division L is to check whether $\mathcal{B}$ is already involutive. The trouble is that we do not know a priori where obstructions to involution might lie. As these multi indices must somehow be related to the non-multiplicative indices of the elements of $\mathcal{B}$, the multi indices $\nu + 1_j$ with $\nu \in \mathcal{B}$ and $j \in \overline{N}_{L,\mathcal{B}}(\nu)$ represent a natural starting point.

Definition 4.1.1. A finite set $\mathcal{B} \subset \mathbb{N}_0^n$ of multi indices is *locally involutive* for the involutive division L, if $\nu + 1_j \in \langle \mathcal{B} \rangle_L$ for every non-multiplicative index $j \in \overline{N}_{L,\mathcal{B}}(\nu)$ of every multi index $\nu \in \mathcal{B}$.

In contrast to involution, local involution can easily be verified effectively, as it requires only a finite number of checks. However, while involution obviously implies local involution, the converse does not hold, as the following rather bizarre, globally defined division due to Gerdt and Blinkov [156, Ex. 4.8] demonstrates.

Example 4.1.2. We consider in $\mathbb{N}_0^3$ the involutive division L defined by the following assignment of multiplicative indices. With the exception of four multi indices all elements of $\mathbb{N}_0^3$ do not have any multiplicative indices. These four exceptions are: $N_L([0,0,0]) = \{1,2,3\}$, $N_L([1,0,0]) = \{1,3\}$, $N_L([0,1,0]) = \{1,2\}$, and $N_L([0,0,1]) = \{2,3\}$. It is straightforward to show that this assignment indeed defines a global involutive division.

Now we take the set $\mathcal{B} = \{[1,0,0],[0,1,0],[0,0,1]\}$. It is not involutive, as the multi index $[1,1,1] \in \langle\mathcal{B}\rangle$ is not contained in the involutive span $\langle\mathcal{B}\rangle_L$. But it is locally involutive, as the three multi indices $[1,1,0]$, $[0,1,1]$ and $[1,0,1]$ obtained by taking the elements of $\mathcal{B}$ and adding their respective non-multiplicative index are contained in $\langle\mathcal{B}\rangle_L$. ◁

Definition 4.1.3. Let L be an involutive division and $\mathcal{B} \subset \mathbb{N}_0^n$ a finite set. Let furthermore $(\nu^{(1)},\dots,\nu^{(t)})$ be a finite sequence of elements of $\mathcal{B}$ where every multi index $\nu^{(k)}$ with $k < t$ has a non-multiplicative index $j_k \in \overline{N}_{L,\mathcal{B}}(\nu^{(k)})$ such that $\nu^{(k+1)} \mid_{L,\mathcal{B}} \nu^{(k)} + 1_{j_k}$. The division L is *continuous*, if any such sequence consists only of distinct elements, i. e. if $\nu^{(k)} \neq \nu^{(\ell)}$ for all $k \neq \ell$.

It is easy to see that the division of Example 4.1.2 is not continuous. We will show now that for continuous divisions local involution always implies involution. This makes these divisions very convenient for concrete computations: if obstructions to involution exist, some of them must be of the form $\nu + 1_j$ with $\nu \in \mathcal{B}$ and index $j \in \overline{N}_{L,\mathcal{B}}(\nu)$ and all others are "multiples" of these.

Proposition 4.1.4. *If the involutive division L is continuous, then any locally involutive set $\mathcal{B} \subset \mathbb{N}_0^n$ is weakly involutive.*

Proof. Let the set Σ contain those obstructions to involution that are of minimal length. We claim that for a continuous division L all multi indices $\sigma \in \Sigma$ are of the form $\nu + 1_j$ with $\nu \in \mathcal{B}$ and $j \in \overline{N}_{L,\mathcal{B}}(\nu)$. This assertion immediately implies our proposition: since for a locally involutive set all such multi indices are contained in $\langle\mathcal{B}\rangle_L$, we must have $\Sigma = \emptyset$ and thus $\langle\mathcal{B}\rangle = \langle\mathcal{B}\rangle_L$.

In order to prove our claim, we choose a $\sigma \in \Sigma$ for which no $\nu \in \mathcal{B}$ exists with $\sigma = \nu + 1_j$. We collect in $\mathcal{B}_\sigma$ all divisors $\nu \in \mathcal{B}$ of σ of maximal length. Let $\nu^{(1)}$ be an element of $\mathcal{B}_\sigma$; by assumption, the multi index $\mu^{(1)} = \sigma - \nu^{(1)}$ satisfies $|\mu^{(1)}| > 1$ and at least one non-multiplicative index $j_1 \in \overline{N}_{L,\mathcal{B}}(\nu^{(1)})$ exists with $\mu^{(1)}_{j_1} > 0$. By the definition of Σ, we have $\nu^{(1)} + 1_{j_1} \in \langle\mathcal{B}\rangle_L$. Thus a multi index $\nu^{(2)} \in \mathcal{B}$ exists with $\nu^{(2)} \mid_{L,\mathcal{B}} \nu^{(1)} + 1_{j_1}$. This implies $\nu^{(2)} \mid \sigma$ and we set $\mu^{(2)} = \sigma - \nu^{(2)}$. By the definition of the set $\mathcal{B}_\sigma$, we have $|\nu^{(2)}| \leq |\nu^{(1)}|$. Hence $\nu^{(2)} + 1_j \in \langle\mathcal{B}\rangle_L$ for all j.

Let us choose a non-multiplicative index $j_2 \in \overline{N}_{L,\mathcal{B}}(\nu^{(2)})$ with $\mu^{(2)}_{j_2} > 0$. Such an index must exist, as otherwise $\sigma \in \langle\mathcal{B}\rangle_L$. By the same arguments as above, a multi index $\nu^{(3)} \in \mathcal{B}$ exists with $\nu^{(3)} \mid_{L,\mathcal{B}} \nu^{(2)} + 1_{j_2}$ and $|\nu^{(3)}| \leq |\nu^{(2)}|$. We can iterate this process and produce an infinite sequence $(\nu^{(1)},\nu^{(2)},\dots)$ where each multi index satisfies $\nu^{(i)} \in \mathcal{B}$ and $\nu^{(i+1)} \mid_{L,\mathcal{B}} \nu^{(i)} + 1_{j_i}$ with $j_i \in \overline{N}_{L,\mathcal{B}}(\nu^{(i)})$. As $\mathcal{B}$ is a finite set, the elements of the sequence cannot be all different. This observation contradicts our assumption that L is a continuous division: by taking a sufficiently large part of this sequence we obtain a finite sequence with all properties mentioned in Definition 4.1.3 but containing some identical elements. Hence a multi index $\nu \in \mathcal{B}$ must exist such that $\sigma = \nu + 1_j$. □

Continuity is not only sufficient for the equivalence of local involution and involution but also useful for proving other properties. The main point is that it excludes the existence of infinite reduction chains. The two involutive divisions introduced in Examples 3.1.4 and 3.1.7 are both continuous.

Lemma 4.1.5. *The Janet and the Pommaret division are continuous.*

Proof. Let $\mathcal{B} \subseteq \mathbb{N}_0^n$ be a finite set and $(\nu^{(i)}, \dots, \nu^{(t)})$ a finite sequence where $\nu^{(i+1)} |_{L,\mathcal{B}} \nu^{(i)} + 1_j$ with $j \in \overline{N}_{L,\mathcal{B}}(\nu^{(i)})$ for $1 \leq i < t$.

We claim that for $L = J$, the Janet division, $\nu^{(i+1)} \succ_{\text{lex}} \nu^{(i)}$ implying that the sequence is strictly ascending and thus cannot contain any identical entries. Let us set $k = \max\{i \mid \mu_i \neq \nu_i\}$. Then $j \leq k$, as otherwise $j \in N_{J,\mathcal{B}}(\nu^{(i+1)})$ would entail that also $j \in N_{J,\mathcal{B}}(\nu^{(i)})$ contradicting our assumption that j is non-multiplicative for the multi index $\nu^{(i)}$. But $j < k$ is also not possible, as then $\nu_k^{(i+1)} < \nu_k^{(i)}$ and so k could not be multiplicative for $\nu^{(i+1)}$. There remains as only possibility $j = k$. In this case we must have $\nu_j^{(i+1)} = \nu_j^{(i)} + 1$ (otherwise j could not be multiplicative for $\nu^{(i+1)}$). Thus we conclude that $\nu^{(i+1)} \succ_{\text{lex}} \nu^{(i)}$ and the Janet division is continuous.

The proof for $L = P$, the Pommaret division, is slightly more subtle.[1] The condition $j \in \overline{N}_P(\nu^{(i)})$ implies that $\operatorname{cls}(\nu^{(i)} + 1_j) = \operatorname{cls}\nu^{(i)}$ and if $\nu^{(i+1)} |_P \nu^{(i)} + 1_j$, then $\operatorname{cls}\nu^{(i+1)} \geq \operatorname{cls}\nu^{(i)}$, i. e. the class of the elements of the sequence is monotonously increasing. If $\operatorname{cls}\nu^{(i+1)} = \operatorname{cls}\nu^{(i)} = k$, then the involutive divisibility requires that $\nu_k^{(i+1)} \leq \nu_k^{(i)}$, i. e. among the elements of the sequence of the same class the corresponding entry is monotonously decreasing. And if finally $\nu_k^{(i+1)} = \nu_k^{(i)}$, then we must have $\nu^{(i+1)} = \nu^{(i)} + 1_j$, i. e. the length of these elements is strictly increasing. Hence we may conclude that all elements of the sequence are different and the Pommaret division is continuous. □

Remark 4.1.6. In Remark 3.1.13 we discussed that for a global division a weak involutive basis of the sum $\mathcal{I}_1 + \mathcal{I}_2$ of two monoid ideals is obtained by simply taking the union of (weak) involutive bases of $\mathcal{I}_1$ and $\mathcal{I}_2$. As a more theoretical application of the concept of continuity, we prove now a similar statement for the product $\mathcal{I}_1 \cdot \mathcal{I}_2$ and the intersection $\mathcal{I}_1 \cap \mathcal{I}_2$ in the special case of the Pommaret division. Let $\mathcal{N}_1$ be a (weak) Pommaret basis of $\mathcal{I}_1$ and $\mathcal{N}_2$ of $\mathcal{I}_2$. We claim that the set

[1] It is tempting to tackle the lemma for the Pommaret division in the same manner as for the Janet division using $\prec_{\text{revlex}}$ instead of $\prec_{\text{lex}}$; in fact, such a "proof" can be found in the literature. Unfortunately, it is not correct, as $\prec_{\text{revlex}}$ is not a term order (see Example A.1.7): if we have $\nu^{(i+1)} = \nu^{(i)} + 1_j$, then $\nu^{(i+1)} \prec_{\text{revlex}} \nu^{(i)}$ although the latter multi index is a divisor of the former one! Thus the sequences considered in Definition 4.1.3 are in general not strictly ascending with respect to $\prec_{\text{revlex}}$.

$\mathcal{N} = \{\mu + \nu \mid \mu \in \mathcal{N}_1, \nu \in \mathcal{N}_2\}$ is a weak Pommaret basis of $\mathcal{I}_1 \cdot \mathcal{I}_2$ and that the set $\hat{\mathcal{N}} = \{\operatorname{lcm}(\mu, \nu) \mid \mu \in \mathcal{N}_1, \nu \in \mathcal{N}_2\}$ is a weak Pommaret basis of $\mathcal{I}_1 \cap \mathcal{I}_2$.

By Proposition 4.1.4, it suffices to show that both $\mathcal{N}$ and $\hat{\mathcal{N}}$ are locally involutive for the Pommaret division. Thus take a generator $\mu + \nu \in \mathcal{N}$, assuming for definiteness that $\operatorname{cls}\mu \leq \operatorname{cls}\nu$, and a non-multiplicative index $j_1 > \operatorname{cls}(\mu + \nu) = \operatorname{cls}\mu$ of it. Then j_1 is also non-multiplicative for $\mu \in \mathcal{N}_1$ and the Pommaret basis $\mathcal{N}_1$ contains a multi index $\mu^{(1)}$ which involutively divides $\mu + 1_{j_1}$. If now the generator $\mu^{(1)} + \nu \in \mathcal{N}$ is an involutive divisor of $\mu + \nu + 1_{j_1}$, then we are done.

Otherwise, an index $k_1 > \operatorname{cls}\nu$ exists such that $(\mu - \mu^{(1)})_{k_1} > 0$. In this case the Pommaret basis $\mathcal{N}_2$ contains a multi index $\nu^{(1)}$ which involutively divides $\nu + 1_{k_1}$. Again we are done, if $\mu^{(1)} + \nu^{(1)} \in \mathcal{N}$ is now an involutive divisor of $\mu + \nu + 1_{j_1}$. Otherwise, there are two possibilities. An index $j_2 > \operatorname{cls}\mu^{(1)}$ could exist such that $(\mu + \nu + 1_{j_1} - \mu^{(1)} + \nu^{(1)})_{j_2} > 0$ which entails the existence of a further generator $\mu^{(2)} \in \mathcal{N}_1$ which involutively divides the multi index $\mu^{(1)} + 1_{j_2}$. Or there could exist an index $k_2 > \operatorname{cls}\nu^{(1)}$ such that $(\mu + \nu + 1_{j_1} - \mu^{(1)} + \nu^{(1)})_{k_2} > 0$ which implies that there is a multi index $\nu^{(2)} \in \mathcal{N}_2$ involutively dividing $\nu^{(1)} + 1_{k_2}$.

Continuing in this manner, one easily sees that we build up two sequences $(\mu, \mu^{(1)}, \mu^{(2)}, \dots) \subseteq \mathcal{N}_1$ and $(\nu, \nu^{(1)}, \nu^{(2)}, \dots) \subseteq \mathcal{N}_2$ as in the definition of a continuous division. Since both Pommaret bases are finite by definition and the Pommaret division is continuous by Lemma 4.1.5, no sequence may become infinite and the above described process must stop with an involutive divisor of $\mu + \nu + 1_{j_1}$. Hence the set $\mathcal{N}$ is locally involutive and a weak Pommaret basis of $\mathcal{I}_1 \cdot \mathcal{I}_2$. The proof for $\hat{\mathcal{N}}$ and $\mathcal{I}_1 \cap \mathcal{I}_2$ goes completely analogously replacing at appropriate places the sum of two multi indices by their least common multiple. ◁

Definition 4.1.7. Let $\mathcal{B} \subset \mathbb{N}_0^n$ be a finite set of multi indices and choose a multi index $\nu \in \mathcal{B}$ and a non-multiplicative index $j \in \overline{N}_{L,\mathcal{B}}(\nu)$ such that

(i) $\nu + 1_j \notin \langle \mathcal{B} \rangle_L$.
(ii) If there exists a multi index $\mu \in \mathcal{B}$ and $k \in \overline{N}_{L,\mathcal{B}}(\mu)$ such that $\mu + 1_k \mid \nu + 1_j$ but $\mu + 1_k \neq \nu + 1_j$, then $\mu + 1_k \in \langle \mathcal{B} \rangle_L$.

The continuous division L is *constructive*, if for any such set $\mathcal{B}$ and any such multi index $\nu + 1_j$ no multi index $\rho \in \langle \mathcal{B} \rangle_L$ with $\nu + 1_j \in \mathcal{C}_{L,\mathcal{B}\cup\{\rho\}}(\rho)$ exists.

In words, constructivity can roughly be explained as follows. The conditions imposed on ν and j ensure a kind of minimality: no proper divisor of $\nu + 1_j$ is of the form $\mu + 1_k$ for a $\mu \in \mathcal{B}$ and not contained in the involutive span $\langle \mathcal{B} \rangle_L$. The conclusion implies that it is useless to add multi indices to $\mathcal{B}$ that lie in some involutive cone, as none of them can be an involutive divisor of $\nu + 1_j$. An efficient completion algorithm for a constructive division should consider only non-multiplicative indices. Because of the restriction to multi indices of the form $\nu + 1_j$, we require that a constructive division is also continuous.

Lemma 4.1.8. *Any globally defined division (and thus the Pommaret division) is constructive. The Janet division is constructive, too.*

Proof. For a globally defined division the proof is very simple. For any multi index $\rho \in \langle \mathcal{B} \rangle_L$ there exists a multi index $\mu \in \mathcal{B}$ such that $\rho \in \mathcal{C}_L(\mu)$. As in the global case the multiplicative indices are independent of the reference set, we must have by the definition of an involutive division that $\mathcal{C}_L(\rho) \subseteq \mathcal{C}_L(\mu)$. Hence adding such a multi index to $\mathcal{B}$ cannot change the involutive span and if $\nu + 1_j \notin \langle \mathcal{B} \rangle_L$, then also $\nu + 1_j \notin \langle \mathcal{B} \cup \{\rho\} \rangle_L$. But this fact implies the constructivity of L.

The proof of the constructivity of the Janet division is more involved. The basic idea is to show that if it were not constructive, it also could not be continuous. Let $\mathcal{B}$, ν, j be as described in Definition 4.1.7. Assume for a contradiction that a multi index $\rho \in \langle \mathcal{B} \rangle_J$ exists with $\nu + 1_j \in \mathcal{C}_{J,\mathcal{B}\cup\{\rho\}}(\rho)$. We write $\rho = \nu^{(1)} + \mu$ for a multi index $\nu^{(1)} \in \mathcal{B}$ with $\rho \in \mathcal{C}_{J,\mathcal{B}}(\nu^{(1)})$. As $\nu + 1_j \notin \langle \mathcal{B} \rangle_J$, we must have $|\mu| > 0$. Consider the multi index $\lambda = \nu + 1_j - \rho$ and let m, l be the maximal indices such that $\mu_m > 0$ and $\lambda_l > 0$, respectively.

We claim that $j > \max\{m, l\}$. Indeed, if $j \leq m$, then $\nu_m^{(1)} < \nu_m$ and, by definition of the Janet division, this implies that $m \notin N_{J,\mathcal{B}}(\nu^{(1)})$, a contradiction. Similarly, we cannot have $j < l$, as then $l \notin N_{J,\mathcal{B}\cup\{\rho\}}(\rho)$. Finally, $j = l$ is not possible. As we know already that $j > m$, we have in this case that $\rho_i = \nu_i^{(1)} = \nu_i$ for all $i > j$ and $\rho_j \leq \nu_j$. Hence $j \in \overline{N}_{J,\mathcal{B}\cup\{\rho\}}(\nu)$ implying $j \in \overline{N}_{J,\mathcal{B}\cup\{\rho\}}(\rho)$, a contradiction.

Now we construct a sequence as in Definition 4.1.3 of a continuous division. Choose an index $j_1 \in \overline{N}_{J,\mathcal{B}}(\nu^{(1)})$ with $\lambda_{j_1} > 0$. Such an index exists, as otherwise $\nu + 1_j \in \mathcal{C}_{J,\mathcal{B}}(\nu^{(1)}) \subseteq \langle \mathcal{B} \rangle_J$. So we may write $\nu + 1_j = (\nu^{(1)} + 1_{j_1}) + \mu + \lambda - 1_{j_1}$. Because of $|\mu| > 0$, the multi index $\nu^{(1)} + 1_{j_1}$ is a proper divisor of $\nu + 1_j$ and according to our assumptions $\nu^{(2)} \in \mathcal{B}$ exists with $\nu^{(1)} + 1_{j_1} \in \mathcal{C}_{J,\mathcal{B}}(\nu^{(2)})$.

By the same arguments as above, we can find an index $j_2 \in \overline{N}_{J,\mathcal{B}}(\nu^{(2)})$ such that $(\mu + \lambda - 1_{j_1})_{j_2} > 0$ and a multi index $\nu^{(3)} \in \mathcal{B}$ such that $\nu^{(2)} + 1_{j_2} \in \mathcal{C}_{J,\mathcal{B}}(\nu^{(3)})$. Thus we can iterate this construction and produce an infinite sequence $(\nu^{(1)}, \nu^{(2)}, \dots)$ where $\nu^{(i+1)} \mid_{J,\mathcal{B}} \nu^{(i)} + 1_{j_i}$ with $j_i \in \overline{N}_{J,\mathcal{B}}(\nu^{(i)})$. By the continuity of the Janet division all members of the sequence must be different. But every multi index $\nu^{(i)}$ is a divisor of $\nu + 1_j$, so obviously only finitely many of them can be different. Thus the sequence must terminate which only happens, if $\nu + 1_j \in \mathcal{C}_{J,\mathcal{B}}(\nu^{(i)})$ for some i contradicting our assumptions. □

4.2 Computation of Involutive Bases

We present now an algorithm for determining involutive bases of left ideals in a polynomial algebra of solvable type $(\mathcal{P}, \star, \prec)$. Note that it includes simultaneously as special cases the completion of polynomial ideals and of linear systems of differential equations (expressed with linear differential operators). As mentioned above, for arbitrary involutive divisions, nobody has so far been able to find a reasonable algorithm. But if we assume that the division is constructive, then a very simple algorithm exists, the basic ideas of which go back to Janet. As in the last chapter, we start with the "monomial" case, i. e. we consider the problem how we can complete a given finite set $\mathcal{B} \subset \mathbb{N}_0^n$ of multi indices to an involutive set with respect to some involutive division L, and extend later to finite polynomial sets $\mathcal{F} \subset \mathcal{P}$.

Algorithm 4.1 Involutive basis in $(\mathbb{N}_0^n, +)$

Input: a finite set $\mathcal{B} \subset \mathbb{N}_0^n$, an involutive division L
Output: an involutive completion $\bar{\mathcal{B}}$ of $\mathcal{B}$

```
 1: B̄ ← B
 2: loop
 3:     S ← { ν + 1_j | ν ∈ B̄, j ∈ N̄_{L,B̄}(ν), ν + 1_j ∉ ⟨B̄⟩_L }
 4:     if S = ∅ then
 5:         return B̄
 6:     else
 7:         choose μ ∈ S such that S does not contain a proper divisor of it
 8:         B̄ ← B̄ ∪ {μ}
 9:     end if
10: end loop
```

The strategy behind Algorithm 4.1 is fairly natural given the results of the last section. It collects in a set $\mathcal{S}$ all obstructions to local involution. For a continuous division L, the basis $\mathcal{B}$ is involutive, if and only if this set vanishes. Furthermore, for a constructive division L it does not make sense to add elements of the involutive span $\langle \mathcal{B} \rangle_L$ to $\mathcal{B}$ in order to complete. Thus we add in Line /8/ an element of $\mathcal{S}$ which is minimal in the sense that the set $\mathcal{S}$ does not contain a proper divisor of it.

Proposition 4.2.1. *Let the finite set $\mathcal{B} \subset \mathbb{N}_0^n$ possess an involutive completion with respect to the constructive division L. Then Algorithm 4.1 terminates with an involutive completion $\bar{\mathcal{B}}$ of $\mathcal{B}$.*

Proof. If Algorithm 4.1 terminates, its correctness is obvious under the made assumptions. The criterion for its termination, $\mathcal{S} = \emptyset$, is equivalent to local involution of the set $\bar{\mathcal{B}}$. By Proposition 4.1.4, local involution implies for a continuous division involution. Thus the result $\bar{\mathcal{B}}$ is an involutive completion of the input set $\mathcal{B}$, as by construction $\mathcal{B} \subseteq \bar{\mathcal{B}} \subset \langle \mathcal{B} \rangle$.

If the set $\mathcal{B}$ is already involutive, then Algorithm 4.1 leaves it unchanged and thus obviously terminates. Let us assume that $\mathcal{B}$ is not yet involutive. In the first iteration of the `loop` a multi index of the form $\mu = \nu + 1_j$ is added to $\mathcal{B}$. It is not contained in $\langle \mathcal{B} \rangle_L$ and $\mathcal{S}$ does not contain a proper divisor of it. If $\mathcal{B}_L$ is an arbitrary involutive completion of $\mathcal{B}$ for the division L, then it must contain a multi index $\lambda \notin \mathcal{B}$ such that $\lambda \,|_{L,\mathcal{B}_L}\, \mu$. We claim that $\lambda = \mu$.

Assume that $\lambda \neq \mu$. Since $\mathcal{B}_L \subset \langle \mathcal{B} \rangle$, the multi index λ must lie in the cone of a generator $\nu^{(1)} \in \mathcal{B}$. We will show that $\lambda \in \langle \mathcal{B} \rangle_L$ because of the continuity of L, contradicting the constructivity of L. If $\nu^{(1)} \,|_{L,\mathcal{B}}\, \lambda$, we are done. Otherwise, we write $\lambda = \nu^{(1)} + \rho^{(1)}$ for some multi index $\rho^{(1)} \in \mathbb{N}_0^n$. By construction, a non-multiplicative index $j_1 \in \overline{N}_{L,\mathcal{B}}(\nu^{(1)})$ with $\rho^{(1)}_{j_1} > 0$ must exist. Consider the multi index $\nu^{(1)}+1_{j_1}$. Because of $\nu^{(1)}+1_{j_1} \mid \lambda$, the multi index $\nu^{(1)}+1_{j_1}$ is a proper divisor of μ. Since the set $\mathcal{S}$ does not contain any proper divisor of μ, we must have $\nu^{(1)}+1_{j_1} \in \langle\mathcal{B}\rangle_L$. Thus a multi index $\nu^{(2)} \in \mathcal{B}$ exists such that $\nu^{(2)} \mid_{L,\mathcal{B}} \nu^{(1)}+1_{j_1}$.

By iteration of this argument we obtain a sequence $(\nu^{(1)},\nu^{(2)},\dots)$ where each element $\nu^{(i)} \in \mathcal{B}$ is a divisor of λ and where $\nu^{(i+1)} \mid_{L,\mathcal{B}} \nu^{(i)}+1_{j_i}$ with a non-multiplicative index $j_i \in \overline{N}_{L,\mathcal{B}}(\nu^{(i)})$. This sequence cannot become infinite for a continuous division, as λ has only finitely many different divisors and all the multi indices $\nu^{(i)}$ in arbitrary finite pieces of the sequence must be different. But the sequence will only stop, if some $\nu^{(i)} \in \mathcal{B}$ exists such that $\nu^{(i)} \mid_{L,\mathcal{B}} \lambda$ and hence we conclude that $\lambda \in \langle\mathcal{B}\rangle_L$.

Thus *every* involutive completion $\mathcal{B}_L$ of the set $\mathcal{B}$ must contain μ. In the next iteration of the `loop`, Algorithm 4.1 treats the enlarged set $\mathcal{B}_1 = \mathcal{B}\cup\{\mu\}$. It follows from our considerations above that any involutive completion $\mathcal{B}_L$ of $\mathcal{B}$ is also an involutive completion of $\mathcal{B}_1$ and hence we may apply the same argument again. As an involutive completion $\mathcal{B}_L$ is by definition a finite set, we must reach after a finite number k of iterations an involutive basis $\mathcal{B}_k$ of $\langle\mathcal{B}\rangle$. □

Note the crucial difference between this result and Theorem B.4.15 on the termination of the Buchberger Algorithm B.3. In the latter case, one shows the termination for arbitrary input, i. e. the theorem provides a constructive proof for the existence of Gröbner bases. Here we are only able to prove the termination under the assumption that a finite involutive basis exists; the existence has to be shown separately. For example, Lemma 3.1.19 guarantees us the existence of a Janet basis for every monoid ideal.

By Proposition 3.1.12, any weak involutive basis becomes strongly involutive after a simple elimination of redundant elements. Thus we obtain an algorithm for constructing a strong involutive basis of $\langle\mathcal{B}\rangle$ by adding an involutive autoreduction as last step to Algorithm 4.1. Alternatively, we could perform the autoreduction as first step. Indeed, if the input set $\mathcal{B}$ is involutively autoreduced, then all intermediate sets $\bar{\mathcal{B}}$ constructed by Algorithm 4.1 are involutively autoreduced, too. This fact follows from the filter axiom in Definition 3.1.1 of an involutive division requiring that involutive cones may only shrink, if we add elements to the set $\mathcal{B}$.

Remark 4.2.2. While we just stated that it suffices to perform an involutive autoreduction as either first or last step in Algorithm 4.1, we now analyse for later use what happens, if we involutively autoreduce $\bar{\mathcal{B}}$ every time a new element has been added to it. The termination argument given in the proof of Proposition 4.2.1 does not remain valid after this modification and we must provide an alternative proof.

Let again $\mathcal{B}_L = \{\mu^{(1)},\dots,\mu^{(r)}\}$ be an involutive completion of the input set $\mathcal{B}$. If we denote by $\bar{\mathcal{B}}_i$ the value of $\bar{\mathcal{B}}$ after the ith iteration of the `loop`, then it was shown in the proof of Proposition 4.2.1 that $\mathcal{B}_L$ is also a weak involutive completion of any set $\bar{\mathcal{B}}_i$. As by definition $\mathcal{B}_L$ is finite and each $\bar{\mathcal{B}}_i$ is a subset of it, the only possibility for non-termination is the appearance of a cycle, i. e. the existence of values k_0, ℓ such that $\bar{\mathcal{B}}_{k+\ell} = \bar{\mathcal{B}}_k$ for all $k \geq k_0$.

Assume that in some iteration of the `loop` the multi index $\mu^{(k)}$ is added to $\bar{\mathcal{B}}$ and that in the subsequent involutive autoreduction some elements of $\bar{\mathcal{B}}$ are eliminated (this is necessary for having a cycle). The first step in the autoreduction must be that some multi index $\mu^{(\ell)}$ is eliminated, because $\mu^{(k)}$ is an involutive divisor of it. Indeed, by Condition (ii) in Definition 3.1.1, any other reduction would have been possible already before the insertion of $\mu^{(k)}$ and thus the previous involutive autoreduction would not have been finished.

As $\mu^{(k)}$ has been added to $\bar{\mathcal{B}}$, a multi index $\mu^{(a_1)} \in \mathcal{B}$ such that $\mu^{(k)} = \mu^{(a_1)} + \rho$ exists. Furthermore, we know that $\mu^{(\ell)} = \mu^{(k)} + \tilde{\sigma}$ for some multi index $\tilde{\sigma}$ with $|\tilde{\sigma}| > 0$ and thus $\mu^{(\ell)} = \mu^{(a_1)} + \sigma$ with $\sigma = \tilde{\sigma} + \rho$ and $|\sigma| > 1$. As we are in a cycle, the multi index $\mu^{(\ell)}$ must have been added to $\bar{\mathcal{B}}$ in a previous iteration of the `loop`, say when analysing $\bar{\mathcal{B}}_i$. Thus $\mu^{(\ell)}$ cannot be involutively divisible by $\mu^{(a_1)}$ and we must have $\sigma_{j_1} > 0$ for a non-multiplicative index $j_1 \in \overline{N}_{L,\bar{\mathcal{B}}_i}(\mu^{(a_1)})$. As $|\sigma| > 1$, we cannot have $\mu^{(a_1)} + 1_{j_1} = \mu^{(\ell)}$ and therefore $\mu^{(a_1)} + 1_{j_1}$ is a proper divisor of $\mu^{(\ell)}$. Hence the set $\bar{\mathcal{B}}_i$ must contain an involutive divisor $\mu^{(a_2)}$ of $\mu^{(a_1)} + 1_{j_1}$, as otherwise this multi index would have been added to $\bar{\mathcal{B}}$ instead of $\mu^{(\ell)}$.

Obviously, $\mu^{(a_2)} \mid \mu^{(k)}$ and, writing $\mu^{(k)} = \mu^{(a_2)} + \pi$, we conclude by the same reasoning as above that $\pi_{j_2} > 0$ for a non-multiplicative index $j_2 \in \overline{N}_{L,\bar{\mathcal{B}}_i}(\mu^{(a_2)})$. Iteration of this argument yields an infinite sequence $\bigl(\mu^{(a_1)}, \mu^{(a_2)}, \dots\bigr)$ as in Definition 4.1.3 of a continuous division. However, since the division L is continuous and the completion $\mathcal{B}_L$ a finite set, we arrive at a contradiction. Thus even with additional involutive autoreductions after each step Algorithm 4.1 terminates. ◁

In principle, our description of Algorithm 4.1 is not complete, as we did not specify how one should choose the multi index μ in Line /7/, if several choices are possible. One would expect that different involutive completions are obtained for different choices. However, an interesting aspect of our proof of Proposition 4.2.1 is that it shows that this is not the case. The choice affects only the order in which multi indices are added but not which or how many multi indices are added during the completion. A simple method for choosing μ consists of taking an arbitrary term order $\prec$ (which could be changed in each iteration) and setting $\mu = \min_\prec \mathcal{S}$.

Corollary 4.2.3. *If Algorithm 4.1 terminates, then its output $\bar{\mathcal{B}}$ is independent of how the multi index μ is chosen in Line /7/. Furthermore, if $\mathcal{B}_L$ is any involutive completion of $\mathcal{B}$ with respect to the division L, then $\bar{\mathcal{B}} \subseteq \mathcal{B}_L$.*

Proof. Consider the set $\mathcal{L}(\mathcal{B})$ of all involutive completions of $\mathcal{B}$ with respect to the division L and introduce

$$\hat{\mathcal{B}} = \bigcap_{\mathcal{B}_L \in \mathcal{L}(\mathcal{B})} \mathcal{B}_L\,. \tag{4.1}$$

We claim that Algorithm 4.1 determines the thus defined set $\hat{\mathcal{B}}$ independent of the way in which μ is chosen in Line /7/. Obviously, this fact implies our corollary.

Our proof of Proposition 4.2.1 showed that all the multi indices added in the course of Algorithm 4.1 are contained in *every* involutive completion of $\mathcal{B}$. Thus all these multi indices are elements of $\hat{\mathcal{B}}$. As our algorithm terminates with an involutive completion, its output is an element of $\mathcal{L}(\mathcal{B})$, too. Hence the output must be $\hat{\mathcal{B}}$. □

Corollary 4.2.4. *If the monoid ideal $\mathcal{I} \subseteq \mathbb{N}_0^n$ possesses an involutive basis for the constructive division L, then $\mathcal{I}$ has a unique minimal involutive basis.*

Proof. If we apply Algorithm 4.1 to the unique minimal basis $\mathcal{B}$ of $\mathcal{I}$ in the usual sense, then it follows trivially from Corollary 4.2.3 that the output is a minimal involutive basis of $\mathcal{I}$ and that no other involutive basis of $\mathcal{I}$ can be minimal, as it is an involutive completion of $\mathcal{B}$. □

An obvious way to compute an involutive basis for an ideal $\mathcal{I}$ in a polynomial algebra $(\mathcal{P}, \star, \prec)$ of solvable type follows immediately from Remark 3.4.12: we determine first a Gröbner basis $\mathcal{G}$ of $\mathcal{I}$ and then with Algorithm 4.1 an involutive completion of the monomial set $\mathrm{le}_\prec \mathcal{G}$. In fact, a similar method is proposed by Sturmfels and White [439] for the construction of Stanley decompositions which implicitly corresponds to the computation of an involutive basis (see Section 5.1). However, we prefer to extend the ideas behind Algorithm 4.1 to a direct completion algorithm for polynomial ideals, as we believe that such an approach is more efficient for most ideals. As a first step, we need subalgorithms for two important tasks: *involutive normal forms* and *involutive head autoreductions*.

The design of an algorithm $\mathtt{NormalForm}_{L,\prec}(g, \mathcal{H})$ determining an involutive normal form of the polynomial g with respect to the finite set $\mathcal{H} \subset \mathcal{P}$ is trivial. We may use the standard Algorithm B.1 for normal forms in the Gröbner theory, if we replace in Lines /1/, /5/ and /7/ of it the ordinary divisibility relation by its involutive version. Obviously, this modification does not affect the termination. Actually, for our purposes it is not necessary to compute a full normal form; we may stop as soon as we have obtained a polynomial that is not involutively head reducible.

Algorithm 4.2 Involutive head autoreduction

Input: a finite set $\mathcal{F} \subset \mathcal{P}$, an involutive division L
Output: an involutively head autoreduced set $\mathcal{H}$ with $\langle \mathcal{H} \rangle = \langle \mathcal{F} \rangle$
1: $\mathcal{H} \leftarrow \mathcal{F}$
2: **while** $\exists h \in \mathcal{H}, f \in \mathcal{H} \setminus \{h\} : \mathrm{le}_\prec f \,|_{L, \mathrm{le}_\prec \mathcal{H}}\, \mathrm{le}_\prec h$ **do**
3: choose such a pair (h, f)
4: $\mu \leftarrow \mathrm{le}_\prec h - \mathrm{le}_\prec f;\quad c \leftarrow \mathrm{lc}_\prec h / \mathrm{lc}_\prec (x^\mu \star f)$
5: $\mathcal{H} \leftarrow \mathcal{H} \setminus \{h\};\quad \bar{h} \leftarrow h - c x^\mu \star f$
6: **if** $\bar{h} \neq 0$ **then**
7: $\mathcal{H} \leftarrow \mathcal{H} \cup \{\bar{h}\}$
8: **end if**
9: **end while**
10: **return** $\mathcal{H}$

A simple realisation of an algorithm $\mathtt{InvHeadAutoReduce}_{L,\prec}(\mathcal{F})$ for an involutive head autoreduction is given by Algorithm 4.2. It uses exactly the same strategy as the autoreduction Algorithm B.2 with obvious adaptions. Its correctness and its termination follows by the same arguments (any term order is a well-order by Lemma A.1.6). For a full autoreduction instead of only a head autoreduction, one must consider in Line /2/ the full support of the polynomial h and not only its leading term. In this case it is tempting to replace the second part of Line /5/ by $\bar{h} \leftarrow \texttt{NormalForm}_{L,\prec}(h,\bar{\mathcal{H}})$ in analogy to Algorithm 4.2. However, this is correct only for global divisions, as then the multiplicative variables are determined with respect to the smaller set $\bar{\mathcal{H}}$ and not the full basis $\mathcal{H}$ (cf. Remark 3.4.9).

Note that in the determination of the coefficient $c \in \Bbbk$ in Line /6/ we must divide by the leading coefficient of the product $x^\mu \star f$. In general, it differs from the leading coefficient of f, as one can see from (3.8). However, for the two cases of greatest interest for us, ordinary polynomials and linear differential operators, we may take $\operatorname{lc}_\prec f$ instead.

We call $\texttt{MonomialComplete}_{L,\prec}$ the "lift" of the monomial completion Algorithm 4.1 to polynomials obtained by replacing Line /3/ of it by

$$\mathcal{S} \leftarrow \left\{x \star h \mid h \in \bar{\mathcal{H}},\, x \in \overline{X}_{L,\operatorname{le}_\prec \bar{\mathcal{H}},\prec}(h),\, \operatorname{le}_\prec(x \star h) \notin \langle \operatorname{le}_\prec \bar{\mathcal{H}} \rangle_L\right\}$$

and adapting all other lines correspondingly. Applying it to a finite involutively head autoreduced set $\mathcal{H} \subset \mathcal{P}$ yields a set $\bar{\mathcal{H}} \supseteq \mathcal{H}$ consisting of $\mathcal{H}$ and some multiples of its elements such that $\operatorname{le}_\prec \bar{\mathcal{H}}$ is an involutive basis of $\langle \operatorname{le}_\prec \mathcal{H} \rangle$.

Based on these three subalgorithms, we present now the simple Algorithm 4.3 for the computation of involutive bases in $\mathcal{P}$. The current basis is first involutively head autoreduced in Line /3/ and then its set of leading exponents is completed to an involutive basis in Line /4/. Finally, we collect all non-vanishing involutive normal forms of non-multiplicative multiples of our generators in the set $\mathcal{S}$; we stop and return the current basis as soon as it is empty.

Algorithm 4.3 Involutive basis in $(\mathcal{P},\star,\prec)$ ("monomial" form)

Input: a finite set $\mathcal{F} \subset \mathcal{P}$, an involutive division L
Output: an involutive basis $\mathcal{H}$ of $\mathcal{I} = \langle \mathcal{F} \rangle$ with respect to L and $\prec$

1: $\mathcal{H} \leftarrow \mathcal{F}$; $\mathcal{S} \leftarrow \emptyset$
2: **repeat**
3: $\quad\mathcal{H} \leftarrow \texttt{InvHeadAutoReduce}_{L,\prec}(\mathcal{H} \cup \mathcal{S})$
4: $\quad\mathcal{H} \leftarrow \texttt{MonomialComplete}_{L,\prec}(\mathcal{H})$
5: $\quad\mathcal{S} \leftarrow \{\texttt{NormalForm}_{L,\prec}(x \star h, \mathcal{H}) \mid h \in \mathcal{H}, x \in \overline{X}_{L,\mathcal{H},\prec}(h)\} \setminus \{0\}$
6: **until** $\mathcal{S} = \emptyset$
7: **return** $\mathcal{H}$

In order to prove the termination of Algorithm 4.3, we need an extension of the notion of local involution and of Proposition 4.1.4 from monomial to polynomial sets. Since it will allow us to simplify some proofs in Section 4.4, we use this occasion for a slight generalisation.

Definition 4.2.5. A finite set $\mathcal{F} \subset \mathcal{P}$ is called *involutive up to the multi index* λ (or *partially involutive*) for the division L, if for every generator $f \in \mathcal{F}$ and every term $t \in \mathbb{T}$ such that $\operatorname{le}_\prec(t \star f) \prec \lambda$ the involutive normal form of $t \star f$ with respect to $\mathcal{F}$ vanishes, i. e. we have $t \star f \in \langle \mathcal{F} \rangle_{L,\prec}$. The set $\mathcal{F}$ is *locally involutive (up to the* *multi index* $\lambda \in \mathbb{N}_0^n$) for the division L, if $x^j \star f \in \langle \mathcal{F} \rangle_{L,\prec}$ for any non-multiplicative variable $x^j \in \overline{X}_{L,\mathcal{F},\prec}(f)$ of any polynomial $f \in \mathcal{F}$ (such that $\mathrm{le}_\prec(x^j \star f) \prec \lambda$).

Lemma 4.2.6. *Let the finite set $\mathcal{F} \subset \mathcal{P}$ be involutively head autoreduced and involutive up to the multi index $\lambda \in \mathbb{N}_0^n$. If the polynomial $g \in \langle \mathcal{F} \rangle$ possesses a representation of the form $g = \sum_{f \in \mathcal{F}} Q_f \star f$ where the coefficients $Q_f \in \mathcal{P}$ are such that $\mathrm{le}_\prec(Q_f \star f) \prec \lambda$ for all generators f, then the involutive normal form of g with respect to $\mathcal{F}$ vanishes. Furthermore, for all polynomials $g \in \mathcal{P}$ such that $\mathrm{le}_\prec g \prec \lambda$ the ordinary and the involutive normal form with respect to $\mathcal{F}$ coincide.*

Proof. Let the polynomial $g \in \langle \mathcal{F} \rangle$ be expressible as $g = \sum_{f \in \mathcal{F}} Q_f \star f$ where $\mathrm{le}_\prec(Q_f \star f) \prec \lambda$ for all generators f. The right hand side consists of summands of the form $t \star f$ with $t \in \mathbb{T}$ and $\mathrm{le}_\prec(t \star f) \prec \lambda$. By the definition of partial involution, this fact implies that the involutive normal form with respect to $\mathcal{F}$ of all these summands vanishes. Hence a representation $g = \sum_{f \in \mathcal{F}} P_f \star f$ exists where $P_f \in \mathbb{k}[X_{L,\mathcal{F},\prec}(f)]$, i. e. all coefficients are multiplicative for the respective generator f. Since the set $\mathcal{F}$ is assumed to be involutively head autoreduced, we may conclude by the same reasoning as in the proof of Lemma 3.4.14 that this is only possible if $\mathrm{le}_\prec(P_f \star f) \preceq \mathrm{le}_\prec g$ and hence the involutive normal form of g vanishes.

The proof of the second assertion is similar. The ordinary and the involutive normal form only differ, if we perform some non-multiplicative reductions during the computation of the ordinary normal form. But any reduction consists of the subtraction of a polynomial $ct \star f$ where $c \in \mathbb{k}$, $t \in \mathbb{T}$, $f \in \mathcal{F}$ and $\mathrm{le}_\prec(ct \star f) \preceq \mathrm{le}_\prec g$. By the definition of partial involution, under the made assumptions the same effect can always be achieved with a sequence of involutive reductions and hence the two normal forms must coincide (by Proposition 3.4.15 involutive normal forms with respect to involutively head autoreduced sets are unique and thus the order of the reductions does not matter). □

Proposition 4.2.7. *Given a continuous division L, any finite involutively head autoreduced set $\mathcal{F} \subset \mathcal{P}$ that is locally involutive is also involutive.*

Proof. For notational simplicity, we consider only the case of "full" local involution, i. e. without the restriction by some multi index λ. It is trivial that the proof remains true with such a restriction.

We claim that for any term x^μ and for any $f_1 \in \mathcal{F}$ a polynomial $h \in \mathcal{F}$ exists such that $\mathrm{le}_\prec(x^\mu \star f_1) \in \mathcal{C}_{L,\mathrm{le}_\prec \mathcal{F}}(\mathrm{le}_\prec h)$. If this claim is true, we may take arbitrary linear combinations (with coefficients in $\mathbb{k}$) of such products and the leading terms cannot cancel, as $\mathcal{F}$ is involutively head autoreduced. This property immediately implies that the leading exponent of any polynomial in $\langle \mathcal{F} \rangle$ lies in the monoid ideal $\langle \mathrm{le}_\prec \mathcal{F} \rangle_L$ and the proposition holds by the definition of an involutive basis.

If x^μ contains only variables that are multiplicative for f_1, nothing is to be shown. Otherwise we choose a non-multiplicative index $j_1 \in \overline{N}_{L,\mathrm{le}_\prec \mathcal{F}}(\mathrm{le}_\prec f_1)$ such that $\mu_{j_1} > 0$. Because of the assumed local involution of the set $\mathcal{F}$, for each generator $f \in \mathcal{F}$ a polynomial $P_f^{(1)} \in \mathbb{k}[X_{L,\mathcal{F},\prec}(f)]$ exists such that $x^{j_1} \star f_1 = \sum_{f \in \mathcal{F}} P_f^{(1)} \star f$. Let the leading term on the right hand side be $\mathrm{lt}_\prec(P_{f_2}^{(1)} \star f_2)$. If the term $x^{\mu - 1_{j_1}}$ contains only variables that are multiplicative for $\mathrm{le}_\prec f_2$, we are done.

Otherwise we choose a non-multiplicative index $j_2 \in \overline{N}_{L,\mathrm{le}_\prec \mathcal{F}}(\mathrm{le}_\prec f_2)$ such that $(\mu - 1_{j_1})_{j_2} > 0$. Applying again the local involution of $\mathcal{F}$, we find for each $f \in \mathcal{F}$ a polynomial $P_f^{(2)} \in \Bbbk[X_{L,\mathcal{F},\prec}(f)]$ such that $x^{j_2} \star f_2 = \sum_{f\in\mathcal{F}} P_f^{(2)} \star f$. Let the leading term on the right hand side be $\mathrm{lt}_\prec(P_{f_3}^{(2)} \star f_3)$ and introduce the multi index $\nu = \mathrm{le}_\prec(x^{j_1} \star f_1) - \mathrm{le}_\prec f_2$. As the multiplication $\star$ in a polynomial algebra of solvable type respects the term order $\prec$, we have $\mathrm{le}_\prec(x^\mu \star f_1) = \mathrm{le}_\prec(x^{\mu+\nu-1_{j_1}-1_{j_2}} \star f_3)$. If all variables on the right hand side are multiplicative, we are done.

If this is not the case, we must iterate: we choose a non-multiplicative index j_3 and decompose $x^{j_3} \star f_3$ via the local involution into multiplicative products obtaining a new polynomial f_4 and so on. This process yields a sequence $\bigl(\nu^{(1)} = \mathrm{le}_\prec f_1, \nu^{(2)} = \mathrm{le}_\prec f_2, \dots\bigr)$ where all multi indices $\nu^{(k)}$ are elements of the finite set $\mathrm{le}_\prec \mathcal{F}$ and where to each $\nu^{(k)}$ a non-multiplicative index $j_k \in \overline{N}_{L,\mathrm{le}_\prec \mathcal{F}}(\nu^{(k)})$ exists with $\nu^{(k+1)} |_{L,\mathrm{le}_\prec \mathcal{F}} \nu^{(k)} + 1_{j_k}$. By the definition of a continuous division, this sequence cannot become infinite and the above described process must terminate. But this observation implies the existence of a polynomial $h \in \mathcal{F}$ such that $\mathrm{le}_\prec(x^\mu \star f_1) \in \mathcal{C}_{L,\mathrm{le}_\prec \mathcal{F}}(h)$. □

For proving the termination of Algorithm 4.3, we do not only need to show that the termination condition of the `repeat`-loop is satisfied after a finite number of iterations, but we must also ensure that the monomial completions in Line /4/ always terminate. Since we do not know much about the monoid ideals $\langle \mathrm{le}_\prec \mathcal{H} \rangle$ in intermediate iterations of the loop, the simplest solution consists of assuming that the chosen division L is Noetherian.

Theorem 4.2.8. *Let L be a constructive Noetherian division and $(\mathcal{P}, \star, \prec)$ a polynomial algebra of solvable type. Then Algorithm 4.3 terminates for any finite input set $\mathcal{F} \subset \mathcal{P}$ with an involutive basis $\mathcal{H}$ of $\mathcal{I} = \langle \mathcal{F} \rangle$.*

Proof. The correctness is obvious by Proposition 4.2.7. The termination is also rather trivial. As the division L is assumed to be Noetherian, in Line /4/ the set $\mathrm{le}_\prec \mathcal{H}$ possesses a finite involutive completion and $\texttt{MonomialComplete}_{L,\prec}$ terminates by Proposition 4.2.1. Since after Line /4/ this set is furthermore an involutive basis of the monoid ideal generated by it, the leading exponent of any polynomial $g \in \mathcal{S}$ cannot be contained in $\langle \mathrm{le}_\prec \mathcal{H} \rangle$. Hence these monoid ideals form a strictly ascending sequence which cannot be infinite by Dickson's Lemma A.1.2. □

Example 4.2.9. If the division L is not Noetherian, then it may happen that, even when the ideal $\mathcal{I} = \langle \mathcal{F} \rangle$ does possess a finite involutive basis with respect to L, Algorithm 4.3 does not terminate for the input $\mathcal{F}$. The problem is that the existence of an involutive basis for $\mathrm{le}_\prec \mathcal{I}$ does not imply that all subideals of it have also an involutive basis. In such a case it may happen that at some stage of Algorithm 4.3 we encounter a basis $\mathcal{H}$ such that $\langle \mathrm{le}_\prec \mathcal{H} \rangle$ does not possess an involutive basis and then the monomial completion in Line /4/ will not terminate.

One of the simplest instance where this termination problem occurs is not for an ideal but for a submodule of the free $\Bbbk[x,y]$-module with basis $\{\mathbf{e}_1, \mathbf{e}_2\}$. Consider the set $\mathcal{F} = \{y^2\mathbf{e}_1, xy\mathbf{e}_1 + \mathbf{e}_2, x\mathbf{e}_2\}$ with the Pommaret division and any term order for which $xy\mathbf{e}_1 \succ \mathbf{e}_2$. One easily sees that the monoid module $\langle \mathrm{le}_\prec \mathcal{F} \rangle$ does not possess a finite Pommaret basis, as its $\mathbf{e}_2$-component contains only elements of class 1. Hence Algorithm 4.3 will not terminate with $\mathcal{F}$ as input. However, adding the generator $y\mathbf{e}_2$ (the S-"polynomial" of the first two generators) makes $\mathcal{F}$ to a reduced Gröbner basis of $\langle \mathcal{F} \rangle$ which is simultaneously a minimal Pommaret basis.

Another simple example is provided by the set $\mathcal{F} = \{z^2 - 1, yz - 1, x\} \subset \Bbbk[x, y, z]$ together with the lexicographic term order $\prec_{\text{lex}}$. Algorithm 4.3 will iterate infinitely adding all monomials of the form xy^k with $k \geq 1$. Nevertheless, a finite Pommaret basis of $\langle \mathcal{F} \rangle$ for $\prec_{\text{lex}}$ exists and is given by $\mathcal{H} = \{z - y, y^2 - 1, xy, x\}$. $\triangleleft$

Addendum: Right and Two-Sided Ideals

In this Addendum we briefly discuss the relation between left and right involutive bases and the computation of bases for two-sided ideals. We use now the following notations: the left ideal generated by $\mathcal{F} \subset \mathcal{P}$ is denoted by $\langle \mathcal{F} \rangle^{(l)}$, the right ideal by $\langle \mathcal{F} \rangle^{(r)}$ and the two-sided ideal by $\langle\langle \mathcal{F} \rangle\rangle$ and corresponding notations for the left, right and two-sided involutive span.

Recall from Example 3.3.14 that even with a coefficient field $\Bbbk$ it is not guaranteed that the ring $\mathcal{P}$ is also right Noetherian and hence generally the existence of right Gröbner bases for right ideals is not clear. However, we also noted there that the ring $\mathcal{P}$ is always right Noetherian, if we assume that the maps $\rho_i : \Bbbk \to \Bbbk$ in (3.9a) are automorphisms. In this Addendum we will always make this assumption.

From a computational point of view, the theory of right ideals is almost identical to the corresponding theory for left ideals. The left-right asymmetry in our definition of polynomial algebras of solvable type leads only to one complication. Suppose that we want to perform a right reduction of a term ax^ν with respect to another term cx^μ with $\mu \mid \nu$. This requires to find a coefficient $b \in \Bbbk$ such that $\mathrm{lc}_\prec(cx^\mu \star bx^{\nu-\mu}) = c\rho_\mu(b)r_{\mu,\nu-\mu} = a$. Since, according to our assumption, all the maps ρ_μ are automorphisms, such a b always exists.

Lemma 4.2.10. *Let $(\mathcal{P}, \star, \prec)$ be an arbitrary polynomial algebra of solvable type where all the maps ρ_μ appearing in the commutation relations* (3.6a) *are automorphisms. A polynomial $f \in \mathcal{P}$ is (involutively) left reducible modulo a finite set $\mathcal{F} \subset \mathcal{P}$ (with respect to an involutive division L), if and only if it is (involutively) right reducible (with respect to L).*

Proof. Because of the made assumptions on the maps ρ_μ, reducibility depends solely on the leading exponents. □

Proposition 4.2.11. *Let $\mathcal{H}_l \subset \mathcal{P}$ be a monic, involutively left autoreduced, minimal left involutive set and $\mathcal{H}_r \subset \mathcal{P}$ a monic, involutively right autoreduced, minimal right involutive set for an involutive division L. If $\langle \mathcal{H}_l \rangle^{(l)} = \langle \mathcal{H}_r \rangle^{(r)} = \mathcal{I}$, then we also have $\mathcal{H}_l = \mathcal{H}_r$.*

Proof. By definition of a minimal basis, the sets $\mathrm{le}_{\prec}\mathcal{H}_l$ and $\mathrm{le}_{\prec}\mathcal{H}_r$ must both be minimal involutive bases of the monoid ideal $\mathrm{le}_{\prec}\mathcal{I}$ and thus are identical. Suppose $(\mathcal{H}_l \setminus \mathcal{H}_r) \cup (\mathcal{H}_r \setminus \mathcal{H}_l) \neq \emptyset$ and let f be an element of this set with minimal leading exponent with respect to $\prec$. Without loss of generality, we assume that $f \in \mathcal{H}_l \setminus \mathcal{H}_r$. Because of the condition $\langle \mathcal{H}_l \rangle^{(l)} = \langle \mathcal{H}_r \rangle^{(r)}$, we have $f \in \langle \mathcal{H}_r \rangle^{(r)}_{L,\prec}$. Thus the (by Proposition 3.4.15 unique) involutive right normal form of f with respect to $\mathcal{H}_r$ is zero. This observation implies in particular that the polynomial f is involutively right reducible with respect to some generator $h \in \mathcal{H}_r$ with $\mathrm{le}_{\prec} h \preceq \mathrm{le}_{\prec} f$.

If $\mathrm{le}_{\prec} h \prec \mathrm{le}_{\prec} f$, then $h \in \mathcal{H}_l$, too, as f was chosen as a minimal element of the symmetric difference of $\mathcal{H}_l$ and $\mathcal{H}_r$. Hence, by Lemma 4.2.10, f is also left involutively reducible with respect to h (because of $\mathrm{le}_{\prec}\mathcal{H}_l = \mathrm{le}_{\prec}\mathcal{H}_r$ the multiplicative variables of h are the same in both cases). But this fact contradicts the assumption that $\mathcal{H}_l$ is involutively left autoreduced.

If $\mathrm{le}_{\prec} h = \mathrm{le}_{\prec} f = \mu$, then we consider the difference $g = f - h \in \mathcal{I}$: both the left involutive normal form of g with respect to $\mathcal{H}_l$ and the right involutive normal form with respect to $\mathcal{H}_r$ must vanish. By construction, we have $\mathrm{le}_{\prec} g \prec \mu$ and $\operatorname{supp} g \subseteq (\operatorname{supp} f \cup \operatorname{supp} h) \setminus \{\mu\}$. Since both $\mathcal{H}_l$ and $\mathcal{H}_r$ are assumed to be involutively autoreduced, no term in this set is involutively reducible by $\mathrm{le}_{\prec}\mathcal{H}_l = \mathrm{le}_{\prec}\mathcal{H}_r$ and hence we must have $\operatorname{supp} g = \emptyset$, i. e. $g = 0$, a contradiction. □

Directly deriving a theory of two-sided involutive bases along the lines of Section 3.4 fails, as two-sided linear combinations are rather unwieldy objects. A general polynomial $f \in \langle\langle \mathcal{H} \rangle\rangle$ for some finite set $\mathcal{H} \subset \mathcal{P}$ is of the form

$$f = \sum_{h \in \mathcal{H}} \sum_{i=1}^{n_h} \ell_i \star h \star r_i \tag{4.2}$$

with polynomials $\ell_i, r_i \in \mathcal{P}$. The definition of a unique involutive standard representation would require control over the numbers n_h which seems rather difficult. Therefore we will take another approach and construct left involutive bases even for two-sided ideals.

Example 4.2.12. Consider the universal enveloping algebra $\mathfrak{U}(\mathfrak{sl}(2))$. It is isomorphic to the polynomial algebra $\Bbbk[x^1, x^2, x^3]$ where the multiplication is defined by

$$\begin{aligned} x^1 \star x^2 &= x^1x^2\,, & x^2 \star x^1 &= x^1x^2 - x^3\,,\\ x^1 \star x^3 &= x^1x^3\,, & x^3 \star x^1 &= x^1x^3 + 2x^1\,,\\ x^2 \star x^3 &= x^2x^3\,, & x^3 \star x^2 &= x^2x^3 - 2x^2 \end{aligned} \tag{4.3}$$

and which is of solvable type for any degree compatible order. Consider the two-sided principal ideal $\mathcal{I} = \langle\langle x^2 \rangle\rangle$. We have $f = (x^3)^2 = x^1 \star x^2 \star x^3 - x^3 \star x^2 \star x^1 \in \mathcal{I}$, but f is contained neither in the left nor in the right ideal generated by x^2 and it is not possible to write f as a linear combination of x^2 with less than two summands (its leading term is not divisible by x^2). In fact, one can show (e. g. with Algorithm 4.4 below) that $\mathcal{I} = \langle x^1, x^2, x^3 \rangle^{(l)}$. ◁

Proposition 4.2.13. *Let $\mathcal{H} \subset (\mathcal{P}, \star, \prec)$ be a finite set and L an involutive division. Then the following five statements are equivalent.*

(i) *$\mathcal{H}$ is a left involutive basis and $\langle \mathcal{H} \rangle^{(l)} = \langle\langle \mathcal{H} \rangle\rangle$.*
(ii) *$\mathcal{H}$ is a right involutive basis and $\langle \mathcal{H} \rangle^{(r)} = \langle\langle \mathcal{H} \rangle\rangle$.*
(iii) *$\mathcal{H}$ is a left involutive basis of $\langle \mathcal{H} \rangle^{(l)}$ and $h \star x^i, h \star c \in \langle \mathcal{H} \rangle^{(l)}$ for all generators $h \in \mathcal{H}$, all variables x^i and all coefficients $c \in \Bbbk$.*
(iv) *$\mathcal{H}$ is a right involutive basis of $\langle \mathcal{H} \rangle^{(r)}$ and $x^i \star h, c \star h \in \langle \mathcal{H} \rangle^{(r)}$ for all generators $h \in \mathcal{H}$, all variables x^i and all coefficients $c \in \Bbbk$.*
(v) *A unique generator $h \in \mathcal{H}$ exists for every polynomial $f \in \langle\langle \mathcal{H} \rangle\rangle$ such that $\mathrm{le}_\prec h \,|_{L,\mathrm{le}_\prec \mathcal{H}}\, \mathrm{le}_\prec f$.*

Proof. We begin with the equivalence of the first two statements. (i) implies that $\langle \mathcal{H} \rangle^{(l)}_{L,\prec} = \langle \mathcal{H} \rangle^{(l)} = \langle\langle \mathcal{H} \rangle\rangle$ and hence $\langle \mathcal{H} \rangle^{(r)} \subseteq \langle \mathcal{H} \rangle^{(l)}$. The same argument as in the proof of Proposition 4.2.11 shows that we have in fact an equality and thus $\langle \mathcal{H} \rangle^{(r)}_{L,\prec} = \langle \mathcal{H} \rangle^{(r)} = \langle\langle \mathcal{H} \rangle\rangle$, i. e. (ii). The converse goes analogously.

Next we consider the equivalence of (i) and (iii); the equivalence of (ii) and (iv) follows by the same argument. (iii) is a trivial consequence of (i). For the converse, note that (iii) implies that $f \star (ct) \in \langle \mathcal{H} \rangle^{(l)}$ for all $f \in \langle \mathcal{H} \rangle^{(l)}$, all terms $t \in \mathbb{T}$ and all constants $c \in \Bbbk$. Indeed, as in the proof of Proposition 3.2.3 we may rewrite the monomial ct as a polynomial in the "terms" $x^{i_1} \star x^{i_2} \star \cdots \star x^{i_q}$ with $i_1 \leq i_2 \leq \cdots \leq i_q$ and then apply repeatedly our assumptions. Obviously, this entails (i).

The equivalence of (i) or (ii), respectively, with (v) is a trivial consequence of the definition of an involutive basis. □

We would like to exploit Statement (iii) for the construction of a left involutive basis for the two-sided ideal $\langle\langle \mathcal{F} \rangle\rangle$. However, if the field $\Bbbk$ is infinite, then it contains an infinite number of conditions and thus is not effective. In order to proceed in this case, we must make one further assumption about our polynomial algebra $\mathcal{P}$. Let $\Bbbk_0 = \{c \in \Bbbk \mid \forall f \in \mathcal{P} : c \star f = f \star c\}$ be the constant part of the centre of $\mathcal{P}$.

Lemma 4.2.14. *$\Bbbk_0$ is a subfield of $\Bbbk$.*

Proof. It is obvious that $\Bbbk_0$ is a subring. Thus there only remains to show that with $c \in \Bbbk_0^\times$ we have $c^{-1} \in \Bbbk_0$, too. If $c \in \Bbbk_0$, then $x^i \star c = cx^i$, i. e. $\rho_i(c) = c$ and $h_i(c) = 0$, for all $1 \leq i \leq n$. Now on one hand $x^i \star (c^{-1} \star c) = x^i$ and on the other hand

$$(x^i \star c^{-1}) \star c = \rho_i(c^{-1})\rho_i(c)x^i + ch_i(c^{-1}) \tag{4.4}$$

($h_i(c^{-1}) \star c = ch_i(c^{-1})$ since $c \in \Bbbk_0$). The associativity of $\star$ implies now that $\rho_i(c^{-1}) = c^{-1}$ and $h_i(c^{-1}) = 0$. Hence c^{-1} commutes with all variables x^i and it is easy to see that this entails $c^{-1} \in \Bbbk_0$. □

We make now the assumption that either $\Bbbk^\times = \{c_1, \dots, c_\ell\}$ is finite or that the extension $\Bbbk/\Bbbk_0$ is finite, i. e. that $\Bbbk$ is a finite-dimensional vector space over $\Bbbk_0$ with basis $\{c_1, \dots, c_\ell\}$. In the latter case, it is easy to see that it suffices in (iii) to require that only all products $h \star c_j$ lie in $\langle \mathcal{H} \rangle^{(l)}$, as for $c = \sum_{j=1}^{\ell} \lambda_j c_j$ with $\lambda_j \in \Bbbk_0$ we have $h \star c = \sum_{j=1}^{\ell} \lambda_j (h \star c_j)$.

These considerations lead to the simple Algorithm 4.4 below. It first constructs in Line /1/ a left involutive basis $\mathcal{H}$ of the left ideal $\langle \mathcal{F} \rangle^{(l)}$ (using e. g. Algorithm 4.3). The `while` loop in Lines /2–19/ extends the set $\mathcal{H}$ to a left generating set of the two-sided ideal $\langle\langle \mathcal{F} \rangle\rangle$ according to our simplified version of statement (iii) in Proposition 4.2.13. Finally, we complete in Line /20/ this set to an involutive basis. Note that in Line /1/ it is not really necessary to compute a left involutive basis; any left Gröbner basis would suffice as well. Similarly, an ordinary left normal form could be used in Lines /6/ and /12/, respectively; the use of $\texttt{InvLeftNormalForm}_{L,\prec}$ anticipates the final involutive basis computation in Line /20/.

Algorithm 4.4 Left Involutive basis for two-sided ideal

Input: finite set $\mathcal{F} \subset \mathcal{P}$, involutive division L
Output: left involutive basis $\mathcal{H}$ of $\langle\langle \mathcal{F} \rangle\rangle$

1: $\mathcal{H} \leftarrow \texttt{LeftInvBasis}_{L,\prec}(\mathcal{F})$; $\mathcal{S} \leftarrow \mathcal{H}$
2: **while** $\mathcal{S} \neq \emptyset$ **do**
3:     $\mathcal{T} \leftarrow \emptyset$
4:     **for all** $f \in \mathcal{S}$ **do**
5:         **for** i **from** 1 **to** n **do**
6:             $h \leftarrow \texttt{InvLeftNormalForm}_{L,\prec}(f \star x^i, \mathcal{H})$
7:             **if** $h \neq 0$ **then**
8:                 $\mathcal{H} \leftarrow \mathcal{H} \cup \{h\}$; $\mathcal{T} \leftarrow \mathcal{T} \cup \{h\}$
9:             **end if**
10:         **end for**
11:         **for** j **from** 1 **to** ℓ **do**
12:             $h \leftarrow \texttt{InvLeftNormalForm}_{L,\prec}(f \star c_j, \mathcal{H})$
13:             **if** $h \neq 0$ **then**
14:                 $\mathcal{H} \leftarrow \mathcal{H} \cup \{h\}$; $\mathcal{T} \leftarrow \mathcal{T} \cup \{h\}$
15:             **end if**
16:         **end for**
17:     **end for**
18:     $\mathcal{S} \leftarrow \mathcal{T}$
19: **end while**
20: **return** $\texttt{LeftInvBasis}_{L,\prec}(\mathcal{H})$

The termination of the `while` loop follows from the fact that $\mathcal{P}$ is Noetherian and hence a finite generating set of $\langle\langle \mathcal{F} \rangle\rangle$ exists. In principle, we perform here a simple breadth-first search for it. The termination of the involutive bases computations in Lines /1/ and /20/, respectively, depends on the conditions discussed above. Thus the termination is guaranteed, if the division L is constructive and Noetherian.

4.3 Pommaret Bases and δ-Regularity

We saw already in Example 3.1.16 that not every monoid ideal in $\mathbb{N}_0^n$ possesses a finite Pommaret basis: the Pommaret division is not Noetherian. Obviously, this fact also implies that there are polynomial ideals $\mathcal{I} \subseteq \mathcal{P} = \Bbbk[x^1,\dots,x^n]$ without a finite Pommaret basis for any given term order. However, we will show that this problem may be considered as solely a question of choosing "good" variables $\mathbf{x}$ and that generic variables are "good." For notational simplicity, we will restrict in this section to the special case of the commutative polynomial ring and to homogeneous ideals. However, many results remain true in more general situations.

More precisely, assuming that our starting point is not an ideal in the polynomial ring $\mathcal{P}$ but an ideal in the symmetric algebra $S\mathcal{V}$ over an n-dimensional $\Bbbk$-linear space $\mathcal{V}$, we will prove that there always exists a basis $\{x^1,\dots,x^n\}$ of $\mathcal{V}$ such that the induced isomorphism to $\mathcal{P}$ yields an ideal possessing a Pommaret basis. For this purpose, we take in the sequel the following point of view: term orders are defined for exponent vectors, i. e. on the monoid $\mathbb{N}_0^n$; choosing a different basis of $\mathcal{V}$, i. e. performing a linear change of variables $\tilde{\mathbf{x}} = A\mathbf{x}$, leads to new exponent vectors in each polynomial which are then sorted according to the same term order as before.

Definition 4.3.1. The variables $\mathbf{x}$ are called *δ-regular* for the ideal $\mathcal{I} \subseteq \mathcal{P}$ and the term order $\prec$, if $\mathcal{I}$ possesses a finite Pommaret basis for $\prec$.

Given Definition 3.4.1 of an involutive basis, δ-regularity is obviously concerned with the existence of a Pommaret basis for the monoid ideal $\operatorname{le}_\prec \mathcal{I}$. A coordinate transformation generally yields a new leading ideal which may or may not possess a Pommaret basis. We will show in this section that to every polynomial ideal $\mathcal{I}$ variables $\mathbf{x}$ exist such that $\mathcal{I}$ has a finite Pommaret basis. It will even turn out that Pommaret bases exist for almost all possible choices of $\mathbf{x}$. Besides showing the mere existence of δ-regular variables, we will develop an effective approach to determining them for any ideal $\mathcal{I} \subseteq \mathcal{P}$.

We begin by proving two useful technical lemmata on homogeneous ideals where we can exploit the natural grading of $\mathcal{P}$ by total degree. The number $\max_{h\in\mathcal{H}} \deg h$ is called the *degree* of the finite set $\mathcal{H} \subset \mathcal{P}$ and denoted by $\deg \mathcal{H}$.

Lemma 4.3.2. *Let the finite set $\mathcal{H}$ be a homogeneous Pommaret basis of the homogeneous ideal $\mathcal{I} \subseteq \mathcal{P}$. For any degree $q \geq \deg \mathcal{H}$, a Pommaret basis of the truncation $\mathcal{I}_{\geq q} = \bigoplus_{p\geq q} \mathcal{I}_p$ is given by*

$$\mathcal{H}_q = \left\{x^\mu h \mid h \in \mathcal{H},\ |\mu| + \deg h = q,\ \forall j > \operatorname{cls} h : \mu_j = 0\right\}. \tag{4.5}$$

Conversely, if $\mathcal{I}_{\geq q}$ possesses a Pommaret basis, then so does $\mathcal{I}$.

Proof. By the conditions in the definition of the set $\mathcal{H}_q$, any polynomial $h \in \mathcal{H}$ is multiplied by terms x^μ containing only variables multiplicative for it. Thus trivially $\operatorname{cls}(x^\mu h) = \operatorname{cls} \mu$. Furthermore, $\mathcal{H}_q$ is involutively head autoreduced, as $\mathcal{H}$ is. Now let $f \in \mathcal{I}_{\geq q}$ be an arbitrary homogeneous polynomial. As $\mathcal{H}$ is a Pommaret basis of $\mathcal{I}$, it has an involutive standard representation $f = \sum_{h\in\mathcal{H}} P_h h$ with

$P_h \in \Bbbk[x^1,\dots,x^{\operatorname{cls} h}]$. Hence f can be written as a linear combination of polynomials $x^\nu h$ where $|\nu| = \deg f - \deg h \geq q - \deg h$ and where x^ν contains only multiplicative variables. We decompose $\nu = \mu + \rho$ with $|\mu| = q - \deg h$ and $\rho_j = 0$ for all $j > \operatorname{cls}\mu$. Thus we may write $x^\nu h = x^\rho(x^\mu h)$ where $x^\mu h \in \mathcal{H}_q$ and x^ρ contains only variables multiplicative for it. But this observation trivially implies the existence of an involutive standard representation $f = \sum_{h' \in \mathcal{H}_q} P_{h'} h'$ with $P_{h'} \in \Bbbk[x^1,\dots,x^{\operatorname{cls} h'}]$ and thus $\mathcal{H}_q$ is a Pommaret basis of $\mathcal{I}_{\geq q}$ by Theorem 3.4.4.

The converse is also very simple. Let $\mathcal{H}_q$ be a finite Pommaret basis of the truncated ideal $\mathcal{I}_{\geq q}$ and $\mathcal{H}_p$ head autoreduced $\Bbbk$-linear bases of the components $\mathcal{I}_p$ for $0 \leq p < q$. If we set $\mathcal{H} = \bigcup_{p=0}^{q} \mathcal{H}_p$, then $\operatorname{le}_\prec \mathcal{H}$ is obviously a weak Pommaret basis of the full monoid ideal $\operatorname{le}_\prec \mathcal{I}$ and by Proposition 3.4.7 an involutive head autoreduction yields a strong basis. □

Lemma 4.3.3. *Using the same notations as in Lemma 4.3.2, write $\mathcal{B} = \operatorname{le}_\prec \mathcal{H}_q$. If $\nu \in \mathcal{B}$ with $\operatorname{cls}\nu = k$, then $\nu - 1_k + 1_j \in \mathcal{B}$ for all $j > k$. Conversely, let $\mathcal{B} \subseteq (\mathbb{N}_0^n)_q$ be a set of multi indices of degree q. If for each element $\nu \in \mathcal{B}$ with $\operatorname{cls}\nu = k$ and each non-multiplicative index $k < j \leq n$ of it the multi index $\nu - 1_k + 1_j$ is also contained in $\mathcal{B}$, then the set $\mathcal{B}$ is involutive for the Pommaret division.*

Proof. As $\mathcal{B}$ is an involutive basis of $\operatorname{le}_\prec \mathcal{I}_{\geq q}$, it must contain a multi index μ with $\mu \mid_P \nu + 1_j$. Obviously, $\operatorname{cls}(\nu + 1_j) = k$ and thus $\operatorname{cls}\mu \geq k$. Because of $|\mu| = |\nu|$, the only possibility is $\mu = \nu + 1_j - 1_k$. The converse is trivial, as each non-multiplicative multiple of $\nu \in \mathcal{B}$ is of the form $\nu + 1_j$ with $j > k = \operatorname{cls}\nu$ and hence has $\nu - 1_k + 1_j$ as an involutive divisor in $\mathcal{B}$. □

As in concrete computations one always represents an ideal $\mathcal{I} \subseteq \mathcal{P}$ by some finite generating set $\mathcal{F} \subset \mathcal{I}$, we also introduce a notion of regularity for such sets. Assume that the given set $\mathcal{F}$ is involutively head autoreduced with respect to an involutive division L and a term order $\prec$. In general, $\mathcal{F}$ is not an involutive basis of $\mathcal{I}$, but its involutive span $\langle \mathcal{F} \rangle_{L,\prec}$ is only a proper subset of $\mathcal{I}$.

Consider now a $\Bbbk$-linear change of variables $\tilde{\mathbf{x}} = A\mathbf{x}$ defined by a regular matrix $A \in \Bbbk^{n \times n}$. It transforms each $f \in \mathcal{P}$ into a polynomial $\tilde{f} \in \tilde{\mathcal{P}} = \Bbbk[\tilde{x}^1,\dots,\tilde{x}^n]$ of the same degree. Thus a given set $\mathcal{F} \subset \mathcal{P}$ is transformed into a set $\tilde{\mathcal{F}} \subset \tilde{\mathcal{P}}$ which generally is no longer involutively head autoreduced. Performing an involutive head autoreduction yields a set $\tilde{\mathcal{F}}^\triangle$. Again, $\tilde{\mathcal{F}}^\triangle$ will in general not be an involutive basis of the transformed ideal $\tilde{\mathcal{I}} \subseteq \tilde{\mathcal{P}}$.

Since we are dealing with homogeneous polynomials, we can use Hilbert functions to measure the size not only of ideals but also of involutive spans. Recall that the Hilbert function of the ideal $\mathcal{I}$ is given by $h_{\mathcal{I}}(r) = \dim_\Bbbk \mathcal{I}_r$ for all integers $r \geq 0$. For a finite, involutively head autoreduced set $\mathcal{F}$ we define similarly $h_{\mathcal{F},L,\prec}(r) = \dim_\Bbbk (\langle \mathcal{F} \rangle_{L,\prec})_r$. Obviously, we always find $h_{\mathcal{F},L,\prec}(r) \leq h_{\mathcal{I}}(r)$ with equality holding for all $r \geq 0$, if and only if $\mathcal{F}$ is an involutive basis. The same is true for the Hilbert function $h_{\tilde{\mathcal{F}}^\triangle,L,\prec}$ defined by the transformed basis $\tilde{\mathcal{F}}^\triangle$.

According to Lemma 3.4.14, an involutively head autoreduced set $\mathcal{F}$ defines the direct sum decomposition (3.31) of its involutive span $\langle \mathcal{F} \rangle_{L,\prec}$. This observation allows us to provide a simple explicit formula for the Hilbert function:

$$h_{\mathcal{F},L,\prec}(r) = \sum_{f\in\mathcal{F}} \binom{r-q_f+k_f-1}{r-q_f} \tag{4.6}$$

where $q_f = \deg f$ and k_f denotes the number of multiplicative variables of f (for $r < q_f$ we understand that the binomial coefficient is zero). Indeed, by (A.4a) the binomial coefficient in (4.6) gives the number of multiplicative multiples of f of degree r and thus the contribution of f to the involutive span at this degree.

Obviously, a $\Bbbk$-linear change of coordinates does not affect the Hilbert function of an ideal and thus we find $h_{\mathcal{I}} = h_{\tilde{\mathcal{I}}}$. However, this is not true for the Hilbert functions of the involutive spans $\langle\mathcal{F}\rangle_{L,\prec}$ and $\langle\tilde{\mathcal{F}}^{\triangle}\rangle_{L,\prec}$, respectively. We may now measure the effect of the made coordinate transformation by comparing the asymptotic behaviour of $h_{\mathcal{F},L,\prec}$ and $h_{\tilde{\mathcal{F}}^{\triangle},L,\prec}$.

Definition 4.3.4. Let the finite set $\mathcal{F} \subset \mathcal{P}$ of homogeneous polynomials be involutively head autoreduced with respect to the Pommaret division and a term order $\prec$. The given coordinates $\mathbf{x}$ are called *asymptotically regular* for $\mathcal{F}$ and $\prec$, if after any linear change of coordinates $\tilde{\mathbf{x}} = A\mathbf{x}$ the inequality $h_{\mathcal{F},P,\prec}(r) \geq h_{\tilde{\mathcal{F}}^{\triangle},P,\prec}(r)$ holds for all sufficiently large values $r \gg 0$.

Example 4.3.5. Let us reconsider Example 3.1.16 where $\mathcal{F} = \{xy\} \subset \Bbbk[x,y]$ with the degree reverse lexicographic order. Independent of how we order the variables, the class of xy is 1. Hence we have $h_{\mathcal{F},P,\prec}(r) = 1$ for all $r > 1$. After the change of coordinates $x = \tilde{x} + \tilde{y}$ and $y = \tilde{y}$, we obtain the set $\tilde{\mathcal{F}} = \{\tilde{y}^2 + \tilde{x}\tilde{y}\} \subset \Bbbk[\tilde{x},\tilde{y}]$. Its leading term is $\tilde{y}^2$ which is of class 2 implying that $h_{\tilde{\mathcal{F}},P,\prec}(r) = \binom{r-1}{r-2} = r-1$ for $r > 1$. Thus the original coordinates are not asymptotically regular for $\mathcal{F}$ and we know from Example 3.1.16 that they are also not δ-regular for the ideal $\mathcal{I} = \langle\mathcal{F}\rangle$. $\triangleleft$

Note that, given variables $\mathbf{x}$, generally asymptotic regularity for a finite set $\mathcal{F}$ according to Definition 4.3.4 and δ-regularity for the ideal $\mathcal{I} = \langle\mathcal{F}\rangle$ according to Definition 4.3.1 are independent properties. For a concrete instance where the two notions differ see Example 4.2.9 where one easily checks that the used coordinates are δ-regular for the whole submodule but not asymptotically regular for the given generating set, as any transformation of the form $x = \bar{x} + a\bar{y}$ with $a \neq 0$ will increase the Hilbert function. The main point is that δ-regularity for the ideal $\mathcal{I}$ is concerned with the monoid ideal $\mathrm{le}_\prec \mathcal{I}$ whereas asymptotic regularity for the set $\mathcal{F}$ depends on the ideal $\langle \mathrm{le}_\prec \mathcal{F}\rangle \subseteq \mathrm{le}_\prec \mathcal{I}$. However, in some cases the two regularity concepts are related. A simple instance is given by the following result.

Proposition 4.3.6. *Let the coordinates* $\mathbf{x}$ *be* δ*-regular for the ideal* $\mathcal{I} \subseteq \mathcal{P}$ *and the term order* $\prec$*. If the set* $\mathcal{H}$ *is a Pommaret basis of* $\mathcal{I}$ *for* $\prec$*, then the coordinates* $\mathbf{x}$ *are asymptotically regular for* $\mathcal{H}$ *and* $\prec$*.*

Proof. If the set $\mathcal{H}$ is a Pommaret basis of the ideal $\mathcal{I}$, then the two Hilbert functions $h_{\mathcal{I}}$ and $h_{\mathcal{H},P,\prec}$ coincide. As for any generating set $\mathcal{F}$ of $\mathcal{I}$ in any coordinate system the inequality $h_{\mathcal{F},P,\prec}(r) \leq h_{\mathcal{I}}(r)$ trivially holds for all $r \geq 0$, our coordinates are indeed asymptotically regular. □

δ-Regularity of the used coordinates $\mathbf{x}$ represents a trivial necessary condition for the existence of Pommaret bases for an ideal $\mathcal{I} \subseteq \mathcal{P}$. From an algorithmic point of view, their asymptotic regularity for the current basis $\mathcal{H}$ is equally important for the effective construction of a Pommaret basis by the completion Algorithm 4.3 (and its optimised variants which will be discussed in the next section). Even if the used coordinates $\mathbf{x}$ are δ-regular for the ideal $\mathcal{I}$, it may still happen that the algorithm does not terminate, as it tries to construct an infinite Pommaret basis for the monoid ideal $\langle \operatorname{le}_\prec \mathcal{H} \rangle$ (recall again Example 4.2.9).

Remark 4.3.7. For Definition 4.3.4 of asymptotic regularity for a finite set $\mathcal{F}$, the behaviour at lower degrees is irrelevant and it suffices to consider the involutive span of $\mathcal{F}$ only for degrees beyond $q = \deg \mathcal{F}$. Thus we can proceed as in Lemma 4.3.2 and replace $\mathcal{F}$ by the set $\mathcal{F}_q$ defined in analogy to (4.5), i. e. we consider all multiplicative multiples of degree q of elements of $\mathcal{F}$. If we perform a coordinate transformation and a subsequent involutive head autoreduction, then we obtain a set $\tilde{\mathcal{F}}_q^{\triangle}$ where again all elements are of degree q.

It is now very easy to decide which Hilbert function becomes asymptotically larger. Let $\beta_q^{(k)}$ denote the number of generators in $\mathcal{F}_q$ which are of class k and similarly $\tilde{\beta}_q^{(k)}$ for the set $\tilde{\mathcal{F}}_q^{\triangle}$. In our special case, it follows immediately from (4.6) that both Hilbert functions are actually polynomials for $r \geq q$. Furthermore, an expansion of the binomial coefficients in (4.6) shows that if we write the Hilbert function in the form $h_{\mathcal{F},P,\prec}(q+r) = \sum_{k=0}^{n-1} h_i r^i$, then each coefficient h_i is determined by a linear combination of $\beta_q^{(i+1)}, \dots, \beta_q^{(n)}$ with positive coefficients.[2] Thus we must simply compare first $\beta_q^{(n)}$ and $\tilde{\beta}_q^{(n)}$, then $\beta_q^{(n-1)}$ and $\tilde{\beta}_q^{(n-1)}$, and so on, until for the first time one coordinate system leads to a larger value; the corresponding Hilbert function is asymptotically larger. ◁

Choosing an arbitrary reference coordinate system $\hat{\mathbf{x}}$, we may identify every system of coordinates $\mathbf{x}$ with the unique regular transformation matrix $A \in \Bbbk^{n\times n}$ for which $\mathbf{x} = A\hat{\mathbf{x}}$. The next result says that asymptotic regularity for a given set $\mathcal{F}$ of polynomials is a generic property in the sense of the Zariski topology, i. e. almost all coordinates are asymptotically regular for $\mathcal{F}$.

Proposition 4.3.8. *The coordinate systems $\mathbf{x}$ which are asymptotically singular for a finite involutively head autoreduced set $\mathcal{F} \subset \mathcal{P}$ and a term order $\prec$ form a Zariski closed proper subset of $\Bbbk^{n\times n}$.*

Proof. By the considerations in Remark 4.3.7, it suffices to treat the case that all elements of $\mathcal{F}$ possess the same degree. Let us perform a coordinate transformation $\bar{\mathbf{x}} = A\mathbf{x}$ with an undetermined matrix A, i. e. we consider its entries as parameters. It obviously leads to an asymptotically regular coordinate system, as each polynomial in $\tilde{\mathcal{F}}^{\triangle}$ will get its maximally possible class. Asymptotically singular coordinates

[2] Exactly the same computation as we will perform in the proof of Proposition 8.2.6 shows that the precise relation is $h_i = \sum_{k=i}^{n-1} \beta_q^{(k+1)} s_{k-i}^{(k)}(0)/k!$ where the coefficients $s_{k-i}^{(k)}(0)$ are modified Stirling numbers (see Appendix A.4).

are defined by the vanishing of certain (leading) coefficients. These coefficients are polynomials in the entries of A. Thus the set of all asymptotically singular coordinate systems corresponds to the zero set of a polynomial ideal. □

Our next goal is an effective criterion for recognising that coordinates are asymptotically singular for a given set $\mathcal{F}$ and a class respecting term order. The basic idea consists of comparing the multiplicative variables assigned by the Pommaret and the Janet division, respectively. The definitions of these two divisions are apparently quite different. Somewhat surprisingly, they nevertheless yield very similar sets of multiplicative indices.

Proposition 4.3.9. *Let the finite set $\mathcal{B} \subset \mathbb{N}_0^n$ of multi indices be involutively autoreduced with respect to the Pommaret division. Then for all multi indices $\nu \in \mathcal{B}$ the inclusion $N_P(\nu) \subseteq N_{J,\mathcal{B}}(\nu)$ holds.*

Proof. Let $\nu \in \mathcal{B}$ with $\operatorname{cls}\nu = k$. Thus $\nu = [0,\dots,0,\nu_k,\dots,\nu_n]$ with $\nu_k > 0$ and $N_P(\nu) = \{1,\dots,k\}$. We must show that the indices $1,\dots,k$ are also multiplicative for ν with respect to the Janet division. In order to decide whether $k \in N_{J,\mathcal{B}}(\nu)$, we study the set $(\nu_{k+1},\dots,\nu_n) \subseteq \mathcal{B}$. If it contained a multi index μ with $\mu_k > \nu_k$, the set $\mathcal{B}$ would not be involutively autoreduced with respect to the Pommaret division, as $\nu \,|_P\, \mu$. Thus k is multiplicative for the Janet division, too. The same argument can be applied for $k-1,\dots,1$ where sets of the form $(0,\dots,0,\nu_k,\dots,\nu_n)$ with an increasing number of zeros are considered. Hence $N_P(\nu) \subseteq N_{J,\mathcal{B}}(\nu)$. □

Thus, if $\mathcal{F} \subset \mathcal{P}$ is a finite, Pommaret head autoreduced set of polynomials, then its involutive spans with respect to the Pommaret and Janet division, respectively, always satisfy the inclusion $\langle\mathcal{F}\rangle_{P,\prec} \subseteq \langle\mathcal{F}\rangle_{J,\prec} \subseteq \langle\mathcal{F}\rangle$. For later use, we note two simple corollaries to Proposition 4.3.9. Recall that by Remark 3.1.15 any finite set $\mathcal{B} \subset \mathbb{N}_0^n$ is involutively autoreduced with respect to the Janet division. We first prove that an involutive autoreduction of $\mathcal{B}$ with respect to the *Pommaret* division can make its *Janet* span only larger but not smaller.

Corollary 4.3.10. *Let $\mathcal{B} \subset \mathbb{N}_0^n$ be an arbitrary finite set of multi indices and set $\mathcal{B}_P = \mathcal{B} \setminus \{\nu \in \mathcal{B} \mid \exists \mu \in \mathcal{B} : \mu \,|_P\, \nu\}$, i. e. we eliminate all multi indices possessing a Pommaret divisor in $\mathcal{B}$. Then $\langle\mathcal{B}\rangle_J \subseteq \langle\mathcal{B}_P\rangle_J$.*

Proof. If $\mu^{(1)} \,|_P\, \mu^{(2)}$ and $\mu^{(2)} \,|_P\, \nu$, then trivially $\mu^{(1)} \,|_P\, \nu$. Thus for each eliminated multi index $\nu \in \mathcal{B} \setminus \mathcal{B}_P$ another multi index $\mu \in \mathcal{B}_P$ exists with $\mu \,|_P\, \nu$. Let $\operatorname{cls}\mu = k$. By the proposition above $\{1,\dots,k\} \subseteq N_{J,\mathcal{B}_P}(\mu)$. Assume that an index $j > k$ exists with $j \in N_{J,\mathcal{B}}(\nu)$. By definition of the Pommaret division, $\mu_i = \nu_i$ for all $i > k$. Thus $\mu \in (\nu_{j+1},\dots,\nu_n)$ and $j \in N_{J,\mathcal{B}}(\mu)$. As by the second condition on an involutive division $N_{J,\mathcal{B}}(\mu) \subseteq N_{J,\mathcal{B}_P}(\mu)$ for all $\mu \in \mathcal{B}_P$, we conclude that $j \in N_{J,\mathcal{B}_P}(\mu)$ and $\mathcal{C}_{J,\mathcal{B}}(\nu) \subset \mathcal{C}_{J,\mathcal{B}_P}(\mu)$. But this immediately implies $\langle\mathcal{B}\rangle_J \subseteq \langle\mathcal{B}_P\rangle_J$. □

The next corollary implies that any Pommaret basis is also a Janet basis (with respect to the same term order $\prec$). Thus if the set $\mathcal{H}$ is a Pommaret basis, then $X_{P,\prec}(h) = X_{J,\mathcal{H},\prec}(h)$ for all polynomials $h \in \mathcal{H}$.

Corollary 4.3.11. *Let the finite set $\mathcal{H} \subset \mathcal{P}$ be involutive with respect to the Pommaret division (and some term order). Then $\mathcal{H}$ is also involutive with respect to the Janet division (and the same term order).*

Proof. Proposition 4.3.9 trivially implies that $\mathcal{H}$ is at least weakly involutive with respect to the Janet division. For the Janet division any weakly involutive set is strongly involutive, provided that no two elements have the same leading terms (recall Example 3.4.13). But in a Pommaret basis this cannot happen. □

We next show that for an asymptotically regular coordinate system and a class respecting term order the inclusions in Proposition 4.3.9 must be equalities. In other words, if a variable x^ℓ exists which is multiplicative for an element of $\mathcal{F}$ with respect to the Janet division but non-multiplicative with respect to the Pommaret division, then the used coordinates are asymptotically singular for this set $\mathcal{F}$. Our proof is constructive in the sense that it shows us explicitly how to find coordinates leading to a larger Hilbert function.

Theorem 4.3.12. *Let the finite set $\mathcal{F} \subset \mathcal{P}$ be involutively head autoreduced with respect to the Pommaret division and a class respecting term order $\prec$ and let the coefficient field $\Bbbk$ be infinite. If the set $\mathcal{F}$ possesses more multiplicative variables for the Janet division than for the Pommaret division, then the used coordinates $\mathbf{x}$ are asymptotically singular for it.*

Proof. By the proposition above, we have $X_{P,\prec}(f) \subseteq X_{J,\mathcal{F},\prec}(f)$ for all $f \in \mathcal{F}$. Assume that for a polynomial $h \in \mathcal{F}$ the strict inclusion $X_{P,\prec}(h) \subset X_{J,\mathcal{F},\prec}(h)$ holds. Thus a variable $x^\ell \in X_{J,\mathcal{F},\prec}(h)$ with $\ell > k = \operatorname{cls} h$ exists. If we define the set $\mathcal{F}_q$ for $q = \deg \mathcal{F}$ as in Remark 4.3.7, then $\mathcal{F}_q$ contains in particular the generator $(x^k)^{q-\deg h} h$ which is still of class k. It is easy to see that the variable x^ℓ is also Janet multiplicative for this generator. Hence we may again assume without loss of generality that all elements of $\mathcal{F}$ are of the same degree q.

We perform now the linear change of variables $x^i = \tilde{x}^i$ for $i \neq k$ and $x^k = \tilde{x}^k + a\tilde{x}^\ell$ with a yet arbitrary parameter $a \in \Bbbk \setminus \{0\}$. It induces the following transformation of the terms $x^\mu \in \mathbb{T}$:

$$x^\mu = \sum_{j=0}^{\mu_k} \binom{\mu_k}{j} a^j \tilde{x}^{\mu - j_k + j_\ell} . \tag{4.7}$$

Note that $\tilde{x}^\mu$ appears on the right hand side as the only term whose coefficient does not depend on a and that all other terms are greater with respect to our class respecting term order (and their coefficients are different powers of a). Let $\operatorname{le}_\prec h = \mu$. Thus $\mu = [0,\dots,0,\mu_k,\dots,\mu_n]$ with $\mu_k > 0$. Consider the multi index $\nu = \mu - (\mu_k)_k + (\mu_k)_\ell$; obviously, $\operatorname{cls} \nu > k$. Applying our transformation to the polynomial h leads to a polynomial $\tilde{h}$ containing the term $\tilde{x}^\nu$. Note that ν cannot be an element of $\operatorname{le}_\prec \mathcal{F}$. Indeed, if it were, it would be an element of the same set $(\mu_{\ell+1},\dots,\mu_n)$ as μ. But this contradicts our assumption that ℓ is multiplicative for the multi index μ with respect to the Janet division, as by construction $\nu_\ell > \mu_\ell$.

Transforming all polynomials $f \in \mathcal{F}$ yields the set $\tilde{\mathcal{F}}$ on which we perform an involutive head autoreduction in order to obtain the set $\tilde{\mathcal{F}}^\triangle$. Under our assumption on the size of the ground field $\Bbbk$, we can choose the parameter a such that after the transformation each polynomial $\tilde{f} \in \tilde{\mathcal{F}}$ has at least the same class as the corresponding polynomial $f \in \mathcal{F}$, as our term order respects classes. This fact is a simple consequence of (4.7): cancellations of terms may occur only, if the parameter a is a zero of some polynomial (possibly one for each member of $\mathcal{F}$) with a degree not higher than $\deg \mathcal{F}$.

We know already that for the polynomial h considered above the transformation leads to a polynomial $\tilde{h}$ of greater class. We consider now all polynomials $f \in \mathcal{F}$ with $\mathrm{cls}\, f > k$ and h. After the change of variables all the transformed polynomials will thus have a class greater than k. Because of the special form of our transformation, the old leading exponent always remains in the support of each transformed polynomial and if exponents appear which are greater for our term order, then they are always accompanied by a coefficient depending on a. Furthermore, we noted above that ν was not contained in $\mathrm{le}_{\prec} \mathcal{F}$. As all our generators are of the same degree q, an involutive head autoreduction amounts to a simple Gaussian elimination. For a generic choice of the parameter a, it follows from our considerations above that even after the involutive head autoreduction each generator has at least the same class as in the original set $\mathcal{F}$ (and at least one a higher class).

Taking the remaining members of $\mathcal{F}$ into account may only increase the number of elements in $\tilde{\mathcal{F}}^{\triangle}$ possessing a class greater than k. But this fact implies that at least one of the values $\tilde{\beta}_q^{(k+1)}, \dots, \tilde{\beta}_q^{(n)}$ is larger than the corresponding value for $\mathcal{F}$. By Remark 4.3.7, the Hilbert function of $\tilde{\mathcal{F}}^{\triangle}$ is then asymptotically greater than the one of $\mathcal{F}$ and our coordinates are not asymptotically regular. □

Corollary 4.3.13. *If the coordinates* $\mathbf{x}$ *are asymptotically regular for the finite Pommaret head autoreduced set* $\mathcal{F} \subset \mathcal{P}$*, then with respect to a class respecting term order* $\prec$ *we have* $X_{P,\prec}(f) = X_{J,\mathcal{F},\prec}(f)$ *for all generators* $f \in \mathcal{F}$*.*

It is important to note that this corollary provides us only with a necessary but *not* with a sufficient criterion for asymptotic regularity of the coordinates $\mathbf{x}$. In other words, even if the Janet and the Pommaret division yield the same multiplicative variables for a given Pommaret head autoreduced set $\mathcal{F} \subset \mathcal{P}$, this fact does not imply that the used coordinates are asymptotically regular for $\mathcal{F}$.

Example 4.3.14. Let $\mathcal{F} = \{\underline{z^2} - y^2 - 2x^2,\ \underline{xz} + xy,\ \underline{yz} + y^2 + x^2\}$. The underlined terms are the leading ones for the degree reverse lexicographic order. One easily checks that the Janet and the Pommaret division yield the same multiplicative variables. If we perform the transformation $\tilde{x} = z$, $\tilde{y} = y + z$ and $\tilde{z} = x$, then after an autoreduction we obtain the set $\tilde{\mathcal{F}}^{\triangle} = \{\underline{\tilde{z}^2} - \tilde{x}\tilde{y},\ \underline{\tilde{y}\tilde{z}},\ \underline{\tilde{y}^2}\}$. Again the Janet and the Pommaret division lead to the same multiplicative variables, but the Hilbert function $h_{\mathcal{F},P,\prec}$ is asymptotically smaller than $h_{\tilde{\mathcal{F}}^{\triangle},P,\prec}$, as we find $\beta_2^{(2)} = 1 < 2 = \tilde{\beta}_2^{(2)}$. Thus the coordinates (x, y, z) are not asymptotically regular for $\mathcal{F}$.

The explanation of this phenomenon is very simple. Obviously our criterion depends only on the leading terms of the set $\mathcal{F}$. In other words, it analyses the monomial ideal $\langle \mathrm{lt}_{\prec} \mathcal{F} \rangle$. Here $\langle \mathrm{lt}_{\prec} \mathcal{F} \rangle = \langle xz, yz, z^2 \rangle$ and one easily verifies that the used generating set is already a Pommaret basis. However, for $\mathcal{I} = \langle \mathcal{F} \rangle$ the leading ideal is $\operatorname{lt}_\prec \mathcal{I} = \langle x^3, xz, yz, z^2 \rangle$ (one obtains a Janet basis for $\mathcal{I}$ by adding the polynomial x^3 to $\mathcal{F}$) and obviously it does not possess a finite Pommaret basis, as such a basis would have to contain all monomials $x^3 y^k$ with $k \in \mathbb{N}$ (or we exploit our criterion noting that y is a Janet but not a Pommaret multiplicative variable for x^3). Thus we have the opposite situation compared to Example 4.2.9: there $\operatorname{lt}_\prec \mathcal{I}$ had a finite Pommaret basis but $\langle \operatorname{lt}_\prec \mathcal{F} \rangle$ not; here it is the other way round. We will show later (Proposition 5.3.7) that whenever $\langle \operatorname{lt}_\prec \mathcal{F} \rangle$ does not possess a finite Pommaret basis, then $\mathcal{F}$ possesses more Janet than Pommaret multiplicative variables. $\triangleleft$

Eventually, we return to the problem of the existence of a finite Pommaret basis for every ideal $\mathcal{I} \subseteq \mathcal{P}$. As the Pommaret division is not Noetherian, Theorem 4.2.8 cannot be directly applied. However, with a little trick exploiting our above results on the relationship between the Pommaret and the Janet division we can achieve our goal at least for infinite fields.

Theorem 4.3.15. *Let $\prec$ be an arbitrary term order and $\Bbbk$ an infinite field. Then every polynomial ideal $\mathcal{I} \subseteq \mathcal{P}$ possesses a finite Pommaret basis for $\prec$ in suitably chosen variables* $\mathbf{x}$.

Proof. As a first step we show that every ideal has a *Pommaret* head autoreduced *Janet* basis. Indeed, let us apply the completion Algorithm 4.3 for the Janet division with a slight modification: we perform the involutive head autoreductions in the Line /3/ with respect to the Pommaret instead of the Janet division. It is obvious that if the algorithm terminates, the result is a basis with the wanted properties.

The Janet division is Noetherian by Lemma 3.1.19. Thus without our modification the termination is obvious. With respect to the Janet division every set of multi indices is involutively autoreduced. Hence a Janet head autoreduction only takes care that no two elements of a set have the same leading exponents (recall Example 3.4.13). But when we form the union $\mathcal{H} \cup \mathcal{S}$ in Line /3/, we add to $\mathcal{H}$ only polynomials that are in involutive normal form with respect to $\mathcal{H}$ so that no involutive head reductions are possible. As the Pommaret head autoreduction may only lead to a larger monoid ideal $\langle \operatorname{le}_\prec \mathcal{H} \rangle$, the Noetherian argument in the proof of Theorem 4.2.8 remains valid even after our modification.

According to Corollary 4.3.10, the Pommaret head autoreductions may only increase the Janet spans $\langle \operatorname{le}_\prec \mathcal{H} \rangle_J$. Thus the termination of Algorithm 4.1 invoked in Line /4/ is not affected by our modification and the modified algorithm still terminates for arbitrary input.

Let us now work in an undetermined coordinate system; i.e. we perform a coordinate transformation $\bar{\mathbf{x}} = A\mathbf{x}$ with an undetermined matrix A as in the proof of Proposition 4.3.8. By the considerations above, the modified algorithm will terminate and thus treats only a finite number of bases $\mathcal{H}_i$. According to Proposition 4.3.8, the coordinate systems that are asymptotically singular for at least one of them form a Zariski closed set. Thus generic coordinates are asymptotically regular for all sets $\operatorname{lt}_\prec \mathcal{H}_i$ and by Corollary 4.3.13[3] their Janet and their Pommaret spans coincide. But

[3] Note that it is not relevant here that the corollary assumes the use of a class respecting term order, since our argument deals only with monomial sets.

this observation implies that the result of the modified algorithm is not only a Janet but also a Pommaret basis. □

The argument at the end of this proof immediately implies the following analogue to Proposition 4.3.8.

Corollary 4.3.16. *The coordinate systems* $\mathbf{x}$ *which are* δ*-singular for a given ideal* $\mathcal{I} \subseteq \mathcal{P}$ *and a term order* $\prec$ *form a Zariski closed proper subset of* $\Bbbk^{n\times n}$.

Remark 4.3.17. Combining our completion Algorithm 4.3 and the criterion for asymptotic singularity provided by Theorem 4.3.12, we can effectively determine δ-regular coordinates for any ideal $\mathcal{I} \subseteq \mathcal{P}$. Our approach is based on the observation that if our given coordinate system $\mathbf{x}$ is not δ-regular for $\mathcal{I}$, then any attempt to compute a Pommaret basis of $\mathcal{I}$ with Algorithm 4.3 will sooner or later lead to a basis $\mathcal{H}$ of $\mathcal{I}$ for which the coordinates $\mathbf{x}$ are not asymptotically regular. Indeed, by the considerations in the proof of Theorem 4.3.15, the completion with respect to the Janet division (using Pommaret head autoreductions) terminates. Thus either the result is simultaneously a Pommaret basis of $\mathcal{I}$ (and the given coordinates $\mathbf{x}$ are already δ-regular for $\mathcal{I}$) or at some stage we encounter a basis $\mathcal{H}$ of $\mathcal{I}$ possessing more Janet than Pommaret multiplicative variables implying by Theorem 4.3.12 that the coordinates $\mathbf{x}$ are not asymptotically regular for $\mathcal{H}$.

There are (at least) two possibilities to exploit this observation in concrete computations. The first one consists of following Algorithm 4.3 with the Pommaret division and checking before each iteration in the subalgorithm `MonomialComplete` whether there are more Janet than Pommaret multiplicative variables. If this is the case, then we perform coordinate transformations of the form used in the proof of Theorem 4.3.12 until the Janet and the Pommaret division yield the same multiplicative variables. Then we continue with the completion. Alternatively, we compute a Pommaret head autoreduced Janet basis (which always exists by the considerations above) and check whether it is simultaneously a Pommaret basis. If this is the case, we again conclude that our coordinates $\mathbf{x}$ are δ-regular. Otherwise, we perform coordinate transformations as above and start again.

It is easy to provide for each approach examples where it fares better than the other one. The main disadvantage of the first approach is that it may perform transformations even if the coordinates $\mathbf{x}$ are δ-regular for the ideal $\mathcal{I}$. Such redundant transformations will always occur, if we encounter during the completion a basis $\mathcal{H}$ such that the coordinates $\mathbf{x}$ are not δ-regular for the monoid ideal $\langle \operatorname{le}_\prec \mathcal{H}\rangle$ (this assertion follows from Proposition 5.3.7). As one can see from Example 4.2.9, sometimes the transformations are indeed necessary for the termination of the completion but sometimes they just make the computations more expensive.

In the second approach this problem does not appear, as we only check at the very end whether we actually have got a Pommaret basis. Thus we consider only $\operatorname{le}_\prec \mathcal{I}$ and not already some subideal contained in it. If the original coordinates $\mathbf{x}$ are δ-regular, then no transformation at all will be performed and we clearly fare better than with the first approach. On the other hand, if the coordinates $\mathbf{x}$ are not δ-regular, then we will not notice this fact before the end of the Janet completion. It will follow from our results in the next chapter that in such a situation, a Janet basis of $\mathcal{I}$ will typically be larger than a Pommaret basis in δ-regular coordinates; in particular, generally the Janet basis will contain elements of unnecessary high degree. Thus in such situations the first approach will typically fare better, as it avoids a number of unnecessary normal form computations.

Note that at this point we are not able to prove that either strategy will lead to a Pommaret basis after a *finite* number of coordinate transformations. With the help of the theory developed in the next chapter, we will be able to provide a proof at least for the most important case of a class respecting order (Remark 5.5.25). The basic problem is that we do not know a bound for the degree of either a Janet or a Pommaret basis. It is clear that both every completion step and every coordinate transformation asymptotically increase the Hilbert function $h_{\mathcal{H},P,\prec}$ of the current basis $\mathcal{H}$. However, without a bound on the degree of the appearing bases, this information is not sufficient to conclude that either approach terminates in a finite number of steps. ◁

Example 4.3.18. Let us apply Algorithm 4.5 to the Pommaret completion of the set $\mathcal{F} = \{\underline{z^2} - y^2 - 2x^2,\ \underline{xz} + xy,\ \underline{yz} + y^2 + x^2\}$ (with respect to the degree reverse lexicographic order). We have seen in Example 4.3.14 that the coordinates are not δ-regular for $\mathcal{I}$, although the Janet and the Pommaret span of $\mathcal{F}$ coincide. According to our algorithm we must first analyse the polynomial $y(xz+xy)$. Its involutive normal form with respect to $\mathcal{F}$ is $-\underline{x^3}$. If we determine the multiplicative variables for the enlarged set, they do not change for the old elements. For the new polynomial the Janet division yields $\{x,y\}$. As y is obviously not multiplicative for the Pommaret division, our criterion tells us that the coordinates are not asymptotically regular for the enlarged basis and that the Pommaret completion may not terminate. Indeed, here it is easy to see that no polynomial of the form x^3y^k with $k > 0$ is involutively reducible and thus no finite Pommaret basis can exist for $\langle\mathcal{F}\rangle$.

In this example, the Janet completion (with or without Pommaret autoreductions) ends with the addition of this single obstruction to involution and we obtain as Janet basis of $\langle\mathcal{F}\rangle$ the set

$$\mathcal{F}_J = \{\underline{z^2} - y^2 - 2x^2,\ \underline{xz} + xy,\ \underline{yz} + y^2 + x^2,\ \underline{x^3}\}\ . \tag{4.8}$$

In Example 4.3.14 we showed that the transformation $\tilde{x} = z$, $\tilde{y} = y + z$ and $\tilde{z} = x$ yields after an autoreduction the set $\tilde{\mathcal{F}}^{\triangle} = \{\underline{\tilde{z}^2} - \tilde{x}\tilde{y},\ \underline{\tilde{y}\tilde{z}} - \tilde{x},\ \underline{\tilde{y}^2} - \tilde{z}\}$. One easily checks that it is a Pommaret and thus also a Janet basis. This example clearly demonstrates that the Janet division also "feels" δ-singularity in the sense that in such coordinates it typically leads to larger bases of higher degree. ◁

Remark 4.3.19. In Theorems 4.3.12 and 4.3.15 we assumed that we are working over an infinite field. A closer look at the proofs reveals that we could relax this assumption to "sufficiently large" where the required size of $\Bbbk$ is essentially determined by the degree and the size of the considered set $\mathcal{F}$. Thus in the case of a finite field, it may be necessary to enlarge $\Bbbk$ in order to guarantee the existence of a Pommaret basis. This problem is similar to the situation when one tries to put a zero-dimensional ideal in normal x^n-position [267]. ◁

4.4 Construction of Minimal Bases and Optimisations

Algorithm 4.3 is simple and easy to analyse theoretically. However, in general it is not particularly efficient. As a first improvement, we present now an algorithm that merges the computations in Lines /4/ and /5/ of Algorithm 4.3 and thus requires in general less normal form computations. The new Algorithm 4.5 directly generalises the strategy of the monomial Algorithm 4.1. We multiply each element of the current basis by its non-multiplicative variables. Then we look whether the result is already contained in the involutive span of the basis; if not, it is added. The check whether an element lies in the involutive span is performed via an involutive normal form computation: since we are always dealing with an involutively head autoreduced basis, the involutive normal form of a member of the ideal vanishes, if and only if it is an element of the involutive span.

Algorithm 4.5 Involutive basis in $(\mathcal{P},\star,\prec)$ (improved form)

Input: a finite set $\mathcal{F}\subset\mathcal{P}$, an involutive division L
Output: an involutive basis $\mathcal{H}$ of $\mathcal{I}=\langle\mathcal{F}\rangle$ with respect to L and $\prec$
1: $\mathcal{H}\leftarrow \texttt{InvHeadAutoReduce}_{L,\prec}(\mathcal{F})$
2: **loop**
3: $\quad\mathcal{S}\leftarrow\{x_j\star h\mid h\in\mathcal{H}, x_j\in\overline{X}_{L,\mathcal{H},\prec}(h), x_j\star h\notin\langle\mathcal{H}\rangle_{L,\prec}\}$
4: $\quad$**if** $\mathcal{S}=\emptyset$ **then**
5: $\quad\quad$**return** $\mathcal{H}$
6: $\quad$**else**
7: $\quad\quad$choose $\bar{g}\in\mathcal{S}$ such that $\operatorname{le}_\prec\bar{g}=\min_\prec\mathcal{S}$
8: $\quad\quad g\leftarrow\texttt{NormalForm}_{L,\prec}(\bar{g},\mathcal{H})$
9: $\quad\quad\mathcal{H}\leftarrow\texttt{InvHeadAutoReduce}_{L,\prec}(\mathcal{H}\cup\{g\})$
10: $\quad$**end if**
11: **end loop**

The manner in which we choose in Line /7/ the next polynomial $\bar{g}$ to be treated (we briefly write $\min_\prec\mathcal{S}$ for the minimal leading exponent of an element of $\mathcal{S}$) corresponds to what is called the normal selection strategy in the theory of Gröbner bases. There, this strategy is known to work very well with degree compatible term orders but not so well for other orders like the purely lexicographic one. Whereas for Gröbner bases the selection strategy concerns only the efficiency, we will see below that here the use of this strategy is important for our termination proof.

Theorem 4.4.1. *Let L be a constructive Noetherian division and $(\mathcal{P},\star,\prec)$ a polynomial algebra of solvable type. Then Algorithm 4.5 terminates for any finite input set $\mathcal{F}\subset\mathcal{P}$ with an involutive basis $\mathcal{H}$ of $\mathcal{I}=\langle\mathcal{F}\rangle$.*

Proof. We begin by proving the *correctness* of the algorithm under the assumption that it terminates. The relation $\mathcal{I}=\langle\mathcal{H}\rangle$ remains valid throughout, although $\mathcal{H}$ changes. But the only changes are the addition of further elements of $\mathcal{I}$ and involutive head autoreductions; both operations do not affect the ideal generated by $\mathcal{H}$.

When the algorithm terminates, we have $\mathcal{S} = \emptyset$ and thus the output $\mathcal{H}$ is locally involutive and by Proposition 4.2.7 involutive.

There remains the problem of *termination*. Algorithm 4.5 produces a sequence $(\mathcal{H}_1, \mathcal{H}_2, \dots)$ with $\langle \mathcal{H}_i \rangle = \mathcal{I}$. The set $\mathcal{H}_{i+1}$ is determined from $\mathcal{H}_i$ in Line /9/. We distinguish two cases, namely whether or not during the computation of the involutive normal form in Line /8/ the leading exponent changes. If $\mathrm{le}_\prec \bar{g} = \mathrm{le}_\prec g$, then trivially $\langle \mathrm{le}_\prec \mathcal{H}_i \rangle = \langle \mathrm{le}_\prec \mathcal{H}_{i+1} \rangle$, as $\mathrm{le}_\prec g = \mathrm{le}_\prec h + 1_j$ for some $h \in \mathcal{H}_i$. Otherwise we claim that $\langle \mathrm{le}_\prec \mathcal{H}_i \rangle \subsetneq \langle \mathrm{le}_\prec \mathcal{H}_{i+1} \rangle$.

Because of the normal selection strategy, the basis $\mathcal{H}_i$ is involutive up to the leading exponent μ of the polynomial $\min_\prec \mathcal{S}$. If $\mathrm{le}_\prec g \prec \mu$, then Lemma 4.2.6 asserts that the ordinary and the involutive normal form of g with respect to $\mathcal{H}_i$ coincide. By construction, g is in involutive normal form, hence we must have $\mathrm{le}_\prec g \notin \langle \mathrm{le}_\prec \mathcal{H}_i \rangle$ and consequently $\langle \mathrm{le}_\prec \mathcal{H}_i \rangle \subsetneq \langle \mathrm{le}_\prec \mathcal{H}_{i+1} \rangle$, as otherwise g would be reducible.

So the `loop` in Algorithm 4.5 generates an ascending chain of monoid ideals $\langle \mathrm{le}_\prec \mathcal{H}_1 \rangle \subseteq \langle \mathrm{le}_\prec \mathcal{H}_2 \rangle \subseteq \cdots \subseteq \mathrm{le}_\prec \mathcal{I}$. As $\mathbb{N}_0^n$ is Noetherian, the chain must become stationary at some index N. It follows from the considerations above that in all iterations of the `loop` after the Nth one we find $\mathrm{le}_\prec \bar{g} = \mathrm{le}_\prec g$ in Line /8/. At this stage Algorithm 4.5 reduces to an involutive completion of the exponent set $\mathrm{le}_\prec \mathcal{H}_N$ using Algorithm 4.1—but with additional involutive autoreductions after each appearance of a new element. Indeed, we choose the polynomial $\bar{g}$ in Line /7/ such that $\mathrm{le}_\prec \bar{g}$ is a possible choice for the multi index μ Algorithm 4.1 adds in Line /8/. Since we assume that our division is Noetherian, this modified algorithm terminates by Remark 4.2.2 with an involutive basis of $\langle \mathrm{le}_\prec \mathcal{H}_N \rangle$. Hence Algorithm 4.5 terminates, too, and our correctness proof above implies that in fact $\langle \mathrm{le}_\prec \mathcal{H}_N \rangle = \mathrm{le}_\prec \mathcal{I}$. □

Remark 4.4.2. The problem of non-termination even if a finite involutive basis exists remains unchanged by the above optimisation of Algorithm 4.3. However, variations of Theorem 4.4.1 hold also for divisions which are not Noetherian. For example, we could assume that all subideals of $\mathrm{le}_\prec \mathcal{I}$ possess an involutive basis. Alternatively, we could restrict to term orders of type ω. Then it suffices to assume that $\mathrm{le}_\prec \mathcal{I}$ has an involutive basis. Indeed, now it is not possible that Algorithm 4.5 iterates endlessly within $\mathrm{le}_\prec \mathcal{H}_i$, as sooner or later an element $\bar{g}$ must be selected in Line /7/ such that $\mathrm{le}_\prec g \notin \mathrm{le}_\prec \mathcal{H}_i$. ◁

Next, we consider the special case of term orders of type ω and show that for them our algorithm always determines in a finite number of steps a *Gröbner basis* of the given ideal, even if it does not terminate.

Proposition 4.4.3. *Let the term order $\prec$ be of type ω. Then Algorithm 4.5 determines for any finite input set $\mathcal{F} \subset \mathcal{P}$ in a finite number of steps a Gröbner basis of the ideal $\mathcal{I} = \langle \mathcal{F} \rangle$.*

Proof. In the proof of Theorem 4.4.1 we introduced the intermediate set $\mathcal{H}_N$ such that $\langle \mathcal{H}_{N+\ell} \rangle = \langle \mathcal{H}_N \rangle$ for all $\ell > 0$. We claim that $\mathcal{H}_N$ is a Gröbner basis of $\mathcal{I}$.

Let $f \in \mathcal{I}$ be an arbitrary element of the ideal. As $\mathcal{H}_N$ is a basis of $\mathcal{I}$, we find for each $h \in \mathcal{H}_N$ a polynomial $g_h \in \mathcal{P}$ such that

$$f = \sum_{h\in\mathcal{H}_N} g_h \star h\,. \tag{4.9}$$

By Proposition B.4.8, $\mathcal{H}_N$ is a Gröbner basis, if and only if we can choose the non-vanishing coefficients g_h so that $\operatorname{lt}_\prec(g_h \star h) \preceq \operatorname{lt}_\prec f$, i.e. if and only if (4.9) is a standard representation. Assume that for f no standard representation exists and let $\mu = \max_{h\in\mathcal{H}_N}\{\operatorname{le}_\prec g_h + \operatorname{le}_\prec h\} \succ \operatorname{lt}_\prec f$. If we denote by $\bar{\mathcal{H}}_N$ the subset of all generators $\bar{h} \in \mathcal{H}_N$ for which we have $\operatorname{le}_\prec g_{\bar{h}} + \operatorname{le}_\prec \bar{h} = \mu$, then the identity $\sum_{\bar{h}\in\bar{\mathcal{H}}_N} \operatorname{lm}_\prec(g_{\bar{h}} \star \bar{h}) = 0$ must hold. This obviously requires that the set $\bar{\mathcal{H}}_N$ has more than one element. As we are dealing with an involutively head autoreduced set, the involutive cones $\mathcal{C}_{L,\operatorname{le}_\prec \mathcal{H}_N}(\operatorname{le}_\prec \bar{h})$ are disjoint and at least one polynomial $\bar{h} \in \bar{\mathcal{H}}_N$ exists such that some non-multiplicative variable $x^j \in \overline{X}_{L,\mathcal{H}_N}(\bar{h})$ divides $\operatorname{lt}_\prec g_{\bar{h}}$.

As $\prec$ is of type ω, after a finite number of steps the non-multiplicative product $x^j \star \bar{h}$ is analysed in Algorithm 4.5. We know already that at this stage of the algorithm the leading term does not change during the involutive normal form computation. Thus for some $n_1 \geq 0$ the set $\mathcal{H}_{N+n_1}$ contains an element $\bar{h}'$ with $\operatorname{le}_\prec \bar{h}' = \operatorname{le}_\prec(x^j \star \bar{h})$. Let $\nu = \operatorname{le}_\prec g_{\bar{h}}$, $x^{\nu-1_j} \star x^j = cx^\nu + r_1$ and $\bar{h}' = dx^j \star \bar{h} + r_2$. Then we may rewrite

$$g_{\bar{h}} \star \bar{h} = \frac{\operatorname{lc}_\prec g_{\bar{h}}}{cd}\left[x^{\nu-1_j} \star (\bar{h}' - r_2) - dr_1 \star \bar{h}\right] + \left(g_{\bar{h}} - \operatorname{lm}_\prec g_{\bar{h}}\right) \star \bar{h}\,. \tag{4.10}$$

The leading term on the right hand side is $\operatorname{lt}_\prec(x^{\nu-1_j} \star \bar{h}')$. If $\operatorname{supp} x^{\nu-1_j}$ contains a non-multiplicative variable $x^k \in \overline{X}_{L,\mathcal{H}_{N+n_1}}(\bar{h}')$, then we can repeat the argument obtaining an element $\bar{h}'' \in \mathcal{H}_{N+n_1+n_2}$ such that $\operatorname{le}_\prec \bar{h}'' = \operatorname{le}_\prec(x^k \star \bar{h}')$ and so on.

Obviously, this process terminates after a finite number of steps—even if we do it for each $\bar{h} \in \bar{\mathcal{H}}_N$. Thus after ℓ further iterations (for some finite number $\ell \geq 0$) we obtain a set $\mathcal{H}_{N+\ell}$ such that, after applying all the found relations (4.10), f can be expressed in the form $f = \sum_{h\in\mathcal{H}_{N+\ell}} \tilde{g}_h \star h$ where still $\mu = \max_{h\in\mathcal{H}_{N+\ell}}\{\operatorname{le}_\prec \tilde{g}_h + \operatorname{le}_\prec h\}$. Denote again by $\bar{\mathcal{H}}_{N+\ell} \subseteq \mathcal{H}_{N+\ell}$ the set of all those polynomials $\bar{h}$ for which this maximum is achieved.

By construction, no term $\operatorname{lt}_\prec \tilde{g}_{\bar{h}}$ with $\bar{h} \in \bar{\mathcal{H}}_{N+\ell}$ contains a non-multiplicative variable for $\bar{h}$. Hence, $|\bar{\mathcal{H}}_{N+\ell}| = 1$, since $\mathcal{H}_{N+\ell}$ is also involutively head autoreduced. But this fact contradicts our assumption that $\mu \succ \operatorname{lt}_\prec f$. Thus every polynomial $f \in \mathcal{I}$ possesses a standard representation already with respect to $\mathcal{H}_N$ and this set is a Gröbner basis implying that $\langle \operatorname{le}_\prec \mathcal{H}_N\rangle = \operatorname{le}_\prec \mathcal{I}$. □

Algorithm 4.5 will in general not produce a minimal involutive basis. As a trivial counterexample consider the monomial ideal generated by $\mathcal{F} = \{x, x^2\} \subset \Bbbk[x]$. Obviously, the generator x^2 is here redundant. But with respect to the Janet division $\mathcal{F}$ is involutively autoreduced and thus Algorithm 4.5 will return $\mathcal{F}$ unchanged as Janet basis of $\mathcal{I} = \langle\mathcal{F}\rangle$ whereas the minimal Janet basis is of course $\{x\}$.

In contrast to the monomial case, now it does not suffice to make some minor changes in order to reach a minimal involutive basis. In particular, it does not suffice to simply apply Algorithm 4.5 to a minimal basis in the ordinary sense. The probably simplest method is given by Algorithm 4.6.

Algorithm 4.6 Minimal involutive basis in $(\mathcal{P},\star,\prec)$

Input: a finite set $\mathcal{F}\subset\mathcal{P}$, an involutive division L
Output: a minimal involutive basis $\mathcal{H}$ of $\mathcal{I}=\langle\mathcal{F}\rangle$ with respect to L and $\prec$

1: $\mathcal{H}\leftarrow\emptyset$; $\quad\mathcal{Q}\leftarrow\mathcal{F}$
2: **repeat**
3: $\quad g\leftarrow 0$
4: $\quad$ **while** $(\mathcal{Q}\neq\emptyset)\wedge(g=0)$ **do**
5: $\quad\quad$ choose $f\in\mathcal{Q}$ such that $\mathrm{le}_\prec f=\min_\prec\mathcal{Q}$
6: $\quad\quad$ $\mathcal{Q}\leftarrow\mathcal{Q}\setminus\{f\}$; $\quad g\leftarrow\mathtt{NormalForm}_{L,\prec}(f,\mathcal{H})$
7: $\quad$ **end while**
8: $\quad$ **if** $g\neq 0$ **then**
9: $\quad\quad$ $\mathcal{H}'\leftarrow\{h\in\mathcal{H}\mid \mathrm{le}_\prec g\prec\mathrm{le}_\prec h\}$; $\quad\mathcal{H}\leftarrow(\mathcal{H}\cup\{g\})\setminus\mathcal{H}'$
10: $\quad\quad$ $\mathcal{Q}\leftarrow\mathcal{Q}\cup\mathcal{H}'\cup\{x\star g\mid x\in\overline{X}_{L,\mathcal{H},\prec}(g)\}$
11: $\quad$ **end if**
12: **until** $\mathcal{Q}=\emptyset$
13: **return** $\mathcal{H}$

The basic idea of this algorithm is fairly simple. We now work with two sets $\mathcal{H}$ and $\mathcal{Q}$ which together always define a basis of the given ideal $\mathcal{I}=\langle\mathcal{F}\rangle$ and for which always $\max_\prec\mathcal{H}\preceq\min_\prec\mathcal{Q}$ holds (when both sets are non-empty). The set $\mathcal{H}$ will eventually be the minimal involutive basis of $\mathcal{I}$. The elements of $\mathcal{Q}$ are potential members of this basis (including in particular all non-multiplicative multiples of the generators $h\in\mathcal{H}$) which still must be checked for involutive reducibility. The decisive difference to Algorithm 4.5 is that the check in Line /6/ is done with respect to the set $\mathcal{H}$ only. Furthermore, $\mathcal{H}$ is built up in such a way that no "internal reduction" can be overlooked: whenever a new element is added to it in Line /9/, all previous elements which have a greater leading exponent are moved back to the set $\mathcal{Q}$ and thus later again checked for involutive reducibility. Following our usual strategy of checking for local involution, we also augment in Line /10/ the set $\mathcal{Q}$ by all products of the new basis member g with its non-multiplicative variables.

This modification has the effect that now input sets like $\mathcal{F}=\{x,x^2\}$ for the Janet division can be reduced to a minimal involutive basis, although they define already an involutive basis. Indeed, after one iteration we have $\mathcal{H}=\{x\}$ and $\mathcal{Q}=\{x^2\}$. If we now compute in Line /6/ the involutive normal form of $f=x^2$ with respect to $\mathcal{H}$, then it trivially vanishes and the output is the minimal involutive basis $\{x\}$.

Theorem 4.4.4. *Under the assumptions of Theorem 4.4.1, Algorithm 4.6 terminates with a minimal involutive basis of $\mathcal{I}=\langle\mathcal{F}\rangle$.*

Proof. It is trivial to see that upon termination of the algorithm the set $\mathcal{H}$ is locally involutive and generates the whole ideal $\mathcal{I}$. Hence it is an involutive basis of $\mathcal{I}$. Thus there only remains to show that the algorithm terminates under the made assumptions and that the output is a minimal involutive basis.

The proof of the *termination* of Algorithm 4.6 requires only slight variations of the arguments given in the proof of Theorem 4.4.1. We consider now the monoid ideals $\mathcal{J}_i=\langle\mathrm{le}_\prec\mathcal{H}_i\rangle+\langle\mathrm{le}_\prec\mathcal{Q}_i\rangle$ where again the index i refers to the value after the ith iteration of the `repeat`-loop. If we have $\mathrm{le}_\prec g=\mathrm{le}_\prec f$ in Line /6/, then trivially

$\mathcal{J}_i = \mathcal{J}_{i+1}$. Otherwise, $\mathrm{le}_\prec g$ is not divisible by the leading exponent of any element of $\mathcal{Q}_i$ by the choice of f and it follows again from our use of the normal selection strategy and from Lemma 4.2.6 that $\mathrm{le}_\prec g \notin \langle \mathrm{le}_\prec \mathcal{H}_i \rangle$. Thus in this case $\mathcal{J}_i \subsetneq \mathcal{J}_{i+1}$.

As the ascending chain $\mathcal{J}_0 \subseteq \mathcal{J}_1 \subseteq \mathcal{J}_2 \subseteq \cdots$ of monoid ideals must become stationary, after a finite number N of iterations we always have $\mathrm{le}_\prec g = \mathrm{le}_\prec f$. No subsequent iteration removes an element from the set $\mathcal{H}$. Thus now even the monoid ideals $\langle \mathrm{le}_\prec \mathcal{H}_i \rangle$ with $i > N$ define an ascending chain and as soon as it becomes stationary, Algorithm 4.6 reduces to the monomial completion Algorithm 4.1 which terminates under the made assumptions.

For proving the *minimality* of $\mathcal{H}$, let us assume that $\mathcal{H}$ was not a minimal basis. By definition, this means that $\mathrm{le}_\prec \mathcal{H}$ is not the unique minimal involutive basis $\mathcal{B}_L$ of $\mathrm{le}_\prec \mathcal{I}$ but that $\mathcal{B}_L \subsetneq \mathrm{le}_\prec \mathcal{H}$. Hence the set $\mathrm{le}_\prec \mathcal{H}$ contains multi indices which are not in $\mathcal{B}_L$; among these let $\mathrm{le}_\prec h$ be the minimal one. We introduce the two sets $\mathcal{B}' = \{\nu \in \mathcal{B}_L \mid \nu \prec \mathrm{le}_\prec h\}$ and $\mathcal{H}' = \{h' \in \mathcal{H} \mid \mathrm{le}_\prec h' \in \mathcal{B}'\}$.

Consider the moment when h was finally entered into $\mathcal{H}$ (h might have already been entered at earlier times but was removed again, when polynomials with smaller leading exponents appeared). Obviously, at this time we had $\mathcal{H} = \mathcal{H}'$ in the algorithm. The second condition in Definition 3.1.1 of an involutive division implies that $N_{L,\mathcal{B}_L}(\mathrm{le}_\prec h') \subseteq N_{L,\mathcal{B}'}(\mathrm{le}_\prec h')$ for all $h' \in \mathcal{H}'$. As $\mathcal{B}_L$ is the minimal involutive basis of $\mathrm{le}_\prec \mathcal{I}$, the multi index $\mathrm{le}_\prec h$ has an involutive divisor in $\mathcal{B}_L$. The above inclusion implies that $\mathrm{le}_\prec h$ must also have an involutive divisor in $\mathrm{le}_\prec \mathcal{H}'$ contradicting that the polynomial h was in involutive normal form with respect to $\mathcal{H}'$ when it was added. Hence $\mathcal{H}$ is a minimal basis. □

Algorithm 4.6 allows us in principle the construction of a minimal involutive basis for any ideal and division satisfying the made assumptions. While it surely fares better than similarly simple versions of Buchberger's Algorithm B.3 for Gröbner bases, it still will fail on larger examples for complexity reasons. Although this problem is not a central topic of this book, we discuss at least a few optimisations omitting some redundant computations.

Example 4.4.5. Let us consider in $\Bbbk[x,y,z]$ the ideal generated by the three polynomials $f_1 = \underline{x} + p_1$, $f_2 = \underline{y^2} + p_2$ and $f_3 = \underline{z^2} + p_3$. Here p_1, p_2, p_3 are some polynomials subject to the sole constraint that the underlined terms are indeed the leading terms for the chosen term order, say the degree reverse lexicographic order. For both the Pommaret and the Janet division the first steps in the completion will be to add the involutive normal forms of the generators $f_4 = \underline{xy} + yp_1$ and $f_5 = \underline{xz} + zp_1$. Obviously, no involutive head reductions can be performed and for simplicity we assume that involutive reductions of lower terms are not possible either.

For the Pommaret division, z is non-multiplicative for f_4 and y is non-multiplicative for f_5. Obviously $zf_4 = yf_5 = yzf_1$ and one of these non-multiplicative products is redundant, as it completely suffices to compute the involutive normal form of only one. For the Janet division, y is multiplicative for f_5. Our algorithm computes zf_4 and then notices that its involutive normal form vanishes due to the involutive reduction with yf_5. Thus for both divisions the algorithm performs unnecessary computations the outcome of which we can predict in advance. ◁

This example presented two trivial cases of redundant computations. In the theory of Gröbner bases more complex situations are covered by the second Buchberger criterion (Proposition B.4.19). In the next chapter the proof of Theorem 5.4.4 will show that many instances of it are automatically taken into account by our strategy of considering only products with non-multiplicative variables in the completion. For this reason, the importance of criteria for detecting unnecessary reductions is much less pronounced as in the classical theory of Gröbner bases. Small to medium size examples can be reasonably handled with the simple Algorithm 4.5; only larger problems require further optimisations.

In Line /3/ of Algorithm 4.5 we multiply each generator in the current basis with all its non-multiplicative variables and compute the involutive normal form of the result in order to determine the set $\mathcal{S}$. Of course, in a concrete implementation one would not explicitly determine the full set $\mathcal{S}$. Instead one would first perform the multiplication that yields a minimal leading term, then the one that yields the second smallest leading term, and so on until we encounter a product with a non-vanishing involutive normal form. Nevertheless, the same product (and its involutive normal form) may still be computed repeatedly which is often quite expensive.

A simple strategy allows us to avoid these redundant computations. We store for each generator $h \in \mathcal{H}$ a set Y of all variables with which we have already multiplied it. In later iterations of the main loop of our algorithm we ignore these variables, even if they are still non-multiplicative for h. While this optimisation is trivially permitted for global divisions, its correctness is less clear for general divisions. Basically, it is again a consequence of our use of the normal selection strategy to analyse always the non-multiplicative product with the smallest leading term that renders this optimisation correct.

Assume that we analysed at some stage the non-multiplicative product $x^j \star h$. Thus we determined its involutive normal form. In general, this computation entails that we performed some involutive reductions. Now it may happen for a not globally defined division that at a later stage of the algorithm, these reductions are no longer involutive, as some multiplicative variables have become non-multiplicative. Hence, if we analysed the product again, we would get a different result.

However, under such circumstances our selection strategy will never choose this particular non-multiplicative product $x^j \star h$. Indeed, the leading exponents of all polynomials subtracted from $x^j \star h$ during the involutive normal form computation are trivially less than or at most equal to $\mathrm{le}_{\prec}(x^j \star h)$. Hence, if one of the reductions has become non-multiplicative, this fact entails that we have now at least one further non-multiplicative product $x^k \star g$ such that $\mathrm{le}_{\prec}(x^k \star g) \preceq \mathrm{le}_{\prec}(x^j \star h)$ which should be analysed first. At the time the polynomial $x^j \star h$ possesses again the smallest leading exponent among all non-multiplicative products, we can redo all former involutive reductions (or perform equivalent ones), as all possible obstructions have already been removed in the meantime. This follows from the same kind of continuity argument that was used for proving Theorem 4.4.1.

In order to avoid such repeated non-multiplicative products, we work in Algorithm 4.7 below with triples (f, μ, Y): $f \in \mathcal{P}$ is a polynomial; the multi index μ satisfies $\mu \preceq \mathrm{le}_{\prec} f$; the set Y contains all variables with which the polynomial f (or some predecessor of it) has already been multiplied. The multi index μ obtains a value which is less than $\mathrm{le}_\prec f$, if f arises as a non-multiplicative product and if in the subsequent normal form computation the leading exponent is not changed (see Line /32/). In this case μ is set equal to the multi index contained in the predecessor triple; thus one might say that it encodes the "history" of the generator f. Given a set $\mathcal{S}$ of such triples, we write briefly $\mathrm{le}_\prec \mathcal{S}$ for the set of the leading exponents of the polynomials f appearing in triples $(f,\mu,Y) \in \mathcal{S}$ and $\min_\prec \mathcal{S}$ for the minimal exponent among these.

Next we discuss the involutive form of Buchberger's second criterion. As usually, it states that certain normal form computations can be avoided, as we can be predict that they yield zero. In order to formulate it in precisely the form in which we apply it, we must already refer to our final completion Algorithm 4.7. There the criterion is realised by the subalgorithm `Criterion`$(g,\nu,\mathcal{H})$ which takes as argument a polynomial g, a multi index ν and a set $\mathcal{H}$ of triples (f,μ,Y). It returns `True`, if $\mathcal{H}$ contains a triple (h,μ,Y) such that $\mathrm{le}_\prec h$ is an involutive divisor of $\mathrm{le}_\prec g$ with respect to $\mathrm{le}_\prec \mathcal{H}$ and $\mathrm{lcm}\,(\nu,\mu) \prec \mathrm{le}_\prec g$. Note that the design of Algorithm 4.7 ensures that the occurring sets $\mathcal{H}$ are always locally involutive up to a sufficiently large multi index so that the following proposition guarantees the correctness of the criterion.

Proposition 4.4.6. *Assume that we obtain in the course of the Algorithm 4.7 within our current basis $\mathcal{F}$ a triple (g,ν,Z) and a subset $\mathcal{H} \subseteq \mathcal{F}$ satisfying:*

(i) *The set $\mathcal{H}$ is involutively head autoreduced and locally involutive up to* $\mathrm{le}_\prec g$.
(ii) *There is a triple $(h,\mu,Y) \in \mathcal{H}$ with* $\mathrm{le}_\prec h \,|_{L,\mathrm{le}_\prec \mathcal{H}}\, \mathrm{le}_\prec g$ *and* $\mathrm{lcm}\,(\mu,\nu) \prec \mathrm{le}_\prec g$.

Then the involutive normal form of g with respect to $\mathcal{H}$ vanishes.

Proof. We prove this proposition by reducing it to the standard form of Buchberger's second criterion.[4] Note that the assumption on $\mathrm{lcm}\,(\mu,\nu)$ implies that either h or g (or both) arose during the algorithm as the non-multiplicative product of some other member of a previous basis. Let us assume for simplicity that h comes from some polynomial h_0; the other case goes completely analogously. We consider now the S-polynomials $S_\prec(h_0,h)$ and $S_\prec(h_0,g)$. Since $\mathrm{le}_\prec h_0 \mid \mathrm{le}_\prec h$, it also divides $\mathrm{lcm}\,(\mathrm{le}_\prec h, \mathrm{le}_\prec g) = \mathrm{le}_\prec g$. It follows that both S-polynomials have a leading exponent which is less than $\mathrm{le}_\prec g$.

The key for proving our claim is the following observation. Within our completion algorithms we compute repeatedly the involutive normal form of a non-multiplicative product $x \star f$. Let us assume that in the first step we reduce involutively its leading term with respect to some polynomial $\bar{f}$. Then we obtain as intermediate result the S-polynomial $S_\prec(f,\bar{f})$. Thus in a somewhat hidden manner our

[4] In Appendix B.4 Buchberger's second criterion is discussed only for the commutative polynomial ring. Kredel [264, Lemma 4.5.8] showed that it remains valid for arbitrary polynomial algebras of solvable type. Note that this is not the case for Buchberger's first criterion: consider for example in the Weyl algebra $\mathcal{W}_1 = \Bbbk[x,\partial]$ the two polynomials $f = x$ and $g = \partial$ [264, Ex. 4.5.11]; obviously, their leading exponents are coprime, but their S-polynomial $S_\prec(f,g) = -1$ is not reducible by either f or g.

Algorithm 4.7 Minimal involutive basis in $(\mathcal{P}, \star, \prec)$ (optimised form)

Input: a finite set $\mathcal{F} \subset \mathcal{P}$, an involutive division L
Output: a minimal involutive basis $\mathcal{H}$ of $\mathcal{I} = \langle \mathcal{F} \rangle$ with respect to L and $\prec$

$\mathcal{F} \leftarrow \texttt{InvHeadAutoReduce}_{L,\prec}(\mathcal{F})$
$\mathcal{Q} \leftarrow \{(f, \mathrm{le}_\prec f, \emptyset) \mid f \in \mathcal{F}\}$; $\quad \mathcal{H} \leftarrow \emptyset$
repeat
  $g \leftarrow 0$
  while $(\mathcal{Q} \neq \emptyset) \wedge (g = 0)$ **do**
    choose $(f, \mu, Y) \in \mathcal{Q}$ such that $\mathrm{le}_\prec f = \min_\prec \mathcal{Q}$; $\quad \mathcal{Q} \leftarrow \mathcal{Q} \setminus \{(f, \mu, Y)\}$
    if $\neg\texttt{Criterion}(f, \mu, \mathcal{H})$ **then**
      $g \leftarrow \texttt{NormalForm}_{L,\prec}(f, \mathcal{H})$
    end if
  end while
  if $g \neq 0$ **then**
    if $\mathrm{le}_\prec g = \mathrm{le}_\prec f$ **then**
      $\mathcal{H} \leftarrow \mathcal{H} \cup \{(g, \mu, Y)\}$
    else
      $\mathcal{H}' \leftarrow \{(h, \mu, Y) \in \mathcal{H} \mid \mathrm{le}_\prec h \succ \mathrm{le}_\prec g\}$
      $\mathcal{H} \leftarrow (\mathcal{H} \cup \{(g, \mathrm{le}_\prec g, \emptyset)\}) \setminus \mathcal{H}'$; $\quad \mathcal{Q} \leftarrow \mathcal{Q} \cup \mathcal{H}'$
    end if
  end if
  repeat
    $\mathcal{S} \leftarrow \{x \star h \mid (h, \mu, Y) \in \mathcal{H},\ x \in \overline{X}_{L,\mathcal{H},\prec}(h) \setminus Y\}$
    if $\mathcal{Q} \neq \emptyset$ **then**
      $\mathcal{S} \leftarrow \{f \in \mathcal{S} \mid \mathrm{le}_\prec f \prec \min_\prec \mathcal{Q}\}$
    end if
    $g \leftarrow 0$
    while $(\mathcal{S} \neq \emptyset) \wedge (g = 0)$ **do**
      choose $(h, \mu, Y) \in \mathcal{H}, x \in \overline{X}_{L,\mathcal{H},\prec}(h) \setminus Y$ such that $\mathrm{le}_\prec (x \star h) = \min_\prec \mathcal{S}$
      $\mathcal{H} \leftarrow (\mathcal{H} \setminus \{(h, \mu, Y)\}) \cup \{(h, \mu, Y \cup \{x\})\}$; $\quad \mathcal{S} \leftarrow \mathcal{S} \setminus \{x \star h\}$
      if $\neg\texttt{Criterion}(x \star h, \mu, \mathcal{H})$ **then**
        $g \leftarrow \texttt{NormalForm}_{L,\prec}(x \star h, \mathcal{H})$
        if $g \neq 0$ **then**
          if $\mathrm{le}_\prec g = \mathrm{le}_\prec (x \star h)$ **then**
            $\mathcal{H} \leftarrow \mathcal{H} \cup \{(g, \mu, \emptyset)\}$
          else
            $\mathcal{H} \leftarrow \mathcal{H} \cup \{(g, \mathrm{le}_\prec g, \emptyset)\}$
          end if
          $\mathcal{H}' \leftarrow \{(h, \mu, Y) \in \mathcal{H} \mid \mathrm{le}_\prec h \succ \mathrm{le}_\prec g\}$
          $\mathcal{H} \leftarrow \mathcal{H} \setminus \mathcal{H}'$; $\quad \mathcal{Q} \leftarrow \mathcal{Q} \cup \mathcal{H}'$
        end if
      end if
    end while
  until $\mathcal{S} = \emptyset$
until $\mathcal{Q} = \emptyset$
return $\{h \mid (h, \mu, Y) \in \mathcal{H}\}$

algorithms also proceed by analysing some S-polynomials—just like the familiar Buchberger Algorithm B.3. Our proof of Theorem 5.4.4 in the next chapter shows that it suffices to consider this particular subset of all S-polynomials, as all other S-polynomials are then automatically covered by Buchberger's second criterion.

According to our first assumption, the set $\mathcal{H}$ is involutively head autoreduced and locally involutive up to $\mathrm{le}_{\prec} g$. It follows then from Proposition 3.4.16 that for all polynomials f with $\mathrm{le}_{\prec} f \prec \mathrm{le}_{\prec} g$ the involutive and the ordinary normal form with respect to $\mathcal{H}$ coincide. Thus by construction both $S_{\prec}(h_0, h)$ and $S_{\prec}(h_0, g)$ and hence again by Buchberger's second criterion also $S_{\prec}(h, g)$ have a vanishing involutive normal form. But since $\mathrm{le}_{\prec} h$ is an involutive divisor of $\mathrm{le}_{\prec} g$, the involutive normal forms of $S_{\prec}(h, g)$ and g are identical. □

Theorem 4.4.7. *Under the assumptions of Theorem 4.4.1, Algorithm 4.7 terminates with a minimal involutive basis of* $\mathcal{I} = \langle \mathcal{F} \rangle$.

Proof. As Algorithm 4.7 follows the same basic strategy as Algorithm 4.6, we can take over many parts of the proof of Theorem 4.4.4. This includes in particular the proof of the minimality of the computed basis. The correctness of the criterion for avoiding the treatment of certain non-multiplicative products is asserted by Proposition 4.4.6 and we also showed that it is not necessary to repeat non-multiplicative prolongations. Thus the correctness of the algorithm is immediate.

For proving the termination we must only show that the arguments given in the proof of Theorem 4.4.4 remain valid, although there exists now a second place where the sets $\mathcal{H}$ and $\mathcal{Q}$ are modified, namely in Lines /30–38/. Again it is obvious that the ideal $\mathcal{J}_i$ remains unchanged, if $\mathrm{le}_{\prec} g = \mathrm{le}_{\prec}(x \star h)$ in Line /29/. Otherwise, $\mathrm{le}_{\prec} g \notin \langle \mathrm{le}_{\prec} \mathcal{Q}_i \rangle$ by construction and $\mathrm{le}_{\prec} g \notin \langle \mathrm{le}_{\prec} \mathcal{H}_i \rangle$ by Lemma 4.2.6, so that $\mathcal{J}_i \subsetneq \mathcal{J}_{i+1}$ and we obtain as before an ascending chain of monoid ideals. □

Algorithm 4.7 can be used as starting point for concrete implementations. Tests have demonstrated that it exhibits a quite good behaviour in practice and in particular that it is highly competitive to traditional Gröbner bases implementations based on variants of the classical Buchberger Algorithm B.3. Hence even if one is not interested in an involutive basis but only in the reduced Gröbner basis of the given ideal, it can be useful to apply Algorithm 4.7. In this case, the output as described above is not really the result one wants, as it still has to be autoreduced. However, with the help of the information contained in the triples (f, μ, Y), this problem can be solved very elegantly.

So far we have used the multi indices μ only for applying the involutive form of Buchberger's second criterion. If one takes a closer look at the assignment of this entry, one sees that $\mu \neq \mathrm{le}_{\prec} f$, if and only if f arises as a non-multiplicative multiple of another generator and if during the subsequent involutive normal form computation the leading exponent does *not* change. Thus even after dropping all such polynomials, we still have a Gröbner basis. In fact, because of our use of the normal selection strategy it will even be the *reduced* Gröbner basis of $\mathcal{I}$ (after a normalisation of the leading coefficients).

In order to prove this assertion we must show that the remaining leading exponents form the minimal basis of $\mathrm{le}_{\prec} \mathcal{I}$. Whenever a new polynomial g is added to the set $\mathcal{H}$ and its leading exponent has changed in the preceeding involutive normal form computation, then $\mathrm{le}_\prec g$ does not possess an involutive divisor in $\mathrm{le}_\prec \mathcal{H}$. We claim now that in fact it has no divisor at all. The key observation is that we always check all "small" non-multiplicative multiples of the elements of $\mathcal{H}$ *before* we look at the next member of $\mathcal{Q}$ (the outer appearance of Algorithm 4.7 seems to indicate the opposite, but note that in Line /2/ we initialise $\mathcal{H}$ with the empty set).

Assume that in the involutive normal form computation in Line /8/ or /29/, respectively, the leading coefficient has changed but nevertheless a generator $h^{(1)} \in \mathcal{H}$ exists with $\mu^{(1)} = \mathrm{le}_\prec h^{(1)} \mid \nu = \mathrm{le}_\prec g$. Since $\mu^{(1)}$ obviously cannot be an involutive divisor, there must be a non-multiplicative variable $x^{(1)} \in \overline{X}_{L,\mathcal{H},\prec}(h^{(1)})$ such that $\mathrm{le}_\prec (x^{(1)} \star h^{(1)}) = \mu^{(1)} + 1_{j^{(1)}}$ still divides ν. But this fact implies that the non-multiplicative multiple $x^{(1)} \star h^{(1)}$ was previously checked and its involutive normal form turned out to vanish. Hence $\mathcal{H}$ must contain a generator $h^{(2)}$ such that $\mu^{(2)} = \mathrm{le}_\prec h^{(2)}$ is an involutive divisor of $\mu^{(1)} + 1_{j^{(1)}} \mid \nu$. As $\mu^{(2)}$ cannot be an involutive divisor of ν either, we can continue in this manner and obtain as result a sequence $(\mu^{(1)}, \mu^{(2)}, \dots)$ as in Definition 4.1.3 of a continuous division. Thus the assumption that $\nu \in \langle \mathrm{le}_\prec \mathcal{H} \rangle$ contradicts the continuity of the division L.

Remark 4.4.8. In the last section we discussed in Remark 4.3.17 the construction of δ-regular coordinates $\mathbf{x}$ for an ideal $\mathcal{I} \subseteq \mathcal{P}$ (and thus of a Pommaret basis for $\mathcal{I}$) in the context of the "monomial" completion Algorithm 4.3. It is, however, trivial to see that the discussion applies with obvious adaptions to any of the improved algorithms derived in this section. ◁

4.5 Semigroup Orders

For a number of applications it is of interest to compute involutive or Gröbner bases not only with respect to monoid orders but also for *semigroup orders* (cf. Definition A.1.1); these applications include in particular computations in local rings which are often needed in algebraic geometry. According to Lemma A.1.6, a semigroup order is a well-order, if and only if it is a monoid order. Hence for semigroup orders which are not a monoid order, infinite strictly descending sequences exist and our argument that the normal form Algorithm B.1 (or its involutive variant, respectively) always terminates breaks down. Indeed, it is easy to provide an example of a non-terminating normal form computation: take the univariate polynomial ring $\mathbb{k}[x]$ with the unique semigroup order satisfying $x \prec 1$; if we try to compute the normal form of $g = x$ with respect to $f = \underline{1} - x$, then we obtain as intermediate results the infinite sequence x, x^2, x^3,... Thus all arguments and proofs based on the termination of normal form computations become invalid. In particular, we can no longer guarantee the termination of Algorithm 4.5 or one of its optimised variants.

Example 4.5.1. Let us consider the Weyl algebra $\mathbb{W}_n$. The only difference to the ordinary commutative polynomial ring is the commutation relation $\partial_i \star x^i = x^i \partial_i + 1$

for all $1 \leq i \leq n$. While $\mathcal{W}_n$ is trivially of solvable type for any monoid order, a semigroup order respects the multiplication $\star$, if and only if $1 \prec x^i\partial_i$ for all $1 \leq i \leq n$. Saito et al [391] call such orders *multiplicative monomial orders*.

An important class of semigroup orders is defined via real weight vectors. Let $(\xi,\zeta) \in \mathbb{R}^{2n}$ be such that all entries of the vector $\xi + \zeta \in \mathbb{R}^n$ are non-negative and let $\prec$ be an arbitrary monoid order. Then we set $x^\mu\partial^\nu \prec_{(\xi,\zeta)} x^\sigma\partial^\tau$, if either $\mu\cdot\xi + \nu\cdot\zeta < \sigma\cdot\xi + \tau\cdot\zeta$ or $\mu\cdot\xi + \nu\cdot\zeta = \sigma\cdot\xi + \tau\cdot\zeta$ and $x^\mu\partial^\nu \prec x^\sigma\partial^\tau$ where the dot denotes the standard scalar product in $\mathbb{R}^n$. It is easy to see that this definition yields a monoid order, if and only if both ξ and ζ contain only non-negative entries. A special case are the orders with weight vectors $(\xi,-\xi)$ arising from the action of the algebraic torus $(\mathbb{k}^\times)^n$ on $\mathcal{W}_n$. They have numerous applications in the theory of $\mathcal{D}$-modules [391] and obviously such an order is never a monoid order. ◁

Because of the weaker properties of semigroup orders, we must slightly modify our definitions of (weak) involutive or Gröbner bases. The proof of Theorem 3.4.4 (and consequently also the one of Corollary 3.4.5 showing that a weak involutive basis of an ideal $\mathcal{I}$ is indeed a basis of $\mathcal{I}$) requires normal form computations and thus is no longer valid. In fact, it is not difficult to provide counterexamples. The same problem occurs for Gröbner bases. Therefore we must explicitly include the basis property as a further condition in our definition.

Definition 4.5.2. Let $(\mathcal{P},\star,\prec)$ be a polynomial algebra of solvable type where $\prec$ is a semigroup order and $\mathcal{I} \subseteq \mathcal{P}$ a left ideal. A *Gröbner basis* of $\mathcal{I}$ is a finite set $\mathcal{G}$ such that $\langle\mathcal{G}\rangle = \mathcal{I}$ and $\langle \mathrm{le}_\prec \mathcal{G}\rangle = \mathrm{le}_\prec \mathcal{I}$. The set $\mathcal{G}$ is a *weak involutive basis* of $\mathcal{I}$ for the involutive division L, if in addition the leading exponents $\mathrm{le}_\prec \mathcal{G}$ form a weakly involutive set for the division L. It is a *(strong) involutive basis*, if $\mathrm{le}_\prec \mathcal{G}$ is even an involutively head autoreduced involutive set.

In the case of Gröbner bases, a classical trick to circumvent the problem of non-termination of normal form computations consists of homogenising the input. Thus let $(\mathcal{P},\star,\prec)$ be a polynomial algebra of solvable type for a semigroup order $\prec$. We set $\tilde{\mathcal{P}} = \mathbb{k}[x^1,\dots,x^{n+1}]$; for convenience we often write t instead of x^{n+1}. We extend the multiplication $\star$ of $\mathcal{P}$ to $\tilde{\mathcal{P}}$ by defining that t commutes with all other variables and the elements of $\mathbb{k}$. The *homogenisation* of a polynomial $f = \sum c_\mu x^\mu \in \mathcal{P}$ of degree q is defined as $f^{(h)} = \sum c_\mu x^\mu t^{q-|\mu|}$. Conversely, for a polynomial $\tilde{f} \in \tilde{\mathcal{P}}$ we denote its projection to the original algebra $\mathcal{P}$ as $f = \tilde{f}|_{t=1}$.

We denote by $\tilde{\mathbb{T}}$ the set of terms in $\tilde{\mathcal{P}}$; obviously, it is in one-to-one correspondence to the multi indices in $\mathbb{N}_0^{n+1}$. We use in the sequel the following convention. Multi indices in $\mathbb{N}_0^{n+1}$ always carry a tilde: $\tilde{\mu} = [\mu_1,\dots,\mu_{n+1}]$. The projection from $\mathbb{N}_0^{n+1}$ to $\mathbb{N}_0^n$ defined by dropping the last entry (i. e. the exponent of the homogenisation variable t) is signalled by omitting the tilde; thus $\mu = [\mu_1,\dots,\mu_n]$. For subsets $\tilde{\mathcal{B}} \subset \mathbb{N}_0^{n+1}$ we also simply write $\mathcal{B} = \{\nu \mid \tilde{\nu} \in \tilde{\mathcal{B}}\} \subset \mathbb{N}_0^n$. Note that generally $|\mathcal{B}| \leq |\tilde{\mathcal{B}}|$, as some multi indices in $\tilde{\mathcal{B}}$ may differ only in their last entry.

We start by lifting the semigroup order $\prec$ on $\mathbb{T}$ to a monoid order $\prec_h$ on $\tilde{\mathbb{T}}$ defined by setting $x^{\tilde{\mu}} \prec_h x^{\tilde{\nu}}$, if either $|\tilde{\mu}| < |\tilde{\nu}|$ or both $|\tilde{\mu}| = |\tilde{\nu}|$ and $x^\mu \prec x^\nu$. As a degree compatible order, $\prec_h$ is trivially a monoid order and one easily verifies that $(\tilde{\mathcal{P}}, \star, \prec_h)$ is again a polynomial algebra of solvable type.

Let $\mathcal{F} \subset \mathcal{P}$ be a finite set and $\mathcal{I} = \langle \mathcal{F} \rangle$ the ideal generated by it. We first note that generally the ideal $\tilde{\mathcal{I}} = \langle \mathcal{F}^{(h)} \rangle \subseteq \tilde{\mathcal{P}}$ generated by the homogenisation $\mathcal{F}^{(h)} = \{f^{(h)} \mid f \in \mathcal{F}\}$ of $\mathcal{F}$ is only a subset of the homogenised ideal $\mathcal{I}^{(h)} = \langle f^{(h)} \mid f \in \mathcal{I} \rangle$. Indeed, if $g = \sum_{f \in \mathcal{F}} P_f \star f$ is an arbitrary element of $\mathcal{I}$, then a homogenisation leads to a relation of the form $t^k g^{(h)} = \sum_{f \in \mathcal{F}} P_f^{(h)} \star f^{(h)}$ for some $k \geq 0$, and we can only conclude that $t^k g^{(h)} \in \tilde{\mathcal{I}}$ but no statement about $g^{(h)}$ itself is possible. Since $\prec_h$ is a term order, it is possible to apply the Buchberger Algorithm B.3 to the homogenised set $\mathcal{F}^{(h)}$ for the construction of a Gröbner basis $\tilde{\mathcal{G}}$ of the ideal $\tilde{\mathcal{I}}$ with respect to the lifted order $\prec_h$.

Lemma 4.5.3. *The dehomogenisation $\mathcal{G} \subset \mathcal{P}$ of $\tilde{\mathcal{G}} \subset \tilde{\mathcal{P}}$ is a Gröbner basis of the ideal $\mathcal{I}$ with respect to the semigroup order $\prec$.*

Proof. Let $\tilde{\mathcal{G}} = \{\tilde{g}_1, \ldots, \tilde{g}_r\}$ and let $f \in \mathcal{I}$ be an arbitrary element. By the considerations above, $\tilde{f} = (x^{n+1})^k f^{(h)} \in \tilde{\mathcal{I}}$ for some $k \geq 0$. As the set $\tilde{\mathcal{G}}$ is assumed to be a Gröbner basis of $\tilde{\mathcal{I}}$, there exists a standard representation $\tilde{f} = \sum_{i=1}^r \tilde{P}_i \star \tilde{g}_i$ with coefficients $\tilde{P}_i \in \tilde{\mathcal{P}}$ such that $\operatorname{lt}_{\prec_h}(\tilde{P}_i \star \tilde{g}_i) \preceq_h \operatorname{lt}_{\prec_h} \tilde{f}$. Dehomogenisation leads to an equation $f = \sum_{i=1}^r P_i \star g_i$ (note that generally some polynomials g_i may coincide!) and hence $\langle \mathcal{G} \rangle = \mathcal{I}$. Since all appearing polynomials in $\tilde{\mathcal{P}}$ are homogeneous, it follows from the definition of $\prec_h$ that the inequality $\operatorname{lt}_{\prec}(P_i \star g_i) \preceq \operatorname{lt}_{\prec} f$ holds. Thus there must exist a generator $g_i \in \mathcal{G}$ such that $\operatorname{lt}_{\prec} g_i \mid \operatorname{lt}_{\prec} f$, i. e. we also have $\langle \operatorname{lt}_{\prec} \mathcal{G} \rangle = \operatorname{lt}_{\prec} \mathcal{I}$ and $\mathcal{G}$ is a Gröbner basis of $\mathcal{I}$. □

While this simple approach yields thus indeed a Gröbner basis $\mathcal{G}$ with respect to the semigroup order $\prec$, we cannot expect that $\mathcal{G}$ is *reduced*. Furthermore, a subsequent autoreduction of $\mathcal{G}$ is generally not possible, as the reduction process requires possibly infinite normal form computations. Hence the existence of a reduced Gröbner basis of $\mathcal{I}$ cannot be guaranteed.

We extend now this approach to involutive bases. Here we encounter the additional difficulty that we must lift not only the order $\prec$ but also the used involutive division L and we must show that properties like being Noetherian or continuous are preserved by the lift which is non-trivial. For lifting the involutive division L, we proceed somewhat similarly to the definition of the Janet division: the homogenisation variable t is multiplicative only for terms which have maximal degree in t (which also explains why we number the homogenisation variable x^{n+1} instead of the more common x^0).

Proposition 4.5.4. *Let L be an involutive division on $\mathbb{N}_0^n$. For a finite set $\tilde{\mathcal{B}} \subset \mathbb{N}_0^{n+1}$ and a multi index $\tilde{\mu} \in \tilde{\mathcal{B}}$, we define $N_{\tilde{L},\tilde{\mathcal{B}}}(\tilde{\mu})$ by saying that (i) $n+1 \in N_{\tilde{L},\tilde{\mathcal{B}}}(\tilde{\mu})$, if and only if $\mu_{n+1} = \max_{\tilde{\nu} \in \tilde{\mathcal{B}}}\{\nu_{n+1}\}$ and (ii) $i \in N_{\tilde{L},\tilde{\mathcal{B}}}(\tilde{\mu})$ for $1 \leq i \leq n$, if and only if $i \in N_{L,\mathcal{B}}(\mu)$. This determines an involutive division $\tilde{L}$ on $\mathbb{N}_0^{n+1}$.*

Proof. Both conditions for an involutive division are easily verified. For the first one, let $\tilde{\rho} \in \mathcal{C}_{\tilde{L},\tilde{\mathcal{B}}}(\tilde{\mu}) \cap \mathcal{C}_{\tilde{L},\tilde{\mathcal{B}}}(\tilde{\nu})$ with $\tilde{\mu}, \tilde{\nu} \in \tilde{\mathcal{B}}$. If $\rho_{n+1} = \mu_{n+1} = \nu_{n+1}$, the last entry can be ignored, and the properties of the division L imply the desired result.

If $\rho_{n+1} = \mu_{n+1} > \nu_{n+1}$, the index $n+1$ must be multiplicative for $\tilde{\nu}$ contradicting $\mu_{n+1} > \nu_{n+1}$. If ρ_{n+1} is greater than both μ_{n+1} and ν_{n+1}, the index 0 must be multiplicative for both implying $\mu_{n+1} = \nu_{n+1}$. In this case we may again ignore the last entry and invoke the properties of L.

For the second condition we note that whether a multiplicative index $i \leq n$ becomes non-multiplicative for some element $\tilde{\nu} \in \tilde{\mathcal{B}}$ after adding a new multi index to $\tilde{\mathcal{B}}$ is independent of the last entry and thus only determined by the involutive division L. If the new multi index has a higher last entry than all elements of $\tilde{\mathcal{B}}$, then $n+1$ becomes non-multiplicative for all elements in $\tilde{\mathcal{B}}$ but this is permitted. □

Next we check to what extent the properties of L are inherited by the lifted division $\tilde{L}$. Given the similarity of the definition of $\tilde{L}$ and the Janet division, it is not surprising that we may reuse many ideas from proofs for the latter one.

Proposition 4.5.5. *If L is a Noetherian division, then so is $\tilde{L}$.*

Proof. Let $\tilde{\mathcal{B}} \subset \mathbb{N}_0^{n+1}$ be an arbitrary finite subset. In order to prove the existence of a finite $\tilde{L}$-completion of $\tilde{\mathcal{B}}$, we first take a finite L-completion $\hat{\mathcal{B}} \subset \mathbb{N}_0^n$ of $\mathcal{B}$ which always exists, as by assumption the division L is Noetherian. Next, we define a finite subset $\tilde{\mathcal{B}}' \subset \langle\tilde{\mathcal{B}}\rangle$ by setting

$$\tilde{\mathcal{B}}' = \Bigl\{\, \tilde{\mu} \in \mathbb{N}_0^{n+1} \mid \mu \in \hat{\mathcal{B}} \wedge \mu_{n+1} \leq \max_{\tilde{\nu}\in\tilde{\mathcal{B}}} \nu_{n+1} \,\Bigr\} \cap \langle\tilde{\mathcal{B}}\rangle\,. \tag{4.11}$$

We claim that this set $\tilde{\mathcal{B}}'$ is an $\tilde{L}$-completion of $\tilde{\mathcal{B}}$. By construction, we have both $\tilde{\mathcal{B}}' \subset \langle\tilde{\mathcal{B}}\rangle$ and $\tilde{\mathcal{B}} \subseteq \tilde{\mathcal{B}}'$, so that we must only show that $\tilde{\mathcal{B}}'$ is involutive.

Let $\tilde{\mu} \in \langle\tilde{\mathcal{B}}'\rangle$ be arbitrary. By construction of $\tilde{\mathcal{B}}'$, we can find $\tilde{\nu} \in \tilde{\mathcal{B}}'$ with $\nu\,|_{L,\hat{\mathcal{B}}}\,\mu$. Moreover, the definition of $\tilde{\mathcal{B}}'$ guarantees that we can choose $\tilde{\nu}$ in such a way that either $\nu_{n+1} = \mu_{n+1}$ or $\nu_{n+1} = \max_{\tilde{\rho}\in\tilde{\mathcal{B}}'} \rho_{n+1} < \mu_{n+1}$ holds. In the former case, we trivially have $\tilde{\nu}\,|_{\tilde{L},\tilde{\mathcal{B}}'}\,\tilde{\mu}$; in the latter case we have $n+1 \in N_{\tilde{L},\tilde{\mathcal{B}}}(\tilde{\nu})$ (see the proof of Proposition 4.5.4). Thus in either case $\tilde{\mu} \in \langle\tilde{\mathcal{B}}'\rangle_{\tilde{L}}$. □

Proposition 4.5.6. *If L is a continuous division, then so is $\tilde{L}$.*

Proof. Let $(\tilde{\nu}^{(1)},\dots,\tilde{\nu}^{(r)})$ with $\tilde{\nu}^{(i)} \in \tilde{\mathcal{B}} \subset \mathbb{N}_0^{n+1}$ be a finite sequence as described in Definition 4.1.3 of a continuous division. We first note that the integer sequence $(\nu_{n+1}^{(1)},\dots,\nu_{n+1}^{(r)})$ is then monotonically increasing. Indeed, if $\nu_{n+1}^{(i)}$ is not maximal among the entries μ_{n+1} for $\tilde{\mu} \in \tilde{\mathcal{B}}$, no multiplicative divisor of $\tilde{\nu}^{(i)} + 1_j$ in $\tilde{\mathcal{B}}$ can have a smaller last entry: if $\nu_{n+1}^{(i)}$ is maximal, the index $n+1$ is multiplicative for $\tilde{\nu}^{(i)}$ and any involutive divisor in $\tilde{\mathcal{B}}$ must also be maximal in the last entry. Thus it suffices to look at those parts of the sequence where equality in the last entries holds. But there the inequality of the multi indices $\tilde{\nu}^{(i)}$ follows from the continuity of the underlying division L. □

Unfortunately, it is much harder to show that constructivity is preserved. So far, a proof is known only for globally defined divisions (and thus in particular for the Pommaret division) and the Janet division.

Proposition 4.5.7. *If the continuous division L is either globally defined or the Janet division J, then the lifted division $\tilde{L}$ is constructive.*

Proof. We select a finite set $\tilde{\mathcal{B}} \subset \mathbb{N}_0^{n+1}$, a multi index $\tilde{\nu} \in \tilde{\mathcal{B}}$ and a non-multiplicative index j of $\tilde{\nu}$ such that the conditions in Definition 4.1.7 of a constructive division are satisfied. Assume that a multi index $\tilde{\rho} \in \tilde{\mathcal{B}}$ exists such that $\tilde{\nu} + 1_j = \tilde{\rho} + \tilde{\sigma} + \tilde{\tau}$ with $\tilde{\rho} + \tilde{\sigma} \in \mathcal{C}_{\tilde{L},\tilde{\mathcal{B}}}(\tilde{\rho})$ and $\tilde{\rho} + \tilde{\sigma} + \tilde{\tau} \in \mathcal{C}_{\tilde{L},\tilde{\mathcal{B}}\cup\{\tilde{\rho}+\tilde{\sigma}\}}(\tilde{\rho} + \tilde{\sigma})$. If such a $\tilde{\rho}$ existed, the division $\tilde{L}$ would not be constructive.

Let L be a globally defined division. If $j = n+1$, then the equality $\nu_{n+1} + 1 = \rho_{n+1} + \sigma_{n+1} + \tau_{n+1}$ implies that $\sigma_{n+1} = \tau_{n+1} = 0$. Indeed, $\sigma_{n+1} > 0$ entails $\rho_{n+1} \geq \nu_{n+1} \geq \rho_{n+1} + \sigma_{n+1} > \rho_{n+1}$, as then $n+1$ must be multiplicative for $\tilde{\rho}$, and the assumption $\sigma_{n+1} = 0$ and $\tau_{n+1} > 0$ leads to a similar contradiction. If $j \leq n$, the argumentation is simple. A global division is always constructive, as adding further elements to $\mathcal{B}$ does not change the multiplicative indices. But the same holds for the indices $k \leq n$ in the lifted division $\tilde{L}$. Thus under the above conditions $\tilde{\nu} + 1_j \in \langle\tilde{\mathcal{B}}\rangle_{\tilde{L}}$ contradicting the made assumptions.

For L the Janet division J, we construct a sequence $(\tilde{\rho}^{(1)}, \tilde{\rho}^{(2)}, \dots)$ of elements $\tilde{\rho}^{(k)} \in \tilde{\mathcal{B}}$ as follows. We start with $\tilde{\rho}^{(1)} = \tilde{\rho}$ and $\tilde{\phi}^{(1)} = \tilde{\sigma} + \tilde{\tau}$. In each step, we select first a non-multiplicative index j_k for $\tilde{\rho}^{(k)}$ such that $\tilde{\phi}^{(k)}_{j_k} > 0$ and then a multiplicative divisor $\tilde{\rho}^{(k+1)} \in \tilde{\mathcal{B}}$ of $\tilde{\rho}^{(k)} + 1_{j_k}$:

$$\begin{aligned}\tilde{\nu} + 1_j &= \tilde{\rho}^{(k)} + \tilde{\phi}^{(k)} = \tilde{\rho}^{(k)} + 1_{j_k} + \tilde{\phi}^{(k)} - 1_{j_k} \\ &= \tilde{\rho}^{(k+1)} + \underbrace{\tilde{\sigma}^{(k+1)} + \tilde{\phi}^{(k)} - 1_{j_k}}_{=\tilde{\phi}^{(k+1)}}\end{aligned} \tag{4.12}$$

If for some k all indices j_k such that $\tilde{\phi}^{(k)}_{j_k} > 0$ were multiplicative for $\tilde{\rho}^{(k)}$, we would have obviously a contradiction, as then $\tilde{\nu} + 1_j \in \langle\tilde{\mathcal{B}}\rangle_{\tilde{L}}$. There remains to show that an involutive divisor $\tilde{\rho}^{(k+1)} \in \tilde{\mathcal{B}}$ of $\tilde{\rho}^{(k)} + 1_{j_k}$ always exists. By the second condition in the definition of constructivity, this will be the case, if $\tilde{\rho}^{(k)} + 1_{j_k}$ is a proper divisor of $\tilde{\nu} + 1_j$. Thus we must show that in each step $|\tilde{\phi}^{(k)}|$ is greater than one (note that $|\tilde{\sigma}|$ and $|\tilde{\tau}|$ cannot be zero).

For this purpose, we next prove some inequalities between the elements of our sequence for the lexicographic order on $\mathbb{N}_0^{n+1}$.

$\tilde{\mu} \prec_{\text{lex}} \tilde{\rho}^{(1)} = \tilde{\rho}$: Let s and t, respectively, be the position of the last non-vanishing entry in the multi indices $\tilde{\sigma}$ and $\tilde{\tau}$. Assume first that $j = n+1$. Then $\mu_{n+1} + 1 = \rho_{n+1} + \sigma_{n+1} + \tau_{n+1}$ entails that for $s = n+1$ we have $\rho_{n+1} \leq \mu_{n+1}$ and that for $s \leq n$, $t = n+1$ we have $\rho_{n+1} + \sigma_{n+1} \leq \mu_{n+1}$. Both possibilities contradict the definition of the lifted division $\tilde{L}$, as then $n+1$ cannot be multiplicative for $\tilde{\rho}$ or $\tilde{\rho} + \tilde{\sigma}$, respectively. If $j = n+1$ and $s \leq n$, $t \leq n$, then $\rho_{n+1} = \mu_{n+1} + 1$ and thus $\tilde{\mu} \prec_{\text{lex}} \tilde{\rho}$. For $j \leq n$, we claim that $j > \max\{s,t\}$, from which the inequality immediately follows. But this relation in turn follows from the properties of the Janet division in the same manner as in the proof of Lemma 4.1.8.

$\tilde{\rho}^{(k)} \prec_{\text{lex}} \tilde{\rho}^{(k+1)}$: According to (4.12), we have $\tilde{\rho}^{(k)} + 1_{j_k} = \tilde{\rho}^{(k+1)} + \tilde{\sigma}^{(k+1)}$. If $j_k \leq n$, then $\tilde{\rho}_{n+1}^{(k+1)} = \tilde{\rho}_{n+1}^{(k)}$ and the claimed inequality follows by the same considerations as in the proof of Lemma 4.1.5. For $j_k = n+1$, we first consider the case where $\tilde{\sigma}_{n+1}^{(k+1)} \neq 0$. This means that $n+1$ is multiplicative for $\tilde{\rho}^{(k+1)}$ but non-multiplicative for $\tilde{\rho}^{(k)}$, although $\tilde{\rho}_{n+1}^{(k)} \geq \tilde{\rho}_{n+1}^{(k+1)}$ which is not possible. Thus $\tilde{\sigma}_{n+1}^{(k+1)} = 0$, $\tilde{\rho}_{n+1}^{(k+1)} > \tilde{\rho}_{n+1}^{(k)}$ and therefore $\tilde{\rho}^{(k)} \prec_{\text{lex}} \tilde{\rho}^{(k+1)}$.

Now suppose that $\tilde{\nu} + 1_j = \tilde{\rho}^{(k)} + \tilde{\phi}^{(k)}$ with $|\tilde{\phi}^{(k)}| = 1$. We may assume without loss of generality that we have chosen $\tilde{\nu}$ maximal among all multi indices $\tilde{\mu}$ with $\tilde{\nu} + 1_j = \tilde{\mu} + 1_\ell$ where ℓ is non-multiplicative for $\tilde{\mu}$. This yields the contradiction $\tilde{\nu} \prec_{\text{lex}} \tilde{\rho}^{(1)} \prec_{\text{lex}} \cdots \prec_{\text{lex}} \tilde{\rho}^{(k)} \preceq_{\text{lex}} \tilde{\nu}$.

Hence we always find an involutive divisor $\tilde{\rho}^{(k+1)}$ allowing us to extend the sequence. But if the sequence $(\tilde{\rho}^{(1)}, \tilde{\rho}^{(2)}, \dots)$ is infinite, its elements cannot be all distinct, as $\tilde{\nu} + 1_j$ has only finitely many divisors, and this observation contradicts the continuity of the lifted Janet division $\tilde{J}$. □

Based on these results, Algorithm 4.3 (or the improved form, Algorithm 4.5) can be extended to semigroup orders. Given a finite set $\mathcal{F} \in \mathcal{P}$, we first determine its homogenisation $\mathcal{F}^{(h)} \in \tilde{\mathcal{P}}$ and then compute an involutive basis of $\langle F^{(h)} \rangle$ with respect to $\tilde{L}$ and $\prec_h$. What remains to be done is first to show that the existence of a finite involutive basis is preserved under the lifting to $\tilde{\mathcal{P}}$ and then to study the properties of the dehomogenised basis.

Proposition 4.5.8. *If the left ideal $\mathcal{I} = \langle \mathcal{F} \rangle \subseteq \mathcal{P}$ possesses an involutive basis with respect to the Noetherian division L and the semigroup order $\prec$, then the lifted left ideal $\tilde{\mathcal{I}} = \langle \mathcal{F}^{(h)} \rangle \subseteq \tilde{\mathcal{P}}$ generated by the homogenisations of the elements in the finite set $\mathcal{F}$ possesses an involutive basis with respect to the lifted division $\tilde{L}$ and the lifted monoid order $\prec_h$.*

Proof. By Theorem 3.3.13, the ideal $\tilde{\mathcal{I}} \subseteq \tilde{\mathcal{P}}$ possesses a Gröbner basis $\tilde{\mathcal{G}}$ with respect to the monoid order $\prec_h$. By Proposition 4.5.5, a finite $\tilde{L}$-completion $\tilde{\mathcal{B}}$ of the set $\text{le}_{\prec_h} \tilde{\mathcal{G}}$ exists. Moreover, by Definition 3.3.12 of a Gröbner basis, the monoid ideals $\langle \text{le}_{\prec_h} \tilde{\mathcal{G}} \rangle$ and $\text{le}_{\prec_h} \tilde{\mathcal{I}}$ coincide. Thus $\tilde{\mathcal{B}}$ is an involutive basis of $\text{le}_{\prec_h} \tilde{\mathcal{I}}$ with respect to the lifted division $\tilde{L}$ and an involutive basis $\tilde{\mathcal{H}}$ of the ideal $\tilde{\mathcal{I}}$ with respect to $\tilde{L}$ is given by

$$\tilde{\mathcal{H}} = \left\{ x^{\tilde{\mu}} \star \tilde{g} \mid \tilde{g} \in \tilde{\mathcal{G}} \wedge \text{le}_{\prec_h}(x^{\tilde{\mu}} \star \tilde{g}) \in \tilde{\mathcal{B}} \right\}. \tag{4.13}$$

This set is obviously finite. □

Hence the lifting leads to a situation where we can apply Theorem 4.4.1 about the correctness and the termination of our basic completion Algorithm 4.5. However, the dehomogenisation of the strong involutive basis computed in $\tilde{\mathcal{P}}$ does not necessarily lead to a *strong* involutive basis in $\mathcal{P}$. But we will always obtain at least a weak involutive basis and thus in particular a Gröbner basis.

Theorem 4.5.9. *Let $\tilde{\mathcal{H}}$ be a strong involutive basis of the left ideal $\tilde{\mathcal{I}} \subseteq \tilde{\mathcal{P}}$ with respect to $\tilde{L}$ and $\prec_h$. Then its dehomogenisation*[5] *$\mathcal{H}$ is a weak involutive basis of the left ideal $\mathcal{I} \subseteq \mathcal{P}$ with respect to L and $\prec$.*

Proof. For any $f \in \mathcal{I}$ an integer $k \geq 0$ exists such that $\tilde{f} = t^k f^{(h)} \in \tilde{\mathcal{I}}$. The polynomial $\tilde{f}$ possesses a unique involutive standard representation

$$\tilde{f} = \sum_{\tilde{h} \in \tilde{\mathcal{H}}} \tilde{P}_{\tilde{h}} \star \tilde{h} \tag{4.14}$$

with $\tilde{P}_{\tilde{h}} \in \Bbbk\big[X_{\tilde{L},\mathrm{le}_{\prec_h}\tilde{\mathcal{H}}}(\tilde{h})\big]$ and $\mathrm{le}_{\prec_h}(\tilde{P}_{\tilde{h}} \star \tilde{h}) \preceq_h \mathrm{le}_{\prec_h} \tilde{f}$. Setting $t = 1$ in (4.14) yields a representation of the polynomial f with respect to the dehomogenised basis $\mathcal{H}$ of the form $f = \sum_{h \in \mathcal{H}} P_h \star h$ where $P_h \in \Bbbk\big[X_{L,\mathrm{le}_\prec \mathcal{H}}(h)\big]$ by the definition of the lifted division $\tilde{L}$. This fact obviously implies that $\langle \mathcal{H} \rangle = \mathcal{I}$. By the definition of the lifted order $\prec_h$ and the homogeneity of the lifted polynomials, we have furthermore that $\mathrm{le}_\prec(P_h \star h) \preceq \mathrm{le}_\prec f$ and hence that $\mathrm{le}_\prec \mathcal{H}$ is a weak involutive basis of $\mathrm{le}_\prec \mathcal{I}$ by Theorem 3.4.4 (note that in the proof of the direction we are using here no normal form arguments are required and it hence remains valid). Since all conditions of Definition 4.5.2 are satisfied, the set $\mathcal{H}$ is therefore indeed a weak involutive basis of the ideal $\mathcal{I}$ for the division L and the semigroup order $\prec$. □

Example 4.5.10. An important class of left ideals in the Weyl algebra $\mathcal{W}_n$ are the Gelfand–Kapranov–Zelevinsky (GKZ) systems describing hypergeometric functions [152]. Any such system can be described by a $d \times n$ integer matrix A of rank d and a vector $\beta \in \Bbbk^d$. Its generators are

$$f_i = \sum_{j=1}^{n} A_{ij} x^j \partial_j - \beta_i\,, \qquad \text{for } i = 1, \dots, d$$

and in addition for every pair of multi indices $\mu, \nu \in \mathbb{N}_0^n$ such that $A\mu = A\nu$ the generator $f_{\mu\nu} = \partial^\mu - \partial^\nu$. The generators f_i express certain homogeneity relations of the solutions of the corresponding linear system of partial differential equations; the operators $f_{\mu\nu}$ generate the toric ideal associated with A (Sturmfels [437] gives concrete algorithms for their construction).

If we take as a concrete example

$$A = \begin{pmatrix} 2 & 1 & 0 \\ 0 & 1 & 2 \end{pmatrix}, \qquad \beta = \begin{pmatrix} -1 \\ 1 \end{pmatrix}, \tag{4.15}$$

then the corresponding GKZ system in $\mathcal{W}_3 = \Bbbk[x, y, z, \partial_x, \partial_y, \partial_z]$ is generated by the three operators

$$f_1 = 2x\partial_x + y\partial_y + 1\,, \quad f_2 = 2z\partial_z + y\partial_y - 1\,, \quad f = \partial_x\partial_z - \partial_y^2\,. \tag{4.16}$$

[5] Note that the dehomogenised basis $\mathcal{H}$ is in general smaller than $\tilde{\mathcal{H}}$, as some elements of $\tilde{\mathcal{H}}$ may differ only in powers of the homogenisation variable t.

The solution space of the induced linear system of partial differential equations is spanned by the two familiar expressions $(-y \pm \sqrt{y^2 - 4xz})/2x$, i. e. the solutions of the quadratic equation $x\lambda^2 + y\lambda + z = 0$.

The construction of formal series solutions of such hypergeometric systems requires Gröbner bases with respect to non-term orders defined by weight vectors of the form $(-\xi, \xi) \in \mathbb{R}^{2n}$ [391]. We take $\xi = (1,0,0)$ and refine by the degree reverse lexicographic term order with $y \prec x \prec z \prec \partial_y \prec \partial_x \prec \partial_z$ (note the permutation of x and y). As involutive division we use the Janet division corresponding to the permutation of the variables used in the term order. The involutive completion algorithm yields then the following basis with eight generators:

$$\begin{aligned} H = \Big\{\ & \underline{2x\partial_x} + y\partial_y + 1,\ \underline{2z\partial_z} + y\partial_y - 1,\ \underline{\partial_x\partial_z} - \partial_y^2, \\ & \underline{y\partial_y\partial_z} + \partial_z + 2x\partial_y^2,\ \underline{2z\partial_y\partial_z} + y\partial_y^2,\ \underline{y\partial_x\partial_y} - \partial_x + 2z\partial_y^2, \\ & \underline{2x\partial_x\partial_y} + y\partial_y^2 + 2\partial_y,\ \underline{y^2\partial_y^2} + y\partial_y - 1 - 4xz\partial_y^2\ \Big\}. \end{aligned} \tag{4.17}$$

As one can see from the underlined leading terms, we have actually obtained a strong Janet basis. In this case, a reduced Gröbner basis exists and consists of six generators, namely one can drop the fifth and the seventh generator in the basis above which are just ∂_y times the first and the second generator, respectively. ◁

Remark 4.5.11. For the Pommaret division P the situation is considerably simpler. There is no need to define a lifted division $\tilde{P}$ according to Proposition 4.5.4. Instead we simply use the standard Pommaret division on $\mathbb{N}_0^{n+1}$. This approach implies that for all multi indices $\tilde{\mu} \in \mathbb{N}_0^{n+1}$ with $\mu \neq 0$ the equality $N_P(\tilde{\mu}) = N_P(\mu)$ holds, as obviously $n+1$ is multiplicative only for multi indices of the form $\tilde{\mu} = \ell_{n+1}$, i. e. for which $\mu = 0$. One easily sees that the above proof of Theorem 4.5.9 is not affected by this change of the division used in $\mathbb{N}_0^{n+1}$ and hence remains true. ◁

It is not a shortcoming of our proof that in general we do not get a strong involutive basis, but actually some ideals do not possess strong involutive bases. In particular, there is no point in invoking Proposition 3.4.7 for obtaining a strong basis. While we may surely obtain by elimination a subset $\mathcal{H}' \subseteq \mathcal{H}$ such that $\mathrm{le}_\prec \mathcal{H}$ is a strong involutive basis of $\mathrm{le}_\prec \mathcal{H}$, in general $\langle \mathcal{H}' \rangle \subsetneq \mathcal{I}$.

Example 4.5.12. Consider in the Weyl algebra $\mathcal{W}_2 = \Bbbk[x, y, \partial_x, \partial_y]$ the left ideal generated by the set $\mathcal{F} = \{\underline{1} + x + y, \underline{\partial_y} - \partial_x\}$. We take the semigroup order induced by the weight vector $(-1,-1,1,1)$ and refined by a term order for which $\partial_y \succ \partial_x \succ y \succ x$. Then the underlined terms are the leading ones. One easily checks that $\mathcal{F}$ is a Gröbner basis for this order. Furthermore, all variables are multiplicative for each generator with respect to the Pommaret division and thus $\mathcal{F}$ is a weak Pommaret basis, too.

Obviously, the set $\mathcal{F}$ is neither a reduced Gröbner basis nor a strong Pommaret basis, as 1 is a (multiplicative) divisor of ∂_y. However, it is easy to see that the left ideal $\mathcal{I} = \langle \mathcal{F} \rangle$ does not possess a reduced Gröbner basis or a strong Pommaret basis.

Indeed, we have $\text{le}_{\prec}\mathcal{I} = \mathbb{N}_0^4$ and thus such a basis had to consist of only a single generator; but $\mathcal{I}$ is not a principal ideal. $\lhd$

A special situation arises for the Janet division. Recall from Remark 3.1.15 that any finite set $\mathcal{B} \subset \mathbb{N}_0^n$ is automatically involutively autoreduced with respect to the Janet division. Thus any weak Janet basis is a strong basis, if all generators have different leading exponents. If we follow the above outlined strategy of applying Algorithm 4.5 to a homogenised basis and then dehomogenising the result, we cannot generally expect this condition to be satisfied. However, with only a minor modification of the algorithm we can achieve this goal.

Theorem 4.5.13. *Let $(\mathcal{P}, \star, \prec)$ be a polynomial algebra of solvable type where $\prec$ is an arbitrary semigroup order. Then every left ideal $\mathcal{I} \subseteq \mathcal{P}$ possesses a strong Janet basis for $\prec$.*

Proof. Assume that at some stage of Algorithm 4.5 the basis $\tilde{\mathcal{H}}$ contains two polynomials $\tilde{f}$ and $\tilde{g}$ such that $\text{le}_{\prec_h}(\tilde{g}) = \text{le}_{\prec_h}(\tilde{f}) + 1_{n+1}$, i. e. the two leading exponents differ only in the last entry. If $\tilde{g} = t\tilde{f}$, we will find $f = g$ after dehomogenisation and no obstruction to a strong basis appears. Otherwise, the homogenisation variable t is non-multiplicative for $\tilde{f}$ by definition of the lifted Janet division $\tilde{J}$. Thus at some later stage the algorithm must consider the non-multiplicative product $t\tilde{f}$ (if it were already treated, $\tilde{\mathcal{H}}$ would not be involutively head autoreduced).

In the usual algorithm, we then determine the involutive normal form of the polynomial $t\tilde{f}$; the first step of this computation is to replace $t\tilde{f}$ by $t\tilde{f} - \tilde{g}$. Alternatively, we may proceed as follows. The polynomial $\tilde{g}$ is removed from the basis $\tilde{\mathcal{H}}$ and replaced by $t\tilde{f}$. Then we continue by analysing the involutive normal form of $\tilde{g}$ with respect to the new basis. Note that this modification concerns only the situation that a multiplication by t has been performed and that the basis $\tilde{\mathcal{H}}$ contains already an element with the same leading exponent as the obtained polynomial.

If the final output $\tilde{\mathcal{H}}$ of the thus modified completion algorithm contains two polynomials $\tilde{f}$ and $\tilde{g}$ such that $\text{le}_{\prec_h}\tilde{g}$ and $\text{le}_{\prec_h}\tilde{f}$ differ only in the last entry, then either $\tilde{g} = t^k\tilde{f}$ or $\tilde{f} = t^k\tilde{g}$ for some exponent $k \in \mathbb{N}$. Thus the dehomogenisation yields a basis $\mathcal{H}$ where all elements possess different leading terms and $\mathcal{H}$ is a strong Janet basis. Looking at the proof of Theorem 4.4.1, it is easy to see that this modification does not affect the correctness and the termination of the algorithm. As the Janet division is Noetherian, these considerations prove together with Proposition 4.5.5 the assertion. $\square$

Note that our modification only achieves its goal, if we really restrict in Algorithm 4.5 to head reductions. Otherwise some other terms than the leading term in $t\tilde{f}$ might be reducible but not the corresponding terms in $\tilde{f}$. Then we could still find after dehomogenisation two generators with the same leading term.

Example 4.5.14. Let us consider in the Weyl algebra $\mathcal{W}_3 = \Bbbk[x, y, z, \partial_x, \partial_y, \partial_z]$ the left ideal generated by the set $\mathcal{F} = \{\partial_z - y\partial_x, \partial_y\}$ corresponding to the partial differential equation (2.57) considered in Example 2.3.9. If we apply the usual involutive completion Algorithm 4.5 (to the homogenisation $\mathcal{F}^{(h)}$), we obtain for the weight vector $(-1,0,0,1,0,0)$ refined by the degree reverse lexicographic order and the Janet division a weak basis with nine generators:

$$\mathcal{H}_1 = \{\, \partial_x,\ \partial_y,\ \partial_z,\ \partial_x\partial_z,\ \partial_y\partial_z,\ y\partial_x,\ y\partial_x + \partial_z,\ y\partial_x\partial_z,\ y\partial_x\partial_z + \partial_z^2 \,\}\,. \tag{4.18}$$

As one easily sees from the last four generators, it is not a strong basis.

Applying the modified algorithm for the Janet division yields the following basis with only seven generators:

$$\mathcal{H}_2 = \{\, \partial_x + \partial_y\partial_z,\ \partial_y,\ \partial_z,\ \partial_x\partial_z,\ \partial_y\partial_z,\ y\partial_x + \partial_z,\ y\partial_x\partial_z + \partial_z^2 \,\}\,. \tag{4.19}$$

Obviously, we now have a strong basis, as all leading terms are different.

This example also demonstrates the profound effect of the homogenisation. It follows from the results in Example 2.3.9 that a strong Janet or Pommaret basis of $\langle\mathcal{F}\rangle$ is simply given by $\mathcal{H} = \{\partial_x, \partial_y, \partial_z\}$ which is simultaneously a reduced Gröbner basis. In the homogeneous ideal $\langle\mathcal{F}^{(h)}\rangle$ many reductions are not possible because the terms contain different powers of t. However, this effect is a general problem of all approaches to Gröbner bases for semigroup orders using homogenisation and not specific for the involutive approach.

In this particular case, one could have applied the involutive completion algorithm directly to the original set $\mathcal{F}$ and it would have terminated with the minimal basis $\mathcal{H}$, although we are not using a monoid order. Unfortunately, it is not clear how to predict when infinite reduction chains appear in normal form computations with respect to such orders, so that one does not know in advance whether one may dispense with the homogenisation. ◁

One computational disadvantage of the outlined approach is that the intermediate basis $\tilde{\mathcal{H}}$ in the homogenised algebra $\tilde{\mathcal{P}}$ is often much larger than the final basis $\mathcal{H}$ in the original algebra $\mathcal{P}$, as upon dehomogenisation generators may become identical. Furthermore, we have seen that it is quite difficult to prove the constructivity of the lifted involutive division L_h which limits the applicability of this technique. Finally, for most divisions we are not able to determine strong involutive bases. There exists an alternative approach which simultaneously avoids all these problems. It is based on two key ideas: we modify the normal form algorithm and we work over a ring of fractions of $\mathcal{P}$.

The source of all problems in computations with semigroup orders which are not monoid orders is that they are no longer well-orders (cf. Lemma A.1.6) and hence normal form computations with the classical algorithms do not necessarily terminate. We introduce now an alternative normal form algorithm which always terminates with a *Mora normal form*. For technical reasons, we do this first in a non-involutive, homogenised form.

Algorithm 4.8 differs from a standard normal form algorithm applied to the homogenised input in two respects. We are allowed to reduce with respect to intermediate results (see Line /5/) and the output is only a "weak" normal form in a sense made precise by the following result.

Algorithm 4.8 Homogenised Mora normal form

Input: polynomial $f \in \mathcal{P}$, finite set $\mathcal{G} \subset \mathcal{P}$
Output: Mora normal form h of f with respect to $\mathcal{G}$

$\tilde{h} \leftarrow f^{(h)}$; $\tilde{\mathcal{G}} \leftarrow \mathcal{G}^{(h)}$
while $(\tilde{h} \neq 0) \wedge (\mathcal{S}_h = \{\tilde{g} \in \tilde{\mathcal{G}} \mid \exists k \in \mathbb{N}_0 : \mathrm{le}_{\prec_h} \tilde{g} \mid \mathrm{le}_{\prec_h}(t^k \tilde{h})\} \neq \emptyset)$ **do**
  choose $\tilde{g} \in \mathcal{S}_h$ with minimal k
  if $k > 0$ **then**
    $\tilde{\mathcal{G}} \leftarrow \tilde{\mathcal{G}} \cup \{\tilde{h}\}$
  end if
  $\mu \leftarrow \mathrm{le}_{\prec_h}(t^k \tilde{h}) - \mathrm{le}_{\prec_h} \tilde{g}$; $\tilde{h} \leftarrow t^k \tilde{h} - \frac{\mathrm{lc}_{\prec_h}(t^k \tilde{h})}{\mathrm{lc}_{\prec_h}(x^\mu \star \tilde{g})} x^\mu \star \tilde{g}$
  choose maximal ℓ such that $t^\ell \mid \tilde{h}$
  $\tilde{h} \leftarrow \tilde{h}/t^\ell$
end while
return h

Proposition 4.5.15. *Algorithm 4.8 always terminates. The output h for the input f and $\mathcal{G}$ satisfies the following two properties: (i) there exists a polynomial $u \in \mathcal{P}$ with $\mathrm{le}_\prec u = 0$ such that the difference $u \star f - h$ possesses a standard representation $u \star f - h = \sum_{g \in \mathcal{G}} P_g \star g$ and (ii) none of the leading exponents $\mathrm{le}_\prec g$ divides $\mathrm{le}_\prec h$. If $\prec$ is a monoid order, then $u = 1$ and h is a normal form in the usual sense.*

Proof. Denote by $\tilde{h}_i$ the value of $\tilde{h}$ after the ith iteration of the `while`-loop and similarly for all other variables in Algorithm 4.8. We show by an induction the existence of polynomials u_j with $\mathrm{le}_\prec u_j = 0$ such that after dehomogenisation we always have a standard representation

$$u_j \star f - h_j = \sum_{g \in \mathcal{G}} P_g^{(j)} \star g\,. \tag{4.20}$$

The case $j = 0$ is trivial: $u_0 = 1$ and $P_g^{(0)} = 0$ for all $g \in \mathcal{G}$.

Assume that $i > 0$ and that (4.20) holds for all $0 \leq j < i$. The values h_i and h_{i-1} are related by

$$h_i = h_{i-1} - m_i \star g_i \tag{4.21}$$

for some monomial m_i. We furthermore have $\mathrm{le}_{\prec_h}(t^{\ell_i}\tilde{h}_i) \prec_h \mathrm{le}_{\prec_h}(t^{k_i}\tilde{h}_{i-1})$ and since all occurring polynomials are homogeneous, this estimate implies by the definition of the lifted order $\prec_h$ that $\mathrm{le}_\prec h_i \prec \mathrm{le}_\prec h_{i-1}$.

We solve (4.21) of h_{i-1} and enter the result into the standard representation (4.20) evaluated for $j = i-1$ leading to

$$u_{i-1} \star f - h_i = \sum_{g \in \mathcal{G}} P_g^{(i-1)} \star g + m_i \star g_i\,. \tag{4.22}$$

We distinguish now two cases: $g_i \in \mathcal{G}$ and $g_i \in \mathcal{G}_{i-1} \setminus \mathcal{G}$. In the first case, $g_i = g' \in \mathcal{G}$ and we simply set $u_i = u_{i-1}$ and $P_g^{(i)} = P_g^{(i-1)} + \delta_{gg'} m_i$. In the second case, $g_i = h_r$ for some $0 \leq r < i-1$. We evaluate (4.20) for $j = r$ and solve it for h_r. Entering the result into (4.22) and sorting yields for $j = i$ a standard representation (4.20) with $u_i = u_{i-1} - m_i \star u_r$ and $P_g^{(i)} = P_g^{(i-1)} - m_i \star P_g^{(r)}$.

Since by construction $\mathrm{le}_{\prec_h}(t^{k_i}\tilde{h}_{i-1}) = \mathrm{le}_{\prec_h}(\tilde{m}_i \star \tilde{h}_r)$, we obtain the inequality $\mathrm{le}_\prec h_r \succ \mathrm{le}_\prec h_{i-1} = \mathrm{le}_\prec (m_i \star h_r)$ which obviously implies $\mathrm{le}_\prec m_i \prec 0$. Hence in the second case we still have $\mathrm{le}_\prec u_i = \mathrm{le}_\prec u_{i-1} = 0$ as required. This observations completes the induction.

By Lemma A.1.6, a monoid order refines the natural partial order by divisibility. Thus $x^\mu \mid x^\nu$ implies $x^\mu \prec x^\nu$ and we always have $\mathcal{S}_h \subseteq \mathcal{G}^{(h)}$. Thus Algorithm 4.8 becomes a standard normal form algorithm and above we are always in the first case so that indeed $u = 1$.

Finally, we consider the question of termination. By construction, the monomial ideals $\langle \mathrm{le}_{\prec_h} \tilde{\mathcal{G}}_i \rangle$ with $i \geq 0$ form an ascending chain. Since $\mathbb{N}_0^{n+1}$ is a Noetherian monoid, an integer $I \geq 0$ exists such that $\langle \mathrm{le}_{\prec_h} \tilde{\mathcal{G}}_i \rangle = \langle \mathrm{le}_{\prec_h} \tilde{\mathcal{G}}_I \rangle$ for all $i \geq I$. Hence after the Ith iteration the set $\tilde{\mathcal{G}}$ does not change anymore and we always find $k_i = 0$. This implies furthermore that for $i \geq I$ we always have $\mathrm{le}_{\prec_h} \tilde{h}_{i+1} \prec_h \mathrm{le}_{\prec_h} \tilde{h}_i$ and since the lifted order $\prec_h$ is a well-order, the computation must terminate after a finite number of iterations. □

Remark 4.5.16. The homogenised version of the Mora normal form algorithm makes the termination proof easier. It is possible to formulate the algorithm completely within the original polynomial ring $\mathcal{P}$. We will do so below, when we adapt it to involutive normal form computations. ◁

So far it has remained unclear why we call the output h of Algorithm 4.8 a weak normal form of f, as obviously it presents only a normal form of the product $u \star f$. As we will show now, we can move to a ring of fractions over $\mathcal{P}$ where all elements u with $\mathrm{le}_\prec u = 0$ are units. In this larger ring it indeed makes sense to consider h as a weak normal form.

Proposition 4.5.17. *Let $(\mathcal{P}, \star, \prec)$ be a polynomial algebra of solvable type where $\prec$ is an arbitrary semigroup order. Then the subset*

$$\mathcal{S}_\prec = \{ f \in \mathcal{P} \mid \mathrm{le}_\prec f = 0 \} \tag{4.23}$$

is multiplicatively closed and the left localisation $\mathcal{P}_\prec = \mathcal{S}_\prec^{-1} \star \mathcal{P}$ exists.

Proof. Obviously, $1 \in \mathcal{S}_\prec$ and each element of $\mathcal{S}_\prec$ is of the form $c(1+f)$ with $c \in \mathbb{k}$ and $\mathrm{le}_\prec f \prec 0$. Since we are working over a field, we ignore the constant c in the sequel. If $1+f$ and $1+g$ are two elements in $\mathcal{S}_\prec$, then the compatibility of the order $\prec$ with the multiplication $\star$ ensures that their product is of the form $(1+f) \star (1+g) = 1+h$ with $\mathrm{le}_\prec h \prec 0$. Hence $\mathcal{S}_\prec$ is multiplicatively closed.

As polynomial algebras of solvable type do not possess zero divisors, the left Ore condition for the existence of the left localisation $\mathcal{S}_\prec^{-1} \star \mathcal{P}$ requires that for all $f \in \mathcal{S}_\prec$ and $g \in \mathcal{P}$ the intersection $(\mathcal{P} \star f) \cap (\mathcal{S}_\prec \star g)$ is not empty. But this fact can be shown using minor modifications of our proof of Proposition 3.2.10.

We first choose coefficients $r_0, s_0 \in \mathcal{R}$ such that in $\bar{h}_1 = r_0 g \star f - s_0 f \star g$ the leading terms cancel, i. e. we have $\mathrm{le}_\prec \bar{h}_1 \prec \mathrm{le}_\prec f + \mathrm{le}_\prec g = \mathrm{le}_\prec g$. Then we compute with Algorithm 4.8 a weak normal form h_1 of $\bar{h}_1$ with respect to $\mathcal{F}_0 = \{f, g\}$. This yields a standard representation $u_1 \star \bar{h}_1 - h_1 = \phi_0 \star f + \psi_0 \star g$ where $\operatorname{le}_\prec u_1 = 0$ by Proposition 4.5.15. Assume now that $\operatorname{le}_\prec \psi_0 \succeq 0$. Then we arrive at the contradiction $\operatorname{le}_\prec(\psi_0 \star g) \succeq \operatorname{le}_\prec g \succ \operatorname{le}_\prec \bar{h}_1 = \operatorname{le}_\prec(u_1 \star \bar{h}_1)$. Thus we have $\operatorname{le}_\prec \psi_0 \prec 0$. If $h_1 = 0$, then $(u \star r_0 g - \phi_0) \star f = (u \star s_0 f + \psi_0) \star g$ and by the considerations above on the leading exponents $u \star s_0 f + \psi_0 \in \mathcal{S}_\prec$ so that indeed $(\mathcal{P} \star f) \cap (\mathcal{S}_\prec \star g) \neq \emptyset$.

If $h_1 \neq 0$, we proceed as in the proof of Proposition 3.2.10. We introduce $\mathcal{F}_1 = \mathcal{F}_0 \cup \{h_1\}$ and choose $r_1, s_1 \in \mathcal{R}$ such that in $\bar{h}_2 = r_1 h_1 \star f - s_1 f \star h_1$ the leading terms cancel. If we compute a weak normal form h_2 of $\bar{h}_2$, then we obtain a standard representation $u_2 \star \bar{h}_2 - h_2 = \phi_1 \star f + \psi_1 \star g + \rho_1 \star h_1$ where again $\operatorname{le}_\prec u_2 = 0$. The properties of a standard representation imply that $\operatorname{le}_\prec \psi_1 + \operatorname{le}_\prec g \preceq \operatorname{le}_\prec \bar{h}_2$ and $\operatorname{le}_\prec \rho_1 + \operatorname{le}_\prec h_1 \preceq \operatorname{le}_\prec \bar{h}_2$. Together with the inequalities $\operatorname{le}_\prec \bar{h}_2 \prec \operatorname{le}_\prec f + \operatorname{le}_\prec h_1 = \operatorname{le}_\prec h_1 \prec \operatorname{le}_\prec g$ this entails that both $\operatorname{le}_\prec \psi_1 \prec 0$ and $\operatorname{le}_\prec \rho_1 \prec 0$. Thus if $h_2 = 0$, then we have again found $\phi \in \mathcal{P}$ and $\psi \in \mathcal{S}_\prec$ such that $\phi \star f = \psi \star g$. If $h_2 \neq 0$, similar inequalities in the subsequent iterations ensure that we always find $\psi \in \mathcal{S}_\prec$. □

Remark 4.5.18. Using the geometric series $(1-y)^{-1} = 1 + y + y^2 + \cdots$, one easily sees that in the commutative case there is a canonical injection $\mathcal{P}_\prec \hookrightarrow \Bbbk[[x^1, \dots, x^n]]$ into the ring of formal power series. Hence in this case we may also think of the elements of the localisation $\mathcal{P}_\prec$ as such series instead of fractions. ◁

As any localisation of a Noetherian ring is again Noetherian, the ring $\mathcal{P}_\prec$ is Noetherian, if $\mathcal{P}$ is so. One sees immediately that the units in $\mathcal{P}_\prec$ are all those fractions where not only the denominator but also the numerator is contained in the set $\mathcal{S}_\prec$. Given an ideal $\mathcal{I} \subseteq \mathcal{P}_\prec$, we may always assume without loss of generality that its generated by a set $\mathcal{F} \subset \mathcal{P}$ of polynomials, as multiplication of a generator by a unit does not change the span. Hence in all computations we will exclusively work with polynomials and not with fractions.

Since all elements of $\mathcal{S}_\prec$ are units in $\mathcal{P}_\prec$, we may extend the notions of leading term, monomial or exponent: if $f \in \mathcal{P}_\prec$, then we can choose a unit $u \in \mathcal{S}_\prec$ with $\operatorname{lc}_\prec u = 1$ such that $u \star f \in \mathcal{P}$ is a polynomial; now we define $\operatorname{le}_\prec f = \operatorname{le}_\prec(u \star f)$ etc. One easily verifies that this definition is independent of the choice of u.

It is straightforward to develop a complete theory of Gröbner bases over $\mathcal{P}_\prec$. Definition 4.5.2 of Gröbner (and involutive) bases can be extended without changes from the ring $\mathcal{P}$ to $\mathcal{P}_\prec$. Furthermore, Theorem 3.3.13 on the existence of Gröbner bases generalises to $\mathcal{P}_\prec$, as its proof is only based on the leading exponents and a simple normal form argument remaining valid due to our considerations above.

However, the design of an involutive version of the Mora normal form faces a problem. Obviously, we want an analogous result to Proposition 4.5.15—now with an involutive standard representation for $u \star f - h$. This extension raises the question how multiplicative variables are assigned to the intermediate results by which the set $\hat{\mathcal{G}}$ is augmented in the course of the computation.

If we look at the proof of Proposition 4.5.15, we see that the reduction with respect to such an intermediate result affects the coefficients P_g of the final standard representation in form of a summand $m_i \star P_g^{(r)}$. Thus we must ensure that such terms are always multiplicative for the generator g. As such terms are products, it is obvious that for arbitrary polynomial algebras of solvable type this cannot be done: even if both m_i and $P_g^{(r)}$ lie in a subset $\Bbbk[X']$ for some restricted set of variables $X' \subset X = \{x^1, \dots, x^n\}$, due to the non-commutativity their product will generally contain further variables from $X \setminus X'$.

For a general involutive division one way out of this problem is to require that for any $X' \subset X$ the subset $\Bbbk[X'] \subset \mathcal{P}$ is actually a subring (which is trivially satisfied by the ordinary polynomial ring or rings of linear differential operators). If more is known about the used involutive division one may weaken this condition. For example, the Pommaret division always yields sets of multiplicative variables of the form $\{x^1, \dots, x^k\}$ and it suffices to require that $\mathcal{P}$ is an iterated solvable algebra in the sense of Definition 3.3.1.

Assuming that this point is settled, the proof of Proposition 4.5.15 motivates immediately the following strategy for the computation of an involutive Mora normal form (employed in Algorithm 4.9). To each member g of the set $\hat{\mathcal{G}}$ with respect to which we reduce we assign (once and forever) a set $N[g]$ of multiplicative indices. We write $\operatorname{le}_\prec g \,|_N \operatorname{le}_\prec h$, if the multi index $\operatorname{le}_\prec h$ lies in the restricted cone of $\operatorname{le}_\prec g$ defined by $N[g]$. The set $\mathcal{S}$ collects all generators $g \in \mathcal{G}$ which have already been used for reductions and the set $\mathcal{N}$ is the intersection of the corresponding sets of multiplicative indices. If a new element h is added to $\hat{\mathcal{G}}$, it is assigned as multiplicative indices the current value of $\mathcal{N}$. It should be stressed that this assignment does *not* correspond to some involutive division in the sense of Definition 3.1.1 and in particular we cannot expect that $N[g] = N_{L,\hat{\mathcal{G}},\prec}(g)$ for the given division L!

As already mentioned in Remark 4.5.16, it is possible to formulate the Mora normal form completely without homogenisation. The effect of a homogenisation can be simulated with the help of the *écart* of a polynomial: for a given semigroup order $\prec$ we set $\operatorname{écart} f = \deg f - \deg \operatorname{lt}_\prec f$. Obviously, this definition implies that $\operatorname{écart} f = \deg_t \operatorname{lt}_{\prec_h} f^{(h)}$ and therefore the écart allows us to recognise when it is necessary in Algorithm 4.8 to multiply $\tilde{h}$ by a power of t.

Proposition 4.5.19. *Algorithm 4.9 always terminates. Let $(\mathcal{P} = \Bbbk[X], \star, \prec)$ be a polynomial algebra of solvable type (for an arbitrary semigroup order $\prec$) such that $\Bbbk[X']$ is a subring of $\mathcal{P}$ for any subset $X' \subset X$. Then the output h is a* weak involutive normal form *of the input f with respect to the set $\mathcal{G}$ in the sense that there exists a polynomial $u \in \mathcal{P}$ with $\operatorname{le}_\prec u = 0$ such that the difference $u \star f - h$ possesses an involutive standard representation*

$$u \star f - h = \sum_{g \in \mathcal{G}} P_g \star g \tag{4.24}$$

and none of the leading exponents $\operatorname{le}_\prec g$ involutively divides $\operatorname{le}_\prec h$. If $\prec$ is a monoid order, then $u = 1$ and h is an involutive normal form of f in the usual sense.

Proof. Let us first ignore the involutive aspects of Algorithm 4.9 and show that the reformulation using the écart is correct. The value k appearing in Lines /3/ and /4/ of Algorithm 4.8 is obviously just $\operatorname{écart} g - \operatorname{écart} h$, hence the selection of the next

Algorithm 4.9 Involutive Mora normal form

Input: polynomial $f \in \mathcal{P}$, finite set $\mathcal{G} \subset \mathcal{P}$, involutive division L
Output: involutive Mora normal form h of f with respect to $\mathcal{G}$

$h \leftarrow f$; $\hat{\mathcal{G}} \leftarrow \mathcal{G}$
for all $g \in \mathcal{G}$ **do**
  $N[g] \leftarrow N_{L,\mathrm{le}_\prec \mathcal{G}}(\mathrm{le}_\prec g)$
end for
$\mathcal{N} \leftarrow \{1,\dots,n\}$; $\mathcal{S} \leftarrow \emptyset$
while $(h \neq 0) \wedge (\mathcal{S}_h = \{g \in \hat{\mathcal{G}} : \mathrm{le}_\prec g \,|_{\mathcal{N}}\, \mathrm{le}_\prec h\} \neq \emptyset)$ **do**
  choose $g \in \mathcal{S}_h$ with écart g minimal
  if $(g \in \mathcal{G}) \wedge (g \notin \mathcal{S})$ **then**
    $\mathcal{S} \leftarrow \mathcal{S} \cup \{g\}$; $\mathcal{N} \leftarrow \mathcal{N} \cap N[g]$
  end if
  if écart $g >$ écart h **then**
    $\hat{\mathcal{G}} \leftarrow \hat{\mathcal{G}} \cup \{h\}$; $N[h] \leftarrow \mathcal{N}$
  end if
  $\mu \leftarrow \mathrm{le}_\prec h - \mathrm{le}_\prec g$; $h \leftarrow h - \frac{\mathrm{lc}_\prec h}{\mathrm{lc}_\prec (x^\mu \star g)} x^\mu \star g$
end while
return h

generator with respect to which we reduce and the addition of new elements to $\hat{\mathcal{G}}$ are done in both algorithms by the same criteria. The "normalisation" in Line /9/ of Algorithm 4.8 ensures that this remains true in all iterations of the `while` loop.

Thus upon dehomogenisation Algorithm 4.8 becomes the non-involutive form of Algorithm 4.9 and the termination of the latter follows immediately from the termination of the former. Restricting the reductions to involutive ones obviously does not affect the termination.

The same arguments as in the proof of Proposition 4.5.15 prove the existence of a unit u and a standard representation of $u \star f - h$. There only remains to show that under the made assumption on $\mathcal{P}$ the standard representation is an involutive one. But this is straightforward. We perform the same case distinction as in the proof of Proposition 4.5.15. If $g_i \in \mathcal{G}$, then m_i is multiplicative for it and all coefficients $P_g^{(i)}$ remain trivially multiplicative for the corresponding generator g. If $g_i \in \mathcal{G}_{i-1} \setminus \mathcal{G}$, i. e. $g_i = h_r$ for some $0 \leq r < i-1$, the monomial m_i is multiplicative for h_r. By our assignment of multiplicative variables, this means that m_i is multiplicative for any $g \in \mathcal{G}$ with $P_g^{(r)} \neq 0$. Since we assume that computing the product $m_i \star P_g^{(r)}$ does not introduce any new variables, we find again that all coefficients $P_g^{(i)}$ remain multiplicative for the corresponding generator g. Hence in the final step we obtain indeed an involutive standard representation of $u \star f - h$. □

Even if the set $\mathcal{G}$ is involutively head autoreduced, we cannot conclude in analogy to Proposition 3.4.15 that the involutive Mora normal form is unique, as we only consider the leading term in Algorithm 4.9 and hence the lower terms in h may still be involutive divisible by the leading term of some generator $g \in \mathcal{G}$. However, Theorem 3.4.4 on the existence of involutive standard representations remains valid.

Theorem 4.5.20. *Let* $(\mathcal{P} = \Bbbk[\mathcal{X}], \star, \prec)$ *be a polynomial algebra of solvable type (for a semigroup order* $\prec$*) such that* $\Bbbk[\mathcal{X}']$ *is a subring of* $\mathcal{P}$ *for any subset* $\mathcal{X}' \subset \mathcal{X}$*. Furthermore, let L be a constructive Noetherian division. For a finite set* $\mathcal{F} \subset \mathcal{P}$ *of polynomials let* $\mathcal{I} = \langle \mathcal{F} \rangle$ *be the left ideal generated by it in the localisation* $\mathcal{P}_{\prec}$*. If we apply Algorithm 4.5 with the involutive Mora normal form instead of the usual one to the set* $\mathcal{F}$*, then it terminates with an involutive basis of the ideal* $\mathcal{I}$*.*

Proof. The termination of Algorithm 4.5 under the made assumptions was shown in Proposition 4.2.7 and Theorem 4.4.1. One easily verifies that their proofs are not affected by the substitution of the normal form algorithm, as they rely mainly on Theorem 3.4.4 and on the fact that the leading term of the normal form is not involutively divisible by the leading term of any generator. Both properties remain valid for the Mora normal form. □

The adaption of this result to the Pommaret division proceeds along the same lines as the proof of Theorem 4.3.15. Note again that in this case it suffices to require that $\mathcal{P}$ is an iterated solvable algebra.

Remark 4.5.21. Note that Theorem 4.5.20 guarantees the existence of *strong* involutive bases. Due to the extension to $\mathcal{P}_{\prec}$, Example 4.5.12 is no longer a valid counterexample. As the first generator in $\mathcal{F}$ is now a unit, we find that $\langle \mathcal{F} \rangle = \mathcal{P}_{\prec}$ and $\{1\}$ is a trivial strong Pommaret basis. ◁

Example 4.5.22. We continue Example 4.5.14. Following the approach given by Theorem 4.5.20, we immediately compute as Janet basis of $\langle \mathcal{F} \rangle$ (over $\mathcal{P}_{\prec}$) the minimal basis $\mathcal{H}_3 = \{\partial_x, \partial_y, \partial_z\}$. Obviously, it is considerably smaller than the bases obtained with the classical homogenisation approach (over $\mathcal{P}$). This effect becomes even more profound, if we look at the sizes of the bases in the homogenised Weyl algebra: both $\tilde{\mathcal{H}}_1$ and $\tilde{\mathcal{H}}_2$ consist of 21 generators. ◁

4.6 Involutive Bases over Rings

Finally, we consider the general case that $\mathcal{P} = \mathcal{R}[x_1, \dots, x_n]$ is a polynomial algebra of solvable type over a Noetherian ring $\mathcal{R}$. We will follow the standard approach and assume in the sequel that two operations can be effectively performed in $\mathcal{R}$:

(i) Given elements $s, r_1, \dots, r_k \in \mathcal{R}$, we can decide whether $s \in \langle r_1, \dots, r_k \rangle_{\mathcal{R}}$ (the left ideal in $\mathcal{R}$ generated by $r_1, \dots, r_k$).
(ii) Given elements $r_1, \dots, r_k \in \mathcal{R}$, we can construct a finite basis of the module $\operatorname{Syz}(r_1, \dots, r_k)$ of left syzygies $s_1 r_1 + \dots + s_k r_k = 0$.

In this case one often says that *linear equations are solvable* in $\mathcal{R}$.

The first operation is obviously necessary for the algorithmic reduction of polynomials with respect to a set $\mathcal{F} \subset \mathcal{P}$. The necessity of the second operation will become evident later. Compared with the commutative case, reduction is a more complicated process, in particular due to the possibility that in the commutation relations (3.6) for the multiplication in $\mathcal{P}$ the maps ρ_μ may be different from the identity on $\mathcal{R}$ and the coefficients $r_{\mu\nu}$ unequal one.

Let $\mathcal{G} \subset \mathcal{P}$ be a finite set. We introduce for any polynomial $f \in \mathcal{P}$ the two sets $\mathcal{G}_f = \{g \in \mathcal{G} \mid \ \mathrm{le}_\prec g \mid \mathrm{le}_\prec f\}$ and

$$\bar{\mathcal{G}}_f = \{x^\mu \star g \mid g \in \mathcal{G}_f \wedge \mu = \mathrm{le}_\prec f - \mathrm{le}_\prec g \wedge \mathrm{le}_\prec (x^\mu \star g) = \mathrm{le}_\prec f\} \tag{4.25}$$

The last condition in the definition of the set $\bar{\mathcal{G}}_f$ is redundant only, if the coefficient ring $\mathcal{R}$ is a domain. Otherwise it may happen that $|\bar{\mathcal{G}}_f| < |\mathcal{G}_f|$, namely if $\rho_\mu(r)r_{\mu\nu} = 0$ for $\mathrm{lm}_\prec g = rx^\nu$. The polynomial f is *head reducible* with respect to the set $\mathcal{G}$, if $\mathrm{lc}_\prec g \in \langle \mathrm{lc}_\prec \bar{\mathcal{G}}_f \rangle_\mathcal{R}$ (note that we use $\bar{\mathcal{G}}_f$ here so that the reduction comes only from the leading terms and is not due to some zero divisors as leading coefficients). *Involutive head reducibility* is defined analogously via two sets $\mathcal{G}_{f,L}$ and $\bar{\mathcal{G}}_{f,L}$ where only involutive divisors with respect to an involutive division L on $\mathbb{N}_0^n$ are taken into account, i. e.

$$\mathcal{G}_{f,L} = \{g \in \mathcal{G} \mid \mathrm{le}_\prec f \in \mathcal{C}_{L,\mathrm{le}_\prec \mathcal{G}}(\mathrm{le}_\prec g)\}\,. \tag{4.26}$$

Thus the set $\mathcal{G}$ is *involutively head autoreduced*, if $\mathrm{lc}_\prec g \notin \langle \mathrm{lc}_\prec (\bar{\mathcal{G}}_{g,L} \setminus \{g\}) \rangle_\mathcal{R}$ for all polynomials $g \in \mathcal{G}$. This is now a much weaker notion than before; in particular, Lemma 3.4.14 is no longer valid.

Definition 4.6.1. Let $\mathcal{I} \subseteq \mathcal{P}$ be a left ideal in the polynomial algebra $(\mathcal{P}, \star, \prec)$ of solvable type over a coefficient ring $\mathcal{R}$ in which linear equations can be solved. A finite set $\mathcal{G} \subset \mathcal{P}$ is a *Gröbner basis* of $\mathcal{I}$, if for every polynomial $f \in \mathcal{I}$ the condition $\mathrm{lc}_\prec f \in \langle \mathrm{lc}_\prec \bar{\mathcal{G}}_f \rangle$ is satisfied. The set $\mathcal{G}$ is a *weak involutive basis* for the involutive division L, if for every polynomial $f \in \mathcal{I}$ the condition $\mathrm{lc}_\prec f \in \langle \mathrm{lc}_\prec \bar{\mathcal{G}}_{f,L} \rangle$ is satisfied. A weak involutive basis is a *strong involutive basis*, if every set $\bar{\mathcal{G}}_{f,L}$ contains precisely one element.

It is easy to see that the characterisation of (weak) involutive bases via the existence of involutive standard representations (Theorem 3.4.4) remains valid. Indeed, only the first part of the proof requires a minor change: the polynomial f_1 is now of the form $f_1 = f - \sum_{h \in \mathcal{H}_{f,L}} r_h h$ where the coefficients $r_h \in \mathcal{R}$ are chosen such that the leading monomials cancel, i. e. $\mathrm{le}_\prec f_1 \prec \mathrm{le}_\prec f$.

Clearly, a necessary condition for the existence of Gröbner and thus of (weak) involutive bases for arbitrary left ideals $\mathcal{I} \subset \mathcal{P}$ is that the algebra $\mathcal{P}$ is a (left) Noetherian ring. As we have seen in Section 3.3, this assumption becomes non-trivial, if the coefficient ring $\mathcal{R}$ is not a field. In this section, we will assume throughout that $\mathcal{P}$ is a polynomial algebra of solvable type over a left Noetherian ring $\mathcal{R}$ with centred commutation relations (cf. Definition 3.3.4) so that Theorem 3.3.7 asserts that $\mathcal{P}$ is left Noetherian, too.[6] A very useful side effect of this assumption is that the scalars appearing in the commutation relations (3.6) are units and thus not zero divisors which is important for some arguments.

[6] The case of iterated polynomial algebras of solvable type (cf. Definition 3.3.1) will be considered in an Addendum to Section 5.4, after we have developed a syzygy theory for involutive bases.

Example 4.6.2. As in the previous section, we cannot generally expect strong involutive bases to exist. As a simple concrete example, also demonstrating the need of the second assumption on $\mathcal{R}$, we consider in $\Bbbk[x,y][z]$ (with the ordinary multiplication) the ideal $\mathcal{I}$ generated by the set $\mathcal{F} = \{x^2z - 1, y^2z + 1\}$. Obviously, both generators have the same leading exponent $[1]$; nevertheless none is reducible by the other one due to the relative primeness of the leading coefficients x^2 and y^2, respectively. Furthermore, the syzygy $\mathbf{S} = x^2\mathbf{e}_2 - y^2\mathbf{e}_1 \in \Bbbk[x,y]^2$ connecting the leading coefficients leads to the polynomial $x^2 + y^2 \in \mathcal{I}$. It is easy to see that a Gröbner and a weak Janet basis of $\mathcal{I}$ is obtained by adding it to $\mathcal{F}$. A strong Janet basis does not exist, as none of these generators may be removed from the basis. $\triangleleft$

This example shows that simply applying our completion Algorithm 4.5 (or some variant of it) will generally not suffice. Obviously, with respect to the Janet division z is multiplicative for both elements of $\mathcal{F}$ so that no non-multiplicative variables exist and thus it is not possible to generate the missing generator by multiplication with a non-multiplicative variable. We must substitute in the completion the involutive head autoreduction by a more comprehensive operation.[7]

Definition 4.6.3. Let $\mathcal{F} \subset \mathcal{P}$ be a finite set and L an involutive division. We consider for each polynomial $f \in \mathcal{F}$ the syzygies $\sum_{\bar{f}\in\bar{\mathcal{F}}_{f,L}} s_{\bar{f}} \operatorname{lc}_{\prec} \bar{f} = 0$ connecting the leading coefficients of the elements of $\bar{\mathcal{F}}_{f,L}$. The set $\mathcal{F}$ is *involutively $\mathcal{R}$-saturated* for the division L, if for any such syzygy $\mathbf{S}$ the polynomial $\sum_{\bar{f}\in\bar{\mathcal{F}}_{f,L}} s_{\bar{f}}\bar{f}$ possesses an involutive standard representation with respect to $\mathcal{F}$.

For checking involutive $\mathcal{R}$-saturation, it obviously suffices to consider a finite basis of each of the finitely many syzygy modules $\operatorname{Syz}(\operatorname{lc}_{\prec} \bar{\mathcal{F}}_{f,L})$ so that such a check can easily be performed effectively. An element $f \in \mathcal{F}$ is involutively head reducible by the other elements of $\mathcal{F}$, if and only if $\operatorname{Syz}(\operatorname{lc}_{\prec} \bar{\mathcal{F}}_{f,L})$ contains a syzygy with $s_f = 1$. For this reason it is easy to combine an involutive $\mathcal{R}$-saturation with an involutive head autoreduction leading to Algorithm 4.10.

The `for` loop in Lines /5-13/ takes care of the involutive head autoreduction (`HeadReduce`$_{L,\prec}(f,\mathcal{H})$ involutively head reduces f with respect to the set $\mathcal{H}\setminus\{f\}$ but with multiplicative variables determined with respect to the full set $\mathcal{H}$—cf. Remark 3.4.9). The `for` loop in Lines /17-22/ checks the involutive $\mathcal{R}$-saturation. Each iteration of the outer `while` loop analyses from the remaining polynomials (collected in $\mathcal{S}$) those with the highest leading exponent. The set $\mathcal{S}$ is reset to the full basis, whenever a new element has been put into $\mathcal{H}$; this ensures that all new

[7] In the case of commutative variables over a coefficient field, it is not difficult to show that for any finite set $\mathcal{F}$ the syzygy module $\operatorname{Syz}(\operatorname{lm}_{\prec}\mathcal{F})$ of the leading *monomials* can be spanned by binomial generators corresponding to the S-polynomials in the Buchberger Algorithm B.3. In Section 5.4 we will show that in any such syzygy at least one component contains a non-multiplicative variable, so that implicitly Algorithm 4.5 also runs over a generating set of this syzygy module. When we move on to coefficient rings, it is well-known that additional, more complicated syzygies coming from the coefficients must be considered. For these we can no longer assume that one component contains a non-multiplicative variable. Hence *partially* we must follow the same approach as in the generalisation of the Buchberger algorithm and this leads to the notion of $\mathcal{R}$-saturation where some syzygies not reachable via non-multiplicative variables are explicitly considered.

reduction possibilities are taken into account. In Line /15/ it does not matter which element $f \in \mathcal{S}_\nu$ we choose, as the set $\mathcal{H}'_{f,L}$ depends only on $\operatorname{le}_\prec f$ and all elements of $\mathcal{S}_\nu$ possess by construction the same leading exponent ν.

Algorithm 4.10 Involutive $\mathcal{R}$-saturation (and head autoreduction)

Input: finite set $\mathcal{F} \subset \mathcal{P}$, involutive division L on $\mathbb{N}_0^n$
Output: involutively $\mathcal{R}$-saturated and head autoreduced set $\mathcal{H}$ with $\langle\mathcal{H}\rangle = \langle\mathcal{F}\rangle$
1: $\mathcal{H} \leftarrow \mathcal{F}$; $\mathcal{S} \leftarrow \mathcal{F}$
2: **while** $\mathcal{S} \neq \emptyset$ **do**
3: $\quad \nu \leftarrow \max_\prec \operatorname{le}_\prec \mathcal{S}$; $\mathcal{S}_\nu \leftarrow \{f \in \mathcal{H} \mid \operatorname{le}_\prec f = \nu\}$
4: $\quad \mathcal{S} \leftarrow \mathcal{S} \setminus \mathcal{S}_\nu$; $\mathcal{H}' \leftarrow \mathcal{H}$
5: $\quad$ **for all** $f \in \mathcal{S}_\nu$ **do**
6: $\qquad h \leftarrow \texttt{HeadReduce}_{L,\prec}(f, \mathcal{H})$
7: $\qquad$ **if** $f \neq h$ **then**
8: $\qquad\quad \mathcal{S}_\nu \leftarrow \mathcal{S}_\nu \setminus \{f\}$; $\mathcal{H}' \leftarrow \mathcal{H}' \setminus \{f\}$
9: $\qquad\quad$ **if** $h \neq 0$ **then**
10: $\qquad\qquad \mathcal{H}' \leftarrow \mathcal{H}' \cup \{h\}$
11: $\qquad\quad$ **end if**
12: $\qquad$ **end if**
13: $\quad$ **end for**
14: $\quad$ **if** $\mathcal{S}_\nu \neq \emptyset$ **then**
15: $\qquad$ choose $f \in \mathcal{S}_\nu$ and determine the set $\bar{\mathcal{H}}'_{f,L}$
16: $\qquad$ compute basis $\mathcal{B}$ of $\operatorname{Syz}(\operatorname{lc}_\prec \bar{\mathcal{H}}'_{f,L})$
17: $\qquad$ **for all** $\mathbf{S} = \sum_{\bar{f} \in \bar{\mathcal{H}}'_{f,L}} s_{\bar{f}} \mathbf{e}_{\bar{f}} \in \mathcal{B}$ **do**
18: $\qquad\quad h \leftarrow \texttt{NormalForm}_{L,\prec}(\sum_{\bar{f} \in \bar{\mathcal{H}}'_{f,L}} s_{\bar{f}} \bar{f}, \mathcal{H}')$
19: $\qquad\quad$ **if** $h \neq 0$ **then**
20: $\qquad\qquad \mathcal{H}' \leftarrow \mathcal{H}' \cup \{h\}$
21: $\qquad\quad$ **end if**
22: $\qquad$ **end for**
23: $\quad$ **end if**
24: $\quad$ **if** $\mathcal{H}' \neq \mathcal{H}$ **then**
25: $\qquad \mathcal{H} \leftarrow \mathcal{H}'$; $\mathcal{S} \leftarrow \mathcal{H}$
26: $\quad$ **end if**
27: **end while**
28: **return** $\mathcal{H}$

Proposition 4.6.4. *Under the made assumptions about the polynomial algebra $\mathcal{P}$, Algorithm 4.10 terminates for any finite input set $\mathcal{F} \subset \mathcal{P}$ with an involutively $\mathcal{R}$-saturated and head autoreduced set $\mathcal{H}$ such that $\langle\mathcal{H}\rangle = \langle\mathcal{F}\rangle$.*

Proof. The correctness of the algorithm is trivial. The termination follows from the fact that both $\mathcal{R}$ and $\mathbb{N}_0^n$ are Noetherian. Whenever we add a new polynomial h to the set $\mathcal{H}'$, we have either that $\operatorname{le}_\prec h \notin \langle \operatorname{le}_\prec \mathcal{H}' \rangle_{\mathbb{N}_0^n}$ or $\operatorname{lc}_\prec h \notin \langle \operatorname{lc}_\prec \mathcal{H}'_{h,L} \rangle_{\mathcal{R}}$. As neither in $\mathbb{N}_0^n$ nor in $\mathcal{R}$ infinite ascending chains of ideals are possible, the algorithm must terminate after a finite number of steps. □

An obvious idea is now to substitute in the completion Algorithm 4.5 the involutive head autoreduction by an involutive $\mathcal{R}$-saturation. Recall that Proposition 4.2.7

was the crucial step for proving the correctness of Algorithm 4.5. Our next goal is thus to show that under the made assumptions for involutively $\mathcal{R}$-saturated sets local involution implies weak involution.

Proposition 4.6.5. *Under the made assumptions about the polynomial algebra $\mathcal{P}$, a finite, involutively $\mathcal{R}$-saturated set $\mathcal{F} \subset \mathcal{P}$ is weakly involutive, if and only if it is locally involutive.*

Proof. We first note that Proposition 4.2.7 remains true under the made assumptions. Its proof only requires a few trivial modifications, as all appearing coefficients (for example, when we rewrite $x^\mu \to x^{\mu-1_j} \star x_j$) are units in the case of centred commutation relations and thus we may proceed as for a field. Hence if $\mathcal{F}$ is locally involutive, then $\mathcal{I} = \langle\mathcal{F}\rangle = \langle\mathcal{F}\rangle_{L,\prec}$ implying that any polynomial $g \in \mathcal{I}$ may be written in the form $g = \sum_{f\in\mathcal{F}} P_f \star f$ with $P_f \in \mathcal{R}[X_{L,\mathcal{F},\prec}(f)]$. Furthermore, it follows from this proof that for centred commutation relations we may assume that the polynomials P_f satisfy $\mathrm{le}_\prec(P_f \star f) = \mathrm{le}_\prec P_f + \mathrm{le}_\prec f$. We are done, if we can show that they can be chosen such that additionally $\mathrm{le}_\prec(P_f \star f) \preceq \mathrm{le}_\prec g$, i. e. such that we obtain an involutive standard representation of g.

If the representation coming out of the proof of Proposition 4.2.7 already satisfies this condition on the leading exponents, nothing has to be done. Otherwise we set $\nu = \max_\prec\{\mathrm{le}_\prec(P_f \star f) \mid f \in \mathcal{F}\}$ and $\mathcal{F}_\nu = \{f \in \mathcal{F} \mid \mathrm{le}_\prec(P_f \star f) = \nu\}$. As by construction $\nu \in \bigcap_{f\in\mathcal{F}_\nu} \mathcal{C}_{L,\mathrm{le}_\prec\mathcal{F}}(\mathrm{le}_\prec f)$, the properties of an involutive division imply that we can write $\mathcal{F}_\nu = \{f_1,\dots,f_k\}$ with $\mathrm{le}_\prec f_1 \mid \mathrm{le}_\prec f_2 \mid \cdots \mid \mathrm{le}_\prec f_k$ and hence $\mathcal{F}_\nu \subseteq \mathcal{F}_{f_k,L}$. Since we have assumed that $\mathrm{le}_\prec(P_f \star f) = \mathrm{le}_\prec P_f + \mathrm{le}_\prec f$, we even find $\mathcal{F}_\nu \subseteq \bar{\mathcal{F}}_{f_k,L}$.

By construction, the equality $\sum_{f\in\mathcal{F}_\nu} \mathrm{lc}_\prec(P_f \star f) = 0$ holds. If we now write $\mathrm{lm}_\prec f = r_f x^{\nu_f}$ and $\mathrm{lm}_\prec P_f = s_f x^{\mu_f}$, then we obtain under the made assumptions: $\mathrm{lc}_\prec(P_f \star f) = s_f \rho_{\mu_f}(r_f) r_{\mu_f\nu_f} = \big[s_f \bar{\rho}_{\mu_f}(r_f) r_{\mu_f\nu_f}\big] r_f$ and hence the above equality corresponds to a syzygy of the set $\mathrm{lc}_\prec \mathcal{F}_{f_k,L}$. As the set $\mathcal{F}$ is involutively $\mathcal{R}$-saturated, there exists an involutive standard representation

$$\sum_{i=1}^{k} \big[s_{f_i} \bar{\rho}_{\mu_{f_i}}(r_{f_i}) r_{\mu_{f_i}\nu_{f_i}}\big] \bar{f}_i = \sum_{f\in\mathcal{F}} Q_f \star f \tag{4.27}$$

with $Q_f \in \Bbbk[X_{L,\mathcal{F},\prec}(f)]$ and $\mathrm{le}_\prec(Q_f \star f) = \mathrm{le}_\prec Q_f + \mathrm{le}_\prec f \prec \nu_{f_k}$.

Introducing now the polynomials $Q'_f = Q_f - \big[s_f \bar{\rho}_{\mu_f}(r_f) r_{\mu_f\nu_f}\big] x^{\nu_{f_k} - \nu_f}$ for $f \in \mathcal{F}_\nu$ and $Q'_f = Q_f$ otherwise, we get the syzygy $\sum_{f\in\mathcal{F}} Q'_f \star f = 0$. If we set furthermore $P'_f = P_f - c_f^{-1} x^{\nu-\nu_{f_k}} \star Q'_f$ with $c_f = \bar{\rho}_{\nu-\nu_{f_k}}\big(s_f \bar{\rho}_{\mu_f}(r_f) r_{\mu_f\nu_f}\big) \bar{\rho}_{\mu_f}(r_f) r_{\mu_f\nu_f}$, then, by construction, $g = \sum_{f\in\mathcal{F}} P'_f \star f$ is another involutive representation of the polynomial g with $\nu' = \max_\prec\{\mathrm{le}_\prec(P'_f \star f) \mid f \in \mathcal{F}\} \prec \nu$.

Repeating this procedure for a finite number of times obviously yields an involutive standard representation of the polynomial g. As g was an arbitrary element of the ideal $\mathcal{I} = \langle\mathcal{F}\rangle$, this implies that $\mathcal{F}$ is indeed weakly involutive. □

Theorem 4.6.6. *Let $\mathcal{P}$ be a polynomial algebra of solvable type satisfying the made assumptions. If the subalgorithm* $\texttt{InvHeadAutoReduce}_{L,\prec}$ *is substituted in Algorithm 4.5 by Algorithm 4.10, then the completion will terminate with a weak involutive basis of $\mathcal{I} = \langle\mathcal{F}\rangle$ for any finite input set $\mathcal{F} \subset \mathcal{P}$ such that the monoid ideal $\mathrm{le}_\prec \mathcal{I}$ possesses a weak involutive basis.*

Proof. The correctness of the modified algorithm follows immediately from Proposition 4.6.5. For the termination we may use the same argument as in the proof of Theorem 4.4.1, as it depends only on the leading exponents. □

4.7 Notes

The notions of continuous and constructive divisions were introduced by Gerdt and Blinkov [156]. Obviously, both are rather technical concepts and especially constructivity is often difficult to prove. They are designed exactly in such a way that our proof of Proposition 4.2.1 works. Combined they ensure that no infinite chains can appear in Algorithm 4.1. Despite its complicated definition, it is obvious that constructivity is a necessary property for the correctness of our algorithms. Otherwise we would be forced to consider also products with multiplicative variables. Concerning continuity, the situation is less clear. While it is sufficient for Proposition 4.1.4 on the equivalence of local involution and weak involution to hold, it is not known whether it is also necessary.

As already mentioned in the Notes of Chapter 3, Apel [20] introduced his own concept of an involutive division. His divisions are defined with respect to only one subset $\mathcal{N} \subseteq \mathbb{N}_0^n$. As during the completion the set $\mathcal{H}$ is changed in every step, we always need a new division for the new set $\mathrm{le}_\prec \mathcal{H}$ and Apel discusses some heuristic strategies for choosing the divisions.[8] Whether this additional freedom can be systematically exploited to achieve a higher efficiency is an open question. A definite advantage of his approach is that it becomes almost trivial to prove the existence of an involutive basis for arbitrary ideals.

However, one should note the following difference. In Apel's approach it is not even possible to formulate the task to determine an involutive basis for a prescribed involutive division, as we do not know the relevant set $\mathrm{le}_\prec \mathcal{H}$ before the completion is over. Thus this approach is only useful, if the goal is to somehow get some direct sum decomposition or if the involutive theory is only used as an alternative for Buchberger's algorithm to compute Gröbner bases; the role of the involutive division is then mainly that of a technical trick to improve efficiency. Up to now, this cannot really be considered as a disadvantage, as not much is known about special properties of the decompositions obtained for different divisions (with the exception

[8] One should note that Apel admits the case $\mathcal{N} = \mathbb{N}_0^n$, i. e. divisions defined on the set of all multi indices. For such divisions, there is no need to change the divisions within a computation and in fact one can formulate the classical involutive divisions like the Pommaret or the Janet division in his framework. In this case his theory largely coincides with ours.

of Pommaret bases, as we will see in the next chapter). But one might expect that in the future involutive divisions will gain a similar importance as term orders and that for different tasks one is interested in different ones.

The basic ideas underlying the completion Algorithm 4.5 are already contained in the Janet–Riquier Theory of differential equations from the early 20th century (compare also the Notes of Chapter 3). Our formulation is essentially due to Blinkov, Gerdt and Zharkov [156, 157, 489]. The simple completion Algorithm 4.3 has been "folklore"; it is explicitly spelled out for the Janet division in [354]. The optimised Algorithm 4.6 is also due to Gerdt and Blinkov [157]. In recent years, they have spent, together with further collaborators, much effort on making the algorithm more efficient and on providing fast implementations. Results of extensive benchmarks are contained in [160] and on their web site http://invo.jinr.ru. The involutive form of Buchberger's second criterion was also first given by Gerdt and Blinkov [156, Thm. 8.1]. We presented a slightly more general formulation with a simpler proof making use of our results on syzygies in the next chapter.

Our proof of Theorem 4.4.1 asserting the correctness and termination of the improved completion Algorithm 4.5 makes essential use of the normal selection strategy. Since it is well-known that this strategy has problems with some term orders, one wonders whether this restriction is really necessary and, indeed, one can show that more refined optimisations can circumvent it [23, 80, 155].

The completion approaches discussed in this chapter are not the only ones. Since Gröbner bases are by now a standard tool in commutative algebra and algebraic geometry, a number of algorithms for their construction have been designed and their improvement is a very active field of research. In a direct comparison of Buchberger's Algorithm B.3 for the construction of Gröbner bases with the involutive approach (recall that any involutive basis is automatically a Gröbner basis) it turns out that the latter one is highly competitive as long as the involutive basis is not considerably larger than the Gröbner basis (see [160] for concrete benchmarks).

Some simple heuristic explanations of this observation are possible (further arguments are given in [20]). The involutive Algorithm 4.5 implicitly also computes S-polynomials but less than a naive implementation of the Buchberger algorithm, as Buchberger's second criterion is not completely but to a large extent automatically contained in the strategy of multiplying only by non-multiplicative variables. This fact follows from our proof of Theorem 5.4.4 in the next chapter.

The strategy underlying Algorithm 4.5 also has certain similarities with the *Hilbert driven Buchberger algorithm* introduced by Traverso [457]. In this variant of the Buchberger algorithm one assumes that one knows already the Hilbert function of the ideal and this knowledge is used to avoid further unnecessary reductions. In concrete computations, this idea has turned out to be highly effective. Unfortunately, the determination of the Hilbert function requires usually a Gröbner basis. Thus the main application of this technique consists of transforming a Gröbner basis with respect to one term order into one with respect to another term order.

In this context, our algorithm can be understood as follows. As we will see in the next chapter, given an involutively head autoreduced basis of the considered ideal, we can always make a conjecture about its Hilbert function based on the assignment of multiplicative variables (we made this idea explicit in our treatment of δ-regularity of a finite set $\mathcal{F}$ in Section 4.3). The multiplication of the elements by their non-multiplicative variables checks whether the conjectured function is already the true Hilbert function or whether it still yields too small values (as the involutive span of a basis is generally only a subset of the full ideal, we always approach the true Hilbert function from below). The algorithm stops, as soon as the true Hilbert function has been reached. Note that in contrast to the Hilbert driven Buchberger algorithm no a priori knowledge of the Hilbert function is required.

Another interesting aspect about the involutive approach to the determination of Gröbner bases is the following one. A classical problem in computer algebra is called *intermediate expression swell*: often intermediate expressions are much larger than the final result. In the case of Gröbner bases, this phenomenon concerns in particular the size of the coefficients of the appearing polynomials.

Arnold [25, Ex. 1.1] provided a particularly striking example. Consider the ideal $\mathcal{I} \subset \mathbb{Q}[x,y,z]$ generated by the following four polynomials:[9]

$$\begin{aligned} f_1 &= 8y^2z^2 + 5y^3z + 3xz^3 + xyz^2 \,, \\ f_2 &= z^5 + 2x^2y^3 + 13x^3y^2 + 5x^4y \,, \\ f_3 &= 8z^3 + 12y^3 + x^2z + 3 \,, \\ f_4 &= 7y^4z^2 + 18x^2y^3z + x^3y^3 \,. \end{aligned} \tag{4.28}$$

The reduced Gröbner basis of $\mathcal{I}$ with respect to the degree reverse lexicographic term order is very small:

$$g_1 = z\,, \quad g_2 = y^3 + 1/4\,, \quad g_3 = x^2\,. \tag{4.29}$$

However, Arnold reports that during the determination of this basis with the computer algebra system MACAULAY2 intermediate polynomials appear possessing coefficients with about 80.000 digits! Obviously, even elementary arithmetic operations with such larger numbers are quite time consuming.

By contrast, Gerdt (private communication) reports the following results for the involutive approach. The Janet basis for this example is obtained by adding to the reduced Gröbner basis the two monomials $g_4 = x^2y$ and $g_5 = x^2y^2$. The largest coefficients of intermediate polynomials appearing during its computation along the lines of the algorithms discussed in this chapter have roughly 400 digits, i. e. are smaller by several orders of magnitude compared with the Buchberger algorithm. Thus, at least for this example, the approach via involutive divisions shows a much smoother behaviour. So far it is unclear to what extent this represents a general phenomenon and how it could be explained theoretically.

Proposition 4.4.3 says that even if Algorithm 4.5 does not terminate, it yields in a finite number of steps a Gröbner basis for term orders of type ω. In the given form, this result is only of theoretical interest, as it is unclear how one should detect

[9] In order to be consistent with our conventions for the standard term orders, x and z are swapped compared with [25].

the Gröbner basis. Using standard criteria would destroy all potential advantages of our algorithm. For the special case of Pommaret bases, Apel [19] found a simple criterion, so that a variant of Algorithm 4.5 can be used for the construction of Gröbner bases independent of the existence of a finite involutive basis.

δ-regularity is often considered a purely technical nuisance which, however, is not true. Its origin are intrinsic algebraic properties of the ideal $\mathcal{I}$ considered: as we will see in Section 5.3 and Chapter 6, it is related to a Noether normalisation and to the associated prime ideals of $\mathcal{P}/\mathcal{I}$. In the theory of differential equations, we recover here the classical notion of characteristics: if the hyperplane $x^n = 0$ is a characteristic surface, then the coordinates are δ-singular for the symbol module (see Section 7.5).

Several solutions to the problem of δ-regularity have been proposed in the literature, mainly in the context of differential equations. Most of them are based on the proof of Proposition 4.3.8. Pommaret [356] advocates the brute force method of computing in a generic coordinate system. As it requires the introduction of n^2 additional parameters which are regarded as algebraically independent transcendentals over $\Bbbk$, this approach is only of theoretical value. Even for small examples computations in generic coordinates very quickly blow up. An advantage is that one obtains an explicit characterisation of all δ-singular coordinate systems which is sometimes of interest (see e. g. [439, Proposition 5.3])

Another possibility consists of relying on Proposition 4.3.8 and performing a random coordinate transformation. This approach is, for example, used by Hartley and Tucker [194] in their implementation of the Cartan–Kähler theory (and also by Sturmfels and White [439] in their above mentioned method for the construction of Rees decompositions). As we have seen, it leads with probability 1 to a δ-regular coordinate system. Besides the possibility that we still can get a δ-singular system and thus have to verify the δ-regularity, the main disadvantage of this approach is that it destroys any sparsity that might be present in the original basis $\mathcal{F}$. In particular, in applications to differential equations the elements of $\mathcal{F}$ are very often sparse, i. e. not every derivative u^α_μ up to a given length q of the multi indices μ does actually appear. Of course, such sparsity presents a great advantage in any concrete computations which one would like to preserve.

The approach presented in Section 4.3 was developed in [198] inspired by the work of Gerdt [154] on the relation between Pommaret and Janet bases (however, the treatment given there is not completely correct; we followed here the improved version from [410]). The coordinate transformation indicated by the criterion for δ-singularity of Theorem 4.3.12 preserves comparatively much of the sparsity of $\mathcal{F}$, as it only affects terms containing x^ℓ. However, it may of course happen that many transformations are necessary and then not much has been gained.

The basic idea to compute Gröbner bases with respect to semigroup orders via homogenisation is due to Lazard [283]. Mora [328] developed the normal form carrying now his name in the context of computing tangent cones which requires the determination of Gröbner bases for a special kind of semigroup orders. He introduced the notion of the *écart* (French for difference or deviation) of a polynomial f

as the difference between the lowest and the highest degree of a term in f (or, alternatively, we may write $\text{écart}\, f = \deg_{x_{n+1}} f^{(h)}$).

The generalisation of his ideas that we used has been first proposed by Greuel and Pfister [184] (see also [185, Sect. 1.6]) and later independently by Gräbe [176, 177]; an extensive treatment is contained in the textbook [99, Chapt. 4]. They noticed that Mora's approach remains valid for arbitrary semigroup orders, if one modifies the definition of the écart to $\text{écart}\, f = \deg f - \deg \text{lt}_{\prec} f$.

The extension to involutive bases was first discussed in [199] for the special case of the Weyl algebra. There only Lazard's approach via homogenisation was considered; the more efficient approach via the Mora normal form comes from [409].

The computation of Gröbner bases in polynomial rings over a coefficient ring have been studied in [164, 459] (see [5, Chapt. 4] for a more extensive textbook discussion); for PBW extensions (recall Example 3.2.6) a theory of Gröbner bases was developed by Giesbrecht et al [165]. The extension of involutive bases to this more general situation was apparently first considered in [409].

The strategy to determine left bases even for two-sided ideals goes back at least to Kandry-Rody and Weispfenning [249] and is by now fairly standard (see e. g. [63, 264, 286]). An alternative approach via bimodules was recently presented in [148]. The material in the Addendum to Section 4.2 may be considered as an involutive version of [249, Sects. 4/5] (see also [264, Sect. 4.11]) and comes again from [409].

The extension of the here discussed completion algorithms to nonlinear differential equations will be discussed in Chapter 7. For the moment we only note that a special situation arises for quasi-linear equations. Since they are linear in the derivatives of highest order, all reductions can be performed effectively as linear operations. However, we cannot expect the equation to remain quasi-linear during the completion, as some of the arising integrability conditions may be fully nonlinear. Provided all appearing equations are indeed quasi-linear, Algorithm 4.5 can be used without problems and produces the correct result. A simple example where this happens are the Navier–Stokes equations.

Chapter 5
Structure Analysis of Polynomial Modules

By the help of God and with His precious assistance, I say that Algebra is a scientific art.

Omar Khayyam

We now apply the theory of involutive bases developed in Chapter 3 to the structure analysis of modules over the commutative polynomial ring, i. e. we do some classical commutative algebra (the question to what extent the results presented here also hold for more general polynomial rings of solvable type is still open and will not be discussed here). The basic observation is that the Pommaret basis with respect to the degree lexicographic term order provides us with an easy access to several interesting invariants; more precisely, this basis is to a considerable extent determined by the structure of the module. As this particular type of basis will also play an important role in the analysis of differential equations, this fact simultaneously allows us an algebraic interpretation of many aspects of the theory of differential equations.

We start in the first section by analysing combinatorial or Stanley decompositions: a polynomial module is written as a direct sum of free modules over polynomial rings in a restricted set of variables. It is a natural consequence of our definition of involutive bases (and in fact the main motivation for it) that any involutive basis of a submodule of a free polynomial module immediately induces such a decomposition. For more general modules the situation is more complicated. Assuming that we deal with a finitely generated module, we can present it as the quotient of a free polynomial module by a submodule and then construct a complementary decomposition to a Gröbner basis of the submodule.

The next section is concerned with the determination of the (Krull) dimension and the depth. It is a trivial application of arbitrary Stanley decompositions that one can directly read off the Hilbert series and thus the dimension; this observation is an immediate consequence of their combinatorial nature. Pommaret bases lead to a particular type of decompositions named after Rees. With their help also regular sequences and thus the depth are easily obtained. As a simple corollary, we obtain a new proof of the Hironaka criterion for Cohen–Macaulay algebras.

In the third section we will relate the question of δ-regularity to the existence of Noether normalisations. Special emphasis will be put on monomial ideals where the existence of a Pommaret basis is equivalent to simultaneous Noether normalisations of both the ideal itself and all its primary components. Furthermore, in an

W.M. Seiler, *Involution*, Algorithms and Computation in Mathematics 24,
DOI 10.1007/978-3-642-01287-7_5, © Springer-Verlag Berlin Heidelberg 2010

Addendum we will discuss how one can extract the standard pairs of a monomial ideal from any complementary decomposition of the ideal.

The topic of Section 5.4 is the syzygy theory of involutive bases. We prove an involutive version of the classical Schreyer Theorem: the syzygies obtained from analysing the non-multiplicative products of the generators form a Gröbner basis of the first syzygy module. For certain divisions, including the Janet and the Pommaret division, one even obtains again an involutive basis. Thus an immediate iteration allows for the straightforward construction of a whole syzygy resolution.

In the special case of the Pommaret division, the arising free resolution is highly structured. Large parts of the last two sections are concerned with exploiting these structures for further statements about the underlying module. For example, we will see that, albeit the resolution is generally not minimal, it is always of minimal length so that we can immediately read off the projective dimension of our module. As a trivial corollary we will recover the Auslander–Buchsbaum formula relating the projective dimension with the depth. A further, very important result is that the degree of a Pommaret basis with respect to the degree reverse lexicographic term order always equals the Castelnuovo–Mumford regularity of the module.

5.1 Combinatorial Decompositions

The distinctive feature underlying the definition of (strong) involutive bases is the direct sum decomposition induced by them. In this section we study this aspect in more details. All results apply to arbitrary finitely generated polynomial modules. But for notational simplicity, we restrict to graded $\mathbb{k}$-algebras $\mathcal{A} = \mathcal{P}/\mathcal{I}$ for some homogeneous ideal $\mathcal{I} \subseteq \mathcal{P}$ where $\mathcal{P} = \mathbb{k}[x^1, \dots, x^n]$ is the ordinary polynomial ring. If we speak of a basis of the ideal $\mathcal{I}$, we always assume that it is homogeneous, too.

Recall that the main motivation of Buchberger for the introduction of Gröbner bases was to be able to compute effectively in such factor spaces. Indeed given a Gröbner basis $\mathcal{G}$ of the ideal $\mathcal{I}$, the normal form with respect to $\mathcal{G}$ distinguishes a unique representative in each equivalence class. Our goal is to show that Pommaret bases yield in addition a lot of structural information about the algebra $\mathcal{A}$. More precisely, we want to compute fundamental invariants like the Hilbert function, the depth or the minimal resolution. Our basic tools are combinatorial decompositions of $\mathcal{A}$ into a direct sum of polynomial rings with a restricted number of variables.

Definition 5.1.1. A *Stanley decomposition* of the graded $\mathbb{k}$-algebra $\mathcal{A} = \mathcal{P}/\mathcal{I}$ consists of an isomorphism as graded $\mathbb{k}$-linear spaces

$$\mathcal{A} \cong \bigoplus_{t \in \mathcal{T}} \mathbb{k}[X_t] \cdot t \tag{5.1}$$

with a finite set $\mathcal{T} \subset \mathbb{T}$ and sets $X_t \subseteq \{x^1, \dots, x^n\}$. The elements of X_t are the *multiplicative variables* of the generator t.

Remark 5.1.2. We may contrast Stanley decompositions with Noether normalisations (cf. Definition B.1.23). In a Stanley decomposition we want to express $\mathcal{A}$ as a direct sum of free polynomial modules. The price we have to pay for the directness is that we must use in general different polynomial rings for the different generators. In a Noether normalisation we want that all appearing modules are free over the same polynomial ring $\mathcal{R}$. But the price for this requirement is that we cannot achieve a direct sum composition. Common to both constructions is that they are finite: in Definition 5.1.1 the set $\mathcal{T}$ of generators must be finite; the crucial property of a Noether normalisation is that it leads to a finitely generated $\mathcal{R}$-module. $\triangleleft$

Corollary 3.4.5 asserts that given any involutive basis of the ideal $\mathcal{I}$ we immediately obtain a similar decomposition of $\mathcal{I}$ itself. Here, however, we are interested in decomposing the quotient space $\mathcal{A} = \mathcal{P}/\mathcal{I}$ and no obvious way seems to exist how an involutive basis of the ideal $\mathcal{I}$ should yield a *complementary decomposition*, i. e. a decomposition of $\mathcal{A}$.

Vasconcelos [468, p. 23] calls Stanley decompositions *"an approach that is not greatly useful computationally but it is often nice theoretically"*. One reason for his assessment is surely that the classical algorithms for their construction work only for monomial ideals $\mathcal{I}$. Thus for a general ideal $\mathcal{I}$ one must first compute a Gröbner basis of it for some term order $\prec$ and then, exploiting that as vector spaces $\mathcal{P}/\mathcal{I} \cong \mathcal{P}/\operatorname{lt}_\prec \mathcal{I}$, one determines a Stanley decomposition of the latter algebra. Its existence is guaranteed by the following result which we formulate again for multi indices, i. e. the monoid $\mathbb{N}_0^n$.

Proposition 5.1.3. *Let $\mathcal{I} \subseteq \mathbb{N}_0^n$ be a monoid ideal and denote by $\overline{\mathcal{I}} = \mathbb{N}_0^n \setminus \mathcal{I}$ its complementary set. Then there exists a finite set $\overline{\mathcal{B}} \subset \overline{\mathcal{I}}$ and for each multi index $\nu \in \overline{\mathcal{B}}$ a set $N_\nu \subseteq \{1,\dots,n\}$ defining the disjoint decomposition*

$$\overline{\mathcal{I}} = \biguplus_{\nu \in \overline{\mathcal{B}}} \left(\nu + \mathbb{N}^n_{N_\nu}\right) . \tag{5.2}$$

Proof. The assertion is trivial for the zero ideal, so we exclude this case and proceed by an induction on the size n of the multi indices. The case $n = 1$ is trivial, as any monoid ideal in $\mathbb{N}_0$ is of the form $\mathcal{I} = \langle [q] \rangle$ for some value $q \in \mathbb{N}_0$ and hence $\overline{\mathcal{I}} = \{[0],[1],\dots,[q-1]\}$ which is of the form (5.2) with $N_\nu = \emptyset$ for all ν.

Suppose now that the assertion is true for $n-1$. We want to show that it holds for n, too. But before we must introduce some notations. Primed letters always refer to multi indices $\nu' \in \mathbb{N}_0^{n-1}$; unprimed ones to multi indices $\nu \in \mathbb{N}_0^n$. For a given $\nu \in \mathbb{N}_0^n$, we define $\nu' = [\nu_1,\dots,\nu_{n-1}] \in \mathbb{N}_0^{n-1}$, i. e. we simply drop the last entry. Finally, we write $[\nu',q]$ for the multi index $[\nu_1,\dots,\nu_{n-1},q] \in \mathbb{N}_0^n$.

Given a monoid ideal $\mathcal{I} \subseteq \mathbb{N}_0^n$, we define for all values $q \in \mathbb{N}_0$ the monoid ideals $\mathcal{I}'_q = \{\nu' \in \mathbb{N}_0^{n-1} \mid [\nu',q] \in \mathcal{I}\}$ in $\mathbb{N}_0^{n-1}$. Thus we slice $\mathcal{I}$ by intersecting it with the hyperplanes $\nu_n = \mathit{Const}$. Obviously, for $q < r$ we have $\mathcal{I}'_q \subseteq \mathcal{I}'_r$ so that these ideals form an ascending chain. As $\mathbb{N}_0^{n-1}$ is a Noetherian monoid, an integer $q_0 \in \mathbb{N}$ exists such that $\mathcal{I}'_q = \mathcal{I}'_{q_0}$ for all $q \geq q_0$.

According to our inductive hypothesis, the complements $\overline{\mathcal{I}}'_q$ of each of these monoid ideals for $0 \leq q \leq q_0$ can be disjointly decomposed in the form

$$\overline{\mathcal{I}}'_q = \biguplus_{\nu' \in \overline{\mathcal{B}}_q^{(n-1)}} \left(\nu' + \mathbb{N}^{n-1}_{N_{\nu'}}\right) \tag{5.3}$$

with some subsets $\overline{\mathcal{B}}_q^{(n-1)} \subset \mathbb{N}_0^{n-1}$ and $N_{\nu'} \subseteq \{1,\dots,n-1\}$. We introduce now the sets $\overline{\mathcal{B}}_q^{(n)} = \big\{[\nu',q] \mid \nu' \in \overline{\mathcal{B}}_q^{(n-1)}\big\}$ and $\overline{\mathcal{B}} = \bigcup_{q=0}^{q_0} \overline{\mathcal{B}}_q^{(n)}$. Furthermore, we choose $N_{[\nu',q]} = N_{\nu'}$ for $q < q_0$ and $N_{[\nu',q]} = N_{\nu'} \cup \{n\}$ for $q = q_0$.

Setting $\mathcal{U} = \bigcup_{\nu \in \overline{\mathcal{B}}} (\nu + \mathbb{N}^n_{N_\nu})$, the union is obviously disjoint. We claim now that $\mathcal{U} = \overline{\mathcal{I}}$ which implies our proposition. Let $\mu \in \mathcal{U}$, then by construction we have that $\mu' \in \overline{\mathcal{I}}'_q$ for some $0 \leq q \leq q_0$ and hence by definition of the ideals $\mathcal{I}'_q$, the multi index μ cannot be an element of $\mathcal{I}$. Thus $\mathcal{U} \subseteq \overline{\mathcal{I}}$. Conversely, $\mu \in \overline{\mathcal{I}}$ implies $\mu' \in \overline{\mathcal{I}}'_{\mu_n}$. If $\mu_n \geq q_0$, then $\mathcal{I}'_{\mu_n} = \mathcal{I}'_{q_0}$ and a multi index $\nu' \in \overline{\mathcal{B}}_{q_0}^{(n-1)}$ exists such that $\mu' \in \nu' + \mathbb{N}^{n-1}_{N_{\nu'}}$. Thus $\mu \in \nu + \mathbb{N}^n_\nu$ with $\nu = [\nu', q_0]$. If $\mu_n < q_0$, then a multi index $\nu' \in \overline{\mathcal{B}}_{\mu_n}^{(n-1)}$ exists such that $\mu' \in \nu' + \mathbb{N}^{n-1}_{N_{\nu'}}$ and in this case $\mu \in \nu + \mathbb{N}^n_\nu$ with $\nu = [\nu', \mu_n]$. Thus we have $\overline{\mathcal{I}} \subseteq \mathcal{U}$ and consequently $\mathcal{U} = \overline{\mathcal{I}}$ as claimed. □

This proof is not completely constructive, as the number q_0 is defined by a Noetherian argument. But one easily sees that we may set $q_0 = \max_{\nu \in \mathcal{B}} \nu_n$ for the minimal basis $\mathcal{B}$ of the monoid ideal $\mathcal{I}$. Now one straightforwardly transforms our proof above into the recursive Algorithm 5.1 for the construction of a complementary decomposition; the only point is to see that the set $\mathcal{B}'_q$ constructed in Line /6/ is indeed a minimal basis of the monoid ideal $\mathcal{I}'_q$, but this fact is trivial to verify.

Algorithm 5.1 Complementary decomposition (from minimal basis)

Input: minimal basis $\mathcal{B}$ of monoid ideal $\mathcal{I} \subset \mathbb{N}_0^n$
Output: finite complementary decomposition $\overline{\mathcal{B}}$
1: **if** n=1 **then** $\{$in this case $\mathcal{B} = \{\nu\}\}$
2: $\quad q_0 \leftarrow \nu_1$; $\overline{\mathcal{B}} \leftarrow \big\{([0],\emptyset),\dots,([q_0-1],\emptyset)\big\}$
3: **else**
4: $\quad q_0 \leftarrow \max_{\nu \in \mathcal{B}} \nu_n$; $\overline{\mathcal{B}} \leftarrow \emptyset$
5: $\quad$ **for** q **from** 0 **to** q_0 **do**
6: $\quad\quad \mathcal{B}'_q \leftarrow \{\nu' \in \mathbb{N}_0^{n-1} \mid \nu \in \mathcal{B},\ \nu_n \leq q\}$
7: $\quad\quad \overline{\mathcal{B}}'_q \leftarrow$ `ComplementaryDecomposition`$(\mathcal{B}'_q)$
8: $\quad\quad$ **if** $q < q_0$ **then**
9: $\quad\quad\quad \overline{\mathcal{B}} \leftarrow \overline{\mathcal{B}} \cup \big\{([\nu',q],N_{\nu'}) \mid (\nu',N_{\nu'}) \in \overline{\mathcal{B}}'_q\big\}$
10: $\quad\quad$ **else**
11: $\quad\quad\quad \overline{\mathcal{B}} \leftarrow \overline{\mathcal{B}} \cup \big\{([\nu',q],N_{\nu'} \cup \{n\}) \mid (\nu',N_{\nu'}) \in \overline{\mathcal{B}}'_q\big\}$
12: $\quad\quad$ **end if**
13: $\quad$ **end for**
14: **end if**
15: **return** $\overline{\mathcal{B}}$

The complementary decomposition (5.2) is not unique and different decompositions may use sets $\overline{\mathcal{B}}$ of different sizes. The elements in the sets N_ν are again called *multiplicative indices*. If a Janet basis of the monoid ideal $\mathcal{I}$ is known, an alternative, non-recursive algorithm exists. It provides us with a decomposition where the multiplicative indices are again chosen according to the Janet division.

Proposition 5.1.4. *Let $\mathcal{B}_J$ be a Janet basis of the monoid ideal $\mathcal{I} \subseteq \mathbb{N}_0^n$. Then the set $\overline{\mathcal{B}} \subset \mathbb{N}_0^n$ in the decomposition* (5.2) *may be chosen such that for all $\nu \in \overline{\mathcal{B}}$ the equality $N_\nu = N_{J,\mathcal{B}_J \cup \{\nu\}}(\nu)$ holds.*

Proof. We explicitly construct such a set $\overline{\mathcal{B}}$. Let $q_n = \max_{\mu \in \mathcal{B}_J} \mu_n$. Then we put into $\overline{\mathcal{B}}$ all multi indices $\nu = [0,\dots,0,q]$ such that $0 \leq q < q_n$ and such that $\mathcal{B}_J$ does not contain any multi index μ with $\mu_n = q$. We set $N_\nu = \{1,\dots,n-1\}$ which obviously is just $N_{J,\mathcal{B}_J \cup \{\nu\}}(\nu)$ according to the definition of the Janet division given in Example 3.1.4.

Let $(d_k,\dots,d_n) \subseteq \mathcal{B}_J$ be one of the subsets (3.2) considered in the assignment of the multiplicative variables to the elements of $\mathcal{B}_J$ according to the definition of the Janet division and set $q_{k-1} = \max_{\mu \in (d_k,\dots,d_n)} \mu_{k-1}$. We enlarge the set $\overline{\mathcal{B}}$ by all those multi indices $\nu = [0,\dots,0,q,d_k,\dots,d_n]$ where $0 \leq q < q_{k-1}$ and where $(d_k,\dots,d_n)$ does not contain any multi index μ with $\mu_{k-1} = q$. The corresponding sets N_ν of multiplicative indices contain the indices $1,\dots,k-2$ and all those indices greater than k that are multiplicative for the elements of $(d_k,\dots,d_n)$. Again it is easy to verify that this entails $N_\nu = N_{J,\mathcal{B}_J \cup \{\nu\}}(\nu)$.

We claim that the thus constructed set $\overline{\mathcal{B}}$ (together with the sets N_ν) defines a disjoint decomposition (5.2) of the complementary set $\overline{\mathcal{I}}$. Like in the proof of the previous proposition we proceed by an induction on n and use again the notations introduced there. The case $n = 1$ is trivial. Suppose that our assertion is true for $n-1$. Let $\rho \in \mathbb{N}_0^n$ be an arbitrary multi index and let $q_1 < q_2 < \cdots < q_r$ represent all values that occur for the entry μ_n in the multi indices $\mu \in \mathcal{B}_J$. We distinguish three cases depending on the value of ρ_n: (i) $\rho_n < q_r$ but $\rho_n \notin \{q_1,\dots,q_{r-1}\}$, (ii) $\rho_n \in \{q_1,\dots,q_{r-1}\}$, and (iii) $\rho_n \geq q_r$.

In the first case, the set $\overline{\mathcal{B}}$ contains the multi index $\nu = [0,\dots,0,\rho_n]$ and, as $N_\nu = \{1,\dots,n-1\}$, obviously $\rho \in \nu + \mathbb{N}^n_{N_\nu}$. It is easy to see that ρ cannot be an element of $\mathcal{I}$ or of the cone $\bar{\nu} + \mathbb{N}^n_{N_{\bar{\nu}}}$ of any other multi index $\bar{\nu} \in \overline{\mathcal{B}}$.

In the second case, we proceed by "slicing" both $\mathcal{B}_J$ and $\overline{\mathcal{B}}$ at the degree ρ_n. This operation leads to the two sets $\mathcal{B}'_J = \{\mu' \in \mathbb{N}_0^{n-1} \mid \mu \in \mathcal{B}_J,\ \mu_n = \rho_n\} \subset \mathbb{N}_0^{n-1}$ and $\overline{\mathcal{B}}' = \{\nu' \in \mathbb{N}_0^{n-1} \mid \nu \in \overline{\mathcal{B}},\ \nu_n = \rho_n\} \subset \mathbb{N}_0^{n-1}$, respectively. If we compute the sets $N_{J,\mathcal{B}'_J}(\mu')$ of multiplicative indices for the elements $\mu' \in \mathcal{B}'_J$, it is straightforward to verify that they are just $N_{J,\mathcal{B}_J}(\mu)$, as μ_n is not maximal and $\mu \in (\rho_n)$, if and only if $\mu' \in \mathcal{B}'_J$. Furthermore, $\mathcal{B}'_J$ is a Janet basis of the monoid ideal $\mathcal{I}'_{\rho_n}$. If we apply the procedure above to this ideal in $\mathbb{N}_0^{n-1}$, we obtain $\overline{\mathcal{B}}'$ as complementary basis and the sets of multiplicative variables remain unchanged.

By our inductive hypothesis, $\overline{\mathcal{B}}'$ defines a disjoint decomposition of the sought form of the set $\overline{\mathcal{I}}'_{\rho_n}$. If $\rho \in \overline{\mathcal{I}}$, then $\rho' \in \overline{\mathcal{I}}'_{\rho_n}$ and a multi index $\nu' \in \overline{\mathcal{B}}'$ exists such that

$\rho' \in \nu' + \mathbb{N}^{n-1}_{N_{\nu'}}$. By construction, $\nu = [\nu', \rho_n] \in \overline{\mathcal{B}}$ and $\rho \in \nu + \mathbb{N}^n_{N_\nu}$. If $\rho \in \mathcal{I}$, then $\rho' \in \mathcal{I}'_{\rho_n}$, and it is not possible to find a multi index $\nu' \in \overline{\mathcal{B}}'$ such that $\rho' \in \nu' + \mathbb{N}^{n-1}_{N_{\nu'}}$. Hence $\rho \notin \nu + \mathbb{N}^n_{N_\nu}$ for any $\nu \in \overline{\mathcal{B}}$.

The third case is very similar to the second one. For the definition of $\mathcal{B}'_J$ and $\overline{\mathcal{B}}'$ we consider only multi indices where the value of the last entry is q_r. Note that for all multi indices $\nu \in \overline{\mathcal{B}}$ that contribute to $\overline{\mathcal{B}}'$ the index n is multiplicative. If $\rho \in \overline{\mathcal{I}}$, then $\rho' \in \overline{\mathcal{I}}'_{\rho_n}$ and a multi index $\nu' \in \overline{\mathcal{B}}'$ exists such that $\rho' \in \nu' + \mathbb{N}^{n-1}_{N_{\nu'}}$ and we conclude as above that $\rho \in \nu + \mathbb{N}^n_{N_\nu}$ for $\nu = [\nu', q_r] \in \overline{\mathcal{B}}$. Again it is obvious that for a multi index $\rho \in \mathcal{I}$, it is not possible to find such a $\nu \in \overline{\mathcal{B}}$. □

In a more formal language this proof leads to Algorithm 5.2. The two outer `for`-loops iterate over all non-empty subsets $(d_{k+1},\dots,d_n) \subseteq \mathcal{B}_J$ (recall that $() = \mathcal{B}_J$ for $k = n$), i. e. they traverse the full Janet tree of $\mathcal{B}_J$ with k denoting the current depth in the tree. Thus in practice one would use a tree algorithm for a breadth-first transversal and the inner `for`-loop is an iteration over the "gaps" between the values at the children of the current node. In Line /4/ it does not matter which multi index $\mu \in (d_{k+1},\dots,d_n)$ one chooses, as we need in Line /5/ only the multiplicative indices greater than k which are the same for all such μ.

Algorithm 5.2 Complementary decomposition (from Janet basis)

Input: Janet basis $\mathcal{B}_J$ of monoid ideal $\mathcal{I} \subset \mathbb{N}^n_0$
Output: finite complementary decomposition $\overline{\mathcal{B}}$
1: $\overline{\mathcal{B}} \leftarrow \emptyset$
2: **for** k **from** n **downto** 1 **do**
3:     **for all** $\emptyset \neq (d_{k+1},\dots,d_n) \subseteq \mathcal{B}_J$ **do**
4:         choose arbitrary $\mu \in (d_{k+1},\dots,d_n)$
5:         $N \leftarrow \{1,\dots,k-1\} \cup \{i \in N_{J,\mathcal{B}_J}(\mu) \mid i > k\}$
6:         **for** i **from** 0 **to** $\max_{\mu \in (d_{k+1},\dots,d_n)} \{\mu_k\}$ **do**
7:             **if** $\nexists \mu \in (d_{k+1},\dots,d_n) : \mu_k = i$ **then**
8:                 $\overline{\mathcal{B}} \leftarrow \overline{\mathcal{B}} \cup \{([0,\dots,0,i,d_{k+1},\dots,n],N)\}$
9:             **end if**
10:         **end for**
11:     **end for**
12: **end for**
13: **return** $\overline{\mathcal{B}}$

We have formulated Proposition 5.1.4 only for the case that the given set $\mathcal{B}_J$ is a Janet basis, but our proof yields an even stronger result. Let $\mathcal{B}_J$ be an *arbitrary* subset of $\mathbb{N}^n_0$ and construct the corresponding set $\overline{\mathcal{B}}$. Then we may substitute everywhere in the proof the full ideal $\mathcal{I}$ by the involutive span $\langle \mathcal{B}_J \rangle_J \subseteq \mathcal{I}$ and still obtain that any multi index $\rho \in \mathbb{N}^n_0$ either lies in $\langle \mathcal{B}_J \rangle_J$ or in exactly one of the cones defined by $\overline{\mathcal{B}}$ and the sets N_ν.

Example 5.1.5. Consider the simple example $\mathcal{B}_J = \{[1,1]\}$. Our construction yields the two multi indices $\nu^{(1)} = [0,0]$ with $N_{\nu^{(1)}} = \{1\}$ and $\nu^{(2)} = [0,1]$ with $N_{\nu^{(2)}} = \{2\}$.

Obviously, the corresponding cones disjointly decompose the complementary set $\overline{\mathcal{I}} = \{[i,j] \in \mathbb{N}_0^2 \mid ij = 0\}$. ◁

According to Corollary 4.3.11, we may apply the same construction to Pommaret bases. But as the Pommaret division is globally defined, it is very easy to provide an alternative decomposition depending only on the degree q of a Pommaret basis of the ideal $\mathcal{I}$ (we will see later in Section 5.5 that this degree is in fact an important invariant of $\mathcal{I}$, namely its Castelnuovo–Mumford regularity). In general, this decomposition is larger than the one obtained with the Janet approach, but it is sometimes useful for theoretical applications.

Proposition 5.1.6. *The monoid ideal $\mathcal{I} \subseteq \mathbb{N}_0^n$ has a Pommaret basis of degree q, if and only if the sets $\overline{\mathcal{B}_0} = \{\nu \in \bar{\mathcal{I}} \mid |\nu| < q\}$ and $\overline{\mathcal{B}_1} = \{\nu \in \bar{\mathcal{I}} \mid |\nu| = q\}$ yield the following disjoint decomposition*

$$\bar{\mathcal{I}} = \overline{\mathcal{B}_0} \uplus \biguplus_{\nu \in \overline{\mathcal{B}_1}} \mathcal{C}_P(\nu)\,. \tag{5.4}$$

Proof. The definition of the Pommaret division implies the identity

$$(\mathbb{N}_0^n)_{\geq q} = \biguplus_{|\nu|=q} \mathcal{C}_P(\nu) \tag{5.5}$$

from which one direction trivially follows. Here $(\mathbb{N}_0^n)_{\geq q}$ denotes the set of all multi indices of length greater than or equal to q. By Definition 3.1.1 of an involutive division, the union on the right hand side is disjoint.

For the converse, we claim that if (5.4) holds, then the set $\mathcal{H} = \{\mu \in \mathcal{I}_q\}$ is a Pommaret basis of the monoid ideal $\mathcal{I}_{\geq q}$; our assertion follows then immediately by Lemma 4.3.2. We choose an arbitrary multi index $\mu \in \mathcal{H}$ and a non-multiplicative index $k = \operatorname{cls}\mu < j \leq n$ of it. If we can show that $\mu + 1_j \in \langle\mathcal{H}\rangle_P$, then $\mathcal{H}$ is involutive by Proposition 4.2.7. But this fact is trivial: we have $\mu + 1_j \in \mathcal{C}_P(\mu - 1_k + 1_j)$ and $\mu - 1_k + 1_j \in \mathcal{I}_q$, as otherwise $\mu + 1_j \notin \mathcal{I}$ contradicting (5.4). □

Example 5.1.7. The decomposition (5.4) is usually redundant. Considering for $\overline{\mathcal{B}}_1$ only multi indices of length q makes the formulation much easier but is not optimal. Consider the trivial example $\mathcal{B}_P = \{[0,1]\}$. According to Proposition 5.1.6, we should set $\overline{\mathcal{B}}_0 = \{[0,0]\}$ and $\overline{\mathcal{B}}_1 = \{[1,0]\}$. But obviously $\overline{\mathcal{I}} = [0,0] + \mathbb{N}^2_{\{1\}}$. Applying Algorithm 5.2 yields directly this more compact form. ◁

Remark 5.1.8. For monomial ideals a refinement of complementary decompositions is provided by *Stanley filtrations*. This notion requires an ordering of the generators of the decomposition and in our multi index notation we can formulate it as follows. Assume that we have $\mathcal{B} = \{\nu^{(1)}, \dots, \nu^{(r)}\}$ in the complementary decomposition (5.2). Defining $\mathcal{I}_k = \mathcal{I} + \langle \nu^{(k+1)}, \dots, \nu^{(r)}\rangle$ for all $0 \leq k \leq r$, we speak of a Stanley filtration, if always

$$\overline{\mathcal{I}_k} = \biguplus_{j=1}^{k} \left(\nu^{(j)} + \mathbb{N}^n_{N_{\nu^{(j)}}}\right)\,, \tag{5.6}$$

i. e., if we have a filtration

$$\emptyset = \overline{\mathcal{I}_0} \subset \overline{\mathcal{I}_1} \subset \cdots \subset \overline{\mathcal{I}_r} = \overline{\mathcal{I}} \,. \tag{5.7}$$

It is not difficult to find a complementary decomposition not admitting an ordering of its generators that makes it to a Stanley filtration. We will now prove that the decompositions determined by Algorithm 5.2 from a Janet basis always induce a Stanley filtration, if we order the generators according to the lexicographic order.

Let again $\mathcal{B}_J$ be a Janet basis of the monoid ideal $\mathcal{I}$ and $\overline{\mathcal{B}}$ the output of Algorithm 5.2. We first note that if we set $\mathcal{B}'_J = \mathcal{B}_J \cup \{\mu\}$ and $\overline{\mathcal{B}'} = \overline{\mathcal{B}} \setminus \{\mu\}$ for any multi index $\mu \in \overline{\mathcal{B}}$, then applying Algorithm 5.2 to $\mathcal{B}'_J$ yields $\overline{\mathcal{B}'}$ and for no element of any of these sets the multiplicative variables change. Indeed, it is easy to see that none of the values q_k appearing in the proof of Proposition 5.1.4 changes and that for $\mu = [0,\dots,0,q,d_{k+1},\dots,d_n]$ the subset $(q,d_{k+1},\dots,d_n)$ was empty for $\mathcal{B}_J$; hence the assignment of multiplicative variables according to the Janet division is not affected by adding μ to $\mathcal{B}_J$.

Now assume that we choose for μ the maximal element of $\overline{\mathcal{B}}$ with respect to the lexicographic order. We claim that then for any $\nu \in \overline{\mathcal{B}'}$ the intersection $\mathcal{C}(\mu) \cap \mathcal{C}_{J,\mathcal{B}'\cup\{\nu\}}(\nu)$ is empty: obviously, this observation implies the complementary decomposition

$$\overline{\mathcal{I} + \langle\mu\rangle} = \biguplus_{\nu \in \overline{\mathcal{B}'}} \left(\nu + \mathbb{N}^n_{N_\nu}\right) \tag{5.8}$$

(and that $\mathcal{B}'_J$ is a Janet basis of $\mathcal{I} + \langle\mu\rangle$); hence by iteration that we have a Stanley filtration. Take an arbitrary $\nu \in \overline{\mathcal{B}'}$; since $\nu \prec_{\text{lex}} \mu$, we must have $\mu_i > \nu_i$, if the last non-vanishing entry of $\nu - \mu$ is at position i. As then $\mu, \nu \in (\mu_{i+1},\dots,\mu_n)$, it follows that the index i is not Janet multiplicative for ν with respect to the set $\mathcal{B}' \cup \{\nu\}$ which implies our claim.

A similar result holds for the decomposition (5.4) obtained via a Pommaret basis. This time, we get a Stanley filtration, if we order the generators according to the degree reverse lexicographic order. Thus we first consider the elements of $\overline{\mathcal{B}_1}$ where one easily sees that for two multi indices $\mu, \nu \in \overline{\mathcal{B}_1}$ with $\nu \prec_{\text{degrevlex}} \mu$ again the intersection $\mathcal{C}(\mu) \cap \mathcal{C}_P(\nu)$ is always empty. Since the elements of $\overline{\mathcal{B}_0}$ do not possess any multiplicative variables in the decomposition (5.4), we may take for them any order refining the partial order induced by divisibility—a condition satisfied by any term order according to Lemma A.1.6. ◁

If $\mathcal{J} \subseteq \mathcal{P}$ is a polynomial ideal possessing a Pommaret basis for some term order $\prec$, then applying Proposition 5.1.6 to $\mathcal{I} = \text{le}_\prec \mathcal{J}$ yields a Stanley decomposition of a special type: all sets X_t are of the form $X_t = \{x^1, x^2, \dots, x^{\text{cls}\,t}\}$ where the number $\text{cls}\,t$ is called the *class*[1] of the generator t. One speaks then of a *Rees decomposition* of $\mathcal{A} = \mathcal{P}/\mathcal{J}$. It is no coincidence that we use here the same terminology as in the definition of the Pommaret division: if $t = x^\mu$ with $\mu \in \bar{\mathcal{B}}_1$, then indeed its class is just $\text{cls}\,\mu$ in the usual sense. Given a polynomial $f \in \mathcal{P}$ and a term order $\prec$, we write

[1] Some authors prefer the term *level*.

in the sequel briefly $\operatorname{cls} f$ instead of $\operatorname{cls}(\operatorname{le}_{\prec} f)$. Elimination of the redundancy in the decomposition (5.4) leads to the following result.

Corollary 5.1.9. *Let $\mathcal{I} \subseteq \mathcal{P}$ be a polynomial ideal having for some term order $\prec$ a Pommaret basis $\mathcal{H}$ where $\min_{h\in\mathcal{H}} \operatorname{cls} h = d$. Then the factor algebra $\mathcal{P}/\mathcal{I}$ possesses a Rees decomposition where the minimal class of a generator is $d-1$.*

Proof. Obviously, it suffices to consider the monomial case and formulate the proof therefore in the multi index language of Proposition 5.1.6. Furthermore, for $d = 1$ there is nothing to be shown so that we assume from now on $d > 1$. Our starting point is the decomposition (5.4). For each $\nu \in \bar{\mathcal{B}}_1$ with $\operatorname{cls}\nu = k < d$ we introduce the multi index $\tilde{\nu} = \nu - (\nu_k)_k$, i. e. $\tilde{\nu}$ arises from ν by simply setting the kth entry to zero. Obviously, the k-dimensional cone $C_\nu = \tilde{\nu} + \mathbb{N}^n_{\{1,\dots,k\}}$ is still completely contained in the complement $\bar{\mathcal{I}}$ and we have $\mathcal{C}_P(\nu) \subset C_\nu$.

Replacing in (5.4) for any such ν the cone $\mathcal{C}_P(\nu)$ by C_ν, we still have a decomposition of $\bar{\mathcal{I}}$, but no longer a disjoint one. We show first that in the thus obtained decomposition all cones C with $0 < \dim C < d-1$ can be dropped without loss. For $k < d-1$ we consider the multi index $\mu = \tilde{\nu} + (\nu_k)_{k+1}$. Since $|\mu| = q$ and $\operatorname{cls}\mu = k+1$ by construction, $\mu \in \bar{\mathcal{B}}_1$ under the made assumptions. Furthermore, $\tilde{\mu} = \mu - (\mu_{k+1})_{k+1}$ is a divisor of $\tilde{\nu}$ ($\tilde{\mu}$ and $\tilde{\nu}$ differ at most in their $(k+1)$st entries and $\tilde{\mu}_{k+1} = 0$) and thus the inclusion $C_\nu \subset C_\mu = \tilde{\mu} + \mathbb{N}^n_{\{1,\dots,k+1\}}$ holds.

The remaining cones with $\dim C \geq d-1$ are all disjoint. This is trivially true for all cones with $\dim C \geq d$, as these have not been changed. For the other ones, we note that if μ and ν are two multi indices with $\operatorname{cls}\mu = \operatorname{cls}\nu = d-1$ and $|\mu| = |\nu| = q$, then they must differ at some position ℓ with $\ell \geq d$. But this fact implies that the cones C_μ and C_ν are disjoint.

Thus there only remains to study the zero-dimensional cones consisting of the multi indices $\nu \in \bar{\mathcal{B}}_0$. If we set $\ell = q - |\nu|$, then $\mu = \nu + \ell_1 \in \bar{\mathcal{B}}_1$, since we assumed $d > 1$, and trivially $\nu \in C_\mu = (\mu - (\mu_1)_1) + \mathbb{N}^n_{\{1\}}$. By our considerations above, the cone C_μ and thus also the multi index ν is contained in some $(d-1)$-dimensional cone. Therefore we may also drop all zero-dimensional cones and obtain a Rees decomposition where all cones are at least $(d-1)$-dimensional. □

Slightly generalising the notion of Rees decompositions, we speak of a *quasi-Rees decomposition*, if there exists a term $\bar{t} \in \mathcal{T}$ such that $\bigcup_{t\in\mathcal{T}} X_t = X_{\bar{t}}$, i. e. there exists a unique maximal set of multiplicative variables containing all other sets of multiplicative variables. Obviously, every Rees decomposition is a quasi-Rees decomposition, but not vice versa. We will see in the next two sections that such decompositions possess special properties.

5.2 Dimension and Depth

We have seen that given a Gröbner basis of the homogeneous ideal $\mathcal{I}$, it is possible to compute a Stanley decomposition of the graded algebra $\mathcal{A} = \mathcal{P}/\mathcal{I}$. As a first and fairly simple application we discuss now the determination of the (Krull) dimension and the depth as two different measures for the size of an algebra (or more generally of a module). While for the dimension any Stanley decompositions can be used (in fact, this question was the original motivation for their introduction), it will turn out that for the depth we need a Rees decomposition and thus a Pommaret basis of $\mathcal{I}$ is here better than any other Gröbner bases.

Proposition 5.2.1. *Let the graded* $\Bbbk$*-algebra* $\mathcal{A} = \mathcal{P}/\mathcal{I}$ *possess the Stanley decomposition* (5.1). *Then its Hilbert series is*

$$\mathcal{H}_{\mathcal{A}}(\lambda) = \sum_{t\in\mathcal{T}} \frac{\lambda^{q_t}}{(1-\lambda)^{k_t}} \tag{5.9}$$

where $q_t = \deg t$ *and* $k_t = |X_t|$. *Thus the (Krull) dimension of the graded algebra* $\mathcal{A}$ *is given by* $\dim\mathcal{A} = D = \max_{t\in\mathcal{T}} k_t$ *and the multiplicity* $\operatorname{mult}\mathcal{A}$ *by the number of terms* $t \in \mathcal{T}$ *with* $k_t = D$.

Proof. The Hilbert series of a polynomial ring $\Bbbk[y^1,\dots,y^d]$ in d variables is given by $\mathcal{H}(\lambda) = 1/(1-\lambda)^d$, since both the Taylor coefficient of order q of this function and the number of monomials of degree q in the ring given by (A.4a) is $\binom{d+q-1}{q}$. If the generator t has degree q_t, we must shift all coefficients by q_t which corresponds to a multiplication with λ^{q_t}. Finally, we simply add the contributions of all generators, as the right hand side of (5.1) is a direct sum, and arrive at (5.9). The statements about the dimension and the multiplicity are trivial corollaries (see the discussion following Theorem B.1.22 and in particular (B.10) relating the Hilbert series and the Hilbert polynomial). □

Given a Pommaret basis of the ideal $\mathcal{I}$, we can provide an alternative characterisation of $\dim\mathcal{A}$ which is useful for many applications. It may be considered as a strengthening of the following observation for arbitrary quasi-Rees decompositions.

Lemma 5.2.2. *Let* $\mathcal{G}$ *be a Gröbner basis of the homogeneous ideal* $\mathcal{I} \subseteq \mathcal{P}$ *for some term order* $\prec$ *and assume that the finite set* $\mathcal{T} \subset \mathbb{T}$ *defines a quasi-Rees decomposition of* $\mathcal{A} = \mathcal{P}/\mathcal{I}$ *with the maximal set* $X_{\bar{t}}$ *of multiplicative variables for some term* $\bar{t} \in \mathcal{T}$. *If* $q = 1 + \max_{t\in\mathcal{T}} \deg t$, *then* $\langle\mathcal{G}, X_{\bar{t}}\rangle_q = \mathcal{P}_q$ *and no smaller set of variables or other set of variables of the same size has this property.*

Proof. Assume first that the term $x^\mu \in \mathcal{P}_q \setminus \operatorname{lt}_\prec\mathcal{I}$ is not contained in the leading ideal. By definition of the degree q, we have $|\mu| > \deg t$ for all $t \in \mathcal{T}$. Hence x^μ must be properly contained in some cone of the quasi-Rees decomposition and can be written as a product mt with some $t \in \mathcal{T}$ and a term m in the variables $X_t \subseteq X_{\bar{t}}$ with $\deg m > 0$. This presentation implies $x^\mu \in \langle X_{\bar{t}}\rangle$.

If the term $x^\mu \in \mathcal{P}_q$ lies in $\operatorname{lt}_\prec\mathcal{I}$, we compute its normal form with respect to the Gröbner basis $\mathcal{G}$. If this normal form vanishes, then $x^\mu \in \langle\mathcal{G}\rangle$. Otherwise, it is a $\Bbbk$-linear combination of terms in $\mathcal{P}_q \setminus \operatorname{lt}_\prec\mathcal{I}$ and thus lies by the considerations above in $\langle X_{\bar{t}}\rangle$. Hence we may conclude that all terms of degree q lie in $\langle\mathcal{G}, X_{\bar{t}}\rangle$.

No other set $\bar{X} \subset X$ of variables with $\bar{X} \neq X_t$ and $|\bar{X}| \leq |X_{\bar{t}}|$ can possess this property, as $\mathrm{lt}_\prec \mathcal{I} \cap \Bbbk[X_{\bar{t}}] = \{0\}$ and hence we can always find a term $x^\mu \in (\Bbbk[X_{\bar{t}}])_q$ not contained in $\langle \mathcal{G}, \bar{X} \rangle$. Indeed, assume that a $x^\nu \in \mathrm{lt}_\prec \mathcal{I} \cap \Bbbk[X_{\bar{t}}]$ existed. Then obviously $\bar{t} \cdot x^\nu \in \mathrm{lt}_\prec \mathcal{I}$, contradicting the fact that $\mathcal{T}$ defines a complementary decomposition with multiplicative variables $X_{\bar{t}}$ for $\bar{t}$. □

Proposition 5.2.3. *Let $\mathcal{H}$ be a Pommaret basis of the homogeneous ideal $\mathcal{I} \subseteq \mathcal{P}$ for some term order $\prec$. If $\deg \mathcal{H} = \max_{h \in \mathcal{H}} \deg h = q$, then the dimension D of the factor algebra $\mathcal{A} = \mathcal{P}/\mathcal{I}$ satisfies*

$$D = \min\left\{ i \mid \langle \mathcal{H}, x_1, \dots, x_i \rangle_q = \mathcal{P}_q \right\}. \tag{5.10}$$

Proof. The Hilbert polynomials of $\mathcal{A}$ and the truncation $\mathcal{A}_{\geq q}$ coincide. Thus it suffice to consider the latter algebra. By Lemma 4.3.2, a Pommaret basis of $\mathcal{I}_{\geq q}$ is given by the set $\mathcal{H}_q$ determined in (4.5). If D is the smallest number such that $\langle \mathcal{H}_q, x^1, \dots, x^D \rangle_q = \mathcal{P}_q$, then all multi indices $\nu \in \mathbb{N}_0^n$ with $|\nu| = q$ and $\mathrm{cls}\,\nu > D$ lie in $\mathrm{le}_\prec \mathcal{H}_q$ but a multi index $\mu \in \mathbb{N}_0^n$ exists with $|\mu| = q$, $\mathrm{cls}\,\mu = D$ and $\mu \neq \mathrm{le}_\prec \mathcal{H}_q$. By Proposition 5.1.6, this observation entails that μ is a generator of class D of the decomposition (5.4) and that the decomposition does not contain a generator of higher class. Hence $\dim \mathcal{A} = D$ by Proposition 5.2.1. □

A subset $X_\mathcal{I} \subseteq \{x^1, \dots, x^n\}$ of variables is called *independent modulo* a given ideal $\mathcal{I} \subseteq \mathcal{P}$, if $\mathcal{I} \cap \Bbbk[X_\mathcal{I}] = \{0\}$, i. e. if $\mathcal{I}$ does not contain any non-zero polynomial depending only on the variables in $X_\mathcal{I}$.[2] A subset $X_\mathcal{I} \subseteq \{x^1, \dots, x^n\}$ is *strongly independent* modulo $\mathcal{I}$ for a given term order $\prec$, if $\mathrm{lt}_\prec \mathcal{I} \cap \Bbbk[X_\mathcal{I}] = \{0\}$. Such a set is automatically also independent modulo $\mathcal{I}$, because for any non-zero polynomial $f \in \mathcal{I} \cap \Bbbk[X_\mathcal{I}]$ in particular the leading term must lie in $\Bbbk[X_\mathcal{I}]$. However, the converse is not true, as the following trivial counterexample shows: consider $\mathcal{P} = \Bbbk[x, y]$ and $\mathcal{I} = \langle y - x \rangle \subset \mathcal{P}$ for any term order $\prec$ such that $x \prec y$; then the set $\{y\}$ is independent but not strongly independent modulo $\mathcal{I}$.

One can show that $\dim \mathcal{A}$ is the maximal size of either a strongly independent or an independent set. It is comparatively easy to establish this fact for independent sets, but less simple for strongly independent sets, as for them seemingly the maximal number of elements depends on the chosen term order. A further complication arises from the fact that in general maximal independent sets of different sizes exist. As a simple concrete example we consider $\mathcal{I} = \langle xz + z, yz + z \rangle \subset \Bbbk[x, y, z]$. Here both $\{x, y\}$ and $\{z\}$ are maximal independent sets modulo $\mathcal{I}$.

Strong independence modulo $\mathcal{I}$ with respect to a term order $\prec$ is easy to verify effectively with the help of a Gröbner basis $\mathcal{G}$ of $\mathcal{I}$ for $\prec$: it follows immediately from Definition B.4.1 of a Gröbner basis that the set $X_\mathcal{I}$ is strongly independent, if and only if it satisfies $\mathrm{lt}_\prec \mathcal{G} \cap \Bbbk[X_\mathcal{I}] = \emptyset$. It is now a combinatorial (and thus sometimes quite expensive) exercise to determine effectively all maximal strongly

[2] One easily shows that for the special case that $\mathcal{I} \subset \mathcal{P}$ is a prime ideal, this notion of independence coincides with the familiar concept of algebraic independence in field theory: the set $X_\mathcal{I}$ is independent modulo $\mathcal{I}$, if and only if the equivalence classes of its elements in $\mathcal{P}/\mathcal{I}$ are algebraically independent over the quotient field of $\mathcal{P}/\mathcal{I}$.

independent sets modulo $\mathcal{I}$ and to compute their maximal size and hence $\dim \mathcal{A}$. The situation becomes much simpler, if $\mathcal{A} = \mathcal{P}/\mathcal{I}$ admits a quasi-Rees decomposition, as we will show now that in this case a unique maximal strongly independent set modulo $\mathcal{I}$ exists. This observation is based on the following result.

Lemma 5.2.4. *Let $\mathcal{I} \subset \mathcal{P}$ be an ideal and $\prec$ a term order. Assume that the finite set $\mathcal{T} \subset \mathbb{T}$ defines a quasi-Rees decomposition of the algebra $\mathcal{A}' = \mathcal{P}/\operatorname{lt}_{\prec}\mathcal{I}$ with the maximal set $X_{\bar{t}}$ of multiplicative variables for some term $\bar{t} \in \mathcal{T}$. Then a variable x^i is not contained in $X_{\bar{t}}$, if and only if the minimal basis of $\operatorname{lt}_{\prec}\mathcal{I}$ contains an element of the form $(x^i)^{e_i}$ for some exponent $e_i \in \mathbb{N}$.*

Proof. Assume first that $x^i \notin X_{\bar{t}}$. By definition of a quasi-Rees decomposition, then $x^i \notin X_t$ for all $t \in \mathcal{T}$. Since $\mathcal{T}$ is a finite set, only finitely many terms of the form $t = (x^i)^{k_t}$ can be contained in it. If we choose k greater than all these values k_t, then $(x^i)^k \in \operatorname{lt}_{\prec}\mathcal{I}$ and the minimal basis of $\operatorname{lt}_{\prec}\mathcal{I}$ must contains an element $(x^i)^{e_i}$.

For the converse, assume that $(x^i)^{e_i}$ lies in the minimal basis of $\operatorname{lt}_{\prec}\mathcal{I}$. Then for any $t \in \mathcal{T}$ the term $t \cdot (x^i)^{e_i}$ lies in $\operatorname{lt}_{\prec}\mathcal{I}$ and thus x^i cannot be an element of X_t by definition of a complementary decomposition. □

Proposition 5.2.5. *Under the assumptions of Lemma 5.2.4, the set $X_{\bar{t}}$ is the unique maximal strongly independent set modulo the ideal $\mathcal{I}$.*

Proof. We showed already in the proof of Lemma 5.2.2 that $\operatorname{lt}_{\prec}\mathcal{I} \cap \mathbb{k}[X_{\bar{t}}] = \{0\}$, i. e. that the set $X_{\bar{t}}$ is strongly independent modulo $\mathcal{I}$. It follows from Lemma 5.2.4 that no variable $x^i \notin X_{\bar{t}}$ can be contained in a strongly independent set modulo $\mathcal{I}$. Hence any such set must be a subset of $X_{\bar{t}}$. □

Corollary 5.2.6. *Let the chosen coordinates $\mathbf{x}$ be δ-regular for the ideal $\mathcal{I} \subset \mathcal{P}$, i. e. $\mathcal{I}$ possesses a Pommaret basis $\mathcal{H}$. Then $\{x^1, \dots, x^D\}$ with $D = \dim \mathcal{A}$ is the unique maximal strongly independent set modulo the ideal $\mathcal{I}$.*

Our next application is the construction of regular sequences for a homogeneous ideal $\mathcal{I} \subseteq \mathcal{P}$ where it will again turn out that δ-regular coordinates are special. Recall that the maximal length of such a sequence yields the depth of $\mathcal{I}$ (see Appendix B.1). While the definition allows for arbitrary polynomials in regular sequences, it suffices for computing the depth to consider only polynomials $f_i \in \mathcal{P}_1$ of degree 1. This fact follows, for example, from [439, Lemma 4.1]. For this reason, our proof of the following result restricts to this case.

Proposition 5.2.7. *Let $\mathcal{I} \subseteq \mathcal{P}$ be a homogeneous ideal and $\mathcal{H}$ a homogeneous Pommaret basis of it for a class respecting term order $\prec$ and set $d = \min_{h \in \mathcal{H}} \operatorname{cls} h$. Then the monomials $(x^1, \dots, x^d)$ form a maximal regular sequence and* $\operatorname{depth} \mathcal{I} = d$.

Proof. A Pommaret basis $\mathcal{H}$ induces a Rees decomposition of $\mathcal{I}$ of the form

$$\mathcal{I} = \bigoplus_{h \in \mathcal{H}} \mathbb{k}[x^1, \dots, x^{\operatorname{cls} h}] \cdot h \,. \tag{5.11}$$

If $d = \min_{h\in\mathcal{H}} \operatorname{cls} h$ denotes the minimal class of a generator in $\mathcal{H}$, then (5.11) trivially implies that the sequence $(x^1,\dots,x^d)$ is regular.

Let us try to extend this sequence by a variable x^k with $k > d$. We introduce the subset $\mathcal{H}_d = \{h \in \mathcal{H} \mid \operatorname{cls} h = d\}$ and choose an element $\bar{h} \in \mathcal{H}_d$ of maximal degree: according to the characterisation of class respecting term orders given in Lemma A.1.8, we have $\bar{h} \in \langle x^1,\dots,x^d\rangle$. Now consider the product $x^k\bar{h}$; by construction, the variable x^k is non-multiplicative for $\bar{h}$. Thus for each generator $h \in \mathcal{H}$ a coefficient $P_h \in \Bbbk[x^1,\dots,x^{\operatorname{cls} h}]$ exists such that $x^k\bar{h} = \sum_{h\in\mathcal{H}} P_h h$. No polynomial $h \in \mathcal{P}$ with $\operatorname{cls} h > d$ can lie in $\langle x^1,\dots,x^d\rangle$ (obviously $\operatorname{lt}_\prec h \notin \langle x^1,\dots,x^d\rangle$). As the leading terms cannot cancel in the sum, $P_h \in \langle x^1,\dots,x^d\rangle$ for all $h \in \mathcal{H}\setminus\mathcal{H}_d$. Thus $x^k\bar{h} = \sum_{h\in\mathcal{H}_d} c_h h + g$ with $c_h \in \Bbbk$ and $g \in \langle x^1,\dots,x^d\rangle\mathcal{I}$. As $\mathcal{I}$ is a homogeneous ideal and as the degree of $\bar{h}$ is maximal in $\mathcal{H}_d$, all constants c_h must vanish.

It is not possible that $\bar{h} \in \langle x^1,\dots,x^d\rangle\mathcal{I}$, as otherwise $\bar{h}$ would be involutively head reducible by some other element of $\mathcal{H}$. Hence we have shown that any variable x^k with $k > d$ is a zero divisor in $\mathcal{I}/\langle x^1,\dots,x^d\rangle\mathcal{I}$ and the $\mathcal{I}$-regular sequence $(x^1,\dots,x^d)$ cannot be extended by any x^k with $k > d$.

Finally, assume that the forms $y^1,\dots,y^{d+1} \in \mathcal{P}_1$ define a regular sequence of length $d+1$. We extend them to a basis $\{y^1,\dots,y^n\}$ of $\mathcal{P}_1$ and perform a coordinate transformation $\mathbf{x} \mapsto \mathbf{y}$. Our basis $\mathcal{H}$ transforms into a set $\mathcal{H}_{\mathbf{y}}$ and after an involutive head autoreduction we obtain a set $\mathcal{H}_{\mathbf{y}}^{\triangle}$. In general, the new coordinates $\mathbf{y}$ are not asymptotically regular for the latter set. But there exist coordinates $\tilde{\mathbf{y}}$ of the special form $\tilde{y}^k = y^k + \sum_{i=1}^{k-1} a_i^k y^i$ with $a_i^k \in \Bbbk$ such that if we transform $\mathcal{H}$ to them and perform afterwards an involutive head autoreduction, then they are asymptotically regular for the obtained set $\tilde{\mathcal{H}}_{\tilde{\mathbf{y}}}^{\triangle} \subset \tilde{\mathcal{P}} = \Bbbk[\tilde{\mathbf{y}}]$.

By Definition 4.3.4 of asymptotic regularity, the involutive spans of the two sets $\mathcal{H}$ and $\tilde{\mathcal{H}}_{\tilde{\mathbf{y}}}^{\triangle}$ possess asymptotically the same Hilbert function. Since $\mathcal{H}$ is assumed to be a Pommaret basis of $\mathcal{I}$, this function is also simultaneously the Hilbert function $h_{\mathcal{I}}$ of the ideal $\mathcal{I}$ implying that the involutive span of $\tilde{\mathcal{H}}_{\tilde{\mathbf{y}}}^{\triangle}$ is the full transformed ideal $\tilde{\mathcal{I}} = \langle\tilde{\mathcal{H}}_{\tilde{\mathbf{y}}}^{\triangle}\rangle \subseteq \tilde{\mathcal{P}}$, i. e. $\tilde{\mathcal{H}}_{\tilde{\mathbf{y}}}^{\triangle}$ is a Pommaret basis of $\tilde{\mathcal{I}}$. Thus $\min_{\tilde{h}\in\tilde{\mathcal{H}}_{\tilde{\mathbf{y}}}^{\triangle}} \operatorname{cls} \tilde{h} = d$ and, by the same argument as above, $\tilde{y}^{d+1}$ is a zero divisor in $\tilde{\mathcal{I}}/\langle\tilde{y}^1,\dots,\tilde{y}^d\rangle\tilde{\mathcal{I}}$. Because of the special form of the transformation $\mathbf{y} \mapsto \tilde{\mathbf{y}}$, we have—considering everything as forms in $\mathcal{P}_1$—the equality $\langle\tilde{y}^1,\dots,\tilde{y}^d\rangle = \langle y^1,\dots,y^d\rangle$ and y^{d+1} must be a zero divisor in $\mathcal{I}/\langle y^1,\dots,y^d\rangle\mathcal{I}$. But this contradicts the assumption that $(y^1,\dots,y^{d+1})$ is a regular sequence and thus indeed $\operatorname{depth}\mathcal{I} = d$. □

With essentially the same arguments, one can directly show that under the assumptions of the proposition $(x^1,\dots,x^{d-1})$ form a maximal regular sequence for the algebra $\mathcal{A} = \mathcal{P}/\mathcal{I}$ and hence $\operatorname{depth}\mathcal{A} = \operatorname{depth}\mathcal{I} - 1 = d - 1$. Using a bit more commutative algebra than we introduced in Appendix B, one may alternatively exploit the trivial exact sequence $0 \to \mathcal{I} \to \mathcal{P} \to \mathcal{A} \to 0$ for a proof of these facts.

Remark 5.2.8. One may wonder to what extent this result really requires the Pommaret division. Given an arbitrary involutive basis $\mathcal{H}$ of $\mathcal{I}$, we may introduce the set $X_{\mathcal{I}} = \bigcap_{h\in\mathcal{H}} X_{L,\mathcal{H},\prec}(h)$; obviously, for a Pommaret basis $\mathcal{X}_{\mathcal{I}} = \{x^1,\dots,x^d\}$ with $d = \min_{h\in\mathcal{H}} \operatorname{cls} h$. Again it is trivial to conclude from the induced decomposition that any sequence formed by elements of $\mathcal{X}_{\mathcal{I}}$ is regular. But in general we cannot claim that these are *maximal* regular sequences. Thus only a lower bound for the depth is obtained.

As a simple example consider the ideal $\mathcal{I} \subset \Bbbk[w,x,y,z]$ generated by the three polynomials $f_1 = z^2 - xy$, $f_2 = yz - wx$ and $f_3 = y^2 - wz$. If we set $x_1 = w$, $x_2 = x$, $x_3 = y$ and $x_4 = z$, then it is straightforward to check that the set $\mathcal{F} = \{f_1, f_2, f_3\}$ is a Pommaret basis of $\mathcal{I}$ with respect to the degree reverse lexicographic order. By Proposition 5.2.7, (w,x,y) is a maximal regular sequence and $\operatorname{depth}\mathcal{I} = 3$.

If we set $x_1 = w$, $x_2 = z$, $x_3 = y$ and $x_4 = x$, then no finite Pommaret basis exists; these coordinates are not δ-regular. In order to obtain a Janet basis $\mathcal{F}_J$ of $\mathcal{I}$ (for the degree reverse lexicographic order with respect to the new ordering of the variables), we must enlarge $\mathcal{F}$ by $f_4 = z^3 - wx^2$ and $f_5 = yz^3 - wx^2z$. We find $X_{\mathcal{I}} = \{w,x\}$, as

$$\begin{aligned} X_{J,\mathcal{F}_J,\prec_{\text{degrevlex}}}(f_1) = X_{J,\mathcal{F}_J,\prec_{\text{degrevlex}}}(f_2) &= \{w,x\}\,, \\ X_{J,\mathcal{F}_J,\prec_{\text{degrevlex}}}(f_3) &= \{w,z,y,x\}\,, \\ X_{J,\mathcal{F}_J,\prec_{\text{degrevlex}}}(f_4) = X_{J,\mathcal{F}_J,\prec_{\text{degrevlex}}}(f_5) &= \{w,z,x\}\,. \end{aligned} \tag{5.12}$$

Thus $X_{\mathcal{I}}$ can be extended to a maximal regular sequence by adding y. However, the Janet basis gives no indications, why y should be added. One could also conjecture that the minimal number of multiplicative variables for a generator gives the depth. But again this is clearly not correct for the above Janet basis. Thus no obvious way seems to exist to deduce $\operatorname{depth}\mathcal{I}$ from $\mathcal{F}_J$. ◁

Let again $\mathcal{H}$ be a Pommaret basis of $\mathcal{I}$ for a class respecting term order. It follows trivially from the Rees decomposition (5.11) that

$$\min\left\{i \mid \langle \mathcal{H}, x_1, \dots, x_i \rangle_q = \mathcal{P}_q\right\} \geq d-1\,. \tag{5.13}$$

Thus, as a trivial corollary of Propositions 5.2.3 and 5.2.7, we can conclude that for any graded algebra $\mathcal{A} = \mathcal{P}/\mathcal{I}$ the inequality $\operatorname{depth}\mathcal{A} \leq \dim\mathcal{A}$ holds. In the limit case $\operatorname{depth}\mathcal{A} = \dim\mathcal{A}$, the algebra is by definition *Cohen–Macaulay*. Thus we obtain the following characterisation of algebras of this type.

Theorem 5.2.9. *Let the set $\mathcal{H}$ be a homogeneous Pommaret basis of the homogeneous ideal $\mathcal{I} \subseteq \mathcal{P}$ for a class respecting term order $\prec$. The algebra $\mathcal{A} = \mathcal{P}/\mathcal{I}$ is Cohen–Macaulay, if and only if $\langle \mathcal{H}, x^1, \dots, x^{d-1} \rangle_q = \mathcal{P}_q$ where $d = \min_{h\in\mathcal{H}} \operatorname{cls} h$ and $q = \deg\mathcal{H}$.*

Proof. By the results above, $\operatorname{depth}\mathcal{A} = \dim\mathcal{A} = d-1$. □

An alternative characterisation, which is more useful for computations, is based on the existence of a special kind of Rees decomposition; one sometimes speaks of a *Hironaka decomposition*.

Corollary 5.2.10 (Hironaka). *The algebra $\mathcal{A} = \mathcal{P}/\mathcal{I}$ is Cohen–Macaulay, if and only if a Rees decomposition of $\mathcal{A}$ exists where all generators have the same class.*

Proof. One direction is trivial. If such a special decomposition exists with d the common class of all generators, then obviously both the dimension and the depth of $\mathcal{A}$ is d and thus $\mathcal{A}$ is Cohen–Macaulay.

For the converse, assume that the algebra $\mathcal{A}$ is Cohen–Macaulay and that we have $\dim\mathcal{A} = \operatorname{depth}\mathcal{A} = d$. According to Theorem 4.3.15, $\mathcal{I}$ possesses in suitably chosen variables $\mathbf{x}$ a Pommaret basis $\mathcal{H}$ with respect to the degree reverse lexicographic order. By our results above, this implies that $\min_{h\in\mathcal{H}} \operatorname{cls} h = d+1$. We introduce the set $\overline{\mathcal{B}} = \{\nu \in \overline{\langle \operatorname{le}_\prec \mathcal{H}\rangle} \mid \operatorname{cls}\nu > d\}$ (recall that in Appendix A.1 we defined $\operatorname{cls}[0,\dots,0] = n$ so that $[0,\dots,0] \in \overline{\mathcal{B}}$ whenever $\mathcal{I} \neq \mathcal{P}$). The set $\overline{\mathcal{B}}$ is finite, as all its elements satisfy $|\nu| < \deg\mathcal{H}$ by Proposition 5.2.3 (in other words, the set $\operatorname{lt}_\prec \mathcal{H}$ generates an Artinian ideal in the subring $\mathbb{k}[x^{d+1},\dots,x^n]$). Now we claim that

$$\mathcal{A} \cong \bigoplus_{\nu\in\overline{\mathcal{B}}} \mathbb{k}[x^1,\dots,x^d]\cdot x^\nu\,. \tag{5.14}$$

More precisely, we claim that (5.14) is the result of applying Algorithm 5.2 for the computation of a complementary decomposition from a Janet basis to the Pommaret basis $\mathcal{H}$ (by Corollary 4.3.11 any Pommaret basis is simultaneously a Janet basis). Consider an element $\nu \in \overline{\mathcal{B}}$; obviously, it is of the form $\nu = [0,\dots,0,\nu_{d+1},\dots,\nu_n]$ with $\sum_{i=d+1}^n \nu_i < q = \deg\mathcal{H}$. If we set $q' = q - \sum_{i=d+2}^n \nu_i$, then $\nu + (q' - \nu_{d+1})_{d+1} \in \langle \operatorname{le}_\prec \mathcal{H}\rangle$ by Theorem 5.2.9 implying the existence of a multi index $\nu' \in \operatorname{le}_\prec \mathcal{H}$ of the form $\nu' = [0,\dots,0,\nu'_{d+1},\nu_{d+2},\dots,\nu_n]$ with $\nu_{d+1} < \nu'_{d+1} \leq q'$. Hence the set $(\nu_{d+2},\dots,\nu_n)$ is considered in the assignment of multiplicative indices for the elements of $\operatorname{le}_\prec \mathcal{H}$ with respect to the Janet division and it consists only of the multi index ν', as $\mathcal{H}$ is involutively head autoreduced (with respect to the Pommaret division). This observation entails that Algorithm 5.2 chooses ν as an element of $\overline{\mathcal{B}}$ and assigns to it the multiplicative indices $1,\dots,d$.

Algorithm 5.2 cannot lead to a larger set $\overline{\mathcal{B}}$, as any further multi index would be of class less than or equal to d and thus contained in $\mathbb{k}[x^1,\dots,x^d]\cdot 1$. But as the sets are disjoint, this cannot happen and we get the decomposition (5.14). □

Example 5.2.11. Consider the ideal $\mathcal{I} = \langle z^3, yz^2 - xz^2, y^2 - xy\rangle \subset \mathbb{k}[x,y,z]$. A Pommaret basis of $\mathcal{I}$ for the degree reverse lexicographic order is given by the set $\mathcal{H} = \{z^3, yz^2 - xz^2, y^2z - xyz, y^2 - xy\}$ and thus both the depth and the dimension of $\mathcal{P}/\mathcal{I}$ is 1. Applying Algorithm 5.2, we obtain $\mathcal{T} = \{1, y, z, yz, z^2\}$ (which is also the complementary set of $\langle z^3, yz^2, y^2z, y^2\rangle \subset \mathbb{k}[y,z]$) and the Hironaka decomposition of $\mathcal{P}/\mathcal{I}$ is given by

$$\mathcal{P}/\mathcal{I} \cong \mathbb{k}[x] \oplus \mathbb{k}[x]\cdot y \oplus \mathbb{k}[x]\cdot z \oplus \mathbb{k}[x]\cdot yz \oplus \mathbb{k}[x]\cdot z^2\,. \tag{5.15}$$

The term z^2 comes from the set $(2) = \{[0,1,2]\}$. The two terms yz and z arise from $(1) = \{[0,2,1]\}$ and finally both 1 and y stem from $(0) = \{[0,2,0]\}$. The set $(3) = \{[0,0,3]\}$ does not contribute to the decomposition. ◁

5.3 Noether Normalisation and Primary Decomposition

An important operation in commutative algebra, e. g. in the context of dimension theory, is the determination of a Noether normalisation for an algebra $\mathcal{A} = \mathcal{P}/\mathcal{I}$ with an ideal $\mathcal{I} \subseteq \mathcal{P}$. As a simple consequence of our results in the previous section, we show now that any complementary quasi-Rees decomposition of $\mathcal{I}$ induces a Noether normalisation of $\mathcal{A}$.

Proposition 5.3.1. *Under the assumptions of Lemma 5.2.4, the restriction of the canonical projection $\pi : \mathcal{P} \to \mathcal{A}$ to $\Bbbk[X_{\bar{t}}]$ defines a Noether normalisation for $\mathcal{A}$.*

Proof. By Proposition 5.2.5, the set $X_{\bar{t}}$ is strongly independent modulo $\mathcal{I}$ and thus also independent modulo $\mathcal{I}$, i. e. $\mathcal{I} \cap \Bbbk[X_{\bar{t}}] = \{0\}$ implying that the restriction of π to $\Bbbk[X_{\bar{t}}]$ is injective. Furthermore, it follows immediately from the definition of a complementary quasi-Rees decomposition that the algebra $\mathcal{A}$ is finitely generated as a module over the ring $\Bbbk[X_{\bar{t}}]$. □

Remark 5.3.2. Recall from Lemma 5.2.4 that for any variable $x^i \notin X_{\bar{t}}$ the minimal basis of $\operatorname{lt}_\prec \mathcal{I}$ contains an element of the form $(x^i)^{e_i}$ for some exponent $e_i \in \mathbb{N}$. Thus any Gröbner basis of $\mathcal{I}$ for the chosen term order $\prec$ must contain an element $g_i \in \mathcal{G}$ with $\operatorname{lt}_\prec g_i = (x^i)^{e_i}$. Assume now that $\prec$ is the lexicographic order. Then g_i must be of the form $g_i = (x^i)^{e_i} + \sum_{j=0}^{e_i-1} P_{i,j}(x^i)^j$ with polynomials $P_{i,j} \in \Bbbk[x^1, \dots, x^{i-1}]$. Thus in this case we even obtain a general Noether normalisation. ◁

Since any Pommaret basis of an ideal $\mathcal{I}$ yields even a Rees decomposition, the existence of a Noether normalisation for every affine algebra $\mathcal{A} = \mathcal{P}/\mathcal{I}$ follows immediately from Theorem 4.3.15 asserting the existence of a Pommaret basis for every ideal $\mathcal{I}$ in appropriately chosen coordinates. Comparing with classical existence proofs, one sees that using δ-regular coordinates $\mathbf{x}$ ensures that the ideal $\mathcal{I}$ is in Noether position. However, the converse is generally not true: even if the variables $\mathbf{x}$ are chosen in such a way that $\Bbbk[x^1, \dots, x^D]$ defines a Noether normalisation of the algebra $\mathcal{A}$, this fact is not sufficient for concluding that the ideal $\mathcal{I}$ possesses a Pommaret basis. As we will show now, the existence of a Pommaret basis is equivalent to a stronger property. Since a Noether normalisation of $\mathcal{P}/\mathcal{I}$ automatically induces one of $\mathcal{P}/\operatorname{lt}_\prec \mathcal{I}$, it suffices to consider monomial ideals.

Definition 5.3.3. A monomial ideal $\mathcal{I} \subseteq \mathcal{P}$ is called *quasi-stable*, if it possesses a finite monomial Pommaret basis.

It makes indeed sense to provide a special name for this class of monomial ideals (the reason for the terminology "quasi-stable" will become apparent in the next section), as in the context of monomial ideals Theorem 4.3.15 is not that useful, since the transformation to δ-regular coordinates usually yields an ideal which is no longer monomial. Hence it is something particular, if a monomial ideal has a monomial Pommaret basis. Recall from Example 3.1.16 that any Artinian monomial ideal is quasi-stable.

We now give a number of equivalent characterisations of quasi-stable ideals via ideal quotients and saturations which are independent of the theory of Pommaret bases and which demonstrate how natural the notion of quasi-stability is. They will provide us with a further criterion for δ-regularity and also lead to a simple description of an irredundant primary decomposition of such ideals.

Proposition 5.3.4. *Let $\mathcal{I} \subseteq \mathcal{P}$ be a monomial ideal with $\dim \mathcal{P}/\mathcal{I} = D$. Then the following six statements are equivalent.*

(i) *$\mathcal{I}$ is quasi-stable.*
(ii) *The variable x^1 is not a zero divisor for $\mathcal{P}/\mathcal{I}^{\text{sat}}$ and for all $1 \leq j < D$ the variable x^{j+1} is not a zero divisor for $\mathcal{P}/\langle \mathcal{I}, x^1, \dots, x^j \rangle^{\text{sat}}$.*
(iii) *We have $\mathcal{I} : \langle x^1 \rangle^\infty \subseteq \mathcal{I} : \langle x^2 \rangle^\infty \subseteq \cdots \subseteq \mathcal{I} : \langle x^D \rangle^\infty$ and for all $D < j \leq n$ an exponent $k_j \geq 1$ exists such that $(x^j)^{k_j} \in \mathcal{I}$.*
(iv) *For all $1 \leq j \leq n$ the equality $\mathcal{I} : \langle x^j \rangle^\infty = \mathcal{I} : \langle x^j, \dots, x^n \rangle^\infty$ holds.*
(v) *For every associated prime ideal $\mathfrak{p} \in \operatorname{Ass}(\mathcal{P}/\mathcal{I})$ an integer $1 \leq j \leq n$ exists such that $\mathfrak{p} = \langle x^j, \dots, x^n \rangle$.*
(vi) *If $x^\mu \in \mathcal{I}$ and $\mu_i > 0$ for some $1 \leq i < n$, then for each $0 < r \leq \mu_i$ and $i < j \leq n$ an integer $s \geq 0$ exists such that $x^{\mu - r_i + s_j} \in \mathcal{I}$.*

Proof. We begin with showing the equivalence of (i) and (iii). Assume first that the ideal $\mathcal{I}$ is quasi-stable with Pommaret basis $\mathcal{H}$. The existence of a term $(x^j)^{k_j} \in \mathcal{I}$ for all $D < j \leq n$ follows then immediately from Proposition 5.2.3. Now consider a term $x^\mu \in \mathcal{I} : \langle x^k \rangle^\infty \setminus \mathcal{I}$ for some $1 \leq k \leq n$. By definition of a colon ideal, there exists an integer ℓ such that $(x^k)^\ell x^\mu \in \mathcal{I}$ and hence a generator $x^\nu \in \mathcal{H}$ such that $x^\nu \mid_P (x^k)^\ell x^\mu$. If $\operatorname{cls} \nu > k$, then ν would also be an involutive divisor of μ contradicting the assumption $x^\mu \notin \mathcal{I}$. Thus we find $\operatorname{cls} \nu \leq k$ and $\nu_k > \mu_k$.

Next we consider for arbitrary exponents $m > 0$ the terms $(x^{k+1})^m x^\nu \in \mathcal{I}$. For each m a generator $x^{\rho^{(m)}} \in \mathcal{H}$ exists which involutively divides $(x^{k+1})^m x^\nu$. By the same reasoning as above, $\operatorname{cls} x^{\rho^{(m)}} > k+1$ is not possible for an involutively autoreduced basis $\mathcal{H}$ yielding the estimate $\operatorname{cls} \nu \leq \operatorname{cls} x^{\rho^{(m)}} \leq k+1$.

We claim now that there exists an integer m_0 such that $\rho^{(m)} = \rho^{(m_0)}$ for all $m \geq m_0$ and $\operatorname{cls} x^{\rho^{(m_0)}} = k+1$. Indeed, if $\operatorname{cls} x^{\rho^{(m)}} < k+1$, then we must have $\rho^{(m)}_{k+1} = \nu_{k+1} + m$, since x^{k+1} is not multiplicative for $x^{\rho^{(m)}}$. Hence $x^{\rho^{(m)}}$ cannot be an involutive divisor of $(x^{k+1})^{m+1} x^\nu$ and $\rho^{(m+1)} \notin \{\rho^{(1)}, \dots, \rho^{(m)}\}$. As the Pommaret basis $\mathcal{H}$ is a finite set, $\operatorname{cls} x^{\rho^{(m_0)}} = k+1$ for some value $m_0 > 0$. Hence x^{k+1} is multiplicative for $x^{\rho^{(m_0)}}$ and $x^{\rho^{(m_0)}}$ is an involutive divisor of $(x^{k+1})^m x^\nu$ for all values $m \geq m_0$.

By construction, the generator $x^{\rho^{(m_0)}}$ is also an involutive divisor of $(x^{k+1})^{m_0} x^\mu$, as the variable x_k is multiplicative for it. Hence this term must lie in $\mathcal{I}$ and consequently x^μ is contained in $\mathcal{I} : \langle x^{k+1} \rangle^\infty$. Thus we can conclude that the inclusion $\mathcal{I} : \langle x^k \rangle^\infty \subseteq \mathcal{I} : \langle x^{k+1} \rangle^\infty$ holds. This proves (iii).

For the converse assume that (iii) holds and let $\mathcal{B}$ be the minimal basis of the ideal $\mathcal{I}$. Let $x^\mu \in \mathcal{B}$ be an arbitrary term of class k. Then $x^\mu / x^k \in \mathcal{I} : \langle x^k \rangle^\infty$. By assumption, this means that $x^\mu / x^k \in \mathcal{I} : \langle x^\ell \rangle^\infty$ for any non-multiplicative index ℓ, too. Hence for each term $x^\mu \in \mathcal{B}$ and for each value $\operatorname{cls}(x^\mu) < \ell \leq n$ there exists an integer $q_{\mu,\ell}$ such that $(x^\ell)^{q_{\mu,\ell}} x^\mu / x^k \notin \mathcal{I}$ but $(x^\ell)^{q_{\mu,\ell}+1} x^\mu / x^k \in \mathcal{I}$. For the values $1 \leq \ell \leq \operatorname{cls} x^\mu$ we set $q_{\mu,\ell} = 0$. Observe that if $x^\nu \in \mathcal{B}$ is a minimal generator dividing $(x^\ell)^{q_{\mu,\ell}+1} x^\mu / x^k$, then $x^\nu \prec_{\text{invlex}} x^\mu$, since $\operatorname{cls}(x^\nu) \geq \operatorname{cls}(x^\mu)$ and $\nu_k < \mu_k$.

Consider now the set

$$\mathcal{H} = \left\{ x^{\mu+\rho} \mid x^\mu \in \mathcal{B},\ \forall 1 \leq \ell \leq n : 0 \leq \rho_\ell \leq q_{\mu,\ell} \right\}. \tag{5.16}$$

We claim that it is a weak involutive completion of $\mathcal{B}$ and thus a weak Pommaret basis of $\mathcal{I}$. In order to prove this assertion, we must show that each term $x^\lambda \in \mathcal{I}$ lies in the involutive cone of a member of $\mathcal{H}$.

As the term x^λ is assumed to be an element of $\mathcal{I}$, we can factor it in the form $x^\lambda = x^{\sigma^{(1)}} x^{\rho^{(1)}} x^{\mu^{(1)}}$ where $x^{\mu^{(1)}} \in \mathcal{B}$ is a minimal generator, $x^{\sigma^{(1)}}$ contains only multiplicative variables for $x^{\mu^{(1)}}$ and $x^{\rho^{(1)}}$ only non-multiplicative ones. If $x^{\mu^{(1)}+\rho^{(1)}} \in \mathcal{H}$, then we are done, as obviously $\operatorname{cls}\left(x^{\mu^{(1)}+\rho^{(1)}}\right) = \operatorname{cls}\left(x^{\mu^{(1)}}\right)$ and hence all variables contained in $x^{\sigma^{(1)}}$ are multiplicative for $x^{\mu^{(1)}+\rho^{(1)}}$, too.

Otherwise there must exist at least one non-multiplicative variables x^ℓ such that $\rho^{(1)}_\ell > q_{\mu^{(1)},\ell}$. Any minimal generator $x^{\mu^{(2)}} \in \mathcal{B}$ dividing $(x^\ell)^{q_{\mu^{(1)},\ell}+1} x^{\mu^{(1)}} / x^k$ is also a divisor of x^λ and we find a second factorisation $x^\lambda = x^{\sigma^{(2)}} x^{\rho^{(2)}} x^{\mu^{(2)}}$ where again $x^{\sigma^{(2)}}$ consists only of multiplicative and $x^{\rho^{(2)}}$ only of non-multiplicative variables for the term $x^{\mu^{(2)}}$. If $x^{\mu^{(2)}+\rho^{(2)}} \in \mathcal{H}$, then we are done by the same argument as above; otherwise we iterate.

According to the observation above, the sequence $(x^{\mu^{(1)}}, x^{\mu^{(2)}}, \dots)$ of minimal generators constructed this way is strictly descending with respect to the inverse lexicographic order. However, the minimal basis $\mathcal{B}$ is a finite set and the iteration cannot go on infinitely. As the iteration only stops, if there exists an involutive cone containing x^λ, the involutive span of $\mathcal{H}$ is indeed $\mathcal{I}$ and the ideal $\mathcal{I}$ quasi-stable.

Next we consider the implication "(ii) $\Rightarrow$ (iii)". One easily verifies that (ii) implies $\mathcal{I}^{\text{sat}} \cap \Bbbk[x^1, \dots, x^D] = \{0\}$ and hence $\{x^1, \dots, x^D\}$ is an independent set modulo the ideal $\mathcal{I}$ (trivially $\mathcal{I} \subseteq \mathcal{I}^{\text{sat}}$). Since $D = \dim \mathcal{P}/\mathcal{I}$, it must in fact be a maximal independent set. Hence $\dim \mathcal{P}/\langle \mathcal{I}, x^1, \dots, x^D \rangle = 0$ which is equivalent to the second assertion in (iii).

For the first assertion we note that trivially $\mathcal{I} : \langle x^k \rangle^\infty \supseteq \mathcal{I}^{\text{sat}}$ for any $1 \leq k \leq n$. As by assumption x^1 is not a zero divisor on $\mathcal{P}/\mathcal{I}^{\text{sat}}$, we must have $\mathcal{I} : \langle x^1 \rangle^\infty = \mathcal{I}^{\text{sat}}$ and consequently $\mathcal{I} : \langle x^1 \rangle^\infty \subseteq \mathcal{I} : \langle x^2 \rangle^\infty$. Suppose now that we have already shown $\mathcal{I} : \langle x^1 \rangle^\infty \subseteq \cdots \subseteq \mathcal{I} : \langle x^j \rangle^\infty$ for some $1 < j < D$ and that $x^\mu \in \mathcal{I} : \langle x^j \rangle^\infty$. Obviously, then $(x^j)^\ell x^\mu \in \mathcal{I} \subseteq \langle \mathcal{I}, x^1, \dots, x^{j-1} \rangle$ for some $\ell > 0$. As x^j is by assumption a non zero divisor on $\mathcal{P}/\langle \mathcal{I}, x^1, \dots, x^{j-1} \rangle^{\text{sat}}$, this observation entails that $x^\mu \in \langle \mathcal{I}, x^1, \dots, x^{j-1} \rangle^{\text{sat}} \subseteq \mathcal{I} : \langle x^{j+1} \rangle^\infty$ and we can prove (iii) by induction.

The implication "(iii) $\Rightarrow$ (iv)" is trivial to prove: since by assumption $\mathcal{I} : \langle x^1 \rangle^\infty \subseteq \mathcal{I} : \langle x^2 \rangle^\infty \subseteq \cdots \subseteq \mathcal{I} : \langle x^D \rangle^\infty \subseteq \mathcal{P} = \mathcal{I} : \langle x^{D+1} \rangle^\infty = \cdots = \mathcal{I} : \langle x^n \rangle^\infty$, we have the inclusion $\mathcal{I} : \langle x^j \rangle^\infty \subseteq \mathcal{I} : \langle x^k \rangle^\infty$ whenever $j < k$ entailing our claim.

Now we claim that (iv) entails $\langle\mathcal{I},x^1,\dots,x^{j-1}\rangle^{\text{sat}} = \langle\mathcal{I},x^1,\dots,x^{j-1}\rangle : \langle x^j\rangle^\infty$ which in turn immediately implies (ii). If $x^\mu \in \langle\mathcal{I},x^1,\dots,x^{j-1}\rangle : \langle x^j\rangle^\infty$, then we have either $x^\mu \in \langle\mathcal{I},x^1,\dots,x^{j-1}\rangle$ or $x^\mu \in \mathcal{I} : \langle x^j\rangle^\infty$. Since we have by assumption $\mathcal{I} : \langle x^j\rangle^\infty = \mathcal{I} : \langle x^j,\dots,x^n\rangle^\infty \subseteq \langle\mathcal{I},x^1,\dots,x^{j-1}\rangle^{\text{sat}}$, our claim follows (the converse inclusion is trivial).

The equivalence of (i) and (vi) is a consequence of the Lemmata 4.3.2 and 4.3.3. Note that it suffices to consider $r=1$, as the general case follows by an easy induction. For the direction "(i) ⇒ (vi)" we choose s so large that $|\mu|+s=q-1>\deg\mathcal{H}$ where $\mathcal{H}$ is as usual the Pommaret basis of $\mathcal{I}$. Now (vi) is obtained by applying Lemma 4.3.3 to the Pommaret basis $\mathcal{H}_q$ of $\mathcal{I}_{\geq q}$. For the converse we choose a value q so large that for each possible choice of μ, i, j and r an s exists such that $\mu - r_i + s_j \in \mathcal{I}_q$. Then the basis $\mathcal{B}_q$ of $\mathcal{I}_q$ satisfies the assumptions of the reverse direction of Lemma 4.3.3 implying (i).

Finally, we discuss the equivalence of (iv) and (v). Suppose first that (iv) holds. Any associated prime $\mathfrak{p}$ of a monomial ideal $\mathcal{I}$ is itself a monomial ideal and thus minimally generated by variables (see Remark B.1.18). Assume that $\mathfrak{p} = \mathcal{I} : f$ for a polynomial $f \in \mathcal{P}$ is an associated prime ideal of $\mathcal{I}$ and $x^j \in \mathfrak{p}$. Thus $f \in \mathcal{I} : \langle x^j\rangle^\infty = \mathcal{I} : \langle x^j,\dots,x^n\rangle^\infty$ trivially implying that $\langle x^j,\dots,x^n\rangle \subseteq \mathfrak{p}$ and hence (v).

For the converse let $\mathcal{I} = \mathfrak{q}_1 \cap\cdots\cap \mathfrak{q}_t$ be an irredundant monomial primary decomposition (which always exists by Theorem B.1.17 and Remark B.1.18). If $\mathcal{J} \subseteq \mathcal{P}$ is any ideal, then $\mathcal{I} : \mathcal{J}^\infty = \bigcap_{k=1}^t \mathfrak{q}_k : \mathcal{J}^\infty$. Hence it suffices to show that (iv) holds for every primary component $\mathfrak{q}_k$. By (v), any associated prime ideal $\mathfrak{p}_k = \sqrt{\mathfrak{q}_k}$ is of the form $\mathfrak{p}_k = \langle x^i,\dots,x^n\rangle$. Thus the ideal $\mathfrak{q}_k$ contains for each $i \leq j \leq n$ a term $(x^j)^{\ell_j}$ so that $\mathfrak{q}_k : \langle x^j\rangle^\infty = \mathfrak{q}_k : \langle x^j,\dots,x^n\rangle^\infty = \mathcal{P}$ and for each $1 \leq j < i$ we find $\mathfrak{q}_k : \langle x^j\rangle^\infty = \mathfrak{q}_k : \langle x^j,\dots,x^n\rangle^\infty = \mathfrak{q}_k$ implying that indeed (iv) holds. □

It is easy to see that quasi-stability is preserved under basic operations with monomial ideals like taking their sum or product.

Lemma 5.3.5. *Let $\mathcal{I}_1,\mathcal{I}_2 \subseteq \mathcal{P}$ be two quasi-stable ideals. Then their sum $\mathcal{I}_1+\mathcal{I}_2$, their product $\mathcal{I}_1\cdot\mathcal{I}_2$ and their intersection $\mathcal{I}_1\cap\mathcal{I}_2$ are quasi-stable, too. If $\mathcal{I} \subseteq \mathcal{P}$ is a quasi-stable ideal, then the quotient $\mathcal{I} : \mathcal{J}$ is again quasi-stable for arbitrary monomial ideals $\mathcal{J} \subseteq \mathcal{P}$.*

Proof. For the sum $\mathcal{I}_1+\mathcal{I}_2$ the claim follows immediately from Remark 3.1.13 which asserts that the union $\mathcal{H}_1\cup\mathcal{H}_2$ of (weak) Pommaret bases $\mathcal{H}_k$ of $\mathcal{I}_k$ is a weak Pommaret basis of the sum $\mathcal{I}_1+\mathcal{I}_2$. Similarly, the case of both the product $\mathcal{I}_1\cdot\mathcal{I}_2$ and the intersection $\mathcal{I}_1\cap\mathcal{I}_2$ was settled in Remark 4.1.6 where for both ideals weak Pommaret bases were constructed.

For the last assertion we use Part (vi) of Proposition 5.3.4. If $\mathcal{J}$ is minimally generated by the monomials $m_1,\dots,m_r$, then $\mathcal{I} : \mathcal{J} = \bigcap_{k=1}^r \mathcal{I} : m_k$ and thus it suffice to consider the case that $\mathcal{J}$ is a principal ideal with a generator $x^\nu \in \mathbb{T}$. Assume that $x^\mu \in \mathcal{I} : x^\nu$ and that $\mu_i > 0$. Since $x^{\mu+\nu}$ lies in the quasi-stable ideal $\mathcal{I}$, we find for each $0 < r \leq \mu_i$ and $i < j \leq n$ an integer $s \geq 0$ such that $x^{\mu+\nu-r_i+s_j} \in \mathcal{I}$. As $r \leq \mu_i$, this observation trivially implies that $x^{\mu-r_i+s_j} \in \mathcal{I} : x^\nu$. □

Remark 5.3.6. Above we actually proved that (iii) may be replaced by the equivalent statement $\mathcal{I} : \langle x^1 \rangle^\infty \subseteq \mathcal{I} : \langle x^2 \rangle^\infty \subseteq \cdots \subseteq \mathcal{I} : \langle x^n \rangle^\infty$ which does not require a priori knowledge of D (the dimension D arises then trivially as the smallest value k such that $\mathcal{I} : \langle x^k \rangle^\infty = \mathcal{P}$, i., e, for which $\mathcal{I}$ contains a minimal generator $(x^k)^\ell$ for some exponent $\ell > 0$). In this formulation it is straightforward to verify (iii) effectively: bases of the colon ideals $\mathcal{I} : \langle x^k \rangle^\infty$ are easily obtained by setting $x^k = 1$ in any basis of $\mathcal{I}$ and for monomial ideals it is trivial to check inclusion, as one must only compare their minimal bases.

We furthermore note that if we have for some value $1 \leq k \leq n$ an ascending chain $\mathcal{I} : \langle x^1 \rangle^\infty \subseteq \mathcal{I} : \langle x^2 \rangle^\infty \subseteq \cdots \subseteq \mathcal{I} : \langle x^k \rangle^\infty$, then for each $1 \leq j \leq k$ the minimal basis $\mathcal{B}_j$ of $\mathcal{I} : \langle x^j \rangle^\infty$ lies in $\Bbbk[x^{j+1}, \dots, x^n]$. Indeed, no element of $\mathcal{B}_j$ can depend on x^j. Now assume that $x^\nu \in \mathcal{B}_j$ satisfies $\operatorname{cls} \nu = \ell < j$. Then $(x^j)^m x^\nu$ is a minimal generator of $\mathcal{I}$ for some suitable exponent $m \in \mathbb{N}_0$. This in turn implies that $(x^j)^m x^\nu / (x^\ell)^{\nu_\ell} \in \mathcal{I} : \langle x^\ell \rangle^\infty \subseteq \mathcal{I} : \langle x^j \rangle^\infty$ and hence $x^\nu / (x^\ell)^{\nu_\ell} \in \mathcal{I} : \langle x^j \rangle^\infty$ which contradicts our assumption that x^ν was a minimal generator. ◁

Part (iii) of Proposition 5.3.4 provides us with a simple effective criterion for δ-regularity, as it is straightforward to determine the required saturations and verify wether they form an ascending chain. With its help, we can furthermore show that δ-singular coordinates for a (polynomial) ideal $\mathcal{I} \subseteq \mathcal{P}$ cannot be asymptotically regular for a Pommaret (head)autoreduced Gröbner basis $\mathcal{G}$ of $\mathcal{I}$. Since in this case $\operatorname{lt}_\prec \mathcal{I} = \langle \operatorname{lt}_\prec \mathcal{G} \rangle$ is not quasi-stable, this assertion is an immediate consequence of the following result and Theorem 4.3.12.

Proposition 5.3.7. *Let $\mathcal{I} \subseteq \mathcal{P}$ be a monomial ideal and the set $\mathcal{B} \subset \mathcal{I}$ a finite monomial basis of it. If $\mathcal{I}$ is not quasi-stable, then at least for one generator contained in the basis $\mathcal{B}$ a variable exists which is Janet but not Pommaret multiplicative.*

Proof. As the ideal $\mathcal{I}$ is not quasi-stable, there exists a minimal value $1 \leq k < n$ such that $\mathcal{I} : \langle x^k \rangle^\infty \nsubseteq \mathcal{I} : \langle x^{k+1} \rangle^\infty$. Let x^μ be a minimal generator of $\mathcal{I} : \langle x^k \rangle^\infty$ which is not contained in $\mathcal{I} : \langle x^{k+1} \rangle^\infty$. Then for a suitable exponent $m \in \mathbb{N}_0$ the term $x^{\bar\mu} = (x^k)^m x^\mu$ is a minimal generator of $\mathcal{I}$ and hence contained in $\mathcal{B}$.

We claim now that $\mathcal{B}$ contains a generator for which x^{k+1} is Janet but not Pommaret multiplicative. If $x^{k+1} \in X_{J,\mathcal{B}}(x^{\bar\mu})$, then we are done, as according to Remark 5.3.6 $\operatorname{cls} \bar\mu = k$ and hence $x^{k+1} \notin X_P(x^{\bar\mu})$. Otherwise $\mathcal{B}$ contains a term x^ν such that $\nu_\ell = \mu_\ell$ for $k+1 < \ell \leq n$ and $\nu_{k+1} > \mu_{k+1}$. If several generators with this property exist in $\mathcal{B}$, we choose one for which ν_{k+1} takes a maximal value so that we have $x^{k+1} \in X_{J,\mathcal{B}}(x^\nu)$ by definition of the Janet division. If $\operatorname{cls} \nu < k+1$, we are again done, as then $x^{k+1} \notin X_P(x^\nu)$. Now assume that $\operatorname{cls} \nu = k+1$ and consider the term $x^\rho = x^\nu / (x^{k+1})^{\nu_{k+1}}$. Obviously, $x^\rho \in \mathcal{I} : \langle x^{k+1} \rangle^\infty$ contradicting our assumption $x^\mu \notin \mathcal{I} : \langle x^{k+1} \rangle^\infty$ since $x^\rho \mid x^\mu$. Hence this case cannot arise. □

We mentioned above that while δ-regular coordinates ensure that $\mathcal{I}$ is in Noether position the converse is not true. Based on Condition (v) in Proposition 5.3.4, we can now formulate a converse for monomial ideals. It shows that a Pommaret basis of a monomial ideal induces not only a Noether normalisation of the ideal itself but simultaneously of all its primary components.

Corollary 5.3.8. *Let $\mathcal{I} \subseteq \mathcal{P}$ be a monomial ideal with* $\dim \mathcal{P}/\mathcal{I} = D$. *Furthermore, let $\mathcal{I} = \mathfrak{q}_1 \cap \cdots \cap \mathfrak{q}_r$ be an irredundant monomial primary decomposition with $D_j = \dim \mathcal{P}/\mathfrak{q}_j$ for $1 \leq j \leq r$. The ideal $\mathcal{I}$ is quasi-stable, if and only if $\Bbbk[x^1,\dots,x^D]$ defines a Noether normalisation of $\mathcal{P}/\mathcal{I}$ and $\Bbbk[x^1,\dots,x^{D_j}]$ one of $\mathcal{P}/\mathfrak{q}_j$ for each primary component $\mathfrak{q}_j$.*

Proof. By assumption, each $\mathfrak{q}_j$ is a monomial primary ideal. This implies that $\Bbbk[x^1,\dots,x^{D_j}]$ defines a Noether normalisation of $\mathcal{P}/\mathfrak{q}_j$, if and only if the associated prime ideal is $\sqrt{\mathfrak{q}_j} = \langle x^{D_j+1},\dots,x^n\rangle$. Now the assertion follows immediately from Condition (v) in Proposition 5.3.4. □

We may also exploit Proposition 5.3.4 for actually deriving an irredundant primary decomposition $\mathcal{I} = \mathfrak{q}_1 \cap \cdots \cap \mathfrak{q}_t$ with monomial ideals $\mathfrak{q}_j$ for an arbitrary quasi-stable ideal $\mathcal{I}$. We first note that the proof of the implication "(v) ⇒ (iv)" in Proposition 5.3.4 has some simple consequences for the form of such a decomposition $\mathcal{I} = \mathfrak{q}_1 \cap \cdots \cap \mathfrak{q}_t$ of a quasi-stable ideal $\mathcal{I}$. Let again $D = \dim \mathcal{P}/\mathcal{I}$. Then $\mathfrak{p} = \langle x^{D+1},\dots,x^n\rangle$ is the unique minimal prime ideal associated to $\mathcal{I}$ and the corresponding unique primary component is given by $\mathcal{I} : \langle x^D\rangle^\infty$ (if $D = 0$, then obviously $\mathcal{I}$ is already a primary ideal). More generally, we find for any $1 \leq k \leq D$ that

$$\mathcal{I} : \langle x^k\rangle^\infty = \bigcap_{\mathfrak{p}_j \subseteq \langle x^{k+1},\dots,x^n\rangle} \mathfrak{q}_j \tag{5.17}$$

where $\mathfrak{p}_j = \sqrt{\mathfrak{q}_j}$ is the corresponding associated prime ideal (note that these immediately implies that $\mathcal{I} : \langle x^k\rangle^\infty$ is quasi-stable, too). Based on these observations, an irredundant primary decomposition can be constructed by working backwards through the chain $\mathcal{I} \subseteq \mathcal{I} : \langle x^1\rangle^\infty \subseteq \mathcal{I} : \langle x^2\rangle^\infty \subseteq \cdots \subseteq \mathcal{I} : \langle x^n\rangle^\infty$.

Let $d = \operatorname{depth} \mathcal{P}/\mathcal{I}$, i. e. $d+1$ is the minimal class of a generator in the Pommaret basis $\mathcal{H}$ of $\mathcal{I}$ according to Proposition 5.2.7.[3] For $1 \leq k \leq D$ we set $s_k = \min\{s \mid \mathcal{I} : \langle x^k\rangle^s = \mathcal{I} : \langle x^k\rangle^{s+1}\}$, i. e. s_k is the highest x^k-degree of a minimal generator of $\mathcal{I}$. Finally, we introduce the ideals

$$\mathcal{J}_k = \mathcal{I} + \langle (x^{k+1})^{s_{k+1}},\dots,(x^D)^{s_D}\rangle \tag{5.18}$$

and their saturations

$$\mathfrak{q}_k = \mathcal{J}_k : \langle x^k\rangle^\infty = \mathcal{I} : \langle x^k\rangle^\infty + \langle (x^{k+1})^{s_{k+1}},\dots,(x^D)^{s_D}\rangle\,. \tag{5.19}$$

It is easy to see that all the ideals $\mathcal{J}_k$ are again quasi-stable provided the original ideal $\mathcal{I}$ is quasi-stable (this follows immediately from Proposition 5.3.4 and the fact that in this case $(\mathcal{I} : \langle x^i\rangle^\infty) : \langle x^j\rangle^\infty = \mathcal{I} : \langle x^j\rangle^\infty$ for $i < j$). For notational simplicity we formally define $\mathcal{I} : \langle x^0\rangle^\infty = \mathcal{I}$ and $\mathfrak{q}_0 = \mathcal{J}_0 = \mathcal{I} + \langle (x^1)^{s_1},\dots,(x^D)^{s_D}\rangle$. Since obviously $\dim \mathcal{P}/\mathcal{J}_k = k$ for $0 \leq k \leq D$, it follows that $\mathfrak{q}_k$ is $\langle x^{k+1},\dots,x^n\rangle$-primary.

---

[3] Note that for determining the depth d in the case of a quasi-stable ideal, it is not necessary to compute the Pommaret basis: since multiplication with a non-multiplicative variable never decreases the class, $d+1$ is also the minimal class of a minimal generator.

Proposition 5.3.9. *Let $\mathcal{I} \subseteq \mathcal{P}$ be a quasi-stable ideal. Then $\mathcal{I} = \bigcap_{k=d}^{D} \mathfrak{q}_k$ defines a primary decomposition of it. Eliminating all those primary ideals $\mathfrak{q}_k$ for which $\mathcal{I} : \langle x^k \rangle^\infty = \mathcal{I} : \langle x^{k+1} \rangle^\infty$ makes it an irredundant decomposition.*

Proof. We first show that the equality $\mathcal{I} : \langle x^k \rangle^\infty = \bigcap_{\ell=k}^{D} \mathfrak{q}_\ell$ holds or equivalently that $\mathcal{I} : \langle x^k \rangle^\infty = \mathfrak{q}_k \cap (\mathcal{I} : \langle x^{k+1} \rangle^\infty)$ for $0 \leq k \leq n$; for $k = d$ this represents the first statement of the proposition, since obviously $\mathcal{I} : \langle x^0 \rangle^\infty = \cdots = \mathcal{I} : \langle x^d \rangle^\infty = \mathcal{I}$. By definition of the value s_{k+1}, we have that (see e. g. [185, Lemma 3.3.6])

$$\mathcal{I} : \langle x^k \rangle^\infty = \big(\mathcal{I} : \langle x^k \rangle^\infty + \langle (x^{k+1})^{s_{k+1}} \rangle\big) \cap \big((\mathcal{I} : \langle x^k \rangle^\infty) : \langle x^{k+1} \rangle^\infty\big) . \tag{5.20}$$

The second factor obviously equals $\mathcal{I} : \langle x^{k+1} \rangle^\infty$. To the first one we apply the same construction and decompose

$$\begin{aligned} \mathcal{I} : \langle x^k \rangle^\infty + \langle (x^{k+1})^{s_{k+1}} \rangle &= \\ &= \big(\mathcal{I} : \langle x^k \rangle^\infty + \langle (x^{k+1})^{s_{k+1}}, (x^{k+2})^{s_{k+2}} \rangle\big) \cap \\ &\quad \big((\mathcal{I} : \langle x^k \rangle^\infty + \langle (x^{k+1})^{s_{k+1}} \rangle) : \langle x^{k+2} \rangle^\infty\big) \\ &= \big(\mathcal{I} : \langle x^k \rangle^\infty + \langle (x^{k+1})^{s_{k+1}}, (x^{k+2})^{s_{k+2}} \rangle\big) \cap \\ &\quad \big(\mathcal{I} : \langle x^{k+2} \rangle^\infty + \langle (x^{k+1})^{s_{k+1}} \rangle\big) . \end{aligned} \tag{5.21}$$

Continuing in this manner, we arrive at a decomposition

$$\mathcal{I} : \langle x^k \rangle^\infty = \mathfrak{q}_k \cap \cdots \cap (\mathcal{I} : \langle x^{k+1} \rangle^\infty) \tag{5.22}$$

where the dots on the right hand side represent further ideals which are of the form $\mathcal{I} : \langle x^\ell \rangle^\infty + \langle (x^{k+1})^{s_{k+1}}, \ldots, (x^{\ell-1})^{s_{\ell-1}} \rangle$ with $\ell \geq k+2$. Since we assume that $\mathcal{I}$ is quasi-stable, the colon ideal $\mathcal{I} : \langle x^{k+1} \rangle^\infty$ is contained in each of them and we may omit them which proves our claim.

In the thus obtained primary decomposition of $\mathcal{I}$ the radicals of all appearing primary ideals are pairwise different. Furthermore, it is obvious that $\mathfrak{q}_k$ is redundant whenever $\mathcal{I} : \langle x^k \rangle^\infty = \mathcal{I} : \langle x^{k+1} \rangle^\infty$. Thus there only remains to prove that all the other primary ideals $\mathfrak{q}_k$ are indeed necessary. Assume that $\mathcal{I} : \langle x^k \rangle^\infty \subsetneq \mathcal{I} : \langle x^{k+1} \rangle^\infty$ (which is in particular the case for $k < d$). Then there exists a minimal generator x^μ of $\mathcal{I} : \langle x^{k+1} \rangle^\infty$ which is not contained in $\mathcal{I} : \langle x^k \rangle^\infty$. Consider the monomial $(x^k)^{s_k} x^\mu$. It cannot lie in $\mathcal{I} : \langle x^k \rangle^\infty$, as otherwise already $x^\mu \in \mathcal{I} : \langle x^k \rangle^\infty$, and thus it also cannot be contained in $\mathfrak{q}_k$ (since we showed above that $\mathcal{I} : \langle x^k \rangle^\infty = \mathfrak{q}_k \cap (\mathcal{I} : \langle x^{k+1} \rangle^\infty)$). On the other hand we find that $(x^k)^{s_k} x^\mu \in \mathfrak{q}_\ell$ for all $\ell > k$ since then $\mathcal{I} : \langle x^{k+1} \rangle^\infty \subseteq \mathfrak{q}_\ell$ and for all $\ell < k$ since then $\langle (x^k)^{s_k} \rangle \subseteq \mathfrak{q}_\ell$. Hence $\mathfrak{q}_k$ is not redundant. □

We remarked already above that the ideals $\mathcal{I} : \langle x^k \rangle^\infty$ are again quasi-stable. It is straightforward to obtain Pommaret bases for them. We disjointly decompose the monomial Pommaret basis $\mathcal{H} = \mathcal{H}_1 \cup \cdots \cup \mathcal{H}_n$ where $\mathcal{H}_k$ contains all generators of class k. Furthermore, we write $\mathcal{H}'_k$ for the set obtained by setting $x^k = 1$ in each generator in $\mathcal{H}_k$.

Lemma 5.3.10. *For any $1 \leq k \leq n$ the set $\mathcal{H}' = \mathcal{H}'_k \cup \bigcup_{\ell=k+1}^n \mathcal{H}_\ell$ is a weak Pommaret basis of the colon ideal $\mathcal{I} : \langle x^k \rangle^\infty$.*

Proof. We first show that $\mathcal{H}'$ is an involutive set. By definition of the Pommaret division, it is obvious that the subset $\bigcup_{\ell=k+1}^n \mathcal{H}_\ell$ is involutive. Thus there only remains to consider the non-multiplicative products of the members of $\mathcal{H}'_k$. Take $x^\mu \in \mathcal{H}'_k$ and let x^ℓ be a non-multiplicative variable for it. Obviously, there exists an $m > 0$ such that $(x^k)^m x^\mu \in \mathcal{H}_k$ and hence a generator $x^\nu \in \bigcup_{\ell=k}^n \mathcal{H}_\ell$ such that $x^\ell (x^k)^m x^\mu$ lies in the involutive cone $\mathcal{C}_P(x^\nu)$. Writing $x^\ell (x^k)^m x^\mu = x^{\rho+\nu}$, we distinguish two cases. If $\operatorname{cls}\nu > k$, then $\rho_k = m$ and we can divide by $(x^k)^m$ in order to obtain an involutive standard representation of $x^\ell x^\mu$ with respect to $\mathcal{H}'$. If $\operatorname{cls}\nu = k$, then the multi index ρ is of the form r_k, i. e. only the kth entry is different from zero, and we even find that $x^\ell x^\mu = x^\nu/(x^k)^r \in \mathcal{H}'_k$.

Thus there only remains to prove that $\mathcal{H}'$ is actually a generating set for $\mathcal{I} : \langle x^k \rangle^\infty$. For this we first note that the Pommaret basis of a quasi-stable ideal contains a generator of class k only, if there is a minimal generator of class k, as applying the monomial completion Algorithm 4.1 to the minimal basis adds only non-multiplicative multiples of the minimal generators (and these are trivially of the same class). By Remark 5.3.6, all minimal generators of $\mathcal{I} : \langle x^k \rangle^\infty$ have at least class $k+1$. Thus setting $x^k = 1$ in any member of $\bigcup_{\ell=1}^{k-1} \mathcal{H}_\ell$ can never produce a minimal generator of $\mathcal{I} : \langle x^k \rangle^\infty$ and thus $\mathcal{H}'$ is a weak involutive completion of the minimal basis of $\mathcal{I} : \langle x^k \rangle^\infty$. According to Proposition 3.1.12, an involutive autoreduction yields a strong basis. □

The ideals $\langle (x^{k+1})^{s_{k+1}}, \dots, (x^D)^{s_D} \rangle$ are trivially irreducible and for $k \geq d$ exactly of the form that they possess a Pommaret basis as discussed in Example 3.1.16. There we also provided an explicit Pommaret basis for such an ideal. Since according to Remark 3.1.13 the union of two (weak) Pommaret bases of two monomial ideals $\mathcal{I}_1$, $\mathcal{I}_2$ yields a weak Pommaret basis of $\mathcal{I}_1 + \mathcal{I}_2$, we obtain this way easily weak Pommaret bases for all primary ideals $\mathfrak{q}_k$ appearing in the irredundant decomposition of Proposition 5.3.9.

Thus the crucial information for obtaining an irredundant primary decomposition of a quasi-stable ideal $\mathcal{I}$ is where "jumps" are located, i. e. where $\mathcal{I} : \langle x^k \rangle^\infty \subsetneq \mathcal{I} : \langle x^{k+1} \rangle^\infty$. Since these ideals are quasi-stable, the positions of the jumps are determined by their depths. A chain with all the jumps is obtained by the following simple recipe: set $\mathcal{I}_0 = \mathcal{I}$ and define $\mathcal{I}_{k+1} = \mathcal{I}_k : \langle x^{d_k} \rangle^\infty$ where $d_k = \operatorname{depth}\mathcal{I}_k$. This leads to the *sequential chain* of $\mathcal{I}$:

$$\mathcal{I}_0 = \mathcal{I} \subsetneq \mathcal{I}_1 \subsetneq \cdots \subsetneq \mathcal{I}_r = \mathcal{P} . \tag{5.23}$$

Remark 5.3.11. With the help of the sequential chain one can easily show that any quasi-stable ideal is *sequentially Cohen–Macaulay* (following Stanley [433], the $\Bbbk$-algebra $\mathcal{A} = \mathcal{P}/\mathcal{I}$ is called sequentially Cohen–Macaulay, if a chain of ideals $\mathcal{I}_0 = \mathcal{I} \subset \mathcal{I}_1 \subset \cdots \subset \mathcal{I}_r = \mathcal{P}$ exists such that all quotients $\mathcal{I}_{k+1}/\mathcal{I}_k$ are Cohen–Macaulay and their dimensions are ascending: $\dim(\mathcal{I}_k/\mathcal{I}_{k-1}) < \dim(\mathcal{I}_{k+1}/\mathcal{I}_k)$).

Indeed, consider the elimination ideal $\mathcal{J}_k = \mathcal{I}_k \cap \Bbbk[x^{d_k}, \dots, x^n]$. By Remark 5.3.6, $\mathcal{J}_k$ has the same minimal generators as $\mathcal{I}_k$; furthermore, by Proposition 5.3.4 (iv)

$\mathcal{J}_k^{\text{sat}} = \mathcal{J}_k : \langle x^{d_k}\rangle^\infty$. Hence, going back to the full polynomial ring $\mathcal{P}$, we find that $\mathcal{I}_{k+1} = \langle \mathcal{J}_k^{\text{sat}}\rangle_{\mathcal{P}}$ and

$$\mathcal{I}_{k+1}/\mathcal{I}_k \cong (\mathcal{J}_k^{\text{sat}}/\mathcal{J}_k)[x^1,\dots,x^{d_k-1}]\,. \tag{5.24}$$

Since the factor ring $\mathcal{J}_k^{\text{sat}}/\mathcal{J}_k$ is trivially finite (as a $\Bbbk$-linear space), the quotient $\mathcal{I}_{k+1}/\mathcal{I}_k$ is thus a (d_k-1)-dimensional Cohen–Macaulay module.

As another simple application of the sequential chain we characterise those quasi-stable ideals which have the *saturated chain property*. An ideal $\mathcal{I} \subset \mathcal{P}$ has the saturated chain property, if to every non-minimal associated prime ideal $\mathfrak{p} \in \operatorname{Ass}\mathcal{I}$ a further associated prime $\mathfrak{p}' \subset \mathfrak{p}$ exists with $\dim \mathcal{P}/\mathfrak{p}' = \dim \mathcal{P}/\mathfrak{p} + 1$. By Proposition 5.3.9, this is the case for a quasi-stable ideal, if and only if no jumps occur in the sequential chain, i. e. if always $\operatorname{depth}\mathcal{J}_{k+1} = \operatorname{depth}\mathcal{J}_k + 1$ for $k > 0$. $\triangleleft$

Addendum: Standard Pairs

Another kind of combinatorial decomposition, albeit not a disjoint one, can be obtained via the standard pairs. We will now show how they are related to Stanley decompositions. Consider pairs (ν, N_ν) where $\nu \in \mathbb{N}_0^n$ is a multi index and $N_\nu \subseteq \{1,\dots,n\}$ a set of associated indices. Such a pair is called *admissible*, if $\operatorname{supp}\nu \cap N_\nu = \emptyset$, i. e. if $\nu_i = 0$ for all $i \in N_\nu$. On the set of admissible pairs one defines a partial order: $(\nu, N_\nu) \leq (\mu, N_\mu)$ if and only if the restricted cone $\mu + \mathbb{N}^n_{N_\mu}$ is completely contained in $\nu + \mathbb{N}^n_{N_\nu}$. Obviously, this containment is equivalent to $\nu \mid \mu$ and any index i such that either $\mu_i > \nu_i$ or $i \in N_\mu$ also lies in N_ν.

Definition 5.3.12. Let $\mathcal{I} \subseteq \mathbb{N}_0^n$ be an arbitrary monoid ideal. An admissible pair (ν, N_ν) is called *standard* for $\mathcal{I}$, if $\nu + \mathbb{N}^n_{N_\nu} \cap \mathcal{I} = \emptyset$ and (ν, N_ν) is minimal with respect to $\leq$ among all admissible pairs with this property. We denote the set of all standard pairs of the ideal $\mathcal{I}$ by $\mathcal{S}_{\mathcal{I}}$.

Proposition 5.3.13. *Let $\mathcal{I} \subseteq \mathbb{N}_0^n$ be an arbitrary monoid ideal. Then the complementary set $\overline{\mathcal{I}}$ of $\mathcal{I}$ can be written in the form*

$$\overline{\mathcal{I}} = \bigcup_{(\nu,N_\nu)\in\mathcal{S}_{\mathcal{I}}} (\nu + \mathbb{N}^n_{N_\nu}) \tag{5.25}$$

and $\mathcal{I}$ can be decomposed as

$$\mathcal{I} = \bigcap_{(\nu,N_\nu)\in\mathcal{S}_{\mathcal{I}}} \langle (\nu_i+1)_i \mid i \notin N_\nu\rangle\,. \tag{5.26}$$

Proof. For the first assertion assume that $\mu \in \overline{\mathcal{I}}$. Then the pair $(\mu, \emptyset)$ is trivially admissible and hence either $(\mu,\emptyset) \in \mathcal{S}_{\mathcal{I}}$ or the set $\mathcal{S}_{\mathcal{I}}$ contains a standard pair (ν, N_ν) such that $(\nu, N_\nu) \leq (\mu, \emptyset)$. In both cases the multi index μ is part of the union on the right hand side of (5.25). The second assertion is a corollary of the first one: a multi index μ is contained in the ideal $\mathcal{I}$, if and only if it is not contained in any of the restricted cones $\nu + \mathbb{N}^n_{N_\nu}$ defined by the standard pairs. But the condition $\mu \notin \nu + \mathbb{N}^n_{N_\nu}$ is obviously equivalent to $\mu \in \langle (\nu_i + 1)_i \mid i \notin N_\nu \rangle$. □

The ideals on the right hand side of (5.26) are obviously irreducible and thus primary (cf. Remark B.1.18), so that (5.26) actually represents a primary decomposition of the ideal $\mathcal{I}$. In general, this decomposition is highly redundant. Let $N \subseteq \{1,\dots,n\}$ be an arbitrary subset and consider all standard pairs (ν, N_ν) with $N_\nu = N$. Obviously, among these only the ones with multi indices ν which are maximal with respect to divisibility are relevant for the decomposition (5.26) and in fact restricting to the corresponding ideals yields the irredundant irreducible decomposition of $\mathcal{I}$ (which is unique according to [322, Thm. 5.27]). Their intersection defines a possible choice for the primary component for the prime ideal $\mathfrak{p}_N = \langle x^i \mid i \notin N \rangle$ (as a trivial corollary of these considerations the standard pairs immediately yield the set $\operatorname{Ass}(\mathcal{P}/\mathcal{I})$ of associated prime ideals, as it consists of all prime ideals $\mathfrak{p}_N$ such that a standard pair (ν, N) exists).

We now show that $\mathcal{S}_\mathcal{I}$ may be extracted from any complementary decomposition using the simple Algorithm 5.3. Thus once a Janet basis of $\mathcal{I}$ is known, we can immediately use Janet's Algorithm 5.2 for the construction of a complementary decomposition and then obtain the standard pairs.

Algorithm 5.3 Standard pairs

Input: finite complementary decomposition $\mathcal{T}_\mathcal{I}$ of monoid ideal $\mathcal{I} \subseteq \mathbb{N}^n_0$
Output: set $\mathcal{S}_\mathcal{I}$ of standard pairs of $\mathcal{I}$

$\overline{\mathcal{S}}_\mathcal{I} \leftarrow \emptyset$
for all $(\nu, N_\nu) \in \mathcal{T}_\mathcal{I}$ **do**
  $\overline{N}_\nu \leftarrow \operatorname{supp} \nu \cap N_\nu$
  if $\overline{N}_\nu = \emptyset$ **then**
    $\overline{\mathcal{S}}_\mathcal{I} \leftarrow \overline{\mathcal{S}}_\mathcal{I} \cup \{(\nu, N_\nu)\}$
  else
    $\bar{\nu} \leftarrow \nu - \sum_{i \in \overline{N}_\nu} (\nu_i)_i$
    $\overline{\mathcal{S}}_\mathcal{I} \leftarrow \overline{\mathcal{S}}_\mathcal{I} \cup \{(\bar{\nu}, N_\nu)\}$
  end if
end for
return set of minimal elements of $\overline{\mathcal{S}}_\mathcal{I}$ with respect to $\preceq$

Algorithm 5.3 performs the following trivial operation. Let the finite input set $\mathcal{T}_\mathcal{I} = \{(\nu, N_\nu) \mid \nu \in \mathbb{N}^n_0, N_\nu \subseteq \{1,\dots,n\}\}$ define a complementary decomposition of the monoid ideal $\mathcal{I}$. If the pair $(\nu, N_\nu) \in \mathcal{T}_\mathcal{I}$ is not admissible, then we substitute it by the pair $(\bar{\nu}, N_\nu)$ where $\bar{\nu}_i = 0$ for all $i \in N_\nu$ and $\bar{\nu}_i = \nu_i$ else. Obviously, this new pair is admissible and the thus obtained set $\overline{\mathcal{S}}_\mathcal{I}$ still defines a (no longer disjoint) decomposition of the complementary set $\overline{\mathcal{I}}$. Finally, we eliminate all pairs in $\overline{\mathcal{S}}_\mathcal{I}$ which are not minimal with respect to the partial order $\preceq$.

Proposition 5.3.14. *Let $\mathcal{T}_{\mathcal{I}}$ be a finite complementary decomposition of the monoid ideal $\mathcal{I} \subseteq \mathbb{N}_0^n$. Then Algorithm 5.3 correctly computes with $\mathcal{T}_{\mathcal{I}}$ as input the set $\mathcal{S}_{\mathcal{I}}$ of standard pairs of $\mathcal{I}$.*

Proof. It is trivial to see that the set $\overline{\mathcal{S}}_{\mathcal{I}}$ computed by Algorithm 5.3 contains only admissible pairs and that $\nu + \mathbb{N}^n_{N_\nu} \subseteq \overline{\mathcal{I}}$ for any pair $(\nu, N_\nu) \in \overline{\mathcal{S}}_{\mathcal{I}}$. Thus there only remains to show that all standard pairs are contained in $\overline{\mathcal{S}}_{\mathcal{I}}$.

Let (μ, N_μ) be an admissible pair such that $\mu + \mathbb{N}^n_{N_\mu} \subseteq \overline{\mathcal{I}}$. Since the union of the cones $\nu + \mathbb{N}^n_{N_\nu}$ with $(\nu, N_\nu) \in \overline{\mathcal{S}}_{\mathcal{I}}$ still covers $\overline{\mathcal{I}}$, the finiteness of $\overline{\mathcal{S}}_{\mathcal{I}}$ implies the existence of a multi index $\overline{\mu} \in \mu + \mathbb{N}^n_{N_\mu}$ and a pair $(\nu, N_\nu) \in \overline{\mathcal{S}}_{\mathcal{I}}$ such that $\overline{\mu} + \mathbb{N}^n_{N_\mu} \subseteq \nu + \mathbb{N}^n_{N_\nu}$ (obviously, it is not possible to cover $\mu + \mathbb{N}^n_{N_\mu}$ with a finite number of lower-dimensional cones). As both (μ, N_μ) and (ν, N_ν) are admissible pairs, this entails that in fact $(\nu, N_\nu) \leq (\mu, N_\mu)$. Hence either $(\mu, N_\mu) \in \overline{\mathcal{S}}_{\mathcal{I}}$ or it is not a standard pair. □

Remark 5.3.15. If we use the decomposition (5.4) derived from a Pommaret basis of degree q, then the determination of the set $\overline{\mathcal{S}}_{\mathcal{I}}$ is completely trivial. For all pairs $(\nu, N_\nu) \in \mathcal{T}_{\mathcal{I}}$ with $|\nu| < q$ we have $N_\nu = \emptyset$ and hence they are admissible. For all other pairs we find by definition of the Pommaret division that $\operatorname{supp} \nu \cap N_\nu = \{k\}$ with $k = \operatorname{cls} \nu$ and hence none of them is admissible. In Line /7/ Algorithm 5.3 simply sets the first non-vanishing entry of such multi indices to zero. ◁

Example 5.3.16. Let us continue with the ideal studied in Example 5.2.11. It follows from the Hironaka decomposition (5.15) that a complementary decomposition of the monoid ideal $\operatorname{le}_\prec \mathcal{I} = \langle [0,0,3], [0,1,2], [0,2,0] \rangle$ is given by

$$\begin{aligned} \mathcal{S}_{\mathcal{I}} = \Big\{ &\big([0,0,0],\{1\}\big), \big([0,1,0],\{1\}\big), \big([0,0,1],\{1\}\big), \\ &\big([0,1,1],\{1\}\big), \big([0,0,2],\{1\}\big) \Big\} \end{aligned} \tag{5.27}$$

and one easily verifies that these are all standard pairs.

If we exploit the Pommaret basis given in Example 5.2.11, then the complementary decomposition constructed via Proposition 5.1.6 is much larger. Besides many multi indices without any multiplicative indices, we obtain the following six multi indices for which 1 is the sole multiplicative index: $[3,0,0]$, $[2,1,0]$, $[2,0,1]$, $[1,1,1]$, $[1,2,0]$ and $[1,0,2]$. After setting the first entry to zero, we find precisely the multi indices appearing in (5.27) plus the multi index $[0,2,0]$. As $\big([0,1,0],\{1\}\big) < \big([0,2,0],\{1\}\big)$, the latter pair is not minimal. The same holds for all pairs corresponding to the multi indices without multiplicative indices and hence we also arrive at (5.27). ◁

Remark 5.3.17. Our considerations above also provide us with a simple direct proof of the implication "(i) ⇒ (v)" in Proposition 5.3.4. If the ideal $\mathcal{I}$ is quasi-stable, then $\mathcal{I}$ admits a complementary Rees decomposition according to Proposition 5.1.6. Together with Proposition 5.3.14 this observation trivially implies that all associated prime ideals are of the form $\mathfrak{p} = \langle x^j, \dots, x^n \rangle$. ◁

5.4 Syzygies and Free Resolutions

Appendix B.4 shows that Gröbner bases are very useful in syzygy theory: if we determine a Gröbner basis with the Buchberger Algorithm B.3, then the Schreyer Theorem B.4.27 gives us directly a Gröbner basis of the first syzygy module (with respect to a special term order). In this section we consider the use of involutive bases for syzygy computations. It will turn out that in particular Pommaret bases allow us to make statements not only about the first syzygy module but about a whole syzygy resolution.

In this section we cannot avoid the explicit use of (free) $\mathcal{P}$-modules and we employ the vector notation introduced in Appendix B.4. Let $\mathcal{H} \subset \mathcal{P}^m$ be a finite subset of the free module $\mathcal{P}^m$ for some $m \in \mathbb{N}$, $\prec$ a term order on the set of vector terms $\mathbb{T}^m$ and L an involutive division on $\mathbb{N}_0^n$. We divide $\mathcal{H}$ into the m disjoint subsets $\mathcal{H}_\alpha = \{\mathbf{h} \in \mathcal{H} \mid \operatorname{lt}_\prec \mathbf{h} = t\mathbf{e}_\alpha,\ t \in \mathbb{T}\}$ according to the position of the leading term. This partitioning leads naturally to m sets of multi indices $\mathcal{B}_\alpha = \{\mu \in \mathbb{N}_0^n \mid x^\mu \mathbf{e}_\alpha \in \operatorname{lt}_\prec \mathcal{H}_\alpha\}$. If we assign now to each generator $\mathbf{h} \in \mathcal{H}_\alpha$ the multiplicative variables $X_{L,\mathcal{H},\prec}(\mathbf{h}) = \{x^i \mid i \in N_{L,\mathcal{B}_\alpha}(\operatorname{le}_\prec \mathbf{h})\}$, then an obvious generalisation of (3.31) defines the involutive span $\langle \mathcal{H} \rangle_{L,\prec}$ of the set $\mathcal{H}$. Thus for the assignment of the multiplicative variables we compare only those generators which have their leading terms in the same component. The definition of (weak and strong) involutive bases for submodules $\mathcal{U} \subseteq \mathcal{P}^m$ proceeds now in complete analogy to the ideal case. In particular, all the theoretical results in Section 3.4 and all the algorithms presented in Chapter 4 can be extended in a straightforward manner.

Let $\mathcal{U} \subseteq \mathcal{P}^m$ be a submodule and $\mathcal{H} = \{\mathbf{h}_1, \dots, \mathbf{h}_s\}$ an involutive basis of $\mathcal{U}$, i. e. we have $\langle \mathcal{H} \rangle_{L,\prec} = \mathcal{U}$ and $\mathcal{H}$ is involutively head autoreduced. Take an arbitrary element $\mathbf{h}_\alpha \in \mathcal{H}$ and choose an arbitrary non-multiplicative variable $x^k \in \overline{X}_{L,\mathcal{H},\prec}(\mathbf{h}_\alpha)$. By Theorem 3.4.4, we can determine with an involutive normal form algorithm for each generator $\mathbf{h}_\beta \in \mathcal{H}$ a unique polynomial $P_\beta^{(\alpha;k)} \in \Bbbk[X_{L,\mathcal{H},\prec}(\mathbf{h}_\beta)]$ such that $x^k \mathbf{h}_\alpha = \sum_{\beta=1}^s P_\beta^{(\alpha;k)} \mathbf{h}_\beta$. The corresponding syzygy is

$$\mathbf{S}_{\alpha;k} = x^k \mathbf{e}_\alpha - \sum_{\beta=1}^{s} P_\beta^{(\alpha;k)} \mathbf{e}_\beta \in \mathcal{P}^s\,. \tag{5.28}$$

We denote the set of all thus obtained syzygies by

$$\mathcal{H}_{\mathrm{Syz}} = \left\{ \mathbf{S}_{\alpha;k} \mid 1 \leq \alpha \leq s;\ x^k \in \overline{X}_{L,\mathcal{H},\prec}(\mathbf{h}_\alpha) \right\}\,. \tag{5.29}$$

Lemma 5.4.1. *Let $\mathbf{S} = \sum_{\beta=1}^s S_\beta \mathbf{e}_\beta \in \operatorname{Syz}(\mathcal{H})$ be an arbitrary syzygy of the finite set $\mathcal{H} \subset \mathcal{P}^m$. Then $S_\beta \in \Bbbk[X_{L,\mathcal{H},\prec}(\mathbf{h}_\beta)]$ for all $1 \leq \beta \leq s$, if and only if $\mathbf{S} = 0$.*

Proof. By definition of a syzygy, $\sum_{\beta=1}^s S_\beta \mathbf{h}_\beta = 0$. As the basis $\mathcal{H}$ is involutively head autoreduced, each element $\mathbf{f} \in \langle \mathcal{H} \rangle$ possesses a unique involutive standard representation by Theorem 3.4.4. In particular, this holds for $0 \in \langle \mathcal{H} \rangle$. Thus either we have $S_\beta \notin \Bbbk[X_{L,\mathcal{H},\prec}(\mathbf{h}_\beta)]$ for at least one β or $\mathbf{S} = 0$. □

We recall from Example B.4.21 the definition of the term order $\prec_{\mathcal{F}}$ on $\mathbb{T}^s$ induced by a finite set $\mathcal{F} = \{\mathbf{f}_1, \dots, \mathbf{f}_s\} \subset \mathcal{P}^m$ and a term order $\prec$ on $\mathbb{T}^m$: given two terms $\mathbf{s} = s\mathbf{e}_\sigma$ and $\mathbf{t} = t\mathbf{e}_\tau$, we set $\mathbf{s} \prec_{\mathcal{F}} \mathbf{t}$, if either $\mathrm{lt}_\prec(s\mathbf{f}_\sigma) \prec \mathrm{lt}_\prec(t\mathbf{f}_\tau)$ or $\mathrm{lt}_\prec(s\mathbf{f}_\sigma) = \mathrm{lt}_\prec(t\mathbf{f}_\tau)$ and $\tau < \sigma$.

Corollary 5.4.2. *If $\mathcal{H} \subset \mathcal{P}$ is an involutive basis, then the set $\mathcal{H}_{\mathrm{Syz}}$ generates the syzygy module* $\mathrm{Syz}(\mathcal{H})$.

Proof. Let $\mathbf{S} = \sum_{\beta=1}^{s} S_\beta \mathbf{e}_\beta$ by an arbitrary non-vanishing syzygy in $\mathrm{Syz}(\mathcal{H})$. By Lemma 5.4.1, at least one of the coefficients S_β must contain a term x^μ with a non-multiplicative variable $x^j \in \overline{X}_{L,\mathcal{H},\prec}(\mathbf{h}_\beta)$. Let $cx^\mu \mathbf{e}_\beta$ be the maximal such term with respect to the term order $\prec_{\mathcal{H}}$ and j the maximal non-multiplicative index with $\mu_j > 0$. Then we eliminate this term by computing $\mathbf{S}' = \mathbf{S} - cx^{\mu - 1_j}\mathbf{S}_{\beta;j}$. If $\mathbf{S}' \neq 0$, we iterate. Since all new terms introduced by the subtraction are smaller than the eliminated term with respect to $\prec_{\mathcal{H}}$, we must reach zero after a finite number of steps. Thus this computation leads to a representation of $\mathbf{S}$ as a linear combination of elements of $\mathcal{H}_{\mathrm{Syz}}$. □

Let $\mathcal{H} = \{\mathbf{h}_1, \cdots, \mathbf{h}_s\}$ be an involutive basis and thus a Gröbner basis for the term order $\prec$. Without loss of generality we may assume that $\mathcal{H}$ is a monic basis. Set $\mathbf{t}_\alpha = \mathrm{lt}_\prec \mathbf{h}_\alpha$ and $\mathbf{t}_{\alpha\beta} = \mathrm{lcm}(\mathbf{t}_\alpha, \mathbf{t}_\beta)$. We have for every S-polynomial a standard representation $\mathbf{S}_\prec(\mathbf{h}_\alpha, \mathbf{h}_\beta) = \sum_{\gamma=1}^{s} f_{\alpha\beta\gamma}\mathbf{h}_\gamma$ where the polynomials $f_{\alpha\beta\gamma} \in \mathcal{P}$ satisfy $\mathrm{lt}_\prec\big(\mathbf{S}_\prec(\mathbf{h}_\alpha, \mathbf{h}_\beta)\big) \succeq \mathrm{lt}_\prec(f_{\alpha\beta\gamma}\mathbf{h}_\gamma)$ for $1 \leq \gamma \leq s$. Setting $\mathbf{f}_{\alpha\beta} = \sum_{\gamma=1}^{s} f_{\alpha\beta\gamma}\mathbf{e}_\gamma$, we introduce for $\alpha \neq \beta$ the syzygy

$$\mathbf{S}_{\alpha\beta} = \frac{\mathbf{t}_{\alpha\beta}}{\mathbf{t}_\alpha}\mathbf{e}_\alpha - \frac{\mathbf{t}_{\alpha\beta}}{\mathbf{t}_\beta}\mathbf{e}_\beta - \mathbf{f}_{\alpha\beta} \,. \tag{5.30}$$

According to the Schreyer Theorem B.4.27, the set

$$\mathcal{H}_{\mathrm{Schreyer}} = \{\mathbf{S}_{\alpha\beta} \mid 1 \leq \alpha < \beta \leq s\} \tag{5.31}$$

of all these syzygies is a Gröbner basis of the first syzygy module $\mathrm{Syz}(\mathcal{H})$ with respect to the induced term order $\prec_{\mathcal{H}}$.

We denote by $\tilde{\mathbf{S}}_{\alpha\beta} = \frac{\mathbf{t}_{\alpha\beta}}{\mathbf{t}_\alpha}\mathbf{e}_\alpha - \frac{\mathbf{t}_{\alpha\beta}}{\mathbf{t}_\beta}\mathbf{e}_\beta$ the syzygy of the leading terms corresponding to $\mathbf{S}_{\alpha\beta}$ and if $\mathcal{S} \subseteq \mathcal{H}_{\mathrm{Schreyer}}$ is a set of syzygies, $\tilde{\mathcal{S}}$ contains the corresponding syzygies of the leading terms.

Lemma 5.4.3. *Let the subset $\mathcal{S} \subseteq \mathcal{H}_{\mathrm{Schreyer}}$ be such that $\tilde{\mathcal{S}}$ generates* $\mathrm{Syz}(\mathrm{lt}_\prec \mathcal{H})$. *Then the subset $\mathcal{S}$ also generates* $\mathrm{Syz}(\mathcal{H})$. *Assume furthermore that the three pairwise distinct indices α, β, γ are such that*[4] *$\mathbf{S}_{\alpha\beta}, \mathbf{S}_{\beta\gamma}, \mathbf{S}_{\alpha\gamma} \in \mathcal{S}$ and $\mathbf{t}_\gamma \mid \mathbf{t}_{\alpha\beta}$. Then the smaller set $\mathcal{S} \setminus \{\mathbf{S}_{\alpha\beta}\}$ still generates* $\mathrm{Syz}(\mathcal{H})$.

Proof. It is a classical result in the theory of Gröbner bases that $\tilde{\mathcal{S}} \setminus \{\tilde{\mathbf{S}}_{\alpha\beta}\}$ still generates $\mathrm{Syz}(\mathrm{lt}_\prec \mathcal{H})$. Actually, this fact is the basic property underlying Buchberger's

[4] If $\alpha > \beta$, then we understand that $\mathbf{S}_{\beta\alpha} \in \mathcal{S}$ etc.

second criterion (see Proposition B.4.19) for avoiding redundant S-polynomials. Thus it suffices to show the first assertion; the second one is a simple corollary.

Let $\mathbf{R} = \sum_{\alpha=1}^{s} R_\alpha \mathbf{e}_\alpha \in \mathrm{Syz}(\mathcal{H})$ be an arbitrary syzygy of the full generators and set $\mathbf{t_R} = \max_\prec \{\mathrm{lt}_\prec (R_\alpha \mathbf{h}_\alpha) \mid 1 \leq \alpha \leq s\}$. Then

$$\tilde{\mathbf{R}} = \sum_{\mathrm{lt}_\prec (R_\alpha \mathbf{h}_\alpha) = \mathbf{t}_R} \mathrm{lt}_\prec (R_\alpha \mathbf{h}_\alpha) \in \mathrm{Syz}(\mathrm{lt}_\prec \mathcal{H}) \,. \tag{5.32}$$

According to our assumption $\tilde{\mathcal{S}}$ is a generating set of $\mathrm{Syz}(\mathrm{lt}_\prec \mathcal{H})$, so that we may write $\tilde{\mathbf{R}} = \sum_{\tilde{\mathbf{S}} \in \tilde{\mathcal{S}}} P_{\tilde{\mathbf{S}}} \tilde{\mathbf{S}}$ for some coefficients $P_{\tilde{\mathbf{S}}} \in \mathcal{P}$. Let us now consider the syzygy $\mathbf{R}' = \mathbf{R} - \sum_{\mathbf{S} \in \mathcal{S}} P_{\tilde{\mathbf{S}}} \mathbf{S}$. Obviously, $\mathbf{t}_{\mathbf{R}'} \prec \mathbf{t_R}$. By iteration, we thus obtain in a finite number of steps a representation $\mathbf{R} = \sum_{\mathbf{S} \in \mathcal{S}} Q_{\mathbf{S}} \mathbf{S}$ and thus the set $\mathcal{S}$ generates the module $\mathrm{Syz}(\mathcal{H})$. □

As a consequence of this simple lemma, we can now show that for each involutive basis $\mathcal{H}$ the set $\mathcal{H}_{\mathrm{Syz}}$ given by (5.29) defines a Gröbner basis of the first syzygy module $\mathrm{Syz}(\mathcal{H})$ for the term order $\prec_\mathcal{H}$. Note that this basis $\mathcal{H}_{\mathrm{Syz}}$ arises automatically as a by-product during the determination of the involutive basis with Algorithm 4.5. This fact is completely analogously to the automatic determination of the Gröbner basis $\mathcal{H}_{\mathrm{Schreyer}}$ with the Buchberger Algorithm B.3.

Theorem 5.4.4. *Let $\mathcal{H}$ be an involutive basis for the involutive division L and the term order $\prec$. Then the set $\mathcal{H}_{\mathrm{Syz}}$ is a Gröbner basis of the syzygy module* $\mathrm{Syz}(\mathcal{H})$ *for the term order $\prec_\mathcal{H}$.*

Proof. Without loss of generality, we may assume that $\mathcal{H}$ is a monic basis, i. e. all leading coefficients are 1. Let $\mathbf{S}_{\alpha;k} \in \mathcal{H}_{\mathrm{Syz}}$. As $\mathcal{H}$ is an involutive basis, the unique polynomials $P_\beta^{(\alpha;k)}$ in (5.28) satisfy $\mathrm{lt}_\prec (P_\beta^{(\alpha;k)} \mathbf{h}_\beta) \preceq \mathrm{lt}_\prec (x^k \mathbf{h}_\alpha)$ and there exists only one index $\bar{\beta}$ such that $\mathrm{lt}_\prec (P_{\bar{\beta}}^{(\alpha;k)} \mathbf{h}_{\bar{\beta}}) = \mathrm{lt}_\prec (x^k \mathbf{h}_\alpha)$. It is easy to see that we have $\mathbf{S}_{\alpha;k} = \mathbf{S}_{\alpha\bar{\beta}}$. Thus $\mathcal{H}_{\mathrm{Syz}} \subseteq \mathcal{H}_{\mathrm{Schreyer}}$.

Let $\mathbf{S}_{\alpha\beta} \in \mathcal{H}_{\mathrm{Schreyer}} \setminus \mathcal{H}_{\mathrm{Syz}}$ be an arbitrary syzygy. We first prove that the set $\mathcal{H}_{\mathrm{Schreyer}} \setminus \{\mathbf{S}_{\alpha\beta}\}$ still generates $\mathrm{Syz}(\mathcal{H})$. Any syzygy in $\mathcal{H}_{\mathrm{Schreyer}}$ has the form $\mathbf{S}_{\alpha\beta} = x^\mu \mathbf{e}_\alpha - x^\nu \mathbf{e}_\beta + \mathbf{R}_{\alpha\beta}$. By construction, one of the monomials x^μ and x^ν must contain a non-multiplicative variable x^k for $\mathbf{h}_\alpha$ or $\mathbf{h}_\beta$, respectively. Without loss of generality, we assume that $x^k \in \overline{X}_{L,\mathcal{H},\prec}(\mathbf{h}_\alpha)$ and $\mu_k > 0$. Then $\mathcal{H}_{\mathrm{Syz}}$ contains the syzygy $\mathbf{S}_{\alpha;k}$ and, as shown above, a unique index $\gamma \neq \beta$ exists such that $\mathbf{S}_{\alpha;k} = \mathbf{S}_{\alpha\gamma}$.

Let $\mathbf{S}_{\alpha\gamma} = x^k \mathbf{e}_\alpha - x^\rho \mathbf{e}_\gamma + \mathbf{R}_{\alpha\gamma}$. By construction, $x^\rho \mathbf{t}_\gamma = x^k \mathbf{t}_\alpha$ divides $x^\mu \mathbf{t}_\alpha = \mathbf{t}_{\alpha\beta}$. Thus $\mathbf{t}_\gamma \mid \mathbf{t}_{\alpha\beta}$ and by Lemma 5.4.3 the set $\mathcal{H}_{\mathrm{Schreyer}} \setminus \{\mathbf{S}_{\alpha\beta}\}$ still generates $\mathrm{Syz}(\mathcal{H})$. If we try to iterate this argument, then we encounter the following problem. In order to be able to eliminate $\mathbf{S}_{\alpha\beta}$ we need both $\mathbf{S}_{\alpha\gamma}$ and $\mathbf{S}_{\beta\gamma}$ in the remaining set. For $\mathbf{S}_{\alpha\gamma} \in \mathcal{H}_{\mathrm{Syz}}$, this is always guaranteed. But we know nothing about $\mathbf{S}_{\beta\gamma}$ and, if it is not an element of $\mathcal{H}_{\mathrm{Syz}}$, it could have been removed in an earlier iteration.

We claim that with respect to the term order $\prec_\mathcal{H}$ the term $\mathrm{lt}_{\prec_\mathcal{H}} \mathbf{S}_{\alpha\beta}$ is greater than both $\mathrm{lt}_{\prec_\mathcal{H}} \mathbf{S}_{\alpha\gamma}$ and $\mathrm{lt}_{\prec_\mathcal{H}} \mathbf{S}_{\beta\gamma}$. Without loss of generality, we may assume for simplicity that $\alpha < \beta < \gamma$, as the syzygies $\mathbf{S}_{\alpha\beta}$ and $\mathbf{S}_{\beta\alpha}$ differ only by a sign. Thus

$\text{lt}_{\prec_\mathcal{H}} \mathbf{S}_{\alpha\beta} = \frac{\mathbf{t}_{\alpha\beta}}{\mathbf{t}_\alpha}\mathbf{e}_\alpha$ and similarly for $\mathbf{S}_{\alpha\gamma}$ and $\mathbf{S}_{\beta\gamma}$. Furthermore, $\mathbf{t}_\gamma \mid \mathbf{t}_{\alpha\beta}$ trivially implies $\mathbf{t}_{\alpha\gamma} \mid \mathbf{t}_{\alpha\beta}$ and hence $\mathbf{t}_{\alpha\gamma} \prec \mathbf{t}_{\alpha\beta}$ for any term order $\prec$. Obviously, the same holds for $\mathbf{t}_{\beta\gamma}$. Now a straightforward application of the definition of the induced term order $\prec_\mathcal{H}$ proves our claim.

Thus if in each step we remove the syzygy $\mathbf{S}_{\alpha\beta} \in \mathcal{H}_{\text{Schreyer}} \setminus \mathcal{H}_{\text{Syz}}$ whose leading term is maximal with respect to the term order $\prec_\mathcal{H}$, then it can never happen that the syzygy $\mathbf{S}_{\beta\gamma}$ required for the application of Lemma 5.4.3 has already been eliminated at an earlier stage and $\mathcal{H}_{\text{Syz}}$ is a generating set of $\text{Syz}(\mathcal{H})$.

It is a simple corollary of Schreyer's Theorem B.4.27 that $\mathcal{H}_{\text{Syz}}$ is even a Gröbner basis of $\text{Syz}(\mathcal{H})$. Indeed, we know that $\mathcal{H}_{\text{Schreyer}}$ is a Gröbner basis of $\text{Syz}(\mathcal{H})$ for the term order $\prec_\mathcal{H}$ and it follows from our considerations above that whenever we remove a syzygy $\mathbf{S}_{\alpha\beta}$ we still have in the remaining set at least one syzygy whose leading term divides $\text{lt}_{\prec_\mathcal{H}} \mathbf{S}_{\alpha\beta}$. Thus we find

$$\langle \text{lt}_{\prec_\mathcal{H}}(\mathcal{H}_{\text{Syz}})\rangle = \langle \text{lt}_{\prec_\mathcal{H}}(\mathcal{H}_{\text{Schreyer}})\rangle = \text{lt}_{\prec_\mathcal{H}} \text{Syz}(\mathcal{H}) \tag{5.33}$$

which proves our assertion. □

This result is in so far not completely satisfying, as it only yields a Gröbner and not an involutive basis of the syzygy module. The latter seems to be hard to achieve for arbitrary involutive divisions L. However, for some divisions it is possible with a little effort. The (technical) point is that for the order $\prec_\mathcal{H}$ the ordering of the generators in the basis $\mathcal{H}$ is important and we must choose the right one.

In order to obtain a good ordering, we associate a directed graph with each involutive basis $\mathcal{H}$. Its vertices are the generators in $\mathcal{H}$. If $x^j \in \overline{X}_{L,\mathcal{H},\prec}(\mathbf{h})$ for an element $\mathbf{h} \in \mathcal{H}$, then the involutive basis $\mathcal{H}$ must contain a unique generator $\bar{\mathbf{h}}$ such that $\text{le}_\prec \bar{\mathbf{h}}$ is an involutive divisor of $\text{le}_\prec(x^j\mathbf{h})$. In this case we include a directed edge from $\mathbf{h}$ to $\bar{\mathbf{h}}$. The thus defined graph is called the *L-graph* of $\mathcal{H}$.

Lemma 5.4.5. *If the division L is continuous, then the L-graph of any involutive set $\mathcal{H} \subset \mathcal{P}$ is acyclic.*

Proof. The leading exponents of the vertices of a path in an L-graph define a sequence as in Definition 4.1.3 of a continuous division. If the path is a cycle, then the sequence contains identical elements contradicting the continuity of L. □

We order the elements of $\mathcal{H}$ as follows: whenever the L-graph of $\mathcal{H}$ contains a path from $\mathbf{h}_\alpha$ to $\mathbf{h}_\beta$, then we must have $\alpha < \beta$. Any ordering satisfying this condition is called an *L-ordering* of $\mathcal{H}$. Note that by Lemma 5.4.5 for a continuous division L-orderings always exist (although they are in general not unique).

For the Pommaret division P it is easy to describe explicitly a P-ordering without using the P-graph: we require that if either $\text{cls}\,\mathbf{h}_\alpha < \text{cls}\,\mathbf{h}_\beta$ or $\text{cls}\,\mathbf{h}_\alpha = \text{cls}\,\mathbf{h}_\beta = k$ and and the last non-vanishing entry of $\text{le}_\prec \mathbf{h}_\alpha - \text{le}_\prec \mathbf{h}_\beta$ is negative, then we must have $\alpha < \beta$. Thus we sort the generators $\mathbf{h}_\alpha$ first by their class and within each class lexicographically. It is straightforward to verify that this defines indeed a P-ordering.

Example 5.4.6. Let us consider the ideal $\mathcal{I} \subset \Bbbk[x,y,z]$ generated by the six polynomials $h_1 = x^2$, $h_2 = xy$, $h_3 = xz - y$, $h_4 = y^2$, $h_5 = yz - y$ and $h_6 = z^2 - z + x$. One easily verifies that they form a Pommaret basis $\mathcal{H}$ for the degree reverse lexicographic order. The corresponding P-graph has the following form

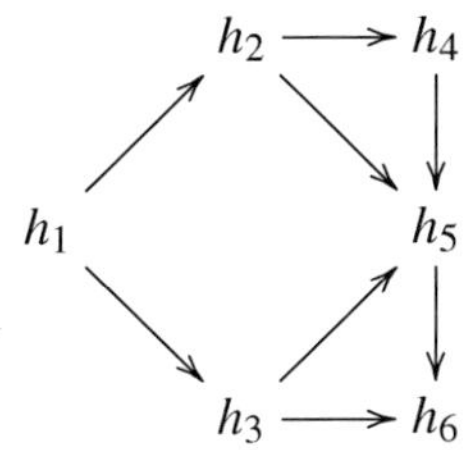

(5.34)

One clearly sees that the generators are already P-ordered, namely according to the description above. ◁

The decisive observation about an L-ordering is that we can now easily determine the leading terms of all syzygies $\mathcal{S}_{\alpha;k} \in \mathcal{H}_{\mathrm{Syz}}$ for the term order $\prec_{\mathcal{H}}$.

Lemma 5.4.7. *Let the elements of the involutive basis $\mathcal{H} \subset \mathcal{P}$ be ordered according to an L-ordering. Then the syzygies $\mathcal{S}_{\alpha;k}$ satisfy $\operatorname{lt}_{\prec_{\mathcal{H}}} \mathcal{S}_{\alpha;k} = x^k \mathbf{e}_\alpha$.*

Proof. By the properties of the unique involutive standard representation, we have in (5.28) $\operatorname{lt}_{\prec}(P_\beta^{(\alpha;k)} \mathbf{h}_\beta) \preceq \operatorname{lt}_{\prec}(x^k \mathbf{h}_\alpha)$ for all β and only one index $\bar{\beta}$ exists for which $\operatorname{lt}_{\prec}(P_{\bar{\beta}}^{(\alpha;k)} \mathbf{h}_{\bar{\beta}}) = \operatorname{lt}_{\prec}(x^k \mathbf{h}_\alpha)$. Thus $\operatorname{le}_{\prec} \mathbf{h}_{\bar{\beta}}$ is an involutive divisor of $\operatorname{le}_{\prec}(x^k \mathbf{h}_\alpha)$ and the L-graph of $\mathcal{H}$ contains an edge from $\mathbf{h}_\alpha$ to $\mathbf{h}_{\bar{\beta}}$. In an L-ordering, this fact implies $\alpha < \bar{\beta}$. Now the assertion follows from the definition of the term order $\prec_{\mathcal{H}}$. □

There remains the problem of controlling the multiplicative variables associated to these leading terms by the involutive division L. For arbitrary divisions it does not seem possible to make any statement. Thus we simply define a class of involutive divisions with the desired properties.

Definition 5.4.8. An involutive division L is of *Schreyer type* for the term order $\prec$, if for any involutive set $\mathcal{H} \subseteq \mathcal{P}$ with respect to L and $\prec$ all sets $\overline{X}_{L,\mathcal{H},\prec}(\mathbf{h})$ with $\mathbf{h} \in \mathcal{H}$ are again involutive.

Lemma 5.4.9. *Both the Janet and the Pommaret division are of Schreyer type for any term order $\prec$.*

Proof. For the Janet division any set of variables, i. e. monomials of degree one, is involutive. Indeed, let $\mathcal{F}$ be such a set and $x^k \in \mathcal{F}$, then

$$X_{J,\mathcal{F}}(x^k) = \{x^i \mid x^i \notin \mathcal{F} \vee i \leq k\} \tag{5.35}$$

which immediately implies the assertion. For the Pommaret division sets of non-multiplicative variables are always of the form $\mathcal{F} = \{x^k, x^{k+1}, \ldots, x^n\}$ and such a set is trivially involutive with respect to the Pommaret division. □

A simple example of an involutive division which is not of Schreyer type is provided by the Thomas division (cf. Example 3.1.3), as one easily sees that no set consisting only of variables can be involutive for it.

Theorem 5.4.10. *Let L be a continuous involutive division of Schreyer type for the term order $\prec$ and $\mathcal{H}$ an L-ordered involutive basis of the polynomial submodule $\mathcal{U} \subseteq \mathcal{P}^m$ with respect to L and $\prec$. Then $\mathcal{H}_{\mathrm{Syz}}$ is an involutive basis of $\mathrm{Syz}(\mathcal{H})$ with respect to L and the term order $\prec_{\mathcal{H}}$.*

Proof. By Lemma 5.4.7, the leading term of $\mathbf{S}_{\alpha;k} \in \mathcal{H}_{\mathrm{Syz}}$ is $x^k\mathbf{e}_\alpha$ and we have one such generator for each non-multiplicative variable $x^k \in \overline{X}_{L,\mathcal{H},\prec}(\mathbf{h}_\alpha)$. Since we assume that L is of Schreyer type, these leading terms form an involutive set. As we know already from Theorem 5.4.4 that $\mathcal{H}_{\mathrm{Syz}}$ is a Gröbner basis of $\mathrm{Syz}(\mathcal{H})$, the assertion follows trivially. □

Example 5.4.11. We continue with the Pommaret basis $\mathcal{H}$ of Example 5.4.6. Since all assumptions of Theorem 5.4.10 are satisfied for it, the eight syzygies

$$\mathbf{S}_{1;3} = z\mathbf{e}_1 - x\mathbf{e}_3 - \mathbf{e}_2\,, \qquad \mathbf{S}_{2;3} = z\mathbf{e}_2 - x\mathbf{e}_5 - \mathbf{e}_2\,, \tag{5.36a}$$

$$\mathbf{S}_{3;3} = z\mathbf{e}_3 - x\mathbf{e}_6 + \mathbf{e}_5 - \mathbf{e}_3 + \mathbf{e}_1\,, \qquad \mathbf{S}_{4;3} = z\mathbf{e}_4 - y\mathbf{e}_5 - \mathbf{e}_4\,, \tag{5.36b}$$

$$\mathbf{S}_{5;3} = z\mathbf{e}_5 - y\mathbf{e}_6 + \mathbf{e}_2\,, \qquad \mathbf{S}_{1;2} = y\mathbf{e}_1 - x\mathbf{e}_2\,, \tag{5.36c}$$

$$\mathbf{S}_{2;2} = y\mathbf{e}_2 - x\mathbf{e}_4\,, \qquad \mathbf{S}_{3;2} = y\mathbf{e}_3 - x\mathbf{e}_5 + \mathbf{e}_4 - \mathbf{e}_2 \tag{5.36d}$$

form a Pommaret basis of the syzygy module $\mathrm{Syz}(\mathcal{H})$ with respect to the induced term order $\prec_{\mathcal{H}}$. Indeed, as

$$z\mathbf{S}_{1;2} = y\mathbf{S}_{1;3} - x\mathbf{S}_{2;3} + x\mathbf{S}_{4;2} + \mathbf{S}_{2;2}\,, \tag{5.37a}$$

$$z\mathbf{S}_{2;2} = y\mathbf{S}_{2;3} - x\mathbf{S}_{4;3} + \mathbf{S}_{2;2}\,, \tag{5.37b}$$

$$z\mathbf{S}_{3;2} = y\mathbf{S}_{3;3} - x\mathbf{S}_{5;3} - \mathbf{S}_{2;3} + \mathbf{S}_{4;3} + \mathbf{S}_{3;2} - \mathbf{S}_{1;2}\,, \tag{5.37c}$$

all products of the generators with their non-multiplicative variables possess an involutive standard representation. ◁

Since Theorem 5.4.10 yields again an involutive basis, we may iterate its application and construct this way a syzygy resolution for any polynomial submodule $\mathcal{U} \subseteq \mathcal{P}^m$ given an involutive basis of it for an involutive division of Schreyer type (obviously, we can thus determine resolutions for arbitrary finitely presented modules $\mathcal{M} \cong \mathcal{P}^m/\mathcal{U}$). Note, however, that after the first step this process requires to compute explicitly involutive standard representations. We specialise now to Pommaret bases, as for them one can even make a number of statements about the size of the resolution without actually performing these computations. In particular, we immediately obtain a stronger form of Hilbert's Syzygy Theorem B.4.29.

Theorem 5.4.12. *Let $\mathcal{H}$ be a Pommaret basis of the polynomial submodule $\mathcal{U} \subseteq \mathcal{P}^m$. Denote by $\beta_0^{(k)}$ the number of generators $\mathbf{h} \in \mathcal{H}$ such that $\mathrm{cls}\,(\mathrm{le}_\prec \mathbf{h}) = k$ and set as usual $d = \min\{k \mid \beta_0^{(k)} > 0\}$. Then $\mathcal{U}$ possesses a free resolution*

$$0 \longrightarrow \mathcal{P}^{t_{n-d}} \longrightarrow \cdots \longrightarrow \mathcal{P}^{t_1} \longrightarrow \mathcal{U} \longrightarrow 0 \tag{5.38}$$

where the ranks of the free modules are given by

$$t_i = \sum_{k=1}^{n-i} \binom{n-k}{i} \beta_0^{(k)} . \tag{5.39}$$

Proof. By Theorem 5.4.10, $\mathcal{H}_{\text{Syz}}$ is a Pommaret basis of $\text{Syz}(\mathcal{H})$. Applying Theorem 5.4.10 again, we can construct a Pommaret basis of the second syzygy module $\text{Syz}^2(\mathcal{H})$ and so on. It follows from the definition (5.28) of the generators $\mathbf{S}_{\alpha;k}$ that $\text{cls}\,\mathbf{S}_{\alpha;k} = k > \text{cls}\,\mathbf{h}_\alpha$. If d is the minimal class appearing in $\mathcal{H}$, the minimal class in $\mathcal{H}_{\text{Syz}}$ is $d+1$. This observation yields the length of the resolution (5.38), as a Pommaret basis with $d = n$ generates a free module.

The ranks of the modules follow from a rather straightforward combinatorial calculation. Let $\beta_i^{(k)}$ denote the number of generators of class k of the ith syzygy module $\text{Syz}^i(\mathcal{H})$. By definition of the generators $\mathbf{S}_{\alpha;k}$, we find $\beta_i^{(k)} = \sum_{j=1}^{k-1} \beta_{i-1}^{(j)}$, as each generator of class less than k in the Pommaret basis of $\text{Syz}^{i-1}(\mathcal{H})$ contributes one generator of class k in the basis of $\text{Syz}^i(\mathcal{H})$. A simple induction allows us to express the values $\beta_i^{(k)}$ in terms of the values $\beta_0^{(k)}$:

$$\beta_i^{(k)} = \sum_{j=1}^{k-i} \binom{k-j-1}{i-1} \beta_0^{(j)} . \tag{5.40}$$

The ranks of the modules in (5.38) are given by $t_i = \sum_{k=1}^{n} \beta_i^{(k)}$; entering (5.40) yields via a classical identity for binomial coefficients (5.39). □

Remark 5.4.13. Theorem 5.4.12 remains valid for any involutive basis $\mathcal{H}$ with respect to a continuous division of Schreyer type, if we define $\beta_0^{(k)}$ (respectively $\beta_i^{(k)}$ in the proof) as the number of generators with k multiplicative variables, since Theorem 5.4.10 holds for any such basis. Indeed, after the first step we always analyse monomial sets of the form $\{x_{i_1}, x_{i_2}, \dots, x_{i_{n-k}}\}$ (the non-multiplicative variables of a generator with k multiplicative variables) with $i_1 < i_2 < \cdots < i_{n-k}$. By assumption, these sets are involutive and this is only possible, if one of the generators possesses n multiplicative variables, another one $n-1$ and so on until the last generator which has only $n-k$ multiplicative variables (this fact follows for example from Proposition 5.2.1 on the form of the Hilbert series of such an ideal). Hence the basic recursion relation $\beta_i^{(k)} = \sum_{j=1}^{k-1} \beta_{i-1}^{(j)}$ and all subsequent combinatorial computations remain valid for any division of Schreyer type.

For the Janet division one may give a simple direct proof of the corresponding statement. Indeed, as for the Pommaret division, it is here possible to determine explicitly the multiplicative variables for any syzygy: if h_α is a generator in the Janet basis $\mathcal{H}$ with the non-multiplicative variables $\overline{X}_{J,\mathcal{H},\prec}(h_\alpha) = \{x_{i_1}, x_{i_2}, \dots, x_{i_{n-k}}\}$ where again $i_1 < i_2 < \cdots < i_{n-k}$, then

$$X_{J,\mathcal{H}_{\mathrm{Syz}},\prec}(\mathbf{S}_{\alpha;i_j}) = \{x_1,\dots,x_n\} \setminus \{x_{i_{j+1}}, x_{i_{j+2}},\dots,x_{i_{n-k}}\}\,, \tag{5.41}$$

as one readily verifies. ◁

We may explicitly write the resolution (5.38) as a complex. Let $\mathcal{W}$ be a free $\mathcal{P}$-module with basis $\{w_1,\dots,w_s\}$, i.e. its rank is given by the size of the Pommaret basis $\mathcal{H}$. Let $\mathcal{V}$ be a further free $\mathcal{P}$-module with basis $\{v_1,\dots,v_n\}$, i.e. its rank is determined by the number of variables in the polynomial ring $\mathcal{P}$, and denote by $\Lambda\mathcal{V}$ the exterior algebra over $\mathcal{V}$. We set $\mathcal{C}_i = \mathcal{W}\otimes_{\mathcal{P}}\Lambda^i\mathcal{V}$ for $0\leq i\leq n$. If $\mathbf{k}=(k_1,\dots,k_i)$ is a strictly ascending sequence of integers with $1\leq k_1<k_2<\cdots<k_i\leq n$ and $v_{\mathbf{k}}$ denotes the wedge product $v_{k_1}\wedge\cdots\wedge v_{k_i}$, then a basis of this free $\mathcal{P}$-module is given by the set of all tensor products $w_\alpha\otimes v_{\mathbf{k}}$. Finally, we introduce the free submodule $\mathcal{S}_i\subset\mathcal{C}_i$ generated by all those basis elements where $k_1>\operatorname{cls}\mathbf{h}_\alpha$. Note that the rank of $\mathcal{S}_i$ is precisely r_i as defined by (5.39).

We denote the elements of the Pommaret basis of $\mathrm{Syz}^i(\mathcal{H})$ by $\mathbf{S}_{\alpha;\mathbf{k}}$ with the inequalities $\operatorname{cls}\mathbf{h}_\alpha<k_1<\cdots<k_i$. An involutive normal form computation determines for every non-multiplicative index $n\geq k_{i+1}>k_i=\operatorname{cls}\mathbf{S}_{\alpha;\mathbf{k}}$ unique polynomials $P^{(\alpha;\mathbf{k},k_{i+1})}_{\beta;\boldsymbol{\ell}}\in\Bbbk[x^1,\dots,x^{\ell_i}]$ such that

$$x^{k_{i+1}}\mathbf{S}_{\alpha;\mathbf{k}} = \sum_{\beta=1}^{s}\sum_{\boldsymbol{\ell}} P^{(\alpha;\mathbf{k},k_{i+1})}_{\beta;\boldsymbol{\ell}}\mathbf{S}_{\beta;\boldsymbol{\ell}} \tag{5.42}$$

where the second sum is over all integer sequences $\boldsymbol{\ell}=(\ell_1,\dots,\ell_i)$ satisfying $\operatorname{cls}\mathbf{h}_\beta<\ell_1<\cdots<\ell_i\leq n$. We define the $\mathcal{P}$-module homomorphisms $\varepsilon:\mathcal{S}_0\to\mathcal{U}$ and $\delta:\mathcal{S}_{i+1}\to\mathcal{S}_i$ by $\varepsilon(w_\alpha)=\mathbf{h}_\alpha$ and

$$\delta(w_\alpha\otimes v_{\mathbf{k},k_{i+1}}) = x^{k_{i+1}}w_\alpha\otimes v_{\mathbf{k}} - \sum_{\beta,\boldsymbol{\ell}} P^{(\alpha;\mathbf{k},k_{i+1})}_{\beta;\boldsymbol{\ell}} w_\beta\otimes v_{\boldsymbol{\ell}}\,. \tag{5.43}$$

We extend the differential δ to a map $\mathcal{C}_{i+1}\to\mathcal{C}_i$ as follows. If $k_i\leq\operatorname{cls}\mathbf{h}_\alpha$, then we set $\delta(w_\alpha\otimes v_{\mathbf{k}})=0$. Otherwise let j be the smallest value such that $k_j>\operatorname{cls}\mathbf{h}_\alpha$ and set (by slight abuse of notation)

$$\delta(w_\alpha\otimes v_{k_1}\wedge\cdots\wedge v_{k_i}) = v_{k_1}\wedge\cdots\wedge v_{k_{j-1}}\wedge\delta(w_\alpha\otimes v_{k_j}\wedge\cdots\wedge v_{k_i})\,. \tag{5.44}$$

Thus the factor $v_{k_1}\wedge\cdots\wedge v_{k_{j-1}}$ remains unchanged and does not affect the differential. This definition makes, by construction, $(\mathcal{C}_*,\delta)$ to a complex and $(\mathcal{S}_*,\delta)$ to an exact sequence which—augmented by the map $\varepsilon:\mathcal{S}_0\to\mathcal{U}$—is isomorphic to the syzygy resolution (5.38).

Example 5.4.14. We continue with the ideal of Example 5.4.6 and 5.4.11, respectively. As here $d=1$, the resolution has length 2 in this case. Using the notation introduced above, the module $\mathcal{S}_0$ is then generated by $\{w_1,\dots,w_6\}$, the module $\mathcal{S}_1$ by the eight elements $\{w_1\otimes v_3,\dots,w_5\otimes v_3,w_1\otimes v_2,\dots,w_3\otimes v_2\}$ (the first three generators in the Pommaret basis $\mathcal{H}$ are of class 1, the next two of class 2 and the final one of class 3) and the module $\mathcal{S}_2$ by $\{w_1\otimes v_2\wedge v_3,\dots,w_3\otimes v_2\wedge v_3\}$ corresponding to the three first syzygies of class 2. It follows from the expressions (5.36) and

(5.37), respectively, for the first and second syzygies that the differential δ is here defined by the relations

$$\begin{aligned}
\delta(w_1 \otimes v_3) &= zw_1 - xw_3 - w_2 , && (5.45a)\\
\delta(w_2 \otimes v_3) &= zw_2 - xw_5 - w_2 , && (5.45b)\\
\delta(w_3 \otimes v_3) &= zw_3 - xw_6 + w_5 - w_3 + w_1 , && (5.45c)\\
\delta(w_4 \otimes v_3) &= zw_4 - yw_5 - w_4 , && (5.45d)\\
\delta(w_5 \otimes v_3) &= zw_5 - yw_6 + w_2 , && (5.45e)\\
\delta(w_3 \otimes v_2) &= yw_3 - xw_5 + w_4 - w_2 , && (5.45f)\\
\delta(w_2 \otimes v_2) &= yw_2 - xw_4 , && (5.45g)\\
\delta(w_1 \otimes v_2) &= yw_1 - xw_2 , && (5.45h)\\
\delta(w_1 \otimes v_2 \wedge v_3) &= zw_1 \otimes v_2 - yw_1 \otimes v_3 + xw_2 \otimes v_3 - && (5.45i)\\
&\qquad xw_3 \otimes v_2 - w_2 \otimes v_2 . \\
\delta(w_2 \otimes v_2 \wedge v_3) &= zw_2 \otimes v_2 - yw_2 \otimes v_3 + xw_4 \otimes v_3 - w_2 \otimes v_2 , && (5.45j)\\
\delta(w_3 \otimes v_2 \wedge v_3) &= zw_3 \otimes v_2 - yw_3 \otimes v_3 + xw_5 \otimes v_3 + && (5.45k)\\
&\qquad w_2 \otimes v_3 - w_4 \otimes v_3 - w_3 \otimes v_2 + w_1 \otimes v_2 ,
\end{aligned}$$

It represents a straightforward albeit rather tedious task to verify explicitly the exactness of the thus constructed complex $(\mathcal{S}_*, \delta)$. ◁

In the case that $m = 1$ and thus $\mathcal{U}$ is actually an ideal in $\mathcal{P}$, it is tempting to try to equip the complex $(\mathcal{C}_*, \delta)$ with the structure of a differential ring. We first introduce a multiplication $\times$ on $\mathcal{W}$. If h_α and h_β are two elements of the Pommaret basis $\mathcal{H}$, then their product possesses a unique involutive standard representation $h_\alpha h_\beta = \sum_{\gamma=1}^{s} P_{\alpha\beta\gamma} h_\gamma$ and we define

$$w_\alpha \times w_\beta = \sum_{\gamma=1}^{s} P_{\alpha\beta\gamma} w_\gamma \tag{5.46}$$

and continue $\mathcal{P}$-linearly on $\mathcal{W}$. This multiplication can be extended to the whole complex $\mathcal{C}_*$ by defining for arbitrary elements $w, \bar{w} \in \mathcal{W}$ and $\omega, \bar{\omega} \in \Lambda\mathcal{V}$

$$(w \otimes \omega) \times (\bar{w} \otimes \bar{\omega}) = (w \times \bar{w}) \otimes (\omega \wedge \bar{\omega}) . \tag{5.47}$$

The distributivity of the thus introduced multiplication $\times$ is obvious from its definition. For obtaining a differential ring, the product $\times$ must furthermore be associative and satisfy the graded Leibniz rule $\delta(a \times b) = \delta(a) \times b + (-1)^{|a|} a \times \delta(b)$ where $|a|$ denotes the form degree of the element a. While in general both conditions are not met, a number of special situations exist where this construction indeed turns $(\mathcal{C}_*, \delta)$ into a differential ring.

Let us first consider the question of the associativity of the product $\times$. For our purposes, it suffices to study it at the level of the module $\mathcal{W}$ where we obtain the following two results, if we multiply in different orders:

$$w_\alpha \times (w_\beta \times w_\gamma) = \sum_{\delta,\varepsilon=1}^{s} P_{\beta\gamma\delta} P_{\alpha\delta\varepsilon} w_\varepsilon \,, \tag{5.48a}$$

$$(w_\alpha \times w_\beta) \times w_\gamma = \sum_{\delta,\varepsilon=1}^{s} P_{\alpha\beta\delta} P_{\gamma\delta\varepsilon} w_\varepsilon \,. \tag{5.48b}$$

One easily checks that both $\sum_{\delta,\varepsilon=1}^{s} P_{\beta\gamma\delta}P_{\alpha\delta\varepsilon}h_\varepsilon$ and $\sum_{\delta,\varepsilon=1}^{s} P_{\alpha\beta\delta}P_{\gamma\delta\varepsilon}h_\varepsilon$ are standard representations of the product $h_\alpha h_\beta h_\gamma$ for the Pommaret basis $\mathcal{H}$. However, we cannot conclude that they are involutive standard representations, as we do not know whether $P_{\beta\gamma\delta}$ and $P_{\alpha\beta\delta}$, respectively, are multiplicative for h_ε. If this were the case, the associativity of $\times$ would immediately follow from the uniqueness of involutive standard representations.

For the graded Leibniz rule the situation is similar but a bit more involved. In the next section, we will consider it in more details for the monomial case. In the end, the discussion boils down to analysing standard representations for products of the form $x^k h_\alpha h_\beta$. Again, we have two different ways for constructing them and a sufficient condition for the satisfaction of the Leibniz rule is that both approaches lead always to the unique involutive standard representation.

Example 5.4.15. We consider again the ideal $\mathcal{I} \subset \Bbbk[x,y,z]$ generated by $h_1 = y^2 - z$, $h_2 = yz - x$ and $h_3 = z^2 - xy$. We showed in Example 3.4.11 that these polynomials form a Pommaret basis of $\mathcal{I}$ for the degree reverse lexicographic term order. The Pommaret basis of the first syzygy module consists of $\mathbf{S}_{1;3} = z\mathbf{e}_1 - y\mathbf{e}_2 + \mathbf{e}_3$ and $\mathbf{S}_{2;3} = z\mathbf{e}_2 - y\mathbf{e}_3 - x\mathbf{e}_1$. As both generators are of class 3, they span a free module and the resolution stops here.

In a straightforward calculation one obtains for the multiplication $\times$ the following defining relations:

$$w_1^2 = w_3 - yw_2 + y^2 w_1 \,, \quad w_1 \times w_2 = -yw_3 + y^2 w_2 - xw_1 \,, \tag{5.49a}$$

$$w_1 \times w_3 = (y^2 - z)w_3 \,, \quad w_2^2 = y^2 w_3 - xw_2 + xyw_1 \,, \tag{5.49b}$$

$$w_2 \times w_3 = (yz - x)w_3 \,, \quad w_3^2 = (z^2 - xy)w_3 \,. \tag{5.49c}$$

Note that all coefficients of w_1 and w_2 are contained in $\Bbbk[x,y]$ and are thus multiplicative for all generators. This fact immediately implies that our multiplication is associative, as any way to evaluate the product $w_\alpha \times w_\beta \times w_\gamma$ leads to the unique involutive standard representation of $h_\alpha h_\beta h_\gamma$.

In the only non-multiplicative products $zh_2 = yh_3 + xh_1$ and $zh_1 = yh_2 + h_3$ all coefficients on the right hand sides lie in the subring $\Bbbk[x,y]$, too, and by the same line of reasoning δ and $\times$ satisfy the graded Leibniz rule so that our complex $(\mathcal{C}_*, \delta)$ has indeed become a differential ring. ◁

The situation is not always as favourable as in this particular example. Already the next example shows that in general we cannot expect that the above introduced multiplication $\times$ makes the complex $(\mathcal{C}_*, \delta)$ to a differential ring (in fact, not even to an associative ring).

Example 5.4.16. Let us continue with the ideal of Examples 5.4.6, 5.4.11 and 5.4.14. Evaluation of the defining relation (5.46) is particularly simple for the products of the form $w_i \times w_6 = h_i w_6$, as all variables are multiplicative for the generator h_6. Two further products are $w_5^2 = y^2 w_6 - y w_5 - x w_4$ and $w_3 \times w_5 = xy w_6 - y w_5 - x w_2$. In a straightforward computation one finds

$$(w_3 \times w_5) \times w_5 - w_3 \times w_5^2 = x^2 w_4 - xy w_2 \,, \tag{5.50}$$

so that the multiplication is not associative. Note that the difference corresponds to the syzygy $x^2 h_4 - xy h_2 = 0$. This is not surprising, as it encodes the difference between two standard representations of $h_3 h_5^2$. The reason for the non-associativity lies in the coefficient y of w_5 in the power w_5^2; it is non-multiplicative for h_2 and the generator w_2 appears in the product $w_3 \times w_5$. Hence computing $w_3 \times w_5^2$ does not lead to an involutive standard representation of $h_3 h_5^2$ whereas the alternative product $(w_3 \times w_5) \times w_5$ does. ◁

In the special case of monomial modules, stronger results can be obtained. In particular, it is possible to obtain a closed form of the differential (5.43) solely by analysing the Pommaret basis $\mathcal{H}$. However, in order to have a Pommaret basis available, we must always assume in the sequel that we are dealing with a quasi-stable submodule $\mathcal{U} \subseteq \mathcal{P}^m$. Let $\mathcal{H} = \{\mathbf{h}_1, \dots, \mathbf{h}_s\}$ with $\mathbf{h}_\alpha \in \mathbb{T}^m$ be its monomial Pommaret basis (by Proposition 3.1.21, it is unique). Furthermore, we introduce the function $\Delta(\alpha, k)$ determining the unique generator in the Pommaret basis $\mathcal{H}$ such that $x^k \mathbf{h}_\alpha = t_{\alpha,k} \mathbf{h}_{\Delta(\alpha,k)}$ with a term $t_{\alpha,k} \in \Bbbk[X_P(\mathbf{h}_{\Delta(\alpha,k)})]$.

Lemma 5.4.17. *The function Δ and the terms $t_{\alpha,k}$ satisfy the following relations.*

(i) *The inequality* $\operatorname{cls} \mathbf{h}_\alpha \leq \operatorname{cls} \mathbf{h}_{\Delta(\alpha,k)} \leq k$ *holds for all non-multiplicative indices* $k > \operatorname{cls} \mathbf{h}_\alpha$.

(ii) *Let* $k_2 > k_1 > \operatorname{cls} \mathbf{h}_\alpha$ *be non-multiplicative indices. If* $\operatorname{cls} \mathbf{h}_{\Delta(\alpha,k_2)} \geq k_1$, *then* $\Delta(\Delta(\alpha,k_1),k_2) = \Delta(\alpha,k_2)$ *and* $x^{k_1} t_{\alpha,k_2} = t_{\alpha,k_1} t_{\Delta(\alpha,k_1),k_2}$. *Otherwise we have* $\Delta(\Delta(\alpha,k_1),k_2) = \Delta(\Delta(\alpha,k_2),k_1)$ *and* $t_{\alpha,k_1} t_{\Delta(\alpha,k_1),k_2} = t_{\alpha,k_2} t_{\Delta(\alpha,k_2),k_1}$.

Proof. Part (i) is trivial. The inequality $\operatorname{cls} \mathbf{h}_\alpha \leq \operatorname{cls} \mathbf{h}_{\Delta(\alpha,k)}$ follows from the definition of Δ and the Pommaret division. If $\operatorname{cls} \mathbf{h}_{\Delta(\alpha,k)} > k$, then $\mathbf{h}_{\Delta(\alpha,k)}$ would be an involutive divisor of $\mathbf{h}_\alpha$ which contradicts the fact that any involutive basis is involutively head autoreduced.

For Part (ii) we compute the involutive standard representation of the product $x^{k_1} x^{k_2} \mathbf{h}_\alpha$. There are two ways to do it. We may either write

$$x^{k_1} x^{k_2} \mathbf{h}_\alpha = x^{k_2} t_{\alpha,k_1} \mathbf{h}_{\Delta(\alpha,k_1)} = t_{\alpha,k_1} t_{\Delta(\alpha,k_1),k_2} \mathbf{h}_{\Delta(\Delta(\alpha,k_1),k_2)} \,, \tag{5.51}$$

which is an involutive standard representation by Part (i), or start with

$$x^{k_1} x^{k_2} \mathbf{h}_\alpha = x^{k_1} t_{\alpha,k_2} \mathbf{h}_{\Delta(\alpha,k_2)} \tag{5.52}$$

which requires a case distinction. If $\operatorname{cls} \mathbf{h}_{\Delta(\alpha,k_2)} \geq k_1$, we have already an involutive standard representation and its uniqueness implies our claim. Otherwise multiplicatively rewriting $x^{k_1}\mathbf{h}_{\Delta(\alpha,k_2)} = t_{\Delta(\alpha,k_2),k_1}\mathbf{h}_{\Delta(\Delta(\alpha,k_2),k_1)}$ yields an involutive standard representation and our assertion follows from its uniqueness. □

Using this lemma, we can now provide a closed form for the differential δ which does not require involutive normal form computations in the syzygy modules $\mathrm{Syz}^i(\mathcal{H})$ (which are of course expensive to perform) but is solely based on information already computed during the determination of $\mathcal{H}$. For its proof we must introduce some additional notations and conventions. If again $\mathbf{k} = (k_1,\dots,k_i)$ is an integer sequence with $1 \leq k_1 < \cdots < k_i \leq n$, then we write $\mathbf{k}_j$ for the same sequence of indices but with k_j eliminated. Its first entry is denoted by $(\mathbf{k}_j)_1$; hence $(\mathbf{k}_j)_1 = k_1$ for $j > 1$ and $(\mathbf{k}_j)_1 = k_2$ for $j = 1$. The syzygy $\mathbf{S}_{\alpha;\mathbf{k}}$ is only defined for $\operatorname{cls}\mathbf{h}_\alpha < k_1$. We extend this notation by setting $\mathbf{S}_{\alpha;\mathbf{k}} = 0$ for $\operatorname{cls}\mathbf{h}_\alpha \geq k_1$. This convention will simplify some sums in the sequel.

Theorem 5.4.18. *Let $\mathcal{U} \subseteq \mathcal{P}^m$ be a quasi-stable submodule and $\mathbf{k} = (k_1,\dots,k_i)$. Then the differential δ of the complex $\mathcal{C}_*$ may be written as*

$$\delta(w_\alpha \otimes v_{\mathbf{k}}) = \sum_{j=1}^{i} (-1)^{i-j}\left(x^{k_j} w_\alpha - t_{\alpha,k_j} w_{\Delta(\alpha,k_j)}\right) \otimes v_{\mathbf{k}_j} \,. \tag{5.53}$$

Proof. Note that all summands where k_j is multiplicative for $\mathbf{h}_\alpha$ vanish. This observation implies trivially (5.44), so that we can restrict to the case that $\operatorname{cls}\mathbf{h}_\alpha < k_1$. Then our theorem is equivalent to

$$\mathbf{S}_{\alpha;\mathbf{k}} = \sum_{j=1}^{i} (-1)^{i-j}\left(x^{k_j}\mathbf{S}_{\alpha;\mathbf{k}_j} - t_{\alpha,k_j}\mathbf{S}_{\Delta(\alpha,k_j);\mathbf{k}_j}\right) \,. \tag{5.54}$$

Some of the terms $\mathbf{S}_{\Delta(\alpha,k);\mathbf{k}_j}$ might vanish by our above introduced convention. The equality (5.54) is trivial for $i = 1$ (with $\mathbf{S}_\alpha = \mathbf{h}_\alpha$) and a simple corollary of Part (ii) of Lemma 5.4.17 for $i = 2$.

For $i > 2$ things become messy. We proceed by induction on i. In our approach, the syzygy $\mathbf{S}_{\alpha;\mathbf{k}}$ arises from the non-multiplicative product $x^{k_i}\mathbf{S}_{\alpha;\mathbf{k}_i}$. Thus we must compute now the involutive normal form of this product. By our induction hypothesis we may write

$$x^{k_i}\mathbf{S}_{\alpha;\mathbf{k}_i} = \sum_{j=1}^{i-1} (-1)^{i-1-j}\left(x^{k_j}x^{k_i}\mathbf{S}_{\alpha;\mathbf{k}_{ji}} - x^{k_i}t_{\alpha,k_j}\mathbf{S}_{\Delta(\alpha,k_j);\mathbf{k}_{ji}}\right) \,. \tag{5.55}$$

As x^{k_i} is always non-multiplicative, using again the induction hypothesis, each summand may be replaced by the corresponding syzygy—but only at the expense of the introduction of many additional terms. The main task in the proof will be to show that most of them cancel. However, the cancellations occur in a rather complicated manner with several cases, so that no simple way for proving (5.54) seems to exist. We obtain the following lengthy expression:

$$
\begin{aligned}
x^{k_i}\mathbf{S}_{\alpha;\mathbf{k}_i} &= \sum_{j=1}^{i-1}(-1)^{i-1-j}\Big[x^{k_j}\underset{①}{\mathbf{S}_{\alpha;\mathbf{k}_j}} - t_{\alpha,k_j}\underset{②}{\mathbf{S}_{\Delta(\alpha,k_j);\mathbf{k}_j}}\Big] \\
&+ \sum_{j=1}^{i-1}x^{k_j}\left[\sum_{\ell=1}^{j-1}(-1)^{\ell+j+1}x^{k_\ell}\underset{③}{\mathbf{S}_{\alpha;\mathbf{k}_{\ell j}}} - \sum_{\ell=j+1}^{i-1}(-1)^{\ell+j+1}x^{k_\ell}\underset{④}{\mathbf{S}_{\alpha;\mathbf{k}_{j\ell}}}\right] \\
&- \sum_{j=1}^{i-1}\left[\sum_{\ell=1}^{j-1}(-1)^{\ell+j+1}x^{k_j}t_{\alpha,k_\ell}\underset{⑤}{\mathbf{S}_{\Delta(\alpha,k_\ell);\mathbf{k}_{\ell j}}} - \right. \\
&\qquad \left.\sum_{\ell=j+1}^{i-1}(-1)^{\ell+j+1}x^{k_j}t_{\alpha,k_\ell}\underset{⑥}{\mathbf{S}_{\Delta(\alpha,k_\ell);\mathbf{k}_{j\ell}}}\right] \\
&+ \sum_{j=1}^{i-2}(-1)^{i-1-j}x^{k_j}t_{\alpha,k_i}\underset{⑦}{\mathbf{S}_{\Delta(\alpha,k_i);\mathbf{k}_{ji}}} + x^{k_{i-1}}t_{\alpha,k_i}\underset{⑧}{\mathbf{S}_{\Delta(\alpha,k_i);\mathbf{k}_{i-1,i}}} \\
&- \sum_{j=1}^{i-1}t_{\alpha,k_j}\left[\sum_{\ell=1}^{j-1}(-1)^{\ell+j+1}x^{k_\ell}\underset{⑨}{\mathbf{S}_{\Delta(\alpha,k_j);\mathbf{k}_{\ell j}}} - \right. \\
&\qquad \left.\sum_{\ell=j+1}^{i-1}(-1)^{\ell+j+1}x^{k_\ell}\underset{⑩}{\mathbf{S}_{\Delta(\alpha,k_j);\mathbf{k}_{j\ell}}}\right] \\
&+ \sum_{j=1}^{i-1}t_{\alpha,k_j}\left[\sum_{\ell=1}^{j-1}(-1)^{\ell+j+1}t_{\Delta(\alpha,k_j),k_\ell}\underset{⑪}{\mathbf{S}_{\Delta(\Delta(\alpha,k_j),k_\ell);\mathbf{k}_{\ell j}}} - \right. \\
&\qquad \left.\sum_{\ell=j+1}^{i-1}(-1)^{\ell+j+1}t_{\Delta(\alpha,k_j),k_\ell}\underset{⑫}{\mathbf{S}_{\Delta(\Delta(\alpha,k_j),k_\ell);\mathbf{k}_{j\ell}}}\right] \\
&- \sum_{j=1}^{i-1}(-1)^{i-1-j}t_{\alpha,k_j}t_{\Delta(\alpha,k_j),k_i}\underset{⑬}{\mathbf{S}_{\Delta(\Delta(\alpha,k_j),k_i);\mathbf{k}_{ji}}}\,.
\end{aligned}
\tag{5.56}
$$

Note that the terms ⑦, ⑧ and ⑬, respectively, correspond to the special case $\ell = i$ (and $j = i-1$) in the sums ⑥ and ⑫, respectively. We list them separately, as they must be treated differently. The existence of any summand where the coefficient contains a term $t_{\cdot,\cdot}$ is bound on conditions.

With the exception of the coefficient $x^{k_{i-1}}$ in the term ⑧, all coefficients are already multiplicative. Thus this term must be further expanded using the induction hypothesis for the last time:

$$
\begin{aligned}
x^{k_{i-1}}t_{\alpha,k_i}\mathbf{S}_{\Delta(\alpha,k_i);\mathbf{k}_{i-1,i}} &= t_{\alpha,k_i}\underset{⑭}{\mathbf{S}_{\Delta(\alpha,k_i);\mathbf{k}_i}} \\
&- \sum_{j=1}^{i-2}(-1)^{i-1-j}x^{k_j}t_{\alpha,k_i}\underset{⑮}{\mathbf{S}_{\Delta(\alpha,k_i);\mathbf{k}_{ji}}} \\
&+ \sum_{j=1}^{i-1}(-1)^{i-1-j}t_{\alpha,k_i}t_{\Delta(\alpha,k_i),k_j}\underset{⑯}{\mathbf{S}_{\Delta(\Delta(\alpha,k_i),k_j);\mathbf{k}_{ji}}}\,.
\end{aligned}
\tag{5.57}
$$

The left hand side of (5.56) and the terms ①, ② and ⑭ represent the syzygy $\mathbf{S}_{\alpha,\mathbf{k}}$ we are looking for. We must thus show that all remaining terms vanish. In order to simplify the discussion of the double sums, we swap j and ℓ in ③, ⑤, ⑨ and ⑪ so that everywhere $j < \ell$. It is now easy to see that ③ and ④ cancel; each summand of ③ also appears in ④ but with the opposite sign. Note that the same argument does not apply to ⑪ and ⑫, as the existence of these terms is bound to different conditions!

For the other cancellations, we must distinguish several cases depending on the classes of the generators in the Pommaret basis $\mathcal{H}$. We first study the double sums and thus assume that $1 \leq j < i$.

- If $\operatorname{cls}\mathbf{h}_{\Delta(\alpha,k_j)} < (\mathbf{k}_j)_1$, then the terms ⑤ and ⑩ are both present and cancel each other. We now make a second case distinction on the basis of $\mathbf{h}_{\Delta(\alpha,k_\ell)}$.
  - If $\operatorname{cls}\mathbf{h}_{\Delta(\alpha,k_\ell)} < (\mathbf{k}_j)_1$, then the terms ⑥ and ⑨ are also present and cancel each other. Furthermore, both ⑪ and ⑫ exist and cancel due to the second case of Part (ii) of Lemma 5.4.17.
  - If $\operatorname{cls}\mathbf{h}_{\Delta(\alpha,k_\ell)} \geq (\mathbf{k}_j)_1$, then none of the four terms ⑥, ⑨, ⑪ and ⑫ appears. For the latter two terms, this fact is a consequence of the first case of Part (ii) of Lemma 5.4.17.
- If $\operatorname{cls}\mathbf{h}_{\Delta(\alpha,k_j)} \geq (\mathbf{k}_j)_1$, then neither ⑤ nor ⑩ nor ⑫ exists. For the remaining double sums, we must again consider the class of $\mathbf{h}_{\Delta(\alpha,k_\ell)}$.
  - If $\operatorname{cls}\mathbf{h}_{\Delta(\alpha,k_\ell)} < (\mathbf{k}_j)_1$, then the terms ⑥ and ⑨ exist and cancel each other. By contrast, the term ⑪ does not exist, as Lemma 5.4.17 implies the inequalities $\operatorname{cls}\mathbf{h}_{\Delta(\Delta(\alpha,k_\ell),k_j)} = \operatorname{cls}\mathbf{h}_{\Delta(\Delta(\alpha,k_j),k_\ell)} \geq \operatorname{cls}\mathbf{h}_{\Delta(\alpha,k_j)} \geq (\mathbf{k}_j)_1$.
  - If $\operatorname{cls}\mathbf{h}_{\Delta(\alpha,k_\ell)} \geq (\mathbf{k}_j)_1$, then neither ⑥ nor ⑨ exist and the term ⑪ is not present either; this time the application of Part (ii) of Lemma 5.4.17 yields the chain of inequalities $\operatorname{cls}\mathbf{h}_{\Delta(\Delta(\alpha,k_\ell),k_j)} \geq \operatorname{cls}\mathbf{h}_{\Delta(\alpha,k_\ell)} \geq (\mathbf{k}_j)_1$.

For the remaining terms everything depends on the class of $\mathbf{h}_{\Delta(\alpha,k_i)}$ controlling in particular the existence of the term ⑧.

- If $\operatorname{cls}\mathbf{h}_{\Delta(\alpha,k_i)} < k_1 \leq (\mathbf{k}_j)_1$, then the term ⑧ exists and generates the terms ⑮ and ⑯. Under this condition, the term ⑦ is present, too, and because of Part (ii) of Lemma 5.4.17 it cancels ⑮. Again by Part (ii) of Lemma 5.4.17, the conditions for the existence of ⑬ and ⑯ are identical and they cancel each other.
- If $\operatorname{cls}\mathbf{h}_{\Delta(\alpha,k_i)} \geq k_1$, then ⑧ and consequently ⑮ and ⑯ are not present. The analysis of ⑦ and ⑬ requires a further case distinction.
  - Under the made assumption, the case $\operatorname{cls}\mathbf{h}_{\Delta(\alpha,k_i)} < (\mathbf{k}_j)_1$ occurs only for $j = 1$ as otherwise $(\mathbf{k}_j)_1 = k_1$. Because of Part (ii) of Lemma 5.4.17, the terms ⑦ and ⑬ exist for $j = 1$ and cancel each other.
  - If $\operatorname{cls}\mathbf{h}_{\Delta(\alpha,k_i)} \geq (\mathbf{k}_j)_1$, then the term ⑦ does not appear. The term ⑬ is also not present, but there are now two different possibilities: depending on which case of Part (ii) of Lemma 5.4.17 applies, we either find $\operatorname{cls}\mathbf{h}_{\Delta(\Delta(\alpha,k_j),k_i)} = \operatorname{cls}\mathbf{h}_{\Delta(\alpha,k_i)}$ or $\operatorname{cls}\mathbf{h}_{\Delta(\Delta(\alpha,k_j),k_i)} = \operatorname{cls}\mathbf{h}_{\Delta(\Delta(\alpha,k_i),k_j)} \geq \operatorname{cls}\mathbf{h}_{\Delta(\alpha,k_i)}$; in any case the class is too high.

Thus we have shown that indeed all terms vanish with the exception of ①, ② and ⑭ which are needed for the syzygy $\mathbf{S}_{\alpha,\mathbf{k}}$. This proves our claim. □

Remark 5.4.19. Above we introduced for any involutive basis $\mathcal{H}$ with respect to a division L its L-graph. We augment now this graph by weights for the edges. Recall that we have a directed edge from $\mathbf{h}$ to $\bar{\mathbf{h}}$, if $\mathrm{le}_{\prec}\,\bar{\mathbf{h}}$ is an involutive divisor of $\mathrm{le}_{\prec}\,(x_k\mathbf{h})$ for some non-multiplicative variable $x^k \in \overline{X}_{\mathcal{H},L,\prec}(\mathbf{h})$. If $\mathrm{le}_{\prec}\,(x_k\mathbf{h}) = \mathrm{le}_{\prec}\,\bar{\mathbf{h}} + \mu$, then we assign the weight x^μ to this edge. For a monomial Pommaret basis the corresponding P-graph has then a directed edge from $\mathbf{h}_\alpha$ to $\mathbf{h}_{\Delta(\alpha,k)}$ with weight $t_{\alpha,k}$ for every non-multiplicative variable $x^k \in \overline{X}_P(\mathbf{h}_\alpha)$. Thus we may say that by Theorem 5.4.18 the whole complex $(\mathcal{C}_*, \delta)$ (and the isomorphic syzygy resolution of $\langle\mathcal{H}\rangle$) is encoded in the weighted P-graph of $\mathcal{H}$. ◁

As above, we introduce for monomial ideals, i. e. for $m = 1$, the product $\times$. The right hand side of its defining equation (5.46) simplifies for a monomial basis $\mathcal{H}$ to

$$w_\alpha \times w_\beta = m_{\alpha,\beta} w_{\Gamma(\alpha,\beta)} \tag{5.58}$$

where the function $\Gamma(\alpha,\beta)$ determines the unique generator $h_{\Gamma(\alpha,\beta)} \in \mathcal{H}$ such that we can write $h_\alpha h_\beta = m_{\alpha,\beta} h_{\Gamma(\alpha,\beta)}$ with a term $m_{\alpha,\beta} \in \Bbbk[X_P(h_{\Gamma(\alpha,\beta)})]$. In analogy to Lemma 5.4.17, we obtain now the following result.

Lemma 5.4.20. *The function Γ and the terms $m_{\alpha,\beta}$ satisfy the following relations.*

(i) $\operatorname{cls} h_{\Gamma(\alpha,\beta)} \geq \max\{\operatorname{cls} h_\alpha, \operatorname{cls} h_\beta\}$.
(ii) $\Gamma\bigl(\Gamma(\alpha,\beta),\gamma\bigr) = \Gamma\bigl(\alpha,\Gamma(\beta,\gamma)\bigr)$ *and* $m_{\alpha,\beta} m_{\Gamma(\alpha,\beta),\gamma} = m_{\beta,\gamma} m_{\Gamma(\beta,\gamma),\alpha}$.
(iii) $\Gamma\bigl(\Delta(\alpha,k),\beta\bigr) = \Delta\bigl(\Gamma(\alpha,\beta),k\bigr)$ *and* $t_{\alpha,k} m_{\Delta(\alpha,k),\beta} = t_{\Gamma(\alpha,\beta),k} m_{\alpha,\beta}$.

Proof. Part (i) is obvious from the definition of the function Γ. Part (ii) and (iii), respectively, follow from the analysis of the two different ways to compute the involutive standard representation of $h_\alpha h_\beta h_\gamma$ and $x^k h_\alpha h_\beta$, respectively. We omit the details, as they are completely analogous to the proof of Lemma 5.4.17. □

Theorem 5.4.21. *Let $\mathcal{H}$ be the Pommaret basis of the quasi-stable monomial ideal $\mathcal{I} \subseteq \mathcal{P}$. Then the product $\times$ defined by* (5.58) *makes the complex $(\mathcal{C}_*, \delta)$ to a graded differential ring.*

Proof. This is a straightforward consequence of Lemma 5.4.20. Writing out the relations to be checked, one easily finds that Part (ii) ensures the associativity of $\times$ and Part (iii) the satisfaction of the graded Leibniz rule. □

Addendum: Iterated Polynomial Algebras of Solvable Type

In Section 4.6 we studied involutive bases in polynomial algebras of solvable type over rings. We had to substitute the notion of an involutively head autoreduced set by the more comprehensive concept of an involutively $\mathcal{R}$-saturated set. In a certain sense, this approach was not completely satisfying, as we had to resort to classical Gröbner techniques, namely computing normal forms of ideal elements arising from syzygies. Using the syzygy theory developed in this section, we provide now an alternative approach for the special case that the coefficient ring $\mathcal{R}$ is again a polynomial algebra of solvable type (over a field). It is obvious that in this case left ideal membership in $\mathcal{R}$ can be decided algorithmically and by Theorem 5.4.4 it is also possible to construct algorithmically a basis of the syzygy module.

Remark 5.4.22. Throughout this chapter we have only considered the ordinary commutative polynomial ring, whereas now we return to general polynomial algebras of solvable type (over a field). However, it is easy to see that all the arguments in the proof of the involutive Schreyer Theorem 5.4.4 depend only on normal form computations and on considerations concerning the leading exponents. The same holds for the classical Schreyer theorem, as one may easily check (see also [285, 286] for a non-commutative version). Thus the in the sequel crucial Theorem 5.4.4 remains valid in the general case of non-commutative polynomial algebras. ◁

In this addendum, we use the following notations: $\mathcal{R} = (\Bbbk[y_1,\dots,y_m],\star,\prec_y)$ and $\mathcal{P} = (\mathcal{R}[x_1,\dots,x_n],\star,\prec_x)$. Furthermore, we are given an involutive division L_y on $\mathbb{N}_0^m$ and a division L_x on $\mathbb{N}_0^n$. For simplicity, we always assume in the sequel that at least L_y is Noetherian. In order to obtain a reasonable theory, we make similar assumptions as in Section 4.6: both $\mathcal{R}$ and $\mathcal{P}$ are solvable algebras with centred commutation relations so that both are (left) Noetherian by Theorem 3.3.7.

We now propose an alternative algorithm for the involutive $\mathcal{R}$-saturation. Until Line /13/ it is identical with Algorithm 4.10; afterwards we perform an involutive completion and multiply in Line /17/ each polynomial in $\bar{\mathcal{H}}'_{f,L_x}$ by the non-multiplicative variables of its leading coefficient. In the determination of involutive normal forms, we may multiply each polynomial $h' \in \mathcal{H}'$ only by monomials rx^μ such that $x^\mu \in \mathcal{R}[X_{L_x,\mathcal{H}',\prec_x}(h')]$ *and* $r \in \Bbbk\left[Y_{L_y,\mathrm{lc}_{\prec_x}(\bar{\mathcal{H}}'_{h',L_x}),\prec_y}(\mathrm{lc}_{\prec_x} h')\right]$.

Proposition 5.4.23. *Let L_y be a Noetherian constructive division. Algorithm 5.4 terminates for any finite input set $\mathcal{F} \subset \mathcal{P}$ with an involutively $\mathcal{R}$-saturated and head autoreduced set $\mathcal{H}$ such that $\langle\mathcal{H}\rangle = \langle\mathcal{F}\rangle$. Furthermore, the sets $\mathrm{lc}_{\prec_x}\bar{\mathcal{H}}_{h,L_x}$ form weak L_y-involutive bases of the $\mathcal{R}$-ideals generated by them for each generator $h \in \mathcal{H}$.*

Proof. The termination criterion in Line /26/ is equivalent to local involution of all the sets $\mathrm{lc}_{\prec_x}\bar{\mathcal{H}}'_{f,L_x}$. Under the made assumptions on the division L_y and because of the fact that $\mathcal{P}$ is Noetherian, the termination of the algorithm and the assertion about these sets is obvious. In general, we only obtain weak involutive bases, as no involutive head autoreductions of these sets are performed. The correctness is a consequence of Theorem 5.4.4: by analysing all non-multiplicative products we have taken into account a whole basis of the syzygy module $\mathrm{Syz}(\mathrm{lc}_{\prec_x}\bar{\mathcal{H}}'_{f,L_x})$. Thus the output $\mathcal{H}$ is indeed involutively $\mathcal{R}$-saturated. □

Theorem 5.4.24. *Let the polynomial ring $\mathcal{P}$ satisfy the made assumptions and let L_x be a Noetherian constructive division. If in Algorithm 4.5 the subalgorithm*

Algorithm 5.4 Involutive $\mathcal{R}$-saturation (iterated case)

Input: finite set $\mathcal{F} \subset \mathcal{P}$, involutive divisions L_y on $\mathbb{N}_0^m$ and L_x on $\mathbb{N}_0^n$
Output: involutively $\mathcal{R}$-saturated and head autoreduced set $\mathcal{H}$ with $\langle \mathcal{H} \rangle = \langle \mathcal{F} \rangle$

1: $\mathcal{H} \leftarrow \mathcal{F}$; $\mathcal{S} \leftarrow \mathcal{F}$
2: **while** $S \neq \emptyset$ **do**
3: $\quad v \leftarrow \max_{\prec_x} \mathrm{le}_{\prec_x} \mathcal{S}$; $\mathcal{S}_v \leftarrow \{f \in \mathcal{H} \mid \mathrm{le}_{\prec_x} f = v\}$
4: $\quad \mathcal{S} \leftarrow \mathcal{S} \setminus \mathcal{S}_v$; $\mathcal{H}' \leftarrow \mathcal{H}$
5: $\quad$ **for all** $f \in \mathcal{S}_v$ **do**
6: $\quad\quad h \leftarrow \texttt{HeadReduce}_{L_x,\prec_x}(f, \mathcal{H})$
7: $\quad\quad$ **if** $f \neq h$ **then**
8: $\quad\quad\quad \mathcal{S}_v \leftarrow \mathcal{S}_v \setminus \{f\}$; $\mathcal{H}' \leftarrow \mathcal{H}' \setminus \{f\}$
9: $\quad\quad\quad$ **if** $h \neq 0$ **then**
10: $\quad\quad\quad\quad \mathcal{H}' \leftarrow \mathcal{H}' \cup \{h\}$
11: $\quad\quad\quad$ **end if**
12: $\quad\quad$ **end if**
13: $\quad$ **end for**
14: $\quad$ **if** $\mathcal{S}_v \neq \emptyset$ **then**
15: $\quad\quad$ choose $f \in \mathcal{S}_v$ and determine the set $\bar{\mathcal{H}}'_{f,L_x}$
16: $\quad\quad$ **repeat**
17: $\quad\quad\quad \mathcal{T} \leftarrow \{y_j \star \bar{f} \mid \bar{f} \in \bar{\mathcal{H}}'_{f,L_x}, y_j \in \overline{Y}_{L_y,\mathrm{lc}_{\prec_x}(\bar{\mathcal{H}}'_{f,L_x}),\prec_y}(\mathrm{lc}_{\prec_x} \bar{f})\}$
18: $\quad\quad\quad$ **repeat**
19: $\quad\quad\quad\quad$ choose $h' \in \mathcal{T}$ such that $\mathrm{le}_{\prec_y}(\mathrm{lc}_{\prec_x} h')$ is minimal
20: $\quad\quad\quad\quad \mathcal{T} \leftarrow \mathcal{T} \setminus \{h'\}$
21: $\quad\quad\quad\quad h \leftarrow \texttt{NormalForm}_{L_x,\prec_x,L_y,\prec_y}(h', \mathcal{H}')$
22: $\quad\quad\quad\quad$ **if** $h \neq 0$ **then**
23: $\quad\quad\quad\quad\quad \mathcal{H}' \leftarrow \mathcal{H}' \cup \{h\}$
24: $\quad\quad\quad\quad$ **end if**
25: $\quad\quad\quad$ **until** $\mathcal{T} = \emptyset \vee h \neq 0$
26: $\quad\quad$ **until** $\mathcal{T} = \emptyset \wedge h = 0$
27: $\quad$ **end if**
28: $\quad$ **if** $\mathcal{H}' \neq \mathcal{H}$ **then**
29: $\quad\quad \mathcal{H} \leftarrow \mathcal{H}'$; $\mathcal{S} \leftarrow \mathcal{H}$
30: $\quad$ **end if**
31: **end while**
32: **return** $\mathcal{H}$

$\texttt{InvHeadAutoReduce}_{L_x,\prec_x}$ *is substituted by Algorithm 5.4, then the completion will terminate with a weak involutive basis of* $\mathcal{I} = \langle \mathcal{F} \rangle$ *for any finite input set* $\mathcal{F} \subset \mathcal{P}$. *Furthermore, the sets* $\mathrm{lc}_{\prec_x} \bar{\mathcal{H}}_{h,L_x}$ *form strong* L_y*-involutive bases of the* $\mathcal{R}$*-ideals generated by them for each* $h \in \mathcal{H}$.

Proof. The proof of the termination and of the correctness of the algorithm is as in Section 4.6. The only new claim is that the sets $\mathrm{lc}_{\prec_x} \bar{\mathcal{H}}_{h,L_x}$ are strongly L_y-involutive. This is a simple consequence of the fact that under the made assumption on the product in $\mathcal{P}$ the loop in Lines /5-13/ of Algorithm 5.4 leads to an involutive head autoreduction of these sets. Hence we indeed obtain strong involutive bases. □

Corollary 5.4.25. *If* L_x *is the Janet division, then every polynomial* $f \in \mathcal{I}$ *possesses a unique involutive standard representation* $f = \sum_{h \in \mathcal{H}} P_h \star h$ *where the coefficients satisfy* $P_h \in \Bbbk[Y_{L_y,\mathrm{lc}_{\prec_x}(\bar{\mathcal{H}}_{h,L_x}),\prec_y}(\mathrm{lc}_{\prec_x} h)][X_{L_x,\mathcal{H},\prec_x}(h)]$.

Proof. For the Janet division the only obstruction for $\mathcal{H}$ being a strong involutive basis is that some elements of it may have the same leading exponents. More precisely, for any $h \in \mathcal{H}$ we have $\mathcal{H}_{h,L_x} = \{h' \in \mathcal{H} \mid \mathrm{le}_{\prec_x} h' = \mathrm{le}_{\prec_x} h\}$. This immediately implies furthermore $\bar{\mathcal{H}}_{h,L_x} = \mathcal{H}_{h,L_x}$. By Theorem 5.4.24 the sets $\mathrm{lc}_{\prec_x} \bar{\mathcal{H}}_{h,L_x}$ form a strong L_y-involutive basis of the ideals they generate. Hence the claimed representation must be unique. □

5.5 Minimal Resolutions and Castelnuovo–Mumford Regularity

Recall that for a graded polynomial module $\mathcal{M}$ a graded free resolution is *minimal*, if all entries of the matrices corresponding to the maps $\phi_i : \mathcal{P}^{r_i} \to \mathcal{P}^{r_{i-1}}$ are of positive degree, i. e. no constant coefficients appear (cf. Definition B.2.32). If $\mathcal{U} \subseteq \mathcal{P}^m$ is a graded submodule, as we will assume throughout this section, then the free resolution (5.38) of $\mathcal{U}$, derived in the last section from a Pommaret basis, is obviously graded, too. However, in general, it is not minimal. In this section we will exploit the structures present in (5.38) to deduce several statements about the minimal resolution of $\mathcal{U}$ without explicitly computing it. In particular, we will show that (5.38) is always of minimal length and that the degree of a Pommaret basis for the degree reverse lexicographic order equals the Castelnuovo–Mumford regularity of $\mathcal{U}$. But our first goal consists of finding conditions under which the resolution (5.38) is minimal. A simple criterion for minimality that can be directly checked on the Pommaret basis $\mathcal{H}$ of $\mathcal{U}$ is provided by the next result.

Lemma 5.5.1. *The resolution* (5.38) *is minimal, if and only if all first syzygies* $\mathbf{S}_{\alpha;k}$ *are free of constant terms.*

Proof. One direction is of course trivial. Since (5.38) was obtained by iterating Theorem 5.4.10, it suffices for proving the converse to show that under the made assumption all second syzygies $\mathbf{S}_{\alpha;k_1,k_2}$ are free of constant terms, too. But this is easy to see: we have $\mathbf{S}_{\alpha;k_1,k_2} = x^{k_2}\mathbf{e}_{\alpha;k_1} - x^{k_1}\mathbf{e}_{\alpha;k_2} + \sum_{\gamma,\ell} c_{\gamma;\ell}\mathbf{e}_{\gamma,\ell}$ where every non-vanishing coefficient $c_{\gamma;\ell}$ is divisible by a coefficient $P_\beta^{(\alpha;k)}$ with $k = k_1$ or $k = k_2$ appearing in the first syzygy $\mathbf{S}_{\alpha;k}$ and thus is of positive degree. □

Theorem 5.5.2. *If the resolution* (5.38) *is minimal, then the submodule* $\mathcal{U} \subseteq \mathcal{P}^m$ *is componentwise linear.*[5]

Proof. Let $\mathcal{H}$ be the Pommaret basis of $\mathcal{U}$ and $d \geq 0$ an arbitrary degree. As in Lemma 4.3.2 it is easy to see that the set

$$\mathcal{G}_d = \{x^\mu h \mid h \in \mathcal{H},\ |\mu| + \deg h = d,\ \forall j > \mathrm{cls}\, h : \mu_j = 0\} \tag{5.59}$$

defines a $\Bbbk$-linear basis of the homogeneous component $\mathcal{U}_d$ and thus generates the module $\mathcal{U}_{\langle d\rangle} = \langle \mathcal{U}_d \rangle$. Consider now a product $x^j \bar{g}$ for some generator $\bar{g} = x^\mu \bar{h} \in \mathcal{G}_d$

[5] See Remark B.2.34 for a definition of this notion.

where $j > k = \operatorname{cls}\bar{g}$ so that x^j is non-multiplicative for $\bar{g}$. We must distinguish two cases. If $j \leq \operatorname{cls}\bar{h}$, then $\mathcal{G}_d$ also contains the generator $\tilde{g} = x^{\mu - 1_k + 1_j}\bar{h}$ and we have $x^j\bar{g} = x^k\tilde{g}$ where the latter product is multiplicative.

Otherwise, the variable x^j is non-multiplicative for the generator $\bar{h}$, too, and our resolution contains a first syzygy corresponding to an involutive standard representation $x^j\bar{h} = \sum_{h\in\mathcal{H}} P_h h$. If $|\mu| > 0$, then we can lift this equation to a standard representation $x^j\bar{g} = \sum_{h\in\mathcal{H}} P_h x^\mu h$. However, in general it will no longer be an involutive one, as the term x^μ may depend on variables which are non-multiplicative for some generators $h \in \mathcal{H}$. In this case, we must rewrite the right hand side using further first syzygies from (5.38). It is not difficult to see that after a finite number of such steps we also arrive at an involutive standard representation

$$x^j\bar{g} = \sum_{h\in\mathcal{H}} P_h h \tag{5.60}$$

where for notational simplicity we still denote the coefficients by P_h.

Assume now that the resolution (5.38) is minimal. Obviously, all first syzygies and thus also the coefficients P_h in (5.60) are then free of constant terms. But this observation implies that we can transform (5.60) into an involutive standard representation $x^j\bar{g} = \sum_{g\in\mathcal{G}_d} Q_g g$ with respect to $\mathcal{G}_d$ and hence this set is a Pommaret basis of the module $\mathcal{U}_{\langle d\rangle}$ by Proposition 4.2.7. As all elements of $\mathcal{G}_d$ are of degree d, it follows immediately from the form of (5.38) evaluated for $\mathcal{G}_d$ that $\mathcal{U}_{\langle d\rangle}$ has a linear resolution and thus $\mathcal{U}$ is componentwise linear. □

Example 5.5.3. The converse of Theorem 5.5.2 is not true, as the following trivial counterexample demonstrates. Consider the monomial ideal $\mathcal{I} = \langle x, y^2\rangle \subset \Bbbk[x,y]$. It is componentwise linear: $\mathcal{I}_{\langle 1\rangle} = \langle x\rangle$ is as principal ideal a free module; all ideals $\mathcal{I}_{\langle d\rangle}$ for $d > 1$ are simply generated by all monomials of degree d and thus possess trivially a linear resolution. For the natural ordering of variables $x_1 = x$ and $x_2 = y$, the Pommaret basis of $\mathcal{I}$ is $\mathcal{H} = \{x, xy, y^2\}$ and since the arising resolution contains the first syzygy $y\mathbf{e}_1 - \mathbf{e}_2$, it is not minimal. Comparing with the proof above, we see that $\mathcal{G}_1 = \{x\}$ is not a Pommaret basis of $\mathcal{I}_{\langle 1\rangle}$ (actually, $\mathcal{I}_{\langle 1\rangle}$ does not even possess a finite Pommaret basis, as it is not quasi-stable).

Note, however, that the situation changes, if we swap the ordering of the variables to $x_1 = y$ and $x_2 = x$. Now the minimal basis $\mathcal{H}' = \{x, y^2\}$ is already a Pommaret basis of $\mathcal{I}$ and the corresponding resolution is trivially minimal, as the only syzygy is $y^2\mathbf{e}_1 - x\mathbf{e}_2$. We will see later (Theorem 5.5.26) that this observation is no accident but that *generically* the resolution (5.38) is minimal for componentwise linear modules (and thus generically (5.39) yields an explicit formula for the Betti numbers of componentwise linear modules). ◁

For quasi-stable monomial modules $\mathcal{U}$ a simple combinatorial characterisation exists when our resolution is minimal. We will also provide a simple alternative characterisation via Pommaret bases.

Definition 5.5.4. A (possibly infinite) set $\mathcal{N} \subseteq \mathbb{N}_0^n$ is called *stable*, if for each multi index $\nu \in \mathcal{N}$ all multi indices $\nu - 1_k + 1_j$ with $k = \operatorname{cls}\nu < j \leq n$ are also contained in the set $\mathcal{N}$. A monomial submodule $\mathcal{U} \subseteq \mathcal{P}^m$ is called *stable*, if each of the sets $\mathcal{N}_\alpha = \{\mu \mid x^\mu \mathbf{e}_\alpha \in \mathcal{U}\} \subseteq \mathbb{N}_0^n$ with $1 \leq \alpha \leq m$ is stable.

Remark 5.5.5. The stable modules are of considerable interest, as they contain as a subset the *Borel fixed modules*, i. e. modules which remain invariant under linear coordinate transformations $\tilde{\mathbf{x}} = A\mathbf{x}$ with a matrix A contained in the Borel group of the invertible lower triangular matrices.[6] Indeed, one can show that (for a coefficient field of characteristic 0) a module is Borel fixed, if and only if it can be generated by a set $\mathcal{S}$ of monomials such that whenever $x^\nu \mathbf{e}_j \in \mathcal{S}$ then also $x^{\nu - 1_k + 1_j}\mathbf{e}_j \in \mathcal{S}$ for all $\operatorname{cls}\nu \leq k < j \leq n$ [125, Theorem 15.23]. Generically, the leading terms of any polynomial module form a Borel fixed module [125, Theorem 15.20][146]. Note that while the definition of stability is obviously independent of the characteristic of the coefficient field, the same is not true for the notion of a Borel fixed module. ◁

Every monomial submodule has a unique minimal basis. For stable submodules it must coincide with the Pommaret basis. This fact represents a very simple and effective criterion of stability. Furthermore, it shows that any stable submodule is quasi-stable and thus explains the terminology introduced in Definition 5.3.3.

Proposition 5.5.6. *Let $\mathcal{U} \subseteq \mathcal{P}^m$ be a monomial submodule. Then $\mathcal{U}$ is stable, if and only if its minimal basis $\mathcal{H}$ is simultaneously a Pommaret basis.*

Proof. Let us assume first that $\mathcal{U}$ is stable; we have to show that $\langle\mathcal{H}\rangle_P = \mathcal{U}$. For every term $\mathbf{s} \in \mathcal{U}$ a unique term $\mathbf{t}_1 \in \mathcal{H}$ exists such that $\mathbf{s} = s_1\mathbf{t}_1$ for some term $s_1 \in \mathbb{T}$. If $s_1 \in \mathbb{k}[X_P(\mathbf{t}_1)]$, we are done. Otherwise there exists an index $j > k = \operatorname{cls}\mathbf{t}_1$ such that $x_j \mid s_1$ and we rewrite

$$\mathbf{s} = \left(\frac{x_k}{x_j}s_1\right)\left(\frac{x_j}{x_k}\mathbf{t}_1\right) . \tag{5.61}$$

Since, by assumption, $\mathcal{U}$ is stable, $(x_j/x_k)\mathbf{t}_1 \in \mathcal{U}$. Thus a term $\mathbf{t}_2 \in \mathcal{H}$ exists such that $(x_j/x_k)\mathbf{t}_1 = s_2\mathbf{t}_2$ for some term $s_2 \in \mathbb{T}$. We are done, if $\mathbf{t}_2 \mid_P \mathbf{s}$. Otherwise we iterate this construction. By the continuity of the Pommaret division, this process cannot go on infinitely, i. e. after a finite number of steps we must reach a term $\mathbf{t}_N \in \mathcal{H}$ such that $\mathbf{t}_N \mid_P \mathbf{s}$ and thus $\mathbf{s} \in \langle\mathcal{H}\rangle_P$.

For the converse, assume that the minimal basis $\mathcal{H}$ is a Pommaret basis. Then a unique element $\mathbf{t} \in \mathcal{H}$ exists for each $\mathbf{s} \in \mathcal{U}$ such that $\mathbf{t} \mid_P \mathbf{s}$. We must show that with $k = \operatorname{cls}\mathbf{s} \leq \operatorname{cls}\mathbf{t}$ for all $i > k$ the terms $(x_i/x_k)\mathbf{s}$ are also elements of $\mathcal{U}$. We distinguish two cases. If $\mathbf{s} = \mathbf{t}$, a $\bar{\mathbf{t}} \in \mathcal{H}$ exists with $\bar{\mathbf{t}} \mid_P (x_i\mathbf{t})$. As $\mathcal{H}$ is a minimal basis, it cannot be that $\bar{\mathbf{t}} = x_i\mathbf{t}$. Instead we must have that $\bar{\mathbf{t}} \mid (x_i/x_k)\mathbf{t} = (x_i/x_k)\mathbf{s}$ and we are done. If $\mathbf{s} \neq \mathbf{t}$, we write $\mathbf{s} = s\mathbf{t}$ with $s \in \mathbb{T}$. If $k < \operatorname{cls}\mathbf{t}$, then $x_k \mid s$ which implies that we can divide by x_k and thus $(x_i/x_k)\mathbf{s} \in \mathcal{U}$. Otherwise, $\operatorname{cls}\mathbf{s} = \operatorname{cls}\mathbf{t}$ and we know from the first case that $(x_i/x_k)\mathbf{t} \in \mathcal{U}$. But $(x_i/x_k)\mathbf{s} = (x_i/x_k)s\mathbf{t}$. □

As an immediate corollary, we obtain the following, perhaps surprising result about the relation between the minimal Pommaret basis and the reduced Gröbner basis of a polynomial module.

---

[6] Classically, the Borel group consists of upper triangular matrices. In our "inverse" conventions we must take lower triangular matrices.

Theorem 5.5.7. *Let $\mathcal{U} \subseteq \mathcal{P}^m$ be a graded submodule in generic position and $\mathcal{G}$ the reduced Gröbner basis of $\mathcal{U}$ for an arbitrary term order $\prec$. If* $\operatorname{char} \Bbbk = 0$*, then $\mathcal{G}$ is also the minimal Pommaret basis of $\mathcal{I}$ for $\prec$.*

Proof. The generic initial module of any module is Borel fixed which implies in particular that it is stable for $\operatorname{char} \Bbbk = 0$ (Remark 5.5.5). By definition of a reduced Gröbner basis, $\operatorname{le}_\prec \mathcal{G}$ is the minimal basis of $\operatorname{le}_\prec \mathcal{U}$. But as $\operatorname{le}_\prec \mathcal{U}$ is generically stable, this fact implies that $\operatorname{le}_\prec \mathcal{G}$ is a Pommaret basis of $\operatorname{le}_\prec \mathcal{U}$ and thus $\mathcal{G}$ is a Pommaret basis of $\mathcal{U}$. As the Pommaret division is global, the set $\mathcal{G}$ is automatically the unique minimal basis. □

Theorem 5.5.8. *Let $\mathcal{U} \subseteq \mathcal{P}^m$ be a quasi-stable submodule. Then the syzygy resolution given by* (5.38) *is minimal, if and only if $\mathcal{U}$ is stable. Hence any stable submodule is componentwise linear.*

Proof. According to Lemma 5.5.1, the resolution (5.38) is minimal, if and only if all first syzygies are free of constant terms. For a monomial module this is the case, if and only if already the minimal basis is the Pommaret basis, since otherwise the Pommaret basis contains generators $\mathbf{h}_1$, $\mathbf{h}_2$ related by $\mathbf{h}_2 = x^j \mathbf{h}_1$ for some non-multiplicative variable x^j leading to a first syzygy $x^j \mathbf{e}_1 - \mathbf{e}_2$ with a constant term. Now our claim follows from Proposition 5.5.6. □

Example 5.5.9. One might be tempted to conjecture that this result extended to polynomial modules, i. e. that the resolution (5.38) was minimal for all polynomial modules $\mathcal{U}$ with a stable leading module $\operatorname{lt}_\prec \mathcal{U}$. Unfortunately, this is not true. Consider the homogeneous ideal $\mathcal{I} \subset \Bbbk[x,y,z]$ generated by $h_1 = z^2 + xy$, $h_2 = yz - xz$, $h_3 = y^2 + xz$, $h_4 = x^2 z$ and $h_5 = x^2 y$. One easily checks that these elements form a Pommaret basis $\mathcal{H}$ for the degree reverse lexicographic term order and that $\operatorname{lt}_\prec \mathcal{U}$ is a stable ideal. A Pommaret basis of $\operatorname{Syz}(\mathcal{H})$ is given by

$$\mathbf{S}_{2;3} = z\mathbf{e}_2 + (x-y)\mathbf{e}_1 + x\mathbf{e}_3 - \mathbf{e}_4 - \mathbf{e}_5 \,, \tag{5.62a}$$
$$\mathbf{S}_{3;3} = z\mathbf{e}_3 - x\mathbf{e}_1 - (x+y)\mathbf{e}_2 - \mathbf{e}_4 + \mathbf{e}_5 \,, \tag{5.62b}$$
$$\mathbf{S}_{4;3} = z\mathbf{e}_4 - x^2\mathbf{e}_1 + x\mathbf{e}_5 \,, \tag{5.62c}$$
$$\mathbf{S}_{5;3} = z\mathbf{e}_5 - x^2\mathbf{e}_2 - x\mathbf{e}_4 \,, \tag{5.62d}$$
$$\mathbf{S}_{4;2} = (y-x)\mathbf{e}_4 - x^2\mathbf{e}_2 \,, \tag{5.62e}$$
$$\mathbf{S}_{5;2} = y\mathbf{e}_5 - x^2\mathbf{e}_3 + x\mathbf{e}_4 \,. \tag{5.62f}$$

As the first two generators show, the resolution (5.38) is not minimal. ◁

Remark 5.5.10. It follows from Lemma 4.3.3, that if $\mathcal{H}$ is a Pommaret basis of the submodule $\mathcal{U} \subseteq \mathcal{P}^m$ of degree q, then $(\operatorname{lt}_\prec \mathcal{U})_{\geq q}$ is a stable monomial submodule, as $\operatorname{lt}_\prec \mathcal{H}_q$ is obviously its minimal basis and simultaneously a Pommaret basis. ◁

We return now to general polynomial submodules $\mathcal{U} \subseteq \mathcal{P}^m$. Like for any graded free resolution, it is a standard task to reduce (5.38) to the minimal resolution using just some linear algebra (see Appendix B.4 for a detailed discussion). However, even in the monomial case it seems highly non-trivial to find a closed form description of the outcome of the minimisation process. Nevertheless, certain statements are possible. In particular, we show now that (5.38) is always of minimal length and hence we can trivially read off a Pommaret basis the projective dimension of $\mathcal{U}$.

Theorem 5.5.11. *Let $\mathcal{U} \subseteq \mathcal{P}^m$ be a graded submodule and $\mathcal{H}$ its Pommaret basis for a class respecting term order. If we set $d = \min_{\mathbf{h}\in\mathcal{H}} \operatorname{cls}\mathbf{h}$, then the projective dimension of $\mathcal{U}$ is* $\operatorname{proj\,dim}\mathcal{U} = n - d$.

Proof. Consider the resolution (5.38) which is of length $n-d$. The last map in it is defined by the syzygies $\mathbf{S}_{\alpha;(d+1,\dots,n)}$ originating in the generators $\mathbf{h}_\alpha \in \mathcal{H}$ with $\operatorname{cls}\mathbf{h}_\alpha = d$. Choose now among these generators an element $\mathbf{h}_\gamma$ of maximal degree (recall that the same choice was crucial in the proof of Proposition 5.2.7). The syzygy $\mathbf{S}_{\gamma;(d+1,\dots,n)}$ cannot contain a constant coefficient, as the coefficients of all vectors $\mathbf{e}_{\beta;\mathbf{k}}$ where the last entry of $\mathbf{k}$ is n must be contained in $\langle x_1,\dots,x_{n-1}\rangle$ and the coefficients of the vectors $\mathbf{e}_{\alpha;(d+1,\dots,n-1)}$ cannot be constant for degree reasons.

If we start now a minimisation process at the end of the resolution, then it will never introduce a constant term into the syzygy $\mathbf{S}_{\gamma;(d+1,\dots,n)}$ and thus it will never be eliminated. It is also not possible that it is reduced to zero, as the last map in a free resolution is obviously injective. This observation implies that the last term of the resolution will not vanish during the minimisation and the length of the minimal resolution, i. e. $\operatorname{proj\,dim}\mathcal{U}$, is still $n-d$. □

The (graded) *Auslander–Buchsbaum formula* arises as a trivial corollary of this theorem and Proposition 5.2.7 on the depth. In contrast to other proofs, our approach is constructive in the sense that we automatically have an explicit regular sequence of maximal length and a (partially) explicit free resolution of minimal length.

Corollary 5.5.12 (Auslander–Buchsbaum). *If $\mathcal{U} \subseteq \mathcal{P}^m$ is a graded submodule with $\mathcal{P} = \Bbbk[x^1,\dots,x^n]$, then* $\operatorname{depth}\mathcal{U} + \operatorname{proj\,dim}\mathcal{U} = n$.

As for a monomial module we do not need a term order, we obtain as a further simple corollary the following relation between the projective dimensions of a polynomial module $\mathcal{U}$ and its leading module $\operatorname{lt}_\prec \mathcal{U}$, respectively.

Corollary 5.5.13. *Let $\mathcal{U} \subseteq \mathcal{P}^m$ be a graded submodule and $\prec$ any term order for which $\mathcal{U}$ possesses a Pommaret basis $\mathcal{H}$. Then* $\operatorname{proj\,dim}\mathcal{U} \leq \operatorname{proj\,dim}(\operatorname{lt}_\prec \mathcal{U})$. *If $\prec$ is a class respecting term order, then we even have equality.*

Proof. Let $\mathcal{H}$ be the Pommaret basis of the polynomial submodule $\mathcal{U}$ for the term order $\prec$ and set $d = \min_{\mathbf{h}\in\mathcal{H}} \operatorname{cls}(\operatorname{lt}_\prec \mathbf{h})$. Then it follows immediately from Theorem 5.5.11 that $\operatorname{proj\,dim}(\operatorname{lt}_\prec \mathcal{U}) = n-d$. On the other hand, Theorem 5.4.12 guarantees the existence of the free resolution (5.38) of length $n-d$ for $\mathcal{U}$ so that this value is an upper bound for $\operatorname{proj\,dim}\mathcal{U}$. For a class respecting term order we have equality by Theorem 5.5.11. □

For notational simplicity, we restrict now again to ideals instead of submodules. In many situations it is of interest to obtain an estimate on the degree of an ideal basis, in particular for the complexity analysis of completion algorithms. Up to date, no simple answer is known to this question. Somewhat surprisingly, the stronger problem of bounding not only the degree of a basis of $\mathcal{I}$ but also of its syzygies can be treated effectively.

Definition 5.5.14. Let $\mathcal{I} \subseteq \mathcal{P}$ be a homogeneous ideal. $\mathcal{I}$ is called *q-regular*, if its ith syzygy module $\mathrm{Syz}^i(\mathcal{I})$ can be generated by elements of degree less than or equal to $q+i$. The *Castelnuovo–Mumford regularity* $\operatorname{reg}\mathcal{I}$ is the least value q for which the ideal $\mathcal{I}$ is q-regular.

As discussed in more detail in Remark B.4.20, among other applications the Castelnuovo–Mumford regularity $\operatorname{reg}\mathcal{I}$ is an important measure for the complexity of Gröbner bases computations. We now show that $\operatorname{reg}\mathcal{I}$ is trivially determined by a Pommaret basis with respect to the degree reverse lexicographic order and then provide several alternative characterisations of it.

Theorem 5.5.15. *Let $\mathcal{I} \subseteq \mathcal{P}$ be a homogeneous ideal. The Castelnuovo–Mumford regularity of $\mathcal{I}$ is $\operatorname{reg}\mathcal{I} = q$, if and only if $\mathcal{I}$ has in some coordinates a Pommaret basis of degree q for the degree reverse lexicographic order.*

Proof. Let $\mathbf{x}$ be some δ-regular coordinates for the ideal $\mathcal{I}$ so that it possess a Pommaret basis $\mathcal{H}$ of degree q with respect to the degree reverse lexicographic order in these coordinates. Then the ith module of the syzygy resolution (5.38) induced by the basis $\mathcal{H}$ is obviously generated by elements of degree less than or equal to $q+i$. Thus we have the trivial estimate $\operatorname{reg}\mathcal{I} \leq q$ and there only remains to show that it is in fact an equality.

For this purpose, consider a generator $h_\gamma \in \mathcal{H}$ of degree q which is of minimal class among all elements of this maximal degree q in $\mathcal{H}$. If $\operatorname{cls} h_\gamma = n$, then h_γ cannot be removed from $\mathcal{H}$ without loosing the basis property, as the leading term of no other generator of class n can divide $\operatorname{lt}_\prec h_\gamma$ and, since we are using the degree reverse lexicographic order, all other generators do not contain any terms of class n. Hence we trivially find $\operatorname{reg}\mathcal{I} = q$ in this case.

If $\operatorname{cls} h_\gamma = n-i$ for some $i > 0$, then the resolution (5.38) contains at the ith position the syzygy $\mathbf{S}_{\gamma;(n-i+1,\dots,n)}$ of degree $q+i$. Assume now that we minimise the resolution step by step starting at the end. We claim that the syzygy $\mathbf{S}_{\gamma;(n-i+1,\dots,n)}$ is not eliminated during this process.

There are two possibilities how $\mathbf{S}_{\gamma;(n-i+1,\dots,n)}$ could be removed during the minimisation. The first one is that a syzygy at the next level of the resolution contained the term $\mathbf{e}_{\gamma;(n-i+1,\dots,n)}$ with a constant coefficient. Any such syzygy is of the form (5.42) with $\operatorname{cls} h_\alpha < n-i$ and $\operatorname{cls} h_\alpha < k_1 < \cdots < k_i < n$ and its leading term is $x_{k_{i+1}}\mathbf{e}_{\alpha;\mathbf{k}}$ with $k_{i+1} > k_i$. However, since $\operatorname{cls}(x_{k_1}\cdots x_{k_{i+1}}h_\alpha) < n-i$ and $\operatorname{cls}(x_{n-i+1}\cdots x_n h_\gamma) = n-i$, it follows from our use of the degree reverse lexicographic order (since we assume that everything is homogeneous, both polynomials have the same degree) and the definition of the induced term orders, that the term $\mathbf{e}_{\gamma;(n-i+1,\dots,n)}$ is greater than the leading term $x_{k_{i+1}}\mathbf{e}_{\alpha;\mathbf{k}}$ of any syzygy $\mathbf{S}_{\alpha;(k_1,\dots,k_{i+1})}$ at the level $i+1$ and thus cannot appear.

The second possibility is that $\mathbf{S}_{\gamma;(n-i+1,\dots,n)}$ itself contained a constant coefficient at some vector $\mathbf{e}_{\beta;\ell}$. However, this required $\deg h_\beta = \deg h_\gamma + 1$ which is again a contradiction.[7] As the minimisation process never introduces new constant coefficients, the syzygy $\mathbf{S}_{\gamma;(n-i+1,\dots,n)}$ may only be modified but not eliminated. Furthermore, the modifications cannot make $\mathbf{S}_{\gamma;(n-i+1,\dots,n)}$ to the zero syzygy, as otherwise a basis vector of the next level was in the kernel of the differential. However, this is not possible, as we assume that the tail of the resolution is already minimised and by the exactness of the sequence any kernel member must be a linear combination of syzygies. Hence the final minimal resolution will contain at the ith position a generator of degree $q+i$ and $\operatorname{reg}\mathcal{I} = q$. □

To some extent this result was to be expected. We know from Theorem 5.5.7 that generically the reduced Gröbner basis is also a Pommaret basis and, according to Bayer and Stillman [34], this basis has for the degree reverse lexicographic order generically the degree $\operatorname{reg}\mathcal{I}$. Thus the only surprise is that Theorem 5.5.15 does not require that the leading ideal is stable and the Pommaret basis $\mathcal{H}$ is not necessarily a reduced Gröbner basis (however, if the ideal $\mathcal{I}$ has a Pommaret basis of degree q, then the truncated ideal $(\operatorname{le}_\prec \mathcal{I})_{\geq q}$ is stable by Remark 5.5.10 and thus the set $\mathcal{H}_q$ defined by (4.5) is the reduced Gröbner basis of $\mathcal{I}_{\geq q}$).

Note furthermore that Theorem 5.5.15 implies a quite remarkable fact: in arbitrary coordinates $\mathbf{x}$ the ideal $\mathcal{I}$ either does not possess a finite Pommaret basis for the degree reverse lexicographic order or the basis is of degree $\operatorname{reg}\mathcal{I}$. Hence using Pommaret bases for this particular order, it becomes trivial to determine the Castelnuovo–Mumford regularity: it is just the degree of the basis.

Remark 5.5.16. The proof of Theorem 5.5.15 also provides us with information about the positions at which the maximal degree is attained in the minimal resolution. We only have to look for all elements of maximal degree in the Pommaret basis; their classes correspond to these positions. ◁

Remark 5.5.17. Recall from Remark 5.4.13 that Theorem 5.4.12 remains valid for any involutive basis $\mathcal{H}$ with respect to a continuous division of Schreyer type (with an obvious modification of the definition of the numbers $\beta_0^{(k)}$) and that it is independent of the used term order. It follows immediately from the form of the resolution (5.38), i. e. from the form of the maps in it given by the respective involutive bases according to Theorem 5.4.10, that the estimate $\operatorname{reg}\mathcal{I} \leq \deg\mathcal{H}$ holds and thus any such basis provides us with a bound for the Castelnuovo–Mumford regularity.

This observation also implies that an involutive basis with respect to a division of Schreyer type and an arbitrary term order can never be of lower degree than the Pommaret basis for the degree reverse lexicographic order. The latter one is thus in this sense optimal. As a concrete example consider again the ideal mentioned in

[7] For later use we note the following fact about this argument. If $\mathbf{e}_{\beta;\ell}$ is a constant term in the syzygy $\mathbf{S}_{\gamma;(n-i+1,\dots,n)}$, then it must be smaller than the leading term and hence $\operatorname{lt}_\prec(x_{\ell_1}\cdots x_{\ell_i}h_\beta) \prec \operatorname{lt}_\prec(x_{k_1}\cdots x_{k_{i+1}}h_\alpha)$ implying that $\operatorname{cls} h_\beta \leq \operatorname{cls} h_\gamma$. Thus it suffices, if h_γ is of maximal degree among all generators $h_\beta \in \mathcal{H}$ with $\operatorname{cls} h_\beta \leq \operatorname{cls} h_\gamma$. For the special case that h_γ is of minimal class, we exploited this observation already in the proof of Theorem 5.5.11.

Remark 5.2.8: in "good" coordinates a Pommaret basis of degree 2 exists for it and after a simple permutation of the variables its Janet basis is of degree 4. ◁

In analogy to the proof of Corollary 5.5.13 comparing the projective dimensions of a polynomial submodule $\mathcal{U} \subseteq \mathcal{P}^m$ and of its leading module $\operatorname{lt}_\prec \mathcal{U}$ with respect to an arbitrary term order $\prec$, we may derive a similar estimate for the Castelnuovo–Mumford regularity.

Corollary 5.5.18. *Let $\mathcal{I} \subseteq \mathcal{P}$ be a homogeneous ideal and $\prec$ an arbitrary term order such that a Pommaret basis $\mathcal{H}$ of $\mathcal{I}$ exists. Then* $\operatorname{reg}\mathcal{I} \leq \operatorname{reg}(\operatorname{lt}_\prec \mathcal{I}) = \deg\mathcal{H}$*. If $\prec$ is the degree reverse lexicographic order, then even* $\operatorname{reg}\mathcal{I} = \operatorname{reg}(\operatorname{lt}_\prec \mathcal{I})$.

Proof. It follows from Theorem 5.5.15 that $\operatorname{reg}(\operatorname{lt}_\prec \mathcal{I}) = \deg\mathcal{H}$. On the other hand, the form of the resolution (5.38) implies trivially that $\operatorname{reg}\mathcal{I} \leq \deg\mathcal{H}$. For the degree reverse lexicographic order Theorem 5.5.15 entails that $\operatorname{reg}\mathcal{I} = \deg\mathcal{H}$, too. □

Specialising again to monomial ideals, a combination of the above results with Remark 5.5.10 and Proposition 5.5.6 immediately implies the following result.

Proposition 5.5.19. *Let $\mathcal{I} \subseteq \mathcal{P}$ be a quasi-stable ideal generated in degrees less than or equal to q. Then $\mathcal{I}$ is q-regular, if and only if the truncation $\mathcal{I}_{\geq q}$ is stable.*

Remark 5.5.20. A refinement of the Castelnuovo–Mumford regularity $\operatorname{reg}\mathcal{I}$ is provided by the *extremal Betti numbers*. Recall that the (graded) Betti number β_{ij} of the ideal $\mathcal{I}$ is defined as the number of minimal generators of degree $i+j$ of the ith module in the minimal free resolution of $\mathcal{I}$ (thus $\operatorname{reg}\mathcal{I}$ is the maximal value j such that $\beta_{i,i+j} > 0$ for some i). A Betti number $\beta_{ij} > 0$ is called extremal, if $\beta_{k\ell} = 0$ for all $k \geq i$ and $\ell > j$. There always exists a least one extremal Betti number: if we take the maximal value i for which $\beta_{i,i+\operatorname{reg}\mathcal{I}} > 0$, then $\beta_{i,i+\operatorname{reg}\mathcal{I}}$ is extremal. In general, there may exist further extremal Betti numbers. One can show that both their positions and their values coincide for any ideal and its generic initial ideal with respect to the degree reverse lexicographic order.

Our proof of Theorem 5.5.15 allows us to make the same statement for the ordinary initial ideal for $\prec_{\text{degrevlex}}$—provided the coordinates are δ-regular. Furthermore, it shows that the extremal Betti numbers of $\mathcal{I}$ can be immediately read off the Pommaret basis $\mathcal{H}$ of $\mathcal{I}$. Finally, if we introduce "pseudo-Betti numbers" for the (in general non-minimal) resolution (5.38), then the positions and values of the extremal ones coincide with the true extremal Betti numbers of $\mathcal{I}$.

Take the generator h_γ used in the proof of Theorem 5.5.15. If $\operatorname{cls} h_\gamma = n - i_1$ and $\deg h_\gamma = q_1$, then the considerations in the proof imply immediately that β_{i_1,q_1+i_1} is an extremal Betti number and its value is given by the number of generators of degree q_1 and class $n - i_1$ in the Pommaret basis $\mathcal{H}$. If $i_1 = \operatorname{depth}\mathcal{I}$, then this is the only extremal Betti number. Otherwise, let q_2 be the maximal degree of a generator $h \in \mathcal{H}$ with $\operatorname{cls} h < n - i_1$ and assume that $n - i_2$ is the minimal class of such a generator. Then the arguments used in the proof of Theorem 5.5.15 show that β_{i_2,q_2+i_2} is also an extremal Betti number and that its value is given by the number of generators of degree q_2 and class $n - i_2$ in the Pommaret basis $\mathcal{H}$. Continuing in this manner, we obtain all extremal Betti numbers. Since all our considerations depend only on the leading terms of the generators, we find exactly the same situation for the leading ideal $\operatorname{lt}_{\prec} \mathcal{I}$. ◁

We provide now some alternative characterisations of q-regularity. The first one may be considered as a more abstract rephrasing of the idea that for every product of a generator with one of its non-multiplicative variables an involutive standard representation should exist. In the next chapter we will rediscover this result in yet another formulation (see Remark 6.3.6).

Theorem 5.5.21. *Let $\mathcal{I} \subseteq \mathcal{P}$ be a homogeneous ideal which can be generated by elements of degree less than or equal to q. Then $\mathcal{I}$ is q-regular, if and only if for some value $0 \leq d \leq n$ linear forms $y^1, \dots, y^d \in \mathcal{P}_1$ exist such that*

$$\big(\langle \mathcal{I}, y^1, \dots, y^{j-1}\rangle : y^j\big)_q = \langle \mathcal{I}, y^1, \dots, y^{j-1}\rangle_q\,, \quad 1 \leq j \leq d\,, \tag{5.63a}$$

$$\langle \mathcal{I}, y^1, \dots, y^d\rangle_q = \mathcal{P}_q\,. \tag{5.63b}$$

Proof. Assume first that the conditions (5.63) are satisfied for some linear forms $y^1, \dots, y^d \in \mathcal{P}_1$ and choose variables $\mathbf{x}$ such that $x^i = y^i$ for $1 \leq i \leq d$. Let the finite set $\mathcal{H}_q$ be a basis of $\mathcal{I}_q$ as a vector space in triangular form with respect to the degree reverse lexicographic order, i. e. $\operatorname{lt}_{\prec} h_1 \neq \operatorname{lt}_{\prec} h_2$ for all $h_1, h_2 \in \mathcal{H}_q$. We claim that $\mathcal{H}_q$ is a Pommaret basis of the truncation $\mathcal{I}_{\geq q}$ implying that the full ideal $\mathcal{I}$ possesses a Pommaret basis of degree $q' \leq q$ and hence by Theorem 5.5.15 that $\operatorname{reg} \mathcal{I} \leq q$.

Let us write $\mathcal{H}_q = \{h_{k,\ell} \mid 1 \leq k \leq n,\ 1 \leq \ell \leq \ell_k\}$ where $\operatorname{cls} h_{k,\ell} = k$. A basis of the vector space $\langle \mathcal{I}, x^1, \dots, x^j\rangle_q$ is then given by all $h_{k,\ell}$ with $k > j$ and all terms in $\langle x^1, \dots, x^j\rangle_q$. We will now show that

$$\mathcal{H}_{q+1} = \{x^j h_{k,\ell} \mid 1 \leq j \leq k,\ 1 \leq k \leq n,\ 1 \leq \ell \leq \ell_k\} \tag{5.64}$$

is a basis of $\mathcal{I}_{q+1}$ as a vector space. This implies that $\mathcal{H}_q$ is locally involutive for the Pommaret division and thus involutive by Proposition 4.2.7. Since $\mathcal{I}$ is generated in degrees less than or equal to q, we have furthermore $\langle \mathcal{H}_q\rangle = \mathcal{I}_{\geq q}$ so that indeed $\mathcal{H}_q$ is a Pommaret basis of the ideal $\mathcal{I}_{\geq q}$.

Let $f \in \mathcal{I}_{q+1}$ and $\operatorname{cls} f = j$. By the properties of the degree reverse lexicographic order, this implies that $f = x^j \hat{f} + g$ with $\hat{f} \in \big(\Bbbk[x^j, \dots, x^n] \setminus \{0\}\big)_q$ and $g \in \big(\langle x^1, \dots, x^{j-1}\rangle\big)_{q+1}$ (cf. Lemma A.1.8). We distinguish two cases. The condition (5.63b) implies that $\big(\langle \mathcal{I}, x^1, \dots, x^d\rangle\big)_q = \mathcal{P}_q$. Thus if $j > d$, we may write $\hat{f} = \sum_{k=d+1}^{n} \sum_{\ell=1}^{\ell_k} c_{k,\ell} h_{k,\ell} + \hat{g}$ with $c_{k,\ell} \in \Bbbk$ and $\hat{g} \in \big(\langle x^1, \dots, x^d\rangle\big)_q$. We introduce $\hat{f}_0 = \sum_{k=j}^{n} \sum_{\ell=1}^{\ell_k} c_{k,\ell} h_{k,\ell}$ and $\hat{f}_1 = \sum_{k=d+1}^{j-1} \sum_{\ell=1}^{\ell_k} c_{k,\ell} h_{k,\ell} + \hat{g}$. Obviously, we have now $\hat{f} \in \big(\langle \mathcal{I}, x^1, \dots, x^{j-1}\rangle : x^j\big)_q$. If $j \leq d$, then the condition (5.63a) implies that actually $\hat{f} \in \langle \mathcal{I}, x^1, \dots, x^{j-1}\rangle_q$. Hence in this case we may decompose $\hat{f} = \hat{f}_0 + \hat{f}_1$ with $\hat{f}_0 = \sum_{k=j}^{n} \sum_{\ell=1}^{\ell_k} c_{k,\ell} h_{k,\ell}$ and $\hat{f}_1 \in \big(\langle x^1, \dots, x^{j-1}\rangle\big)_q$.

It is trivial that $\langle \mathcal{H}_{q+1}\rangle_{\Bbbk} \subseteq \mathcal{I}_{q+1}$ (here the linear span over $\Bbbk$ is considered and not over $\mathcal{P}$). We show by an induction over j that $\mathcal{I}_{q+1} \subseteq \langle \mathcal{H}_{q+1}\rangle_{\Bbbk}$. If $j = 1$, then $f = x^1 \hat{f}$

with $\hat{f} \in \mathcal{I}_q$. Thus $f \in \langle \mathcal{H}_{q+1} \rangle_{\Bbbk}$. If $j > 1$, then we write $f = f_0 + f_1$ with $f_0 = x^j \hat{f}_0$ and $f_1 = x^j \hat{f}_1 + g$ where the polynomials $\hat{f}_0$ and $\hat{f}_1$ have been defined above. By construction, $f_0 \in \langle \mathcal{H}_{q+1} \rangle_{\Bbbk}$, as x^j is multiplicative for all generators contained in $\hat{f}_0$, and $f_1 \in \mathcal{I}_{q+1}$ with $\operatorname{cls} f_1 < j$. According to our inductive hypothesis this fact implies that $f_1 \in \langle \mathcal{H}_{q+1} \rangle_{\Bbbk}$, too. Hence $\langle \mathcal{H}_{q+1} \rangle_{\Bbbk} = \mathcal{I}_{q+1}$.

Assume conversely that the ideal $\mathcal{I}$ is q-regular. Then, by Theorem 5.5.15, it possesses a Pommaret basis $\mathcal{H}$ of degree $\operatorname{reg} \mathcal{I} \leq q$ with respect to the degree reverse lexicographic order. We claim that by choosing $y^i = x^i$ for $1 \leq i \leq d = \dim \mathcal{P}/\mathcal{I}$ the conditions (5.63) are satisfied. For the second equality (5.63b), this fact follows immediately from Proposition 5.2.3 which shows that it actually holds already at degree $\operatorname{reg} \mathcal{I} \leq q$.

For the equality (5.63a) take a polynomial $f \in \big(\langle \mathcal{I}, x^1, \dots, x^{j-1} \rangle : x^j \big)_q$. By definition, we have then $x^j f \in \langle \mathcal{I}, x^1, \dots, x^{j-1} \rangle$. If $f \in \langle x^1, \dots, x^{j-1} \rangle$, then there is nothing to prove. Otherwise, a polynomial $g \in \langle x^1, \dots, x^{j-1} \rangle$ exists such that $x^j f - g \in \mathcal{I}$ and it obviously satisfies $\operatorname{cls}(x^j f - g) = j$. If we introduce the set $\mathcal{H}_{\geq j} = \{ h \in \mathcal{H} \mid \operatorname{cls} h \geq j \}$, the involutive standard representation of $x^j f - g$ induces an equation $x^j f = \sum_{h \in \mathcal{H}_{\geq j}} P_h h + \bar{g}$ where $\bar{g} \in x^j \langle x^1, \dots, x^{j-1} \rangle$ and $P_h \in \langle x^j \rangle$ (this is trivial if $\operatorname{cls} h > j$ and follows from $\deg h \leq q$ if $\operatorname{cls} h = j$). Thus we can divide by x^j and find that already $f \in \langle \mathcal{I}, x^1, \dots, x^{j-1} \rangle_q$. □

One can prove that in *generic* coordinates it is not possible to find a Gröbner basis of degree less than $\operatorname{reg} \mathcal{I}$ and that this estimate is sharp, as it is realised by bases with respect to the degree reverse lexicographic order. The restriction to the generic case is here essential, as for instance most monomial ideals are trivial counterexamples. Unfortunately, no effective criterion is known for deciding whether or not a given ideal is in generic position.

Example 5.5.22. Because of the above mentioned non-effective genericity condition, Theorem 5.5.21 is only of limited use for computing the Castelnuovo–Mumford regularity. Consider the homogeneous ideal

$$\mathcal{I} = \langle z^8 - wxy^6,\ y^7 - x^6 z,\ yz^7 - wx^7 \rangle \subset \Bbbk[w,x,y,z] \,. \tag{5.65}$$

The given basis of degree 8 is already a reduced Gröbner basis for the degree reverse lexicographic order. If we perform a simple permutation of the variables and consider $\mathcal{I}$ as an ideal in $\Bbbk[w,y,x,z]$, then we obtain for the same term order a reduced Gröbner basis of degree 50:

$$\begin{aligned} \{ y^7 - x^6 z,\ yz^7 - wx^7,\ z^8 - wxy^6,\ y^8 z^6 - wx^{13}, & \\ y^{15} z^5 - wx^{19},\ y^{22} z^4 - wx^{25},\ y^{29} z^3 - wx^{31}, & \\ y^{36} z^2 - wx^{37},\ y^{43} z - wx^{43},\ y^{50} - wx^{49} \} \,. & \end{aligned} \tag{5.66}$$

Unfortunately, neither coordinate system is generic for $\mathcal{I}$: as $\operatorname{reg} \mathcal{I} = 13$, one yields a basis of too low degree and the other one of too high degree.

Using Theorem 5.5.15, the determination of the Castelnuovo–Mumford regularity is trivial, as the first coordinate system is δ-regular. A Pommaret basis of $\mathcal{I}$

for the degree reverse lexicographic order is obtained by adding the polynomials $z^k(y^7 - x^6 z)$ with $1 \leq k \leq 6$ to the first basis and thus its degree is indeed 13. ◁

Example 5.5.23. Consider the ideal $\mathcal{I} = \langle x^q, y^q \rangle$ for an integer $q > 1$. Its Pommaret basis is $\mathcal{H} = \{x^q y^r \mid 0 \leq r < q\} \cup \{y^q\}$ implying that $\operatorname{reg} \mathcal{I} = 2q - 1$. Thus the ideal is not in general position and one may check that most non-trivial linear coordinate transformations lead to a basis whose completion yields a Gröbner basis of degree at least $2q - 1$. ◁

Another characterisation of q-regularity is based on the existence of a special kind of resolution for the truncation $\mathcal{I}_{\geq q}$.

Theorem 5.5.24. *The homogeneous ideal $\mathcal{I} \subseteq \mathcal{P}$ is q-regular, if and only if its truncation $\mathcal{I}_{\geq q}$ admits a linear resolution.*

Proof. If the ideal $\mathcal{I}$ is q-regular, then by Theorem 5.5.15 it possesses in suitable coordinates a Pommaret basis $\mathcal{H}$ of degree $\operatorname{reg} \mathcal{I} \leq q$. The set $\mathcal{H}_q$ defined by (4.5) is a Pommaret basis of the truncated ideal $\mathcal{I}_{\geq q}$ according to Lemma 4.3.2. Now it follows immediately from Theorem 5.4.10 that $\mathcal{I}_{\geq q}$ possesses a linear free resolution, as all syzygies in the resolution (5.38) derived from the basis $\mathcal{H}_q$ are necessarily homogeneous of degree 1.

The converse is trivial. The existence of a linear resolution for $\mathcal{I}_{\geq q}$ immediately implies that $\operatorname{reg} \mathcal{I}_{\geq q} = q$. Hence $\mathcal{I}_{\geq q}$ possesses a Pommaret basis of degree q by Theorem 5.5.15 entailing the existence of a Pommaret basis for $\mathcal{I}$ of degree $q' \leq q$. Hence, again by Theorem 5.5.15, $\operatorname{reg} \mathcal{I} = q' \leq q$. □

Remark 5.5.25. We are now finally in a position where we can finish the discussion started in Remark 4.3.17 on the effective construction of Pommaret bases. There we were not able to prove that after a finite number of coordinate transformations based on our criterion for asymptotic singularity (Theorem 4.3.12) one always arrives at a δ-regular coordinate system for a given homogeneous ideal $\mathcal{I} \subseteq \mathcal{P}$. Recall that our main problem in Remark 4.3.17 was that we do not have a bound for the degrees of either Pommaret or Janet bases of $\mathcal{I}$. Our results above do not provide us with such a bound, but it still turns out that we can prove the termination of our approach by studying what happens at the finite degree $q = \operatorname{reg} \mathcal{I}$.

We assume from now on that we are working with a class respecting order and with an infinite field $\Bbbk$. By the considerations in the proof above, our coordinates $\mathbf{x}$ are δ-regular, if and only if an involutively head autoreduced, $\Bbbk$-linear basis of $\mathcal{I}_q$ is also a Pommaret basis of $\mathcal{I}_{\geq q}$. Denote, as in Remark 4.3.7, by $\beta_q^{(k)}$ the number of elements of class k in such a basis. There we already noted that these numbers are invariants of $\mathcal{I}$, as they are in a one-to-one correspondence with the coefficients of the Hilbert polynomial $H_{\mathcal{I}}$.

Consider now some basis $\mathcal{H}$ arising during the completion process. It induces a subset $\mathcal{H}_q \subset \mathcal{I}_q$ by taking all Pommaret multiplicative multiples of elements up to degree q; let $\tilde{\beta}_q^{(k)}$ be the number of members of class k in it. If $\mathcal{H}$ is not a Pommaret basis, then a comparison of the values $\beta_q^{(k)}$ and $\tilde{\beta}_q^{(k)}$ starting with $k = n$ will sooner or later lead to a smaller value $\tilde{\beta}_q^{(k)}$; more precisely, we have the inequality $\sum_{k=1}^{n} k\tilde{\beta}_q^{(k)} \leq \sum_{k=1}^{n} k\beta_q^{(k)}$ with equality holding, if and only if $\mathcal{H}$ is a Pommaret basis.

Each completion step which adds an element of degree q or less increases the value of the sum $\sum_{k=1}^{n} k\tilde{\beta}_q^{(k)}$. Consider now the effect of a coordinate transformation of the form used in the proof of Theorem 4.3.12. All new terms arising on the right hand side of (4.7) are greater than the original one with respect to any class respecting term order. Thus in general we can expect that after such a transformation at least some leading terms of the new set $\mathcal{H}_q$ are greater than before. In fact, by the same argument as in the proof of Theorem 4.3.12, we even can be sure that after a finite number of transformations this will indeed be the case. But this observation implies that after a finite number of transformations the sum $\sum_{k=1}^{n} k\tilde{\beta}_q^{(k)}$ must increase and eventually we must obtain after a finite number of completion steps and coordinate transformations the right value for this sum implying that we have obtained δ-regular coordinates and a Pommaret basis. ◁

As a further corollary to Theorem 5.5.24, we provide a converse to Theorem 5.5.2 generalising Theorem 5.5.8 from the monomial case to polynomial submodules.

Theorem 5.5.26. *Let $\mathcal{U} \subseteq \mathcal{P}^m$ be a componentwise linear submodule with Pommaret basis $\mathcal{H}$. Then generically the resolution* (5.38) *is minimal. It is minimal, if and only if $\mathcal{H}$ is a minimal basis of $\mathcal{U}$.*

Proof. As Example 5.5.3 demonstrated, the problem is that generally the sets $\mathcal{G}_d$ defined by (5.59) are not Pommaret bases of the modules $\mathcal{U}_{\langle d\rangle}$ for all $d \geq 0$. According to Theorems 5.5.24 and 5.5.15, this is trivially the case for all degrees $d \geq q = \deg\mathcal{H}$, since for them $\mathcal{U}_{\langle d\rangle} = \mathcal{U}_{\geq d}$. Thus it suffices to consider the finitely many modules $\mathcal{U}_{\langle 0\rangle}, \dots, \mathcal{U}_{\langle q\rangle}$. By Corollary 4.3.16, generic coordinate systems are simultaneously δ-regular for all these modules (and then also for the whole submodule $\mathcal{U}$).

Let $\mathcal{H}$ be the Pommaret basis of $\mathcal{U}$ in such a coordinate system and consider the corresponding sets $\mathcal{G}_d$ for $0 \leq d \leq q$. By construction, $\mathcal{U}_{\langle d\rangle} = \langle \mathcal{G}_d \rangle$. According to our assumption, all modules $\mathcal{U}_{\langle d\rangle}$ possess linear resolutions and thus $\operatorname{reg}\mathcal{U}_{\langle d\rangle} = d$. Hence the Pommaret basis of $\mathcal{U}_{\langle d\rangle}$ is of degree d by Theorem 5.5.15, which is only possible, if it is already given by $\mathcal{G}_d$. If all sets $\mathcal{G}_d$ are involutive, then no first syzygy of the Pommaret basis $\mathcal{H}$ can contain a constant term and it follows from Lemma 5.5.1 that the resolution (5.38) is minimal. Furthermore, in this case $\mathcal{H}$ must be a minimal generating set of $\mathcal{U}$. Indeed, it is trivial that the elements of $\mathcal{H}$ of lowest degree are minimal generators and since no element of a higher degree d can be contained in a module $\mathcal{U}_{\langle d'\rangle}$ for any $d' < d$, it must also be a minimal generator. □

Remark 5.5.27. These considerations in the proof above can be exploited for effectively deciding whether a given submodule $\mathcal{U} \subseteq \mathcal{P}^m$ is componentwise linear. We compute a Pommaret basis $\mathcal{H}$ for $\mathcal{U}$, changing to δ-regular coordinates if necessary. If the resolution (5.38) determined by $\mathcal{H}$ is minimal, then $\mathcal{U}$ is componentwise linear by Theorem 5.5.2 (the minimality of the resolution is trivial to check with Lemma 5.5.1). Otherwise, there are first syzygies in (5.38) containing a constant term. Let $\mathbf{S}_{\alpha;\ell}$ be one of minimal degree. If $\deg \mathbf{h}_\alpha = d$, then obviously all modules $\mathcal{U}_{\langle d' \rangle}$ for degrees $d' < d$ possess linear resolutions (coming from their Pommaret bases $\mathcal{G}_d$). For analysing the module $\mathcal{U}_{\langle d \rangle}$, we take the corresponding set $\mathcal{G}_d$ and complete it to a Pommaret basis $\mathcal{H}_d$ (potentially performing further coordinate transformation). If $\deg \mathcal{H}_d = d$, then we recompute the Pommaret basis $\mathcal{H}$ of $\mathcal{U}$ in the new coordinates, which trivially are still δ-regular for $\mathcal{U}$ and all modules $\mathcal{U}_{\langle d' \rangle}$ with $d' < d$, and check again for minimality. In the case that obstructions in some degree $d < \bar{d} < \operatorname{reg} \mathcal{U}$ appear, we continue in the same manner. After a finite number of steps we either obtain a minimal resolution and $\mathcal{U}$ is componentwise linear or we find a degree d such that the module $\mathcal{U}_{\langle d \rangle}$ does not possess a linear resolution. $\triangleleft$

The Castelnuovo–Mumford regularity is closely related to certain saturations. Thus we discuss next how one can effectively determine the saturation $\mathcal{I}^{\text{sat}} = \mathcal{I} : \mathcal{P}_+^\infty$ of a homogeneous ideal $\mathcal{I} \subseteq \mathcal{P}$ using Pommaret bases.

Proposition 5.5.28. *Let $\mathcal{H}$ be a Pommaret basis of the homogeneous ideal $\mathcal{I}$ for the degree reverse lexicographic order. We introduce the sets $\mathcal{H}_1 = \{h \in \mathcal{H} \mid \operatorname{cls} h = 1\}$ and $\bar{\mathcal{H}}_1 = \{h/(x^1)^{\deg_{x^1} \operatorname{lt}_\prec h} \mid h \in \mathcal{H}_1\}$. Then $\bar{\mathcal{H}} = (\mathcal{H} \setminus \mathcal{H}_1) \cup \bar{\mathcal{H}}_1$ is a weak Pommaret basis of the saturation $\mathcal{I}^{\text{sat}}$.*

Proof. By the definition of the reverse lexicographic order, the leading term $\operatorname{lt}_\prec h$ has the lowest x^1-degree of all terms in a generator $h \in \mathcal{H}_1$. Thus the set $\bar{\mathcal{H}}_1$ is well-defined and does not contain a generator of class 1 anymore.

We first show that indeed $\bar{\mathcal{H}}_1 \subset \mathcal{I}^{\text{sat}}$. Introduce $d_1 = \max_{h \in \mathcal{H}_1} \{\deg_{x^1} \operatorname{lt}_\prec h\}$ and $\Delta = d_1 + \max_{h \in \mathcal{H}_1} \{\deg h\} - \min_{h \in \mathcal{H}_1} \{\deg h\}$. We claim that $\bar{h} \cdot \mathcal{P}_\Delta \subset \mathcal{I}$ for every generator $\bar{h} \in \bar{\mathcal{H}}_1$. Thus let $x^\mu \in \mathcal{P}_\Delta$ and choose $k \in \mathbb{N}_0$ such that $(x^1)^k \bar{h} \in \mathcal{H}_1$; obviously, we have $k \leq d_1$. Since the polynomial $x^\mu (x^1)^k \bar{h}$ lies in $\mathcal{I}$, it possesses an involutive standard representation of the form

$$x^\mu (x^1)^k \bar{h} = \sum_{h \in \mathcal{H} \setminus \mathcal{H}_1} P_h h + \sum_{h \in \mathcal{H}_1} Q_h h \tag{5.67}$$

with $P_h \in \mathbb{k}[x^1, \dots, x^{\operatorname{cls} h}]$ and $Q_h \in \mathbb{k}[x^1]$.

The left hand side of this equation is contained in $\langle (x^1)^k \rangle$ and thus also the right hand side. Analysing an involutive normal form computation leading to the representation (5.67), one immediately sees that this implies that all coefficients P_h (since here $\operatorname{cls} h > 1$) and all summands $Q_h h$ lie in $\langle (x^1)^k \rangle$. As a first consequence of this representation we observe that for any monomial x^μ (not necessarily of degree Δ) we may divide (5.67) by $(x^1)^k$ and then obtain an involutive standard representation of $x^\mu \bar{h}$ with respect to the set $\bar{\mathcal{H}}$; hence this set is indeed weakly involutive for the Pommaret division and the given term order.

If $x^\mu \in \mathcal{P}_\Delta$, then we find for any $h \in \mathcal{H}_1$ that $|\deg \bar{h} - \deg h| \leq \Delta$ and hence $\deg Q_h = \deg\left(x^\mu (x^1)^k \bar{h}\right) - \deg h \geq k$. Since $Q_h \in \mathbb{k}[x^1]$, this implies that under the made assumption on x^μ already the coefficient Q_h lies in $\langle (x^1)^k \rangle$ so that the product $x^\mu \bar{h}$ possesses an involutive standard representation with respect to $\mathcal{H}$ and thus is contained in the ideal $\mathcal{I}$ as claimed.

Now we show that every polynomial $f \in \mathcal{I}^{\text{sat}}$ may be decomposed into an element of $\mathcal{I}$ and a linear combination of elements of $\bar{\mathcal{H}}_1$. We may write $f = \tilde{f} + g$ where $\tilde{f}$ is the involutive normal form of f with respect to $\mathcal{H}$ and $g \in \mathcal{I}$. If $\tilde{f} = 0$, then already $f \in \mathcal{I}$ and nothing is to be shown. Hence we assume that $\tilde{f} \neq 0$. By definition of the saturation $\mathcal{I}^{\text{sat}}$, there exists a $k \in \mathbb{N}_0$ such that $\tilde{f} \cdot \mathcal{P}_k \subset \mathcal{I}$, hence in particular $(x^1)^k \tilde{f} \in \mathcal{I}$. This implies that $\text{lt}_\prec\big((x^1)^k \tilde{f}\big) \in \langle \text{lt}_\prec \mathcal{H} \rangle_P$. Therefore a unique generator $h \in \mathcal{H}$ exists with $\text{lt}_\prec h \,|_P\, \text{lt}_\prec\big((x^1)^k \tilde{f}\big)$.

So let $\text{lt}_\prec\big((x^1)^k \tilde{f}\big) = x^\mu \,\text{lt}_\prec h$ and assume first that $\text{cls}\, h > 1$. Since the term on the left hand side is contained in $\langle (x^1)^k \rangle$, we must have $\mu_1 \geq k$ so that we can divide by $(x^1)^k$. But this implies that already $\text{lt}_\prec \tilde{f} \in \langle \text{lt}_\prec \mathcal{H} \rangle_P$ contradicting our assumption that $\tilde{f}$ is in involutive normal form. Hence we must have $\text{cls}\, h = 1$ and by the same argument as above $\mu_1 < k$.

Division by $(x^1)^k$ shows that $\text{lt}_\prec \tilde{f} \in \langle \text{lt}_\prec \bar{\mathcal{H}}_1 \rangle_P$. Performing the corresponding involutive reduction leads to a new element $f_1 \in \mathcal{I}^{\text{sat}}$. We compute again its involutive normal form $\tilde{f}_1$ and apply the same argument as above, if $\tilde{f}_1 \neq 0$. After a finite number of such reductions we obtain an involutive standard representation of f with respect to the set $\bar{\mathcal{H}}$ proving our assertion. □

By Proposition 3.4.7, an involutive head autoreduction of the set $\bar{\mathcal{H}}$ yields a strong Pommaret basis for the saturation $\mathcal{I}^{\text{sat}}$. As a trivial consequence of the considerations in the proof above we find that in δ-regular coordinates $\mathcal{I}^{\text{sat}}$ is simply given by the quotient $\mathcal{I} : \langle x^1 \rangle^\infty$. This observation in turn implies immediately that for degrees $q \geq \deg \mathcal{H}_1$ we have $\mathcal{I}_q = \mathcal{I}_q^{\text{sat}}$ and that $\deg \mathcal{H}_1$ is the lowest value having this property, i. e. the satiety $\text{sat}\, \mathcal{I}$.

Corollary 5.5.29. *Let the set $\mathcal{H}$ be a Pommaret basis of the homogeneous ideal $\mathcal{I} \subseteq \mathcal{P}$ for the degree reverse lexicographic order. Then $\mathcal{I}$ is saturated, if and only if $\mathcal{H}_1 = \emptyset$. If $\mathcal{I}$ is not saturated, then* $\text{sat}\, \mathcal{I} = \deg \mathcal{H}_1$. *Independent of the existence of a Pommaret basis, we have for any homogeneous generating set $\mathcal{F}$ of the socle $\mathcal{I} : \mathcal{P}_+$ the equality*

$$\text{sat}\, \mathcal{I} = 1 + \max\{\deg f \mid f \in \mathcal{F} \wedge f \notin \mathcal{I}\}\,. \tag{5.68}$$

Proof. Except of the last statement, everything has already been proven in the discussion above. For its proof we may assume without loss of generality that the coordinates are δ-regular so that a Pommaret basis $\mathcal{H}$ of $\mathcal{I}$ exists, as all quantities appearing in (5.68) are invariant under linear coordinate transformations.

Let $\tilde{h}$ be an element of $\mathcal{H}_1$ of maximal degree. We claim that $\tilde{h}/x^1 \in \mathcal{I} : \mathcal{P}_+$. Indeed, if we analyse for any $1 < \ell \leq n$ the involutive standard representation of $x^\ell \tilde{h}$, then all coefficients of generators $h \in \mathcal{H} \setminus \mathcal{H}_1$ are trivially contained in $\langle x^1 \rangle$ and for the coefficients of elements $h \in \mathcal{H}_1$ the same holds for degree reasons. Hence we can divide by x^1 and find that $x^\ell \tilde{h}/x^1 \in \mathcal{I}$ for all $1 \leq \ell \leq n$. Since $\mathcal{I} : \mathcal{P}_+ \subseteq \mathcal{I} : \langle x^1 \rangle$, all generators $f \in \mathcal{F}$ with $f \notin \mathcal{I}$ must be linear combinations of polynomials of the form $\tilde{h}/x^1$ with $\tilde{h} \in \mathcal{H}_1$ and the third statement follows from the second one. □

Remark 5.5.30. Specialising to *monomial* ideals, Corollary 5.5.29 immediately implies that for ideals with $\mathcal{I} : \mathcal{P}_+ = \mathcal{I} : \langle x^1 \rangle$ the satiety $\text{sat}\, \mathcal{I}$ is the maximal degree of a minimal generator of $\mathcal{I}$ divisible by x^1. If the ideal $\mathcal{I}$ possesses a Pommaret basis $\mathcal{H}$, this statement also follows from the fact that under the made assumption all elements of $\mathcal{H}_1$ are minimal generators.

Indeed, assume to the contrary that $\mathcal{H}_1$ contains two elements $h_1 \neq h_2$ such that $h_1 \mid h_2$. Obviously, this implies $\deg_{x^1} h_1 = \deg_{x^1} h_2$ and a non-multiplicative index $1 < \ell \leq n$ exists such that $x^\ell h_1 \mid h_2$. Without loss of generality, we may assume that $h_2 = x^\ell h_1$. But this immediately entails that $x^\ell h_1/x^1 = h_2/x^1 \notin \mathcal{I}$ and hence $h_1/x^1 \in (\mathcal{I} : \langle x^1 \rangle) \setminus (\mathcal{I} : \mathcal{P}_+)$. ◁

A first trivial consequence of our results is the following well-known formula relating Castelnuovo-Mumford regularity and saturation.

Corollary 5.5.31. *Let $\mathcal{I} \subseteq \mathcal{P}$ be an ideal. Then* $\operatorname{reg}\mathcal{I} = \max\{\operatorname{sat}\mathcal{I}, \operatorname{reg}\mathcal{I}^{\text{sat}}\}$.

Proof. Without loss of generality, we may assume that we use δ-regular coordinates so that the ideal $\mathcal{I}$ possesses a Pommaret basis $\mathcal{H}$ with respect to the degree reverse lexicographic order. Now the statement follows immediately from Proposition 5.5.28 and Corollary 5.5.29. □

For monomial ideals in generic position one can determine $\operatorname{reg}\mathcal{I}$ with the help of certain evaluations. Let $d = \dim(\mathcal{P}/\mathcal{I})$. We introduce for $j = 0, \dots, d$ the polynomial subrings $\mathcal{P}^{(j)} = \Bbbk[x^{j+1}, \dots, x^n]$ and within them the ideals $\mathcal{I}^{(j)} = \mathcal{I} \cap \mathcal{P}^{(j)}$ and their saturations $\tilde{\mathcal{I}}^{(j)} = \mathcal{I}^{(j)} : \langle x^{j+1} \rangle^\infty$. Obviously, a basis of $\mathcal{I}^{(j)}$ is obtained by setting $x^1 = \cdots = x^j = 0$ in a basis of $\mathcal{I}$ and for a basis of $\tilde{\mathcal{I}}^{(j)}$ we must additionally set $x^{j+1} = 1$. Now we define the numbers

$$c_j = \sup\{q \mid (\tilde{\mathcal{I}}^{(j)}/\mathcal{I}^{(j)})_q \neq 0\}\,, \qquad 0 \leq j < D\,, \tag{5.69a}$$

$$c_D = \sup\{q \mid (\mathcal{P}^{(d)}/\mathcal{I}^{(d)})_q \neq 0\}\,. \tag{5.69b}$$

One can show that whenever none of these numbers is infinite (which is the case in generic coordinates), then their maximum is given by $\operatorname{reg}\mathcal{I}$. We prove now that this is indeed the case for δ-regular coordinates and express the numbers c_j as satieties.

Theorem 5.5.32. *The numbers $c_0, \dots, c_D$ are all finite, if and only if the monomial ideal $\mathcal{I} \subseteq \mathcal{P}$ is quasi-stable. In this case $c_j = \operatorname{sat}\mathcal{I}^{(j)}$ for $0 \leq j \leq D$ and*

$$\max\{c_0, \dots, c_D\} = \operatorname{reg}\mathcal{I}\,. \tag{5.70}$$

If $d = \operatorname{depth}\mathcal{I}$, then it suffices to consider $c_d, \dots, c_D$.

Proof. We assume first that $\mathcal{I}$ is quasi-stable and thus possesses a Pommaret basis which we write $\mathcal{H} = \{h_{k,\ell} \mid 1 \leq k \leq n,\ 1 \leq \ell \leq \ell_k\}$ where $\operatorname{cls} h_{k,\ell} = k$. One easily verifies that the subset $\mathcal{H}^{(j)} = \{h_{k,\ell} \in \mathcal{H} \mid k > j\}$ is the Pommaret basis of the ideal $\mathcal{I}^{(j)}$. If we set $a_{k,\ell} = \deg_{x^k} h_{k,\ell}$, then the Pommaret basis of $\tilde{\mathcal{I}}^{(j)}$ is $\tilde{\mathcal{H}}^{(j)} = \mathcal{H}^{(j+1)} \cup \{h_{j+1,\ell}/(x^{j+1})^{a_{j+1,\ell}} \mid 1 \leq \ell \leq \ell_{j+1}\}$. This fact immediately implies that $c_j = \max\{\deg h_{j+1,\ell} \mid 1 \leq \ell \leq \ell_{j+1}\}$. By construction, $\dim(\mathcal{P}^{(D)}/\mathcal{I}^{(D)}) = 0$ and Proposition 5.2.3 entails that for $\hat{q} = \deg\mathcal{H}^{(D)}$ the equality $\mathcal{I}^{(D)}_{\hat{q}} = \mathcal{P}^{(D)}_{\hat{q}}$ holds.

Hence we find $c_D = \hat{q}$ (it is not possible that $c_D < \hat{q}$, as otherwise the set $\mathcal{H}$ was not involutively autoreduced).

Thus we find that $\max\{c_0,\dots,c_D\} = \deg\mathcal{H}$ and Theorem 5.5.15 yields (5.70). Furthermore, it follows immediately from Corollary 5.5.29 and Proposition 5.2.3, respectively, that $c_j = \operatorname{sat}\mathcal{I}^{(j)}$ for $0 \leq j \leq D$. Finally, Proposition 5.2.7 entails that the values $c_0,\dots,c_{d-1}$ vanish.

Now assume that the ideal $\mathcal{I}$ was not quasi-stable. By Part (ii) of Proposition 5.3.4 this entails that for some $0 \leq j < D$ the variable x^{j+1} is a zero divisor in the ring $\mathcal{P}/\langle\mathcal{I},x^1,\dots,x^j\rangle^{\mathrm{sat}} \cong \mathcal{P}^{(j)}/(\mathcal{I}^{(j)})^{\mathrm{sat}}$. Thus a polynomial $f \notin (\mathcal{I}^{(j)})^{\mathrm{sat}}$ exists for which $x^{j+1}f \in (\mathcal{I}^{(j)})^{\mathrm{sat}}$ which means that we can find for any sufficiently large degree $q \gg 0$ a polynomial $g \in \mathcal{P}^{(j)}$ with $\deg g = q - \deg f$ such that $fg \notin \mathcal{I}^{(j)}$ but $x^{j+1}fg \in \mathcal{I}^{(j)}$. The equivalence class of fg is now a non-vanishing element of $(\tilde{\mathcal{I}}^{(j)}/\mathcal{I}^{(j)})_q$ so that for a not quasi-stable ideal $\mathcal{I}$ at least one c_j is not finite. □

One direction of the proof uses the same idea as the one given above for Theorem 5.5.15: the Castelnuovo–Mumford regularity is determined by the basis members of maximal degree and their classes give us the positions in the minimal resolution where it is attained (recall Remark 5.5.16; here these are just the indices j for which c_j is maximal). However, while Theorem 5.5.15 holds for arbitrary homogeneous ideals, the evaluation approach can only be applied to monomial ideals.

Proposition 5.5.33. *Let $\mathcal{I} \subseteq \mathcal{P}$ be a quasi-stable ideal and $\mathcal{I} = \mathcal{J}_1 \cap \cdots \cap \mathcal{J}_r$ its unique irredundant decomposition into irreducible monomial ideals. Then the equality* $\operatorname{reg}\mathcal{I} = \max\{\operatorname{reg}\mathcal{J}_1,\dots,\operatorname{reg}\mathcal{J}_r\}$ *holds.*

Proof. We first note that the Castelnuovo-Mumford regularity of an irreducible monomial ideal $\mathcal{J} = \langle (x^{i_1})^{\ell_1},\dots,(x^{i_k})^{\ell_k}\rangle$ is easily determined using the considerations in Example 3.1.16. There we showed that any such ideal becomes quasi-stable after a simple renumbering of the variables and explicitly gave its Pommaret basis. Up to the renumbering, the unique element of maximal degree in this Pommaret basis is the term $(x^{i_1})^{\ell_1}(x^{i_2})^{\ell_2-1}\cdots(x^{i_k})^{\ell_k-1}$ and thus it follows from Theorem 5.5.15 that $\operatorname{reg}\mathcal{J} = \sum_{j=1}^{k}\ell_j - k + 1$.

Recall from Proposition 5.3.13 that an irreducible decomposition can be constructed via standard pairs. As discussed in the Addendum to Section 5.3, the decomposition (5.26) is in general redundant; among all standard pairs (ν, N_ν) with $N_\nu = N$ for a given set N only those exponents ν which are maximal with respect to divisibility appear in the irredundant decomposition and thus are relevant.

If we now determine the standard pairs of $\mathcal{I}$ from a Pommaret basis according to Remark 5.3.15, then we must distinguish two cases. We have first the standard pairs coming from the terms x^μ of degree $q = \deg\mathcal{H}$ not lying in $\mathcal{I}$. They are of the form $\bigl(x^\nu,\{x^1,\dots,x^k\}\bigr)$ where $k = \operatorname{cls}\mu$ and $x^\nu = x^\mu/(x^k)^{\mu_k}$. By Proposition 5.3.13, each such standard pair leads to the irreducible ideal $\mathcal{J} = \langle (x^\ell)^{\nu_\ell+1} \mid k < \ell \leq n\rangle$. By the remarks above, $\operatorname{reg}\mathcal{J} = |\nu| + 1 \leq |\mu| = q = \operatorname{reg}\mathcal{I}$.

The other standard pairs come from the terms $x^\nu \notin \mathcal{I}$ with $|\nu| < q$. It is easy to see that among these the relevant ones correspond one-to-one to the "end points" of the monomial completion process: we call an element of the Pommaret basis $\mathcal{H}$

of $\mathcal{I}$ an end point, if each non-multiplicative multiple of it has a *proper* involutive divisor in the basis (and thus one branch of the completion process ends with this element[8]). If the generator $x^\mu \in \mathcal{H}$ is such an end point, then the corresponding standard pair consists of the monomial $x^\nu = x^\mu / x^k$ where $k = \operatorname{cls}\mu$ and the empty set and it yields the irreducible ideal $\mathcal{J} = \langle (x^\ell)^{\nu_\ell + 1} \mid 1 \leq \ell \leq n \rangle$. Thus we find again $\operatorname{reg}\mathcal{J} = |\nu| + 1 = |\mu| \leq q = \operatorname{reg}\mathcal{I}$.

These considerations prove the estimate $\operatorname{reg}\mathcal{I} \geq \max\{\operatorname{reg}\mathcal{J}_1, \dots, \operatorname{reg}\mathcal{J}_r\}$. The claimed equality follows from the observation that any element of degree q in $\mathcal{H}$ must trivially be an end point of the completion process and the corresponding standard pair yields an irreducible ideal $\mathcal{J}$ with $\operatorname{reg}\mathcal{J} = q$. □

Yet another characterisation of $\operatorname{reg}\mathcal{I}$ using the homological nature of this invariant will be given in the next chapter where we will relate it to the Koszul homology of the factor module $\mathcal{P}/\mathcal{I}$.

The question of bounding the Castelnuovo–Mumford regularity of a homogeneous ideal $\mathcal{I}$ in terms of the degree q of an arbitrary generating set has attracted quite some interest. It is well-known that a doubly exponential bound exists for $\operatorname{reg}\mathcal{I}$ and that this bound is indeed sharp (cf. our discussion of the complexity of Gröbner basis computations in Appendix B.4).

For monomial ideals $\mathcal{I}$ the situation is much more favourable. The Taylor resolution provides us with an explicit resolution for any such ideal supported by the lcm-lattice of the given basis (thus the generators in the kth term of the resolution are determined by the least common multiples of k generators of $\mathcal{I}$).[9] It immediately implies for such ideals the existence of a *linear* bound

$$\operatorname{reg}\mathcal{I} \leq n(q-1) + 1 \tag{5.71}$$

where n is again the number of variables. Indeed, the degrees of the generators of its kth term are trivially bounded by kq and, by Hilbert's Syzygy Theorem B.4.29, it suffices to consider the first n terms which yields the above bound. For a quasi-stable ideal $\mathcal{I}$, a simple corollary of Proposition 5.5.33 yields an improved bound using the minimal generators.

Corollary 5.5.34. *Let the monomials $m_1, \dots, m_r$ be the minimal generators of the quasi-stable ideal $\mathcal{I} \subseteq \Bbbk[x^1, \dots, x^n]$. Setting furthermore $x^\lambda = \operatorname{lcm}(m_1, \dots, m_r)$ and $d = \min\{\operatorname{cls} m_1, \dots, \operatorname{cls} m_r\}$ (i. e. $\operatorname{depth}\mathcal{I} = d$), the Castelnuovo–Mumford regularity of $\mathcal{I}$ satisfies the estimate*

$$\operatorname{reg}\mathcal{I} \leq |\lambda| + d - n \tag{5.72}$$

and this bound is sharp.

Proof. Applying repeatedly the rule $\langle \mathcal{F}, t_1 t_2 \rangle = \langle \mathcal{F}, t_1 \rangle \cap \langle \mathcal{F}, t_2 \rangle$ for arbitrary generating sets $\mathcal{F}$ and coprime monomials t_1, t_2, one determines an irreducible decomposition of $\mathcal{I}$. Obviously, in the worst case one of the obtained irreducible ideals is

[8] Note that an end point may very well be a member of the minimal basis of $\mathcal{I}$!

[9] See the Notes at the end of this chapter for more details.

$\mathcal{J} = \langle (x^d)^{\lambda_d}, \dots, (x^n)^{\lambda_n} \rangle$. As we already know that $\operatorname{reg}\mathcal{J} = |\lambda| + d - n$, this value bounds $\operatorname{reg}\mathcal{I}$ by Proposition 5.5.33. □

Remark 5.5.35. An alternative direct proof of the corollary goes as follows. Let $\mathcal{H}$ be the Pommaret basis of $\mathcal{I}$. We claim that each generator $x^\mu \in \mathcal{H}$ with $\operatorname{cls}\mu = k$ satisfies $\mu_k \leq \lambda_k$ and $\mu_j < \lambda_j$ for all $j > k$. The estimate for μ_k is obvious, as it follows immediately from our completion algorithm that there is a minimal generator $x^\nu \mid x^\mu$ with $\nu_k = \mu_k$.

Assume for a contradiction that the Pommaret basis $\mathcal{H}$ contains a generator x^μ where $\mu_j > \lambda_j$ for some $j > \operatorname{cls}\mu$. If several such generators exist for the same value j, choose one for which μ_j is maximal. Obviously, j is non-multiplicative for the term x^μ and hence the multiple $x^j x^\mu$ must contain an involutive divisor $x^\nu \in \mathcal{H}$. Because of our maximality assumption $\nu_j \leq \mu_j$ and hence j must be multiplicative for x^ν so that $\operatorname{cls}\nu \geq j$. But this fact trivially implies that $x^\nu \mid_P x^\mu$ contradicting that the basis $\mathcal{H}$ is by definition involutively autoreduced.

Now the assertion of the proposition follows immediately: under the made assumptions we find $\operatorname{cls}\lambda = d$ and in the worst case $\mathcal{H}$ contains the generator $(x^d)^{\lambda_d}(x^{d+1})^{\lambda_{d+1}-1}\cdots(x^n)^{\lambda_n-1}$ which is of degree $|\lambda| + d - n$. ◁

Remark 5.5.36. The same arguments together with Proposition 5.5.28 also yield immediately a bound for the satiety of a quasi-stable ideal $\mathcal{I}$. As already mentioned above, a quasi-stable ideal is not saturated, if and only if $d = 1$. In this case, we have trivially $\operatorname{sat}\mathcal{I} \leq |\lambda| + 1 - n$. Again the bound is sharp, as shown by exactly the same class of irreducible ideals as considered in the proof above. ◁

Finally, we recall that, given two quasi-stable ideals $\mathcal{I}, \mathcal{J} \subseteq \mathcal{P}$ and their respective Pommaret bases, we explicitly constructed in Remarks 3.1.13 and 4.1.6, respectively, weak Pommaret bases for the sum $\mathcal{I} + \mathcal{J}$, the product $\mathcal{I} \cdot \mathcal{J}$ and the intersection $\mathcal{I} \cap \mathcal{J}$. Together with Theorem 5.5.15 the respective form of these weak bases lead to the following result.

Proposition 5.5.37. *Let $\mathcal{I}, \mathcal{J} \subseteq \mathcal{P}$ be two quasi-stable ideals. Then the following three estimates hold:*

$$\operatorname{reg}(\mathcal{I} + \mathcal{J}) \leq \max\{\operatorname{reg}\mathcal{I}, \operatorname{reg}\mathcal{J}\}\,, \tag{5.73a}$$

$$\operatorname{reg}(\mathcal{I} \cdot \mathcal{J}) \leq \operatorname{reg}\mathcal{I} + \operatorname{reg}\mathcal{J}\,, \tag{5.73b}$$

$$\operatorname{reg}(\mathcal{I} \cap \mathcal{J}) \leq \max\{\operatorname{reg}\mathcal{I}, \operatorname{reg}\mathcal{J}\}\,. \tag{5.73c}$$

Proof. The first two estimates follow immediately from the weak Pommaret bases given in the above mentioned remarks and Theorem 5.5.15. The last estimate is a simple corollary of Proposition 5.5.33. □

5.6 Notes

Combinatorial decompositions of polynomial modules and involutive bases appeared already in the *Janet–Riquier Theory* of systems of partial differential equations developed in the late 19th, early 20th century by a number of authors (see e. g. [233, 235, 381, 454, 458]). In this theory, the term *passive* introduced by Méray and Riquier [321] is more common than *involutive*; the latter one seems to go back to Lie. The idea of assigning multiplicative variables is due to Janet.

The results of this theory remained more or less unknown in commutative algebra until fairly recently. Rees [371] introduced much later the special decomposition now carrying his name. The concept was generalised by Stanley [431, 432] who was interested in a simple way to determine Hilbert functions; Proposition 5.2.1 is due to him. In recent years, a number of authors have picked up these ideas and applied them in various fields; see e. g. [32, 44, 151, 436, 439] and references therein. In the context of a complexity analysis of Gröbner bases, Dubé [114] reinvented combinatorial decompositions (apparently unaware of all the previous work on this topic). Stanley filtrations were introduced by Maclagan and Smith [300]; they also provide a concrete example [300, Ex. 3.8] of a Stanley decomposition not admitting an ordering that makes it to a Stanley filtration.

Within an algebraic context, concrete algorithms for the construction of combinatorial decompositions were proposed by various authors (in the Notes of Chapter 7 one can find some references to implementations of the Janet–Riquier Theory of partial differential equations which at least implicitly also determine such decompositions; complementary decompositions arise here in the context of determining formally well-posed initial value problems—see Section 9.3). Sturmfels and White [439] presented a simple recursive algorithm to determine a Stanley decomposition for arbitrary monomial modules which is essentially equivalent to our proof of Proposition 5.1.3. Thus its extension to polynomial modules $\mathcal{P}^m/\mathcal{N}$ requires that one must first compute a Gröbner basis of $\mathcal{N}$ for some term order $\prec$ in order to obtain a basis of $\operatorname{lt}_\prec \mathcal{N}$. They further developed a special method for determining Rees decompositions. However, it is rather expensive, as it needs repeated computations of Gröbner bases. Our approach via Pommaret bases is more direct and efficient; the precise relation between the two approaches is detailed in [410]. Both Dubé [114] and Maclagan and Smith [300] mention algorithms for the construction of combinatorial decompositions for monomial ideals (in the latter case the algorithm even yields Stanley filtrations). Robertz [385] presented an algorithm for the construction of a quasi-Rees decomposition using Janet bases[10] in the context of determining a Noether normalisation. Its basic idea is very similar to our approach to the construction of δ-regular coordinates.

Proposition 5.1.3 on the decomposition of the complementary set is a classical result in commutative algebra [98, pp. 417–418] and is sometimes used to prove the existence of the Hilbert polynomial. We slightly modified the proof in order to obtain

[10] Although this fact is not mentioned in [385], the algorithm works for arbitrary Noetherian involutive divisions.

a disjoint decomposition. Gerdt [153, Lemma 24] proposed a version that relates the form of the cones to an involutive division. However, his proof is not correct, as the there constructed set $\mathcal{U}$ (in the notation of our proof of Proposition 5.1.3) may also contain elements of $\mathcal{I}$. Indeed, it is trivial to provide examples for this effect using the Janet division.

Proposition 5.1.4 was already proven by Janet [235, Section 15]—again in the context of initial value problems for overdetermined differential equations. In fact, our proof is essentially the one given by him. Complementary decompositions were constructed already before by Riquier [381], Chapt. V, §§79–88. Janet was the first to consider disjoint decompositions of both the ideal $\mathcal{I}$ and its factor algebra $\mathcal{P}/\mathcal{I}$. Oberst and Pauer [340, Sect. 2.2] provide an algorithm for the simultaneous construction of both decompositions; their determination of the complementary decomposition is essentially equivalent to our Algorithm 5.1.

The combination of Corollary 5.1.9 and Proposition 5.2.7 allows us to make some statements about the *Stanley conjecture*. It concerns the minimal number of multiplicative variables for a generator in a Stanley decomposition. Following Apel [21, Def. 1] and Herzog et al [212] we call this number the *Stanley depth* of the decomposition and for an ideal $\mathcal{I} \subseteq \mathcal{P}$ the Stanley depth of the algebra $\mathcal{A} = \mathcal{P}/\mathcal{I}$, written $\operatorname{sdepth}\mathcal{A}$, is defined as the maximal Stanley depth of a complementary decomposition for $\mathcal{I}$. In its simplest form the Stanley conjecture claims that we always have the inequality $\operatorname{sdepth}\mathcal{A} \geq \operatorname{depth}\mathcal{A}$. Obviously, Corollary 5.1.9 together with Proposition 5.2.7 (plus the existence Theorem 4.3.15 for Pommaret bases) shows that this inequality holds for arbitrary ideals.

The rigorous formulation of the Stanley conjecture [432, Conj. 5.1] concerns monomial ideals and requires that all generators in the decomposition are again monomials. Furthermore, no coordinate transformation is allowed. Then our results only allow us to conclude that the Stanley conjecture is true for all quasi-stable ideals. Some further results on this question have been achieved by Apel [21, 22] with the help of a slightly different notion of involutive bases.[11]

The notion of a standard pair was introduced by Sturmfels et al [440]. They used it to derive bounds on the geometric and the arithmetic degree, respectively, of a homogeneous ideal; in particular, the arithmetic degree is simply given by the number of standard pairs [440, Lemma 3.3]. We cannot give here an introduction into these degree concepts; we refer instead to [468, Section 9.1] or [33]. Hoşten and Smith [219] present two algorithms for the direct construction of the set $\mathcal{S}_{\mathcal{I}}$ of all standard pairs given the minimal basis of $\mathcal{I}$.

The concept of an independent set of variables modulo an ideal $\mathcal{I}$ goes back to Gröbner [186, Def. 1] (see also [188, Section 131]). This approach to the dimension of an ideal has been taken up again by Kredel and Weispfenning [265] using Gröbner bases (see also [36, Sections 6.3 and 9.3]) who introduced the notion of a strongly independent set and also provided a recursive algorithm for the effective

[11] Apel [21] considers the Stanley conjecture for the ideal $\mathcal{I} \subseteq \mathcal{P}$ itself instead of the factor algebra $\mathcal{A} = \mathcal{P}/\mathcal{I}$. Here it follows immediately from the definition of a Pommaret basis and Proposition 5.2.7 that the Stanley conjecture is true in its weak form for arbitrary polynomial ideals and in its strong form for all quasi-stable ideals.

determination of all maximal strongly independent sets and thus the dimension of $\mathcal{A} = \mathcal{P}/\mathcal{I}$. The relation to Pommaret bases and the special role of δ-regular coordinates was first exhibited in [410]. The terminology *Hironaka decomposition* seems to have been coined by Sturmfels [436, Sect. 2.3].

The last statement in Corollary 5.5.29 relating the satiety of a homogeneous ideal $\mathcal{I} \subseteq \mathcal{P}$ to its socle is originally due to Bermejo and Gimenez [41, Prop. 2.1]. They provided a direct proof without Pommaret bases. The characterisation of $\operatorname{sat}\mathcal{I}$ for monomial ideals via the equality $\mathcal{I} : \mathcal{P}_+ = \mathcal{I} : \langle x^1 \rangle$ also stems from Bermejo and Gimenez [41, Cor. 2.4]; it generalises a classical result about Borel fixed ideals [182, Cor. 2.10] where this equality always holds. The here presented approach via Pommaret bases is taken from [410].

A number of results in Section 5.3 are also due to Bermejo and Gimenez [41]. They introduced the notion of a *monomial ideal of nested type* defined by Condition (v) of Proposition 5.3.4 and showed that it is equivalently characterised by the Conditions (ii), (iii) or (iv) [41, Proposition 3.2]. The proof that every quasi-stable ideal is of nested type and vice versa, i. e. Condition (i) in Proposition 5.3.4, comes from [410]. Condition (vi) and the observation that quasi-stable ideals are sequentially Cohen–Macaulay are due to Herzog et al [211, Proposition 2.2, Corollary 2.5] (these authors call quasi-stable ideals *ideals of Borel type*; Caviglia and Sbarra [79] use the terminology *weakly stable ideals*).

The observation that these conditions are related to the existence of simultaneous Noether normalisations for all primary components (Corollary 5.3.8) and the implications of Proposition 5.3.4 on the form of an irredundant primary decomposition of a quasi-stable ideal are again due to Bermejo and Gimenez [41, Proposition 3.6, Remark 3.3]. The construction of an irredundant primary decomposition for a quasi-stable ideal is taken from [200]. That a Pommaret basis induces a Noether normalisation for an arbitrary polynomial ideal was first demonstrated in [410].

The classical approach to the construction of Noether normalisations consists of performing random coordinate transformations (see e. g. [185, Sect. 3.4]) which is of course very unpleasant for all subsequent computations. Another well-known method is due to Logar [293]. Recently, two new algorithms were developed by Hashemi [196] and Robertz [385], respectively. The first one uses implicitly similar ideas as are underlying Pommaret bases. The second one uses Janet bases and was already mentioned above. It performs similar coordinate changes as in our approach to the construction of δ-regular coordinates, but stops as soon as a Noether normalisation is reached.

The characterisation of stable monomial modules in terms of Pommaret bases is due to Mall [305] (but see also the remark after Lemma 1.2 in [128]). Eliahou and Kervaire [128] provide an explicit representation of the minimal free resolution of such ideals. It is based on some manipulations of a Koszul complex. Unfortunately, it is not possible to extend their results to the polynomial case. The main reason is very simple. Assume that we are given a stable monomial module with basis $\{\mathbf{t}_1, \dots, \mathbf{t}_r\}$. If we multiply the generator $\mathbf{t}_\alpha$ with a term s, there exists a unique generator $\mathbf{t}_\beta$ such that $s\mathbf{t}_\alpha \in \mathcal{C}_P(\mathbf{t}_\beta)$. The estimate $\operatorname{cls}\mathbf{t}_\beta \geq \operatorname{cls}\mathbf{t}_\alpha$ is completely trivial but essential for the construction of the differential of Eliahou and Kervaire. In the polynomial case with a basis $\{\mathbf{h}_1,\dots,\mathbf{h}_r\} \subset \mathcal{P}^m$, the standard representation of a product $s\mathbf{h}_\alpha$ consists of a sum involving in general many generators $\mathbf{h}_\beta$. In particular, it is no longer possible to provide a lower bound for the classes of the $\mathbf{h}_\beta$. This prevents the construction of a complex along the lines of Eliahou and Kervaire.

Based on the Schreyer Theorem B.4.27, any Gröbner basis may be used as starting point for the construction of a syzygy resolution. In many textbooks such a resolution is used for a proof of Hilbert's Syzygy Theorem B.4.29. However, the special structures induced by a Pommaret basis does not only yield a stronger estimate on the length of the minimal resolution (in fact, we have seen that it always yields a resolution of minimal length) but even a closed formula for the ranks of the modules contained in the resolution. At the moment, it is still an open question to what extent this result may be exploited for obtaining tight estimates on the Betti numbers or even for computing them directly.

Componentwise linear ideals were defined by Herzog and Hibi [210]. An example of such ideals are *Gotzmann ideals* (first introduced in the same article): a homogeneous ideal $\mathcal{I} \subseteq \mathcal{P}$ is Gotzmann, if the unique lexicographic ideal $\mathcal{I}^{\text{lex}}$ which has the same Hilbert function as $\mathcal{I}$ also has the same number of minimal generators as $\mathcal{I}$ (see [334] for more information). The fact that the resolution coming from a Pommaret basis is minimal only, if the ideal is componentwise linear, and that generically also the converse is true was first shown in [410].

The involutive Schreyer Theorem also stems from [410] (originally only for Pommaret bases, later in the form presented here). The idea of an L-graph is a generalisation of a construction by Plesken and Robertz [354] for the special case of the Janet division. This article also contains a Schreyer Theorem for Janet bases.

We used the Taylor resolution for deriving the linear bound (5.71) for the regularity of a monomial ideal. This resolution was the first explicit resolution for arbitrary monomial ideals and presented by Taylor [452] in her Ph.D. thesis.

Let $\{m_1,\dots,m_r\}$ be the minimal basis of the monomial ideal $\mathcal{I}$ and $\mathcal{V}$ an arbitrary $\mathbb{k}$-linear space of dimension r with basis $\{v_1,\dots,v_r\}$. If $\mathbf{k} = (k_1,\dots,k_q)$ is a sequence of integers with $1 \leq k_1 < k_2 < \cdots < k_q \leq r$, we set $m_{\mathbf{k}} = \operatorname{lcm}(m_{k_1},\dots,m_{k_q})$ and $v_{\mathbf{k}} = v_{k_1} \wedge \cdots \wedge v_{k_q}$. The $\mathcal{P}$-module $\mathcal{T}_q = \mathcal{P} \otimes \Lambda^q \mathcal{V}$ is then freely generated by all wedge products $v_{\mathbf{k}}$. Finally, we introduce on the algebra $\mathcal{T} = \mathcal{P} \otimes \Lambda \mathcal{V}$ a $\mathcal{P}$-linear differential δ by requiring that

$$\delta v_{\mathbf{k}} = \sum_{\ell=1}^{q} (-1)^{\ell-1} \frac{m_{\mathbf{k}}}{m_{\mathbf{k}_\ell}} v_{\mathbf{k}_\ell}\,, \tag{5.74}$$

where $\mathbf{k}_\ell$ denotes the sequence $\mathbf{k}$ with the entry k_ℓ removed. One can show that $(\mathcal{T},\delta)$ is a complex representing a free resolution of the ideal $\mathcal{I}$. Obviously, the length of the resolution is given by the number r of monomials. This implies immediately that the resolution is rarely minimal (several characterisations of the case that $\mathcal{T}$ defines a minimal resolution have been given by Fröberg [143]). Note that the ordering of the monomials m_i in the minimal basis has no real influence on the result: the arising resolutions are trivially isomorphic.

Lyubeznik [297] proved later in his Ph.D. thesis that in fact already a subcomplex $\mathcal{L} \subseteq \mathcal{T}$ defines a free resolution of $\mathcal{I}$. Let $\mathbf{k}$ again be an integer sequence; we denote for $1 \leq i < r$ by $\mathbf{k}_{>i}$ the subsequence of all entries $k_j > i$. If we eliminate from the basis of the Taylor complex $\mathcal{T}$ all generators $v_{\mathbf{k}}$ where for at least one $1 \leq i < r$ the monomial m_i divides $m_{\mathbf{k}_{>i}}$, then the remaining part $\mathcal{L}$ is still a complex defining a resolution of $\mathcal{I}$. However, even this smaller resolution is rarely minimal.[12]

A derivation of the Taylor resolution via the Schreyer Theorem B.4.27 was presented in [407]. There it is also shown that the Lyubeznik resolution $\mathcal{L}$ may be considered as a simple consequence of Buchberger's second criterion for the avoidance of redundant S-polynomials (Proposition B.4.19). Furthermore, the explicit contracting homotopy for the Taylor complex given by Fröberg [143] is related there to normal form computations with respect to a Gröbner basis.

A number of different approaches to the Castelnuovo–Mumford regularity exists; a few notes on the history of this concept can be found in [125, pp. 514–515]. Often it is defined using local cohomology; we used the more constructive approach via syzygies. Characterisations and the effective determination of the Castelnuovo–Mumford regularity has attracted some interest in recent time starting mainly with the work of Bayer and Stillman [34]. Theorem 5.5.21 is due to them and they also emphasised the special role of the degree reverse lexicographic term order by showing that generically no Gröbner basis has a lower degree than the one for this order. The fact that *any* Pommaret basis with respect to this order has the Castelnuovo–Mumford regularity as degree was proven in [410].

Theorem 5.5.24 on the existence of a linear resolution for the truncation of $\mathcal{I}$ was discovered by Eisenbud and Goto [126] (in fact, it was already noted by Serre in his letter appended to [189] who called it a "curiosité"). Trung [460] introduced the evaluation based approach to monomial ideals underlying Theorem 5.5.32. The approach of Bermejo and Gimenez [41] to the computation of $\operatorname{reg}\mathcal{I}$ is essentially equivalent to this result; they also were the first to prove Proposition 5.5.33 [41, Cor. 17]. Proposition 5.5.19 was first proven by Eisenbud et al [127, Prop. 10] for the special case of a Borel fixed ideal. In all cases originally different proofs were given; the here presented proofs via Pommaret bases are again taken from [410]. The concept of extremal Betti numbers is due to Bayer et al [35]. They also proved that both their positions and their values coincide for any ideal and its generic initial ideal with respect to the degree reverse lexicographic order [35, Thm. 1.6].

For a monomial ideal $\mathcal{I} \subseteq \mathcal{P}$, Maclagan and Smith [300, Rem. 4.2] deduce a bound on the Castelnuovo–Mumford regularity $\operatorname{reg}(\mathcal{P}/\mathcal{I})$ from any Stanley filtration of $\mathcal{P}/\mathcal{I}$, namely the maximal degree of a generator (actually, the main result of Maclagan and Smith [300, Thm. 4.1] is a bound on a multigraded generalisation of $\operatorname{reg}(\mathcal{P}/\mathcal{I})$). It follows then from Remark 5.1.8 that the maximal degree of any Janet basis of $\mathcal{I}$ provides such a bound. Note, however, that their results do not give any indications when this bound may be sharp.

A basic reference for the question of bounding the Castelnuovo–Mumford regularity is the survey article by Bayer and Mumford [33]. The general doubly

[12] Explicit minimal resolutions are only known for some special classes of monomial ideals like stable ideals. Further classes can be found in [322].

exponential bound was already very early derived by Hermann [204]. Only much later Mayr and Meyer [318] provided explicit examples that proved that this bound was sharp. Our proof of the bound (5.72) for the case of quasi-stable ideals comes from [410]. It also follows immediately from the results of Bermejo and Gimenez [41]. Yet another proof was given by Hashemi [195]. The estimates in Proposition 5.5.37 were proven in an alternative way by Cimpoeaş [91, 90].

If one insists on having an estimate involving only the maximal degree q of the minimal generators and the depth, then Corollary 5.5.34 yields immediately the following estimate for quasi-stable ideals, variations of which appear in [10, 79, 89] (as usual, n denotes here the number of variables and d the depth of $\mathcal{I}$):

$$q \leq \operatorname{reg}\mathcal{I} \leq (n-d+1)(q-1)+1\,. \tag{5.75}$$

Indeed, under the made assumptions the length of the exponent λ of the least common multiple of the minimal generators is bounded by $|\lambda| \leq (n-d+1)q$. Note that both bounds in (5.75) are sharp: the upper bound is realised by the irreducible ideal $\mathcal{I} = \langle (x^1)^q, \dots, (x^n)^q \rangle$; the lower bound is attained, if the ideal $\mathcal{I}$ is even stable, as then Proposition 5.5.6 implies that $\operatorname{reg}\mathcal{I} = q$ independent of $\operatorname{depth}\mathcal{I}$.

Eisenbud et al [127] presented a variation of the estimate (5.75). They introduced the notion of s-stability as a generalisation of stability: let $s \geq 1$ be an integer; a monomial ideal $\mathcal{I}$ is *s-stable*, if for every monomial $x^\mu \in \mathcal{I}$ and for every index $n \geq j > \operatorname{cls}\mu = k$ an exponent $1 \leq e \leq s$ exists such that $x^{\mu - e_k + e_j} \in \mathcal{I}$. Then it is easy to see that for an s-stable ideal $\mathcal{I} \subseteq \mathcal{P}$ generated in degrees less than or equal to q the estimate

$$\operatorname{reg}\mathcal{I} \leq q + (n-1)(s-1) \tag{5.76}$$

holds, as $\mathcal{I}_{\geq q+(n-1)(s-1)}$ is stable (thus any s-stable ideal is trivially quasi-stable). However, in general this bound is an overestimate, as it based on the assumption that the ideal $\mathcal{I}$ possesses a minimal generator of class 1 and degree q which must be multiplied by $(x^2)^{s-1}(x^3)^{s-1}\cdots(x^n)^{s-1}$ in order to reach a stable set.

The results in this chapter also demonstrate that the notion of δ-regularity is related to many genericity questions in commutative algebra and algebraic geometry. Many statements that are only generically true do hold in δ-regular coordinates. In particular, in δ-regular coordinates many properties of an affine algebra $\mathcal{A} = \mathcal{P}/\mathcal{I}$ may already be read off the monomial algebra $\mathcal{A}' = \mathcal{P}/\operatorname{lt}_\prec \mathcal{I}$ where $\prec$ is the degree reverse lexicographic order.

It follows immediately from Proposition 5.2.7 that $\operatorname{depth}\mathcal{A} = \operatorname{depth}\mathcal{A}'$ and that $(x^1, \dots, x^d)$ represents a maximal regular sequence for both algebras. As in the homogeneous case it is also trivial that $\dim\mathcal{A} = \dim\mathcal{A}'$, we see that the algebra $\mathcal{A}$ is Cohen–Macaulay if and only if $\mathcal{A}'$ is so. Similarly, it is an easy consequence of Theorem 5.5.11 that the projective dimensions coincide, $\operatorname{proj}\dim\mathcal{A} = \operatorname{proj}\dim\mathcal{A}'$, and of Theorem 5.5.15 that the same holds for the Castelnuovo–Mumford regularity: $\operatorname{reg}\mathcal{A} = \operatorname{reg}\mathcal{A}'$. An exception are the Betti numbers where Example 5.5.9 shows that even in δ-regular coordinates $\mathcal{A}$ and $\mathcal{A}'$ may have different ones.

These equalities are of course not new; they can already be found in [34] (some even in earlier references). However, one should note an important difference:

Bayer and Stillman [34] work with the *generic initial ideal*, whereas we assume δ-regularity of the coordinates. These are two different genericity concepts, as even in δ-regular coordinates $\mathrm{lt}_{\prec}\mathcal{I}$ is not necessarily the generic initial ideal (in contrast to the former, the latter is always Borel fixed).

When we proved in Corollary 5.5.13 and 5.5.18, respectively, the two inequalities $\operatorname{proj\,dim}\mathcal{A} \leq \operatorname{proj\,dim}\mathcal{A}'$ and $\operatorname{reg}\mathcal{I} \leq \operatorname{reg}(\mathrm{lt}_{\prec}\mathcal{I})$ for arbitrary term orders $\prec$, we had to assume the existence of a Pommaret basis of $\mathcal{I}$ for $\prec$. It is well-known that these inequalities remain true, even if we drop this assumption (see for example the discussions in [33, 34, 56]). We presented here alternative proofs because of their great simplicity and they cover at least the generic case.

In contrast to the two previous chapters, we restricted now in the applications of Pommaret bases to the ordinary polynomial ring. Indeed, many of the results presented here do not hold in arbitrary polynomial algebras of solvable type. For example, in all our proofs concerning the depth or the Castelnuovo–Mumford regularity of a module we implicitly assumed that if $f \in \langle x^i \rangle$, then each term in $\operatorname{supp} f$ is divisible by x^i. For the universal enveloping algebra $\mathfrak{U}\bigl(\mathfrak{so}(3)\bigr)$, this is obviously not the case according to (3.16). In fact, we even find that $\langle x^1, x^2 \rangle = \langle x^1, x^2, x^3 \rangle$, since $x^3 = x^1 \star x^2 - x^2 \star x^1$.

The definition of a Stanley decomposition for a module over a polynomial algebra of solvable type is identical to the definition in the case of the ordinary polynomial ring, as it is solely based on an isomorphism as $\Bbbk$-linear spaces. Thus the multiplication does not matter and it is important to note that on the right hand side of the decomposition (5.1) the $\cdot$ is *not* replaced by a $\star$. If the multiplication $\star$ preserves the natural grading of the polynomial ring, Proposition 5.2.1 remains valid. Thus involutive bases are a valuable tool for computing Hilbert functions even in the non-commutative case and therefore allow us to compute the *Gelfand–Kirillov dimension* [319, Sect. 8.1.11] of quotient rings as the degree of the Hilbert polynomial. Some examples for such computations in the context of quantum groups (however, using Gröbner instead of involutive bases) may be found in [61].

For syzygies the situation is more complicated. The proof of Theorem 5.4.4 is independent of the precise form of the multiplication and thus we may conclude that we can always construct at least a Gröbner basis of the syzygy module. Our proof of Theorem 5.4.10 relies mainly on normal form arguments that generalise. A minimal requirement is that the term order $\prec_{\mathcal{H}}$ respects the multiplication $\star$, as otherwise the theorem does not even make sense. Furthermore, we must be careful with all arguments involving multiplicative variables. We need that if x_i and x_j are both multiplicative for a generator, then $x_i \star x_j = c_{ij} x_i x_j + h_{ij}$ must also contain only multiplicative variables which will surely happen, if the polynomial h_{ij} depends only on variables x_k with $k \leq \max\{i, j\}$. This is for example the case for linear differential operators (but again not for $\mathfrak{U}\bigl(\mathfrak{so}(3)\bigr)$), so that we may conclude that Theorem 5.4.10 and its consequences remain true for the Weyl algebra and other rings of differential operators. In particular, this observation implies that the global dimension of these rings is n, as for the commutative polynomial ring.

Chapter 6
Involution II: Homological Theory

Nothing proves more clearly that the mind seeks truth, and nothing reflects more glory upon it, than the delight it takes, sometimes in spite of itself, in the driest and thorniest researches of algebra.
Bernard de Fontenelle

A reader with some experience in commutative algebra has probably noticed that almost all the invariants of polynomial modules which we computed in the last chapter with the help of Pommaret bases are actually of a homological origin. Hence one suspects that this special type of involutive bases must be closely related to some homological constructions. The main purpose of this chapter is to demonstrate that such a relationship indeed exists.

The first section introduces the two relevant complexes: the polynomial de Rham and the Koszul complex, respectively, over a finite-dimensional vector space. In particular, we study the duality between them and prove that they are exact in positive degree (and hence define a (co)resolution). Then we restrict these complexes to certain subcomplexes. In the case of the Koszul complex, this restriction is simply achieved by tensoring it with some polynomial module leading to the familiar Koszul homology of the module.

For the polynomial de Rham complex one introduces classically the notion of a symbolic system. As we will see, such a system is nothing but a polynomial *co*module, if one exploits the fact that the de Rham complex can be equally well defined over the symmetric coalgebra. Taking this point of view, the Spencer cohomology of a symbolic system is obtained completely dually to the Koszul homology by cotensoring the polynomial de Rham complex with a comodule. Finally, we define a (co)module to be involutive at a certain degree, if the corresponding bigraded (co)homology vanishes beyond this degree.

The second section is concerned with the derivation of effective criteria for an involutive (co)module. This leads to a homological version of the famous Cartan test used in the theory of exterior differential systems. As this test requires the choice of a basis for the underlying vector space, we meet again the problem of δ- or quasiregularity. And again we can show that generic bases are quasiregular, this time based on a standard theorem about associated prime ideals.

The final section discusses the relation of these homological constructions to Pommaret bases in a polynomial ring $\mathcal{P}$. We show that if the ideal $\mathcal{I} \subseteq \mathcal{P}$ has a Pommaret basis of degree q, then the module $\mathcal{P}/\mathcal{I}$ becomes involutive at degree q and hence that the notion of involution is essentially the same as the

W.M. Seiler, *Involution*, Algorithms and Computation in Mathematics 24,
DOI 10.1007/978-3-642-01287-7_6, © Springer-Verlag Berlin Heidelberg 2010

notion of Castelnuovo–Mumford regularity. Furthermore, we show that the concept of quasiregularity appearing in the dual Cartan test for $\mathcal{P}/\mathcal{I}$ is equivalent to δ-regularity in the sense of existence of a Pommaret basis of $\mathcal{I}$. These results are basically corollaries to the simple complementary decomposition (5.4) defined by a Pommaret basis. In the monomial case we will even see that the kernels appearing in the dual Cartan test induce the complete Pommaret basis so that the latter may be considered as an intrinsic object.

6.1 Spencer Cohomology and Koszul Homology

Let $\mathcal{V}$ be a finite-dimensional vector space (over a field $\Bbbk$) with $\dim\mathcal{V} = n$. With the help of the symmetric algebra $S\mathcal{V}$ and the exterior algebra $\Lambda\mathcal{V}$, we introduce two natural complexes of vector spaces based on the product spaces $S_q\mathcal{V}\otimes\Lambda_p\mathcal{V}$. Any element of these spaces may be written as a $\Bbbk$-linear sum of separable elements, i. e. elements of the form $w_1\cdots w_q\otimes v_1\wedge\cdots\wedge v_p$ with vectors $w_i, v_j\in\mathcal{V}$. By convention, we set $S_j\mathcal{V} = \Lambda_j\mathcal{V} = 0$ for values $j < 0$.

Definition 6.1.1. For any integer $r\geq 0$ the cochain complex

$$\begin{array}{c} 0 \longrightarrow S_r\mathcal{V} \xrightarrow{\delta} S_{r-1}\mathcal{V}\otimes\mathcal{V} \xrightarrow{\delta} S_{r-2}\mathcal{V}\otimes\Lambda_2\mathcal{V} \xrightarrow{\delta} \cdots \\ \cdots \xrightarrow{\delta} S_{r-n}\mathcal{V}\otimes\Lambda_n\mathcal{V} \longrightarrow 0 \end{array} \tag{6.1}$$

where the differential δ is defined by

$$\delta(w_1\cdots w_q\otimes v_1\wedge\cdots\wedge v_p) = \sum_{i=1}^{q} w_1\cdots\widehat{w_i}\cdots w_q\otimes w_i\wedge v_1\wedge\cdots\wedge v_p \tag{6.2}$$

(the hat $\widehat{\ }$ signals that the corresponding factor is omitted) is called the *polynomial de Rham complex* $R_r(\mathcal{V})$ at degree r over $\mathcal{V}$. The *Koszul complex* $K_r(\mathcal{V})$ at degree r over $\mathcal{V}$ is the chain complex

$$\begin{array}{c} 0 \longrightarrow S_{r-n}\mathcal{V}\otimes\Lambda_n\mathcal{V} \xrightarrow{\partial} S_{r-n+1}\mathcal{V}\otimes\Lambda_{n-1}\mathcal{V} \xrightarrow{\partial} \cdots \\ \cdots \xrightarrow{\partial} S_r\mathcal{V} \longrightarrow 0 \end{array} \tag{6.3}$$

where now the differential ∂ is defined as

$$\partial(w_1\cdots w_q\otimes v_1\wedge\cdots\wedge v_p) = \sum_{i=1}^{p}(-1)^{i+1} w_1\cdots w_q v_i\otimes v_1\wedge\cdots\wedge\widehat{v_i}\wedge\cdots\wedge v_p\,. \tag{6.4}$$

It is trivial to verify that, due to the skew-symmetry of the wedge product, these differentials satisfy $\delta^2 = 0$ and $\partial^2 = 0$, respectively, so that we are indeed dealing with complexes.

Let $\{x^1,\dots,x^n\}$ be a basis of $\mathcal{V}$. Then a basis of the vector space $S_q\mathcal{V}$ is given by all terms x^μ with μ a multi index of length q. For a basis of the vector space $\Lambda_p\mathcal{V}$ we use the following convention: let I be a *sorted* repeated index of length p, i.e. $I=(i_1,\dots,i_p)$ with $1\leq i_1<i_2<\cdots<i_p\leq n$; then we write x^I for $x^{i_1}\wedge\cdots\wedge x^{i_p}$ and the set of all such "terms" provides a basis of $\Lambda_p\mathcal{V}$. With respect to these bases, we obtain the following expressions for the above differentials:

$$\delta(x^\mu\otimes x^I)=\sum_{i=1}^{n}\operatorname{sgn}\bigl(\{i\}\cup I\bigr)\mu_i x^{\mu-1_i}\otimes x^{\{i\}\cup I} \tag{6.5}$$

and

$$\partial(x^\mu\otimes x^I)=\sum_{j=1}^{p}(-1)^{j+1}x^{\mu+1_{i_j}}\otimes x^{I\setminus\{i_j\}}\,. \tag{6.6}$$

Formally, (6.5) looks like the exterior derivative applied to a differential p-form with polynomial coefficients. This observation explains the name "polynomial de Rham complex" for (6.1) and in principle one should use the usual symbol d for the differential but the notation δ has become standard.

Remark 6.1.2. While the de Rham differential δ indeed needs the algebra structure of the exterior algebra $\Lambda\mathcal{V}$, it exploits only the vector space structure of the symmetric algebra $S\mathcal{V}$. Thus we may substitute the symmetric algebra $S\mathcal{V}$ by the symmetric *coalgebra* $\mathfrak{S}\mathcal{V}$ and define δ on the components of the free $\mathfrak{S}\mathcal{V}$-comodule $\mathfrak{S}\mathcal{V}\otimes\Lambda\mathcal{V}$, since both are identical as vector spaces (cf. Appendix B.3). It is not difficult to verify that the differential δ is a comodule morphism. In fact, we will see later that in our context this comodule interpretation is even more natural. For the Koszul differential ∂ we have the opposite situation: we need the algebra $S\mathcal{V}$ but only the vector space $\Lambda\mathcal{V}$. However, this fact will not become relevant for us. $\triangleleft$

Lemma 6.1.3. *We have* $\bigl(\delta\circ\partial+\partial\circ\delta\bigr)(\omega)=(p+q)\omega$ *for all* $\omega\in S_q\mathcal{V}\otimes\Lambda_p\mathcal{V}$.

Proof. For $\omega=w_1\cdots w_q\otimes v_1\wedge\cdots\wedge v_p$ one readily computes that

$$(\partial\circ\delta)(\omega)=q\omega+\sum_{i=1}^{q}\sum_{j=1}^{p}(-1)^j w_1\cdots\widehat{w_i}\cdots w_q v_j\otimes w_i\wedge v_1\wedge\cdots\wedge\widehat{v_j}\wedge\cdots\wedge v_p \tag{6.7}$$

and similarly

$$(\delta\circ\partial)(\omega)=p\omega+\sum_{j=1}^{p}\sum_{i=1}^{q}(-1)^{j+1}w_1\cdots\widehat{w_i}\cdots w_q v_j\otimes w_i\wedge v_1\wedge\cdots\wedge\widehat{v_j}\wedge\cdots\wedge v_p \tag{6.8}$$

which immediately implies our claim. $\square$

Proposition 6.1.4. *The two complexes* $R_q(\mathcal{V})$ *and* $K_q(\mathcal{V})$ *are exact for all values* $q>0$. *For* $q=0$ *both complexes are of the form* $0\rightarrow\Bbbk\rightarrow 0$.

Proof. This assertion is an immediate consequence of Lemma 6.1.3. It implies that for any degree $q>0$ the map ∂ induces a contracting homotopy for $R_q(\mathcal{V})$ and conversely δ for $K_q(\mathcal{V})$ connecting the respective identity and zero maps. According to Remark B.2.26, this observation entails exactness. □

We may consider the two complexes $R_q(\mathcal{V})$ and $K_q(\mathcal{V})$ as homogeneous components of complexes $R(\mathcal{V})$ and $K(\mathcal{V})$ over the $S\mathcal{V}$-modules $S\mathcal{V}\otimes\Lambda_i\mathcal{V}$. Because of the grading of $S\mathcal{V}$, both the cohomology $H^\bullet\bigl(R(\mathcal{V}),\delta\bigr)$ and the homology $H_\bullet\bigl(K(\mathcal{V}),\partial\bigr)$ are bigraded; we denote by $H^{q,p}\bigl(R(\mathcal{V}),\delta\bigr)$ and $H_{q,p}\bigl(K(\mathcal{V}),\partial\bigr)$ the respective (co)homology module at $S_q\mathcal{V}\otimes\Lambda_p\mathcal{V}$. By Proposition 6.1.4, all these modules vanish except for $H^{0,0}\bigl(R(\mathcal{V}),\delta\bigr)\cong H_{0,0}\bigl(K(\mathcal{V}),\partial\bigr)\cong\Bbbk$ (for the polynomial de Rham complex, Proposition 6.1.4 is known as the *formal Poincaré Lemma*, as one may consider it as a special case of the Poincaré Lemma for general differential forms on a manifold—cf. Remark C.2.9). This observation implies that the Koszul complex $K(\mathcal{V})$ defines a free resolution and the polynomial de Rham complex $R(\mathcal{V})$ a free coresolution of the underlying field $\Bbbk$.

The polynomial de Rham and the Koszul complex are related by duality. As detailed in Appendix B.2, we may introduce for any complex of $\mathcal{R}$-modules its dual complex by applying the functor $\mathrm{Hom}_{\mathcal{R}}(\cdot,\mathcal{R})$. In the case of finite-dimensional vector spaces, Proposition B.2.28 asserts that the homology of the dual complex is the dual space of the cohomology of the original complex.

Remark 6.1.5. We obtain a canonical isomorphism $S_q(\mathcal{V}^*)\cong(S_q\mathcal{V})^*$ by interpreting the separable element $\phi_1\cdots\phi_q\in S_q(\mathcal{V}^*)$ as the linear map on $S_q\mathcal{V}$ defined by

$$(\phi_1\cdots\phi_q)(v_1\cdots v_q)=\sum_{\pi\in\mathcal{S}_q}\prod_{i=1}^{q}\phi_i(v_{\pi(i)}) \tag{6.9}$$

where $\mathcal{S}_q$ denotes the symmetric group of all permutations of $1,\dots,q$. The same construction can be applied to exterior products and thus we can extend to a canonical isomorphism $S_q(\mathcal{V}^*)\otimes\Lambda_p(\mathcal{V}^*)\cong(S_q\mathcal{V}\otimes\Lambda_p\mathcal{V})^*$.

At the level of bases, this isomorphism takes the following form. We denote again by $\{x^1,\dots,x^n\}$ a basis of $\mathcal{V}$ and by $\{y_1,\dots,y_n\}$ the corresponding dual basis of $\mathcal{V}^*$. Then the monomials x^μ with $|\mu|=q$ form a basis of $S_q\mathcal{V}$ and similarly the monomials $y_\mu=y_1^{\mu_1}\cdots y_n^{\mu_n}$ with $|\mu|=q$ form a basis of $S_q(\mathcal{V}^*)$. However, these two bases are *not* dual to each other, since according to (6.9) $y_\mu(x^\nu)=\mu!\delta_\mu^\nu$. Thus the dual basis consists of the *divided powers* $\frac{y_\mu}{\mu!}$. For the exterior algebra no such combinatorial factor arises, as the evaluation of the expression corresponding to the right hand side of (6.9) on basis vectors yields only one non-vanishing summand.

Another way to see that the dualisation leads to the divided powers is based on the coalgebra approach of Remark 6.1.2. If we substitute in the definition of the polynomial de Rham complex the symmetric algebra $S\mathcal{V}$ by the symmetric coalgebra $\mathfrak{S}\mathcal{V}$, then the dual algebra is $S(\mathcal{V}^*)$ and the evaluation of the polynomial convolution product (B.67) leads to (6.9). ◁

Proposition 6.1.6. $\bigl(R(\mathcal{V})^*,\delta^*\bigr)$ *is isomorphic to* $\bigl(K(\mathcal{V}^*),\partial\bigr)$.

Proof. There only remains to show that ∂ is indeed the pull-back of δ. Choosing the above described dual bases, this is a straightforward computation. By definition of the pull-back,

$$\delta^*\Big(\frac{y^\mu}{\mu!}\otimes y^I\Big)(x^\nu\otimes x^J)=\begin{cases}\nu_j\,\mathrm{sgn}\,(\{j\}\cup J) & \text{if } \exists j:\left\{\begin{array}{l}\mu=\nu-1_j\\ I=\{j\}\cup J\end{array}\right\},\\ 0 & \text{otherwise}\,.\end{cases} \tag{6.10}$$

Note that $\nu_j=\frac{\nu!}{\mu!}$ if $\mu=\nu-1_j$; hence we find that

$$\delta^*(y^\mu\otimes y^I)=\sum_{j=1}^{p}(-1)^{j+1}y^{\mu+1_{i_j}}\otimes y^{I\setminus\{i_j\}}\,. \tag{6.11}$$

Comparison with (6.6) yields the desired result. □

For reasons that will become apparent in the next chapter when we apply the here developed theory to differential equations, we prefer to consider the Koszul complex over $\mathcal{V}$ and the polynomial de Rham complex over its dual space $\mathcal{V}^*$. Thus we will always use $R(\mathcal{V}^*)$ and $K(\mathcal{V})$. If $\mathcal{U}$ is a further finite-dimensional $\Bbbk$-linear space with $\dim\mathcal{U}=m$, then we may extend to the tensor product complex $R(\mathcal{V}^*)\otimes\mathcal{U}$ and dually to $K(\mathcal{V})\otimes\mathcal{U}^*$. Everything we have done so far remains valid with trivial modifications, as the differentials of the complexes are essentially unaffected by this operation: one must only add a factor $u\in\mathcal{U}$ (or $v\in\mathcal{U}^*$, respectively) to each equation and consider all our computations above as componentwise.

Definition 6.1.7. Let $\mathcal{N}_q\subseteq S_q(\mathcal{V}^*)\otimes\mathcal{U}$ be a linear subspace. Its *prolongation* is the following linear subspace[1]

$$\mathcal{N}_{q,1}=\big\{f\in S_{q+1}(\mathcal{V}^*)\otimes\mathcal{U}\mid \delta(f)\in\mathcal{N}_q\otimes\mathcal{V}^*\big\}\,. \tag{6.12}$$

A sequence of linear subspaces $\big(\mathcal{N}_q\subseteq S_q(\mathcal{V}^*)\otimes\mathcal{U}\big)_{q\in\mathbb{N}_0}$ is called a *symbolic system* over $S(\mathcal{V}^*)$, if $\mathcal{N}_{q+1}\subseteq\mathcal{N}_{q,1}$ for all $q\in\mathbb{N}_0$.

We may equivalently introduce the prolongation as

$$\mathcal{N}_{q,1}=(\mathcal{V}^*\otimes\mathcal{N}_q)\cap\big(S_{q+1}(\mathcal{V}^*)\otimes\mathcal{U}\big) \tag{6.13}$$

where the intersection is understood to take place in $\mathcal{V}^*\otimes\big(S_q(\mathcal{V}^*)\otimes\mathcal{U}\big)$. This equivalence follows immediately from the definition of the differential δ. The extension to higher prolongations $\mathcal{N}_{q,r}\subseteq S_{q+r}(\mathcal{V}^*)\otimes\mathcal{U}$ proceeds then either by induction, $\mathcal{N}_{q,r+1}=(\mathcal{N}_{q,r})_{,1}$ for all $r\in\mathbb{N}$, or alternatively by generalising (6.13) to $\mathcal{N}_{q,r}=\big(\bigotimes_{i=1}^r\mathcal{V}^*\otimes\mathcal{N}_q\big)\cap\big(S_{q+r}(\mathcal{V}^*)\otimes\mathcal{U}\big)$ where the intersection is now understood to take place in $\bigotimes_{i=1}^r\mathcal{V}^*\otimes\big(S_q(\mathcal{V}^*)\otimes\mathcal{U}\big)$.

The notion of a symbolic system is fairly classical in the formal theory of differential equations (see Proposition 7.1.15). The next result shows, however, that if we

[1] In the literature the prolongation is usually denoted by $\mathcal{N}_q^{(1)}$. However, this convention clashes with our notation for projections and also with a notation introduced in the next section. For this reason we use the non-standard $\mathcal{N}_{q,1}$.

take the coalgebra point of view of the polynomial de Rham complex mentioned in Remark 6.1.2, then a symbolic system correspond to a simple algebraic structure.

Lemma 6.1.8. *Let $(\mathcal{N}_q)_{q\in\mathbb{N}_0}$ be a symbolic system. Then $\mathcal{N} = \bigoplus_{q=0}^{\infty}\mathcal{N}_q$ is a graded (right) subcomodule of the free $\mathfrak{S}\mathcal{V}$-comodule $\mathfrak{S}(\mathcal{V}^*)\otimes\mathcal{U}$. Conversely, the sequence $(\mathcal{N}_q)_{q\in\mathbb{N}_0}$ of the homogeneous components of any graded subcomodule $\mathcal{N}\subseteq\mathfrak{S}(\mathcal{V}^*)\otimes\mathcal{U}$ is a symbolic system.*

Proof. Let $(\mathcal{N}_q)_{q\in\mathbb{N}_0}$ be a symbolic system and $f\in\mathcal{N}_q$ a homogeneous element of degree q. Choosing a basis $\{y_1,\dots,y_n\}$ of $\mathcal{V}^*$ and identifying $S(\mathcal{V}^*)$ with the ring $\mathcal{P}=\Bbbk[y_1,\dots,y_n]$, we may consider f as polynomial. Now $\delta f\in\mathcal{N}_{q-1}\otimes\mathcal{V}^*$ implies $\partial f/\partial y_i\in\mathcal{N}_{q-1}$ for all $1\leq i\leq n$, since our differential δ is just the exterior derivative. Using induction, we thus find that $\partial^{|\mu|}f/\partial y^{\mu}\in\mathcal{N}_{q-r}$ for all μ with $|\mu|=r$. By the definition (B.65) of the polynomial coproduct Δ, this condition is equivalent to $\Delta(f)\in\mathcal{N}\otimes\mathfrak{S}(\mathcal{V}^*)$ and hence $\mathcal{N}$ is a subcomodule. For the converse, we must only revert every step of this argument to find that $\Delta(f)\in\mathcal{N}\otimes\mathfrak{S}(\mathcal{V}^*)$ implies that $\mathcal{N}_q\subseteq\mathcal{N}_{q-1,1}$ for all $q>0$. □

The proof above gives yet another description of the prolongation completely independent of any complex. If $\mathcal{P}=\Bbbk[y_1,\dots,y_n]$ is a polynomial ring in n variables and $\mathcal{N}_q\subseteq\mathcal{P}_q$ a homogeneous subspace, then its rth prolongation is given by

$$\mathcal{N}_{q,r}=\left\{f\in\mathcal{P}_{q+r}\mid\forall\mu\in(\mathbb{N}_0^n)_r:\frac{\partial f}{\partial y^{\mu}}\in\mathcal{N}_q\right\}. \tag{6.14}$$

Example 6.1.9. Let $\mathcal{V}$ be a two-dimensional vector space with basis $\{x,y\}$ and dual basis $\{\bar{x},\bar{y}\}$. The subspaces $\mathcal{N}_0=\Bbbk$, $\mathcal{N}_1=\mathcal{V}^*$ and $\mathcal{N}_q=\langle\bar{x}^q\rangle\subset S_q(\mathcal{V}^*)$ for $q\geq 2$ define a symbolic system where $\mathcal{N}_{q,1}=\mathcal{N}_{q+1}$ for all $q\geq 2$. Indeed, if $k+\ell=q$ with $\ell>0$, then $\delta(\bar{x}^k\bar{y}^\ell)=\bar{x}^{k-1}\bar{y}^\ell\otimes\bar{x}+\bar{x}^k\bar{y}^{\ell-1}\otimes\bar{y}$ so that the result lies in $\mathcal{N}_{q-1}\otimes\mathcal{V}^*$ only for $\ell=0$. We will see later that this symbolic system is associated with partial differential equations of the form $u_{yy}=F(\mathbf{x},u^{(1)})$, $u_{xy}=G(\mathbf{x},u^{(1)})$ for one unknown function u depending on $\mathbf{x}=(x,y)$.

Another simple symbolic system over the same dual space $\mathcal{V}^*$ is obtained by setting $\mathcal{N}_0=\Bbbk$, $\mathcal{N}_1=\mathcal{V}^*$, $\mathcal{N}_2=S_2(\mathcal{V}^*)$, $\mathcal{N}_3=\langle\bar{x}^2\bar{y},\bar{x}\bar{y}^2\rangle$, $\mathcal{N}_4=\langle\bar{x}^2\bar{y}^2\rangle$ and $\mathcal{N}_q=0$ for all $q\geq 5$. This system is related to partial differential equations of the form $u_{yyy}=F(\mathbf{x},u^{(2)})$, $u_{xxx}=G(\mathbf{x},u^{(2)})$. One can show that any such equation has a finite-dimensional solution space and this fact is reflected by the vanishing of the associated symbolic system beyond a certain degree. ◁

From now on, we will not distinguish between a symbolic system $(\mathcal{N}_q)_{q\in\mathbb{N}_0}$ and the corresponding subcomodule $\mathcal{N}\subseteq\mathfrak{S}(\mathcal{V}^*)\otimes\mathcal{U}$. We are mainly interested in subcomodules $\mathcal{N}$ where almost all components $\mathcal{N}_q$ are different from zero. It follows immediately from the rule (B.68) for cogeneration in $\mathfrak{S}(\mathcal{V}^*)$ that such comodules cannot be finitely cogenerated. However, the duality between $\mathfrak{S}(\mathcal{V}^*)$ and $S\mathcal{V}$ leads at least to the following result.

Corollary 6.1.10. *Let $(\mathcal{N}_q)_{q\in\mathbb{N}_0}$ be an arbitrary symbolic system. There exists an integer $q_0\geq 0$ such that $\mathcal{N}_{q+1}=\mathcal{N}_{q,1}$ for all $q\geq q_0$.*

Proof. By Proposition B.3.6, the annihilator $\mathcal{N}^0 \subseteq S\mathcal{V}\otimes\mathcal{U}^*$ is a submodule of a free module over the symmetric algebra $S\mathcal{V}$. Now $\mathcal{N}_{q+1} \subsetneq \mathcal{N}_{q,1}$ implies that any minimal basis of $\mathcal{N}^0$ contains at least one generator of degree $q+1$. Since, by Hilbert's Basis Theorem B.1.13, any polynomial ring in a finite number of variables and hence also the isomorphic symmetric algebra $S\mathcal{V}$ is Noetherian, an upper bound q_0 for such values q must exist. □

By this corollary, we may consider any symbolic system $\mathcal{N}$ as a kind of finitely cogenerated "differential comodule": since the truncated comodule $\mathcal{N}_{\leq q_0}$ is a finite-dimensional vector space, it is obviously finitely cogenerated and by repeated prolongations of the component $\mathcal{N}_{q_0}$ we obtain the remainder of the comodule $\mathcal{N}$. Thus we may conclude that every symbolic system $\mathcal{N}$ is uniquely determined by a finite number of generators.

Definition 6.1.11. Let $\mathcal{N}$ be a (right) graded comodule over the symmetric coalgebra $\mathcal{C} = \mathfrak{S}(\mathcal{V}^*)$. Its *Spencer complex* $\big(R(\mathcal{N}),\delta\big)$ is the cotensor[2] product complex $\mathcal{N}\boxtimes_{\mathcal{C}} R(\mathcal{V}^*)$. The *Spencer cohomology* of $\mathcal{N}$ is the corresponding bigraded cohomology; the cohomology group at $\mathcal{N}_q\otimes\Lambda_p(\mathcal{V}^*)$ is denoted by $H^{q,p}(\mathcal{N})$.

Since $\mathcal{N}\boxtimes_{\mathcal{C}}\mathcal{C}\cong\mathcal{N}$ for any $\mathcal{C}$-comodule $\mathcal{N}$, the components of the complex $\mathcal{N}\boxtimes_{\mathcal{C}} R(\mathcal{V}^*)$ are indeed just the vector spaces $\mathcal{N}_q\otimes\Lambda_p(\mathcal{V}^*)$. We will be mainly interested in the special case that $\mathcal{N}$ is a subcomodule of a free comodule $\mathcal{C}\otimes\mathcal{U}$ and then the differential in the Spencer complex $R(\mathcal{N})$ is simply the restriction of the differential δ in the polynomial de Rham complex $R(\mathcal{V}^*)$ to the subspaces $\mathcal{N}_q\otimes\Lambda_p(\mathcal{V}^*)\subseteq\mathfrak{S}_q(\mathcal{V}^*)\otimes\Lambda_p(\mathcal{V}^*)\otimes\mathcal{U}$; this observation explains why we keep the notation δ for the differential. One can also verify by direct computation that this restriction makes sense whenever the sequence $(\mathcal{N}_q)_{q\in\mathbb{N}_0}$ defines a symbolic system (this requires basically the same computation as the one showing the equivalence of the two definitions (6.12) and (6.13) of the prolongation); in fact, this restriction is the classical approach to define the Spencer complex.

Remark 6.1.12. It is important to emphasise here that the Spencer cohomology is *bigraded*. Ignoring the polynomial degree and considering only the exterior degree, we obtain the comodules $H^p(\mathcal{N}) = \bigoplus_{q=0}^{\infty} H^{q,p}(\mathcal{N})$. For these, another point of view is possible. According to Remark B.3.5 any free comodule is injective and we have exactly the situation of the definition of cotorsion groups (see Appendix B.3): we take an injective coresolution (of $\mathbb{k}$) and cotensor it with a comodule $\mathcal{N}$. Thus we may consider the Spencer cohomology as the right derived functor of $\mathcal{N}\boxtimes_{\mathcal{C}}\cdot$ and identify $H^p(\mathcal{N}) = \operatorname{Cotor}^p_{\mathcal{C}}(\mathcal{N},\mathbb{k})$.

As discussed in Appendix B.2 for arbitrary derived functors, the definition of the groups $\operatorname{Cotor}^p_{\mathcal{C}}(\mathcal{N},\mathbb{k})$ is independent of the coresolution used for its computation or, more precisely, the results obtained with different coresolutions are isomorphic to each other. However, given some other way to explicitly determine $\operatorname{Cotor}^p_{\mathcal{C}}(\mathcal{N},\mathbb{k})$, say via a coresolution of $\mathcal{N}$, it may be a non-trivial task to recover the bigrading of the Spencer cohomology. ◁

[2] For the definition of the cotensor product see (B.62).

Lemma 6.1.13. *Let $\mathcal{N} \subseteq \mathfrak{S}(\mathcal{V}^*) \otimes \mathcal{U}$ be an arbitrary symbolic system. Then we have $H^{q,0}(\mathcal{N}) = 0$ and $\dim H^{q-1,1}(\mathcal{N}) = \dim \left(\mathcal{N}_{q-1,1}/\mathcal{N}_q\right)$ for all $q > 0$.*

Proof. The first claim follows immediately from the formal Poincaré Lemma (i. e. Proposition 6.1.4). For the second claim we consider a non-vanishing element $f \in \mathcal{N}_{q-1,1} \setminus \mathcal{N}_q$. Then $g = \delta f \in \ker \delta\big|_{\mathcal{N}_{q-1} \otimes \mathcal{V}^*}$ and, again because of the formal Poincaré Lemma, $g \neq 0$. However, by construction, $g \notin \operatorname{im} \delta\big|_{\mathcal{N}_q}$ and hence we find that $0 \neq [g] \in H^{q-1,1}(\mathcal{N})$. This observation implies immediately the inequality $\dim H^{q-1,1}(\mathcal{N}) \geq \dim \left(\mathcal{N}_{q-1,1}/\mathcal{N}_q\right)$. Conversely, consider an arbitrary non-vanishing cohomology class $[g] \in H^{q-1,1}(\mathcal{N})$. Again by the formal Poincaré Lemma, an element $f \in \mathfrak{S}_q(\mathcal{V}^*) \otimes \mathcal{U}$ exists such that $g = \delta f$ and, by definition of the prolongation, $f \in \mathcal{N}_{q-1,1} \setminus \mathcal{N}_q$. Thus we also have the opposite inequality $\dim H^{q-1,1}(\mathcal{N}) \leq \dim \left(\mathcal{N}_{q-1,1}/\mathcal{N}_q\right)$. □

Together with Corollary 6.1.10, this result implies that $H^{q,1}(\mathcal{N}) = 0$ for a sufficiently high degree q. We will see below that a similar finiteness result holds in fact for all exterior degrees p. Dualisation of Definition 6.1.11 leads to the following classical construction in commutative algebra with a polynomial module.

Definition 6.1.14. Let $\mathcal{M}$ be a graded module over the algebra $\mathcal{P} = S\mathcal{V}$. Its *Koszul complex* $\left(K(\mathcal{M}), \partial\right)$ is the tensor product complex $\mathcal{M} \otimes_{\mathcal{P}} K(\mathcal{V})$. The *Koszul homology* of $\mathcal{M}$ is the corresponding bigraded homology; the homology group at $\mathcal{M}_q \otimes \Lambda_p \mathcal{V}$ is denoted by $H_{q,p}(\mathcal{M})$.

Remark 6.1.15. We observed already above that the Koszul complex defines a free resolution of the field $\Bbbk$. Hence, as for the Spencer cohomology, we may take another point of view and consider the Koszul homology as the right derived functor of $\mathcal{M} \otimes_{\mathcal{P}} \cdot$. According to Definition B.2.36 of the torsion groups, this observation leads to the identification $H_p(\mathcal{M}) = \bigoplus_{q=0}^{\infty} H_{q,p}(\mathcal{M}) = \operatorname{Tor}_p^{\mathcal{P}}(\mathcal{M}, \Bbbk)$ where we consider $\Bbbk$ as a $\mathcal{P}$-module as in Remark B.1.9. But again this interpretation ignores the natural bigrading of the Koszul complex $K(\mathcal{M})$.

An alternative way to compute $\operatorname{Tor}_p^{\mathcal{P}}(\mathcal{M}, \Bbbk)$ consists of using a free resolution of the module $\mathcal{M}$. If $\mathcal{C} \to \mathcal{M} \to 0$ is such a resolution, then the Koszul homology $H_\bullet(\mathcal{M})$ is isomorphic to the homology of the tensor product complex $\mathcal{C} \otimes_{\mathcal{P}} \Bbbk$. Each component in $\mathcal{C}$ is of the form $\mathcal{P}^m$ and therefore $\mathcal{P}^m \otimes_{\mathcal{P}} \Bbbk = \Bbbk^m$. Now assume that we actually have a *minimal* resolution. In this case all differentials in $\mathcal{C}$ possess a positive degree and it follows from the $\mathcal{P}$-action on $\Bbbk$ that the induced differential on the complex $\mathcal{C} \otimes_{\mathcal{P}} \Bbbk$ is the zero map. Hence we find that $H_\bullet(\mathcal{M}) \cong \mathcal{C} \otimes_{\mathcal{P}} \Bbbk$. Since a minimal resolution is necessarily graded, we even recover the bigrading of the Koszul complex and find that the dimensions $\dim H_{q,p}(\mathcal{M})$ are just the bigraded Betti numbers $\beta_{q,p}\mathcal{M}$. In this sense we may say that the Koszul homology corresponds to a minimal free resolution. ◁

Lemma 6.1.16. *Let $\mathcal{M}$ be a graded $\mathcal{P}$-module. Then $H_{q,0}(\mathcal{M}) = \mathcal{M}_q/(\mathcal{V} \cdot \mathcal{M}_{q-1})$ and thus $\dim H_{q,0}(\mathcal{M})$ gives the number of generators of degree q in any minimal basis of $\mathcal{M}$. Furthermore,*

$$H_{q,n}(\mathcal{M}) \cong \{m \in \mathcal{M}_q \mid \operatorname{Ann}(m) = S_+\mathcal{V}\} . \tag{6.15}$$

Proof. The first assertion follows trivially from the definition of the Koszul homology. Elements of $H_{q,n}(\mathcal{M})$ are represented by cycles in $\mathcal{M}_q \otimes \Lambda_n\mathcal{V}$. If $\{x^1,\dots,x^n\}$ is a basis of $\mathcal{V}$, these are forms $\omega = m \otimes x^1 \wedge \cdots \wedge x^n$ and the condition $\partial\omega = 0$ is equivalent to $x^i m = 0$ for $1 \le i \le n$. □

Lemma 6.1.17. *Let $\mathcal{M}$ be a graded $\mathcal{P}$-module. Multiplication by an arbitrary element of $S_+\mathcal{V}$ induces the zero map on the Koszul homology $H_\bullet(\mathcal{M})$.*

Proof. We first observe that if $\omega \in \mathcal{M}_q \otimes \Lambda_p\mathcal{V}$ is a cycle, i. e. $\partial\omega = 0$, then for any $v \in \mathcal{V}$ the form $v\omega$ is a boundary, i. e. $v\omega \in \operatorname{im}\partial$. Indeed,

$$\partial\bigl(v \wedge \omega\bigr) = -v \wedge (\partial\omega) + v\omega = v\omega . \tag{6.16}$$

Since ∂ is $S\mathcal{V}$-linear, this observation remains true, if we take for v an arbitrary element of $S_+\mathcal{V}$, i. e. a polynomial without constant term. □

By Proposition B.3.6, each subcomodule $\mathcal{N} \subseteq \mathfrak{S}(\mathcal{V}^*) \otimes \mathcal{U}$ induces a submodule $\mathcal{M} = \mathcal{N}^0 \subseteq S\mathcal{V} \otimes \mathcal{U}^*$, its annihilator. Conversely, the annihilator of any submodule $\mathcal{M} \subseteq S\mathcal{V} \otimes \mathcal{U}^*$ defines a subcomodule $\mathcal{N} = \mathcal{M}^0 \in \mathfrak{S}(\mathcal{V}^*) \otimes \mathcal{U}$. In view of the duality between the polynomial de Rham and the Koszul complex, we expect a simple relation between the Spencer cohomology $H^\bullet(\mathcal{N})$ of the comodule $\mathcal{N}$ and the Koszul homology $H_\bullet(\mathcal{N}^0)$ of its annihilator $\mathcal{N}^0$.

Such a relation is easily obtained with the help of the $S\mathcal{V}$-module $\mathcal{N}^*$ dual to $\mathcal{N}$. If we take the dual $\pi^* : \bigl((\mathfrak{S}(\mathcal{V}^*) \otimes \mathcal{U})/\mathcal{N}\bigr)^* \to \bigl(\mathfrak{S}(\mathcal{V}^*) \otimes \mathcal{U}\bigr)^* = S\mathcal{V} \otimes \mathcal{U}^*$ of the projection $\pi : \mathfrak{S}(\mathcal{V}^*) \otimes \mathcal{U} \to \bigl(\mathfrak{S}(\mathcal{V}^*) \otimes \mathcal{U}\bigr)/\mathcal{N}$, then $\operatorname{im}\pi^* = \mathcal{N}^0$. Like for any map, we have for π the canonical isomorphism $\operatorname{coker}(\pi^*) \cong (\ker\pi)^* = \mathcal{N}^*$ and hence may identify $\mathcal{N}^*$ with the factor module $(S\mathcal{V} \otimes \mathcal{U}^*)/\mathcal{N}^0$.

Proposition 6.1.18. *Let $\mathcal{N} \subseteq \mathfrak{S}(\mathcal{V}^*) \otimes \mathcal{U}$ be a symbolic system. Then for all $q \ge 0$ and $1 \le p \le n$*

$$\bigl(H^{q,p}(\mathcal{N})\bigr)^* \cong H_{q,p}(\mathcal{N}^*) \cong H_{q+1,p-1}(\mathcal{N}^0) \tag{6.17}$$

where the second isomorphism is induced by the Koszul differential ∂.

Proof. The first isomorphism follows from Proposition 6.1.6. For the second one we note that the considerations above lead to the short exact sequence

$$0 \longrightarrow \mathcal{N}^0 \stackrel{\iota}{\hookrightarrow} S\mathcal{V} \otimes \mathcal{U}^* \stackrel{\pi}{\longrightarrow} \mathcal{N}^* \longrightarrow 0 \tag{6.18}$$

where the first map is the natural inclusion and the second one the canonical projection. Tensoring with the vector space $\Lambda_p\mathcal{V}$ is a flat functor (cf. Remark B.2.16) and hence does not affect the exactness so that we obtain a short exact sequence of the corresponding Koszul complexes:

$$0 \longrightarrow K(\mathcal{N}^0) \hookrightarrow K(\mathcal{V}) \otimes \mathcal{U}^* \longrightarrow K(\mathcal{N}^*) \longrightarrow 0 . \tag{6.19}$$

Since $K(\mathcal{V}) \otimes \mathcal{U}^*$ is exact in positive exterior degree, the long exact homological sequence (see Appendix B.2 for a discussion) for (6.19) yields an isomorphism $H_p(\mathcal{N}^*) \to H_{p-1}(\mathcal{N}^0)$. Furthermore, as the maps in (6.18) are just the natural inclusion and the restriction to the comodule $\mathcal{N}$, respectively, it follows straightforwardly from the construction of the connecting homomorphism that this isomorphism is induced by the Koszul differential ∂. Hence, taking the bigrading into account, we obtain an isomorphism $H_{q,p}(\mathcal{N}^*) \to H_{q+1,p-1}(\mathcal{N}^0)$. □

Remark 6.1.19. From a computational point of view, it is often more convenient to work with the annihilator $\mathcal{N}^0$ instead of the dual $\mathcal{N}^*$. Our proof of Proposition 6.1.18 gave the isomorphism only in one direction. However, Lemma 6.1.3 allows us to derive an explicit expression for the inverse map. Let $\omega \in \mathcal{N}_q^* \otimes \Lambda_p\mathcal{V}$ be a cycle and $\tilde{\omega} \in S_q\mathcal{V} \otimes \Lambda_p\mathcal{V} \otimes \mathcal{U}^*$ an arbitrary form such that $\pi(\tilde{\omega}) = \omega$. Then $\partial\omega = 0$ implies that $\bar{\omega} = \partial\tilde{\omega} \in \mathcal{N}_{q+1}^0 \otimes \Lambda_{p-1}\mathcal{V}$ and the isomorphism used in the proof above maps $[\omega] \mapsto [\bar{\omega}]$. By Lemma 6.1.3, $\delta\bar{\omega} = (p+q)\tilde{\omega} - \partial(\delta\tilde{\omega})$ and hence $\left[\frac{1}{p+q}\delta\bar{\omega}\right] = [\tilde{\omega}]$. But this observation implies immediately that the inverse of our isomorphism is given by the map $[\bar{\omega}] \mapsto \left[\frac{1}{p+q}\pi(\delta\bar{\omega})\right]$. ◁

Remark 6.1.20. It follows from (the proof of) Proposition 6.1.18 and Lemma 6.1.16 that for modules of the form $\mathcal{M} = (S\mathcal{V} \otimes \mathcal{U}^*)/\mathcal{N}^0$ the exterior degree 0 part of the Koszul homology is given by $H_{0,0}(\mathcal{M}) \cong \mathbb{k}^m$ with $m = \dim\mathcal{U}$ and $H_{q,0} = 0$ for $q > 0$. In exterior degree 1 we have the isomorphism $H_{q,1}(\mathcal{M}) \cong \mathcal{N}_q^0/(\mathcal{V} \cdot \mathcal{N}_{q-1}^0)$ so that now $\dim H_{q,1}(\mathcal{M})$ equals the number of minimal generators of $\mathcal{N}^0$ of degree q. ◁

For us, one of the most important properties of the Spencer cohomology is the following finiteness result which obviously requires the bigrading.

Theorem 6.1.21. *Let $\mathcal{N} \subseteq \mathfrak{S}(\mathcal{V}^*) \otimes \mathcal{U}$ be a symbolic system. There exists an integer $q_0 \geq 0$ such that $H^{q,p}(\mathcal{N}) = 0$ for all $q \geq q_0$ and $0 \leq p \leq n$. Dually, let $\mathcal{M}$ be a finitely generated graded polynomial module. There exists an integer $q_0 \geq 0$ such that $H_{q,p}(\mathcal{M}) = 0$ for all $q \geq q_0$ and $0 \leq p \leq n$.*

Proof. By the duality of Proposition 6.1.18, it suffices to show the assertion for the polynomial module $\mathcal{M}$. Here it is a simple consequence of Lemma 6.1.17 above. The cycles in $\mathcal{M} \otimes \Lambda_p\mathcal{V}$ form a finitely generated $S\mathcal{V}$-module. Thus there exists an integer $q_0 \geq 0$ such that the polynomial degree of all elements in a finite generating set of it is less than q_0. All cycles of higher polynomial degree can then be written as linear combinations of these generating ones with polynomial coefficients without constant terms. According to Lemma 6.1.17, they are therefore boundaries. Hence indeed $H_{q,p}(\mathcal{M}) = 0$ for all $q \geq q_0$. □

Definition 6.1.22. The *degree of involution* of the $\mathfrak{S}(\mathcal{V}^*)$-comodule $\mathcal{N}$ is the smallest value q_0 such that $H^{q,p}(\mathcal{N}) = 0$ for all $q \geq q_0$ and $0 \leq p \leq n = \dim\mathcal{V}$. More generally, we say that $\mathcal{N}$ is *s-acyclic* at degree q_0 for an integer $0 \leq s \leq n$, if $H^{q,p}(\mathcal{N}) = 0$ for all $q \geq q_0$ and $0 \leq p \leq s$. A comodule $\mathcal{N}$ that is n-acyclic at a degree q_0 is called *involutive* at degree q_0. Dually, we call an $S\mathcal{V}$-module $\mathcal{M}$ *involutive* at degree q_0, if its Koszul homology vanishes beyond degree q_0: $H_{q,p}(\mathcal{M}) = 0$ for all $q \geq q_0$ and $0 \leq p \leq n$.

With this terminology we may reformulate Lemma 6.1.13 as follows: if the symbolic system $\mathcal{N}$ is such that its annihilator $\mathcal{N}^0$ can be generated in degree less than or equal to q_0, then $\mathcal{N}$ is 1-acyclic at degree q_0, and if conversely q_0 is the smallest degree at which $\mathcal{N}$ is 1-acyclic, then any generating set of $\mathcal{N}^0$ contains an element of degree q_0. We will see later in Section 7.2 that 2-acyclicity is relevant for checking formal integrability. It follows trivially from the definition that if $\mathcal{N}$ is involutive at some degree q_0, then it is also involutive at any higher degree $q \geq q_0$.

Remark 6.1.23. According to Remark 6.1.15, the Koszul homology of a module $\mathcal{M}$ is equivalent to its minimal free resolution. This fact implies immediately that the degree of involution of $\mathcal{M}$ is nothing but $\operatorname{reg}\mathcal{M}+1$, where $\operatorname{reg}\mathcal{M}$ denotes again the Castelnuovo–Mumford regularity of the module $\mathcal{M}$. In the case that we are dealing with a factor module $\mathcal{M} = (S\mathcal{V} \otimes \mathcal{U}^*)/\mathcal{N}^0$, it follows easily from Proposition 6.1.18 that the degree of involution of $\mathcal{M}$ equals $\operatorname{reg}\mathcal{N}^0$. ◁

For complexity considerations, it is of interest to bound for a given comodule $\mathcal{N}$ or module $\mathcal{M}$, respectively, its degree of involution. In our applications to differential equations we will be mainly concerned with the special case that $\mathcal{M}$ is a submodule of a free $S\mathcal{V}$-module of rank m generated by elements of degree q. For this situation, a bound $\bar{q}$ exists depending only on the values of n, m and q. It may be expressed as a nested recursion relation:

$$\begin{aligned} \bar{q}(n,m,q) &= \bar{q}\left(n, m\binom{q+n-1}{n}, 1\right), \\ \bar{q}(n,m,1) &= m\binom{\bar{q}(n-1,m,1)+n}{n-1} + \bar{q}(n-1,m,1) + 1, \\ \bar{q}(0,m,1) &= 0. \end{aligned} \tag{6.20}$$

Table 6.1 shows $\bar{q}(n,m,1)$ for different values of m and n. One sees that the values rapidly explode, if the dimension n of the underlying vector space $\mathcal{V}$ increases. The situation is still worse for modules generated in higher order. It seems to be an open question whether this bound is sharp, i. e. whether or not for some modules the degree of involution is really that high. Fortunately, $\bar{q}(n,m,q)$ yields usually a coarse over-estimate of the actual degree of involution.

$n\backslash m$	1	2	3	4
1	2	3	4	5
2	7	14	23	34
3	53	287	999	2.699
4	29.314	8.129.858	503.006.503	13.151.182.504

Table 6.1 $\bar{q}(n,m,1)$ for different values of m and n.

Example 6.1.24. Consider for $m = 1$ the monomial ideal $I = \langle x^q, y^q \rangle \subset \mathcal{P} = \Bbbk[x,y]$ in $n = 2$ variables for some value $q > 0$. For $q = 2$ we encountered this ideal already in Example 3.1.10; for $q = 3$ it is the annihilator $\mathcal{N}^0$ of the second symbolic system $\mathcal{N}$ in Example 6.1.9. A trivial computation yields that the only non-vanishing Koszul homology modules are $H_{q,0}(\mathcal{I}) = \langle [x^q], [y^q] \rangle$ and $H_{2q-1,1}(\mathcal{I}) = \langle [x^q y^{q-1} \otimes y - x^{q-1} y^q \otimes x^1] \rangle$. Hence the degree of involution of $\mathcal{I}$ is $2q-1$. By contrast, evaluation of the bound (6.20) yields

$$\bar{q}(2,1,q) = \frac{1}{4}q^4 + \frac{1}{2}q^3 + \frac{9}{4}q^2 + 2q + 2\,, \tag{6.21}$$

i. e. a polynomial in q of degree 4. ◁

6.2 Cartan's Test

In this section we study some explicit criteria for deciding whether or not a given (co)module is involutive which do not require to determine the full (co)homology of the (co)module. As such criteria appeared first (though only in an implicit manner) in Cartan's theory of exterior differential systems, they are collectively known under the name *Cartan's test*. We start with a symbolic system $\mathcal{N} \subseteq \mathfrak{S}(\mathcal{V}^*) \otimes \mathcal{U}$. As before, let $\{x^1, \dots, x^n\}$ be an ordered basis of the underlying vector space $\mathcal{V}$ and $\{y_1, \dots, y_n\}$ the corresponding dual basis of $\mathcal{V}^*$. Then we may introduce for any index $0 \le k \le n$ the following subspaces of the homogeneous component $\mathcal{N}_q$ of degree q:

$$\begin{aligned} \mathcal{N}_q^{(k)} &= \Big\{ f \in \mathcal{N}_q \mid f(x^i, v_2, \dots, v_q) = 0\,, \\ &\qquad\qquad \forall 1 \le i \le k\,,\ \forall v_2, \dots, v_q \in \mathcal{V} \Big\} \\ &= \Big\{ f \in \mathcal{N}_q \mid \frac{\partial f}{\partial y_i} = 0\ \forall 1 \le i \le k \Big\}\,. \end{aligned} \tag{6.22}$$

In the first line we interpreted elements of $\mathcal{N}_q$ as multilinear maps on $\mathcal{V}$ and in the last line we considered them as polynomials in the dual "variables" $y_1, \dots, y_n$.

Obviously, these subspaces define a filtration

$$0 = \mathcal{N}_q^{(n)} \subseteq \mathcal{N}_q^{(n-1)} \subseteq \cdots \subseteq \mathcal{N}_q^{(1)} \subseteq \mathcal{N}_q^{(0)} = \mathcal{N}_q\,. \tag{6.23}$$

Obviously, this filtration (and in particular the dimensions of the involved subspaces) depends on the chosen basis for $\mathcal{V}^*$. It turns out that certain bases are distinguished, as they show a special behaviour with respect to these dimensions. We speak here again of δ-regularity; the relationship to the notion of δ-regularity introduced in Section 4.3 will be clarified in the next section.

Definition 6.2.1. Let $\mathcal{N} \subseteq \mathfrak{S}(\mathcal{V}^*) \otimes \mathcal{U}$ be a symbolic system. With respect to a given basis $\{y_1, \dots, y_n\}$ of the dual space $\mathcal{V}^*$, we define the *Cartan characters* of the homogeneous component $\mathcal{N}_q$ as

$$\alpha_q^{(k)} = \dim \mathcal{N}_q^{(k-1)} - \dim \mathcal{N}_q^{(k)}\,, \qquad 1 \leq k \leq n\,. \tag{6.24}$$

The basis $\{y_1, \dots, y_n\}$ is called *δ-regular* for $\mathcal{N}_q$, if the sum $\sum_{k=1}^{n} k\alpha_q^{(k)}$ attains a minimal value.

Because of the coefficient k appearing in the above weighted sum of the Cartan characters, we may describe δ-regularity in words as follows. We first select all bases of the dual space $\mathcal{V}^*$ which lead to a minimal value of the highest Cartan character $\alpha_q^{(n)}$. Among these we choose those which yield a minimal value for the second highest Cartan character $\alpha_q^{(n-1)}$ and so on. The bases which remain at the end of this process are the δ-regular ones.

We know from the proof of Lemma 6.1.8 that the differentiation with respect to any dual variable y_k maps the homogeneous component $\mathcal{N}_{q+1}$ into the component $\mathcal{N}_q$. It follows trivially from the definition of the subspaces $\mathcal{N}_q^{(k)}$ that we may consider the restrictions $\partial_{y_k,q} : \mathcal{N}_{q+1}^{(k-1)} \to \mathcal{N}_q^{(k-1)}$.

Proposition 6.2.2. *Let $\mathcal{N} \subseteq \mathfrak{S}(\mathcal{V}^*) \otimes \mathcal{U}$ be a symbolic system and $\{y_1, \dots, y_n\}$ a basis of $\mathcal{V}^*$. Then we have for any degree $q \geq 0$ the inequality*

$$\dim \mathcal{N}_{q+1} \leq \sum_{k=0}^{n-1} \dim \mathcal{N}_q^{(k)} = \sum_{k=1}^{n} k\alpha_q^{(k)}\,. \tag{6.25}$$

Equality holds, if and only if the restricted maps $\partial_{y_k,q} : \mathcal{N}_{q+1}^{(k-1)} \to \mathcal{N}_q^{(k-1)}$ are surjective for all values $1 \leq k \leq n$.

Proof. By definition of the subspaces $\mathcal{N}_q^{(k)}$, we have the exact sequences

$$0 \longrightarrow \mathcal{N}_{q+1}^{(k)} \hookrightarrow \mathcal{N}_{q+1}^{(k-1)} \xrightarrow{\partial_{y_k,q}} \mathcal{N}_q^{(k-1)} \tag{6.26}$$

implying the inequalities $\dim \mathcal{N}_{q+1}^{(k-1)} - \dim \mathcal{N}_{q+1}^{(k)} \leq \dim \mathcal{N}_q^{(k-1)}$. Summing over all values $0 \leq k \leq n$ yields immediately the inequality (6.25). Equality in (6.25) is obtained, if and only if in all these dimension relations equality holds. But this is the case, if and only if all the maps $\partial_{y_k,q}$ are surjective. □

Proposition 6.2.3. *The symbolic system $\mathcal{N} \subseteq \mathfrak{S}(\mathcal{V}^*) \otimes \mathcal{U}$ is involutive at some degree $q_0 \geq 0$, if and only if a basis $\{y_1, \dots, y_n\}$ of $\mathcal{V}^*$ can be chosen such that the maps $\partial_{y_k,q} : \mathcal{N}_{q+1}^{(k-1)} \to \mathcal{N}_q^{(k-1)}$ are surjective for all $q \geq q_0$ and all $1 \leq k \leq n$.*

Proof. We prove only one direction; the converse will follow from our subsequent dual considerations for the Koszul homology of $\mathcal{N}^*$ (see Remark 6.2.14). Let us take an arbitrary cycle $\omega \in \mathcal{N}_q \otimes \Lambda_p \mathcal{V}^*$ with $1 \leq p \leq n$; we want to show that if in

(6.25) equality holds, then a form $\eta \in \mathcal{N}_{q+1} \otimes \Lambda_{p-1}\mathcal{V}^*$ exists with $\omega = \delta\eta$ implying that the cohomology group $H^{q,p}(\mathcal{N})$ vanishes.

We do this in an iterative process, assuming first that the exterior part of ω depends only on $y_k, y_{k+1}, \dots, y_n$. Then we may decompose $\omega = \omega_1 + y_k \wedge \omega_2$ where the exterior parts of both ω_1 and ω_2 depend only on $y_{k+1}, \dots, y_n$. Since ω is a cycle, we have $\delta\omega = \delta\omega_1 - y_k \wedge \delta\omega_2 = 0$. Consider now in this equation those terms where the exterior part is of the form $y_\ell \wedge y_k \wedge \cdots$ with $\ell < k$. Such terms occur only in the second summand and hence we must have $\partial\omega_2/\partial y_l = 0$ for all $1 \leq \ell < k$. This observation implies $\omega_2 \in \mathcal{N}_q^{(k-1)} \otimes \Lambda_{p-1}\mathcal{V}^*$.

By assumption, $\partial_{y_k,q} : \mathcal{N}_{q+1}^{(k-1)} \to \mathcal{N}_q^{(k-1)}$ is surjective so that there exists a form $\eta^{(k)} \in \mathcal{N}_{q+1}^{(k-1)} \otimes \Lambda_{p-1}\mathcal{V}^*$ such that $\partial_{y_k}\eta^{(k)} = \omega_2$. Hence the exterior part of the form $\omega^{(k)} = \omega - \delta\eta^{(k)}$ depends only on $y_{k+1}, \dots, y_n$ and we can iterate. Thus, starting with $k = 1$, we successively obtain $\omega = \delta\bigl(\eta^{(1)} + \cdots + \eta^{(n-1)}\bigr)$. □

While Proposition 6.2.3 is nice from a theoretical point of view, it is not very useful computationally, as we must check infinitely many conditions, namely one for each degree $q \geq q_0$ (thus we are in the same situation as for deciding formal integrability directly on the basis of Definition 2.3.15). Under modest assumptions on the symbolic system $\mathcal{N}$, it suffices to consider only the one degree q_0 and then we obtain an effective criterion for involution representing an algebraic reformulation of the classical Cartan test in the theory of exterior differential systems. It uses only linear algebra with the two finite-dimensional components $\mathcal{N}_{q_0}$ and $\mathcal{N}_{q_0+1}$ (note, however, that the test can only be applied in δ-regular bases). In particular, it is not necessary to determine explicitly any Spencer cohomology group. In the context of differential equations, this observation will later translate into the fact that it is easier to check involution than formal integrability.

Theorem 6.2.4 (Cartan's Test). *Let $\mathcal{N} \subseteq \mathfrak{S}(\mathcal{V}^*) \otimes \mathcal{U}$ be a symbolic system such that $\mathcal{N}_{q,1} = \mathcal{N}_{q+1}$ for all $q \geq q_0$. Then $\mathcal{N}$ is involutive at the degree q_0, if and only if a basis $\{y_1, \dots, y_n\}$ of $\mathcal{V}^*$ exists such that we have equality in* (6.25) *for $q = q_0$.*

Essentially, we will prove this theorem by reducing it to Proposition 4.2.7 asserting that for a continuous division like the Pommaret division local involution implies involution. In order to make our presentation independent of the rather technical material in Chapter 4, we first explicitly formulate the special case of Proposition 4.2.7 that we will need in the sequel and provide a simple direct proof of it.

Proposition 6.2.5. *Let $\mathcal{H}_q \subset \mathcal{P}_q$ be a finite triangular set of homogeneous polynomials of degree q which is locally involutive for the Pommaret division and a term order $\prec$. Then the set*

$$\mathcal{H}_{q+1} = \bigl\{x^i h \mid h \in \mathcal{H}_q,\ x^i \in X_{P,\prec}(h)\bigr\} \subset \mathcal{P}_{q+1} \tag{6.27}$$

is also triangular and locally involutive (by induction this assertion obviously implies that $\mathcal{H}_q$ is a Pommaret basis of $\langle\mathcal{H}_q\rangle$ for the term order $\prec$).

Proof. It is trivial to see that $\mathcal{H}_{q+1}$ is again triangular (all leading terms are different). For showing that it is also locally involutive, we consider an element $x^i h \in \mathcal{H}_{q+1}$. By construction, $\operatorname{cls}(x^i h) = i \leq \operatorname{cls} h$. We must show that for any non-multiplicative index $i < j \leq n$ the polynomial $x^j(x^i h)$ is expressible as a linear combination of polynomials $x^k \bar{h}$ where $\bar{h} \in \mathcal{H}_{q+1}$ and $x^k \in X_{P,\prec}(\bar{h})$. In the case that $i < j \leq \operatorname{cls} h$, this is trivial, as we may choose $\bar{h} = x^j h$ and $k = i$.

Otherwise x^j is non-multiplicative for h and since $\mathcal{H}_q$ is assumed to be locally involutive, the polynomial $x^j h$ can be written as a $\mathbb{k}$-linear combination of elements of $\mathcal{H}_{q+1}$. For exactly one summand $\bar{h}$ in this linear combination $\operatorname{lt} \bar{h} = \operatorname{lt}(x^j h)$ and hence $x^i \in X_{P,\prec}(\bar{h})$. If x^i is also multiplicative for all other summands, we are done. If the variable x^i is non-multiplicative for some summand $\bar{h}' \in \mathcal{H}_{q+1}$, then we analyse the product $x^i \bar{h}'$ in the same manner writing $\bar{h}' = x^k h'$ for some $h' \in \mathcal{H}_q$. Since $\operatorname{lt} \bar{h}' \prec \operatorname{lt}(x^j h)$, this process terminates after a finite number of steps leading to an involutive standard representation of $x^j(x^i h)$. □

In order to make contact with the Cartan test, we reformulate the condition that the finite set $\mathcal{H}_q \subset \mathcal{P}_q$ is triangular for the special case that the leading terms are determined with respect to the degree reverse lexicographic order. If we denote by $\beta_q^{(k)}$ the number of generators in $\mathcal{H}_q$ where the leading exponent is of class k (cf. Remark 4.3.7 and Lemma 8.2.1), then obviously the set $\mathcal{H}_{q+1}$ contains $\sum_{k=1}^n k\beta_q^{(k)}$ elements. Thus, if we denote the ideal $\langle \mathcal{H}_q \rangle$ generated by $\mathcal{H}_q$ by $\mathcal{N}^0$, then the inequality

$$\dim \mathcal{N}^0_{q+1} \geq \sum_{k=1}^{n} k\beta_q^{(k)} \tag{6.28}$$

always holds and $\mathcal{H}_q$ is locally involutive, if and only if we have equality. We will show now that the similarity between (6.25) and (6.28) is not only formal but that in fact the two inequalities are equivalent.

Proof (of Theorem 6.2.4). It follows immediately from Proposition 6.1.18 that the degrees of involution of the given symbolic system $\mathcal{N}$ and of the dual factor module $\mathcal{M} = (S\mathcal{V} \otimes \mathcal{U}^*)/\mathcal{N}^0 \cong \mathcal{N}^*$, respectively, coincide and that according to Remark 6.1.23 this degree of involution is the Castelnuovo–Mumford regularity of the annihilator $\mathcal{N}^0 \subseteq S\mathcal{V} \otimes \mathcal{U}^*$. As Theorem 5.5.15 asserts that $\operatorname{reg} \mathcal{N}^0$ equals the degree of any Pommaret basis of the submodule $\mathcal{N}^0$ for the degree reverse lexicographic order, it suffices to show that equality in (6.25) for $q = q_0$ is equivalent to the existence of a Pommaret basis of degree q_0 for $\mathcal{N}^0$.

For notational simplicity, we assume in the sequel that $\dim \mathcal{U} = 1$ so that we can ignore $\mathcal{U}$ and consider the annihilator $\mathcal{N}^0$ as an ideal in $S\mathcal{V}$. We first choose a basis $\{x^1, \dots, x^n\}$ of $\mathcal{V}$ that is δ-regular for this ideal in the sense of Definition 4.3.1 and then a basis $\mathcal{H}_{q_0} \subset \mathbb{k}[x^1, \dots, x^n]$ of the vector space $\mathcal{N}^0_{q_0}$ which is triangular with respect to the degree reverse lexicographic order $\prec$. If we set $(\mathbb{N}_0^n)_{q_0} \setminus \operatorname{le}_\prec \mathcal{H}_{q_0} = \{\nu^{(1)}, \dots, \nu^{(s)}\}$ where $s = \operatorname{codim} \mathcal{H}_{q_0} = \dim \mathcal{N}_{q_0}$, then by elementary linear algebra a basis $\mathcal{G}_{q_0} = \{g_1, \dots, g_s\} \subset \mathbb{k}[y_1, \dots, y_n]$ of the vector space $\mathcal{N}_{q_0}$ exists such that $\min_\prec \operatorname{supp} g_i = y^{\nu^{(i)}}$ (here $\{y_1, \dots, y_n\}$ denotes again the dual basis to $\{x^1, \dots, x^n\}$).

By the defining property of the degree reverse lexicographic term order (cf. Lemma A.1.8), this condition implies that the subspace $\mathcal{N}_{q_0}^{(k-1)}$ is generated by all polynomials g_i with $\operatorname{cls} \nu^{(i)} \geq k$. Hence the Cartan characters of $\mathcal{N}_{q_0}$ are

$$\alpha_{q_0}^{(k)} = \binom{n-k+q_0-1}{q_0-1} - \beta_{q_0}^{(k)}, \tag{6.29}$$

since by (A.4b) the appearing binomial coefficient gives the number of terms of degree q_0 and class k. A well-known identity for binomial coefficients shows now that the inequalities (6.25) and (6.28) are equivalent (obviously we have here $\dim \mathcal{N}_{q_0} + \dim \mathcal{N}_{q_0}^0 = \dim \mathcal{P}_{q_0} = \binom{n+q_0-1}{q_0}$). In particular, if and only if equality holds, then $\mathcal{H}_{q_0}$ is a Pommaret basis of $\mathcal{N}_{\geq q_0}^0$ by Proposition 6.2.5 and thus $\operatorname{reg} \mathcal{N}^0 \leq q_0$ implying our claim. □

Example 6.2.6. Let us consider over a three-dimensional vector space $\mathcal{V}$ with basis $\{x,y,z\}$ and dual basis $\{\bar{x},\bar{y},\bar{z}\}$ the symbolic system $\mathcal{N} \subset \mathfrak{S}(\mathcal{V}^*)$ defined by $\mathcal{N}_0 = \Bbbk$, $\mathcal{N}_1 = \mathcal{V}^*$, $\mathcal{N}_2 = \langle \bar{x}^2, \bar{x}\bar{y}, \bar{x}\bar{z}, \bar{y}^2 \rangle$ and $\mathcal{N}_q = \mathcal{N}_{q-1,1}$ for $q \geq 3$. One easily verifies that here $\mathcal{N}_2^{(1)} = \langle \bar{y}^2 \rangle$ and $\mathcal{N}_2^{(2)} = \mathcal{N}_2^{(3)} = 0$ and therefore the only non-vanishing Cartan characters of $\mathcal{N}$ are $\alpha_2^{(1)} = 3$ and $\alpha_2^{(2)} = 1$. Furthermore, $\mathcal{N}_3 = \langle \bar{x}^3, \bar{x}^2\bar{y}, \bar{x}^2\bar{z}, \overline{xy}^2, \bar{y}^3 \rangle$. Since $\alpha_2^{(1)} + 2\alpha_2^{(2)} = 5 = \dim \mathcal{N}_3$, the symbolic system $\mathcal{N}$ passes the Cartan test and is involutive at degree $q = 2$. One also immediately sees that the map $\partial_{\bar{x},2} : \mathcal{N}_3 \to \mathcal{N}_2$ is indeed surjective and that the map $\partial_{\bar{y},2} : \mathcal{N}_3^{(1)} = \langle \bar{y}^3 \rangle \to \mathcal{N}_2^{(1)}$ is even bijective (there is no need to consider also $\partial_{\bar{z}}$, since both $\mathcal{N}_2^{(2)}$ and $\mathcal{N}_3^{(2)}$ vanish). ◁

Example 6.2.7. For an instance where the Cartan test is not passed, we return to the second symbolic system $\mathcal{N}$ in Example 6.1.9. Since it vanishes from degree 5 on, it is trivially involutive at degree 5. It is clear that the map $\partial_{\bar{x},4} : \mathcal{N}_5 \to \mathcal{N}_4$ cannot be surjective and also $\alpha_4^{(1)} = 1 > \dim \mathcal{N}_5 = 0$. Hence the symbolic system $\mathcal{N}$ is not involutive at degree $q = 4$. ◁

The essence of the proof of Proposition 6.1.6 is that differentiation with respect to y_k is dual to multiplication with x^k. Hence when we now study the dualisation of the considerations above, it is not surprising that the multiplication with elements $v \in \mathcal{V}$ is central. We start with a simple observation extending Lemma 6.1.16.

Lemma 6.2.8. *Let $\mathcal{M}$ be a finitely generated graded $S\mathcal{V}$-module and $q > 0$ an arbitrary degree. Then the following statements are equivalent.*

(i) $H_{r,n}(\mathcal{M}) = 0$ *for all* $r \geq q$.
(ii) *If* $\operatorname{Ann}(m) = S_+\mathcal{V}$ *for an* $m \in \mathcal{M}$, *then* $m \in \mathcal{M}_{<q}$.
(iii) *The existence of an element* $v \in \mathcal{V}$ *with* $v \cdot m = 0$ *entails* $m \in \mathcal{M}_{<q}$.
(iv) *For all* $v \in \mathcal{V}$ *except the elements of a finite number of proper subspaces of* $\mathcal{V}$ *the equation* $v \cdot m = 0$ *entails* $m \in \mathcal{M}_{<q}$.

Proof. The equivalence of (i) and (ii) follows immediately from Lemma 6.1.16. Furthermore, it is trivial that (iv) implies (iii) implies (ii). Hence there only remains to show that (iv) is a consequence of (ii).

Assume that (ii) holds and let $\mathcal{A} = \{m \in \mathcal{M}_{<q} \mid \operatorname{Ann}(m) = S_+\mathcal{V}\}$. We choose a complement $\mathcal{K}$ such that $\mathcal{M}_{<q} = \mathcal{A} \oplus \mathcal{K}$ and set $\overline{\mathcal{M}} = \mathcal{K} \oplus \bigoplus_{r\geq q} \mathcal{M}_r$. Because of (ii) no element of $\overline{\mathcal{M}} \setminus \{0\}$ is annihilated by $S_+\mathcal{V}$ and hence $S_+\mathcal{V}$ is not an associated prime ideal of the module $\overline{\mathcal{M}}$. By Theorem B.1.15, the set $\operatorname{Ass}\overline{\mathcal{M}}$ of all associated prime ideals of $\overline{\mathcal{M}}$ contains only finitely many elements. The intersection of any of these with $\mathcal{V}$ yields a proper subspace, since we just showed that $S_+\mathcal{V}$ is not among them. If we choose $v \in \mathcal{V}$ such that it is not contained in any of these subspaces, then $v \cdot m = 0$ entails $m \in \mathcal{M}_{<q}$. □

The property of v in Part (iii) will become so important in the sequel that we provide a special name for it. It is closely related to the idea of a regular sequence except that for the latter it is not permitted that the multiplication with v has a non-trivial kernel whereas here we only restrict the degree of the kernel.

Definition 6.2.9. A vector $v \in \mathcal{V}$ is called *quasi-regular* at degree q for the module $\mathcal{M}$, if $v \cdot m = 0$ entails $m \in \mathcal{M}_{<q}$. A finite sequence $(v_1, \dots, v_k)$ of elements of $\mathcal{V}$ is *quasi-regular* at degree q for the module $\mathcal{M}$, if each v_i is quasi-regular at degree q for the factor module $\mathcal{M}/\langle v_1, \dots, v_{i-1}\rangle\mathcal{M}$.

Obviously, if a vector $v \in \mathcal{V}$ is quasi-regular at degree q, it is also quasi-regular at any higher degree $r > q$. Furthermore, the vectors in a quasi-regular sequence $(v_1, \dots, v_k)$ are linearly independent. Thus such a sequence of length $n = \dim \mathcal{V}$ defines a basis of the vector space $\mathcal{V}$. In the sequel, we will abbreviate the statement in Part (iv) of Lemma 6.2.8 by saying that a *generic* vector $v \in \mathcal{V}$ is quasi-regular; this is the same notion of genericity (based on the Zariski topology) as in Proposition 4.3.8 and the subsequent discussion.

Lemma 6.2.10. *Let $v \in \mathcal{V}$ be quasi-regular at degree q. Then for each $r \geq q$ and $1 \leq p \leq n$ there is a short exact sequence*

$$0 \longrightarrow H_{r,p}(\mathcal{M}) \xrightarrow{\alpha} H_{r,p}(\mathcal{M}/v\mathcal{M}) \xrightarrow{\beta} H_{r,p-1}(\mathcal{M}) \longrightarrow 0 \tag{6.30}$$

and the multiplication with v is injective on $\mathcal{M}_{\geq q}$.

Proof. As above we decompose $\mathcal{M} = \mathcal{A} \oplus \overline{\mathcal{M}}$. Since $\mathcal{A} \subseteq \mathcal{M}_{<q}$, we have the equality $H_{r,p}(\mathcal{M}) = H_{r,p}(\overline{\mathcal{M}})$ for all $r \geq q$ and, because of $v \cdot \mathcal{A} = 0$, similarly $H_{r,p}(\mathcal{M}/v\mathcal{M}) = H_{r,p}(\overline{\mathcal{M}}/v\overline{\mathcal{M}})$ for all $r \geq q$.

It follows trivially from the definition of quasi-regularity that multiplication with v is injective on $\mathcal{M}_{\geq q}$. In fact, it is injective on $\overline{\mathcal{M}}$. Indeed, suppose that $v \cdot m = 0$ for some homogeneous element $m \in \overline{\mathcal{M}}$. Let us assume first that $m \in \mathcal{M}_{q-1}$. Then $v \cdot (w \cdot m) = 0$ for all $w \in \mathcal{V}$ and since $w \cdot m \in \mathcal{M}_q$, this is only possible, if $w \cdot m = 0$ and thus $\operatorname{Ann}(m) = S_+\mathcal{V}$ implying $m \in \mathcal{A}$. Iterating this argument, we conclude that m cannot be contained in $\overline{\mathcal{M}}_{q-2}$ either and so on. Hence $m \in \mathcal{A}$.

Because of the injectivity, the sequence

$$0 \longrightarrow \mathcal{M} \xrightarrow{v} \mathcal{M} \xrightarrow{\pi} \mathcal{M}/v\mathcal{M} \longrightarrow 0 \tag{6.31}$$

of graded modules is exact at all degrees $r \geq q$. Tensoring (6.31) with the vector space $\Lambda\mathcal{V}$ yields a similar sequence for the corresponding Koszul complexes $K(\mathcal{M})$ and $K(\mathcal{M}/v\mathcal{M})$, respectively, with the same exactness properties. Now we consider the associated long exact homological sequence

$$\begin{aligned} \cdots \longrightarrow H_{r-1,p}(\mathcal{M}) &\xrightarrow{H(v)} H_{r,p}(\mathcal{M}) \xrightarrow{H(\pi)} \\ \longrightarrow H_{r,p}(\mathcal{M}/v\mathcal{M}) &\xrightarrow{\beta} H_{r,p-1}(\mathcal{M}) \xrightarrow{H(v)} \cdots . \end{aligned} \tag{6.32}$$

Since, by Lemma 6.1.17, $H(v)$ is the zero map, it decomposes into the desired short exact sequences with $\alpha = H(\pi)$ and $\beta\big([\omega]\big) = \big[\frac{1}{v} \cdot \partial(\omega)\big]$. □

Based on these two lemmata, we can relate the notion of quasi-regularity with the Koszul homology of $\mathcal{M}$. As a corollary we will then obtain the dual form of Cartan's test for involution.

Proposition 6.2.11. *Let $\mathcal{M}$ be a finitely generated graded $S\mathcal{V}$-module and the sequence $(v_1,\dots,v_k)$ quasi-regular for it at degree q. Then $H_{r,p}(\mathcal{M}) = 0$ for all $r \geq q$ and $n-k < p \leq n$. If we set $\mathcal{M}^{(i)} = \mathcal{M}/\langle v_1,\dots,v_i\rangle\mathcal{M}$, then*

$$H_{r,n-k}(\mathcal{M}) \cong H_{r,n-k+1}\big(\mathcal{M}^{(1)}\big) \cong \cdots \cong H_{r,n}\big(\mathcal{M}^{(k)}\big) \tag{6.33}$$

for all $r \geq q$.

Proof. We proceed by induction over the length k of the sequence. For $k = 1$, it follows from Lemma 6.2.8 that $H_{r,n}(\mathcal{M}) = 0$ for all $r \geq q$. Entering this result into the short exact sequence (6.30) of Lemma 6.2.10 leads immediately to an isomorphism $H_{r,n-1}(\mathcal{M}) \cong H_{r,n}(\mathcal{M}^{(1)})$.

Assume that both assertions hold for any quasi-regular sequence with less than k vectors. Then we know that $H_{r,p}(\mathcal{M}) = 0$ for all $r \geq q$ and $n-k+1 < p \leq n$ and that $H_{r,n-k+1}(\mathcal{M}) \cong H_{r,n}(\mathcal{M}^{(k-1)})$. Since the vector v_k is quasi-regular at degree q for $\mathcal{M}^{(k-1)}$, the latter homology group vanishes by Lemma 6.2.8 proving the first assertion. Applying the induction hypothesis to $\mathcal{M}^{(i-1)}$ and the quasi-regular sequence $(v_i,\dots,v_k)$ shows that $H_{r,n-k+i}(\mathcal{M}^{(i-1)}) = 0$. Now we use again the exact sequence of Lemma 6.2.10 to conclude that $H_{r,n-k+i}(\mathcal{M}^{(i)}) \cong H_{r,n-k+i-1}(\mathcal{M}^{(i-1)})$. This proves the second assertion. □

Proposition 6.2.12. *Let $\mathcal{M}$ be a graded $S\mathcal{V}$-module finitely generated in degree less than $q > 0$. The module $\mathcal{M}$ is involutive at degree q, if and only if a basis $\{x^1,\dots,x^n\}$ of $\mathcal{V}$ exists such that the maps*

$$\mu_{k,r} : \mathcal{M}_r/\langle x^1,\dots,x^{k-1}\rangle\mathcal{M}_{r-1} \longrightarrow \mathcal{M}_{r+1}/\langle x^1,\dots,x^{k-1}\rangle\mathcal{M}_r \tag{6.34}$$

induced by the multiplication with x^k are injective for all $r \geq q$ and $1 \leq k \leq n$.

Proof. We first note that the statement that $\mathcal{M}$ is generated in degree less than q is equivalent to $H_{r,0}(\mathcal{M}) = 0$ for all $r \geq q$ by Lemma 6.1.16.

If $\mathcal{M}$ is involutive at degree q, then $H_{r,n}(\mathcal{M}) = 0$ for all $r \geq q$ and Lemma 6.2.8 implies that a generic vector $x^1 \in \mathcal{V}$ is quasi-regular at degree q. Now we proceed by iteration. Setting $\mathcal{M}^{(k)} = \mathcal{M}/\langle x^1, \cdots, x^k \rangle \mathcal{M}$, we find by Lemma 6.2.10 that $H_{r,n}(\mathcal{M}^{(k)}) = H_{r,n-k}(\mathcal{M}) = 0$ for all $r \geq q$. Thus we may again apply Lemma 6.2.8 in order to show that for any $1 \leq k < n$ every sequence $(x^1, \dots, x^k)$ which is quasi-regular at degree q can be further extended by a generic vector $x^{k+1} \in \mathcal{V}$. As already remarked above, a quasi-regular sequence of length n defines a basis of $\mathcal{V}$. Now the injectivity of the maps $\mu_{k,r}$ follows from Lemma 6.2.10.

For the converse, we note that if all the maps $\mu_{k,r}$ are injective, then obviously $(x^1, \dots, x^n)$ defines an quasi-regular sequence of length n at degree q. Now the vanishing of all homology groups $H_{r,p}(\mathcal{M})$ with $r \geq q$ and $1 \leq p \leq n$ follows from Proposition 6.2.11 and $\mathcal{M}$ is involutive. □

Again we face the problem that this proposition requires an infinite number of checks and thus cannot be applied effectively. However, for finitely presented modules it suffices to consider only the components $\mathcal{M}_q$ and $\mathcal{M}_{q+1}$. This fact leads to the following dual formulation of the Cartan test (Theorem 6.2.4).

Theorem 6.2.13 (Dual Cartan Test). *Let $\mathcal{N}^0 \subseteq S\mathcal{V} \otimes \mathcal{U}^*$ be a homogeneous submodule of the free $S\mathcal{V}$-module $S\mathcal{V} \otimes \mathcal{U}^*$ finitely generated in degree less than $q > 0$. Then the factor module $\mathcal{M} = (S\mathcal{V} \otimes \mathcal{U}^*)/\mathcal{N}^0$ is involutive at degree q, if and only if a basis $\{x^1, \dots, x^n\}$ of $\mathcal{V}$ exists such that the maps*

$$\mu_{k,q} : \mathcal{M}_q/\langle x^1, \dots, x^{k-1} \rangle \mathcal{M}_{q-1} \longrightarrow \mathcal{M}_{q+1}/\langle x^1, \dots, x^{k-1} \rangle \mathcal{M}_q \tag{6.35}$$

induced by the multiplication with x^k are injective for all $1 \leq k \leq n$.

Remark 6.2.14. Let $\mathcal{N} \subseteq \mathfrak{S}(\mathcal{V}^*) \otimes \mathcal{U}$ be a symbolic system and consider the dual $S\mathcal{V}$-module $\mathcal{M} = \mathcal{N}^* \cong (S\mathcal{V} \otimes \mathcal{U}^*)/\mathcal{N}^0$. Let furthermore $\{x^1, \dots, x^n\}$ be a basis of $\mathcal{V}$ and $\{y_1, \dots, y_n\}$ the dual basis of $\mathcal{V}^*$. Then we find that $\mu_1^* = \partial_{y_1}$ and hence that $(\mathcal{N}^{(1)})^* = (\ker \partial_{y_1})^* \cong \operatorname{coker} \mu_1 = \mathcal{M}^{(1)}$. Iteration of this argument yields $(\mathcal{N}^{(k)})^* \cong \mathcal{M}^{(k)}$ for all $1 \leq k \leq n$ (considering always ∂_{y_k} as a map on $\mathcal{N}^{(k-1)}$ so that $\mathcal{N}^{(k)} = \ker \partial_{y_k}$ and μ_k as a map on $\mathcal{M}^{(k-1)}$ so that $\mathcal{M}^{(k)} = \operatorname{coker} \mu_k$). We also have $\mu_k = \partial_{y_k}^*$ and hence obtain the isomorphisms $(\ker \mu_k)^* \cong \operatorname{coker} \partial_{y_k}^*$ (again considering the maps on the appropriate domains of definition). Thus injectivity of all the maps μ_k is equivalent to surjectivity of all the maps ∂_{y_k}. Hence applying Proposition 6.2.12 to $\mathcal{M}$ proves dually Proposition 6.2.3 for $\mathcal{N}$ and similarly for the Theorems 6.2.4 and 6.2.13. Furthermore, it is obvious that if the basis $\{x^1, \dots, x^n\}$ is quasi-regular at degree q, then the dual basis $\{y_1, \dots, y_n\}$ is δ-regular for $\mathcal{N}_q$. The converse does not necessarily hold, as δ-regularity is a much weaker condition than quasi-regularity (the latter implies involution via the dual Cartan test; the former is only a necessary condition for applying the Cartan test). ◁

Example 6.2.15. For the symbolic system $\mathcal{N}$ of Example 6.2.6 the annihilator $\mathcal{N}^0$ is the ideal $\mathcal{I} \subset \mathcal{P} = S\mathcal{V}$ generated by the monomials yz and z^2. We apply now the dual Cartan test to the factor module $\mathcal{M} = \mathcal{P}/\mathcal{I}$. For the two relevant components we obtain after a trivial computation that $\mathcal{M}_2 \cong \langle x^2, xy, xz, y^2\rangle \cong \mathcal{N}_2$ and $\mathcal{M}_3 \cong \langle x^3, x^2y, x^2z, xy^2, y^3\rangle \cong \mathcal{N}_3$. Similarly, we find that the non-vanishing factor modules required for the dual Cartan test are given by $\mathcal{M}_2^{(1)} \cong \langle y^2\rangle \cong \mathcal{N}_2^{(1)}$ and $\mathcal{M}_3^{(1)} \cong \langle y^3\rangle \cong \mathcal{N}_3^{(1)}$. It is now trivial to see that the map $\mu_x : \mathcal{M}_2 \to \mathcal{M}_3$ induced by the multiplication with x is injective and that the map $\mu_y : \mathcal{M}_2^{(1)} \to \mathcal{M}_3^{(1)}$ induced by the multiplication with y is even bijective. Hence according to the dual Cartan test the module $\mathcal{M}$ is involutive at degree 2. ◁

Remark 6.2.16. Another way to formulate the assumptions of Theorem 6.2.13 is to require that $\mathcal{M}$ is a finitely generated graded $S\mathcal{V}$-module such that the homology groups $H_{r,0}(\mathcal{M})$ and $H_{r,1}(\mathcal{M})$ vanish for all $r \geq q$. This fact follows immediately from Remark 6.1.20 and implies that for submodules of a free submodule, we must also see in which degrees their minimal syzygies live.

As a concrete example consider the monomial ideal $\mathcal{I} = \langle x^3, y^3\rangle \subset \Bbbk[x,y]$ generated in degree 3, i. e. the annihilator of the second symbolic system $\mathcal{N}$ in Example 6.1.9. It is trivial that $\mu_x : \mathcal{I}_4 \to \mathcal{I}_5$ is injective. For the map μ_y we note that $\mathcal{I}_4/x\mathcal{I}_3 \cong \langle x^3y, x^4\rangle$ and thus it is again easy to see that μ_y is injective in degree 4.

If $\mathcal{I}$ were a factor module, it would now follow from Theorem 6.2.13 that the map μ_y was also injective at all higher degree. However, using the identification $\mathcal{I}_5/x\mathcal{I}_4 \cong \langle x^3y^2, y^5\rangle$, we find here for $\mu_y : \mathcal{I}_5/x\mathcal{I}_4 \to \mathcal{I}_6/x\mathcal{I}_5$ that $\mu_y([x^3y^2]) = [x^3y^3] = 0$ so that the map is not injective at degree 5. This phenomenon is easily understood by considering the syzygies. $\mathrm{Syz}(\mathcal{I}) \cong H_1(\mathcal{I})$ is generated by the single element $y^3\mathbf{e}_1 - x^3\mathbf{e}_2$ of degree 6 (and hence Theorem 6.2.13 holds for $\mathcal{I}$ only from this degree on). As its coefficients are of degree 3, nothing happens with the map μ_y before we encounter $\mathcal{I}_6$ and then the equation $\mu_y([x^3y^2]) = 0$ is a trivial consequence of this syzygy. ◁

6.3 Pommaret Bases and Homology

Finally, we study the relationship between Pommaret bases and the homological constructions of the last two sections. We assume throughout that a fixed basis $\{x^1, \dots, x^n\}$ of $\mathcal{V}$ has been chosen so that we may identify $S\mathcal{V} \cong \Bbbk[x^1, \dots, x^n] = \mathcal{P}$. For simplicity, we restrict to homogeneous ideals $\mathcal{I} \subseteq \mathcal{P}$. We only consider Pommaret bases for the degree reverse lexicographic term order $\prec_{\text{degrevlex}}$, as Theorem 5.5.15 (together with Remark 5.5.17) implies that for any other term order the corresponding Pommaret basis (if it exists) cannot be of lower degree.

It turns out that this relationship takes its simplest form, if we compare the Pommaret basis of the ideal $\mathcal{I}$ and the Koszul homology of its factor algebra $\mathcal{P}/\mathcal{I}$. We have seen in the last chapters that everything relevant for involutive bases can be read off the leading ideal. Therefore, we show first that at least for our chosen term order quasi-regularity is also already decided by the leading ideal.

Lemma 6.3.1. *Let $\mathcal{I} \subseteq \mathcal{P}$ be a homogeneous ideal and $\prec$ the degree reverse lexicographic order. The sequence $(x^1,\dots,x^n)$ is quasi-regular at degree q for the factor module $\mathcal{P}/\mathcal{I}$, if and only if it is quasi-regular at degree q for $\mathcal{P}/\operatorname{lt}_\prec \mathcal{I}$.*

Proof. Let $\mathcal{G}$ be a Gröbner basis of $\mathcal{I}$ for $\prec$. Recall from Remark B.4.10 that the normal form with respect to the basis $\mathcal{G}$ defines an isomorphism between the vector spaces $\mathcal{M} = \mathcal{P}/\mathcal{I}$ and $\mathcal{M}' = \mathcal{P}/\operatorname{lt}_\prec \mathcal{I}$. One direction is now trivial, as an obvious necessary condition for $m = [f] \in \mathcal{M}$ to satisfy $x^1 \cdot m = 0$ is that $x^1 \cdot [\operatorname{lt}_\prec f] = 0$ in the module $\mathcal{M}'$. Hence quasi-regularity of x^1 for $\mathcal{M}'$ implies quasi-regularity of x^1 for $\mathcal{M}$ and by iteration the same holds true for the whole sequence (note that so far we could have used any term order).

For the converse let $r \geq q$ be an arbitrary degree. Because of the mentioned isomorphism, we may choose for the homogeneous component $\mathcal{M}_r$ a $\Bbbk$-linear basis where each member is represented by a term (the standard terms of Remark B.4.10), i. e. the representatives simultaneously induce a basis of the component $\mathcal{M}'_r$. Let x^μ be one of these terms. As we assume that x^1 is quasi-regular at degree q for the module $\mathcal{M}$, we must have $x^1 \cdot [x^\mu] \neq 0$ in $\mathcal{M}$. Suppose now that $x^1 \cdot [x^\mu] = 0$ in $\mathcal{M}'$ so that x^1 was not quasi-regular at degree q for $\mathcal{M}'$.

Thus $x^{\mu+1_1} \in \operatorname{lt}_\prec \mathcal{I}$. Since $\operatorname{lt}_\prec \mathcal{I} = \langle \operatorname{lt}_\prec \mathcal{G}\rangle$ by Definition B.4.1 of a Gröbner basis, the set $\mathcal{G}$ must contain a polynomial g with $\operatorname{lt}_\prec g \mid x^{\mu+1_1}$. Because of the assumption $x^\mu \notin \operatorname{lt}_\prec \mathcal{I}$, we must have $\operatorname{cls}(\operatorname{lt}_\prec g) = 1$. According to Lemma A.1.8, this fact implies that every term in g is of class 1. Iteration of this argument shows that the normal form of $x^{\mu+1_1}$ with respect to $\mathcal{G}$ can be written as $x^1 f$ with $f \in \mathcal{P}_r$ and $\operatorname{lt}_\prec f \prec x^\mu$. Consider now the polynomial $\bar{f} = x^\mu - f \in \mathcal{P}_r \setminus \{0\}$. As it consists entirely of terms not contained in $\operatorname{lt}_\prec \mathcal{I}$, we have $[\bar{f}] \neq 0$ in $\mathcal{M}_r$. However, by construction $x^1 \cdot [\bar{f}] = 0$ contradicting the injectivity of multiplication by x^1 on $\mathcal{M}_r$.

For the remaining elements of the sequence $(x^1,\dots,x^n)$ we note for each value $1 \leq k < n$ the trivial isomorphism $\mathcal{M}^{(k)} = \mathcal{M}/\langle x^1,\dots,x^k\rangle\mathcal{M} \cong \mathcal{P}^{(k)}/\mathcal{I}^{(k)}$ where $\mathcal{P}^{(k)} = \Bbbk[x^{k+1},\dots,x^n]$ and $\mathcal{I}^{(k)} = \mathcal{I} \cap \mathcal{P}^{(k)}$. It implies that we may iterate the arguments above so that indeed quasi-regularity of $(x^1,\dots,x^n)$ for $\mathcal{M}'$ is equivalent to quasi-regularity of the sequence for $\mathcal{M}'$. □

The deeper reason that this lemma holds only for the degree reverse lexicographic order is given by Remark 5.5.17. In general, we have only the inequality $\operatorname{reg}(\operatorname{lt}_\prec \mathcal{I}) \geq \operatorname{reg} \mathcal{I}$ and if it is strict, then a sequence may be quasi-regular for $\mathcal{P}/\mathcal{I}$ at any degree $\operatorname{reg} \mathcal{I} \leq q < \operatorname{reg}(\operatorname{lt}_\prec \mathcal{I})$, but it cannot be quasi-regular for $\mathcal{P}/\operatorname{lt}_\prec \mathcal{I}$ at such a degree by the results below.

Theorem 6.3.2. *The basis $\{x^1,\dots,x^n\}$ is δ-regular for the homogeneous ideal $\mathcal{I} \subseteq \mathcal{P}$ in the sense that $\mathcal{I}$ possesses a Pommaret basis $\mathcal{H}$ for the degree reverse lexicographic term order with $\deg \mathcal{H} = q$, if and only if the sequence $(x^1,\dots,x^n)$ is quasi-regular for the factor module $\mathcal{P}/\mathcal{I}$ at degree q but not at any lower degree.*

Proof. It suffices to consider monomial ideals $\mathcal{I}$: for Pommaret bases it is obvious from their definition that a basis is δ-regular for $\mathcal{I}$, if and only if it is so for $\operatorname{lt}_\prec \mathcal{I}$; by Lemma 6.3.1 the same is true for quasi-regularity.

Let us first assume that the basis $\{x^1,\dots,x^n\}$ is δ-regular in the described sense. By Proposition 5.1.6, the leading terms $\operatorname{lt}_{\prec}\mathcal{H}$ induce a complementary decomposition of the form (5.4) of $\mathcal{M}=\mathcal{P}/\mathcal{I}$ where all generators are of degree q or less. Thus, if $\mathcal{M}_q\neq 0$ (otherwise there is nothing to show), then we can choose a vector space basis of it as part of the complementary decomposition and the variable x^1 is multiplicative for all its members. But this observation immediately implies that multiplication with x^1 is injective from degree q on, so that x^1 is quasi-regular for the module $\mathcal{M}$ at degree q.

For the remaining elements of the basis $\{x^1,\dots,x^n\}$ we proceed as in the proof of Lemma 6.3.1 and use the isomorphisms $\mathcal{M}^{(k)}\cong\mathcal{P}^{(k)}/\mathcal{I}^{(k)}$. One easily verifies that a Pommaret basis of $\mathcal{I}^{(k)}$ is obtained by setting $x^1=\cdots=x^k=0$ in the partial basis $\mathcal{H}^{(k)}=\{h\in\mathcal{H}\mid \operatorname{cls}h>k\}$. Thus we can again iterate for each value $1<k\leq n$ the argument above so that indeed $(x^1,\dots,x^n)$ is a quasi-regular sequence for the module $\mathcal{M}$ at degree q.

For the converse, we first show that quasi-regularity of the sequence $(x^1,\dots,x^n)$ implies the existence of a Rees decomposition for $\mathcal{M}$. Exploiting again the isomorphism $\mathcal{M}^{(k)}\cong\mathcal{P}^{(k)}/\mathcal{I}^{(k)}$, one easily sees that a vector space basis of $\mathcal{M}_q^{(k)}$ is induced by all terms $x^\mu\notin\mathcal{I}$ with $|\mu|=q$ and $\operatorname{cls}\mu\geq k$. By the definition of quasi-regularity, multiplication with x^k is injective on $\mathcal{M}^{(k)}$, hence we take $\{x^1,\dots,x^k\}$ as multiplicative variables for such a term (which is exactly the assignment used in the Rees decomposition induced by a Pommaret basis according to Proposition 5.1.6).

We claim now that this assignment yields a Rees decomposition of $\mathcal{M}_{\geq q}$ (and hence induces one of $\mathcal{M}$, since we only have to add all terms $x^\mu\notin\mathcal{I}$ such that $|\mu|<q$ without any multiplicative variables). There only remains to prove that our decomposition indeed covers all of $\mathcal{M}_{\geq q}$. But this fact is trivial. If $x^\mu\notin\mathcal{I}$ is an arbitrary term with $|\mu|=q+1$ and $\operatorname{cls}\mu=k$, then we can write $x^\mu=x^k\cdot x^{\mu-1_k}$. Obviously, $x^\mu\notin\mathcal{I}$ implies $x^{\mu-1_k}\notin\mathcal{I}$ and $\operatorname{cls}(\mu-1_k)\geq k$ so that x^k is multiplicative for it. Hence all of $\mathcal{M}_{q+1}$ is covered and an easy induction shows that we have indeed a decomposition of $\mathcal{M}_{\geq q}$.

Proposition 5.1.6 entails now that the ideal $\mathcal{I}$ possesses a *weak* Pommaret basis of degree q. As the reduction to a strong basis along the lines of the proof of Proposition 3.1.12 can only decrease the degree, we conclude that $\mathcal{I}$ has a strong Pommaret basis of degree at most q. If the degree of the basis actually decreased, then, by the converse statement already proven, $(x^1,\dots,x^n)$ would be a quasi-regular sequence for $\mathcal{M}$ at a lower degree than q contradicting our assumptions.

The same "reverse" argument shows that if the ideal $\mathcal{I}$ has a Pommaret basis of degree q, then the sequence $(x^1,\dots,x^n)$ cannot be quasi-regular for $\mathcal{M}$ at any degree less than q, as otherwise a Pommaret basis of lower degree would exist which is not possible by the discussion following Theorem 5.5.15. □

For monomial ideals $\mathcal{I}\subseteq\mathcal{P}$ a much stronger statement is possible. Using again the isomorphisms $\mathcal{M}^{(k)}\cong\mathcal{P}^{(k)}/\mathcal{I}^{(k)}$, we may identify elements of $\mathcal{M}^{(k)}$ with linear combinations of the terms $x^\nu\notin\mathcal{I}$ satisfying $\operatorname{cls}x^\nu>k$. Finally, if we denote as before by $\mu_k:\mathcal{M}^{(k-1)}\to\mathcal{M}^{(k-1)}$ the map induced by multiplication with x^k, then we obtain the following simple relationship between the kernels of the maps μ_k and the Pommaret basis of $\mathcal{I}$ showing that the latter can in fact be defined intrinsically.

Proposition 6.3.3. *Let $\mathcal{I} \subseteq \mathcal{P}$ be a quasi-stable ideal and $\mathcal{H}$ its Pommaret basis. If we define $\mathcal{H}_k = \{x^\nu \in \mathcal{H} \mid \operatorname{cls}\nu = k\}$ for any $1 \leq k \leq n$, then the set $\{x^{\nu-1_k} \mid x^\nu \in \mathcal{H}_k\}$ is a basis of $\ker \mu_k$.*

Proof. Assume that $x^\nu \in \mathcal{H}_k$. Then $x^{\nu-1_k} \notin \mathcal{I}$, as otherwise the Pommaret basis $\mathcal{H}$ was not involutively autoreduced, and hence we find $x^{\nu-1_k} \in \ker \mu_k$.

Conversely, suppose that $x^\nu \in \ker \mu_k$. Obviously, this implies $x^{\nu+1_k} \in \mathcal{I}$ and the Pommaret basis $\mathcal{H}$ must contain an involutive divisor of $x^{\nu+1_k}$. If this divisor were not $x^{\nu+1_k}$ itself, the term x^ν would have to be an element of $\mathcal{I}$ which is obviously not possible. Since $x^\nu \in \ker \mu_k$ entails $\operatorname{cls}(\nu + 1_k) = k$, we thus find $x^{\nu+1_k} \in \mathcal{H}_k$. □

We noted already in Remark 6.1.23 that the degree of involution is essentially just the Castelnuovo–Mumford regularity. There we used the equivalence of the Koszul homology to the minimal free resolution. Based on Theorems 5.5.15 and 6.3.2, we can also give a simple direct proof.

Corollary 6.3.4. *Let $\mathcal{I} \subseteq \mathcal{P}$ be a homogeneous ideal. Then the factor module $\mathcal{M} = \mathcal{P}/\mathcal{I}$ is involutive at degree q but not at any lower degree, if and only if the Castelnuovo–Mumford regularity of $\mathcal{I}$ takes the value $\operatorname{reg}\mathcal{I} = q$.*

Proof. By Theorem 5.5.15, $\operatorname{reg}\mathcal{I} = q$, if and only if $\mathcal{I}$ possesses in suitable variables $x^1, \dots, x^n$ a Pommaret basis $\mathcal{H}$ with $\deg \mathcal{H} = q$. According to Theorem 6.3.2, the sequence $(x^1, \dots, x^n)$ is then quasi-regular for $\mathcal{M}$ at degree q but not at any lower degree, so that by the dual Cartan test (Theorem 6.2.13) the module $\mathcal{M}$ is involutive at degree q but not at any lower degree. □

Remark 6.3.5. In fact, the reasoning behind this proof allows us to determine easily from a Pommaret basis $\mathcal{H}$ of $\mathcal{I}$ the dimensions of the highest non-vanishing homology groups. We follow the ideas used in the proof of Theorem 5.5.24 on the existence of a linear resolution of the truncated ideal $\mathcal{I}_{\geq q}$: if we form the set $\mathcal{H}_q \subset \mathcal{P}_q$ according to (4.5) by taking the multiplicative multiples of the elements of $\mathcal{H}$ of lower degree, then $\mathcal{H}_q$ is a Pommaret basis of the truncated ideal $\mathcal{I}_{\geq q}$ and applying Theorem 5.4.12 yields the minimal free resolution of $\mathcal{I}_{\geq q}$. Since the Koszul homology groups $H_{q,p}(\mathcal{I})$ depend only on the truncation $\mathcal{I}_{\geq q}$, Proposition 6.1.18 and the equivalence of the minimal free resolution to the Koszul homology entail for any value $0 \leq p \leq n$ isomorphisms

$$H_{q,p}(\mathcal{I}) \cong H_{q-1,p+1}(\mathcal{P}/\mathcal{I}) \cong \operatorname{Syz}^p(\mathcal{H}_q)\,. \tag{6.36}$$

Finally, according to Theorem 5.4.12, (5.39) gives us the dimensions of these vector spaces, i. e. the corresponding bigraded Betti numbers:

$$\beta_{q,p}(\mathcal{I}) = \beta_{q-1,p+1}(\mathcal{P}/\mathcal{I}) = \sum_{k=1}^{n-p} \binom{n-k}{p} \beta_q^{(k)}\,, \tag{6.37}$$

where $\beta_q^{(k)}$ denotes the number of elements of class k in the basis $\mathcal{H}_q$. Note that in the case that $\mathcal{I} = \mathcal{I}_{\geq q}$, i. e. the whole ideal $\mathcal{I}$ is generated in degree q, these vector spaces are the only non-vanishing homology groups. While from an algebraic point of view, this situation is very special, it actually is quite common in the context of differential equations, as we will see in the next chapter. ◁

Remark 6.3.6. In view of Corollary 6.3.4, it is not so surprising to see that the characterisation of the Castelnuovo–Mumford regularity provided by Theorem 5.5.21 and the dual Cartan test in Theorem 6.2.13 are equivalent. Consider a homogeneous ideal $\mathcal{I} \subseteq \mathcal{P}$ for which the basis $\{x^1, \dots, x^n\}$ of $\mathcal{V}$ is δ-regular and assume that for some degree $q \geq 0$ the condition (5.63a) of Theorem 5.5.21 is violated for some $1 \leq j \leq D = \dim(\mathcal{P}/\mathcal{I})$. Thus there exists a polynomial $f \in \mathcal{P}_{q-1}$ such that $f \notin \langle \mathcal{I}, x^1, \dots, x^{j-1} \rangle$ but $x^j f$ is contained in this ideal. If we introduce $\mathcal{M}^{(j)} = \mathcal{P}/\langle \mathcal{I}, x^1, \dots, x^j \rangle$, then obviously the equivalence class $[f]$ lies in the kernel of the map $\mu_j : \mathcal{M}^{(j-1)} \to \mathcal{M}^{(j-1)}$ induced by multiplication with the variable x^j. Since for $\mathcal{M} = \mathcal{P}/\mathcal{I}$ we have the trivial isomorphism $\mathcal{M}^{(j)} \cong \mathcal{M}/\langle x^1, \dots, x^j \rangle \mathcal{M}$, the conditions of Theorem 6.2.13 are not satisfied for $\mathcal{M}$ either. Conversely, any representative of a non-trivial element of $\ker \mu_j$ of degree q provides us with such an f. There is no need to consider a $j > D$, since we know from Proposition 5.2.3 that $\mathcal{M}^{(D)}_{\geq \operatorname{reg} \mathcal{I}} = 0$. ◁

Remark 6.3.7. The results above also provide us with a simple proof of the characterisation (ii) of a quasi-stable ideal in Proposition 5.3.4. If the ideal $\mathcal{I}$ is quasi-stable, then the coordinates $x^1, \dots, x^n$ are δ-regular for it, hence by Theorem 6.3.2 they form a quasi-regular sequence for $\mathcal{P}/\mathcal{I}$ at a suitably chosen degree. By Proposition 5.5.28, we have that $\mathcal{I}^{\text{sat}} = \mathcal{I} : \langle x^1 \rangle^\infty$ and hence multiplication by x^1 is injective on $\mathcal{P}/\mathcal{I}^{\text{sat}}$. As obviously $\mathcal{P}/\langle \mathcal{I}, x^1, \dots, x^j \rangle^{\text{sat}} \cong \mathcal{P}^{(j)}/(\mathcal{I}^{(j)})^{\text{sat}}$, we can apply the same argument also for all $1 \leq j < D$.

Conversely, if $x^1 \cdot f \in \mathcal{I}$ for a polynomial $f \in \mathcal{P} \setminus \mathcal{I}$, then $f \in \mathcal{I}^{\text{sat}} \setminus \mathcal{I}$ and hence $\deg f < \operatorname{sat} \mathcal{I}$. Thus x^1 is quasi-regular for $\mathcal{P}/\mathcal{I}$ at the degree $\operatorname{sat} \mathcal{I}$. Using again the isomorphisms $\mathcal{P}/\langle \mathcal{I}, x^1, \dots, x^j \rangle^{\text{sat}} \cong \mathcal{P}^{(j)}/(\mathcal{I}^{(j)})^{\text{sat}}$, we can apply the same argument for all $1 \leq j < D$, so that $(x^1, \dots, x^D)$ is a quasi-regular sequence for $\mathcal{P}/\mathcal{I}$ at a sufficiently high degree.

As already discussed in the proof of Proposition 5.3.4, the characterisation (ii) of Proposition 5.3.4 implies that the set $\{x^1, \dots, x^D\}$ is maximally independent modulo the saturation $\mathcal{I}^{\text{sat}}$. Hence $\dim \mathcal{P}/\langle \mathcal{I}, x^1, \dots, x^D \rangle = 0$ entailing that $(x^1, \dots, x^n)$ is a quasi-regular sequence for $\mathcal{P}/\mathcal{I}$ at a sufficiently high degree. By Theorem 6.3.2, the ideal $\mathcal{I}$ is thus quasi-stable. ◁

As an application, we finally show that our proof of Proposition 5.2.7 determining the depth of an ideal or module from its Pommaret basis is actually of a homological nature. The following theorem provides a characterisation of the depth via Koszul homology which is sometimes used as definition of $\operatorname{depth} \mathcal{M}$ (cf. Remark B.2.39). Note that, taking into account the relation between the minimal free resolution of a module and its Koszul homology discussed in Remark 6.1.15, it also trivially implies the Auslander–Buchsbaum formula (Corollary 5.5.12).

Theorem 6.3.8. *Let $\mathcal{M}$ be a polynomial module. Then* $\operatorname{depth}\mathcal{M} = d$, *if and only if* $H_{n-d}(\mathcal{M}) \neq 0$ *and* $H_{n-d+1}(\mathcal{M}) = \cdots = H_n(\mathcal{M}) = 0$.

Proof. For notational simplicity, we give the proof only for the case of an ideal $\mathcal{I} \subseteq \mathcal{P}$. The extension to modules is straightforward. We use the same notations as in the proof of Proposition 5.2.7: $\mathcal{H}$ is the Pommaret basis of $\mathcal{I}$ with respect to the degree reverse lexicographic order, $d = \min_{h\in\mathcal{H}} \operatorname{cls} h$ (and thus $\operatorname{depth}\mathcal{I} = d$) and $\mathcal{H}_d = \{h \in \mathcal{H} \mid \operatorname{cls} h = d\}$. We consider again an element $\bar{h} \in \mathcal{H}_d$ of maximal degree and show now that it induces a non-zero element of $H_{n-d}(\mathcal{I})$.

By Lemma A.1.8, $\bar{h} \in \langle x^1, \dots, x^d\rangle$ and thus it possesses a unique representation $\bar{h} = x^1\bar{h}^{(1)} + \cdots + x^d\bar{h}^{(d)}$ with $\bar{h}^{(i)} \in \Bbbk[x^i, \dots, x^n]$. The polynomial $\bar{h}^{(d)}$ cannot lie in $\mathcal{I}$, as otherwise there must exist an $h \in \mathcal{H}$ with $\operatorname{lt}_\prec h \,|_P\, \operatorname{lt}_\prec \bar{h}^{(d)} \,|_P\, \operatorname{lt}_\prec \bar{h}$ contradicting the fact that any Pommaret basis is involutively autoreduced. We showed already in the proof of Proposition 5.2.7 that for any $d < k \leq n$ polynomials $P_h \in \langle x^1, \dots, x^d\rangle$ exist such that $x^k\bar{h} = \sum_{h\in\mathcal{H}} P_h h$. As above, each of these coefficients may be uniquely decomposed $P_h = x^1P_h^{(1)} + \cdots + x^dP_h^{(d)}$ with $P_h^{(i)} \in \Bbbk[x^i, \dots, x^n]$. Because of the uniqueness of these decompositions we find that $x^k\bar{h}^{(i)} = \sum_{h\in\mathcal{H}} P_h^{(i)} h$ and thus $x^k\bar{h}^{(i)} \in \mathcal{I}$ for any $d < k \leq n$.

Let $I = (i_1, \dots, i_{d-1})$ be a repeated index with $i_1 < i_2 < \cdots, i_{d-1}$. Then its complement $\bar{I} = \{1, \dots, n\} \setminus I$ is a repeated index of length $n - d + 1$ and we may represent any element $\bar{\omega} \in \mathcal{P} \otimes \Lambda_{n-d+1}\mathcal{V}$ in the form $\bar{\omega} = \sum_{|I|=d-1} \bar{f}_I \mathrm{d}x^{\bar{I}}$. We consider now in particular all repeated indices with $i_{d-1} \leq d$. For each of them a unique value $i \in \{1, \dots, d\}$ exists such that $i \notin I$ and we set $\bar{f}_I = (-1)^{d-i}\bar{h}^{(i)}$. For all remaining coefficients we only assume that $\bar{f}_I \in \mathcal{I}$. Then, by our considerations above, the thus chosen form $\bar{\omega}$ is not contained in $\mathcal{I} \otimes \Lambda_{n-d+1}\mathcal{V}$.

We claim that $\omega = \partial\bar{\omega} \in \mathcal{I} \otimes \Lambda_{n-d}\mathcal{V}$. If we write $\omega = \sum_{|I|=d} f_I \mathrm{d}x^{\bar{I}}$, then by definition of the Koszul differential $f_I = \sum_{j=1}^{d} (-1)^j x^{i_j} \bar{f}_{I\setminus\{i_j\}}$. Let us first assume that $i_d > d$. It follows from our choice of $\bar{\omega}$ that $f_{I\setminus\{i_j\}} \in \mathcal{I}$ for all $j < d$ and that always $x^{i_d} f_{I\setminus\{i_d\}} \in \mathcal{I}$ implying trivially that $f_I \in \mathcal{I}$. If $i_d = d$, one easily verifies that we have chosen $\bar{\omega}$ precisely such that $f_I = \bar{h} \in \mathcal{I}$. Hence our claim is proven.

If we can show that it is not possible to choose a form $\tilde{\omega} \in \mathcal{P} \otimes \Lambda_{n-d+2}\mathcal{V}$ such that $\bar{\omega} + \partial\tilde{\omega} \in \mathcal{I} \otimes \Lambda_{n-d+1}\mathcal{V}$, then we have constructed a non-zero element $[\omega] \in H_{n-d}(\mathcal{I})$. But this is easy to achieve by considering in particular the coefficient $\bar{f}_{(1,2,\dots,d-1)} = \bar{h}^{(d)} \notin \mathcal{I}$. The corresponding coefficient of the form $\partial\tilde{\omega}$ is given by $\sum_{j=1}^{d-1} (-1)^j x^j \tilde{f}_{(1,2,\dots,d-1)\setminus\{j\}} \in \langle x^1, \dots, x^{d-1}\rangle$. As noted above, we have $\bar{h}^{(d)} \in \Bbbk[x^d, \dots, x^n]$ so that it is not possible to eliminate it in this manner and hence no form $\bar{\omega} + \partial\tilde{\omega}$ can be contained in $\mathcal{I} \otimes \Lambda_{n-d+1}\mathcal{V}$.

There remains to show that $H_{n-d+1}(\mathcal{I}) = \cdots = H_n(\mathcal{I}) = 0$ under our assumptions. $H_n(\mathcal{I}) = 0$ follows immediately from Lemma 6.1.16. Consider now a cycle $\omega \in \mathcal{I} \otimes \Lambda_{n-k}\mathcal{V}$ with $0 < k < d$. Since the Koszul complex $K(\mathcal{P})$ is exact, a form $\bar{\omega} \in \mathcal{P} \otimes \Lambda_{n-k+1}\mathcal{V}$ exists with $\partial\bar{\omega} = \omega$. For all I we have by assumption $j_I = \sum_{j=1}^{d} (-1)^j x^{i_j} \bar{f}_{I\setminus\{i_j\}} \in \mathcal{I}$; our goal is to show that (modulo $\operatorname{im}\partial$) we can always choose $\bar{\omega}$ such that all coefficients $\bar{f}_J \in \mathcal{I}$, too.

Without loss of generality, we may assume that all coefficients $\bar{f}_J$ are in normal form with respect to the Pommaret basis $\mathcal{H}$, as the difference is trivially contained in $\mathcal{I}$. In addition, we may assume that $\operatorname{lt}_\prec f_I = \operatorname{lt}_\prec(x^{i_j}\bar{f}_{I\setminus\{i_j\}})$ for some value j. Indeed, it is easy to see that cancellations between such leading terms can always be eliminated by subtracting a suitable form $\partial\tilde{\omega}$ from $\bar{\omega}$.

We begin with those repeated indices $I = (i_1,\dots,i_k)$ for which all indices $i_j < d = \min_{h\in\mathcal{H}} \operatorname{cls} h$. In this case $\operatorname{lt}_\prec f_I \in \langle \operatorname{lt}_\prec \mathcal{H}\rangle_P = \operatorname{lt}_\prec \mathcal{I}$ implies that already $\operatorname{lt}_\prec \bar{f}_{I\setminus\{i_j\}} \in \operatorname{lt}_\prec \mathcal{I}$ for the above j. But unless $\bar{f}_{I\setminus\{i_j\}} = 0$, this observation contradicts our assumption that all f_J are in normal form and thus do not contain any terms from $\operatorname{lt}_\prec \mathcal{I}$. Therefore all $\bar{f}_J$ where all entries of J are less than d must vanish.

We continue with those repeated indices $I = (i_1,\dots,i_k)$ for which only one index $i_\ell > d$. Then, by our considerations above, $\bar{f}_{I\setminus\{i_\ell\}} = 0$ and hence $\operatorname{lt}_\prec f_I = \operatorname{lt}_\prec(x^{i_j}\bar{f}_{I\setminus\{i_j\}})$ for some value $j \neq \ell$. Thus $i_j < d$ and the same argument as above implies that all such $\bar{f}_{I\setminus\{i_j\}} = 0$. A trivial induction proves now that in fact all $\bar{f}_J = 0$ and therefore $\bar{\omega} \in \mathcal{I} \otimes \Lambda_{n-k+1}\mathcal{V}$. □

6.4 Notes

The cohomological approach to involution was pioneered by Spencer [427] and collaborators [167, 168, 365]; later discussions can be found e. g. in [58, 84, 246, 302, 304, 307, 115, 263, 356]. One should mention that the Spencer cohomology appeared first not in the context of differential equations but in deformation theory [426]. Pommaret [356] explicitly determines at least the size of the cohomology groups for a number of concrete examples. Although the cohomology appears at many places, one may probably say that so far it has not been much applied; most of the references only use it for defining an involutive symbol. The comodule point of view was developed in [279]; this article also contains a discussion of possible algorithms for the explicit construction of the Spencer cohomology.

Instead of the direct definition (6.2), the differential δ of the polynomial de Rham complex may also be introduced as the composition of several natural maps relating the full tensor algebra with the symmetric and exterior algebra, respectively, regarded as quotient algebras. More precisely, one defines first the map

$$\Delta_{q,r} : S_{q+r}\mathcal{V} \longrightarrow \bigotimes^{q+r}\mathcal{V} \cong \bigotimes^{q}\mathcal{V} \otimes \bigotimes^{r}\mathcal{V} \longrightarrow S_q\mathcal{V} \otimes S_r\mathcal{V} \tag{6.38}$$

where the first arrow represents the natural inclusion and the second arrow the canonical projection (alternatively, we may consider $\Delta_{q,r}$ as the dual map to the symmetric multiplication $S_q\mathcal{V}^* \otimes S_r\mathcal{V}^* \to S_{q+r}\mathcal{V}^*$ over the dual space $\mathcal{V}^*$). The Spencer differential δ arises then as the composition

$$\delta : S_{q+1}\mathcal{V} \otimes \Lambda_p\mathcal{V} \xrightarrow{\Delta_{q,1}\otimes \mathrm{id}} S_q\mathcal{V} \otimes \mathcal{V} \otimes \Lambda_p\mathcal{V} \xrightarrow{\mathrm{id}\otimes\wedge} S_q\mathcal{V} \otimes \Lambda_p\mathcal{V}\,. \tag{6.39}$$

Obviously, this construction somewhat hides the fact that δ is nothing but the exterior derivative restricted to forms with polynomial coefficients, i. e. the relation to the classical de Rham complex.

Some of the ideas underlying this construction (and thus the Spencer cohomology) already appeared in the work of Johnson [242, 243] on an algebraic formulation of the concept of prolongation in Cartan's theory of exterior differential systems where they are also used for giving a new proof of the Cartan–Kuranishi Theorem asserting that every exterior system is either inconsistent or becomes involutive after a finite number of prolongations (see Section 7.4 for the partial differential equations version of this result). The bound (6.20) on the number of prolongations required is due to Sweeney [447, Corollary 7.7].

While the notion of prolongation (of a symbol) arises naturally in the context of differential equations, it is also of interest in pure algebra. Sidman and Sullivant [417] recently showed (generalising earlier results by Landsberg and Manivel [280]) that prolongations provide direct access to the lowest degree part of the ideals of secant varieties. As a side product they furthermore proved that already in the monomial case directly computing prolongations is *NP*-hard.

Mansfield [306] gives a detailed discussion of the polynomial de Rham complex and various associated constructions in local bases. In particular, she explicitly proves that the differential ∂ of the Koszul complex provides a contracting homotopy. However, she misses the precise formulation of the duality between $R(\mathcal{V})$ and $K(\mathcal{V}^*)$, as she does not perform a proper dualisation. She uses the scalar product induced by the standard coordinate basis of $S_q\mathcal{V}\otimes\Lambda_p\mathcal{V}$ to introduce an adjoint of δ differing from the Koszul differential ∂ by combinatorial coefficients. This is solely due to the fact that she does not go to the dual spaces and thus does not use the dual bases. The combinatorial factors contained in the definition of the dual bases cancel the coefficients appearing in her adjoint.

The Koszul complex is a fundamental tool in commutative algebra and discussed in almost any textbook on this subject. Some discussions of computational aspects and, in particular, of applications in the geometry of projective varieties are given by Green [181] (see also [125, Sect. 17.5]). One should note that different conventions exist, but the arising complexes are trivially isomorphic. The one used here seems to be the most common one, but see [125, Chapt. 17] for an alternative. Apparently, the duality between the Spencer cohomology and the Koszul homology was first noted by Singer and Sternberg [419] who attributed it to private discussions with Grothendieck and Mumford (see also [365]). An independent proof was later given by Ruiz [390].

It is also possible to give a more abstract description of the Koszul differential ∂ as the composition of natural maps; the construction proceeds dually to the one for the Spencer differential δ. We take the map $\iota_{p,r}:\Lambda_{p+r}\mathcal{V}\rightarrow\bigotimes^{p+r}\mathcal{V}\rightarrow\Lambda_p\mathcal{V}\otimes\Lambda_r\mathcal{V}$ dual to the wedge product $\wedge:\Lambda_p\mathcal{V}^*\otimes\Lambda_r\mathcal{V}^*\rightarrow\Lambda_{p+r}\mathcal{V}^*$ over the dual space $\mathcal{V}^*$. Denoting the multiplication in the symmetric algebra simply by $\cdot$, we obtain now the Koszul differential ∂ as the composition

$$\partial : S_q\mathcal{V}\otimes\Lambda_p\mathcal{V} \xrightarrow{\text{id}\otimes\iota_{1,p-1}} S_q\mathcal{V}\otimes\mathcal{V}\otimes\Lambda_{p-1}\mathcal{V} \xrightarrow{\cdot\otimes\text{id}} S_{q+1}\mathcal{V}\otimes\Lambda_{p-1}\mathcal{V}\,. \tag{6.40}$$

The fact that this differential defines a complex follows immediately from the observation that the map $(\text{id}\otimes\iota_{1,p-2})\circ\iota_{1,p-1}$ takes its values in $S_2\mathcal{V}\otimes\Lambda_{p-2}\mathcal{V}$ and hence must be the zero map.

The relationship between the Koszul homology and the depth of a module given in Theorem 6.3.8 is discussed e. g. by Serre [414, Sect. IV.A.2/4] where it is (via isomorphic extension groups) used to define the notion of depth. Our proof via Pommaret bases appeared first in [408].

The Cartan test has its origin in the Cartan–Kähler theory of exterior differential systems where involution is defined for a subset $\mathcal{N}_q\subseteq S\mathcal{V}^*\otimes\mathcal{U}$ as equality in (6.25) with $\mathcal{N}_{q+1}=\mathcal{N}_{q,1}$. In this approach the problem is not obtaining a finite criterion for involution (as it was for us in Section 6.2), but in proving that if $\mathcal{N}_q$ is involutive, then the same is true for $\mathcal{N}_{q+1}$. However, basically these two problems are equivalent and our proof of Theorem 6.2.4 is in principle a straightforward adaption of a classical proof of Janet [234].

The modern homological reformulation of the Cartan test is essentially due to Matsushima [311, 312]. The infinite form of the dual Cartan test (Proposition 6.2.12) was developed by Serre in a letter appended to [189]. Quillen [365, App., Prop. 8] provided a bit later the effective version (Theorem 6.2.13); a slightly different proof was given by Malgrange [303, Prop. 1.2]. The relation between the two notions of δ-regularity and quasi-regularity (Theorem 6.3.2) was first exhibited in [408].

Rather surprisingly, the simple fact that the degree of involution and the Castelnuovo–Mumford regularity coincide (Remark 6.1.23 or Corollary 6.3.4, respectively) has been overlooked for a long time. It is implicitly contained in [406] and explicitly mentioned for the first time only fairly recently by Malgrange [303]. It seems that in the differential equations community almost exclusively the Spencer cohomology was considered without noticing the extensive theory on Koszul homology and the Castelnuovo–Mumford regularity developed by algebraists; conversely, the concept of involution (in the sense of this book) has remained unknown to most algebraists. This becomes particulary evident in Remark 6.3.6 showing that the criterion for q-regularity of Bayer and Stillman [34] (Theorem 5.5.21) is equivalent to the dual Cartan test (Theorem 6.2.13).

Chapter 7
Involution III: Differential Theory

I consider that I understand an equation when I can predict the properties of its solutions, without actually solving it.
Paul A.M. Dirac

In this chapter we return again to the study of differential equations. We will now combine the geometric theory introduced in Chapter 2 with the algebraic and homological constructions of the last four chapters in order to arrive finally at the notion of an involutive equation. The key is the (geometric) symbol $\mathcal{N}_q \subseteq V\pi^q_{q-1}$ of an equation $\mathcal{R}_q \subseteq J_q\pi$ which we define in the first section. The fundamental identification $V\pi^q_{q-1} \cong S_q(T^*\mathcal{X}) \underset{J_{q-1}\pi}{\otimes} V\pi$ discussed in Section 2.2 builds a bridge between the geometric and the algebraic side of formal theory.

Intrinsically, the symbol defines via prolongation at each point on $\mathcal{R}_q$ a subcomodule of a free comodule over the symmetric coalgebra, i. e. a symbolic system, to which we may apply the homological theory of Chapter 6. Dually, the symbol equations generate a submodule of a free module over the symmetric algebra. In local coordinates we may identify the symmetric algebra with the polynomial algebra and thus are exactly in the situation treated in Chapter 5.

Either way, the algebraic theory leads naturally to the notion of an involutive symbol. For readers not so familiar with abstract algebra, we will repeat some results from the previous chapters in differential equations terminology. In particular, we will discuss a concrete criterion for an involutive symbol which is useful in coordinate computations. However, it is valid only in δ-regular coordinates so that it must be applied with some care.

In the second section we finally define involutive differential equations: formally integrable equations with an involutive symbol. Somewhat surprisingly, in local coordinates—assumed to be δ-regular—involution is easier to verify than formal integrability: a first hint that involution is a stronger notion. We also introduce the Cartan normal form of an involutive equation, a useful tool in many situations.

Again the natural question arises what we should do, if we encounter an equation which is not involutive, and again the answer is to complete it to an equivalent involutive one. In contrast to the algebraic situation of Chapter 4, completion is now a more complicated operation which, in particular for nonlinear equations, is not always fully algorithmic. We consider first in Section 7.3 the special case of ordinary differential equations where the completion process has a very intuitive geometric

W.M. Seiler, *Involution*, Algorithms and Computation in Mathematics 24,
DOI 10.1007/978-3-642-01287-7_7, © Springer-Verlag Berlin Heidelberg 2010

description. In the next section we prove the Cartan–Kuranishi Theorem asserting the existence of an equivalent involutive equation for any regular partial differential equation. Although our proof is constructive, it does not provide us with an algorithm in the strict sense. We will show later in Section 10.7 how for linear systems the Cartan–Kuranishi approach may be combined with the algebraic algorithms of Chapter 4 in order to obtain an effective algorithm.

As a first application of the notion of involution we revisit in Section 7.5 the principal symbol and provide a rigorous definition of under- and overdetermined equations. It turns out that the classical counting rules comparing the number of equations and unknown functions, respectively, are not always valid. Furthermore, we indicate how the principal symbol allows us a partial solution of the problem of δ-regularity (actually, for many systems appearing in applications this partial solution completely suffices).

A full solution of the question of δ-regularity is possible via a combinatorial extension of the notion of a principal symbol which we will discuss in the last section of this chapter. This is a rather technical matter and in practice the approach of Section 4.3 is often easier; thus readers not particularly interested in this topic are advised to skip this section. The main advantage of the new approach is that it does not require to apply a coordinate transformation, as it determines the correct indices of a symbol indirectly (by simulating the effect of a partial transformation).

7.1 (Geometric) Symbol and Principal Symbol

In our introduction of the hierarchy of jet bundles over a fibred manifold $\pi:\mathcal{E}\to\mathcal{X}$ in Chapter 2 we stressed several times the fact that the qth order jet bundle $J_q\pi$ is an affine bundle over the jet bundle $J_{q-1}\pi$ of order $q-1$ modelled on the vector bundle $S_q(T^*\mathcal{X})\otimes V\pi$; in fact our intrinsic approach to jet bundles in Section 2.2 was built on this property. The reason for this emphasis will become apparent now. It allows us to introduce a very useful geometric object, namely the symbol of a differential equation. Many properties of a differential equation can be read off its symbol and it effectively reduces some problems to questions in linear algebra.

Definition 7.1.1. Let $\mathcal{R}_q\subseteq J_q\pi$ be a differential equation of order q. The *(geometric) symbol* $\mathcal{N}_q$ of $\mathcal{R}_q$ is a distribution over the manifold $\mathcal{R}_q$ where the component at a point $\rho\in\mathcal{R}_q$ is given by

$$(\mathcal{N}_q)_\rho = T_\rho\mathcal{R}_q\cap V_\rho\pi^q_{q-1} = V_\rho\left(\pi^q_{q-1}|_{\mathcal{R}_q}\right)\,. \tag{7.1}$$

Thus the geometric symbol is the vertical part of the tangent space of the submanifold $\mathcal{R}_q$ with respect to the fibration π^q_{q-1} (here we identify $T\mathcal{R}_q$ with a subspace of $T(J_q\pi)$). If the equation $\mathcal{R}_q$ is globally described by a map $\Phi:J_q\pi\to\mathcal{E}'$ with a vector bundle $\pi':\mathcal{E}'\to\mathcal{X}$, then we introduce the *symbol map* $\sigma:V\pi^q_{q-1}\to T\mathcal{E}'$ as a restriction of the tangent map, $\sigma=T\Phi|_{V\pi^q_{q-1}}$, and find $\mathcal{N}_q=\ker\sigma$.

Locally, we obtain the following picture. Let $(\mathbf{x},\mathbf{u}^{(q)})$ be local coordinates on $J_q\pi$ in a neighbourhood of ρ. We must first determine $T_\rho\mathcal{R}_q$ as a subspace of $T_\rho(J_q\pi)$. Let $(\mathbf{x},\mathbf{u}^{(q)};\dot{\mathbf{x}},\dot{\mathbf{u}}^{(q)})$ be the induced coordinates on $T_\rho(J_q\pi)$; then every vector $v\in T_\rho(J_q\pi)$ has the form $v=\dot{x}^i\partial_{x^i}+\dot{u}^\alpha_\mu\partial_{u^\alpha_\mu}$. Assuming that a local representation of the submanifold $\mathcal{R}_q$ is given by the equations $\Phi^\tau(\mathbf{x},\mathbf{u}^{(q)})=0$ with $\tau=1,\dots,t$, its tangent space $T_\rho\mathcal{R}_q$ consists of all vectors $v\in T_\rho(J_q\pi)$ such that $\mathrm{d}\Phi^\tau(v)=v\Phi^\tau=0$ (see Remark C.2.8). This yields the linear system

$$\sum_{i=1}^{n}\frac{\partial\Phi^\tau}{\partial x^i}(\rho)\,\dot{x}^i+\sum_{\substack{1\leq\alpha\leq m\\0\leq|\mu|\leq q}}\frac{\partial\Phi^\tau}{\partial u^\alpha_\mu}(\rho)\,\dot{u}^\alpha_\mu=0 \tag{7.2}$$

for the coefficients $\dot{x}^i$, $\dot{u}^\alpha_\mu$ of the vector v.

The symbol $(\mathcal{N}_q)_\rho$ is by definition the vertical part of this tangent space with respect to the fibration π^q_{q-1}. Thus we are only interested in those solutions of (7.2) where $\dot{\mathbf{x}}=\dot{\mathbf{u}}^{(q-1)}=0$ and locally $(\mathcal{N}_q)_\rho$ can be described as the solution space of the following system of linear equations:

$$(\mathcal{N}_q)_\rho:\left\{\sum_{\substack{1\leq\alpha\leq m\\|\mu|=q}}\frac{\partial\Phi^\tau}{\partial u^\alpha_\mu}(\rho)\,\dot{u}^\alpha_\mu=0\,,\qquad\tau=1,\dots,t\right. \tag{7.3}$$

(note that the summation is now only over the multi indices of length q). Eq. (7.3) is a linear system with real coefficients, as the derivatives $\partial\Phi^\tau/\partial u^\alpha_\mu$ are evaluated at the point $\rho\in\mathcal{R}_q$. If there are n independent and m dependent variables, the coefficient matrix of (7.3) has t rows (labelled by τ) and $m\binom{n+q-1}{n-1}$ columns (labelled by α and μ). We call it *symbol matrix* and denote it by $M_q(\rho)$. It also represents the matrix of the symbol map σ in local coordinates.

Obviously, $\dim(\mathcal{N}_q)_\rho$—or equivalently $\operatorname{rank}M_q(\rho)$—might vary with the point $\rho\in\mathcal{R}_q$, i. e. we cannot generally expect that $\mathcal{N}_q$ is a distribution of constant rank. Only if the dimension remains constant over $\mathcal{R}_q$, the symbol $\mathcal{N}_q$ is actually a vector subbundle of $V(\pi^q_{q-1}|_{\mathcal{R}_q})$. In particular, $\mathcal{N}_q$ is a vector bundle, if $\mathcal{R}_q$ is globally defined as the zero set of a fibred map $\Phi:J_q\pi\to\mathcal{E}'$. In line with our usual regularity assumptions, we avoid a discussion of the problems associated with jumps in the dimensions $\dim(\mathcal{N}_q)_\rho$.

The symbol is most easily understood for linear differential equations. The symbol matrix is then simply the highest-order or principal part of the system (considered as algebraic equations). For non-linear systems we perform via the Jacobian a brute force linearisation at the point ρ in order to obtain $M_q(\rho)$.

Remark 7.1.2. It is easy to see that a first-order partial differential equation $\mathcal{R}_1$ is of finite type in the sense introduced in Remark 2.3.6, if and only if the dimension of its symbol $\mathcal{N}_1$ is zero. This observation leads to the general definition that any differential equation with a vanishing symbol is said to be of finite type (the reason for this terminology will become apparent in Remark 8.2.4). ◁

Of course, not only the original equation $\mathcal{R}_q$ has a symbol $\mathcal{N}_q$, but also every prolongation $\mathcal{R}_{q+r} \subseteq J_{q+r}\pi$ of it possesses a symbol $\mathcal{N}_{q+r} \subseteq T(J_{q+r}\pi)$. It follows easily from the coordinate expression (2.14) of the formal derivative that for obtaining a local representation of the prolonged symbol $(\mathcal{N}_{q+r})_{\hat{\rho}}$ at a point $\hat{\rho} \in \mathcal{R}_{q+r}$ lying over $\rho \in \mathcal{R}_q$, there is no need to explicitly compute a local representation of the prolonged differential equation $\mathcal{R}_{q+r}$. We can directly derive it from a local representation of the symbol $(\mathcal{N}_q)_\rho$ at order q.

Given the form of the symbol equations (7.3) and the local representation (2.54) of the prolonged equation $\mathcal{R}_{q+r}$, we need for determining the symbol $(\mathcal{N}_{q+1})_{\hat{\rho}}$ only the partial derivatives $\partial D_i\Phi^\tau/\partial u^\alpha_\nu(\hat{\rho})$ with $1 \leq i \leq n$ and $|\nu| = q+1$, i. e. the highest-order part of the formal derivative $D_i\Phi^\tau$. According to (2.14), it is given by $\partial D_i\Phi^\tau/\partial u^\alpha_\nu(\hat{\rho}) = \partial\Phi^\tau/\partial u^\alpha_{\nu-1_i}(\rho)$ and thus depends indeed only on the projection $\rho = \pi^{q+1}_q(\hat{\rho})$ (if $\nu_i = 0$, the derivative vanishes). Thus a local representation of the first prolongation $(\mathcal{N}_{q+1})_{\hat{\rho}}$ is

$$(\mathcal{N}_{q+1})_{\hat{\rho}} : \begin{cases} \displaystyle\sum_{\substack{1\leq\alpha\leq m\\ |\mu|=q}} \frac{\partial\Phi^\tau}{\partial u^\alpha_\mu}(\rho)\, \dot{u}^\alpha_{\mu+1_i} = 0\,, & \begin{array}{l}\tau = 1,\dots,t\,,\\ i = 1,\dots,n\,.\end{array}\end{cases} \tag{7.4}$$

One might be tempted to think that the linear systems (7.3) and (7.4) have the same coefficient matrices. This is of course not true. The symbol matrix $M_{q+1}(\hat{\rho})$ in (7.4) has nt rows and $m\binom{n+q}{n-1}$ columns and is thus much larger.[1] Because of the particular way in which the index i appears in (7.4), one may say that the matrix $M_{q+1}(\hat{\rho})$ emerges from $M_q(\rho)$ by a kind of combinatorial blow up. For each row in $M_q(\rho)$ we obtain n rows in $M_{q+1}(\hat{\rho})$ and similarly for the columns. These n rows contain exactly the same entries but distributed in different patterns.

Example 7.1.3. We consider a scalar second-order equation in two independent variables x, y with the local representation

$$\mathcal{R}_2 : \Phi(x,y,u,u_x,u_y,u_{xx},u_{xy},u_{yy}) = 0\,. \tag{7.5}$$

If we define the three functions $a(\rho) = \partial\Phi/\partial u_{yy}(\rho)$, $b(\rho) = \partial\Phi/\partial u_{xy}(\rho)$ and $c(\rho) = \partial\Phi/\partial u_{xx}(\rho)$, each $\mathcal{R}_2 \to \mathbb{R}$, then the symbol matrix at a point $\rho \in \mathcal{R}_2$ contains only a single row:

$$M_2(\rho) = \big(a(\rho)\ b(\rho)\ c(\rho)\big) \tag{7.6}$$

[1] Strictly speaking, the matrix $M_{q+1}(\hat{\rho})$ has even t rows more, containing only zeros. These correspond to the t equations defining $\mathcal{R}_q$ which are of course also part of the local representations of any prolongation. However, these equations are trivially of lower order and do not affect the prolonged symbol $(\mathcal{N}_{q+1})_{\hat{\rho}}$. Hence we ignore them.

(we usually order the columns according to the degree reverse lexicographic order). Thus $(\mathcal{N}_2)_\rho$ consists of all vertical vectors $v = \dot{u}_{yy}\partial_{u_{yy}} + \dot{u}_{xy}\partial_{u_{xy}} + \dot{u}_{xx}\partial_{u_{xx}} \in T_\rho\mathcal{R}_q$ with $a(\rho)\dot{u}_{yy} + b(\rho)\dot{u}_{xy} + c(\rho)\dot{u}_{xx} = 0$. If no point $\rho \in \mathcal{R}_2$ exists where all three coefficient functions a, b, c simultaneously vanish, then the symbol $\mathcal{N}_2$ defines a two-dimensional vector bundle over $\mathcal{R}_2$.

For the prolonged symbol $(\mathcal{N}_3)_{\hat{\rho}}$ at any point $\hat{\rho} \in \mathcal{R}_3$ such that $\pi^3_2(\hat{\rho}) = \rho$ we obtain the 2×4 symbol matrix

$$M_3(\hat{\rho}) = \begin{pmatrix} a(\rho) & b(\rho) & c(\rho) & 0 \\ 0 & a(\rho) & b(\rho) & c(\rho) \end{pmatrix} . \tag{7.7}$$

Thus $(\mathcal{N}_3)_{\hat{\rho}}$ consists of all vectors $v = \dot{u}_{yyy}\partial_{u_{yyy}} + \dot{u}_{xyy}\partial_{u_{xyy}} + \dot{u}_{xxy}\partial_{u_{xxy}} + \dot{u}_{xxx}\partial_{u_{xxx}}$ where the coefficients satisfy the equations $a(\rho)\dot{u}_{yyy} + b(\rho)\dot{u}_{xyy} + c(\rho)\dot{u}_{xxy} = 0$ and $a(\rho)\dot{u}_{xyy} + b(\rho)\dot{u}_{xxy} + c(\rho)\dot{u}_{xxx} = 0$. If the symbol $\mathcal{N}_2$ is a vector bundle, then the prolonged symbol $\mathcal{N}_3$ is, too. $\triangleleft$

The considerations above are straightforwardly extended by iteration to higher prolongations $\mathcal{R}_{q+r}$ and we obtain as local representation for the symbol at any point $\hat{\rho} \in \mathcal{R}_{q+r}$ lying in the fibre $(\pi^{q+r}_q)^{-1}(\rho)$ over $\rho \in \mathcal{R}_q$

$$(\mathcal{N}_{q+r})_{\hat{\rho}} : \begin{cases} \displaystyle\sum_{\substack{1\leq\alpha\leq m \\ |\mu|=q}} \frac{\partial\Phi^\tau}{\partial u^\alpha_\mu}(\rho)\,\dot{u}^\alpha_{\mu+\nu} = 0\,, & \begin{array}{l}\tau = 1,\dots,t\,, \\ |\nu| = r\,.\end{array} \end{cases} \tag{7.8}$$

The size of the corresponding symbol matrices $M_{q+r}(\hat{\rho})$ grows rapidly. Because of the formula (2.32) for the dimension of the vertical bundle, one easily calculates that M_{q+r} has $t\binom{n+r-1}{n-1}$ rows and $m\binom{n+q+r-1}{n-1}$ columns (again ignoring all equations which are trivially of lower order).

We stop now with explicitly mentioning the point ρ at which a symbol is considered except if it is of importance. In fact, if not explicitly stated otherwise, we make the following assumption.

Blanket Assumption 7.1.4. *All properties of the symbol are independent of the considered point. In particular, all considered symbols are vector bundles.*

Remark 7.1.5. The matrices M_{q+r} of the prolonged symbols $\mathcal{N}_{q+r}$ implicitly appeared already in Section 2.3 during the construction of a formal power series solution of the form (2.65). The rank of the symbol matrix M_q tells us how many Taylor coefficients of order q (nothing is said about the coefficients of lower order!) are determined by (2.66), i. e. are principal. The matrices M_{q+r} of the prolonged symbols appear, if we interpret (2.68) as an inhomogeneous linear system for the coefficients of order $q+r$: its matrix is precisely M_{q+r}. Thus the symbol $\mathcal{N}_{q+r}$ is the solution space of the corresponding homogeneous system and its dimension measures the number of possible solutions for the coefficients of order $q+r$, i. e. equals the number of parametric coefficients. In Section 8.2 we will exploit this observation for showing how the size of the (usually infinite-dimensional) formal solution space can be measured with some finite quantities. ◁

In Section 2.3 we saw that this simple algorithm for the construction of formal power series solutions only works, if we are dealing with a formally integrable equation. The symbol helps us in deciding whether or not a given equation is formally integrable. It indicates whether during the prolongation of a differential equation integrability conditions may have occurred and it allows us to construct them using only linear algebra. The proof of the next theorem shows this explicitly.

Theorem 7.1.6. *Let $\mathcal{R}_q \subseteq J_q\pi$ be an arbitrary differential equation. If the prolonged symbol $\mathcal{N}_{q+1}$ is a vector bundle, then the dimension of the prolonged equation $\mathcal{R}_{q+1}$ and its projection $\mathcal{R}_q^{(1)} = \pi_q^{q+1}(\mathcal{R}_{q+1s})$, respectively, are related by*

$$\dim \mathcal{R}_q^{(1)} = \dim \mathcal{R}_{q+1} - \dim \mathcal{N}_{q+1} \,. \tag{7.9}$$

Proof. The dimension of a submanifold can be determined by counting the number of independent equations needed to describe it locally, i. e. by the rank of a Jacobian. Let $\mathcal{R}_q$ have the local representation $\boldsymbol{\Phi}^\tau(\mathbf{x},\mathbf{u}^{(q)}) = 0$. The part of the Jacobian of the corresponding local representation of the prolonged equation $\mathcal{R}_{q+1}$ relevant for the determination of its fibre dimension can be divided into four blocks:

$$\left(\begin{array}{|c|c|}\hline \dfrac{\partial D_i\Phi^\tau}{\partial u^\alpha_\mu}\,,\ |\mu| = q+1 & \dfrac{\partial D_i\Phi^\tau}{\partial u^\alpha_\mu}\,, 0 \le |\mu| \le q \\ \hline 0 & \dfrac{\partial \Phi^\tau}{\partial u^\alpha_\mu}\,,\ 0 \le |\mu| \le q \\ \hline \end{array}\right) . \tag{7.10}$$

The lower part stems from the local representation of $\mathcal{R}_q$; its rank is $\operatorname{codim}\mathcal{R}_q$. The upper part comes from the prolongation; the left block is just the symbol matrix M_{q+1}. We must distinguish two cases. If M_{q+1} has maximal (row) rank, then it is not possible to construct by linear operations a row in the upper part with only zeros in the left block and thus we cannot extract any lower-order equations from the equations $D_i\Phi^\tau = 0$. This observation implies that $\mathcal{R}_q^{(1)} = \mathcal{R}_q$, as the restriction of the projection π_q^{q+1} to $\mathcal{R}_{q+1}$ is surjective. Furthermore, we find in this case $\dim\mathcal{R}_{q+1} = \dim\mathcal{R}_q + \dim\mathcal{N}_{q+1}$ and thus our claim holds.

If the matrix M_{q+1} does not have maximal rank, then one can construct by linear operations a row in the upper part with only zeros in the left block. The question is how this row looks like in the right block. If it is linearly independent of the rows in the lower part, then we have found an integrability condition. It is constructed by performing exactly the same linear operations on the full equations $D_i\Phi^\tau = 0$. Otherwise, our local representation of $\mathcal{R}_{q+1}$ is redundant and this redundancy leads to *identities* between the equations of the system. Comparing ranks yields the claim for one point, but since $\mathcal{N}_{q+1}$ is a vector bundle, it holds everywhere. □

Remark 7.1.7. The above mentioned identities appear naturally in every overdetermined system. In the case of linear equations, they lead to *compatibility conditions*, if we add a right hand side: equations that the right hand sides must satisfy, if a solution is to exist. A simple classical example may explain this phenomenon. Consider the differential equation $\mathcal{R}_1$ locally described by

$$\mathcal{R}_1 : \begin{cases} \phi(x,y,u,u_x,u_y) = u_x = 0\,, \\ \psi(x,y,u,u_x,u_y) = u_y = 0\,, \end{cases} \tag{7.11}$$

(or more generally by $\nabla u = 0$ in arbitrary dimensions). If we differentiate each equation in the system with respect to both independent variables, we obtain a local representation of $\mathcal{R}_2$ containing twice the equations $u_{xy} = 0$ which leads to the identity $D_y\phi - D_x\psi = 0$.

Let us add some right hand sides, i. e. we consider the inhomogeneous system $u_x = F(x,y)$ and $u_y = G(x,y)$; a solution u of it is called a *potential* for the two-dimensional vector field $(F,G)^t$. The effect of the above identity consists of imposing on the right hand sides F, G the well-known compatibility condition $F_y = G_x$ which is obviously a necessary (and, as we will later show in Section 10.5, also a sufficient) condition for the existence of a potential.

The compatibility conditions form again a linear system with the same independent variables and the right hand sides as dependent variables.[2] If we consider in the ring of linear differential operators the module $\mathcal{M}$ generated by the operators defining the left hand sides of our linear system, then the compatibility conditions generate the syzygy module $\operatorname{Syz}(\mathcal{M})$. This point of view will be studied in much more details in Section 10.5.

In some respect the distinction between integrability and compatibility conditions is artificial and merely a question of the chosen point of view. If we consider in our example the functions F, G not as some arbitrary right hand sides but as additional dependent variables in the inhomogeneous system above, then the equation $F_y = G_x$ becomes an ordinary integrability condition. Indeed, in the literature the terms "integrability condition" and "compatibility condition" are often used synonymously. However, in this book the latter one always refers to right hand sides.

The restriction to right hand sides is also important to ensure that the compatibility conditions are again linear. Of course, one could consider a general linear system depending parametrically on some functions $\mathbf{F}$ and then study under which conditions on $\mathbf{F}$ the system is formally integrable. Because of the linearity, these conditions will always form separate equations (i. e. containing only $\mathbf{F}$ as dependent variables). However, in general, they will cease to form a linear system. Nevertheless, such considerations are sometimes very useful. For instance, in the theory of (completely) integrable systems one has the notion of a *Lax pair* which corresponds exactly to such a situation. Consider the linear system

---

[2] In principle, one can add right hand sides also to a nonlinear system. However, then one does not obtain a separate system for the right hand sides, i. e. the compatibility conditions will depend on both the old dependent variables and the right hand sides. Therefore the concept of a compatibility condition is not particularly useful is this case.

$$u_{xx} + (\lambda - F)u = 0\,, \qquad u_t - 2(F + 2\lambda)u_x + F_x u = 0 \tag{7.12}$$

where λ is a real parameter and F a yet arbitrary function of x, t. In a straightforward computation one can show that the system is (formally) solvable, if and only if the function F is a solution of the non-linear differential equation

$$F_t - 6FF_x + F_{xxx} = 0 \tag{7.13}$$

which is nothing but the famous Korteweg–de Vries equation. The fact that it can be derived from a linear problem is in particular important for its integration by the inverse scattering method. While it is common in the literature to call (7.13) a compatibility condition for the Lax pair (7.12), we stress again that this is not correct in our terminology. ◁

In the classical theory of partial differential equations a different notion of symbol appears which should not be confused with the geometric symbol introduced above: the classical symbol is *not* an intrinsic object. Our notion of symbol is closely related to what is traditionally called the principal symbol which is again an intrinsically defined object.

Assume we are given a one-form $\chi \in T^*\mathcal{X}$. For notational simplicity, we write briefly $V\pi$ instead of the pull-back bundle $(\pi_0^q)^*(\tau_{\mathcal{E}}|_{V\pi}) : (\pi_0^q)^*(V\pi) \to J_q\pi$ where $\tau_{\mathcal{E}} : T\mathcal{E} \to \mathcal{E}$ is the tangent bundle projection. The fundamental identification ε_q allows us to define a map $\iota_{\chi,q} : V\pi \to V\pi^q_{q-1}$ by setting $\iota_{\chi,q}(v) = \varepsilon_q^{-1}(\chi^q \otimes v)$. Here χ^q denotes the q-fold symmetric product of χ. In local coordinates, we write $\chi = \chi_i \mathrm{d}x^i$ and obtain $\iota_{\chi,q}(v^\alpha \partial_{u^\alpha}) = \chi_\mu v^\alpha \partial_{u^\alpha_\mu}$ where μ runs over all multi indices of length q and $\chi_\mu = \chi_1^{\mu_1} \cdots \chi_n^{\mu_n}$. Now we can define the principal symbol in terms of the symbol map σ.

Definition 7.1.8. Let $\chi \in T^*\mathcal{X}$ be a one-form over $\mathcal{X}$ and $\mathcal{R}_q \subseteq J_q\pi$ a differential equation globally described by the map $\Phi : J_q\pi \to \mathcal{E}'$. The *principal symbol* of $\mathcal{R}_q$ is the linear map $\tau_\chi : V\pi \to T\mathcal{E}'$ defined by $\tau_\chi = \sigma \circ \iota_{\chi,q}$.

Remark 7.1.9. Over a given point $x \in \mathcal{X}$, a slightly different point of view will later be useful for a generalisation. Let $\mathcal{Y} \subseteq \mathcal{X}$ be a one-dimensional submanifold with $x \in \mathcal{Y}$ such that its cotangent bunde $T^*\mathcal{Y} \subseteq T^*\mathcal{X}$ is generated by χ (i. e. Y is an integal manifold through x for the codistribution generated by χ). Then we may identify the pull-back bundle $(\pi_0^q)^*(V\pi)$ with $S_q(T^*\mathcal{Y}) \otimes V\pi$. Now the map $\iota_{\chi,q}$ simply becomes the inverse fundamental identification. ◁

The one-form χ plays the role of a parameter in the principal symbol τ_χ. For many purposes, it is of great importance how the properties of the linear map τ_χ change with χ. Locally, we associate a matrix $T[\chi]$ with the linear map τ_χ:

$$T^\tau_\alpha[\chi] = \sum_{|\mu|=q} \frac{\partial \Phi^\tau}{\partial u^\alpha_\mu} \chi^\mu\,. \tag{7.14}$$

If $\dim \mathcal{E} = m$ and $\dim \mathcal{E}' = t$ (i. e. for a system of t equations in m dependent variables), it has t rows and m columns. Its entries are homogeneous polynomials of degree q in the coefficients of the one-form χ. We may think of $T[\chi]$ as a kind of "contraction" of the symbol matrix M_q of $\mathcal{R}_q$. Both matrices have the same number of rows. The column with index α of $T[\chi]$ is a linear combination of all columns in M_q corresponding to a variable $\dot{u}^\alpha_\mu$ with the coefficients given by the product χ_μ. Hence the principal symbol contains the same information as the geometric symbol, only encoded in a slightly different way: the role of the many different columns in M_q is taken over by the dependency on the parameter χ.

Let us assume that the functions Φ^τ appearing in the chosen local representation of our differential equation $\mathcal{R}_q$ are elements of some differential field $\mathbb{F}$. Then the entries of the matrix $T[\chi]$ are polynomials in $\mathcal{P} = \mathbb{F}[\chi_1,\dots,\chi_n]$ and the rows of $T[\chi]$ may be considered as elements of the free polynomial module $\mathcal{P}^m$. We call the graded submodule $\mathcal{T} \subseteq \mathcal{P}^m$ generated by them the *principal symbol module* of the differential equation $\mathcal{R}_q$. This submodule contains in some sense the principal symbols of all prolongations of $\mathcal{R}_q$.

Lemma 7.1.10. *If $\mathcal{T}$ is the principal symbol module of the differential equation $\mathcal{R}_q$, then for any $r \geq 0$ the principal symbol module of the prolonged equation $\mathcal{R}_{q+r}$ is the truncation $\mathcal{T}_{\geq q+r}$.*

Proof. Let the equations $\Phi^\tau = 0$ with $1 \leq \tau \leq t$ form a local representation of $\mathcal{R}_q$ and $\mathbf{t}_\tau \in \mathcal{P}^m$ denote the τth row of the matrix $T[\chi]$. For the principal symbol only the highest-order part of the given differential equation is relevant. Thus for the principal symbol of the prolongation $\mathcal{R}_{q+r}$ we need to consider only the equations $D_\nu\Phi^\tau = 0$ with $|\nu| = r$. As one can easily see, such an equation contributes to the principal symbol module of $\mathcal{R}_{q+r}$ the generator $\chi^\nu\mathbf{t}_\tau$. This observation immediately implies our claim. □

Example 7.1.11. Consider again the second-order equation $\mathcal{R}_2$ described by (7.5). Its symbol matrix is the 1×3 matrix shown in (7.6). The matrix $T[\chi]$ of its principal symbol is 1×1 with the single entry $f = a(\rho)\chi_y^2 + b(\rho)\chi_x\chi_y + c(\rho)\chi_x^2$ arising by adding up the columns of M_2 with suitable monomials in the variables χ as coefficients. Hence the principal symbol module is here the homogeneous ideal $\mathcal{T} = \langle f \rangle \subset \mathbb{R}[\chi_x,\chi_y]$. The principal symbol of the prolonged equation $\mathcal{R}_3$ is described by the 2×1 matrix $\begin{pmatrix} \chi_x f \\ \chi_y f \end{pmatrix}$. Obviously, its two entries form an $\mathbb{R}$-linear basis of the homogeneous component $\mathcal{T}_3$. ◁

Remark 7.1.12. The principal symbol module $\mathcal{T}$ allows us to relate the construction of integrability conditions to syzygy computations. Let $\mathbf{s} \in \mathrm{Syz}(\mathcal{T}) \subseteq \mathcal{P}^t$ be a syzygy of the rows of $T[\chi]$. The substitution $\chi_i \to D_i$ (where D_i denotes as usual the formal derivative with respect to x^i) transforms each component s_τ of $\mathbf{s}$ into a linear differential operator $\hat{s}_\tau$. Then the equation $\Psi = \hat{s}_\tau\Phi^\tau = 0$ is a linear combination of differential consequences of the differential equation $\mathcal{R}_q$ in which, by construction, the highest-order terms cancel (this is just the syzygy property). In fact, this approach represents nothing but the rigorous mathematical formulation of "taking a (generalised) cross-derivative."

If r is the maximum of the degrees of the components s_τ, then generically Ψ is a function defined on the jet bundle J_{q+r-1}. Note that the question whether or not the equation $\Psi = 0$ represents an integrability condition can*not* be decided with the principal symbol. In order to answer it, we must check whether the function Ψ is algebraically independent of the induced local representation of the prolonged equation $\mathcal{R}_{q+r-1}$ and the answer will generally depend on the full equations and not only on their principal part. Thus one may say that the algebraic theory of the principal symbol tells us where *potentially* integrability conditions are hidden, but the actual verification still requires differential computations.

As a simple concrete example let us take up the linear differential equation $\mathcal{R}_1$ of Example 2.3.9. It was locally defined by the system $u_z + yu_x = 0 = u_y$. Because of the linearity, we may use here as underlying differential field $\mathbb{F} = \mathbb{R}(x,y,z)$, the rational functions in the independent variables. Now the principal symbol module of $\mathcal{R}_1$ is the ideal $\mathcal{T} = \langle \chi_z + y\chi_x, \chi_y \rangle \subset \mathbb{F}[\chi_x, \chi_y, \chi_z]$. Its syzygy module $\mathrm{Syz}(\mathcal{T})$ is trivially generated by $\mathbf{s} = \chi_y\mathbf{e}_1 - (\chi_z + y\chi_x)\mathbf{e}_2$. Hence a potential integrability condition arises by requiring that the "generalised cross-derivative" corresponding to $\mathbf{s}$ vanishes:

$$D_y(u_z + yu_x) - (D_z + yD_x)u_y = 0\,. \tag{7.15}$$

Evaluating the formal derivatives yields the equation $u_x = 0$, i. e. exactly the integrability condition we found in Example 2.3.9.

One should note that this approach gives us only those integrability conditions that come from cross-derivatives. As lower-order equations do not contribute to the principal symbol, integrability conditions of the second kind cannot be determined this way. In our discussion of elliptic systems in Section 10.3 we will introduce the *reduced principal symbol*. With its help one can algebraically find all potential integrability conditions.

Since the polynomial ring $\mathcal{P}$ is Noetherian, the syzygy module $\mathrm{Syz}(\mathcal{T}) \subseteq \mathcal{P}^t$ is finitely generated. Obviously, it suffices to check the potential integrability conditions coming from an arbitrary generating set of $\mathrm{Syz}(\mathcal{T})$, as all other syzygies lead trivially to algebraically dependent equations. We will show later in this chapter that this observation will permit us to derive a finite criterion for formal integrability (Theorem 7.2.8). ◁

The principal symbol also underlies the notion of a characteristic. This concept is for example crucial for the classification of differential equations into elliptic or hyperbolic systems. While this topic will not play an important role in this book (see however Sections 10.3 and 10.4), it will later turn out that characteristics are closely related to the problem of δ-regularity.

Definition 7.1.13. A non-zero one-form $\chi \in T^*\mathcal{X}$ is *characteristic* for a differential equation $\mathcal{R}_q$ at a given point $\rho \in \mathcal{R}_q$, if the principal symbol τ_χ is not injective at ρ. All characteristic one-forms at ρ form the *characteristic variety* $C_\rho \subseteq \mathbb{R}^n$.

The set C_ρ is indeed a variety in the strict algebraic meaning of the word. The condition that the principal symbol τ_χ is not injective at a point ρ can be expressed as the vanishing of certain subdeterminants of the matrix $T[\chi]$, i. e. as the vanishing of (homogeneous) polynomials in the variables χ_i. For *semi-linear* equations, where the symbol $(\mathcal{N}_q)_\rho$ depends only on $x = \pi^q(\rho) \in \mathcal{X}$, one can define characteristic one-forms over points $x \in \mathcal{X}$ in the base space.

Example 7.1.14. In many textbooks on differential equations one does not consider characteristic one-forms but *characteristic surfaces*. We exhibit now—for notational simplicity only for linear first-order systems—the relation between the two approaches. Consider

$$\mathcal{R}_1 : \left\{ \sum_{\alpha=1}^{m} \sum_{i=1}^{n} A_\alpha^{\tau i}(\mathbf{x}) u_i^\alpha + \sum_{\alpha=1}^{m} B_\alpha^\tau(\mathbf{x}) u^\alpha + C^\tau(x) = 0 \,, \qquad 1 \leq \tau \leq p \right. \tag{7.16}$$

and the $(n-1)$-dimensional hypersurface $\Sigma \subset \mathcal{X}$ described by the equation

$$\Sigma = \left\{ x \in \mathcal{X} \mid x^n = \phi(x^1, \dots, x^{n-1}) \right\} \tag{7.17}$$

(note that we have singled out the independent variable x^n). Such a surface is called characteristic, if prescribing the value of $\mathbf{u}$ on Σ does not suffice to compute all first-order derivatives of $\mathbf{u}$ on Σ. Obviously, this condition concerns only the derivatives $(\partial\mathbf{u}/\partial x^n)\big|_\Sigma$ with respect to x^n; all other ones follow easily from our "initial values".

Thus assume that $\mathbf{u}\big|_\Sigma = \mathbf{f}(x^1, \dots, x^{n-1})$. Using the chain rule we obtain

$$\left.\frac{\partial \mathbf{u}}{\partial x^i}\right|_\Sigma + \left.\frac{\partial \phi}{\partial x^i}\frac{\partial \mathbf{u}}{\partial x^n}\right|_\Sigma = \frac{\partial \mathbf{f}}{\partial x^i} \,, \qquad 1 \leq i < n \,. \tag{7.18}$$

Substituting these relations into (7.16) and restricting to Σ yields a linear system for the derivatives $(\partial\mathbf{u}/\partial x^n)\big|_\Sigma$ whose matrix T is

$$T_\alpha^\tau = A_\alpha^{\tau n} - \sum_{i=1}^{n-1} A_\alpha^{\tau i} \frac{\partial \phi}{\partial x^i} \,. \tag{7.19}$$

If $\operatorname{rank} T < m$, i. e. if the associated linear map is not injective, then it is *not* possible to determine all derivatives on the hypersurface Σ and it is characteristic. The matrix of the principal symbol of our linear system (7.16) is $T_\alpha^\tau[\chi] = A_\alpha^{\tau i}\chi_i$ which coincides with the matrix (7.19) for the special choice $\chi = \chi_\phi = \mathrm{d}x^n - (\partial\phi/\partial x^i)\mathrm{d}x^i$. Hence the hypersurface Σ is characteristic, if and only if this particular one-form is characteristic. More abstractly, we may say, using a terminology introduced in Appendix C.2, that Σ is characteristic, if and only if its conormal space $N^*\Sigma$ (which in our particular case is generated by the single one-form χ_ϕ) consists entirely of characteristic one-forms. ◁

Now we want to apply the algebraic methods developed in the last four chapters for the analysis of the (geometric) symbol and in particular for the definition of an involutive symbol. Thus we must first identify corresponding algebraic structures, i. e. either a symbolic system or dually a polynomial module. This identification takes place pointwise; more precisely, we find a symbolic system or a polynomial module, respectively, at each point $\rho \in \mathcal{R}_q$.

Another important point is that while all our definitions are intrinsic, their concrete application usually happens in local coordinates. Hence we meet again the question of δ- or quasi-regularity. As an effective method to deal with it has already been developed in Section 4.3, we assume throughout this section that our coordinates are δ-regular.

The intrinsic approach to involutive symbols is based on the Spencer cohomology introduced in Section 6.1. Because of its coordinate independence it represents the natural way to introduce the notion of an involutive symbols. Once we have given this definition, we will show how Pommaret bases provide us with a simple effective criterion for involution. Afterwards we will present yet another formulation of this criterion which does not require any algebra (albeit a proper understanding will not be possible without at least some ideas of the underlying algebraic theory).

Recall that the fundamental identification provides us at each point $\rho \in \mathcal{R}_q$ with an isomorphism ε_q between the vertical space $V_\rho \pi^q_{q-1}$ and the vector space $S_q(T^*_x\mathcal{X}) \otimes V_\xi \pi$ where $\xi = \pi^q_0(\rho)$ and $x = \pi(\xi)$. Hence we may identify $(\mathcal{N}_q)_\rho$ with a subspace of the latter vector space.

In local coordinates ε_q is given by (2.21); hence its main effect is the introduction of some combinatorial factors which can be absorbed in the choice of an appropriate basis. More precisely, we recover here the discussion in Remark 6.1.5. If we use as usual $\{\partial_{x^1}, \dots, \partial_{x^n}\}$ as basis of the tangent space $T_x\mathcal{X}$ and the dual basis $\{\mathrm{d}x^1, \dots, \mathrm{d}x^n\}$ for the cotangent space $T^*_x\mathcal{X}$, then the "terms" $\partial_{x^\mu} = \partial^{\mu_1}_{x^1} \cdots \partial^{\mu_n}_{x^n}$ with $|\mu| = q$ form a basis of $S_q(T_x\mathcal{X})$ whereas the dual basis of $S_q(T^*_x\mathcal{X})$ is given by the "divided powers" $\frac{1}{\mu!}\mathrm{d}x^\mu$. If we express an element $f \in S_q(T^*_x\mathcal{X}) \otimes V_\xi \pi$ in this basis as $f = \frac{1}{\mu!} f^\alpha_\mu \mathrm{d}x^\mu \otimes \partial_{u^\alpha}$ where μ runs over all multi indices with $|\mu| = q$, then the symbol $(\mathcal{N}_q)_\rho$ consists of all such f satisfying the linear system of equations

$$\sum_{\substack{1 \leq \alpha \leq m \\ |\mu| = q}} \frac{\partial \Phi^\tau}{\partial u^\alpha_\mu}(\rho) f^\alpha_\mu = 0\,, \qquad \tau = 1, \dots, t \tag{7.20}$$

which is the same linear system as (7.3) defining the symbol as subspace of $V_\rho \pi^q_{q-1}$.

Proposition 7.1.15. *Let $\mathcal{R}_q \subseteq J_q\pi$ be a differential equation and $\bigl(\rho_r \in \mathcal{R}_r\bigr)_{r \geq q}$ a sequence of points such that $\pi^r_q(\rho_r) = \rho_q$. With $\xi = \pi^q_0(\rho_q)$ and $x = \pi^q(\rho_q)$, we set $\mathcal{N}_r = \mathfrak{S}_r(T^*_x\mathcal{X}) \otimes V_\xi \pi$ for $0 \leq r < q$. Then the sequence $\bigl((\mathcal{N}_r)_{\rho_r}\bigr)_{r \in \mathbb{N}_0}$ defines a symbolic system in $\mathfrak{S}(T^*_x\mathcal{X}) \otimes V_\xi \pi$ which satisfies $\mathcal{N}_{r+1} = \mathcal{N}_{r,1}$ for all $r \geq q$.*

Proof. According to Definition 6.1.7 of a symbolic system, we must verify that $\mathcal{N}_{r+1} \subseteq \mathcal{N}_{r,1}$ for all $r \geq q$. For notational simplicity, we do this only for $r = q$. Thus let $f = \frac{1}{\nu!} f^\alpha_\nu \mathrm{d}x^\nu \otimes \partial_{u^\alpha}$ where ν runs over all multi indices with $|\nu| = q+1$ be an arbitrary element of the prolonged space $\mathcal{N}_{q,1}$. By definition of the prolongation, this membership is equivalent to $\delta(f) \in \mathcal{N}_q \otimes T^*_x\mathcal{X}$ and hence we must have $\frac{\nu_i}{\nu!} f^\alpha_\nu \mathrm{d}x^{\nu - 1_i} \otimes \partial_{u^\alpha} \in \mathcal{N}_q$ for every $1 \leq i \leq n$. In other words, the coefficients f^α_ν satisfy the linear system of equations

$$\sum_{\substack{1\leq\alpha\leq m\\ |\nu|=q+1,\ \nu_i>0}} \frac{\partial\Phi^\tau}{\partial u^\alpha_{\nu-1_i}} f^\alpha_\nu = 0\,, \qquad \begin{cases} \tau=1,\dots,t\,;\\ i=1,\dots,n\,. \end{cases} \tag{7.21}$$

A comparison with (7.4) shows that these are just the equations describing the prolonged symbol $\mathcal{N}_{q+1}$. Hence we even have $\mathcal{N}_{r+1} = \mathcal{N}_{r,1}$ for all $r \geq q$ (which also shows why it makes sense to call $\mathcal{N}_{r,1}$ a prolongation of $\mathcal{N}_r$). □

With the help of this result, one can now easily verify that the symbolic systems given in Example 6.1.9 are associated to the there mentioned differential equations. In the formulation of Proposition 7.1.15 we used a sequence of points $\rho_r \in \mathcal{R}_r$ with $\pi^r_q(\rho_r) = \rho_q$ in order to consider the symbols $(\mathcal{N}_r)_{\rho_r}$. Obviously, such a sequence does not necessarily exists (unless we are dealing with a formally integrable equation). However, the proof shows us that in fact the obtained symbolic system is independent of the choice of these points, as we may simply set $\mathcal{N}_{r+1} = \mathcal{N}_{r,1}$ for all $r \geq q$. Hence at each point $\rho \in \mathcal{R}_q$ the symbol $(\mathcal{N}_q)_\rho$ induces a whole symbolic system which, according to Lemma 6.1.8, we may alternatively consider as a subcomodule $\mathcal{N}[\rho] \subseteq \mathfrak{S}(T^*_x\mathcal{X}) \otimes V_\xi\pi$ which we call the *symbol comodule* of the differential equation $\mathcal{R}_q$ at the point $\rho \in \mathcal{R}_q$. Following our Blanket Assumption 7.1.4, we assume that the properties of the symbol comodule are independent of the chosen point $\rho \in \mathcal{R}_q$ and therefore mostly omit the explicit reference to ρ.

Remark 7.1.16. In Proposition 7.1.15 and consequently in the definition of the symbol comodule $\mathcal{N}$ we set the lower-order components $\mathcal{N}_r$ for $0 \leq r < q$ to the full degree r component of the free comodule $\mathfrak{S}(T^*_x\mathcal{X}) \otimes V_\xi\pi$. A more precise approach would be to consider instead the symbols of the projected equations $\mathcal{R}^{(q-r)}_r$ (in order to obtain the proper annihilator according to the previous remark, one should then also add their principal symbols to the generating set of the module $\mathcal{M}$). However, for the subsequent involution analysis it only matters what happens from degree q on which is not affected by such changes in lower order. Hence we stick to this simpler approach. ◁

Definition 7.1.17. The symbol $\mathcal{N}_q$ of the differential equation $\mathcal{R}_q \subseteq J_q\pi$ of order q is *involutive*, if the comodule $\mathcal{N}$ is involutive at degree q.

Remark 7.1.18. We can also relate the principal symbol τ_χ to constructions from the last chapter. If we evaluate the matrix $T[\chi]$ defined by (7.14) at the point $\rho \in \mathcal{R}_q$, then its entries are homogeneous polynomials of degree q in $\mathcal{P} = \mathbb{R}[\chi_1,\dots,\chi_n]$ and the rows may be considered as elements of the free polynomial module $\mathcal{P}^m$ generating a submodule $\mathcal{M}[\rho]$. It follows immediately from the local coordinate expressions that $\mathcal{M}[\rho]$ is isomorphic to the annihilator $\mathcal{N}[\rho]^0$ of the symbol comodule. Thus by Proposition 6.1.18 the comodule $\mathcal{N}[\rho]$ is involutive at degree q, if and only if the factor module $\mathcal{P}^m/\mathcal{M}[\rho]$ is involutive at degree q (the quotient "undoes" the index shift in (6.17)). ◁

Choosing local coordinates $(\mathbf{x}, \mathbf{u}^{(q)})$ in a neighbourhood of the point $\rho \in \mathcal{R}_q$, then we can apply the Cartan test (Theorem 6.2.4) for deciding involution of the symbol comodule. Recall that this test requires only linear algebra computations with the two symbols $\mathcal{N}_q$ and $\mathcal{N}_{q+1}$ and thus is easily performed effectively.

In practice, one uses a dual approach with a Pommaret basis. By Proposition B.3.6, the annihilator $\mathcal{N}^0 \subseteq S(T_x\mathcal{X}) \otimes V_\xi\pi$ of the symbol comodule $\mathcal{N}$ is an $S(T_x\mathcal{X})$-submodule, the *symbol module*. As already mentioned above, by Proposition 6.1.18, we may equivalently study either the Spencer cohomology of the comodule $\mathcal{N}$ or the Koszul homology of the module $\bigl(S(T_x\mathcal{X}) \otimes V_\xi\pi\bigr)/\mathcal{N}^0$, as the two are isomorphic. Computationally, the latter approach is often more convenient.

If in our chosen local coordinates and bases the symbol $\mathcal{N}_q$ is defined by (7.20), then the submodule $\mathcal{N}^0$ is generated by the "polynomials"

$$\sum_{\substack{1\leq\alpha\leq m\\ |\mu|=q}} \frac{\partial\Phi^\tau}{\partial u^\alpha_\mu}\,\partial^\mu_x \otimes \partial_{u^\alpha}\,, \qquad \tau = 1,\dots,t\,, \tag{7.22}$$

i. e. the left hand sides of the equations in (7.20) or (7.3). Identifying $S(T_x\mathcal{X})$ with the polynomial ring $\mathcal{P} = \mathbb{R}[\partial_{x^1},\dots,\partial_{x^n}]$ (perhaps one should emphasise that ∂_{x^i} is here only the name of a variable in $\mathcal{P}$ but not a differential operator), one readily recognises in $\mathcal{N}^0$ the polynomial module generated by the rows of the matrix $T[\chi]$ of the principal symbol which appeared in Remark 7.1.12. We may now apply the theory of Pommaret bases to the submodule $\mathcal{N}^0$. Then the following result follows immediately from Theorem 6.3.2.

Proposition 7.1.19. *The symbol $\mathcal{N}_q$ of the differential equation $\mathcal{R}_q \subseteq J_q\pi$ is involutive, if and only if in suitable local coordinates an involutive head autoreduction transforms the generators* (7.22) *into a Pommaret basis of the symbol module $\mathcal{N}^0$ for a class respecting term order.*

Remark 7.1.20. The symbol of an *ordinary* differential equation is always involutive, as in this case the annihilator $\mathcal{N}^0$ is a module over a polynomial ring in only one variable and any set of homogeneous elements with the same degree trivially forms an involutive basis. ◁

Remark 7.1.21. When constructing a Pommaret basis of the module $\mathcal{N}^0$ we encounter the problem of δ-regularity; applying the Cartan test to the factor module $\bigl(S(T_x\mathcal{X}) \otimes V_\xi\pi\bigr)/\mathcal{N}^0$ is only valid in a quasi-regular basis. According to Theorem 6.3.2, both notions are equivalent, so that also in this respect it does not matter which approach we take. ◁

Proposition 7.1.19 transforms the Cartan test into an easily applicable effective criterion for an involutive symbol. In the remainder of this section, we express the approach taken so far in a less algebraic language in order to recover some results frequently mentioned in the literature. For most computational purposes this reformulation is completely sufficient. The main point is that in our special situation checking whether or not an involutive head autoreduction transforms the generators (7.22) into a Pommaret basis of the symbol module $\mathcal{N}^0$ for a class respecting term order amounts to some simple operations on the symbol matrix M_q.

Recall that the columns of M_q correspond to the unknowns $\dot{u}^\alpha_\mu$ (or alternatively to the basis vectors $\frac{1}{\mu!}dx^\mu \otimes \partial_{u^\alpha}$ of $S_q(T_x\mathcal{X}) \otimes V_\xi\pi$); we sort them according to a class respecting term order. In fact, it suffices, if we take care that a column corresponding to an unknown $\dot{u}^\alpha_\mu$ is always to the left of a column corresponding to the unknown $\dot{u}^\beta_\nu$, if $\operatorname{cls}\mu > \operatorname{cls}\nu$. Now an involutive head autoreduction is equivalent to determining a row echelon form $M_q^\triangle$ of M_q using only row operations. We are interested in where the first non-vanishing entry of each row in $M_q^\triangle$ sits; the unknown $\dot{u}^\alpha_\mu$ corresponding to this column is called the *leader* of the row.

Definition 7.1.22. Let $\beta_q^{(k)}$ be the number of leaders that are of class k. These numbers are the *indices* of the symbol $\mathcal{N}_q$.

The problem of δ-regularity concerns this definition. The class of a derivative is not invariant under coordinate transformations. In different coordinate systems we may thus obtain different values for the indices. δ-regular coordinates are distinguished by the fact that the sum $\sum_{k=1}^n k\beta_q^{(k)}$ takes its maximal value. This property is already familiar in the context of Hilbert regularity, as we also get it, if we apply Remark 4.3.7 to the set of "polynomials" corresponding to the rows in $M_q^\triangle$ with respect to the Pommaret division and a class respecting term order.

Remark 7.1.23. Obviously, a necessary condition for δ-regularity is that the highest index $\beta_q^{(n)}$ takes its maximal value. If we restrict for simplicity to first order, then a maximal number of equations must be solvable for an x^n-derivative. Using the classical terminology introduced in Example 7.1.14, we may say that in a δ-regular coordinate system the surface $x^n = 0$ must not be characteristic. This observation demonstrates again that δ-regularity is not just a technical nuisance, but that it is related to important intrinsic properties of the differential equation. ◁

Proposition 7.1.24. *The symbol $\mathcal{N}_q$ with the indices $\beta_q^{(k)}$ is involutive, if and only if the matrix M_{q+1} of the prolonged symbol $\mathcal{N}_{q+1}$ satisfies*

$$\operatorname{rank} M_{q+1} = \sum_{k=1}^{n} k\beta_q^{(k)} . \tag{7.23}$$

Proof. We show that the new criterion (7.23) is equivalent to Proposition 7.1.19. One direction is an immediate consequence of the definition of a Pommaret basis. If the rows of the matrix $M_q^\triangle$ define such a basis, all elements of degree $q+1$ in $\mathcal{N}^0$ are obtained via multiplicative prolongations and their number is given by the right hand side of (7.23). For the converse, we simply note that the equality (7.23) implies that the basis defined by the rows of $M_q^\triangle$ is locally involutive. Hence, by Proposition 4.2.7, this basis is in fact a Pommaret basis. □

The leaders of an involutive symbol matrix in row echelon form cannot take arbitrary values. The following result is a reformulation of Lemma 4.3.3 showing that the leading exponents define a stable monoid ideal. We therefore omit a proof.

Lemma 7.1.25. *Let $\mathcal{N}_q$ be an involutive symbol and $M_q^{\triangle}$ its matrix in row echelon form where the columns have been ordered according to a class respecting term order. If $\dot{u}^{\alpha}_{\mu}$ is a leader of $M_q^{\triangle}$ of class k, then $\dot{u}^{\alpha}_{\mu-1_k+1_j}$ is also a leader for all values $j > k$.*

Example 7.1.26. We apply our criterion to the analysis of the symbol of the Einstein equations (2.96). In arbitrary coordinates it is rather unpleasant to write down the defining equations. So we exploit that involution is a local property: if we want to check involution at a point $\rho \in \mathcal{R}_2$ in the fibre over $x_0 \in \mathcal{X}$, we may use the locally geodesic coordinates mentioned in Example 2.4.4.

The arising form (2.97) of Einstein's equations permits an easy construction of the symbol $\mathcal{N}_2$ at ρ. As no lower-order terms are present, it makes no difference whether we consider the symbol equations or the differential equations. Our first task is to determine the leaders. We distinguish four cases.

$i \neq n$ and $j \neq n$: The leader is $\partial_{x^n x^n} g_{ij}$.
$i = n$ and $j < n-1$: The leader is $\partial_{x^n x^{n-1}} g_{n-1,j}$.
$i = n$ and $j = n-1$: The leader is $\partial_{x^n x^{n-1}} g_{11}$.
$i = n$ and $j = n$: The leader is $\partial_{x^{n-1} x^{n-1}} g_{11}$.

In the last case several substitutions of other equations are necessary for obtaining an autoreduced form. The other cases are rather obvious. One easily verifies that the leaders satisfy Lemma 7.1.25.

The indices of the Einstein equations are therefore

$$\beta_2^{(n)} = \frac{n(n-1)}{2}\,, \quad \beta_2^{(n-1)} = n\,, \quad \beta_2^{(n-k)} = 0 \ \text{ for } 1 < k < n\,. \tag{7.24}$$

We mentioned already in Example 2.4.4 that for the first prolongation one may act as if (2.97) were indeed a differential equation and not just some relations valid only in the fibre over x_0. So the prolonged symbol is again easily constructed. The rank of its matrix evaluates to $\frac{1}{2}n^2(n+1) - n$ which is just $n\beta_2^{(n)} + (n-1)\beta_2^{(n-1)}$. Hence, by Proposition 7.1.24, the Einstein equations have an involutive symbol. Note that this fact entails that locally geodesic coordinates are δ-regular, as in δ-singular coordinates the criterion of Proposition 7.1.24 always fails. ◁

Remark 7.1.27. In cases like the last example, it is in fact trivial to determine the Spencer cohomology of the symbol comodule $\mathcal{N}$. Assume that all equations in a local representation of the involutive differential equation $\mathcal{R}_q$ are of the same order (which is not uncommon in mathematical physics, as the above example of Einstein's equations shows). According to Proposition 7.1.19, the principal parts of the equations define after an involutive head autoreduction a Pommaret basis $\mathcal{H}_q$ of the symbol module $\mathcal{N}^0$ (even in the strict sense of Remark 7.1.16). We discussed already in Remark 6.3.5 how $\mathcal{H}_q$ and the corresponding *minimal* resolution (5.38) allow us to determine the Koszul homology of $\mathcal{N}^0$; we even obtained an explicit formula (6.37) for the bigraded Betti numbers. The duality (6.17) thus yields an isomorphism for the only non-vanishing groups[3] $H^{q-1,p}(\mathcal{N}) \cong \mathrm{Syz}^{p-1}(\mathcal{H}_q)$ where $1 \leq p \leq 1 + \operatorname{proj}\dim \mathcal{N}^0$ and their dimensions are given by

$$\dim H^{q-1,p}(\mathcal{N}) = \sum_{k=1}^{n-p+1} \binom{n-k}{p-1} \beta_q^{(k)}, \tag{7.25}$$

with the indices $\beta_q^{(k)}$ of the symbol $\mathcal{N}_q$.

The Einstein equations (2.96) contain only equations of class n and $n-1$, so that the projective dimension of its symbol module $\mathcal{N}^0$ is 1 and the only two non-vanishing Spencer cohomology groups are $H^{1,1}(\mathcal{N})$ and $H^{1,2}(\mathcal{N})$. For their dimensions we find from (7.24):

$$\dim H^{1,1}(\mathcal{N}) = \beta_2^{(n)} + \beta_2^{(n-1)} = \frac{n(n+1)}{2}, \tag{7.26a}$$

$$\dim H^{1,2}(\mathcal{N}) = \beta_2^{(n-1)} = n. \tag{7.26b}$$

Note that the first dimension is just the number m of dependent variables (expressing that Einstein's equations are square) and the second dimension gives us the number of the contracted Bianchi identities. For testing whether the Einstein equations are formally integrable, these identities are crucial. We will show in the next section (Theorem 7.2.8) that not only in this particular case but generally the second Spencer cohomology groups $H^{\bullet,2}(\mathcal{N})$ are decisive for deciding formal integrability; more precisely, their dimensions give the number of identities to be checked. ◁

Corollary 7.1.28. *Let $\mathcal{N}_1$ be the involutive symbol of a first-order equation in n independent and m dependent variables. Then its indices form an ascending sequence*

$$0 \leq \beta_1^{(1)} \leq \beta_1^{(2)} \leq \cdots \leq \beta_1^{(n)} \leq m. \tag{7.27}$$

Proof. This is an immediate consequence of Lemma 7.1.25. If $\dot{u}_k^\alpha$ is a leader of class k, then for any $j > k$ we have a leader $\dot{u}_j^\alpha$ of class j. Hence $\beta_1^{(k)} \leq \beta_1^{(j)}$ for $k < j$. □

This corollary is only true for the symbol of a *first-order* equation. In higher order we cannot expect a similar result to hold, as the following trivial counterexample demonstrates: for the symbol $\mathcal{N}_2$ of the second-order finite type equation $\mathcal{R}_2$ defined by $u_{yy} = u_{xy} = u_{xx} = 0$ the indices are $\beta_2^{(1)} = 2 > \beta_2^{(2)} = 1$.

Remark 7.1.29. A special situation arises, if $m = 1$, i. e. if there is only one dependent variable, as then *any* first-order symbol $\mathcal{N}_1$ is involutive. Indeed, the symbol module $\mathcal{N}^0$ is now an ideal in $\mathcal{P}$ generated by linear polynomials. Using some linear algebra, we may always assume that all generators have different leading terms (with respect to the degree reverse lexicographic order). Because of the linearity, this trivially implies that all leading terms are relatively prime. By Buchberger's first criterion (Proposition B.4.18) all S-polynomials reduce to zero and our generating set

[3] We exclude from our discussion the trivially non-vanishing group $H^{0,0}(\mathcal{N}) \cong V\pi$ whose dimension is for any differential equation just the number m of dependent variables.

is a Gröbner basis. Furthermore, it follows immediately from Lemma 7.1.25 that the leading terms involutively generate the leading ideal and hence we have a Pommaret basis of $\mathcal{N}^0$ or equivalently $\mathcal{N}_1$ is involutive.

This observation is the deeper reason for a classification suggested by Drach (see Chapter [435, Chapt. 5]). Using the techniques described in Appendix A.3, we may transform any differential equation $\mathcal{R}_q$ into one with only one dependent variable. If we first rewrite $\mathcal{R}_q$ as a first-order equation, then the transformed equation will be of second order. Only in special circumstances one can derive a first-order equation in one dependent variables. Thus from a theoretical point of view we may distinguish two basic classes of differential equations: first-order and second-order equations, respectively, in one dependent variable. The first class is much simpler, as its symbol is always involutive (like for ordinary differential equations).

A further special property of the ideal $\mathcal{N}^0$ is that it is a Cohen–Macaulay module. This fact follows trivially from Theorem 5.2.9. ◁

Example 7.1.30. Let us analyse the symbol of Maxwell's equations (2.85). We order the independent variables as (x,y,z,t) so that time derivatives have the highest class. With this ordering (2.85) yields the symbol matrix directly in row echelon form. In fact, as no lower-order terms are present in the equations, it makes no real difference whether we study the symbol or the differential equation. Thus we do not introduce any extra notations for the variables in the symbol equations.

The rows of maximal class, i. e. of class 4, correspond to the six evolution equations in (2.85a); the remaining two constraint equations (2.85b) are of class 3, as each can by solved for a z-derivative. Thus we obtain for the indices

$$\beta_1^{(4)} = 6\,, \quad \beta_1^{(3)} = 2\,, \quad \beta_1^{(2)} = \beta_1^{(1)} = 0\,. \tag{7.28}$$

They form indeed an ascending sequence of the form (7.27). Only for the z-components of the fields $\mathbf{E}$ and $\mathbf{B}$ several leaders appear, namely $E_z^{(z)}$ and $E_t^{(z)}$ (and correspondingly for $B^{(z)}$). This observation agrees with Lemma 7.1.25.

In order to check whether the symbol is involutive, we must determine what happens in the non-multiplicative prolongations. In our case, this means that we must differentiate the constraint equations (2.85b) with respect to t:

$$\partial_t(\nabla\cdot\mathbf{E}) = \nabla\cdot\mathbf{E}_t = \nabla\cdot(\nabla\times\mathbf{B}) = 0 \tag{7.29}$$

where in the first step we used multiplicative prolongations of equations in (2.85a) and the last step is a well-known result in vector analysis. The same result is obtained with the equations for the magnetic field $\mathbf{B}$. Thus we can rewrite any non-multiplicative prolongation as a linear combination of multiplicative ones and the symbol is involutive. Following Remark 7.1.27, we obtain for the two non-vanishing Spencer cohomology groups the following dimensions:

$$\dim H^{1,1}(\mathcal{N}) = \beta_1^{(4)} + \beta_1^{(3)} = 8\,, \tag{7.30a}$$

$$\dim H^{1,2}(\mathcal{N}) = \beta_2^{(n-1)} = 2\,. \tag{7.30b}$$

Again the first dimension is just the number of differential equations in our system and the second dimension the number of compatibility conditions (one for each Gauss law). ◁

7.2 Involutive Differential Equations

Now we finally introduce the notion of an involutive differential equation. It combines the geometric concept of formal integrability with the algebraic ideas underlying involutive symbols. We will see in the subsequent chapters that this combination indeed makes sense and provides us with a very powerful framework for analysing many different aspects of differential equations.

Definition 7.2.1. The differential equation $\mathcal{R}_q$ is called *involutive*, if it is formally integrable and if its symbol $\mathcal{N}_q$ is involutive.

In the literature, the term "involution" is often used in a rather imprecise manner. In particular, involution is sometimes taken as a synonym for formal integrability. While Definition 7.2.1 obviously implies that an involutive equation is also formally integrable, the converse is generally not true: involution is a stronger concept than formal integrability.

Example 7.2.2. The probably simplest example of a differential equation which is formally integrable but *not* involutive is the following second-order equation in two independent and one dependent variable:

$$\mathcal{R}_2 : \begin{cases} u_{yy} = 0\,, \\ u_{xx} = 0\,. \end{cases} \tag{7.31}$$

We showed in Example 7.2.4 that its symbol $\mathcal{N}_2$ is not involutive. But $\mathcal{R}_2$ is trivially formally integrable, as all right hand sides are zero. Since the prolonged symbol $\mathcal{N}_3$ is involutive, $\mathcal{R}_3$ is an involutive equation. ◁

Recall that Definition 2.3.15 of formal integrability imposes an infinite number of conditions (the surjectivity of the induced projections $\hat{\pi}^{q+r+1}_{q+r} : \mathcal{R}_{q+r+1} \rightarrow \mathcal{R}_{q+r}$ for all $r \geq 0$) and thus cannot be effectively verified. We will expose in the sequel a finite criterion for an involutive differential equation. It requires some further result on involutive symbols.

Recall from the last section that the definition of the indices $\beta_q^{(k)}$ depends on a row echelon form of the symbol matrix. If we compare with the construction of formal power series solutions in Section 2.3, then the computation of the row echelon form corresponds to a special choice of the principal derivatives: the derivative u^α_μ is principal, if and only if the unknown $\dot{u}^\alpha_\mu$ is a leader. Thus the index $\beta_q^{(k)}$ determines the number of principal derivatives of order q and class k.

Let us consider this observation at the level of the differential equation and not only the symbol. In the neighbourhood of a point $\rho \in \mathcal{R}_q$ we solve each equation in a local representation for a derivative. We can do this in such a way that each equation is solved for a different derivative and that this derivative is the leading derivative of the equation with respect to a class respecting term order. If the class of a leading derivative is k, we assign the multiplicative variables $x^1,\dots,x^k$ to the corresponding equation (just as we are used from the Pommaret division). Taking a local representation in triangular form of the prolonged equation $\mathcal{R}_{q+1}$ in the neighbourhood of a point in $\left(\pi_q^{q+1}\right)^{-1}(\rho)$, we find of course that all equations of order $q+1$ in it are algebraically dependent of the formal derivatives of the equations of order q in the representation of $\mathcal{R}_q$. If the symbol is involutive, we do not need all formal derivatives of these equations.

Proposition 7.2.3. *The symbol $\mathcal{N}_q$ of the differential equation $\mathcal{R}_q$ is involutive, if and only if all equations of order $q+1$ in a local representation of the prolonged equation $\mathcal{R}_{q+1}$ are algebraically dependent of the formal derivatives of the equations of order q in a local representation of $\mathcal{R}_q$ with respect to their multiplicative variables only.*

Proof. Obviously, we may prepare our local representation of $\mathcal{R}_q$ in such a way that applying (7.3) yields the symbol directly in row echelon form. All equations obtained by prolonging each equation only with respect to its multiplicative variables are independent, as they possess distinct leading derivatives. Since there are $\beta_q^{(k)}$ equations of order q and class k in our local representation of $\mathcal{R}_q$, we get this way at least $\sum_{k=1}^{n} k\beta_q^{(k)}$ algebraically independent equations of order $q+1$ in the local representation of $\mathcal{R}_{q+1}$. By assumption, $\operatorname{rank}\mathcal{N}_{q+1}$ equals this number. Hence the local representation of $\mathcal{R}_{q+1}$ cannot contain more independent equations of order $q+1$ (it may well contain further independent equations, but of lower order, i. e. they are integrability conditions—cf. the proof of Theorem 7.1.6). □

Example 7.2.4. We reconsider Example 3.1.10 in form of the second-order differential equation $\mathcal{R}_2$ defined by the system $u_{xx} = u_{yy} = 0$. Its symbol is

$$\mathcal{N}_2 : \begin{cases} \dot{u}_{yy} = 0\,, \\ \dot{u}_{xx} = 0\,. \end{cases} \tag{7.32}$$

If we order $x^1 = x$ and $x^2 = y$, we see that the first equation is of class 2 and the second one of class 1. The symbol matrix is already in row echelon form and for the second equation y is a non-multiplicative variable.

For the indices of $\mathcal{N}_2$ we find $\beta_2^{(2)} = \beta_2^{(1)} = 1$. The prolonged symbol $\mathcal{N}_3$ is defined by the equations

$$\mathcal{N}_3 : \begin{cases} \dot{u}_{yyy} = 0\,, & \dot{u}_{xyy} = 0\,, \\ \dot{u}_{xxy} = 0\,, & \dot{u}_{xxx} = 0\,. \end{cases} \tag{7.33}$$

Thus we get that $\operatorname{rank} M_3 = 4 \neq \beta_2^{(1)} + 2\beta_2^{(2)} = 3$ and according to our definition $\mathcal{N}_2$ is *not* involutive. By contrast, the symbol $\mathcal{N}_3$ is involutive, as $\dim \mathcal{N}_3 = 0$: a vanishing symbol is trivially involutive.

At the level of the differential equation we get the following picture. Prolonging the equation $u_{xx} = 0$ with respect to its non-multiplicative variable yields the equation $u_{xxy} = 0$. It is easy to see that this equation is algebraically independent of all multiplicative prolongations. Note that the equation $u_{xxy} = 0$ is *not* an integrability condition; one easily checks that $\mathcal{R}_q^{(1)} = \mathcal{R}_q$ for all $q \geq 2$. As in Chapter 3 we call such an equation an *obstruction to involution*.

It is straightforward to extend this example to higher order. Consider the differential equation $\mathcal{R}_q$ defined by the system $u_{x\cdots x} = u_{y\cdots y} = 0$ with $q \geq 2$. Its symbol module is the ideal $\langle x^q, y^q \rangle$ already mentioned in Example 6.1.24. One easily verifies that we obtain only after $q-1$ prolongations an involutive symbol, as in all intermediate systems we encounter an obstruction to involution of the form $u_{x\cdots xy\cdots y} = 0$ containing q differentiations with respect to x and $1 \leq \ell < q$ differentiations with respect to y. ◁

Proposition 7.2.3 permits us to make Algorithm 2.1 for the construction of formal power series solutions more efficient, as it provides us with a simple systematic method for computing minimal local representations of prolonged equations $\mathcal{R}_{q+r}$. Algorithm 7.1 depends on the following assumption: we know a local representation of $\mathcal{R}_q$ where each equation is solved for a different principal derivative (below we will call such a representation a Cartan normal form). Because of this assumption we obtain a unique solution, as the arising algebraic equations are all linear in the principal derivatives.

Algorithm 2.1 gives us the general power series solution depending on some parameters (note that Algorithm 2.1 does not specify a separation into principal and parametric coefficients and in fact does not even need one; it just asks for a parametric representation of the solution space of the arising algebraic equations). In Algorithm 7.1 we take now a different approach, as our assumption on the local representation of $\mathcal{R}_q$ implies a split into a set Π of parametric and a set Δ of principal derivatives. We "simulate" the solution of an initial value problem by requiring that a map $\delta : \Pi \to \mathbb{R}$ is given that provides us with values for any parametric derivative (we will discuss inSection 9.3 how such a map can be obtained from more common ways of specifying initial data).

The `for` loop in Lines /9-13/ computes the equations that must be added to a local representation of $\mathcal{R}_s$ in order to obtain one for $\mathcal{R}_{s+1}$ and immediately solves these equations for the principal coefficients of order $s+1$. For the prolongation it suffices by Proposition 7.2.3 to take all equations of maximal order and to differentiate them formally with respect to their multiplicative variables only; this approach avoids the redundancies of Algorithm 2.1.

Finally, we prove the following two somewhat technical results. Every prolongation of an involutive symbol is again involutive. In principle, this is obvious from the properties of a Pommaret basis; however, we give here a direct computational proof. The second result concerns a kind of commutativity of projection and prolongation

Algorithm 7.1 Power series solution of involutive differential equation

Input: involutive differential equation $\mathcal{R}_q$ defined by a local representation in solved form $u^\alpha_\mu = \phi^\alpha_\mu(\mathbf{x}, \tilde{\mathbf{u}}^{(|\mu|)})$, expansion point x_0, initial data map $\delta : \Pi \to \mathbb{R}$, truncation order $t \geq q$
Output: unique power series solution of $\mathcal{R}_q$ up to order t

for all parametric derivatives u^α_μ with $|\mu| \leq t$ **do**
  $a^\alpha_\mu \leftarrow \delta(u^\alpha_\mu)$
end for
for all principal derivatives u^α_μ with $|\mu| \leq q$ **do**
  $a^\alpha_\mu \leftarrow \phi^\alpha_\mu(x_0, \mathbf{a}^{(|\mu|)})$
end for
$s \leftarrow q$
repeat
  for all equations $u^\alpha_\mu = \phi^\alpha_\mu$ with $|\mu| = s$ **do**
    for i **from** 1 **to** $\operatorname{cls}\mu$ **do**
      add equation $u^\alpha_{\mu+1_i} = \phi^\alpha_{\mu+1_i} = D_i\phi^\alpha_\mu$
      $a^\alpha_{\mu+1_i} \leftarrow \phi^\alpha_{\mu+1_i}(x_0, \mathbf{a}^{(s+1)})$
    end for
  end for
  $s \leftarrow s+1$
until $s = t$
return $\mathbf{a}^{(t)}$

for equations with an involutive symbol. Both results will be needed in the next section for proving a simple criterion for involutive differential equations.

Proposition 7.2.5. *Let the symbol $\mathcal{N}_q$ of the equation $\mathcal{R}_q$ be involutive.*

(i) *The prolonged symbol $\mathcal{N}_{q+1}$ is involutive, too.*
(ii) *The equality $(\mathcal{R}^{(1)}_q)_{+1} = \mathcal{R}^{(1)}_{q+1}$ holds.*

Proof. We start with Part (i); Part (ii) will be a corollary. Without loss of generality, we assume that the symbol matrix M_q is in row echelon form with the columns ordered according to a class respecting term order. Let $\dot{u}^\alpha_\mu$ be the leaders of M_q. By Proposition 7.2.3, the leaders of M_{q+1} are $\dot{u}^\alpha_{\mu+1_\ell}$ with $1 \leq \ell \leq \operatorname{cls}\mu$. The class of such a leader is ℓ.

In order to prove Part (i), we must show that prolonging a row in M_{q+1} with respect to a non-multiplicative variable leads to a linearly dependent equation. Thus set $k = \operatorname{cls}\mu$ and assume $\ell \leq k$. Prolonging the rows of M_{q+1} with respect to all independent variables leads to rows with leaders of the form $\dot{u}^\alpha_{\mu+1_\ell+1_j}$ for $1 \leq j \leq n$. Three cases must be distinguished:

$j \leq \ell$: These are the prolongations with respect to multiplicative variables. All obtained rows are linearly independent, as their leaders are different.

$\ell < j \leq k$: Since $j \leq k$, the variable x^j is multiplicative for the row in M_q with the leader $\dot{u}^\alpha_\mu$. Thus we can first prolong this row with respect to x^j and then the result with respect to x^ℓ and obtain the same result. As now each prolongation is only with respect to a multiplicative variable, the resulting row is already contained in the rows of the previous case.

$j > k$: Let us denote the row in M_{q+2} by r. As in the previous case we interchange the prolongations, thus prolong first with respect to x^j and then with respect to x^ℓ to obtain r. Since $j > k$, $\dot{u}^\alpha_{\mu-1_k+1_j}$ is by Lemma 7.1.25 also a leader in M_q. Its class is greater than or equal k; thus x^k is a multiplicative variable for the corresponding row. By prolonging it with respect to x^k, we see that $\dot{u}^\alpha_{\mu+1_j}$ is among the leaders of the matrix M_{q+1}. Since x^ℓ is multiplicative for this equation, we have found a row contained in the first case with leader $\dot{u}^\alpha_{\mu+1_\ell+1_j}$.
If we compare the two rows with leader $\dot{u}^\alpha_{\mu+1_\ell+1_j}$, we see that they are both prolongations with respect to x^ℓ of rows with the leader $\dot{u}^\alpha_{\mu+1_j}$. One of these was obtained by prolonging a row of M_q with respect to a non-multiplicative variable. Since the symbol $\mathcal{N}_q$ is involutive, this row is not contained in the set of independent rows we use for $\mathcal{N}_{q+1}$ but is a linear combination of those. Thus r can be written as x^ℓ-prolongation of this linear combination. If x^ℓ is a non-multiplicative variable for some of the rows involved, we apply the same analysis again. Since the considered leaders become smaller and smaller, this process eventually terminates and then we have expressed r as a linear combination of rows contained in the first case.

Hence we have shown that all independent rows in M_{q+2} can be obtained by prolonging the rows of M_{q+1} only with respect to their multiplicative variables. But this fact implies that the prolonged symbol $\mathcal{N}_{q+1}$ is involutive.

Part (ii) may look trivial at first sight, but it is not. On the left hand side only the integrability conditions that arise during the projection from order $q+1$ to order q are taken into account; then one prolongs again to order $q+1$. On the right hand side, however, we project from order $q+2$ to order $q+1$. Hence we have to show that all integrability conditions which occur during this projection are prolongations of integrability conditions of the projection of order $q+1$ to q.

In order to achieve this goal we make basically the same considerations as in the proof of Part (i) but now with the full equations instead of only with their highest-order part. Let $\Phi^\tau = 0$ be an equation of class k in a local description of $\mathcal{R}_q$. We have to analyse all equations of the form $D_\ell D_j \Phi^\tau = 0$. We distinguish two cases assuming always $\ell \leq j$:

$j \leq k$: In this case $D_\ell D_j \Phi^\tau = 0$ contains only prolongations with respect to multiplicative variables. By Proposition 7.2.3, these are the only independent equations of order $q+2$ in our local description of $\mathcal{R}_{q+2}$.

$k < j$: This implies that $D_j \Phi^\tau = 0$ was a prolongation with respect to a non-multiplicative variable. By Proposition 7.2.3 we can express it as a linear combination of prolongations of equations of $\mathcal{R}_q$ with respect to multiplicative variables and perhaps some integrability conditions. Following again the argument of the third case above, we find that $D_\ell D_j \Phi^\tau = 0$ is a linear combination of the equations contained in the first case and of prolonged integrability conditions.

Note that it is not necessary here to distinguish between $\ell \leq k$ and $\ell > k$, as we are allowed to use all prolongations to construct $(\mathcal{R}^{(1)}_q)_{+1}$. □

Part (ii) says that for a differential equation with involutive symbol we obtain the same, if we either first prolong it twice and then project once or alternately prolong, project and prolong. Denoting prolongations by one order by ρ and projections by one order by π, this may concisely be written as $\rho \circ \pi \circ \rho = \pi \circ \rho^2$ (it is crucial that we start on both sides with a prolongation; we may not omit the trailing ρ).

Theorem 7.2.6. *$\mathcal{R}_q \subseteq J_q\pi$ is an involutive differential equation, if and only if its symbol $\mathcal{N}_q$ is involutive and $\mathcal{R}_q^{(1)} = \mathcal{R}_q$.*

Proof. The necessity of the two given conditions for involution is obvious from Definition 7.2.1, as they are weaker than the conditions there. The interesting point is their sufficiency representing a simple corollary of Proposition 7.2.5, which can be applied since the symbol $\mathcal{N}_q$ is assumed to be involutive. By Part (ii) of it and our assumption on $\mathcal{R}_q^{(1)}$, we obtain $\mathcal{R}_{q+1}^{(1)} = (\mathcal{R}_q^{(1)})_{+1} = \mathcal{R}_{q+1}$. Part (i) implies that also all prolonged symbols $\mathcal{N}_{q+r}$ are involutive, so that we can apply this argument inductively at any order. Hence $\mathcal{R}_{q+r}^{(1)} = \mathcal{R}_{q+r}$ for all $r \geq 0$ and the differential equation $\mathcal{R}_q$ is formally integrable. □

Now we can understand the deeper meaning of the at first sight rather technical Proposition 7.2.5. If the symbol $\mathcal{N}_q$ is involutive and at some order of prolongation integrability conditions are hidden, then at least some of these conditions must arise already during the *first* prolongation (here we meet again the idea of local involution discussed in Section 4.1); the remaining ones are prolongations of these. This observation is obviously a strong result and forms the basis of any concrete computation with involutive equations. In particular, the Cartan–Kuranishi completion to involution, which we will discuss in Section 7.4, depends crucially on it.

Recall that checking whether or not the symbol $\mathcal{N}_q$ is involutive via the Cartan test (Theorem 6.2.4 or its alternative formulation Proposition 7.1.24) requires only (linear) computations in order q and $q+1$. Obviously, (except for the linearity) the same is true for verifying the equality $\mathcal{R}_q^{(1)} = \mathcal{R}_q$. Hence Theorem 7.2.6 represents indeed a finite criterion for involution.

Example 7.2.7. Theorem 7.2.6 permits us to conclude that the Einstein equations are involutive. We showed already in Example 2.4.4 the equality $\mathcal{R}_2^{(1)} = \mathcal{R}_2$ and in Example 7.1.26 that the symbol $\mathcal{N}_2$ is involutive. Thus all conditions of the theorem are satisfied. ◁

A closer look at the homological theory in Chapter 6 yields in fact a finite criterion for formal integrability independent of involution. However, it requires the computation of certain Spencer cohomology modules (or at least of their dimensions) which is avoided by the Cartan test. Hence in practice it is often easier to verify involution than formal integrability.

Theorem 7.2.8. *Let the symbolic system $\mathcal{N}$ induced by the symbol $\mathcal{N}_q$ of the differential equation $\mathcal{R}_q$ be 2-acyclic at degree $q+r$ for some integer $r \geq 0$. Then $\mathcal{R}_q$ is formally integrable, if and only if the equality $\mathcal{R}_{q+r'}^{(1)} = \mathcal{R}_{q+r'}$ holds for all values $0 \leq r' \leq \max\{0, r-1\}$.*

Proof. One direction is trivial. If the differential equations $\mathcal{R}_q$ is formally integrable, then we even have $\mathcal{R}^{(1)}_{q+r'} = \mathcal{R}_{q+r'}$ for all $r' \geq 0$. Furthermore, by Theorem 6.1.21 the Spencer cohomology of any symbolic system is finite and hence $\mathcal{N}$ must become 2-acyclic at some degree $q+r$.

For the converse, we first note that the symbolic system $\mathcal{N}$ is trivially 1-acyclic at degree q by Lemma 6.1.13. Our assumption says that in addition the Spencer cohomology modules $H^{q+s,2}(\mathcal{N})$ vanish for all $s \geq r$. According to Proposition 6.1.18, the dual Koszul homology modules $H_{q+s,1}(\mathcal{N}^0)$ of the symbol module $\mathcal{N}^0$ must then also vanish for all $s > r$.

Recall from Remark 6.1.15 that the Koszul homology corresponds to a minimal free resolution of the annihilator $\mathcal{N}^0$ and hence our assumption tells us that the maximal degree of a minimal generator of the first syzygy module $\mathrm{Syz}(\mathcal{N}^0)$ is $q+r$. In Remark 7.1.12 we have seen that the syzygies of $\mathcal{N}^0$ are related to those integrability conditions arising from generalised cross-derivatives between the highest-order equations. If now the equality $\mathcal{R}^{(1)}_{q+r'} = \mathcal{R}_{q+r'}$ holds for all $0 \leq r' \leq \max\{0, r-1\}$, then none of these cross-derivatives can produce an integrability condition (for the computation of $\mathcal{R}^{(1)}_{q+r-1}$ we need $\mathcal{R}_{q+r}$, so that indeed all syzygies are covered). Furthermore, no integrability conditions of the second kind can arise from lower-order equations, as we always consider at least $r' = 0$. Hence the differential equation $\mathcal{R}_q$ is formally integrable. □

Thus, even for deciding formal integrability, it suffices to check a finite number of projections $\pi^{q+r+1}_{q+r} : \mathcal{R}_{q+r+1} \to \mathcal{R}_{q+r}$ for surjectivity. From a computational point of view, we have learned even more. The proof above says that a (preferably minimal) generating set of the Koszul homology module $H_1(\mathcal{N}^0)$ shows us exactly which generalised cross-derivatives may produce integrability conditions. Hence in a concrete computation there is no need to calculate local representations of all prolonged equations $\mathcal{R}_{q+r'}$ but we can directly treat the relevant cross-derivatives. Of course, it cannot be decided solely on the basis of the symbol $\mathcal{N}_q$ whether or not these integrability conditions vanish modulo the equations describing $\mathcal{R}_q$, as this depends on lower-order terms.

Remark 7.2.9. For linear differential equations we can directly link involution in the sense of Definition 7.2.1 with Pommaret bases. Let $\mathbb{F}$ be a differential field containing $\mathbb{R}$ as its field of constants. By making an appropriate choice of a fibred manifold $\pi : \mathcal{E} \to \mathcal{X}$ (e. g. $\pi : \mathbb{R}^n \times \mathbb{R}^m \to \mathbb{R}^n$), we can identify elements of the free module $\mathcal{D}^m$ where $\mathcal{D} = \mathbb{F}[\partial_1, \dots, \partial_n]$ with linear differential operators in the geometric sense of Remark 2.3.2.

Assume now that we know a Pommaret basis $\mathcal{H}$ of a submodule $\mathcal{M} \subseteq \mathcal{D}^m$ for a degree compatible and class respecting term order $\prec$ and let q be the maximal order of an element of $\mathcal{H}$. In general, the basis $\mathcal{H}$ also contains operators of lower order than q. Hence, if we directly translate $\mathcal{H}$ into a differential equation $\mathcal{R}_q \subseteq J_q\pi$, then $\mathcal{R}_q$ will not be formally integrable because of the presence of integrability conditions of the second kind. But if we first add all multiplicative prolongations of the generators of $\mathcal{H}$ up to order q, then it is trivial to see that the corresponding equation $\mathcal{R}_q$ will be involutive in the sense of Definition 7.2.1.

Conversely, let $\mathcal{R}_q$ be an involutive differential equation possessing a linear local representation. Then we can translate this representation into a finite set $\mathcal{F} \subset \mathcal{D}^m$ of linear differential operators. Now it is easy to see that an involutive head autoreduction with respect to a degree compatible and class respecting term order transforms the set $\mathcal{F}$ into a Pommaret bases of $\langle \mathcal{F} \rangle$.

It follows from our proof of Theorem 7.2.8 that similar statements hold for formally integrable equations. They are in the same manner related to Gröbner bases with respect to degree compatible term orders. Indeed, the S-polynomials of a Gröbner basis induce a generating set of the first syzygy module by Theorem B.4.26 and according to Theorem 7.2.8 it suffices to analyse such a generating set for proving formal integrability. ◁

Involutive differential equations possess local representations of a special form: the *Cartan normal form*. In order to avoid excessive notation, we present it only for a first-order equation $\mathcal{R}_1$ (as shown in Appendix A.3, every differential equation can be transformed into an equivalent first-order one). The two main points about the Cartan normal form are that it yields the symbol matrix immediately in row echelon form and that each equation in the system is solved for its principal derivative (which requires some mild regularity assumptions). Thus it looks as follows:

$$u_n^\alpha = \phi_n^\alpha(\mathbf{x}, \mathbf{u}, u_j^\gamma, u_n^\delta)\,, \qquad \begin{cases} 1 \leq \alpha \leq \beta_1^{(n)}\,, \\ 1 \leq j < n\,, \\ \beta_1^{(n)} < \delta \leq m\,, \end{cases} \tag{7.34a}$$

$$u_{n-1}^\alpha = \phi_{n-1}^\alpha(\mathbf{x}, \mathbf{u}, u_j^\gamma, u_{n-1}^\delta)\,, \qquad \begin{cases} 1 \leq \alpha \leq \beta_1^{(n-1)}\,, \\ 1 \leq j < n-1\,, \\ \beta_1^{(n-1)} < \delta \leq m\,, \end{cases} \tag{7.34b}$$

$$\vdots$$

$$u_1^\alpha = \phi_1^\alpha(\mathbf{x}, \mathbf{u}, u_1^\delta)\,, \qquad \begin{cases} 1 \leq \alpha \leq \beta_1^{(1)}\,, \\ \beta_1^{(1)} < \delta \leq m\,, \end{cases} \tag{7.34c}$$

$$u^\alpha = \phi^\alpha(\mathbf{x}, \mathbf{u})\,, \qquad \begin{cases} 1 \leq \alpha \leq \beta_0 \leq m;, \\ \beta_0 < \beta \leq m\,. \end{cases} \tag{7.34d}$$

Important is here the "fine-print," i. e. the ranges of the indices.[4] The first subsystem (7.34a) comprises all equations of class n, the second one (7.34b) all of class $n-1$ and so on. The derivatives on the right hand side have at most the same class as the one on the left hand side. The last subsystem (7.34d) collects all algebraic constraints; their number is denoted by β_0. In general, the subsystems may be empty from a certain class on, as $0 \leq \beta_1^{(1)} \leq \cdots \leq \beta_1^{(n)} \leq m$ by Corollary 7.1.28.

[4] Note that e. g. the ranges given in (7.34a) do not imply that always $\beta_1^{(n)} < m$. They should rather be understood in the sense that if $\beta_1^{(n)} = m$, then no derivatives u_n^δ may appear on the right hand side and correspondingly for the other subsystems.

These purely structural aspects of the normal form (7.34) do not yet capture that our differential equation is involutive; they only express that we have chosen a local representation in triangular form solved for the leading derivatives. Under mild regularity assumptions, any differential equation can be brought into such a form (for fully nonlinear equations one must usually make some case distinctions). The most important point about the Cartan normal form is that involution implies a number of relations between prolongations of the individual equations in (7.34). First of all, involution of the symbol $\mathcal{N}_1$ requires by Proposition 7.2.3 that any non-multiplicative prolongation can be written as a linear combination of multiplicative ones. Thus functions $A_{ij}^{\beta}(\mathbf{x},\mathbf{u}^{(1)})$, $B_i^{\beta}(\mathbf{x},\mathbf{u}^{(1)})$, and $C^{\beta}(\mathbf{x},\mathbf{u})$ must exist such that whenever $1 \leq \ell < k \leq n$

$$\begin{aligned} D_k(u_\ell^\alpha - \phi_\ell^\alpha) = \sum_{i=1}^{k}\sum_{\beta=1}^{\beta_1^{(i)}} \left\{ \sum_{j=1}^{i} A_{ij}^{\beta} D_j(u_i^\beta - \phi_i^\beta) + B_i^\beta (u_i^\beta - \phi_i^\beta) \right\} + \\ + \sum_{\beta=1}^{\beta_0} C^\beta (u^\beta - \phi^\beta)\,, \end{aligned} \tag{7.35}$$

as the variable x^k is non-multiplicative for all equations solved for an x^ℓ-derivative. It follows easily from an analysis of the appearing derivatives that no equations of class higher than k are needed on the right hand side. Strictly speaking, involution of the symbol $\mathcal{N}_1$ requires only that in (7.35) the second-order derivatives cancel, i. e. that functions A_{ij}^{β} exist so that the difference of the left and the right hand side is at most first order. The coefficient functions B_i^β and C^β take already care that the performed non-multiplicative prolongation does not lead to an integrability condition, i. e. that the difference actually vanishes.

Formal integrability of the differential equation $\mathcal{R}_1$ requires furthermore that the prolongation of the algebraic equations in (7.34d) does not lead to new equations, i. e. that no integrability conditions of the second kind appear. This requirement will be satisfied, if functions $\bar{C}^\beta(\mathbf{x},\mathbf{u})$ exist such that for all values $1 \leq k \leq n$

$$\phi_k^\alpha - \sum_{\beta=\beta_0+1}^{m} \frac{\partial \phi^\alpha}{\partial u^\beta}\phi_k^\beta = \frac{\partial \phi^\alpha}{\partial x^k} + \sum_{\beta=1}^{\beta_0} \bar{C}^\beta \left(u^\beta - \phi^\beta\right) . \tag{7.36}$$

Note that involution of the differential equation $\mathcal{R}_1$ requires not only that $\beta_0 \leq \beta_1^{(1)}$ but in fact that even on the right hand side of (7.34d) always $\beta \leq \beta_1^{(1)}$. Otherwise, some derivatives in the prolonged algebraic equations could not be eliminated using the differential equations in (7.34).

Remark 7.2.10. From a computational point of view, we may summarise these considerations as follows. Given a system in the triangular form (7.34), it is straightforward to check effectively whether or not it is involutive. We compute of each equation the formal derivative with respect to each of its non-multiplicative variables (recall that we defined the class of the zero multi index as zero so that for the algebraic equations in (7.34d) all variables are non-multiplicative). Then we reduce the obtained equations with respect to our system (7.34) and all its multiplicative prolongations (at least the reductions with respect to the differential equations are easy to perform effectively, as our system is in solved form and thus we simply substitute the principal derivatives by the corresponding right hand sides). The system will be involutive, if and only if every such computation yields zero, as then we have identities of the form (7.35) and (7.36), respectively.

Note that this recipe does not even require to determine explicitly the symbol $\mathcal{N}_1$. In the case that the system is not involutive, it is still easy to reconstruct whether the problem comes from the symbol or from the existence of integrability conditions. If a reduction does not lead to zero, we must only look at the order of the remaining equation: the existence of a second-order equation indicates by Proposition 7.2.3 that the symbol $\mathcal{N}_1$ is not involutive; any remainder of lower order represents a non-trivial integrability condition.

This observation is the basis of the completion algorithm for linear systems which we will develop in Section 10.7. For nonlinear systems the problem is less the check for involution; as we showed above, it can to a large extent be performed effectively under the made assumptions. The real problem is that if the system is not formally integrable, then it might be difficult to solve the arising integrability conditions for their leading derivatives. Thus generally we are not able to analyse effectively the projected equations $\mathcal{R}_1^{(1)}$. ◁

Remark 7.2.11. Making some further simplifying assumptions, we can even explicitly follow the recipe presented in Remark 7.2.10 and determine the conditions for involution of the symbol $\mathcal{N}_1$ and formal integrability of $\mathcal{R}_1$ in closed form. We assume firstly that no algebraic equations (7.34d) are present and secondly that nowhere on the right hand side a principal derivative appears (thus in the ith subsystem of the normal form (7.34) on the right hand side only the derivatives u_j^γ with $1 \leq j \leq i$ and $\beta_1^{(j)} < \gamma \leq m$ appear). The first assumption essentially requires that we are able to solve explicitly the present algebraic equations and to use the result for eliminating some dependent variables. The second assumption can always be achieved by an autoreduction.

If we set $\Phi_i^\alpha = u_i^\alpha - \phi_i^\alpha$ and choose a non-multiplicative index $i < j \leq n$, then, expressing the formal derivative D_j via the contact vector fields $C_j^{(1)}$ and C_γ^k introduced in (2.11), we can write the corresponding prolongation in the form

$$D_j\Phi_i^\alpha = u_{ij}^\alpha - C_j^{(1)}(\phi_i^\alpha) - \sum_{k=1}^{i} \sum_{\gamma=\beta_1^{(k)}+1}^{m} u_{kj}^\gamma C_\gamma^k(\phi_i^\alpha)\,. \tag{7.37}$$

Each of the arising second-order derivatives can be eliminated using a multiplicative prolongation. Performing these eliminations leads after a tedious but in principle straightforward computation to the following lengthy expression (all the gory details of the required computations for deriving this expression are explicitly derived in the thesis of Fesser [131]):

$$
\begin{aligned}
&D_j\Phi_i^\alpha - D_i\Phi_j^\alpha + \sum_{k=1}^{i}\sum_{\gamma=\beta_1^{(k)}+1}^{m} C_\gamma^k(\phi_h^\alpha)D_k\Phi_j^\gamma \\
&= C_i^{(1)}(\phi_j^\alpha) - C_j^{(1)}(\phi_i^\alpha) - \sum_{k=1}^{i}\sum_{\gamma=\beta_1^{(k)}+1}^{m} C_\gamma^k(\phi_k^\alpha)C_k^{(1)}(\phi_j^\gamma) \\
&\quad - \sum_{k=1}^{i-1}\sum_{\delta=\beta_1^{(k)}+1}^{m} u_{kk}^\delta\left[\sum_{\gamma=\beta_1^{(k)}+1}^{\beta_1^{(j)}} C_\gamma^k(\phi_k^\alpha)C_\delta^k(\phi_j^\gamma)\right] \\
&\quad - \sum_{1\le k<\ell<i}\left\{\sum_{\delta=\beta_1^{(k)}+1}^{\beta_1^{(\ell)}} u_{k\ell}^\delta\left[\sum_{\gamma=\beta_1^{(\ell)}+1}^{\beta_1^{(j)}} C_\gamma^\ell(\phi_\ell^\alpha)C_\delta^k(\phi_j^\gamma)\right] + \right. \\
&\qquad\qquad \sum_{\delta=\beta_1^{(\ell)}+1}^{m} u_{k\ell}^\delta\left[\sum_{\gamma=\beta_1^{(k)}+1}^{\beta_1^{(j)}} C_\gamma^k(\phi_k^\alpha)C_\delta^\ell(\phi_j^\gamma) + \right. \\
&\qquad\qquad\qquad \left.\left.\sum_{\gamma=\beta_1^{(\ell)}+1}^{\beta_1^{(j)}} C_\gamma^\ell(\phi_\ell^\alpha)C_\delta^k(\phi_j^\gamma)\right]\right\} \\
&\quad - \sum_{k=1}^{i-1}\left\{\sum_{\delta=\beta_1^{(k)}+1}^{\beta_1^{(i)}} u_{ki}^\delta\left[-C_\delta^k(\phi_k^\alpha) + \sum_{\gamma=\beta_1^{(i)}+1}^{\beta_1^{(j)}} C_\gamma^i(\phi_i^\alpha)C_\delta^k(\phi_j^\gamma)\right] + \right. \\
&\qquad\qquad \sum_{\delta=\beta_1^{(i)}+1}^{m} u_{ki}^\delta\left[-C_\delta^k(\phi_k^\alpha) + \sum_{\gamma=\beta_1^{(i)}+1}^{\beta_1^{(j)}} C_\gamma^i(\phi_i^\alpha)C_\delta^k(\phi_j^\gamma) + \right. \\
&\qquad\qquad\qquad \left.\left.\sum_{\gamma=\beta_1^{(k)}+1}^{\beta_1^{(j)}} C_\gamma^k(\phi_k^\alpha)C_\delta^i(\phi_j^\gamma)\right]\right\} \\
&\quad - \sum_{k=1}^{i-1}\sum_{\delta=\beta_1^{(j)}+1}^{m} u_{kj}^\delta\left[C_\delta^k(\phi_k^\alpha) + \sum_{\gamma=\beta_1^{(k)}+1}^{\beta_1^{(j)}} C_\gamma^k(\phi_k^\alpha)C_\delta^j(\phi_j^\gamma)\right] \\
&\quad - \sum_{k=i}^{j-1}\sum_{\delta=\beta_1^{(k)}+1}^{m} u_{ik}^\delta\left[-C_\delta^k(\phi_k^\alpha) + \sum_{\gamma=\beta_1^{(i)}+1}^{\beta_1^{(j)}} C_\gamma^i(\phi_i^\alpha)C_\delta^k(\phi_j^\gamma)\right] \\
&\quad - \sum_{\substack{k=1\\ i+1\le\ell<j}}^{i-1}\sum_{\delta=\beta_1^{(\ell)}}^{m} u_{k\ell}^\delta\left[\sum_{\gamma=\beta_1^{(k)}+1}^{\beta_1^{(j)}} C_\gamma^k(\phi_k^\alpha)C_\delta^\ell(\phi_j^\gamma)\right] \\
&\quad - \sum_{\delta=\beta_1^{(j)}+1}^{m} u_{ij}^\delta\left[C_\delta^i(\phi_i^\alpha) - C_\delta^j(\phi_j^\alpha) + \sum_{\gamma=\beta_1^{(i)}+1}^{\beta_1^{(j)}} C_\gamma^i(\phi_i^\alpha)C_\delta^j(\phi_j^\gamma)\right].
\end{aligned}
\tag{7.38}
$$

None of the second-order derivatives appearing here can be obtained as a multiplicative prolongation of a principal derivative in (7.34). Hence all the expressions in square brackets represent obstructions to involution and the symbol $\mathcal{N}_1$ is involutive, if and only if they vanish. The terms in the first line of the right hand side are of first order. Assuming that all the obstructions to involution vanish, their sum represents the remaining integrability condition. Since no principal derivatives are contained in it, no further simplification is possible and involution of $\mathcal{R}_1$ requires that it vanishes, too. ◁

Example 7.2.12. Linear homogeneous first-order equations in one dependent variable represent a classical problem which found much interest in the nineteenth century. Most textbooks on differentia l equations from the early twentieth century discuss it at length (see e. g. [68]). Its solution leads to what is nowadays called the *Frobenius Theorem* (Theorem C.3.3) in differential geometry.

In local coordinates we are dealing with the system

$$\mathcal{R}_1 : \left\{ \sum_{i=1}^{n} a_\tau^i(\mathbf{x}) u_i = 0 \,, \qquad 1 \leq \tau \leq t \,. \right. \tag{7.39}$$

Note that no terms of order zero appear. We want to study under what conditions $\mathcal{R}_1$ is involutive. In order to facilitate the analysis of the symbol, we transform the system into a triangular form. By taking appropriate linear combinations we solve for some derivatives. We assume that these are those of maximal class; otherwise we renumber the independent variables. So we arrive at a system of the form

$$\mathcal{R}_1 : \left\{ u_{n+1-\tau} + \sum_{i=1}^{n-t} \bar{a}_\tau^i(\mathbf{x}) u_i = 0 \,, \qquad 1 \leq \tau \leq t \,. \right. \tag{7.40}$$

If (7.39) is involutive, then this is its Cartan normal form.

Now we must check whether relations of the form (7.35) hold; so we consider the non-multiplicative prolongations of the equations in (7.40). For the equation with index τ these are the prolongations with respect to the variables $x^{n+1-\sigma}$ for $1 \leq \sigma < \tau$. They lead to second-order equations

$$u_{n+1-\sigma,n+1-\tau} + \sum_{i=1}^{n-t} \bar{a}_\tau^i(\mathbf{x}) u_{i,n+1-\sigma} + \sum_{i=1}^{n-t} \frac{\partial \bar{a}_\tau^i(\mathbf{x})}{\partial x^{n+1-\sigma}} u_i = 0 \,. \tag{7.41}$$

In order for $\mathcal{R}_1$ to be involutive, we must be able to express (7.41) as a linear combination of the remaining prolongations. The first term can be cancelled by subtracting the prolongation of the equation with index σ with respect to the multiplicative variable $x^{n+1-\tau}$: it is obtained by interchanging σ and τ in (7.41).

Next we study the remaining second-order derivatives in the first sum in (7.41) (and in the equation with σ and τ interchanged, respectively). Each of them can be cancelled individually by subtracting (respectively adding) the prolongation of the equation with index τ (respectively σ) with respect to the multiplicative variable x^i

multiplied by $\bar{a}^i_\tau(\mathbf{x})$ (respectively $\bar{a}^i_\sigma(\mathbf{x})$). One easily checks that hereby all appearing derivatives u_{ij} with $1 \leq i, j \leq n-t$ cancel.

Thus we are able to eliminate all second-order derivatives implying that the symbol $\mathcal{N}_1$ is involutive. But if we perform all these eliminations, we obtain as potential integrability condition for each pair $1 \leq \sigma < \tau \leq t$ a first-order differential equation:

$$\sum_{i=1}^{n-t} \left[\frac{\partial \bar{a}^i_\tau(\mathbf{x})}{\partial x^{n+1-\sigma}} - \frac{\partial \bar{a}^i_\sigma(\mathbf{x})}{\partial x^{n+1-\tau}} + \sum_{j=1}^{n-t} \bar{a}^j_\sigma(\mathbf{x}) \frac{\partial \bar{a}^i_\tau(\mathbf{x})}{\partial x^j} - \sum_{j=1}^{n-t} \bar{a}^j_\tau(\mathbf{x}) \frac{\partial \bar{a}^i_\sigma(\mathbf{x})}{\partial x^j} \right] u_i = 0 \,. \tag{7.42}$$

As it does not contain any derivative u_i with $i > n-t$, it cannot be expressed as a linear combination of equations in (7.40): $\mathcal{R}_1$ is formally integrable and thus involutive, if and only if in (7.42) the coefficients in the bracket vanishes.

This result possesses a simple geometric interpretation. We introduce the following set of vector fields

$$\bar{X}_\tau = \partial_{x^{n+1-\tau}} + \sum_{i=1}^{n-t} \bar{a}^i_\tau(\mathbf{x}) \partial_{x^i} \,, \qquad 1 \leq \tau \leq t \,. \tag{7.43}$$

They allow us to express (7.40) succinctly in the form $\bar{X}_\tau u = 0$. Comparing with (C.7), it is not difficult to see that the above integrability conditions are equivalent to the vanishing of the Lie brackets of the fields $\bar{X}_\tau$:

$$[\bar{X}_\tau, \bar{X}_\sigma] = 0 \,, \qquad 1 \leq \sigma < \tau \leq t \,. \tag{7.44}$$

Thus we are in the same situation as in Example 2.3.17 where also the integrability conditions could be expressed as the vanishing of Lie brackets.

We may write the original system (7.39) in terms of the vector fields

$$X_\tau = \sum_{i=1}^{n} a^i_\tau(\mathbf{x}) \partial_{x^i} \,, \qquad 1 \leq \tau \leq t \,. \tag{7.45}$$

They are linear combinations of the vector fields $\bar{X}_\tau$ – but with $\mathfrak{X}(\mathcal{E})$ considered as $\mathcal{F}(\mathcal{E})$-module (see Appendix C.2). The Lie bracket (C.7) is bilinear only with respect to linear combinations with real coefficients. Thus we cannot conclude that for an involutive system (7.39) the vector fields X_τ must commute. But it is easy to see that the commutator of two of the vector fields must be an $\mathcal{F}(\mathcal{E})$-linear combination of the X_τ, i. e. functions $C^\rho_{\sigma\tau}(\mathbf{x})$ must exist such that

$$[X_\sigma, X_\tau] = \sum_{\rho=1}^{t} C^\rho_{\sigma\tau}(\mathbf{x}) X_\rho \,, \tag{7.46}$$

i. e. the distribution $\mathcal{D}$ spanned by these vector fields is closed under the Lie bracket. Thus the equation $\mathcal{R}_1$ is involutive, if and only if the distribution $\mathcal{D}$ is involutive.

In the classical language, a linear system (7.39) such that the vector fields X_τ satisfy (7.46) is called *complete* or *involutive* (explaining the origin of this term); this terminology goes back at least to Clebsch [93]. If the system possesses the triangular form (7.40), one sometimes speaks of a *Jacobian system*. A generalisation of the theory of complete systems to nonlinear equations is possible using the Mayer bracket (recall Example 2.3.12).

Finally, let us note that any complete system can be integrated solving only *ordinary* differential equations. If it consists of t equations, then it possesses $n-t$ functionally independent solutions $\psi_k(\mathbf{x})$ with $1 \leq k \leq n-t$ and its general solution is $\Psi\big(\psi_1(\mathbf{x}), \dots, \psi_{n-t}(\mathbf{x})\big)$ where Ψ is an arbitrary function.

Indeed, our proof of the Frobenius Theorem C.3.3 is constructive and requires only the repeated use of the Rectification Lemma C.2.3. The proof of the latter in turn was based on the solution of a system of ordinary differential equations (for the exponentiation of the vector field). Explicitly performing these computations thus requires to solve t systems of ordinary differential equations. Mayer [315] proposed a more efficient approach where one needs to integrate only one system consisting of $n-t$ equations (our presentation follows [137, §§34,35,41] to where we also refer for the proof of the correctness of the method). It assumes that we are dealing with the Jacobian system defined by the vector fields (7.43).

Consider the one-forms

$$\omega^k = \mathrm{d}x^k - \sum_{\tau=n-t+1}^{n} \bar{a}^k_\tau(\mathbf{x})\mathrm{d}x^\tau\,, \qquad 1 \leq k \leq n-t\,. \tag{7.47}$$

They generate a Pfaffian system which is the annihilator $\mathcal{D}^0$ of the distribution $\mathcal{D}$ spanned by vector fields (7.43). We may think of it as the (generalised) characteristic system for our partial differential equation, as at any point $x_0 \in \mathcal{X}$ the forms $\omega^k\big|_{x_0}$ generate the characteristic variety C_{x_0} of our equation.

The direct integration of this Pfaffian system requires the solution of as many systems of ordinary differential equations as the original linear system. The approach of Mayer consists of a clever coordinate transformation $\mathbf{x} \mapsto \mathbf{y}$ after which we can reduce (7.47) immediately to a single system of $n-t$ ordinary differential equations. The simplest form of the transformation is

$$x^i = \begin{cases} y^i & \text{if } 1 \leq i \leq n-t \text{ or } i = n\,, \\ \alpha^i + (y^n - \alpha^n)y^i & \text{if } n-t < i < n\,. \end{cases} \tag{7.48}$$

Here the α^i are arbitrary constants.

The transformed system is still in Jacobian form $u_{n+1-\tau} + \sum_{i=1}^{n-t} b^i_\tau(\mathbf{y})u_i = 0$ with coefficients $b^i_1 = \bar{a}^i_1 + \sum_{\tau=2}^{t} \bar{a}^i_\tau y^\tau$ and $b^i_\tau = (y^n - \alpha^n)\bar{a}^i_\tau$ for $\tau > 1$. The "trick" of Mayer consists of the observation that essentially it suffices to consider only the first equation of the system treating the variables $y^{n-t+1}, \dots, y^{n-1}$ as constants. Its characteristic system can be written in the form

$$\frac{dy^i}{dy^n} = b_1^i(\mathbf{y}) , \qquad 1 \le i \le n-t . \tag{7.49}$$

Assume that we are able to find $n-t$ first integrals $\tilde{\psi}_k(\mathbf{y})$ of it (all the time treating $y^{n-t+1}, \dots, y^{n-1}$ as constants). Then reverting the change of coordinates in the equation $\tilde{\psi}_k(\mathbf{y}) = \tilde{\psi}_k(\boldsymbol{\alpha})$ yields $n-t$ first integrals $\psi_{k(\mathbf{x})}$ of our original system so that its general solution is $u = F\bigl(\psi_1(\mathbf{x}), \dots, \psi_{n-t}(\mathbf{x})\bigr)$. Here $\alpha^{n-t+1}, \dots, \alpha^n$ are the constants already appearing in (7.48) and the remaining α^i are arbitrary.

We demonstrate the procedure for the following Jacobian system

$$\mathcal{R}_1 : \begin{cases} u_6 = \dfrac{x_5 - x_3}{x_5 - x_4} u_1 + \dfrac{x_3 - x_4}{x_5 - x_4} u_2 , & u_5 = \dfrac{x_5 - x_2}{x_5 - x_4} u_1 + \dfrac{x_2 - x_4}{x_5 - x_4} u_2 , \\ u_4 = \dfrac{x_5 - x_1}{x_5 - x_4} u_1 + \dfrac{x_1 - x_4}{x_5 - x_4} u_2 , & u_3 = \dfrac{x_5 - x_6}{x_5 - x_4} u_1 + \dfrac{x_6 - x_4}{x_5 - x_4} u_2 . \end{cases} \tag{7.50}$$

Application of the Mayer method yields then a system of two ordinary differential equations, $dy^1/dy^6 = b_1^1$ and $dy^2/dy^6 = b_1^2$, which we consider under the assumption that y^3, y^4 and y^5 are constant. Since $b_1^1 + b_1^2 = -(1 + y^3 + y^4 + y^5)$ is constant, we have found a first integral

$$\tilde{\psi}_1(\mathbf{y}) = y^1 + y^2 + (1 + y^3 + y^4 + y^5) y_6 . \tag{7.51}$$

Reverting the coordinate transformation in the equation $\tilde{\psi}_1(\mathbf{y}) = \tilde{\psi}_1(\boldsymbol{\alpha})$ shows that a first integral of (7.50) is given by $\psi_1(\mathbf{x}) = x^1 + x^2 + x^3 + x^4 + x^5 + x^6$. Some further computations yield as an additional first integral $\psi_2 = x^1x^4 + x^2x^5 + x^3x^6$. Hence the general solution of the Jacobian system (7.50) is given by

$$u(\mathbf{x}) = F(x^1 + x^2 + x^3 + x^4 + x^5 + x^6, x^1x^4 + x^2x^5 + x^3x^6) \tag{7.52}$$

with an arbitrary function F of two arguments. ◁

The final proposition in this section should by now be rather obvious to any reader who has understood the meaning of involution. We mention it here only for historical reasons, as it generalises an old result of Finzi [135] (cf. also [342, Sect. 2.6] for a discussion) on systems which have the same number of equations and dependent variables.

Proposition 7.2.13. *A differential equation $\mathcal{R}_q$ possesses identities or integrability conditions, if and only if either $\beta_q^{(n-1)} > 0$ or $\mathcal{R}_{q-1}^{(1)} \subsetneq J_{q-1}\pi$.*

Proof. The second case, $\mathcal{R}_{q-1}^{(1)} \subsetneq J_{q-1}\pi$, is trivial. It means that any local representation of $\mathcal{R}_q$ contains not only equations of order q but also of lower order. Obviously, this fact implies the existence of either identities or integrability conditions arising from the prolongation of these lower order equations. Thus we exclude this possibility and concentrate on the first case.

If we have $\beta_q^{(n-1)} > 0$, then any local representation of $\mathcal{R}_q$ contains equations with non-multiplicative variables, as only equations of class n are without. Thus, according to Proposition 7.2.3, the system must possess either identities or integrability conditions.

For the converse we note that $\beta_q^{(n-1)} = 0$ implies that all indices $\beta_q^{(k)}$ with $k < n$ must vanish; this can be shown analogously to Corollary 7.1.28. Thus a local representation of $\mathcal{R}_q$ exists where all equations are of class n. Hence all their prolongations are algebraically independent and neither identities nor integrability conditions can arise from cross-derivatives. □

7.3 Completion of Ordinary Differential Equations

We turn now our attention to the problem of completing a given differential equation to an equivalent involutive one. We begin with ordinary differential equations; partial differential equations will be treated in the next section. We first restrict to *autonomous* systems where the independent variable x does not occur explicitly. This restriction is made only for the simplicity of the presentation and to make the theory more intuitive. The general case will be treated at the end of the section.

Autonomous systems are traditionally described within the tangent bundle (see Appendix C.1). So let $\mathcal{U}$ be an m-dimensional manifold. The right hand side of an autonomous system in the solved form

$$\mathbf{u}' = \boldsymbol{\phi}(\mathbf{u}) \tag{7.53}$$

is interpreted as a vector field $X \in \mathfrak{X}(\mathcal{U})$. Systems of the more general form

$$\boldsymbol{\Phi}\left(\mathbf{u}, \mathbf{u}'\right) = 0\,, \tag{7.54}$$

where the number of equations may differ from $\dim \mathcal{U}$, appear in many applications and are known in the literature under different names. A classical one is *implicit equation*; in numerical analysis the name *differential algebraic equation* has become very popular in recent times; in engineering applications one often speaks of a *descriptor form* in contrast to the *state space form* (7.53). Geometrically, such a system corresponds (under some regularity assumptions) to a submanifold $\mathcal{S}^{(0)} \subseteq T\mathcal{U}$.

Assuming as usual that $\boldsymbol{\Phi}$ is a smooth function and that the rank of the Jacobian $\partial\boldsymbol{\Phi}/\partial\mathbf{u}'$ is constant, we transform (7.54) into *semi-explicit form*: we split the vector $\mathbf{u}$ into two parts $\mathbf{v}$ and $\mathbf{w}$ (in general of different dimensions) such that $\operatorname{rank}(\partial\boldsymbol{\Phi}/\partial\mathbf{v}') = \operatorname{rank}(\partial\boldsymbol{\Phi}/\partial\mathbf{u}')$; by the Implicit Function Theorem we can then solve for the derivative $\mathbf{v}'$ and obtain

$$\mathbf{v}' = \boldsymbol{\phi}(\mathbf{v}, \mathbf{w}, \mathbf{w}')\,, \tag{7.55a}$$

$$0 = \boldsymbol{\psi}(\mathbf{v}, \mathbf{w})\,. \tag{7.55b}$$

This form also explains the name *differential algebraic equation*; in semi-explicit form such a system comprises a mixture of differential and algebraic equations. The algebraic equations are often called *constraints*. Note that so far we do not assume that we are dealing with an involutive equation.

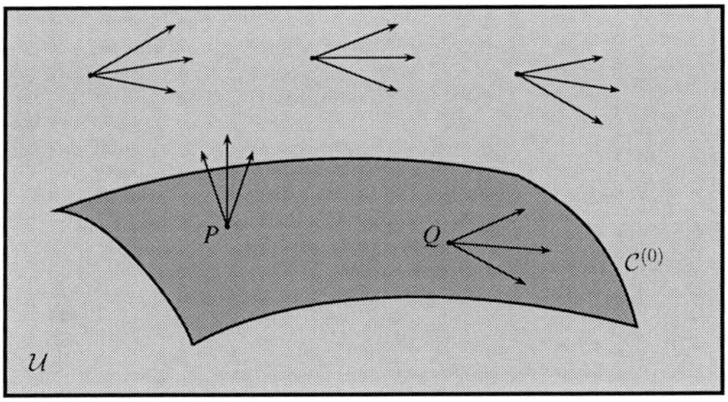

Fig. 7.1 Vector field pencil and submanifold

Thus we are not given a simple vector field on a manifold but a whole *pencil* of vector fields on $\mathcal{U}$ and a *submanifold* $\mathcal{C}^{(0)} = \tau_{\mathcal{U}}(\mathcal{S}^{(0)}) \subseteq \mathcal{U}$, the *constraint manifold* obtained via the tangent bundle projection (see Figure 7.1). In the semi-explicit form, the submanifold is described by the constraint equations (7.55b) and we have a pencil $\mathcal{P}^{(0)}$ of vector fields (or alternatively a parametrised vector field), as (7.55a) does not uniquely determine all components of a vector field (the $\mathbf{w}$ components remain arbitrary and may be considered as parameters).

Example 7.3.1. As a trivial example we may consider the following linear system in semi-explicit form defined over $\mathcal{U} = \mathbb{R}^3$ with $\mathbf{u} = (u, v, w)$:

$$v' = u - 2\,, \quad w' = -v\,, \quad v = 3\,. \tag{7.56}$$

Here the constraint manifold $\mathcal{C}^{(0)}$ is the plane $v = 3$ and the pencil $\mathcal{P}^{(0)}$ contains all vector fields $X_a \in \mathfrak{X}(\mathcal{U})$ of the form $X_a(\mathbf{u}) = \big(a(\mathbf{u}), u-2, -v\big)^t$ where $a \in \mathcal{F}(\mathcal{U})$ is an arbitrary function. ◁

A solution of the differential algebraic equation must define an integral curve to one of the vector fields in the pencil $\mathcal{P}^{(0)}$ and this curve must lie in the constraint manifold $\mathcal{C}^{(0)}$. Obviously, these two conditions can be simultaneously satisfied only, if the vector field is tangential to $\mathcal{C}^{(0)}$. As shown in Figure 7.1, two possibilities arise. It may happen at some points $P \in \mathcal{C}^{(0)}$ that all vectors contained in $\mathcal{P}^{(0)}$ are transversal to $\mathcal{C}^{(0)}$; such points can never lie on a curve defined by a solution of the differential algebraic equation and must be removed. Such eliminations lead to a smaller constraint manifold $\mathcal{C}^{(1)} \subseteq \mathcal{C}^{(0)}$ (as usual, we assume here for simplicity that the remaining points do indeed form a manifold). At all points $Q \in \mathcal{C}^{(1)}$ the pencil $\mathcal{P}^{(0)}$ contains at least one vector tangential to $\mathcal{C}^{(0)}$. Removal of all the transversal vectors leads to a shrunk pencil $\mathcal{P}^{(1)} \subseteq \mathcal{P}^{(0)}$.

If $\mathcal{C}^{(1)}$ is a proper submanifold of $\mathcal{C}^{(0)}$, we must again check the tangency, as vectors tangential to $\mathcal{C}^{(0)}$ are not necessarily tangential to $\mathcal{C}^{(1)}$. Thus we obtain a decreasing sequence of submanifolds $\mathcal{C}^{(0)} \supset \mathcal{C}^{(1)} \supset \cdots \supset \mathcal{C}^{(f)}$ and accompanying vector field pencils $\mathcal{P}^{(0)} \supseteq \mathcal{P}^{(1)} \supseteq \cdots \supseteq \mathcal{P}^{(f)}$. This completion process stops as soon as $\mathcal{C}^{(i)} = \mathcal{C}^{(i+1)}$; depending on whether or not $\mathcal{P}^{(i)} = \mathcal{P}^{(i+1)}$ we have $f = i$ or $f = i+1$. At the end we obtain a final submanifold $\mathcal{C}^{(f)} \subseteq \mathcal{U}$ and a final pencil $\mathcal{P}^{(f)}$ of vector fields tangential to it.

In a more constructive manner we describe this process by two sequences of submanifolds $\mathcal{C}^{(i)} \subseteq \mathcal{U}$ and $\mathcal{S}^{(i)} \subseteq T\mathcal{U}$. The constraint manifold $\mathcal{C}^{(i)}$ is always obtained via the tangent bundle projection: $\mathcal{C}^{(i)} = \tau_{\mathcal{U}}(\mathcal{S}^{(i)})$. The new differential equation $\mathcal{S}^{(i+1)}$ arises as the intersection of the old one with the tangent space of its constraint manifold: $\mathcal{S}^{(i+1)} = \mathcal{S}^{(i)} \cap T\mathcal{C}^{(i)}$. The vector field pencils $\mathcal{P}^{(i)}$ are uniquely defined only on the constraint manifolds $\mathcal{C}^{(i)}$; there they simply span the submanifolds $\mathcal{S}^{(i)}$. Off the submanifolds $\mathcal{C}^{(i)}$, the vector fields contained in $\mathcal{P}^{(i)}$ may be continued arbitrarily.

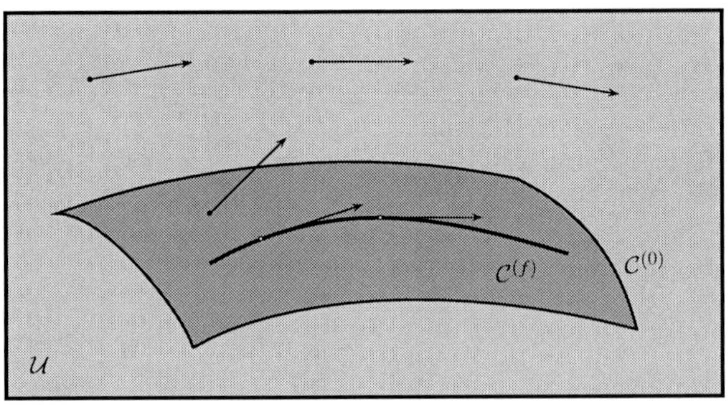

Fig. 7.2 Completed system

In most applications the final pencil consists of a single vector field (see Figure 7.2); otherwise we are dealing with an underdetermined system. Such a field is called an *underlying vector field*. Note again that it is uniquely determined only *on* the final constraint manifold $\mathcal{C}^{(f)}$. If we add to an underlying vector field any field vanishing on $\mathcal{C}^{(f)}$, we obtain another underlying vector field. It is well possible that $\mathcal{C}^{(f)}$ is an empty submanifold. In this case the original system was *inconsistent*: it does not possess any solution, as the conditions it poses cannot be fulfilled simultaneously.

The final constraint manifold $\mathcal{C}^{(f)}$ is obviously an *invariant manifold* for any vector field contained in the final pencil $\mathcal{P}^{(f)}$. Indeed, in dynamical systems theory an invariant manifold for a vector field has the property that any solution for initial data on it stays within the manifold for all times and this property was the guiding principle in our discussion above (we will formulate it later in a more general form in Proposition 9.1.1).

Example 7.3.2. We continue Example 7.3.1. As the constraint manifold $\mathcal{C}^{(0)}$ is parallel to the u-w plane, a vector field is tangential to it, if and only if its v-component vanishes on $\mathcal{C}^{(0)}$. As the parameter a does not appear in this component, the tangency cannot be achieved by shrinking the pencil $\mathcal{P}^{(0)}$. Instead, we have to remove all points $P \in \mathcal{C}^{(0)}$ which are not simultaneously on the plane $u = 2$. Thus the manifold $\mathcal{C}^{(1)}$ is the straight line $u = 2$, $v = 3$. On it all vector fields contained in the pencil $\mathcal{P}^{(0)}$ are tangential to $\mathcal{C}^{(0)}$, so that there is no need to shrink the pencil and we have $\mathcal{P}^{(1)} = \mathcal{P}^{(0)}$.

As the constraint manifold has become smaller, we have to check again the tangency. Vector fields tangential to $\mathcal{C}^{(1)}$ must possess a u-component which vanishes on $\mathcal{C}^{(1)}$. This property can be achieved solely by shrinking the pencil $\mathcal{P}^{(1)}$ to those vector fields X_a where the function a vanishes on $\mathcal{C}^{(1)}$. The constraint manifold remains unchanged, so that the completion stops here.

Thus we have $f = 2$; the final constraint manifold $\mathcal{C}^{(2)}$ is the straight line $u = 2$, $v = 3$ and one possible choice for the final pencil $\mathcal{P}^{(2)}$ is the underlying vector field $X_1(\mathbf{u}) = (0,0,-3)^t$. The field $X_2(\mathbf{u}) = (2-u, v-3, u-v-2)^t$ represents also a valid choice, as the two fields X_1 and X_2 coincide on the final constraint manifold $\mathcal{C}^{(2)}$. ◁

The extension of these considerations to *non-autonomous systems* requires essentially not more than replacing everywhere the tangent by the jet bundle. Thus we assume now that we are given a fibred manifold $\pi : \mathcal{E} \to \mathcal{X}$ over a one-dimensional base space $\mathcal{X}$; for simplicity we take $\mathcal{X} = \mathbb{R}$. Our starting point is a differential equation $\mathcal{R}_1 \subseteq J_1\pi$. Instead of the sequence $\mathcal{S}^{(i)}$ we consider the sequence $\mathcal{R}_1^{(i)} \subseteq J_1\pi$ constructed by $\mathcal{R}_1^{(i+1)} = \pi_1^2(\mathcal{R}_2^{(i)})$. The completion stops as soon as the equality $\mathcal{R}_1^{(i+1)} = \mathcal{R}_1^{(i)}$ holds.

At first sight, this procedure looks very different compared to the completion process for autonomous systems. The definition of $\mathcal{R}_1^{(i+1)}$ requires as first step a prolongation to second order and then we project back to first order. In the autonomous case, we reverse the order of the operations: first we project to the constraint manifold, then we "prolong" by considering its tangent space. But for *ordinary* differential equations we have the following result.

Proposition 7.3.3. *Let $\mathcal{R}_q \subseteq J_q\pi$ be an ordinary differential equation. If we consider its projection $\mathcal{R}_{q-1}^{(1)} \subseteq J_{q-1}\pi$ together with the corresponding natural projection $\hat{\pi}^{q-1} : \mathcal{R}_{q-1}^{(1)} \to \mathcal{X}$, then $\mathcal{R}_q^{(1)} = \mathcal{R}_q \cap J_1\hat{\pi}^{q-1}$ (here we consider $J_1\hat{\pi}^{q-1}$ as a submanifold of $J_q\pi$).*

Proof. For notational simplicity we give the proof for the case of a first-order equation: $q = 1$. The extension to the general case is trivial. Let a local representation of the differential equation $\mathcal{R}_1$ be

$$\mathcal{R}_1 : \begin{cases} \mathbf{v}' = \boldsymbol{\phi}(x, \mathbf{v}, \mathbf{w}, \mathbf{w}') , \\ 0 = \boldsymbol{\psi}(x, \mathbf{v}, \mathbf{w}) . \end{cases} \tag{7.57}$$

This form represents the obvious generalisation of the semi-explicit form (7.55) to non-autonomous systems. We need a local description of $\mathcal{R}_1^{(1)}$. Because of the semi-explicit form, it is easy to see that the prolongation of the equations containing $\mathbf{v}'$ leads to second-order equations solved for the derivatives $\mathbf{v}''$. Thus no cancellations of second-order derivatives are possible and these equations drop out when we project back to first order.

A local representation of $\mathcal{R}_1^{(1)}$ is obtained by adding to the equations already contained in (7.57) the new equations

$$\frac{\partial \boldsymbol{\psi}}{\partial x} + \frac{\partial \boldsymbol{\psi}}{\partial \mathbf{v}}\mathbf{v}' + \frac{\partial \boldsymbol{\psi}}{\partial \mathbf{w}}\mathbf{w}' = 0 \tag{7.58}$$

obtained by prolonging the algebraic equations in (7.57). Of course, in general this approach leads to a redundant description, as these new equations are not necessarily independent of the original ones. But this fact is of no importance here.

It is trivial to see that the same equations also define a local representation of $\mathcal{R}_1 \cap J_1\hat{\pi}^1$. The constraint manifold $\mathcal{R}_0^{(1)}$ is locally described by the algebraic equations in (7.57) and we must simply add their prolongations (7.58) to the original equations for the intersection. □

Hence we may reverse the order of prolongation and projection for ordinary differential equations. Obviously, any smooth solution of the original equation $\mathcal{R}_1$ is also a solution of the completed differential equation $\mathcal{R}_1^{(f)}$, as it automatically satisfies all the integrability conditions added during the completion process, and vice versa. There only remains to show that the final equation $\mathcal{R}_1^{(f)}$ is indeed involutive.

Corollary 7.3.4. *The final differential equation $\mathcal{R}_1^{(f)}$ is involutive.*

Proof. As we are dealing with ordinary differential equations, involution is equivalent to formal integrability. By the previous proposition and our construction it is obvious that $\mathcal{R}_1^{(f)} = \mathcal{R}_1^{(f+\ell)}$ for any $\ell > 0$. But this fact implies that it is a formally integrable equation. □

A closer analysis of (7.58) clarifies the two geometric effects described in the autonomous case: shrinking of either the constraint manifold or the vector field pencil. The derivatives $\mathbf{v}'$ in (7.58) can be eliminated with the help of the differential part of (7.57) leading to

$$\frac{\partial \boldsymbol{\psi}}{\partial x} + \frac{\partial \boldsymbol{\psi}}{\partial \mathbf{v}}\boldsymbol{\phi} + \frac{\partial \boldsymbol{\psi}}{\partial \mathbf{w}}\mathbf{w}' = 0\,. \tag{7.59}$$

We find additional *differential* equations, if the Jacobian of this equation with respect to $\mathbf{w}'$ does not vanish identically on $\mathcal{R}_1$; hence the rank of the matrix $(\partial\boldsymbol{\psi}/\partial\mathbf{v})(\partial\boldsymbol{\phi}/\partial\mathbf{w}') + (\partial\boldsymbol{\psi}/\partial\mathbf{w})$ determines their number. Such new differential equations correspond to a shrinking of the pencil. If the Jacobian does not possess maximal rank, we can additionally derive some *algebraic* equations from (7.58). If they are independent of the constraints in (7.57), we have found further integrability conditions leading to a smaller constraint manifold.

Obviously, the completion process always terminates, as the proof of Proposition 7.3.3 shows that $\dim \mathcal{R}_1^{(i+1)} < \dim \mathcal{R}_1^{(i)}$ until the process stops (we have additional equations). As $J_1\pi$ is a finite-dimensional manifold, we cannot have an infinite decreasing sequence. One should note that the assumption of a regular equation is crucial for this proof. It is equivalent to the assumption of constant rank of all appearing Jacobians which is necessary to allow us at each step the distinction into algebraic and differential equations.

In the geometric language we have used so far in this section, the completion process appears to be very simple. However, in concrete computations it becomes notoriously subtle. The considerations in the proof of Proposition 7.3.3 lead immediately to the informal "Algorithm" 7.2.

Algorithm 7.2 Completion of ordinary differential equation

Input: system of ordinary differential equations
Output: involutive completion

```
1: separate into algebraic and differential equations
2: repeat
3:     differentiate (new) algebraic equations
4:     remove all functionally dependent equations
5:     separate new equations into algebraic and differential ones
6: until no new algebraic equations appear
7: return final system
```

One great problem for any concrete realisation of this "algorithm" is that we must check in Line /4/ whether some smooth functions are functionally independent. Generally, this question is undecidable. In practice, the problem is less severe, as explicitly given constraints are usually not only smooth but even analytic. The real problem is that we must decide the independence not on the full jet bundle $J_1\pi$ but only on the differential equation $\mathcal{R}_1^{(i)}$, i. e. modulo a local representation of this submanifold. Fully effective methods are known only for representations which are polynomial in all variables. For the special case of the Dirac algorithm for Hamiltonian systems discussed below, Gerdt and Gogilidze [158] presented a completely algorithmic solution based on Gröbner bases for elimination orders.

A further problem arises in the following manner. Assume that we obtained at some stage a new algebraic equation $\phi = 0$ where the left hand side is of the particular form $\phi = \psi^n$ for some other function ψ and an exponent $n > 1$. From a geometric point of view, this special form makes no difference, as the equations $\phi = 0$ and $\psi = 0$ obviously describe the same zero set. But according to (2.15), prolongations require the differential. As $d\phi = n\psi d\psi$, the differential of ϕ trivially vanishes on the constraint manifold $\mathcal{R}_0^{(i)}$ and is useless for prolongations. Thus if we continue to work with ϕ, then we might overlook some constraints.[5]

If all constraints are polynomial, then this question represents a very classical topic in algebraic geometry: the ideal $\mathcal{I}$ generated by the constraints is not *radical*. Since by Hilbert's Nullstellensatz varieties are in a one-to-one correspondence with radical ideals (and $\mathcal{V}(\mathcal{I}) = \mathcal{V}(\sqrt{\mathcal{I}})$), we must take care to determine at each iteration of the algorithm a basis of the radical ideal corresponding to the current constraint

[5] In the physical literature one sometimes calls constraints like ϕ *ineffective*.

manifold. Again, the theory of Gröbner bases yields an algorithmic solution of this problem, albeit such computations are often prohibitively expensive.

Addendum: Constrained Hamiltonian Systems

Long before the expression differential algebraic equation was even invented, theoretical physicists began already to study such equations under the heading *constrained dynamics*. They have been mainly concerned with the special case of Hamiltonian systems. As the corresponding theory was initiated by Dirac [107, 108] (see also [109]), it is often called *Dirac theory*. An intrinsic geometric version was developed by Gotay et al [173].

The classical coordinate formulation of the Dirac theory goes as follows. We assume that $\mathcal{U}$ is a symplectic manifold with the symplectic two-form ω. The dynamics is described by the Hamiltonian vector field X_H for some Hamiltonian $H \in \mathcal{F}(\mathcal{U})$. In concrete applications, H usually arises from the Legendre transformation of a Lagrangian L. If L is degenerate, then H is not uniquely defined. In this case, the Hamiltonian vector field X_H may be naturally considered as a vector field pencil, as in our considerations above. Finally, we assume that the dynamics is constrained to a submanifold $\mathcal{C}^{(0)} \subseteq \mathcal{U}$ given as the zero set of some *primary constraints* $\phi^\tau \in \mathcal{F}(\mathcal{U})$ with $1 \leq \tau \leq r$.

According to our considerations above, we must study at which points $P \in \mathcal{C}^{(0)}$ the condition $X_H\big|_P \in T_P\mathcal{C}^{(0)}$ is satisfied. This condition may be equivalently formulated as $d\phi^\tau(X_H) = 0$ for $1 \leq \tau \leq r$ (see Remark C.2.8). It follows from Definition C.6.3 of a Hamiltonian vector field and from the representation (C.46) of the Poisson bracket on a symplectic manifold, respectively, that we find for any function $\phi \in \mathcal{F}(\mathcal{U})$ the equality

$$d\phi(X_H) = (\iota_{X_\phi}\omega)(X_H) = \omega(X_\phi, X_H) = \{\phi, H\} . \tag{7.60}$$

Thus the tangency condition is satisfied at those points P where not only the given primary constraints ϕ^τ but in addition also the *secondary constraints* $\psi^\tau = \{\phi^\tau, H\}$ vanish (if X_H is actually a vector field pencil, these conditions may also be satisfied by a suitable restriction of the pencil) and the latter ones define a new constraint manifold $\mathcal{C}^{(1)} \subseteq \mathcal{C}^{(0)}$. Note that the secondary constraints lead to a smaller constraint manifold only, if they are algebraically independent of the primary ones.

If $\mathcal{C}^{(1)}$ is a proper submanifold of $\mathcal{C}^{(0)}$, we must again study the tangency of the Hamiltonian vector field X_H. By the same reasoning as above, this analysis amounts to requiring that $d\psi^\tau(X_H) = \{\psi^\tau, H\} = 0$ which may yield new tertiary constraints. Iteration leads to the famous *Dirac algorithm*: we always compute the Poisson brackets of the current constraint generation with the Hamiltonian until no new constraints arise. As we have seen, there is nothing special behind this algorithm; it is based on a simple reformulation of the tangency condition that is only possible, if we are working on a symplectic manifold.

Dirac's derivation did not use the tangency condition but an equivalent reasoning. If the dynamics is constrained to the manifold $\mathcal{C}^{(0)}$, then the functions ϕ^τ must be constant (actually vanish) along trajectories. This observation obviously entails that their variation along the trajectories, $\dot{\phi}^\tau$, must vanish. By (C.51), this variation is given by $\dot{\phi}^\tau = \{\phi^\tau, H\}$. Thus we arrive at the same result as above.

The geometric formulation of Gotay et al [173] is obtained by expressing the tangency conditions in yet another way, namely,

$$\mathcal{C}^{(i+1)} = \{p \in \mathcal{C}^{(i)} \mid \mathrm{d}H\big((T\mathcal{C}^{(i)})^\perp\big)(p) = 0\} . \tag{7.61}$$

Here $(T\mathcal{C}^{(i)})^\perp$ denotes the symplectic complement of $T\mathcal{C}^{(i)}$ in the full tangent bundle $T\mathcal{U}\big|_{\mathcal{C}^{(i)}}$. Using the annihilator, the tangency condition may thus be expressed more concisely as $\mathrm{d}H \in \big((T\mathcal{C}^{(i)})^\perp\big)^0$. This fact can be seen as follows. Let us denote by $(T\mathcal{C}^{(i)})^\flat$ the image of the musical isomorphism $\flat$ restricted to $T\mathcal{C}^{(i)}$, i. e. $(T\mathcal{C}^{(i)})^\flat = \{\iota_Y\omega \mid Y \in T\mathcal{C}^{(i)}\} \subseteq T^*\mathcal{U}\big|_{\mathcal{C}^{(i)}}$. Then

$$\big((T\mathcal{C}^{(i)})^\flat\big)^0 = \{Z \in T\mathcal{U}\big|_{\mathcal{C}^{(i)}} \mid \omega(Y,Z) = 0\ \forall Y \in T\mathcal{C}^{(i)}\} = (T\mathcal{C}^{(i)})^\perp \tag{7.62}$$

implying that $\big((T\mathcal{C}^{(i)})^\perp\big)^0 = (T\mathcal{C}^{(i)})^\flat$. Hence if the one-form $\mathrm{d}H$ is contained in this space, then by the definition of $\flat$ the corresponding Hamiltonian vector field is tangential to $\mathcal{C}^{(i)}$. Conversely, if the field $X_H\big|_{\mathcal{C}^{(i)}}$ is tangential to the constraint manifold, then for all $Y \in (T\mathcal{C}^{(i)})^\perp$ the equality $\omega(X_H, Y) = \mathrm{d}H(Y) = 0$ holds and thus we find $\mathrm{d}H \in \big((T\mathcal{C}^{(i)})^\perp\big)^0$.

Example 7.3.5. As a concrete example we consider a *rigid rotator* in d dimensions. This is a particle of mass m moving in a d-dimensional Euclidean space $\mathcal{Q}$; it is not subject to any external forces but constrained to stay on the surface of the unit sphere. We denote the spatial coordinates in $\mathcal{Q}$ by $\mathbf{q}$ and the corresponding momenta in $T^*\mathcal{Q}$ by $\mathbf{p} = m\dot{\mathbf{q}}$. The Hamiltonian of a free particle is simply $H_0 = \frac{1}{2m}\mathbf{p}^2$ and our primary constraint is given by $\phi = \frac{1}{2}(\mathbf{q}^2 - 1)$. Thus we start our analysis with the Hamiltonian $H = \frac{1}{2}\mathbf{p}^2 + \lambda\phi$; here λ is a yet arbitrary function (a "multiplier") on our phase space $T^*\mathcal{Q}$ expressing the fact that H is well-defined only on the constraint surface $\mathcal{C}$ described by the equation $\phi = 0$.

According to our discussion above, we must require that the Hamiltonian vector field X_H (note that X_H is actually a pencil because of the arbitrariness of λ) is tangent to $\mathcal{C}$ and this condition is equivalent to $\{\phi, H\}|_{\mathcal{C}} = 0$. As we are using here the canonical Poisson bracket (C.48), we find that the function $\frac{1}{m}\mathbf{qp} + \{\phi, \lambda\}\phi$ must vanish on $\mathcal{C}$. Obviously, we may neglect the last term and take $\psi = \mathbf{qp}$ as secondary constraint. It is trivial to see that it expresses nothing but the fact that the velocity $\dot{\mathbf{q}} = \mathbf{p}/m$ must always be tangential to the unit sphere. Thus $\mathcal{C}^{(1)}$ is the submanifold of $T^*\mathcal{Q}$ described by the equations $\phi = \psi = 0$.

In the next step of the Dirac algorithm we have to evaluate the Poisson bracket $\{\psi, H\} = \frac{1}{m}\mathbf{p}^2 - \lambda\mathbf{q}^2 + \{\psi, \lambda\}\phi$. We may again neglect the last term, as it trivially vanishes on $\mathcal{C}^{(1)}$, and find this time that the Hamiltonian vector field X_H will be tangent to $\mathcal{C}^{(1)}$, if we choose $\lambda = \frac{1}{m}\frac{\mathbf{p}}{\mathbf{q}}$. As the constraint manifold did not shrink any further, the Dirac algorithm stops here.

The dynamics of a rigid rotor is therefore described by the Hamiltonian vector field X_{H_f} belonging to the final Hamiltonian $H_f = \frac{\mathbf{p}^2}{2m}(2 - \frac{1}{\mathbf{q}^2})$. Note that on the final constraint manifold $\mathcal{C}^{(1)}$ the Hamiltonians H_0 and H_f coincide. However, the same is not true for the corresponding Hamiltonian vector fields, as their computation requires also normal differentiations. From a physical point of view the additional terms in X_H represent *constraint forces*, i. e. those forces that ensure that our particle stays on the constraint manifold. If we assume e. g. that our particle is connected to the origin by a rod, then these forces correspond to the tension in the rod. ◁

In Example 2.4.6 we intrinsically formulated non-autonomous Hamiltonian systems using cosymplectic geometry. Now we present the corresponding modification of the Dirac theory. We assume again that the total space of our fibred manifold is a cosymplectic manifold $(\mathcal{E}, \omega, \eta)$—the extended phase space—with a transversal Reeb vector field R.

According to Example 2.4.6, the dynamics of an unconstrained system defined by a Hamiltonian H is given by the evolution vector field E_H. For the analysis of a constrained system an alternative formulation is often advantageous. We can combine the cosymplectic structure and the Hamiltonian by introducing the 2-form $\omega_H = \omega + \mathrm{d}H \wedge \eta$. One easily checks that $(\mathcal{E}, \omega_H, \eta)$ is again a cosymplectic manifold with Reeb vector field $R_H = E_H$.

Now we impose on this system some constraints, i. e. we require that the dynamics takes place on a prescribed fibred submanifold $\mathcal{C} \subseteq \mathcal{E}$. According to our considerations above, this condition entails that solutions can only exist at points where the Reeb vector field R_H is tangent to $\mathcal{C}$. This leads to a sequence of submanifolds

$$\mathcal{C}^{(i+1)} = \{\xi \in \mathcal{C}^{(i)} \mid R_h|_\xi \in T_\xi \mathcal{C}^{(i)}\} \tag{7.63}$$

with $\mathcal{C}^{(1)} = \mathcal{C}$. Using the cosymplectic complement, we may reformulate this tangency condition in analogy to (7.61) as

$$\mathcal{C}^{(i+1)} = \{\xi \in \mathcal{C}^{(i)} \mid \eta_\xi\big((T_\xi \mathcal{C}^{(i)})^\perp\big) = 0\}\,. \tag{7.64}$$

As the proof of this equality is very similar to the computations above, we skip it and refer to [83] instead.

As discussed in Example 2.4.6, the contact map allows us to associate with the Reeb vector field R_H a section $\gamma : \mathcal{E} \to J_1\pi$ whose image defines a Hamiltonian differential equation $\bar{\mathcal{R}}_1$. The constrained system may then be realised as

$$\mathcal{R}_1 = \{(\xi, \lambda) \in \bar{\mathcal{R}}_1 \mid \xi \in \mathcal{C}\}\,. \tag{7.65}$$

Here we use our intrinsic Definition 2.2.2 of the first-order jet bundle $J_1\pi$ and represent any point $\rho \in J_1\pi$ as a pair (ξ, λ) with $\xi \in \mathcal{E}$ and $\lambda \in T_x^*\mathcal{X} \otimes T_\xi \mathcal{E}$. Obviously, $\mathcal{R}_0^{(1)} = \mathcal{C}$. As above, the completion process leads to a sequence of submanifolds

$\mathcal{R}_1^{(i)}$ until $\mathcal{R}_1^{(i+1)} = \mathcal{R}_1^{(i)}$. We claim now that $\mathcal{R}_0^{(i)} = \mathcal{C}^{(i)}$, i. e. our two alternative descriptions of the completion algorithm are equivalent.

Indeed, this equivalence is a simple consequence of Definition 2.2.2. The basic idea is to describe the constraint manifolds $\mathcal{R}_0^{(i)}$ in a similar manner as (7.63). By construction, the equality

$$\Gamma_1\big(\gamma(\xi), T_\xi\pi(R_H|_\xi)\big) = R_H|_\xi \tag{7.66}$$

holds. Thus the differential equation $\bar{\mathcal{R}}_1$ defined by the section γ is an underlying differential equation. In fact, it would suffice, if (7.66) held only for all $\xi \in \mathcal{C}$, i. e. we could take any section such that $\mathcal{R}_1 \subseteq \operatorname{Im}\gamma$. It is now easy to see that in close analogy to (7.63)

$$\mathcal{R}_0^{(i+1)} = \{\xi \in \mathcal{R}_0^{(i)} \mid \gamma(\xi) \in J_1\mathcal{R}_0^{(i)}\}\,. \tag{7.67}$$

Assume that $\gamma(\xi) = (\xi, \lambda) \in J_1\mathcal{R}_0^{(i)}$ which means that $\lambda \in T^*_{\pi(\xi)}\mathcal{X} \otimes T_\xi\mathcal{R}_0^{(i)}$ according to our definition of the jet bundle. Applying (7.66) shows that this property is equivalent to $R_H|_\xi \in T_\xi\mathcal{R}_0^{(i)}$, as for such a λ the contact map Γ_1 maps to the tangent space $T_\xi\mathcal{R}_0^{(i)}$. Hence $\xi \in \mathcal{R}_0^{(i)}$ implies $\xi \in \mathcal{C}^{(i)}$ and vice versa.

7.4 Cartan–Kuranishi Completion

In the previous section we considered only ordinary differential equations, where in particular for autonomous system the completion process is very transparent and intuitive. Now we derive in an intrinsic manner a rather general completion method that can—in principle—handle any *regular* system of partial differential equations. While it is still fairly simple, it is less intuitive. Furthermore, we will see that some of the required operations cannot always be performed effectively.

Recall from Section 2.3 that a differential equation $\mathcal{R}_q$ is regular, if all its prolongations and subsequent projections lead to fibred submanifolds. Furthermore, in accordance with our Blanket Assumption 7.1.4, we assume that the indices of all symbols are constant. In particular, if integrability conditions arise, it is generally not possible to check in advance whether a given differential equation satisfies these assumptions; they can only be verified during the completion. A trivial exception are linear equations, where only the zeros and singularities of the coefficient functions must be analysed more closely.

Theorem 7.4.1 (Cartan–Kuranishi). *Let $\mathcal{R}_q \subseteq J_q\pi$ be a regular differential equation. Then two integers $r, s \geq 0$ exist such that $\mathcal{R}_{q+r}^{(s)}$ is involutive.*

Proof. We give a constructive proof leading directly to the promised general completion method. It is based on the simple Algorithm 7.3 which consists of two nested loops: in the inner one we prolong until an involutive symbol is reached; in the outer one we check whether one further prolongation and subsequent projection yields integrability conditions. If we can show that this algorithm always terminates with an involutive equation of the form $\mathcal{R}^{(s)}_{q+r}$, we are done. Thus there are two points to prove: the correctness and the termination.

Algorithm 7.3 Cartan–Kuranishi completion

Input: regular differential equation $\mathcal{R}_q \subseteq J_q\pi$
Output: equivalent involutive equation $\mathcal{R}^{(s)}_{q+r} \subseteq J_{q+r}\pi$

```
r ← 0; s ← 0
repeat
    while N^(s)_{q+r} not involutive do
        r ← r + 1
    end while
    intConds ← R^(s+1)_{q+r} ⊊ R^(s)_{q+r}
    if intConds then
        s ← s + 1
    end if
until ¬intConds
return R^(s)_{q+r}
```

The *correctness* of Algorithm 7.3 is simple. Assume that it terminates with some values r and s. By construction, the symbol $\mathcal{N}^{(s)}_{q+r}$ is involutive at the end of the inner loop. At the end of the outer loop the boolean variable *intConds* is false implying that $\mathcal{R}^{(s+1)}_{q+r} = \mathcal{R}^{(s)}_{q+r}$. Thus both conditions of Theorem 7.2.6 are satisfied and the differential equation $\mathcal{R}^{(s)}_{q+r}$ is involutive.

For the *termination* we must consider the two loops in Algorithm 7.3. The inner loop terminates because of Theorem 6.1.21: if r_0, s_0 are the values of the counters at the beginning of the loop and $\mathcal{N}$ the symbol comodule of the corresponding differential equation $\mathcal{R}^{(s_0)}_{q+r_0}$, then this theorem guarantees the existence of a finite value $r_1 \geq r_0$ such that the Spencer cohomology $H^\bullet(\mathcal{N})$ vanishes from degree $q+r_1$ on; but this fact is equivalent to involution of the symbol $\mathcal{N}^{(s_0)}_{q+r_1}$.

Alternatively, we may work in local coordinates. At a point $\rho \in \mathcal{R}^{(s_0)}_{q+r_0}$ the inner loop corresponds to the determination of a finite Pommaret basis for the symbol module. According to Theorem 4.4.1 (together with the results of Section 4.3), such a basis always exists and thus the loop always terminates—provided our chosen coordinates are δ-regular.

The termination of the outer loop follows from a Noetherian argument. Consider the polynomial module $\mathcal{M}$ finitely generated by the symbol modules of $\mathcal{R}_q$ and all its projections $\mathcal{R}^{(q-q')}_{q'}$ with $0 \leq q' < q$, respectively (thus we take the annihilator of the symbol comodule constructed in the more refined manner of Remark 7.1.16 where also the lower-order equations are taken into account). In local coordinates, this approach corresponds to choosing a local representation of $\mathcal{R}_q$ with a Jacobian in triangular form and taking the principal part of each equation as a generator.

If integrability conditions occur at the end of the outer loop, i. e. the corresponding projection is not surjective, then the module $\mathcal{M}'$ correspondingly constructed for $\mathcal{R}_{q+r}^{(s+1)}$ is larger than $\mathcal{M}$. Indeed, choosing again a local representation in triangular form, the principal parts of the new equations stemming from the integrability conditions cannot lie in $\mathcal{M}$. Hence adding these new generators leads to a strictly larger (but still finitely generated) polynomial module $\mathcal{M}' \supset \mathcal{M}$.

Therefore the outer loop of Algorithm 7.3 leads to an ascending chain of submodules of a free polynomial module. As polynomial modules are Noetherian, this chain must become stationary for some value of s. But this fact implies that in the iteration of the outer loop corresponding to this value *no* integrability condition was found. Hence the algorithm terminates at this iteration. □

If we apply Algorithm 7.3 to a concrete differential equation, then it obviously will alternate prolongations and projections. Thus, strictly speaking, it works with differential equations $((((\mathcal{R}_{q+r_1})^{(1)})_{+r_2})^{(1)})\cdots)_{+r_s}$. However, it always projects equations with an involutive symbol. Thus, by Part (ii) of Proposition 7.2.5, these equations coincide with $\mathcal{R}_{q+r_1+\cdots+r_s}^{(s)}$.

Remark 7.4.2. The involutive completed equation $\mathcal{R}_{q+r}^{(s)}$ is *equivalent* to the original differential equation $\mathcal{R}_q$ in the sense that both possess the same formal solution space. In fact, both equations have even the same smooth solutions. Differences may show up only for solutions of $\mathcal{R}_q$ with a finite differentiability: solutions of $\mathcal{R}_q$ which are only k times differentiable for $q \leq k < q+r$ cannot be strong solutions of $\mathcal{R}_{q+r}^{(s)}$, as they lack sufficiently many derivatives. But we may expect that they are at least weak solutions in a suitable sense. ◁

Example 7.4.3. We demonstrate the working of Algorithm 7.3 for a classical linear second-order equation due to Janet [235, Ex. 47]:

$$\mathcal{R}_2 : \begin{cases} u_{zz} + y u_{xx} = 0\,, \\ u_{yy} = 0\,. \end{cases} \tag{7.68}$$

Thus we have one dependent variable and three independent variables. Obviously, the system is overdetermined.

According to our algorithm, we prolong until the symbol becomes involutive. One readily checks that the indices of the symbol $\mathcal{N}_2$ are $\beta_2^{(3)} = \beta_2^{(2)} = 1$ and $\beta_2^{(1)} = 0$. The rank of the symbol matrix M_3 is $6 > 3\beta_2^{(3)} + 2\beta_2^{(2)} = 5$ (the non-multiplicative prolongation $u_{yyz} = 0$ is necessary to generate $\mathcal{N}_3$), thus the symbol $\mathcal{N}_2$ is not involutive and we must prolong. In a lengthy computation one shows that $\mathcal{N}_3$ is involutive and the inner loop terminates with $r = 1$. Now we must check for the appearance of integrability conditions. One finds $\dim \mathcal{R}_3^{(1)} = 11 < \dim \mathcal{R}_3 = 12$. Thus $\mathcal{R}_3$ is not involutive, as one condition is hidden. Indeed,

$$D_{yy}(u_{zz} + y u_{xx}) - (D_{zz} + y D_{xx}) u_{yy} = u_{xxy} = 0\,. \tag{7.69}$$

Thus after the first iteration of the outer loop $r = s = 1$ and we continue with $\mathcal{R}_3^{(1)}$. The calculations now become more and more tedious, as the number of equations needed in local representations of either the differential equations or their symbols rapidly grows. So we can only sketch the progress of the computation.

$\mathcal{N}_3^{(1)}$ is again not involutive, since setting up $\mathcal{N}_4^{(1)}$ requires the non-multiplicative prolongation $u_{xxyz} = 0$. The symbol $\mathcal{N}_4^{(1)}$ is involutive and at the end of the inner loop $r = 2$. But again we find an integrability condition,

$$D_{xxy}(u_{zz} + yu_{xx}) - (D_{zz} + yD_{xx})u_{xxy} = u_{xxxx} = 0\,, \tag{7.70}$$

and $\mathcal{R}_4^{(1)}$ is not involutive. Thus now $s = 2$ and we continue with $\mathcal{R}_4^{(2)}$.

Again, the symbol $\mathcal{N}_4^{(2)}$ is not involutive because of the non-multiplicative prolongation $u_{xxxxz} = 0$. Thus we prolong once more in order to obtain $\mathcal{R}_5^{(2)}$. Finally, the algorithm terminates with $r = 3$ and $s = 2$, as the symbol $\mathcal{N}_5^{(2)}$ is involutive and no integrability conditions appear during the prolongation so that the differential equation $\mathcal{R}_5^{(2)}$ is involutive, too. Thus, $\mathcal{R}_4^{(2)}$ is another example of an equation which is formally integrable but *not* involutive. However, we can detect this fact only in hindsight, after the algorithm has terminated.

In fact, involution of the symbol $\mathcal{N}_5^{(2)}$ is trivial, as it vanishes. Thus we are dealing here with an equation of finite type. Its solution space is twelve-dimensional as $\dim \mathcal{R}_5^{(2)} = 12$. Indeed, a direct integration of $\mathcal{R}_5^{(2)}$ yields the solution

$$\begin{aligned} u(x,y,z) = &\frac{1}{3}a_1x^3z - a_2xyz^2 + a_3x^3 + a_2x^2z + a_4xyz - a_1yz^2 + \\ &a_5x^2 + a_6xy + a_7xz + a_8yz + a_9x + a_{10}y + a_{11}z + a_{12}\,. \end{aligned} \tag{7.71}$$

(Note that this integration is greatly simplified by the knowledge of the hidden integrability conditions. This effect is typical and explains why completion has found so much interest in Lie symmetry theory where one often must solve large overdetermined systems for obtaining the symmetry generators.)

As one can see, the straightforward application of Algorithm 7.3 may become quite tedious even for such small equations. In our particular example one needs a total of 132 prolongations in order to get local representations for all appearing equations. Furthermore, one must compute row echelon forms of symbol matrices and Jacobians; the largest is an 86×84 matrix. We will show later in Section 10.7 that by a clever organisation of the computations using ideas from the algebraic theory of involutive bases one can completely avoid the determination of row echelon forms and reduce the number of prolongations by about 80%. ◁

Algorithm 7.3 does not really represent an algorithm but only a "method", i. e. it is not clear whether all steps can be performed effectively. This question concerns in particular the test for inequality in Line /6/. To decide whether or not a symbol is involutive requires basically only some linear algebra and is thus fully algorithmic.

However, even there it may happen for non-linear systems that not all necessary operations with the coefficients of the symbol can be performed effectively.

The test in Line /6/ requires to decide whether certain equations are algebraically independent. Depending on the type of the equation this question can be highly non-trivial. For *linear* equations it is again only a matter of some linear algebra. For *polynomial* equations it can—at least in principle—be done using Gröbner bases (see Appendix B.4), although the computations suffer potentially from a very high complexity. For arbitrary non-linear equations no general criterion is known. Depending on the class of functions considered the situation may become even worse. While analytic functions are either independent or not, smooth functions may be independent in some domains and dependent in others.

Finally, we compare Algorithm 7.3 with the geometric completion procedure for ordinary differential equations introduced in the last section. As any ordinary differential equation has an involutive symbol, the inner loop becomes superfluous. The algorithm reduces to a simple *prolong–project*, *prolong–project*, ... But we have seen in the proof of Proposition 7.3.3 that it is possible to swap the operations and use *project–prolong* cycles. The result is the same, but the modified algorithm is computationally more efficient, as less equations must be prolonged.

One should stress that this modification is allowed only for *ordinary* differential equations, because for them no integrability conditions arise from cross-derivatives which always take place in higher order. Even if we are dealing with a first-order partial differential equation in only one dependent variable, we must use *prolong–project* cycles, although such an equation always has an involutive symbol according to the Drach classification in Remark 7.1.29. Otherwise we risk to overlook hidden integrability conditions.

Remark 7.4.4. Recall from Remark 7.1.12 that integrability conditions are related to syzygies of the principal symbol module. Furthermore, by (our proof of) Theorem 7.2.8 we can also use these syzygies for deciding formal integrability. These considerations lead to a different view of the computations underlying any effective realisation of Algorithm 7.3 and also highlight some of the differences between formal integrability and involution.

In Remark 6.1.23 we noted that the degree of involution of a polynomial module is nothing but its Castelnuovo–Mumford regularity. Thus assume that the polynomial module $\mathcal{M}$ which we constructed in the termination proof of Algorithm 7.3 becomes involutive at the degree $q+r$. Then the truncation $\mathcal{M}_{\geq q+r}$ is nothing but the annihilator $(\mathcal{N}^{(s)}_{q+r})^0$ of the current symbol and, according to Theorem 5.5.24, it possesses a linear resolution. This observation implies in particular that the first syzygy module can be generated by syzygies of degree one. Given the connection between syzygies and integrability conditions, any integrability condition must therefore show up already in the next prolongation—a fact we proved in a different manner in Theorem 7.2.6.

Going back to Example 7.4.3, one can easily verify these considerations. The polynomial module $\mathcal{M}$ associated to the original equation $\mathcal{R}_2$ is here the ideal $\langle \chi_3^2 + y\chi_1^2, \chi_2^2 \rangle$. One easily sees that $\mathrm{Syz}(\mathcal{M})$ is a principal module with the single generator $\mathbf{S}_1 = \chi_2^2\mathbf{e}_1 + (\chi_3^2 + y\chi_1^2)\mathbf{e}_2$. As it is of degree 2, the resolution of $\mathcal{M}_{\geq 2} = \mathcal{M}$ is not linear and the symbol $\mathcal{N}_2$ cannot be involutive. At the level of the construction of integrability conditions, it is now obvious that we must prolong twice in Algorithm 7.3 before we can find the first condition $u_{xxy} = 0$ (one prolongation is done in the inner loop; the other one is hidden in the computation of the boolean variable `intConds`). Note that the construction of this condition in (7.69) follows immediately from the syzygy above.

Continuing now in the same manner with the differential equation $\mathcal{R}_3^{(1)}$, we must now analyse the polynomial module $\mathcal{M}$ obtained by adding the generator $\chi_1^2\chi_2$. This leads to two additional minimal generators for $\mathrm{Syz}(\mathcal{M})$, namely $\mathbf{S}_2 = \chi_1^2\mathbf{e}_2 - \chi_2\mathbf{e}_3$ and $\mathbf{S}_3 = \chi_1^2\chi_2\mathbf{e}_1 - (\chi_3^2 + y\chi_1^2)\mathbf{e}_3$. One easily sees that $\mathbf{S}_2$ does not lead to an integrability condition (but to an identity), whereas $\mathbf{S}_3$ yields (7.70). Again the minimal generators are of higher order, so that $\mathcal{M}_{\geq 3}$ does not possess a linear resolution and in the execution of Algorithm 7.3 we had again to prolong twice to find this integrability condition, as $\mathcal{N}_3^{(1)}$ is not involutive.

Thus we are now dealing with the differential equation $\mathcal{R}_4^{(2)}$ and we must add the minimal generator χ_1^4 in order to obtain the modified polynomial module $\mathcal{M}$. For its syzygy module $\mathrm{Syz}(\mathcal{M})$ we need again two further minimal generators, namely $\mathbf{S}_4 = \chi_1^2\mathbf{e}_3 - \chi_2\mathbf{e}_4$ and $\mathbf{S}_5 = (\chi_3^2 + y\chi_1^2)\mathbf{e}_4 - \chi_1^4\mathbf{e}_1$. One easily checks than none of them leads to a new integrability condition so that $\mathcal{R}_4^{(2)}$ is a formally integrable equation according to Theorem 7.2.8. Thus, if we are just interested in integrability conditions, then we can stop the completion process at this stage. But if we go for involution, then it follows again from the degrees of the syzygies that $\mathcal{N}_4^{(2)}$ cannot be involutive and that further prolongations are required. The number of prolongations needed cannot be deduced from the syzygy, as it depends on the full syzygy resolution (or equivalently the full Koszul homology).

Compared with the direct execution of Algorithm 7.3, one sees that this approach requires much less computations, as we need only those prolonged equations which are actually part of a generalised cross-derivative leading to an integrability condition. However, the above description is slightly misleading, as we simply wrote down the minimal generators of the respective syzygy modules. In practice, these must be determined using some form of Gröbner basis (for example, an involutive basis). Indeed, the efficient algebraic realisation of the Cartan–Kuranishi Algorithm 7.3 for linear systems which we will present in Section 10.7 implicitly does exactly this, although there syzygies will not be mentioned explicitly. $\triangleleft$

7.5 The Principal Symbol Revisited

In this section we prove a result on the relation of the highest index $\beta_q^{(n)}$ and the principal symbol of a differential equation. An extension of this approach to a fully constructive solution of the problem of δ-regularity will be given in the next section.

Here we will use the mentioned relation for a rigorous definition of under- and overdetermined equations.

Theorem 7.5.1. *Let $\mathcal{R}_q$ be a differential equation in n independent variables with principal symbol τ_χ. In different coordinate systems we may obtain different values for the index $\beta_q^{(n)}$ of the geometric symbol $\mathcal{N}_q$ at the point $\rho \in \mathcal{R}_q$ with $\pi(\rho) = x$; let $\tilde{\beta}_q^{(n)}$ be the maximum of these values. Then at the point ρ*

$$\max_{\chi \in T_x^* \mathcal{X}} \operatorname{rank} \tau_\chi = \tilde{\beta}_q^{(n)} . \tag{7.72}$$

Proof. By the definition of $\tilde{\beta}_q^{(n)}$, we find in a neighbourhood of ρ a local description of $\mathcal{R}_q$ where the first $\tilde{\beta}_q^{(n)}$ equations $\Phi^\tau(\mathbf{x}, \mathbf{u}^{(q)}) = 0$ satisfy

$$\frac{\partial \Phi^\tau}{\partial p^\alpha_{[0,\dots,0,q]}} = \delta^\tau_\alpha . \tag{7.73}$$

Setting $\chi = \chi_i \mathrm{d}x^i\big|_x$ has the following effect on the matrix $T[\chi]$ of the principal symbol τ_χ at the point ρ: the diagonal elements of the top left $\tilde{\beta}_q^{(n)} \times \tilde{\beta}_q^{(n)}$ submatrix of $T[\chi]$ are of the form $(\chi_n)^q + \dots$ and the monomial $(\chi_n)^q$ appears nowhere else in the matrix. If we choose $\chi = \mathrm{d}x^n\big|_x$, all entries vanish with the exception of these diagonal elements which become one. Hence the rank of τ_χ at ρ is $\tilde{\beta}_q^{(n)}$ for this particular choice of χ.

Assume that for some one-form $\chi \in T_x^* \mathcal{X}$ the rank of τ_χ was greater than $\tilde{\beta}_q^{(n)}$. We choose in a neighbourhood of the point x a new coordinate system $\tilde{\mathbf{x}}$ given by $\tilde{\mathbf{x}} = \boldsymbol{\psi}(\mathbf{x})$ where the component ψ^n satisfies $\mathrm{d}\psi^n\big|_x = \chi$. Obviously, such coordinates always exist, as this is only a condition at the single point x. For the dependency of the transformed local representation $\tilde{\Phi}^\tau = 0$ on the derivatives of class n we obtain

$$\frac{\partial \tilde{\Phi}^\tau}{\partial \tilde{p}^\alpha_{[0,\dots,0,q]}} = \frac{\partial \Phi^\tau}{\partial p^\alpha_\nu} \prod_{i=1}^n \left(\frac{\partial \psi^n}{\partial x^i} \right)^{\nu_i} . \tag{7.74}$$

At the considered point x we can evaluate the derivatives of ψ^n and the right hand side becomes the matrix $T[\chi]$ of the principal symbol τ_χ. By assumption, its rank is greater than $\tilde{\beta}_q^{(n)}$. Thus we have found a local coordinate system where more than $\tilde{\beta}_q^{(n)}$ equations are of class n contradicting the definition of $\tilde{\beta}_q^{(n)}$. □

Recall that by definition of δ-regular coordinates, they must always yield the value $\tilde{\beta}_q^{(n)}$ for the index $\beta_q^{(n)}$. Thus for such coordinates the one-form $\chi = \mathrm{d}x^n$ must lead to a principal symbol of maximal rank. One-forms that lead to a lower rank are sometimes called *systatic*.

Corollary 7.5.2. *If a one-form $\chi \in T^* \mathcal{X}$ exists such that the principal symbol τ_χ of the differential equation $\mathcal{R}_q$ is surjective, then $\mathcal{R}_q$ is involutive.*

Proof. The principal symbol τ_χ can only be surjective, if the number t of (independent) equations in any local representation of the equation $\mathcal{R}_q$ is less than or equal to the number m of dependent variables and if all the equations are of order q. The rank of τ_χ is then t and by Theorem 7.5.1 all equations are of class n. But this observation trivially implies that $\mathcal{N}_q$ is involutive and that no integrability conditions arise; hence $\mathcal{R}_q$ is involutive. □

Example 7.5.3. Despite the computational complexity of determining the rank of parametrised matrices, Theorem 7.5.1 is sometimes quite useful for checking the δ-regularity of a coordinate system. We demonstrate this fact for the Yang–Mills equations (2.93). As they are semi-linear, there is no need for a linearisation and the principal part of the equation $\Phi^a_\ell = 0$ is simply $\eta^{ij}(\partial_{x^i x^j} A^a_\ell - \partial_{x^i x^\ell} A^a_j)$. Thus we obtain $\beta_2^{(n)} = d(n-1)$, as for all equations $\Phi^a_\ell = 0$ with an index $\ell \neq n$ we may take $\partial_{x^n x^n} A^a_\ell$ as principal derivative of class n. In the d remaining equations with $\ell = n$ we choose $\partial_{x^{n-1} x^n} A^a_{n-1}$ as principal derivative of class $n-1$. Thus $\beta_2^{(n-1)} = d$ and all other indices vanish.

Now the question arises whether some of the equations we have classified as being of class $n-1$ could be "promoted" to class n by a coordinate transformation, i. e. whether our coordinates are δ-regular. As only $\beta_2^{(n)}$ and $\beta_2^{(n-1)}$ are greater than zero, we are done once we know whether $\beta_2^{(n)} = \tilde{\beta}_2^{(n)}$ and by Theorem 7.5.1 this can be decided with the principal symbol. For the entries of its matrix we obtain

$$T^{ak}_{b\ell}[\chi] = \delta^a_b \eta^{ij} \chi_i \left(\delta^k_\ell \chi_j - \delta^k_j \chi_\ell \right) \tag{7.75}$$

with $1 \leq a, b \leq d$ and $1 \leq k, \ell \leq n$. The indices a, ℓ label the rows (i. e. the equations $\Phi^a_\ell = 0$ in the system) and the indices b, k the columns (i. e. the fields A^b_k).

It is easy to see that these entries satisfy for any one-form χ the d identities $\eta^{m\ell} \chi_m T^{ak}_{b\ell}[\chi] = 0$, as then the expression in the parentheses vanishes—note that we have here just the principal parts of the Noether identities (2.94). This observation implies that the rank of the matrix $T[\chi]$ can never be larger than $d(n-1)$. But from our analysis above, we already know that for $\chi = \mathrm{d}x^n$ we obtain this value as rank. Thus $d(n-1)$ is indeed the maximal rank and our coordinates are δ-regular.

The form of these identities immediately implies that the symbol $\mathcal{N}_2$ of the Yang–Mills equations is involutive, as a multiplication with a component of χ corresponds to a differentiation with respect to the corresponding independent variable. The principal part of the prolonged equation $D_k \Phi^a_\ell$ is of the form $\eta^{ij}(\partial_{x^i x^j x^k} A^a_\ell - \partial_{x^i x^\ell x^k} A^a_j)$ and thus consists of dn^2 equations. Contraction with $\eta^{k\ell}$ yields obviously zero and we have at least d linear dependencies between rows of the symbol matrix M_3 (corresponding to the d identities above) and its maximal rank is $d(n^2-1)$. On the other hand, the number of multiplicative variables of $\mathcal{N}_2$ is a lower bound for this rank and in our case $n\beta_2^{(n)} + (n-1)\beta_2^{(n-1)} = d(n^2-1)$. Hence the symbol $\mathcal{N}_2$ is involutive by Proposition 7.1.24. We already determined in Example 2.4.3 that $\mathcal{R}_2^{(1)} = \mathcal{R}_2$. Hence the Yang–Mills equations are involutive.

Note that for this result the concrete form of neither the metric η^{ij} nor the structure constants C^a_{bc} plays a role! In particular, it is independent of the signature of the metric and remains valid in curved space-times. In the analysis of the symbol the structure constants do not appear at all; the formal integrability requires only that they satisfy the Jacobi identity. This observation also implies that the dimensions of the non-vanishing Spencer cohomology groups are independent of both metric and structure constants. Following Remark 7.1.27 we find:

$$\dim H^{1,1}(\mathcal{N}) = \beta_2^{(n)} + \beta_2^{(n-1)} = dn\,, \tag{7.76a}$$

$$\dim H^{1,2}(\mathcal{N}) = \beta_2^{(n-1)} = d\,. \tag{7.76b}$$

Again the first dimension is just the number of differential equations in our system (which here coincides with the number of dependent variables) and the second dimension the number of (Noether) identities. ◁

Example 7.5.4. In the same manner one may analyse the Einstein equations and give an alternative proof that their symbol is involutive. In locally geodesic coordinates the matrix of the principal symbol has the entries

$$\begin{aligned} T^{ij}_{k\ell}[\chi] = \chi_m \Big[&\eta^{jm}(\delta^i_\ell \chi_k + \delta^i_k \chi_\ell) + \eta^{im}(\delta^j_\ell \chi_k + \delta^j_k \chi_\ell) - \\ &2\eta^{ij}\chi_k\chi_\ell - \eta^{mn}\chi_m\chi_n(\delta^i_k\delta^j_\ell + \delta^i_\ell\delta^j_k)\Big]\,. \end{aligned} \tag{7.77}$$

Due to the symmetries of the metric and the Ricci tensor, T is a square matrix with $n(n+1)/2$ rows and columns; the former ones are labelled by the indices k, ℓ, the latter ones by i, j. One easily verifies the identities

$$\eta^{m\ell}\left(T^{ij}_{km}[\chi]\chi_\ell - \frac{1}{2}T^{ij}_{m\ell}[\chi]\chi_k\right) = 0\,. \tag{7.78}$$

They imply that the maximal rank of T cannot exceed $n(n-1)/2$. But we know already from Example 7.1.26 that in locally geodesic coordinates the index $\beta_2^{(n)}$ has this value. Thus no further identities exist and the coordinates are δ-regular.

We may again interpret the identities (7.78) as relations between the rows of the prolonged symbol matrix and a count similar to the previous example proves involution of the symbol. Note that this time (7.78) represents the principal symbol of the contracted Bianchi identity. Thus our calculations in locally geodesic coordinates yield an indirect proof of them. ◁

It is really surprising that one has difficulties finding in the literature rigorous definitions for such basic notions as under- and overdeterminacy. Often one uses the simple counting familiar from linear algebra: how many equations and how many unknown functions are there? Unfortunately, this method easily leads to wrong results, as the next example shows. The main point is that here a definition and a criterion are confused. In linear algebra, a linear system of equations is called underdetermined, if some components of the solution vector are not restricted by the system, i. e. may be chosen arbitrarily. It is trivial to show that this happens, if and only if the system comprises less independent equations than unknowns. For differential equations the situation is more complicated, as the "only if" is no longer true.

Example 7.5.5. We consider the $U(1)$-Yang–Mills equations on a two-dimensional space-time (i. e. electrodynamics in one space dimension):

$$\mathcal{R}_2 : \begin{cases} u_{tt} - v_{xt} = 0 \,, \\ u_{xt} - v_{xx} = 0 \,. \end{cases} \tag{7.79}$$

Obviously, we have a system with two equations and two unknown functions $u(x,t)$ and $v(x,t)$ (corresponding to A_2 and A_1). Thus one might expect that it should be a well-determined system; however, (7.79) is underdetermined! This can be seen as follows. One easily checks that the system (7.79) is invariant under the transformation[6] $u \longrightarrow u + \partial_x \Lambda$ and $v \longrightarrow v + \partial_t \Lambda$ where Λ is an arbitrary function of the independent variable x and t. Thus we may choose either u or v arbitrarily and still find a solution of (7.79). This behaviour represents precisely what we expect of an under- and not of a well-determined system.

In fact, all Yang–Mills equations in any dimension of space-time are underdetermined, as they remain invariant under the transformation

$$\bar{A}_i^a = A_i^a + \partial_{x^i} \Lambda^a + C_{bc}^a A_i^b \Lambda^c \tag{7.80}$$

where the number of arbitrary functions $\Lambda^a(\mathbf{x})$ equals the dimension of the underlying Lie group $\mathcal{G}$. In differential forms language we may write $\bar{\mathbf{A}} = \mathbf{A} + \mathrm{d}\boldsymbol{\Lambda}$ where the function $\boldsymbol{\Lambda}$ takes its values in the Lie algebra $\mathfrak{g}$ of $\mathcal{G}$. ◁

With the help of the principal symbol we can give rigorous definitions of under-, over-, and well-determined equations. They reduce the problem to linear algebra. In concrete computations with local coordinates the classification requires only the analysis of a parametrised matrix, essentially the determination of its rank.

Definition 7.5.6. We call the involutive differential equation $\mathcal{R}_q \subseteq J_q\pi$ *underdetermined* at a point $\rho \in \mathcal{R}_q$, if no one-form $\chi \in T^*_{\pi^q(\rho)}\mathcal{X}$ exists such that the principal symbol τ_χ of $\mathcal{R}_q$ is injective at ρ. If a one-form χ exists such that τ_χ is bijective at the point ρ, the differential equation $\mathcal{R}_q$ is there *well-determined* or *normal*. In any other case $\mathcal{R}_q$ is called *overdetermined* at ρ.

These definitions are rather abstract, but we will see in Section 8.2 that they indeed capture the intuitive idea of under- or overdeterminacy, respectively. In particular, it will turn out that a differential equation is underdetermined, if and only if some of the unknown functions $\mathbf{u}$ appearing in it may be chosen arbitrarily and one can still find a solution of it (as in Example 7.5.5 above). For practical purposes, we

[6] This transformation represents a simple example of a *gauge symmetry*; more on this topic is contained in Section 8.3.

provide now simple criteria in terms of the highest index $\beta_q^{(n)}$ of the symbol $\mathcal{N}_q$ (of course computed in δ-regular coordinates).

Proposition 7.5.7. *An involutive differential equation $\mathcal{R}_q$ in n independent and m dependent variables is underdetermined, if and only if $\beta_q^{(n)} < m$. It is normal, if and only if $\beta_q^{(n)} = m$, all other indices vanish and $\mathcal{R}_{q-1}^{(1)} = J_{q-1}\pi$.*

Proof. By definition, $\mathcal{R}_q$ is underdetermined, if and only if no one-form χ exists such that rank $\tau_\chi = m$. By Theorem 7.5.1, this happens if and only if $\beta_q^{(n)} < m$. Similarly, we get the value of $\beta_q^{(n)}$ for a normal equation. It follows from the bijectivity of the principal symbol that in this case all other indices vanish, as a local representation cannot contain any further equations. Indeed, they would be of lower order and thus contribute zero rows to the matrix of the principal symbol. Conversely, the condition on the projection $\mathcal{R}_{q-1}^{(1)}$ ensures that no lower-order equations are present and the assumed values of the indices $\beta_q^{(k)}$ imply that at least for one one-form $\chi \in T^*\mathcal{X}$ the matrix $T[\chi]$ is regular. □

Based on this result it is very easy to characterise the local form of normal equations. In the case of an ordinary differential equation, normality excludes the presence of lower-order equations. Thus a first-order normal ordinary differential equation possesses local representations of the form $\dot{\mathbf{u}} = \mathbf{f}(x, \mathbf{u})$, i. e. each equation is solved for a derivative and we have one equation for each derivative. In the case of a partial differential equation we rename x^n as t. Then a normal first-order equation takes locally the *Cauchy–Kovalevskaya form* $\mathbf{u}_t = \mathbf{f}(\mathbf{x}, t, \mathbf{u}, \mathbf{u_x})$: each equation is solved for a t-derivative and we have one equation for each t-derivative. Thus, as expected, normality requires more than just the same number of equations and unknown functions; a distinguished independent variable t must exist such that we can write the system in the above special form.

Recalling Definition 7.1.13 of a characteristic one-form, we see that for underdetermined equations any one-form $\chi \in T^*\mathcal{X}$ is characteristic. For all other equations characteristic and systatic are equivalent and we may say that a necessary condition for a δ-regular coordinate system is that dx^n is not characteristic. Indeed, if δ-singular coordinates appear in applications, the singularity is mostly due to the use of characteristic coordinates. A trivial example is the wave equation in the form $u_{xy} = 0$; it represents just the differential equation version of Example 3.1.16.

Example 7.5.8. We continue with the two-dimensional $U(1)$-Yang–Mills equations (Example 7.5.5). The matrix of the principal symbol is

$$T[\chi] = \begin{pmatrix} (\chi_1)^2 & -\chi_1\chi_2 \\ \chi_1\chi_2 & -(\chi_2)^2 \end{pmatrix} . \tag{7.81}$$

If we multiply the first row by χ_2 and the second by χ_1, we obtain the same. Thus an upper bound for the rank of T is 1 and it takes indeed this value for every non-vanishing one-form χ. Hence $\tilde{\beta}_2^{(2)} = 1$ and it is not possible to find a coordinate system in which both equations of (7.79) are of class 2. The principal symbol is never injective and every one-form is characteristic. Thus $\mathcal{R}_2$ is indeed underdetermined, although there are as many equations as dependent variables. ◁

Note that Definition 7.5.6 applies only to *involutive* equations. Otherwise, no reasonable definitions of under- or overdeterminacy can be given, as hidden integrability conditions may change the character of the equation. We must be sure that all equations possibly affecting the principal symbol have been exhibited.

Example 7.5.9. We augment (7.79) by the first-order equation $v_t - u_x = 0$ (this condition is an example of a gauge fixing—see Section 8.3—and usually called the *Lorenz gauge*). The arising equation $\bar{\mathcal{R}}_2$ is no longer formally integrable. If we nevertheless applied Definition 7.5.6, it would still classify the equation as underdetermined: the addition of a lower-order equation only adds a row of zeros to the matrix of the principal symbol. A local representation of $\bar{\mathcal{R}}_2^{(1)}$ is

$$\bar{\mathcal{R}}_2^{(1)} : \begin{cases} u_{tt} - u_{xx} = 0\,, \quad u_{xt} - v_{xx} = 0\,, \\ v_{tt} - v_{xx} = 0\,, \quad v_{xt} - u_{xx} = 0\,, \\ \qquad v_t - u_x = 0\,. \end{cases} \tag{7.82}$$

It is not difficult to show that this differential equation is involutive. For the matrix of its principal symbol we obtain

$$T[\chi] = \begin{pmatrix} (\chi_1)^2 - (\chi_2)^2 & 0 \\ \chi_1\chi_2 & -(\chi_2)^2 \\ 0 & (\chi_1)^2 - (\chi_2)^2 \\ -(\chi_2)^2 & \chi_1\chi_2 \\ 0 & 0 \end{pmatrix} \tag{7.83}$$

and it is obviously injective for any one-form χ with $\chi_1 \neq \chi_2$. Furthermore, the principal symbol cannot be bijective and thus we are dealing with an overdetermined equation. In fact, the basic idea behind such a gauge fixing consists of reducing the arbitrariness in the solutions of a differential equation. ◁

Remark 7.5.10. In the literature one sometimes finds for linear systems another definition of overdeterminacy: the equation locally described as the kernel of the linear differential operator L_0 is overdetermined, if a linear differential operator $L_1 \neq 0$ exists such that $L_1 \cdot L_0 = 0$. The underlying idea is that a linear overdetermined system always possesses compatibility conditions; this fact follows immediately from Definition 7.5.6 and Proposition 7.2.13. Nevertheless, the two definitions of overdeterminacy are not equivalent, as the following trivial example of a linear system with three unknown functions in three independent variables demonstrates:

$$\mathcal{R}_1 : \begin{cases} u_z - v_z = 0\,, \\ w_z = 0\,, \\ w_y = 0\,. \end{cases} \tag{7.84}$$

Obviously, this system is underdetermined, as the matrix of its principal symbol

$$T[\chi] = \begin{pmatrix} \chi_3 & -\chi_3 & 0 \\ 0 & 0 & \chi_3 \\ 0 & 0 & \chi_2 \end{pmatrix} \tag{7.85}$$

has at most rank 2. Nevertheless, if we set $L_1 = (0\ \partial_y\ \partial_z)$, then $L_1 \cdot L_0 = 0$ for the linear differential operator L_0 corresponding to our system and thus according to the mentioned definition the system would be classified as overdetermined.

This discrepancy has a simple resolution. A closer look at (7.84) shows that the system naturally decouples into a subsystem for u, v and a second one for w. The first subsystem is trivially underdetermined, as we may for example choose v arbitrarily and still find solutions. It is equally obvious that the second subsystem is overdetermined in any reasonable sense. The question is now how to classify the combined system. Our definition gives precedence to underdeterminacy: whenever we can choose some unknown functions arbitrarily in the solution, we call the differential equation underdetermined (which is very natural from an intuitive point of view). Indeed, according to Definition 7.5.6 an equation is overdetermined only, if it is neither under- nor well-determined. By contrast, the above mentioned approach prefers overdeterminacy and classifies an equation as overdetermined as soon as it possesses compatibility conditions. ◁

7.6 δ-Regularity and Extended Principal Symbols

In the last section we saw that the principal symbol permits us to determine the highest index $\beta_q^{(n)}$ in an intrinsic manner. Now we want to discuss how this method can be extended to all indices $\beta_q^{(k)}$ and thus provide an alternative solution to the problem of δ-regularity. It should be mentioned that for most concrete computations the approach of Section 4.3 is probably simpler and more efficient. The main advantage of the new solution is that it does not require to perform coordinate transformations on the whole differential equation. Instead we study in an indirect manner the effect of such a transformation on the symbol.

δ-Regularity concerns Definition 7.1.22 of the indices $\beta_q^{(k)}$ of a symbol, as it is obviously coordinate dependent: in different coordinate systems we will generally obtain different values for them. Following the notation introduced in Theorem 7.5.1, we denote by $\tilde{\beta}_q^{(k)}$ those values for which all sums $\sum_{k=i}^{n} \tilde{\beta}_q^{(k)}$ with $i = 1, \dots, n$ attain their maximum. If we take into account that in a symbol all equations are of the same order, then by Remark 4.3.7 we are lead to the following "differential" version of δ-regularity.[7]

[7] If we compare with our discussion in Section 4.3, then we should actually speak here of *asymptotic* and not of δ-regularity, as our approach is closely related to Definition 4.3.4. However, it is costumary in the literature on differential equations to use the terminology δ-regularity. In the decisive case of an involutive symbol the two concepts coincide anyway.

Definition 7.6.1. Let $\mathcal{N}_q$ be the symbol of the differential equation $\mathcal{R}_q$ and let $\beta_q^{(k)}$ be the number of rows of class k in the symbol matrix M_q in a given coordinate system. Then the coordinate system is called *δ-regular* for $\mathcal{R}_q$, if $\beta_q^{(k)} = \tilde{\beta}_q^{(k)}$.

Note that this definition depends crucially on the equation under consideration. Especially after a projection, when new equations are added, the coordinate system must be checked again, even if it was δ-regular before (prolongations are of course harmless). We further see that even for differential equations it suffices to consider only linear changes of the independent variables. For the analysis of the symbol only one aspect is important: which derivatives of order q do occur after the transformation. But this question is answered by the Jacobian. Thus the linearised transformation contains already all relevant information. By the same argument as in Proposition 4.3.8, generic coordinate systems are δ-regular.

We will now introduce extended principal symbols $\tau_{[\chi^{(k)},\dots,\chi^{(n)}]}$ depending on up to n one-forms $\chi^{(\ell)} \in T^*\mathcal{X}$ such that for δ-regular coordinates $\mathbf{x}$ their matrices have maximal rank, if we choose $\chi^{(\ell)} = dx^\ell$, and such that conversely, if $\chi^{(1)},\dots,\chi^{(n)}$ are a set of one-forms which lead to maximal rank of all these matrices, then any coordinates $\bar{\mathbf{x}}$ satisfying $d\bar{x}^k = \chi^{(k)}$ are δ-regular. Furthermore, the maximal ranks will directly determine the indices $\tilde{\beta}_q^{(1)},\dots,\tilde{\beta}_q^{(n)}$.

We will follow the point of view given in Remark 7.1.9. Let $\chi^{(k)},\dots,\chi^{(n)} \in T^*\mathcal{X}$ be $n-k+1$ linearly independent one-forms. Since we are later only interested in studying the forms at a single point, we may assume without loss of generality that they define an integrable codistribution and that through any point $x \in \mathcal{X}$ an $(n-k+1)$-dimensional integral submanifold $\mathcal{Y} \subseteq \mathcal{X}$ for this codistribution exists. Considering $S_q(T^*\mathcal{Y})$ as a subspace of $S_q(T^*\mathcal{X})$, we can now generalise Definition 7.1.8 of the principal symbol as follows.

Definition 7.6.2. Let $\mathcal{Y}$ and $\chi^{(k)},\dots,\chi^{(n)}$ be as above. Let furthermore the differential equation $\mathcal{R}_q \subseteq J_q\pi$ be globally described by the map $\Phi : J_q\pi \to \mathcal{E}'$. The *k-extended principal symbol* is then the linear map

$$\tau_{[\chi^{(k)},\dots,\chi^{(n)}]} = \sigma \circ \varepsilon_q^{-1} : S_q(T^*\mathcal{Y}) \otimes V\pi \longrightarrow T\mathcal{E}' . \tag{7.86}$$

Any vector in $S_q(T^*\mathcal{Y}) \subseteq S_q(T^*\mathcal{X})$ can be expressed as a linear combination of the elements $\chi^{(\nu)}$ where $\nu \in \mathbb{N}_0^n$ is a multi index with $|\nu| = q$ and $\operatorname{cls}\nu \geq k$ and where $\chi^{(\nu)} = (\chi^{(k)})^{\nu_k}\cdots(\chi^{(n)})^{\nu_n}$. If we express these symmetric products with respect to the standard basis dx^μ of $S_q(T^*\mathcal{X})$, then we obtain $\chi^{(\nu)} = C^\nu_\mu dx^\mu$ where

$$C^\nu_\mu = \sum_{s\in\mathcal{S}_\nu}\prod_{i=1}^{q}\chi^{(s_i)}_{m_i} , \qquad m = r(\mu) . \tag{7.87}$$

Therefore the matrix of the k-extended principal symbol is

$$T^{\tau\nu}_\alpha[\chi^{(k)},\dots,\chi^{(n)}] = \sum_{|\mu|=q}\left(\frac{\partial\phi^\tau}{\partial u^\alpha_\mu}\right)C^\nu_\mu , \qquad \text{with } \operatorname{cls}\nu \geq k\,,\ |\nu| = q\,. \tag{7.88}$$

It has p rows indexed by τ and mr_k columns indexed by α and ν where p is again the dimension of $\mathcal{E}'$ and $r_k = \dim S_q(T^*\mathcal{Y}) \otimes V\mathcal{E} = \binom{n-k+q}{n-k}$. The n-extended principal symbol is the usual one as defined in Section 7.1. If $\dim \mathcal{Y} = n$, we may assume that $\mathcal{Y} = \mathcal{X}$ and the 1-extended principal symbol as the largest one has the same size as the geometric symbol. Indeed, we will later see that we can identify it with the geometric symbol in a generic coordinate system.

Lemma 7.6.3. *The matrix of the k-extended principal symbol contains the matrices of all ℓ-extended symbols with $\ell > k$ as submatrices.*

Proof. We may view the matrix $T[\chi^{(k)}, \dots, \chi^{(n)}]$ as a horizontal concatenation of r_k matrices with p rows and m columns, each generated in a similar manner as the principal symbol $T[\chi]$ but with different mappings ι_χ. Our claim follows from the fact that in (7.88) all multi-indices with $\operatorname{cls} \nu \geq k$ are taken into account. This includes of course all multi-indices with $\operatorname{cls} \nu \geq \ell$ for $\ell > k$. The definition (7.87) of the combinatorial factor C^ν_μ shows that for such multi-indices the entries of the k-extended principal symbol do not depend on $\chi^{(k)}, \dots, \chi^{(\ell-1)}$. Thus we obtain all columns of the ℓ-extended principal symbol. □

Theorem 7.6.4. *Let $\mathcal{R}_q$ be a differential equation with indices $\tilde{\beta}_q^{(1)}, \dots, \tilde{\beta}_q^{(n)}$. Then for $1 \leq k \leq n$*

$$\max_{\chi^{(k)}, \dots, \chi^{(n)}} \operatorname{rank} \tau_{[\chi^{(k)}, \dots, \chi^{(n)}]} = \sum_{i=k}^{n} \tilde{\beta}_q^{(i)} . \tag{7.89}$$

Proof. The proof is a straightforward generalisation of the one of Theorem 7.5.1. In a δ-regular coordinate system there exists a local representation of $\mathcal{R}_q$ such that the first $\tilde{\beta}_q^{(n)}$ equations are solved for derivatives of class n, the next $\tilde{\beta}_q^{(n-1)}$ equations are solved for derivatives of class $n-1$ and so on (e. g. the Cartan normal form of $\mathcal{R}_q$). We order the columns of the k-extended principal symbol in such a way that the first m columns represent the n-extended principal symbol, the first mr_{n-1} columns the $(n-1)$-extended one and so on. This is always possible by Lemma 7.6.3.

Choosing $\chi^{(k)} = dx^k, \dots, \chi^{(n)} = dx^n$, we can easily evaluate the factors C^ν_μ. They vanish whenever $\mu \neq \nu$ and are one otherwise. Thus we have eliminated all contributions to the k-extended principal symbol by derivatives whose class is less than k. If the principal derivative of equation τ is u^α_μ, then in row τ the column indexed by α and μ contains the first non-vanishing entry. Since our local representation was chosen in such a way that the symbol is in solved form, all principal derivatives are different and we have at least $\sum_{i=k}^{n} \tilde{\beta}_q^{(i)}$ linearly independent rows.

On the other hand, assume that there are one-forms $\chi^{(k)}, \dots, \chi^{(n)}$ such that the rank of the k-extended principal symbol is larger than the claimed value. Then we apply a coordinate transformation $\bar{x}^j = \psi^j(x)$ where the functions ψ^j satisfy at an arbitrary but fixed point $x_0 \in \mathcal{X}$

$$\frac{\partial \psi^j}{\partial x^i}(x_0) = \chi^{(j)}_i , \quad j = k, \dots, n . \tag{7.90}$$

As a maximal rank of the k-extended principal symbol can only be achieved, if the one-forms $\chi^{(k)},\dots,\chi^{(n)}$ are linearly independent, such a transformation always exists. Its effect on the symbol is determined by the highest-order part of the transformation law for the derivatives of order q. It is given by

$$u^\alpha_\mu = \sum_{|\nu|=q} \bar{u}^\alpha_\nu \left(\sum_{s\in\mathcal{S}_\nu} \prod_{i=1}^{g} \frac{\partial\psi^{s_i}}{\partial x^{m_i}} \right), \qquad |\mu| = |\nu| = q,\ m = r(\mu)\,, \tag{7.91}$$

where the bar denotes derivatives with respect to the new coordinates $\bar{\mathbf{x}}$. Since we are interested in the values of the indices $\beta_q^{(k)},\dots,\beta_q^{(n)}$ in this new coordinate system, we only consider the following derivatives

$$\frac{\partial\bar{\Phi}^\tau}{\partial\bar{u}^\alpha_\nu} = \sum_{|\mu|=q} \left(\frac{\partial\Phi^\tau}{\partial u^\alpha_\mu} \right) \frac{\partial u^\alpha_\mu}{\partial\bar{u}^\alpha_\nu}\,, \qquad \operatorname{cls}\nu \geq k\,. \tag{7.92}$$

At the chosen point x_0 we can evaluate the derivatives of the function ψ^j occurring on the right hand side by plugging in (7.91). Comparing (7.92) and (7.88), we immediately recognise the k-extended principal symbol on the right hand side. Thus we have found a coordinate system in which we have more equations of class greater than or equal to k than given by $\sum_{i=k}^n \tilde{\beta}_q^{(i)}$. But this fact contradicts the definition of the indices $\tilde{\beta}_q^{(k)},\dots,\tilde{\beta}_q^{(n)}$. □

Definition 7.6.5. The linearly independent one-forms $\chi^{(1)},\dots,\chi^{(n)} \in T^*\mathcal{X}$ form a *non-systatic basis* of $T^*\mathcal{X}$ for the differential equation $\mathcal{R}_q$, if

$$\operatorname{rank}\tau_{[\chi^{(k)},\dots,\chi^{(n)}]} = \sum_{i=k}^{n} \tilde{\beta}_q^{(i)}\,, \qquad k = 1,\dots,n\,. \tag{7.93}$$

Coordinates $\mathbf{x}$ such that $dx^i = \chi^{(i)}$ are called an *associated coordinate system.*

Different associated coordinate systems differ only by constants, namely the choice of an origin. Note that in this definition only those $\chi^{(k)}$ where $\tilde{\beta}_q^{(k)} > 0$ really matter. For the remaining ones one can choose any one-forms which complete to a basis of $T^*\mathcal{X}$. Theorem 7.6.4 implies now immediately the following result.

Corollary 7.6.6. *A local coordinate system on $\mathcal{X}$ is δ-regular for the differential equation $\mathcal{R}_q$, if and only if it is associated to a non-systatic basis of $T^*\mathcal{X}$.*

From (7.91) and (7.92) we also see that if we write down the symbol in a generic coordinate system, i.e. in coordinates $\bar{x}^j = \sum a^j_i x^i$ with undetermined coefficients a^j_i, then we get the 1-extended principal symbol after identifying a^j_i with $\chi_i^{(j)}$. Thus the extended principal symbols allow us to introduce such generic coordinates step by step. Since we often must change only some of the coordinates, this observation can save considerable time in concrete computations, as one rarely has to go until the 1-extended principal symbol. Note furthermore that Lemma 7.6.3 allows us to organise the computation in an efficient way: if we calculate the rank of the k-extended principal symbol as column rank, we can keep the results from the $(k+1)$-extended principal symbol as we only add further columns to the matrix.

Example 7.6.7. We demonstrate the use of the extended principal symbol with a differential equation stemming from a concrete application [247]. The analysis of moving pseudo-spherical surfaces in $\mathbb{R}^3$ leads to the *Bianchi equation*, a third-order equation of the form:

$$u_{xyt} - u_{yt}u_x \cot u + u_{xt}u_y \tan u = 0\,, \tag{7.94a}$$

$$\left(\frac{u_{xt}}{\cos u}\right)_x - K(K\sin u)_t - \frac{u_{yt}u_y}{\sin u} = 0\,, \tag{7.94b}$$

$$\left(\frac{u_{yt}}{\sin u}\right)_y + K(K\cos u)_t - \frac{u_{xt}u_x}{\cos u} = 0\,, \tag{7.94c}$$

where K is a given function of t. Obviously the used coordinate system is δ-singular no matter how we order the coordinates, for there is no derivative of class 3. We will use the ordering $x^1 = t$, $x^2 = x$ and $x^3 = y$. Since each equation in (7.94) contains only one third-order derivative, the symbol takes after some trivial manipulations the simple form

$$v_{xyt} = 0\,, \qquad v_{xxt} = 0\,, \qquad v_{yyt} = 0\,. \tag{7.95}$$

Now the matrix of the 3-extended principal symbol is readily computed:

$$T[\chi^{(3)}] = \begin{pmatrix} \chi_1^{(3)}\chi_2^{(3)}\chi_3^{(3)} \\ \chi_1^{(3)}(\chi_2^{(3)})^2 \\ \chi_1^{(3)}(\chi_3^{(3)})^2 \end{pmatrix}. \tag{7.96}$$

We find two families of characteristic one-forms

$$\omega_1 = \alpha\, dx^1\,, \qquad \omega_2 = \beta\, dx^2 + \gamma dx^3\,. \tag{7.97}$$

This result implies especially that the three coordinate forms dx^1, dx^2, dx^3 are all characteristic. Thus it is not surprising that the coordinate system is not δ-regular. The maximal rank of the 3-extended principal symbol is 1 and obtained e.g. with the choice $\chi^{(3)} = dx^1 + dx^3$.

Next we compute the 2-extended principal symbol. There exist four third-order derivatives of class 2, namely $[0,0,3]$, $[0,1,2]$, $[0,2,1]$ and $[0,3,0]$. Hence its matrix has four columns:

$$T[\chi^{(2)},\chi^{(3)}] = \begin{pmatrix} \chi_1^{(3)}\chi_2^{(3)}\chi_3^{(3)} & \chi_1^{(2)}\chi_2^{(3)}\chi_3^{(3)}+\chi_1^{(3)}\chi_2^{(2)}\chi_3^{(3)}+\chi_1^{(3)}\chi_2^{(3)}\chi_3^{(2)} & \cdots \\ \chi_1^{(3)}(\chi_2^{(3)})^2 & \chi_1^{(2)}(\chi_2^{(3)})^2+2\chi_1^{(3)}\chi_2^{(3)}\chi_2^{(2)} & \cdots \\ \chi_1^{(3)}(\chi_3^{(3)})^2 & \chi_1^{(2)}(\chi_3^{(3)})^2+2\chi_1^{(3)}\chi_3^{(3)}\chi_3^{(2)} & \cdots \end{pmatrix}$$

$$\begin{pmatrix} \cdots & \chi_1^{(2)}\chi_2^{(2)}\chi_3^{(3)}+\chi_1^{(2)}\chi_2^{(3)}\chi_3^{(2)}+\chi_1^{(3)}\chi_2^{(2)}\chi_3^{(2)} & \chi_1^{(2)}\chi_2^{(2)}\chi_3^{(2)} \\ \cdots & \chi_1^{(3)}(\chi_2^{(2)})^2+2\chi_1^{(2)}\chi_2^{(2)}\chi_2^{(3)} & \chi_1^{(2)}(\chi_2^{(2)})^2 \\ \cdots & \chi_1^{(3)}(\chi_3^{(2)})^2+2\chi_1^{(2)}\chi_3^{(2)}\chi_3^{(3)} & \chi_1^{(2)}(\chi_3^{(2)})^2 \end{pmatrix} . \tag{7.98}$$

One could compute the maximal rank of this matrix using, say, Gaussian elimination, but it is much easier to substitute the above mentioned non-characteristic one-form $\chi^{(3)} = \mathrm{d}x^1 + \mathrm{d}x^3$. This yields the following, much simpler matrix:

$$T[\chi^{(2)},dx^1+dx^3] = \begin{pmatrix} 0 & \chi_2^{(2)} & \chi_1^{(2)}\chi_2^{(2)}+\chi_2^{(2)}\chi_3^{(2)} & \chi_1^{(2)}\chi_2^{(2)}\chi_3^{(2)} \\ 0 & 0 & (\chi_2^{(2)})^2 & \chi_1^{(2)}(\chi_2^{(2)})^2 \\ 1 & \chi_1^{(2)}+2\chi_2^{(2)} & (\chi_3^{(2)})^2+2\chi_1^{(2)}\chi_3^{(2)} & \chi_1^{(2)}(\chi_3^{(2)})^2 \end{pmatrix} . \tag{7.99}$$

One sees at once that the rank of this matrix is 3 whenever $\chi_2^{(2)} \neq 0$. Thus we get $\tilde{\beta}_3^{(3)} = 1$ and $\tilde{\beta}_3^{(2)} = 2$ and we have also found a non-systatic basis of $T^*\mathcal{X}$

$$\chi^{(3)} = \mathrm{d}x^1 + \mathrm{d}x^3\,, \qquad \chi^{(2)} = \mathrm{d}x^2\,, \qquad \chi^{(1)} = \mathrm{d}x^1\,. \tag{7.100}$$

An associated δ-regular coordinate system is given by $\bar{x}^3 = y+t$, $\bar{x}^2 = x$ and $\bar{x}^1 = t$. This yields the following form for the symbol

$$\bar{v}_{333} + \bar{v}_{331} = 0\,, \qquad \bar{v}_{332} + \bar{v}_{321} = 0\,, \qquad \bar{v}_{322} + \bar{v}_{221} = 0\,, \tag{7.101}$$

where one easily can directly read off the indices. ◁

7.7 Notes

The principal symbol is a very classical and fundamental concept in the theory of partial differential equations. It is discussed in any advanced textbook. Often it is introduced via Fourier analysis and forms the basis of the classification into elliptic and hyperbolic equations. If one computes the Fourier transform of a linear differential equation, one obtains a polynomial called the symbol of the equation. The principal symbol is its highest degree part. Note, however, that the full symbol is not intrinsically defined but only the principal symbol. For this reason the classical full symbol does not appear in the geometric theory of differential equations.

A classical method to derive the principal symbol in an intrinsic manner goes as follows [453, Chapt. 2, Sect. 9]. We consider for notational simplicity only a scalar linear differential operator $L = \sum_{|\mu|\leq q} L_\mu(\mathbf{x})D^\mu$ of order q acting on one unknown function u. Its principal symbol is in our notation the homogeneous polynomial $T[\chi] = \sum_{\mu=q} L_\mu(\mathbf{x})\chi^\mu$ parametrised by a one-form $\chi \in T^*\mathcal{X}$. Given some real-valued function $\psi \in \mathcal{F}(\mathcal{X})$, we have

$$T[\mathrm{d}\psi]u = \lim_{\lambda\to\infty} \lambda^{-q} e^{-i\lambda\psi} L\big(e^{i\lambda\psi}u\big)\,. \tag{7.102}$$

Indeed, $L\big(e^{i\lambda\psi}u\big) = \big(\lambda^q T[\mathrm{d}\psi]u + r(\mathbf{x},\lambda)\big)e^{i\lambda\psi}$ where r is a polynomial in λ of degree less than q, as one easily verifies by a straightforward differentiation.

The geometric symbol is a natural object given the internal structures of the jet bundles. Its definition is in a certain formal analogy to the definition of the graded ring associated with a filtered ring. Indeed, thinking of jets as truncated power series, it becomes obvious that we must find homogeneous polynomials (or in an intrinsic language the symmetric algebra) within the jet hierarchy.

We remarked already at the end of the last chapter that many of the ideas around involution have their origin in the theory of differential equations (although we presented the theory first in algebraic form). But the Janet–Riquier theory mentioned there was not the only approach to deal with general systems of differential equations. A completely intrinsic theory based on a representation by an exterior ideal of differential forms was developed by Cartan [74] and Kähler [248]; a modern presentation of their work is contained in [58, 231]. The Cartan–Kähler Theory is in particular very natural for many problems in differential geometry.

Within Cartan–Kähler Theory, the distinction between formally integrable and involutive equations does not appear so clearly as in our approach. In fact, the term formal integrability is usually not employed at all. This fact leads sometimes to confusion and one can find in the literature erroneous claims that our notion of formal integrability was equivalent to Cartan's notion of an involutive exterior system. We will see repeatedly in the following chapters that formal integrability is not sufficient for a deeper analysis; for many purposes one really needs the more advanced concept of an involutive differential equation.

We mentioned in the Notes to Chapter 2 that the theory of exterior differential systems does not require a fibred manifold as basis, i. e. there is no need for a distinction into dependent and independent variables. However, the notion of an involutive exterior system is defined with respect to *independence forms*, a set of n linearly independent one-forms. In the case of a fibred manifold $\pi : \mathcal{E} \to \mathcal{X}$ one simply chooses a basis of $\pi^*(T^*\mathcal{X})$ as independence forms and then the two notions of involution become equivalent. But it is worth while noting that the Cartan–Kähler Theory can handle more general situations.

The finite criterion for formal integrability given by Theorem 7.2.8 is due to Goldschmidt [168] who gave a direct proof without using the duality between Spencer cohomology and Koszul homology. Our proof via syzygies is of a more computational nature. In Example 2.3.17 we saw that for differential equations of finite type, integrability conditions may be considered as obstructions to the flatness of the underlying connection. Hence they correspond to the curvature of this connection. Generalising this idea, Goldschmidt [168] introduced a notion of curvature for arbitrary differential equations $\mathcal{R}_q$; it is also discussed by Pommaret [356].

We described the completion process for ordinary differential equations first in terms of the tangent bundle, as in this form it has been rediscovered many times; see e. g. [320, 332, 368, 372, 448]. In some of these works the authors also treat non-autonomous systems within the tangent bundle. This requires slightly awkward constructions. Usually, the time variable is included in the configuration space, so that one deals with the manifold $\mathcal{U} \times \mathbb{R}$. Then one defines the "restricted" tangent bundle as the set of all those vectors in $T(\mathcal{U} \times \mathbb{R})$ which have a 1 in the last component. The jet bundle offers a much more natural framework for such systems and the restricted tangent bundle is easily recovered with the contact map. Our intrinsic proof of the equivalence of the tangent and the jet bundle approach to the completion of Hamiltonian equations was given in [134].

In our discussion we assumed (as throughout this book) that the differential equation is indeed represented by a smooth manifold. Already in the case of equations with polynomial nonlinearities this assumption requires that we exclude all singularities. Pritchard [362] gave a formulation of the completion process for implicit autonomous ordinary differential equations described by polynomial equations based on ideas from commutative algebra and algebraic geometry. The tangent bundle is then replaced by the tangent variety, but the basic idea that completion consists of imposing tangency conditions remains valid.

In the theory of differential algebraic equations, the number of tangency conditions required for the completion is known as the *(differentiation) index* of the equation. A discussion of various index concepts (also for partial differential equations) from the point of view of the formal theory is contained in [402].

In the Addendum to Section 7.3 we discussed only one aspect of the Dirac theory of constrained Hamiltonian systems, namely the Dirac algorithm, i. e. the reformulation of the geometric completion process in terms of Poisson brackets with the Hamiltonian. The full Dirac theory goes beyond completion, as it also introduces a classification of the obtained constraints. Once the completion has been done, one distinguishes *first-class constraints* possessing vanishing Poisson brackets with all other constraints and *second-class constraints* which are the remaining ones.

From a geometric point of view, this classification has the following meaning. If we consider the submanifold $\mathcal{S}$ of the full phase space $\mathcal{U}$ given by the zero set of the second-class constraints only, then the symplectic two-form ω on $\mathcal{U}$ induces a symplectic two-form on $\mathcal{S}$, i. e. $\mathcal{S}$ is again a symplectic manifold. The Poisson bracket corresponding to the induced symplectic structure is usually called the *Dirac bracket*.[8] By contrast, the submanifold defined by the first-class constraints does not carry a symplectic structure, as the two-form induced by ω is always degenerate. First-class constraints are the infinitesimal generators of *gauge symmetries*, i. e. they define transformations of the phase space under which the given Hamiltonian system remains invariant. A *gauge fixing* consists of the addition of further constraints that

[8] Strictly speaking, the Dirac bracket defines a degenerate Poisson bracket on the full phase space which coincides on the constraint manifold with the induced bracket. For an intrinsic geometric discussion see [424]; some numerical applications can be found in [403].

render the first-class ones second-class; thus they remove the arbitrariness due to the gauge symmetry.

Dirac's motivation for the study of constrained Hamiltonian systems was their *quantisation*. Most classical field theories like electromagnetism or relativity are not normal due to the presence of gauge symmetries, so that a priori it is not clear how their quantisation should proceed. As a discussion of the physical aspects of constrained systems is outside the scope of this book, we refer for an extensive treatment to the textbooks [201, 443].

Dirac's work started also a rather unfortunate "tradition" in the physical literature on constrained dynamics: very often results are rigorously stated and proved only in the finite-dimensional case, i. e. for ordinary differential equations, but then applied without hesitations to field theories. In [134, 411] one can find concrete examples of the type of problems that may arise in such an approach. In particular, a straightforward generalisation of Dirac's approach to field theories leads to an algorithm that potentially overlooks hidden constraints. Furthermore, we have seen in Section 7.4 that the completion of partial differential equations leads generally to an increase of the order; but a Hamiltonian formulation should always be of first-order. It is probably only fair to say that we still have only a moderate understanding of constrained Hamiltonian field theories.

We considered in the Addendum only constrained Hamiltonian systems. Indeed, it is common in physics to study constrained systems in the Hamiltonian formalism, as it forms the starting point for the quantisation of systems. But of course one may use also other formalisms; e. g. a Lagrangian discussion is given by Sudarshan and Mukunda [441]. From the point of view of exhibiting all hidden constraints, one can always resort to the general completion procedure described in Section 7.3 which can be applied independent of any particular formalism used for setting up the equations of motion.

The Cartan–Kuranishi Theorem has its origin in the work of Cartan on exterior differential systems. Cartan recognised that in general prolongations are necessary in the completion; however, he was not able to prove that every consistent exterior system becomes involutive after a *finite* number of prolongations. The first complete proof of this fact is due to Kuranishi [274] (again in the language of exterior systems). Within the formal theory of differential equations one has a stronger separation between the geometric and the algebraic aspects of the completion and the question of how many prolongations are needed can be treated completely algebraically in the framework of the Spencer cohomology of the symbol comodule. As shown in Section 6.1, the finiteness becomes then trivial due to the duality to the Koszul homology.

For many applications one would like to have a simple algorithm to determine directly all integrability conditions. We saw in Example 7.4.3 that their knowledge significantly simplifies the explicit integration of the differential equation. In principle, one could design such an algorithm on the basis of the second Spencer homology groups $H^{r,2}(\mathcal{N})$. However, to our knowledge nobody has done this so far. All approaches to the completion of differential equations construct more than just the integrability conditions (e. g. also all obstructions to involution) and they differ in what this "more" is. Most of them have been studied in the context of Lie symmetry theory [48, 342] for the treatment of the determining systems, which are typically large overdetermined linear systems of partial differential equations. Hence a number of references to theoretical works and implementations can be found in the review articles of Hereman [202, 203].

We mention here only three approaches popular in computer algebra. The probably first implementation of the *Janet–Riquier Theory* in a computer algebra system is due to Topunov [456]. Reid [373] developed a *standard form*, also inspired by the Janet–Riquier Theory, and implemented it in MAPLE. The approach is fully algorithmic only for linear equations, but this suffices for applications in symmetry theory. In the more recent work [374], he speaks of a "reduced involutive form" which is not identical with our notion of involution.

Given the importance of Gröbner bases for the study of polynomial ideals, it is not surprising that similar concepts were developed for differential equations. The easiest case are obviously linear equations where one can straightforwardly extend the classical commutative theory as we have seen in Chapter 3. For more general equations one needs differential algebra, where one studies differential ideals in the ring of differential polynomials, i. e. ideals which are closed under differentiation. In this context, Carrà Ferro [69] gave the first definition of a *differential Gröbner basis* followed later by Ollivier [341]. Around the same time, Mansfield [306] introduced an alternative and inequivalent notion of differential Gröbner basis.

However, one must be careful with the terminology "differential Gröbner basis." As shown in Appendix B.4, Gröbner bases can be characterised and thus defined in many equivalent ways. If one tries to extend these alternative definitions to the differential case, they sometimes become inequivalent. For example, Mansfield's bases cannot solve the differential ideal membership problem and thus one may argue whether they should be called Gröbner bases. A recent survey of the theory with many references was compiled by Carrà Ferro [70].

One reason for these problems is the fact that while we can still use a term order (in differential algebra it is more common to speak of a *ranking*) for selecting in each equation a leading derivative, it will generally appear in some power so that we must resort to *pseudo-reductions* (see e. g. [323]). This effect makes it harder to interpret the outcome. In particular, it is possible that one leaves the differential ideal spanned by the original system (one works then in some saturation of it), as equations must be multiplied with differential polynomials in order to make reductions possible. In the language of differential algebra one obtains a *coherent autoreduced set* and the corresponding form of the Buchberger algorithm is called *Kolchin–Ritt algorithm* [258, 383] (it would be straightforward to design an involutive version of it replacing S-polynomials by non-multiplicative prolongations).

Another reason is that the leading coefficient—the *initial* in the language of differential algebra—is now generally also a non-trivial differential polynomial with a zero set (the same is true for the *separant*, i. e. the derivative of a differential polynomial with respect to its leading derivative). On such zero sets our pseudo-reductions become incorrect so that the algorithms must make case distinctions (the situation is similar to the analysis of parametric linear systems of equations [420] or the theory of comprehensive Gröbner bases [477]). In a less computational language, this means that we must augment the theory with structural decompositions of the ideal (or more precisely of its radical): the given ideal is written as the intersection of finitely many "better behaved" ideals for which then bases are computed (one often speaks of *triangulation-decomposition algorithms*). Thus the result of such a computation is not a single basis but several bases, each consisting of equations and inequalities. This approach was pioneered by Boulier et al [51]; a more recent discussion with some improvements can be found in [228, 421].

From a geometric point of view, the need for such decompositions can be understood as follows. In our discussion of the Cartan–Kuranishi completion we implicitly assumed that many objects behave uniformly over the treated differential equations. This assumption concerns in particular the ranks of certain Jacobians or the indices of the corresponding symbols. For a fully non-linear differential equation such a uniform behaviour is, however, exceptional. Thus one is forced to make case distinctions depending on whether or not certain expressions vanish. Therefore in reality the geometric completion process also does not end with a single involutive equation but with a finite number of involutive equations each locally described by a system comprising also inequalities.[9]

Finally, we must mention as third approach the Cartan–Kähler Theory of exterior differential systems [58, 74, 248, 326]. Its first implementation is probably due to Arajs et al [24]. An extensive implementation of it has later been provided by Hartley and Tucker [194] (see also [192, 193]).

A high worst case complexity features as a common problem for all these approaches. As already mentioned, this fact does not come very surprising, as most of them comprise Gröbner basis computations as a special case. Thus the complexity results obtained for Gröbner bases (see Appendix B.4) represent a lower bound for the completion in the differential case. Besides this fairly simple observation not much is known about the complexity of all the mentioned completion algorithms.

Our definition of an underdetermined equation is more precise than the one given by Pommaret [356, Definition 6.6]. He defines an equation to be underdetermined, if a one-form χ exists such that the principal symbol τ_χ is surjective. While this condition covers the elementary case that any local representation comprises less equations than unknown functions, it fails for Example 7.5.5: the rank of the matrix $T[\chi]$ given by (7.81) is at most one, thus τ_χ can never be surjective.

Our approach generalises the definitions of Olver [342, Definition 2.86] for systems with the same number of equations and unknown functions. They are in so far not satisfactory, as he does not require the equation to be involutive. Thus it may happen that the addition of an integrability condition renders an underdetermined equation overdetermined or vice versa. It should be stressed again that this classification makes sense only for involutive equations!

[9] The same observation underlies the notion of a *generically involutive system* recently introduced by Malgrange [304].

Chapter 8
The Size of the Formal Solution Space

Knowing what is big and what is small is more important than being able to solve partial differential equations.

Stan Ulam

The results in this chapter should be used with some care. We introduce measures for the size of the formal solution space of a differential equation. First of all, one cannot stress strongly enough that these considerations only concerns *formal* solutions, i. e. formal power series satisfying the given equation. In contrast to Chapter 9, we do not bother here about the convergence of these series and thus the results do not imply any statement on the existence of strong solutions. So one might say that we perform in this chapter only combinatorial games. Nevertheless, there are situations where the results can be quite useful. This should be evident from the fact that scientists like Cartan and Einstein actively participated in these "games".

In the first section we briefly discuss the concept of a "general" solution. This word is often used but hardly anybody cares to give a rigorous definition. We do not give one either, as this turns out to be rather difficult and of doubtful value. So we only discuss some of the arising problems in order to motivate why we afterwards restrict to the formal solution space where the situation is much simpler.

Section 8.2 introduces the main tools of the formal theory for measuring the size of the formal solution space: the Cartan characters and the Hilbert function. We exhibit the connection between them and show how they can be explicitly determined for an involutive equation. For a classical analyst these concepts may appear to be rather abstract measures. Therefore we also discuss how—in certain circumstances—they can be translated into statements about how many functions are needed for parametrising the solution space.

In applications to physics, one is interested in subtracting the effect of gauge symmetries, as for a physicist solutions related by a gauge transformation represent the same physical state. This goal is easily achieved using a pseudogroup approach. In Section 8.3 we analyse how the Cartan characters and the Hilbert function change under certain operations on the differential equation. This analysis leads to the rather general concept of a differential relation. In an Addendum of a more "historical" nature, we show how Einstein's approach, based on what is called the strength of a differential equation, fits into our framework. It turns out that the machinery of formal theory is more powerful and flexible than his approach, as the strength corresponds to just one of the Cartan characters.

W.M. Seiler, *Involution*, Algorithms and Computation in Mathematics 24,
DOI 10.1007/978-3-642-01287-7_8, © Springer-Verlag Berlin Heidelberg 2010

8.1 General Solutions

The notion of a "general solution" or a "general integral" is notoriously difficult. Although it is intuitively clear what it is supposed to mean, it is almost impossible to give a rigorous definition. From the point of view of the functional analytic approach that has dominated the theory of partial differential equations for many decades, this fact is not surprising as obviously the answer decisively depends on the considered function spaces. The classical idea of a parameter dependent expression which yields all solutions through suitable specialisations of the parameters becomes rather meaningless, if one is interested in solutions living in some abstract function spaces.

Example 8.1.1. If one admits distributions as solutions, already for ordinary differential equations the classical counting rules fail. As a trivial example we may consider the following first-order equation $xu' = 0$ [379]. The classical point of view is that the solution space of a scalar linear first-order equation is a one-dimensional vector space. In our case a basis of this space is obviously given by $u(x) = 1$. However, because of the singularity of the coefficient at $x = 0$, the equation admits besides this *strong* solution further *weak* solutions in a distributional sense, for example $u(x) = H(x)$ where $H(x)$ denotes the Heaviside or step function defined by $H(x) = 0$ for $x < 0$ and $H(x) = 1$ for $x \geq 0$. Thus this first-order equation possesses an at least two-dimensional solution space![1] $\lhd$

This appearance of additional distributional solutions is easily explicable in terms of the symbol. In an intrinsic geometric manner, we may characterise the point $x = 0$ by the fact that there the dimension of the symbol increases. Indeed, for our example the symbol matrix is simply the scalar x and its rank drops whenever $x = 0$. At this point the behaviour of the differential equation changes abruptly, thus it is not surprising that we also find solutions with abrupt changes.

Another problem with the concept of a general solution is the existence of *singular integrals*. They appear only in non-linear equations. The word "singular" does not refer to singularities in the solution but to the fact that such solutions cannot be obtained from a general solution; typically they arise as envelopes of families of solutions. Extensive discussions of the relations between singular and general integrals are contained in most old textbooks on differential equations. We restrict ourselves here to give a few examples of ordinary differential equations with singular solutions. A slightly deeper discussion is contained in Section 9.1.

Example 8.1.2. A *Clairaut equation* is of the form $u = xu' + f(u')$ for some function $f : \mathbb{R} \to \mathbb{R}$ whose second derivative f'' vanishes nowhere. Such equations were first introduced by Clairaut [92] in the study of the motion of rectangular wedges. One readily checks that $u(x) = cx + f(c)$ defines a one-parameter family of solutions which one would classically call the general solution. However, in addition to this one-parameter family, we find a singular integral geometrically represented

[1] An algebraic approach to handle such situations will be presented in Section 10.5.

by the curve $x(\tau) = -f'(\tau)$ and $u(\tau) = -\tau f'(\tau) + f(\tau)$. The parameter τ is here nothing but the value of the derivative u'. Because of our assumption that the second derivative f'' vanishes nowhere, this curve is indeed the graph of a function $u(x)$.

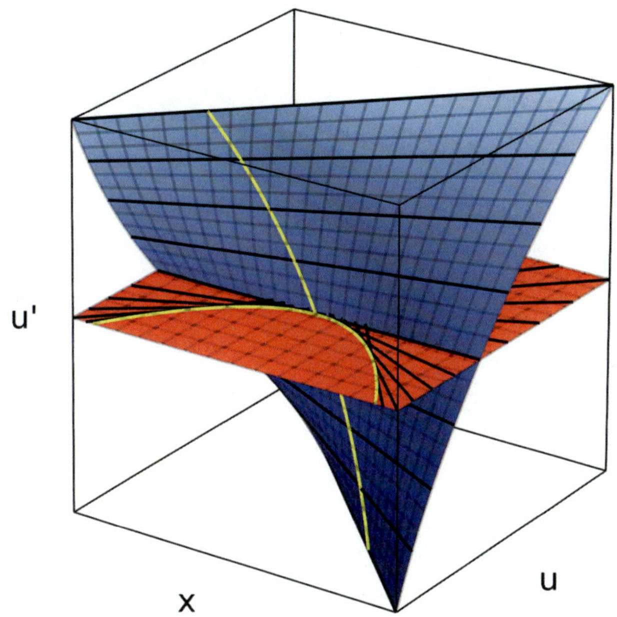

Fig. 8.1 Singular solution of a Clairaut equation

Figure 8.1 shows the manifold $\mathcal{R}_1$ corresponding to the Clairaut equation obtained for the choice $f(s) = -\frac{1}{4}s^2$. The singular solution is here a parabola and one can easily see in the x-u plane that it is indeed the envelope of the straight lines defined by the general solution. Note that this enveloping property holds only in the x-u plane; in the jet bundle the prolongation of the singular solution intersects the prolongation of all other solutions in precisely one point.

Due to the fact that Clairaut equations are generally nonlinear in the derivative u', it is not possible to represent the manifold $\mathcal{R}_1$ globally by an equation $u' = \Phi(x,u)$. In our case, to most points $\rho \in \mathcal{R}_0^{(1)} \subset \mathcal{E}$ there are two points $\hat{\rho} \in \mathcal{R}_1$ such that $\pi_0^1(\hat{\rho}) = \rho$: if $\rho = (x,u)$, then $u' = 2(x \pm \sqrt{x^2 - u})$. This is shown in Figure 8.2 for two choices of ρ. An exception are those points ρ lying on the singular solution; for them a unique $\hat{\rho}$ exists (lying on the prolongation of the singular integral). Indeed, for our choice of the function f, the singular integral is given by $u(x) = x^2$ so that for those points the square root vanishes and in Figure 8.2 the prolongation of the singular integral gives the "fold" of the manifold $\mathcal{R}_1$.

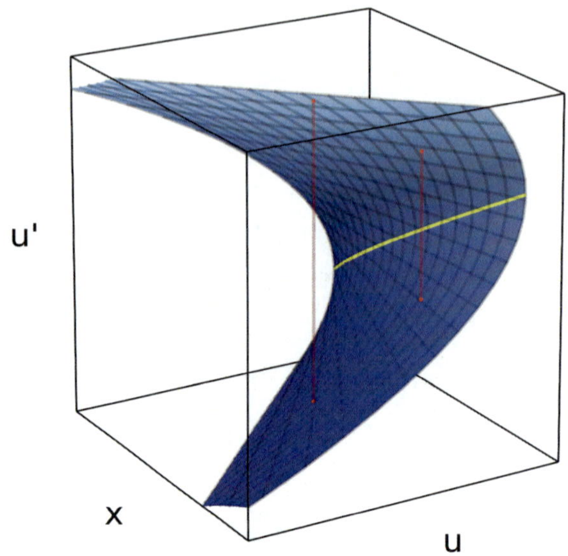

Fig. 8.2 Nonuniqueness in a Clairaut equation

It is straightforward to extend this example to a partial differential equation: take $u = x^i u_i + f(u_{(1)})$ where we assume that the Hessian determinant of the function $f : \mathbb{R}^n \to \mathbb{R}$ vanishes nowhere. Again the general solution is given by a family of straight lines, $u(\mathbf{x}) = c_i x^i + f(\mathbf{c})$, depending now on n constants $c_1, \dots, c_n$. The singular integral corresponds to the surface

$$x^i(\boldsymbol{\tau}) = -\frac{\partial f}{\partial u_i}(\boldsymbol{\tau}), \qquad u(\boldsymbol{\tau}) = -\tau_i \frac{\partial f}{\partial u_i}(\boldsymbol{\tau}) + f(\boldsymbol{\tau}) \tag{8.1}$$

where again the parameters represent the values of the derivatives, $u_i(\boldsymbol{\tau}) = \tau_i$, and where our assumption on the Hessian guarantees that we are indeed dealing with the graph of a function $u(\mathbf{x})$. $\triangleleft$

Singular integrals also only arise, if the dimension of the symbol is not constant. In the case of the Clairaut equations in Example 8.1.2, the symbol matrix is again scalar and its entry is $f'(u') + x$. Thus its rank drops at all points $\hat{\rho} \in \mathcal{R}_1$ where this expression vanishes. Singular solutions appear, if the augmented system consisting of the original system plus the conditions for a rank drop of the symbol is consistent. In our case, we obtain the system $u = xu' - \frac{1}{4}(u')^2$ and $u' = 2x$. Obviously, it is consistent and possess as sole solution our singular integral.

Example 8.1.3. It follows from this observation that in larger systems we may also find whole families of singular integrals. The general solution of the following first-order system in two unknown functions u and v [139, p. 159]

$$(u')^2 + xu' + v' - u = 0\,, \tag{8.2a}$$
$$u'v' + xv' - v = 0 \tag{8.2b}$$

is the two-parameter family $u(x) = ax + a^2 + b$ and $v(x) = bx + ab$ of linear functions. However, there exists an additional one-parameter family of singular integrals, namely $u(x) = -\frac{1}{4}(x+c)^2 + c^2$ and $v(x) = -\frac{1}{4}c(x-c)^2$. Obviously, it is impossible to choose the parameters a, b in such a way that these solutions are part of the general solution.

Again we easily obtain these singular integrals via an analysis of the symbol. The determinant of the symbol matrix is $D = 2(u')^2 + 3xu' - v' + x^2$ and one easily checks that D vanishes on the singular integrals and is different from zero on all other solutions. ◁

Within our geometric approach to differential equations one may use a slightly different ansatz that avoids some of these problems, at least for smooth solutions. The downside is that we can only define *a* and not *the* general solution. Furthermore, the definition is not very constructive. The basic idea is simple and rather old. Consider the inverse problem: given an expression containing some arbitrary constants and functions, we wish to determine differential equations for which this expression is a solution. For the "smallest" such differential equation our expression is a general solution. Following Hermann [208] we can make this idea more precise in our geometric framework.

Definition 8.1.4. A *general solution* of the differential equation $\mathcal{R}_q \subseteq J_q\pi$ is a subset $\Theta \subseteq \Gamma_{loc}(\pi)$ such that the following two conditions are satisfied.

(i) For all $r \geq 0$ we have $j_{q+r}(\Theta) \subseteq \mathcal{R}_{q+r}$.
(ii) For no $r \geq 0$ a submanifold $\bar{\mathcal{R}}_{q+r} \subset \mathcal{R}_{q+r}$ of lower dimension exists such that $j_{q+r}(\Theta) \subseteq \bar{\mathcal{R}}_{q+r}$.

Θ is a *local general solution* at a point $x \in \mathcal{X}$, if all sections $\sigma \in \Theta$ are defined in neighbourhood of x.

A natural question is how one may describe such subsets $\Theta \subseteq \Gamma_{loc}(\pi)$. The intuitive idea is that they stem from an expression parametrised by some arbitrary constants and functions. Again Hermann [208] showed how this idea may be formulated more rigorously in a geometric manner. Let $\mathcal{X}_\lambda$ be manifolds and $\mathcal{V}_\lambda$ vector spaces for $1 \leq \lambda \leq \ell$ and let $\mathcal{W}$ be a further manifold. All manifolds and spaces are assumed to be finite-dimensional. Then a (local) *closed form representation* of a (local) general solution Θ is a smooth map

$$s : \mathcal{C}^\infty(\mathcal{X}_1, \mathcal{V}_1) \times \cdots \times \mathcal{C}^\infty(\mathcal{X}_\ell, \mathcal{V}_\ell) \times \mathcal{W} \to \Gamma_{loc}(\pi) \tag{8.3}$$

such that $\operatorname{im} s = \Theta$. Here $\dim \mathcal{W}$ corresponds to the number of arbitrary constants in the general solution and each factor $\mathcal{C}^\infty(\mathcal{X}_\lambda, \mathcal{V}_\lambda)$ represents $\dim \mathcal{V}_\lambda$ functions depending on $\dim \mathcal{X}_\lambda$ arguments.

Example 8.1.5. We apply these ideas to the wave equation $u_{tt} - u_{xx} = 0$. As it is linear, no singular integrals appear and it is simple to derive closed form expressions for its general solution. Using e. g. the method of characteristics we obtain the classical superposition formula

$$s_1 : \begin{cases} \mathcal{C}^\infty(\mathbb{R}, \mathbb{R}^2) & \longrightarrow & \Gamma_{loc}(\pi) \\ \begin{pmatrix} f \\ g \end{pmatrix} & \longmapsto & f(x-t) + g(x+t) \end{cases} . \tag{8.4}$$

Alternatively, we take the D'Alembert solution of the initial value problem

$$s_2 : \begin{cases} \mathcal{C}^\infty(\mathbb{R}, \mathbb{R}^2) & \longrightarrow & \Gamma_{loc}(\pi) \\ \begin{pmatrix} f \\ g \end{pmatrix} & \longmapsto & \frac{1}{2}\big[f(x-t) + f(x+t)\big] + \frac{1}{2}\int_{x-t}^{x+t} g(\tau)d\tau \end{cases} . \tag{8.5}$$

In both cases, the right hand side is to be interpreted as a section of the trivial fibre bundle $\pi : \mathbb{R}^2 \times \mathbb{R} \to \mathbb{R}^2$. ◁

8.2 Cartan Characters and Hilbert Function

We are interested in measuring the size of the formal solution space. In particular, we would like to be able to compare two differential equations. A simple possibility would be to count the parametric derivatives. However, for most partial differential equations the result would be infinity which does not help much in comparing different equations. Intuitively it is clear that although most partial differential equations have an infinite-dimensional solution space, there are differences in the size. Let us for example assume that we could find a closed form representation like (8.3). Then for some equations more arbitrary functions or manifolds $\mathcal{Y}_\lambda$ of higher dimensions are needed than for others.

The basic trick to capture these ideas in a rigorous manner consists of exploiting the natural grading of derivatives. Instead of simply counting all parametric derivatives, we count them order by order, as this yields at each order a finite number. Just comparing these numbers for one fixed order does not give us much information, but studying the asymptotic growth turns out to be very useful. Obviously, this approach is in complete analogy to the introduction of the Hilbert function in commutative algebra (Definition B.1.21).

In Section 7.1 we introduced the indices $\beta_q^{(k)}$ of the symbol $\mathcal{N}_q$ of a differential equation $\mathcal{R}_q$. If the equation $\mathcal{R}_q$ is involutive, $\beta_q^{(k)}$ corresponds to the number of principal derivatives of class k and order q. With a simple combinatorial calculation we can determine from $\beta_q^{(k)}$ the number of the remaining derivatives of class k and order q which are of course parametric. It turns out that we have already encountered these numbers, namely they are the Cartan characters $\alpha_q^{(k)}$ defined in Section 6.2.

Lemma 8.2.1. *Let $\mathcal{R}_q \subseteq J_q\pi$ be a differential equation of order q in n independent and m dependent variables. Then the Cartan characters $\alpha_q^{(k)}$ and the indices $\beta_q^{(k)}$ of its symbol $\mathcal{N}_q$ are related by*

$$\alpha_q^{(k)} + \beta_q^{(k)} = m\binom{q+n-k-1}{q-1}, \qquad 1 \leq k \leq n\,. \tag{8.6}$$

Proof. If we identify a parametric derivative u_μ^α of order q with the basis vector $\frac{1}{\mu!}dx^\mu \otimes \partial_{u^\alpha}$ of $S_q(T^*\mathcal{X}) \otimes V\pi$, then it follows from the definition (6.22) of the subspace $\mathcal{N}_q^{(k)}$ that as a vector space it is spanned by all parametric derivatives of order q and class greater than k. The Cartan character $\alpha_q^{(k)}$ is defined as the difference of the dimensions of $\mathcal{N}_q^{(k)}$ and $\mathcal{N}_q^{(k-1)}$ and thus indeed the number of parametric derivatives of order q and class k. By (A.4b), the right hand side of (8.6) simply counts the total number of derivatives of order q and class k and $\beta_q^{(k)}$ gives the number of principal derivatives among these. □

The Cartan characters $\alpha_q^{(1)}, \dots, \alpha_q^{(n)}$ of an involutive differential equation form a descending sequence. Note that the corresponding property for the indices $\beta_q^{(k)}$, Corollary 7.1.28, holds only for first-order equations.

Proposition 8.2.2. *The Cartan characters $\alpha_q^{(1)}, \dots, \alpha_q^{(n)}$ of an involutive differential equation $\mathcal{R}_q$ in n independent variables satisfy*

$$\alpha_q^{(1)} \geq \alpha_q^{(2)} \geq \cdots \geq \alpha_q^{(n)} \geq 0\,. \tag{8.7}$$

Proof. For $q = 1$ this follows immediately from Corollary 7.1.28, as in this case $\alpha_1^{(k)} = m - \beta_1^{(k)}$. For a higher-order equation we simply exploit the fact (shown in Proposition A.3.1), that it can be transformed into an equivalent involutive first-order equation with exactly the same Cartan characters. □

One of the key properties of an involutive symbol is that one can compute the indices and the Cartan characters of its prolongations without explicitly determining their equations. Taking an algebraic point of view, this fact may be considered as a simple consequence of our combinatorial interpretation of involutive bases.

Proposition 8.2.3. *Let $\mathcal{N}_q$ be an involutive symbol with Cartan characters $\alpha_q^{(k)}$ and indices $\beta_q^{(k)}$. Then the Cartan characters $\alpha_{q+r}^{(k)}$ and indices $\beta_{q+r}^{(k)}$ of the prolonged symbols $\mathcal{N}_{q+r}$ are given by*

$$\alpha_{q+r}^{(k)} = \sum_{i=k}^{n} \binom{r+i-k-1}{r-1} \alpha_q^{(i)}\,, \tag{8.8a}$$

$$\beta_{q+r}^{(k)} = \sum_{i=k}^{n} \binom{r+i-k-1}{r-1} \beta_q^{(i)}\,. \tag{8.8b}$$

Proof. It suffices to prove (8.8b), as (8.8a) is then an immediate consequence of Lemma 8.2.1. We use an induction over the number r of prolongations. Formally differentiating an equation of class k or higher with respect to x^k yields an equation of class k. By Proposition 7.2.3, the arising equations are the only independent ones of order $q+1$ in the prolongation (and hence the only ones that matter for $\mathcal{N}_{q+1}$), as the symbol $\mathcal{N}_q$ is assumed to be involutive. Thus $\beta_{q+1}^{(k)} = \beta_q^{(k)} + \cdots + \beta_q^{(n)}$ and we have proven (8.8b) for $r = 1$.

Let us assume that (8.8b) holds for all values $1 \leq s < r$. Since according to Proposition 7.2.5 the prolongation of an involutive symbol is again involutive, we can apply the formula for $r = 1$ to the case $s = r-1$ and get $\beta_{q+r}^{(k)} = \beta_{q+r-1}^{(k)} + \cdots + \beta_{q+r-1}^{(n)}$. The induction hypothesis yields then

$$\beta_{q+r}^{(k)} = \sum_{i=k}^{n} \left[\sum_{l=k}^{i} \binom{r+l-k-2}{r-2} \right] \beta_q^{(i)} = \sum_{i=k}^{n} \binom{r+i-k-1}{r-1} \beta_q^{(i)}\,, \tag{8.9}$$

where the last equality follows from an identity for binomial coefficients. □

Remark 8.2.4. Based on this result, we can finally explain the terminology *differential equation of finite type* introduced in Remark 7.1.2. By definition, the geometric symbol $\mathcal{N}_q$ of such an equation is zero-dimensional. This fact implies immediately that all its Cartan characters $\alpha_q^{(k)}$ vanish since $\dim \mathcal{N}_q = \sum_{k=1}^{n} \alpha_q^{(k)}$. By (8.8a) we find that also any Cartan character $\alpha_{q+r}^{(k)}$ at higher orders vanish and consequently all prolonged symbols $\mathcal{N}_{q+r}$ are zero-dimensional, too. Hence such an equation possesses only a finite number of parametric derivatives and its formal solution space is finite-dimensional. ◁

We proceed now in analogy to commutative algebra. Recall that there Hilbert functions are used to measure the size of graded modules (cf. Definition B.1.21). The ring of formal power series also possesses a natural grading. Although the formal solution space of a differential equation $\mathcal{R}_q$ is in general not a submodule, we may define in a similar manner a Hilbert function for it. As we have seen in Section 2.3, the formal solution space is parametrised by the parametric coefficients; hence the following definition is very natural.

Definition 8.2.5. The *Hilbert function* $h(r)$ of a differential equation $\mathcal{R}_q$ gives the number of parametric derivatives of order r.

If the differential equation $\mathcal{R}_q$ is involutive, then it is straightforward to express its Hilbert function in a purely geometric way:

$$h(r) = \begin{cases} \dim \mathcal{R}_0^{(q)} & \text{for } r = 0\,, \\ \dim \mathcal{R}_r^{(q-r)} - \dim \mathcal{R}_{r-1}^{(q-r+1)} & \text{for } 0 < r \leq q\,, \\ \dim \mathcal{R}_r - \dim \mathcal{R}_{r-1} & \text{for } r > q\,. \end{cases} \tag{8.10}$$

But obviously this form is not very satisfactory: neither does it provide any real insight nor is it useful for an explicit computation of $h(r)$. Again it is necessary to invoke algebraic tools.

For an involutive equation $\mathcal{R}_q$ the Hilbert function $h(r)$ of the differential equation coincides with the Hilbert function $h_{\mathcal{N}}(r)$ of the symbol comodule $\mathcal{N}$ for all $r \geq q$. Indeed, recall from Remark 7.1.5 that the principal coefficients of order $r \geq q$ arise as the solutions of an inhomogeneous linear systems of equations and that the component $\mathcal{N}_r$ is nothing but the solution space of the corresponding homogeneous system. Hence its dimension determines the number of parametric coefficients and we find $h(r) = \dim \mathcal{N}_r$ for all $r \geq q$ (this fact also follows from Theorem 7.1.6 taking into account that $\mathcal{R}_q$ is assumed to be formally integrable).

If we take for the lower-order components of the comodule $\mathcal{N}$ the more precise approach of Remark 7.1.16, i. e. use the symbols of the projected equations $\mathcal{R}_r^{(q-r)}$, then $h(r)$ is exactly the Hilbert function of $\mathcal{N}$. Thus algebraically it is a standard task to determine the Hilbert function of differential equation explicitly. In practice, the easiest approach consists in determining a complementary decomposition of the symbol module $\mathcal{N}^0$ (using the methods discussed in Section 5.1) from which the Hilbert function is easy to read off (Proposition 5.2.1).

As it is thus possible to interpret the Hilbert function of a differential equation as the Hilbert function of a module, we can apply Theorem B.1.22 asserting that it becomes asymptotically a polynomial and introduce the *Hilbert polynomial*[2] of a differential equation. As the following result shows, the coefficients of the Hilbert polynomial are in one-to-one correspondence with the Cartan characters.

Proposition 8.2.6. *The Hilbert polynomial $H(r)$ of the involutive differential equation $\mathcal{R}_q \subseteq J_q\pi$ in n independent variables with Cartan characters $\alpha_q^{(1)}, \dots, \alpha_q^{(n)}$ can be written in the form*

$$H(q+r) = \sum_{i=0}^{n-1} \left(\sum_{k=i}^{n-1} \frac{\alpha_q^{(k+1)}}{k!} s_{k-i}^{(k)}(0) \right) r^i \tag{8.11}$$

where the combinatorial coefficients $s_{k-i}^{(k)}(0)$ are modified Stirling numbers (see Appendix A.4 for their definition and basic properties).

Conversely, if the Hilbert polynomial of the differential equation $\mathcal{R}_q \subseteq J_q\pi$ has the form $H(q+r) = \sum_{i=0}^{n-1} h_i r^i$, then its Cartan characters can be computed recursively starting with $k = n$ by

[2] In the context of differential algebra, one often uses the term *differential dimension polynomial* [258, Sect. III.5], [259, Sect. 5.2].

$$\alpha_q^{(k)} = (k-1)!\,h_{k-1} - \sum_{i=k+1}^{n} \frac{(k-1)!}{(i-1)!} s_{i-k}^{(i-1)}(0)\alpha_q^{(i)} . \tag{8.12}$$

Proof. As mentioned above, $H(r) = \dim \mathcal{N}_r$ for all values $r \geq q$. It follows trivially from the definition of the indices and the Cartan characters, respectively, that we have $\dim \mathcal{N}_{q+r} = \sum_{i=1}^{n} \alpha_{q+r}^{(i)}$. Together with (8.8a) this yields

$$\begin{aligned} H(q+r) &= \sum_{i=1}^{n}\sum_{k=i}^{n} \binom{r+k-i-1}{r-1} \alpha_q^{(k)} \\ &= \sum_{k=1}^{n}\sum_{i=1}^{k} \binom{r+k-i-1}{r-1} \alpha_q^{(k)} = \sum_{k=1}^{n} \binom{r+k-1}{r} \alpha_q^{(k)} . \end{aligned} \tag{8.13}$$

Here we used in the last step the identity (A.44). Now we can enter the defining relation (A.39) of the modified Stirling numbers and obtain after a simple index shift (8.11). It defines obviously a one-to-one correspondence between the coefficients h_i of the Hilbert polynomial and the Cartan characters $\alpha_q^{(k)}$. A straightforward inversion yields (8.12). □

Of course, this result represents nothing but a reformulation of Proposition 5.2.1 for the special case of a Rees decomposition induced by a Pommaret basis where all generators are of the same degree. Note that the Hilbert regularity of $\mathcal{R}_q$, i. e. the order from which on $h(r)$ and $H(r)$ coincide, is generally less than q, but the simple form (8.11) for the Hilbert polynomial can be obtained only at the order q.

Example 8.2.7. We computed in Example 7.1.26 the indices (7.24) of the Einstein equations in n dimensions. Applying (8.6) yields for their Cartan characters

$$\alpha_2^{(n)} = n\,, \quad \alpha_2^{(n-1)} = n^2\,, \quad \alpha_2^{(n-k)} = \frac{n(n+1)(n-k+1)}{2} \quad \text{for } 1 < k < n\,. \tag{8.14}$$

In arbitrary dimensions there does not seem to exist a simple expression for the Hilbert polynomial. Evaluation of the modified Stirling numbers arising in (8.11) yields for the leading terms

$$H(2+r) = \frac{n}{(n-1)!} r^{n-1} + \frac{3}{2}\frac{n^2}{(n-2)!} r^{n-2} + \cdots . \tag{8.15}$$

If we specialise to the physically most relevant case $n = 4$, then we obtain as Cartan characters $\alpha_2^{(4)} = 4$, $\alpha_2^{(3)} = 16$, $\alpha_2^{(2)} = 30$, $\alpha_2^{(1)} = 40$ and

$$H(2+r) = \frac{2}{3} r^3 + 12r^2 + \frac{184}{3} r + 90 \tag{8.16}$$

for the Hilbert polynomial. ◁

In some sense the Cartan characters (or the Hilbert function) are the best measures for the size of the formal solution space available, as they do not require any assumptions about the form of a general solution. The main caveat is that the Cartan characters are local quantities defined only at a point $\rho \in \mathcal{R}_q$. In particular, if singular integrals exist, the Cartan characters will vary over $\mathcal{R}_q$. As the question of the existence of global solutions is beyond the scope of this book anyway, we must live with this restriction (or resort to Blanket Assumption 7.1.4 which excludes variations in the values of the Cartan characters).

For the remainder of this section we want to relate the Cartan characters to numbers that can be more easily interpreted, namely to the number of arbitrary functions appearing in a closed form solution like (8.3). However, such an interpretation can be given only under some restrictive assumptions that are often *not* satisfied in applications. The main assumption is that the map s in (8.3) is purely algebraic, i. e. does not involve any differentiations or integrations. Thus we exclude cases as the D'Alembert formula (8.5). More precisely, we require that our solutions can be written in the form

$$\mathbf{u}(\mathbf{x}) = \boldsymbol{\Psi}\Bigl(\dots, F_\lambda\bigl(\phi_1(\mathbf{x}), \dots, \phi_k(\mathbf{x})\bigr), \dots\Bigr)\,. \tag{8.17}$$

Here F_λ represents the arbitrary functions. The one shown depends on k arguments which are defined by the fixed functions ϕ_i (these functions define the manifold $\mathcal{Y}_\lambda$). Note that no explicit dependency on the independent variables $\mathbf{x}$ is permitted; they appear only via the functions ϕ_i.

Example 8.2.8. The nonlinear equation $uu_{xy} - u_x u_y = 0$ admits such an algebraic representation of its general solution. It contains two arbitrary functions F_1 and F_2 depending on one argument and is given by $u(x,y) = \Psi\bigl(F_1(\phi_1(x,y)), F_2(\phi_2(x,y))\bigr)$ with $\Psi(a,b) = ab$, $\phi_1(x,y) = x$ and $\phi_2(x,y) = y$. Thus our differential equation describes all those functions of x and y that can be separated multiplicatively: $u(x,y) = F_1(x)F_2(y)$. ◁

Example 8.2.9. It is a fairly strong assumption that such an algebraic representation of the general solution exists. Hilbert [215] introduced the concept of a differential equation possessing a *resolution without integrals* ("integrallose Auflösung"): its general solution is expressible by a finite number of functions and their derivatives up to a finite order. Thus this concept is more general than our algebraic representations, as derivatives are allowed. Nevertheless, Hilbert [215] showed in a somewhat lengthy calculation that the underdetermined ordinary differential equation $u' = (v'')^2$—nowadays called *Cartan–Hilbert equation*—does not possess a resolution without integrals (Cartan [72] provided soon afterwards a more elegant proof via the Goursat normal form of a Pfaffian system). ◁

Our goal is to determine the number f_k of arbitrary functions in (8.17) that depend on k arguments for $1 \leq k \leq n$. It turns out that if such a representation exists, these numbers are unique. Following [308], we introduce the following shorthand

$$\begin{bmatrix} k \\ r \end{bmatrix} = \binom{k+r-1}{r}\,. \tag{8.18}$$

Then a smooth function of k arguments possesses $\left[\begin{smallmatrix}k\\r\end{smallmatrix}\right]$ Taylor coefficients of order r (cf. (2.32) for $m=1$).

We expand (8.17) in a Taylor series. In order to avoid degenerate cases, we choose the expansion point $x_0 \in \mathcal{X}$ such that if $\partial\phi_j/\partial x^i \not\equiv 0$ for some $1 \leq i \leq n$, then $\left(\partial\phi_j/\partial x^i\right)(x_0) \neq 0$. The right hand side of (8.17) depends thus on

$$T(r) = \sum_{k=1}^{n} f_k \begin{bmatrix} k \\ r \end{bmatrix} \tag{8.19}$$

arbitrary Taylor coefficients of order r. If (8.17) is indeed the general solution of the differential equation $\mathcal{R}_q$, these coefficients must be in a one-to-one correspondence with the parametric derivatives of $\mathcal{R}_q$. At least for orders $r \geq q$, it is straightforward to give a necessary condition for the existence of such a correspondence. We write both the Hilbert polynomial $H(r)$ of $\mathcal{R}_q$ and $T(r)$ as explicit polynomials in r and equate the coefficients. This yields a triangular system for the numbers f_k which is easily solved by back-substitution. We phrase the result in form of a proposition.

Proposition 8.2.10. *If the involutive differential equation $\mathcal{R}_q$ with Cartan characters $\alpha_q^{(k)}$ possesses an algebraic representation of its general solution in the form* (8.17)*, then this representation contains f_k arbitrary functions depending on k arguments where the numbers f_k are determined by the recursion relation*

$$f_n = \alpha_q^{(n)}\,, \tag{8.20a}$$

$$f_k = \alpha_q^{(k)} + \sum_{i=k+1}^{n} \frac{(k-1)!}{(n-1)!}\left(s_{i-k}^{(i-1)}(0)\alpha_q^{(i)} - s_{i-k}^{(i-1)}(q)f_i\right)\,. \tag{8.20b}$$

Such an algebraic representation can exist only, if the solution of this recursion relation contains only non-negative integers.[3]

Now we see why Definition 7.5.6 of an underdetermined equation makes sense. By Proposition 7.5.7, we get for the highest index of an underdetermined equation $\beta_q^{(n)} < m$ and thus $\alpha_q^{(n)} > 0$. Hence any algebraic representation of the general solution of such an equation depends on at least one arbitrary function of n arguments. Obviously, we may simply choose $\mathbf{x}$ for these arguments and at least one component of $\mathbf{u}$ remains completely unrestricted by the differential equation.

Example 8.2.11. We consider again Maxwell's equations (2.85). We know from Example 2.4.1 that they are formally integrable and from Example 7.1.30 that their symbol is involutive. We determined already that $\beta_1^{(4)} = 6$, $\beta_1^{(3)} = 2$ and all other indices vanish. This yields the Cartan characters

[3] One can show by explicit computations that even for fairly large values of n, the recursion relation (8.20) always yields integer solutions f_k. Thus it is tempting to conjecture that this remains true for all $n \in \mathbb{N}$. However, one can also show that this fact is highly non-trivial, as the modified Stirling number $s_{i-k}^{(i-1)}(q)$ is in general not divisible by the denominator $(n-1)!$.

$$\alpha_1^{(4)} = 0\,, \quad \alpha_1^{(3)} = 4\,, \quad \alpha_1^{(2)} = \alpha_1^{(1)} = 6\,. \tag{8.21}$$

The Hilbert polynomial is quadratic, $H(1+r) = 2r^2 + 12r + 16$, and for the numbers of arbitrary functions in an algebraic representation of the general solution we compute from (8.20)

$$f_4 = 0\,, \quad f_3 = 4\,, \quad f_2 = 2\,, \quad f_1 = 0\,. \tag{8.22}$$

In this particular case, these numbers are easy to interpret. Maxwell's equations are obviously not underdetermined. For each of the x- and y-components of the fields we have one evolution equation; thus we may prescribe in an initial value problem arbitrary values on a three-dimensional hypersurface. For each of the z-components we have two equations. As we will discuss in more detail in Chapter 9, this observation implies that, at least formally, we may prescribe initial data for them only on a two-dimensional surface. ◁

For first-order equations the situation becomes very simple. For $q = 1$ we can solve the recursion relation (8.20) in closed form and we can even show that the solution consists always of non-negative integers.

Corollary 8.2.12. *For an involutive first-order equation $\mathcal{R}_1$ the solution of the recursion relation* (8.20) *is given by*

$$f_n = \alpha_1^{(n)} = m - \beta_1^{(n)}\,, \tag{8.23a}$$

$$f_k = \alpha_1^{(k)} - \alpha_1^{(k+1)} = \beta_1^{(k+1)} - \beta_1^{(k)}\,, \quad 1 \le k < n\,. \tag{8.23b}$$

It contains only non-negative integers.

Proof. The proof is by a descending induction. The case $k = n$ is obvious. If we assume that (8.23b) holds for $k+1$, we get for k that

$$\begin{aligned} f_k = \alpha_1^{(k)} - \alpha_1^{(k+1)} + \\ (k-1)! \sum_{i=k+2}^{n-1} i!\,\alpha_1^{(i+1)} \Big(i s_{i-k}^{(i-1)}(1) + s_{i-k+1}^{(i)}(0) - s_{i-k+1}^{(i)}(1) \Big)\,. \end{aligned} \tag{8.24}$$

By Lemma A.4.3 the expression in the bracket vanishes for all values of i and k; this observation proves (8.23b) for k. It is obvious that all values f_k are integers and Proposition 8.2.2 implies that they are non-negative. □

One might be tempted to think that the general solutions of most partial differential equations should possess such algebraic representations. In fact, for higher-order equations it is a rather rare phenomenon. In particular, even in the important special case of normal equations no algebraic representations exist.

Example 8.2.13. Consider a normal differential equation $\mathcal{R}_q$ of order $q > 1$ in n independent and m dependent variables. By Proposition 7.5.7 we find that $\beta_q^{(n)} = m$ and all other indices vanish. Thus its Cartan characters are

$$\alpha_q^{(n)} = 0\,, \qquad \alpha_q^{(k)} = m\binom{q+k-1}{q-1}\,, \qquad 1 \leq k < n\,. \tag{8.25}$$

Evaluating (8.20) yields $f_n = 0$, $f_{n-1} = mq$ and $f_{n-2} = mq(1-q)/2$. Thus for $q > 1$ and $n > 2$ we always get a negative value for f_{n-2}. ◁

This simple example shows that the notion of an algebraic representation is only of limited use. One can consider more general representations involving differentiations and integrations of the arbitrary functions [399], but one obtains in general no unique values for the numbers f_k. Furthermore, several technical assumptions are required. So the value of such a formal counting of arbitrary functions is somewhat dubious. We have presented it here only because it helps to get a better intuition of how the Cartan characters and the Hilbert polynomial measure the size of the formal solution space. These abstract tools do not depend on any particular representation of the general solution and thus can always be applied without problems.

It is no coincidence that for first-order equations one always obtains reasonable values for the numbers f_k. As we will see in Chapter 9, these numbers are related to a concrete initial value problem. Solving this problem with a power series ansatz yields then indeed at each order a one-to-one correspondence between the Taylor coefficients of the initial data and the parametric coefficients of the solution.

For higher-order equations the situation becomes more difficult, as in general it is no longer so easy to relate the orders of the coefficients of the data and of the solution.[4] Only one thing is clear: the asymptotic growth of the number of parametric coefficients is dominated by the arbitrary functions with the highest number of arguments. The number k_0 of their arguments determines the degree of the Hilbert polynomial, $\deg H = k_0 - 1$ and is equal to the highest index k_0 such that $\alpha_q^{(k_0)} > 0$. One calls k_0 the *Cartan genus* $g[\mathcal{R}_q]$ and the Cartan character $\alpha_q^{(k_0)}$ the *index of generality* $e[\mathcal{R}_q]$ of the differential equation.[5] It is not difficult to see that any representation of the general solution depends on precisely $\alpha_q^{(k_0)}$ arbitrary functions of k_0 arguments. Given the fact that the Hilbert function of an equation $\mathcal{R}_q$ coincides with the one of its symbol comodule $\mathcal{N}$, we see that the Cartan genus corresponds to the (Krull) dimension $\dim\mathcal{N}$ and Proposition 8.2.6 implies that the index of generality equals the multiplicity $\operatorname{mult}\mathcal{N}$.

For any reasonable physical field theory one expects $g[\mathcal{R}_q] = n-1$, i. e. the Cartan genus is the number of spatial variables, as such theories always describe evolutionary processes and in an initial value problem one would like to prescribe the values of all fields at the initial time $t = 0$. The index of generality $e[\mathcal{R}_q]$ gives the number of *degrees of freedom*. For simple field theories like the Klein–Gordon equation $u_{tt} - \Delta u + m^2 u = 0$ where m is a real parameter (the mass of the field), but also for Maxwell's equations, this interpretation obviously makes sense.

---

[4] Of course, a simple solution is to rewrite the equation as an equivalent first-order equation as described in Appendix A.3 and then to analyse this equation.

[5] In the context of differential algebra, the degree of the Hilbert polynomial, i. e. $k_0 - 1$, is called the *differential type* and its leading coefficient, i. e. $\alpha_q^{(k_0)}/k_0!$ the *typical differential dimension* [258, Sect. III.5], [259, Sect. 5.2].

However, if we look at the results obtained in Example 8.2.7 for Einstein's equation, then we see that here the Cartan genus is n, i. e. we are dealing with an underdetermined equation. The same problem appears for the Yang–Mills equations. As we will see in the next section, it is due to the presence of a gauge symmetry. In order to obtain physically reasonable values for the Cartan genus and the index of generality, we must "subtract" the effect of the gauge freedom.

8.3 Differential Relations and Gauge Symmetries

One can act with various kinds of transformations on a differential equation. In the simplest case, the transformation consists just of a change of variables in a local representation of the equation and, obviously, it does not affect the Hilbert function or the Cartan characters, since these are geometric invariants. However, more "brutal" transformations may well change these measures for the size of the formal solution space. A simple example is the reduction to a first-order equation as discussed in Appendix A.3; depending on the way the reduction is performed, the Cartan characters may change.

In this section we provide a rather general framework for such transformations: differential relations. These are slightly more complicated objects than differential equations; in a certain sense they contain two equations simultaneously. Using this concept, we can derive an equation connecting the Hilbert functions of differentially related differential equations.

Definition 8.3.1. Let $\pi : \mathcal{E} \to \mathcal{X}$ and $\bar{\pi} : \bar{\mathcal{E}} \to \mathcal{X}$ be two fibred manifolds over the same base space $\mathcal{X}$. A *differential relation* of order $s, \bar{s}$ between π and $\bar{\pi}$ is a submanifold $\mathcal{T}_{s,\bar{s}} \subseteq J_s\pi \underset{\mathcal{X}}{\times} J_{\bar{s}}\bar{\pi}$ such that $\mathcal{T}_{s,\bar{s}}$ is a fibred submanifold with respect to the two fibrations induced by π and $\bar{\pi}$, namely $\pi^s \underset{\mathcal{X}}{\times} \mathrm{id} : J_s\pi \underset{\mathcal{X}}{\times} J_{\bar{s}}\bar{\pi} \to J_{\bar{s}}\bar{\pi}$ and $\mathrm{id} \underset{\mathcal{X}}{\times} \bar{\pi}^{\bar{r}} : J_s\pi \underset{\mathcal{X}}{\times} J_{\bar{s}}\bar{\pi} \to J_s\pi$.

The defining properties of a differential relation allow us to consider for a given section $\sigma : \mathcal{X} \to \mathcal{E}$ the submanifold $\mathcal{T}_{\bar{s}}[\sigma] = \mathcal{T}_{s,\bar{s}} \cap \left(\operatorname{im} j_s\sigma \times J_{\bar{s}}\bar{\pi}\right)$ as a differential equation in $J_{\bar{s}}\bar{\pi}$ and similarly for any fixed section $\bar{\sigma} : \mathcal{X} \to \bar{\mathcal{E}}$ the submanifold $\mathcal{T}_s[\bar{\sigma}] = \mathcal{T}_{s,\bar{s}} \cap \left(J_s\pi \times \operatorname{im} j_{\bar{s}}\bar{\sigma}\right)$ as a differential equation in $J_s\pi$.

Definition 8.3.2. Let $\mathcal{R}_q \subseteq J_q\pi$ and $\bar{\mathcal{R}}_{\bar{q}} \subseteq J_{\bar{q}}\bar{\pi}$ be two differential equations. They are *differentially related* via $\mathcal{T}_{s,\bar{s}} \subseteq J_s\pi \underset{\mathcal{X}}{\times} J_{\bar{s}}\bar{\pi}$, if the following condition is satisfied: if the local section $\sigma \in \Gamma_{loc}(\pi)$ is a solution of $\mathcal{R}_q$, then every solution of $\mathcal{T}_{\bar{s}}[\sigma]$ solves the equation $\bar{\mathcal{R}}_q$, and conversely if $\bar{\sigma} \in \Gamma_{loc}(\bar{\pi})$ is a solution of $\bar{\mathcal{R}}_q$, then every solution of $\mathcal{T}_s[\bar{\sigma}]$ solves the equation $\mathcal{R}_q$.

Example 8.3.3. Bäcklund transformations [353, 386] represent a classical instance of differential relations. Two additional conditions are imposed on them: (i) both $\mathcal{T}_s[\bar{\sigma}]$ and $\mathcal{T}_{\bar{s}}[\sigma]$ are solvable, if and only if σ and $\bar{\sigma}$ are solutions of $\mathcal{R}_q$ and $\bar{\mathcal{R}}_{\bar{q}}$, respectively; (ii) both $\mathcal{T}_s[\bar{\sigma}]$ and $\mathcal{T}_{\bar{s}}[\sigma]$ are equations of finite type. These two conditions allow for a geometric interpretation of Bäcklund transformations as what is often called a zero-curvature representation based on Remark 2.3.6; we do not pursue here this point of view but refer to [52].

Bäcklund [31] introduced this type of transformations for the sine-Gordon equation $u_{xt} = \sin u$. Here we even find an auto-Bäcklund transformation mapping the equation into itself. It is usually written in the following form:

$$\mathcal{T}_{1,1}: \begin{cases} \left(\dfrac{u-v}{2}\right)_t = \dfrac{1}{a}\sin\left(\dfrac{u+v}{2}\right), \\[2ex] \left(\dfrac{u+v}{2}\right)_x = a\sin\left(\dfrac{u-v}{2}\right) \end{cases} \tag{8.26}$$

where $a \neq 0$ is an arbitrary real parameter. Obviously, given v this is an equation of finite type for u and vice versa. In order to check whether these finite type equations are formally integrable, we differentiate the first equation with respect to t and the second one with respect to x and enter (8.26) into the result. With the help of some elementary trigonometric identities we find that it is necessary and sufficient for the formal integrability that both u and v satisfy the sine-Gordon equation.

Bäcklund transformations play a very important role in the theory of soliton equations. In particular, they are related to the inverse scattering transform and allow for the iterative construction of N-soliton solutions [1, Section 3.1]. For example, considering the transformation (8.26) for the trivial solution $u(x,t) \equiv 0$ leads to $v_x = -2a\sin(v/2)$ and $v_t = -(2/a)\sin(v/2)$. Combining these two equations yields the simple linear equation $v_x - a^2 v_t = 0$ whose general solution is trivially given by $v(x,t) = F(t + a^2 x)$ with a yet arbitrary function F. Entering it into the first equation of our system shows that F must satisfy the ordinary differential equation $aF' = -2\sin(F/2)$. This is a separable equation and an elementary integration gives $F(\tau) = 4\arctan\big(\exp(c - \tau/a)\big)$ where c is an arbitrary integration constant. Thus, starting from the trivial solution $u(x,t) \equiv 0$, we have constructed the single kink solution

$$v(x,t) = 4\arctan\big(\exp(ax + \frac{t}{a} + c)\big) \tag{8.27}$$

of the sine-Gordon equation. Iteration yields solutions with more kinks.

A further, rather trivial example, well-known in complex analysis, is provided by the Cauchy–Riemann equations

$$\mathcal{T}_{1,1}: \{\, u_x = v_y\,, \quad u_y = -v_x\,, \tag{8.28}$$

defining an auto-Bäcklund transformation for the Laplace equation $u_{xx} + u_{yy} = 0$, as one easily verifies.

Finally, we consider a Bäcklund transformation between two different equations. In characteristic coordinates the Liouville equation is given by $u_{xt} = e^u$. Via

$$\mathcal{T}_{1,1}: \begin{cases} u_t - v_t = \sqrt{2}e^{(u+v)/2}, \\ u_x + v_x = \sqrt{2}e^{(u-v)/2} \end{cases} \tag{8.29}$$

it is differentially related to the wave equation in characteristic coordinates $v_{xt} = 0$. A trivial solution of the latter one is $v(x,t) \equiv 0$. Entering it into (8.29) and integrating the arising simple system for u yields the non-trivial one-parameter family

$$u(x,t) = -2\ln\Big(C - \frac{1}{\sqrt{2}}(x+t)\Big) \tag{8.30}$$

of solutions of the Liouville equation. ◁

If two equations $\mathcal{R}_q$ and $\bar{\mathcal{R}}_{\bar{q}}$ are differentially related by a relation $\mathcal{T}_{s,\bar{s}}$, then we can associate to each solution of one of them some solutions of the other one. But generally we do not obtain a one-to-one correspondence. In the case of the Bäcklund transformation (8.26), $v(x,t)$ solves an equation of finite type and each solution $u(x,t)$ of the sine-Gordon equation leads via the transformation to a one-parameter family of new solutions. Our goal is now to compare the Hilbert functions of two such equations. For simplicity we will assume that all differential equations involved are involutive.

Let $h(r)$ and $\bar{h}(r)$ be the Hilbert functions of the two equations $\mathcal{R}_q$ and $\bar{\mathcal{R}}_{\bar{q}}$, respectively. We introduce furthermore two different Hilbert functions, $g(r)$ and $\bar{g}(r)$, for the differential relation $\mathcal{T}_{s,\bar{s}}$: for the former one we consider an equation $\mathcal{T}_s[\bar{\sigma}]$ and for the second one an equation $\mathcal{T}_{\bar{s}}[\sigma]$. We assume here and in the sequel for simplicity that the Hilbert functions of these equations do not depend on the chosen sections σ and $\bar{\sigma}$, respectively.

Proposition 8.3.4. *The four above defined Hilbert functions satisfy for all arguments $r \geq \max\{s,\bar{s}\}$ the equality*

$$h(r) - g(r) = \bar{h}(r+\bar{s}-s) - \bar{g}(r+\bar{s}-s). \tag{8.31}$$

Proof. The formal power series solution of $\mathcal{R}_q$ contains $h(r)$ parametric coefficients of order r. Via the differential relation $\mathcal{T}_{s,\bar{s}}$, a family of solutions of $\bar{\mathcal{R}}_{\bar{q}}$ depending on $\bar{g}(r)$ parametric coefficients of order r corresponds to each solution σ of $\mathcal{R}_q$. However, each of these solutions of $\bar{\mathcal{R}}_{\bar{q}}$ can be obtained not only by starting with the section σ but also by starting with any solution in a whole family parameterised at order s by $g(s)$ coefficients (obtained by applying the relation $\mathcal{T}_{s,\bar{s}}$ "backwards"). Since for $r \geq \max\{s,\bar{s}\}$ a coefficient of σ of order r corresponds via $\mathcal{T}_{s,\bar{s}}$ to coefficients of order $r+\bar{s}-s$ of $\bar{\sigma}$, the formal power series solution of $\bar{\mathcal{R}}_{\bar{q}}$ contains $\bar{h}(r+\bar{s}-s) = h(r) - g(r) + \bar{g}(r+\bar{s}-s)$ parametric coefficients of order $r+\bar{s}-s$. □

For lower values of r, the equality (8.31) does not need to hold. This is due to the fact that in general we cannot say that the coefficients of σ of order $r \leq \max\{s,\bar{s}\}$ are related to coefficients of $\bar{\sigma}$ of order $r+\bar{s}-s$. We only know that the coefficients of σ *up to order* s correspond to the coefficients of $\bar{\sigma}$ *up to order* $\bar{s}$. Only when the relation $\mathcal{T}_{s,\bar{s}}$ has a special form, we may be able to make stronger statements.

In particular, if both equations $\mathcal{T}_s[\bar{\sigma}]$ and $\mathcal{T}_{\bar{s}}[\sigma]$ are of Cauchy–Kovalevskaya form, then (8.31) holds for the full Hilbert functions.

Example 8.3.5. We exhibited the relation between Maxwell's equations and the $U(1)$ Yang–Mills equations in Example 2.4.3. Maxwell's equations use as dependent variables the electric and magnetic fields **E** and **B** which may be combined into the field strength tensor F given by (2.95); the Yang–Mills equations are based on the vector potential A. As already mentioned, in the language of differential forms F may be considered as the exterior derivative of A. At the level of the components the two sets of fields are related by the equality

$$F_{ij} = \frac{\partial A_i}{\partial x^j} - \frac{\partial A_j}{\partial x^i}\,, \tag{8.32}$$

representing nothing but a differential relation $\mathcal{T}_{1,0}$. For a given potential A, the field strength F is uniquely determined; hence the corresponding Hilbert function is simply $\bar{g}(r) = 0$. For given field strength F, (8.32) represents a first-order differential equation in the potential A; its Hilbert function is easily determined to be

$$g(0) = 4\,, \quad g(1) = 10\,, \quad g(2+r) = \frac{1}{6}r^3 + \frac{5}{2}r^2 + \frac{37}{3}r + 20\,. \tag{8.33}$$

The Hilbert functions of the two differential equations are given by

$$h(0) = 4\,, \quad h(1) = 16\,, \quad h(2+r) = \frac{1}{6}r^3 + \frac{9}{2}r^2 + \frac{73}{3}r + 36 \tag{8.34}$$

for the second-order Yang–Mills equations and

$$\bar{h}(0) = 6\,, \quad \bar{h}(1+r) = 2r^2 + 12r + 16 \tag{8.35}$$

for the first-order Maxwell equations, respectively. As predicted by our theory, these Hilbert functions satisfy the equality (8.31) for all values $r \geq 0$. ◁

A somewhat similar situation arises in the formal analysis of gauge theories. Such theories are of great importance in physics; today all fundamental interactions are modelled by field theories invariant under gauge symmetries. In fact, this idea is the guiding principle in modern elementary particle physics. In this section we use a very broad definition of the term "gauge symmetry" which is much more general than the one appearing in theoretical physics. In particular, when a symmetry is to be promoted to a gauge symmetry depends generally on the physical interpretation and cannot be decided on the basis of purely mathematical considerations.

We assume that we are given a differential equation $\mathcal{R}_q \subseteq J_q\pi$, the field equations in physical applications. We do not distinguish whether these are ordinary or partial differential equations and we do not assume any special structure of $\mathcal{R}_q$ like that they are of a Lagrangian or Hamiltonian nature. Our considerations are valid for any kind of formalism to set up the field equations.

For us symmetry transformations are (local) diffeomorphisms $\gamma : \mathcal{E} \to \mathcal{E}$ where $\mathcal{E}$ denotes for the moment some manifold (not necessarily fibred) with local coordinates $\mathbf{u}$. We are particularly interested in whole (Lie) groups of transformations. Classically, one would work with an explicit parametrisation of such a group. For example, the group of global affine transformations of $\mathcal{E}$ could be described by the set of all maps $\mathbf{u} \mapsto \bar{\mathbf{u}} = A\mathbf{u} + \mathbf{b}$ with an arbitrary regular matrix A and an arbitrary vector $\mathbf{b}$. However, not for all relevant groups it is possible to provide explicit parametrisations (consider for example for an orientable manifold the group of volume preserving transformations) and such a representation is also not convenient for our purposes. We will instead use a pseudogroup approach and describe our transformations as the solutions of a differential equation.

We may consider any diffeomorphism $\gamma: \mathcal{E} \to \mathcal{E}$ as a section of the trivial bundle $\mathrm{pr}_1 : \mathcal{E} \times \mathcal{E} \to \mathcal{E}$. For simplicity of notation, we will in the sequel always identify sections $\mathcal{E} \to \mathcal{E} \times \mathcal{E}$ with maps $\mathcal{E} \to \mathcal{E}$ and use whatever point of view is more convenient. As for any fibred manifold, we can form for any order $s \geq 0$ the jet bundle $J_s\mathrm{pr}_1$ over this trivial bundle. However, not every section of pr_1 corresponds to a diffeomorphism. We therefore define the open subbundles $I_s\mathrm{pr}_1$ consisting only of the jets of those local sections that are invertible (thus $I_s\mathrm{pr}_1$ is the complement of the submanifold described by the vanishing of the Jacobian determinant).

Definition 8.3.6. A *Lie pseudogroup* is the solution space of a differential equation $\mathcal{G}_s \subseteq I_s\mathrm{pr}_1 \subset J_s\mathrm{pr}_1$, the corresponding *finite Lie equation*, whose solutions satisfy the following three conditions.

(i) The identity $\mathrm{id}_{\mathcal{E}}$ is a solution of $\mathcal{G}_s$.
(ii) If the two sections $\gamma_1 : \mathcal{U}_1 \subseteq \mathcal{E} \to \mathcal{E}$ and $\gamma_2 : \mathcal{U}_2 \subseteq \mathcal{E} \to \mathcal{E}$ are both local solutions of $\mathcal{G}_s$ with $\gamma_1(\mathcal{U}_1) \cap \mathcal{U}_2 \neq \emptyset$, then the section $\gamma_2 \circ \gamma_1$ defined on some subset $\mathcal{U}_3 \subseteq \mathcal{U}_1$ with $\gamma_1(\mathcal{U}_3) \subseteq \mathcal{U}_2$ is also a local solution of $\mathcal{G}_s$.
(iii) If the section $\gamma: \mathcal{U} \subseteq \mathcal{E} \to \mathcal{E}$ is a local solution of $\mathcal{G}_s$, then the inverse section $\gamma^{-1} : \gamma(\mathcal{U}) \to \mathcal{U}$ is also a local solution of $\mathcal{G}_s$.

Obviously, this definition just rephrases the usual axioms of a local transformation group: the identity is contained in it and composition or inversion of group elements yields—where defined—again group elements. The associativity follows trivially from the properties of the composition of maps. For the above mentioned group of affine transformations a pseudogroup description would be simply given by the second-order Lie equation

$$\mathcal{G}_2 : \left\{ \frac{\partial^2 \bar{\mathbf{u}}}{\partial \mathbf{u}^2} = 0 \right. . \tag{8.36}$$

If we consider only the subgroup of homogeneous transformations, i. e. if we require $\mathbf{b} = 0$, then we could use a first-order description $\frac{\partial \bar{\mathbf{u}}}{\partial \mathbf{u}}\mathbf{u} - \bar{\mathbf{u}} = 0$. The group of volume preserving transformations is also described by a first-order Lie equation, namely $\det(\partial\bar{\mathbf{u}}/\partial\mathbf{u}) = 1$. In general, given an explicit parametrisation of a group, we can always derive a corresponding finite Lie equation by eliminating the parameters with the help of differentiations.

In our case, $\mathcal{E}$ is not an arbitrary manifold but the total space of a fibred manifold $\pi : \mathcal{E} \to \mathcal{X}$. A *gauge transformation* γ preserves the fibration; in other words, it is a fibred morphism and induces a map $\bar{\gamma}: \mathcal{X} \to \mathcal{X}$ such that $\pi \circ \gamma = \bar{\gamma} \circ \pi$. Thus if we use our usual adapted coordinates $(\mathbf{x}, \mathbf{u})$ on the fibred manifold $\mathcal{E}$, then a gauge transformation is of the form

$$\bar{\mathbf{x}} = \bar{\gamma}(\mathbf{x}) , \quad \bar{\mathbf{u}} = \hat{\gamma}(\mathbf{x}, \mathbf{u}) . \tag{8.37}$$

For simplicity, we will mostly assume that $\bar{\gamma}$ is the identity map on $\mathcal{X}$.

Assume we are given a finite Lie equation $\mathcal{G}_s$ and some solution γ of it. If $\sigma \in \Gamma_{loc}(\pi)$ is an arbitrary section of the fibred manifold π, then in general the concatenation $\gamma \circ \sigma$ is no longer a section. However, by definition of a Lie pseudogroup, the identity map is contained in it. If our solution γ is sufficiently close to it, then $\gamma \circ \sigma$ will again be a section. $\mathcal{G}_s$ describes now a *symmetry* group of the differential equation $\mathcal{R}_q$, if it maps solutions into solutions, i. e. if the section σ is a solution of $\mathcal{R}_q$, then $\gamma \circ \sigma$ must again be a solution—provided it is still a section.

Thus we have now obtained the above mentioned very general mathematical definition of a gauge symmetry; it is nothing but a fibre preserving symmetry. In physical applications, gauge symmetries have a more narrow meaning which, however, incorporates already physical interpretations.[6] Thus no mathematical criterion exists when a fibre preserving symmetry should be called a gauge symmetry in the physical sense. In physics, gauge symmetries represent a certain redundancy in the mathematical description of a theory. Solutions related by a gauge symmetry are identified, as one says that they correspond to the same physical state and hence cannot be distinguished by measurements.

These considerations imply that a physicist is not so much interested in the size of the full solution space but only in the size modulo the action of the gauge group, i. e. in the size of the "state space." We determine now this "gauge corrected" size purely formally from the two Hilbert functions of the field equation $\mathcal{R}_q$ and the Lie equation $\mathcal{G}_s$, respectively.

Proposition 8.3.7. *Let $h(r)$ be the Hilbert function of the involutive differential equation $\mathcal{R}_q$ and $g(r)$ the Hilbert function of the involutive Lie equation $\mathcal{G}_s$. Then the size of the state space is measured by the gauge corrected Hilbert function*

$$\bar{h}(r) = h(r) - g(r) . \tag{8.38}$$

Proof. We proceed as in the proof of Proposition 8.3.4. By definition of the Hilbert function h, the formal power series solution of the field equations $\mathcal{R}_q$ depends on $h(r)$ parametric coefficients of order r. Using the gauge invariance, $g(r)$ of these coefficients can be given arbitrary values. Thus $\bar{h}(r)$ parametric coefficients remain to distinguish physical states. □

The gauge corrected Hilbert function leads of course immediately to a gauge corrected Hilbert polynomial $\bar{H}(r)$. With its help we may introduce a gauge corrected

[6] Mathematically, physical gauge theories are described by a more specialised machinery, namely the formalism of principal fibre bundles. For our purposes the here presented simple approach is sufficient and easier to handle.

Cartan genus $\bar{g}[\mathcal{R}_q]$ and index of generality $\bar{e}[\mathcal{R}_q]$, respectively (more generally, we can introduce via Proposition 8.2.6 gauge corrected Cartan characters $\bar{\alpha}_q^{(k)}$). Under the assumption that $\bar{g}[\mathcal{R}_q] = n-1$, the number of true degrees of freedom of the gauge theory is then given by $\bar{e}[\mathcal{R}_q]$.

Example 8.3.8. Let us consider again the $U(1)$ Yang–Mills equations. For them Maxwell's equations represent a gauge-invariant form, so that we can explicitly compare the result of a gauge correction with the values obtained directly from Maxwell's equations.

As already mentioned in Example 7.5.5, the $U(1)$ Yang–Mills equations are invariant under the gauge transformation $\bar{A}_i = A_i + \partial_{x^i}\Lambda$ (or $\bar{A} = A + d\Lambda$ in differential forms language) with an arbitrary function Λ. These transformations build the solution space of the following Lie equation

$$\mathcal{G}_1 : \left\{ \frac{\partial \bar{A}_i}{\partial x^j} - \frac{\partial \bar{A}_j}{\partial x^i} = 0\,, \quad \frac{\partial \bar{A}_i}{\partial A_j} = \delta_i^j\,. \right. \tag{8.39}$$

The differential equation $\mathcal{G}_1$ is involutive and in a four-dimensional space-time its Hilbert function is ($\dim I_1 \mathrm{pr}_1 = 36$ here and (8.39) consists of 22 first-order equations and of no algebraic equations)

$$g(0) = 4\,, \quad g(1) = 10\,, \quad g(2+r) = \frac{1}{6}r^3 + \frac{5}{2}r^2 + \frac{37}{3}r + 20\,. \tag{8.40}$$

According to Proposition 8.3.7, we must subtract it from the Hilbert function (8.34) of the $U(1)$ Yang–Mills equations. Thus the gauge corrected Hilbert function is

$$\bar{h}(0) = 0 \quad \bar{h}(1) = 6\,, \quad \bar{h}(2+r) = 2r^2 + 12r + 16\,. \tag{8.41}$$

Up to a shift of the argument by one which is necessary because of the lower order of Maxwell's equations, we find indeed the Hilbert function (8.35) of Maxwell's equations. It is no coincidence that (8.40) is identical with (8.33): the gauge symmetry is sitting in the differential relation $\mathcal{T}_{1,0}$; if A and $\bar{A}$ are connected by a gauge transformation, then they yield the same field strength F.

The gauge corrected Hilbert function corresponds to the following gauge corrected Cartan characters for the Yang–Mills equations:

$$\bar{\alpha}_2^{(4)} = 0\,, \quad \bar{\alpha}_2^{(3)} = 4\,, \quad \bar{\alpha}_2^{(2)} = \bar{\alpha}_2^{(1)} = 6\,. \tag{8.42}$$

Of course, they are identical with the characters (8.21) of Maxwell's equations. Since the highest character vanishes after the gauge correction, the underdeterminacy of the Yang–Mills equations is solely due to the gauge freedom and the gauge corrected Cartan genus is $\bar{g}[\mathcal{R}_2] = 3$, as it should be for a reasonable gauge theory.

The Hilbert function of an arbitrary Yang–Mills theory with a d-dimensional gauge group over a four-dimensional space-time is simply d times (8.34), respectively (8.41) after the gauge correction. This observation does not come surprising, since the Hilbert function is determined by the symbol where the structure constants do not appear. The principal part of an arbitrary Yang–Mills theory consists of d copies of the principal part of the $U(1)$ Yang–Mills theory; no coupling exists between terms belonging to different generators of the Lie algebra. ◁

Gauge symmetries can be *fixed* by adding auxiliary equations such that the augmented differential equation is no longer invariant. Obviously, this modification reduces the solution space. However, the gauge fixing conditions must be chosen so that we still have at least one solution in each orbit of the gauge group. Otherwise we have not removed gauge freedom but parts of the state space. If we are able to completely fix the gauge, then we obtain a differential equation such that each of these solutions lies on a different orbit. Thus its solution space can be bijectively mapped to the state space. The Hilbert function of this equation is then $\bar{h}(s)$. Using (8.12), we can also determine in advance its Cartan characters.

Example 8.3.9. The Lorenz gauge of Example 7.5.9 can be extended to general Yang–Mills theories. It does not eliminate all the arbitrariness due to the gauge group, but at least it manages to render the originally underdetermined equation overdetermined. The gauge fixing condition is $\eta^{\mu\nu}\partial_\mu A^a_\nu = 0$. As it is of lower order, the augmented equation is not formally integrable but becomes involutive after one prolongation with subsequent projection. It is straightforward to determine its Cartan characters:

$$\tilde{\alpha}_2^{(4)} = 0\,, \quad \tilde{\alpha}_2^{(3)} = 4d\,, \quad \tilde{\alpha}_2^{(2)} = 12d\,, \quad \tilde{\alpha}_2^{(1)} = 16d\,. \tag{8.43}$$

Comparing with (8.42) we find that while the two higher characters have the same value as the gauge corrected ones the two lower ones are larger. Thus the gauge fixed system still contains a residual gauge freedom. ◁

Example 8.3.10. Finally, we consider again Einstein's equations. Here we must "cheat" a little bit, as a proper mathematical description of its gauge symmetries would require the introduction of frame bundles. Again we take a simpler approach by considering (somewhat incorrectly) coordinate transformations $\mathbf{x} \mapsto \bar{\mathbf{x}}$ on the space-time manifold as gauge transformations. The dependent variables in Einstein's equations are the components of the metric and transform according to

$$g_{ij} = \frac{\partial \bar{x}^k}{\partial x^i}\frac{\partial \bar{x}^l}{\partial x^j}\bar{g}_{kl}\,, \tag{8.44}$$

i.e. here $\bar{\gamma}$ is not only different from the identity, but all arbitrariness of the gauge transformation is contained in it. Because of this special situation, there is no need to write down some finite Lie equation $\mathcal{G}_s$, as we can immediately deduce its Hilbert polynomial $G(r)$: it follows from (8.44) that the number of parametric coefficients of order r in an expansion of the new dependent variables $\bar{g}$ is given by the number of Taylor coefficients of order $r+1$ (as (8.44) is algebraic in $\bar{g}$ but first-order in $\bar{\mathbf{x}}$) of n functions of n variables (namely $\bar{\mathbf{x}}(\mathbf{x})$), hence

$$G(2+r) = n\binom{n+r+2}{r+3} \tag{8.45}$$

(we determine $G(2+r)$ in order to facilitate the comparison with the Hilbert polynomial H of the Einstein equations which we computed in Example 8.2.7 also in the form $H(2+r)$, as Einstein's equations are second-order.)

If we apply the identity (A.39) and evaluate the arising modified Stirling numbers, then we obtain

$$G(2+r) = \frac{n}{(n-1)!}\sum_{k=0}^{n-1} s_{n-k-1}^{(n-1)}(3)\,r^k = \frac{n}{(n-1)!}r^{n-1} + \frac{1}{2}\frac{n(n+6)}{(n-2)!}r^{n-2} + \cdots . \tag{8.46}$$

As the leading term of the difference $H(2+r) - G(2+r)$ is given by $\frac{n(n-3)}{(n-2)!}r^{n-2}$, we obtain for the two highest gauge corrected Cartan characters

$$\bar{\alpha}_2^{(n)} = 0\,, \quad \bar{\alpha}_2^{(n-1)} = n(n-3)\,. \tag{8.47}$$

So again we see that all underdeterminacy is solely due to the gauge symmetry, as the gauge corrected Cartan genus is $n-1$, and Einstein's equation possess $n(n-3)$ degrees of freedom.

In $n = 4$ dimensions we obtain $G(2+r) = \frac{2}{3}r^3 + 10r^2 + \frac{148}{3}r + 80$ and thus for the gauge corrected Cartan characters

$$\bar{\alpha}_2^{(4)} = 0\,, \quad \bar{\alpha}_2^{(3)} = 4\,, \quad \bar{\alpha}_2^{(2)} = 6\,, \quad \bar{\alpha}_2^{(1)} = 0\,. \tag{8.48}$$

If we compare with the corresponding results for Maxwell's equations (8.21) (or the gauge corrected characters (8.42) for the $U(1)$ Yang–Mills equations), then we see that only the last one differs; in particular, both theories possess the same number of degrees of freedom.

One possible gauge fixing for the Einstein equations are *harmonic coordinates*. They correspond to adding the first-order equations

$$\frac{1}{\sqrt{-g}}\partial_i\bigl(\sqrt{-g}g^{ij}\bigr) = 0 \tag{8.49}$$

where $g = \det g^{ij}$. If we use the same trick as in Example 2.4.4 and express these conditions in locally geodesic coordinates, then they become $\partial_i g^{ij}(x_0) = 0$. As above, we must prolong once and project back in order to obtain an involutive system. In a straightforward computation one determines its Cartan characters to

$$\alpha_2^{(n)} = 0\,, \quad \alpha_2^{(n-1)} = n(n-1)\,, \quad \alpha_2^{(n-k)} = \frac{n^2(k+1)+n(k-1)}{2}\,. \tag{8.50}$$

Again we observe that although the system is no longer underdetermined, the gauge fixing is not complete, as already the first non-vanishing Cartan character is too large. If we look at the values (8.48) obtained for the gauge corrected Cartan characters in $n = 4$ dimensions, then we see that they are not descending: $\bar{\alpha}_2^{(1)} < \bar{\alpha}_2^{(2)}$. However, according to Proposition 8.2.2, the Cartan characters of any differential equation form a descending sequence. Hence we conclude that for Einstein's equations in $n = 4$ dimensions, it is not possible to achieve a complete gauge fixing with the help of differential equations. ◁

A non-physical example of a gauge symmetry and its fixing appears in Appendix A.3 in the discussion of the Drach transformation.

Addendum: Einstein's Strength

Einstein [124] introduced an alternative measure for the size of the formal solution space of a differential equation $\mathcal{R}_q$: the *strength*. We show now how his approach fits into our framework. Let $\mathcal{R}_q$ be an involutive differential equation in n independent variables. The starting point is the function

$$Z(r) = h(r) / \begin{bmatrix} n \\ r \end{bmatrix} \tag{8.51}$$

where h is the Hilbert function of the equation $\mathcal{R}_q$. For values $r \geq q$ (at least) the function Z is rational in its argument r: its denominator is of degree $n-1$ and the numerator is then the Hilbert polynomial (8.11) whose degree is at most $n-1$. The function Z describes the growth of the number of parametric coefficients relative to the growth of the number of Taylor coefficients of an analytic function of n variables. We expand it in powers of $1/r$:

$$Z(q+r) = Z_q^{(0)} + Z_q^{(1)}/r + O(1/r^2)\,. \tag{8.52}$$

Definition 8.3.11. $Z_q^{(0)}$ is the *compatibility coefficient*, $Z_q^{(1)}$ the *strength*[7] of the differential equation $\mathcal{R}_q$. An equation with $Z_q^{(0)} = 0$ is called *absolutely compatible.*

As the next proposition shows (but this fact also easily follows from our interpretation of the function Z), an equation which is not absolutely compatible is necessarily underdetermined. Einstein actually defined the strength only for absolutely compatible equations, but obviously Definition 8.3.11 makes sense for arbitrary involutive differential equations.

Proposition 8.3.12. *The compatibility coefficient and the strength of an involutive differential equation $\mathcal{R}_q$ in n independent variables with Cartan characters $\alpha_q^{(k)}$ are*

---

[7] Actually, Einstein called $Z_q^{(1)}$ the *coefficient of freedom*. He only titled his whole discussion of the subject with the word "strength," as he considered a differential equation as strong, if it imposes many restrictions on its solutions, i. e. if its solution space is small. Later, other authors started to call $Z_q^{(1)}$ strength, although this terminology is quite contradictory: the stronger a differential equation in the sense of Einstein, the smaller its strength!

$$Z_q^{(0)} = \alpha_q^{(n)} , \tag{8.53a}$$

$$Z_q^{(1)} = (n-1)\Big(\frac{1}{2}n\,\alpha_q^{(n)} + \alpha_q^{(n-1)}\Big) . \tag{8.53b}$$

Proof. Proposition 8.2.6 provides the explicit form (8.11) of the Hilbert polynomial in terms of the Cartan characters. Together with (A.39), it yields readily the compatibility coefficient $Z_q^{(0)}$ by taking the limit $r \to \infty$. Subtracting the result from (8.52), multiplying by r and taking again the limit $r \to \infty$ gives the formula for the strength $Z_q^{(1)}$. □

In the case of gauge theories, Einstein used the gauge corrected Hilbert function for the definition of Z. Then the gauge corrected Cartan characters appear in Proposition 8.3.12. As mentioned above, we always expect for a reasonable physical theory that all underdeterminacy stems from the gauge freedom. After the gauge correction, the field equations of such a theory are then absolutely compatible and we get $Z_q^{(1)} = (n-1)\bar{\alpha}_q^{(n-1)} = (n-1)\bar{e}[\mathcal{R}_q]$. Thus up to a constant factor $n-1$ depending solely on the dimension of space-time, the strength coincides with the gauge corrected index of generality and thus the true number of degrees of freedom (we assume here that $\bar{g}[\mathcal{R}_q] = n-1$ which is the case for all classical examples of field theories).

In any case, we may conclude that Einstein's approach provides us with less information than a full involution analysis of the symbol. The compatibility coefficient and the strength correspond only to the two highest Cartan characters; the lower characters are ignored. One could probably recover them by determining further coefficients in the expansion (8.52). But it is clear that analysing the symbol yields the characters much faster and easier.

8.4 Notes

The notions of "general" and "singular" integrals have disappeared from most modern textbooks on differential equations. Older books discuss them often at considerable length; in Forsyth's treatise the topic appears in at least three different volumes: [138, Chapts. VIII, XIII] discusses ordinary differential equations; [140, Chapt. V] studies first-order partial differential equations and [141, Chapt. XII] higher-order partial differential equations. Our approach is based on ideas of Ampère [16], who was apparently the first to propose the definition that a general solution uniquely determines the differential equation, and follows a modern reformulation given by Hermann [208]. Within differential algebra, i. e. for differential equations with at most polynomial non-linearities, a rigorous definition of general and singular solutions exists. Apparently Ritt [383] was the first to introduce it; more recent work in this direction is due to Hubert [226, 227].

According to Ince [230], App. A5, the first singular integral was already discovered by Taylor [451, p. 26] in 1715 in the analysis of the equation

$$(1+x^2)^2(u')^2 = 4u^3 - 4u^2 \,. \tag{8.54}$$

Apparently, this fact is not very well-known, as the discovery of singular integrals is usually attributed to Clairaut [92] (1734) whose family of equations discussed in Example 8.1.2 still represents the prototype of equations with such solutions.

The here used notion of a differential relation was introduced in [404]. It generalises the classical concept of a Bäcklund transformation and allows for rather "rough" transformations of a differential equation. Concerning the gauge correction of Cartan characters, [399] presented a first approach based on an explicit representation of the gauge transformation; the more flexible approach via Lie pseudogroups is again from [404]. This article also discusses the use of the gauge corrected Cartan characters for counting the degrees of freedom of a gauge theory. The formal analysis of gauge fixing conditions was first proposed in [401].

In his quest for a unified field theory, Einstein showed considerable interest in formal measures for the size of the solution space. He [124, p. 133] describes his motivation in the following way:

> Given certain field variables and a system of field equations for them, the latter will not in general determine the field completely. There still remain certain free data for a solution of the field equation. The smaller the number of free data consistent with the system of field equations, the "stronger" is the system. It is clear that in the absence of any other viewpoint from which to select the equations, one will prefer the "stronger" system to a less strong one.

Around 1930 he discussed this question in his correspondence with Cartan [75]. Cartan tried to explain Einstein his theory of involutive systems; in fact, one letter is one of few places where Cartan explained his theory in terms of partial differential equations and not of exterior differential systems. But for whatever reasons, Einstein seems to have never used this approach.

Many years later, he developed his own notion, the strength, in an appendix of his book on the meaning of relativity [124, pp. 133–139]. This appendix on a generalised theory of gravitation was introduced in the third edition and completely rewritten for the fifth edition under the title "Relativistic Theory of the Non-Symmetric Field". The second version starts with the following words:

> This discussion is of intrinsic interest quite apart from the particular theory presented here. For a deeper understanding of our problem, however, it is indispensable.

Of course, Einstein did not use the Hilbert function for defining the strength. He gave a much more cumbersome recipe for determining it. In order to obtain the numerator of (8.51), he counted the number of field equations, the number of identities satisfied by them, the number of identities of these identities and so on. Obviously, this process is prone to errors, in particular when applied to an equation which is not involutive. Indeed, one can find several examples in the literature where wrong results for the strength were obtained. By contrast, the determination of the Cartan characters via the symbol is trivial.

The idea of the strength of a differential equation did not spark much interest. Only few authors applied it to various physical field theories [64, 220, 308, 313, 314,

335, 397, 418]. Penney [350] used it for metaphysical speculations on the dimensionality of the world. He conjectured that the world is four-dimensional, because only in four dimensions the Maxwell and the Einstein equations possess the same strength. In fact, already Einstein [124, p. 139] remarked:

> It is surprising that the gravitational equations for empty space determine their field just as strongly as do Maxwell's equations in the case of the electromagnetic field.

However, if one computes the full set of gauge corrected Cartan characters for both theories, then one sees that even in four dimensions only the higher ones are identical; the last one and thus the size of the formal solution spaces of the two theories are different. Hence Einstein's comment is not really correct. Furthermore, no compelling physical reason exists why the two theories should have formal solution spaces of the same size and one can seriously argue about the importance of *formal* solutions for physics.

It seems that Sué [442] was the first who tried to relate Einstein's and Cartan's approach, especially for gauge theories. But he compared a strength which was corrected for the gauge freedom with Cartan characters which were not corrected, so some of his conclusions are misleading. The relation (8.53) between strength and compatibility coefficient on one side and the two highest Cartan characters on the other side was first derived in [399].

Finally, we note that recently Levin [287] extended the concept of strength to linear systems of *difference* equations with constant coefficients and showed how the strength may be computed via Gröbner bases.

Chapter 9
Existence and Uniqueness of Solutions

In order to solve this differential equation you look at it till a solution occurs to you.

George Pólya

A fundamental issue in the theory of differential equations consists of proving the existence and uniqueness of solutions. Before we discuss partial differential equations, we analyse in Section 9.1 the situation for general systems of ordinary differential equations, often called differential algebraic equations. For involutive equations it is here straightforward to extend the classical existence and uniqueness theorem. The formal theory also provides us with a natural geometric approach to the treatment of certain types of singularities.

Traditionally, (first-order) ordinary differential equations are studied via vector fields on the manifold $\mathcal{E}$ (actually, one usually restricts to the autonomous case assuming that $\mathcal{E} = \mathcal{X} \times \mathcal{U}$ and considers vector fields on $\mathcal{U}$). However, for a unified treatment of many singular phenomena it turns out to be much more useful to associate with the equation a vector field (or more precisely a distribution) in the first jet bundle $J_1\pi$ arising very naturally from the contact structure. We will not develop a general theory of singularities but study a number of situations that have attracted much interest in the literature.

In local coordinates, one may say that the study of power series solutions underlies much of the formal theory. Hence, it is not surprising that results on *analytic* solutions of partial differential equations are fairly straightforward to obtain. In Section 9.2 we recall the famous Cauchy–Kovalevskaya Theorem for normal systems. The main point of the proof consists of showing that the easily obtained formal power series solution of the usual initial value problem actually converges.

While for normal systems it is more or less obvious how many initial conditions we must impose in order to obtain a unique solution, this question becomes much less clear for overdetermined systems. If one insists on prescribing the initial data in the customary way on a hypersurface, then the data cannot be chosen arbitrarily but must satisfy certain consistency conditions. In Section 9.3 we introduce the notion of a formally well-posed initial value problem possessing a unique formal power series solution. The definition is based on the combinatorial decompositions of polynomial modules introduced in Section 5.1.

The term "initial" data should not be taken too literally here, as it is well possible to obtain "initial" conditions where some dependent functions are prescribed on,

W.M. Seiler, *Involution*, Algorithms and Computation in Mathematics 24, 357
DOI 10.1007/978-3-642-01287-7_9, © Springer-Verlag Berlin Heidelberg 2010

say, the hypersurface $x = 0$ and others on $y = 0$. Of course, this effect is familiar from characteristic initial value problems for hyperbolic equations. However, here it is in general not related to characteristics but simply stems from the properties of the chosen term order.

Given an involutive system in δ-regular coordinates where the linearised principal part yields a Pommaret basis for the symbol module, we can directly write down the corresponding formally well-posed initial value problem. In it the initial data are prescribed on a flag of submanifolds, so that we always recover the intuitive idea of an initial value problem. In the analytic category we may now generalise the Cauchy–Kovalevskaya Theorem for normal systems to an existence and uniqueness theorem for arbitrary involutive systems, the Cartan–Kähler Theorem. We will see again why formal integrability is *not* sufficient for analysing differential equations but that one needs involution in order to ensure uniqueness.

From an application point of view the restriction to analytic solutions is very severe. It represents the price we have to pay for the Cartan–Kähler Theorem requiring essentially no other assumptions than involution. At such a level of generality, no stronger results can be expected. However, one may argue that our proof of the Cartan–Kähler Theorem is actually more important than the theorem itself. In situations where more is known about the structure of the system stronger statements are possible. We will see later two simple instances of this effect: in Section 10.2 the classical Holmgren Theorem asserting the uniqueness of continuously differentiable solutions is extended from normal systems to arbitrary linear involutive systems and in Section 10.4 an existence and uniqueness theorem for smooth solutions of hyperbolic equations with elliptic constraints are proven.

The final two sections of this chapter cover the Vessiot theory. We used this approach already in our treatment of ordinary differential equations in the first section; now we discuss its extension to partial differential equations. The material we present here concerns mainly an alternative (dual) approach to the Cartan–Kähler Theory of exterior differential systems. However, it should be pointed out that the Vessiot theory leads to important geometric structures associated with any differential equation. These structures have found many applications, e. g. in the integration of hyperbolic equations (the famous Darboux method). Furthermore, they relate more closely the formal theory with the Cartan–Kähler Theory and allow us to apply directly (i. e. at the level of differential equations and without a transformation to an exterior system) many techniques developed there.

As we present in this chapter mainly local results, we will work most of the time in local coordinates and therefore speak of systems.

9.1 Ordinary Differential Equations

In order to simplify the discussion of the existence and uniqueness of solutions of an involutive first-order ordinary differential equation, we specialise the Cartan normal form (7.34) to the case of only one independent variable. It induces a decomposition of the vector of unknown functions into three components $\mathbf{u}$, $\mathbf{v}$ and $\mathbf{w}$ (not necessarily of the same dimension) and (7.34) may be written as

$$\mathbf{u}' = \boldsymbol{\phi}(x,\mathbf{u},\mathbf{v},\mathbf{w},\mathbf{w}')\,, \tag{9.1a}$$

$$\mathbf{v}' = \boldsymbol{\psi}(x,\mathbf{u},\mathbf{v},\mathbf{w},\mathbf{w}')\,, \tag{9.1b}$$

$$\mathbf{u} = \boldsymbol{\chi}(x,\mathbf{v},\mathbf{w})\,. \tag{9.1c}$$

If the equation $\mathcal{R}_1$ is not underdetermined, then the component $\mathbf{w}$ is empty. Note that here, in contrast to (7.34), we further normalised by eliminating the principal derivatives $\mathbf{u}'$ and $\mathbf{v}'$ from the right hand sides.

Involution imposes restrictions on the right hand sides $\boldsymbol{\phi}$, $\boldsymbol{\psi}$, $\boldsymbol{\chi}$. The conditions (7.35) are empty here, as they stem from cross-derivatives which are not possible for ordinary differential equations. The conditions (7.36) require the existence of a matrix $\mathbf{A}(x,\mathbf{u},\mathbf{v},\mathbf{w},\mathbf{w}')$ of smooth functions such that

$$\frac{\partial\boldsymbol{\chi}}{\partial x} + \frac{\partial\boldsymbol{\chi}}{\partial\mathbf{v}}\boldsymbol{\psi} + \frac{\partial\boldsymbol{\chi}}{\partial\mathbf{w}}\mathbf{w}' - \boldsymbol{\phi} = \mathbf{A}(\mathbf{u}-\boldsymbol{\chi})\,. \tag{9.2}$$

Proposition 9.1.1. *Assume that the system* (9.1) *is involutive and not underdetermined. Then the function* $\mathbf{u}-\boldsymbol{\chi}(x,\mathbf{v},\mathbf{w})$ *is a weak invariant of the normal ordinary differential equation* (9.1a,9.1b).

Proof. This is a trivial reformulation of (9.2). □

The terminology "weak" invariant refers to the fact that the left hand side of the identity (9.2) does not necessarily vanish everywhere but only on the constraint manifold $\mathcal{C} = \mathcal{R}_0^{(1)} \subseteq \mathcal{E}$ defined by the algebraic equations $\mathbf{u} = \boldsymbol{\chi}$ which leads via Hadamard's Lemma C.1.5 to the right hand side of (9.2). For a strong invariant one requires that they vanish on the whole manifold $\mathcal{E}$. Another way to formulate Proposition 9.1.1 is to say that $\mathcal{C}$ is an *invariant manifold* of the ordinary differential equation (9.1a,9.1b) i. e. if we choose initial data lying on $\mathcal{C}$, then the whole arising solution will stay on $\mathcal{C}$.

Proposition 9.1.1 makes the existence and uniqueness theory for an involutive ordinary differential equation $\mathcal{R}_1$ almost trivial. We can reduce the problem to a normal ordinary differential equation living on some submanifold $\mathcal{C} \subseteq \mathcal{E}$ defined by the algebraic part (9.1c) and apply the standard existence and uniqueness theorem of Picard–Lindelöf [13, Theorem 7.4]. For simplicity, we assume in the sequel that $\mathcal{X} \subseteq \mathbb{R}$ which is not a very restrictive assumption, as any connected one-dimensional manifold is diffeomorphic to either a submanifold of $\mathbb{R}$ or $\mathcal{S}^1$.

Theorem 9.1.2. *Let* (9.1) *be an involutive and not underdetermined system. Furthermore, let* $[a,b] \subset \mathcal{X} \subseteq \mathbb{R}$ *be a closed interval with a point* $\xi \in (a,b)$ *in it. Assume that* $\boldsymbol{\phi}$ *and* $\boldsymbol{\psi}$ *are continuous functions on* $\pi^{-1}([a,b]) \cap \mathcal{R}_0^{(1)}$ *and satisfy there a Lipschitz condition in the variables* $(\mathbf{u},\mathbf{v})$. *Then the initial value problem for* (9.1) *with initial data* $\big(\xi,\mathbf{u}(\xi),\mathbf{v}(\xi)\big) \in \mathcal{R}_0^{(1)}$, *i. e. with* $\mathbf{u}(\xi) = \boldsymbol{\chi}\big(\xi,\mathbf{v}(\xi)\big)$, *possesses a unique solution in some interval* $[\xi-r,\xi+r] \subseteq [a,b]$ *with* $r>0$.

Remark 9.1.3. Theorem 9.1.2 generalises the usual form of the Picard–Lindelöf Theorem to involutive equations, as it speaks only about the local existence and uniqueness of solutions. With a little more effort it is not difficult to show that if the function $\boldsymbol{\phi}(x,\mathbf{u})$ is Lipschitz continuous in a domain Ω, then any solution of $\mathbf{u}' = \boldsymbol{\phi}(x,\mathbf{u})$ can always be extended until it reaches the boundary $\partial\Omega$ (note that this boundary may be at infinity!)—see e. g. [13, Theorem 7.6] for a precise statement and a proof. It is straightforward to generalise this stronger form, too. ◁

Of course, one must emphasise that we obtain such a trivial theory only because of our standard assumption that we are dealing with a *regular* differential equation. It entails that the ranks of certain Jacobians are constant and that we may indeed rewrite any local representation in the form (9.1). Many ordinary differential equations appearing in applications do not satisfy this assumption and their existence and uniqueness theory is more complicated, as we will see below.

In order to obtain a geometric formulation of the existence and uniqueness theory, we exploit the contact structure of the jet bundle $J_1\pi$. The contact distribution $\mathcal{C}_1 \subset TJ_1\pi$ is $(m+1)$-dimensional and locally generated by the vector fields $\{\partial_x + (u')^\alpha\partial_{u^\alpha},\ \partial_{(u')^\alpha}\}$ (as one easily sees by specialising (2.11) to $q = n = 1$). Given an ordinary differential equation $\mathcal{R}_1 \subseteq J_1\pi$, we introduce on it the distribution $\mathcal{V}[\mathcal{R}_1] = T\mathcal{R}_1 \cap \mathcal{C}_1|_{\mathcal{R}_1}$ which we call the *Vessiot distribution* of the equation (in Section 9.5 we will generalise this construction to partial differential equations).

Example 9.1.4. If the given differential equation $\mathcal{R}_1$ is normal, then its Vessiot distribution $\mathcal{V}[\mathcal{R}_1]$ is always one-dimensional and it is straightforward to write down a generating vector field. Let $\mathcal{R}_1$ be locally represented by the system $\mathbf{u}' = \boldsymbol{\phi}(x,\mathbf{u})$. A general contact vector field $X = a\big(\partial_x + (u')^\alpha\partial_{u^\alpha}\big) + b^\alpha\partial_{(u')^\alpha}$ is contained in $T\mathcal{R}_1$, if it annihilates the one-forms $\mathrm{d}(u')^\alpha - \frac{\partial\phi^\alpha}{\partial u^\beta}\mathrm{d}u^\beta - \frac{\partial\phi^\alpha}{\partial x}\mathrm{d}x$. As on $\mathcal{R}_1$ we may always replace $\mathbf{u}'$ by $\boldsymbol{\phi}$, this condition yields for the generator

$$X = \partial_x + \phi^\alpha\partial_{u^\alpha} + \Big(\frac{\partial\phi^\alpha}{\partial x} + \phi^\beta\frac{\partial\phi^\alpha}{\partial u^\beta}\Big)\partial_{(u')^\alpha}\,. \tag{9.3}$$

Obviously, application of the projection $T\pi^1_0 : TJ_1\pi \to T\mathcal{E}$ yields the familiar evolution vector field $\partial_x + \phi^\alpha\partial_{u^\alpha}$ corresponding to our differential equation. The coefficient in the direction $(u')^\alpha$ is also easily understood: prolonging our equation shows that it is just the value of the second-order derivative $(u'')^\alpha$. ◁

As for ordinary differential equations $\dim\mathcal{X} = 1$, we may consider the image of a section $\sigma : \mathcal{X} \to \mathcal{E}$ as a curve in $\mathcal{E}$ parametrised by $\mathcal{X}$. Now we obtain the following geometric interpretation of the distribution $\mathcal{V}[\mathcal{R}_1]$.

Lemma 9.1.5. *The local section $\sigma \in \Gamma_{loc}(\pi)$ is a solution of the differential equation $\mathcal{R}_1$, if and only if $T_\rho(\operatorname{im} j_1\sigma) \subseteq \mathcal{V}_\rho[\mathcal{R}_1]$ for every point ρ on the image of the prolonged section $j_1\sigma \in \Gamma_{loc}(\pi^1)$.*

Proof. By the definition of a solution, the curve $\operatorname{im} j_1\sigma$ lies completely in $\mathcal{R}_1$ and thus its tangent space $T(\operatorname{im} j_1\sigma)$ in $T\mathcal{R}_1$. On the other hand, by Proposition 2.2.4, the tangent space of the image of a prolonged section is always a subspace of the contact distribution $\mathcal{C}_1$. □

As a corollary of this lemma, we may easily characterise geometrically the existence of unique solutions for initial value problems. The differential equation $\mathcal{R}_1$ possesses a unique local solution passing through a prescribed point $\rho \in \mathcal{R}_1$, if and only if the vector space $\mathcal{V}_\rho[\mathcal{R}_1]$ is one-dimensional and transversal to the fibration $\pi^1 : J_1\pi \to \mathcal{X}$. Obviously, a necessary condition for uniqueness is that $\mathcal{R}_1$ is not underdetermined. A one-dimensional distribution is trivially involutive and thus integrable. Hence a unique integral curve to $\mathcal{V}[\mathcal{R}_1]$ exists passing through ρ. Since $\mathcal{V}_\rho[\mathcal{R}_1]$ is transversal, it follows from the properties of the contact distribution that this integral curve is in fact the image of a prolonged section $j_1\sigma$ and by Lemma 9.1.5 the thus determined section σ is a solution of $\mathcal{R}_1$.

Theorem 9.1.6. *Let the regular ordinary differential equation $\mathcal{R}_1$ be involutive and not underdetermined. Then the distribution $\mathcal{V}[\mathcal{R}_1]$ is everywhere one-dimensional and transversal so that the initial value problem for $\mathcal{R}_1$ is locally uniquely solvable.*

Proof. As the differential equation $\mathcal{R}_1$ is assumed to be regular and not underdetermined, we can locally transform it into the semi-explicit form $\mathbf{u}' = \boldsymbol{\phi}(x,\mathbf{u})$, $\boldsymbol{\chi}(x,\mathbf{u}) = 0$. Any vector field in $T\mathcal{R}_1$ must thus annihilate the following one-forms:

$$\mathrm{d}(u')^\alpha - \frac{\partial \phi^\alpha}{\partial u^\beta}\mathrm{d}u^\beta - \frac{\partial \phi^\alpha}{\partial x}\mathrm{d}x\,, \qquad \frac{\partial \chi^\gamma}{\partial u^\beta}\mathrm{d}u^\beta + \frac{\partial \chi^\gamma}{\partial x}\mathrm{d}x\,. \tag{9.4}$$

The distribution $\mathcal{V}[\mathcal{R}_1]$ is generated by the contact vector fields $X \in \mathcal{C}_1$ satisfying this condition. The ansatz $X = a\big(\partial_x + (u')^\alpha\partial_{u^\alpha}\big) + b^\alpha\partial_{(u')^\alpha}$ yields the following linear system for the coefficient functions a, b^α:

$$b^\alpha = a\left(\frac{\partial \phi^\alpha}{\partial x} + (u')^\beta\frac{\partial \phi^\alpha}{\partial u^\beta}\right), \qquad 0 = a\left(\frac{\partial \chi^\gamma}{\partial x} + (u')^\beta\frac{\partial \chi^\gamma}{\partial u^\beta}\right). \tag{9.5}$$

As we consider these equations only *on* $\mathcal{R}_1$, the second subsystem vanishes for an involutive equation (this follows immediately from a trivial adaption of (9.2)). Thus the linear system (9.5) has a one-dimensional solution space. Furthermore, one easily sees from the first subsystem that for every non-trivial solution $a \neq 0$.

Hence, we may conclude that the distribution $\mathcal{V}[\mathcal{R}_1]$ is indeed one-dimensional and transversal to the fibration $\pi^1 : J_1\pi \to \mathcal{X}$. Locally, it is thus generated by a single transversal vector field and every integral curve of it is the image of the prolongation of a solution σ of our differential equation $\mathcal{R}_1$. □

For notational simplicity, we formulated this theorem for first-order equations. But obviously all the considerations extend straightforwardly to higher-order equations; there is not need to rewrite them as first-order equations, as it is usually done. For a regular involutive and not underdetermined equation $\mathcal{R}_q \subset J_q\pi$ we may always introduce the one-dimensional distribution $\mathcal{V}[\mathcal{R}_q] = T\mathcal{R}_q \cap \mathcal{C}_q|_{\mathcal{R}_q}$ the integral curves of which are the images of prolongations $j_q\sigma$ of solutions σ of $\mathcal{R}_q$.

Remark 9.1.7. One should always keep in mind that the initial value problem for an implicit differential equation $\mathcal{R}_1$ consists of prescribing a point $\rho \in \mathcal{R}_1$, i. e. on the jet bundle and not only on the constraint manifold $\mathcal{R}_0^{(1)}$. Indeed, assume for example that we are given a differential equation of the form $(u')^k = \phi(x,u)$ with an exponent $k > 1$. Then prescribing values (x_0, u_0) does not suffice for obtaining a unique solution, as there are k possible values u_0' such that $(x_0, u_0, u_0') \in \mathcal{R}_1$ leading to k different solutions. Prescribing a point $\rho \in \mathcal{R}_1$ corresponds to choosing one of these possible values and thus yields a unique solution. ◁

So far we have always made the blanket assumption that we are in a completely regular situation. Now we will admit some singularities. In fact, it will turn out that the geometric theory developed above gives us with the Vessiot distribution a very useful tool for studying certain types of singularities where the classical existence and uniqueness results fail. However, in most cases a purely formal analysis does not suffice but must be complemented by additional analytic considerations.

We will not attempt to develop here a general theory of singularities of ordinary differential equations, as this topic would require a book of its own. Based on a few examples, we will only exhibit some typical phenomena and show how they may be tackled using the above geometric ideas and the Vessiot distribution. In particular, we want to show that a proper analysis of the singularities of a qth order nonlinear equation can only be performed on the jet bundle $J_q\pi$ and not already on $\mathcal{E}$.

We first remark that in our context the word "singular" may have for a differential equation $\mathcal{R}_q$ (at least) the following different meanings: (i) $\mathcal{R}_q$ is not everywhere a smooth manifold; (ii) a solution of $\mathcal{R}_q$ cannot be described as a smooth section; (iii) singular solutions in the sense of Section 8.1 exist (i. e. solutions which are not contained in the general one). We will ignore here the first possibility, although it is of great practical importance, as nonlinear differential equations typically lead to subvarieties and not to submanifolds of the jet bundle.

Concerning the second point, we introduce now a more general concept of solutions. So far we have defined solutions as sections of the underlying fibred manifold $\pi : \mathcal{E} \to \mathcal{X}$. In the case of ordinary differential equations, we have thus always worked with curves parametrised by the base space $\mathcal{X}$, as we considered them as images of prolonged sections. Now we drop this restriction and allow arbitrary curves.

Definition 9.1.8. A *generalised solution* of the ordinary differential equation $\mathcal{R}_q$ is an integral curve of its Vessiot distribution $\mathcal{V}[\mathcal{R}_q]$.

For the following considerations it is crucial to realise that a generalised solution is a curve in the jet bundle $J_q\pi$ and not in $\mathcal{E}$. If a generalised solution stems from a solution in the usual sense, then this curve projects onto the image of a section $\sigma : \mathcal{X} \to \mathcal{E}$, i. e. the generalised solution is just the prolongation of a classical solution. This will be the case, whenever the Vessiot distribution is transversal along the generalised solution. In general, however, the projection to $\mathcal{E}$ of a generalised solution can be a rather complicated object.

Definition 9.1.9. Let $\mathcal{R}_q \subseteq J_q\pi$ be an involutive differential equation which is not underdetermined. A point $\rho \in \mathcal{R}_q$ is called[1]

(i) *regular*, if $\mathcal{V}_\rho[\mathcal{R}_q]$ is one-dimensional and transversal,
(ii) *regular singular*, if $\mathcal{V}_\rho[\mathcal{R}_q]$ is one-dimensional but not transversal,
(iii) *irregular singular* or *s-singular* if $\dim \mathcal{V}_\rho[\mathcal{R}_q] = s+1$ with $s > 0$.

We exclude here again the case of an underdetermined equation, as for it any point is trivially irregular singular. At a regular point we are in the situation treated by Theorem 9.1.2, i. e. at such a point we find exactly one classical solution. We will show below that generically there are *two* classical solutions at regular singular points. In the irregular singular case several possibilities exist. For example we could have a singular integral which leads again to the existence of two classical solutions. But it is also possible that infinitely many classical solutions may start or end at such points (one then sometimes speaks of a *funnel*).

Proposition 9.1.10. *Let the system $\boldsymbol{\Phi}(x,\mathbf{u},\mathbf{u}') = 0$ for m unknown functions $\mathbf{u}(x)$ be involutive. A point $\rho = (\bar{x},\bar{\mathbf{u}},\bar{\mathbf{u}}')$ is regular, if and only if*

$$\operatorname{rank}\left(\frac{\partial\boldsymbol{\Phi}}{\partial\mathbf{u}'}\right)_\rho = m\,. \tag{9.6}$$

The point ρ is regular singular, if and only if it is not regular and

$$\operatorname{rank}\left(\frac{\partial\boldsymbol{\Phi}}{\partial\mathbf{u}'}\;\middle|\;\frac{\partial\boldsymbol{\Phi}}{\partial x} + \mathbf{u}'\frac{\partial\boldsymbol{\Phi}}{\partial\mathbf{u}}\right)_\rho = m\,. \tag{9.7}$$

Proof. This claim is a simple consequence of the local computation of the Vessiot distribution $\mathcal{V}[\mathcal{R}_1]$. If we make as usual the ansatz $X = a(\partial_x + \mathbf{u}'\partial_{\mathbf{u}}) + \mathbf{b}\partial_{\mathbf{u}'}$, then $\mathcal{V}[\mathcal{R}_1]$ is determined by the linear system

$$\mathbf{b}\frac{\partial\boldsymbol{\Phi}}{\partial\mathbf{u}'} + a\left(\frac{\partial\boldsymbol{\Phi}}{\partial x} + \mathbf{u}'\frac{\partial\boldsymbol{\Phi}}{\partial\mathbf{u}}\right) = 0\,. \tag{9.8}$$

Obviously, we obtain therefore a one-dimensional Vessiot distribution, if and only if (9.7) is satisfied. Transversality of $\mathcal{V}[\mathcal{R}_1]$ is equivalent to $a \neq 0$ which is the case if already the submatrix $\partial\boldsymbol{\Phi}/\partial\mathbf{u}'$ has maximal rank. □

Remark 9.1.11. The case of a regular point can be easily formulated in an intrinsic manner via the symbol: a point $\rho \in \mathcal{R}_q \subseteq J_q\pi$ is regular, if and only if $\dim \mathcal{N}_q|_\rho = 0$. Indeed, the condition (9.6) ensures that the symbol equations have maximal rank and hence the symbol $\mathcal{N}_q$ vanishes. More generally, we conclude that the symbol matrix

[1] This terminology follows Arnold [27], §4. Rabier [367] objects that it may be confused with similar notions in the Fuchs–Frobenius theory of linear ordinary differential equations in the complex plane (see e. g. [364]). While this is certainly true, it is equally true that from a geometric point of view this terminology appears very natural, as the classification is based on whether the Vessiot distribution is regular or singular at ρ.

is always part of the matrix of the linear system determining the Vessiot distribution and that its properties decide the transversality of $\mathcal{V}[\mathcal{R}_1]$. ◁

Example 9.1.12. All the different types of singular points may already appear in a scalar implicit equation, if it cannot be solved for the highest-order derivative. Let us consider the first-order differential equation

$$\mathcal{R}_1 \,:\, (u')^2 + u^2 + x^2 - 1 = 0\,. \tag{9.9}$$

Obviously, the submanifold $\mathcal{R}_1$ is nothing but the unit sphere in $J_1\pi$. Strictly speaking, $\mathcal{R}_1$ is not a differential equation in the sense of Definition 2.3.1, as at the equator $u' = 0$ the projection $\pi^1 : \mathcal{R}_1 \to \mathcal{X}$ is not a submersion. In fact, if we take $\mathcal{X} = \mathbb{R}$, then it is not even surjective. We will simply ignore this point; in particular, as it disappears as soon as one considers the equation over the complex numbers $\mathbb{C}$.

One easily verifies with the help of Proposition 9.1.10 that all points off the equator $u' = 0$ are regular and that on the equator all points are regular singular except for the two 1-singular points $(0, \pm 1, 0)$. Indeed, except for these two irregular singularities the Vessiot distribution $\mathcal{V}[\mathcal{R}_1]$ is globally spanned by the vector field $X = u'\partial_x + (u')^2\partial_u - (x + uu')\partial_{u'}$. Figure 9.1 shows the direction defined by $\mathcal{V}[\mathcal{R}_1]$ at some points on $\mathcal{R}_1$; the singular points are marked red.

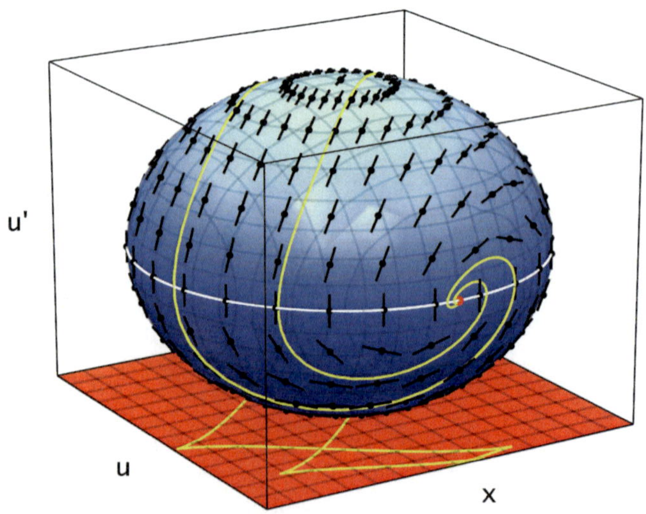

Fig. 9.1 Vessiot distribution $\mathcal{V}[\mathcal{R}_1]$ and solutions of the differential equation (9.9)

Locally, it is possible to solve for u': the upper hemisphere corresponds to the branch with the positive square root; the lower hemisphere to the one with the negative square root. As Figure 9.1 demonstrates, $\mathcal{V}[\mathcal{R}_1]$ possesses integral curves crossing the equator (in fact, they cross it repeatedly, as they spiral into the singular points). This behaviour means that they jump from one branch of the square root to the other one. If we project such a generalised solution into the x-u plane, then singularities in the form of cusps arise at the location of these jumps. This effect is not surprising, as at the equator vector field X is not transversal. Only the segments between two singularities define images of sections $\sigma : \mathcal{X} \to \mathcal{E}$ and thus classical solutions; the full projected curves correspond to multivalued sections.

We can summarise the existence and uniqueness theory of this differential equation as follows. Through every regular point we have exactly one classical solution. It can be extended until its prolongation reaches a singular point. At every regular singular point *two* classical solutions either start or end (depending on the sign of the x-coordinate). At the two irregular singular points infinitely many solutions start and end so that they are funnels. In fact, every generalised solution connects these two points. Note that for a generalised solution it does not makes sense to speak about start or end, as it is an integral curve of a one-dimensional distribution and hence no direction is distinguished as forward. ◁

As already mentioned above in Remark 9.1.3, it is a classical result in the theory of explicit ordinary differential equations $\mathbf{u}' = \mathbf{f}(x, \mathbf{u})$ that every solution can be extended until it reaches the boundary of the domain of definition of $\mathbf{f}$ (which is possibly at infinity in the case of a blow-up). As Example 9.1.12 demonstrates, the situation is more complicated for implicit equations, as a solution may also break down at a singular point.

Theorem 9.1.13. *Let $\mathcal{R}_1$ be a first-order ordinary differential equation such that everywhere $\dim \mathcal{V}[\mathcal{R}_1] = 1$ (i. e. $\mathcal{R}_1$ contains no irregular singular points). If $\rho \in \mathcal{R}_1$ is a regular point, then there exists a unique solution $\sigma \in \Gamma_{loc}(\mathcal{E})$ with $\rho \in \operatorname{im} j_1\sigma$. This solution can be extended until $\operatorname{im} j_1\sigma$ reaches either the boundary of $\mathcal{R}_1$ or a regular singular point. If $\rho \in \mathcal{R}_1$ is a regular singular point, then generically two solutions $\sigma_1, \sigma_2 \in \Gamma_{loc}(\mathcal{E})$ with $\rho \in \operatorname{im} j_1\sigma_i$ exist. Again each can be extended until $\operatorname{im} j_1\sigma_i$ reaches either the boundary of $\mathcal{R}_1$ or a regular singular point.*

Proof. The first assertion is simple. Under our assumptions the Vessiot distribution is locally generated by a smooth vector field to which we can apply the standard existence and uniqueness theorem of Picard–Lindelöf. Hence any generalised solution can be extended until it reaches the boundary of $\mathcal{R}_1$. However, its projection ceases to define a classical solution, as soon as it reaches a singular point.

If a generalised solution goes through a regular singular point, then generically the ∂_x-component of its tangent vector changes sign (for instance this is always the case in Example 9.1.12). Thus upon projection to $\mathcal{E}$ we obtain two classical solutions which both either start or end at this point. □

Remark 9.1.14. In the above theorem we deliberately formulated the case of a regular singular initial point rather vaguely by making only a generic statement. One could now start to study in much more detail the different possibilities or to give analytic criteria for them. As such an analysis becomes rapidly highly technical, we omit it. Our main point here is simply to demonstrate how useful and natural it is to consider the Vessiot distribution of the given differential equation. ◁

In the example above we had two isolated irregular singular points. It may also happen that we have a whole manifold of such singularities. This is for example the case when singular solutions in the sense of Section 8.1 exist.

Example 9.1.15. We consider again the Clairaut equation $u = xu' + f(u')$ already studied in Example 8.1.2. Now we can systematically derive the observations made there. Using our standard ansatz $X = a(\partial_x + u'\partial_u) + b\partial_{u'}$ for the generators of the Vessiot distribution $\mathcal{V}[\mathcal{R}_1]$, we find that the coefficient functions a, b must satisfy the condition $\bigl(x + f'(u')\bigr)b = 0$. Hence all points on $\mathcal{R}_1$ with $x + f'(u') \neq 0$ are regular; the remaining points form an irregular singular submanifold $\bar{\mathcal{R}}_1 \subset \mathcal{R}_1$.

At the regular points the Vessiot distribution $\mathcal{V}[\mathcal{R}_1]$ is globally generated by the vector field $X = \partial_x + u'\partial_u$. The integral curves of X are of the form $x \mapsto (x, px + f(p), p)$ with a parameter p and obviously describe the general solution found in Example 8.1.2. In contrast to Example 9.1.12, we do not find infinitely many solutions through the irregular singular points. The main difference is that here $\bar{\mathcal{R}}_1$ may again be interpreted as a differential equation (necessarily overdetermined). The Vessiot distribution $\mathcal{V}[\bar{\mathcal{R}}_1]$ is determined by the equation $a + f''(u')b = 0$ and thus always one-dimensional. All points on $\bar{\mathcal{R}}_1$ with $f''(u') \neq 0$ are regular. $\mathcal{V}[\bar{\mathcal{R}}_1]$ is generated by the vector field $Y = f''(u')\bigl(\partial_x + u'\partial_u\bigr) + \partial_{u'}$. Since the manifold $\bar{\mathcal{R}}_1$ is one-dimensional, it is the single integral curve of Y and it can be parametrised in the form $\tau \mapsto \bigl(-f'(\tau), f(\tau) - \tau f'(\tau), \tau\bigr)$. Obviously, this curve is nothing but the prolongation of the singular solution found in Example 8.1.2.

Thus we can summarise the situation as follows. The analysis of the Vessiot distribution $\mathcal{V}[\mathcal{R}_1]$ leads naturally to a decomposition of $\mathcal{R}_1$ into two parts: $\mathcal{R}_1 \setminus \bar{\mathcal{R}}_1$ and $\bar{\mathcal{R}}_1$. Considered as individual differential equations, both parts consist only of regular points (assuming for simplicity that $f''(u')$ does not vanish anywhere on $\bar{\mathcal{R}}_1$) and we can apply Theorem 9.1.13. Taking a slightly different point of view, we see that the single vector field generating $\mathcal{V}[\mathcal{R}_1 \setminus \bar{\mathcal{R}}_1]$ can be analytically continued to the whole submanifold $\mathcal{R}_1$. Thus, although $\mathcal{V}_\rho[\mathcal{R}_1]$ is two-dimensional at every point $\rho \in \bar{\mathcal{R}}_1$, not all vectors in it appear as tangents to generalised solutions. Only two directions in this plane are relevant: the continuation of the distribution $\mathcal{V}[\mathcal{R}_1 \setminus \bar{\mathcal{R}}_1]$ to ρ (i. e. $\langle X_\rho \rangle$) and $\mathcal{V}_\rho[\bar{\mathcal{R}}_1] = \langle Y_\rho \rangle$. This observation implies that there are exactly two solutions through every point $\rho \in \bar{\mathcal{R}}_1$: one member of the general solution and the singular solution.[2]

Note that this effect becomes visible only in the jet bundle $J_1\pi$ and not already on $\mathcal{E}$, since the projections satisfy $T_\rho \pi^1_0(X_\rho) = T_\rho \pi^1_0(Y_\rho)$. This identity expresses the fact that here the singular solution is the envelope of the general solution and thus the corresponding curves have a first-order contact at ρ. But as one can clearly see in Figure 8.1, the corresponding generalised solutions have a clean intersection, since the $\partial_{u'}$-components of X_ρ and Y_ρ are different (0 and $f''(u')$, respectively). ◁

---

[2] From an analytic point of view, we have infinitely many solutions through any point $\rho \in \bar{\mathcal{R}}_1$: follow a piece of the singular solution of arbitrary length and then continue along some member of the general solution. But at ρ we have only two possible directions into which we can move.

Example 9.1.16. We consider the fully nonlinear differential equation

$$\mathcal{R}_1 : x(u')^2 - 2uu' - x = 0 . \tag{9.10}$$

Obviously, it is quadratic in the derivative u' and hence for every choice of initial conditions $u(x_0) = u_0$ (except for $x_0 = u_0 = 0$) two different solutions exist. In this respect, the equation is similar to the Clairaut equations studied in Example 8.1.2. However, in contrast to these, (9.10) does not possess singular solutions. Its symbol vanishes whenever $xu' = u$ and the submanifold in $J_1\pi$ described by this equation and (9.10) is the u'-axis and thus does not represent a valid differential equation, as it is not a fibred submanifold. This observation explains why we must exclude the initial condition $x_0 = u_0 = 0$; for it no solution exists.

It is not difficult to solve the equation (9.10) explicitly: its solutions are the parabolas $u(x) = \alpha x^2 - 1/(4\alpha)$ with $\alpha \neq 0$. If we resolve (9.10) algebraically for the derivative u', we obtain $u' = (u \pm \sqrt{x^2 + u^2})/x$. Depending on the sign of u the limit $x \to 0$ exists only for one branch. But this suffices, as one easily checks that the parabolas with $\alpha > 0$ correspond to the branch with the positive square root, whereas $\alpha < 0$ yields the negative square root. The solutions and their 1-jets are shown in Figure 9.2. The parabolas for positive and negative, respectively, values of α are shown in different colours. They enclose the point $x = u = 0$ and one can see that the intersection of the u'-axis and $\mathcal{R}_1$ is the open interval $(-1, 1)$.

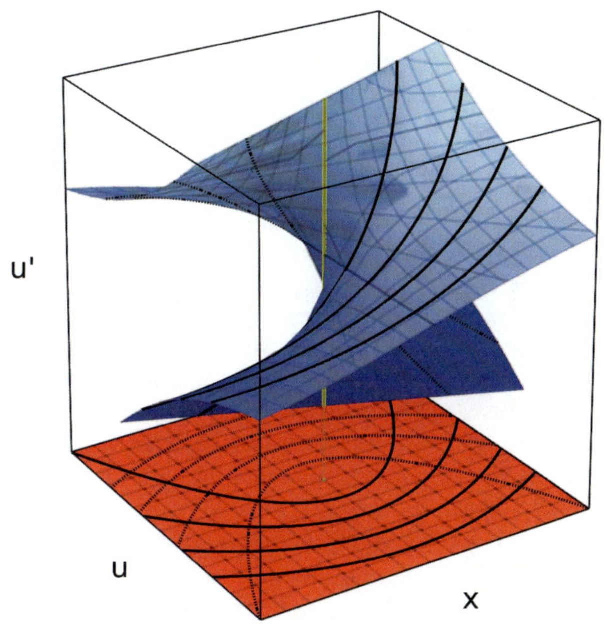

Fig. 9.2 Solutions of (9.10)

From a numerical point of view, the integration of (9.10) is not completely straightforward, as we have at each point $(x,u) \in \mathcal{E}$ two solutions for u' corresponding to the two intersecting parabolas. In this particular case, this is not so bad, as we may easily solve for u' and simply follow one branch. However, in larger examples where no symbolic solution for u' is possible and its value must be determined numerically, we obviously face the problem of choosing the right solution, if several exist. This could for example be achieved with homotopy techniques [345].

If we use the Vessiot theory, the numerical solution of (9.10) is straightforward. The Vessiot distribution $\mathcal{V}[\mathcal{R}_1]$ is everywhere one-dimensional, as it is spanned by the vector field $X = 2(xu' - u)\partial_x + 2(xu' - u)u'\partial_u + \bigl((u')^2 + 1\bigr)\partial_{u'}$. For some points on the manifold $\mathcal{R}_1$ the direction defined by it is shown in Figure 9.2. We choose a starting point $(x_0, u_0, u'_0) \in \mathcal{R}_1$ and then follow the uniquely determined integral curve of the field X through this point. As Figure 9.2 clearly shows, these integral curves never intersect.

Although the Vessiot distribution is everywhere regular, the points on the intersection of $\mathcal{R}_1$ with the u'-axis are singular in the following sense: the projection of the Vessiot distribution into the x-u plane yields the zero vector. Thus although this intersection is an integral curve of $\mathcal{V}[\mathcal{R}_1]$ and thus a generalised solution, it does not project onto a classical solution, as it is not transversal to the fibration $\pi^1_0 : J_1\pi \rightarrow \mathcal{E}$. All other generalised solutions are simply prolonged classical solutions. ◁

We restrict now to the special case of an implicit quasi-linear equation $\mathcal{R}_1$ locally represented by the system $A(x,\mathbf{u})\mathbf{u}' = \mathbf{f}(x,\mathbf{u})$ where the values of A are $m \times m$-matrices and and those of the right hand side $\mathbf{f}$ m-dimensional vectors. Thus we are dealing with a square system. Their analysis has received considerable attention in the literature (see e. g. [86, 87, 369, 370, 376, 462, 463]), in particular in the context of the numerical integration of such systems, and lead to the following notions. A point $\xi = (x,\mathbf{u}) \in \mathcal{E}$ is called an *s-singular point* of the system, if $\dim\ker A(\xi) = s$ (note that this is a different notation of s-singularity than the one introduced in Definition 9.1.9!). If furthermore $\mathbf{f}(\xi) \in \operatorname{im} A(\xi)$, one speaks of a *geometric* s-singular point, otherwise of an *algebraic* one.

From the point of view of the formal theory not much can be said about geometric s-singular points. If the equation $\mathcal{R}_1$ is involutive, then it possesses at such a point an s-dimensional formal solution space, i. e. the system is underdetermined. If $\mathcal{R}_1$ is not yet involutive, no predictions can be made about the possible effect of the completion on these points. This observation is generally valid: a reasonable singularity analysis is only possible for involutive equations. Many of the rather cumbersome calculations in the literature have their origin in the fact that one tries to analyse non-involutive equations.

By contrast, it is clear that an algebraic s-singular point is not contained in the constraint manifold $\mathcal{R}^{(1)}_0$. Hence no solution can exist at these points. Nevertheless, one frequently observes that already in the neighbourhood of such points the numerical integration of $\mathcal{R}_1$ encounters difficulties. In the literature one often speaks about *impasse points*, although the rigorous definition of this term often differs from author to author.

Example 9.1.17. Instead of trying to develop a general theory for this class of differential equations, we consider in more detail the equation $\mathcal{R}_1$ in two dependent variables u, v defined by the simple system $u' = 1$, $u = g(v)$ with some smooth function g. A number of examples in the above cited works have this particular structure. One sees immediately that the system is not formally integrable: prolongation of the algebraic equation leads to the integrability condition $\bigl(g(v)\bigr)' = g'(v)v' = 1$.

Thus the differential part of the projected equation $\mathcal{R}_1^{(1)}$ is

$$\begin{pmatrix} 1 & 0 \\ 0 & g'(v) \end{pmatrix} \begin{pmatrix} u' \\ v' \end{pmatrix} = \begin{pmatrix} 1 \\ 1 \end{pmatrix} \tag{9.11}$$

and all points $(x,u,v) \in \mathcal{E}$ with $g'(v) = 0$ are algebraic 1-singular points for it and not contained in $\mathcal{R}_0^{(2)}$. Furthermore, we see that if a solution approaches such a point, then its derivative v' must diverge which leads to the above mentioned problems in the numerical integration.

We compute now the Vessiot distribution $\mathcal{V}[\mathcal{R}_1^{(1)}]$. A general contact vector field is of the form $X = a(\partial_x + u'\partial_u + v'\partial_v) + b_1\partial_{u'} + b_2\partial_{v'}$ and lies in $T\mathcal{R}_1^{(1)}$, if its coefficients satisfy the linear system $b_1 = 0$ and $g''(v)(v')^2 a + g'(v)b_2 = 0$. Since $v' = 1/g'(v)$ on $\mathcal{R}_1^{(1)}$, a generator of the Vessiot distribution is

$$X = \bigl(g'(v)\bigr)^3\partial_x + \bigl(g'(v)\bigr)^3\partial_u + \bigl(g'(v)\bigr)^2\partial_v - g''(v)\partial_{v'}\,. \tag{9.12}$$

Thus all points on $\mathcal{R}_1^{(1)}$ are regular and we can apply without problems our existence and uniqueness Theorem 9.1.13. In principle, the formal analysis of this class of systems ends here. However, using additional analytic considerations, we can even resolve some of the problems classically encountered.

By definition, the Vessiot distribution $\mathcal{V}[\mathcal{R}_1^{(1)}]$ lives on the submanifold $\mathcal{R}_1^{(1)}$. However, we can readily extend the generator X to the whole jet bundle $J_1\pi$. The thus obtained vector field $\tilde{X} \in \mathfrak{X}(J_1\pi)$ has no singularities except that it vanishes at points where $g'(v) = g''(v) = 0$. Hence outside of these points, the field $\tilde{X}$ possesses smooth integral curves which we may consider as generalised solutions.

Actually, even these remaining singularities of $\tilde{X}$ can be removed. Because of the quasi-linearity, there is no real need to work with a distribution in $J_1\pi$, if we are only interested in a solution curve in $\mathcal{E}$. The vector field $\tilde{X}$ can be projected to $\mathcal{E}$, as its components do not depend on the derivatives. On $\mathcal{E}$ the vector field $Y = g'(v)(\partial_x + \partial_u) + \partial_v$ generates the same distribution as the projected field $(\pi_0^1)_*\tilde{X}$ but has no singularities at all.

The integral curves of either $\tilde{X}$ or Y are easily determined numerically; the algebraic 1-singular points of $\mathcal{R}_1$ possess no special meaning in this process. The origin of the numerical problems encountered in classical approaches to the integration of $\mathcal{R}_1$ in the neighbourhood of these points lies in the fact that Y ceases there to be transversal and thus we are in the same situation as at regular singular points: the integral curves of Y can no longer be interpreted as images of sections.

If a system of the type studied here arises in the mathematical modelling of a physical problem, then one must discuss whether any reasonable interpretation can be attached to these generalised solutions. In fact, this is probably rarely possible and one must say that the model breaks down at such points. Nevertheless, our approach via the vector field Y at least allows us to approach these points without any numerical problem. $\triangleleft$

9.2 The Cauchy–Kovalevskaya Theorem

With respect to the class of differential equations to which it can be applied, the Cauchy–Kovalevskaya Theorem represents probably the most general existence and uniqueness result we know for the initial value problem for normal equations. Its only structural restriction is that we are dealing with a non-characteristic problem. However, the downside of this structural generality is that it applies only to *analytic* differential equations, *analytic* initial data and makes only statements about *analytic* solutions. These restrictions are very severe, in fact, too severe for most applications.

Nevertheless, the Cauchy–Kovalevskaya Theorem is the natural continuation of the work we have done so far. In Section 2.3 we discussed the construction of formal power series solutions. Now we show that under certain conditions these series actually converge. We prove the theorem only for quasi-linear first-order equations, as according to Remark A.3.4 we can always reduce the initial value problem for a general normal non-linear equation to this special case by a quasi-linearisation.

It follows from Proposition 7.5.7 that in δ-regular coordinates any normal quasi-linear first-order system may be brought into the form

$$\mathbf{u}_{x^n} = \sum_{i=1}^{n-1} C_i(\mathbf{x},\mathbf{u})\,\mathbf{u}_{x^i} + \mathbf{c}(\mathbf{x},\mathbf{u}) \tag{9.13a}$$

where the C_i are $m \times m$ matrices with entries $c^{\alpha}_{i\beta}$ and $\mathbf{c}$ is an m-dimensional vector with components c^{α}. We consider the homogeneous initial conditions

$$\mathbf{u}(x^1,\dots,x^{n-1},0) = 0\,. \tag{9.13b}$$

The extension to general (analytic) initial data $\mathbf{f}$ is trivial; one only has to perform the coordinate transformation $\mathbf{v} = \mathbf{u} - \mathbf{f}$.

Note that the structure of (9.13a) implies that the hyperplane $x^n = 0$ on which the initial data are prescribed is non-characteristic. More general non-characteristic surfaces may always be mapped by a coordinate transformation into this hyperplane. In physical applications, the "evolution parameter" x^n represents usually the time t.

Theorem 9.2.1 (Cauchy–Kovalevskaya). *Let the functions C_i and $\mathbf{c}$ be real-analytic at the origin. Then the initial value problem* (9.13) *possesses a unique solution $\mathbf{u}(\mathbf{x})$ that is real-analytic at the origin.*

Proof. As a normal equation, (9.13a) is trivially formally integrable and we can apply the order by order construction of a formal power series solution discussed in Section 2.3. It is easy to distinguish here the principal and the parametric coefficients: any coefficient a^α_μ with $\mu_n > 0$ is principal; all others are parametric. The initial conditions (9.13b) determine uniquely all parametric coefficients of the series. It is not difficult to see that for our special initial data any Taylor coefficient a^α_μ can be represented as a polynomial with positive coefficients in the Taylor coefficients of the functions $c^\alpha_{i\beta}$ and c^α expanded at the origin.

Thus the initial value problem (9.13) possesses a unique formal power series solution. There remains to show that it converges in a neighbourhood of the origin. As the finitely many functions $c^\alpha_{i\beta}$ and c^α are analytic at the origin, two constants M and r exist such that they are all contained in $\mathcal{C}_{M,r}(0)$ (see Appendix A.2). Consider the initial value problem

$$\mathbf{U}_{x^n} = \sum_{i=1}^{n-1} \bar{C}_i(\mathbf{x},\mathbf{U})\,\mathbf{U}_{x^i} + \bar{\mathbf{c}}(\mathbf{x},\mathbf{U})\,, \tag{9.14a}$$

$$\mathbf{U}(x^1,\dots,x^{n-1},0) = 0 \tag{9.14b}$$

where all entries $\bar{c}^\alpha_{i\beta}$ of the matrices $\bar{C}_i$ and components $\bar{c}^\alpha$ of the vector $\bar{\mathbf{c}}$ are given by the same function

$$\phi(\mathbf{x},\mathbf{U}) = \frac{Mr}{r-(x^1+\cdots+x^n+U^1+\cdots+U^m)} \tag{9.15}$$

and thus majorise the functions in (9.13a): $\bar{c}^\alpha_{i\beta} \succeq c^\alpha_{i\beta}$ and $\bar{c}^\alpha \succeq c^\alpha$.

The formal power series solution $\mathbf{U}$ of (9.14) majorises the formal power series solution $\mathbf{u}$ of (9.13a) (recall, the Taylor coefficients a^α_μ are determined by polynomials with *positive* coefficients). Hence our proof is completed, if we can show that the series $\mathbf{U}$ converges.

Entering (9.15) into (9.14) results in the differential equation

$$\mathbf{U}_{x^n} = \frac{Mr}{r-(x^1+\cdots+x^n+U^1+\cdots+U^m)}\left(1+\sum_{i=1}^{n-1}\sum_{\alpha=1}^{m} U^\alpha_{x^i}\right) \tag{9.16}$$

which can be solved in closed form. With the help of the simple uniform ansatz $U^\alpha(\mathbf{x}) = V(x^1+\cdots+x^{n-1},x^n)$ for $1\le\alpha\le m$, we can reduce (9.16) to a scalar equation for the single unknown function $V(\sigma,\tau)$. Its solution, computed for example by the method of characteristics, is given by

$$V(\sigma,\tau) = \frac{1}{nm}\left(r-\sigma-\sqrt{(r-\sigma)^2-2nmMr\tau}\right)\,. \tag{9.17}$$

This function is analytic at $\sigma=\tau=0$ and (9.16) possesses a unique solution which is analytic at the origin. Thus the above constructed power series solution of (9.14) converges in a neighbourhood of the origin. □

It should be stressed that the uniqueness of the solution is asserted only in the category of functions that are real-analytic at the origin. The Cauchy–Kovalevskaya Theorem does not exclude the existence of further solutions that are not real-analytic. We will see in Section 10.2 that for linear systems a much stronger uniqueness statement is possible.

Another important point is that Theorem 9.2.1 gives a purely local result; nothing is asserted about the domain of definition of the solution. A lower bound for its radius of convergence is given by the radius of convergence of the function V defined by (9.17). We do not give a precise formula, but obviously the bound depends only on the number of independent and dependent variables, respectively, n and m and on the numbers M, r determined by the coefficients of the equations.

Example 9.2.2. The assumption of analyticity is crucial for the validity of the Cauchy–Kovalevskaya Theorem. In general, it is not possible to relax it, say to smooth functions, as the following classical result of Lewy [288] demonstrates (see also [238]). He proved the existence of a real-valued function $f \in \mathcal{C}^\infty(\mathbb{R}^3)$ such that the following linear equation for a complex-valued unknown $u(x,y,z)$

$$u_x + iu_y - 2i(x+iy)u_z = f(x,y,z) \tag{9.18}$$

possesses no solutions in $\mathcal{C}^1(\Omega)$ for any open set $\Omega \subseteq \mathbb{R}^3$. Thus this equation is formally integrable but not locally solvable in the smooth category.

At that time (1957), the existence of a linear equation without solution came as a big surprise. Nowadays, this phenomenon is much better understood: Hörmander [223, Chapt. VI] proved his famous *Poisson bracket condition* for local solvability. In local coordinates, the principal symbol of such a scalar equation $Lu = f$ in one dependent variable u is a 1×1 matrix, i. e. a function $T(\mathbf{x},\chi)$ polynomial in the coefficients of χ. Recall that $\chi = \chi_i \mathrm{d}x^i$ is a one-form and that the cotangent bundle $T^*\mathcal{X}$ is canonically a symplectic manifold (see Example C.6.2). Thus $\operatorname{Re} T$ and $\operatorname{Im} T$ may be interpreted as smooth functions on a symplectic manifold and we may compute their Poisson bracket in canonical coordinates:

$$C(\mathbf{x},\chi) = \{\operatorname{Re} T, \operatorname{Im} T\} = \frac{\partial \operatorname{Re} T}{\partial x^i}\frac{\partial \operatorname{Im} T}{\partial \chi_i} - \frac{\partial \operatorname{Re} T}{\partial \chi_i}\frac{\partial \operatorname{Im} T}{\partial x^i}\,. \tag{9.19}$$

One can show that if $T(\mathbf{x},\chi) = 0$ but $C(\mathbf{x},\chi) \neq 0$ at a point $x \in \Omega$ and for some one-form $\chi \in T_x^*\mathcal{X}$, then there exists a function $f \in \mathcal{C}^\infty(\Omega)$ such that $Lu = f$ does not possess a solution. Indeed, for Lewy's equation (9.18) we find $C(\mathbf{x},\chi) = -4\chi_z$ and thus for $\chi = -2y\mathrm{d}x + 2x\mathrm{d}y + \mathrm{d}z$ the condition of non-solvability is satisfied. $\lhd$

The way we stated the Cauchy–Kovalevskaya Theorem (which is the usual one in textbooks), it gives *sufficient* conditions for the existence of a unique analytic solution. An obvious—but rarely posed—question is whether these are also *necessary*. Kovalevskaya gave in her thesis [262] an affirmative answer, however, without proof. The essential question is whether it is necessary to require that on the right hand side of (9.13a) the derivatives with respect to x^i for $i < n$ are also of first order.

All other conditions are obviously necessary. If the initial conditions are posed on a characteristic surface, it is not possible to construct a unique power series solution; this fact is the essence of the definition of characteristic. If x^n is non-characteristic and the system cannot be brought into the form (9.13a), it is either under- or overdetermined. In the first case we do not get a unique solution, in the latter one the initial data cannot be specified arbitrarily but must satisfy compatibility conditions. The necessity of the analyticity of the involved functions is trivial.

Example 9.2.3. Kovalevskaya [262, p. 22] provided a simple counterexample to her theorem, if the right hand side of (9.13a) contains higher-order derivatives. Consider the following initial value problem for the heat equation:

$$u_t = u_{xx}\,, \tag{9.20a}$$

$$u(x,0) = 1/(1-x)\,. \tag{9.20b}$$

It is straightforward to construct its unique formal power series solution:

$$u(x,t) = \sum_{i,j=0}^{\infty} \frac{(2i)!}{i!j!} x^j t^i\,. \tag{9.21}$$

However, one can show that this series converges for no $t \neq 0$. Indeed, it is well-known that the solution of the pure initial value problem for the heat equation is given by convolution with the heat kernel

$$u(x,t) = \frac{1}{\sqrt{4\pi t}} \int_{-\infty}^{\infty} \exp\left[-\frac{(x-\xi)^2}{4t}\right] f(\xi)\,d\xi \tag{9.22}$$

for $u(x,0) = f(x)$. Obviously, the integral does not converge for arbitrary analytic initial data f, but we must impose on f some decay conditions for $|x| \to \infty$.

As shown by Tikhonov [455] (our presentation follows [95, Sect. 12.5]), the heat equation also nicely demonstrates the relation between analyticity and uniqueness. For any smooth function $\phi(t)$ the series

$$v(x,t) = \sum_{k=0}^{\infty} \phi^{(k)}(t) \frac{x^{2k}}{(2k)!} \tag{9.23}$$

defines a formal solution of the heat equation.

We consider now specifically the function $\phi(t) = \exp(-1/t^2)$. This function is analytic on $\mathbb{R} \setminus \{0\}$, but at the origin it is only smooth (cf. Appendix A.2). In fact, ϕ is even analytic in $\mathbb{C} \setminus \{0\}$. Thus by Cauchy's Theorem we find that for any simple closed contour Γ such that t is on the inside and the origin on the outside of Γ

$$\phi^{(k)}(t) = \frac{k!}{2\pi i} \oint_{\Gamma} \frac{\phi(z)}{(z-t)^{k+1}} dz\,. \tag{9.24}$$

If we choose $\Gamma = \left\{t + \frac{1}{2} t e^{i\theta} \mid 0 \leq \theta \leq 2\pi\right\}$, then $\operatorname{Re}(1/z^2) \geq 4/(81t^2)$ for all $z \in \Gamma$. This allows us to estimate for $t > 0$

$$|\phi^{(k)}(t)| \leq k!\left(\frac{2}{t}\right)^{k+1} \exp\left(-\frac{4}{81t^2}\right) \tag{9.25}$$

and consequently

$$|v(x,t)| \leq \frac{2}{t}\exp\left(-\frac{4}{81t^2}\right)\sum_{k=0}^{\infty}\frac{2^k k!}{(2k)!}\frac{x^{2k}}{t^k} \leq \frac{2}{t}\exp\left(\frac{x^2}{t}-\frac{4}{81t^2}\right). \tag{9.26}$$

A first consequence of this estimate is that (9.23) defines for our choice of ϕ not only a formal but a smooth solution of the heat equation. Secondly, it implies that $|v(x,t)| \to 0$ for $t \to 0+$ (even uniformly on any finite x-interval) and $v(x,t)$ solves the initial value problem with the initial condition $v(x,0) = 0$. Thus if $u(x,t)$ is any smooth solution for initial data $f(x)$, then $u(x,t) + v(x,t)$ is another smooth solution of the same initial value problem. Tikhonov [455] showed that we can achieve uniqueness by imposing the additional condition that $u(x,t)$ satisfies the estimate

$$|u(x,t)| \leq Me^{ax^2} \tag{9.27}$$

with positive constants a, M. Obviously, $v(x,t)$ violates this condition. ◁

Of course, such a counterexample does not suffice to prove the necessity of the conditions in the Cauchy–Kovalevskaya Theorem. A rigorous proof was achieved comparatively recently by Mizohata [324] for the case of linear systems. However, the techniques used are beyond the scope of this book.

9.3 Formally Well-Posed Initial Value Problems

In the previous section we discussed the initial value problem for normal differential equations. There it is immediately obvious what initial conditions one should impose: all unknown functions $u^\alpha(\mathbf{x})$ are prescribed on the same hypersurface $x^n = 0$. For a non-normal equation this question becomes less trivial. In fact, it is not even completely clear what one should call an "initial value problem." We will take the following simple point of view: in an initial value problem one prescribes certain derivatives $u^\alpha_\mu(\mathbf{x})$ on coordinate planes of the form[3] $x^{i_1} = \cdots x^{i_k} = 0$. We admit that for different derivatives different coordinate planes are used and that the codimensions k of these planes vary. In this section we study the problem of choosing the right form and number of initial conditions on the level of formal power series, i. e. we are only considering formal power series solutions and the initial data take the

[3] Of course, one could use planes described by equations of the form $x^i = c^i$ with a constant c^i, but for simplicity we assume that our local coordinates are chosen in such a way that all planes contain the origin. Note that this excludes the possibility that two different planes are parallel. However, this is done on purpose, as such a situation would correspond more to a boundary than an initial value problem. Our methods are then not applicable, as we need a common point of all planes around which we can expand our solution.

form of formal power series, too. The next section will deal with the extension of the results to analytic solutions and analytic initial data.

Definition 9.3.1. An initial value problem for a differential equation $\mathcal{R}_q$ is *formally well-posed*, if it possesses a unique formal power series solution for arbitrary formal power series as initial data.

Thus in a formally well-posed initial value problem the Taylor coefficients of the initial data are in a one-to-one correspondence with the parametric derivatives of the differential equation. It is easy to see that the initial value problem considered in the Cauchy–Kovalevskaya Theorem 9.2.1 is formally well-posed. Here all derivatives that are only with respect to $x^1,\dots,x^{n-1}$ are parametric. As the initial data prescribed consist of functions depending on these variables, each such derivative is uniquely determined by the corresponding Taylor coefficient of the initial data.

Remark 9.3.2. The concept of well-posedness plays an important role in the theory of differential equations. According to the classical definition of Hadamard a problem is well-posed, if (i) it has a solution, (ii) the solution is unique and (iii) it depends continuously on the initial data. Of course, this description is still rather vague and must be made rigorous by specifying function spaces and topologies. In Definition 9.3.1 we gave a precise formulation of the first two points by restricting to formal power series, but we completely ignored the third point.

This discussion points to a general problem with the Cauchy–Kovalevskaya Theorem: it admits initial value problems that are not well-posed in the classical sense, as a famous example of Hadamard [190, p. 33] demonstrates. Consider for an arbitrary integer $n \in \mathbb{N}$ the following initial value problem for the Laplace equation:

$$u_{tt} + u_{xx} = 0\,, \tag{9.28a}$$
$$u(x,0) = 0\,, \tag{9.28b}$$
$$u_t(x,0) = \sin(nx)/n\,. \tag{9.28c}$$

Its (unique) solution is given by $u(x,t) = \sinh(nt)\sin(nx)/n^2$. If we consider the limit $n \to \infty$, we see that the initial data vanish but that for any $t > 0$ the solution blows up. As $u(x,t) \equiv 0$ is the unique solution for vanishing initial data, solutions of (9.28) cannot depend continuously on the initial data: arbitrarily small changes of the initial data may lead to arbitrarily large changes in the solution.

This effect is typical for elliptic equations and the reason why one usually studies boundary value problems for them. We will ignore in the sequel the question of continuous dependency on the initial data and concentrate solely on the existence and uniqueness of (formal) solutions. ◁

Our goal in this section is to determine in a systematic way formally well-posed initial value problems for involutive systems. We start with (homogeneous) linear systems and for simplicity we further restrict to the case of one dependent variable; in the general case one must simply perform the same analysis for each dependent variable separately. Taking up again the notations of Example 3.2.4, such a system corresponds to a set $\mathcal{F} \subset \mathcal{D}$ where $\mathcal{D} = \mathbb{F}[\partial_1, \dots, \partial_n]$ denotes the ring of linear differential operators with coefficients in some differential field $\mathbb{F}$. However, from an algebraic point of view, the set $\mathcal{F}$ does not really matter but only the (left) ideal $\mathcal{I} = \langle \mathcal{F} \rangle$ generated by it; indeed, $\mathcal{I}$ consists of all equations obtainable from the set $\mathcal{F}$ by differentiating and taking linear combinations. Therefore, we will now sometimes also speak of an initial value problem for an ideal $\mathcal{I} \subseteq \mathcal{D}$.

Example 9.3.3. Before going into technical details, we demonstrate our basic strategy for the construction of formally well-posed initial value problems at a simple example in one dependent and two independent variables where the right initial conditions are easily found by inspection. Consider the differential equation

$$\mathcal{R}_2 : \begin{cases} u_{yy} = \alpha_1 u + \alpha_2 u_x + \alpha_3 u_y \\ u_{xy} = \beta_1 u + \beta_2 u_x + \beta_3 u_y \end{cases} \tag{9.29}$$

where the coefficients α_i, β_i are smooth functions of x, y such that no integrability conditions arise.[4] One easily verifies that under this assumption $\mathcal{R}_2$ is involutive. A natural choice as principal derivatives are u_{yy}, u_{xy} and all their derivatives. Conversely, the parametric derivatives are then u, u_y and all pure x-derivatives.

According to our definition above, all parametric coefficients must be uniquely determined by the initial conditions in a formally well-posed initial value problem. As all pure x-derivatives are parametric, it is obvious that the initial data must comprise a function of x. Thus we take as one initial condition $u(x,0) = f(x)$. This determines already all parametric coefficients in our system with the exception of the one corresponding to u_y. Hence we impose as a second initial condition $u_y(0,0) = g$. This time it is not possible to prescribe an arbitrary function but only a constant. If we required $u_y(x,0) = g(x)$, then we would have two possibilities to determine the principal coefficient corresponding to u_{xy} leading to the consistency condition

$$g'(0) = \beta_1(0,0)f(0) + \beta_2(0,0)f'(0) + \beta_3(0,0)g(0) \tag{9.30}$$

(and similar conditions arising from the higher order derivatives $u_{x\cdots xy}$) so that the initial data f, g could not be chosen arbitrarily contradicting Definition 9.3.1.

It is a typical feature of non-normal systems that even in the non-characteristic initial value problem one cannot prescribe all initial data on one hypersurface as in the Cauchy–Kovalevskaya Theorem. In the case of an underdetermined system the data would not suffice to determine all parametric coefficients, as some unknown functions are completely arbitrary. For an overdetermined system one could not prescribe arbitrary data, as they would have to satisfy consistency conditions. ◁

As a first step towards the construction of formally well-posed initial value problems, we need an algebraic approach to the distinction into principal and parametric derivatives. Given the ideal $\mathcal{I} \subseteq \mathcal{D}$ describing the differential equation, this is easily achieved with the help of a term order and leads to concepts familiar from Chapter 5.

---

[4] One easily verifies that one can choose α_3, β_2 and β_3 arbitrarily and for formal integrability the remaining coefficients must be of the form $\alpha_1 = \beta_2(\beta_2 - \alpha_3)$, $\alpha_2 = 0$ and $\beta_1 = -\beta_2\beta_3$.

Definition 9.3.4. Let $\mathcal{I} \subseteq \mathcal{D}$ be a (left) ideal. For a given term order $\prec$, we call the derivatives ∂^μ with $\mu \in \Delta_\prec(\mathcal{I}) = \mathrm{le}_\prec \mathcal{I}$ *principal*. All remaining derivatives are *parametric* and their exponents form the *complementary set* $\Pi_\prec(\mathcal{I}) = \mathbb{N}_0^n \setminus \Delta_\prec(\mathcal{I})$.

Comparing with the introduction of principal and parametric derivatives (or coefficients, respectively) during the order by order construction of formal power series solutions in Section 2.3, we see that there at each order of prolongation the principal derivatives could be chosen completely independent of the choices at the previous orders, as we treated at each order a separate linear system of equations. Now we select the principal derivatives in a systematic manner based on a term order.

While $\Delta_\prec(\mathcal{I})$ is a monoid ideal, the complementary set $\Pi_\prec(\mathcal{I})$ does in general not possess any algebraic structure.[5] Given an involutive division L and an involutive basis $\mathcal{H}$ of the ideal $\mathcal{I}$ with respect to L and $\prec$, we immediately obtain a disjoint decomposition of $\Delta_\prec(\mathcal{I})$ via the involutive cones of the multi indices in $\mathrm{le}_\prec \mathcal{H}$. The basic idea underlying the construction of formally well-posed initial value problems is to construct a similar disjoint decomposition of the complementary set $\Pi_\prec(\mathcal{I})$. But the question of complementary decompositions was already successfully treated in Section 5.1 (we even provided with the Algorithms 5.1 and 5.2 two methods for their effective construction) and we exploit here the results for obtaining formally well-posed initial value problems.

Given an ideal $\mathcal{I} \subseteq \mathcal{D}$, we decompose the set $\Pi_\prec(\mathcal{I})$ in the form (5.2) and choose an arbitrary point $\mathbf{x}_0 \in \mathcal{X}$ (typically the origin of our local coordinate system). For subsets $N = \{i_1, \dots, i_k\} \subseteq \{1, \dots, n\}$ we introduce two notations: $\mathbf{y} = N(\mathbf{x})$ where $y^i = x^i$ for $i \in N$ and $y^i = x_0^i$ otherwise and $\mathbf{x}_N = (x^{i_1}, \dots, x^{i_k})$. In other words, $N(\mathbf{x})$ denotes a projection of the vector $\mathbf{x}$ where those components of it that do not correspond to indices contained in N are set to the corresponding entry in $\mathbf{x}_0$ and $\mathbf{x}_N$ simply collects those components of $\mathbf{x}$ that correspond to the elements of N. The decomposition (5.2) is defined by a set $\overline{\mathcal{B}} \subseteq \Pi_\prec(\mathcal{I})$ and associated sets $N_\nu \subseteq \{1, \dots, n\}$ for each $\nu \in \overline{\mathcal{B}}$. Based on these data, we prescribe in the neighbourhood of the chosen point $\mathbf{x}_0$ the following initial conditions for our system:

$$(D_\nu u)\big(N_\nu(\mathbf{x})\big) = f_\nu(\mathbf{x}_{N_\nu}) , \qquad \forall \nu \in \overline{\mathcal{B}} . \tag{9.31}$$

Here, the right hand sides f_ν are arbitrary formal power series in their arguments. If $N_\nu = \emptyset$, then f_ν is simply an arbitrary constant.

Theorem 9.3.5. *The initial value problem for the ideal $\mathcal{I}$ defined by the initial conditions* (9.31) *is formally well-posed.*

Proof. We expand the solution of the ideal $\mathcal{I}$ into a formal power series about the point $\mathbf{x}_0$. By definition, its parametric coefficients are in a one-to-one correspondence with the elements of the set $\Pi_\prec(\mathcal{I})$. Let $\mu \in \Pi_\prec(\mathcal{I})$ be the multi index corresponding to a given parametric coefficient. Then exactly one multi index $\nu \in \overline{\mathcal{B}}$

[5] One sometimes says that $\Pi_\prec(\mathcal{I})$ is an *order ideal* in $\mathbb{N}_0^n$ meaning that if an exponent μ is contained in it, then any "divisor" of it lies in $\Pi_\prec(\mathcal{I})$, too. Obviously, this property is characteristic for complements to monoid ideals.

exists in the complementary decomposition underlying (9.31) such that $\mu = \nu + \rho$ with $\rho \in \mathbb{N}^n_{N_\nu}$. Our initial value problem prescribes for this coefficient the unique value $D_\rho f_\nu(\mathbf{x}_0)$ (note that this derivative makes sense, as $\rho \in \mathbb{N}^n_{N_\nu}$ and f_ν depends on all the variables in $\mathbf{x}_{N_\nu}$). Conversely, for every $\rho \in \mathbb{N}^n_{N_\nu}$ the multi index $\mu = \nu + \rho$ lies in $\Pi_\prec(\mathcal{I})$. This implies that we obtain for arbitrary initial data a unique formal power series solution and, by definition, our problem is formally well-posed. □

Thus we see that the construction of a formally well-posed initial value problem for a differential equation $\mathcal{F} \subset \mathcal{D}$ is in fact equivalent to the determination of a Stanley decomposition of the $\mathcal{D}$-module $\mathcal{D}/\mathcal{I}$ where $\mathcal{I} = \langle \mathcal{F} \rangle$. Both operations reduce to the same problem, namely to find a decomposition of the form (5.2) for the complementary set of the monoid ideal $\mathrm{le}_\prec \mathcal{I}$. Note that in order to determine the latter ideal we need at least a Gröbner basis of $\mathcal{I}$.

Example 9.3.6. Given a system in Cauchy–Kovalevskaya form, we must analyse for each dependent variable the monoid ideal $\langle \{[0,\dots,0,1]\} \rangle$. Its complement is obviously given by $[0,\dots,0] + \mathbb{N}^n_{\{1,\dots,n-1\}}$. Thus we are led to the initial value problem

$$u^\alpha(x^1,\dots,x^{n-1},x^n_0) = f^\alpha(x^1,\dots,x^{n-1}) , \qquad 1 \leq \alpha \leq m \tag{9.32}$$

which is used in the Cauchy–Kovalevskaya Theorem.

As a simple example of a system not in Cauchy–Kovalevskaya form we consider the wave equation in characteristic coordinates $u_{xy} = 0$. The corresponding monoid ideal $\mathcal{I} = \langle \{[1,1]\} \rangle$ was already studied in Example 5.1.5. We obtain the classical characteristic initial value problem

$$u(x,y_0) = f(x) , \qquad u_y(x_0,y) = g(y) . \tag{9.33}$$

The second condition is often substituted by $u(x_0,y) = \hat{g}(y)$, but then we must impose the consistency condition $f(0) = \hat{g}(0)$ so that the initial value problem does not possess a solution for arbitrary initial data and thus is not formally well-posed. Our formulation allows us to choose g arbitrarily.

The integration of this initial value problem is of course trivial and we can express the general solution in terms of f and g as

$$u(x,y) = f(x) + \int_{y_0}^{y} g(\xi)\,\mathrm{d}\xi . \tag{9.34}$$

It looks slightly different from the classical form $u(x,y) = f(x) + \hat{g}(y)$ but is of course equivalent (a systematic derivation of this expression will be given in Proposition 9.3.8 below). Our form has the advantage that we have a bijection between the parametric Taylor coefficients of the solution and the coefficients of the functions f and g (as expected since (9.33) is formally well-posed).

The classical form shows a different behaviour, as the zeroth order coefficient of u is given by the sum of the zeroth order coefficients of f and $\hat{g}$. One can derive the classical form from the standard pairs $\big([0,0],\{1\}\big)$ and $\big([0,0],\{2\}\big)$ of the monoid ideal $\langle [1,1] \rangle$. These two pairs induce a complementary decomposition which is not disjoint, as the multi index $[0,0]$ lies in the intersection of the corresponding cones. This redundancy obviously reflects the fact that we cannot uniquely determine the zeroth order coefficients of f and $\hat{g}$ from u. ◁

Remark 9.3.7. In Section 8.1 we introduced closed form representations of general solutions in which manifolds $\mathcal{X}_\ell$ appear. Our initial value problems take planes for these manifolds, namely, $\mathcal{X}_\nu = \{\mathbf{x} \in \mathcal{X} \mid x^i = x_0^i\ \forall i \notin N_\nu\}$. If we construct them via Pommaret bases, these submanifolds form a flag $\{\mathbf{x}_0\} = \mathcal{X}_0 \subset \mathcal{X}_1 \subset \cdots \subset \mathcal{X}_n = \mathcal{X}$ where $\mathcal{X}_k = \{\mathbf{x} \in \mathcal{X} \mid x^i = 0\ \forall i > k\}$. For other decompositions this is not necessarily the case, as we have seen above for the characteristic initial value problem of the wave equation (this observation demonstrates again that δ-regularity is related to characteristics, as Pommaret bases exist only in δ-regular coordinates). ◁

A particularly simple situation arises, if we are dealing with a monomial system, i. e. if the set $\mathcal{F} \subset \mathcal{D}$ is monomial. In this case we can easily solve the above constructed initial value problem and any complementary decomposition of the factor module $\mathcal{D}/\langle\mathcal{F}\rangle$ leads immediately to a closed-form representation of the general solution of the system $\mathcal{F}$.

Proposition 9.3.8. *Let $\mathcal{I} \subseteq \mathcal{D}$ be a monomial ideal. If a disjoint decomposition $\mathcal{T}$ of the complementary set $\Pi(\mathcal{I})$ is defined by the finite set $\overline{\mathcal{B}} \subset \mathbb{N}_0^n$ and associated sets $N_\nu \subseteq \{1,\dots,n\}$ of multiplicative variables for each multi index $\nu \in \overline{\mathcal{B}}$, then the unique smooth solution of the associated initial value problem* (9.31) *is given by*

$$u(\mathbf{x}) = \sum_{\nu\in\overline{\mathcal{B}}} \int_{x_0^{i_q}}^{\zeta_q} \cdots \int_{x_0^{i_1}}^{\zeta_1} f_\nu(\hat{\boldsymbol{\xi}}_\nu)\,\mathrm{d}\xi_1 \cdots \mathrm{d}\xi_q\,. \tag{9.35}$$

Here $(i_1,\dots,i_q) = R(\nu)$ denotes the unique realisation of the multi index ν of length $q = |\nu|$ as repeated index with $i_1 \leq i_2 \leq \cdots \leq i_q$. The upper integration limit ζ_k equals ξ_{k+1}, if $i_k = i_{k+1}$, and x^{i_k} otherwise. Finally, for any index $j \in N_\nu$ the component $\hat{\boldsymbol{\xi}}_\nu^j$ is given by x^j, if $j \notin \operatorname{supp}\nu$, and otherwise by ξ_k where $k = \min\{\ell \mid i_\ell = j\}$.

Example 9.3.9. Before we prove Proposition 9.3.8, we give an example for the closed-form solution (9.35) in order to make the simple idea behind it more transparent. We consider a slight modification of Example 9.3.6, namely the fourth-order equation $u_{xxyy} = 0$. A disjoint decomposition of the complementary set $\Pi(\mathcal{I})$ of parametric derivatives is defined by the four pairs

$$\big([0,0],\{1\}\big),\ \big([0,1],\{1\}\big),\ \big([0,2],\{2\}\big),\ \big([1,2],\{2\}\big)\,. \tag{9.36}$$

Thus we may take as formally well-posed initial value problem for our equation

$$\begin{aligned} u(x,y_0) &= f_{[0,0]}(x)\,, & u_y(x,y_0) &= f_{[0,1]}(x)\,,\\ u_{yy}(x_0,y) &= f_{[0,2]}(y)\,, & u_{xyy}(x_0,y) &= f_{[1,2]}(y)\,. \end{aligned} \tag{9.37}$$

Evaluation of our solution formula (9.35) yields now

$$\begin{aligned} u(x,y) = f_{[0,0]}(x) + \int_{y_0}^{y} f_{[0,1]}(x)\,\mathrm{d}\xi + \\ \int_{y_0}^{y}\int_{y_0}^{\xi_2} f_{[0,2]}(\xi_1)\,\mathrm{d}\xi_1\mathrm{d}\xi_2 + \int_{x_0}^{x}\int_{y_0}^{y}\int_{y_0}^{\xi_2} f_{[1,2]}(\xi_1)\,\mathrm{d}\xi_1\mathrm{d}\xi_2\mathrm{d}\xi_3\,. \end{aligned} \tag{9.38}$$

Two integrations can immediately be performed, as the integrands do not depend on the integration variable. This simplification leads to the final expression

$$\begin{aligned} u(x,y) = f_{[0,0]}(x) + (x-x_0)f_{[0,1]}(x) + \\ \int_{y_0}^{y}\int_{y_0}^{\xi_2} f_{[0,2]}(\xi)\,\mathrm{d}\xi_1\mathrm{d}\xi_2 + (x-x_0)\int_{y_0}^{y}\int_{y_0}^{\xi_2} f_{[1,2]}(\xi_1)\,\mathrm{d}\xi_1\mathrm{d}\xi_2\,. \end{aligned} \tag{9.39}$$

It is easy to see that generally in (9.35) all integrations corresponding to a value i_k which is not contained in N_ν can be explicitly performed. Truly nested integrations appear only, if several entries i_k of $R(\nu)$ are identical and contained in N_ν, as it is here the case for $\nu = [0,2]$ and $\nu = [1,2]$. ◁

Proof (of Proposition 9.3.8). Let $(\bar{\nu}, N_{\bar{\nu}})$ be one of the pairs defining the complementary decomposition. If we compute $D_{\bar{\nu}}u\big(N_{\bar{\nu}}(\mathbf{x})\big)$ for $u(\mathbf{x})$ given by (9.35), then the summand with $\nu = \bar{\nu}$ obviously just yields $f_{\bar{\nu}}(\mathbf{x}_{N_{\bar{\nu}}})$, as the differentiations just undo the integrations. Otherwise, there exists an index i such that $\nu_i \neq \bar{\nu}_i$. If $\nu_i < \bar{\nu}_i$, then $i \notin N_\nu$, as otherwise the decomposition was not disjoint. Hence the integrand does not depend on x^i and since we differentiate more often with respect to x^i than there are integrations over x^i, the corresponding summand vanishes.

If conversely $\nu_i > \bar{\nu}_i$, then $i \notin N_{\bar{\nu}}$ by the same argument as above. Now we have more integrations than differentiations with respect to x^i. Since $i \notin N_{\bar{\nu}}$, the lower and the upper limit of the outer surviving integration become identical upon the restriction to $N_{\bar{\nu}}(\mathbf{x})$. Hence we find again that the corresponding summand vanishes.

We conclude that the function $u(\mathbf{x})$ defined by (9.35) satisfies the initial conditions (9.31). Furthermore, the same argument as in the case $\nu_i < \bar{\nu}_i$ shows that $u(\mathbf{x})$ is a solution of the given differential equation. Indeed, if $\partial^\mu \in \mathcal{I}$, then for any multi index $\nu \in \bar{\mathcal{B}}$ an index $i \notin N_\nu$ exists such that $\mu_i > \nu_i$ and hence $D_\mu u = 0$. Thus (9.35) defines a trivially smooth solution of our initial value problem. We will show in Section 10.2 that for linear systems smooth solutions of formally well-posed initial value problems are unique. □

One easily verifies that (9.35) defines a general solution in the sense of our geometric Definition 8.1.4. The proposition remains correct for any strong solution, i. e. if q is the maximal length of the multi indices $\mu^{(\tau)}$, then we must only assume that at least $u \in \mathcal{C}^q$ entailing that also the functions g_ν on the right hand side of (9.35) are at least in $\mathcal{C}^{q-|\nu|}$.

A closer look at the proof reveals that at least for one direction the differentiability of the functions g_ν is not needed at all: for showing that (9.35) solves our system, it was sufficient to differentiate each summand with respect to a variable on which the corresponding function g_ν does not depend. Thus (9.35) may be considered as a (weak) solution of $\mathcal{I}$ for more or less arbitrary functions g_ν.

If we do not insist on the uniqueness of the representation (9.35), then we may also use the standard pairs (ν, N_ν) of the monoid ideal $\mathrm{le}_\prec \mathcal{I}$. As shown in Proposition 5.3.13, they also induce a decomposition of the complementary set. As this decomposition is not disjoint, different choices of the functions g_ν may now lead to the same solution u (cf. Example 9.3.6).

For arbitrary linear systems we cannot hope to obtain such a closed form representation of the solution, but we can at least provide the simple Algorithm 9.1 for the construction of any Taylor coefficient of the formal power series solution of a formally well-posed initial value problem. It requires as input a Gröbner basis of the ideal $\mathcal{I} \subseteq \mathcal{D}$ for some term order $\prec$ with leading ideal $\Delta_\prec(\mathcal{I})$ and a corresponding complementary decomposition $\mathcal{T}$ of $\Pi_\prec(\mathcal{I})$ together with the necessary initial data for defining the initial value problem (9.31).

Algorithm 9.1 Taylor coefficient of formal solution (linear version)

Input: Gröbner basis $\mathcal{G}$ of $\mathcal{I}$ for term order $\prec$, complementary decomposition $\mathcal{T}$ with corresponding initial data for each $(\nu, N_\nu) \in \mathcal{T}$, expansion point $x_0 \in \mathcal{X}$, multi index $\rho \in \mathbb{N}_0^n$

Output: Taylor coefficient a_ρ of unique formal power series solution of initial value problem corresponding to given complementary decomposition

$f \leftarrow \mathtt{NormalForm}_\prec(\partial^\rho, \mathcal{G})$
for all $\mu \in \operatorname{supp} f$ **do**
  find unique $(\nu, N_\nu) \in \mathcal{T}$ such that $\mu \in \nu + \mathbb{N}^n_{N_\nu}$
  $a_\mu \leftarrow \partial^{\mu-\nu} f_\nu(x_0)$
end for
write $f = \sum_\mu c_\mu(\mathbf{x})\partial^\mu$
return $\sum_\mu c_\mu(x_0) a_\mu$

Example 9.3.10. If we denote by $\mathfrak{m}_\partial = \langle \partial_1, \dots, \partial_n \rangle$ the homogeneous maximal ideal in $\mathcal{D}$, then a particularly simple situation arises for ideals $\mathcal{I} \subseteq \mathcal{D}$ such that $\mathfrak{m}_\partial^s \subseteq \mathcal{I}$ for some exponent $s \in \mathbb{N}$ (in the case that $\mathcal{D}$ is commutative, i. e. for linear differential operators with constant coefficients, this condition means that $\mathcal{I}$ is $\mathfrak{m}_\partial$-primary and we can take for s the satiety $\operatorname{sat}\mathcal{I}$). Obviously, the solution space of the differential equation corresponding to $\mathfrak{m}_\partial^s$ consists only of polynomials of degree less than s. Hence our given system possesses also only polynomial solutions. More precisely, under the made assumption the set $\mathcal{N} = \{\mu \in \mathbb{N}_0^n \mid \partial^\mu \notin \mathcal{I}\}$ is finite and it is easy to see that the support of any polynomial solution contains only terms x^μ with $\mu \in \mathcal{N}$.

Furthermore, such an ideal $\mathcal{I}$ is trivially zero-dimensional and hence we are dealing with a differential equation of finite type. Thus $\Pi_\prec(\mathcal{I})$ is a finite set with a trivial disjoint decomposition: $\mathcal{T} = \{(\nu, \emptyset) \mid \nu \in \Pi_\prec(\mathcal{I})\}$. We choose an expansion point $\mathbf{x}_0 \in \mathcal{X}$ (a natural choice would be of course $\mathbf{x}_0 = 0$). We can construct for each multi index $\mu \in \Pi_\prec(\mathcal{I})$ the polynomial solution h_μ for the initial data $f_\nu(\mathbf{x}_0) = \delta_{\nu\mu}$ by applying Algorithm 9.1 for each multi index $\rho \in \mathcal{N}$. Then the set $\{h_\mu \mid \mu \in \Pi_\prec(\mathcal{I})\}$ is a basis of the solution space of our differential equation. ◁

Finally, we study the extension of these results to systems with arbitrary nonlinearities. The only available notion of involution for such systems is the intrinsic

Definition 7.2.1 within our geometric approach. Thus we start with an involutive equation $\mathcal{R}_q \subseteq J_q\pi$ and construct in local coordinates an initial value problem in the neighbourhood of a point $\rho \in \mathcal{R}_q$. We prepare again a local representation in Cartan normal form: each equation is solved for a different derivative which is the largest derivative in the equation with respect to the degree reverse lexicographic order.

Consider again the symbol module $\mathcal{N}^0$ (as we have restricted for simplicity to the case of only one dependent variables, $\mathcal{N}^0$ is in fact an ideal); obviously only the equations of order q contribute to it. As we assume that $\mathcal{R}_q$ is involutive, these equations induce by Definition 7.1.17 of an involutive symbol a Pommaret basis of it and their leading derivatives yield a Pommaret basis of $\mathrm{le}_\prec \mathcal{N}^0$. Let u_μ be the leading derivative of an equation whose order is less than q. As $\mathcal{R}_q$ is formally integrable, we may choose our local representation such that $u_{\mu+1_i}$ for any $1 \leq i \leq n$ is the leading derivative of some equation, too.

Let $\Delta_\prec(\mathcal{R}_q) \subseteq \mathbb{N}_0^n$ be the monoid ideal generated by the multi indices of the leading derivatives;[6] obviously, $\big(\Delta_\prec(\mathcal{R}_q)\big)_{\geq q} = \mathrm{le}_\prec \mathcal{N}^0$. Thus our generating set is a weak Pommaret basis of $\Delta_\prec(\mathcal{R}_q)$ and, by Proposition 3.1.12, an involutive autoreduction yields a strong Pommaret basis. But now we are in the same situation as above for linear systems: we have a monoid ideal defined by the principal derivatives and must decompose its complement in the form (5.2). Then we impose the initial conditions (9.31) and show as in Theorem 9.3.5 that this yields a formally well-posed initial value problem in the neighbourhood of ρ.

Remark 9.3.11. The considerations above lead straightforwardly to Algorithm 9.2 for determining any Taylor coefficient in the formal power series solution of an involutive differential equation $\mathcal{R}_q$. It assumes that the Cartan normal form of $\mathcal{R}_q$ is given by equations $u_\mu = \phi_\mu\big(\mathbf{x}, \tilde{u}^{(|\mu|)}\big)$ where the principal derivatives u_μ induce a Pommaret basis of the monoid ideal $\mathrm{le}_\prec \mathcal{N}^0$ and $\tilde{u}$ denotes the parametric derivatives only and that we know already a complementary decomposition $\mathcal{T}$ of $\mathrm{le}_\prec \mathcal{N}^0$ together with the necessary initial data for the corresponding formally well-posed initial value problem (9.31). If we make the ansatz $u^\alpha(\mathbf{x}) = \sum_{\mu \in \mathbb{N}_0^n} \frac{a_\mu}{\mu!} x^\mu$ for the unique power series solution of this problem, then the algorithm determines the coefficient a_ρ for any multi index $\rho \in \mathbb{N}_0^n$.

The algorithm first checks in Line /1/ whether a_ρ is a parametric coefficient. This is the case, if and only if the complementary decomposition $\mathcal{T}$ contains a (necessarily unique) pair (ν, N_ν) such that $\rho \in \nu + \mathbb{N}^n_{N_n}$. Then a_ρ is simply the corresponding Taylor coefficient of the initial datum f_ν and thus can be computed as $a_\rho = \partial^{\rho-\nu} f_\nu(0)$.

If the coefficient a_ρ is principal, then—by the properties of a Pommaret basis—we find exactly one equation $u_\mu = \phi_\mu\big(\mathbf{x}, \tilde{u}^{(|\mu|)}\big)$ such that ρ is a multiplicative multiple of μ, i. e. $\rho \in \mathcal{C}_P(\mu)$. Prolonging this equation yields $a_\rho = \big(D^{\rho-\mu}\phi_\mu\big)\big(0, c^{(|\rho|)}\big)$. Note that here on the right hand side further coefficients a_σ up to order $|\rho|$ may appear which we must then compute recursively in order to obtain a_ρ. Since by construction always $\sigma \prec_{\mathrm{degrevlex}} \rho$, the recursion always terminates. ◁

[6] Despite our notation, this monoid ideal is of course not intrinsically defined but depends on the chosen local representation.

Algorithm 9.2 Taylor coefficient of formal solution (general version)

Input: Cartan normal form $u_\mu = \phi_\mu$ of involutive differential equation $\mathcal{R}_q$, complementary decomposition $\mathcal{T}$ of monoid ideal $\Delta_\prec$ with corresponding initial data f_ν for each $(\nu, N_\nu) \in \mathcal{T}$, expansion point $x_0 \in \mathcal{X}$, multi index $\rho \in \mathbb{N}_0^n$

Output: Taylor coefficient a_ρ of unique formal power series solution of initial value problem corresponding to given complementary decomposition

```
if ρ ∈ ν + N^n_{N_ν} for some (ν, N_ν) ∈ T then {a_ρ parametric coefficient}
    return ∂^{ρ−ν} f_ν(x_0)
else {a_ρ principal coefficient}
    find unique equation u_μ = φ_μ such that ρ ∈ C_P(μ)
    return D^{ρ−μ} φ_μ(x_0, a^{(|ρ|)})
end if
```

Before ending this section, we use the obtained results to characterise those involutive differential equations $\mathcal{R}_q$ for which the symbol module $\mathcal{N}^0$ possesses a complementary Hironaka decomposition, i. e. for which $\mathcal{P}^m/\mathcal{N}^0$ is a Cohen–Macaulay module. It turns out that in a certain sense this property is equivalent to normality; thus such equations have a particularly simple structure.

Proposition 9.3.12. *Let $\mathcal{N}^0 \subseteq \mathcal{P}^m$ be the symbol module in the more precise sense of Remark 7.1.16 (i. e. including contributions from the lower-order equations) of the involutive differential equation $\mathcal{R}_q \subseteq J_q\pi$ in m dependent variables. The factor module $\mathcal{M} = \mathcal{P}^m/\mathcal{N}^0$ is a Cohen–Macaulay module with $\dim \mathcal{M} = \operatorname{depth} \mathcal{M} = d$, if and only if $\mathcal{R}_q$ is equivalent to a normal first-order equation in $d+1$ independent and $\operatorname{mult} \mathcal{M}$ dependent variables in the sense that a bijection between the formal solution spaces of the two equations exists.*

Proof. According to Corollary 5.2.10, the factor module $\mathcal{M}$ is Cohen–Macaulay, if and only if it possesses a Hironaka decomposition. This decomposition consists then of $\operatorname{mult} \mathcal{M}$ cones of dimension d. If we take as local representation of $\mathcal{R}_q$ a Cartan normal form, then the principal parts of the equations define a weak Pommaret basis of $\mathcal{N}^0$ as discussed above. The monomial submodule generated by the leading terms of this basis has as complementary decomposition exactly the above Hironaka decomposition. Thus the corresponding formally well-posed initial value problem prescribes $\operatorname{mult} \mathcal{M}$ functions of the d variables $x^1, \dots, x^d$. But these are exactly the initial data for a normal first-order equation with $\operatorname{mult} \mathcal{M}$ dependent and $d+1$ independent variables. □

This proof is rather formal and exploits the fact that two differential equations are already equivalent in the used sense, if it is possible to find for both initial value problems with the same initial data. However, one can make this proof much more concrete, as it is possible to construct explicitly the equivalent normal equation and implicitly the bijection. Since this construction is very similar to the proof of the Cartan–Kähler Theorem, we will postpone it until the end of the next section. Here we close the discussion by recalling that according to the considerations in the proof of Corollary 5.2.10 the factor module $\mathcal{M}$ is Cohen–Macaulay of depth d, if and only if all derivatives of degree q and class greater than d are principal for $\mathcal{R}_q$. Thus this property is very easy to decide, as soon as one has brought the symbol matrix of $\mathcal{N}_q$ in row echelon form.

9.4 The Cartan–Kähler Theorem

The Cartan–Kähler Theorem extends the Cauchy–Kovalevskaya Theorem to arbitrary involutive systems. This fact implies that it inherits all its restrictions. So again the practical applicability of the theorem is rather limited. However, the proof is quite enlightening. It does not only highlight the importance of involution in contrast to mere formal integrability; it also demonstrates a fairly general methodology to extend results on normal systems to involutive ones.

For simplicity, we formulate it for an involutive first-order system in Cartan normal form without any algebraic equations. Thus we assume that we are given an involutive system of the following form (cf. (7.34)):

$$u_n^\alpha = \phi_n^\alpha(x^i, u^\beta, u_j^\gamma, u_n^\delta)\,, \qquad \begin{cases} 1\leq\alpha\leq\beta_1^{(n)}\,, \\ 1\leq j<n\,, \\ \beta_1^{(n)}<\delta\leq m\,, \end{cases} \tag{9.40a}$$

$$u_{n-1}^\alpha = \phi_{n-1}^\alpha(x^i, u^\beta, u_j^\gamma, u_{n-1}^\delta)\,, \qquad \begin{cases} 1\leq\alpha\leq\beta_1^{(n-1)}\,, \\ 1\leq j<n-1\,, \\ \beta_1^{(n-1)}<\delta\leq m\,, \end{cases} \tag{9.40b}$$

$$\vdots$$

$$u_1^\alpha = \phi_1^\alpha(x^i, u^\beta, u_1^\delta)\,, \qquad \begin{cases} 1\leq\alpha\leq\beta_1^{(1)}\,, \\ \beta_1^{(1)}<\delta\leq m\,. \end{cases} \tag{9.40c}$$

Since we assume that the system (9.40) is involutive, the rows of its symbol matrix yield a Pommaret basis (and hence, by Corollary 4.3.11, also a Janet basis) of the symbol module $\mathcal{N}^0$. By Definition 3.4.1 of an involutive basis, their leading terms define a Pommaret basis of the monoid module generated by them. Following the considerations in the last section and computing a complementary decomposition for this monoid module using Algorithm 5.2, a formally well-posed initial value problem for the system (9.40) leading to a unique formal solution is then obtained, if we prescribe the initial conditions

$$\begin{aligned}
&u^{\alpha}(x^1,\dots,x^n)=f^{\alpha}(x^1,\dots,x^n)\,, && \beta_1^{(n)}<\alpha\le m\,, && \text{(9.41a)}\\
&u^{\alpha}(x^1,\dots,x^{n-1},0)=f^{\alpha}(x^1,\dots,x^{n-1})\,, && \beta_1^{(n-1)}<\alpha\le\beta_1^{(n)}\,, && \text{(9.41b)}\\
&\vdots\\
&u^{\alpha}(x^1,0,\dots,0)=f^{\alpha}(x^1)\,, && \beta_1^{(1)}<\alpha\le\beta_1^{(2)}\,, && \text{(9.41c)}\\
&u^{\alpha}(0,\dots,0)=f^{\alpha}\,, && 1\le\alpha\le\beta_1^{(1)}\,. && \text{(9.41d)}
\end{aligned}$$

Recall from the discussion of (7.34) that the given ranges for the indices are to be understood in the sense that if for example $\beta_1^{(n)}=m$, then no derivatives u_n^{δ} appear in (9.40a) and the conditions (9.41a) are empty etc.

Again one sees that the system is underdetermined, if $\beta_1^{(n)}<m$, as then some of the dependent variables u^{α} are not restricted at all by the system (9.40) and must therefore be completely prescribed by the initial conditions. One might say that these variables have more the character of parameters than of unknowns.

Theorem 9.4.1 (Cartan–Kähler). *Let the functions ϕ_k^{α} and f^{α} be real-analytic at the origin and let the system* (9.40) *be involutive. Then it possesses one and only one solution that is analytic at the origin and satisfies the initial conditions* (9.41).

We first prove a lemma representing the core of the proof of the Cartan–Kähler Theorem which we will present afterwards. This lemma is also of independent interest, as it shows an important property of involutive systems.

Lemma 9.4.2. *Under the assumptions of Theorem 9.4.1, consider the subsystem consisting of the equations of class k in* (9.40),

$$u_k^{\alpha}=\phi_k^{\alpha}(x^i,u^{\beta},u_j^{\gamma},u_k^{\delta})\,,\qquad \begin{cases}1\le\alpha\le\beta_1^{(k)}\,,\\ 1\le j\le k\,,\\ \beta_1^{(k)}<\delta\le m\,,\end{cases}\tag{9.42}$$

and set $x^i=0$ for $i>k$ in it. Let $\mathbf{U}(x^1,\dots,x^k)$ be an analytic solution of the thus restricted system (9.42) *for the initial conditions*

$$\begin{aligned}
&u^{\alpha}(x^1,\dots,x^{k-1},0)=f^{\alpha}(x^1,\dots,x^{k-1})\,, && 1\le\alpha\le\beta_1^{(k)}\,,\\
&u^{\alpha}(x^1,\dots,x^{k-1},x^k)=f^{\alpha}(x^1,\dots,x^{k-1},x^k,0,\dots,0)\,, && \beta_1^{(k)}<\alpha\le m\,.
\end{aligned}$$

(Thus the functions u^{α} with $\alpha>\beta_1^{(k)}$ appear only as parameters). If the initial data $\mathbf{f}(\mathbf{x})$ are such that for $x^k=\cdots=x^n=0$ they satisfy all equations in (9.40) *of class lower than k, then the solution $\mathbf{U}(x^1,\dots,x^k)$ satisfies these equations also for $x^k\neq 0$.*

Proof. We must analyse what happens, if we enter the given solution $\mathbf{U}(x^1,\dots,x^k)$ into the equations of lower class. Therefore we introduce the (by assumption analytic) residuals stemming from the equations of class ℓ

$$\Delta_\ell^\alpha(x^1,\dots,x^k) = \frac{\partial U^\alpha}{\partial x^\ell}(x^1,\dots,x^k) - \phi_\ell^\alpha\Bigl(x^1,\dots,x^k,0,\dots,0,\mathbf{U}(x^1,\dots,x^k),\frac{\partial \mathbf{U}}{\partial \mathbf{x}}(x^1,\dots,x^k)\Bigr)$$

for all classes $\ell < k$ and all indices $1 \leq \alpha \leq \beta_1^{(\ell)}$. We exploit the fact that the system (9.40) is involutive in order to derive some partial differential equations for the functions Δ_ℓ^α. Adapting (7.35) to the special case at hand, we see that involution of (9.40) is equivalent to the existence of functions $A_{ij}^\beta(\mathbf{x},\mathbf{u}^{(1)})$ and $B_i^\beta(\mathbf{x},\mathbf{u}^{(1)})$ such that whenever $1 \leq \ell < k \leq n$

$$D_k(u_\ell^\alpha - \phi_\ell^\alpha) = \sum_{i=1}^{k}\sum_{\beta=1}^{\beta_1^{(i)}}\Bigl\{\sum_{j=1}^{i} A_{ij}^\beta D_j(u_i^\beta - \phi_i^\beta) + B_i^\beta(u_i^\beta - \phi_i^\beta)\Bigr\}. \tag{9.43}$$

Entering $\mathbf{U}(x^1,\dots,x^k)$ into these relations and setting $x^i = 0$ for $i > k$ implies that the residuals Δ_ℓ^α must satisfy the linear system

$$\frac{\partial \Delta_\ell^\alpha}{\partial x^k} = \sum_{i=1}^{k-1}\sum_{\beta=1}^{\beta_1^{(i)}}\Bigl\{\sum_{j=1}^{i} \bar{A}_{ij}^\beta \frac{\partial \Delta_i^\beta}{\partial x^j} + \bar{B}_i^\beta \Delta_i^\beta\Bigr\}, \quad 1 \leq \alpha \leq \beta_1^{(k)}, \tag{9.44}$$

where the coefficient functions

$$\bar{A}_{ij}^\beta(x^1,\dots,x^k) = A_{ij}^\beta\Bigl(x^1,\dots,x^k,0,\dots,0,\mathbf{U}(x^1,\dots,x^k),\frac{\partial \mathbf{U}}{\partial \mathbf{x}}(x^1,\dots,x^k)\Bigr),$$
$$\bar{B}_i^\beta(x^1,\dots,x^k) = B_i^\beta\Bigl(x^1,\dots,x^k,0,\dots,0,\mathbf{U}(x^1,\dots,x^k),\frac{\partial \mathbf{U}}{\partial \mathbf{x}}(x^1,\dots,x^k)\Bigr)$$

are analytic, as A_{ij}^β, B_i^β and $\mathbf{U}$ are all analytic. On the right hand side of (9.44) the terms with $i = k$ have disappeared, as by assumption $\mathbf{U}$ is a solution of (9.42). This makes (9.44) an analytic *normal* system in the independent variables $x^1,\dots,x^k$. Obviously, $\Delta_\ell^\alpha \equiv 0$ is an analytic solution of it and, by the Cauchy–Kovalevskaya Theorem 9.2.1, this solution is unique. Thus all residuals vanish and the given solution $\mathbf{U}$ also satisfies the equations of lower class. □

Lemma 9.4.2 expresses a kind of conservation property of involutive systems: if the equations of lower class are satisfied for $x^k = 0$, then they are also satisfied at all values of x^k for which a solution exists. For the special case of Maxwell's equations (2.85), this result is very classical. As already mentioned in Example 2.4.1, if a solution of the evolutionary part of Maxwell's equations satisfies the Gauss laws at the time $t = 0$, it satisfies them at all times. Using a well-known identity from vector analysis, this fact is proven in almost any textbook on electrodynamics. But as we have seen, it actually represents a very general property of involutive systems.

We may also consider Lemma 9.4.2 as a generalisation of Proposition 9.1.1 to partial differential equations. It states that the subsystems of lower class are a kind of "weak invariants" for the "flow" generated by the subsystem of class k. In fact, this point of view becomes very useful within the formalism of semigroups [349, 379] where partial differential equations are considered as ordinary differential equations on infinite-dimensional spaces.

Proof (of the Cartan–Kähler Theorem 9.4.1). We follow the strategy given by Pommaret [356]: The solution is constructed step by step; at each step a normal system appears to which we apply the Cauchy–Kovalevskaya Theorem 9.2.1. Finally, we use repeatedly Lemma 9.4.2.

We start with the subsystem (9.40c) comprising all equations of class 1:

$$u_1^\alpha = \phi_1^\alpha(x^i, u^\beta, u_1^\gamma)\,, \qquad \begin{cases} 1\leq\alpha\leq\beta_1^{(1)}\,, \\ \beta_1^{(1)}<\gamma\leq m\,. \end{cases} \tag{9.45a}$$

If we set $x^i = 0$ for $i > 1$, the initial conditions (9.41) allow us to substitute for u^β and u_1^β with $\beta > \beta_1^{(1)}$ the functions $\bar f^\beta(x^1) = f^\beta(x^1, 0, \dots, 0)$ and their x^1-derivatives. Now (9.45a) becomes a normal system (even of ordinary differential equations) in the independent variable x^1 and we may apply the Cauchy–Kovalevskaya Theorem for the last set of initial conditions in (9.41)

$$u^\alpha(0,\dots,0) = f^\alpha \in \mathbb{R}\,, \qquad 1 \leq \alpha \leq \beta_1^{(1)}\,. \tag{9.45b}$$

It guarantees for $1 \leq \alpha \leq \beta_1^{(1)}$ the existence of a unique analytic solution $U_1^\alpha(x^1)$ of the initial value problem (9.45).

In the next step, we consider the equations of class 2 in (9.40)

$$u_2^\alpha = \phi_2^\alpha(x^i, u^\beta, u_1^\gamma, u_2^\delta)\,, \qquad \begin{cases} 1\leq\alpha\leq\beta_1^{(2)}\,, \\ \beta_1^{(2)}<\delta\leq m\,. \end{cases} \tag{9.46a}$$

This time we set $x^i = 0$ only for $i > 2$. Then we substitute u^β, u_1^β and u_2^β with $\beta > \beta_1^{(2)}$ by the functions $\bar f^\beta(x^1, x^2) = f^\beta(x^1, x^2, 0, \dots, 0)$ and their derivatives with respect to x^1 and x^2. This makes (9.46a) a normal system in the two independent variables x^1 and x^2. We prescribe the initial conditions

$$u^\alpha(x^1, 0) = \begin{cases} f^\alpha(x^1) & \text{for } \beta_1^{(1)} < \alpha \leq \beta_1^{(2)}\,, \\ U_1^\alpha(x^1) & \text{for } 1 \leq \alpha \leq \beta_1^{(1)}\,. \end{cases} \tag{9.46b}$$

Thus for the components u^α which we considered already in the first step we take as initial data the solution $U_1^\alpha(x^1)$ obtained above. For the remaining components we use the initial data in (9.41).

We may again apply the Cauchy–Kovalevskaya Theorem, as all appearing functions are analytic. It guarantees us the existence of a unique analytic solution $U_2^\alpha(x^1, x^2)$ with $1 \leq \alpha \leq \beta_1^{(2)}$ of the initial value problem (9.46). However, it is not obvious that these functions also satisfy (9.45a) with $x^i = 0$ for $i > 2$; we only know they do, if in addition $x^2 = 0$. But this situation is covered by Lemma 9.4.2: it asserts that the functions $U_2^\alpha(x^1,x^2)$ also satisfy the equations of lower class, i. e. in our case (9.45a), for all values of the variable x^2.

We iterate this process. In step k we consider the equations of class k, i. e. (9.42). Setting $x^i = 0$ for $i > k$, we substitute u^β and u_j^β with $\beta > \beta_1^{(k)}$ by the functions $\bar{f}^\beta(x^1,\dots,x^k) = f^\beta(x^1,\dots,x^k,0,\dots,0)$ and their derivatives. Then (9.42) becomes a normal system in the independent variables $x^1,\dots,x^k$. As initial conditions we use

$$u^\alpha(x^1,\dots,x^{k-1},0) = \begin{cases} f^\alpha(x^1,\dots,x^{k-1}) & \text{for } \beta_1^{(k-1)} < \alpha \le \beta_1^{(k)}\,, \\ U_{k-1}^\alpha(x^1,\dots,x^{k-1}) & \text{for } 1 \le \alpha \le \beta_1^{(k-1)}\,. \end{cases} \tag{9.47}$$

Here $U_{k-1}^\alpha(x^1,\dots,x^{k-1})$ represents the solution of the previous step. By the Cauchy–Kovalevskaya Theorem the initial value problem (9.42, 9.47) possesses a unique analytic solution $U_k^\alpha(x^1,\dots,x^k)$ with $1 \le \alpha \le \beta_1^{(k)}$ which by Lemma 9.4.2 also satisfies all equations of lower class.

The solution $U_n^\alpha(x^1,\dots,x^n)$ with $1 \le \alpha \le \beta_1^{(n)}$ obtained in step n is then the sought analytic solution of the complete original initial value problem (9.40, 9.41), as the remaining components of **u** are given by the initial conditions. The uniqueness of the solution follows from the construction. □

Corollary 9.4.3. *Every analytic and formally integrable differential equation is locally solvable in the analytic category.*

Proof. Recall from Remark 2.3.16 that a differential equation $\mathcal{R}_q$ is locally solvable, if for every point $\rho \in \mathcal{R}_q$ a local solution σ exists such that $\rho \in \operatorname{im} j_q\sigma$. If $\mathcal{R}_q$ is a formally integrable equation, then some finite prolongation $\mathcal{R}_{q+r}$ of it is involutive and, because of the assumed formal integrability, we can find to every point $\rho \in \mathcal{R}_q$ a point $\hat{\rho} \in \mathcal{R}_{q+r}$ such that $\pi_q^{q+r}(\hat{\rho}) = \rho$, i. e. $\hat{\rho}$ lies in the fibre over ρ. Without loss of generality, we may assume that in suitable local coordinates, $\pi^{q+r}(\hat{\rho}) = 0$. Consider now for the involutive equation $\mathcal{R}_{q+r}$ the initial value problem treated in the Cartan–Kähler Theorem. Obviously, we can choose the initial data in such a way that the corresponding unique analytic solution σ satisfies $\hat{\rho} \in \operatorname{im} j_{q+r}\sigma$ and thus $\rho \in \operatorname{im} j_q\sigma$; in fact, this can even be achieved with polynomial initial data. Note that σ is *not* uniquely determined by ρ, as in general many different choices for both the point $\hat{\rho}$ and the initial data are possible. □

One may wonder whether in the Cartan–Kähler Theorem one really needs the assumption that the system (9.40) is involutive; perhaps formal integrability suffices? In fact, the step by step construction used in the proof above is in principle possible for *any* system in triangular form. But it is unclear whether the thus constructed solution also satisfies all equations of lower class. Our proof of Lemma 9.4.2 depends crucially on the involution of the system. The construction of the normal system (9.44) requires that the non-multiplicative prolongations of the equations with class less than k with respect to x^k are expressible as linear combinations of multiplicative prolongations. For proving the Cartan–Kähler Theorem, we had to apply

Lemma 9.4.2 for all classes k appearing in (9.40) and hence had to consider *all possible* non-multiplicative prolongations. According to Proposition 7.2.3, this proof will therefore only work, if the symbol $\mathcal{N}_q$ is involutive. The following example demonstrates explicitly, what happens if obstructions to involution are present.

Example 9.4.4. We consider the first-order equation

$$\mathcal{R}_1 : \begin{cases} v_t = w_x \, , \\ w_t = 0 \, , \\ v_x = 0 \, . \end{cases} \tag{9.48}$$

It arises when we rewrite the second-order equation (7.31) as a first-order equation by setting $w = u_t$ and $v = u_x$ (compare Appendix A.3). Thus we already know that $\mathcal{R}_1$ is formally integrable but not involutive: the non-multiplicative prolongation $D_t(v_x)$ cannot be reduced to zero and we obtain the obstruction to involution $w_{xx} = 0$.

The crucial step in the proof above applied to $\mathcal{R}_1$ would be to show that any solution of the normal subsystem comprising the first two equations of (9.48) that satisfies the third equation for $t = 0$ also satisfies it for all values of t. If $\mathcal{R}_1$ were involutive, this would be the case by Lemma 9.4.2. But as we are only dealing with a formally integrable equation, it is easy to construct a counterexample.

Consider the functions

$$v(t,x) = \alpha'(x)t + c \, , \quad w(t,x) = \alpha(x) \tag{9.49}$$

with an arbitrary constant $c \in \mathbb{R}$ and an arbitrary real-analytic function $\alpha(x)$. They solve the first two equations of (9.48). Furthermore, for $t = 0$, they also solve the last equation. But obviously this is not the case for any other value of t, if we choose $\alpha(x)$ such that $\alpha''(x) \neq 0$. Only for $\alpha(x) = ax + b$ with constants $a, b \in \mathbb{R}$ we obtain a solution of $\mathcal{R}_1$, as only for this choice of α the equation $w_{xx} = 0$ is satisfied. $\triangleleft$

In fact, this example can easily be generalised to a proof that involution is not only sufficient for Lemma 9.4.2 to be applicable at each step of the proof of Theorem 9.4.1 but also necessary. Formal integrability is trivially necessary; thus we restrict to systems which are formally integrable but not involutive. Without loss of generality, we may further assume that in the Cartan normal form (9.40) the subsystem of class n is normal and that an obstruction to involution arises from the prolongation of an equation of lower class with respect to x^n, as this is precisely the situation in which we apply Lemma 9.4.2 in the proof of Theorem 9.4.1.

As the subsystem of class n is normal, we can eliminate any x^n-derivative in the obstruction to involution. Thus it may be considered as a differential equation in $x^1, \dots, x^{n-1}$ which, by construction, is independent of the equations of lower class present in our system. In our example above this was the equation $w_{xx} = 0$. According to the assumptions of Lemma 9.4.2, we choose initial data for $x^n = 0$ satisfying all equations of lower class. But we can always find initial data that do *not* satisfy our obstruction to involution. If we solve the subsystem of class n for such initial data, for $x^n \neq 0$ the solution will *not* satisfy that equation of lower class that generated the obstruction to involution, as its x^n-prolongation does not vanish.

Thus, if (9.40) is only formally integrable but not involutive, then it does not possess a solution for arbitrary choices of the initial data in (9.41). Hence involution is a necessary condition in Theorem 9.4.1, as otherwise our initial value problem is not even formally well-posed. Furthermore, these considerations show that it is a characteristic property of involutive systems that Lemma 9.4.2 holds for $1 \leq k \leq n$.

The Cartan–Kähler Theorem inherits all the restrictions and disadvantages of the Cauchy–Kovalevskaya Theorem, as its proof is based on the latter. For arbitrary differential equations there seems to be no way to avoid this problem, as no other existence and uniqueness theorem for arbitrary normal equations exists. But if more is known about the structure of the subsystems considered in the proof above, one might be able to use other existence and uniqueness theorems and to obtain stronger results still following the basic strategy of our proof.

If we want to go beyond analytic solutions, a crucial step will be the extension of Lemma 9.4.2. We used the Cauchy–Kovalevskaya Theorem in its proof. But in fact all we need is a proof of the uniqueness of the zero solution of (9.44). Note that this is always a *linear* system, even if (9.40) is non-linear. Thus one could think of applying the Holmgren Theorem (see Section 10.2) instead of the Cauchy–Kovalevskaya Theorem which would even yield that the zero solution is unique among all C^1 solutions. However, this approach still requires that we are dealing with analytic equations. In order to be able to relax that condition one needs more information about the structure of the system (9.40). For a certain class of linear systems we will demonstrate this possibility in Section 10.4.

Finally, we continue the discussion—started at the end of the last section—of involutive differential equations $\mathcal{R}_q$ for which the factor module $\mathcal{M} = \mathcal{P}^m/\mathcal{N}^0$ is Cohen–Macaulay with $\dim \mathcal{M} = \operatorname{depth} \mathcal{M} = d$. Assume first that we are dealing with a first-order equation $\mathcal{R}_1$ without an algebraic part. In this case $\mathcal{N}^0$ is generated by the rows of the symbol matrix M_1 and $d+1$ is the lowest class of an equation in the Cartan normal form of $\mathcal{R}_1$. In our proof of the Cartan–Kähler Theorem we considered the subsystems of constant class. If $\mathcal{M}$ is a Cohen–Macaulay module, each of them becomes normal, as soon as we set $x^{k+1},\dots,x^n$ to constant values where k is the class of the subsystem. No further substitutions are necessary, as all initial conditions comprise only functions of $x^1,\dots,x^d$.

In the first step of the proof we study the subsystem of class $d+1$, as no equations of lower class are present. Its initial conditions are identical with the initial conditions for the full equation $\mathcal{R}_1$. Thus we may take it for the normal system referred to in Proposition 9.3.12. The bijection between the formal solution spaces of the full system and this normal system is implicitly contained in the proof of the Cartan–Kähler Theorem. We extend the solution $\mathbf{U}(x^1,\dots,x^{d+1})$ of the subsystem to the solution $\mathbf{u}(x^1,\dots,x^n)$ of the full system following step by step the proof. At each step the initial conditions are just the solutions of the previous step; since $\mathcal{M}$ is a Cohen–Macaulay module, no additional conditions appear. In other words, in (9.47) no functions f^α appear, as $\beta_1^{(k-1)} = \beta_1^{(k)} = m$. Thus the extension is uniquely determined by the subsystems of higher class; their solutions define the bijection.

The discussion of a higher-order equation $\mathcal{R}_q$ is similar, only the construction of the required normal systems is slightly different. The starting point is now the

Pommaret basis of $\mathcal{N}^0$. If $\mathcal{M}$ is Cohen–Macaulay, then all its leading terms must be contained in $\mathbb{k}[x^{d+1},\dots,x^n]^m$ and generated there a zero-dimensional module. Its complementary set consists of $\operatorname{mult}\mathcal{M}$ terms and we introduce for each of them a new unknown function. Then, using the relations between the derivatives corresponding to these terms and the original differential equation, we can construct a differential equation for the x^{d+1}-derivative of each new unknown function. This way we obtain the equivalent normal first-order equation of Proposition 9.3.12. The bijection between its solution space and the one of our original differential equation is obtained by solving a series of normal differential equation similar to the proof of the Cartan–Kähler Theorem in the next section. Instead of providing complicated formulae, we demonstrate the construction for a concrete example.

Example 9.4.5. The linear differential equation naturally corresponding to the ideal $\mathcal{I} \subset \mathcal{P} = \mathbb{k}[x,y,z]$ treated in Example 5.2.11 has the local representation

$$\mathcal{R}_3 : \begin{cases} u_{zzz} = 0\,, \\ u_{yzz} = u_{xzz}\,, \\ u_{yy} = u_{xy}\,. \end{cases} \tag{9.50}$$

$\mathcal{R}_3$ is not involutive, as we still have to take into account the prolongations of the lower-order equation $u_{yy} = u_{xy}$. However, one easily verifies that $\mathcal{R}_3^{(1)}$ is involutive. Indeed, we noted in Example 5.2.11 that a Pommaret basis of $\mathcal{I}$ is obtained after addition of the generator $y^2z - xyz$ and hence the symbol $\mathcal{N}_3^{(1)}$ is involutive.

The ideal $\mathcal{I}$ is the symbol module of $\mathcal{R}_3^{(1)}$ (or actually also of $\mathcal{R}_3$) in the more precise sense of Remark 7.1.16. In Example 5.2.11 we already determined that the factor module $\mathcal{M} = \mathcal{P}/\mathcal{I}$ is Cohen–Macaulay (with $\dim\mathcal{M} = \operatorname{depth}\mathcal{M} = 1$) and that the corresponding Hironaka decomposition is given by (5.15). It is not difficult to see that $\operatorname{mult}\mathcal{M} = 5$ here. Note furthermore that all leading terms of the Pommaret basis are contained in the subring $\mathbb{k}[y,z]$. It follows from the decomposition (5.15) that a formally well-posed initial value problem for $\mathcal{R}_3^{(1)}$ is given by the conditions:

$$\begin{aligned} u(x,y_0,z_0) &= f_1(x)\,, \\ u_y(x,y_0,z_0) &= f_2(x)\,, \\ u_z(x,y_0,z_0) &= f_3(x)\,, \\ u_{yz}(x,y_0,z_0) &= f_4(x)\,, \\ u_{zz}(x,y_0,z_0) &= f_5(x)\,. \end{aligned} \tag{9.51}$$

In order to derive the equivalent normal equation, we introduce the new dependent variables

$$v^1 = u\,, \quad v^2 = u_y\,, \quad v^3 = u_z\,, \quad v^4 = u_{yz}\,, \quad v^5 = u_{zz} \tag{9.52}$$

corresponding to the generators of (5.15) and consider them first as functions of the independent variables x, y only. The relations between these functions and the system (9.50) lead to the following normal system:

$$v^1_y = v^2\,, \quad v^2_y = v^2_x\,, \quad v^3_y = v^4\,, \quad v^4_y = v^4_x\,, \quad v^5_y = v^5_x\,. \tag{9.53}$$

Its natural initial conditions are equivalent to (9.51), namely $v^\alpha(x, y_0) = f_\alpha(x)$. The bijection between the solution spaces of (9.53) and (9.50), respectively, is determined by the solution of a further normal system which again arises from the relations between the $\mathbf{v}$ and the system (9.50):

$$v^1_z = v^3\,, \quad v^2_z = v^4\,, \quad v^3_z = v^5\,, \quad v^4_z = v^5_x\,, \quad v^5_z = 0 \tag{9.54}$$

where we now consider the dependent variables as functions of all three independent variables and where we take the solutions of (9.53) as initial data. The v^1-component of a solution of the second system (9.54) is then a solution of the original differential equation (9.50) and conversely every solution of (9.50) induces a solution of the normal system (9.53) as the initial data of the corresponding solution of the second system (9.54). The equivalence is reflected in the fact that the construction of the two systems (9.53) and (9.54) required all relations between the functions $\mathbf{v}$ and all equations contained in (9.50).

If we had more independent variables, then each further variable would lead to one more normal system constructed in the same manner as (9.54). Each system takes as initial data the solution of the previous one and adds the dependence on one more independent variable to the unknown functions $\mathbf{v}$. ◁

9.5 The Vessiot Distribution

Obviously, it is difficult to solve differential equations. An idea going back at least to Cartan is to consider first *infinitesimal solutions* or *integral elements*; in Section 2.3 we have seen that their construction requires essentially only linear algebra. We will not follow the approach of Cartan which is based on differential forms and leads to the theory of exterior differential systems. Instead we will develop in this and the next section a dual vector field approach originally proposed by Vessiot [470].

The emerging theory may be considered as a generalisation of the Frobenius Theorem C.3.3. Recall from Example 2.3.17 that this theorem essentially provides a complete solvability theory for any differential equation $\mathcal{R}_1$ of finite type: an involutive equation of finite type defines a flat connection on $\pi : \mathcal{E} \rightarrow \mathcal{X}$ and the images of its solutions σ are the integral manifolds of the corresponding horizontal bundle.

For a general differential equation $\mathcal{R}_q$ the geometry is less simple, as the equation does not correspond to a connection. However, it will turn out that in a certain sense the equation $\mathcal{R}_q$ can be "covered" by (infinitely many) flat connections and to each of these we may apply the Frobenius Theorem. More precisely, we will find the following situation. The embedding $\iota : \mathcal{R}_q \hookrightarrow J_q\pi$ induces naturally a distribution $\mathcal{V}[\mathcal{R}_q] \subseteq T\mathcal{R}_q$ on the submanifold $\mathcal{R}_q$ tangent to any prolonged solution and we will search for n-dimensional transversal (with respect to the fibration $\hat{\pi}^q : \mathcal{R}_q \rightarrow \mathcal{X}$) *involutive* subdistributions of it. Such a subdistribution may be interpreted as the horizontal bundle of a flat connection for $\hat{\pi}^q$. Obviously, it possesses n-dimensional integral manifolds and the properties of $\mathcal{V}[\mathcal{R}_q]$ will ensure that they are automatically of the form $\operatorname{im} j_q\sigma$ for a solution σ of $\mathcal{R}_q$.

For the simple case of an ordinary differential equation we applied this idea already in Section 9.1: we constructed solutions as the projection of integral curves of a one-dimensional distribution $\mathcal{V}[\mathcal{R}_q]$ on $\mathcal{R}_q$. As such a distribution is trivially involutive, there was no need to analyse subdistributions.

In Sections 2.1 and 2.2 we introduced the contact structure of the jet bundle. Recall that, by Proposition 2.1.6 (or 2.2.7, respectively), if $\rho = j_q\sigma(x) \in J_q\pi$ for some point $x \in \mathcal{X}$ and section $\sigma \in \Gamma_{loc}(\pi)$, then the tangent space $T_\rho(\operatorname{im} j_q\sigma)$ to the image of the prolonged section $j_q\sigma$ is a subspace of the contact distribution $\mathcal{C}_q\big|_\rho$. If the section σ is a solution of $\mathcal{R}_q$, it furthermore satisfies by definition $\operatorname{im} j_q\sigma \subseteq \mathcal{R}_q$ and hence $T(\operatorname{im} j_q\sigma) \subseteq T\mathcal{R}_q$. These considerations (already used in the proof of Lemma 9.1.5) motivate the following definition.

Definition 9.5.1. Let $\mathcal{R}_q \subseteq J_q\pi$ be a differential equation and $\iota : \mathcal{R}_q \hookrightarrow J_q\pi$ the corresponding natural inclusion map. The *Vessiot distribution* of $\mathcal{R}_q$ is the distribution $\mathcal{V}[\mathcal{R}_q] \subseteq T\mathcal{R}_q$ defined by

$$T\iota\big(\mathcal{V}[\mathcal{R}_q]\big) = T\iota(T\mathcal{R}_q) \cap \mathcal{C}_q\big|_{\mathcal{R}_q} . \tag{9.55}$$

Eq. (9.55) defines the Vessiot distribution $\mathcal{V}[\mathcal{R}_q]$ uniquely, as the inclusion map ι is trivially injective. Here we have explicitly written the tangent map instead of identifying as usual $T\mathcal{R}_q$ with the subspace $T\iota(T\mathcal{R}_q) \subseteq (TJ_q\pi)\big|_{\mathcal{R}_q}$. The reason for this rigour is the same as in the discussion of integral elements in Section 2.3 and will become apparent in Example 9.5.4 below. Note that $\mathcal{V}[\mathcal{R}_q]$ is defined only on the submanifold $\mathcal{R}_q \subseteq J_q\pi$. In general, we cannot expect that it is of constant rank (as seen in Section 9.1, this is not even the case for ordinary differential equations). However, in line with the other regularity assumptions we have made throughout this book, we will ignore this possibility in the sequel.

Blanket Assumption 9.5.2. *The Vessiot distribution $\mathcal{V}[\mathcal{R}_q]$ is of constant rank on the differential equation $\mathcal{R}_q$.*

The above definition of the Vessiot distribution is not the one usually found in the literature (see, however, [296, Sect. 1.8]) which is based on pulling-back the contact codistribution $\mathcal{C}_q^0$. But it is very easy to prove the equivalence of the two approaches.

Proposition 9.5.3. *We have the equality*

$$\mathcal{V}[\mathcal{R}_q] = (\iota^*\mathcal{C}_q^0)^0 . \tag{9.56}$$

Proof. Let $\omega \in \mathcal{C}_q^0$ be a contact form and $X \in \mathfrak{X}(\mathcal{R}_q)$ a vector field on $\mathcal{R}_q$. Then, by the definition of the pull-back and the push-forward, respectively, $\iota^*\omega(X) = \omega(\iota_*X)$ (cf. (C.15)). This fact implies that $X \in (\iota^*\mathcal{C}_q^0)^0$ is equivalent to $\iota_*X \in \mathcal{C}_q\big|_{\mathcal{R}_q}$ and consequently $T\iota\big((\iota^*\mathcal{C}_q^0)^0\big) = T\iota(T\mathcal{R}_q) \cap \mathcal{C}_q\big|_{\mathcal{R}_q}$. The claimed equality follows now from the injectivity of $T\iota$. □

Example 9.5.4. Assuming that we are given an explicit parametrisation of the submanifold $\mathcal{R}_q \subseteq J_q\pi$, the inclusion map ι is of course trivial in the corresponding local coordinates. However, the local expressions of the pull-back ι^* and the tangent map $T\iota$, respectively, are *not* trivial, so that one must take care whether one works in coordinates on $\mathcal{R}_q$ or in coordinates on $J_q\pi$, i. e. whether one deals with a vector field $X \in \mathfrak{X}(\mathcal{R}_q)$ or with its push-forward $\iota_* X \in \mathfrak{X}(J_q\pi)$.

We demonstrate the explicit determination of the Vessiot distribution in both coordinate systems for the simple differential equation

$$\mathcal{R}_1 \,:\, u_y = F(u_x)\,. \tag{9.57}$$

Here we use our standard coordinates on $J_1\pi$: (x,y,u,u_x,u_y). As our system is explicitly solved for u_y, a natural choice for coordinates on $\mathcal{R}_1$ is $(\overline{x},\overline{y},\overline{u},\overline{u_x})$. We bar these coordinates in order to distinguish them clearly from the corresponding coordinates on $J_1\pi$. We will follow this convention throughout this and the next section.

The tangent space $T\mathcal{R}_1$ is spanned by the fields $\partial_{\overline{x}}$, $\partial_{\overline{y}}$, $\partial_{\overline{u}}$, $\partial_{\overline{u_x}}$. If we compute their push-forwards, then we obtain simply the corresponding fields for the unbarred variables with one exception:

$$\iota_*\partial_{\overline{u_x}} = \partial_{u_x} + F'(u_x)\partial_{u_y}\,. \tag{9.58}$$

Dually, the cotangent space $T^*(J_1\pi)$ is spanned by the forms $\mathrm{d}x$, $\mathrm{d}y$, $\mathrm{d}u$, $\mathrm{d}u_x$, $\mathrm{d}u_y$. Their pull-backs are the corresponding forms for the barred variables except for

$$\iota^*(\mathrm{d}u_y) = F'(u_x)\mathrm{d}\overline{u_x}\,. \tag{9.59}$$

The contact distribution $\mathcal{C}_1$ on $J_1\pi$ is by definition the annihilator of the contact form $\omega = \mathrm{d}u - u_x\mathrm{d}x - u_y\mathrm{d}y$ and the space $T\iota(T\mathcal{R}_1)$ is the annihilator of the one-form $\eta = \mathrm{d}u_y - F'(u_x)\mathrm{d}u_x$. A straightforward computation yields that the intersection of the two spaces is spanned by the vector fields

$$\iota_*X_1 = \partial_x + u_x\partial_u\,, \quad \iota_*X_2 = \partial_y + u_y\partial_u\,, \quad \iota_*X_3 = \partial_{u_x} + F'(u_x)\partial_{u_y}\,. \tag{9.60}$$

The pull-back of the contact form is according to our considerations above simply $\iota^*\omega = \mathrm{d}\overline{u} - \overline{u_x}\mathrm{d}\overline{x} - F(\overline{u_x})\mathrm{d}\overline{y}$. Thus, if we did not carefully distinguish between the coordinates on $\mathcal{R}_1$ and $J_1\pi$, one could think that we simply *restricted* ω to the submanifold $\mathcal{R}_1$. Indeed, it has become customary in the literature to say that the Vessiot distribution is obtained via a "restriction" of the contact codistribution. This is at best an abuse of language but in principle plainly wrong!

The main reason for this confusion is that as long as our system contains only equations solved for derivatives of maximal order, the contact forms seemingly do not notice the pull-back. If our system contained an additional lower-order equation, say, $u = G(x)$, then we would find $\iota^*(\mathrm{d}u) = G'(x)\mathrm{d}\overline{x}$ and the pull-back $\iota^*\omega$ would look very differently from the restriction $\omega|_{\mathcal{R}_1}$.

The annihilator of $\iota^*\omega$ and thus the Vessiot distribution $\mathcal{V}[\mathcal{R}_1]$ of the differential equation $\mathcal{R}_1$ is spanned by the three vector fields

$$\overline{X}_1 = \partial_{\overline{x}} + \overline{u_x}\partial_{\overline{u}}\,, \quad \overline{X}_2 = \partial_{\overline{y}} + F(\overline{u_x})\partial_{\overline{u}}\,, \quad \overline{X}_3 = \partial_{\overline{u_x}}\,. \tag{9.61}$$

Computing the push-forwards (and taking into account that $u_y = F(u_x)$ on the submanifold $\mathcal{R}_1$), we find that indeed $\overline{X}_k = X_k$. ◁

For an implicit differential system, we generally are not able to determine explicitly the pull-back of the contact forms, as no closed form expression of the inclusion map $\iota : \mathcal{R}_q \hookrightarrow J_q\pi$ is available. In such a situation we must work in $T(J_q\pi)$ and stick to the space $T\iota(T\mathcal{R}_q) \cap \mathcal{C}_q|_{\mathcal{R}_q}$ a basis of which can always be computed by simply solving a linear system.

Proposition 2.1.6 provides us with a basis of the contact distribution, namely the contact fields $C_i^{(q)}$ and C_α^μ, respectively, defined by (2.11). Thus any vector field in $\mathcal{V}[\mathcal{R}_q]$ is of the form $\iota_* X = a^i C_i^{(q)} + b_\mu^\alpha C_\alpha^\mu$ with coefficients $a^i, b_\mu^\alpha \in \mathcal{F}(\mathcal{R}_q)$. If a local representation of the differential equation $\mathcal{R}_q$ is given by $\boldsymbol{\Phi} = 0$, then according to Remark C.2.8 we must have $d\boldsymbol{\Phi}(\iota_* X) = \iota_* X(\boldsymbol{\Phi}) = 0$. Hence the coefficients solve the homogeneous linear system

$$C_i^{(q)}(\boldsymbol{\Phi})a^i + C_\alpha^\mu(\boldsymbol{\Phi})b_\mu^\alpha = 0\,. \tag{9.62}$$

Remark 9.5.5. Setting $a^i = 0$ in (9.62), we obtain the linear system $C_\alpha^\mu(\boldsymbol{\Phi})b_\mu^\alpha = 0$ which, by the definition (2.11b) of the contact fields C_α^μ, coincides with the symbol equations (7.3). Hence the symbol $\mathcal{N}_q$ of the differential equation $\mathcal{R}_q$ lies in the Vessiot distribution $\mathcal{V}[\mathcal{R}_q]$. Furthermore, it contains all the vertical vectors in $\mathcal{V}[\mathcal{R}_q]$, since any solution of the linear system (9.62) where at least one coefficient a^i does not vanish is transversal (both with respect to π_{q-1}^q and π^q!). ◁

Remark 9.5.6. Determining the Vessiot distribution $\mathcal{V}[\mathcal{R}_q]$ is closely related to prolonging the differential equation $\mathcal{R}_q$. Indeed, it requires essentially the same computations. Given the above local representation of $\mathcal{R}_q$, the prolongation yields for $1 \le i \le n$ the following additional equations for $\mathcal{R}_{q+1}$:

$$D_i\Phi^\tau = C_i^{(q)}(\Phi^\tau) + C_\alpha^\mu(\Phi^\tau)u_{\mu+1_i}^\alpha = 0\,. \tag{9.63}$$

In the context of computing formal power series solutions (cf. Section 2.3) these equations are considered as an inhomogeneous linear system for the Taylor coefficients of order $q+1$ (compare (9.63) with (2.67)). Taking this point of view, we may call (9.62) the "projective" version of (9.63). In fact for $n = 1$, i. e. for ordinary differential equations, this is even true in a rigorous sense and in Section 9.1 we have seen that this point of view is highly useful for the analysis of singular solutions. ◁

Provided that we know an explicit parametrisation of $\mathcal{R}_q$, the approach via the pull-back of the contact codistribution is computationally simpler. First of all, one works on a lower dimensional manifold and therefore the linear system to be solved for the determination of $\mathcal{V}[\mathcal{R}_q]$ is smaller. Secondly, no derivatives of the right hand sides of the equations are required.

This second effect did not become very apparent in Example 9.5.4, as for simplicity we assumed that F was only a function of u_x. If F were a function of all remaining jet variables, the push-forward would effect all basis vector fields. In this case, the vector fields $\iota_* X_k$ would have much more complicated coordinate expressions whereas the fields $\overline{X}_k$ would not change.

The Vessiot distribution $\mathcal{V}[\mathcal{R}_q]$ of a *partial* differential equation $\mathcal{R}_q$ is generally not involutive (see Remark 9.5.8 below for an exception). But it may possess involutive subdistributions; more precisely, in the sequel we are interested in all n-dimensional involutive transversal[7] subdistributions of $\mathcal{V}[\mathcal{R}_q]$. The reason is contained in the following statement.

Proposition 9.5.7. *Let $\sigma \in \Gamma_{loc}(\pi)$ be a solution of the differential equation $\mathcal{R}_q$. Then the tangent bundle $T(\operatorname{im} j_q\sigma)$ is an n-dimensional involutive transversal subdistribution of $\mathcal{V}[\mathcal{R}_q]\big|_{\operatorname{im} j_q\sigma}$. Conversely, let $\mathcal{U} \subseteq \mathcal{V}[\mathcal{R}_q]$ be an n-dimensional transversal involutive subdistribution. Then any integral manifold of $\mathcal{U}$ is locally of the form* $\operatorname{im} j_q\sigma$ *with a solution $\sigma \in \Gamma_{loc}(\pi)$ of $\mathcal{R}_q$.*

Proof. The first assertion is trivial and just restates the considerations that motivated above the Definition 9.5.1 of the Vessiot distribution. The converse is equally easy. The Frobenius Theorem C.3.3 guarantees the existence of n-dimensional integral manifolds. Since by definition $T\iota\bigl(\mathcal{V}[\mathcal{R}_q]\bigr)$ is contained in the contact distribution, any integral manifold of it is of the form $\operatorname{im} j_q\sigma$ for some section σ by Proposition 2.1.6 and since we are on $\mathcal{R}_q$ this section trivially satisfies $\operatorname{im} j_q\sigma \subseteq \mathcal{R}_q$. Hence any integral manifold of $\mathcal{U}$ corresponds to a (local) solution of $\mathcal{R}_q$. □

Remark 9.5.8. For ordinary differential equations this relationship between solutions and the Vessiot distribution was already the topic of Lemma 9.1.5. In this special case, there is generally no need to consider subdistributions, as according to Theorem 9.1.6 the Vessiot distribution is always one-dimensional for a regular, not underdetermined equation. These results can be easily extended to partial differential equations of finite type.

If we assume for simplicity that no lower-order equations are present, then a differential equation $\mathcal{R}_q$ of finite type may be represented as the image of a global section $\gamma : J_{q-1}\pi \rightarrow J_q\pi$. By definition, the symbol $\mathcal{N}_q$ of such an equation vanishes. Consequently, the Vessiot distribution $\mathcal{V}[\mathcal{R}_q]$ does not contain any vertical vectors. As it is furthermore n-dimensional, it may be interpreted as the horizontal bundle of a connection on the fibred manifold $\hat{\pi}^q : \mathcal{R}_q \rightarrow \mathcal{X}$. On the other hand, by Remark 2.2.3, such a global section γ induces a connection on $\pi^{q-1} : J_{q-1}\pi \rightarrow \mathcal{X}$; let $\mathcal{H}[\gamma]$ be the corresponding horizontal bundle.

In our special case $\mathcal{R}_q = \operatorname{im}\gamma$ is diffeomorphic to $J_{q-1}\pi$ and it is straightforward to verify that the diffeomorphism γ preserves the given connection, i. e. $\gamma_* : \mathcal{H}[\gamma] \rightarrow \mathcal{V}[\mathcal{R}_q]$ is an isomorphism between the respective horizontal bundles (with $T\pi^q_{q-1}$ as inverse). We know already from Example 2.3.17 that $\mathcal{R}_q$ is formally

---

[7] Recall that, by definition, $\mathcal{R}_q$ is a *fibred* submanifold of $\pi^q : J_q\pi \rightarrow \mathcal{X}$. Hence it makes sense to speak of transversal distributions on $\mathcal{R}_q$, namely those that are transversal with respect to the restricted fibration $\hat{\pi}^q : \mathcal{R}_q \rightarrow \mathcal{X}$.

integrable, if and only if the distribution $\mathcal{H}[\gamma]$ is involutive. As according to (C.9) the push-forward preserves Lie brackets, the Vessiot distribution $\mathcal{V}[\mathcal{R}_q]$ will also be involutive, if and only if $\mathcal{R}_q$ is formally integrable. Thus for (regular) differential equations of finite type, the Vessiot theory essentially coincides with the geometric theory developed in Example 2.3.17; we only work "one level higher", namely with the Vessiot distribution $\mathcal{V}[\mathcal{R}_q]$ (defined on $\mathcal{R}_q \subset J_q\pi$) instead of the horizontal bundle $\mathcal{H}[\gamma]$ (defined on $J_{q-1}\pi$). ◁

We could now take any basis $\{X_1, \dots, X_r\}$ of the Vessiot distribution $\mathcal{V}[\mathcal{R}_q]$ and analyse when n fields $U_i = a_i^j X_j$ with coefficients $a_i^j \in \mathcal{F}(\mathcal{R}_q)$ define an n-dimensional transversal involutive subdistribution. However, choosing a basis adapted to the geometry of the jet bundle leads to simpler structure equations.

Recall from Remark 2.2.9 that the contact distribution on $J_q\pi$ may be decomposed as $\mathcal{C}_q = V\pi^q_{q-1} \oplus \mathcal{H}_q$ where $V\pi^q_{q-1}$ is as usual the vertical bundle of the fibration $\pi^q_{q-1} : J_q\pi \to J_{q-1}\pi$ and $\mathcal{H}_q$ some complement defining a connection on the fibration $\pi^q : J_q\pi \to \mathcal{X}$. Since the symbol $\mathcal{N}_q$ of the differential equation $\mathcal{R}_q$ is defined as the intersection $V\pi^q_{q-1}|_{\mathcal{R}_q} \cap T\mathcal{R}_q$, we obtain a similar decomposition of the Vessiot distribution,

$$\mathcal{V}[\mathcal{R}_q] = \mathcal{N}_q \oplus \mathcal{H}\,, \tag{9.64}$$

with some complement $\mathcal{H}$ which again is not uniquely determined.

It follows from Remark 9.5.5 that such a complement $\mathcal{H}$ is always transversal to the fibration $\hat{\pi}^q : \mathcal{R}_q \to \mathcal{X}$ and hence at most n-dimensional. If $\mathcal{H}$ is indeed n-dimensional, then it induces a connection for $\hat{\pi}^q : \mathcal{R}_q \to \mathcal{X}$ which we call a *Vessiot connection* of $\mathcal{R}_q$. The involutive subdistributions $\mathcal{U}$ we are looking for must be such n-dimensional complements $\mathcal{H}$ and a major step towards solving $\mathcal{R}_q$ consists of finding all its *flat* Vessiot connections.

Example 9.5.9. Consider the trivial differential equation $\mathcal{R}_1$ defined by $u_t = 1 - u_x$ and $u = x$. Obviously, this equation has exactly one solution, namely $u(x,t) = x$. If we pull-back the contact form $\omega = \mathrm{d}u - u_x\mathrm{d}x - u_t\mathrm{d}t$, then we find (in the coordinates $(\overline{x}, \overline{t}, \overline{u_x})$ on $\mathcal{R}_1$) that $\iota^*\omega = (1 - \overline{u_x})(\mathrm{d}\overline{x} - \mathrm{d}\overline{t})$ and hence that the Vessiot distribution $\mathcal{V}[\mathcal{R}_1]$ is spanned by the two vector fields $X = \partial_{\overline{x}} + \partial_{\overline{t}}$ and $Y = \partial_{\overline{u_x}}$. The latter vector field spans the symbol $\mathcal{N}_1$. Obviously, any complement $\mathcal{H}$ to $\mathcal{N}_1$ is only one-dimensional and cannot be the horizontal space of a connection.

The explanation for this phenomenon is simple. $\mathcal{R}_1$ is not formally integrable, as the integrability condition $u_t = 0$ is hidden. Adding it and computing the Vessiot distribution of the projected equation $\mathcal{R}_1^{(1)}$, we obtain this time $\iota^*\omega = 0$ so that $\mathcal{V}[\mathcal{R}_1^{(1)}] = \langle \partial_{\overline{x}}, \partial_{\overline{t}} \rangle$. Since the symbol $\mathcal{N}_1^{(1)}$ vanishes, the only choice for a complement is $\mathcal{H} = \mathcal{V}[\mathcal{R}_1^{(1)}]$ which is obviously two-dimensional (and involutive). ◁

Thus, if a system is not formally integrable, then it may well happen that the decomposition (9.64) does not lead to an n-dimensional complement. A closer analysis reveals that this phenomenon appears only, if we have integrability conditions of the second kind, i. e. those arising from the prolongation of lower-order equations and not from generalised cross-derivatives.

Proposition 9.5.10. *Let $\mathcal{R}_q \subseteq J_q\pi$ be a differential equation in n independent variables with symbol $\mathcal{N}_q$. Its Vessiot distribution possesses locally a decomposition $\mathcal{V}[\mathcal{R}_q] = \mathcal{N}_q \oplus \mathcal{H}$ where the complement $\mathcal{H}$ is n-dimensional, if and only if $\mathcal{R}_q$ does not lead to integrability conditions of the second kind.*

Proof. Assume that $\mathcal{R}_q$ has a local representation of the form

$$\mathcal{R}_q : \begin{cases} \boldsymbol{\Phi}(\mathbf{x}, \mathbf{u}^{(q)}) &= 0\,, \\ \boldsymbol{\Psi}(\mathbf{x}, \mathbf{u}^{(q-1)}) &= 0\,, \end{cases} \tag{9.65}$$

where the Jacobian $\partial\boldsymbol{\Phi}/\partial\mathbf{u}_{(q)}$ has maximal (row) rank, so that it is not possible to extract further lower-order equations from the upper subsystem. If we follow the above described method for computing the Vessiot distribution in $TJ_q\pi$, then (9.62) takes for our system the form

$$C_i^{(q)}(\boldsymbol{\Phi})a^i + C_\alpha^\mu(\boldsymbol{\Phi})b_\mu^\alpha = 0\,, \qquad C_i^{(q)}(\boldsymbol{\Psi})a^i = 0\,. \tag{9.66}$$

According to the made assumptions, the matrix $C_\alpha^\mu(\boldsymbol{\Phi})$ has maximal (row) rank and $C_i^{(q)}(\boldsymbol{\Psi}) = D_i\boldsymbol{\Psi}$. Because of the maximal rank, the left subsystem may always be solved for a subset of the variables b_μ^α. If, and only if, no integrability conditions of the second kind exist, then $D_i\boldsymbol{\Psi} = 0$ on $\mathcal{R}_q$ and thus the right subsystem of (9.66) vanishes. If, and only if, this is the case, (9.66) possesses for each $1 \leq j \leq n$ a solution where $a^j = 1$ and all other a^i vanish. Obviously, the existence of such solutions is equivalent to the existence of an n-dimensional complement $\mathcal{H}$. □

Note that as a by-product this proof shows that in the absence of integrability conditions of the second kind, we may simply ignore lower-order equations during the determination of the Vessiot distribution.

Remark 9.5.11. If one does not care about the distinction between integrability conditions of the first and the second kind and simply requires that $\mathcal{R}_q = \mathcal{R}_q^{(1)}$ (i. e. no integrability conditions at all appear in the first prolongation of $\mathcal{R}_q$), then one can provide a more geometric proof for the existence of an n-dimensional complement (of course, in contrast to Proposition 9.5.10, the converse is not true then).

The assumption $\mathcal{R}_q = \mathcal{R}_q^{(1)}$ implies that to every point $\rho \in \mathcal{R}_q$ at least one point $\hat{\rho} \in \mathcal{R}_{q+1}$ with $\pi_q^{q+1}(\hat{\rho}) = \rho$ exists. We choose such a point and consider $\operatorname{im}\Gamma_{q+1}(\hat{\rho}) \subset T_\rho(J_q\pi)$. By definition of the contact map Γ_{q+1}, this is an n-dimensional transversal subset of $\mathcal{C}_q|_\rho$. Thus there only remains to show that it is also tangential to $\mathcal{R}_q$, as then we can define a complement by $T_\rho\iota(H_\rho) = \operatorname{im}\Gamma_{q+1}(\hat{\rho})$. But this tangency is a trivial consequence of $\hat{\rho} \in \mathcal{R}_{q+1}$; using for example the local coordinates expression (2.35) for the contact map and a local representation $\Phi^\tau = 0$ of $\mathcal{R}_q$, one immediately sees that the vector $v_i = \Gamma_{q+1}(\hat{\rho}, \partial_{x^i}) \in T_\rho(J_q\pi)$ satisfies $d\Phi^\tau|_\rho(v_i) = D_i\Phi^\tau(\hat{\rho}) = 0$ and thus is tangential to $\mathcal{R}_q$ by Remark C.2.8.

Hence it is possible to construct for each point $\rho \in \mathcal{R}_q$ a complement $\mathcal{H}_\rho$ such that $\mathcal{V}_\rho[\mathcal{R}_q] = (\mathcal{N}_q)_\rho \oplus \mathcal{H}_\rho$. There remains to show that these complements can be chosen so that they form a distribution (which by definition is smooth). Our assumption $\mathcal{R}_q = \mathcal{R}_q^{(1)}$ implies that the restricted projection $\hat{\pi}_q^{q+1} : \mathcal{R}_{q+1} \to \mathcal{R}_q$ is a surjective submersion, i. e. it defines a fibred manifold. Thus if we choose a section $\gamma : \mathcal{R}_q \to \mathcal{R}_{q+1}$ and then always take $\hat{\rho} = \gamma(\rho)$, it follows immediately that the corresponding complements $\mathcal{H}_\rho$ define a smooth distribution as required. ◁

Remark 9.5.12. For many differential equations $\mathcal{R}_q$ appearing in applications one can easily find an n-dimensional complement $\mathcal{H}$. Suppose that no integrability conditions of the second kind are hidden and that $\mathcal{R}_q$ can be locally represented by a system where the qth order equations are of the solved form $u_\mu^\alpha = \phi_\mu^\alpha(\mathbf{x}, \mathbf{u}, \tilde{\mathbf{u}}^{(q)})$. with $(\alpha, \mu) \in \mathcal{B}$. We assume that the principal derivatives u_μ^α are all pairwise different and of order q. Then we introduce the set $\mathcal{B}$ of all pairs (α, μ) such that our system contains an equation for u_μ^α and denote by $\tilde{\mathbf{u}}^{(q)}$ all the remaining (parametric) derivatives of order q. We claim that under the made assumptions $\mathcal{V}[\mathcal{R}_q] = \mathcal{N}_q \oplus \mathcal{H}$ where $\mathcal{H}$ is generated by the fields

$$\iota_* X_i = C_i^{(q)} + \sum_{(\alpha,\mu)\in\mathcal{B}} C_i^{(q)}(\phi_\mu^\alpha) C_\alpha^\mu\,, \qquad 1 \leq i \leq n\,. \tag{9.67}$$

For notational simplicity we restrict to a first-order equation $\mathcal{R}_1$. Furthermore, since we have seen above that under the made assumptions lower-order equations do not affect the determination of the Vessiot distribution, we simply assume that none are present. Then we are given an explicit parametrisation of $\mathcal{R}_1$ and we can compute on the submanifold $\mathcal{R}_1$ where a local coordinate system is given by $\overline{\mathbf{x}}$, $\overline{\mathbf{u}}$ and the parametric derivatives $\overline{u}_i^\alpha$ with $(\alpha, i) \notin \mathcal{B}$. Hence the pull-back of the contact forms $\omega^\alpha = \mathrm{d}u^\alpha - u_i^\alpha \mathrm{d}x^i$ is

$$\iota^* \omega^\alpha = \mathrm{d}\overline{u}^\alpha - \sum_{(\alpha,i)\in\mathcal{B}} \phi_i^\alpha \mathrm{d}\overline{x}^i - \sum_{(\alpha,i)\notin\mathcal{B}} \overline{u}_i^\alpha \mathrm{d}\overline{x}^i\,. \tag{9.68}$$

The common annihilator of these m one-forms is generated by the vertical vector fields $\partial_{\overline{u}_i^\alpha}$ with $(\alpha, i) \notin \mathcal{B}$ and the transversal fields

$$X_i = \partial_{\overline{x}^i} + \sum_{(\alpha,i)\in\mathcal{B}} \phi_i^\alpha \partial_{\overline{u}^\alpha} + \sum_{(\alpha,i)\notin\mathcal{B}} \overline{u}_i^\alpha \partial_{\overline{u}^\alpha}\,. \tag{9.69}$$

There remains to compute the push-forward of these fields. One easily verifies that the vertical fields

$$\iota_* \partial_{\overline{u}_i^\alpha} = \partial_{u_i^\alpha} + \sum_{(\beta,j)\in\mathcal{B}} \frac{\partial \phi_j^\beta}{\partial u_i^\alpha} \partial_{u_j^\beta} \tag{9.70}$$

form a basis of the symbol $\mathcal{N}_1$. Indeed, due to the solved form of our system, computing the push-forward amounts to nothing but solving the symbol equations (7.3). For the push-forward of the transversal vector fields $\overline{X}_i$ we first note that

$$\iota_*\partial_{\bar{u}^\alpha} = \partial_{u^\alpha} + \sum_{(\beta,j)\in\mathcal{B}} \frac{\partial\phi_j^\beta}{\partial u^\alpha}\partial_{u_j^\beta}\,, \quad \iota_*\partial_{\bar{x}^i} = \partial_{x^i} + \sum_{(\beta,j)\in\mathcal{B}} \frac{\partial\phi_j^\beta}{\partial x^i}\partial_{u_j^\beta}\,. \tag{9.71}$$

Entering these expressions in (9.69) and noting that on $\mathcal{R}_1$ we may replace u_i^α by ϕ_i^α for all $(\alpha,i)\in\mathcal{B}$, we arrive at (9.67). ◁

Remark 9.5.13. In Remark 2.2.9 we discussed the relation between the contact distributions $\mathcal{C}_q$ and $\mathcal{C}_{q+r}$ of order q and $q+r$, respectively. It implies trivially an analogous relation between the Vessiot distributions of the differential equation $\mathcal{R}_q$ and its prolongation $\mathcal{R}_{q+r}$, respectively. Assuming that $\mathcal{R}_q^{(r)} = \mathcal{R}_q$ so that the restricted projection $\hat{\pi}_q^{q+r}: \mathcal{R}_{q+r} \to \mathcal{R}_q$ is surjective, we have that

$$\mathcal{V}[\mathcal{R}_q] = T\hat{\pi}_q^{q+r}\bigl(\mathcal{V}[\mathcal{R}_{q+r}]\bigr) + \mathcal{N}_q\,. \tag{9.72}$$

If $\gamma: \mathcal{R}_q \to \mathcal{R}_{q+r}$ is a section of the fibration $\hat{\pi}_q^{q+r}$, then we obtain at every point $\rho\in\mathcal{R}_q$ the direct sum decomposition

$$\mathcal{V}_\rho[\mathcal{R}_q] = T_{\gamma(\rho)}\bigl(\mathcal{V}_{\gamma(\rho)}[\mathcal{R}_{q+r}]\bigr) \oplus \mathcal{N}_q\,. \tag{9.73}$$

Note that as in Remark 2.2.9 not every possible complement $\mathcal{H}$ to the symbol $\mathcal{N}_q$ can be obtained in this manner. ◁

We return to the question of determining the structure equations of the Vessiot distribution $\mathcal{V}[\mathcal{R}_q]$, as their knowledge is obviously necessary for our search for flat Vessiot connections of $\mathcal{R}_q$. We will derive them based on the decomposition $\mathcal{V}[\mathcal{R}_q] = \mathcal{N}_q \oplus \mathcal{H}$. Let first the vector fields $Y_1,\dots,Y_r$ generate the r-dimensional symbol $\mathcal{N}_q$. The symbol is by definition the intersection of two involutive distributions and hence itself an involutive distribution.[8] Thus there must exist some coefficients $\Delta_{k\ell}^m \in \mathcal{F}(\mathcal{R}_q)$ such that the structure equations of the symbol $\mathcal{N}_q$ are

$$[Y_k, Y_\ell] = \Delta_{k\ell}^m Y_m\,, \qquad 1\leq k,\ell\leq r\,. \tag{9.74a}$$

By contrast, the complement $\mathcal{H}$ is generally not involutive and hence its structure equations take a more complicated form. Since $T\mathcal{R}_q$ is trivially involutive, we find for the derived Vessiot distribution $\mathcal{V}'[\mathcal{R}_q] \subseteq T\iota(T\mathcal{R}_q)\cap \mathcal{C}_q'|_{\mathcal{R}_q}$. Let $Z_1,\dots,Z_t$ be some vector fields such that $\mathcal{V}'[\mathcal{R}_q] = \mathcal{V}[\mathcal{R}_q]\oplus\langle Z_1,\dots,Z_t\rangle$. In local coordinates, it follows from the commutator relations (2.13) of the contact fields that we may choose the fields Z_a as linear combinations of the fields $\partial_{u_\mu^\alpha}$ with $|\mu| = q-1$, i.e. there are coefficients $\kappa_{a\mu}^\alpha \in \mathcal{F}(\mathcal{R}_q)$ such that $Z_a = \kappa_{a\mu}^\alpha \partial_{u_\mu^\alpha}$. For a generic system, *all* these fields appear as generators of the derived Vessiot distribution; on the other hand, for a formally integrable system of finite type $\mathcal{V}[\mathcal{R}_q]$ is involutive by Remark 9.5.8 and hence $t = 0$.

[8] Beware: *as a distribution* the symbol is always involutive; but this observation is unrelated to the notion of an involutive symbol in the sense of Definition 7.1.17!

If the fields $\{X_1,\dots,X_n\}$ form a basis of the complement $\mathcal{H}$, then the structure equations of $\mathcal{H}$ have generally the form

$$[X_i,X_j] = \Pi^h_{ij}X_h + \tilde{\Delta}^m_{ij}Y_m + \Theta^c_{ij}Z_c\,, \qquad 1\leq i,j\leq n\,, \tag{9.74b}$$

with coefficients $\Pi^h_{ij},\tilde{\Delta}^m_{ij},\Theta^c_{ij}\in\mathcal{F}(\mathcal{R}_q)$. For the structure equations of the full Vessiot distribution $\mathcal{V}[\mathcal{R}_q]$ we finally also need the Lie brackets

$$[X_i,Y_k] = \bar{\Pi}^h_{ik}X_h + \bar{\Delta}^m_{ik}Y_m + \Xi^c_{ik}Z_c\,, \qquad 1\leq i\leq n\,, 1\leq k\leq r \tag{9.74c}$$

where again $\bar{\Pi}^h_{ik},\bar{\Delta}^m_{ik},\Xi^c_{ik}\in\mathcal{F}(\mathcal{R}_q)$.

For the analysis of a concrete differential equation $\mathcal{R}_q$ it suffices to consider the structure equations in the form (9.74).[9] However, for the theoretical considerations in the next section we need further simplifications which are—at least in principle—always possible. For an involutive distribution like $\mathcal{N}_q$ we can always find a basis $\{Y_1,\dots,Y_r\}$ such that $\Delta^m_{k\ell}=0$. Starting with an arbitrary basis, this requires only some linear algebra: in some local coordinates $\mathbf{y}$ on $\mathcal{R}_q$ we bring Y_k into the form $\partial_{y^k}+\tilde{Y}_k$ where all the fields $\tilde{Y}_k$ contain only ∂_{y^a} with $a>r$ (in the classical language of Example 7.2.12 we transform to a Jacobian system). With a similar transformation of the vector fields X_i we can always achieve that also $\Pi^h_{ij}=0$ and $\bar{\Pi}^h_{ik}=0$.

Remark 9.5.14. The situation becomes particularly simple, if in some local coordinates $(\mathbf{x},\mathbf{u}^{(q)})$ on $J_q\pi$ we can solve each equation of the system for a principal derivative (this solved form immediately induces then a coordinate chart on $\mathcal{R}_q$). First, we may now simply choose for the symbol fields $Y_k=\partial_{u^\beta_\nu}$ with u^β_ν running over all parametric derivatives of order q; in this form obviously $[Y_k,Y_\ell]=0$. If we introduce furthermore the vector fields $W_a=\partial_{u^\alpha_\mu}$ with u^α_μ running over all principal derivatives of order q, then the distribution $\mathcal{W}$ spanned by them is trivially involutive and furthermore satisfies $V\pi^q_{q-1}|_{\mathcal{R}_q}=\mathcal{N}_q\oplus\mathcal{W}$; i. e. the vector fields Y_k and W_a define together a basis of $V\pi^q_{q-1}|_{\mathcal{R}_q}$.

There remains to make a good choice for a "reference" complement $\mathcal{H}_0$ (not necessarily involutive). Recall that the contact fields $C^{(q)}_i$ and C^μ_α defined by (2.11) span the contact distribution $\mathcal{C}_q$. Of these fields only the $C^{(q)}_i$ are transversal to the fibration π^q. Hence the basis $\{X_1,\dots,X_n\}$ of any complement $\mathcal{H}$ in (9.64) can be chosen in the form $\iota_*X_i=C^{(q)}_i+\xi^\alpha_{i\mu}C^\mu_\alpha$ with suitable coefficients $\xi^\alpha_{i\mu}\in\mathcal{F}(\mathcal{R}_q)$. But the fields C^μ_α generate the distribution $V\pi^q_{q-1}$ and on $\mathcal{R}_q$ we just found a basis for it better adapted to our purposes. Indeed, it is now natural to choose the vectors X_i in the form $\iota_*X_i=C^{(q)}_i+\xi_i^aW_a$.

Making these special choices in our construction of a basis $\{X_1,\dots,X_n,Y_1,\dots,Y_r\}$ for the Vessiot distribution $\mathcal{V}[\mathcal{R}_q]$ eliminates most of the coefficients in the structure

[9] Compared with the literature on Vessiot's approach, already our use of the decomposition $\mathcal{V}[\mathcal{R}_q]=\mathcal{N}_q\oplus\mathcal{H}$ leads to a considerable simplification in the computations presented in the next section, as we will treat there exclusively linear systems whereas traditionally quadratic systems must be analysed.

equations (9.74) and we finally obtain a much simplified version of them,

$$[X_i, X_j] = \Theta^a_{ij} Z_a \,, \quad [X_i, Y_k] = \Xi^a_{ik} Z_a \,, \quad [Y_k, Y_\ell] = 0 \,, \tag{9.75}$$

as one can easily verify. ◁

Example 9.5.15. Going back to Example 9.5.4, we see that there we chose our basis in the manner described in Remark 9.5.14; we must only rename $Y_1 = X_3$. Furthermore, we have $Z_1 = \partial_u$ and the structure equations are

$$[X_1, X_2] = 0 \,, \quad [X_1, Y_1] = -Z_1 \,, \quad [X_2, Y_1] = -F'(u_x) Z_1 \,. \tag{9.76}$$

Indeed, in practical computations one usually constructs the Vessiot distribution automatically such that its structure equations are of the form (9.75). Note that by chance we have chosen here a reference complement $\mathcal{H}_0$ which defines already a flat Vessiot connection. ◁

Example 9.5.16. More generally, let $\mathcal{R}_1$ be a first-order equation with Cartan normal form (7.34) satisfying the additional simplifying assumptions of Remark 7.2.11, i. e. no algebraic equations are present and no principal derivatives appear on the right hand side of the system. If we assume furthermore that $\mathcal{R}_1^{(1)} = \mathcal{R}_1$, then we may choose—according to Remark 9.5.12—the "reference" complement $\mathcal{H}_0$ as the linear span of the vector fields X_i defined by (9.67); note that this corresponds exactly to the form used in Remark 9.5.14. Choosing also the symbol fields Y_k as described in Remark 9.5.14, we can explicitly evaluate the structure equations (9.75) on $\mathcal{R}_1$.

One remaining problem is that it is generally difficult to make any statement about the size and form of the derived Vessiot distribution $\mathcal{V}'[\mathcal{R}_1]$, i. e. to predict $t = \dim \mathcal{V}'[\mathcal{R}_1] - \dim \mathcal{V}[\mathcal{R}_1]$ and to give some explicit expressions for the fields Z_a. We only know that in our case the fields $Z_a = \kappa^\alpha_a \partial_{u^\alpha}$ span some subdistribution of $\langle \partial_{u^1}, \dots, \partial_{u^m} \rangle$. For this reason, we write the right hand sides of (9.75) in the form $\Theta^\alpha_{ij} \partial_{u^\alpha}$ and $\Xi^\alpha_{ik} \partial_{u^\alpha}$. Obviously, the new and the old coefficients are related by

$$\Theta^\alpha_{ij} = \kappa^\alpha_a \Theta^a_{ij} \,, \qquad \Xi^\alpha_{ik} = \kappa^\alpha_a \Xi^a_{ik} \,. \tag{9.77}$$

As part of a basis of $\mathcal{V}'[\mathcal{R}_1]$ the fields Z_a are linearly independent and thus these relations can be inverted and there exist some coefficients $\bar\kappa^a_\alpha$ such that

$$\Theta^a_{ij} = \bar\kappa^a_\alpha \Theta^\alpha_{ij} \,, \qquad \Xi^a_{ik} = \bar\kappa^a_\alpha \Xi^\alpha_{ik} \,. \tag{9.78}$$

A straightforward but slightly tedious calculation yields for $i < j$ that

$$\Theta^\alpha_{ij} = \begin{cases} 0 & \text{if } (\alpha, i) \notin \mathcal{B} \text{ and } (\alpha, j) \notin \mathcal{B} \,, \\ C^{(1)}_i(\phi^\alpha_j) & \text{if } (\alpha, i) \notin \mathcal{B} \text{ and } (\alpha, j) \in \mathcal{B} \,, \\ C^{(1)}_i(\phi^\alpha_j) - C^{(1)}_j(\phi^\alpha_i) & \text{if } (\alpha, i) \in \mathcal{B} \text{ and } (\alpha, j) \in \mathcal{B} \end{cases} \tag{9.79}$$

and, if $Y_k = \partial_{u^\beta_j}$ with some $(\beta, j) \notin \mathcal{B}$, that

$$\Xi_{ik}^{\alpha} = \begin{cases} 0 & \text{if } (\alpha,i) \notin \mathcal{B} \text{ and } (\alpha,i) \neq (\beta,j)\,, \\ -1 & \text{if } (\alpha,i) \notin \mathcal{B} \text{ and } (\alpha,i) = (\beta,j)\,, \\ -C_{\beta}^{j}(\phi_i^{\alpha}) & \text{if } (\alpha,i) \in \mathcal{B}\,. \end{cases} \tag{9.80}$$

Let as usual m be the fibre dimension of $\mathcal{E}$ and $r = \dim \mathcal{N}_1$. We introduce for $1 \leq i,j \leq n$ the vectors $\boldsymbol{\Theta}_{ij} \in \mathcal{F}(\mathcal{R}_1)^m$ with entries Θ_{ij}^{α} and for $1 \leq i \leq n$ the matrices $\boldsymbol{\Xi}_i \in \mathcal{F}(\mathcal{R}_1)^{r\times m}$ with entries Ξ_{ik}^{α}. In the next section we will show that all relevant information about the formal integrability or involution of the considered differential equation $\mathcal{R}_1$ are encoded in these vectors and matrices. Note that (9.79) implies that for a linear system with constant coefficients all vectors $\boldsymbol{\Theta}_{ij}$ vanish. We will see later that this observation is equivalent to the fact that in such a system no integrability conditions can be hidden.

The matrix $\boldsymbol{\Xi}_i$ has a simple block structure, if one orders the columns and rows in the right way (this observation is the reason why we introduced the $r \times m$ matrices built out of the coefficients Ξ_{ik}^{α} instead of the potentially smaller $r \times t$ matrices built out of the coefficients Ξ_{ik}^{a}; the involved contractions of rows destroys the block structure making it much harder to analyse the corresponding matrices). We order the m rows according to increasing α. Each column corresponds to a symbol field $Y_k = \partial_{u_j^{\beta}}$ with some $(\beta,j) \notin \mathcal{B}$; these we order first by increasing values of j and for the same j by increasing values of β. Using this ordering, one finds

$$\boldsymbol{\Xi}_i = \begin{pmatrix} -C_{\beta_1}^{1}(\phi_i^{\alpha}) & \cdots & -C_{\beta_{i-1}}^{i-1}(\phi_i^{\alpha}) & -C_{\beta_i}^{i}(\phi_i^{\alpha}) & 0\cdots 0 \\ 0 & \cdots & 0 & -\mathbb{1}_{\alpha_1^{(i)}} & 0\cdots 0 \end{pmatrix} \tag{9.81}$$

where for $1 \leq j \leq i$ we have $\beta_1^{(j)} < \beta_j \leq m$. Here $\beta_1^{(j)}$ denotes as usual the jth index of the symbol $\mathcal{N}_1$ and $\alpha_1^{(j)} = m - \beta_1^{(j)}$ the corresponding Cartan character. The unit block plus the ranges for the indices β_j lead trivially to the estimate

$$\alpha_1^{(i)} \leq \operatorname{rank} \boldsymbol{\Xi}_i \leq \sum_{j=1}^{i} \alpha_1^{(j)} \leq m\,. \tag{9.82}$$

The zero block at the right end of the matrix $\boldsymbol{\Xi}_i$ is a consequence of our assumption that no principal derivatives appear on the right hand side of the given differential system: we have $C_{\beta}^{j}(\phi_i^{\alpha}) = 0$ whenever $j > i$. ◁

At the end of Section 2.3 we introduced the notion of integral elements for a differential equation $\mathcal{R}_q$. Proposition 2.3.19 gave a characterisation of them via the ideal $\mathcal{I}[\mathcal{R}_q] = \langle \iota^* \mathcal{C}_q^0 \rangle_{\text{diff}} \subseteq \Omega(\mathcal{R}_q)$. Recall that it was essential to use the *differential* ideal, i. e. to include the exterior derivatives of the pulled-back contact forms. The dual operation to exterior differentiation is the Lie bracket. Thus it is not surprising that the Vessiot distribution provides us with a dual characterisation of integral elements using Lie brackets.

However, there is one caveat that must be taken into account. Proposition 2.3.19 allowed us to characterise integral elements $\mathcal{U}_\rho$ at a *single* point $\rho \in \mathcal{R}_q$. As the

Lie bracket is only defined for vector fields $X \in \mathfrak{X}(\mathcal{R}_q)$ but not individual vectors $v \in T_\rho \mathcal{R}_q$, we must now use whole distributions (defined at least on a neighbourhood of ρ).[10] These considerations lead to the following result.

Proposition 9.5.17. *Let $\mathcal{U} \subseteq \mathcal{V}[\mathcal{R}_q]$ be a transversal subdistribution of the Vessiot distribution of constant rank k. The spaces $\mathcal{U}_\rho$ are k-dimensional integral elements for all points $\rho \in \mathcal{R}_q$, if and only if $[\mathcal{U},\mathcal{U}] \subseteq \mathcal{V}[\mathcal{R}_q]$.*

Proof. This is a straightforward consequence of the explicit formula (C.22) for the exterior derivative. Let $\{\omega_1,\dots,\omega_r\}$ be a basis of the codistribution $\iota^*\mathcal{C}_q^0$. Then $\mathcal{I}[\mathcal{R}_q] = \langle \omega_1,\dots,\omega_r, d\omega_1,\dots,d\omega_r\rangle_{\text{alg}}$. Any vector field $X \in \mathcal{U}$ trivially satisfies $\omega_i(X) = 0$ by Proposition 9.5.3. For arbitrary fields $X_1, X_2 \in \mathcal{U}$, (C.22) yields $d\omega_i(X_1,X_2) = X_1\bigl(\omega_i(X_2)\bigr) - X_2\bigl(\omega_i(X_1)\bigr) + \omega_i\bigl([X_1,X_2]\bigr)$. The first two terms on the right hand side vanish trivially and the remaining equation implies our claim. □

Definition 9.5.18. A subdistribution $\mathcal{U}$ of the Vessiot distribution $\mathcal{V}[\mathcal{R}_q]$ satisfying the assumptions of Proposition 9.5.17 is called an *integral distribution* of the differential equation $\mathcal{R}_q$.

One should not confuse an "integral" distribution with an "integrable" distribution; the name simply reflects that such a distribution consists of integral elements. Even if $\mathcal{U}$ is an integral distribution of rank n, there is no reason why it should be involutive and hence possesses integral manifolds. In other words, infinitesimal solutions are only a first step, as it is not clear whether they can be combined in such a manner that a finite solution $\sigma \in \Gamma_{loc}(\pi)$ emerges for which $\operatorname{im} j_q\sigma$ is an integral manifold. We will study the conditions under which this is possible in more detail in the next section.

Addendum: Generalised Prolongations

In most of what we have done so far a differential equation was represented by a fibred submanifold $\mathcal{R}_q \subseteq J_q\pi$. It is important to note that $\mathcal{R}_q$ as a manifold of its own does not carry enough information; we must consider it as submanifold within the ambient jet bundle $J_q\pi$. This situation has now changed: the relevant data of the embedding $\iota : \mathcal{R}_q \hookrightarrow J_q\pi$ are encoded in the Vessiot distribution $\mathcal{V}[\mathcal{R}_q]$. Thus we may alternatively represent a differential equation by a fibred manifold $\pi : \mathcal{M} \to \mathcal{X}$ together with a distribution $\mathcal{D} \subseteq T\mathcal{M}$ on it and solutions are transversal integral manifolds of $\mathcal{D}$. This point of view leads to a generalised concept of prolongation.

Definition 9.5.19. A *generalised prolongation* of a differential equation given as such a pair $(\pi : \mathcal{M} \to \mathcal{X}, \mathcal{D})$ is a fibred manifold $\tilde{\pi} : \tilde{\mathcal{M}} \to \mathcal{M}$ together with a

[10] In the exterior approach we also had to consider implicitly at least a neighbourhood of ρ, namely to compute the exterior derivatives required for obtaining the differential ideal $\mathcal{I}[\mathcal{R}_q]$. But as soon as this ideal is determined, the remaining computations can be performed at a point.

distribution $\tilde{\mathcal{D}} \subseteq T\tilde{\mathcal{M}}$ such that (i) $T\tilde{\pi}(\tilde{\mathcal{D}}) \subseteq \mathcal{D}$ and (ii) for every transversal integral manifold $\mathcal{N} \subseteq \mathcal{M}$ of $\mathcal{D}$ a transversal integral manifold $\tilde{\mathcal{N}} \subseteq \tilde{\mathcal{M}}$ of $\tilde{\mathcal{D}}$ exists which satisfies $\tilde{\pi}(\tilde{\mathcal{N}}) = \mathcal{N}$.

In Condition (ii) the transversality of $\tilde{\mathcal{N}}$ refers of course to the induced fibration $\pi \circ \tilde{\pi} : \tilde{\mathcal{M}} \to \mathcal{X}$. It follows trivially from Condition (i) that any transversal integral manifold $\tilde{\mathcal{N}}$ of $\tilde{\mathcal{D}}$ projects on a transversal integral manifold of $\mathcal{D}$. The converse requirement that any such manifold $\mathcal{N}$ is obtainable by a projection is highly non-trivial and imposes severe restrictions on $(\tilde{\mathcal{M}}, \tilde{\mathcal{D}})$.

Example 9.5.20. The prototype of a generalised prolongation must of course be our familiar prolongation process. Indeed, let $\mathcal{R}_q \subseteq J_q\pi$ be a differential equation with Vessiot distribution $\mathcal{V}[\mathcal{R}_q]$. Then the restricted projection $\hat{\pi}^{q+1}_q : \mathcal{R}_{q+1} \to \mathcal{R}_q$ together with the Vessiot distribution $\mathcal{V}[\mathcal{R}_{q+1}]$ is a generalised prolongation. Here the second condition is trivial: we know that any transversal integral manifold of $\mathcal{V}[\mathcal{R}_q]$ is of the form $\operatorname{im} j_q\sigma$ with a solution $\sigma \in \Gamma_{loc}(\pi)$ and obviously $\operatorname{im} j_{q+1}\sigma$ is a transversal integral manifold of $\mathcal{V}[\mathcal{R}_{q+1}]$ projecting on $\operatorname{im} j_q\sigma$. In fact, it is the only one with this property.

Concerning Condition (i) we note that trivially $T\hat{\pi}^{q+1}_q : T\mathcal{R}_{q+1} \to T\mathcal{R}_q$. By Proposition 2.1.6, the contact distribution $\mathcal{C}_{q+1}$ is locally generated by the vector fields $C^{(q+1)}_i$ and C^{μ}_{α} with $|\mu| = q+1$, respectively. From (2.11) it is easy to see that $T\pi^{q+1}_q(\mathcal{C}_{q+1}) = \langle C^{(q)}_i \rangle \subset \mathcal{C}_q$. Thus we find $T\hat{\pi}^{q+1}_q\bigl(\mathcal{V}[\mathcal{R}_{q+1}]\bigr) \subseteq \mathcal{V}[\mathcal{R}_q]$ as required by Definition 9.5.19.

The above description still depends on the embedding $\mathcal{R}_q \hookrightarrow J_q\pi$, as our definition of the prolonged equation $\mathcal{R}_{q+1}$ goes through $J_{q+1}\pi$. Using Grassmannians we can completely omit any reference to such ambient spaces. Taking the restricted projection $\hat{\pi}^q : \mathcal{R}_q \to \mathcal{X}$ and considering as in Remark 2.2.5 the transversal Grassmannian $G(\hat{\pi}^q) \cong J_1\hat{\pi}^q$, one easily verifies that

$$\mathcal{R}_{q+1} = \bigl\{(\rho, \mathcal{V}) \in G(\hat{\pi}^q) \mid \mathcal{V} \subseteq \mathcal{V}_\rho[\mathcal{R}_q]\bigr\}\,. \tag{9.83}$$

Indeed, (9.83) is just a reformulation of the alternative definition (2.61) of the prolonged equation $\mathcal{R}_{q+1}$. There we required that ρ, considered as an equivalence class, contains a section σ such that $\mathcal{V} = T_\rho(\operatorname{im} j_q\sigma)$. It is obvious that any $\mathcal{V} \subseteq T_\rho\mathcal{R}_q$ with this property is contained in the Vessiot distribution. Conversely, let $\mathcal{V} \subseteq \mathcal{V}_\rho[\mathcal{R}_q]$ be a transversal n-dimensional subspace. Since $\mathcal{V}$ consists of contact fields, it must be of the form $\mathcal{V} = T_\rho(\operatorname{im} j_q\sigma)$ for some section σ contained in ρ.

Any vector space $\mathcal{V}$ admitted in (9.83) is a possible horizontal complement $\mathcal{H}_\rho$ in the decomposition $\mathcal{V}_\rho[\mathcal{R}_q] = \mathcal{N}_q|_\rho \oplus \mathcal{H}_\rho$. Above we discussed that such a decomposition does not necessarily exist everywhere on the differential equation $\mathcal{R}_q$ but only on $\mathcal{R}^{(1)}_q$. In (9.83) points $\rho \in \mathcal{R}_q$ where no appropriate complement exists are automatically eliminated. ◁

Suppose that a connection with horizontal bundle $\tilde{\mathcal{H}}$ is given on the fibred manifold $\tilde{\pi} : \tilde{\mathcal{M}} \to \mathcal{M}$. The associated horizontal lift $A : \tilde{\mathcal{M}} \times_{\mathcal{M}} T\mathcal{M} \to T\tilde{\mathcal{M}}$ allows us to lift any vector field in $\mathfrak{X}(\mathcal{M})$ to a horizontal field in $\mathfrak{X}(\tilde{\mathcal{M}})$. Thus we may define on $\tilde{\mathcal{M}}$ a distribution $\tilde{\mathcal{D}}$ by setting $\tilde{\mathcal{D}}_{\tilde{p}} = A(\tilde{p}, \mathcal{D}_p)$ for all $p \in \mathcal{M}$ and $\tilde{p} \in \tilde{\pi}^{-1}(p)$. Obviously, $\tilde{\mathcal{D}} \subseteq \tilde{\mathcal{H}}$.

Proposition 9.5.21. *If the derived distribution $\tilde{\mathcal{D}}'$ satisfies $\tilde{\mathcal{D}}' \subseteq \tilde{\mathcal{H}}$, then $(\tilde{\mathcal{M}}, \tilde{\mathcal{D}})$ is a generalised prolongation of $(\mathcal{M}, \mathcal{D})$.*

Proof. Concerning the first condition in Definition 9.5.19 of a generalised prolongation, we note that the horizontal lift is a right inverse to the map $\rho = (\tau_{\tilde{\mathcal{M}}}, T\tilde{\pi})$. This fact implies in particular that $T\tilde{\pi}(\tilde{\mathcal{D}}) = \mathcal{D}$.

For the second condition let $\mathcal{N} \subseteq \mathcal{M}$ be an arbitrary transversal integral manifold of the distribution $\mathcal{D}$ and set $\hat{\mathcal{N}} = \tilde{\pi}^{-1}(\mathcal{N})$. Then our projection $\tilde{\pi}$ restricts to a fibration $\hat{\pi} : \hat{\mathcal{N}} \to \mathcal{N}$. Furthermore, we consider the distribution $\hat{\mathcal{D}}$ on $\hat{\mathcal{N}}$ defined by $\hat{\mathcal{D}}_{\hat{p}} = A(\hat{p}, D_p)$ for all $p \in \mathcal{N}$ and $\hat{p} \in \hat{\pi}^{-1}(p)$. Obviously, $\hat{\mathcal{D}} \subseteq \tilde{\mathcal{D}}|_{\hat{\mathcal{N}}}$. We claim that the distribution $\hat{\mathcal{D}}$ is involutive. Indeed, it follows immediately that $\tilde{\mathcal{H}}|_{\hat{\mathcal{N}}} \cap T\hat{\mathcal{N}} = \hat{\mathcal{D}}$ by construction and since we assumed that $\hat{\mathcal{D}}' \subseteq \tilde{\mathcal{D}}'|_{\hat{\mathcal{N}}} \subseteq \tilde{\mathcal{H}}|_{\hat{\mathcal{N}}}$, we have $\hat{\mathcal{D}}' = \hat{\mathcal{D}}$.

By the Frobenius Theorem C.3.3, the distribution $\hat{\mathcal{D}}$ thus possesses integral manifolds; let $\tilde{\mathcal{N}} \subseteq \tilde{\mathcal{M}}$ be one of them. By construction, $T\tilde{\pi}(\tilde{\mathcal{N}}) = \mathcal{N}$. Hence we are dealing with a transversal integral manifold of the required form. □

In contrast to the classical prolongation discussed in Example 9.5.20, here the lifted integral manifold $\tilde{\mathcal{N}}$ is no longer unique. Any integral manifold of $\hat{\mathcal{D}}$ has the required properties. Note furthermore that the given connection does not need to be flat; in fact, in most applications its horizontal bundle $\tilde{\mathcal{H}}$ will not be integrable. However, the imposed condition $\tilde{\mathcal{D}}' \subseteq \tilde{\mathcal{H}}$ (which may equivalently be formulated as the vanishing of the curvature $\Omega_{\tilde{\mathcal{H}}}$ on $\tilde{\mathcal{D}}'$) ensures that the induced connection on the fibration $\hat{\pi} : \hat{\mathcal{N}} \to \mathcal{N}$ with horizontal bundle $\hat{\mathcal{D}}$ is always flat.

Definition 9.5.22. A connection on the fibred manifold $\tilde{\pi} : \tilde{\mathcal{M}} \to \mathcal{M}$ with a horizontal bundle $\mathcal{H}$ such that $\tilde{\mathcal{D}}' \subseteq \tilde{\mathcal{H}}$ is called an *adapted connection* for the differential equation $(\mathcal{M}, \mathcal{D})$.

Let us assume that $(\mathcal{M}, \mathcal{D})$ is in fact a differential equation $\mathcal{R}_q$ together with its Vessiot distribution $\mathcal{V}[\mathcal{R}_q]$. As usual, we use as local coordinates on $\mathcal{M} = \mathcal{R}_q$ the independent variables $\mathbf{x}$ and all parametric derivatives u^α_μ with $0 \leq |\mu| \leq q$ which we denote collectively by $\tilde{\mathbf{u}}^{(q)}$. Furthermore, let $\mathbf{z}$ be fibre coordinates for the fibration $\tilde{\pi} : \tilde{\mathcal{M}} \to \mathcal{M}$ with fibre dimension $\tilde{m}$. The horizontal bundle of an adapted connection is then locally spanned by vector fields of the form

$$\partial_{x^i} + \Gamma^a_i(\mathbf{x}, \tilde{\mathbf{u}}^{(q)}, \mathbf{z})\partial_{z^a}\,, \quad \partial_{u^\alpha_\mu} + \Gamma^{a\mu}_\alpha(\mathbf{x}, \tilde{\mathbf{u}}^{(q)}, \mathbf{z})\partial_{z^a}\,. \tag{9.84}$$

In applications one is mainly concerned with the special case $\Gamma^{a\mu}_\alpha = 0$ called a *prolongation by pseudo-potentials*. The origin of this terminology lies in the even more specialised case where the remaining coefficients Γ^a_i are independent of the fibre coordinates $\mathbf{z}$. If the section $\sigma \in \Gamma_{loc}(\pi)$ is a solution of $\mathcal{R}_q$, then $\operatorname{im} j_q\sigma$ is a transversal integral manifold of $\mathcal{D} = \mathcal{V}[\mathcal{R}_q]$. It follows from (9.84) and our special assumptions that the horizontal lift of $T \operatorname{im} j_q\sigma$ is spanned by the vector fields

$$\hat{X}_i = \partial_{x^i} + (\Gamma_i^a \circ j_q\sigma)\partial_{z^a} . \tag{9.85}$$

By our considerations in the proof of Proposition 9.5.21, these fields span the horizontal bundle $\hat{\mathcal{H}}$ of a connection of the fibration $\hat{\pi} : \tilde{\pi}^{-1}(\operatorname{im} j_q\sigma) \to \operatorname{im} j_q\sigma$. Furthermore, we have seen that for an adapted connection this induced connection is always flat. Hence the distribution $\hat{\mathcal{H}}$ must be involutive. Computing the Lie brackets between the vector fields $\hat{X}_i$ shows that this is the case if and only if

$$\frac{\partial(\Gamma_i^a \circ j_q\sigma)}{\partial x^k} - \frac{\partial(\Gamma_k^a \circ j_q\sigma)}{\partial x^i} = 0 \tag{9.86}$$

for all $1 \leq i,k \leq n$. But these equations represent the compatibility conditions for the existence of potentials V^a for the functions $\Gamma_i^a \circ j_q\sigma$ (cf. Remark 7.1.7).

We know from Remark 2.3.6 that a connection corresponds to a differential equation of finite type and according to Example 2.3.17 this differential equation is formally integrable, if and only if the connection is flat. Thus returning to the case of a prolongation by pseudo-potentials, we find in the case of an adapted connection that the differential equation

$$\tilde{\mathcal{R}}_1 : \left\{ \frac{\partial z^a}{\partial x^i} = \Gamma_i^a\bigl(j_q\sigma(\mathbf{x}),\mathbf{z}\bigr) , \qquad \begin{matrix} 1\leq a\leq \tilde{m} , \\ 1\leq i\leq n \end{matrix} \right. \tag{9.87}$$

is formally integrable whenever σ is a solution of $\mathcal{R}_q$. Thus in a generalised sense we may consider the differential equation $\mathcal{R}_q$ as a compatibility condition for the system $\partial z^a/\partial x^i = \Gamma_i^a(\mathbf{x},\tilde{\mathbf{u}}^{(q)},\mathbf{z})$. This is particularly of interest, if we are dealing with a linear connection, as then (9.87) becomes a linear equation for $\mathbf{z}$.

Addendum: Symmetry Theory and the Method of Characteristics

Most, if not all, methods to determine at least some explicit solutions of (non-linear) partial differential equations rely on Lie symmetries. Roughly speaking, symmetries are transformations (in the simplest case diffeomorphisms $\mathcal{E} \to \mathcal{E}$) mapping solutions of the equation into solutions. For a general introduction into this theory we refer to [48, 342]. In this Addendum we take a non-standard approach using the Vessiot distribution and show as an application how the well-known method of characteristics can be derived in this manner.

Definition 9.5.23. Let $\mathcal{D} \subseteq T\mathcal{M}$ be a distribution on a manifold $\mathcal{M}$. A vector field $X \in \mathfrak{X}(\mathcal{M})$ is called an *infinitesimal symmetry* of $\mathcal{D}$, if $\mathcal{L}_X\mathcal{D} = [X,\mathcal{D}] \subseteq \mathcal{D}$.

It follows immediately from the Jacobi identity for the Lie bracket that the infinitesimal symmetries of $\mathcal{D}$ form a Lie algebra, i. e. if the vector fields X_1 and X_2 are two infinitesimal symmetries, then the same is true for their commutator $[X_1,X_2]$, as $\bigl[[X_1,X_2],\mathcal{D}\bigr] = \bigl[X_1,[X_2,\mathcal{D}]\bigr] - \bigl[X_2,[X_1,\mathcal{D}]\bigr] \subseteq \mathcal{D}$.

Proposition 9.5.24. *Let $\mathcal{D}$ be a distribution on a manifold $\mathcal{M}$. Then $\mathcal{D}$ is invariant under the flow of a vector field $X \in \mathfrak{X}(\mathcal{M})$, i. e. the tangent map $T\exp(tX)$ restricted to $\mathcal{D}$ is an isomorphism for sufficiently small values of the parameter t, if and only if X is an infinitesimal symmetry of $\mathcal{D}$.*

Proof. This is an immediate consequence of the definition of the Lie derivative of a vector field given by (C.12): we have for an arbitrary vector field $Y \in \mathcal{D}$ that $\mathcal{L}_X Y = \frac{d}{dt}\left[\left(T\exp(tX)\right)^{-1} Y\right]_{t=0}$. Thus we stay within the distribution $\mathcal{D}$, if and only if $\mathcal{D}$ is invariant under the Lie derivative with respect to X. □

Let $\mathcal{N} \subseteq \mathcal{M}$ be a k-dimensional integral manifold of the distribution $\mathcal{D}$. For sufficiently small values of the parameter t, we introduce the set $\mathcal{N}_t = \exp(tX)(\mathcal{N})$; it follows from standard results on the smooth dependency of the solutions of ordinary differential equations on the initial data that $\mathcal{N}_t$ is again a k-dimensional submanifold for t in some open interval $0 \in \mathbb{I} \subseteq \mathbb{R}$. Furthermore, Proposition 9.5.24 implies that $\mathcal{N}_t$ is also an integral manifold of $\mathcal{D}$, if X is an infinitesimal symmetry. Thus considering integral manifolds as "solutions" of a distribution, we find here the classical picture that symmetries map solutions into solutions.

Definition 9.5.25. Let $\mathcal{D}$ be a distribution. A vector field $C \in \mathcal{D}$ is *Cauchy characteristic*, if $\mathcal{L}_C\mathcal{D} = [C, \mathcal{D}] \subseteq \mathcal{D}$. The integral curves of C are called the *Cauchy characteristics* of the distribution $\mathcal{D}$.

Obviously, any Cauchy characteristic vector field is an infinitesimal symmetry but not vice versa, as an infinitesimal symmetry is not necessarily contained in the distribution $\mathcal{D}$. It follows by the same argument as above that the Lie bracket of two Cauchy characteristic vector fields is again Cauchy characteristic; thus these fields form an involutive subdistribution of $\mathcal{D}$.

Proposition 9.5.26. *Let C be a Cauchy characteristic vector field for the distribution $\mathcal{D}$ and $\mathcal{N}$ a k-dimensional integral manifold of $\mathcal{D}$ for which C is everywhere transversal. Then the set $\mathcal{N}_C = \bigcup_{t\in\mathbb{I}} \mathcal{N}_t$ is for a sufficiently small interval $0 \in \mathbb{I} \subseteq \mathbb{R}$ a $(k+1)$-dimensional integral manifold of $\mathcal{D}$.*

Proof. We know already from our considerations above that each subset $\mathcal{N}_t$ is an integral manifold. Since, by definition, the vector field C is contained in the distribution $\mathcal{D}$, the claim follows immediately. □

This result demonstrates the great importance of Cauchy characteristic vector fields. According to Proposition 9.5.24, we may use any infinitesimal symmetry to map an integral manifold $\mathcal{N}$ into new ones; but these new ones are always of the same dimension. If we know a transversal Cauchy characteristic vector field C, then we can extend an integral manifold to a higher-dimensional one and this extension requires essentially only the solution of an ordinary differential equation, namely the determination of the flow of C.

It is a natural thought to take for the manifold $\mathcal{M}$ a differential equation $\mathcal{R}_q$ and as distribution $\mathcal{D}$ its Vessiot distribution $\mathcal{V}[\mathcal{R}_q]$. In fact, this approach would immediately lead to a generalisation of the classical Lie symmetry theory, as one can show that the infinitesimal symmetries of $\mathcal{V}[\mathcal{R}_q]$ correspond to what is called *internal symmetries* in the literature [17]. We will not work out the corresponding theory in this Addendum but only note that from a geometric point of view most reduction methods for partial differential equations consists of enforcing the existence of a Cauchy characteristic vector field for the Vessiot distribution.

A standard approach to solving scalar first-order partial differential equations in one dependent variable is the method of characteristics (see e. g. [238]). We show now that it may be understood as a combination of symmetry theory with symplectic geometry. The basic observation is that the Vessiot distribution of such an equation possesses a Cauchy characteristic vector field. Our presentation combines ideas from [27, §8] and [231, Sect. 6.1].

Let $\pi : \mathcal{E} \to \mathcal{X}$ be a fibred manifold with one-dimensional fibres, i. e. $m = 1$ in our conventions. Then $\dim J_1\pi = 2n+1$ where $n = \dim \mathcal{X}$ and the contact codistribution $\mathcal{C}_1^0$ is one-dimensional with the local generator $\omega = \mathrm{d}u - u_i \mathrm{d}x^i$. Dually, its annihilator, the contact distribution $\mathcal{C}_1$, is locally spanned by the $2n$ vector fields $C_i^{(1)} = \partial_{x^i} + u_i \partial_u$ and $C^i = \partial_{u_i}$ (this is just (2.11) for $q = m = 1$).

By Remark C.3.7, we may associate with any distribution $\mathcal{D}$ on a manifold $\mathcal{M}$ a vector-valued curvature two-form Ω taking its values in the "normal" vector bundle $\mathcal{K} = \mathcal{D}'/\mathcal{D}$ where $\mathcal{D}'$ is the derived distribution. Let us apply this construction to the contact distribution $\mathcal{C}_1$. According to (2.13), the only non-vanishing commutators are $[C_i^{(1)}, C^i] = \partial_u$. Thus here $\dim \mathcal{K} = 1$ and the curvature is now easily seen to be given by the ordinary two-form $\Omega = \mathrm{d}\omega = \mathrm{d}x^i \wedge \mathrm{d}u_i$. Using the local coordinate expression, one readily verifies that Ω is closed and non-degenerate. Hence, the contact distribution $\mathcal{C}_1$ has the structure of a $2n$-dimensional symplectic manifold with the symplectic two-form Ω.[11]

Let $\mathcal{R}_1 \subset J_1\pi$ be a scalar differential equation, i. e. $\operatorname{codim} \mathcal{R}_1 = 1$. Provided that the tangent space $T\mathcal{R}_1$ is not completely contained in the contact distribution $\mathcal{C}_1|_{\mathcal{R}_1}$, the Vessiot distribution $\mathcal{V}[\mathcal{R}_1]$ is $(2n-1)$-dimensional. Because of $\Omega(X,Y) = \omega\big([X,Y]\big)$ for all $X, Y \in \mathcal{C}_1$, any vector field C that is Cauchy characteristic in $\iota_*\big(\mathcal{V}[\mathcal{R}_1]\big)$ satisfies $\Omega(C,X) = 0$ for all $X \in \iota_*\big(\mathcal{V}[\mathcal{R}_1]\big)$ and thus must lie in the symplectic complement of the Vessiot distribution.

Lemma 9.5.27. *Let $(\mathcal{V}, \Omega)$ be a symplectic vector space and $\mathcal{W} \subset \mathcal{V}$ a subspace with* $\operatorname{codim} \mathcal{W} = 1$*. Then the symplectic complement $\mathcal{W}^\perp$ is one-dimensional and satisfies $\mathcal{W}^\perp \subseteq \mathcal{W}$.*

Proof. Let $\dim \mathcal{V} = 2n$. The subspace $\mathcal{W}^\perp$ is described by $2n-1$ linear equations; hence trivially $\dim \mathcal{W}^\perp \geq 1$. We choose some vector $z \in \mathcal{V}$ such that $\mathcal{V} = \mathcal{W} \oplus \langle z \rangle$. If $v \in \mathcal{W}^\perp$, then by definition $\Omega(v,w) = 0$ for all $w \in \mathcal{W}$. As a symplectic form is always non-degenerate, we must have $\Omega(v,z) \neq 0$ for all $v \in \mathcal{W}^\perp$. Assume now that $v_1, v_2 \in \mathcal{W}^\perp$ are two linearly independent vectors and set $c_i = \Omega(v_i, z) \neq 0$. Then

---

[11] There exists a generalisation of these considerations to jet bundles of arbitrary order and with any number of dependent variables. There one does not obtain a classical symplectic structure but a so-called *metasymplectic structure* defined by a vector valued two-form [296].

the linear combination $v = c_2 v_1 - c_1 v_2$ defines a non-vanishing element of $\mathcal{W}^\perp$ with $\Omega(v,z) = 0$. As this contradicts our considerations above, we have $\dim \mathcal{W}^\perp = 1$.

Let v be a generator of $\mathcal{W}^\perp$; we write it as $v = w + cz$ with $w \in \mathcal{W}$ and some constant $c \in \mathbb{R}$. We must have $\Omega(v,z) = \Omega(w,z) \neq 0$ and thus in particular $w \neq 0$ by our considerations above. On the other hand, by definition of the complement $\mathcal{W}^\perp$, we find $\Omega(w,v) = c\Omega(w,z) = 0$. Hence $c = 0$ and $\mathcal{W}^\perp \subseteq \mathcal{W}$. □

We apply this result with $\mathcal{V} = \mathcal{C}_1$ and $\mathcal{W} = \iota_*\big(\mathcal{V}[\mathcal{R}_1]\big)$. It implies the existence of a distinguished direction within the Vessiot distribution. However, if C is a non-vanishing vector field pointing everywhere in this direction, then our reasoning above with the symplectic two-form Ω only shows that $[C,X] \in \mathcal{C}_1$ for all fields $X \in \iota_*\big(\mathcal{V}[\mathcal{R}_1]\big)$. In order to prove that C is Cauchy characteristic for the Vessiot distribution, we must show that in addition $[C,X] \in T\mathcal{R}_1$. But this condition is trivially satisfied, as all considered fields live in $T\mathcal{R}_1$ and the Lie bracket of any two vector fields in $T\mathcal{R}_1$ lies again in $T\mathcal{R}_1$.

The initial value problem for $\mathcal{R}_1$ may be formulated geometrically as follows. Let $\Gamma \hookrightarrow \mathcal{X}$ be an $(n-1)$-dimensional submanifold and $\hat{\pi} : \pi^{-1}(\Gamma) \to \Gamma$ the corresponding restriction of the fibred manifold $\pi : \mathcal{E} \to \mathcal{X}$. As initial data we use a section $\gamma : \Gamma \to \pi^{-1}(\Gamma)$. Now let $\tilde{\gamma} \in \Gamma_{loc}(\pi)$ be a section defined in an open neighbourhood of Γ and satisfying both $\tilde{\gamma}|_\Gamma = \gamma$ and $\mathcal{N}_0 = \operatorname{im}\big((j_1\tilde{\gamma})|_\Gamma\big) \subset \mathcal{R}_1$. Then $\mathcal{N}_0$ is an $(n-1)$-dimensional submanifold of $\mathcal{R}_1$. We call the thus formulated initial value problem *non-characteristic*, if the field C is everywhere transversal to $\mathcal{N}_0$.

Proposition 9.5.28. *Let (Γ,γ) define a non-characteristic initial value problem for the given differential equation $\mathcal{R}_1$. Then this problem possesses a unique solution $\sigma \in \Gamma_{loc}(\pi)$ defined in a neighbourhood of Γ.*

Proof. By construction, $\mathcal{N}_0 \subset \mathcal{R}_1$ is an integral manifold of the Vessiot distribution $\iota_*\big(\mathcal{V}[\mathcal{R}_1]\big)$. Since the Cauchy characteristic field C is assumed to be transversal to $\mathcal{N}_0$, we may apply Proposition 9.5.26 and obtain that $\mathcal{N}_C$ is an n-dimensional integral manifold of the Vessiot distribution. But this means that $\mathcal{N}_C = \operatorname{im} j_1\sigma$ for some solution σ of $\mathcal{R}_1$ and thus that the given Cauchy problem is solvable.

In order to prove the uniqueness of the solution, we first show that the Cauchy characteristic field C is tangent to *every* prolonged solution $j_1\sigma$ of the differential equation $\mathcal{R}_1$. Let $\kappa : \operatorname{im} j_1\sigma \hookrightarrow J_1\pi$ be the canonical inclusion. Since the exterior derivative commutes with the pull-back, the symplectic two-form Ω satisfies $\kappa^*\Omega = d(\kappa^*\omega) = 0$. Now assume that for a given point $\rho \in \operatorname{im} j_1\sigma$ the vector C_ρ was not contained in $\kappa_*\big(T_\rho(\operatorname{im} j_1\sigma)\big)$. Then Ω would vanish on the $(n+1)$-dimensional subspace $\kappa_*\big(T_\rho(\operatorname{im} j_1\sigma)\big) \oplus \langle C_\rho\rangle \subset \mathcal{C}_1\big|_\rho$ which is not possible for a non-degenerate two-form on a $2n$-dimensional space.

This result implies that all prolonged solutions are foliated by characteristics and therefore uniquely determined by any $(n-1)$-dimensional submanifold transversal to C. There only remains to show that our submanifold $\mathcal{N}_0$ is uniquely determined by the initial data (γ,Γ). This follows basically from a simple counting argument. The requirement that the section $\tilde{\gamma}$ coincides with γ on Γ leads to $n-1$ independent conditions on the n jet components of $(j_1\tilde{\gamma})|_\Gamma$. For a non-characteristic problem, we have $C_\rho \notin T_\rho\big(\mathrm{im}\,(j_1\tilde{\gamma})|_\Gamma\big)$ for any point $\rho \in \mathrm{im}\,(j_1\tilde{\gamma})|_\Gamma$ but $C_\rho \in T_\rho(\mathrm{im}\, j_1\tilde{\gamma})$. Hence the requirement $\mathcal{N}_0 \subset \mathcal{R}_1$ provides an independent nth condition so that $(j_1\tilde{\gamma})|_\Gamma$ is uniquely determined by the section γ. □

In local coordinates we obtain the following picture. Let $\Phi(\mathbf{x},\mathbf{u}^{(1)}) = 0$ be a local representation of the equation $\mathcal{R}_1$. Its tangent space $T\mathcal{R}_1$ is the annihilator of the one-form $\mathrm{d}\Phi$ and thus completely contained in the contact distribution $\mathcal{C}_1$, if and only if the two one-forms $\mathrm{d}\Phi$ and ω are linearly dependent. This is the case, if and only if both $\partial\Phi/\partial u_i = 0$ and $\partial\Phi/\partial x^i + u_i(\partial\Phi/\partial u) = 0$. Thus if we assume that $\mathcal{R}_1$ is everywhere truly a first-order equation, then this cannot happen.

A vector field $X = a^i\partial_{x^i} + b\partial_u + c_i\partial_{u_i}$ is contained in $\mathcal{C}_1$, if and only if $b = u_i a^i$. Thus the Vessiot distribution is defined by the linear equations

$$b = u_i a^i\,, \qquad \left(\frac{\partial\Phi}{\partial x^i} + u_i\frac{\partial\Phi}{\partial u}\right)a^i + \frac{\partial\Phi}{\partial u_i}c_i = 0\,. \tag{9.88}$$

If we make the ansatz $C = \bar{a}^i(\partial_{x^i} + u_i\partial_u) + \bar{c}_i\partial_{u_i}$ for the Cauchy characteristic vector fields, then the condition $\Omega(C,X) = 0$ for all $X \in \iota_*\big(\mathcal{V}[\mathcal{R}_1]\big)$ leads to the linear equation $c_i\bar{a}^i - a^i\bar{c}_i = 0$ which must be satisfied for all values a^i, c_i solving the linear system (9.88). This is obviously only then the case, if $\bar{a}^i = \lambda\partial\Phi/\partial u_i$ and $\bar{c}_i = -\lambda\big(\partial\Phi/\partial x^i + u_i)(\partial\Phi/\partial u\big)$ for some function $\lambda \in \mathcal{F}(\mathcal{R}_1)$. Hence all Cauchy characteristic vector fields are multiples of

$$C = \frac{\partial\Phi}{\partial u_i}(\partial_{x^i} + u_i\partial_u) - \left(\frac{\partial\Phi}{\partial x^i} + u_i\frac{\partial\Phi}{\partial u}\right)\partial_{u_i} \tag{9.89}$$

which is exactly the vector field describing the classical characteristic system for a first-order equation $\mathcal{R}_1$.

For the initial value problem we may assume that our local coordinates on $\mathcal{X}$ are chosen in such a way that the hypersurface Γ is defined by $x^n = 0$. For a non-characteristic problem, the field C must then have a non-vanishing ∂_{x^n}-component which entails that the differential equation $\mathcal{R}_1$ has a local representation of the form $u_n = \phi(\mathbf{x},u,u_1,\dots,u_{n-1})$. The initial data γ defines a real-valued function $f(x^1,\dots,x^{n-1})$ and the submanifold $\mathcal{N}_0 \subset \mathcal{R}_1$ consists of all points $(\mathbf{x}_0,\mathbf{u}_0^{(1)}) \in J_1\pi$ such that $x_0^n = 0$, $u_0 = f(x_0^1,\dots,x_0^{n-1})$, $u_{0,i} = \big(\partial f/\partial x^i\big)(x_0^1,\dots,x_0^{n-1})$ for $1 \leq i < n$ and, finally, $u_{0,n} = \phi(\mathbf{x}_0,u_0,u_{0,1},\dots,u_{0,n-1})$. Thus $n-1$ derivatives are prescribed by the initial data and the last one is determined by the differential equation. Obviously, the thus described problem is the usual initial value problem for the Cauchy–Kovalevskaya Theorem. Note, however, that Proposition 9.5.28 is an existence and uniqueness theorem in the *smooth* category!

In introductory textbooks one often finds only simplified versions of the method of characteristics either for homogeneous linear or for quasi-linear equations. They correspond to situations where the above Cauchy characteristic field C is projectable (cf. Remark C.2.2) to a field either on $\mathcal{X}$ or on $\mathcal{E}$. In the linear case, we have $\Phi(\mathbf{x},u,\mathbf{p}) = a^i(\mathbf{x})u_i$ and hence $T\pi^1(C) = a^i(\mathbf{x})\partial_{x^i}$ is a well-defined vector field on the base manifold $\mathcal{X}$. In the quasi-linear case, $\Phi(\mathbf{x},u,\mathbf{p}) = a^i(\mathbf{x},u)u_i - b(\mathbf{x},u)$ and now $T\pi^1_0(C) = a^i(\mathbf{x},u)\partial_{x^i} + b(\mathbf{x},u)\partial_u$ is a well-defined field on the total space $\mathcal{E}$.

Example 9.5.29. We apply the method of characteristics to the *inviscid Burgers equation* $u_t = uu_x$. One readily computes its Cauchy characteristic vector field:

$$C = \partial_t - u\partial_x + (u_t - uu_x)\partial_u + u_t u_x \partial_{u_t} + u_x^2 \partial_{u_x} \,. \tag{9.90}$$

As promised by the remarks above, this field is projectable, as we are dealing with a quasi-linear equation and on $\mathcal{R}_1$ the ∂_u-component vanishes.

We consider the standard initial value problem prescribed on the hypersurface $\Gamma : t = 0$ with the section γ defined by a function $f(x)$. Then the (projected) characteristic system takes the simple form

$$\dot{t} = 1\,, \qquad \dot{x} = -u\,, \qquad \dot{u} = 0\,. \tag{9.91}$$

Because of the first equation, we can use t as parameter for the (projected) characteristic curves and obtain as solution of the characteristic system $u(t) = u_0$ and $x(t) = x_0 - u_0 t$. Thus the solution of our initial value problem is given by the implicit equation $u = f(x+ut)$. It can be solved whenever $tf'(x+ut) \neq 1$.

At the points where $tf'(x+ut) \neq 1$ the solution develops a *shock*. This phenomenon can be easily understood by taking the full characteristic system into account. The solution of the equation $\dot{u}_x = u_x^2$ is given by $u_x(t) = (1/u_{x,0} - t)^{-1}$ with the initial value $u_{x,0} = f'(x_0)$. Hence, the derivative u_x (and similarly u_t) becomes singular whenever $tf'(x+ut) = 1$. ◁

9.6 Flat Vessiot Connections

Recall from the last section that the goal of Vessiot's approach to solving a general partial differential equation $\mathcal{R}_q$ consists of constructing all its flat Vessiot connections, i. e. all n-dimensional transversal involutive subdistributions $\mathcal{U}$ of the Vessiot distribution $\mathcal{V}[\mathcal{R}_q]$. Given a basis $\{X_1,\dots,X_n,Y_1,\dots,Y_r\}$ of $\mathcal{V}[\mathcal{R}_q]$ with structure equations of the form (9.74), it is clear that any such distribution $\mathcal{U}$ can be locally generated by n vector fields $U_i \in \mathfrak{X}(\mathcal{R}_q)$ of the form $U_i = X_i + \zeta_i^k Y_k$ with some smooth coefficients $\zeta_i^k \in \mathcal{F}(\mathcal{R}_q)$.

Before we proceed to the actual determination of such distributions $\mathcal{U}$, we discuss their relation to power series solutions. Let $\sigma \in \Gamma_{loc}(\pi)$ be a smooth solution of $\mathcal{R}_q$ with the property that $\operatorname{im} j_q\sigma$ is an integral manifold of $\mathcal{U}$. Thus, if $\rho = j_q\sigma(x_0)$ for some $x_0 \in \mathcal{X}$, then $T_\rho(\operatorname{im} j_q\sigma) = \mathcal{U}_\rho$. We learned in Section 2.2 that knowing the tangent space of a q-fold prolonged section is equivalent to knowing its $(q+1)$-jet. If in our local coordinates $\sigma(\mathbf{x}) = \big(\mathbf{x},\mathbf{s}(\mathbf{x})\big)$ and the function $\mathbf{s}$ has a Taylor expansion around x_0 of the form $s^\alpha(\mathbf{x}) = \sum_{|\mu|=0}^{\infty} c_\mu^\alpha (\mathbf{x}-\mathbf{x}_0)^\mu/\mu!$, then the coefficients c_μ^α of order $q+1$ must be related to the coefficients ζ_i^k of the vector fields U_i.

In order to exhibit this relationship, we first note that it follows from our discussion of the contact map Γ_{q+1} in Section 2.2 that $T\iota(\mathcal{U}_\rho) = \operatorname{im}\Gamma_{q+1}(\hat{\rho})$ where $\hat{\rho} = j_{q+1}\sigma(x_0)$. Now assume that the fields X_i and Y_k have been chosen in the manner described in Remark 9.5.14. Thus the symbol fields satisfy $\iota_* Y_k = \partial_{u^\beta_\nu} + \xi^\alpha_{k\mu}\partial_{u^\alpha_\mu}$ where u^β_ν is a parametric derivative of order q and the summation goes over all principal derivatives u^α_μ of order q. The basis of our "reference" complement is of the form $\iota_* X_i = C^{(q)}_i + \bar{\xi}^\alpha_{i\mu}\partial_{u^\alpha_\mu}$ where again the summation goes only over the principal derivatives of order q. It follows from the local coordinate form of Γ_{q+1} and of the vector fields U_i, respectively, that $\iota_* U_i|_\rho = \Gamma_{q+1}(\hat{\rho}, \partial_{x^i})$ and this identity implies that

$$c^\beta_{\nu+1_i} = \zeta^k_i\,, \qquad c^\alpha_{\mu+1_i} = \bar{\xi}^\alpha_{i\mu} + \zeta^k_i \xi^\alpha_{k\mu} \tag{9.92}$$

where we again assume that u^β_ν is a parametric and u^α_μ a principal derivative of order q for our differential equation $\mathcal{R}_q$.

Note that in general the identification (9.92) implies certain relations between the coefficients $\boldsymbol{\zeta}$. Assume that index values k, ℓ, i, j exist such that $\iota_* Y_k = \partial_{u^\beta_\nu} + \xi^\alpha_{k\mu}\partial_{u^\alpha_\mu}$ and $\iota_* Y_\ell = \partial_{u^\beta_\rho} + \xi^\alpha_{\ell\mu}\partial_{u^\alpha_\mu}$ with $\nu + 1_i = \rho + 1_j$. In such a situation it follows from the first equation in (9.92) that the equality $\zeta^k_i = \zeta^\ell_j$ must hold. We will derive these equalities below in a more intrinsic manner.

Example 9.6.1. We continue with Example 9.5.4 (and 9.5.15, respectively). Expanding all definitions, we find for the fields U_i that

$$\begin{aligned} \iota_* U_1 &= \partial_x + u_x\partial_u + \zeta^1_1\partial_{u_x} + F'(u_x)\zeta^1_1\partial_{u_y}\,, \\ \iota_* U_2 &= \partial_y + u_y\partial_u + \zeta^1_2\partial_{u_x} + F'(u_x)\zeta^1_2\partial_{u_y}\,. \end{aligned} \tag{9.93}$$

This result implies at once that

$$c_{xx} = \zeta^1_1\,, \qquad c_{xy} = \zeta^1_2 = F'(u_x)\zeta^1_1\,, \qquad c_{yy} = F'(u_x)\zeta^1_2\,. \tag{9.94}$$

Hence the two coefficients ζ^1_1, ζ^1_2 cannot be chosen arbitrarily for a flat Vessiot connection but must satisfy $\zeta^1_2 = F'(u_x)\zeta^1_1$. This observation is consistent with the fact that the formal power series solution of the differential equation $u_y = F(u_x)$ has at every order only one parametric coefficient. ◁

Remark 9.6.2. In Section 2.3 we discussed a simple method for the order by order construction of formal power series solutions of a given differential equation $\mathcal{R}_q$. Once we got beyond order q, we obtained a linear system for all Taylor coefficients of the current order. While resolving this system, we can make a new choice of the principal and parametric coefficients, respectively, independent of the choice we made at the previous order. The situation changed when we analysed the symbol module using involutive bases. As discussed in Section 9.3 (see Definition 9.3.4), we performed then the classification in a systematic manner using a term order.

In the Vessiot theory we do something intermediate. If we follow the recipe described in Remark 9.5.14 for the construction of our basis $\{X_1,\dots,X_n,Y_1,\dots,Y_r\}$ of the Vessiot distribution, we decided on the principal and parametric derivatives at order q. Now we say that the parametric derivatives at order $q+1$ must be chosen among the derivatives of the parametric derivatives at order q. Indeed, each coefficient ζ_i^k corresponds to such a derivative. Of course, in general we obtain this way more coefficients than our differential equation admits parametric coefficients at order $q+1$. For this reason, the coefficients $\boldsymbol{\zeta}$ must satisfy some conditions.[12] In solving these for some of the coefficients we finally decide which of our potentially parametric derivatives are actually taken as principal ones. $\triangleleft$

We finally derive the conditions which the coefficients $\zeta_i^k \in \mathcal{F}(\mathcal{R}_q)$ of our ansatz $U_i = X_i + \zeta_i^k Y_k$ must satisfy in order to lead to an involutive subdistribution $\mathcal{U}$. Using the structure equations (9.74) of the Vessiot distribution $\mathcal{V}[\mathcal{R}_q]$, a straightforward computation yields the following conditions for $1 \leq i < j \leq n$:

$$\begin{aligned}[U_i, U_j] &= \Gamma_{ij}^h U_h + \left(\Theta_{ij}^a - \Xi_{jk}^a \zeta_i^k + \Xi_{ik}^a \zeta_j^k\right) Z_a \\ &\quad + \left(U_i(\zeta_j^k) - U_j(\zeta_i^k) + \tilde{\Delta}_{i\ell}^k \zeta_j^\ell - \tilde{\Delta}_{j\ell}^k \zeta_i^\ell + \Delta_{m\ell}^k \zeta_i^m \zeta_j^\ell - \Gamma_{ij}^\ell \zeta_\ell^k\right) Y_k\end{aligned} \tag{9.95}$$

where the coefficients Γ_{ij}^h are given by

$$\Gamma_{ij}^h = \Pi_{ij}^h + \tilde{\Pi}_{ik}^h \zeta_j^k - \tilde{\Pi}_{jk}^h \zeta_i^k \,. \tag{9.96}$$

If we assume that we can find a basis for the Vessiot distribution $\mathcal{V}[\mathcal{R}_q]$ such that the structure equation take the simpler form (9.75), then the fields U_i define a Jacobian system (in the terminology of Example 7.2.12), as they are obviously in a triangular form. This observation implies that they generate an involutive distribution, if and only if their Lie brackets vanish. (9.95) now simplifies to:

$$[U_i, U_j] = \left(\Theta_{ij}^a - \Xi_{jk}^a \zeta_i^k + \Xi_{ik}^a \zeta_j^k\right) Z_a + \left(U_i(\zeta_j^k) - U_j(\zeta_i^k)\right) Y_k = 0 \,. \tag{9.97}$$

As the vector fields Y_k and Z_a have been chosen linearly independent, their coefficients in (9.95) (or (9.97), respectively) must vanish. Thus we may distinguish two sets of conditions on the unknowns $\boldsymbol{\zeta}$. The first ones, the coefficients of the vector fields Z_a, form a linear system of (algebraic) equations:

$$G_{ij}^a = \Theta_{ij}^a - \Xi_{jk}^a \zeta_i^k + \Xi_{ik}^a \zeta_j^k = 0 \,, \qquad \begin{cases} 1 \leq a \leq t \,, \\ 1 \leq i < j \leq n \,. \end{cases} \tag{9.98a}$$

The second set, the coefficients of the fields Y_k, forms a quasi-linear system of differential equations. If we do not expand the fields U_i, it takes the concise form

[12] Compared with the direct determination of the Taylor coefficients of order $q+1$, these equations form a smaller system for a lower number of variables. Thus one might think that this should give us a more efficient approach for computing the coefficients. But this is not true; the work required for determining the other Taylor coefficients is only hidden in the computation of the fields $X_1, \dots, X_n$.

$$H_{ij}^k = U_i(\zeta_j^k) - U_j(\zeta_i^k) - \Lambda_{ij}^k = 0\,, \qquad \begin{cases} 1\leq k\leq r\,, \\ 1\leq i<j\leq n \end{cases} \tag{9.98b}$$

where the inhomogeneity Λ_{ij}^k vanishes for the case of the simplified conditions (9.97) and otherwise contains a number of linear and quadratic terms in $\boldsymbol{\zeta}$.

Remark 9.6.3. If and only if we choose the coefficients $\boldsymbol{\zeta}$ such that the algebraic equations (9.98a) are satisfied, then all commutators $[U_i, U_j]$ will vanish modulo the Vessiot distribution $\mathcal{V}[\mathcal{R}_q]$. According to Proposition 9.5.17, these equations therefore represent necessary and sufficient conditions for $\mathcal{U}$ to be an integral distribution. Above we noted in our comparison of the unknowns $\boldsymbol{\zeta}$ with the Taylor coefficients of formal power series solutions that certain components of $\boldsymbol{\zeta}$ must coincide. As we will show now, the deeper intrinsic reason for this restriction is exactly this observation about the meaning of (9.98a).

By Definition 2.3.18, an integral element of $\mathcal{R}_q$ lies always in the image of the contact map Γ_{q+1}. This observation permits us to reduce the number of unknowns in our ansatz for $\mathcal{U}$. Assume that we have index values $1 \leq i, j \leq n$, $1 \leq \alpha \leq m$ and $\mu \in \mathbb{N}_0^n$ (with $|\mu| = q-1$) such that both $u^\alpha_{\mu+1_i}$ and $u^\alpha_{\mu+1_j}$ are parametric derivatives (and thus obviously the derivative $u^\alpha_{\mu+1_i+1_j}$ of order $q+1$, too). Then there will exist two corresponding symbol fields $Y_k = \iota_*(\overline{\partial_{u^\alpha_{\mu+1_i}}})$ and $Y_l = \iota_*(\overline{\partial_{u^\alpha_{\mu+1_j}}})$. Now it follows from the coordinate form (2.35) of the contact map that a necessary condition for $\mathcal{U}$ to be an integral distribution is that $\zeta_j^k = \zeta_i^l$. ◁

Proposition 9.6.4. *If the combined conditions* (9.98) *are solvable everywhere on a given differential equation $\mathcal{R}_q \subseteq J_q\pi$, then $\mathcal{R}_q$ is formally integrable.*

Proof. According our usual regularity assumptions, solvability of (9.98) implies that for every point $\rho \in \mathcal{R}_q$ a neighbourhood exists on which a flat Vessiot connection lives. By the Frobenius Theorem C.3.3, this connection possesses an integral manifold containing ρ. As discussed in the last section, this integral manifold is the image of a prolonged solution σ of $\mathcal{R}_q$. Hence we can conclude that $\mathcal{R}_q$ is locally solvable in the smooth category and, as already mentioned in Remark 2.3.16, this fact implies trivially that the differential equation $\mathcal{R}_q$ is formally integrable. □

Example 9.6.5. We consider a first-order equation in one dependent variable

$$\mathcal{R}_1 : \begin{cases} u_n = \phi_n(\mathbf{x}, u, u_1, \dots, u_{n-r-1})\,, \\ \qquad \vdots \\ u_{n-r} = \phi_{n-r}(\mathbf{x}, u, u_1, \dots, u_{n-r-1})\,. \end{cases} \tag{9.99}$$

Recall from Remark 7.1.29 that the symbol of such an equation is always involutive. In order to simplify the notation, we use the following convention for the indices: $1 \leq k, \ell < n-r$ and $n-r \leq a, b \leq n$. Thus we may express our system in the form $u_a = \phi_a(\mathbf{x}, u, u_k)$ and local coordinates on $\mathcal{R}_1$ are $(\overline{\mathbf{x}}, \overline{u}, \overline{u}_k)$.

The pull-back of the contact form $\omega = du - u_i dx^i$ generating the contact codistribution $\mathcal{C}_1^0$ is $\iota^*\omega = d\overline{u} - \overline{u}_k d\overline{x}^k - \phi_a d\overline{x}^a$ and we obtain for the Vessiot distribution

$\mathcal{V}[\mathcal{R}_1] = \langle X_1, \dots, X_n, Y_1, \dots, Y_{n-r-1} \rangle$ where

$$X_k = \partial_{\overline{x}^k} + \overline{u}_k \partial_{\overline{u}} \,, \quad X_a = \partial_{\overline{x}^a} + \phi_a \partial_{\overline{u}} \,, \quad Y_k = \partial_{\overline{u}_k} \,. \tag{9.100}$$

The fields Y_k span the symbol $\mathcal{N}_1$ and the fields X_i our choice of a reference complement $\mathcal{H}_0$. Setting $Z = \partial_{\overline{u}}$, the structure equations of $\mathcal{V}[\mathcal{R}_1]$ are

$$\begin{gathered} [X_k, X_\ell] = 0 \,, \qquad [Y_k, Y_\ell] = 0 \,, \\ [X_k, X_a] = X_k(\phi_a) Z \,, \qquad [X_a, X_b] = \bigl(X_a(\phi_b) - X_b(\phi_a)\bigr) Z \,, \\ [X_k, Y_\ell] = -\delta_{k\ell} Z \,, \qquad [X_a, Y_k] = -Y_k(\phi_a) Z \,. \end{gathered} \tag{9.101}$$

Now we make the above discussed ansatz $U_i = X_i + \zeta_i^k Y_k$ for the generators of a transversal complement $\mathcal{H}$. Modulo the Vessiot distribution $\mathcal{V}[\mathcal{R}_1]$ we obtain for their Lie brackets

$$[U_k, U_\ell] \equiv (\zeta_k^\ell - \zeta_\ell^k) Z \mod \mathcal{V}[\mathcal{R}_1] \,, \tag{9.102a}$$

$$[U_a, U_k] \equiv \bigl(\zeta_a^k - \zeta_k^\ell Y_\ell(\phi_a) - X_k(\phi_a)\bigr) Z \mod \mathcal{V}[\mathcal{R}_1] \,, \tag{9.102b}$$

$$\begin{aligned} [U_a, U_b] \equiv \bigl(&\zeta_a^k Y_k(\phi_b) - \zeta_b^\ell Y_\ell(\phi_a) + \\ &X_a(\phi_b) - X_b(\phi_a)\bigr) Z \mod \mathcal{V}[\mathcal{R}_1] \,. \end{aligned} \tag{9.102c}$$

The algebraic system (9.98a) is now obtained by requiring that all the expressions in parentheses on the right hand sides vanish. Its solution is straightforward. The first subsystem (9.102a) implies the equalities $\zeta_k^\ell = \zeta_\ell^k$. This result was to be expected by the discussion in Remark 9.6.3: both u_k and u_ℓ are parametric derivatives for $\mathcal{R}_1$ and thus we could have made this identification already in our ansatz for the complement. The second subsystem (9.102b) yields that $\zeta_a^k = \zeta_k^\ell Y_\ell(\phi_a) + X_k(\phi_a)$. If we enter these results into the third subsystem (9.102c), then all unknowns ζ drop out and the solvability condition

$$X_a(\phi_b) - X_b(\phi_a) + X_k(\phi_a) Y_k(\phi_b) - X_k(\phi_b) Y_k(\phi_a) = 0 \tag{9.103}$$

arises. Thus in this example the algebraic system (9.98a) has a solution, if and only if this condition is satisfied.

Comparing with Example 2.3.12, one easily verifies that this solvability condition is equivalent to the vanishing of the Mayer or Jacobi bracket $[u_a - \phi_a, u_b - \phi_b]$ on the submanifold $\mathcal{R}_1$ which in turn was a necessary and sufficient condition for the formal integrability of the differential equation $\mathcal{R}_1$. Thus we may conclude that $\mathcal{R}_1$ possesses n-dimensional integral distributions, if and only if it is formally integrable (which in our case is also equivalent to $\mathcal{R}_1$ being involutive).

Note the manner in which the integrability condition (9.103) of $\mathcal{R}_1$ appears: it represents a necessary and sufficient condition for the existence of integral distributions. Thus here involution can be decided solely on the basis of the algebraic system (9.98a) whereas Proposition 9.6.4 requires the solvability of the combined system (9.98). We will see below that this observation does not represent a special property of a very particular class of differential equations but a general feature of the Vessiot theory. ◁

We have now collected all ingredients for applying Vessiot's approach to a concrete differential equation $\mathcal{R}_q$: the flat Vessiot connections arise as solutions of the above conditions (9.98). Explicitly deriving these conditions is in principle straightforward and solving the algebraic subsystem is a fairly trivial exercise, as it is linear. For the differential subsystem one will rarely be able to determine its general solution. But at least we can systematically determine this way all integral distributions of the differential equation $\mathcal{R}_q$.

However, our analysis so far does not permit us to make any statements about the actual solvability of (9.98) and thus about the existence of flat Vessiot connections. We will show next that (9.98)—in fact, already (9.98a)—is solvable in a certain manner, if and only if the differential equation $\mathcal{R}_q$ is involutive. As our proof of this fact is of a fairly computational nature, we will assume in the sequel that we are in the situation of Example 9.5.16: we consider a first-order equation $\mathcal{R}_1$ in Cartan normal form without algebraic equations; every equation is solved for a different principal derivative and no principal derivatives appear on the right hand side. From a theoretical point of view, these assumptions do not represent a serious restriction, as every regular differential equation can be transformed into such an equation.

The conditions (9.98) represent a first-order differential equation $\mathcal{Z}_1$ for the coefficients $\boldsymbol{\zeta}$. As each flat Vessiot connection of the original equation $\mathcal{R}_1$ corresponds to a solution of $\mathcal{Z}_1$, we must study its solvability which basically means that we must check whether or not $\mathcal{Z}_1$ is involutive. If this is the case, then the Cartan–Kähler Theorem 9.4.1 guarantees us that for analytic equations flat Vessiot connections indeed exist. Thus we must perform an involution analysis of $\mathcal{Z}_1$. As a first step we make the following simple observation which actually is independent of our made assumptions, as its proof relies solely on the Jacobi identity.

Lemma 9.6.6. *The differential conditions* (9.98b) *form an involutive system.*

Proof. We first note that the differential part (9.98b) is already more or less in Cartan normal form. The coefficients ζ_i^k are functions of all coordinates on $\mathcal{R}_1$ and we order these in such a way that the original independent variables $x^1,\dots,x^n$ are the highest ones among the new independent variables (thus as usual the x^n-derivatives are those of maximal class). A derivative with respect to x^i only appears in the vector field U_i by the form of our ansatz. Thus the equations of maximal class in (9.98b) are the equations $H_{in}^k = 0$, as only these contain the x^n-derivatives $\partial\zeta_i^k/\partial x^n$. The equations of second highest class are $H_{i,n-1}^k = 0$ and so on.

By the Jacobi identity we have for all $1 \le i < j < k \le n$ that

$$\big[U_i,[U_j,U_k]\big] + \big[U_j,[U_k,U_i]\big] + \big[U_k,[U_i,U_j]\big] = 0\,. \tag{9.104}$$

Evaluation of this identity *on* $\mathcal{Z}_1$ leads to the equation

$$\Bigl(U_i(G^a_{jk}) + U_j(G^a_{ki}) + U_k(G^a_{ij})\Bigr)Z_a + \Bigl(U_i(H^\ell_{jk}) + U_j(H^\ell_{ki}) + U_k(H^\ell_{ij})\Bigr)Y_\ell = 0, \quad (9.105)$$

since all other terms vanish. As the vector fields on the right hand side are linearly independent, the expressions in the parentheses must vanish individually.

Since $i < j < k$, the term $U_k(H^\ell_{ij})$ contains a non-multiplicative prolongation of the equation $H^\ell_{ij} = 0$ and all other terms represent multiplicative prolongations. As these equations cover all non-multiplicative prolongations of equations in (9.98b), we conclude that cross-derivatives do not lead to any integrability conditions, as any non-multiplicative prolongation can be expressed as a linear combination of multiplicative ones. By Remark 7.2.10, this observation entails that (9.98b) is an involutive system. □

Note that this result automatically implies that the symbol of $\mathcal{Z}_1$ is involutive. In order to check the formal integrability of $\mathcal{Z}_1$, we are left with analysing the effect of the algebraic part (9.98a), as we already know that cross-derivatives cannot lead to integrability conditions. However, one easily sees that in general not all of its prolongations are contained in the differential part (9.98b). Hence $\mathcal{Z}_1$ is generally not formally integrable.

The classical approach would now consist of adding these prolongations of the algebraic equations in order to obtain a local representation of $\mathcal{Z}_1^{(1)}$ and to check it for involution. However, it seems to be simpler to solve explicitly the algebraic equations (recall that they are linear) and to enter the result into the differential equations. Therefore we will follow this approach.

Remark 9.6.7. For the analysis of the algebraic equations, we follow the considerations in Example 9.5.16 and use the matrices $\boldsymbol{\Xi}_h$ and the vectors $\boldsymbol{\Theta}_{ij}$ defined there. Recall that they have more rows than really necessary, but that this redundancy leads to a simpler structure; in fact for the class of differential equations we are currently studying, we could even derive the closed formulae (9.79) and (9.80), respectively, for their entries.

On the other hand, since the unknowns ζ_i^k may be understood as labels for the columns of the matrices $\boldsymbol{\Xi}_h$, the identifications discussed in Remark 9.6.3 allow us to work with matrices will less columns. We introduce contracted matrices $\hat{\boldsymbol{\Xi}}_h$ which arise as follows: whenever $\zeta_j^k = \zeta_i^l$ the corresponding columns of any matrix $\boldsymbol{\Xi}_h$ are added. Similarly, we introduce reduced vectors $\hat{\boldsymbol{\zeta}}_h$ where the redundant components are left out. Obviously, these column operations do not affect the vectors $\boldsymbol{\Theta}_{ij}$. ◁

While so far we have considered all algebraic conditions (9.98a) together, we now separate them into subsystems. This idea is already familiar to us: our proof of the Cartan–Kähler Theorem 9.4.1 was based on separating a differential equation into subsystems according to the classes of the individual equations and all versions of Cartan's test in Section 6.2 relied on similar filtrations. This idea is also the very point where the question of δ-regularity always pops up.

The separation into subsystems is equivalent to building up the integral distribution $\langle U_1, \dots, U_n \rangle$ step by step. We first choose arbitrary values for the coefficients $\hat{\boldsymbol{\xi}}_1$. Then we try to determine values for the coefficients $\hat{\boldsymbol{\xi}}_2$ such that the corresponding vector fields U_1, U_2 satisfy $[U_1, U_2] \equiv 0 \bmod \mathcal{V}[\mathcal{R}_1]$. Note that in this computation we consider the coefficients $\hat{\boldsymbol{\xi}}_1$ as parameters and only the coefficients $\hat{\boldsymbol{\xi}}_2$ as variables. More precisely, (taking Remark 9.6.7 into account) we rewrite the conditions $\mathbf{G}_{12} = 0$ in the form

$$\hat{\boldsymbol{\Xi}}_1 \hat{\boldsymbol{\xi}}_2 = \hat{\boldsymbol{\Xi}}_2 \hat{\boldsymbol{\xi}}_1 - \boldsymbol{\Theta}_{12} \tag{9.106}$$

and require that this first subsystem does not impose any restrictions on the coefficients $\hat{\boldsymbol{\xi}}_1$. Obviously, this is the case, if and only if

$$s_1 = \operatorname{rank} \hat{\boldsymbol{\Xi}}_1 = \operatorname{rank} (\hat{\boldsymbol{\Xi}}_1 \ \hat{\boldsymbol{\Xi}}_2) . \tag{9.107}$$

Assuming that this identity holds, the actual solvability of the linear system $\mathbf{G}_{12} = 0$ is equivalent to the augmented rank condition

$$\operatorname{rank} \hat{\boldsymbol{\Xi}}_1 = \operatorname{rank} (\hat{\boldsymbol{\Xi}}_1 \ \hat{\boldsymbol{\Xi}}_2 \ \ -\boldsymbol{\Theta}_{12}) , \tag{9.108}$$

as otherwise the right hand sides lead to an inconsistency.

Assume that in the previous step we succeeded in constructing a two-dimensional integral distribution $\langle U_1, U_2 \rangle$. Then we try next to extend it to a three-dimensional one by finding values $\hat{\boldsymbol{\xi}}_3$ such that the corresponding vector field U_3 satisfies $[U_1, U_3] \equiv 0 \bmod \mathcal{V}[\mathcal{R}_1]$ and $[U_2, U_3] \equiv 0 \bmod \mathcal{V}[\mathcal{R}_1]$. This time we consider all coefficients $\hat{\boldsymbol{\xi}}_1$, $\hat{\boldsymbol{\xi}}_2$ as parameters and only the coefficients $\hat{\boldsymbol{\xi}}_3$ as variables. Hence the conditions $\mathbf{G}_{13} = 0$ and $\mathbf{G}_{23} = 0$ are rewritten in the form

$$\begin{aligned} \hat{\boldsymbol{\Xi}}_1 \hat{\boldsymbol{\xi}}_3 &= \hat{\boldsymbol{\Xi}}_3 \hat{\boldsymbol{\xi}}_1 - \boldsymbol{\Theta}_{13} , \\ \hat{\boldsymbol{\Xi}}_2 \hat{\boldsymbol{\xi}}_3 &= \hat{\boldsymbol{\Xi}}_3 \hat{\boldsymbol{\xi}}_2 - \boldsymbol{\Theta}_{23} \end{aligned} \tag{9.109}$$

and we require that this second subsystem does not impose any restrictions on the coefficients $\hat{\boldsymbol{\xi}}_1$ and $\hat{\boldsymbol{\xi}}_2$. Therefore we must have

$$s_1 + s_2 = \operatorname{rank} \begin{pmatrix} \hat{\boldsymbol{\Xi}}_1 \\ \hat{\boldsymbol{\Xi}}_2 \end{pmatrix} = \operatorname{rank} \begin{pmatrix} \hat{\boldsymbol{\Xi}}_1 & \hat{\boldsymbol{\Xi}}_3 & 0 \\ \hat{\boldsymbol{\Xi}}_2 & 0 & \hat{\boldsymbol{\Xi}}_3 \end{pmatrix} . \tag{9.110}$$

As above, if this condition is satisfied, then the solvability of the linear system $\mathbf{G}_{13} = \mathbf{G}_{23} = 0$ is equivalent to an augmented rank identity:

$$\operatorname{rank} \begin{pmatrix} \hat{\boldsymbol{\Xi}}_1 \\ \hat{\boldsymbol{\Xi}}_2 \end{pmatrix} = \operatorname{rank} \begin{pmatrix} \hat{\boldsymbol{\Xi}}_1 & \hat{\boldsymbol{\Xi}}_3 & 0 & -\boldsymbol{\Theta}_{13} \\ \hat{\boldsymbol{\Xi}}_2 & 0 & \hat{\boldsymbol{\Xi}}_3 & -\boldsymbol{\Theta}_{23} \end{pmatrix} . \tag{9.111}$$

By an obvious iteration, we assume in the ith step that we have already constructed an $(i-1)$-dimensional integral distribution $\langle U_1, \dots, U_{i-1} \rangle$ and search for an extension by a vector field U_i. This approach leads to a rewriting of the conditions

$\mathbf{G}_{1i} = 0, \ldots, \mathbf{G}_{i-1,i} = 0$ in the form

$$\begin{aligned} \hat{\boldsymbol{\Xi}}_1 \hat{\boldsymbol{\zeta}}_i &= \hat{\boldsymbol{\Xi}}_i \hat{\boldsymbol{\zeta}}_1 - \boldsymbol{\Theta}_{1i} , \\ &\vdots \\ \hat{\boldsymbol{\Xi}}_{i-1} \hat{\boldsymbol{\zeta}}_i &= \hat{\boldsymbol{\Xi}}_i \hat{\boldsymbol{\zeta}}_{i-1} - \boldsymbol{\Theta}_{i-1,i} . \end{aligned} \tag{9.112}$$

The requirement that this subsystem does not impose any restrictions on the coefficients $\hat{\boldsymbol{\zeta}}_j$ with $1 \leq j < i$ leads to the rank equality

$$s_1 + \cdots + s_{i-1} = \operatorname{rank} \begin{pmatrix} \hat{\boldsymbol{\Xi}}_1 \\ \hat{\boldsymbol{\Xi}}_2 \\ \vdots \\ \hat{\boldsymbol{\Xi}}_{i-1} \end{pmatrix} = \operatorname{rank} \begin{pmatrix} \hat{\boldsymbol{\Xi}}_1 & \hat{\boldsymbol{\Xi}}_i & & \\ \hat{\boldsymbol{\Xi}}_2 & & \hat{\boldsymbol{\Xi}}_i & 0 \\ \vdots & 0 & & \ddots \\ \hat{\boldsymbol{\Xi}}_{i-1} & & & & \hat{\boldsymbol{\Xi}}_i \end{pmatrix} . \tag{9.113}$$

Assuming this equality, the linear system (9.112) is solvable, if and only if

$$\operatorname{rank} \begin{pmatrix} \hat{\boldsymbol{\Xi}}_1 \\ \hat{\boldsymbol{\Xi}}_2 \\ \vdots \\ \hat{\boldsymbol{\Xi}}_{i-1} \end{pmatrix} = \operatorname{rank} \begin{pmatrix} \hat{\boldsymbol{\Xi}}_1 & \hat{\boldsymbol{\Xi}}_i & & & -\boldsymbol{\Theta}_{1i} \\ \hat{\boldsymbol{\Xi}}_2 & & \hat{\boldsymbol{\Xi}}_i & 0 & -\boldsymbol{\Theta}_{2i} \\ \vdots & 0 & & \ddots & \vdots \\ \hat{\boldsymbol{\Xi}}_{i-1} & & & \hat{\boldsymbol{\Xi}}_i & -\boldsymbol{\Theta}_{i-1,i} \end{pmatrix} . \tag{9.114}$$

Example 9.6.8. Before we proceed with our theoretical analysis, let us demonstrate with a concrete differential equation that in general we cannot expect that the above outlined step-by-step construction of integral distributions works. In order to keep the size of the example reasonably small, we use the second-order equation (it could always be rewritten as a first-order one satisfying the assumptions made above and the phenomenon we want to discuss is independent of this transformation)

$$\mathcal{R}_2 : \begin{cases} u_{xx} = \alpha u , \\ u_{yy} = \beta u \end{cases} \tag{9.115}$$

with two real constants α, β. Note that its symbol $\mathcal{N}_2$ is our standard example of a non-involutive symbol. However, one easily proves that $\mathcal{R}_2$ is formally integrable for arbitrary choices of the constants α, β.

One readily computes that the Vessiot distribution $\mathcal{V}[\mathcal{R}_2]$ is generated by the following three vector fields on $\mathcal{R}_2$:

$$\begin{aligned} X_1 &= \partial_{\overline{x}} + \overline{u_x}\partial_{\overline{u}} + \alpha\overline{u}\partial_{\overline{u_x}} + \overline{u_{xy}}\partial_{\overline{u_y}} , \\ X_2 &= \partial_{\overline{y}} + \overline{u_y}\partial_{\overline{u}} + \overline{u_{xy}}\partial_{\overline{u_x}} + \beta\overline{u}\partial_{\overline{u_y}} , \\ Y_1 &= \partial_{\overline{u_{xy}}} . \end{aligned} \tag{9.116}$$

They yield as structure equations for $\mathcal{V}[\mathcal{R}_2]$

$$[X_1, X_2] = \beta \overline{u_x} \partial_{\overline{u_y}} - \alpha \overline{u_y} \partial_{\overline{u_x}}\,, \quad [X_1, Y_1] = -\partial_{\overline{u_y}}\,, \quad [X_2, Y_1] = -\partial_{\overline{u_x}}\,. \tag{9.117}$$

For a two-dimensional integral distribution $\mathcal{U} \subset \mathcal{V}[\mathcal{R}_2]$ we make as above the ansatz $U_i = X_i + \zeta_i^1 Y_1$ with two coefficients ζ_i^1. As we want to perform a step-by-step construction, we assume that we have chosen some fixed value for ζ_1^1 and try now to determine ζ_2^1 such that $[U_1, U_2] \equiv 0 \bmod \mathcal{V}[\mathcal{R}_2]$. Evaluation of the Lie bracket yields the equation

$$[U_1, U_2] \equiv (\beta \overline{u_x} - \zeta_2^1) \partial_{\overline{u_y}} - (\alpha \overline{u_y} - \zeta_1^1) \partial_{\overline{u_x}} \mod \mathcal{V}[\mathcal{R}_2]\,. \tag{9.118}$$

A necessary condition for the vanishing of the right hand side is that $\zeta_1^1 = \alpha \overline{u_y}$. Hence it is not possible to choose this coefficient arbitrarily, as we assumed, but (9.118) determines *both* functions ζ_i^1 uniquely.

Note that the conditions on the coefficients ζ_i^1 imposed by (9.118) are trivially solvable and hence $\mathcal{R}_2$ possesses an integral distribution (which, in fact, is even a flat Vessiot connection). The problem is that it could not be constructed systematically with the above outlined *step-by-step* process. We made already a similar observation when we discussed the order-by-order construction of formal power series solutions: it is possible only for a formally integrable system, but of course many systems which are not formally integrable nevertheless possess such solutions. ◁

We relate now the rank conditions (9.113) and (9.114) to the formal analysis of the differential equation $\mathcal{R}_1$. Recall that in Remark 7.2.11 we explicitly determined the conditions for a first-order equation in Cartan normal form to be involutive. It will turn out that these conditions show up again in the analysis of our rank conditions. Albeit we will discuss this point in more details below, we stress already now that for the following result it is crucial to use the contracted matrices $\hat{\Xi}_i$ introduced in Remark 9.6.7.

Theorem 9.6.9. *Assume that δ-regular coordinates have been chosen for the given differential equation $\mathcal{R}_1$. The rank condition* (9.113) *is satisfied for all $1 < i \leq n$, if and only if the symbol $\mathcal{N}_1$ is involutive. The augmented rank condition* (9.114) *holds for all $1 < i \leq n$, if and only if the equation $\mathcal{R}_1$ is involutive.*

Proof. In order to prove (9.113), we transform the matrices into row echelon form. Since each matrix $\hat{\Xi}_i$ contains a unit block, there is an obvious way to do this. We describe the transformation as a sequence of block operations. For specifying blocks within the matrix $\hat{\Xi}_i$ we use the following notation: ${}_a^b\big[\hat{\Xi}_h\big]_c^d$ denotes the block consisting of the entries from the ath to the bth row and from the cth to the dth column. As we shall see, the relevant entries in this row echelon form are the coefficients of the second-order derivatives u_{hk}^δ in the explicit expression (7.38) for the obstruction to involution of a first-order system of the assumed form. Therefore their vanishing is equivalent to involution of the symbol $\mathcal{N}_1$.

We start with $i = 2$, i. e. with (9.107). Since $\hat{\Xi}_1$ is a negative unity matrix of $\alpha_1^{(1)}$ rows with a $\beta_1^{(1)} \times \alpha_1^{(1)}$-matrix stacked upon it and only zeros for all other entries, we have rank $\hat{\Xi}_1 = \alpha_1^{(1)}$. Next, we transform the matrix $(\hat{\Xi}_1\ \hat{\Xi}_2)$ into row echelon form using the special structure of the matrices $\hat{\Xi}_i$ as given in (9.81). This is achieved by replacing the blocks in the following way:

$$ {}^{\beta_1^{(1)}}_{1}\bigl[\hat{\Xi}_1\bigr]^{\alpha_1^{(1)}}_{1} \leftarrow {}^{\beta_1^{(1)}}_{1}\bigl[\hat{\Xi}_1\bigr]^{\alpha_1^{(1)}}_{1} + {}^{\beta_1^{(1)}}_{1}\bigl[\hat{\Xi}_1\bigr]^{\alpha_1^{(1)}}_{1} \cdot {}^{m}_{\beta_1^{(1)}+1}\bigl[\hat{\Xi}_1\bigr]^{\alpha_1^{(1)}}_{1}\,, \tag{9.119a} $$

$$ {}^{\beta_1^{(1)}}_{1}\bigl[\hat{\Xi}_2\bigr]^{\alpha_1^{(1)}}_{1} \leftarrow {}^{\beta_1^{(1)}}_{1}\bigl[\hat{\Xi}_2\bigr]^{\alpha_1^{(1)}}_{1} + {}^{\beta_1^{(1)}}_{1}\bigl[\hat{\Xi}_2\bigr]^{\alpha_1^{(1)}}_{1} \cdot {}^{m}_{\beta_1^{(1)}+1}\bigl[\hat{\Xi}_2\bigr]^{\alpha_1^{(1)}}_{1}\,, \tag{9.119b} $$

$$ {}^{\beta_1^{(1)}}_{1}\bigl[\hat{\Xi}_2\bigr]^{\alpha_1^{(1)}+\alpha_1^{(2)}}_{\alpha_1^{(1)}+1} \leftarrow {}^{\beta_1^{(1)}}_{1}\bigl[\hat{\Xi}_2\bigr]^{\alpha_1^{(1)}+\alpha_1^{(2)}}_{\alpha_1^{(1)}+1} + {}^{\beta_1^{(1)}}_{1}\bigl[\hat{\Xi}_1\bigr]^{\alpha_1^{(1)}}_{1} \cdot {}^{m}_{\beta_1^{(1)}+1}\bigl[\hat{\Xi}_2\bigr]^{\alpha_1^{(1)}+\alpha_1^{(2)}}_{\alpha_1^{(1)}+1}\,. \tag{9.119c} $$

If, for the sake of simplicity, we use the same names for the changed blocks, then we find that after this operation

$$ {}^{\beta_1^{(1)}}_{1}\bigl[\hat{\Xi}_2\bigr]^{\alpha_1^{(1)}}_{1} = \Bigl(-C^1_\delta(\phi_2^\alpha) + \sum_{\gamma=\beta_1^{(1)}+1}^{\beta_1^{(2)}} C^1_\gamma(\phi_1^\alpha)C^1_\delta(\phi_2^\gamma)\Bigr)_{\substack{1\le\alpha\le\beta_1^{(1)}\\ \beta_1^{(1)}+1\le\delta\le m}}\,, \tag{9.120a} $$

$$ {}^{\beta_1^{(1)}}_{1}\bigl[\hat{\Xi}_2\bigr]^{\alpha_1^{(1)}+\alpha_1^{(2)}}_{\alpha_1^{(1)}+1} = \Bigl(C^1_\delta(\phi_1^\alpha) - C^2_\delta(\phi_2^\alpha) + \sum_{\gamma=\beta_1^{(1)}+1}^{\beta_1^{(2)}} C^1_\gamma(\phi_1^\alpha)C^2_\delta(\phi_2^\gamma)\Bigr)_{\substack{1\le\alpha\le\beta_1^{(1)}\\ \beta_1^{(2)}+1\le\delta\le m}}\,. \tag{9.120b} $$

A comparison with the obstructions to involution obtained by evaluating (7.38) for $i = 1$ and $j = 2$ shows that all these entries will vanish, if and only if all the obstructions vanish. It follows that then the first $\beta_1^{(1)}$ rows of the matrix $(\hat{\Xi}_1\ \hat{\Xi}_2)$ contain only zeros. The last $\alpha_1^{(1)}$ rows begin with the block $-\mathbb{1}_{\alpha_1^{(1)}}$ and hence $\operatorname{rank}(\hat{\Xi}_1\ \hat{\Xi}_2) = \alpha_1^{(1)} = \operatorname{rank}\hat{\Xi}_1$. Thus we may conclude that the rank condition (9.107) holds, if and only if in our system no non-multiplicative prolongation $D_2\Phi_1^a$ leads to an obstruction of involution.

The claim for the augmented condition (9.108) follows from the explicit expression (9.79) for the entries Θ^α_{ij}. Performing the same computations as above described with the augmented system yields as additional relevant entries of the transformed matrix exactly the integrability conditions arising from (7.38) evaluated for $i = 1$ and $j = 2$. Hence (9.108) holds, if and only if no non-multiplicative prolongation $D_2\Phi_1^\alpha$ yields an integrability condition.

As one might expect from the above considerations for $i = 2$, the analysis of the rank condition (9.113) for some fixed value $2 < i \le n$ will require the non-multiplicative prolongations $D_i\Phi_1^\alpha, D_i\Phi_2^\alpha, \dots D_i\Phi_{i-1}^\alpha$. It follows trivially from the block form (9.81) of the matrices Ξ_i that the rank of the matrix on the left hand side of (9.113) is given by $\sum_{k=1}^{i-1}\alpha_1^{(k)}$.

We skip the very technical details for the general case (they can be found in [131]). We follow the same steps as in the case $i = 1$. The transformation of the matrix on the right hand side of (9.113) can be described using block matrices, and the resulting matrix in row echelon form has as its entries in the rows where no unit block appears the coefficients of the second-order derivatives in (7.38). Thus we may conclude again that satisfaction of (9.113) is equivalent to the fact that in the non-multiplicative prolongations $D_i\Phi_1^\alpha, \dots, D_i\Phi_{i-1}^\alpha$ no obstructions to involution arise. In the case of the augmented conditions (9.114), it follows again from the explicit expression (9.79) for the entries Θ_{ij}^α that the additional relevant entries are identical with the potential integrability conditions produced by the non-multiplicative prolongations $D_i\Phi_1^\alpha, \dots, D_i\Phi_{i-1}^\alpha$. □

Remark 9.6.10. Note that as by-product of this proof we find that the values s_i defined by the ranks of the linear systems (9.112) appearing in our step by step construction of the integral distribution $\mathcal{U}$ are just given by the Cartan characters $\alpha_i^{(1)}$. Thus these important numbers measuring the size of the formal solution space are also automatically determined during the Vessiot approach. ◁

The proof above also makes the reason for the introduction of the contracted matrices $\hat{\Xi}_i$ apparent. As all functions are assumed to be smooth, partial derivatives commute: $u_{ij}^\alpha = u_{ji}^\alpha$. In (7.38) each obstruction to involution actually consists of two parts: one arises as coefficient of u_{ij}^α, the other one as coefficient of u_{ji}^α. While this decomposition does not show in (7.38) because both derivatives are collected into one term, the two parts appear in different columns of the matrices Ξ_i and in general the rank condition (9.113) will not hold, if we replace the contracted matrices $\hat{\Xi}_i$ by the original matrices Ξ_i (see the example below). The effect of the contraction is to combine the two parts in order to obtain the right rank. At the level of the contact map Γ_1 the commutativity of partial derivatives is of course also the reason that certain coefficients in its image and hence some ζ_i^k must coincide.

Theorem 9.6.11. *Assume that δ-regular coordinates have been chosen for the given differential equation $\mathcal{R}_1$ and that furthermore $\mathcal{R}_1$ is analytic. Then the full set of conditions* (9.98) *is solvable.*

Proof. As the proof of this theorem is very technical, we do not give the details (which can be found in [131]), but only briefly explain the underlying strategy. We showed already in Lemma 9.6.6 that the differential part of (9.98) on its own forms an involutive system. Instead of analysing the effect of prolonging the algebraic part, we do the following. In the proof of Theorem 9.6.9 we derived row echelon forms of the linear systems appearing in our step by step approach to solving the algebraic conditions. With their help one can explicitly write down the solution of these conditions. Then one enters the results into the differential conditions. One can show that whenever these substitutions affect the leading derivative of an equation, then the corresponding equation vanishes identically. This fact implies that the remaining system is involutive and thus solvable by the Cartan–Kähler Theorem 9.4.1. □

Example 9.6.12. Consider the first-order differential equation

$$\mathcal{R}_1 : \begin{cases} u_t = v_t = w_t = u_s = 0\,, & v_s = 2u_x + 4u_y\,, \\ w_s = -u_x - 3u_y\,, & u_z = v_x + 2w_x + 3v_y + 4w_y \end{cases} \tag{9.121}$$

with five independent variables x, y, z, s, t and three dependent variables u, v, w. It is formally integrable, and its symbol is involutive with $\dim \mathcal{N}_1 = 8$. Thus $\mathcal{R}_1$ is an involutive equation. For the matrices $\Xi_1, \dots, \Xi_5$ we find that Ξ_5 contains only zeros and the other four matrices are given by

$$\Xi_1 = \begin{pmatrix} -1 & 0 & 0 & 0 & 0 & 0 & 0 & 0 & 0 \\ 0 & -1 & 0 & 0 & 0 & 0 & 0 & 0 & 0 \\ 0 & 0 & -1 & 0 & 0 & 0 & 0 & 0 & 0 \end{pmatrix}, \qquad \Xi_2 = \begin{pmatrix} 0 & 0 & 0 & -1 & 0 & 0 & 0 & 0 \\ 0 & 0 & 0 & 0 & -1 & 0 & 0 & 0 \\ 0 & 0 & 0 & 0 & 0 & -1 & 0 & 0 \end{pmatrix},$$

$$\Xi_3 = \begin{pmatrix} 0 & -1 & -2 & 0 & -3 & -4 & 0 & 0 \\ 0 & 0 & 0 & 0 & 0 & 0 & -1 & 0 \\ 0 & 0 & 0 & 0 & 0 & 0 & 0 & -1 \end{pmatrix}, \qquad \Xi_4 = \begin{pmatrix} 0 & 0 & 0 & 0 & 0 & 0 & 0 & 0 \\ -2 & 0 & 0 & -4 & 0 & 0 & 0 & 0 \\ 1 & 0 & 0 & 3 & 0 & 0 & 0 & 0 \end{pmatrix}. \tag{9.122}$$

For the first two steps in the construction of the fields U_i, the rank conditions are trivially satisfied even for the non-contracted matrices. But not so in the third step where we have in the row echelon form of the arising (9×32)-matrix in the 7th row zero entries throughout except in the 12th column (where we have -2) and in the 17th column (where we have 2). As a consequence, we obtain the equality $\zeta_1^4 = \zeta_2^1$ and the rank condition for this step does not hold. However, since both u_x and u_y are parametric derivatives and in our ordering $Y_1 = \iota_*(\partial_{\overline{u_x}})$ and $Y_4 = \iota_*(\partial_{\overline{u_y}})$, this equality is already taken into account in our reduced ansatz and for the matrices $\hat{\Xi}_i$ the rank condition is satisfied.

The rank condition is first violated when the rank reaches the symbol dimension. From then on, the rank of the left matrix in (9.113) stagnates at $\dim \mathcal{N}_1$ while the rank of the augmented matrix may rise further. The entries breaking the rank condition differ only by their sign, while the corresponding coefficients in (7.38) are collected into one sum which thus vanishes. ◁

9.7 Notes

The geometric approach to the initial value problem for ordinary differential equations presented in Section 9.1 is not only of theoretical interest. While our proof of Theorem 9.1.6 was based on a local representation in semi-explicit form, this is not necessary for numerical computations. Even if a fully non-linear representation is given, the distribution $\mathcal{V}[\mathcal{R}_1]$ always arises as the solution of a linear system. The only advantage of the semi-linear form is that it yields this system in the special form (9.5); but for its numerical solution this structure is not important. For example, both the Vessiot distribution and the generalised solutions in Figure 9.1 were computed solely numerically.

In our treatment of the initial value problem we always required that the initial data correspond to a point $(x_0, \mathbf{u}_0, \mathbf{u}'_0)$ lying *on* the differential equation $\mathcal{R}_1$, i. e. we restricted to *consistent* initial data. Obviously, this condition is necessary for the existence of a strong solution. If one allows for distributional solutions (in the sense of functional analysis), then one may also study the effect of inconsistent initial data.

For simplicity, we demonstrate this effect only for a square linear system with constant coefficients where it is easily possible to obtain closed form solutions. We consider an inhomogeneous system of the form

$$E\mathbf{u}' = A\mathbf{u} + \mathbf{f}(x) \tag{9.123}$$

where $E, A \in \mathbb{R}^{m \times m}$ are square matrices and $\mathbf{f}$ is some sufficiently often differentiable vector-valued function together with an initial condition $\mathbf{u}(0) = \mathbf{u}_0$. If E is regular, then we are dealing with a normal system.

Thus we assume that E is a singular matrix but that the determinant $\det(E - \lambda A)$ is a non-zero polynomial in λ (i. e. E, A define a *regular matrix pencil*). By a classical result of matrix theory [147, Chapt. XII,§§2,7], under the made assumptions regular matrices P, Q exist such that $QEP = \operatorname{diag}(\mathbb{1}_{m_1}, N)$ and $QAP = \operatorname{diag}(\bar{A}, \mathbb{1}_{m_2})$ where N is a nilpotent $(m_2 \times m_2)$ matrix and $\bar{A}$ an arbitrary $(m_1 \times m_1)$ matrix with $m_1 + m_2 = m$. Thus after a linear transformation $\mathbf{u} \mapsto P^{-1}\mathbf{u}$ the above system splits into two subsystems

$$\mathbf{v}' = \bar{A}\mathbf{v} + \mathbf{g}(x)\,, \tag{9.124a}$$

$$N\mathbf{w}' = \mathbf{w} + \mathbf{h}(x) \tag{9.124b}$$

of dimensions m_1 and m_2, respectively, with $(\mathbf{g}, \mathbf{h})^T = Q\mathbf{f}$ and corresponding initial data $\mathbf{v}_0, \mathbf{w}_0$. Obviously, the first subsystem (9.124a) represents a normal system; thus we may restrict our attention to the second one (9.124b).

Let ℓ be the nilpotency index of N, i. e. ℓ is the smallest integer for which $N^\ell = 0$. Then prolonging (9.124b) ℓ times yields after a trivial computation that this subsystem possesses the unique solution

$$\mathbf{w}_{\text{strong}}(x) = -\sum_{k=0}^{\ell-1} N^k \mathbf{h}^{(k)}(x)\,. \tag{9.125}$$

Thus for strong solutions there remains no choice for the initial data: we must take $\mathbf{w}_0 = \mathbf{w}_{\text{strong}}(0)$. However, using for example a Laplace transformation, it is not difficult to show that even for arbitrary choices of $\mathbf{w}_0$ a unique distributional solution of (9.124b) exists, namely

$$\mathbf{w}(x) = \mathbf{w}_{\text{strong}}(x) - \sum_{k=1}^{\ell-1} \delta^{(k-1)}(x) N^k \big(\mathbf{w}_0 - \mathbf{w}_{\text{strong}}(0)\big) \tag{9.126}$$

where δ denotes the Dirac delta distribution. Thus the system responds with a strong singularity to the inconsistency of the initial data.

In control theory systems of the form (9.123) are often called *regular descriptor systems* and hidden behind the inhomogeneity $\mathbf{f}$ are the inputs of the system, i. e. in this context the function $\mathbf{f}$ is not a priori fixed but we can make choices. The second term in (9.126) is then called the *impulse response* of the system. Obviously, such "impulses", i. e. delta distributions in the solution, are unwanted and hence the inputs must be chosen in such a way that the parentheses in (9.126) vanishes.

The terminology "impasse point" stems from electrical engineering where this phenomenon was early recognised in the analysis of nonlinear networks [85]. Subsequently, it received much attention in this domain; notable are in particular the work of Chua and Deng [86, 87] and Reißig [375, 376, 377, 378]. Other relevant contributions are [191, 310, 423, 449]. In the general context of differential algebraic equations one must mention the related works of Rabier and Rheinboldt [367, 369, 370]. For the special case of a scalar implicit ordinary differential equation, the use of the Vessiot distribution was already advocated by Arnold [27, §3] and plays a crucial role in the singularity theory of such equations.

Much of the here presented material on singularities is based on the work of Tuomela [462, 463, 464]. He also pioneered the use of the geometric theory for the numerical solution of overdetermined systems of ordinary differential equations and developed special integrators for the Vessiot distribution exploiting the Riemannian geometry of the submanifold $\mathcal{R}_1 \subseteq J_1\pi$ (for a trivial bundle $\pi : \mathbb{R} \times \mathbb{R}^m \to \mathbb{R}$ and the canonical Riemannian structure of $\mathbb{R}^{2m+1} \cong J_1\pi$).

The use of the Vessiot distribution of partial differential equations as discussed in Sections 9.5 and 9.6 goes back to Vessiot [470]. The fundamental idea of constructing integral elements step by step is, however, due to Cartan [71] who formulated the theory in terms of differential forms. The Vessiot theory represents essentially a dual form of the Cartan–Kähler theory that may be applied directly to partial differential equations without any need to transform them into a differential ideal. Modern accounts of the theory were given by Fackerell [130] and by Stormark [435]. However, none of these references provides a rigorous analysis of when Vessiot's approach succeeds. Such an analysis, leading to our Theorem 9.6.9 and 9.6.11, was first given by Fesser [131] (see also [132, 133]). Applications of Vessiot's theory appear for example in the analysis of hyperbolic equations, see e. g. [469].

Our presentation of the Vessiot theory seems to be the first one that exploits the geometry of the jet bundle for choosing a suitable basis of the Vessiot distribution. This concerns in particular the fact that we may choose for the vertical part always commuting vector fields. Seemingly, it has been overlooked so far and in all references we are aware of the algebraic step of Vessiot's construction leads to a quadratic system in the coefficients of the ansatz. Using a step by step approach corresponding to ours in Section 9.6, one can actually solve it by treating a series of linear systems. But this requires δ-regular coordinates, otherwise one must indeed work with quadratic equations which is never necessary in our approach.

The term "integral distributions" for the result of the algebraic step of Vessiot's construction is non-standard. In the literature, one usually speaks of "involutions" (see for example [435, Sect. 3.2]) As generally these distributions are not yet involutive, we consider this terminology as confusing and inappropriate. Our name appears much more natural given the fact that according to Proposition 2.3.19 such distributions consist of integral elements.

In particular in the Russian literature (see e. g. [263, 271, 296]), the contact distribution $\mathcal{C}_q$ appears often under the name *Cartan distribution* and also $\mathcal{V}[\mathcal{R}_q]$ is called the Cartan distribution of the differential equation $\mathcal{R}_q$. The terminology Vessiot distribution seems to have been proposed by Fackerell [130]. The above references also discuss the decomposition (9.64) of the Vessiot distribution as the direct sum of the symbol $\mathcal{N}_q$ and a transversal complement $\mathcal{H}$; the latter is called a *Cartan connection* and a *Bott connection*, respectively, if its curvature vanishes.

Our approach to generalised prolongations represents essentially a dual version of the treatment by Molino [326]. Prolongations by pseudo-potentials have been applied in particular to *completely integrable systems* (also known as *soliton equations*). Here the knowledge of pseudo-potentials often leads to many other important structures like Bäcklund transformations, Lax pairs or inverse scattering problems. The investigation of this approach was initiated by Wahlquist and Estabrook [471] for the special case of the Korteweg–de Vries equation; Hermann [209] exposed the underlying geometric idea. All these works use exterior differential systems. Fackerell [130] first hinted at a dual description in the context of the Vessiot theory, however, without actually formulating it in detail.

A very important question we did not discuss is how one actually finds generalised prolongations. So far no algorithmic approach is known. For the special case of a prolongation by pseudo-potentials a number of techniques have been developed based on incomplete Lie algebras. We refer to [136, 326, 327, 388] and references therein for more information.

The proof of the Cauchy–Kovalevskaya Theorem we have used is the one found in most textbooks (see e. g. [238, 379]) and essentially goes back to Goursat [175]. Cauchy gave a first proof in 1842 in a series of notes to the French Academy of Science [76, 77, 78]; later Kovalevskaya studied the problem in her thesis [262]. Note that Cauchy's method of majorants proves the convergence of the formal power series solutions indirectly by comparison with a known series. A direct convergence proof turns out to be surprisingly tricky; Shinbrot and Welland [416] provided a relatively simple one.

Several alternative proofs of the Cauchy–Kovalevskaya Theorem exist which do not use power series but functional analytic methods. The basic idea is to convert the initial value problem into a fixed point equation in some Banach space; a nice example for this technique can be found in [473]. These approaches have the advantage that they can also handle more general situations leading to what is often called "abstract" Cauchy–Kovalevskaya Theorems.

The Janet–Riquier Theory (see the Notes of Chapter 3) also provides a generalisation of the Cauchy–Kovalevskaya Theorem: *Riquier's Theorem*. It considers a formally well-posed initial value problem constructed with the help of an involutive basis (classically one speaks of a *passive* and *orthonomic* system). In contrast to our proof of the Cartan–Kähler Theorem, one tries to prove directly the convergence of the corresponding unique formal power series solution (similarly to our proof of the Cauchy–Kovalevskaya Theorem in Section 9.2).

The theorem appeared first in Riquier's fundamental book [381, Chapt. VII,§115]. A modern discussion (in particular with a careful analysis of the notions of an orthonomic system and of a reduced Gröbner basis) and a simplified proof were comparatively recently given by Oberst and Pauer [340], Sect. 3.2.

An involutive basis is always defined with respect to a term order. In differential algebra one often uses the terminology *ranking* instead of term order. So far, Riquier's Theorem has been proven only for special rankings. First of all, the ranking must be *orderly* (corresponding to a degree compatible term order), i. e. derivatives of higher order are always greater than those of lower order. Furthermore, if for some multi indices μ and ν the relation $p^\alpha_\mu \prec p^\alpha_\nu$ holds for one value of α, then it must hold for all values of α (thus in case there is only one unknown function, this assumption is automatically satisfied). Such rankings are called *Riquier rankings*.

Thus we meet again the condition that each equation must be solved for a derivative of maximal order. We noted already that this was a necessary condition for the Cauchy–Kovalevskaya Theorem. Whether or not it is also necessary for Riquier's Theorem seems to be an open question. The fact that the theorem has been proven only for Riquier rankings is often neglected. The Riquier property has the following meaning: using the Drach transformation of Appendix A.3, we may rewrite any system in several dependent variables as a system in one dependent variable at the expense of introducing further independent variables. Only a Riquier ranking induces again a ranking on the new derivatives, otherwise one obtains contradictions.

Lemaire [284] constructed fairly recently an explicit example of a formally well-posed initial value problem for an orderly ranking which is not a Riquier ranking where it is possible to prescribe analytic initial data but the corresponding formal power series solution diverges. He considered the differential system

$$\begin{aligned} u_{xx} &= u_{xy} + u_{yy} + v \,, \\ v_{yy} &= v_{xy} + v_{xx} + u \end{aligned} \tag{9.127}$$

with the initial conditions $u(0,y) = u_x(0,y) = e^y$ and $v(x,0) = v_y(x,0) = e^x$. Note that the system is actually normal, as we could solve both equations either for pure x- or pure y-derivatives of maximal order. However, Lemaire used a ranking leading to the shown principal derivatives. Then the given initial conditions are precisely those obtained by our analysis in Section 9.3. One can show via a comparison with the Fibonacci numbers that the unique formal power series solution of this initial value problem diverges. Thus Riquier's Theorem does not hold here, although all assumptions are satisfied except that the ranking is not of Riquier type.

As already indicated in the Notes to Chapter 5, historically the construction of formally well-posed initial value problems for general systems of differential equations preceeded the computation of complementary decompositions. Riquier [381], Chapt. V, §§79–88 and Janet [235, Sect. 15] considered this question already in 1910 and 1929, respectively (see also [382, Chapt. IX]) and both provided even algorithmic solutions. Much later, this question was again studied from an algorithmic point of view (with an implementation in Maple) by Reid [373]. While implicitly underlying these works, it seems that the rigorous notion of a formally well-posed initial value problem has not been introduced before.

The considerations at the end of Section 9.3 concerning equations with a Cohen–Macaulay symbol module seem to be new despite their simplicity (and the fact that they are very natural in view of the characterisation of Cohen-Macaulay modules by Hironaka decompositions given in Corollary 5.2.10). Further results on such equations have been found by Kruglikov and Lychagin [270].

Chapter 10
Linear Differential Equations

> *We [Kaplansky and Halmos] share a philosophy about linear algebra: we think basis-free, we write basis-free, but when the chips are down we close the office door and compute with matrices like fury.*
>
> *Irving Kaplansky*

Linear differential equations are simpler in many respects. The truth of this statement is already obvious from the fact that their solution spaces possess the structure of a vector space. Thus it is not surprising that some of our previous results may be improved in this special case. In the first section we study how the linearity can be expressed within our geometric framework for differential equations. This topic includes in particular a geometric formulation of the linearisation of an arbitrary equation along one of its solutions.

In the discussion of the Cartan–Kähler Theorem in Section 9.4 we emphasised that the uniqueness statement holds only within the category of analytic functions; it is possible that further solutions with lower regularity exist. In the case of linear systems stronger statements hold. In Section 10.2 we will use our proof of the Cartan–Kähler Theorem to extend the classical Holmgren Theorem on the uniqueness of $\mathcal{C}^1$ solutions from normal equations to arbitrary involutive ones.

A fundamental topic in the theory of partial differential equations is the classification into elliptic and hyperbolic equations. We will study the notion of ellipticity in Section 10.3 for arbitrary involutive equations. One of our main results will be that the approach to ellipticity via weights usually found in the literature is not only insufficient but also unnecessary, if one restricts to involutive equations: if a system is elliptic with weights, then its involutive completion is also elliptic without weights; the converse is not true.

The following section is devoted to hyperbolic equations. Special emphasis is given to a simple class of linear equations: weakly overdetermined systems with a hyperbolic evolution part and elliptic constraints. This class includes in particular Maxwell's equations. We will derive the conditions that such a system is involutive and prove (under some additional assumptions) an existence and uniqueness theorem for smooth solutions. The main point is here less a particularly strong result but the ease with which the technique of the Cartan–Kähler Theorem (cf. Section 9.4) allows us to extend results from normal systems to overdetermined ones.

In Section 10.5 we complement the geometric theory of Section 10.1 with an introduction to the algebraic analysis of linear systems. Of course, for some basic aspects we may simply refer to Chapter 3 where we developed the theory

W.M. Seiler, *Involution*, Algorithms and Computation in Mathematics 24,
DOI 10.1007/978-3-642-01287-7_10, © Springer-Verlag Berlin Heidelberg 2010

of involutive bases immediately for polynomial algebras of solvable type which include both linear differential operators over a coefficient field and the Weyl algebra (i. e. linear differential operators with polynomial coefficients). Here we will show how one can associate naturally which each linear system a module and how this idea leads to a generalised notion of solution. As a first application, we study the question of solvability of linear inhomogeneous systems leading to what is often called the fundamental principle. It turns out that the answer depends on the algebraic properties of the function space in which we look for solutions and leads to rather deep analytic problems. We will therefore give a complete treatment only for the simplest case of formal power series solutions of constant coefficients system where an elementary algebraic argument suffices.

In the following section we study the inverse syzygy problem: when is a given linear differential equation the compatibility condition of another equation? Somewhat surprisingly, this question—which is of great interest in applied fields like mathematical physics or control theory—admits a simple constructive answer, if one studies it in the more general context of finitely generated modules over a coherent ring. It turns out that the solvability of this problem is related to torsionlessness. In an Addendum we will furthermore demonstrate that as a by-product our algorithm can also been used for the construction of certain extension groups.

The topic of completion to involution is taken up again in Section 10.7. We design a completion algorithm for linear systems that combines the geometric approach of the Cartan–Kuranishi Theorem (cf. Section 7.4) with the purely algebraic algorithms of Section 4.2. In general, algebraic algorithms are more efficient than geometric ones, as they avoid redundant prolongations. Here we show how a clever "bookkeeping" allows to maintain this efficiency while nevertheless obtaining the full geometric information of the Cartan–Kuranishi completion.

In the last section, we consider linear systems of finite type with constant coefficients. We present an algorithm for the explicit construction of the general solution. For ordinary differential equations it is of course well-known that the solution is of polynomial-exponential type, but it is much less known that this fact remains true when we proceed to partial differential equations. For systems which are not of finite type we will briefly indicate a similar result, however without proof.

10.1 Elementary Geometric Theory

If we want to speak about a *linear* differential equation, then we need somewhere a linear structure. In Chapter 2 we only made the assumption that $\pi : \mathcal{E} \to \mathcal{X}$ is a fibred manifold. Obviously, this setting does not entail any linear structure in general. In order to define linear differential equations, we must make the stronger assumption that $\pi : \mathcal{E} \to \mathcal{X}$ is furthermore a vector bundle (in the simple case of a trivial bundle $\mathcal{E} = \mathcal{X} \times \mathcal{U}$ this requirement implies that $\mathcal{U}$ is a vector space). As we will now show, the jet bundles $J_q\pi$ inherit this property.

Proposition 10.1.1. *If the fibred manifold $\pi : \mathcal{E} \to \mathcal{X}$ is a vector bundle, then the prolongation $\pi^q : J_q\pi \to \mathcal{X}$ is a vector bundle, too.*

Proof. We denote as in Section 2.2 by $[\sigma]_x^{(q)}$ a class of sections that are equivalent to order q at the point $x \in \mathcal{X}$; the points in $J_q\pi$ may be identified with such classes. If $\mathcal{E}$ is a vector bundle, then we introduce a vector bundle structure on the qth order jet bundle $J_q\pi$ by simply setting $[\sigma]_x^{(q)} + [\sigma']_x^{(q)} = [\sigma + \sigma']_x^{(q)}$ for arbitrary sections $\sigma, \sigma' \in \Gamma_{loc}(\pi)$ and similarly $r \cdot [\sigma]_x^{(q)} = [r \cdot \sigma]_x^{(q)}$ for arbitrary constants $r \in \mathbb{R}$ (as the values of sections live now in vector spaces, the vector bundle structure of $\mathcal{E}$ allows us to define the addition of sections and the multiplication with scalars simply pointwise: $(\sigma + \sigma')(x) = \sigma(x) + \sigma'(x)$ and $(r \cdot \sigma)(x) = r\sigma(x)$). □

A local coordinates proof can be given as follows (for notational simplicity we consider only the first-order case; the extension to higher-order jet bundles is straightforward). Assume that we have two overlapping charts in a local trivialisation of $J_1\pi$. We denote the coordinates in one chart by $(\mathbf{x}, \mathbf{u}^{(1)})$ and in the other one by $(\mathbf{y}, \mathbf{v}^{(1)})$. We must then show that the dependent variables and the derivatives transform linearly, i. e. in the form $\mathbf{v}^{(1)} = A(\mathbf{x})\mathbf{u}^{(1)}$ with an $n \times n$ matrix A. But this property is a simple consequence of the chain rule.

As $\mathcal{E}$ is by assumption a vector bundle, we are restricted to structure-preserving coordinate transformations $\mathbf{y} = \mathbf{g}(\mathbf{x})$ and $\mathbf{v} = F(\mathbf{x})\mathbf{u}$ yielding the following affine transformation law for the derivatives

$$\frac{\partial \mathbf{v}}{\partial \mathbf{y}} = \frac{\partial \mathbf{x}}{\partial \mathbf{y}} \frac{\partial \mathbf{v}}{\partial \mathbf{x}} = \left(\frac{\partial g}{\partial \mathbf{x}}\right)^{-1} \left[F \frac{\partial \mathbf{u}}{\partial \mathbf{x}} + \frac{\partial F}{\partial \mathbf{x}} \mathbf{u}\right]. \tag{10.1}$$

But we are interested in the joint transformation of derivatives and dependent variables and there we obtain

$$\begin{pmatrix} \mathbf{v} \\ \mathbf{v}_{(1)} \end{pmatrix} = \begin{pmatrix} 0 & F \\ \left(\frac{\partial g}{\partial \mathbf{x}}\right)^{-1} F & \left(\frac{\partial g}{\partial \mathbf{x}}\right)^{-1} \frac{\partial F}{\partial \mathbf{x}} \end{pmatrix} \begin{pmatrix} \mathbf{u} \\ \mathbf{u}_{(1)} \end{pmatrix}. \tag{10.2}$$

Obviously, (10.2) is a linear transformation of the sought form.

Our earlier result that $\pi^q_{q-1} : J_q\pi \to J_{q-1}\pi$ is an affine bundle (Proposition 2.1.11) is in some sense a necessary condition for this statement: a linear transformation over $\mathcal{X}$ obviously implies an affine transformation of $J_{q-1}\pi$. However, one should be careful in comparing the two propositions. We are studying different bundle structures with different projections and albeit $\pi^q = \pi^{q-1} \circ \pi^q_{q-1}$, this decomposition says nothing about the properties of the corresponding bundles. A classical counterexample is that the double tangent bundle $T(T\mathcal{X})$ is not a vector bundle over $\mathcal{X}$, even though both $T\mathcal{X} \to \mathcal{X}$ and $T(T\mathcal{X}) \to T\mathcal{X}$ are vector bundles.

Remark 10.1.2. With essentially the same arguments as in the proof of Proposition 10.1.1 one can show that if $\rho : \mathcal{A} \to \mathcal{X}$ is an affine bundle modelled on the vector bundle $\pi : \mathcal{E} \to \mathcal{X}$, then $\rho^q : J_q\rho \to \mathcal{X}$ is again an affine bundle modelled on the vector bundle $\pi^q : J_q\pi \to \mathcal{X}$. ◁

The usual idea of a linear system is that the equations depend linearly on the derivatives and on the dependent variables (but with arbitrary dependency on the independent variables). We may now formulate this concept in an intrinsic manner.

Definition 10.1.3. Let $\pi : \mathcal{E} \to \mathcal{X}$ be a vector bundle. We call a differential equation $\mathcal{R}_q \subseteq J_q\pi$ *linear*, if it is a vector subbundle.

Similar to Remark 2.3.2, we will usually assume that a second vector bundle $\pi' : \mathcal{E}' \to \mathcal{X}$ over the same base space exists such that the differential equation $\mathcal{R}_q$ is the kernel of a global vector bundle morphism $\Phi : J_q\pi \to \mathcal{E}'$. This point of view leads then naturally to the notion of a *linear differential operator* as the linear map $\Delta[\Phi] : \Gamma_{loc}(\pi) \to \Gamma_{loc}(\pi')$ defined by $\Delta[\Phi](\sigma) = (\Phi \circ j_q)(\sigma)$.

Remark 10.1.4. Quasi-linear equations may be approached similarly. Their definition does not require any additional assumptions on the fibred manifold $\pi : \mathcal{E} \to \mathcal{X}$, as it is solely based on the always existing affine structure of $J_q\pi$ over $J_{q-1}\pi$. A differential equation $\mathcal{R}_q \subseteq J_q\pi$ is *quasi-linear*, if the subset $\mathcal{R}_q$ may be considered as an affine subbundle of the affine bundle $\pi^q_{q-1} : J_q\pi \to J_{q-1}\pi$. Indeed, this requirement obviously captures the usual idea of linearity in the highest-order derivatives. We need an affine structure, since quasi-linear equations may contain a "right hand side", i. e. lower-order terms. As quasi-linearity is a much weaker property than linearity, we will not separately study such equations. $\triangleleft$

Let $\pi : \mathcal{E} \to \mathcal{X}$ be a fibred manifold and assume that we know a local solution $\sigma \in \Gamma_{loc}(\pi)$ of some (nonlinear) differential equation $\mathcal{R}_q \subseteq J_q\pi$. For notational simplicity, we pretend that it was actually a global solution and use the full spaces instead of a chart. A frequent operation is the *linearisation* of $\mathcal{R}_q$ about this solution. We will now provide a geometric formulation of this operation.

Denote by $\nu = \pi \circ \tau_{\mathcal{E}} : V\pi \to \mathcal{X}$ the natural projection which gives the vertical bundle of $\pi : \mathcal{E} \to \mathcal{X}$ also the structure of a fibred manifold over $\mathcal{X}$ and thus allows us the construction of the jet bundles $\nu^q : J_q\nu \to \mathcal{X}$. Similarly, we may introduce the natural projection $\hat{\nu}^q = \pi^q \circ \tau_{J_q\pi} : V\pi^q \to \mathcal{X}$ from the vertical space of the jet bundle $\pi^q : J_q\pi \to \mathcal{X}$ to the base space providing us with a further fibration. Note that once we first go to the vertical space and then construct jet bundles over it and once we proceed the other way round by considering the vertical space of a jet bundle. According to the following lemma, we obtain in both cases the same result.

Lemma 10.1.5. *The two fibred manifolds $\hat{\nu}^q : V\pi^q \to \mathcal{X}$ and $\nu^q : J_q\nu \to \mathcal{X}$ are canonically diffeomorphic.*

Proof. By definition, a vertical vector $\mathbf{v} \in V_\rho\pi^q$ is an equivalence class of curves $\gamma : \mathbb{I} \subseteq \mathbb{R} \to (J_q\pi)_x$ living in the fibre over $x = \pi^q(\rho)$ and satisfying $\gamma(0) = \rho$. Instead of such a curve, we may consider a family of local sections $\sigma_t \in \Gamma_{loc}(\pi)$ defined in a neighbourhood of x such that $\gamma(t) = [\sigma_t]^{(q)}_x$ for all $t \in \mathbb{I}$. This point of view allows us to define the map

$$\phi : \begin{cases} V\pi^q & \longrightarrow & J_q\nu \\ \dfrac{d}{dt}\left(\left[\sigma_t\right]_x^{(q)}\right)_{t=0} & \longmapsto & \left[\left(\dfrac{d\sigma_t}{dt}\right)_{t=0}\right]_x^{(q)} \end{cases} \tag{10.3}$$

which is obviously a canonical diffeomorphism. It is also trivial to check that ϕ is independent of the chosen representative γ and of the family σ_t. □

The coordinate picture is as follows. Let $(\mathbf{x},\mathbf{u}^{(q)})$ be our usual local coordinates on the jet bundle $J_q\pi$. Then any vector in $V\pi^q$ is of the form $v^\alpha_\mu\partial_{u^\alpha_\mu}$ (i. e. without $\partial_{\mathbf{x}}$-components). Collecting the coefficients v^α_μ in a vector $\mathbf{v}^{(q)}$, coordinates on $V\pi^q$ are $(\mathbf{x},\mathbf{u}^{(q)},\mathbf{v}^{(q)})$. Similarly, we have as coordinates on $V\pi$ the tuple $(\mathbf{x},\mathbf{u},\mathbf{v})$ where $\mathbf{v}$ contains the coefficients v^α of a vertical vector $v^\alpha\partial_{u^\alpha}$. As we treat here $V\pi^q$ as a fibred manifold over X, both $\mathbf{u}$ and $\mathbf{v}$ must be considered as dependent variables and coordinates on $J_q\nu$ are obviously $(\mathbf{x},\mathbf{u}^{(q)},\mathbf{v}^{(q)})$. Hence the identification of Lemma 10.1.5 is trivial at the level of local coordinates.

We return to our solution $\sigma\in\Gamma_{loc}(\pi)$ of the differential equation $\mathcal{R}_q$; it defines an embedding of $\mathcal{X}$ in $\mathcal{E}$. The linearisation of the manifold $\mathcal{E}$ in a neighbourhood of $\operatorname{im}\sigma$ is given by the tangent space $T\mathcal{E}|_{\operatorname{im}\sigma}$. However, if we study the linearisation of a differential equation, we are only interested in the vertical part of it. The space $\bar{\mathcal{E}} = V\pi|_{\operatorname{im}\sigma}$ is obviously a vector bundle over $\mathcal{X}$ with the same fibre dimension as $\mathcal{E}$; we denote the corresponding projection by $\bar{\pi}$. The same operation may be performed with the prolonged section $j_q\sigma$.

Lemma 10.1.6. *The projection $V\pi^q|_{\operatorname{im} j_q\sigma}\rightarrow\mathcal{X}$ defines a vector bundle canonically isomorphic to $J_q\bar{\pi}\rightarrow\mathcal{X}$.*

Proof. Using Lemma 10.1.5, we have the inclusions

$$V\pi^q|_{\operatorname{im} j_q\sigma}\subset V\pi^q\cong J_q\nu\supset J_q\bar{\pi}\,. \tag{10.4}$$

It follows trivially from the definition (10.3) of the diffeomorphism ϕ that the subspaces on both ends are mapped into each other by ϕ, as we may choose the family of sections σ_t such that $\sigma_0=\sigma$, and that restricted to these subspaces ϕ is in fact a vector bundle isomorphism. □

Definition 10.1.7. Let $\mathcal{R}_q\subseteq J_q\pi$ be a differential equation and $\sigma\in\Gamma_{loc}(\pi)$ a solution of it. The *linearisation* of $\mathcal{R}_q$ along the solution σ is the linear differential equation $R_q=(T\mathcal{R}_q\cap V\pi^q)|_{\operatorname{im} j_q\sigma}\subseteq J_q\bar{\pi}$.

A local representation of R_q is easily obtained. Let $\Phi^\tau(\mathbf{x},\mathbf{u}^{(q)})=0$ be a local representation of the original equation $\mathcal{R}_q\subseteq J_q\pi$ and $Y=v^\alpha_\mu\partial_{u^\alpha_\mu}$ a vertical vector field in $V\pi^q|_{\operatorname{im} j_q\sigma}\cong J_q\bar{\pi}$. Then the field Y takes its values in R_q, if and only if it is everywhere tangent to $\mathcal{R}_q|_{\operatorname{im} j_q\sigma}$, i. e. if and only if $\mathrm{d}\Phi^\tau|_{\operatorname{im} j_q\sigma}(Y)=0$. But this condition is of course equivalent to

$$\left.\frac{\partial \Phi^\tau}{\partial u^\alpha_\mu}\right|_{\operatorname{im} j_q\sigma} v^\alpha_\mu = 0\,. \tag{10.5}$$

Taking as above $(\mathbf{x}, \mathbf{v}^{(q)})$ as coordinates in $J_q\bar{\pi}$, (10.5) is obviously the local representation of a linear differential equation and one immediately recognises the familiar coordinate form of the linearisation about the solution σ.

As $V\pi^q_{q-1}$ may be considered as a subbundle of $V\pi^q$, the definition of R_q implies that the symbol (and all its prolongations) of the linearised equation R_q and of the original equation $\mathcal{R}_q$, respectively, coincide (this fact is also obvious from the local coordinate form—cf. the proof of Theorem 7.1.6 and in particular (7.10)). The same holds for the dimensions of the equations, their prolongations and projections etc. These considerations prove the following result.

Proposition 10.1.8. *The linearised equation R_q is involutive, if and only if the original differential equation $\mathcal{R}_q$ is involutive.*

10.2 The Holmgren Theorem

For linear systems it is possible to eliminate some of the restrictions of the Cartan–Kähler Theorem. Namely, we can prove the uniqueness of $\mathcal{C}^1$ solutions to analytic differential equations. For normal systems, this result is known as *Holmgren's Theorem.* Based on the proof of the Cartan–Kähler Theorem 9.4.1 presented in the last chapter, it can be easily extended to arbitrary involutive systems.

The basic idea is simple; it is more difficult to make it rigorous. We consider first the following *normal* linear homogeneous system with analytic coefficients

$$A^{\alpha\beta}_i(\mathbf{x})u^\alpha_i + B^{\alpha\beta}(\mathbf{x})u^\alpha = 0\,, \qquad 1 \leq \beta \leq m \tag{10.6}$$

in a "lens-shaped" domain $\Omega \subset \mathbb{R}^n$ with a boundary consisting of two non-characteristic hypersurfaces R and S (see Figure 10.1). We furthermore assume that S is analytic. The claim is that the initial value problem for (10.6) with $\mathbf{u}$ prescribed on R has at most one solution in $\mathcal{C}^1(\overline{\Omega})$.

An informal proof can be given as follow. We impose the homogeneous initial conditions $\mathbf{u}|_R \equiv 0$. Then we add the equations in (10.6) with some yet undetermined functions $\bar{\mathbf{u}} \in \mathcal{C}^1(\overline{\Omega})$ as coefficients and integrate over the domain Ω. A partial integration yields

$$\begin{aligned} 0 &= \int_\Omega \left(\bar{u}^\beta A^{\alpha\beta}_i u^\alpha_i + \bar{u}^\beta B^{\alpha\beta} u^\alpha\right) dx \\ &= \int_\Omega \left(\left[-\bar{u}^\beta_i A^{\alpha\beta}_i u^\alpha + \bar{u}^\beta \frac{\partial A^{\alpha\beta}_i}{\partial x^i} u^\alpha\right] + \bar{u}^\beta B^{\alpha\beta} u^\alpha\right) dx + \\ &\quad \int_{\partial\Omega} \bar{u}^\beta A^{\alpha\beta}_i u^\alpha n_i\, dA \end{aligned} \tag{10.7}$$

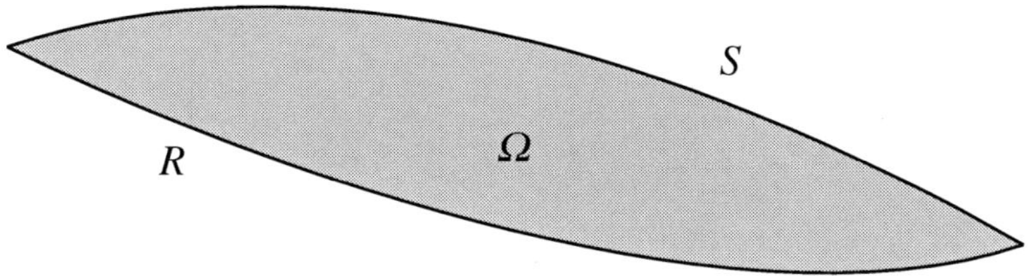

Fig. 10.1 A "lens-shaped" domain

where $\mathbf{n}$ represents the outer unit normal to the hypersurface $\partial\Omega = R \cup S$ and as usual $\bar{u}_i^\beta$ is the x^i-derivative of $\bar{u}^\beta$.

Let us assume that $\bar{\mathbf{u}}$ is a solution of the adjoint system

$$-A_i^{\alpha\beta}\bar{u}_i^\beta + \left[\frac{\partial A_i^{\alpha\beta}}{\partial x^i} + B^{\alpha\beta}\right]\bar{u}^\beta = 0\,, \quad 1 \leq \alpha \leq m \tag{10.8}$$

satisfying the initial conditions $\bar{\mathbf{u}}|_S = \mathbf{f}$ for some arbitrary continuous function $\mathbf{f}$ defined on S. Then (10.7) yields

$$\int_S f^\beta \left(A_i^{\alpha\beta} n_i\right) u^\alpha \, dA = 0\,. \tag{10.9}$$

Since we assumed that S is an analytic non-characteristic hypersurface, the term $\sum_{i=1}^n A_i^{\alpha\beta} n_i$ vanishes nowhere on S. Thus (10.9) can only hold for *all* continuous functions $\mathbf{f}$, if $\mathbf{u}|_S \equiv 0$.

The problem is to show that we always find such functions $\bar{\mathbf{u}}$. As (10.6) is an analytic system, (10.8) is analytic, too. For analytic functions $\mathbf{f}$ we can thus apply the Cauchy–Kovalevskaya Theorem. By the Weierstraß Approximation Theorem any continuous function on a compact subset of $\mathbb{R}^n$ can be uniformly approximated by polynomials, so that it indeed suffices to treat analytic initial data, namely polynomials. However, the Cauchy–Kovalevskaya Theorem is a purely local result and we do not know whether the solution of the initial value problem for (10.8) exists in the whole domain Ω.

This problem forces us to proceed in the following rather technical fashion. We replace the single surface S by a one-parameter family of surfaces S_λ sweeping through the whole domain Ω which is now assumed to be of a somewhat special form. Let $\Omega' \subset \mathbb{R}^n$ be a bounded domain such that the coefficients of (10.6) are analytic in it. Let $\Psi(x)$ be an analytic function defined in Ω' such that $\nabla\Psi$ does not vanish anywhere. We set $Z = \{x \in \Omega' \mid x^n = 0\}$ and assume that it is a non-empty and non-characteristic hypersurface.

We introduce the family of hypersurfaces $S_\lambda = \{x \in \Omega' \mid \Psi(x) = \lambda, x^n \geq 0\}$ with a parameter $\lambda \in [a,b]$. We make the following assumptions (see Figure 10.2):

1. All $\lambda \in [a,b]$ are regular values of Ψ.

2. $\overline{\Omega} = \bigcup_{\lambda\in[a,b]} S_\lambda$.
3. $S_a \subset Z$ consists of a single point.
4. For each $\lambda \in (a,b]$ the hypersurfaces S_λ are non-characteristic and intersect Z transversally.

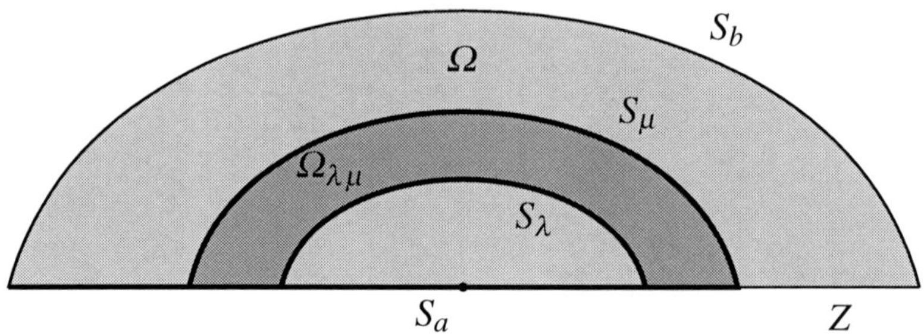

Fig. 10.2 A parametrised family of hypersurfaces

Theorem 10.2.1 (Holmgren). *Let* $\mathbf{u} \in \mathcal{C}^1(\overline{\Omega})$ *be a solution of* (10.6) *such that* $\mathbf{u} \equiv 0$ *on* $Z \cap \partial\Omega$. *Then* $\mathbf{u} \equiv 0$ *in* $\overline{\Omega}$.

Proof. We set $\Lambda = \{\lambda \in [a,b] \mid \mathbf{u}|_{S_\lambda} \equiv 0\}$. The theorem is equivalent to $\Lambda = [a,b]$. By definition Λ is closed; we will show that it is also open in $[a,b]$, as then $\Lambda = [a,b]$.

The functions $A_i^{\alpha\beta}$, $B^{\alpha\beta}$, Ψ are analytic and $\overline{\Omega}$ is compact; thus we can find values M, r independent of $\mathbf{x}$ such that all of them are in $\mathcal{C}_{M,r}(\mathbf{x})$ at each point $\mathbf{x} \in \overline{\Omega}$. If we solve the initial value problem for (10.6) with initial data in $\mathcal{C}_{M,r}$ prescribed on the hypersurface S_λ for $\lambda \in (a,b]$, then the Cauchy–Kovalevskaya Theorem asserts the existence of a unique analytic solution in an ε-neighbourhood of S_λ where ε does not depend on λ.

Polynomials are analytic functions and thus elements of some class $\mathcal{C}_{\bar{M},\bar{r}}$. As polynomials possess only a finite number of non-vanishing derivatives, we can choose an arbitrarily large value for $\bar{r}$ provided we take $\bar{M}$ large enough. As (10.6) is a linear system, we can multiply any solution by a constant and obtain again a solution; this does not change the domain of definition of the solution. By multiplying a polynomial in $\mathcal{C}_{\bar{M},r}$ with $M/\bar{M}$ we obtain a polynomial in $\mathcal{C}_{M,r}$. Thus we can conclude that for any polynomial prescribed as initial data on S_λ the unique analytic solution of (10.8) exists in the whole ε-neighbourhood of S_λ.

Let $\lambda \in [a,b]$ and $\varepsilon > 0$ be given. In the neighbourhood of some point $\mathbf{y} \in S_\lambda$ we can apply the Implicit Function Theorem to solve the parametric equation $\Psi(\mathbf{x}) = \mu$ for some coordinate x^i. This yields a function $x^i = \psi(x^1,\dots,x^{i-1},x^{i+1},\dots,x^n;\mu)$ which is continuous in μ. So we can choose a $\delta(\mathbf{y}) > 0$ such that for all $\mu \in [a,b]$ with $|\lambda-\mu| < \delta(\mathbf{y})$ there exists a point $\mathbf{z} \in S_\mu$ with $||\mathbf{z}-\mathbf{y}|| < \varepsilon$. As S_λ is compact, there exists a constant δ such that $0 < \delta \leq \delta(\mathbf{y})$ for all $\mathbf{y} \in S_\lambda$. But this implies that for all $\mu \in [a,b]$ with $|\lambda - \mu| < \delta$ the hypersurface S_μ lies completely in an ε-neighbourhood of the hypersurface S_λ.

Now take a $\lambda \in \Lambda$ and a $\mu \in [a,b]$ such that $\mu \neq \lambda$ and $|\lambda - \mu| < \delta$. Then we can apply the argument above to the domain $\Omega_{\lambda\mu}$ bounded by the surfaces S_λ, S_μ and Z (recall that above we did not require that R is analytic, thus we may take S_λ plus parts of Z for the surface R). This time we can guarantee that the solution of the adjoint system (10.8) exists in the whole domain. Thus $\mathbf{u}|_{S_\mu} \equiv 0$ and $\mu \in \Lambda$. But this observation implies that Λ is open in $[a,b]$, as obviously $a \in \Lambda$ and thus Λ is not empty. □

Theorem 10.2.1 does not present the strongest possible formulation of the Holmgren Theorem. Using, say, density arguments one can extend this uniqueness statement to considerable more general function spaces; for example, Taylor [453] gives a version valid in Sobolev spaces. Note that we are still forced to require that the system itself is analytic, as we apply the Cauchy–Kovalevskaya Theorem at an intermediate step. But this assumption is much less severe than the restriction to analytic solutions which has now been removed.

Recall that Lemma 9.4.2 was a crucial ingredient of our proof of the Cartan–Kähler Theorem, as it guarantees that the solution obtained in one step is also a solution of all previous subsystems. The for the proof of this statement decisive normal system (9.44) is always linear. Thus provided that its coefficients $\bar{A}^{\beta}_{ij}(\mathbf{x})$ and $\bar{B}^{\beta}_{i}(\mathbf{x})$ are analytic (for example, if A^{β}_{ij} and B^{β}_{i} in (9.43) do not depend on $\mathbf{u}^{(1)}$), we can prove Lemma 9.4.2 for much larger classes of solutions, as we can replace the use of the Cauchy–Kovalevskaya Theorem by the Holmgren Theorem.

Finally, we proceed to extend the Holmgren Theorem to arbitrary involutive systems. As in Section 9.4, we consider a system in the Cartan normal form (9.40), however now assuming that is linear, i. e. we assume that there exist analytic functions $A^{\alpha\delta}_{k}(\mathbf{x})$, $B^{\alpha\gamma}_{kj}(\mathbf{x})$, $C^{\alpha\beta}_{k}(\mathbf{x})$, $D^{\alpha}_{k}(\mathbf{x})$ such that the right hand sides can be written in the form

$$\begin{aligned}\phi^{\alpha}_{k}(\mathbf{x},\mathbf{u},u^{\gamma}_{j},u^{\delta}_{k}) = \sum_{\delta=\beta^{(k)}_{1}+1}^{m} A^{\alpha\delta}_{k}(\mathbf{x})u^{\delta}_{k} + \sum_{\gamma=1}^{m}\sum_{j=1}^{k-1} B^{\alpha\gamma}_{kj}(\mathbf{x})u^{\gamma}_{j} + \\ + \sum_{\beta=1}^{m} C^{\alpha\beta}_{k}(\mathbf{x})u^{\beta} + D^{\alpha}_{k}(\mathbf{x})\,.\end{aligned} \tag{10.10}$$

Theorem 10.2.2. *The initial value problem* (9.40), (9.41) *with the right hand sides given by* (10.10) *possesses at most one solution in* $\mathcal{C}^1(\overline{\Omega})$.

Proof. As for a normal system, it suffices to prove that in the case of a homogeneous system the only solution for vanishing initial data is the zero solution. Thus we study what happens in our proof of the Cartan–Kähler Theorem in this situation. In the first step we consider the normal system (9.45a) with vanishing initial data. Theorem 10.2.1 implies that the unique solution in $\mathcal{C}^1(\overline{\Omega} \cap \{x^2 = \cdots = x^n = 0\})$ is

$U_1^\alpha(x^1) \equiv 0$. Thus in the next step we have again vanishing initial data and can again apply Theorem 10.2.1. In this manner we can go through all steps and obtain finally that $\mathbf{u} \equiv 0$ is the unique solution in $\mathcal{C}^1(\overline{\Omega})$. □

10.3 Elliptic Equations

The classification into elliptic and hyperbolic equations is fundamental for the theory of partial differential equations. Roughly speaking, for the former ones boundary value problems are well-posed whereas the latter ones describe evolutionary problems and thus require the prescription of initial values. While the case of normal equations is extensively treated in any advanced textbook on linear partial differential equations, the extension to non-normal equations encounters some problems. As we will see, the restriction to involutive equations will solve some of them.

For a linear differential equation $Lu = 0$ with constant coefficients one can show that the classification has implications on the form of the support of a fundamental solution (i. e. a solution of the inhomogeneous equation $Lu = \delta$ where δ denotes the Dirac delta distribution). An elliptic equation always possesses a fundamental solution which is everywhere analytic except at the origin, whereas a hyperbolic equation has a fundamental solution the support of which is contained in a cone (see e. g. [224, Thm. 7.1.20] and [225, Thm. 12.5.1], respectively, or the discussion in [102, Chapt. V, §2]). Thus the classification is reflected in properties of the solution space and we expect that it remains invariant under transformations of the equation which do not affect the solution space, say a reduction to first order.

In this section we discuss the definition of ellipticity for general, i. e. possibly under- over overdetermined, equations. It turns out that it is non-trivial to provide such a definition, as integrability conditions may affect the usual criterion. Hyperbolic equations will be the topic of the next section.

Definition 10.3.1. The differential equation $\mathcal{R}_q$ is *elliptic* at a given point $\rho \in \mathcal{R}_q$, if its principal symbol τ_χ is injective at ρ for all non-vanishing one-forms $\chi \in T^*\mathcal{X}$. We call $\mathcal{R}_q$ *elliptic*, it it is elliptic at every point $\rho \in \mathcal{R}_q$.

Thus, according to Definition 7.1.13, we may alternatively define an elliptic equation as one without real characteristic one-forms. This is the formulation usually found in introductory textbooks. As ellipticity is a property defined at points and depending only on the symbol, it suffices, if we study it for linear equations. In fact, it even suffices to restrict to linear systems with constant coefficients and we will do so throughout our treatment of ellipticity. Then it is no longer necessary to distinguish different points on $\mathcal{R}_q$, because if the equation is elliptic at one point, it will be so at any other point, too.

Remark 10.3.2. Note that, by Definition 7.5.6 of an underdetermined equation, such an equation can never be elliptic. This could be changed by permitting in Definition 10.3.1 that τ_χ is alternatively surjective. At the level of the matrix $T[\chi]$ of τ_χ in some local representation this modification corresponds to requiring only that $T[\chi]$ has full rank whereas our stricter definition requires full *column* rank. In some situations it is useful to allow for underdetermined elliptic equations (a simple example is given by a "Gauss law" like $\nabla \cdot \mathbf{u} = 0$). But typically they appear then as subequations of larger equations (indeed such a situation will be studied in the next section) and are not of independent interest. Hence we will assume for the remainder of this section that we are not dealing with an underdetermined equation. ◁

Example 10.3.3. The prototype of an elliptic equation is the *Laplace equation* $\sum_{i=1}^{n} u_{x^i x^i} = 0$. For a one-form $\chi = \chi_i \mathrm{d}x^i \in T^*\mathcal{X}$, its principal symbol τ_χ is locally represented by a 1×1-matrix whose single entry is the polynomial $T[\chi] = \sum_{i=1}^{n} \chi_i^2$. Thus the only real one-form χ for which $T[\chi]$ vanishes is the zero form and the Laplace equation is indeed elliptic.

Let us restrict for simplicity to $n = 2$ and write $x = x^1$ and $y = x^2$. In many textbooks the following reduction to first order is proposed:

$$\mathcal{R}_1 : \begin{cases} u_x - v = 0\,, \\ u_y - w = 0\,, \\ v_x + w_y = 0\,. \end{cases} \tag{10.11}$$

Obviously, $\mathcal{R}_1$ is not involutive, as an integrability condition is hidden: $v_y - w_x = 0$. We obtain an involutive equation, $\mathcal{R}_1^{(1)}$, only after its addition.

The principal symbol τ_χ of $\mathcal{R}_1$ has the following matrix representation:

$$T[\chi] = \begin{pmatrix} \chi_x & 0 & 0 \\ \chi_y & 0 & 0 \\ 0 & \chi_x & \chi_y \end{pmatrix}\,. \tag{10.12}$$

It is easy to see that τ_χ cannot be injective for any one-form χ and thus we obtain the surprising result that the above first-order reduction $\mathcal{R}_1$ of the Laplace equation is not elliptic in the sense of Definition 10.3.1.

However, if we add the above mentioned integrability condition, i. e. if we analyse the involutive equation $\mathcal{R}_1^{(1)}$ (which is the proper first-order reduction[1] of the Laplace equation according to Appendix A.3), then the matrix $T[\chi]$ acquires a further row $(0\ \chi_y\ -\chi_x)$ and one easily verifies that now the principal symbol is injective for any real non-zero form χ. Thus we arrive at the expected result that the reduction to first order preserves ellipticity. ◁

Instead of adding this simple integrability condition, one finds in the literature an alternative ansatz to solve the problem that seemingly $\mathcal{R}_1$ is no longer elliptic. As this approach is not intrinsic, we must first choose some local coordinates and consider a concrete local representation

[1] In fact, the simplest first-order reduction of the Laplace equation is the (trivially elliptic) system $v_x + w_y = v_y - w_x = 0$, as the second equation is just the compatibility condition which guarantees the existence of a function u such that $u_x = v$ and $u_y = w$.

$$\mathcal{R}_q : \left\{ \Phi^\tau(\mathbf{x}, \mathbf{u}^{(q)}) = A_\alpha^{\tau\mu} u_\mu^\alpha = 0 \,, \qquad 1 \le \tau \le t \,. \right. \tag{10.13}$$

Working now in a fixed coordinate system, we do not distinguish in the sequel between the principal symbol τ_χ and its matrix $T[\chi]$, calling the latter also principal symbol. We say that $T[\chi]$ is *elliptic*, if it defines an injective map for any non-vanishing one-form $\chi \in T^*\mathcal{X}$, i. e. if the matrix has always full column rank.

Next we introduce two sets of integer *weights*: one weight s_τ, $1 \le \tau \le t$, for each equation in the given system and one weight t_α, $1 \le \alpha \le m$, for each dependent variable. They must be chosen such that $s_\tau + t_\alpha \ge q_{\tau\alpha}$ where $q_{\tau\alpha}$ is the maximal order of a derivative u_μ^α in the τth equation, i. e. we have $q_{\tau\alpha} = \max\{|\mu| \mid A_\alpha^{\tau\mu} \neq 0\}$.

Definition 10.3.4. Let $w = (s_1, \dots, s_t; t_1, \dots, t_m)$ be a set of weights for the differential system (10.13). Then the corresponding *weighted principal symbol* is the $t \times m$ matrix $T_w[\chi]$ with entries

$$\bigl(T_w[\chi]\bigr)_\alpha^\tau = \sum_{|\mu| = s_\tau + t_\alpha} A_\alpha^{\tau\mu} \chi^\mu \,. \tag{10.14}$$

The system (10.13) is called *DN-elliptic*,[2] if a set w of weights exists such that the weighted principal symbol $T_w[\chi]$ is elliptic.

Note that DN-ellipticity requires only that for *some* choice of w the weighted principal symbol is injective and in general there are different possible choices. In particular, the property of being DN-elliptic is *not* independent of the choice of coordinates. Also it may not be easy to construct effectively suitable weights/coordinates.

For the special choice $w = (0, \dots, 0; q, \dots, q)$ we find $T_w[\chi] = T[\chi]$ and hence any elliptic system is DN-elliptic. The converse is not true: if we return to the first-order form (10.11) for the Laplace equation, then it is DN-elliptic, as its weighted principal symbol $T_w[\chi]$ is injective for the choice $w = (-1, -1, 0; 2, 1, 1)$. This example also clearly demonstrates the simple idea behind the introduction of the weights. The v- and the w-term in the first two equations of (10.11) do not contribute to the usual principal symbol, as they are of lower order; they only become relevant after the cross-derivation leading to the hidden integrability condition. Using the above weights, they are "visible" already in the original system (10.11).

This observation clearly indicates a relation between the completion of (10.13) and ellipticity. Indeed, our main goal in this section is to show that if some local representation of a differential equation $\mathcal{R}_q$ is DN-elliptic, then the involutive completion $\mathcal{R}_{q+r}^{(s)}$ is elliptic in the ordinary sense. Thus, if we restrict the classification to involutive equations, we do not need at all the concept of DN-ellipticity!

Remark 10.3.5. There are two other special choices of weights worth mentioning. If q_τ denotes the order of the τth equation in the system (10.13), then the weights $w = (q_1 - q, \dots, q_t - q; q, \dots, q)$ lead to the *reduced principal symbol* $T_{\text{red}}[\chi]$. We will

[2] The terminology *DN-elliptic* refers to Douglis and Nirenberg [111] who first considered this generalisation of the classical definition of ellipticity.

show later that if our system has an elliptic reduced symbol, then by simply prolonging some of the lower-order equations we obtain an elliptic system in the classical sense. Hence for all computational purposes it suffices to consider $T_{\text{red}}[\chi]$ instead of $T[\chi]$. More generally, we call any weighted principal symbol *reduced*, if all weights t_α have the same value.

Let $\tilde{q}_\alpha$ be the maximal order of a derivative u^α_μ in the whole system (10.13). With the weights $w = (0,\dots,0;\tilde{q}_1,\dots,\tilde{q}_m)$ we obtain what is often called the *Petrovskij (principal) symbol* $T_P[\chi]$. If $T_P[\chi]$ is elliptic, then we call (10.13) *elliptic in the sense of Petrovskij* or for short *P-elliptic*. This notion of ellipticity features prominently in the Russian literature, see e. g. [9]. ◁

Obviously, the weighted symbol $T_w[\chi]$ remains unchanged, if we replace all weights s_τ by $s_\tau + k$ and all weights t_α by $t_\alpha - k$ for some integer $k \in \mathbb{Z}$. Hence we may always suppose that (after some renumbering) $s_1 \leq s_2 \leq \cdots \leq s_t = 0$ and $t_1 \geq t_2 \geq \cdots \geq t_m \geq 0$. Furthermore, we introduce values a, b and the two sets of indices τ_k, $1 \leq k \leq a$, and α_ℓ, $1 \leq \ell \leq b$, such that

$$\begin{aligned} s_1 = \cdots = s_{\tau_1} < s_{\tau_1+1} = \cdots = s_{\tau_1+\tau_2} < \cdots < s_{\tau_1+\cdots+\tau_{a-1}+1} = \cdots = s_k = 0\,, \\ t_1 = \cdots = t_{\alpha_1} > t_{\alpha_1+1} = \cdots = t_{\alpha_1+\alpha_2} > \cdots > t_{\alpha_1+\cdots+\alpha_{b-1}+1} = \cdots = t_m \end{aligned} \tag{10.15}$$

and $\tau_a = t - \sum_{k=1}^{a-1} \tau_k$, $\alpha_b = m - \sum_{\ell=1}^{b-1} \alpha_\ell$. In addition, we set

$$\begin{aligned} \bar{\tau}_0 = 0\,, \quad \bar{\tau}_k = \tau_1 + \cdots + \tau_k\,, \\ \bar{\alpha}_0 = 0\,, \quad \bar{\alpha}_\ell = \alpha_1 + \cdots + \alpha_\ell\,. \end{aligned} \tag{10.16}$$

With these conventions, the weighted principal symbol $T_w[\chi]$ can be decomposed into blocks as follows:

$$T_w[\chi] = \begin{pmatrix} T_{11} & T_{12} & \cdots & T_{1b} \\ T_{21} & T_{22} & \cdots & T_{2b} \\ \vdots & \vdots & \ddots & \vdots \\ T_{a1} & T_{a2} & \cdots & T_{ab} \end{pmatrix}. \tag{10.17}$$

Here the block $T_{k\ell}$ is a $\tau_k \times \alpha_\ell$ matrix and its entries are homogeneous polynomials in χ of degree $\nu_{k\ell} = s_{\bar{\tau}_k} + t_{\bar{\alpha}_\ell}$.

Lemma 10.3.6. *Suppose that the system* (10.13) *has an elliptic weighted principal symbol $T_w[\chi]$ for the weights $w = (s_1,\dots,s_t;t_1,\dots,t_m)$. Consider the system obtained from* (10.13) *by adding all equations obtained by differentiating the τth equation p times with respect to each independent variable. Then its weighted principal symbol is elliptic for the following weights: s_τ is set to zero, the weights for the new equations are $s_\tau + p$, and all other weights remain unchanged.*

Proof. Denote by $(T_w[\chi])^\tau$ the τth row in the weighted principal symbol $T_w[\chi]$. We apply p times the formal derivative D_i to the τth equation and set the weight of the obtained new equation to $s_\tau + p$. In terms of the weighted principal symbol, this corresponds to adding the row $\chi_i^p(T_w[\chi])^\tau$ to the original weighted principal symbol. Hence, doing this for all $1 \leq i \leq n$ and using the described weights, it is easy to see that full column rank of $T_w[\chi]$ entails full column rank of the new weighted principal symbol. □

By repeated application of this lemma for $p = 1$ we obtain immediately the promised result for reduced principal symbols.

Corollary 10.3.7. *Suppose that a reduced principal symbol $T_{\text{red}}[\chi]$ of the qth-order system* (10.13) *is elliptic. If we prolong all lower-order equations to order q, then we obtain an elliptic system in the classical sense.*

Lemma 10.3.8. *Assume that all rows in* (10.13) *are of order q and that the weights are ordered as in* (10.15)*. Assume that for some weights w the weighted principal symbol $T_w[\chi]$ is elliptic. Then*

(i) $s_1 + t_1 = q$ *and* $T_{k1} = 0$ *for* $1 < k \leq a$*;*
(ii) *The block T_{11} is an elliptic principal symbol; i. e.* $\operatorname{rank} T_{11} = \alpha_1$ *and in particular $\tau_1 \geq \alpha_1$;*
(iii) $t_m \geq 0$ *and without loss of generality we may suppose that* $t_m \geq 1$.

Proof. If $s_1 + t_1 < q$, then the first τ_1 equations could not be of order q. If $s_1 + t_1 > q$, then the first block column would be zero, so that $T_w[\chi]$ could not be elliptic. Thus $s_1 + t_1 = q$ and for all $\tau > \tau_1$ we have $s_\tau + t_1 > q$ implying $T_{k1} = 0$ for $1 < k \leq a$. If $\operatorname{rank} T_{11} < \tau_1$, then $T_w[\chi]v = 0$ for some nonzero vector v of the form

$$v = \left(v_1, \dots, v_{\tau_1}, 0, \dots, 0\right)^t . \tag{10.18}$$

As the kernel of an elliptic symbol is trivial, we must have $\tau_1 \geq \alpha_1$. If $t_m < 0$, then the last block column would be zero and $T_w[\chi]$ could not be elliptic.

Suppose that $t_m = 0$ and call the dependent variables $u^{\bar{\alpha}_{b-1}+1}, \dots, u^m$ algebraic, since no derivatives of them appear in the system. Moreover, the first $\bar{\tau}_{a-1}$ equations do not depend on these variables. Hence the first $\bar{\tau}_{a-1}$ equations form a DN-elliptic system with variables $u^1, \dots, u^{\bar{\alpha}_{b-1}}$. Because T_{ab} is of full column rank, the algebraic variables can be solved in terms of other variables. In case $b = 1$, the block T_{a1} is of full column rank and we can again solve the algebraic variables in terms of other variables. Hence we obtain then a system without algebraic variables. □

As a consequence of these results, we may suppose, whenever it is convenient, that the (weighted) principal symbol of our differential equation has the following block form:

$$T_w[\chi] = \begin{pmatrix} T_{11} & T_{12} & \dots & T_{1b} \\ 0 & T_{22} & \dots & T_{2b} \\ \vdots & \vdots & \ddots & \vdots \\ 0 & T_{a2} & \dots & T_{ab} \end{pmatrix} \qquad T[\chi] = \begin{pmatrix} T_{11} & 0 & \dots & 0 \\ T'_{21} & T'_{22} & \dots & T'_{2b} \\ \vdots & \vdots & \ddots & \vdots \\ T'_{a1} & T'_{a2} & \dots & T'_{ab} \end{pmatrix} . \tag{10.19}$$

Proposition 10.3.9. *Assume that at the end of an iteration of the outer loop of the Cartan–Kuranishi Algorithm 7.3 the reduced principal symbol becomes elliptic, then it will remain elliptic until termination.*

Proof. Each iteration of the outer loop of Algorithm 7.3 begins with prolongations. It follows trivially from Lemma 10.3.6 that if the reduced principal symbol is elliptic before a prolongation, it will remain so afterwards. Going from $\mathcal{R}_q$ to $\mathcal{R}_q^{(1)}$, i.e. performing a prolongation with a subsequent projection, adds further rows to the principal symbol (corresponding to the hidden integrability conditions) and thus can only increase the column rank. □

Remark 10.3.10. For simplicity, we formulated Proposition 10.3.9 only for the geometric Cartan–Kuranishi completion. But a similar result holds for any other completion method, say the algebraic Algorithm 4.5 applied to linear differential operators. The addition of obstructions of involution or of integrability conditions can always only increase the column rank. Algebraic operations like autoreductions or normal form computations correspond at the level of the principal symbol to elementary row operations and thus preserve the rank. ◁

In order to proceed, we divide the vector $\mathbf{u}$ of the dependent variables into segments $\mathbf{u}^{(l)}$ according to the indices $\bar{\alpha}_\ell$ defined in (10.16): $\mathbf{u}^{(l)} = \left(u^{\bar{\alpha}_{\ell-1}+1}, \dots, u^{\bar{\alpha}_\ell}\right)^t$. This induces a corresponding segmentation of the coefficient matrix $A_\alpha^{\tau\mu}$ of our linear system (10.13) and hence of the weighted principal symbol. As this segmentation is in accordance with the jumps in the sequence (10.15), the pieces are reduced principal symbols:

$$T_w[\chi] = \left(T_{\mathrm{red},1}[\chi], \dots, T_{\mathrm{red},b}[\chi]\right) . \tag{10.20}$$

This observation will lead to a simple inductive proof of our main result. Recall from Remark 7.1.12 that the construction of integrability conditions is related to syzygies of the rows of $T_{\mathrm{red}}[\chi]$. In order to go on we therefore need the following technical result.

Lemma 10.3.11. *Let $\mathcal{P} = \Bbbk[x^1, \dots, x^n]$ be the ordinary commutative polynomial ring and $B = \left(B_1, C_1\right)$ a polynomial matrix where $B_1 \in \mathcal{P}^{\ell \times m_1}$, $C_1 \in \mathcal{P}^{\ell \times m_2}$ and $m = m_1 + m_2$. Furthermore, let*

$$B' = \begin{pmatrix} B_1 & 0 \\ B_2 & S^t C_1 \end{pmatrix} \tag{10.21}$$

where $S \in \mathcal{P}^{k \times r}$ is a syzygy matrix[3] of B_1^t and $B_2 \in \mathcal{P}^{r \times m_1}$ an arbitrary matrix. For a given vector $\chi \in \Bbbk^n$ we consider the matrix $B(\chi) \in \Bbbk^{\ell \times m}$ obtained by substituting $x^i = \xi^i$ and similarly for B'. If $\ell > m_1$, then $\ker\left(B(\chi)\right) = 0$ for all $\chi \neq 0$ implies $\ker\left(B'(\chi)\right) = 0$ for all $\chi \neq 0$.

[3] The columns of a syzygy matrix S of a polynomial matrix A span the syzygy module of the columns of A; hence in particular $AS = 0$

Proof. The assumption $\ell > m_1$ ensures that the syzygy matrix S does not vanish, as the matrix $B_1^t \in \mathcal{P}^{m_1 \times \ell}$ cannot possess full column rank.

Assume that a non-vanishing vector $\bar{\chi} \in \Bbbk^n$ exists with $\ker(B'(\bar{\chi})) \neq 0$. Then a non-vanishing vector $\mathbf{v} = (\tilde{\mathbf{v}}, \hat{\mathbf{v}}) \in \Bbbk^m$ exists such that $B'(\bar{\chi})\mathbf{v} = 0$ and hence $B_1(\bar{\chi})\tilde{\mathbf{v}} = 0$. Since $\ker(B(\bar{\chi})) = 0$ entails $\ker(B_1(\bar{\chi})) = 0$, it follows that $\tilde{\mathbf{v}} = 0$. Thus we find $S^t(\bar{\chi})C_1(\bar{\chi})\hat{\mathbf{v}} = 0$ and $C_1(\bar{\chi})\hat{\mathbf{v}} \in \ker(S^t(\bar{\chi}))$.

By definition of S, we have an exact sequence $\mathcal{P}^r \xrightarrow{S} \mathcal{P}^\ell \xrightarrow{B_1^t} \mathcal{P}^{m_1}$. Since $\ker(B_1(\chi)) = 0$ for all $\chi \neq 0$, it follows now from Corollary B.4.33 that the sequence $\Bbbk^{m_1} \xrightarrow{B_1(\bar{\chi})} \Bbbk^\ell \xrightarrow{S^t(\bar{\chi})} \Bbbk^r$ is exact, too. Hence $\ker(S^t(\bar{\chi})) = \operatorname{im}(B_1(\bar{\chi}))$ and there exists a vector $\hat{\mathbf{u}} \in \Bbbk^{m_1}$ such that $B_1(\bar{\chi})\hat{\mathbf{u}} + C_1(\bar{\chi})\hat{\mathbf{v}} = 0$. Putting $u = (\hat{u}, \hat{v}) \neq 0$ (since $\hat{\mathbf{v}} \neq 0$) yields $B(\bar{\chi})\mathbf{u} = 0$. But this contradicts our assumption that $\ker(B(\chi)) = 0$ for all vectors $\chi \neq 0$. □

Theorem 10.3.12. *If the differential equation $\mathcal{R}_q$ has a DN-elliptic local representation, then its involutive completion $\mathcal{R}_{q+r}^{(s)}$ is elliptic.*

Proof. We exploit the decomposition (10.20) assuming that $b > 1$. Let S be the syzygy matrix of the $\alpha_1 \times t$ matrix $(T_{\text{red},1}[\chi])^t$. According to Lemma 10.3.8 we have $t > \tau_1 \geq \alpha_1$ and $T_{\text{red},1}[\chi]$ is elliptic. By the same argument as in the proof of Lemma 10.3.11, these facts imply that $S \neq 0$. Let c be the number of the columns of the matrix S and denote the columns by $\mathbf{v}^{(r)}$ with $r = 1, \dots, c$. Since the entries of $(T_{\text{red},1}[\chi])^t$ are homogeneous polynomials, for each r there is some m_r such that the degree of $\mathbf{v}_\tau^{(r)}$ is $m_r - s_\tau$ or $v_\tau^{(r)}$ is zero.

Substituting the formal derivative D_i for the variable χ_i in the matrix S, we construct a differential operator $\hat{S}$. Let us now consider the extended linear system obtained by adding to (10.13) all the equations obtained by applying the operator $\hat{S}^t$ to (10.13). With respect to the weights $w^{(1)}$ obtained by setting $t_\alpha^{(1)} = t_\alpha + 1$ for $\alpha > \bar{\alpha}_1$, $s_{t+r}^{(1)} = m_r - 1$ for $r = 1, \dots, c$ and leaving all other weights unchanged, its weighted principal symbol $T_w^{(1)}[\chi]$ is of the form

$$T_w^{(1)}[\chi] = \begin{pmatrix} T_{\text{red},1}[\chi] & 0 \\ B & S^t\left(T_{\text{red},2}[\chi], \dots, T_{\text{red},b}[\chi]\right) \end{pmatrix} . \tag{10.22}$$

This choice of weights is consistent with the conditions we imposed on admissible weights, since the order of any derivative of variables $\mathbf{u}^{(1)}$ in the rth new equation is always less than or equal to $t_{\bar{\alpha}_1} + m_r - 1$.

Since the weighted principal symbol $T_w[\chi]$ is elliptic and $t > \alpha_1$, all assumptions of Lemma 10.3.11 are satisfied and we conclude that $T_w^{(1)}[\chi]$ is elliptic, too. Thus we can apply the same arguments to the extended system, until we obtain after ν iterations a system such that $t_{\bar{\alpha}_1} = t_{\bar{\alpha}_2}^{(\nu)}$. This process reduces in a finite number of steps a DN-elliptic system with b block columns to an equivalent DN-elliptic system with $b - 1$ block columns.

Continuing in this fashion, we arrive after finitely many steps at a system equivalent to the original one and with $b = 1$ block column, i. e. with an elliptic reduced symbol. By Corollary 10.3.7, this observation suffices to prove our assertion. □

Remark 10.3.13. On the surface, the construction used in the proof above has nothing to do with any of the completion algorithms discussed in this book. Again this observation signals that Theorem 10.3.12 is only a particular formulation of a more general result. Every equation added during the proof is a differential consequence of the original system. In the geometric theory this implies that it vanishes on some prolongation of the involutive completion. In the algebraic theory we find that its involutive normal form with respect to an involutive basis of the system vanishes. Similar remarks holds for other completion approaches. Thus we may conclude that adding sufficiently many integrability conditions to a DN-elliptic system yields a system that is elliptic in the ordinary sense. ◁

Example 10.3.14. We illustrate the fairly technical construction used above in the proof of Theorem 10.3.12 with the following sixth-order equation[4] in four dependent variables (u^1, u^2, u^3, u^4) and two independent variables x^1, x^2:

$$\mathcal{R}_6 : \begin{cases} u^1_{[0,2]} + u^2_{[1,0]} - u^3 = 0\,, \\ u^1_{[1,0]} + u^2_{[2,0]} + u^2_{[0,2]} + u^3 = 0\,, \\ u^1_{[4,0]} + u^2_{[1,2]} + u^3_{[0,2]} + u^4_{[0,1]} = 0\,, \\ u^2_{[3,3]} + u^4_{[4,0]} + u^4_{[0,4]} = 0\,. \end{cases} \tag{10.23}$$

With respect to the weights $w = (-2,-2,0,2;4,4,2,2)$, we obtain as weighted principal symbol

$$T_w[\chi] = \begin{pmatrix} \chi_2^2 & 0 & -1 & 0 \\ 0 & \chi_1^2 + \chi_2^2 & 1 & 0 \\ \chi_1^4 & 0 & \chi_2^2 & 0 \\ 0 & \chi_1^3\chi_2^3 & 0 & \chi_1^4 + \chi_2^4 \end{pmatrix}. \tag{10.24}$$

Our local representation is thus a DN-elliptic system: one easily verifies that $\det\bigl(T_w[\chi]\bigr) = (\chi_1^2 + \chi_2^2)(\chi_1^4 + \chi_2^4)^2$.

Here the decomposition (10.20) splits $T_w[\chi]$ into its first two and its last two columns (which we call $T_{w,1}[\chi]$ and $T_{w,2}[\chi]$, respectively). A standard computation yields for the syzygy matrix of $\bigl(T_{w,1}[\chi]\bigr)^t$

$$S = \begin{pmatrix} \chi_1^4 & 0 \\ 0 & \chi_1^3\chi_2^3 \\ -\chi_2^2 & 0 \\ 0 & -\chi_1^2 - \chi_2^2 \end{pmatrix}. \tag{10.25}$$

Thus in the notation of the proof above we have $m_1 = 2$ and $m_2 = 4$.

[4] Due to the high order of this example, we use here the multi index notation usually reserved for theoretical considerations.

Using the differential operator $\hat{S}$ corresponding to the polynomial matrix S, we obtain as new additional equations:

$$\begin{aligned} &u^2_{[5,0]} - u^2_{[1,4]} - u^3_{[4,0]} - u^3_{[0,4]} - u^4_{[0,3]} = 0\,, \\ &u^1_{[4,3]} + u^3_{[3,3]} - u^4_{[6,0]} - u^4_{[4,2]} - u^4_{[2,4]} - u^4_{[0,6]} = 0\,. \end{aligned} \tag{10.26}$$

The weighted principal symbol of the extended linear system is then

$$T_w^{(1)}[\chi] = \begin{pmatrix} \chi_2^2 & 0 & 0 & 0 \\ 0 & \chi_1^2+\chi_2^2 & 0 & 0 \\ \chi_1^4 & 0 & 0 & 0 \\ 0 & \chi_1^3\chi_2^3 & 0 & 0 \\ 0 & \chi_1^5-\chi_1\chi_2^4 & -\chi_1^4-\chi_2^4 & 0 \\ \chi_1^4\chi_2^3 & 0 & \chi_1^3\chi_1^3 & -(\chi_1^2+\chi_2^2)(\chi_1^4+\chi_2^4) \end{pmatrix}, \tag{10.27}$$

if we use the weights $t_3^{(1)} = t_4^{(1)} = 3$, $s_5^{(1)} = m_1 - 1 = 1$, $s_6^{(1)} = m_2 - 1 = 3$ and all other weights as for $T_w[\chi]$.

Since the values $t_1 = 4$ and $t_3^{(1)} = 3$ are not yet equal, we now compute the syzygy matrix S_1 of $\big(T_w^{(1)}[\chi]\big)^t$ obtaining

$$S_1 = \begin{pmatrix} 0 & 0 & \chi_1^4 & 0 \\ 0 & \chi_1^3-\chi_1\chi_2^2 & 0 & \chi_1\chi_2^5 \\ \chi_2^3 & 0 & -\chi_2^2 & 0 \\ 0 & 0 & 0 & -\chi_1^2-\chi_2^2 \\ 0 & -1 & 0 & \chi_2^3 \\ -1 & 0 & 0 & 0 \end{pmatrix}. \tag{10.28}$$

Thus we deduce that $m_1^{(1)} = 3$, $m_2^{(1)} = 1$, $m_3^{(1)} = 2$ and $m_4^{(1)} = 4$. Applying now the differential operator $\hat{S}_1^t$, we get

$$\hat{S}_1^t A^{(1)} : \begin{cases} u^2_{[1,5]} - u^3_{[3,3]} - u^3_{[0,5]} + u^4_{[6,0]} + u^4_{[4,2]} + u^4_{[2,4]} + u^4_{[0,6]} + u^4_{[0,4]} = 0\,, \\ u^1_{[4,0]} - u^1_{[2,2]} + u^3_{[4,0]} + u^3_{[0,4]} + u^3_{[3,0]} - u^3_{[1,2]} + u^4_{[0,3]} = 0\,, \\ u^2_{[5,0]} - u^2_{[1,4]} - u^3_{[4,0]} - u^3_{[0,4]} - u^4_{[0,3]} = 0\,, \\ u^1_{[2,5]} - u^3_{[4,3]} - u^3_{[0,7]} + u^3_{[1,5]} - u^4_{[6,0]} - u^4_{[4,2]} - u^4_{[2,4]} - 2u^4_{[0,6]} = 0\,. \end{cases} \tag{10.29}$$

The operator $A^{(2)} = \big(A^{(1)}, \hat{S}_1^t A^{(1)}\big)$ has the weighted principal symbol

$$T_w^{(2)}[\chi] = \begin{pmatrix} \chi_2^2 & 0 & 0 & 0 \\ 0 & \chi_1^2+\chi_2^2 & 0 & 0 \\ \chi_1^4 & 0 & 0 & 0 \\ 0 & \chi_1^3\chi_3^3 & 0 & 0 \\ 0 & \chi_1^5-\chi_1\chi_2^4 & 0 & 0 \\ \chi_1^4\chi_2^3 & 0 & 0 & 0 \\ 0 & \chi_1\chi_2^5 & -\chi_1^3\chi_2^3 & (\chi_1^2+\chi_2^2)(\chi_1^4+\chi_2^4) \\ \chi_1^4-\chi_1^2\chi_2^2 & 0 & \chi_1^4+\chi_2^4 & 0 \\ 0 & \chi_1^5-\chi_1\chi_2^4 & 0 & 0 \\ \chi_1^2\chi_2^5 & 0 & -\chi_1^4\chi_2^3-\chi_2^7 & 0 \end{pmatrix}, \tag{10.30}$$

if we set $t_3^{(2)} = t_4^{(2)} = 4$, $s_7^{(2)} = m_1^{(1)} - 1 = 2$, $s_8^{(2)} = m_2^{(1)} - 1 = 0$, $s_9^{(2)} = m_3^{(1)} - 1 = 1$, $s_{10}^{(2)} = m_4^{(1)} - 1 = 3$ and keep all other weights as in $\sigma_w A^{(1)}$.

Since $T_w^{(2)}[\chi] = T_{\text{red}}^{(2)}[\chi]$, the reduced symbol of the operator $A^{(2)}$ is elliptic, i. e. we have transformed the DN-elliptic operator A into an equivalent operator $A^{(2)}$ with an elliptic reduced symbol. By Corollary 10.3.7, the involutive form of the differential operator $A^{(2)}$ is elliptic. ◁

Example 10.3.15. Let us consider the differential equation $\mathcal{R}_1$ defined by the system $\nabla \times \mathbf{u} + \mathbf{u} = 0$ where $\mathbf{u}$ is a three-dimensional vector of unknown functions of three independent variables. One can show that it is *not* DN-elliptic. However, adding the integrability condition $\nabla \cdot \mathbf{u} = 0$ gives for the equation $\mathcal{R}_1^{(1)}$ the principal symbol

$$T[\chi] = \begin{pmatrix} 0 & \chi_3 & -\chi_2 \\ -\chi_3 & 0 & \chi_1 \\ \chi_2 & -\chi_1 & 0 \\ \chi_1 & \chi_2 & \chi_3 \end{pmatrix} \tag{10.31}$$

which is obviously elliptic. Thus we see that the approach via weights is not always sufficient for detecting ellipticity. ◁

It seems that the only sure way for checking whether or not a given differential equation is elliptic consists of first completing it to involution (or something similar as discussed above) and then testing for ellipticity in the ordinary sense. DN-ellipticity provides a shortcut only for the case of an DN-elliptic system, as then Theorem 10.3.12 guarantees us the ellipticity of the completed equation. In the negative case no statements are possible as demonstrated by the example above.

10.4 Hyperbolic Equations

We proceed to the definition of *hyperbolicity*. In contrast to ellipticity, it is not an absolute notion but always defined with respect to a direction, i. e. a distinguished independent variable t exists playing the role of time in evolution problems.

We therefore assume for simplicity that $\mathcal{X} = \Omega \times \mathbb{R}$ with $\Omega \subseteq \mathbb{R}^n$ some domain and slightly change our notation and denote the independent variables by $(x^1, \dots, x^n, t)$, i.e. we have now $n+1$ variables and write t instead of x^{n+1}.

We will define hyperbolicity only for first-order linear systems (with variable coefficients) and always assume that the systems are solved for all t-derivatives, i.e. they are given in the form

$$\mathbf{u}_t + P_0\mathbf{u} = 0\,, \tag{10.32a}$$

$$P_1\mathbf{u} = 0\,. \tag{10.32b}$$

Here P_0, P_1 are arbitrary linear first-order differential operators containing only spatial derivatives ∂_{x^i} (but with possibly t-dependent coefficients). Note that this assumption again excludes underdetermined systems, as we trivially have $\beta_1^{(n+1)} = m$ for (10.32) and hence it cannot be underdetermined by Proposition 7.5.7. The reason for this restriction will become apparent below.

A straightforward computation reveals that a necessary condition for (10.32) being involutive is the existence of a linear first-order differential operator Q containing only spatial derivatives such that

$$QP_1 = \frac{\partial P_1}{\partial t} - P_1 P_0\,. \tag{10.33}$$

Here $\partial P_1/\partial t$ denotes the operator obtained by differentiating all coefficients of P_1 with respect to t. Indeed, if no such Q exists, then the prolongation of (10.32b) with respect to the non-multiplicative variable t yields either obstructions to involution or integrability conditions. If P_1 is a normal operator, then (10.33) is also sufficient for involution. Otherwise further non-multiplicative prolongations (with respect to spatial variables) of equations in the subsystem (10.32b) must be analysed.

As in the case of ellipticity, the definition of a hyperbolic system is based solely on the principal symbol. However, we consider this time only "spatial" one-forms $\chi = \chi_i dx^i \in T^*\Omega$, i.e. without a dt-component. Let $T_0[\chi]$, $T_1[\chi]$ be the principal symbols of the operators P_0, P_1. As the operator $\partial P_1/\partial t$ is of lower order than the other two terms in (10.33), our involution condition implies that $T_0[\chi]$ maps vectors in $\ker T_1[\chi]$ again into $\ker T_1[\chi]$. Based on this observation, we can finally define hyperbolicity.

Definition 10.4.1. Assume that the linear system (10.32) is involutive. Then it is *hyperbolic in t-direction* at a point $(\mathbf{x}, t) \in \mathcal{X}$ of the base space, if for any "spatial" one-form $\chi \in T^*\Omega$ at this point the restriction of $T_0[\chi]$ to $\ker T_1[\chi]$ has only real eigenvalues and an eigenbasis.

Example 10.4.2. Assume that the operator P_1 is elliptic in the sense of Definition 10.3.1. Then obviously $\ker T_1[\chi] = 0$ and the linear system (10.32) is trivially hyperbolic. ◁

Remark 10.4.3. In order to understand the reasoning behind the fairly technical Definition 10.4.1, let us restrict to a homogeneous system with constant coefficients:

$\mathbf{u}_t + A^j\mathbf{u}_{x^j} = 0$, $C^j\mathbf{u}_{x^j} = 0$. Then we look for *normal modes*, i. e. solutions of the special form $\mathbf{u}(\mathbf{x},t) = \exp\left[i(\omega t - \mathbf{k}\cdot\mathbf{x})\right]\mathbf{u}_0$ with a scalar $\omega \in \mathbb{C}$ and a vector $\mathbf{k} \in \mathbb{C}^n$ (here i is not an index but the imaginary unit!). Obviously, a necessary and sufficient condition for the existence of a non-trivial solution of this form is that $k_jC^j = 0$ and that ω is an eigenvalue of the matrix k_jA^j with eigenvector $\mathbf{u}_0$. The condition that for those vectors $\mathbf{k}$ satisfying $k_jC^j = 0$ all eigenvalues of k_jA^j are real ensures that all normal modes are oscillatory and hence do not lead to an instability. If furthermore an eigenbasis of k_jA^j exists, then it follows from elementary Fourier analysis that any solution of our system may be written as a linear combination of normal modes. This fact renders the initial value problem well-posed.

The question of well-posedness is so simple only for this highly restricted class of systems. As soon as one allows for variable coefficients or lower-order terms, the situation becomes more involved. The problem of finding sufficient and necessary conditions for well-posedness of the initial value problem has lead for normal systems to the development of different concepts like strong, strict or symmetric hyperbolicity (see e. g. [266, Sects. 2.4.1, 3.3.1] or [453, Chapt. 6, Sect. 5]). Applied to a normal system, our Definition 10.4.1 actually corresponds to *strict* hyperbolicity and it suffices then also in the variables coefficients case for well-posedness. ◁

Remark 10.4.4. Definition 10.4.1 is formulated only for involutive systems. In principle, this assumption could be relaxed a bit. The linear system (10.32) must be close enough to involution to ensure the existence of an operator Q satisfying (10.33), as without this condition we cannot restrict the principal symbol $T_0[\chi]$ to the subspace $\ker T_1[\chi]$. If then the restriction has only real eigenvalues and an eigenbasis, the same will be true for the involutive completion (we assume here of course that (10.32) is consistent), as integrability conditions enlarge the operator P_1 and thus shrink $\ker T_1[\chi]$. However, the converse is obviously not true: if (10.32) is not yet involutive, the considered $\ker T_1[\chi]$ may simply be too large. ◁

We formulated our definition of a hyperbolic system only for the first-order case. In principle, this suffices, as the methods of Appendix A.3 allow us to rewrite any system as an equivalent first-order system. However, in practice it is often inconvenient to perform an explicit reduction. Analysing the computation leading to (10.33), we note that it is not necessary to assume that the operators P_0, P_1 are first order. Hence Definition 10.4.1 actually applies to any system that is first order in time. This final restriction can also be easily removed.

We consider now a system of the form

$$\frac{\partial^q\mathbf{u}}{\partial t^q} + \sum_{k=0}^{q-1} P_0^{(k)}\frac{\partial^k\mathbf{u}}{\partial t^k} = 0\,, \tag{10.34a}$$

$$\sum_{k=0}^{q-1} P_1^{(k)}\frac{\partial^k\mathbf{u}}{\partial t^k} = 0 \tag{10.34b}$$

where again the differential operators $P_i^{(k)}$ contain only spatial derivatives. For simplicity we assume that the order of $P_i^{(k)}$ is less than or equal to $q-k$ so that we are dealing with a system of order q. Prolonging the constraints (10.34b) with respect to the non-multiplicative variable t yields neither obstructions to involution nor integrability conditions, if and only if a differential operator Q exists such that

$$QP_1^{(k)} = \frac{\partial P_1^{(k)}}{\partial t} + P_1^{(k-1)} - P_1^{(q-1)} P_0^{(k)}\,, \qquad 0 \leq k < q \tag{10.35}$$

(for $k = 0$ we set $P_1^{(-1)} = 0$). In this calculation we have assumed that $P_1^{(q-1)} \neq 0$. If this assumption is not satisfied, then (10.34) cannot be involutive, as prolonging the constraints (10.34b) with respect to t trivially leads to obstructions to involution.

We remarked above that in Definition 10.4.1 it is only important that (10.32) is of first-order in time. Thus we introduce new dependent variables $\mathbf{v}_k = \partial^k \mathbf{u}/\partial t^k$ for $0 \leq k < q$ and obtain as reduced system

$$\frac{\partial \mathbf{v}_{q-1}}{\partial t} + \sum_{k=0}^{q-1} P_0^{(k)} \mathbf{v}_k = 0\,, \tag{10.36a}$$

$$\frac{\partial \mathbf{v}_k}{\partial t} - \mathbf{v}_{k+1} = 0\,, \qquad 0 \leq k \leq q-2\,, \tag{10.36b}$$

$$\sum_{k=0}^{q-1} P_1^{(k)} \mathbf{v}_k = 0\,. \tag{10.36c}$$

It is of the form (10.32) and the two relevant principal symbols are given by

$$T_0[\chi] = \begin{pmatrix} 0 & -1 & 0 & & 0 \\ & & -1 & & \\ & & & \ddots & 0 \\ 0 & & & & -1 \\ T_0^{(0)}[\chi] & & \cdots & & T_0^{(q-1)}[\chi] \end{pmatrix}\,, \tag{10.37a}$$

$$T_1[\chi] = \begin{pmatrix} T_1^{(0)}[\chi] & \cdots & T_1^{(q-1)}[\chi] \end{pmatrix} \tag{10.37b}$$

where $T_i^{(k)}[\chi]$ is the principal symbol of the operator $P_i^{(k)}$. The condition (10.35) ensures that $\ker T_1[\chi]$ is an invariant subspace for $T_0[\chi]$. Thus we may formulate Definition 10.4.1 directly for the higher-order system (10.34) by using the composed principal symbols $T_i[\chi]$ defined by (10.37).

Remark 10.4.5. For square systems, i. e. systems with as many equations as unknown functions, an alternative definition of hyperbolicity exists: such a system is hyperbolic in t-direction, if (i) the one-form $\mathrm{d}t$ is non-characteristic (which ensures that the equation is normal) and (ii) for any non-vanishing form $\chi = \chi_i \mathrm{d}x^i$ the condition $\det T[\lambda \mathrm{d}t + \chi] = 0$ considered as equation for λ possesses only simple real zeros. Note that this definition is independent of the order of the system.

One can show that this approach is equivalent to Definition 10.4.1 upon a reduction to first order. As the proof is quite tedious, we only treat the simplest example, namely a scalar second-order equation $au_{tt} + 2bu_{xt} + cu_{xx} = 0$ (we omit a lower-order part, as it does not affect the result). We obtain

$$\det T[\lambda \mathrm{d}t + \chi \mathrm{d}x] = a\lambda^2 + 2b\lambda\chi + c\chi^2 \tag{10.38}$$

and hence the familiar result that the equation is hyperbolic in t-direction, if and only if $a \neq 0$ and $b^2 > ac$. For the reduction to first order we introduce $v = u_t$ and $w = u_x$. Writing $\mathbf{u} = (u\ v\ w)^t$, the reduced system is

$$\begin{gathered}\mathbf{u}_t + \begin{pmatrix} 0 & 0 & 0 \\ 0 & 2b/a & c/a \\ 0 & -1 & 0 \end{pmatrix} \mathbf{u}_x + \begin{pmatrix} 0 & -1 & 0 \\ 0 & 0 & 0 \\ 0 & 0 & 0 \end{pmatrix} \mathbf{u} = 0\,, \\ \begin{pmatrix} 1 & 0 & 0 \end{pmatrix} \mathbf{u}_x + \begin{pmatrix} 0 & 0 & -1 \end{pmatrix} \mathbf{u} = 0\,.\end{gathered} \tag{10.39}$$

Given this matrix form, one straightforwardly computes that

$$\ker T_1[\chi \mathrm{d}x] = \ker \begin{pmatrix} \chi & 0 & 0 \end{pmatrix} = \left\langle \begin{pmatrix} 0 \\ 1 \\ 0 \end{pmatrix}, \begin{pmatrix} 0 \\ 0 \\ 1 \end{pmatrix} \right\rangle \tag{10.40}$$

and hence

$$T_0[\chi \mathrm{d}x]\big|_{\ker T_1[\chi \mathrm{d}x]} = \begin{pmatrix} 2b/a & c/a \\ -1 & 0 \end{pmatrix}. \tag{10.41}$$

This matrix has real eigenvalues, if and only if $b^2 \geq ac$. However, in the degenerate case $b^2 = ac$ where the two eigenvalues coincide, only a single eigenvector exists. Thus we obtain the same result as above: the reduced system is hyperbolic in t-direction, if and only if $b^2 > ac$.

The same matrix as in (10.41) arises, if we use the above outlined approach where only the time derivatives are reduced to first order. The reduced system is then normal and the relevant principal symbol is described by the transposed matrix of (10.41). ◁

Our final goal in this section will be to prove the existence and uniqueness of *smooth* solutions for a special class of linear systems. By making further structural assumptions we will be able to extend the technique used in the proof of the Cartan–Kähler Theorem 9.4.1 beyond analytic functions, as we no longer need to invoke the Cauchy–Kovalevskaya Theorem.

Definition 10.4.6. An involutive differential equation $\mathcal{R}_q$ in n independent variables is *weakly overdetermined*, if $\beta_q^{(n-1)} > 0$ and $\beta_q^{(k)} = 0$ for $k < n-1$ (i. e. a local representation in δ-regular coordinates contains only equations of class n and $n-1$).

Note that such a weakly overdetermined equation might well be underdetermined in the sense of Definition 7.5.6. In the sequel, we are only interested in equations that can be interpreted in some sense as evolution equations. As above, we slightly change our notation and denote the independent variables by $(x^1, \dots, x^n, t)$, i. e. we have $n+1$ variables and write t instead of x^{n+1}. We study linear systems with variable coefficients of the following form:

$$E\mathbf{u}_t = A^i\mathbf{u}_{x^i} + B\mathbf{u}\,, \tag{10.42a}$$

$$0 = C^i\mathbf{u}_{x^i} + D\mathbf{u}\,. \tag{10.42b}$$

Here $\mathbf{u}$ is as usual the m-dimensional vector of dependent variables. The matrices $E(\mathbf{x},t)$, $A^i(\mathbf{x},t)$ and $B(\mathbf{x},t)$ have $r \leq m$ rows and m columns; the matrices $C^i(\mathbf{x},t)$ and $D(\mathbf{x},t)$ have $s \leq m$ rows. We make the following three assumptions that are crucial for our subsequent analysis.

(i) The hypersurface $t=0$ is non-characteristic.
(ii) $\operatorname{rank} E(\mathbf{x},t) \equiv r$.
(iii) For at least one i we have $\operatorname{rank} C^i(\mathbf{x},t) \equiv s$. Without loss of generality, we assume that this is the case for C^n.

All three assumptions together imply that we are dealing with a weakly overdetermined system in a δ-regular coordinate system. Note that (i) excludes parabolic systems (for any standard definition of parabolicity). (ii) implies that (10.42a) contains all equations of class $n+1$. (iii) ensures that all equations in (10.42b) are of class n and thus, if the system is involutive, it is indeed weakly overdetermined.

Example 10.4.7. An important example of a weakly overdetermined system are *Maxwell's equations*. It is well-known that their evolution part (2.85a), corresponding to (10.42a) forms a hyperbolic system, whereas the constraint equations (2.85b), corresponding to (10.42b), are elliptic. ◁

Our first task is to derive the conditions under which (10.42) represents an involutive system. It requires only a straightforward computation.

Lemma 10.4.8. *The linear system* (10.42) *is involutive, if and only if $s\times r$ matrices $M^i(\mathbf{x},t)$, $N(\mathbf{x},t)$ and $s\times s$ matrices $H^i(\mathbf{x},t)$, $K(\mathbf{x},t)$ exist such that the following equalities hold for all values $1\leq i,j\leq n$:*

$$M^iE = C^i\,, \tag{10.43a}$$

$$H^iC^j + H^jC^i = M^iA^j + M^jA^i\,, \tag{10.43b}$$

$$NE = D - M^kE_{x^k}\,, \tag{10.43c}$$

$$H^iD + KC^i + H^kC^i_{x^k} = M^iB + NA^i + M^kA^i_{x^k} + C^i_t\,, \tag{10.43d}$$

$$KD + H^kD_{x^k} = NB + M^kB_{x^k} + D_t\,. \tag{10.43e}$$

Proof. As the system is weakly overdetermined, non-multiplicative variables appear only in (10.42b), namely t is non-multiplicative for each equation. Thus we must try to express the t-prolongation of (10.42b) as a linear combination of the original equations and their multiplicative prolongations. The matrices M^i, N, H^i, K are nothing but the coefficients of this linear combination. More precisely, a simple calculation shows that the formal equality

$$\partial_t(10.42\text{b}) = M^i\partial_{x^i}(10.42\text{a}) + H^i\partial_{x^i}(10.42\text{b}) + N(10.42\text{a}) + K(10.42\text{b}) \tag{10.44}$$

holds, if and only if the coefficient matrices satisfy the above conditions. □

The first two conditions are necessary for an involutive symbol; the remaining three conditions ensure the formal integrability of the system. Because of the rank assumptions (ii) and (iii), it is not difficult to see that if such matrices M^i, N, H^i, K exist, they are uniquely determined by (10.43). For later use, we derive the compatibility conditions of the linear system (10.42) under the assumption that it is involutive. For this purpose, we add on the right hand side of (10.42a) an inhomogeneity $\boldsymbol{\delta}$ and on the right hand side of (10.42b) an inhomogeneity $-\boldsymbol{\epsilon}$.

Lemma 10.4.9. *The inhomogeneous system possesses a formal solution, if and only if the right hand sides satisfy the compatibility condition*

$$\boldsymbol{\epsilon}_t - H^i\boldsymbol{\epsilon}_{x_i} - K\boldsymbol{\epsilon} = M^i\boldsymbol{\delta}_{x_i} + N\boldsymbol{\delta}\,. \tag{10.45}$$

Proof. (10.45) follows immediately from (10.44). □

In order to obtain an existence and uniqueness theorem for smooth solutions, we need to make some further assumptions on the structure of (10.42). First of all, we want to avoid underdetermined situations. So we assume that $r = m$. This implies together with assumption (ii) that E is a regular matrix and without loss of generality, we take $E(\mathbf{x},t) = \mathbb{1}_m$, the $m \times m$ identity matrix. Then it follows from (10.43a) that $M^i = C^i$.

According to our definition, (10.42a) with $E = \mathbb{1}_m$ is hyperbolic in t-direction at a point $(\mathbf{x},t)$, if for any "spatial" one-form $\chi = \chi_i \mathrm{d}x^i$ (i. e. without a $\mathrm{d}t$-component) the matrix $A_\chi(\mathbf{x},t) = \chi_i A^i(\mathbf{x},t)$ has only real eigenvalues and an eigenbasis. We call it *strongly hyperbolic*, if furthermore there exists for any $A_\chi(\mathbf{x},t)$ a symmetric, positive definite matrix $P_\chi(\mathbf{x},t)$, a *symmetriser*, depending smoothly on $\mathbf{x}$, t and χ such that $P_\chi A_\chi - A_\chi^t P_\chi = 0$. Following Remark 10.3.2, we call (10.42b) *elliptic*, if the matrix $C_\chi(\mathbf{x},t) = \chi_i C^i(\mathbf{x},t)$ defines for all non-vanishing one-forms χ a surjective mapping. This condition implies that *all* matrices C^i must possess everywhere maximal rank: $\operatorname{rank} C^i(\mathbf{x},t) \equiv s$.

Strongly hyperbolic systems possess a well-established existence and uniqueness theory which will form the basis of our results for (10.42). It provides in addition an estimate on the norm of the solution. In the case of a symmetric hyperbolic problem, i. e. $A = A^t$ and we do not need a symmetriser, we take the usual Sobolev norms $\|\mathbf{u}(\mathbf{x})\|_{H^p}$ defined as square root of the sum of the squares of the L^2 norm of $\mathbf{u}(\mathbf{x})$ and all its derivatives up to order p (note we consider here only functions of the spatial variables $\mathbf{x}$).

In the general case we need a kind of weighted Sobolev norm where the "weight" consists of a pseudodifferential operator constructed out of the symmetriser $P_\chi(\mathbf{x},t)$. More precisely, we identify now the one-form χ with a vector $\boldsymbol{\chi} \in \mathbb{R}^n$ whose components are considered as dual variables to $\mathbf{x}$ and introduce the operator

$$\hat{P}(t)v(\mathbf{x}) = \int P_{\chi/|\chi|}(\mathbf{x},t)e^{2\pi i\chi\cdot\mathbf{x}}\hat{v}(\chi)\,\mathrm{d}\chi \tag{10.46}$$

acting on functions depending only on the spatial variables. Here $\hat{v}$ is as usually the Fourier transform of v. Then we define the inner product

$$\big(\mathbf{v}(\mathbf{x}),\mathbf{w}(\mathbf{x})\big)_{P(t)} = \int \mathbf{v}(\mathbf{x})\cdot\big(P(t)\mathbf{w}(\mathbf{x})\big)\,\mathrm{d}\mathbf{x}\,. \tag{10.47}$$

The norm $\|\mathbf{u}(\mathbf{x})\|_{P(t),H^p}$ is then a Sobolev norm where the underlying L^2 norm has been substituted by the norm induced by the inner product (10.47). There are some subtleties involved in the definition of this norm like the domain over which the integrals are taken; in general, one needs a partition of unity on this domain etc; we refer to [266, Sect. 6.2] and references therein for details. Using these norms, we have the following result [266, Theorem 6.2.2].[5]

Theorem 10.4.10. *Let the normal linear system*

$$\mathbf{u}_t = A^i(\mathbf{x},t)\mathbf{u}_{x_i} + B(\mathbf{x},t)\mathbf{u} + \mathbf{F}(\mathbf{x},t) \tag{10.48}$$

have smooth coefficients A^i, B, $\mathbf{F}$ and let it be strongly hyperbolic with the symmetriser P_χ. Then it possesses a unique smooth solution satisfying the smooth initial conditions

$$\mathbf{u}(\mathbf{x},0) = \mathbf{f}(\mathbf{x})\,. \tag{10.49}$$

This solution satisfies at any time $t \in [0,T]$ the estimate

$$\|\mathbf{u}(\cdot,t)\|_{P(t),H^p} \leq K_p\Big[\|\mathbf{f}\|_{P(t),H^p} + \int_0^t \|\mathbf{F}(\cdot,\tau)\|_{P(t),H^p}\,d\tau\Big] \tag{10.50}$$

where the constant K_p depends only on the coefficients of the system, their derivatives up to order p and on T.

Our goal is to extend this theorem to weakly overdetermined systems where the evolutionary part (10.42a) is strongly hyperbolic and the constraints (10.42b) are elliptic. If we could apply Lemma 9.4.2, such a generalisation would be straightforward. However, our equations are not analytic. Thus we must show that the lemma still holds in our more general situation. This can be done with the help of the compatibility condition (10.45) inheriting some properties of (10.42a).

Lemma 10.4.11. *Let the evolutionary system* (10.42a) *with $E = \mathbb{1}_m$ be hyperbolic in t-direction and the constraints* (10.42b) *elliptic. Then the compatibility condition* (10.45)*—considered as a system for $\boldsymbol{\epsilon}$ only—is hyperbolic in t-direction, too.*

Proof. As we assume that the given linear system (10.42) is involutive, the matrices A^i, $M^i = C^i$, H^i satisfy the relations (10.43b). This implies (with obvious notation)

[5] Strictly speaking, the formulation of Theorem 10.4.10 is incomplete. We have not specified a spatial domain Ω in which we consider the differential equation. Kreiss and Lorenz [266] take $\Omega = \mathbb{R}^n$ and assume that all coefficients, initial data and solutions are 1-periodic. The periodicity assumption avoids a specification of the behaviour for $|\mathbf{x}| \to \pm\infty$. But essentially the same result holds, if we assume that all functions are in L^2. Therefore we neglect this issue.

that $H_\chi C_\chi = C_\chi A_\chi$ for any one-form χ. Let $\mathbf{v}$ be an eigenvector of A_χ for the (real) eigenvalue λ. As $H_\chi C_\chi \mathbf{v} = C_\chi A_\chi \mathbf{v} = \lambda C_\chi \mathbf{v}$, we see that $C_\chi \mathbf{v}$ is an eigenvector of H_χ for the eigenvalue λ.

According to our assumptions, all matrices A_χ possess only real eigenvalues and an eigenbasis and all matrices C_χ define surjective mappings. Thus there exists an eigenbasis of H_χ consisting of vectors of the form $C_\chi \mathbf{v}$ where $\mathbf{v}$ is an eigenvector of A_χ. This implies furthermore that all eigenvalues of H_χ are real, as they are also eigenvalues of A_χ. □

Lemma 10.4.12. *Let the assumptions of Lemma 10.4.11 be satisfied and let furthermore* (10.42a) *be strongly hyperbolic with symmetriser $P_\chi(\mathbf{x},t)$. Then the compatibility condition* (10.45) *is strongly hyperbolic, too. As symmetriser we may use the matrix $Q_\chi(\mathbf{x},t)$ defined by*

$$Q_\chi^{-1} = C_\chi P_\chi^{-1} C_\chi^t \,. \tag{10.51}$$

Proof. We must show that Q_χ satisfies the three properties of a symmetriser. Obviously, it inherits the symmetry of P_χ. For the positive definiteness we note that $\mathbf{v}^t Q_\chi^{-1} \mathbf{v} = (C_\chi^t \mathbf{v})^t P_\chi^{-1} (C_\chi^t \mathbf{v})$. As C_χ defines a surjective mapping, its transpose C_χ^t defines an injective mapping. Thus the expression above vanishes only for $\mathbf{v} = 0$. This also implies that Q_χ^{-1} is indeed invertible. Finally, using again (10.43b)

$$\begin{aligned} H_\chi Q_\chi^{-1} - Q_\chi^{-1} H_\chi^t &= H_\chi C_\chi P_\chi^{-1} C_\chi^t - C_\chi P_\chi^{-1} C_\chi^t H_\chi^t \\ &= C_\chi (A_\chi P_\chi^{-1} - P_\chi^{-1} A_\chi^t) C_\chi^t \,. \end{aligned} \tag{10.52}$$

But this last expression vanishes, as P_χ is a symmetriser for A_χ. □

Based on these two technical lemmata, it is now straightforward to extend Theorem 10.4.10 to weakly overdetermined equations. The decisive point in the proof of this generalisation is that Theorem 10.4.10 can also be applied to the compatibility condition (10.45) and implies a uniqueness result for it.

Theorem 10.4.13. *Let* (10.42a) *with $E = \mathbb{1}_m$ be strongly hyperbolic and* (10.42b) *elliptic. If the initial data in* (10.49) *satisfies for $t = 0$ the constraints* (10.42b), *then the initial value problem* (10.42, 10.49) *possesses a unique smooth solution.*

Proof. We first ignore the constraints (10.42b). Then Theorem 10.4.10 guarantees the existence and uniqueness of a smooth solution. Entering this solution into the equations (10.42b) yields residuals $\boldsymbol{\epsilon}$. These residuals must satisfy the compatibility condition (10.45) with $\boldsymbol{\delta} = 0$. As by assumption the initial data satisfy (10.42b) at $t = 0$, our initial conditions for (10.45) are $\boldsymbol{\epsilon}(\mathbf{x},0) = 0$. Obviously, $\boldsymbol{\epsilon} \equiv 0$ is one solution of this initial value problem. According to Lemma 10.4.12, we are dealing with a strongly hyperbolic system, so that we can again apply Theorem 10.4.10. It guarantees that the zero solution is in fact the only solution. Hence there exists a unique smooth solution of the overdetermined problem (10.42, 10.49). □

Our method of proof has as a side effect some implications for the numerical integration of a weakly overdetermined system. A simple approach consists of considering the equations of lower class (10.42b) only as constraints on the initial data and otherwise ignoring them, i. e. one simply solves the initial value problem for (10.42a) with initial data satisfying (10.42b).

For exact solutions this approach works nicely, as we have seen above. For numerical solutions the situation is somewhat different. In general, we must expect a *drift* off the constraints, i. e. after some time the numerical solution ceases to satisfy the equations of lower class (10.42b). This phenomenon is easy to understand on the basis of our results. A numerical solution is only an approximation of an exact solution, hence there appears a non-vanishing residual $\boldsymbol{\delta}$ leading to a non-vanishing right hand side in the compatibility condition (10.45) which determines the evolution of the residual $\boldsymbol{\epsilon}$ in the equations (10.42b). So the size of the drift can be estimated with the help of (10.50):

$$\|\boldsymbol{\epsilon}(\cdot,t)\|_{Q(t),H^p} \leq K_p \int_0^t \|M^i \boldsymbol{\delta}_{x_i}(\cdot,\tau) + N\boldsymbol{\delta}(\cdot,\tau)\|_{Q(t),H^p}\, d\tau \tag{10.53}$$

The problem is aggravated by the fact that the right hand side of the estimate depends not only on the residuals $\boldsymbol{\delta}$ but also on their spatial derivatives. While any reasonable numerical method takes care of controlling the size of $\boldsymbol{\delta}$, it is hardly possible to control the size of the derivatives.

Example 10.4.14. The approach outlined above is fairly popular for numerically solving Maxwell's equations. Here the compatibility conditions are the continuity equations (2.87). It follows from their derivation that for Maxwell's equations the estimate (10.50) depends only on the divergence of $\boldsymbol{\delta}$ and not on $\boldsymbol{\delta}$ itself. Thus a good numerical method for Maxwell's equations should be constructed such that the divergence of the residual vanishes.

Neglecting this effect can lead to numerical problems (see e. g. [217, 236, 333] and references therein). One has shown experimentally that many methods lead to a significant violation of the Gauss laws and it seems that ignoring them yields spurious modes in some computations. The "mimetic" discretisations introduced by Hyman and Shashkov [229] are examples of methods that take care that discrete analogues of the decisive relations are preserved and thus no drift appears. ◁

10.5 Basic Algebraic Analysis

In Section 10.1 we were concerned with a geometric description of linear differential equations. Now we consider them from an algebraic point of view. We introduced the ring of linear differential operators already in Example 3.2.4. As any homogeneous linear differential equation is defined by such an operator, it is natural to study modules over this ring. This idea will automatically lead to a rather abstract generalised notion of solutions.

Let $\mathcal{D} = \mathcal{R}[\partial_1, \dots, \partial_n]$ be a ring of linear differential operators over a differential coefficient ring $\mathcal{R}$ with commuting derivations $\partial_1, \dots, \partial_n$. Many of the algebraic constructions that will follow in the remainder of this chapter are independent of the precise nature of the ring $\mathcal{D}$, i. e. whether or not it is composed of differential operators, but for definiteness and as differential equations are our main interest we will stick to this case. For the moment we only assume that the coefficient ring $\mathcal{R}$ contains a field $\Bbbk$ which is either $\mathbb{R}$ (or some finite extension of $\mathbb{Q}$ for computational purposes) or $\mathbb{C}$, so that a typical choice for $\mathcal{R}$ would be $\Bbbk(x^1,\dots,x^n)$ (which is even a field) or $\Bbbk[x^1,\dots,x^n]$ (which yields the Weyl algebra $\mathcal{W}_n$) and ∂_i represents then of course the usual partial derivative with respect to x^i. A linear differential system of order q with r equations for m unknown functions is defined by a matrix $L \in \mathcal{D}^{r\times m}$:

$$L\mathbf{u} = \sum_{\alpha=1}^{m} L^\beta_\alpha u^\alpha = 0\,, \qquad 1 \leq \beta \leq r \tag{10.54}$$

where $L^\beta_\alpha = \sum_{0\leq|\mu|\leq q} L^\beta_{\alpha\mu} D^\mu \in \mathcal{D}$ with coefficients $L^\beta_{\alpha\mu} \in \mathcal{R}$.

In order to get rid of the dependence on individual matrices, we would like to consider a more intrinsic object, namely a $\mathcal{D}$-module associated with the operator L. If $\mathcal{D}$ were a commutative ring, then this goal could be easily achieved by introducing the $\mathcal{D}$-module homomorphism $\Lambda : \mathcal{D}^m \to \mathcal{D}^r$ with $\Lambda(\mathbf{P}) = L\cdot\mathbf{P}$ and studying either $\operatorname{im}\Lambda$ or $\operatorname{coker}\Lambda$. However, $\mathcal{D}$ is non-commutative and thus Λ as defined is *not* even a left $\mathcal{D}$-module homomorphism (obviously, $\Lambda(Q\mathbf{P}) = L\cdot(Q\mathbf{P}) \neq Q(L\cdot\mathbf{P}) = Q\Lambda(\mathbf{P})$ for arbitrary $Q \in \mathcal{D}$). We resolve this problem by defining Λ via a multiplication with L from the *right*:[6]

$$\Lambda : \begin{cases} \mathcal{D}^{1\times r} & \longrightarrow & \mathcal{D}^{1\times m} \\ \mathbf{P} & \longmapsto & \mathbf{P}\cdot L \end{cases} . \tag{10.55}$$

It is trivial to check that (10.55) indeed defines a left $\mathcal{D}$-module homomorphism. Note that, since we multiply with L from the right, we must treat $\mathbf{P}$ here as a row vector (as indicated by writing $\mathcal{D}^{1\times r}$). Now we may consider the arising cokernel $\mathcal{M} = \operatorname{coker}\Lambda = \mathcal{D}^{1\times m}/\mathcal{D}^{1\times r}L$, i. e. we factor the free module $\mathcal{D}^{1\times m}$ by the left submodule generated by the rows of the matrix L. Obviously, Λ provides us with a finite presentation of $\mathcal{M}$.

We may turn around these considerations and say that a linear differential system *is* a finitely presented left $\mathcal{D}$-module $\mathcal{M}$, as any such module can be written as the cokernel of some homomorphism between two free left $\mathcal{D}$-modules by simply choosing a finite generating set and determining its relations (cf. (B.1)). As we will see below, different presentations of the same module $\mathcal{M}$ correspond to differential equations with isomorphic solution spaces.

In order to be able to speak about solutions, we must first specify some function space in which we look for them. From an abstract point of view we need a

[6] Most textbooks on modules over non-commutative rings insist on using a standard matrix representation where the matrix is to the left of the vector. The simplest solution is then to consider homomorphisms between *right* modules, as for them the matrix must indeed be on the left. For left modules this is only possible, if one works with the *opposite ring* $\mathcal{D}^{\mathrm{op}}$. As a set $\mathcal{D}^{\mathrm{op}} = \mathcal{D}$, but its multiplication $*$ is defined by $f * g = g\cdot f$ where $\cdot$ denotes the multiplication in $\mathcal{D}$. Now obviously any left $\mathcal{D}$-module is a right $\mathcal{D}^{\mathrm{op}}$-module and vice versa. We prefer to avoid the use of either $\mathcal{D}^{\mathrm{op}}$ or right modules and instead simply multiply with L from the right.

(left) $\mathcal{D}$-module $\mathcal{A}$. One natural choice would be $\mathcal{A} = \mathcal{C}^\infty(\Omega, \Bbbk)$, the ring of smooth functions on some domain $\Omega \subseteq \Bbbk^n$ (for $\Bbbk = \mathbb{R}$; for $\Bbbk = \mathbb{C}$ one would consider holomorphic functions), but again much of what we will do is independent of the precise nature of $\mathcal{A}$. Given a matrix-valued linear differential operator $L \in \mathcal{D}^{r\times m}$, we denote by $\ker_{\mathcal{A}} L = \{\mathbf{f} \in \mathcal{A}^m \mid L\mathbf{f} = 0\}$ the solution space of the homogeneous linear system (10.54) with respect to $\mathcal{A}$.

Note that the elements of the $\mathcal{D}$-module $\mathcal{A}$ are not required to be (differentiable) functions in the familiar sense; in fact, the action of $\mathcal{D}$ on $\mathcal{A}$ may be of an arbitrary form not necessarily including a differentiation. Therefore one should speak here of generalised solutions. Important generalisations of classical function spaces include the *micro-* and *hyperfunctions* leading to *microlocal analysis*. We do not go here into further details but refer to the literature [250, 253, 254, 331].

Let $\mathcal{E} = \mathrm{Hom}_{\mathcal{D}}(\mathcal{A}, \mathcal{A})$ be the ring of $\mathcal{D}$-endomorphisms of $\mathcal{A}$. Then $\ker_{\mathcal{A}} L \subseteq \mathcal{A}^m$ possesses a natural structure as an $\mathcal{E}$-module: we simply apply $\phi \in \mathcal{E}$ componentwise. Consider now the space $\mathrm{Hom}_{\mathcal{D}}(\mathcal{M}, \mathcal{A})$. If $\mathcal{D}$ is not a commutative ring (which is generally the case, if we do not deal with systems with constant coefficients), then it is not a $\mathcal{D}$-module, but it is always an $\mathcal{E}$-module with the structure map given by the composition of maps. Central for the algebraic analysis of linear systems is the following simple result known as *Malgrange isomorphism* which allows us to consider generalised solutions as homomorphisms.

Proposition 10.5.1. *Let $L \in \mathcal{D}^{r\times m}$ be a matrix-valued linear differential operator and set $\mathcal{M} = \mathcal{D}^{1\times m}/\mathcal{D}^{1\times r}L$. Then there exists a canonical $\mathcal{E}$-linear isomorphism* $\ker_{\mathcal{A}} L \cong \mathrm{Hom}_{\mathcal{D}}(\mathcal{M}, \mathcal{A})$.

Proof. The definition of $\mathcal{M}$ trivially implies the isomorphism

$$\begin{aligned} \mathrm{Hom}_{\mathcal{D}}(\mathcal{M}, \mathcal{A}) &\cong \bigl\{\phi \in \mathrm{Hom}_{\mathcal{D}}(\mathcal{D}^{1\times m}, \mathcal{A}) \mid \phi(\mathcal{D}^{1\times r}L) = 0\bigr\} \\ &\cong \bigl\{\phi \in \mathrm{Hom}_{\mathcal{D}}(\mathcal{D}^{1\times m}, \mathcal{A}) \mid \forall\beta : \phi(L^\beta) = 0\bigr\}\,. \end{aligned} \tag{10.56}$$

Then we have the obvious isomorphism $\mathrm{Hom}_{\mathcal{D}}(\mathcal{D}^{1\times m}, \mathcal{A}) \cong \mathcal{A}^m$ mapping a homomorphism ϕ into the vector $\mathbf{f}_\phi$ whose components are simply the images of the standard basis: $(\mathbf{f}_\phi)_\alpha = \phi(\mathbf{e}_\alpha) \in \mathcal{A}$ (its inverse maps the vector $\mathbf{f}$ into the homomorphism $\phi_{\mathbf{f}}$ defined by $\phi_{\mathbf{f}}(\mathbf{P}) = \mathbf{P}\cdot\mathbf{f}$). Thus we have $\phi(L^\beta) = L^\beta \cdot \mathbf{f}_\phi$ implying that

$$\mathrm{Hom}_{\mathcal{D}}(\mathcal{M}, \mathcal{A}) \cong \bigl\{\mathbf{f} \in \mathcal{A}^m \mid L\cdot\mathbf{f} = 0\bigr\} = \ker_{\mathcal{A}} L\,. \tag{10.57}$$

It is trivial to verify that all involved maps are $\mathcal{E}$-linear. □

Remark 10.5.2. An alternative proof of Proposition 10.5.1 uses some homological algebra. Since the module $\mathcal{M}$ is defined as the cokernel of the homomorphism Λ introduced in (10.55), we have the exact sequence

$$\mathcal{D}^{1\times r} \xrightarrow{\ \Lambda\ } \mathcal{D}^{1\times m} \xrightarrow{\ \pi\ } \mathcal{M} \longrightarrow 0 \tag{10.58}$$

where π denotes the canonical projection. Since the functor $\mathrm{Hom}_{\mathcal{D}}(\cdot, \mathcal{A})$ is left exact by Proposition B.2.9, the dual sequence

$$0 \longrightarrow \operatorname{Hom}_{\mathcal{D}}(\mathcal{M},\mathcal{A}) \longrightarrow \mathcal{A}^m \xrightarrow{\Lambda^*} \mathcal{A}^r \tag{10.59}$$

is exact, too. Here we exploited again the isomorphism $\operatorname{Hom}_{\mathcal{D}}(\mathcal{D}^{1\times m},\mathcal{A}) \cong \mathcal{A}^m$. It is easy to see that after this identification, the dual map Λ^* is given by applying the matrix operator L. Hence $\operatorname{Hom}_{\mathcal{D}}(\mathcal{M},\mathcal{A})$ is isomorphic to its image in $\mathcal{A}^m$ which by the exactness of the sequence is given by $\ker_{\mathcal{A}} L$. ◁

Remark 10.5.3. One may wonder why we are suddenly able to make statements about the solution space of a differential equation without any assumptions about involution or at least formal integrability. All these notions are hidden behind the definition of the associated module $\mathcal{M}$. In order to obtain a concrete representation of it, we usually need at least a Gröbner basis of the submodule $\mathcal{D}^{1\times r}L \subseteq \mathcal{D}^{1\times m}$, i. e. of the row module of the matrix L. In fact, for many purposes the best representation of the module $\mathcal{M}$ is obtained via a complementary decomposition of this row module, since such a decomposition provides us with an explicit $\mathcal{R}$-linear basis of $\mathcal{M}$ in the case that the coefficient ring $\mathcal{R}$ is actually a field. As discussed in Section 5.1, the determination of such a decomposition is particularly simple, if we know a Pommaret basis (which is always simultaneously a Janet basis) of the row module so that we can apply Algorithm 5.2 (or Proposition 5.1.6). But obviously determining a Pommaret basis of the row module of L is equivalent to completing the linear system $L\mathbf{u} = 0$ to involution (see Section 10.7 for an efficient completion algorithm for linear systems). ◁

We proceed by studying the properties of *inhomogeneous* linear systems. We are particularly interested in the question under which conditions on the right hand sides such systems are solvable. Recall from Remark 7.1.7 that we speak here of *compatibility conditions* and that they correspond to syzygies. Let $L_0 \in \mathcal{D}^{m_1\times m_0}$ be a matrix-valued linear differential operator. We consider the inhomogeneous problem $L_0\mathbf{u} = \mathbf{v}$ where $\mathbf{u} \in \mathcal{A}^{m_0}$ is to be determined and $\mathbf{v} \in \mathcal{A}^{m_1}$ is a given right hand side. As shown in Remark 7.1.7, we cannot expect for an overdetermined operator L_0 that a solution $\mathbf{u}$ exists for every right hand side $\mathbf{v}$. We are interested in necessary and sufficient conditions on $\mathbf{v}$ for the existence of solutions $\mathbf{u}$. It will turn out that deriving necessary conditions is fairly straightforward using the formal theory, while sufficient conditions are much harder to get and depend decisively on the space $\mathcal{A}$ in which we are looking for solutions. We will give such sufficient conditions only for the simple case that we are dealing with a constant coefficients operator L_0 and that $\mathcal{A} = \Bbbk[[x^1,\dots,x^n]]$ is the ring of formal power series.

Definition 10.5.4. The linear differential operator $L_1 \in \mathcal{D}^{m_2\times m_1}$ is a *compatibility operator* for $L_0 \in \mathcal{D}^{m_1\times m_0}$, if $L_1 \cdot L_0 = 0$ and if for any other differential operator $\tilde{L}_1 \in \mathcal{D}^{\tilde{m}_2\times m_1}$ with $\tilde{L}_1 \cdot L_0 = 0$ an operator $\tilde{L} \in \mathcal{D}^{\tilde{m}_2\times m_2}$ exists such that $\tilde{L}_1 = \tilde{L}\cdot L_1$.

In other words, if L_1 is a compatibility operator for L_0, then a necessary condition for the existence of a solution of the inhomogeneous linear system $L_0\mathbf{u} = \mathbf{v}$ for a given right hand side $\mathbf{v}$ is that $L_1\mathbf{v} = 0$. Furthermore, L_1 has the universal property that any other differential operator annihilating L_0 is a multiple of it. Obviously, the first property in Definition 10.5.4 implies that

$$\mathcal{A}^{m_0} \xrightarrow{L_0 \cdot} \mathcal{A}^{m_1} \xrightarrow{L_1 \cdot} \mathcal{A}^{m_2} \tag{10.60}$$

is a complex of $\mathcal{D}$-modules.

Example 10.5.5. Depending on the choice of the space $\mathcal{A}$, some linear operators do not possess a compatibility operator. Consider for example for $\mathcal{R} = \mathcal{A} = \mathcal{C}^\infty(\mathbb{R})$ the trivial operator L_0 defined by multiplication with a smooth function $f \in \mathcal{R}$ satisfying $f(x) = 0$ for all $x \leq 0$ and $f(x) > 0$ for all $x > 0$ (a classical instance of such a function can be found in (A.12)). Assume now that L_1 were a compatibility operator for L_0. Obviously, this property implies that L_1 must vanish on the half line $x \geq 0$. Hence we can form the operator $\tilde{L}_1 = \frac{1}{x} L_1$ which also satisfies $\tilde{L}_1 \cdot L_0 = 0$. However, it is not difficult to see that no differential operator $\tilde{L} \in \mathcal{R}[\partial]$ exists such that $\tilde{L}_1 = \tilde{L} \cdot L_1$ and therefore no compatibility operator exists. ◁

Assuming that a compatibility operator L_1 exists, we may iterate the above construction and search for a compatibility operator of L_1 and so on. The iteration leads to a sequence of the form

$$0 \longrightarrow \ker_{\mathcal{A}} L_0 \longrightarrow \mathcal{A}^{m_0} \xrightarrow{L_0 \cdot} \mathcal{A}^{m_1} \xrightarrow{L_1 \cdot} \mathcal{A}^{m_2} \longrightarrow \cdots . \tag{10.61}$$

One speaks of a *compatibility* or *Janet sequence* for the linear operator L_0 (the latter term is often reserved for the case that the sequence is exact).

Example 10.5.6. A simple but classical instance of a compatibility sequence is formed by the basic operators of (three-dimensional) vector analysis: gradient, curl and divergence. Let us start with the differential equation $\operatorname{grad} u = \nabla u = 0$. If we add a right hand side, we obtain the system

$$u_x = v^1 , \quad u_y = v^2 , \quad u_z = v^3 . \tag{10.62}$$

It is trivial to check that the homogeneous system is involutive. A necessary condition for the existence of solutions of this inhomogeneous system is that the three functions v^α appearing on the right hand side satisfy the differential equation

$$v^2_z - v^3_y = 0 , \quad v^3_x - v^1_z = 0 , \quad v^1_y - v^2_x = 0 , \tag{10.63}$$

in other words, that $\operatorname{curl} \mathbf{v} = \nabla \times \mathbf{v} = 0$. Obviously, this fact corresponds to the well-known identity $\operatorname{curl} \circ \operatorname{grad} = 0$.

Now we may add a right hand side to (10.63), i. e. we consider $\operatorname{curl} \mathbf{v} = \mathbf{w}$. This differential equation is solvable, if and only if $\mathbf{w}$ satisfies

$$w^1_x + w^2_y + w^3_z = 0 . \tag{10.64}$$

Of course, this fact means nothing but $\operatorname{div} \mathbf{w} = \nabla \cdot \mathbf{w} = 0$ and we recover the next classical identity $\operatorname{div} \circ \operatorname{curl} = 0$.

We may represent these identities in the following sequence

$$\begin{aligned} 0 \longrightarrow \mathbb{R} \longrightarrow \mathcal{C}^\infty(\mathbb{R}^3,\mathbb{R}) \xrightarrow{\text{grad}} \mathcal{C}^\infty(\mathbb{R}^3,\mathbb{R}^3) \xrightarrow{\text{curl}} \\ \xrightarrow{\text{curl}} \mathcal{C}^\infty(\mathbb{R}^3,\mathbb{R}^3) \xrightarrow{\text{div}} \mathcal{C}^\infty(\mathbb{R}^3,\mathbb{R}) \longrightarrow 0\,. \end{aligned} \tag{10.65}$$

The term $\mathbb{R}$ at the left end represents the constant functions on $\mathbb{R}^3$ which trivially form the kernel of the gradient.

Readers familiar with differential geometry know that (10.65) is nothing but a special case of the *de Rham complex*. Let $\mathcal{M}$ be an n-dimensional manifold. The vector bundles $\Omega^k(\mathcal{M})$ of differential k-forms (cf. Appendix C.2) form together with the exterior derivative d as differential the complex

$$0 \longrightarrow \mathbb{R} \longrightarrow \Omega^0(\mathcal{M}) \xrightarrow{d} \Omega^1(\mathcal{M}) \xrightarrow{d} \cdots \xrightarrow{d} \Omega^n(\mathcal{M}) \longrightarrow 0\,. \tag{10.66}$$

If we consider global forms, then it is in general not exact and its cohomology contains important topological information on the manifold $\mathcal{M}$. Locally, the sequence (10.66) is exact by the Poincaré Lemma (see Remark C.2.9). ◁

Some obvious questions are when such a compatibility sequence (10.61) exists, how one can construct it and when it will be exact. If the sequence is exact, one calls the operator L_0 a *parametrisation* of L_1, as any element of $\ker_{\mathcal{A}} L_1$ can be represented in the form $L_0\mathbf{u}$ for some $\mathbf{u} \in \mathcal{A}^{m_0}$ (since $\ker_{\mathcal{A}} L_1 = \operatorname{im}_{\mathcal{A}} L_0$). Thus in this case the differential equation $L_1\mathbf{v} = 0$ is also a sufficient condition for the solvability of the inhomogeneous system $L_0\mathbf{u} = \mathbf{v}$; this fact is often called the *fundamental principle*. A natural approach (following the ideas of Remark 10.5.2) consists of computing first a free resolution

$$\cdots \xrightarrow{\cdot L_2} \mathcal{D}^{1\times m_2} \xrightarrow{\cdot L_1} \mathcal{D}^{1\times m_1} \xrightarrow{\cdot L_0} \mathcal{D}^{1\times m_0} \xrightarrow{\pi} \mathcal{M} \longrightarrow 0 \tag{10.67}$$

where as usual $\mathcal{M} = \mathcal{D}^{1\times m_1}/\mathcal{D}^{1\times m_0}L_0$. For many practically relevant coefficient rings $\mathcal{R}$ such a computation can be effectively done using Gröbner (or involutive) bases (cf. B.2 and B.4 or Section 5.4). Once a free resolution has been found, we can dualise it by applying the contravariant functor $\operatorname{Hom}_{\mathcal{D}}(\cdot,\mathcal{A})$ and obtain a complex of the form (10.61). The next result shows that, in the Noetherian case, satisfaction of the fundamental principle is equivalent to the injectivity of the function space $\mathcal{A}$, thus we are dealing here with a purely algebraic property.

Proposition 10.5.7. *Let $\mathcal{D}$ be a Noetherian ring. Then the fundamental principle holds over the $\mathcal{D}$-module $\mathcal{A}$, i. e. a necessary and sufficient condition for the solvability of the inhomogeneous system $L_0\mathbf{u} = \mathbf{v}$ over $\mathcal{A}$ is $L_1\mathbf{v} = 0$, if and only if $\mathcal{A}$ is an injective module.*

Proof. If $\mathcal{A}$ is injective, then the functor $\operatorname{Hom}_{\mathcal{D}}(\cdot,\mathcal{A})$ is exact according to Proposition B.2.11. Thus in this case it follows immediately from the exactness of the free resolution (10.67) that the dual sequence (10.61) is exact, too.

For the converse we apply Baer's Criterion (Proposition B.2.7). Let $\mathcal{I} \subseteq \mathcal{D}$ be an ideal. Since $\mathcal{D}$ is Noetherian, we can find a finite generating set: $\mathcal{I} = \langle Q_1, \dots, Q_p \rangle$. If $\phi : \mathcal{I} \to \mathcal{A}$ is a homomorphism, then we consider the elements $v_i = \phi(Q_i) \in \mathcal{A}$. Baer's Criterion is satisfied (and hence $\mathcal{A}$ injective), if we can find a function $u \in \mathcal{A}$ such that $Q_i u = v_i$ for $1 \leq i \leq p$.

Obviously, such a u is nothing but a solution of an overdetermined linear system. Since we are assuming that the fundamental principle holds over $\mathcal{A}$, a solution will exist if and only if the right hand sides v_i satisfy the compatibility conditions of this system. Let $\mathbf{S}_1, \dots, \mathbf{S}_r \in \mathcal{D}^p$ be a basis of the syzygy module $\mathrm{Syz}(\mathcal{I})$, i. e. we have $\sum_{i=1}^p S_{ji} Q_i = 0$. As ϕ is a homomorphism, we have

$$0 = \phi\Big(\sum_{i=1}^{p} S_{ji} Q_i\Big) = \sum_{i=1}^{p} S_{ji} \phi(Q_i) = \sum_{i=1}^{p} S_{ji} v_i \,. \tag{10.68}$$

Hence the right hand sides v_i satisfy the compatibility conditions and a function $u \in \mathcal{A}$ as required exists. □

Of course, Proposition 10.5.7 does not really answer our questions, but only replaces them with a new one: which $\mathcal{D}$-modules $\mathcal{A}$ are injective? It is highly non-trivial to provide an answer for practically relevant function spaces. We will consider here only the simplest case that the coefficient ring $\mathcal{R}$ is a field $\mathbb{k}$ and that $\mathcal{A} = \mathbb{k}[[x^1, \dots, x^n]]$, i. e. we restrict to formal power series solutions of linear systems with constant coefficients.

Proposition 10.5.8. *The fundamental principle holds for formal power series solutions of linear systems with constant coefficients.*

Proof. The above made assumptions imply that both $\mathcal{D}$ and $\mathcal{A}$ are $\mathbb{k}$-linear spaces. Furthermore, we have a canonical isomorphism between the $\mathbb{k}$-linear dual space $\mathcal{D}^* = \mathrm{Hom}_{\mathbb{k}}(\mathcal{D}, \mathbb{k})$ and the function space $\mathcal{A}$ given by $\phi \mapsto \sum_{\mu \in \mathbb{N}_0^n} \frac{1}{\mu!} \phi(\partial^\mu) x^\mu$. Since for a field $\mathbb{k}$ the functor $\mathrm{Hom}_{\mathbb{k}}(\cdot, \mathbb{k})$ is always exact by Proposition B.2.9, its application to (10.67) provides us with an exact sequence of the form

$$0 \longrightarrow \mathrm{Hom}_{\mathbb{k}}(\mathcal{M}, \mathbb{k}) \longrightarrow \mathrm{Hom}_{\mathbb{k}}(\mathcal{D}^{1 \times m_0}, \mathbb{k}) \longrightarrow \cdots \,. \tag{10.69}$$

Now we can apply Remark B.1.8 according to which we have an isomorphism $\mathrm{Hom}_{\mathbb{k}}(\mathcal{M}, \mathbb{k}) \cong \mathrm{Hom}_{\mathcal{D}}(\mathcal{M}, \mathcal{D}^*)$ and similar for the other modules in (10.69). Entering the above introduced isomorphism $\mathcal{D}^* \cong \mathcal{A}$ yields then an exact compatibility sequence of the form (10.61). □

Example 10.5.9. We return to Example 10.5.6 above. Here $L_0 = \mathrm{grad} = (\partial_x \; \partial_y \; \partial_z)^t$. Obviously, the entries of L_0 form a Pommaret basis of the ideal they generate in $\mathcal{D} = \mathbb{k}[\partial_x, \partial_y, \partial_z]$. Thus we can apply Theorem 5.4.12 in order to determine a free resolution of minimal length (in fact, according to Theorem 5.5.8 we even get the minimal resolution, as our ideal is trivially stable). According to Proposition 10.5.8, we obtain thus the exactness of (10.65) hat the level of formal power series. ◁

Injectivity of $\mathcal{A}$ ensures that exactness of (10.67) implies the exactness of (10.61). The converse is not necessarily true; it requires that the functor $\mathrm{Hom}_{\mathcal{D}}(\cdot,\mathcal{A})$ is faithful. According to Remark B.2.22, this is the case for an injective module $\mathcal{A}$, if and only if $\mathcal{A}$ is in addition a *cogenerator* (see Definition B.2.18). From the point of view of differential equations, this property is of great interest, since we will now show that for injective cogenerators we have a bijective correspondence between finitely generated $\mathcal{D}$-modules and the solutions spaces of linear differential operators. In other words, the solution space defines uniquely (up to isomorphisms) the corresponding $\mathcal{D}$-module. Loosely speaking, the cogenerator property means that the chosen function space $\mathcal{A}$ is large enough.

The action of $\mathcal{D}$ on $\mathcal{A}$ allows us to introduce for each $m \in \mathbb{N}$ a pairing

$$(\cdot,\cdot):\begin{cases}\mathcal{D}^m \times \mathcal{A}^m & \longrightarrow \quad \mathcal{A}\\ \mathbf{P}\times\mathbf{u} & \longmapsto \quad (\mathbf{P},\mathbf{u}) = \sum_{i=1}^m P_i u_i\end{cases} \tag{10.70}$$

which is $\mathcal{D}$-bilinear and $\mathcal{E}$-linear in the second argument. With its help we may define "orthogonal" complements: if $\mathcal{T} \subseteq \mathcal{D}^m$ and $\mathcal{B} \subseteq \mathcal{A}^m$ are arbitrary subsets, then we set $\mathcal{T}^\perp = \{\mathbf{u} \in \mathcal{A}^m \mid (\mathcal{T},\mathbf{u}) = 0\}$ and $\mathcal{B}^\perp = \{\mathbf{P} \in \mathcal{D}^m \mid (\mathbf{P},\mathcal{B}) = 0\}$. Obviously, we have $\mathcal{T}^\perp = \langle\mathcal{T}\rangle_{\mathcal{D}}^\perp$ and $\mathcal{B}^\perp = \langle\mathcal{B}\rangle_{\mathcal{E}}^\perp$, i. e. the complements depend only on the submodules generated by $\mathcal{T}$ and $\mathcal{B}$, respectively.

If we denote by $\mathbb{P}_{\mathcal{D}}(\mathcal{D}^m)$ and $\mathbb{P}_{\mathcal{E}}(\mathcal{A}^m)$ the lattice of all $\mathcal{D}$- respectively $\mathcal{E}$-submodules of $\mathcal{D}^m$ respectively $\mathcal{A}^m$ partially ordered by inclusion, then the operation of taking the complement with respect to the above pairing clearly defines a correspondence $\cdot^\perp : \mathbb{P}_{\mathcal{D}}(\mathcal{D}^m) \leftrightarrow \mathbb{P}_{\mathcal{E}}(\mathcal{A}^m)$, i. e. the maps revert inclusions and we have $\mathcal{T} \subseteq \mathcal{T}^{\perp\perp}$ and $\mathcal{B} \subseteq \mathcal{B}^{\perp\perp}$ for all $\mathcal{T} \in \mathbb{P}_{\mathcal{D}}(\mathcal{D}^m)$ and all $\mathcal{B} \in \mathbb{P}_{\mathcal{E}}(\mathcal{A}^m)$.

Theorem 10.5.10. *If the $\mathcal{D}$-module $\mathcal{A}$ is an injective cogenerator, then the maps $\cdot^\perp$ define a bijective correspondence between the lattice $\mathbb{P}_{\mathcal{D}}(\mathcal{D}^m)$ and the sublattice $\mathcal{K} = \{\mathcal{B} \in \mathbb{P}_{\mathcal{E}}(\mathcal{A}^m) \mid \exists r \in \mathbb{N}, L \in \mathcal{D}^{r\times m} : \mathcal{B} = \ker_{\mathcal{A}} L\}$ consisting of all submodules which are the solution space of some linear differential operator.*

Proof. We first show that under the made assumption on $\mathcal{A}$ the equality $\mathcal{T} = \mathcal{T}^{\perp\perp}$ holds for any submodule $\mathcal{T} \in \mathbb{P}_{\mathcal{D}}(\mathcal{D}^m)$. Since $\mathcal{T} \subseteq \mathcal{T}^{\perp\perp}$, it follows immediately that $\mathcal{T}^\perp = \mathcal{T}^{\perp\perp\perp}$. Consider now the exact sequence

$$0 \longrightarrow \ker\alpha \longrightarrow \mathcal{D}^m/\mathcal{T} \stackrel{\alpha}{\longrightarrow} \mathcal{D}^m/\mathcal{T}^{\perp\perp} \tag{10.71}$$

where α is the canonical map induced by the inclusion $\mathcal{T} \subseteq \mathcal{T}^{\perp\perp}$. Since the functor $\mathrm{Hom}_{\mathcal{D}}(\cdot,\mathcal{A})$ is left exact by Proposition B.2.9, it yields an exact sequence

$$0 \longleftarrow \mathrm{Hom}_{\mathcal{D}}(\ker\alpha,\mathcal{A}) \longleftarrow \mathrm{Hom}_{\mathcal{D}}(\mathcal{D}^m/\mathcal{T},\mathcal{A}) \longleftarrow \mathrm{Hom}_{\mathcal{D}}(\mathcal{D}^m/\mathcal{T}^{\perp\perp},\mathcal{A})\,. \tag{10.72}$$

Similar to Proposition 10.5.1, one can show the existence of an isomorphism $\mathrm{Hom}_{\mathcal{D}}(\mathcal{D}^m/\mathcal{T},\mathcal{A}) \cong \mathcal{T}^\perp$ which maps a homomorphism ϕ to $\phi\circ\pi \in \mathrm{Hom}_{\mathcal{D}}(\mathcal{D}^m,\mathcal{A})$ with the canonical projection $\pi : \mathcal{D}^m \to \mathcal{D}^m/\mathcal{T}$ and then exploits the natural isomorphy $\mathrm{Hom}_{\mathcal{D}}(\mathcal{D}^m,\mathcal{A}) \cong \mathcal{A}^m$. Combining these results yields an isomorphism

$$\mathrm{Hom}_{\mathcal{D}}(\mathcal{D}^m/\mathcal{T},\mathcal{A}) \xrightarrow{\sim} \mathcal{T}^{\perp} = \mathcal{T}^{\perp\perp\perp} \xrightarrow{\sim} \mathrm{Hom}_{\mathcal{D}}(\mathcal{D}^m/\mathcal{T}^{\perp\perp},\mathcal{A})\,. \tag{10.73}$$

By construction, this combined map is nothing but the inverse of $\mathrm{Hom}_{\mathcal{D}}(\alpha,\mathcal{A})$. Thus the latter map is actually an isomorphism and it follows from the exact sequence (10.72) that $\mathrm{Hom}_{\mathcal{D}}(\ker\alpha,\mathcal{A}) = 0$. Since we assume that the module $\mathcal{A}$ is an injective cogenerator, this fact implies by Remark B.2.22 that $\ker\alpha = 0$, too, and thus that α is an isomorphism or equivalently that $\mathcal{T} = \mathcal{T}^{\perp\perp}$.

Now assume that $\mathcal{B} \in \mathcal{K}$. Obviously, the condition $\mathcal{B} = \ker_{\mathcal{A}} L$ is equivalent to $\mathcal{B} = \mathcal{T}^{\perp}$ where $\mathcal{T} = \mathcal{D}^{1\times r}L$ is the row module of the operator L. Thus we also have $\mathcal{B}^{\perp\perp} = \mathcal{T}^{\perp\perp\perp} = \mathcal{T}^{\perp} = \mathcal{B}$ and the two maps $\cdot^{\perp}$ are inverse to each other. □

Corollary 10.5.11. *Let $\mathcal{A}$ be an injective cogenerator. Two differential operators $L_1 \in \mathcal{D}^{r_1\times m}$, $L_2 \in \mathcal{D}^{r_2\times m}$ satisfy $\ker_{\mathcal{A}} L_1 = \ker_{\mathcal{A}} L_2$, if and only if they possess the same row module (or, equivalently, if there are matrices $X \in \mathcal{D}^{r_1\times r_2}$, $Y \in \mathcal{D}^{r_2\times r_1}$ such that $L_1 = YL_2$ and $L_2 = XL_1$).*

Remark 10.5.12. It is highly non-trivial to prove for function spaces of practical relevance that they are injective cogenerators. Important examples of spaces possessing this property are:

(i) the ring $\mathbb{C}\{x^1,\dots,x^n\}$ of convergent power series;
(ii) the ring $\mathcal{O}(\mathbb{C}^n)$ of entire functions;
(iii) the ring of smooth complex-valued functions $\mathcal{C}^\infty(\mathbb{R}^n,\mathbb{C})$;
(iv) the ring of complex-valued distributions $\mathcal{D}'(\mathbb{R}^n,\mathbb{C})$.

The injectivity of all these $\mathcal{D}$-modules (with the obvious action of $\mathcal{D}$ via differentiation) was proven by Ehrenpreis [118, Chapt. 5] and Palamodov [348, Chapt. VII, §8] (here even the more general case of functions defined on an open convex subset Ω of $\mathbb{R}^n$ or $\mathbb{C}^n$, respectively, is considered) already in 1970. The cogenerator property was shown much later (1990) by Oberst [337, Sect. 4]. ◁

10.6 The Inverse Syzygy Problem

In the last section we discussed compatibility operators. Our starting point was a linear differential operator $L_0 \in \mathcal{D}^{m_1\times m_0}$ and we looked for another differential operator $L_1 \in \mathcal{D}^{m_2\times m_1}$ such that the complex (10.60) is exact. Now we turn our attention to the inverse problem: given an operator L_1, decide whether or not it is the compatibility operator of another differential operator L_0. Recall that we called such an operator L_0 a parametrisation of L_1.

In the sequel, we will assume that the function space $\mathcal{A}$ is an injective cogenerator so that the functor $\mathrm{Hom}_{\mathcal{D}}(\cdot,\mathcal{A})$ is exact and faithful. According to the discussion in the last section, an exact compatibility sequence over $\mathcal{A}$ is then dual to a free resolution over $\mathcal{D}$, hence it suffice to treat the question as a syzygy problem for $\mathcal{D}$-modules, as the dualisation reverts arrows.

Besides its obvious theoretical interest, the inverse problem is of considerable importance in practice. First of all, if we can construct a parametrisation L_0, then we have effectively solved the homogeneous equation $L_1\mathbf{u} = 0$, as any solution of it is of the form $\mathbf{u} = L_0\mathbf{v}$ and conversely any function $\mathbf{u} = L_0\mathbf{v}$ is a solution of it. Secondly, the inverse problem is of relevance in many fields like mathematical physics or algebraic systems theory (see the Addendum at the end of this section). In a physical context one usually speaks about *potentials* instead of parametrisations; we will see some concrete examples below.

As it actually makes the theory simpler, we will now take for $\mathcal{D}$ an arbitrary coherent ring (not necessarily of polynomial type). The precise formulation of the inverse syzygy problem goes then as follows: given a (left) $\mathcal{D}$-module homomorphism $\beta : \mathcal{D}^r \to \mathcal{D}^m$ between two free $\mathcal{D}$-modules of finite rank, does there exists a further homomorphism $\gamma : \mathcal{D}^m \to \mathcal{D}^s$ such that $\operatorname{im}\beta = \ker\gamma$? By analogy to the case of linear differential operators, we call such a map γ a *parametrisation* of β. In the sequel we will show how one can effectively decide whether or not a parametrisation exists and how it can be explicitly computed, if it exists.

The basic observation is that a syzygy computation "goes in the wrong direction": it determines a map $\alpha : \mathcal{D}^p \to \mathcal{D}^r$ with $\operatorname{im}\alpha = \ker\beta$. Thus we must revert the directions of the arrows which can be achieved by a *dualisation*, i. e. by applying the functor $\operatorname{Hom}_{\mathcal{D}}(\cdot, \mathcal{D})$. As usual, we denote for a left $\mathcal{D}$-module $\mathcal{M}$ the dual module $\operatorname{Hom}_{\mathcal{D}}(\mathcal{M}, \mathcal{D})$ by $\mathcal{M}^*$ and for a homomorphism $\alpha : \mathcal{M} \to \mathcal{N}$ the dual map by $\alpha^* : \mathcal{N}^* \to \mathcal{M}^*$. By Remark B.1.7, $\mathcal{M}^*$ is always a right $\mathcal{D}$-module.

These considerations lead to the simple Algorithm 10.1. All maps and modules appearing in its course are sketched in the following diagram

$$\begin{array}{ccccc} \mathcal{D}^{\hat{r}} & \xrightarrow{\hat{\beta}} & & & \\ & & \mathcal{D}^m & \xrightarrow{\gamma} & \mathcal{D}^s \\ \mathcal{D}^r & \xrightarrow{\beta} & \updownarrow & & \updownarrow \\ \updownarrow & & & & \\ (\mathcal{D}^r)^* & \xleftarrow{\beta^*} & (\mathcal{D}^m)^* & \xleftarrow{\gamma^*} & (\mathcal{D}^s)^* \end{array} \tag{10.74}$$

where dotted lines denote dualisations. The condition in Line /2/ implies that $\beta^* \circ \gamma^* = (\gamma \circ \beta)^* = 0$ and thus $\gamma \circ \beta = 0$. But although $\operatorname{im}\gamma^* = \ker\beta^*$, we cannot conclude that the equality $\operatorname{im}\beta = \ker\gamma$ holds (according to Proposition B.2.11 the functor $\operatorname{Hom}_{\mathcal{D}}(\cdot, \mathcal{D})$ is not exact, as free $\mathcal{D}$-modules are not necessarily injective). Thus we need Line /4/ in order to verify that the map γ computed in Line /3/ is indeed a parametrisation. Note that for the correctness of Algorithm 10.1 it is indeed necessary that the ring $\mathcal{D}$ is both left and right coherent: we need the left coherence to guarantee that $\ker\gamma$ and the right coherence that $\ker\beta^*$ is finitely generated.

If $\operatorname{im}\beta = \operatorname{im}\hat{\beta}$, then γ is clearly a parametrisation. It is less obvious that no parametrisation exists, if $\operatorname{im}\beta \subsetneq \operatorname{im}\hat{\beta}$. We will characterise the solvability of the inverse syzygy problem via the question whether the cokernel $\operatorname{coker}\beta$ is torsionless and provide simultaneously a correctness proof for Algorithm 10.1. Recall from Remark B.1.7 that a module $\mathcal{M}$ is called torsionless, if the natural homomorphism

Algorithm 10.1 Parametrisation test

Input: map $\beta : \mathcal{D}^r \to \mathcal{D}^m$
Output: if it exists map $\gamma : \mathcal{D}^m \to \mathcal{D}^s$ such that $\operatorname{im}\beta = \ker\gamma$ otherwise `FAIL`

compute dual map β^*
compute $\gamma^* : (\mathcal{D}^s)^* \to (\mathcal{D}^m)^*$ with $\operatorname{im}\gamma^* = \ker\beta^*$
compute dual map $\gamma : \mathcal{D}^m \to \mathcal{D}^s$
compute map $\hat{\beta} : \mathcal{D}^{\hat{r}} \to \mathcal{D}^m$ such that $\operatorname{im}\hat{\beta} = \ker\gamma$
if $\operatorname{im}\hat{\beta} = \operatorname{im}\beta$ **then**
    return γ
else
    return `FAIL`
end if

$\eta_{\mathcal{M}} : \mathcal{M} \to \mathcal{M}^{**}$ connecting $\mathcal{M}$ and its bidual is injective. The two for us relevant properties of a torsionless module are collected in the following lemma.

Lemma 10.6.1. *Let $\phi : \mathcal{M} \to \mathcal{N}$ be a homomorphism of left $\mathcal{D}$-modules such that its dual map $\phi^* : \mathcal{N}^* \to \mathcal{M}^*$ is surjective.*

(i) *If $\mathcal{M}$ is torsionless, then ϕ is injective.*
(ii) *If $\mathcal{N}$ is torsionless, then* $\ker\eta_{\mathcal{M}} = \ker\phi$.

Proof. Consider the following commutative diagram

$$\begin{array}{ccc} \mathcal{M} & \xrightarrow{\phi} & \mathcal{N} \\ {\scriptstyle\eta_{\mathcal{M}}}\downarrow & & \downarrow{\scriptstyle\eta_{\mathcal{N}}} \\ \mathcal{M}^{**} & \xrightarrow[\phi^{**}]{} & \mathcal{N}^{**} \end{array} \tag{10.75}$$

where the bidual map ϕ^{**} is trivially injective, since we assume here that the dual map ϕ^* is surjective.

In the case of the first assertion, $\eta_{\mathcal{M}}$ is injective, too, as the module $\mathcal{M}$ is assumed to be torsionless. Hence the composed map $\phi^{**} \circ \eta_{\mathcal{M}} = \eta_{\mathcal{N}} \circ \phi$ is injective which is only possible, if ϕ is injective.

For the second assertion, we note that the injectivity of ϕ^{**} implies that

$$\ker\eta_{\mathcal{M}} = \ker(\phi^{**} \circ \eta_{\mathcal{M}}) = \ker(\eta_{\mathcal{N}} \circ \phi) = \ker\phi \tag{10.76}$$

where the last equality follows from the assumption that this time the module $\mathcal{N}$ is torsionless and hence $\eta_{\mathcal{N}}$ injective. □

Theorem 10.6.2. *Let $\mathcal{D}$ be a coherent ring and $\beta : \mathcal{D}^r \to \mathcal{D}^m$ a homomorphism of finitely generated free $\mathcal{D}$-modules. If the map $\hat{\beta} : \mathcal{D}^{\hat{r}} \to \mathcal{D}^m$ is constructed by Algorithm 10.1 above, then the following three statements are equivalent.*

(i) *The left $\mathcal{D}$-module* $\operatorname{coker}\beta = \mathcal{D}^m / \operatorname{im}\beta$ *is torsionless.*
(ii) *A left $\mathcal{D}$-module homomorphism $\bar{\gamma} : \mathcal{D}^m \to \mathcal{D}^{\bar{s}}$ exists with* $\operatorname{im}\beta = \ker\bar{\gamma}$.

(iii) *The equality* $\operatorname{im}\beta = \operatorname{im}\hat{\beta}$ *holds.*

Proof. (iii) trivially implies (ii); we simply choose $\bar{\gamma} = \gamma$. Assume that (ii) holds and consider the module homomorphism $\phi : \operatorname{coker}\beta \to \mathcal{D}^{\bar{s}}$ given by $\phi([\mathbf{P}]) = \bar{\gamma}(\mathbf{P})$ where $[\mathbf{P}]$ denotes the equivalence class of $\mathbf{P} \in \mathcal{D}^m$ in $\operatorname{coker}\beta$. The map ϕ is well-defined as $\operatorname{im}\beta = \ker\bar{\gamma}$ and obviously injective. Thus $\operatorname{coker}\beta$ is isomorphic to a submodule of a free module and trivially torsionless so that (ii) implies (i).

There remains to show that (i) implies (iii). Since $\gamma\circ\beta = 0$, the homomorphism γ can be decomposed as $\gamma = \phi\circ\pi$ with $\pi : \mathcal{D}^m \to \operatorname{coker}\beta$ the canonical projection and $\phi : \operatorname{coker}\beta \to \mathcal{D}^s$ the induced map. Dually, we obtain $\gamma^* = \pi^*\circ\phi^*$. By construction, we have on one side $\ker\beta^* = \operatorname{im}\gamma^*$ and on the other side $\ker\beta^* = \operatorname{im}\pi^*$. Hence $\operatorname{im}\gamma^* = \operatorname{im}\pi^*$ and since π^* is trivially injective, ϕ^* must be surjective. Now it follows from the first assertion in Lemma 10.6.1 that ϕ is injective and hence $\operatorname{im}\beta = \ker\pi = \ker(\phi\circ\pi) = \ker\gamma = \operatorname{im}\hat{\beta}$. □

Remark 10.6.3. As a free module, $\mathcal{D}^s$ is trivially torsionless. Using again the decomposition $\gamma = \phi\circ\pi$ from the proof above and the fact that the dual map ϕ^* is surjective, $\ker\eta_{\mathcal{M}} = \ker\phi$ by the second assertion in Lemma 10.6.1. Furthermore, we have the trivial isomorphism $\ker\phi \cong \ker\gamma/\ker\pi = \operatorname{im}\hat{\beta}/\operatorname{im}\beta$. Hence Algorithm 10.1 allows us to determine explicitly $\ker(\eta_{\operatorname{coker}\beta})$ and so provides us with an effective test for torsionlessness of any finitely presented module $\operatorname{coker}(\beta)$. ◁

For the concrete computational realisation of Algorithm 10.1, we use matrix representations of the maps involved. Following the discussion at the beginning of the last section, we consider now β as a map $\mathcal{D}^{1\times r} \to \mathcal{D}^{1\times m}$ defined by a matrix $B \in \mathcal{D}^{r\times m}$ as $\beta(\mathbf{P}) = \mathbf{P}B$. Using the natural isomorphism $(\mathcal{D}^{1\times m})^* \cong \mathcal{D}^m$ identifying each element of the dual module with its image on the standard basis, it is easy to see that we may then consider β^* as the map $\mathcal{D}^m \to \mathcal{D}^r$ given by $\beta^*(\mathbf{Q}) = B\mathbf{Q}$ (as we are now dealing with a right module homomorphism, we must indeed multiply from the left).[7] Thus the first and the third step of Algorithm 10.1 are trivial.

Line /2/ requires to compute the solution space $\ker_{\mathcal{D}}(B\cdot) \subseteq \mathcal{D}^m$ of the linear system of equations $B\mathbf{Q} = 0$ which is equivalent to determining the (right) syzygies of the column module of the matrix B. If $\{\mathbf{Q}_1,\dots,\mathbf{Q}_s\}$ is a generating set of $\ker_{\mathcal{D}}(B\cdot)$, then we may set $C = (\mathbf{Q}_1,\dots,\mathbf{Q}_s)$ thus defining the map γ. Dually, Line /4/ requires to compute a generating set $\{\mathbf{P}_1,\dots,\mathbf{P}_{\hat{r}}\}$ of the solution space $\ker_{\mathcal{D}}(\cdot C) \subseteq \mathcal{D}^{1\times m}$ of the linear system of equations $\mathbf{P}C = 0$, i. e. the determination of the (left) syzygies of the row module of the matrix C. Finally, we must check in Line /5/ whether the row module of B equals $\langle\mathbf{P}_1,\dots,\mathbf{P}_{\hat{r}}\rangle$.

Example 10.6.4. Algorithm 10.1 allows us to derive systematically the relation between the $U(1)$ Yang–Mills equations and the Maxwell equations discussed in Example 2.4.3. We start with the four equations $\mathbf{B}_t = -\nabla\times\mathbf{E}$ and $\nabla\cdot\mathbf{B} = 0$ (i. e. the Bianchi identity $\mathrm{d}F = 0$). Rewriting them as a linear differential operator yields

---

[7] If $\mathcal{D}$ is a commutative ring, then we may of course use the more common realisations $\beta(\mathbf{P}) = B\mathbf{P}$ and $\beta^*(\mathbf{Q}) = B^t\mathbf{Q}$ where now $\mathbf{P}$ is a column vector, too.

$$B = \begin{pmatrix} \partial_t & 0 & 0 & 0 & -\partial_z & \partial_y \\ 0 & \partial_t & 0 & \partial_z & 0 & -\partial_x \\ 0 & 0 & \partial_t & -\partial_y & \partial_x & 0 \\ \partial_x & \partial_y & \partial_z & 0 & 0 & 0 \end{pmatrix} . \tag{10.77}$$

According to the discussion above, we need a generating set for the right syzygies of the columns of B. A straightforward computation yields

$$C = \begin{pmatrix} 0 & -\partial_z & \partial_y & 0 \\ \partial_z & 0 & -\partial_x & 0 \\ -\partial_y & \partial_x & 0 & 0 \\ -\partial_t & 0 & 0 & -\partial_x \\ 0 & -\partial_t & 0 & -\partial_y \\ 0 & 0 & -\partial_t & -\partial_z \end{pmatrix} \tag{10.78}$$

as the matrix representation of γ and a further syzygy computation yields that γ is indeed a parametrisation of β. This result does not come as a surprise, as C obviously encodes the familiar relations $\mathbf{B} = \nabla \times \mathbf{A}$ and $\mathbf{E} = -\nabla \cdot A_4 - \partial_t \mathbf{A}$. Thus we have rediscovered the vector potential.

It is interesting to note that if we omit in the matrix B one of the first three rows, we still obtain the same matrix C. But now $\operatorname{im}\beta \neq \operatorname{im}\hat{\beta}$ so that C does not define a parametrisation and $\hat{\beta}$ is given by the original matrix B. Thus the compatibility analysis with Algorithm 10.1 automatically recovers the dropped equation. ◁

Example 10.6.5. As a further quite famous example we study whether the linearised Einstein equations are the compatibility conditions of some other system. This problem was posed by Wheeler about 1970 and much later solved by Pommaret [358].

The linearisation of the Einstein equations (2.96) around the Minkowski metric, i. e. the ansatz[8] $g_{ij} = \eta_{ij} + h_{ij}$, yields the following linear system

$$\eta^{k\ell}\left[\partial_{x^i x^j} h_{k\ell} + \partial_{x^k x^\ell} h_{ij} - \partial_{x^k x^i} h_{\ell j} - \partial_{x^k x^j} h_{\ell i}\right] = 0 \,. \tag{10.79}$$

By symmetry, we may restrict to $1 \le i \le j \le n$ so that (10.79) contains $n(n+1)/2$ independent equations for the same number of unknowns but the system is still underdetermined. Thus for $n = 4$ we have 10 equations.

Because of its size, we omit the matrix $B \in \mathcal{D}^{10\times 10}$ representing the homomorphism β corresponding to the linearised Einstein equations in four dimensions; its entries follow straightforwardly from (10.79). A syzygy computation yields as matrix representation of the homomorphism γ

$$C = \begin{pmatrix} 2\partial_x & 0 & 0 & 0 & \partial_y & 0 & 0 & \partial_z & 0 & \partial_t \\ 0 & 2\partial_y & 0 & 0 & \partial_x & \partial_z & 0 & 0 & \partial_t & 0 \\ 0 & 0 & 2\partial_z & 0 & 0 & \partial_y & \partial_t & \partial_x & 0 & 0 \\ 0 & 0 & 0 & -2\partial_t & 0 & 0 & -\partial_z & 0 & -\partial_y & -\partial_x \end{pmatrix} . \tag{10.80}$$

[8] Recall from Example 2.4.3 that $\eta^{ii} = 1$ for $1 \le i < n$, $\eta^{nn} = -1$ and $\eta^{ij} = 0$ in all other cases. η_{ij} are the entries of the inverse matrix.

However, a further syzygy computation yields that a minimal generating set of the left syzygy module of the row module of C consists of 20 elements. Thus we have $\operatorname{im}\beta \subsetneq \operatorname{im}\hat{\beta}$ and (10.79) does not admit a parametrisation.

One can obtain in this manner further interesting results in differential geometry; for lack of space we refer again to [358]. We mention here only that the differential operator corresponding to γ^* has a simple geometric interpretation. We may write it as $h_{ij} = \eta_{kj}\partial_{x^i}X^k + \eta_{ik}\partial_{x^j}X^k$ or $h = \mathcal{L}_X\eta$, i. e. h is the Lie derivative of the Minkowski metric with respect to the vector field $X = X^k\partial_{x^k}$. ◁

Example 10.6.6. A special case arises in Algorithm 10.1, if the map β^* is injective. Obviously, we may now take for γ^* the zero map (for an arbitrary value of the rank s). The dual map γ is then of course the zero map, too, and a (trivial) parametrisation of β exists, if and only if β is surjective. Representing as above β by a matrix $B \in \mathcal{D}^{r\times m}$, i. e. $\beta(\mathbf{P}) = \mathbf{P}B$ and $\beta^*(\mathbf{Q}) = B\mathbf{Q}$, injectivity of β^* is equivalent to the matrix B having full column rank, whereas surjectivity of β requires the existence of a left inverse for B, i. e. a matrix X with $XB = \mathbb{1}$.

As a concrete example from physics, we may consider the *Poincaré–Steklov problem* in three dimensions. Given a vector field $\mathbf{v}$ and a function w, it asks for a vector field $\mathbf{u}$ satisfying $\nabla \times \mathbf{u} = \mathbf{v}$ and $\nabla \cdot \mathbf{u} = w$. These equations appear for instance naturally in electrostatics (if we study only time-independent solutions, then Maxwell's equations (2.85) decouple and any static electrical field $\mathbf{E}$ solves a Poincaré–Steklov problem) but also in the theory of the Navier–Stokes equations.

One can show via an involution analysis that only a single compatibility condition exists for this problem, namely the obvious one: $\nabla \cdot \mathbf{v} = 0$. If we study the inverse syzygy problem via Algorithm 10.1, the matrix of the differential operator β is

$$B = \begin{pmatrix} 0 & -\partial_z & \partial_y \\ \partial_z & 0 & -\partial_x \\ -\partial_y & \partial_x & 0 \\ \partial_x & \partial_y & \partial_z \end{pmatrix}. \tag{10.81}$$

Obviously, the columns of this matrix are linearly independent and thus the dual map β^* is injective. However, β is not surjective, as one easily sees that B does not possess a left inverse, and thus no parametrisation exists. ◁

Remark 10.6.7. We discussed the inverse syzygy problem from the view point of differential equations theory emphasising the problem of finding a parametrisation. Thus we started with the homomorphism β and our goal was the construction of a homomorphism γ with $\operatorname{im}\beta = \ker\gamma$. Algebraically this formulation is not the most natural one. Here a better starting point is the left $\mathcal{D}$-module $\mathcal{M} = \operatorname{coker}\beta$. We say that $\mathcal{M}$ is an *nth syzygy*, if a free coresolution

$$0 \longrightarrow \mathcal{M} \longrightarrow \mathcal{F}_1 \longrightarrow \cdots \longrightarrow \mathcal{F}_n \tag{10.82}$$

of length at least n exists. It is not difficult to relate the two formulation. If Algorithm 10.1 provides us with a map $\gamma : \mathcal{D}^m \rightarrow \mathcal{D}^s$ such that $\operatorname{im}\beta = \ker\gamma$, then we obviously have a coresolution of length 1 of the form $0 \rightarrow \mathcal{M} \xrightarrow{\tilde{\gamma}} \mathcal{D}^s$ with

$\tilde{\gamma}([\mathbf{P}]) = \gamma(\mathbf{P})$ so that $\mathcal{M}$ is a first syzygy ($\tilde{\gamma}$ is well-defined, since $\operatorname{im}\beta = \ker\gamma$). Thus Theorem 10.6.2 asserts that $\mathcal{M}$ is a first syzygy, if and only if it is torsionless.

Now we may iterate and apply Algorithm 10.1 to the map γ constructed in the previous step. If we obtain a map $\delta : \mathcal{D}^s \to \mathcal{D}^t$ with $\operatorname{im}\gamma = \ker\delta$, then $\mathcal{M}$ is a second syzygy and so on. Effectively, this procedure amounts to the step by step construction of a free resolution of the module $\operatorname{coker}\beta^*$ (which is actually the Auslander–Bridger dual of $\mathcal{M}$—see the Addendum below) and its subsequent dualisation. However, we must check at each step whether the dual sequence is still exact. If this property is indeed preserved until the nth step, then we say that $\mathcal{M}$ is *n-torsionless* (1-torsionlessness is obviously equivalent to the classical notion of torsionlessness). Thus we may conclude with Theorem 10.6.2 that $\mathcal{M}$ is an nth syzygy, if and only if it is n-torsionless. ◁

In the literature the existence of a parametrisation is usually connected with the question whether $\operatorname{coker}\beta$ is torsion*free* and not torsion*less* as in Theorem 10.6.2 above. Recall from Remark B.1.7 that a torsionless module is always torsionfree but not vice versa. However, if the underlying ring $\mathcal{D}$ satisfies certain conditions, then the converse is true, too. We will now use Algorithm 10.1 to derive an instance where this is the case.

Theorem 10.6.8. *Let $\mathcal{D}$ be a coherent ring which satisfies the left Ore condition for the subset $\mathcal{S}$ of regular elements and whose total ring of left quotients $\mathcal{Q} = \mathcal{S}^{-1}\mathcal{D}$ is right self-injective (see Definition B.2.5). Then a finitely generated left $\mathcal{D}$-module is torsionfree, if and only if it is torsionless.*

Proof. As mentioned above, it suffices to show that under the made assumption any finitely generated torsionfree module is also torsionless. Since $\mathcal{D}$ is coherent, we may consider any finitely generated left $\mathcal{D}$-module as the cokernel $\operatorname{coker}\beta$ of some homomorphism $\beta : \mathcal{D}^m \to \mathcal{D}^r$. Assume now that $\mathcal{M} = \operatorname{coker}\beta$ is torsionfree and determine the maps γ and $\hat{\beta}$ of Algorithm 10.1. According to Example B.2.17, the quotient ring $\mathcal{Q}$ is a flat extension of $\mathcal{D}$. Hence the tensor product functor $\mathcal{Q}\otimes_{\mathcal{D}}\cdot$ is exact and its application yields the equalities

$$\ker(\mathcal{Q}\otimes_{\mathcal{D}}\gamma) = \mathcal{Q}\otimes_{\mathcal{D}}\ker\gamma = \mathcal{Q}\otimes_{\mathcal{D}}\operatorname{im}\hat{\beta} = \operatorname{im}(\mathcal{Q}\otimes_{\mathcal{D}}\hat{\beta}) \tag{10.83}$$

(since by construction $\ker\gamma = \operatorname{im}\hat{\beta}$) and

$$\ker(\mathcal{Q}\otimes_{\mathcal{D}}\beta)^* = \ker(\beta^*\otimes_{\mathcal{D}}\mathcal{Q}) = \operatorname{im}(\gamma^*\otimes_{\mathcal{D}}\mathcal{Q}) = \operatorname{im}(\mathcal{Q}\otimes_{\mathcal{D}}\gamma)^* \tag{10.84}$$

(as $\ker\beta^* = \operatorname{im}\gamma^*$).

By assumption the quotient ring $\mathcal{Q}$ is right self-injective, so that we may dualise the second equality and obtain $\operatorname{im}(\mathcal{Q}\otimes_{\mathcal{D}}\beta) = \ker(\mathcal{Q}\otimes_{\mathcal{D}}\gamma)$ which together with the first equality gives $\operatorname{im}(\mathcal{Q}\otimes_{\mathcal{D}}\beta) = \operatorname{im}(\mathcal{Q}\otimes_{\mathcal{D}}\hat{\beta})$. Thus any element $\mathbf{P}\in\operatorname{im}\hat{\beta}$ can be written in the form $S^{-1}\tilde{\mathbf{P}}$ with $\tilde{\mathbf{P}}\in\operatorname{im}\beta$ and $S\in\mathcal{S}$ a regular element implying that $S[\mathbf{P}] = 0$ in $\mathcal{M}$. As the module $\mathcal{M}$ is assumed to be torsionfree and S is regular, we must have $[\mathbf{P}] = 0$ in $\mathcal{M}$ or equivalently $\mathbf{P}\in\operatorname{im}\beta$. Hence $\operatorname{im}\beta = \operatorname{im}\hat{\beta}$ and, by Theorem 10.6.2, the module $\mathcal{M}$ is torsionless. □

By the same reasoning as in Remark 10.6.3 above, we conclude that under the assumptions of Theorem 10.6.8 Algorithm 10.1 allows us to determine explicitly the torsion submodule $t(\mathcal{M})$ of the module $\mathcal{M} = \operatorname{coker}\beta$, as it is given by $\operatorname{im}\hat{\beta}/\operatorname{im}\beta$.

Remark 10.6.9. Note that the assumptions of the theorem are trivially satisfied, if the ring $\mathcal{D}$ is a left Ore domain, as in this case the quotient ring $\mathcal{Q}$ is actually a skew field and it follows immediately from Baer's Criterion (Proposition B.2.7) that any field is self-injective (alternatively, one could argue that the column and the row rank of a matrix over a skew field are always equal [277]). ◁

Addendum: Computing Extension Groups

Algorithm 10.1 allows us the computation of certain extension groups. We discussed in Remark 10.6.3 that $\operatorname{im}\hat{\beta}/\operatorname{im}\beta = \ker\eta_{\mathcal{M}}$ where $\eta_{\mathcal{M}} : \mathcal{M} \to \mathcal{M}^{**}$ is the natural homomorphism between $\mathcal{M} = \operatorname{coker}\beta$ and its bidual $\mathcal{M}^{**}$. Now we will show that this kernel is actually an extension group. But before we need the notion of the *Auslander–Bridger dual* $D(\mathcal{M})$ of a finitely presented module $\mathcal{M}$: if $\mathcal{M} = \operatorname{coker}\beta$, then we define $D(\mathcal{M}) = \operatorname{coker}\beta^*$, i. e. as the cokernel of the dual map. Note that $D(\mathcal{M})$ is not uniquely determined by $\mathcal{M}$ but depends on the chosen presentation β (but one can show that $D(\mathcal{M})$ is unique up to projective direct summands).

Proposition 10.6.10. *Let $\mathcal{M}$ be a finitely generated left $\mathcal{D}$-module. Then there is an exact sequence*

$$0 \longrightarrow \operatorname{Ext}^1_{\mathcal{D}}(D(\mathcal{M}),\mathcal{D}) \longrightarrow \mathcal{M} \xrightarrow{\eta_{\mathcal{M}}} \mathcal{M}^{**} \longrightarrow \operatorname{Ext}^2_{\mathcal{D}}(D(\mathcal{M}),\mathcal{D}) \longrightarrow 0\,. \tag{10.85}$$

In other words, we have two natural isomorphisms $\ker\eta_{\mathcal{M}} \cong \operatorname{Ext}^1_{\mathcal{D}}(D(\mathcal{M}),\mathcal{D})$ *and* $\operatorname{coker}\eta_{\mathcal{M}} \cong \operatorname{Ext}^2_{\mathcal{D}}(D(\mathcal{M}),\mathcal{D})$.

Proof. As usual, we assume that the module $\mathcal{M}$ is given as a cokernel via an exact sequence $\mathcal{D}^r \xrightarrow{\beta} \mathcal{D}^m \xrightarrow{\pi} \mathcal{M} \longrightarrow 0$. Since the dual module $\mathcal{M}^*$ is obviously finitely generated, too, we have a further exact sequence $\mathcal{D}^s \xrightarrow{\alpha} \mathcal{D}^t \xrightarrow{\rho} \mathcal{M}^* \longrightarrow 0$. Splicing the dual of the second sequence with the first one via the homomorphism $\eta_{\mathcal{M}}$ yields the following complex:

$$\mathcal{D}^r \xrightarrow{\beta} \mathcal{D}^m \xrightarrow{\rho^*\circ\eta_{\mathcal{M}}\circ\pi} (\mathcal{D}^t)^* \xrightarrow{\alpha^*} (\mathcal{D}^s)^*\,. \tag{10.86}$$

Since π and ρ are both surjective (and hence ρ^* injective), we first note that $\operatorname{im}(\rho^*\circ\eta_{\mathcal{M}}\circ\pi) \cong \operatorname{im}\eta_{\mathcal{M}}$ and, as $\ker\alpha^* \cong \mathcal{M}^{**}$, we see that the homology of this complex at $(\mathcal{D}^t)^*$ is isomorphic to $\operatorname{coker}\eta_{\mathcal{M}}$. Furthermore, again by the injectivity of ρ^* and because of $\operatorname{im}\beta = \ker\pi$, we find that the homology of the complex at $\mathcal{D}^m$ is isomorphic to $\ker\eta_{\mathcal{M}}$.

By definition, the Auslander–Bridger dual $D(\mathcal{M})$ is the cokernel of β^*. Thus combining the free presentation of $\mathcal{M}^*$ with the dualised presentation of $\mathcal{M}$ yields the exact sequence

$$\mathcal{D}^s \xrightarrow{\alpha} \mathcal{D}^t \xrightarrow{\pi^* \circ \rho} (\mathcal{D}^m)^* \xrightarrow{\beta^*} (\mathcal{D}^s)^* \longrightarrow D(\mathcal{M}) \longrightarrow 0 \tag{10.87}$$

which we may consider as the beginning of a free resolution of $D(\mathcal{M})$. In order to compute the sought extension modules, we must dualise this sequence. Since one easily verifies that $\pi^{**} = \eta_{\mathcal{M}} \circ \pi$ after the usual identification $(\mathcal{D}^m)^{**} = \mathcal{D}^m$, the relevant part of the dualised sequence is just (10.86) and therefore we have indeed that $\operatorname{Ext}^1_{\mathcal{D}}(D(\mathcal{M}),\mathcal{D}) \cong \ker \eta_{\mathcal{M}}$ and $\operatorname{Ext}^2_{\mathcal{D}}(D(\mathcal{M}),\mathcal{D}) \cong \operatorname{coker} \eta_{\mathcal{M}}$. □

Since according to Remark 10.6.3 $\ker \eta_{\mathcal{M}} = \operatorname{im} \hat{\beta} / \operatorname{im} \beta$, Algorithm 10.1 provides us with an explicit method to determine $\operatorname{Ext}^1_{\mathcal{D}}(D(\mathcal{M}),\mathcal{D})$. It follows from the discussion in Remark 10.6.7 that an iteration allows us to compute the higher extension groups, too; each further group requires essentially two additional syzygy computations. We obtain then the following extension of the diagram (10.74)

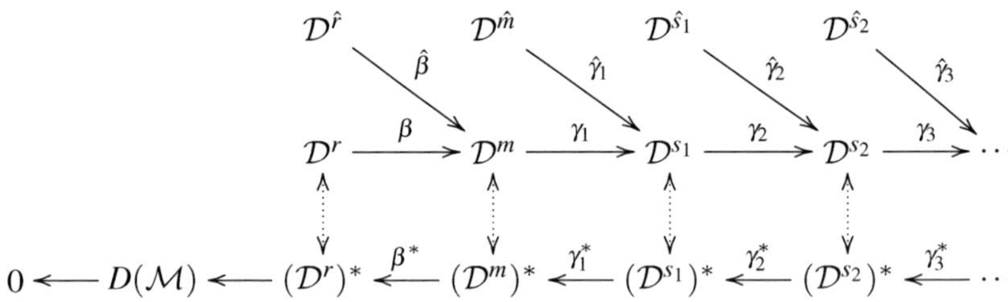

(10.88)

where the bottom row defines a free resolution of $D(\mathcal{M})$ and our results imply now the isomorphisms

$$\operatorname{Ext}^1_{\mathcal{D}}(D(\mathcal{M}),\mathcal{D}) \cong \operatorname{im} \hat{\beta} / \operatorname{im} \beta\,, \quad \operatorname{Ext}^{i+1}_{\mathcal{D}}(D(\mathcal{M}),\mathcal{D}) \cong \operatorname{im} \hat{\gamma}_i / \operatorname{im} \gamma_i\,. \tag{10.89}$$

Remark 10.6.11. Our results so far concern the extension groups of the Auslander–Bridger dual $D(\mathcal{M})$ and, as mentioned above, $D(\mathcal{M})$ is not uniquely determined by the module $\mathcal{M}$ but also depends on the chosen presentation β. The definition of $D(\mathcal{M})$ trivially implies that $D\bigl(D(\mathcal{M})\bigr) = \mathcal{M}$, since for a homomorphism between free modules we may identify $\beta^{**} = \beta$. Hence, by reverting the role of β and β^*, we can use Algorithm 10.1 for computing the extension groups $\operatorname{Ext}^i_{\mathcal{D}}(\mathcal{M},\mathcal{D})$.

Furthermore, the groups $\operatorname{Ext}^i_{\mathcal{D}}(D(\mathcal{M}),\mathcal{D})$ depend only on $\mathcal{M}$ and not on β, as they are trivially related to the extension groups of the dual module $\mathcal{M}^*$. By definition of $D(\mathcal{M})$, we have an exact sequence

$$0 \longrightarrow \mathcal{M}^* \xrightarrow{\pi^*} (\mathcal{D}^m)^* \xrightarrow{\beta^*} (\mathcal{D}^r)^* \longrightarrow D(\mathcal{M}) \longrightarrow 0\,. \tag{10.90}$$

Assume now that $\mathcal{F} \xrightarrow{\alpha} \mathcal{M}^* \longrightarrow 0$ is a free resolution of $\mathcal{M}^*$. Then obviously,

$$\mathcal{F} \xrightarrow{\pi^* \circ \alpha} (\mathcal{D}^m)^* \xrightarrow{\beta^*} (\mathcal{D}^r)^* \longrightarrow D(\mathcal{M}) \longrightarrow 0 \tag{10.91}$$

is a free resolution of $D(\mathcal{M})$ and we obtain the equalities

$$\operatorname{Ext}^{i+2}_{\mathcal{D}}(D(\mathcal{M}), \mathcal{D}) = \operatorname{Ext}^{i}_{\mathcal{D}}(\mathcal{M}^*, \mathcal{D}) \tag{10.92}$$

for all values $i \geq 1$. Together with Proposition 10.6.10, this observation implies that indeed all extension groups $\operatorname{Ext}^{i}_{\mathcal{D}}(D(\mathcal{M}), \mathcal{D})$ depend only on $\mathcal{M}$. Another way to see this fact consists of recalling that the Auslander–Bridger dual $D(\mathcal{M})$ is unique up to projective summands. According to Proposition B.2.37, such projective summands do not affect extension groups. ◁

Addendum: Algebraic Systems Theory

Many of the results in this and in the last section are of great relevance for algebraic systems theory and in fact this application has motivated many theoretical development. Therefore, we now briefly sketch a few basic ideas from this field; for more details we refer to [361, 485, 487].

Classical linear systems theory is concerned with the study of linear ordinary differential equations of the form

$$\mathbf{w}' = P\mathbf{w} + Q\mathbf{v}\,. \tag{10.93}$$

Here the vector $\mathbf{u}$ of the dependent variables is split into two components, the *state variables* $\mathbf{w}$ and the *input* $\mathbf{v}$, and P, Q are two constant matrices of appropriate dimensions. Obviously, (10.93) represents an underdetermined system in the sense of Definition 7.5.6, as in order to obtain a formally well-posed initial value problem we must completely prescribe the component $\mathbf{v}$. The central goal of control theory consists of finding inputs $\mathbf{v}$ such that the state component $\mathbf{w}$ of the solution of (10.93) exhibits some desired properties.

Within the behavioural approach to systems theory, one dispenses with the distinction into state and input variables and defines a *behaviour* as the solution space of a linear differential equation:

$$\mathcal{B} = \{\mathbf{u} \in \mathcal{A}^m \mid B\mathbf{u} = 0\} = \ker_{\mathcal{A}} B\,. \tag{10.94}$$

Here $B \in \mathcal{D}^{r\times m}$ is a matrix of linear differential operators and $\mathcal{A}$ is some function space, the *signal space*, in which the system is considered. If $\mathcal{D}$ is a ring of linear partial differential operators, one speaks of a *multidimensional* or *distributed* system. In abstract systems theory, $\mathcal{D}$ could be any ring and $\mathcal{A}$ any left $\mathcal{D}$-module. While we admit in the sequel non-commutative rings $\mathcal{D}$, we make two simplifying assumptions: $\mathcal{D}$ is a domain and it is both left and right Noetherian (these properties imply that $\mathcal{D}$ satisfies the left and the right Ore condition [319, Theorem 2.1.15] and that $\mathcal{D}$ is left and right coherent). Furthermore, we assume that the signal space $\mathcal{A}$ is an injective cogenerator.

For obvious reasons, one calls (10.94) a *kernel representation* of $\mathcal{B}$. Note that given the matrix B the behaviour $\mathcal{B} = \ker_{\mathcal{A}} B$ is uniquely determined whereas conversely for a given behaviour $\mathcal{B}$ in general many choices for the matrix B are possible. More generally, if $\mathcal{B}_1, \mathcal{B}_2 \subseteq \mathcal{A}^m$ are two behaviours defined by matrices B_1 and B_2, respectively, then $\mathcal{B}_1 \subseteq \mathcal{B}_2$, if and only if a further matrix X exists such that $B_2 = XB_1$. If both B_1 and B_2 have full row rank, then $\mathcal{B}_1 = \mathcal{B}_2$, if and only if X is a unimodular matrix.[9]

Introducing the map $\beta : \mathcal{D}^{1\times r} \to \mathcal{D}^{1\times m}$ with $\beta(\mathbf{P}) = \mathbf{P}B$, we may associate with the behaviour $\mathcal{B}$ the $\mathcal{D}$-module $\mathcal{M} = \operatorname{coker}\beta$ which obviously remains invariant under such multiplications with unimodular matrices. According to Theorem 10.5.10, the maps $\cdot^{\perp}$ define a bijective correspondence between submodules of $\mathcal{D}^m$ and behaviours $\mathcal{B} \subseteq \mathcal{A}^m$; now we may identify the submodule corresponding to a given behaviour $\mathcal{B}$ with $\operatorname{im}\beta$.

Let the matrix $B \in \mathcal{D}^{r\times m}$ has rank $p \leq m$ so that we can choose p columns which are linearly independent over the quotient field $\operatorname{Quot}(\mathcal{D})$. After a suitable renumbering of the components of $\mathbf{u}$, we may decompose $B = (-Q \mid P)$ where the submatrix $P \in \mathcal{D}^{r\times p}$ has still rank p. The corresponding splitting $\mathbf{u} = \binom{\mathbf{v}}{\mathbf{w}}$ allows us to rewrite the system $B\mathbf{u} = 0$ in the form $P\mathbf{w} = Q\mathbf{v}$ which is fairly similar to (10.93), although P, Q are now matrix operators. One speaks of an *input-output structure* for the behaviour $\mathcal{B}$. This terminology is justified by the following result.

Proposition 10.6.12. *Let* $\mathcal{B} = \{\binom{\mathbf{v}}{\mathbf{w}} \in \mathcal{A}^m \mid P\mathbf{w} = Q\mathbf{v}\}$ *be a behaviour with an input-output structure where* $\operatorname{rank} P = p$. *Then to every* $\mathbf{v} \in \mathcal{A}^{m-p}$ *there exists a* $\mathbf{w} \in \mathcal{A}^p$ *such that* $P\mathbf{w} = Q\mathbf{v}$.

Proof. Consider the map $\pi : \mathcal{D}^{1\times r} \to \mathcal{D}^{1\times p}$ defined by right multiplication with P. By a classical syzygy computation, we can determine a homomorphism ρ such that $\ker\pi = \operatorname{im}\rho$. According to the fundamental principle (which is applicable here by Proposition 10.5.7, since the signal space $\mathcal{A}$ is an injective cogenerator), the existence of a solution $\mathbf{w}$ of the equation $P\mathbf{w} = Q\mathbf{v}$ is equivalent to $\mathbf{v}$ satisfying the equation $RQ\mathbf{v} = 0$ where R is the matrix corresponding to ρ. As $\operatorname{rank} B = \operatorname{rank} P$, there exists a matrix $H \in \mathcal{D}^{p\times(m-p)}$ such that $Q = PH$.[10] Thus we obtain the equation $RPH\mathbf{v} = 0$ which holds for any $\mathbf{v}$, since by construction $RP = 0$. □

The αth component of $\mathbf{u}$ is called a *free variable* of the behaviour $\mathcal{B}$, if the natural projection $\pi_\alpha : \mathcal{B} \hookrightarrow \mathcal{A}^m \to \mathcal{A}$ mapping $\mathbf{u} \in \mathcal{B}$ to the component u^α is surjective. The existence of a free variable is obviously equivalent to $B\mathbf{u} = 0$ being an underdetermined system in the sense of Definition 7.5.6. Indeed, the surjectivity of π_α implies that for any arbitrary function $f \in \mathcal{A}$ we can find a solution $\mathbf{u} \in \mathcal{B}$ such that $u^\alpha = f$. Thus, by Proposition 10.6.12, the component $\mathbf{v}$ of an input-output structure consists entirely of free variables.

[9] A square matrix $U \in \mathcal{D}^{r\times r}$ is *unimodular*, if $\det U$ is a unit in $\mathcal{D}$. Obviously, this property is necessary and sufficient for the existence of an inverse $V \in \mathcal{D}^{r\times r}$ with $UV = VU = \mathbb{1}$.

[10] This unique H is called the *transfer matrix* of $\mathcal{B}$ for the chosen input-output structure.

Definition 10.6.13. The behaviour $\mathcal{B}$ is *autonomous*, if no free variables exist.

The considerations above imply that the behaviour $\mathcal{B}$ is autonomous, if and only if the matrix B has full column rank (so that $P = B$) which is equivalent to the dual map β^* being injective (cf. Example 10.6.6). Autonomy of an abstract behaviour possesses a simple algebraic characterisation in terms of torsion.

Theorem 10.6.14. *The behaviour $\mathcal{B}$ is autonomous, if and only if the associated module $\mathcal{M}$ is a torsion module.*

Proof. Assume first that $\mathcal{B}$ is not autonomous. Then for some index value α the sequence $\mathcal{B} \stackrel{\pi_\alpha}{\to} \mathcal{A} \to 0$ is exact. Using the Malgrange isomorphism (Proposition 10.5.1), we may write this sequence as $\operatorname{Hom}_{\mathcal{D}}(\mathcal{M},\mathcal{A}) \to \operatorname{Hom}_{\mathcal{D}}(\mathcal{D},\mathcal{A}) \to 0$. Since we assume that $\mathcal{A}$ is an injective cogenerator, the functor $\operatorname{Hom}_{\mathcal{D}}(\cdot,\mathcal{A})$ is faithful by Remark B.2.22 and the dual sequence $\mathcal{M} \stackrel{\iota_\alpha}{\leftarrow} \mathcal{D} \leftarrow 0$ is exact, too. Here we denote by ι_α the homomorphism induced by $\iota_\alpha(1) = [\mathbf{e}_\alpha]$. Thus for any $L \in \mathcal{D}$ we have $\iota_\alpha(L) = [L\mathbf{e}_\alpha] \neq 0$ because of the injectivity of ι_α. But this fact implies that $[\mathbf{e}_\alpha] \in \mathcal{M}$ cannot be a torsion element.

For the converse, assume now that $\mathcal{M}$ is not a torsion module. We claim that then at least for one index value α, the vector $[\mathbf{e}_\alpha]$ is not a torsion element of $\mathcal{M}$. This claim immediately implies that the behaviour $\mathcal{B}$ cannot be autonomous by simply reverting the argument above ($\operatorname{Hom}_{\mathcal{D}}(\cdot,\mathcal{A})$ is here also an exact functor because of the assumed injectivity of $\mathcal{A}$). Now assume on the contrary that $[\mathbf{e}_\alpha]$ is a torsion element for all values of α. Hence for each α there exists an operator $L_\alpha \in \mathcal{D}$ with $[L_\alpha \mathbf{e}_\alpha] = 0$. Let $[\mathbf{P}] \in \mathcal{M}$ be an arbitrary element of $\mathcal{M}$ with $\mathbf{P} = \sum_\alpha P_\alpha \mathbf{e}_\alpha$. According to the left Ore condition, $\mathcal{D}$ must contain two non-zero elements Q_α, R_α such that $Q_\alpha L_\alpha = R_\alpha P_\alpha$. Now consider a common left multiple R of the thus obtained elements R_α (which exists by the left Ore condition), i. e. $R = S_\alpha R_\alpha$ for each α. We find $R[\mathbf{P}] = [\sum_\alpha R P_\alpha \mathbf{e}_\alpha] = \sum_\alpha S_\alpha [L_\alpha \mathbf{e}_\alpha] = 0$ so that $\mathcal{M}$ is a torsion module in contradiction to our assumption. □

Remark 10.6.15. Since we assume that the underlying ring $\mathcal{D}$ contains no zero divisors, $\mathcal{M}$ is a torsion module, if and only if $\operatorname{Hom}_{\mathcal{D}}(\mathcal{M},\mathcal{D})$ vanishes. According to Part (i) of Proposition B.2.37, we can thus characterise autonomy by the vanishing of the extension group $\operatorname{Ext}^0_{\mathcal{D}}(\mathcal{M},\mathcal{D})$. ◁

A central concept in systems theory is controllability. In the context of the classical theory, it means intuitively that we can "steer" the system from any point of the state space to any other point by a suitable choice of the inputs. For abstract systems, we define it in a rather different manner, but we will show below for a special class of systems that we can recover this intuitive interpretation in a generalised sense.

Definition 10.6.16. The behaviour $\mathcal{B} \subseteq \mathcal{A}^m$ is *controllable*, if it possesses an *image representation*, i. e. there exists a matrix $C \in \mathcal{D}^{r\times s}$ such that

$$\mathcal{B} = \{\mathbf{u} \in \mathcal{A}^m \mid \exists \mathbf{z} \in \mathcal{A}^s : \mathbf{u} = C\mathbf{z}\} = \operatorname{im}_{\mathcal{A}} C\,. \tag{10.95}$$

In the language of physics one may say that a behaviour is controllable, if it admits a *potential* $\mathbf{z}$.

If we compare the image representation (10.95) with the original kernel representation (10.94), then it is obvious that the existence of the former is equivalent to the existence of a parametrisation C for the operator B (this is again just the fundamental principle). Now Theorem 10.6.2 immediately implies the following characterisation of controllability (and Algorithm 10.1 provides us with an effective test for it).

Theorem 10.6.17. *The behaviour $\mathcal{B}$ is controllable, if and only if the associated module $\mathcal{M}$ is torsionless.*

Remark 10.6.18. According to Theorem 10.6.2, the module $\mathcal{M}$ is torsionless, if and only if we find at the end of Algorithm 10.1 that $\operatorname{im}\hat{\beta} = \operatorname{im}\beta$. Furthermore, by Remark 10.6.3, the kernel of the natural homomorphism $\eta_{\mathcal{M}} : \mathcal{M} \to \mathcal{M}^{**}$ is isomorphic to the quotient $\operatorname{im}\hat{\beta}/\operatorname{im}\beta$. Hence, it follows from Proposition 10.6.10 that controllability of the behaviour $\mathcal{B}$ is equivalent to the vanishing of the extension group $\operatorname{Ext}^1_{\mathcal{D}}(D(\mathcal{M}),\mathcal{D})$ where $D(\mathcal{M})$ is again the Auslander-Bridger dual of $\mathcal{M}$. $\triangleleft$

Controllable and autonomous behaviours, respectively, may be considered as two opposite extrema; a randomly selected behaviour will generally be neither controllable nor autonomous. However, we will now show that any behaviour can be written as the sum of an autonomous and a controllable subbehaviour.

Proposition 10.6.19. *Let $\mathcal{B} \subseteq \mathcal{A}^m$ be a behaviour. Then there exists a unique maximal controllable subbehaviour $\mathcal{B}_c \subseteq \mathcal{B}$, the* controllable part *of $\mathcal{B}$.*

Proof. Let $\mathcal{B}$ be given by the kernel representation (10.94). Applying Algorithm 10.1 to the homomorphism β defined by multiplication with B from the right yields a homomorphism $\hat{\beta}$ defined by some matrix $\hat{B}$. We claim that the behaviour $\hat{\mathcal{B}} = \ker_{\mathcal{A}} \hat{B}$ possesses all desired properties. Obviously, by construction, it is indeed controllable and satisfies $\hat{\mathcal{B}} \subseteq \mathcal{B}$.

Let $\mathcal{B}' = \ker_{\mathcal{A}} B' \subseteq \mathcal{B}$ be a further controllable subbehaviour; we must show that $\mathcal{B}' \subseteq \hat{\mathcal{B}}$. Since $\mathcal{B}' \subseteq \mathcal{B}$, a matrix X exists such that $B = XB'$, and since $\mathcal{B}'$ is controllable, a parametrisation C' of B' exists. Because of our assumption that $\mathcal{D}$ is Noetherian, we can use the results obtained in the proof of Theorem 10.6.8: for any $\mathbf{P} \in \operatorname{im}\hat{\beta}$ a non-zero operator $L \in \mathcal{D}$ exists such that $L\mathbf{P} \in \operatorname{im}\beta$. Since we are only dealing with finitely generated modules, there exists a "global" operator $L \in \mathcal{D}$ such that $L\operatorname{im}\hat{\beta} \subseteq \operatorname{im}\beta$. Hence we have $L\hat{B} = YB$ for some matrix Y and therefore find that $L\hat{B}C' = YBC' = XYB'C' = 0$. As $\mathcal{D}$ is a domain, this observation implies $\hat{B}C' = 0$ and thus, since C' a parametrisation of B', that $\hat{B} = ZB'$ for some matrix Z. But the latter equality is equivalent to $\mathcal{B}' \subseteq \hat{\mathcal{B}}$. $\square$

Remark 10.6.20. It is not difficult to characterise $\mathcal{B}_c$ via torsion. Let $\mathcal{M}$ be the differential module corresponding to the behaviour $\mathcal{B}$ and $\mathcal{M}'$ the one for some subbehaviour $\mathcal{B}' \subseteq \mathcal{B}$. Obviously, $\mathcal{M}'$ is isomorphic to a quotient module $\mathcal{M}/\mathcal{N}$ with some submodule $\mathcal{N} \subseteq \mathcal{M}$. According to Theorems 10.6.2 and 10.6.8, $\mathcal{B}'$ is controllable, if and only if $\mathcal{M}'$ is torsionfree. Trivially, $\mathcal{M}' \cong \mathcal{M}/\mathcal{N}$ is torsionfree, if and only if $t(\mathcal{M}) \subseteq \mathcal{N}$ where $t(\mathcal{M})$ is the torsion submodule of $\mathcal{M}$. Thus the module corresponding to the controllable part $\mathcal{B}_c \subseteq \mathcal{B}$ is isomorphic to $\mathcal{M}/t(\mathcal{M})$. $\triangleleft$

Theorem 10.6.21. *Let $\mathcal{B}$ be a behaviour with controllable part $\mathcal{B}_c \subseteq \mathcal{B}$. Then there exists an autonomous subbehaviour $\mathcal{B}_a \subseteq \mathcal{B}$ such that $\mathcal{B} = \mathcal{B}_a + \mathcal{B}_c$.*

Proof. Let $B = (-Q \mid P)$ be an input-output structure for $\mathcal{B}$. Since $\mathcal{B}_c \subseteq \mathcal{B}$ and thus $B = XB_c$ for some matrix X, we obtain an induced input-output structure $B_c = (-Q_c \mid P_c)$ with $\operatorname{rank} P_c = \operatorname{rank} P$. We claim that the (trivially autonomous) subbehaviour $\mathcal{B}_a = \{\binom{\mathbf{v}}{\mathbf{w}} \in \mathcal{B} \mid P\mathbf{w} = 0 \wedge \mathbf{v} = 0\}$ satisfies $\mathcal{B}_a + \mathcal{B}_c = \mathcal{B}$. The inclusion $\mathcal{B}_a + \mathcal{B}_c \subseteq \mathcal{B}$ is obvious. Let $\binom{\mathbf{v}}{\mathbf{w}} \in \mathcal{B}$ be an arbitrary element; according to Proposition 10.6.12, the controllable part $\mathcal{B}_c$ contains an element of the form $\binom{\mathbf{v}}{\tilde{\mathbf{w}}}$. Now we can decompose $\binom{\mathbf{v}}{\mathbf{w}} = \binom{\mathbf{v}}{\tilde{\mathbf{w}}} + \binom{0}{\mathbf{w}-\tilde{\mathbf{w}}}$ where the first summand lives in $\mathcal{B}_c$ and the second one in $\mathcal{B}_a$. □

One should note that, in contrast to the controllable part $\mathcal{B}_c$, the autonomous part $\mathcal{B}_a$ in the above decomposition is far from unique and generally it is not possible to choose it in such a manner that we obtain a direct sum.

For the simplest case, namely that we are considering the smooth solutions of a linear system of partial differential equations with constant coefficients, i. e. for the choice $\mathcal{D} = \mathbb{R}[\partial_1, \dots, \partial_n]$ and $\mathcal{A} = \mathcal{C}^\infty(\mathbb{R}^n)$, autonomy and controllability can be given very concrete interpretations in terms of properties of the solution space. Here we also make contact with classical differential equations theory.

Lemma 10.6.22. *Let $\mathcal{B} \subseteq \mathcal{A}^m$ be an autonomous behaviour. If a solution $\mathbf{u} \in \mathcal{B}$ has compact support, then $\mathbf{u} = 0$.*

Proof. If $\mathbf{u} \in \mathcal{B}$ has compact support, then we may introduce its Fourier transform

$$\hat{\mathbf{u}}(\boldsymbol{\xi}) = \int_{\mathbb{R}^n} \mathbf{u}(\mathbf{x}) e^{-i(\boldsymbol{\xi}\cdot\mathbf{x})} d\mathbf{x} \tag{10.96}$$

satisfying the algebraic equation $B(i\boldsymbol{\xi})\hat{\mathbf{u}}(\boldsymbol{\xi}) = 0$. Since the entries of B are polynomials and $\hat{\mathbf{u}}$ is an analytic function, we may consider this equation over the field of meromorphic functions. According to our discussion above, for an autonomous behaviour the matrix B has full column rank, hence the only solution is $\hat{\mathbf{u}} = 0$ and thus we obtain $\mathbf{u} = 0$ after the inverse Fourier transform. □

Proposition 10.6.23. *Let $\mathcal{B} \subseteq \mathcal{A}^m$ be a behaviour. $\mathcal{B}$ is controllable, if and only if it possesses the following property: if $\Omega_1, \Omega_2 \subset \mathbb{R}^n$ are two arbitrary open subsets with $\overline{\Omega_1} \cap \overline{\Omega_2} = \emptyset$, then to any pair $\mathbf{u}_1, \mathbf{u}_2 \in \mathcal{B}$ there exists a further solution $\mathbf{u} \in \mathcal{B}$ such that $\mathbf{u}|_{\Omega_1} = \mathbf{u}_1|_{\Omega_1}$ and $\mathbf{u}|_{\Omega_2} = \mathbf{u}_2|_{\Omega_2}$.*

Proof. Assume first that $\mathcal{B}$ is controllable so that it possesses an image representation of the form (10.95). Then there exist two functions $\mathbf{z}_1, \mathbf{z}_2 \in \mathcal{A}^s$ such $\mathbf{u}_i = C\mathbf{z}_i$ for $i = 1, 2$. It is a well-known property of our chosen signal space $\mathcal{A}$ that for any pair of open subsets $\Omega_1, \Omega_2 \subset \mathbb{R}^n$ with $\overline{\Omega_1} \cap \overline{\Omega_2} = \emptyset$ a smooth function $\chi \in \mathcal{A}$ exists such that $\chi|_{\Omega_1} \equiv 1$ and $\chi|_{\Omega_2} \equiv 0$. Consider now the function $\mathbf{z} = \chi \cdot \mathbf{z}_1 + (1-\chi) \cdot \mathbf{z}_2$: since $\mathbf{z}|_{\Omega_i} = \mathbf{z}_i$, the function $\mathbf{w} = C\mathbf{z} \in \mathcal{B}$ has the desired properties.

For the converse assume that $\mathcal{B}$ is not controllable and hence its controllable part is a proper subset $\mathcal{B}_c \subsetneq \mathcal{B}$. If we consider an arbitrary solution $0 \neq \mathbf{u} \in \mathcal{B} \setminus \mathcal{B}_c$, then

$B\mathbf{u} = 0$ but $\hat{\mathbf{u}} = B_c\mathbf{u} \neq 0$. Recall from the proof of Proposition 10.6.19 that there exists an operator $L \in \mathcal{D}$ and a matrix Y such that $LB_c = YB$. Hence the components of the function $\hat{\mathbf{u}}$ are contained in the autonomous behaviour $\hat{\mathcal{B}} = \{u \in \mathcal{A} \mid Lu = 0\}$. Now choose some point $x_0 \in \mathbb{R}^n$ such that $\mathbf{u}(x_0) \neq 0$ and consider for two arbitrary radii $0 < r_1 < r_2$ the two open sets $\Omega_1 = \{x \in \mathbb{R}^n \mid \|x - x_0\| < r_1\}$ and $\Omega_2 = \{x \in \mathbb{R}^n \mid \|x - x_0\| > r_2\}$. According to our assumption, the behaviour $\mathcal{B}$ must then contain a solution $\tilde{\mathbf{u}}$ with $\tilde{\mathbf{u}}|_{\Omega_1} = \mathbf{u}|_{\Omega_1}$ and $\tilde{\mathbf{u}}|_{\Omega_2} \equiv 0$ which obviously has compact support and whose components lie in $\hat{\mathcal{B}}$. But this contradicts Lemma 10.6.22. □

Corollary 10.6.24. *Any controllable behaviour $\mathcal{B} \neq 0$ contains a solution $\mathbf{u} \in \mathcal{B}$ with compact support.*

Proof. Let $\mathbf{u} \in \mathcal{B}$ be such that $\mathbf{u}(x_0) \neq 0$ for some point $x_0 \in \mathbb{R}^n$. Consider two open sets Ω_1, Ω_2 as in the previous proof; according to Proposition 10.6.23, the behaviour $\mathcal{B}$ must contain a solution $\tilde{\mathbf{u}}$ such that $\tilde{\mathbf{u}}|_{\Omega_1} = \mathbf{u}|_{\Omega_1}$ and $\tilde{\mathbf{u}}|_{\Omega_2} \equiv 0$ which obviously has compact support. □

Proposition 10.6.25. *Let $\mathcal{B} \subseteq \mathcal{A}^m$ be a behaviour. $\mathcal{B}$ is autonomous, if and only if $\mathcal{B}$ contains no function $\mathbf{u} \neq 0$ with compact support.*

Proof. One direction was proven above in Lemma 10.6.22. For the converse assume that the behaviour $\mathcal{B}$ is not autonomous. It follows from Theorem 10.6.14 that the associated module $\mathcal{M}$ is not a torsion module and hence $t(\mathcal{M}) \subsetneq \mathcal{M}$. The discussion in Remark 10.6.20 implies that the controllable part $\mathcal{B}_c \subseteq \mathcal{B}$ is then non-zero and therefore contains a solution $\mathbf{u} \neq 0$ with compact support by Corollary 10.6.24. □

10.7 Completion to Involution

We stressed in Section 7.4 that the Cartan–Kuranishi completion is more a "method" than an algorithm, as it is not clearly specified how the required operations could be actually performed. We now develop for linear differential equations an effective algebraic realisation of it based on Algorithm 4.5 for involutive bases. A naive realisation of the Cartan–Kuranishi completion is immediate: one explicitly determines local representations for all arising differential equations $\mathcal{R}_{q+r}^{(s)}$ and symbols $\mathcal{N}_{q+r}^{(s)}$. For linear or polynomial equations one can then—at least in principle—perform all required operations algorithmically using Gröbner bases.

Obviously, such a brute force approach is not very efficient; in general, many unnecessary prolongations (which are usually rather expensive to compute) are performed and one has to deal with rather large matrices (recall Example 7.4.3). Pure algebra in form of involutive bases leads to fairly fast algorithms for linear equations, but all geometric information is lost. For example, it is not immediate to determine the number of performed prolongations and projections. We combine the geometric and the algebraic approach into a "hybrid" algorithm which maintains the efficiency of pure algebra and still obtains the full geometric information by clever "book-keeping." The key lies in the simple involution test via the Cartan normal form described in Remark 7.2.10.

In the sequel we assume that the coefficients of our system stem from a differential field $\mathbb{F}$ where all required operations (arithmetics and differentiations) can be performed effectively and which contains the independent variables $\mathbf{x}$. Thus a typical instance of $\mathbb{F}$ would be a finite algebraic extension of the field $\mathbb{Q}(x^1,\dots,x^n)$ of rational functions with rational coefficients.

We assume that the linear system is given in the form

$$\Phi^\tau(\mathbf{x},\mathbf{u}^{(q)}) = A^{\tau\mu}_\alpha(x)u^\alpha_\mu = 0\,, \qquad \tau = 1,\dots,t \tag{10.97}$$

where $0 \leq |\mu| \leq q$. Thus we are only considering homogeneous equations. But of course one could add right hand sides and would then in addition obtain explicitly the compatibility conditions studied in Section 10.5. For the remainder of this section, we concentrate on the left hand sides and call them *rows*. Thus a linear system is for us a set $\mathcal{F}$ of rows; one may also say that we are operating on a matrix representation of the system.

The derivatives are ordered according to the TOP lift of the degree reverse lexicographic order $\prec_{\text{degrevlex}}$; the leading term of a row $f \in \mathcal{F}$ is called *leading derivative* and denoted by[11] $\operatorname{ld} f$. As usual, multiplicative and non-multiplicative variables are assigned to a row according to its leading derivative. Reductions correspond now to elementary row operations and an autoreduction yields a triangular form of the system. We introduce the following notation for the result of a reduction. Let $f = a(x)\operatorname{ld} f + \sum_{u^\alpha_\mu \prec \operatorname{ld} f} A^\mu_\alpha(x)u^\alpha_\mu$ and $g = b(x)\operatorname{ld} g + \sum_{u^\alpha_\mu \prec \operatorname{ld} g} B^\mu_\alpha(x)u^\alpha_\mu$ be two rows of the system $\mathcal{F}$ with $\operatorname{ld} f = \operatorname{ld} g$; then we set

$$\operatorname{rem}(f,g) = \sum_{u^\alpha_\mu \prec \operatorname{ld} f} \left(A^\mu_\alpha(x) - \frac{a(x)}{b(x)}B^\mu_\alpha(x)\right) u^\alpha_\mu\,. \tag{10.98}$$

Note that this operation is purely algebraic; no differentiations are allowed!

In the naive realisation of the Cartan–Kuranishi completion, it is necessary to differentiate every equation in the system with respect to every independent variables in order to obtain a local representation of the prolonged equation. We use the Pommaret division to distinguish multiplicative and non-multiplicative variables. Our goal is to avoid differentiations with respect to multiplicative variables as much as possible. As in Algorithm 4.5 they should appear only in involutive normal form computations. In order to achieve this goal and still be able to decide involution of the appearing symbols and differential equations, we need a careful book-keeping. The necessary tools are introduced in the next definition. It might appear somewhat strange at first sight, but it will prove its usefulness in the sequel.

Definition 10.7.1. Let f denote a row from a linear system $\mathcal{F}$. We can make f into an *indexed row*, either a *single indexed row* $f_{(k)}$ or a *double indexed row* (or *phantom*

---

[11] As we have now fixed one specific term order, we omit for notational simplicity the subscript $\prec$.

row) $f_{(k,\ell)}$ by assigning to it one respectively two non-negative integers k and ℓ where $0 \leq k < \ell$ in the second case. We call k the *initial level* and ℓ the *phantom level* of the indexed row. For a *global level* $\lambda \geq k$, we define the *truncated involutive cone* $\bar{\mathcal{C}}_\lambda$ of an indexed row to be the following set of rows:[12]

$$\bar{\mathcal{C}}_\lambda(f_{(k)}) = \{D_\mu f \mid 0 \leq |\mu| \leq (\lambda - k);\ \forall i > \operatorname{cls} f : \mu_i = 0\}\,, \tag{10.99a}$$

$$\bar{\mathcal{C}}_\lambda(f_{(k,\ell)}) = \{D_\mu f \mid (\lambda - \ell) < |\mu| \leq (\lambda - k);\ \forall i > \operatorname{cls} f : \mu_i = 0\}\,. \tag{10.99b}$$

Thus the truncated involutive cone of an indexed row determined by the integers λ, k and possibly ℓ contains all multiplicative prolongations whose orders satisfy the given conditions (see Figure 10.3 for an illustration). With the help of this notion, we introduce the basic object of our approach, the skeleton of a linear system.

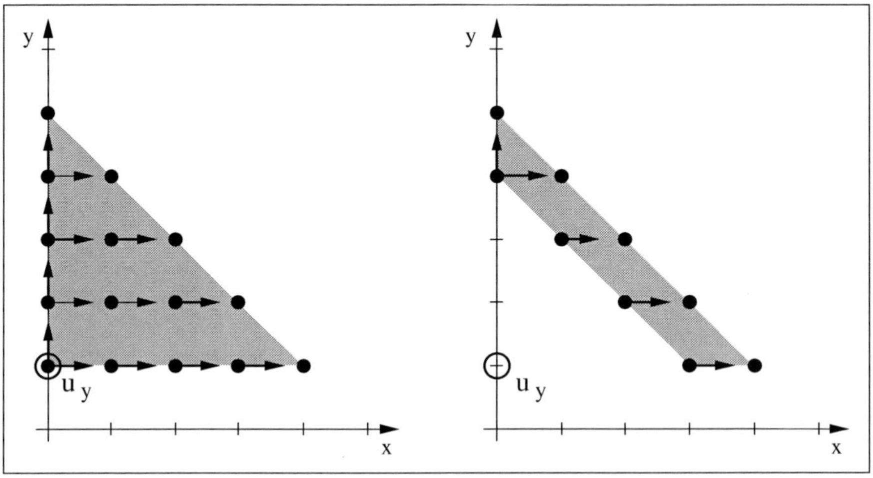

Fig. 10.3 *Left:* involutive truncated cone of a single indexed row $\hat{\mathcal{C}}_4\big((u_y)_{(0)}\big)$. *Right:* involutive truncated cone of a double indexed (phantom) row $\hat{\mathcal{C}}_4\big((u_y)_{(0,2)}\big)$.

Definition 10.7.2. A *skeleton* $\mathcal{S}_\lambda$ with global level λ consists of a set $\mathcal{U}_\lambda = \{f^\tau_{(k_\tau)}\}$ of single indexed rows and of a set $\mathcal{W}_\lambda = \{f^\tau_{(k_\tau, l_\tau)}\}$ of double indexed rows such that for all indices τ the inequalities $k_\tau \leq \lambda$ and $k_\tau < l_\tau \leq \lambda$, respectively, hold. It defines the linear system

$$\bar{\mathcal{S}}_\lambda = \{\bar{\mathcal{C}}_\lambda(f_{(k)}) \mid f_{(k)} \in \mathcal{U}_\lambda\} \cup \{\bar{\mathcal{C}}_\lambda(f_{(k,\ell)}) \mid f_{(k,\ell)} \in \mathcal{W}_\lambda\}\,. \tag{10.100}$$

Thus in order to obtain the full linear system $\bar{\mathcal{S}}_\lambda$ corresponding to a given skeleton $\mathcal{S}_\lambda$, we must simply add "flesh" in form of all permitted multiplicative prolongations of the contained rows and then drop the indices.

[12] For notational simplicity we drop the indices when we are performing any operations with a row; thus an indexed row $f_{(k)}$ should actually be understood as a pair (f, k).

In each iteration of the Cartan–Kuranishi completion, the system is *prolonged* to the next higher order. Since we need autoreduced systems, we must afterwards *triangulate*. We consider now these two operations at the level of skeletons. We may assume that the system $\mathcal{F}$ to be completed is already given in triangular, i. e. (head) autoreduced, form. We turn $\mathcal{F}$ into a skeleton $\mathcal{S}_0$ by simply setting the initial level k_τ of every row f^τ to 0. The global level λ is always equal to the number of the current iteration step; if a new row enters the skeleton in this step, its initial level becomes λ. Double indexed rows arise, if during the triangulation process a row is eliminated which has already produced some multiplicative prolongations. In this case, the phantom level is set to the current global level. Thus the completion algorithm produces a sequence of skeletons

$$\mathcal{S}_0 = \mathcal{S}_0^\triangle \longrightarrow \mathcal{S}_1 \longrightarrow \mathcal{S}_1^\triangle \longrightarrow \mathcal{S}_2 \longrightarrow \mathcal{S}_2^\triangle \longrightarrow \mathcal{S}_3 \longrightarrow \cdots \tag{10.101}$$

where the $^\triangle$ indicates a skeleton in triangular form.

The prolongation Algorithm 10.2 merely computes the non-multiplicative prolongations of those single indexed rows that have been newly added in the previous iteration. Due to the increase of the global level multiplicative prolongations of *all* rows from the skeleton are automatically present in the linear system it defines. No rows are explicitly removed; the new global level causes the elimination of the least-order multiplicative prolongations of all phantom rows in the system; this is shown in Figure 10.4 for a concrete example. We will prove below that these operations yield the desired skeleton for the prolonged system.

Algorithm 10.2 Prolongation of skeleton

Input: skeleton $\mathcal{S}_\lambda = \mathcal{U}_\lambda \cup \mathcal{W}_\lambda$, number n of independent variables
Output: prolonged skeleton $\mathcal{S}_{\lambda+1} = \mathcal{U}_{\lambda+1} \cup \mathcal{W}_{\lambda+1}$
$\mathcal{U}_{\lambda+1} \leftarrow \mathcal{U}_\lambda$; $\mathcal{W}_{\lambda+1} \leftarrow \mathcal{W}_\lambda$
for all $f^\tau_{(\lambda)} \in \mathcal{U}_\lambda$ **do**
  for cls $f^\tau_{(\lambda)} < i \leq n$ **do**
    $\mathcal{U}_{\lambda+1} \leftarrow \mathcal{U}_{\lambda+1} \cup \{(D_i f^\tau)_{(\lambda+1)}\}$
  end for
end for
return $\mathcal{U}_{\lambda+1} \cup \mathcal{W}_{\lambda+1}$

The triangulation Algorithm 10.3 is slightly more complicated, but very similar to the autoreduction mentioned in Section 4.2. It suffices to reduce only the single indexed rows, as the multiplicative prolongations of the phantom rows obviously have different leading derivatives. In the outer `while` loop, we take a row $f^\tau_{(k_\tau)}$ from the set $\mathcal{U}_\lambda$ with maximal leading derivative and try to reduce it modulo the rows contained in the full system $\bar{\mathcal{S}}_\lambda$. Hence we check in Line /4/ whether a reduction with a row lying in a truncated involutive cone is possible (since we consider here both single and double indexed rows, we omitted for notational simplicity the indices, but they are of course necessary for determining the cones according to (10.99)). The actual reduction is performed in the inner `while` loop. If no changes occur, the

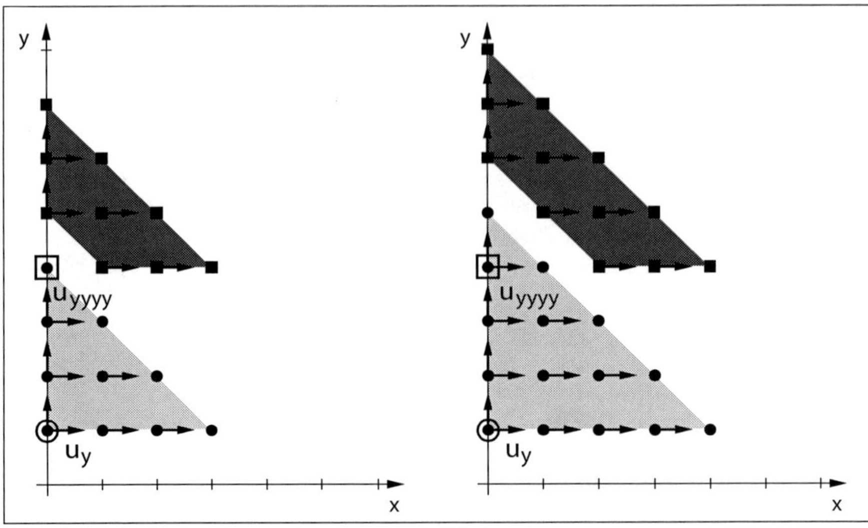

Fig. 10.4 Prolongation of a real and a phantom row, respectively, shown for the skeleton $\mathcal{S}_3 = \{(u_y)_{(0)}, (u_{yyyy})_{(0,3)}\}$ (left) and its prolongation $\mathcal{S}_4$ (right).

row $f^\tau_{(k_\tau)}$ is written back into the skeleton in Line /8/; otherwise the reduced row $\bar{f}$ is added with the initial level λ in Line /11/. If already multiplicative prolongations of the eliminated row $f^\tau_{(k_\tau)}$ have been computed at the current level λ, its removal would cause them to vanish as well. Thus we must keep track of their existence: in Line /14/ we turn $f^\tau_{(k_\tau)}$ into a phantom row by adding the current level as phantom level. This approach assures that in future iterations all relevant multiplicative prolongations are still taken into account.

The correctness of these two algorithms is equivalent to showing that the full system $\bar{\mathcal{S}}^\triangle_\lambda$ defined by the skeleton $\mathcal{S}^\triangle_\lambda$ yields a local representation for the differential equation $\mathcal{R}_{q+\lambda}$, if the initial differential equation was $\mathcal{R}_q$ (represented by the skeleton $\mathcal{S}_0$). As the main step towards this result, we consider the relation between the prolonged rows of $\bar{\mathcal{S}}^\triangle_\lambda$ and rows in $\mathcal{S}_{\lambda+1}$.

Proposition 10.7.3. *Consider the sequence* (10.101) *derived from a skeleton $\mathcal{S}_0$ by repeatedly applying Algorithms 10.2 and 10.3. For some $\lambda \geq 0$, let g be a linear combination of rows of the system $\bar{\mathcal{S}}^\triangle_\lambda$. Then each prolongation of g by one order can be written as a linear combination of rows of the system $\bar{\mathcal{S}}_{\lambda+1}$.*

Proof. We first consider a slightly modified version of the triangulation Algorithm 10.3 where we replace Line /14/ with $\mathcal{U}^\triangle_\lambda \leftarrow \mathcal{U}^\triangle_\lambda \cup \{f^\tau_{(k_\tau)}\}$. This modification implies that no phantom rows occur. So we ignore, for the moment, all references to the set $\mathcal{W}_\lambda$. Any row of which multiplicative prolongations have been computed is contained in all subsequent skeletons. Of course, one thereby loses the triangular form of the systems $\bar{\mathcal{S}}^\triangle_\lambda$.

Algorithm 10.3 Triangulation of skeleton

Input: skeleton $\mathcal{S}_\lambda = \mathcal{U}_\lambda \cup \mathcal{W}_\lambda$

Output: equivalent skeleton $\mathcal{S}_\lambda^\triangle = \mathcal{U}_\lambda^\triangle \cup \mathcal{W}_\lambda^\triangle$ such that $\bar{\mathcal{S}}_\lambda^\triangle$ is in triangular form

$\mathcal{U}_\lambda^\triangle \leftarrow \emptyset;\quad \mathcal{W}_\lambda^\triangle \leftarrow \mathcal{W}_\lambda$
while $\mathcal{U}_\lambda \neq \emptyset$ **do**
  $f^\tau_{(k_\tau)} \leftarrow \max_\prec \mathcal{U}_\lambda;\quad \bar{f} \leftarrow f^\tau;\quad \mathcal{U}_\lambda \leftarrow \mathcal{U}_\lambda \setminus \{f^\tau_{(k_\tau)}\}$
  while $\exists f^\sigma \in \mathcal{S}_\lambda : \operatorname{ld}\bar{f} \in \bar{\mathcal{C}}_\lambda(\operatorname{ld} f^\sigma)$ **do**
    $u^\alpha_\mu \leftarrow \operatorname{ld}\bar{f};\quad u^\alpha_\nu \leftarrow \operatorname{ld} f^\sigma;\quad g \leftarrow D_{\mu-\nu} f^\sigma;\quad \bar{f} \leftarrow \operatorname{rem}(\bar{f}, g)$
  end while
  if $\bar{f} = f^\tau$ **then**
    $\mathcal{U}_\lambda^\triangle \leftarrow \mathcal{U}_\lambda^\triangle \cup \{f^\tau_{(k_\tau)}\}$
  else
    if $\bar{f} \neq 0$ **then**
      $\mathcal{U}_\lambda^\triangle \leftarrow \mathcal{U}_\lambda^\triangle \cup \{\bar{f}_{(\lambda)}\}$
    end if
    if $k_\tau < \lambda$ **then**
      $\mathcal{W}_\lambda^\triangle \leftarrow \mathcal{W}_\lambda^\triangle \cup \{f^\tau_{(k_\tau,\lambda)}\}$
    end if
  end if
end while
return $\mathcal{U}_\lambda^\triangle \cup \mathcal{W}_\lambda^\triangle$

Let the skeleton $\mathcal{S}_\lambda^\triangle$ consist of the single indexed rows $\{f^\tau_{(k_\tau)}\}$ and let

$$g = \sum_{l=1}^{s} a_l(x) D_{\mu_l} f^{\tau_l} \tag{10.102}$$

be a finite linear combination of differential consequences of rows in the triangular system $\bar{\mathcal{S}}_\lambda^\triangle$. The rows in the system $\bar{\mathcal{S}}_\lambda$ are multiplicative prolongations of rows in the corresponding skeleton $\mathcal{S}_\lambda$. We formally differentiate (10.102) with respect to an arbitrary independent variable x^j obtaining

$$D_j g = \sum_{l=1}^{s} \left[a_l(x) D_{\mu_l+1_j} f^{\tau_l} + \frac{\partial a_l(x)}{\partial x^j} D_{\mu_l} f^{\tau_l} \right] . \tag{10.103}$$

We must show that all the rows of $\bar{\mathcal{S}}_\lambda^\triangle$ *and* their prolongations by one order are elements of $\bar{\mathcal{S}}_{\lambda+1}$. By the remark above, this fact is obvious for the unprolonged rows. For the remaining rows, we proceed by an induction over the global level λ. For $\lambda = 0$, Algorithm 10.2 computes all non-multiplicative prolongations (since $k_\tau = \lambda = 0$ for each row $f^\tau_{(k_\tau)}$). The multiplicative prolongations are taken care of by the incrementation of the global level λ. Now assume $\lambda > 0$. Let $h = D_{\mu+1_j} f$ be a prolongation with $D_\mu f \in \bar{\mathcal{S}}_\lambda^\triangle$ and $f_{(k)} \in \mathcal{S}_\lambda^\triangle$. We prove that $h \in \bar{\mathcal{S}}_{\lambda+1}$ by a Noetherian argument. The terminal cases are:

- If x^j is a *multiplicative* variable for f, we are done: all differentiations in $D_{\mu+1_j}$ are multiplicative for f and, as the global level is raised by the prolongation, $h \in \bar{\mathcal{S}}_{\lambda+1}$.
- If $|\mu| = 0$ and x^j is a *non-multiplicative* variable for f, then the prolongation is explicitly computed and therefore contained in the next system.

In all other cases, we set $c = \operatorname{cls} \mu + 1_j$ and rewrite $h = D_c(D_{\mu+1_j-1_c} f)$. Because of $|\mu| \leq (\lambda - k)$, the total derivative in the parentheses is an element of $\bar{\mathcal{S}}_{\lambda'}$ for some index value $\lambda' \leq \lambda$. Thus it is also contained in $\bar{\mathcal{S}}_\lambda$ and we may apply the induction hypothesis in order to write h in the form

$$h = D_c \left[\sum_{l=1}^{s'} b_l(x) D_{\nu_l} f^{\tau_l} \right] . \tag{10.104}$$

Note that for the summand h' possessing the highest leading derivative (its multi index is $\nu = \mu + 1_j - 1_c$) the variable x^c is by construction multiplicative. Hence we find $D_c h' \in \bar{\mathcal{S}}_{\lambda+1}$ as desired. With each of the remaining summands, we repeat the above manipulations. As their leading exponents are all less than $\mu + 1_j - 1_c$, this process must terminate after a finite number of steps. This observation proves the statement above under the made assumption.

Finally, we include the phantom rows, i. e. we restore Line /14/ of Algorithm 10.3 in its original form. It follows from (10.103) that if a row is dependent of the other rows in $\bar{\mathcal{S}}_\lambda$ and therefore eliminated, all its multiplicative prolongations by one order reduce to zero modulo $\bar{\mathcal{S}}_{\lambda+1}$. Therefore it does not matter for the system (however, it does matter for its triangular form) whether we carry out these reductions or not. Non-multiplicative prolongations of phantom rows never need to be computed, as for $f_{(k,\ell)}$ always $k < \ell \leq \lambda$ holds. □

Once this somewhat messy proof is finished, we obtain as corollary the for us crucial theorem on the relation between the sequence (10.101) obtained by alternately prolonging and triangulising the initial skeleton $\mathcal{S}_0$ and the prolongations of the linear differential equation $\mathcal{R}_q$ corresponding to $\mathcal{S}_0$.

Theorem 10.7.4. *Let the system $\mathcal{F}$ be a local representation for the linear differential equation $\mathcal{R}_q$. If we turn $\mathcal{F}$ into a skeleton $\mathcal{S}_0$ as described above and form the sequence* (10.101)*, then for any $\lambda \geq 0$ the system $\bar{\mathcal{S}}_\lambda^\triangle$ is triangulised and represents the equation $\mathcal{R}_{q+\lambda}$.*

Proof. We denote by $\mathcal{F}_0 = \mathcal{F}, \mathcal{F}_1, \mathcal{F}_2, \dots$ the prolongations of the system $\mathcal{F}$. Then the system $\mathcal{F}_i$ locally represents the prolonged differential equation $\mathcal{R}_{q+i}$. If we set $\mathcal{F}_0 = \bar{\mathcal{S}}_0^\triangle$, then we have also equality of the linear hulls: $\langle \mathcal{F}_0 \rangle = \langle \bar{\mathcal{S}}_0^\triangle \rangle$. By repeatedly applying Proposition 10.7.3, we get for $\lambda \geq 0$: $\langle \mathcal{F}_\lambda \rangle = \langle \bar{\mathcal{S}}_\lambda \rangle$. As we perform only elementary row operations during Algorithm 10.3, it follows that $\bar{\mathcal{S}}_\lambda^\triangle$ is a local representation of $\mathcal{R}_{q+\lambda}$.

It remains to show is that the systems $\bar{\mathcal{S}}_\lambda^\triangle$ are really in triangular form. First of all, the rows in $\mathcal{U}_\lambda^\triangle$ obviously possess different leading derivatives. Now suppose that there are two rows $f^1, f^2 \in \bar{\mathcal{S}}_\lambda^\triangle$ with $\operatorname{ld} D_{\mu_1} f^1 = \operatorname{ld} D_{\mu_2} f^2$. Let $f^1_{(k_1,\ell_1)}$ and $f^2_{(k_2,\ell_2)}$ be the corresponding rows in the skeleton $\mathcal{S}_\lambda^\triangle$ from which they have been derived by multiplicative prolongations. If either f^1 or f^2 is a single indexed row, we simply ignore all references made to the phantom level.

By definition of an involutive division, if two involutive cones intersect, one must be entirely contained in the other one. Thus without loss of generality, we may assume $\operatorname{ld} f^2 = \operatorname{ld} D_\nu f^1$ for some multiplicative prolongation D_ν. This assumption implies that in step $k_1 + |\nu|$ the row f^2 was reduced and became a phantom row. In the next $|\mu_2|$ steps, all multiplicative prolongations of f^2 up to this order are reduced. From $|\nu| + |\mu_2| = |\mu_1| \leq (\lambda - k_1)$ it follows that $D_{\mu_2} f^2 \notin \bar{\mathcal{S}}_\lambda^\triangle$, and we have arrived at a contradiction. □

By working with skeletons instead of full systems, we need to compute the multiplicative prolongations only when they are required for a reduction. In each step of the algorithm, we prolong the skeleton by exactly one order and compute a triangular form. However, as we want to realise the Cartan–Kuranishi completion, we have to relate these ideas to the geometric theory.

We have already seen that the skeleton $\mathcal{S}_\lambda^\triangle$ yields a local representation for the differential equation $\mathcal{R}_{q+\lambda}$. In fact, it contains even more information. Projecting into jet bundles of lower order, one can extract from $\mathcal{S}_\lambda^\triangle$ local representations for all differential equations $\mathcal{R}^{(s)}_{q+r}$ with $\lambda = r + s$:

$$\mathcal{S}_\lambda^\triangle : \left\{ \mathcal{R}_{q+\lambda} \overset{\pi^{q+\lambda}_{q+\lambda-1}}{\longrightarrow} \mathcal{R}^{(1)}_{q+\lambda-1} \overset{\pi^{q+\lambda-1}_{q+\lambda-2}}{\longrightarrow} \mathcal{R}^{(2)}_{q+\lambda-2} \overset{\pi^{q+\lambda-2}_{q+\lambda-3}}{\longrightarrow} \cdots \right. . \tag{10.105}$$

Because of the triangular form of the system $\bar{\mathcal{S}}_\lambda^\triangle$, these projections amount to nothing more than dropping the rows of appropriate orders, although we will never do this explicitly. Instead, we always identify the differential equation $\mathcal{R}^{(s)}_{q+r}$ currently considered by the values of the two parameters r and s.

If we juxtapose all the sequences of the form (10.105), then we arrive at the grid pictured in Figure 10.5. Starting with $\mathcal{R}_q$, the Cartan–Kuranishi Algorithm 7.3 moves along the horizontal lines searching for an equation with involutive symbol. If one is found, it is checked for integrability conditions by comparing it with the equation one row below and one column to the right. If the two equations differ, then the process is continued in the row below, i. e. after adding the integrability conditions. Recall that in general $\bigl(\mathcal{R}^{(s)}_{q+r}\bigr)_{+1} \neq \mathcal{R}^{(s)}_{q+r+1}$; to ensure equality we need an involutive symbol (cf. Part (ii) of Proposition 7.2.5). So two tasks remain:

1. decide via the skeleton which equations have an involutive symbol;
2. determine whether two equations $\mathcal{R}^{(s)}_{q+r}$ and $\mathcal{R}^{(s+1)}_{q+r}$ are equal.

The first problem is solved by the following result.

Proposition 10.7.5. *Let $\mathcal{S}_\lambda^\triangle$ be a skeleton for the differential equation $\mathcal{R}_{q+\lambda}$ in a δ-regular coordinate system and let t denote the maximal order of a single indexed*

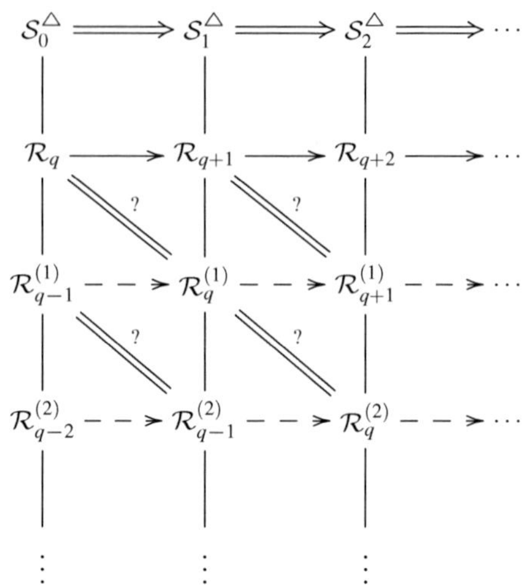

Fig. 10.5 Skeletons and the corresponding equations

row in the prolonged skeleton $\mathcal{S}^{\triangle}_{\lambda+1}$. Then the symbol $\mathcal{N}^{(s)}_{q+r}$ is involutive for all index values r, s such that $r+s=\lambda$ and $q+r\geq t$.

Proof. We choose r,s with $r+s=\lambda$. A local representation for $\mathcal{R}^{(s)}_{q+r}$ is obtained from $\bar{\mathcal{S}}^{\triangle}_{\lambda}$ by dropping all rows of order greater than $q+r$. We prolong this system by one order and compute a triangular form. If no non-multiplicative prolongations of order $q+r+1$ remain, then Proposition 7.2.3 asserts that the symbol $\mathcal{N}^{(s)}_{q+r}$ is involutive. By assumption, even after the prolongation and subsequent triangulation of the full skeleton no single indexed rows of this or a greater order exist in $\mathcal{S}^{\triangle}_{\lambda+1}$. As an independent non-multiplicative prolongation would result in such a row, the involution of the examined symbol can be read off from this skeleton. □

As a useful side effect, this result allows us to carry out additional projections while working inside a given skeleton: we trace a vertical line in Figure 10.5 belonging to some skeleton $\mathcal{S}^{\triangle}_{\lambda}$ downwards as long as the symbols of the respective differential equations remain involutive and continue with the last such equation. This strategy is illustrated in Figure 10.6 in comparison to the generic version of the Cartan–Kuranishi completion (boxed equations possess involutive symbols).

For the second task mentioned above, we note that it suffices to consider the dimensions of the involved manifolds, as one is always a submanifold of the other. As the triangulation guarantees algebraic independence of the rows in a skeleton, the determination of dimensions reduces to a simple counting of equations.

Definition 10.7.6. Let $\mathcal{S}^{\triangle}_{\lambda}$ be a triangular skeleton. We denote by $\#\mathcal{S}^{\triangle}_{\lambda,t}$ the number of rows of order t in the system $\bar{\mathcal{S}}^{\triangle}_{\lambda}$. The *rank vector* of $\mathcal{S}^{\triangle}_{\lambda}$ is the list

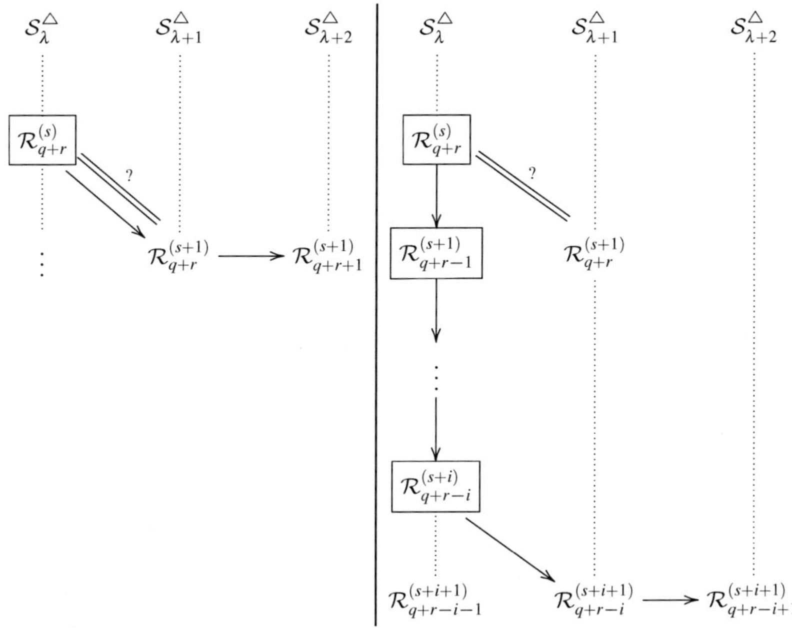

Fig. 10.6 Analysis of symbols and differential systems (left: classical Cartan–Kuranishi algorithm; right: hybrid algorithm).

$$\mathbf{r}_{\mathcal{S}_\lambda^\triangle} = \left[\#\mathcal{S}_{\lambda,0}^\triangle, \dots, \#\mathcal{S}_{\lambda,q+\lambda}^\triangle\right] \tag{10.106}$$

where q is the order of the original skeleton $\mathcal{S}_0$.

The numbers $\#\mathcal{S}_{\lambda,t}^\triangle$ yield straightforwardly the dimensions of all differential equations and symbols represented by $\mathcal{S}_\lambda^\triangle$. For $\lambda = r+s$

$$\dim \mathcal{R}_{q+r}^{(s)} = \dim J_{q+r}\pi - \sum_{i=0}^{q+r} \#\mathcal{S}_{\lambda,i}^\triangle\,, \tag{10.107a}$$

$$\dim \mathcal{N}_{q+r}^{(s)} = \dim S_q(T^*\mathcal{X}) \underset{J_{q-1}\pi}{\otimes} V\mathcal{E} - \#\mathcal{S}_{\lambda,q+r}^\triangle\,. \tag{10.107b}$$

The equations $\mathcal{R}_{q+r}^{(s)}$ and $\mathcal{R}_{q+r}^{(s+1)}$ are identical, i. e. no integrability conditions appear during the next prolongation, if the two rank vectors $\mathbf{r}_{\mathcal{S}_\lambda^\triangle}$ and $\mathbf{r}_{\mathcal{S}_{\lambda+1}^\triangle}$ coincide in their first $q+r$ entries. The higher entries play no role for this comparison.

It is straightforward to modify the Algorithms 10.2 and 10.3 so that they update the rank vector. Computing it for the initial skeleton is trivial. During the prolongation from $\mathcal{S}_\lambda$ to $\mathcal{S}_{\lambda+1}$ the following corrections are necessary for each row $f \in \mathcal{S}_\lambda$:

- If $f_{(k)}$ is a single indexed row of order t with level $k = \lambda$, prolongations with respect to all independent variables are computed. Hence the entry for order $t+1$ is increased by n, the number of independent variables.
- If the initial level of $f_{(k)}$ with order t and class c is less than λ, $\binom{\lambda-k+c}{c-1}$ new rows of order $t+\lambda-k+1$ enter the system.
- If $f_{(k,\ell)}$ is a double indexed row of order t and class c, there are $\binom{\lambda-k+c}{c-1}$ new rows of order $t+\lambda-k+1$, and $\binom{\lambda-\ell+c}{c-1}$ rows of order $t+(\lambda-\ell)+1$ are removed from the system.

For a triangulation, the corrections are even simpler: if a row is reduced to a new row of lower order, the entry in the rank vector for the old order has to be decreased and the one for the new order has to be increased by one.

Combining all these ideas and results we finally arrive at the promised hybrid Algorithm 10.4. In contrast to the Cartan–Kuranishi Algorithm 7.3, it consists of only one loop in which the skeleton is prolonged and triangulised (Line /4/) using the above described algorithms. The analysis of the symbol based on Proposition 10.7.5 takes place in Line /5/. A simple comparison of rank vectors in Line /8/ signals whether integrability conditions have appeared. Note that we compare the entries of the vectors only up to the currently relevant order. The `while` loop in Lines /13–15/ realises the idea sketched in Figure 10.6 and projects as far as possible.

Algorithm 10.4 Hybrid completion of linear systems

Input: skeleton $\mathcal{S}_0^{\triangle}$ of the differential equation $\mathcal{R}_q$
Output: skeleton $\mathcal{S}_\lambda^{\triangle}$ and integers $r, s \in \mathbb{N}$ such that $\mathcal{R}_{q+r}^{(s)}$ is involutive

1: $r \leftarrow 0$; $s \leftarrow 0$; $\lambda \leftarrow 0$; *systemInvolutive* $\leftarrow$ *FALSE*
2: **repeat**
3: $\quad \bar{\mathbf{r}} \leftarrow \mathbf{r}_{\mathcal{S}_\lambda^{\triangle}}$; $\lambda \leftarrow \lambda + 1$
4: $\quad \mathcal{S}_\lambda \leftarrow$ `ProlongSkeleton`$(\mathcal{S}_{\lambda-1}^{\triangle})$; $\mathcal{S}_\lambda^{\triangle} \leftarrow$ `TriangulisekSkeleton`$(\mathcal{S}_\lambda)$
5: $\quad$ **if** $\exists f_{(k)} \in \mathcal{U}_\lambda^{\triangle} : \operatorname{ord} f_{(k)} = q+r+1$ **then** {symbol not involutive}
6: $\qquad r \leftarrow r+1$
7: $\quad$ **else** {$\mathcal{N}_{q+r}^{(s)}$ involutive; check formal integrability of $\mathcal{R}_{q+r}^{(s)}$}
8: $\qquad$ **if** $\forall 0 \le i \le q+r : \mathbf{r}_{\mathcal{S}_\lambda^{\triangle}}[i] = \bar{\mathbf{r}}[i]$ **then**
9: $\qquad\quad$ *systemInvolutive* $\leftarrow$ *TRUE*
10: $\qquad$ **else**
11: $\qquad\quad r \leftarrow r+1$
12: $\qquad$ **end if**
13: $\qquad$ **while** $\neg\big(\exists f_{(k)} \in \mathcal{U}_\lambda^{\triangle} : \operatorname{ord} f_{(k)} = q+r\big)$ **do** {project as far as possible}
14: $\qquad\quad r \leftarrow r-1$; $s \leftarrow s+1$
15: $\qquad$ **end while**
16: $\quad$ **end if**
17: **until** *systemInvolutive*
18: **return** $(r, s, \mathcal{S}_\lambda^{\triangle})$

Strictly speaking, Algorithm 10.4 is *not* a realisation of the Cartan–Kuranishi completion, as the inner `while` loop performs additional projections. A strict realisation would allow only one iteration through this loop. These additional projections are, however, a benefit, as one is interested in those values r, s for which $\mathcal{R}^{(s)}_{q+r}$ is involutive *and* r is minimal. Our new algorithm delivers these values.

Theorem 10.7.7. *If the used coordinates* $\mathbf{x}$ *are* δ*-regular for all appearing symbols, Algorithm 10.4 terminates with an involutive system.*

Proof. The claim follows immediately from the Cartan–Kuranishi Theorem 7.4.1. The only possible problem for the termination is that we might fail to recognise that some symbol is involutive. In this case, Algorithm 10.4 could prolong infinitely often. But as long as the used coordinate system is δ-regular for every appearing symbol, involution of the symbols is easily checked by Proposition 10.7.5. □

Theorem 4.3.12 allows us to verify the condition of Theorem 10.7.7. If the used coordinates are not δ-regular for one of the symbols, then at some stage of the completion a larger Janet span appears. Thus with the help of this simple criterion, we can effectively avoid an infinite computation.

Given the triangulised skeleton $\mathcal{S}^{\triangle}_{\lambda}$ of a differential equation $\mathcal{R}_q$, we refine the above considerations about the dimensions in order to determine the indices $\beta^{(i)}_q$ of its symbol $\mathcal{N}_q$. A row $f_{(k,\ell)} \in \mathcal{S}^{\triangle}_{\lambda}$ of order t and class c obviously contributes to them, if and only if (i) $\lambda - k + t \geq q$ and (ii) $\ell = 0$ or $\lambda - \ell + t < q$. Its contribution $\beta^{(i)}_q(f_{(k,\ell)})$ is then given by

$$\beta^{(i)}_q(f_{(k,\ell)}) = \begin{cases} B(c-i+1, q-t, 1) & \text{for } 1 \leq i \leq c\,, \\ 0 & \text{for } c < i \leq n \end{cases} \tag{10.108}$$

where $B(n,q,c) = \binom{n-c+q-1}{q-1}$ is the number of multi indices of length n, order q and class c. The Cartan characters $\alpha^{(i)}_q$ of the equation are then computed from (8.6).

Example 10.7.8. We demonstrate the working of Algorithm 10.4 for the Janet Example 7.4.3. Our algorithm starts with the skeleton

$$\mathcal{S}_0 = \mathcal{S}^{\triangle}_0 : \begin{cases} (u_{zz} + yu_{xx})_{(0)}\,, \\ (u_{yy})_{(0)}\,. \end{cases} \tag{10.109}$$

The initial rank vector is $\mathbf{r}_{\mathcal{S}^{\triangle}_0} = [0,0,2]$ and the algorithm proceeds as follows.

1. *Iteration*: There is only one non-multiplicative prolongation: u_{yyz}. It cannot be reduced and it is added to the skeleton with initial level 1. Since $\mathcal{S}^{\triangle}_1$ contains a third-order row, the symbol $\mathcal{N}_2$ is not involutive.
2. *Iteration*: u_{yyzz} is the only new non-multiplicative prolongation and it reduces to u_{xxy}. As thus no row of order 4 remains in the skeleton, $\mathcal{N}_3$ is involutive. We must now compare the rank vectors up to order 3: $\bar{\mathcal{S}}^{\triangle}_1$ contains 6 equations of order 3, whereas $\bar{\mathcal{S}}^{\triangle}_2$ contains 7. Thus one integrability condition has appeared (which is of course u_{xxy}) and the algorithm continues with the equation $\mathcal{R}^{(1)}_3$.

3. *Iteration*: The computed non-multiplicative prolongations are u_{xxyy} and u_{xxyz}. Only the first row can be reduced to zero, while the second one obstructs the involution of $\mathcal{N}_3^{(1)}$.
4. *Iteration*: Again there are two non-multiplicative prolongations: u_{xxyyz} can be reduced to zero, whereas u_{xxyzz} is replaced by u_{xxxx}. Due to the lack of an equation of order 5 in $\mathcal{S}_4^{\triangle}$, the symbol $\mathcal{N}_4^{(1)}$ is involutive. The rank vectors that must be compared are: $\mathbf{r}_{\mathcal{S}_3^{\triangle}} = [0,0,2,7,13,17]$ and $\mathbf{r}_{\mathcal{S}_4^{\triangle}} = [0,0,2,7,14,19,24]$. Only the first five components are taken into account, but the difference between the entries 13 and 14 signals the appearance of another integrability condition (u_{xxxx}). The next equation to be examined is $\mathcal{R}_4^{(2)}$.
5. *Iteration*: Of the two non-multiplicative prolongations, u_{xxxxy} is reduced to zero, whereas u_{xxxxz} represents an obstruction to involution for the symbol $\mathcal{N}_4^{(2)}$. The current skeleton is now

$$\mathcal{S}_5^{\triangle} : \begin{cases} (u_{xxxxz})_{(5)}\,, & (u_{yyz})_{(1)}\,, \\ (u_{xxxx})_{(4)}\,, & (u_{zz}+yu_{xx})_{(0)}\,, \\ (u_{xxyz})_{(3)}\,, & (u_{yy})_{(0)}\,, \\ (u_{xxy})_{(2)} & \end{cases} \tag{10.110}$$

with rank vector $\mathbf{r}_{\mathcal{S}_5^{\triangle}} = [0,0,2,7,14,21,26,32]$.
6. *Iteration*: As both non-multiplicative prolongations u_{xxxxyz} and u_{xxxxzz} can be reduced to zero, the symbol $\mathcal{N}_5^{(2)}$ is involutive. The comparison of the rank vectors shows that no new integrability conditions have appeared. Thus the algorithm stops and outputs the skeleton $\mathcal{S}_6^{\triangle}$, which is identical with $\mathcal{S}_5^{\triangle}$ given above, for the differential equation $\mathcal{R}_5^{(2)}$. A local representation is obtained by prolonging all equations in the skeleton multiplicatively up to fifth order.

Thus the path of the Cartan–Kuranishi completion can be summarised in the following sequence of differential equations:

$$\mathcal{R}_2 \to \mathcal{R}_3 \to \mathcal{R}_3^{(1)} \to \mathcal{R}_4^{(1)} \to \mathcal{R}_4^{(2)} \to \mathcal{R}_5^{(2)}\,. \tag{10.111}$$

In order to demonstrate the superior efficiency compared to a direct implementation of the Cartan–Kuranishi completion, we do some statistics. The new algorithm computes 10 non-multiplicative and 18 multiplicative prolongations (for the reductions); 5 of the latter ones are used several times. By contrast, a full determination of all systems requires the prolongation of 132 equations—almost five times more. In addition, the naive approach requires to determine row echelon forms for a number of matrices (Jacobians and symbols), the largest being an 86×84 matrix. ◁

Remark 10.7.9. One may wonder why we need rank vectors, as all integrability conditions in the Janet example above stem from non-multiplicative prolongations that do not reduce to zero. However, recall that there are two types of integrability conditions: those arising from (generalised) cross-derivatives and thus in our formalism from non-multiplicative prolongations and those arising from the prolongation of lower-order equations. For detecting the latter ones—without actually computing them—we need the rank vectors.

As a trivial example consider the equation $\mathcal{R}_2$ locally represented by $u_{yy} = u_{xy} = u_x = 0$. Obviously, the symbol $\mathcal{N}_2$ is involutive. But $\mathcal{R}_2$ is not formally integrable: prolongation of the lower order equation $u_x = 0$ with respect to the *multiplicative* variable x yields the additional equation $u_{xx} = 0$ required to represent $\mathcal{R}_2^{(1)}$. In our algorithm the two skeletons $\mathcal{S}_0^\triangle$ and $\mathcal{S}_1^\triangle$ contain exactly the same rows. But their rank vectors differ: $\mathbf{r}_{\mathcal{S}_0^\triangle} = [0,1,2]$ and $\mathbf{r}_{\mathcal{S}_1^\triangle} = [0,1,3,3]$. The increase in the third entry is due to the integrability condition $u_{xx} = 0$. Algorithm 10.4 never explicitly determines such integrability conditions; they are solely visible through changes in the rank vector. ◁

In the Janet example, phantom rows do not occur and the projections are always only by one order. In the next example, stemming from Lie symmetry theory, both phenomena are present.

Example 10.7.10. Consider the determining system for Lie point symmetries of the heat equation. It consists of nine equations in three dependent variables η, ξ, τ and three independent variables t, x, u:

$$\mathcal{R}_2 : \begin{cases} \tau_u = 0\,, & \tau_{ux} + \xi_u = 0\,, & \eta_{uu} - \xi_{ux} = 0\,, \\ \tau_x = 0\,, & \tau_{xx} + 2\xi_x - \tau_t = 0\,, & \eta_{xx} - \eta_t = 0\,, \\ \tau_{uu} = 0\,, & \eta_{ux} - \frac{1}{2}\xi_{xx} + \frac{1}{2}\xi_t = 0\,, & \xi_{uu} = 0\,. \end{cases} \tag{10.112}$$

As above, we obtain the skeleton $\mathcal{S}_0^\triangle$ by dropping the right hand side of each equation and assigning an initial level of 0. During the triangulation of $\mathcal{S}_1$, the row $\tau_{xx} + 2\xi_x - \tau_t$ can be reduced by the (multiplicative) prolongation with respect to x of the row τ_x. This yields the row $\xi_x - \frac{1}{2}\tau_t$, which is added to $\mathcal{S}_1^\triangle$. However, the system it defines also contains multiplicative prolongations of $\tau_{xx} + 2\xi_x - \tau_t$. Thus dropping the row completely would lose information and lead to wrong values for the dimensions. Instead, its phantom level is set to 1. In the λth iteration, all rows obtained by prolonging λ times the row $\tau_{xx} + 2\xi_x - \tau_t$ are still present in $\bar{\mathcal{S}}_\lambda$ and available for reducing other rows.

The algorithm continues to produce the following sequence of differential equations (note the double projection in the third iteration step):

$$\mathcal{R}_2 \to \mathcal{R}_3 \to \mathcal{R}_3^{(1)} \to \mathcal{R}_2^{(3)} \to \mathcal{R}_2^{(4)} \to \mathcal{R}_3^{(4)}\,. \tag{10.113}$$

The final skeleton is

$$\mathcal{S}_5^\triangle : \begin{cases} \tau_{ttt} = 0\,, & \eta_{xx} - \eta_t = 0\,, & \xi_u = 0\,, \\ \eta_{uu} - \xi_{ux} = 0\,, & \xi_{tt} = 0\,, & \xi_x - \frac{1}{2}\tau_t = 0\,, \\ \eta_{ut} + \frac{1}{2}\xi_{xt} = 0\,, & \tau_u = 0\,, & \tau_x = 0\,, \\ \eta_{ux} - \frac{1}{2}\xi_{xx} + \frac{1}{2}\xi_t = 0\,. \end{cases} \tag{10.114}$$

Since only head reductions have been performed during the algorithm, there exist further possibilities for simplifications which, however, do not affect the leaders of the equations. ◁

10.8 Linear Systems of Finite Type with Constant Coefficients

The natural ring for treating systems with constant coefficients is $\mathcal{D} = \Bbbk[\partial_1, \dots, \partial_n]$ with a field $\Bbbk$ consisting only of constants for the derivations ∂_i. In the sequel we will exclusively consider the case $\Bbbk = \mathbb{C}$, as for some arguments it will be of importance whether or not the field $\Bbbk$ is algebraically closed. Furthermore, we choose as function space $\mathcal{A} = \mathcal{C}^\infty(\Omega, \mathbb{C})$ with some open convex subset $\Omega \subseteq \mathbb{R}^n$. As mentioned in Remark 10.5.12, this space is an injective cogenerator.

Remark 10.8.1. The choice of $\mathcal{A}$ is not so crucial here. All our statements about equations of finite type below will remain true in larger function spaces (say, if we include distributional solutions). Indeed, any differential equation of finite type is trivially elliptic and hence hypoelliptic (see e. g. [225, Thm 11.1.10]) which means that even a distributional solution is automatically smooth. ◁

We will present in this section two approaches to the explicit integration of linear systems with constant coefficients. The first one is based on the theory developed in Section 10.5, i. e. on the differential module associated with the system. The second one exploits that for constants coefficients $\mathcal{D}$ is a commutative ring and applies some standard constructions in commutative algebra like primary decomposition. These two approaches appear very different. However, they are actually closely related by the fact that the zeros of a zero-dimensional polynomial ideal $\mathcal{J}$ are encoded in the eigenvalues and eigenvectors of the multiplication operators on the finite-dimensional vector space $\mathcal{P}/\mathcal{J}$. For more details on this relation we refer to [434] and references therein.

Obviously, the simplest example of a linear system with constant coefficients of finite type is an *ordinary* differential equation $\mathbf{u}' = A\mathbf{u}$ with some matrix $A \in \Bbbk^{m\times m}$ and it is well-known that its general solution is given by $\mathbf{u}(x) = \exp(xA)\mathbf{u}_0$ with an arbitrary vector $\mathbf{u}_0 \in \Bbbk^m$. We will show later that for *partial* differential equations of finite type $L\mathbf{u} = 0$ the general solution is given by a straight-forward generalisation of this expression. The existence of such a closed form solution is not surprising, as we saw already in Example 2.3.17 that the integration of an equation of finite type may always be reduced to the integration of ordinary differential equations via the Frobenius Theorem C.3.3. Although the linearity is preserved during this reduction, we will follow a completely different approach. We begin with the simplest problem of this type: a homogeneous first-order system. Here it is straightforward to obtain a closed form solution of the claimed form.

Lemma 10.8.2. *Let $A_1, \dots, A_n \in \Bbbk^{m\times m}$ be n commuting matrices. Then the unique smooth solution of the linear system*

$$\mathbf{u}_{x^1} = A_1\mathbf{u}\,, \quad \dots\,, \quad \mathbf{u}_{x^n} = A_n\mathbf{u} \tag{10.115}$$

for the initial condition $\mathbf{u}(0) = \mathbf{u}_0$ *is given by*

$$\mathbf{u}(\mathbf{x}) = \exp(x^iA_i)\mathbf{u}_0\,. \tag{10.116}$$

Proof. Let $\mathbf{v}$ be an arbitrary solution of (10.115). We claim that for each $1 \leq j \leq n$, the function $\mathbf{v}_j = \exp\left[-(x^1A_1 + \cdots + x^jA_j)\right]\mathbf{v}$ is independent of $x^1,\dots,x^j$. Obviously, this observation implies our assertion. The proof is by induction over j. For $j = 1$ we compute

$$\frac{\partial \mathbf{v}_1}{\partial x^1} = e^{(-x^1A_1)}\left(\frac{\partial \mathbf{v}}{\partial x^1} - A_1\mathbf{v}_1\right) = 0\,, \tag{10.117}$$

since $\mathbf{v}$ satisfies in particular the first equation in (10.115).

Now we assume that our claim holds for some $j < n$. We introduce the function $\tilde{\mathbf{v}} = \exp\left(-x^{j+1}A_{j+1}\right)\mathbf{v}_j$. By assumption, the matrices A_i commute and hence

$$\exp(-x^{j+1}A_{j+1})\mathbf{v}_j = \exp\left[-(x^1A_1 + \cdots + x^jA_j)\right]\tilde{\mathbf{v}}\,. \tag{10.118}$$

$\tilde{\mathbf{v}}$ (and thus the right hand side) does not depend on x^{j+1} by a similar computation as above and $\mathbf{v}_j$ (and thus the left hand side) not on $x^1,\dots,x^j$ by assumption. As both sides are equal to $\mathbf{v}_{j+1}$, our claim follows. □

Note that a linear system of the form (10.115) is formally integrable, if and only if the matrices A_i commute with each other. Hence this condition does not restrict the applicability of the lemma.

Remark 10.8.3. An alternative approach to Lemma 10.8.2 consists of exploiting the considerations at the end of Example 2.3.17. We are here in the special case that the right hand sides are of the form $\phi_i^\alpha(\mathbf{x},\mathbf{u}) = A_{i\beta}^\alpha u^\beta$. Thus we are lead to the following overdetermined system for one unknown function $v(\mathbf{x},\mathbf{u})$:

$$\frac{\partial v}{\partial x^i} + A_{i\beta}^\alpha u^\beta \frac{\partial v}{\partial u^\beta} = 0\,, \qquad 1 \leq i \leq n\,. \tag{10.119}$$

Using the method of characteristics it is not difficult to see that its general solution is given by $v(\mathbf{x},\mathbf{u}) = F\left(\mathbf{u} - \exp(x^iA_i)\mathbf{u}_0\right)$ with an arbitrary function $F : \mathbb{R}^n \to \mathbb{R}$ and an arbitrary vector $\mathbf{u}_0 \in \mathbb{R}^n$. Thus we obtain the same expression for the solution. ◁

Before we actually integrate linear systems, we study the dimension of their solution space. It turns out that over the coefficient field $\Bbbk$ the associated module $\mathcal{M}$ has the same dimension as $\ker_{\mathcal{A}} L$.

Proposition 10.8.4. *Let* $L \in \mathcal{D}^{r\times m}$ *be a matrix-valued linear differential operator and* $\mathcal{M} = \mathcal{D}^{1\times m}/\mathcal{D}^{1\times r}L$ *the associated* $\mathcal{D}$*-module. The solution space* $\ker_{\mathcal{A}} L$ *is finite-dimensional as a* $\Bbbk$*-linear space, if and only if the module* $\mathcal{M}$ *is also finite-dimensional as a* $\Bbbk$*-linear space. In this case* $\dim_{\Bbbk} \ker_{\mathcal{A}} L = \dim_{\Bbbk} \mathcal{M}$.

Proof. We assume first that $\mathcal{M}$ is finite-dimensional and proceed by an induction over $d = \dim_{\Bbbk} \mathcal{M}$. If $d = 1$, then $\mathcal{M} \cong \Bbbk \cong \mathcal{D}/\mathfrak{m}$ with a maximal ideal $\mathfrak{m} \subset \mathcal{D}$. Since $\Bbbk$ is assumed to be algebraically closed, any maximal ideal is of the form $\mathfrak{m} = \langle \partial_1 - a_1, \dots, \partial_n - a_n \rangle$ with coefficients $a_i \in \Bbbk$ (this is one formulation of Hilbert's Nullstellensatz). Hence, the given system is equivalent to a system of the form (10.115) with $m = 1$ and Lemma 10.8.2 implies that $\dim_{\Bbbk} \ker_{\mathcal{A}} L = 1$, too.

Now assume that our claim was correct for all $\mathcal{D}$-modules $\mathcal{N}$ with $\dim_{\Bbbk} \mathcal{N} < d$. Then choose some proper submodule $0 \neq \mathcal{N} \subset \mathcal{M}$ and consider the associated short exact sequence of $\Bbbk$-linear spaces

$$0 \longrightarrow \mathcal{N} \longrightarrow \mathcal{M} \longrightarrow \mathcal{M}/\mathcal{N} \longrightarrow 0 \tag{10.120}$$

where as usual the maps are the inclusion and the canonical projection. By elementary linear algebra we have that $\dim_{\Bbbk} \mathcal{M} = \dim_{\Bbbk} \mathcal{N} + \dim_{\Bbbk} (\mathcal{M}/\mathcal{N})$.

As the $\mathcal{D}$-module $\mathcal{A}$ is injective, application of the functor $\operatorname{Hom}_{\mathcal{D}}(\cdot, \mathcal{A})$ yields a dual exact sequence of $\Bbbk$-linear spaces

$$0 \longleftarrow \operatorname{Hom}_{\mathcal{D}}(\mathcal{N}, \mathcal{A}) \longleftarrow \operatorname{Hom}_{\mathcal{D}}(\mathcal{M}, \mathcal{A}) \longleftarrow \operatorname{Hom}_{\mathcal{D}}(\mathcal{M}/\mathcal{N}, \mathcal{A}) \longleftarrow 0\,. \tag{10.121}$$

Applying the Malgrange isomorphism (Proposition 10.5.1) and exploiting that both $\mathcal{N}$ and $\mathcal{M}/\mathcal{N}$ are of smaller dimension so that our claim holds for them, we obtain

$$\begin{aligned} \dim_{\Bbbk} \ker_{\mathcal{A}} L &= \dim_{\Bbbk} \operatorname{Hom}_{\mathcal{D}}(\mathcal{N}, \mathcal{A}) + \dim_{\Bbbk} \operatorname{Hom}_{\mathcal{D}}(\mathcal{M}/\mathcal{N}, \mathcal{A}) \\ &= \dim_{\Bbbk} \mathcal{N} + \dim_{\Bbbk} (\mathcal{M}/\mathcal{N}) = \dim_{\Bbbk} \mathcal{M}\,. \end{aligned} \tag{10.122}$$

Hence our claim is also valid for dimension d.

For the converse, we use an indirect proof. Assume that the module $\mathcal{M}$ was infinite-dimensional over $\Bbbk$. Then $\mathcal{M}$ cannot be Artinian and thus the dual object $\operatorname{Hom}_{\mathcal{D}}(\mathcal{M}, \mathcal{A}) \cong \ker_{\mathcal{A}} L$ cannot be Noetherian (any infinite descending sequence of submodules $\mathcal{M} \supset \mathcal{N}_1 \supset \mathcal{N}_2 \supset \cdots$ induces an infinite ascending sequence of submodules of the form $\{\phi \in \operatorname{Hom}_{\mathcal{D}}(\mathcal{M}, \mathcal{A}) \mid \phi|_{\mathcal{N}_i} = 0\}$). But this is only possible, if $\dim_{\Bbbk} \ker_{\mathcal{A}} L = \infty$ contradicting our assumption. □

Using the reduction technique described in Appendix A.3, we can transform any involutive homogeneous linear system of finite type into a first-order system of the form (10.115). However, in general this reduction leads to a considerable increase in the size of the system. As the evaluation of a matrix exponential is fairly expensive, the reduction approach is therefore not well suited for concrete computations. We will now show how algebraic methods allow us always the reduction to a system of the form (10.115), the dimension of which is given by the dimension of the solution space. Obviously, no smaller dimension can be achieved.

In the sequel we take $\Omega \subseteq \mathbb{R}^n$, an open connected set, and $\mathcal{A} = \mathcal{C}^\infty(\Omega, \Bbbk)$. As usual, our linear system on Ω is given by a linear differential operator $L \in \mathcal{D}^{r \times m}$ and we do not make any assumption about the order of L. If the differential operator L defines a differential equation of finite type, then the $\mathcal{D}$-module $\mathcal{M} = \mathcal{D}^{1 \times m}/\mathcal{D}^{1 \times r} L$

is finite dimensional as a vector space over $\mathbb{k}$. Our goal is now to find an equivalent linear system of the form (10.115) so that we can apply Lemma 10.8.2.

Two basic ideas underlie the construction of such an equivalent system. According to Proposition 10.5.1, the solution space $\ker_{\mathcal{A}} L$ over $\mathcal{A}$ is canonically isomorphic to $\mathrm{Hom}_{\mathcal{D}}(\mathcal{M},\mathcal{A})$, which may be considered as a subspace of $\mathrm{Hom}_{\mathbb{k}}(\mathcal{M},\mathcal{A})$ (obviously, $\mathcal{D}$-linearity implies $\mathbb{k}$-linearity but not vice versa). Now elements of $\mathrm{Hom}_{\mathbb{k}}(\mathcal{M},\mathcal{A})$ map an element of $\mathcal{M}$ to a map from Ω to $\mathbb{k}$. The next lemma shows that we may switch the order of evaluation in these nested maps. In order to formulate rigorously this "switching," we need the $\mathbb{k}$-linear dual $\mathcal{M}^* = \mathrm{Hom}_{\mathbb{k}}(\mathcal{M},\mathbb{k})$.

Lemma 10.8.5. *The vector spaces* $\mathcal{C}^\infty(\Omega,\mathcal{M}^*)$ *and* $\mathrm{Hom}_{\mathbb{k}}(\mathcal{M},\mathcal{A})$ *are canonically isomorphic over* $\mathbb{k}$.

Proof. Consider the $\mathbb{k}$-linear map $\Phi : \mathcal{C}^\infty(\Omega,\mathcal{M}^*) \to \mathrm{Hom}_{\mathbb{k}}(\mathcal{M},\mathcal{A})$ defined by setting $\Phi(\mathbf{w})(m)(x) = \mathbf{w}(x)(m)$ for $\mathbf{w} \in \mathcal{C}^\infty(\Omega,\mathcal{M}^*)$, $m \in \mathcal{M}$ and $x \in \Omega$. Similarly, we introduce the $\mathbb{k}$-linear map $\Psi : \mathrm{Hom}_{\mathbb{k}}(\mathcal{M},\mathcal{A}) \to \mathcal{C}^\infty(\Omega,\mathcal{M}^*)$ by setting $\Psi(\phi)(x)(m) = \phi(m)(x)$ for $\phi \in \mathrm{Hom}_{\mathbb{k}}(\mathcal{M},\mathcal{A})$. Both maps are well-defined according to the definitions of the various spaces. Now assume that $\phi = \Phi(\mathbf{w})$. Then $\Psi(\phi)(x)(m) = \Phi(\mathbf{w})(m)(x) = \mathbf{w}(x)(m)$ and hence $\Psi(\phi) = \mathbf{w}$ so that $\Psi \circ \Phi = \mathrm{id}$. Similarly, we get $\Phi \circ \Psi = \mathrm{id}$ by assuming $\mathbf{w} = \Psi(\phi)$ and applying Φ. Hence Φ and Ψ are mutually inverse isomorphisms. □

The second basic observation is that solving a linear system is closely related to studying the multiplication operators on the module $\mathcal{M}$. It follows from Macaulay's Theorem (cf. Remark B.4.10) that the parametric derivatives induce a $\mathbb{k}$-linear basis of $\mathcal{M}$. In our case this basis is finite, as our system is assumed to be of finite type. Hence the—according to Theorem 9.3.5 formally well-posed—initial value problem (9.31) consists of prescribing for each parametric derivative a value at the origin; in other words, we may identify initial value data with elements of the dual space $\mathcal{M}^*$.

Algorithm 9.1 for the construction of arbitrary Taylor coefficients of the general solution of a linear system proceeds essentially by representing the sought derivative as a prolongation of a parametric derivative (which corresponds to repeated multiplications with the variables ∂_i of the polynomial ring $\mathcal{D}$) and then computing a normal form of it with respect to a Gröbner basis. But this procedure is equivalent to giving an explicit representation of the multiplication maps $\partial_i : \mathcal{M} \to \mathcal{M}$ defined by $[\mathcal{P}] \mapsto [\partial_i \cdot \mathcal{P}]$. In our special case, $\mathcal{M}$ is a finite dimensional $\mathbb{k}$-linear space so that we may represent each map ∂_i by a matrix $A_i \in \mathbb{k}^{\ell\times\ell}$ where $\ell = \dim_{\mathbb{k}} \mathcal{M}$ after choosing a basis of $\mathcal{M}$. Since we are dealing with constant coefficients, the operators ∂_i commute and hence the matrices A_i commute, too. We will actually need below the dual maps $\partial_i^* : \mathcal{M}^* \to \mathcal{M}^*$; obviously, with respect to the dual basis they are represented by the transposed matrices A_i^t.

Now we combine these two basic observations. On the space $\mathcal{C}^\infty(\Omega,\mathcal{M}^*)$ we have a natural $\mathcal{D}$-module action by the usual differentiation and we may consider the following linear system with constant coefficients:

$$\frac{\partial \mathbf{w}}{\partial x^1} = A_1^t \mathbf{w}\,, \quad \dots\,, \quad \frac{\partial \mathbf{w}}{\partial x^n} = A_n^t \mathbf{w}\,. \tag{10.123}$$

Obviously, it is of finite type and of the form (10.115). We claim now that this system is equivalent to our original one $L\mathbf{u} = 0$.

Lemma 10.8.6. *There is a canonical isomorphism between the solution space of the linear system* (10.123) *and* $\mathrm{Hom}_{\mathcal{D}}(\mathcal{M}, \mathcal{A})$ *as* $\Bbbk$*-linear spaces.*

Proof. Note that a $\Bbbk$-homomorphism $\phi \in \mathrm{Hom}_{\Bbbk}(\mathcal{M}, \mathcal{A})$ is simultaneously a $\mathcal{D}$-module homomorphism, if and only if $\partial_i \cdot \phi(m) = \phi(\partial_i \cdot m) = \phi(A_i m)$ for all $m \in \mathcal{M}$. If we write $\phi = \Phi(\mathbf{w})$ where Φ is the map introduced in the proof of Lemma 10.8.5, then this condition is equivalent to $\mathbf{w} \in \mathcal{C}^\infty(\Omega, \mathcal{M}^*)$ being a solution of (10.123), as for any $m \in \mathcal{M}$

$$\Phi(\mathbf{w})(A_i m)(x) = \mathbf{w}(x)(A_i m) = A_i^t\big(\mathbf{w}(x)\big)(m) \tag{10.124}$$

by definition of the dual map ∂_i^* and

$$\partial_i \cdot \big[\Phi(\mathbf{w})(m)(x)\big] = \partial_i \cdot \big[\mathbf{w}(x)(m)\big] = \frac{\partial \mathbf{w}}{\partial x^i}(x)(m) \tag{10.125}$$

by definition of the action of $\mathcal{D}$ on $\mathcal{A}$ and $\mathcal{C}^\infty(\Omega, \mathcal{M}^*)$, respectively. □

Since the solution space of our given linear system $L\mathbf{u} = 0$ and the solution space of the first order system (10.123) are isomorphic, we can write down the closed form solution of our system as soon as we have an explicit representation of this isomorphism, as the latter space is determined by Lemma 10.8.2. But such a representation is easily obtained: identifying the standard bases of $\Bbbk^m$ and $\mathcal{D}^m$, we have a canonical homomorphism $\pi : \Bbbk^m \to \mathcal{M}$ and its dual map $\pi^* : \mathcal{M}^* \to \Bbbk^m$. Given a function $\mathbf{w} \in \mathcal{C}^\infty(\Omega, \mathcal{M}^*)$, we thus associate with it the function $\pi^* \circ \mathbf{w} \in \mathcal{A}^m$ and obtain finally the following result.

Theorem 10.8.7. *Every solution of the linear differential equation of finite type described by the* $\mathcal{D}$*-module* $\mathcal{M}$ *is of the form*

$$\mathbf{u}(x) = \pi^* \circ \big(\exp(x^i A_i^t)\lambda\big) \tag{10.126}$$

with $\lambda \in \mathcal{M}^*$ *the corresponding initial data vector.*

As every step in the considerations above was constructive, it is not difficult to formulate Algorithm 10.5 for the construction of the closed form solution of a linear system with constant coefficients of finite type. For the construction of the complementary decomposition in Line /1/ we may use any of the algorithms presented in Section 5.1. The evaluation of the matrix exponential can be done via Jordan normal forms, if one manages to obtain the eigenvalues of the matrices A_i^t; in practice, this step will be the most difficult one and only possible, if $\dim_{\Bbbk} \mathcal{M}$ is small.

Example 10.8.8. We demonstrate Algorithm 10.5 for a concrete linear system of finite type, namely the second-order system

Algorithm 10.5 Solution of linear systems with constant coefficients of finite type

Input: linear system with constant coefficients of finite type $L\mathbf{u} = 0$
Output: general solution in explicit form
1: compute complementary decomposition $\mathcal{M} \cong \bigoplus_{t\in\mathcal{T}} \Bbbk \cdot t$
2: compute matrix representations A_i of the $\Bbbk$-linear maps $\partial_i : \mathcal{M} \to \mathcal{M}$ with respect to the $\Bbbk$-linear basis $\mathcal{T}$
3: compute matrix representation P of $\pi^* : \mathcal{M}^* \to \mathcal{M}^*$
4: evaluate the matrix exponentials $\exp(x^i A_i^t)$
5: **return** $P\big(\exp(x^i A_i^t)\lambda\big)$

$$\mathcal{R}_2 : \begin{cases} u_{yy} - 2u_y + u = 0\,, \\ u_{xx} - u_y = 0\,. \end{cases} \tag{10.127}$$

The symbol of this system coincides with that of the equation studied in Example 7.2.2. One readily checks that it is formally integrable and becomes involutive after one prolongation. Computing the corresponding complementary decomposition, we find that

$$\mathcal{M} \cong \mathbb{R}\cdot\bar{1} \oplus \mathbb{R}\cdot\overline{\partial_x} \oplus \mathbb{R}\cdot\overline{\partial_y} \oplus \mathbb{R}\cdot\overline{\partial_x\partial_y}\,, \tag{10.128}$$

since u, u_x, u_y and u_{xy} are the parametric derivatives of (10.127). Thus as a vector space $\mathcal{M}$ is isomorphic to $\mathbb{R}^4$.

The action of ∂_x and ∂_y, respectively, on the $\mathcal{D}$-module $\mathcal{M}$ is described (with respect to the above used basis) by the matrices

$$A_x = \begin{pmatrix} 0 & 0 & 0 & -1 \\ 1 & 0 & 0 & 0 \\ 0 & 1 & 0 & 2 \\ 0 & 0 & 1 & 0 \end{pmatrix}, \qquad A_y = \begin{pmatrix} 0 & 0 & -1 & 0 \\ 0 & 0 & 0 & -1 \\ 1 & 0 & 2 & 0 \\ 0 & 1 & 0 & 2 \end{pmatrix} \tag{10.129}$$

(this result follows immediately from the observation that modulo the linear system (10.127) $u_{xx} = u_y$, $u_{xxy} = u_{yy} = 2u_y - u$, $u_{yy} = 2u_y - u$ and $u_{xyy} = 2u_{xy} - u_x$). For the same basis, the homomorphism π^* is obviously represented by the matrix

$$P = \begin{pmatrix} 1 & 0 & 0 & 0 \end{pmatrix}\,, \tag{10.130}$$

as we have chosen $\bar{1}$ as our first basis vector.

A vector $\lambda \in \mathcal{M}^* \cong \mathbb{R}^4$ is uniquely determined by its values $(u^{(0)}, u_x^{(0)}, u_y^{(0)}, u_{xy}^{(0)})$ at the four basis vectors $\bar{1}$, $\overline{\partial_x}$, $\overline{\partial_y}$, $\overline{\partial_{xy}}$ (equivalently we may say that these values are prescribed for the parametric derivatives u, u_x, u_y and u_{xy} at some fixed point $x_0 \in \Omega$, for simplicity here taken as the origin). Evaluating the matrix exponentials and applying P to the result, we obtain the following closed form expression for the general solution of (10.127):

$$\begin{aligned} u(x,y) = \frac{1}{4}\Big[\big(-u^{(0)} - u_x^{(0)} + u_y^{(0)} + u_{xy}^{(0)}\big)(x+2y) + \big(2u^{(0)} + 3u_x^{(0)} - u_{xy}^{(0)}\big)\Big]e^{y+x} + \\ \frac{1}{4}\Big[\big(u^{(0)} - u_x^{(0)} - u_y^{(0)} + u_{xy}^{(0)}\big)(x-2y) + \big(2u^{(0)} - 3u_x^{(0)} + u_{xy}^{(0)}\big)\Big]e^{y-x}\,. \end{aligned} \tag{10.131}$$

Note that this representation depends on the basis chosen for $\mathcal{M}$. Using a different complementary decomposition will usually yield a different, but of course equivalent form of the general solution.

A systematic application of the reduction approach would have lead here to a higher dimension. Because of the obstruction to involution $u_{xxy} - 2u_y + u = 0$, we must take as starting point the prolongation to third order and thus introduce five new dependent variables corresponding to the derivatives $u_x, u_y, u_{xx}, u_{xy}, u_{yy}$. Hence this approach would have required to work with six-dimensional matrices.

A closer look at the above obstruction to involution shows that only one third-order derivative effectively appears in it and hence it is not really necessary to introduce new dependent variables for u_{xx} and u_{yy}. This observation yields the following reduced first-order formulation of (10.127):

$$\begin{array}{llll} u_x = v\,, & v_x = w\,, & w_x = z\,, & z_x = 2w - u\,, \\ u_y = w\,, & v_y = z\,, & w_y = 2w - u\,, & z_y = 2z - v\,. \end{array} \tag{10.132}$$

Rewriting it in the matrix formulation corresponding to (10.123) for the vector $\mathbf{w} = (u,v,w,z)^t$ leads exactly to the two matrices A_x, A_y given in (10.129). Hence we obtain exactly the same closed form solution expression (10.131), as the projector P selects the first component of the vector $\mathbf{w}$. Note, however, that already a simple linear transformation of the independent variables would have destroyed this simplification and thus forced us to work in higher dimension. The above described algebraic approach allows us to obtain a system of minimal dimension independent of such considerations. ◁

For our second approach to the integration of linear systems with constant coefficients of finite type, we exploit the obvious isomorphism between the ring $\mathcal{D}$ and the ordinary commutative polynomial ring $\mathcal{P} = \Bbbk[x^1,\dots,x^n]$. Given a linear differential operator $L \in \mathcal{D}$ we can associate with it the polynomial $f_L \in \mathcal{P}$ obtained by simply replacing ∂_i by x^i. Conversely, given an ideal $\mathcal{J} \subseteq \mathcal{P}$ and a point $\boldsymbol{\zeta} \in \Bbbk^n$, we denote by $\mathcal{J}^{(\zeta)} \subseteq \mathcal{D}$ the "shifted" ideal obtained by replacing x^i with $\partial_i + \zeta^i$.

It is easy to see that solving the differential equation $Lu = 0$ and computing the zeros of the polynomial f_L are closely related: the point $\boldsymbol{\zeta} \in \Bbbk^n$ is a zero of f_L, if and only if $u(\mathbf{x}) = \exp(\boldsymbol{\zeta}\cdot\mathbf{x})$ is a solution of $Lu = 0$ (here it is of course important that we assume that $\Bbbk$ is algebraically closed). In fact, in the univariate case it is well-known that the integration of the linear differential equation yields even more precise information, namely the multiplicity of the zeros, as in the case of a multiple root we have additional solutions of the form $x^\ell \exp(\zeta x)$ with exponents ℓ up to the multiplicity of the root. We will show now that this observation carries over to linear partial differential equations of finite type. For simplicity, we restrict to one unknown function u so that we can formulate everything with ideals instead of modules; the extension to the general case is straightforward.

Given an ideal $\mathcal{I} \subseteq \mathcal{D}$ of linear differential operators, we associate with it the polynomial ideal $\mathcal{J} = \{f_L \mid L \in \mathcal{I}\} \subseteq \mathcal{P}$. Obviously, $\mathcal{I}$ corresponds to an equation of finite type, if and only if $\mathcal{J}$ is a zero-dimensional ideal, i. e. the variety $\mathcal{V} \subset \Bbbk^n$ of $\mathcal{J}$ consists only of finitely many points. If $\mathcal{V} = \{\boldsymbol{\zeta}_1, \dots, \boldsymbol{\zeta}_r\}$, then the associated prime ideals of $\mathcal{J}$ are the maximal ideals $\mathfrak{m}_{\boldsymbol{\zeta}_\ell} = \langle x^1 - \zeta_\ell^1, \dots, x^n - \zeta_\ell^n \rangle$. Thus the primary decomposition of $\mathcal{J}$ is of the form $\mathcal{J} = \mathfrak{q}_1 \cap \cdots \mathfrak{q}_r$ with $\sqrt{\mathfrak{q}_\ell} = \mathfrak{m}_{\boldsymbol{\zeta}_\ell}$. Obviously, no embedded prime ideals can exist here so that the decomposition is unique. Since $\mathfrak{q}_\ell$ is $\mathfrak{m}_{\boldsymbol{\zeta}_\ell}$-primary, the "shifted" ideal $\mathfrak{q}_\ell^{(\boldsymbol{\zeta}_\ell)}$ is $\mathfrak{m}_\partial$-primary where $\mathfrak{m}_\partial = \langle \partial_1, \dots, \partial_n \rangle$ denotes the homogeneous maximal ideal in $\mathcal{D}$.

Lemma 10.8.9. *The space* $\ker_{\mathcal{P}} \mathfrak{q}_\ell^{(\boldsymbol{\zeta}_\ell)}$ *of polynomial solutions of the ideal* $\mathfrak{q}_\ell^{(\boldsymbol{\zeta}_\ell)} \subseteq \mathcal{D}$ *and the factor ring* $\mathcal{P}/\mathfrak{q}_\ell$ *are isomorphic over* $\Bbbk$.

Proof. Since $\mathfrak{q}_\ell^{(\boldsymbol{\zeta}_\ell)}$ is $\mathfrak{m}_\partial$-primary, there exists an exponent s (the satiety $\operatorname{sat} \mathfrak{q}_\ell^{(\boldsymbol{\zeta}_\ell)}$) such that $\mathfrak{m}_\partial^s \subseteq \mathfrak{q}_\ell^{(\boldsymbol{\zeta}_\ell)}$. Since trivially $\ker_{\mathcal{P}} \mathfrak{m}_\partial^s = \mathcal{P}_{<s}$, the exponent s provides us with a degree bound for the polynomial solutions of $\mathfrak{q}_\ell^{(\boldsymbol{\zeta}_\ell)}$. Now we may define on the two $\Bbbk$-linear spaces $\mathcal{D}/\mathfrak{m}_\partial^s$ and $\mathcal{P}_{<s}$ (which have obviously the same finite dimension) a pairing by setting $\big([L], f\big) = Lf$. This pairing is well-defined and non-degenerate. Furthermore, by construction, the subspace $\ker_{\mathcal{P}} \mathfrak{q}_\ell^{(\boldsymbol{\zeta}_\ell)} \subseteq \mathcal{P}_{<s}$ is, with respect to this pairing, the "orthogonal" complement of the image of $\mathfrak{q}_\ell^{(\boldsymbol{\zeta}_\ell)}$ in $\mathcal{D}/\mathfrak{m}_\partial^s$ under the canonical projection. But this observation implies that the dimension of $\ker_{\mathcal{P}} \mathfrak{q}_\ell^{(\boldsymbol{\zeta}_\ell)}$ is the same as the dimension of $\mathcal{D}/\mathfrak{q}_\ell^{(\boldsymbol{\zeta}_\ell)} \cong \mathcal{P}/\mathfrak{q}_\ell$. □

Our next result says that for determining the smooth solution space $\ker_{\mathcal{A}} \mathcal{I}$ of our given linear system of finite type it suffices to compute the *polynomial* solutions $\ker_{\mathcal{P}} \mathfrak{q}_\ell^{(\boldsymbol{\zeta}_\ell)}$ of auxiliary systems defined by $\mathfrak{m}_\partial$-primary ideals, as only these show up as coefficients of the exponentials.

Theorem 10.8.10. *Let* $\mathcal{I} \subseteq \mathcal{D}$ *be a zero-dimensional ideal defined by a system of linear partial differential equations with constant coefficients of finite type. If the corresponding polynomial ideal* $\mathcal{J} \subseteq \mathcal{P}$ *possesses an irredundant primary decomposition* $\mathcal{J} = \mathfrak{q}_1 \cap \cdots \cap \mathfrak{q}_r$ *with* $\sqrt{\mathfrak{q}_\ell} = \mathfrak{m}_{\boldsymbol{\zeta}_\ell}$ *where* $\mathcal{V}(\mathcal{J}) = \{\boldsymbol{\zeta}_1, \dots, \boldsymbol{\zeta}_r\}$, *then the solution space* $\ker_{\mathcal{A}} \mathcal{I}$ *is generated over* $\Bbbk$ *by the functions of the form* $h \exp(\boldsymbol{\zeta}_\ell \cdot \mathbf{x})$ *with arbitrary polynomials* $h \in \ker_{\mathcal{P}} \mathfrak{q}_\ell^{(\boldsymbol{\zeta}_\ell)}$.

Proof. One easily verifies that a linear differential operator $L \in \mathcal{D}$ satisfies the equation $L\big(h \exp(\boldsymbol{\zeta}_\ell \cdot \mathbf{x})\big) = 0$, if and only if the "shifted" operator $\hat{L}$ obtained by replacing ∂_i by $\partial_i + \zeta_\ell^i$ satisfies $\hat{L}h = 0$. This observation implies that the given functions indeed lie in $\ker_{\mathcal{A}} \mathcal{I}$. It follows from Lemma 10.8.9 that they generate a vector space of dimension $\sum_{\ell=1}^r \dim_{\Bbbk}(\mathcal{P}/\mathfrak{q}_\ell)$. Since the differential module associated with our linear system $\mathcal{M} = \mathcal{D}/\mathcal{I}$ obviously possesses the same dimension over $\Bbbk$, these functions generate the full solution space by Proposition 10.8.4. □

Thus for the explicit integration of $\mathcal{I}$ there only remains the problem of determining the polynomial solutions of an $\mathfrak{m}_\partial$-primary ideal $\mathfrak{q} \subseteq \mathcal{D}$. But this question was already treated in Example 9.3.10 using Gröbner bases and Algorithm 9.1 (even for the case of variable coefficients). For convenience, we repeat the result in closed form providing an alternative proof.

Let $\prec$ be a term order and $\mathcal{G}$ a Gröbner basis of the ideal $\mathfrak{q}$ with respect to it. We introduce $\mathcal{B} = \mathbb{N}_0^n \setminus \mathrm{le}_\prec \mathfrak{q}$ (the multi indices of the parametric coefficients) and $\mathcal{N} = \{\nu \in \mathbb{N}_0^n \mid \partial^\nu \notin \mathfrak{q}\}$. Since $\mathfrak{q}$ is assumed to be a $\mathfrak{m}_\partial$-primary ideal, both sets are finite (and obviously $\mathcal{B} \subseteq \mathcal{N}$). By definition, the normal form $\mathrm{NF}_{\mathcal{G},\prec}(\partial^\nu)$ vanishes whenever $\nu \notin \mathcal{N}$. For the multi indices $\nu \in \mathcal{N}$ we write

$$\mathrm{NF}_{\mathcal{G},\prec}(\partial^\nu) = \sum_{\mu \in \mathcal{B}} c_{\nu\mu} \partial^\mu \,. \tag{10.133}$$

With the help of the thus defined coefficients $c_{\nu\mu} \in \Bbbk$ we introduce for each multi index $\mu \in \mathcal{B}$ a polynomial $h_\mu \in \mathcal{P}$ by setting

$$h_\mu = \sum_{\nu \in \mathcal{N}} c_{\nu\mu} \frac{x^\nu}{\nu!} \,. \tag{10.134}$$

This polynomial is obtained, if we follow the procedure described in Example 9.3.10. Indeed, if we apply Algorithm 9.1 to $\mathfrak{q}$ with the initial data $f_\rho(\mathbf{x}_0) = \delta_{\rho\mu}$ for all multi indices $\rho \in \mathcal{B}$, then it returns for any $\nu \in \mathcal{N}$ the Taylor coefficient $a_\nu = c_{\nu\mu}$. By the considerations above, all other Taylor coefficients vanish, so that the complete solution is the polynomial h_μ.

Lemma 10.8.11. *The polynomials h_μ form a basis of the solution space of* $\mathfrak{q}$*:*

$$\ker_{\mathcal{A}} \mathfrak{q} = \ker_{\mathcal{P}} \mathfrak{q} = \langle h_\mu \mid \mu \in \mathcal{B} \rangle_{\Bbbk} \,. \tag{10.135}$$

Proof. The equality $\ker_{\mathcal{A}} \mathfrak{q} = \ker_{\mathcal{P}} \mathfrak{q}$ follows from the fact that $\mathfrak{m}_\partial^s \subseteq \mathfrak{q}$ for some exponent $s \in \mathbb{N}$. The above constructed polynomials h_μ are linearly independent over $\Bbbk$, as the term x^μ appears only in h_μ. Furthermore, according to Lemma 10.8.9, their span has the same dimension as $\ker_{\mathcal{P}} \mathfrak{q}$. Hence there only remains to show that the polynomials h_μ are indeed solutions.

Let $L = \sum_\rho L_\rho \partial^\rho \in \mathcal{D}$ be an arbitrary linear differential operator. Then the evaluation of L at the polynomial h_μ yields

$$Lh_\mu = \sum_{\nu \in \mathcal{N}} c_{\nu\mu} \sum_{\rho \leq \nu} L_\rho \frac{x^{\nu-\rho}}{(\nu-\rho)!} = \sum_\sigma \Bigl(\sum_\rho L_\rho c_{\rho+\sigma,\mu}\Bigr) \frac{x^\sigma}{\sigma!} \tag{10.136}$$

where the second equality is obtained by the index shift $\sigma = \nu - \rho$. On the other hand, (10.133) implies that

$$\mathrm{NF}_{\mathcal{G},\prec}(\partial^\sigma L) = \sum_{\mu \in \mathcal{B}} \Bigl(\sum_\rho L_\rho c_{\rho+\sigma,\mu}\Bigr) x^\mu \,. \tag{10.137}$$

If the operator L lies in the ideal $\mathfrak{q}$, then this normal form and hence the expression in the parentheses trivially vanishes so that $Lh_\mu = 0$. □

Example 10.8.12. We treat the same linear system $\mathcal{R}_2$ as in Example 10.8.8 with the above outlined alternative approach. Thus as a first step we need a primary decomposition of the polynomial ideal $\mathcal{J} = \langle y^2 - 2y + 1, x^2 - y \rangle \subset \Bbbk[x,y]$ corresponding to (10.127); it has the form $\mathcal{J} = \mathfrak{q}_1 \cap \mathfrak{q}_2$ where $\mathfrak{q}_1 = \langle y^2 - 2y + 1, 2x - y + 1 \rangle$ is primary to $\mathfrak{p}_1 = \langle y - 1, x - 1 \rangle$ and $\mathfrak{q}_2 = \langle y^2 - 2y + 1, 2x + y + 1 \rangle$ to $\mathfrak{p}_2 = \langle y - 1, x + 1 \rangle$. Obviously, $\mathfrak{p}_1$ is the maximal ideal to the zero $\boldsymbol{\zeta}_1 = (1,1)$ and $\mathfrak{p}_2$ to $\boldsymbol{\zeta}_2 = (-1,1)$.

As a second step we need the polynomial solutions of the shifted primary ideals $\mathfrak{q}_1^{(\zeta_1)} = \langle \partial_y^2, 2\partial_x - \partial_y \rangle$ and $\mathfrak{q}_2^{(\zeta_2)} = \langle \partial_y^2, 2\partial_x + \partial_y \rangle$ in $\Bbbk[\partial_x, \partial_y]$. Their determination proceeds for both ideals almost identical, since in both case we find $\mathcal{B} = \{1, \partial_y\}$ and $\mathcal{N} = \mathcal{B} \cup \{\partial_x\}$. The given bases are already Gröbner bases for any term order where $\partial_x \succ \partial_y$ and the normal forms of ∂_x with respect to them are $\pm\frac{1}{2}\partial_y$. Hence we obtain as generators of the polynomial solution space the functions $h_1 = 1$ and $h_2 = y \pm \frac{1}{2}x$. In conclusion, a basis of the solution space of the linear system (10.127) is

$$\left\{ e^{y+x},\ (y + \tfrac{1}{2}x)e^{y+x},\ e^{y-x},\ (y - \tfrac{1}{2}x)e^{y-x} \right\}, \tag{10.138}$$

which obviously coincides with our results in Example 10.8.8. ◁

For arbitrary linear systems with constant coefficients not necessarily of finite type one can also show that all smooth solutions may be represented with integrals over polynomial-exponential functions. However, the proof of this assertion is much more involved and we cite here only the result.

Theorem 10.8.13. *Let $\mathcal{I} \subseteq \mathcal{D}$ be an ideal of linear differential operator with constant coefficients and $\mathcal{J} \subseteq \mathcal{P}$ the corresponding polynomial ideal. Then there exist finitely many pairs $(\mathcal{V}_\ell, P_\ell)$ where $\mathcal{V}_\ell$ is the variety of an associated prime ideal of $\mathcal{J}$ and $P_\ell \in \Bbbk[x^1, \dots, x^n, \xi_1, \dots, \xi_n]$ is a polynomial in $2n$ variables with the following property: if $\Omega \subset \mathbb{R}^n$ is a compact and convex subset and $u \in \mathcal{C}^\infty(\Omega)$ a smooth solution of $\mathcal{I}$, then there exists on each variety $\mathcal{V}_\ell$ a measure μ_ℓ such that*

$$u(\mathbf{x}) = \sum_\ell \int_{\mathcal{V}_\ell} P_\ell(\mathbf{x}, \boldsymbol{\xi}) \exp(\mathbf{x} \cdot \boldsymbol{\xi}) d\mu_\ell(\boldsymbol{\xi}) . \tag{10.139}$$

Conversely, every such u is a solution of $\mathcal{I}$.

One should emphasise that the varieties $\mathcal{V}_\ell$ are not necessarily different in Theorem 10.8.13. In other words, in general several polynomials P belong to every variety of an associated prime ideal of $\mathcal{J}$. We do not discuss here the explicit construction of the polynomials P_ℓ, but consider Theorem 10.8.13 as a mere structure theorem providing us with information about the kind of functions that may appear as solutions of a linear systems with constant coefficients.

10.9 Notes

Our proof of Holmgren's Theorem follows [379]. The basic trick of using the adjoint system goes back to Holmgren [221]; its proper global formulation with the one-parameter family of surfaces S_λ is due to John [238]. Despite its simplicity, the extension to arbitrary involutive systems seems to have appeared first in [398]. We formulated the Holmgren Theorem only for continuously differentiable solutions. Using, say, density arguments it is not difficult to extend it to larger function spaces. For example, Hörmander [224, Cor. 8.6.9] provides a distributional version.

The definition of ellipticity for general overdetermined differential equations is quite rarely found in the literature; one accessible exception is the encyclopaedia article by Dudnikov and Samborski [116]. The approach via a weighted symbol is due to Agmon et al [7, 8] and Douglis and Nirenberg [111]; it has become standard in the literature. The here presented alternative and the proof of the insufficiency of the weights comes from [273]. Quantifier elimination provides a method for effectively deciding whether or not a given system is DN-elliptic and determining all corresponding sets of weights [412]; however, this method is often prohibitively expensive for non-trivial systems.

Besides the above mentioned encyclopaedia article [116] and the research monograph [450], the question of defining ellipticity for overdetermined equations was taken up only by few authors—see e. g. [96, 216, 363]. Notable are here in particular the results of Cosner [96] who constructed for any equation which is elliptic with weights an equivalent one which is elliptic without weights.

Agmon [6, pp. 63–67] developed a regularity theory for overdetermined elliptic equations in one dependent variable. The Drach transformation of Appendix A.3 allows us to rewrite any overdetermined equation in an arbitrary number of dependent variables as an equivalent one in one dependent variable and in [273] it is shown that ellipticity is preserved by this operation, if we perform the gauge fixing (A.36). Thus one can extend Agmon's results to arbitrary elliptic problems. Of course, it is possible to formulate such results directly without a Drach transformation: [116, 450] give some relevant a priori estimates in terms of Sobolev space norms which show precisely the regularity of the solution in terms of the data. In fact, in these estimates the weights needed for DN-ellipticity acquire a rather natural interpretation. For getting the relevant estimates, one should also specify correct boundary conditions. It turns out that, in addition to the ellipticity of the operator, the boundary operators should satisfy the *Shapiro–Lopatinskij condition*. Its discussion is beyond the scope of this book and we just refer to [9, 116] for definitions and to [272] for a recent treatment of the overdetermined case.

Our definition of hyperbolicity is also taken from the encyclopaedia article by Dudnikov and Samborski [116]. An equivalent definition appears in the work of Johnson [244, 245]. One of the few other references dealing with the overdetermined case is an article by Peradzynski [351]. The notion of a weakly overdetermined system was introduced in [405]. That article studies only systems with constant coefficients where the situation is simpler. In this case one does not need the assumption that the constraints are elliptic, as one may simply apply the Holmgren Theorem for proving that the equations of lower class are satisfied.

The algebraic approach to the analysis of linear systems presented in Section 10.5 originated from the work of Malgrange [301]. It has since become fundamental for algebraic systems theory (see e. g. [337] for an in-depth treatment of the constant coefficients case and [144] for a survey on results for variable coefficients ordinary differential equations). The behavioural approach to systems and control theory was pioneered by Willems [478, 479, 480] in a series of articles on systems described by linear ordinary differential equations. More details can be found in the survey articles [145, 481] and the textbook [355]. The extension to distributed systems, i. e. partial differential equations, was started by Pillai and Shankar [352].

The fundamental principle was first announced by Ehrenpreis [117]. As both Malgrange [301] and Palamodov [348] have also made significant contributions to its proof, one sometimes speaks of the *principle of Ehrenpreis–Malgrange–Palamodov*. As we have seen, from an algebraic point of view it is just a reformulation of the fact that the function space $\mathcal{A}$ in which the differential equation is studied is an injective module. Under the assumption that $\mathcal{A}$ is in addition a cogenerator, we showed in Theorem 10.5.10 the existence of a bijection between finitely generated $\mathcal{D}$-modules and linear differential systems. Oberst [337] extended this bijection to a full categorical duality (for systems with constant coefficients; variable coefficients ordinary differential equations were later treated in [144]) for the case that $\mathcal{A}$ is even a *large* injective cogenerator (which means that every finitely generated $\mathcal{D}$-module can be embedded $\mathcal{D}$-linearly into a finite product $\mathcal{A}^k$). In [337] it is also shown that all the modules listed in Remark 10.5.12 satisfy this property.

On the theoretical side basically everything about the inverse syzygy problem and the notion of a torsionless module was already said by Auslander and Bridger [30].[13] However, they use a fairly heavy machinery (for commutative rings a simplified version was presented by Bruns and Vetter [57, Sect. 16E] from where also our proof of Proposition 10.6.10 is taken). Our presentation follows [488]; in particular the rather simple proof of Theorem 10.6.2 originates from there. It seems that in commutative algebra this problem has not attracted much interest—in contrast to differential equations where it possesses quite some importance for applications. Algorithms for the inverse syzygy problem and other questions in systems theory together with a concrete implementation are described by Chyzak et al [88] for the case that $\mathcal{D}$ is an Ore ring. Damiano et al [101] connected the inverse syzygy problem for linear differential operators with the Hartogs phenomenon of complex analysis allowing for the removability of compact singularities.

In the second Addendum to Section 10.6, we characterised both autonomy and controllability of a behaviour $\mathcal{B}$ by the vanishing of an extension groups (Remarks 10.6.15 and 10.6.18, respectively). If $\mathcal{M}$ be the $\mathcal{D}$-module corresponding to $\mathcal{B}$ and $D(\mathcal{M})$ its Auslander–Bridger dual, then autonomy is equivalent to $\mathrm{Ext}^0_{\mathcal{D}}(\mathcal{M},\mathcal{D}) = 0$ and controllability to $\mathrm{Ext}^1_{\mathcal{D}}(D(\mathcal{M}),\mathcal{D}) = 0$. This observation has motivated the introduction of stronger notions of controllability and autonomy [486] (see also

[13] Note that Auslander and Bridger use the term 1-torsionfree instead of torsionless.

[88, 359]). We call $\mathcal{B}$ *strongly controllable*, if all extension groups $\mathrm{Ext}^i_{\mathcal{D}}(D(\mathcal{M}),\mathcal{D})$ with $i \geq 1$ vanish (for rings $\mathcal{D}$ of finite global dimension, only finitely many groups can be non-zero). Similarly, the behaviour $\mathcal{B}$ is *strongly autonomous*, if all groups $\mathrm{Ext}^i_{\mathcal{D}}(\mathcal{M},\mathcal{D})$ with $i \geq 0$ vanish.

The basic idea of Algorithm 10.1 seems to be due to Oberst [337] who proved its correctness for the commutative case, i. e. for linear differential equations with constant coefficients (see also [360] and [485, Section 1.2]). Pommaret [357] provides a version for variable coefficients using the formally adjoint operator instead of a dualisation. Chyzak et al [88] showed later that the algorithm remains valid over arbitrary Ore algebras.

In all these references, the existence of a parametrisation is related to the question whether the cokernel $\operatorname{coker}\beta$ is torsionfree and the corresponding proofs require the introduction of a quotient field, i. e. that the ring $\mathcal{D}$ is a domain satisfying an Ore condition as discussed in Remark 10.6.9. The proofs are usually based on the fact that for a field $\mathcal{Q}$ the functor $\mathrm{Hom}_{\mathcal{Q}}(\cdot,\mathcal{Q})$ is always exact and then one shows that any syzygy over $\mathcal{Q}$ can be lifted to one over $\mathcal{D}$ by multiplication with the common denominator. Essentially, this approach is equivalent to our proof of Theorem 10.6.8 where we slightly relaxed the assumption by only requiring that $\mathcal{D}$ has a self-injective ring of quotients. The idea that by using the notion of a torsionless (instead of a torsionfree) module one can simultaneously simplify the proof and dispense with the Ore condition was developed in [488].

The integral representation Theorem 10.8.13 for linear systems with constant coefficients is also due to Ehrenpreis [118, Chapt. VII] and Palamodov [348, Chapt. VI]. Simpler presentations of the proof are contained in [45, Chapt. 8] and [222, Sect. 7.7] (without proofs, the topic is also covered in [438, Chapt. 10]). In our formulation of Theorem 10.8.13 we simply wrote about polynomials P_ℓ in $2n$ variables. In fact, it is more common to consider the P_ℓ as elements of the Weyl algebra $\mathcal{W}_n = \mathbb{k}[x^1,\dots,x^n,\partial_1,\dots,\partial_n]$ and to call them *Noetherian operators*. One can show [348, Chapt. IV, §§3,4], [45, Chapt. 8, §4] that to each primary ideal $\mathfrak{q} \subseteq \mathcal{P} = \mathbb{k}[x^1,\dots,x^n]$ a finite number of Noetherian operators $P_\ell \in \mathcal{W}_n$ exists such that a polynomial $f \in \mathcal{P}$ lies in $\mathfrak{q}$, if and only if $P_\ell f \in \sqrt{\mathfrak{q}}$ (i. e. the polynomial $P_\ell f$ vanishes on the variety $\mathcal{V}(\mathfrak{q})$). The effective construction of Noetherian operators is discussed in [100, 339].

For equations of finite type the situation is much less involved. Our presentation of the first explicit integration method follows Lomadze and Zerz [294]; the second one is taken from [438, Chapt. 10] and originated from [338]. It is not so easy to compare the efficacy and the efficiency of the two methods. The first method has troubles, if $\dim_{\mathbb{k}} \mathcal{M}$ gets larger, as then the Jordan normal forms of larger matrices must be computed. The second method requires the determination of a primary decomposition which computer algebra systems can usually compute only over finite extensions of $\mathbb{Q}$ (and the computation of which is rather expensive for larger problems). The bottleneck of both methods is the determination of the zeros of the corresponding polynomial ideal; in the first method these appear as eigenvalues, in the second method they are encoded in the prime ideals underlying the primary decomposition. The proof of Proposition 10.8.4 is due to Oberst [338] (there also more general situations like linear difference operators or operators with coefficients in a not algebraically closed field are considered).

One should note that many of the ideas underlying our treatment of linear systems of finite type go back to Gröbner [187]. This concerns in particular the use of the "shifted" ideals $\mathfrak{q}^{\zeta}$ for the integration of zero-dimensional primary ideals and of matrices representing the multiplication on the factor ring. Obviously, for the whole theory it is only necessary that $\mathcal{D}$ is a commutative polynomial ring. But this remains true if we consider e. g. $D = \mathbb{k}[x^1\partial_1, \dots, x^n\partial_n]$ and thus one can also analyse certain types of linear systems with variable coefficients with the same methods. However, Gröbner [187], §5.2 showed that such systems can always be transformed into a system with constant coefficients.

The hybrid algorithm of Section 10.7 combining geometric and algebraic ideas was developed in the diploma thesis of Hausdorf [197] (see also [198]). He implemented the algorithm in the computer algebra system *MuPAD*. Compared with earlier implementations [396, 400] using a naive realisation of the Cartan–Kuranishi completion, it is much more efficient. The results of Section 4.3 on detecting δ-singular coordinate systems can also be incorporated into the algorithm; for details we refer to [198, Section 5]. For the application to non-linear equations, the same comments as in Chapter 4 for the purely algebraic case apply.

Appendix A
Miscellaneous

Analysis takes back with one hand what it gives with the other. I recoil in fear and loathing from that deplorable evil: continuous functions with no derivatives.

Charles Hermite

The first section of this chapter fixes some notations for multi indices that are widely used in the main text. Except for the distinction into multi and "repeated" indices these are standard notations in multivariate analysis. We also show an important finiteness result for multi indices, namely Dickson's Lemma, and discuss term orders. The next section recalls those properties of (real-)analytic functions that are needed in the proof of the Cauchy–Kovalevskaya Theorem. Then some elementary operations with differential equations like reduction to first order or quasi-linearisation are discussed within the context of the formal theory. Finally, a few simple facts about the modified Stirling numbers appearing in manipulations of Cartan characters and Hilbert functions are presented.

A.1 Multi Indices and Orders

We use mostly the standard multi index notation of multivariate analysis. If we are dealing with n independent variables $\mathbf{x} = (x^1, \dots, x^n)$, a *multi index* is an ordered n-tuple $\mu = [\mu_1, \dots, \mu_n] \in \mathbb{N}_0^n$ of integers (we always use square brackets to denote multi indices). The sum of its entries is called its *length* written $|\mu| = \mu_1 + \cdots + \mu_n$. The *support* of the multi index μ contains the positions of the non-vanishing entries: $\operatorname{supp}\mu = \{i \mid \mu_i > 0\}$.

The factorial of a multi index $\mu \in \mathbb{N}_0^n$ is defined by $\mu! = \mu_1! \cdots \mu_n!$. The expression $(\mathbf{x}-\mathbf{y})^\mu$ is a short hand for $(x^1 - y^1)^{\mu_1} \cdots (x^n - y^n)^{\mu_n}$. For the derivatives of a smooth function $\phi(\mathbf{x})$ we write

$$\frac{\partial^{|\mu|}\phi}{\partial \mathbf{x}^\mu} = \frac{\partial^{|\mu|}\phi}{\partial (x^1)^{\mu_1} \cdots \partial (x^n)^{\mu_n}} \,. \tag{A.1}$$

For notational convenience, we define for the special multi index $0 = [0, \dots, 0]$ that $x^0 = 1$ and $\partial^{|0|}\phi/\partial \mathbf{x}^0 = \phi$. Obviously, $0! = 1$ and $|0| = 0$. Other special multi indices appearing occasionally are

$$\ell_i = [0,\dots,0,\ell,0,\dots,0] \tag{A.2}$$

where $\ell \in \mathbb{N}$ is the ith entry and all other entries vanish.

The addition of multi indices is defined componentwise, i. e.

$$\mu + \nu = [\mu_1 + \nu_1,\dots,\mu_n + \nu_n]\,. \tag{A.3}$$

If we want to increase the ith entry of a multi index μ by one, we can thus simply write $\mu + 1_i$ using (A.2). With this addition, the set $\mathbb{N}_0^n$ of all multi indices acquires the structure of an Abelian monoid with neutral element 0. Within the polynomial ring $\mathcal{P} = \Bbbk[x^1,\dots,x^n]$ the subset $\mathbb{T} = \{x^\mu \mid \mu \in \mathbb{N}_0^n\}$ of all *terms*[1] together with the usual multiplication forms an obviously isomorphic monoid with neutral element 1.

If k is the smallest value such that $\mu_k \neq 0$, we call it the *class* of the multi index μ and write $\operatorname{cls}\mu = k$. In (A.1) we find then no differentiations with respect to the variables $x^1,\dots,x^{k-1}$. In principle, the class of the multi index 0 is undefined, but in many situations it is convenient to set $\operatorname{cls} 0 = 0$.

Some elementary combinatorics leads to the following results:

$$\bigl|\{\mu \in \mathbb{N}_0^n \mid |\mu| = q\}\bigr| = \binom{n+q-1}{q}\,, \tag{A.4a}$$

$$\bigl|\{\mu \in \mathbb{N}_0^n \mid |\mu| = q \wedge \operatorname{cls}\mu = k\}\bigr| = \binom{n-k+q-1}{q-1}\,, \tag{A.4b}$$

$$\bigl|\{\mu \in \mathbb{N}_0^n \mid 0 \leq |\mu| \leq q\}\bigr| = \binom{n+q}{q}\,. \tag{A.4c}$$

A multi index μ of length q and class k is obtained by first setting μ_k to a value $1 \leq q' \leq q$ and then adding an arbitrary multi index in $\mathbb{N}_0^{n-k}$ of length $q - q'$. Thus the second line arises from summing up the first one from 0 to $q-1$ (the possible values of $q - q'$) after replacing n by $n-k$ and using a classical identity for binomial coefficients. In a similar manner, the third line stems from summing up the first one from 0 to q and applying the same identity.

Occasionally we use an alternative multi index notation, which we call for lack of a better name *repeated index* notation. We distinguish it from multi indices by parentheses instead of brackets; furthermore, we take Greek letters like μ or ν for multi indices and capital letters like I or J for repeated indices. A repeated index I of length q contains exactly q entries with values between 1 and n: $I = (i_1,\dots,i_q)$. The analogue to (A.1) is now

$$\frac{\partial^q \phi}{\partial \mathbf{x}^I} = \frac{\partial^q \phi}{\partial x^{i_1}\cdots\partial x^{i_q}}\,. \tag{A.5}$$

We always consider a repeated index as an ordered tuple; thus $(1,2) \neq (2,1)$. Of course, these two indices would yield in (A.5) the same result, as for us partial

[1] There is a notorious dispute in the commutative algebra community whether the elements of $\mathbb{T}$ should be called terms or monomials. We call them terms; a monomial is the product of a term and a coefficient.

derivatives commute, but in other places the order plays a role. For example, if we are working in a non-commutative polynomial ring, then $x^I = x^{i_1} \cdots x^{i_q}$ and now different orderings yield different results. Finally, sort(I) denotes the repeated index obtained from I by sorting the entries in ascending order.

If $\mu \in \mathbb{N}_0^n$ is an arbitrary multi index, then we denote by $\mathcal{S}_\mu$ the set of all possible realisations of it as a repeated index (in combinatorics one would speak here of permutations with repetitions). We further write $R(\mu)$ for the unique element of $\mathcal{S}_\mu$ where all entries are sorted in ascending order. Thus we have for example $\mathcal{S}_{[1,2]} = \{(1,2,2);(2,1,2);(2,2,1)\}$ and $R([1,2]) = (1,2,2)$. It is a standard result in elementary combinatorics that the size of $\mathcal{S}_\mu$ for a multi index with $|\mu| = q$ is given by the polynomial coefficient

$$|\mathcal{S}_\mu| = \binom{q}{\mu} = \frac{q!}{\mu_1! \cdots \mu_n!}\,. \tag{A.6}$$

Repeated indices are particularly useful for representing exterior products. Then we only consider indices $I = (i_1,\dots,i_q)$ with $i_1 < i_2 < \cdots < i_q$, i.e. all entries are different and sorted in ascending order. If I, J are two such repeated indices, then $I \cup J$ denotes the index obtained by first concatenating I and J and then sorting the entries. Obviously, this only yields a valid result, if I and J have no entries in common. We set $\operatorname{sgn}(I \cup J) = \pm 1$ depending on whether an even or odd number of transpositions is required for the sorting. If I and J have entries in common, we set $\operatorname{sgn}(I \cup J) = 0$; this convention is useful to avoid case distinctions in some sums.

On various occasions we need ordered structures; in particular, orders on the monoid $(\mathbb{N}_0^n,+) \cong (\mathbb{T},\cdot)$ are crucial for the definition of Gröbner bases in Appendix B.4. For this reason, we collect here some related notions and for us important properties of them. A good general reference for orders in the context of commutative algebra is [36, Chapt. 4].

Definition A.1.1. A *partial order* on a set $\mathcal{S}$ is a relation $\preceq$ such that for all elements $s_1, s_2, s_3 \in \mathcal{S}$ (i) $s_1 \preceq s_1$ holds, (ii) $s_1 \preceq s_2$ and $s_2 \preceq s_3$ imply $s_1 \preceq s_3$, (iii) $s_1 \preceq s_2$ and $s_2 \preceq s_1$ entail $s_1 = s_2$. If in addition for every pair $s_1, s_2 \in \mathcal{S}$ either $s_1 \preceq s_2$ or $s_2 \preceq s_1$ holds, then $\preceq$ is a *total* (or *linear*) *order*. If every non-empty subset $\mathcal{S}' \subseteq \mathcal{S}$ has a minimal element with respect to $\preceq$ or equivalently $\mathcal{S}$ does not contain infinite strictly descending sequences, then $\preceq$ is called a *well-order*.

A total order $\preceq$ on an Abelian semigroup $(\mathcal{S},\cdot)$ is a *semigroup order*, if it is compatible with the product $\cdot$, i.e. if $s \preceq t$ implies $r \cdot s \preceq r \cdot t$ for all $r,s,t \in \mathcal{S}$ (*monotonicity* of semigroup orders). If $(\mathcal{S},\cdot)$ is even a monoid with neutral element 1, then $\preceq$ is a *monoid order*, if besides being a semigroup order it satisfies $1 \prec s$ for all $s \in \mathcal{S} \setminus \{1\}$. A monoid order on $(\mathbb{T},\cdot) \cong (\mathbb{N}_0^n,+)$ is also called a *term order*,

While strictly speaking only a relation like $\preceq$ (i.e. including equality) can be a total order in the sense of Definition A.1.1, we will follow the popular convention to denote orders by $\prec$. The natural order on $\mathbb{N}_0$ given by the usual $\leq$-relation induces a partial order on $\mathbb{N}_0^n$ defined by $\mu \leq \nu$, if $\mu_i \leq \nu_i$ for all $1 \leq i \leq n$. Identifying the multi index μ and the term x^μ, we often write $\mu \mid \nu$ instead of $\mu \leq \nu$, as obviously $\mu \leq \nu$ is equivalent to $x^\mu \mid x^\nu$.

The following result is usually called *Dickson's Lemma*, since its first appearance in the literature seems to be [106]. However, it has been rediscovered so many times that it is difficult to give proper credit.

Lemma A.1.2 (Dickson). *Let $\mathcal{N} \subseteq \mathbb{N}_0^n$ be an arbitrary set of multi indices. Then it possesses a finite subset $\mathcal{B} \subseteq \mathcal{N}$, a* Dickson basis, *such that for every $\nu \in \mathcal{N}$ a multi index $\mu \in \mathcal{B}$ exists with $\mu \mid \nu$. Choosing for $\mathcal{N}$ a monoid ideal, it follows that $(\mathbb{N}_0^n, +)$ is a Noetherian monoid.*

Proof. We use an induction over the number n of entries of the multi indices. The case $n = 1$ is trivial: we first define $d = \min\{\nu_1 \mid \nu \in \mathcal{N}\}$ and then simply choose $\mathcal{B} = \{[d]\}$. Now let $n > 1$ and assume that our claim is correct for $n-1$. We associate to each $\mu \in \mathbb{N}_0^n$ the multi index $\mu' = [\mu_1, \dots, \mu_{n-1}] \in \mathbb{N}_0^{n-1}$. Then we consider for each $i \in \mathbb{N}_0$ the set $\mathcal{S}_i = \{\mu' \mid \mu \in \mathcal{S} \wedge \mu_n = i\}$. According to our induction hypothesis, each set $\mathcal{S}_i$ possesses a Dickson basis $\mathcal{B}_i$. Furthermore, the union $\bigcup_{i \in \mathbb{N}_0} \mathcal{B}_i$ also has a Dickson basis $\mathcal{B}'$. Since $\mathcal{B}'$ is finite, there exists an index $k \geq 0$ such that $\mathcal{B}' \subseteq \mathcal{B}_0 \cup \cdots \cup \mathcal{B}_k$.

We claim that the set $\mathcal{B} = \{\mu \mid 0 \leq \mu_n \leq k \wedge \mu' \in \mathcal{B}_{\mu_n}\}$ is a Dickson basis for $\mathcal{S}$. So let $\nu \in \mathcal{S}$ be an arbitrary element of $\mathcal{S}$. By definition, $\nu' \in \mathcal{S}_{\nu_n}$ and there exists a multi index $\mu' \in \mathcal{B}_{\nu_n}$ with $\mu' \mid \nu'$. If $\nu_n \leq k$, we set $\mu = [\mu_1, \dots, \mu_{n-1}, \nu_n]$. Otherwise there exists an index $0 \leq i \leq k$ and a multi index $\lambda' \in \mathcal{B}_i$ such that $\lambda' \mid \mu'$. Then we set $\mu = [\lambda_1, \dots, \lambda_{n-1}, i] \in \mathcal{B}_i$. In both cases we find $\mu \in \mathcal{B}$ and $\mu \mid \nu$. □

Let $\prec$ be a partial order on the set $\mathcal{S}$. We introduce for each element $s \in \mathcal{S}$ its *upper set* $\mathcal{U}_s = \{s' \in \mathcal{S} \mid s \prec s'\}$ and say that the set $\mathcal{S}$ has the *König property* with respect to the partial order $\prec$, if both $\mathcal{S}$ and all upper sets $\mathcal{U}_s$ have only finitely many minimal elements. The following result is known as *König's Lemma*, as a similar statement for graphs was first shown by König [260].

Lemma A.1.3 (König). *Let $\mathcal{S}$ be an infinite set that has the König property with respect to the partial order $\prec$. Then $\mathcal{S}$ contains an infinite strictly ascending sequence $s_1 \prec s_2 \prec s_3 \prec \cdots$.*

Proof. Let $\mathcal{M} \subset \mathcal{S}$ be the finite set of minimal elements in $\mathcal{S}$; obviously, this implies $\mathcal{S} = \bigcup_{m \in \mathcal{M}} (\{m\} \cup \mathcal{U}_m)$. Since by assumption $\mathcal{S}$ is infinite, at least one $m \in \mathcal{M}$ must have an infinite upper set $\mathcal{U}_m$. We choose $s_1 = m$ and iterate the argument for $\mathcal{S} = \mathcal{U}_m$ obtaining an element $s_2 \succ s_1$. As we always retain at least one infinite set $\mathcal{U}_{s_k}$, we can iterate infinitely often and construct this way the desired sequence. □

Remark A.1.4. The König property is related to the question of whether a ring (or a monoid) is Noetherian. Let $\mathcal{R}$ be a (left) Noetherian ring and $\mathcal{S}$ a set of (left) ideals of $\mathcal{R}$. If $\mathcal{S}$ has the König property for the partial order induced by set inclusion, then it must be finite, as otherwise König's Lemma A.1.3 would imply the existence of an infinite strictly ascending chain of ideals. ◁

Consider again the polynomial ring $\mathcal{P} = \mathbb{k}[x^1, \dots, x^n]$ over a field $\mathbb{k}$ and the corresponding monoid of terms $\mathbb{T}$. Any polynomial $f \in \mathcal{P} \setminus \{0\}$ can be written in the form $f = \sum_{t \in \mathbb{T}} c_t t$ where only finitely many coefficients $c_t \in \Bbbk$ are different from 0. The finite set

$$\operatorname{supp} f = \{t \in \mathbb{T} \mid c_t \neq 0\} \tag{A.7}$$

containing all terms with a non-zero coefficient is called the *support* of f. For the theory of Gröbner bases considered in Appendix B.4 the following notions are of fundamental importance.

Definition A.1.5. Let $f \in \mathcal{P}$ be a non-zero polynomial. For a given order $\prec$ on $\mathbb{T}$, the set $\operatorname{supp} f$ contains a largest term, the *leading term* $\operatorname{lt}_\prec f$ of the polynomial f. Its coefficient is the *leading coefficient* $\operatorname{lc}_\prec f$ and if $\operatorname{lt}_\prec f = x^\mu$, we call μ the *leading exponent* $\operatorname{le}_\prec f$. Finally, the *leading monomial* is defined by $\operatorname{lm}_\prec f = \operatorname{lc}_\prec f \cdot \operatorname{lt}_\prec f$. If $\mathcal{F} \subseteq \mathcal{P}$ is a set of polynomials, we write $\operatorname{lt}_\prec \mathcal{F}$ for the set $\{\operatorname{lt}_\prec f \mid f \in \mathcal{F}\}$.

In this context, well-orders are particularly important. Many constructive proofs in the theory of Gröbner bases proceed by the following scheme. We apply repeatedly a certain operation to an element of an ordered set obtaining a new element which is smaller with respect to the given order. If the used order is a well-order, this process cannot go on infinitely but must end after a finite number of steps. Usually, the finally obtained element possesses then some particular property. Our next result characterises those semigroup orders on $\mathbb{T}$ which are well-orders.

Lemma A.1.6. *Let $\prec$ be a semigroup order on $(\mathbb{T}, \cdot)$. Then the following three statements are equivalent.*

(i) *$\prec$ is a well-order.*
(ii) *$\prec$ is a monoid order.*
(iii) *If $s \mid t$ and $s \neq t$ for two terms $s, t \in \mathbb{T}$, then $\mu \prec \nu$, i. e. the semigroup order $\prec$ refines the natural partial order on $\mathbb{T}$ defined by divisibility.*

Proof. Suppose that we have $x^i \prec 1$ for some index $1 \leq i \leq n$. Then obviously $(x^i)^k \prec (x^i)^\ell$ whenever $k > \ell$ by the monotonicity of a semigroup order and the subset $\{(x^i)^k \mid k \in \mathbb{N}\} \subset \mathbb{T}$ does not possess a minimal element. Hence for a well-order we must have $1 \prec x^i$ for all $1 \leq i \leq n$. Again the monotonicity of $\prec$ implies that also $1 \prec x^i x^j$ for any $1 \leq j \leq n$. Thus by a simple induction $1 \prec t$ for all $t \in \mathbb{T} \setminus \{1\}$ and (ii) follows from (i).

If $s \mid t$ and $s \neq t$, then a term $r \neq 1$ exists such that $t = rs$. As $1 \prec r$ for a monoid order, $s = 1 \cdot s \prec r \cdot s = t$ by monotonicity and (ii) implies (iii).

Finally, for showing that (iii) implies (i), let $t_1 \succ t_2 \succ t_3 \succ \cdots$ be an infinite descending sequence in $\mathbb{T}$ and set $\mathcal{T} = \{t_1, t_2, \dots\}$. By Dickson's Lemma A.1.2, a finite subset $\mathcal{B} \subseteq \mathcal{T}$ exists which contains for every $t \in \mathcal{T}$ a term $s \in \mathcal{B}$ such that $s \mid t$. Without loss of generality, we may assume that it is of the form $\mathcal{B} = \{t_1, t_2, \dots, t_k\}$ for some $k \geq 1$. Thus there must exist a value $1 \leq j \leq k$ such that $t_j \mid t_{k+1}$ leading to the contradiction $t_j \prec t_{k+1}$. □

Let $\prec$ be a semigroup order on $(\mathbb{T}, \cdot)$. If $s \prec t$ whenever $\deg s < \deg t$, it is called *degree compatible* (clearly, such an order is a monoid order). We say that an order is of *type* ω, if for any two terms $s, t \in \mathbb{T}$ with $s \prec t$ only finitely many terms r_i exist with $s \prec r_1 \prec r_2 \prec \cdots \prec t$. Obviously, any degree compatible order is of type ω.

Example A.1.7. Some important term orders are the following ones. The *lexicographic* order is defined by $x^\mu \prec_{\text{lex}} x^\nu$, if the last non-vanishing entry of $\mu - \nu$ is negative. Thus $(x^2)^2 x^3 \prec_{\text{lex}} x^1(x^3)^2$. With respect to the *reverse lexicographic* order, $x^\mu \prec_{\text{revlex}} x^\nu$, if the first non-vanishing entry of $\mu - \nu$ is positive. Now we have $x^1(x^3)^2 \prec_{\text{revlex}} (x^2)^2 x^3$. However, $\prec_{\text{revlex}}$ is only a semigroup order, as $x^i \prec_{\text{revlex}} 1$ for all $1 \leq i \leq n$. The reverse lexicographic order should not be confused with the *inverse lexicographic* order $\prec_{\text{invlex}}$ which arises from $\prec_{\text{lex}}$ by inverting the order of the variables, i. e. $x^\mu \prec_{\text{invlex}} x^\nu$, if the first non-vanishing entry of $\mu - \nu$ is negative.

The lexicographic order is the classical example of an order which is not of type ω. Indeed, we have for instance $1 \prec_{\text{lex}} x^1 \prec_{\text{lex}} (x^1)^2 \prec_{\text{lex}} \cdots \prec_{\text{lex}} x^2$. Degree compatible versions exist of all these orders. $x^\mu \prec_{\text{deglex}} x^\nu$, if $|\mu| < |\nu|$ or if $|\mu| = |\nu|$ and $x^\mu \prec_{\text{lex}} x^\nu$. Similarly, $x^\mu \prec_{\text{degrevlex}} x^\nu$, if $|\mu| < |\nu|$ or if $|\mu| = |\nu|$ and $x^\mu \prec_{\text{revlex}} x^\nu$. Note that $\prec_{\text{degrevlex}}$ is a *monoid* order, in fact a very important one!

Assume that we split the variables $x^1, \dots, x^n$ into two sets, say $x^1, \dots, x^k$ and $x^{k+1}, \dots, x^n$. For notational simplicity we write $y^1, \dots, y^{n-k}$ for the variables in the second set. Given a term order $\prec_x$ on the submonoid $\mathbb{T}_x$ of terms in $x^1, \dots, x^k$ and term order $\prec_y$ on the submonoid $\mathbb{T}_y$ of terms in $y^1, \dots, y^{n-k}$, we may define a *product* or *block order* $\prec$ on the full monoid $\mathbb{T}$ by setting $x^{\mu_1} y^{\nu_1} \prec x^{\mu_2} y^{\nu_2}$, if $y^{\nu_1} \prec_y y^{\nu_2}$ or $\nu_1 = \nu_2$ and $x^{\mu_1} \prec_x x^{\mu_2}$.

Here we defined the orders inverse to the usual convention in most textbooks on Gröbner bases. The classical forms arise, if one inverts the order of the variables: $x^1, \dots, x^n \mapsto x^n, \dots, x^1$. Our version fits better to the conventions used in differential equations theory, in particular to our definition of the class of a multi index. $\lhd$

For certain applications the lexicographic and the degree reverse lexicographic order, respectively, are particularly important. The reasons lie in the following characterisations of these two term order.

Lemma A.1.8. *Let for a term order $\prec$ the condition $\operatorname{lt}_\prec f \in \Bbbk[x^1, \dots, x^k]$ with an arbitrary value $1 \leq k \leq n$ be equivalent to $f \in \Bbbk[x^1, \dots, x^k]$ for any polynomial $f \in \mathcal{P}$. Then $\prec$ is the lexicographic order $\prec_{\text{lex}}$. If $\prec$ is degree compatible and the condition $\operatorname{lt}_\prec f \in \langle x^1, \dots, x^k \rangle$ is equivalent to $f \in \langle x^1, \dots, x^k \rangle$ for any homogeneous polynomial $f \in \mathcal{P}$, then $\prec$ is the degree reverse lexicographic order $\prec_{\text{degrevlex}}$.*

Proof. We leave the case of the lexicographic order as an exercise; it goes analogously to the for us more important case of the degree reverse lexicographic order. Let x^μ and x^ν be two arbitrary monomials of the same degree such that the first non-vanishing entry of $\mu - \nu$ is $\mu_k - \nu_k$. Without loss of generality, we assume that $\mu_k > \nu_k$. Set $\rho = [\nu_1, \dots, \nu_k, 0, \dots, 0]$ and consider the multi indices $\bar\mu = \mu - \rho$ and $\bar\nu = \nu - \rho$. Obviously, $x^{\bar\mu} \in \langle x^1, \dots, x^k \rangle$ whereas $x^{\bar\nu} \notin \langle x^1, \dots, x^k \rangle$. Considering the homogeneous polynomial $f = x^{\bar\mu} + x^{\bar\nu}$, the assumption of the lemma implies that the leading term of f must be $x^{\bar\nu}$. By the monotonicity of term orders, we conclude that $x^\mu \prec x^\nu$. But by definition, we also have $x^\mu \prec_{\text{degrevlex}} x^\nu$. $\square$

We say that a term order *respects classes*, if for multi indices μ, ν of the same length $\operatorname{cls}\mu < \operatorname{cls}\nu$ implies $x^\mu \prec x^\nu$. It is now easy to see that by Lemma A.1.8 on terms of the same degree any class respecting term order coincides with the degree reverse lexicographic order.

Example A.1.9. For some applications *elimination orders* are useful. The order $\prec$ has the elimination property for the variables $x^{k+1},\dots,x^n$, if $\mathrm{lt}_\prec f \in \Bbbk[x^1,\dots,x^k]$ implies $f \in \Bbbk[x^1,\dots,x^k]$ for any polynomial $f \in \Bbbk[x^1,\dots,x^n]$. Obviously, product orders possess this property. By Lemma A.1.8 the lexicographic order is characterised by the fact that it is an elimination order for any value $1 \le k \le n$. $\lhd$

Addendum: Computing Derivative Trees

Consider the following problem: we are given a function $f(x,y)$ and must compute the derivatives f_{xx} and f_{xy}. Obviously, there are two possibilities. The more efficient one first computes f_x and then the two derivatives we need. The other one consists of first computing f_y and f_{xy} and only afterwards f_x and f_{xx}; hence it needs one derivation more. In this simple example it is trivial to find the optimal way, but if we are given the task to compute, say, six derivatives of up to degree ten of a function of four variables, things are getting more complicated.

In concrete computations with differential equations this problem appears fairly often. One trivial application consists of checking whether a certain function is actually a solution of a given differential equation. Then we must compute all derivatives that appear in the equation. More generally, one may consider arbitrary differential substitutions where a substitution for u^α_μ automatically entails substitutions for all its prolongations $u^\alpha_{\mu+\nu}$.

We may formalise this question as follows. We are given a finite set $\mathcal{N} \subset \mathbb{N}_0^n$ of multi indices and we must compute a connected tree, the *derivative tree*, the root of which is the zero index and which contains all elements of $\mathcal{N}$ as leaves. This is an instance of what computer scientists call the *rectilinear Steiner tree problem*. Unfortunately, one can show that it is *NP*-complete [150].

In our application this fact implies that there is no point in looking for an algorithm to construct explicitly a minimal Steiner tree, as the time needed for the construction will probably be larger than any gains one achieves by avoiding some differentiations. The recursive Algorithm A.1 below is based on a simple divide-and-conquer heuristics.[2] While it does not necessarily find the optimal solution, it comes up with a reasonably good solution very fast.

The key is to introduce a distance function for multi indices. As we consider now also differences of multi indices, we must accommodate for negative entries and define $\|\mu\| = \sum_{i=1}^n |\mu_i|$ for any $\mu \in \mathbb{Z}^n$. Then we define for two arbitrary multi indices $\mu,\nu \in \mathbb{N}_0^n$

[2] The main ideas of this algorithm stem from Sven Helmer (Universität Mannheim). It was first presented in [37].

$$\operatorname{dist}(\mu,\nu) = \begin{cases} 0 & \text{if } \|\mu-\nu\| = \big|\|\mu\| - \|\nu\|\big|\,, \\ \|\mu-\nu\| & \text{else}\,. \end{cases} \tag{A.8}$$

Algorithm A.1 always searches for the "least common multiple" of the given set of multi indices (thinking of them as terms). It is subtracted from all multi indices and returned as next node in the derivative tree. Now the set is divided into two subsets: we first search for two multi indices with a maximal distance from each other and then group every multi index to the one to which it is closer. Finally, we recursively apply the algorithm to these subsets. The results are appended as children to the current node (we represent here the tree as a nested list).

Algorithm A.1 Derivative tree

Input: finite set $\mathcal{N} \subset \mathbb{N}_0^n$
Output: derivative tree $\mathcal{T}$ with elements of $\mathcal{N}$ as leaves

$\alpha \leftarrow \operatorname{lcm}\mathcal{N}$
$\mathcal{N} \leftarrow \{\mu - \alpha \mid \mu \in \mathcal{N}\} \setminus \{0\}$
choose $\mu,\nu \in \mathcal{N}$ with $\operatorname{dist}(\mu,\nu)$ maximal
$\mathcal{N}_\mu \rightarrow \{\rho \in \mathcal{N} \mid \operatorname{dist}(\rho,\mu) < \operatorname{dist}(\rho,\nu)\}$
return $\Big(\alpha, \big(\mathtt{Tree}(\mathcal{N}_\mu), \mathtt{Tree}(\mathcal{N}\setminus\mathcal{N}_\mu)\big)\Big)$

This approach to dividing into two subsets based on the distance explains the case distinction in the definition of $\operatorname{dist}(\mu,\nu)$. If one multi index is a "divisor" of the other, we want of course that the two always remain in the same subset, as the optimal path to the larger one leads through the smaller one. Hence we define their distance as zero.

Example A.1.10. Figure A.1 shows a derivative tree for the set

$$\Big\{[1,0,3,2],[0,2,1,4],[0,4,4,0],[0,3,3,3],[3,0,4,2],[0,5,1,4]\Big\} \subset \mathbb{N}^4 \tag{A.9}$$

found by Algorithm A.1. Thus these six derivative can be computed with at most 26 differentiations. ◁

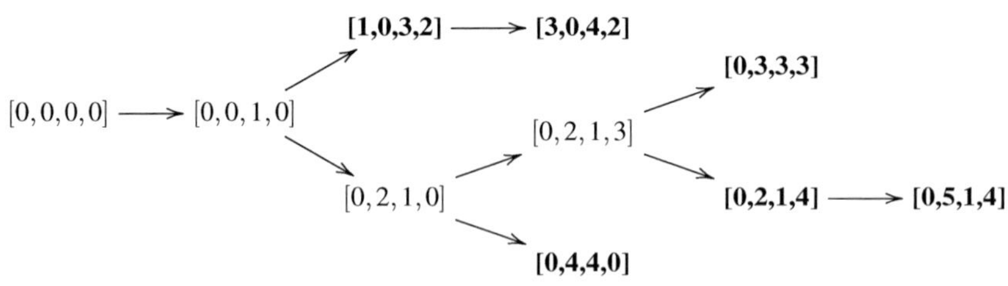

Fig. A.1 Derivative tree for Example A.1.10

A.2 Real-Analytic Functions

Much of the formal theory is based on formal power series. If such a series converges, it defines an analytic function. In this chapter we collect some basic facts about these functions. Throughout we consider real-valued functions defined in some domain $\Omega \subset \mathbb{R}^n$ (i. e. in a connected open set).

Definition A.2.1. The function $f : \Omega \to \mathbb{R}$ is *real-analytic* at the point $y \in \Omega$, if an open neighbourhood $y \in \mathcal{U} \subseteq \Omega$ exists such that for all $x \in \mathcal{U}$ the function f can be represented by a power series

$$f(x) = \sum_{|\mu|=0}^{\infty} a_\mu (x-y)^\mu \,. \tag{A.10}$$

f is real-analytic in the whole domain Ω, if it is real-analytic at every point $y \in \Omega$. The set of all functions that are real-analytic in the domain Ω is denoted by $\mathcal{C}^\omega(\Omega)$.

Any analytic function is smooth but not conversely: $\mathcal{C}^\omega(\Omega) \subsetneq \mathcal{C}^\infty(\Omega)$. By Taylor's Theorem, the coefficients a_μ are thus given by

$$a_\mu = \frac{1}{\mu!}\frac{\partial^{|\mu|} f}{\partial x^\mu}(y) \,, \tag{A.11}$$

i. e. they are determined by the derivatives of f at the point y, and we may say that the analytic functions are those smooth functions for which the Taylor series converges to the function. This condition leads to a strict subset of the smooth functions. Consider the function $f(x) = \exp(-1/x^2)$. It is not difficult to verify that f is smooth everywhere on $\mathbb{R}$. However, f is not analytic at the origin. Indeed, all derivatives of f vanish there so that the power series expansion vanishes in any neighbourhood of the origin. This function f allows us the construction of other functions like

$$\bar{f} : x \mapsto \begin{cases} 0 & x \leq 0 \,, \\ f(x) & x > 0 \,. \end{cases} \tag{A.12}$$

Again it is easy to see that $\bar{f}$ is smooth everywhere on $\mathbb{R}$, but obviously it is not analytic at the origin.

Analyticity is a rather strong property, as the following proposition implies that if we know an analytic function and all its derivatives at one point of Ω, we know it in all Ω. This fact is sometimes called *unique continuation*.

Proposition A.2.2. *Let $f, g \in \mathcal{C}^\omega(\Omega)$ be two functions that are real-analytic in the domain Ω. If there exists a point $y \in \Omega$ such that*

$$\frac{\partial^{|\mu|} f}{\partial x^\mu}(y) = \frac{\partial^{|\mu|} g}{\partial x^\mu}(y) \qquad \forall \mu \in \mathbb{N}_0^n \,, \tag{A.13}$$

then $f \equiv g$ in the whole domain Ω.

Proof. We use a simple topological argument. Consider the set

$$\Sigma = \left\{ z \in \Omega \mid \frac{\partial^{|\mu|} f}{\partial x^{\mu}}(z) = \frac{\partial^{|\mu|} g}{\partial x^{\mu}}(z) \quad \forall \mu \in \mathbb{N}_0^n \right\}. \tag{A.14}$$

It is relatively closed in Ω, as it is defined by equations. On the other hand, Σ must be open, because at any point $x \in \Sigma$ the two functions f and g possess identical power series and hence are identical in an open neighbourhood of y. As, by definition, a domain is a connected open set, we must have $\Sigma = \Omega$. □

Another important and characteristic property of analytic functions is the existence of a bound on their derivatives. For a smooth function f defined in a neighbourhood of an arbitrary but fixed point $y \in \Omega$, we say that $f \in \mathcal{C}_{M,r}(y)$, if its derivatives at y satisfy *Cauchy's estimate*

$$\left| \frac{\partial^{|\mu|} f}{\partial x^{\mu}}(y) \right| \le M \frac{|\mu|!}{r^{|\mu|}} \qquad \forall \mu \in \mathbb{N}^n. \tag{A.15}$$

Proposition A.2.3. *Let $f \in \mathcal{C}^{\infty}(\Omega)$ be a smooth function. Then f is analytic, i. e. $f \in \mathcal{C}^{\omega}(\Omega)$, if and only if for every compact subset $\Sigma \subset \Omega$ two numbers M and r exist such that $f \in \mathcal{C}_{M,r}(x)$ for all points $x \in \Sigma$.*

Proof. Let us first assume that f is analytic. Then it is not difficult to show that for all $x \in \Omega$ its derivatives satisfy in a neighbourhood $U(x)$ the estimate (A.15) with constants $M(x)$ and $r(x)$ (see e. g. [379, Section 2.2.1]). As any compact subset Σ can be covered by a finite number of such neighbourhoods, we can take the maximum of the $M(x)$ and the minimum of the $r(x)$ and obtain the assertion.

For the converse, take an arbitrary point $y \in \Omega$ and choose as compact set Σ a closed ball of radius R with its centre at x (obviously, the radius R must be chosen small enough so that $\Sigma \subset \Omega$). Then, by assumption, there exist numbers M and r such that $f \in \mathcal{C}_{M,r}(y)$. Consider a point $x \in \Sigma$ with $\delta = \sum_{i=1}^{n} |y_i - x_i| < \min(r, R)$ and introduce for $0 \le \tau \le 1$ the function $F(\tau) = f\bigl(y + \tau(x-y)\bigr)$. We apply Taylor's Theorem to $F(\tau)$ and obtain

$$f(x) = F(1) = \sum_{k=0}^{q-1} \frac{1}{k!} \frac{d^k F}{d\tau^k}(1) + \frac{1}{q!} \frac{d^q F}{d\tau^q}(\bar{\tau}) \tag{A.16}$$

for some value $0 \le \bar{\tau} \le 1$. A straightforward computation yields for any order $q \ge 0$ the following estimate of the remainder term

$$\left| \frac{1}{q!} \frac{d^q F}{d\tau^q}(\bar{\tau}) \right| \le M \left(\frac{\delta}{r} \right)^q. \tag{A.17}$$

As by construction $\delta < r$, it tends to zero and we find that f can be represented in a neighbourhood of y by a convergent series of the form (A.10). □

Finally, we recall that an analytic function f is *majorised* in a neighbourhood of the point y by another analytic function g, written $f \preceq g$, if the inequality

$$\left|\frac{\partial^{|\mu|}f}{\partial x^\mu}(y)\right| \leq \left|\frac{\partial^{|\mu|}g}{\partial x^\mu}(y)\right| \tag{A.18}$$

holds for all multi indices $\mu \in \mathbb{N}_0^n$. As a simple example we note that any function in $\mathcal{C}_{M,r}(0)$ is majorised by the function

$$g(x) = \frac{Mr}{r-(x_1+\cdots+x_n)}\,, \tag{A.19}$$

as the Taylor coefficients of g at the origin are just given by the right hand side of Cauchy's estimate (A.15).

A.3 Elementary Transformations of Differential Equations

In this section, we study three elementary operations with differential equations: rewriting a higher-order equation as a first-order one, quasi-linearisation and the transformation to one dependent variable. In particular, we determine in each case what happens, if we apply the operation to an involutive equation. As usual, we assume that $\pi : \mathcal{E} \to \mathcal{X}$ is a fibred manifold and that a higher-order differential equation $\mathcal{R}_q \subseteq J_q\pi$ is given.

Reduction to first order

For many purposes it is useful to rewrite a differential equation of order $q > 1$ as a first-order equation through the introduction of additional dependent variables representing the derivatives of the original dependent variables. Geometrically, this transformation is based on the identification of $J_q\pi$ with a submanifold of $J_1\pi^{q-1}$ via the immersion $\iota_{q,1}$ (see Section 2.2). We define intrinsically $\tilde{\mathcal{R}}_1 = \iota_{q,1}(\mathcal{R}_q) \subseteq J_1\pi^{q-1}$.

In local coordinates we get the following picture. Let $(\mathbf{x}, \mathbf{u}^{(q)})$ be our usual coordinates on $J_q\pi$; on the iterated jet bundle $J_1\pi^{q-1}$ we use $\big(\mathbf{x}, (\mathbf{u}^{(q-1)})^{(1)}\big)$. If the original equation $\mathcal{R}_q$ is locally represented by the system $\Phi^\tau(\mathbf{x}, \mathbf{u}^{(q)}) = 0$, a local representation of its reduction $\tilde{\mathcal{R}}_1$ is

$$\tilde{\mathcal{R}}_1 : \begin{cases} \tilde{\Phi}^\tau\big(\mathbf{x}, (\mathbf{u}^{(q-1)})^{(1)}\big) = 0\,, & \\ u^\alpha_{\mu,i} = u^\alpha_{\mu+1_i}\,, & |\mu| < q-1\,,\ 1 \leq i \leq n\,, \\ u^\alpha_{\mu,i} = u^\alpha_{\mu-1_k+1_i,k}\,, & |\mu| = q-1\,,\ k = \operatorname{cls}\mu < i \leq n\,. \end{cases} \tag{A.20}$$

The second and the third line of this system represent the equations defining the submanifold $\iota_{q,1}(J_q\pi) \subset J_1\pi^{q-1}$. The function $\tilde{\Phi}^\tau$ is not uniquely defined, as in general several possibilities exist to express a higher-order derivative u^α_μ by one of the new coordinates. One way, which will allow us afterwards to compute easily the indices of the symbol $\tilde{\mathcal{N}}_1$, is to use the mapping

$$u^\alpha_\mu \longmapsto \begin{cases} u^\alpha_\mu & \text{for } |\mu| \leq q-1\,, \\ u^\alpha_{\mu-1_k,k} & \text{for } |\mu| = q,\ \operatorname{cls}\mu = k\,. \end{cases} \tag{A.21}$$

Proposition A.3.1. *Let $\tilde{\alpha}_1^{(1)}, \dots, \tilde{\alpha}_1^{(n)}$ be the Cartan characters of the reduced equation $\tilde{\mathcal{R}}_1 \subseteq J_1\pi^{q-1}$ and $\alpha_q^{(1)}, \dots, \alpha_q^{(n)}$ those of the original equation $\mathcal{R}_q \subseteq J_q\pi$. Then $\tilde{\alpha}_1^{(k)} = \alpha_q^{(k)}$ for $1 \leq k \leq n$ and there exists a one-to-one correspondence between the solutions*[3] *of $\tilde{\mathcal{R}}_1$ and $\mathcal{R}_q$. The differential equation $\mathcal{R}_q$ is involutive, if and only if its reduction $\tilde{\mathcal{R}}_1$ is involutive.*

Proof. Due to the substitution rule (A.21) the first line of (A.20) yields $\beta_q^{(k)}$ equations of class k, i.e. the same value as in $\mathcal{R}_q$. The second line represents $\dim J_{q-2}\pi$ equations of class k, and a little combinatorial calculation shows that the last line leads to additionally $\dim\bigl(S_{q-1}(T^*\mathcal{X})\otimes V\mathcal{E}\bigr) - \binom{q+n-k-1}{q-1}$ equations of class k. Adding these values we get

$$\tilde{\beta}_1^{(k)} = \dim J_{q-2}\pi + \dim\bigl(S_{q-1}(T^*\mathcal{X})\otimes V\mathcal{E}\bigr)\,. - \alpha_q^{(k)} \tag{A.22}$$

Since the fibre dimension of $\pi^{q-1} : J_{q-1}\pi \to \mathcal{X}$ corresponds to the number of the dependent variables in the reduced equation $\tilde{\mathcal{R}}_1$, its Cartan characters are:

$$\tilde{\alpha}_1^{(k)} = \dim J_{q-1}\pi - \tilde{\beta}_1^{(k)} = \alpha_q^{(k)}\,. \tag{A.23}$$

If $(\mathbf{x})$ defines a δ-regular coordinate system for $\mathcal{R}_q$, then it does so for $\tilde{\mathcal{R}}_1$, too, and vice versa. The additional equations in (A.20) are in a form that leads to maximal values for the number of equations of class n and so on. Thus only the equations $\tilde{\Phi}^\tau = 0$ could make trouble. But due to our substitution rule (A.21) we see that if we get maximal values for the indices of $\mathcal{N}_q$ we get them also for $\tilde{\mathcal{N}}_1$ and vice versa.

These considerations imply furthermore that involution of the symbol of one of the equations is equivalent to involution of the symbol of the other one. It is easy to see that the second and third line in (A.20) alone form an involutive system and that there arise no integrability conditions between them and the equations $\tilde{\Phi}^\tau = 0$. But if the equations $\Phi^\tau = 0$ generate integrability conditions, then the same happens with the equations $\tilde{\Phi}^\tau = 0$ and vice versa.

The claimed relation between the solution spaces of the differential equations $\mathcal{R}_q$ and $\tilde{\mathcal{R}}_1$, respectively, is obvious. If a section $\sigma : \mathcal{X} \to \mathcal{E}$ is a solution of $\mathcal{R}_q$, then its prolongation $j_{q-1}\sigma : \mathcal{X} \to J_{q-1}\pi$ is a solution of $\tilde{\mathcal{R}}_1$. Conversely, the second and third line of (A.20) ensure that every solution $\tilde{\sigma} : \mathcal{X} \to J_{q-1}\pi$ of $\tilde{\mathcal{R}}_1$ is a section of the form $\tilde{\sigma} = j_{q-1}\sigma$ with $\sigma : \mathcal{X} \to \mathcal{E}$ a solution of $\mathcal{R}_q$. □

[3] As usual we are only considering either smooth or formal solutions.

Quasi-Linearisation

The above mentioned method for a reduction to first order is not the only possible one. In particular, we may apply it to the prolongation $\mathcal{R}_{q+1}$ instead of $\mathcal{R}_q$ itself. Recall that a prolongation always leads to a quasi-linear equation. This fact implies that the arising first-order equation $\bar{\mathcal{R}}_1 \subset J_1\pi^q$ is always quasi-linear; the original equations are transformed into algebraic constraints.

Proposition A.3.2. *Let $\bar{\alpha}_1^{(1)}, \dots, \bar{\alpha}_1^{(n)}$ be the Cartan characters of the quasi-linear equation $\bar{\mathcal{R}}_1 \subseteq J_1\pi^q$ and $\alpha_q^{(1)}, \dots, \alpha_q^{(n)}$ those of the original differential equation $\mathcal{R}_q \subseteq J_q\pi$. Then $\bar{\alpha}_1^{(k)} = \sum_{i=k}^{n} \alpha_q^{(i)}$ for $1 \leq k \leq n$ and there exists a one-to-one correspondence between the solutions of $\bar{\mathcal{R}}_1$ and $\mathcal{R}_q$. If $\mathcal{R}_q$ is involutive, then $\bar{\mathcal{R}}_1$ is involutive, too. If $\bar{\mathcal{R}}_1$ is involutive, then $\mathcal{R}_q$ is formally integrable and the prolongation $\mathcal{R}_{q+1}$ is involutive.*

Proof. Most of the assertions follow immediately from Proposition A.3.1. According to Proposition 8.2.3, the Cartan characters of $\mathcal{R}_{q+1}$ and $\mathcal{R}_q$ are related by $\alpha_{q+1}^{(k)} = \sum_{i=k}^{n} \alpha_q^{(i)}$. If the equation $\mathcal{R}_q$ is involutive, then $\mathcal{R}_{q+1}$ and hence $\bar{\mathcal{R}}_1$ are involutive, too.

For the final assertion, we first note that if $\bar{\mathcal{R}}_1$ is involutive, then by our previous results $\mathcal{R}_{q+1}$ is involutive. Applying our transformation to $\mathcal{R}_q$ yields a differential equation $\bar{\mathcal{R}}_0 \supseteq \bar{\mathcal{R}}_0^{(1)}$. As the transformation of $\mathcal{R}_{q+1}$ yields $\bar{\mathcal{R}}_1$ and as, because of the involution of $\bar{\mathcal{R}}_1$, the equality $(\bar{\mathcal{R}}_0^{(1)})_{+1} = \bar{\mathcal{R}}_1$ holds, we must have $\bar{\mathcal{R}}_0 = \bar{\mathcal{R}}_0^{(1)}$. But this implies that $\mathcal{R}_q^{(1)} = \mathcal{R}_q$. Hence $\mathcal{R}_q$ is at least formally integrable. □

Example A.3.3. We cannot conclude that involution of $\bar{\mathcal{R}}_1$ entails involution of the original equation $\mathcal{R}_q$, as its construction uses a prolongation. Consider again the differential equation of Example 7.2.2:

$$\mathcal{R}_2 : \left\{ u_{yy} = 0 \,, \quad u_{xx} = 0 \,. \right. \tag{A.24}$$

In order to make the notation less cumbersome, we introduce the following variable names in $J_2\pi$: $r = u_{xx}$, $s = u_{xy}$, $t = u_{yy}$, $v = u_x$ and $w = u_y$. The first-order equation $\bar{\mathcal{R}}_1 \subset J_1\pi^2$ has a local representation of the form

$$\bar{\mathcal{R}}_1 : \begin{cases} t_y = 0\,, & t_x = 0\,, \\ s_x = 0\,, & r_x = 0\,, \\ r = 0\,, & t = 0\,, \\ r_y = s_x\,, & s_y = t_x\,, \\ v_y = s\,, & v_x = r\,, \\ w_y = t\,, & w_x = s\,, \\ u_y = w\,, & u_x = v\,. \end{cases} \tag{A.25}$$

The first six equations represent $\mathcal{R}_3$ in the new coordinates; the remaining equations define the submanifold $\iota_{3,1}(J_3\pi) \subset J_1\pi^2$. One easily checks that $\bar{\mathcal{R}}_1$ is involutive.

We know from Example 7.2.2 that $\mathcal{R}_2$ is only formally integrable but not involutive whereas $\mathcal{R}_3$ is involutive. ◁

Remark A.3.4. A special situation arises, if we start with a *normal* first-order differential equation $\mathcal{R}_1$. Such an equation is locally represented by a system of the form $u^\alpha_n = \phi^\alpha(\mathbf{x},\mathbf{u},\tilde{\mathbf{u}}_{(1)})$ where $\tilde{\mathbf{u}}_{(1)}$ denotes all derivatives of class less than n. Applying the above quasi-linearisation process yields a differential equation $\bar{\mathcal{R}}_1$. We consider now only the subsystem of class n of the Cartan normal form of $\bar{\mathcal{R}}_1$:

$$\begin{aligned} u^\alpha_{,n} &= u^\alpha_n\,, \\ u^\alpha_{j,n} &= \begin{cases} u^\alpha_{n,j} & \text{for } 1 \le j < n\,, \\ \dfrac{\partial \phi^\alpha}{\partial x^n} + \dfrac{\partial \phi^\alpha}{\partial u^\beta} u^\beta_n + \dfrac{\partial \phi^\alpha}{\partial u^\beta_j} u^\beta_{n,j} & \text{for } j = n\,. \end{cases} \end{aligned} \tag{A.26}$$

It defines a quasi-linear normal first-order equation. Of course, this equation is *not* equivalent to the original equation $\mathcal{R}_1$, as its local representation is only a subsystem of the equivalent equation $\bar{\mathcal{R}}_1$. Thus its solution space is larger.

Assume now that we consider for $\mathcal{R}_1$ the initial value problem with initial conditions $\mathbf{u}(x^1,\dots,x^{n-1},0) = \mathbf{f}(x^1,\dots,x^{n-1})$. The solution of this problem induces a solution of (A.26) satisfying the initial conditions

$$\begin{aligned} u^\alpha(x^1,\dots,x^{n-1},0) &= f^\alpha(x^1,\dots,x^{n-1})\,, \\ u^\alpha_j(x^1,\dots,x^{n-1},0) &= \begin{cases} \dfrac{\partial f^\alpha}{\partial x^j}(x^1,\dots,x^{n-1}) & \text{for } 1 \le j < n\,, \\ \phi^\alpha\Bigl(x^1,\dots,x^{n-1},0,f^\beta,\dfrac{\partial f^\beta}{\partial x^j}\Bigr) & \text{for } j = n\,. \end{cases} \end{aligned} \tag{A.27}$$

Conversely, any solution of the quasi-linear system (A.26) for initial conditions of this special form yields a solution of the initial value problem for the original equation $\mathcal{R}_1$. Hence it suffices for a uniqueness and existence theory of normal systems to consider only the case of quasi-linear systems. ◁

Transformation to one dependent variable

It is possible to transform any differential equation $\mathcal{R}_q$ in m dependent variables u^α and n independent variables x^i into an equation with only one dependent variable v. As this trick seems to be due to Drach [112], we call it *Drach transformation*. It requires the introduction of m additional independent variables y_α. Furthermore, the order is raised by one and the relation between the transformed and the original equation is more complicated than in the reduction to first order discussed above.

Traditionally, the Drach transformation is described only in local coordinates. We give here an intrinsic geometric picture, too. It requires that $\pi : \mathcal{E} \to \mathcal{X}$ is not only a fibred manifold but a vector bundle. Based on the dual bundle $\pi^* : \mathcal{E}^* \to \mathcal{X}$, we introduce the trivial line bundle $\hat{\pi} : \hat{\mathcal{E}} = \mathcal{E}^* \times \mathbb{R} \to \mathcal{E}^*$. Let $\sigma : \mathcal{X} \to \mathcal{E}$ be a section with $\sigma(x) = \bigl(x, s(x)\bigr)$; we associate with it the unique section $\hat{\sigma} : \mathcal{E}^* \to \hat{\mathcal{E}}$ defined at an arbitrary point $\xi = (x, u^*) \in \mathcal{E}^*$ by $\hat{\sigma}(\xi) = \bigl(\xi, u^*\bigl(s(x)\bigr)\bigr)$. Note that this definition makes sense, as the fibre $\mathcal{E}^*_x$ is the dual space of the fibre $\mathcal{E}_x$ and thus we may consider u^* as a linear functional acting on $\mathcal{E}_x$.

If we use local coordinates (x^i, u^α) on $\mathcal{E}$, the dual coordinates (x^i, y_α) on $\mathcal{E}^*$ and v as fibre coordinate on $\hat{\mathcal{E}}$, then $\hat{\sigma}(x^i, y_\alpha) = \bigl(x^i, y_\alpha; y_\alpha s^\alpha(x^i)\bigr)$ for $\sigma(x^i) = \bigl(x^i, s^\alpha(x^i)\bigr)$. Thus the basic idea of the Drach transformation is very simple: we define a new dependent variable by

$$v = y_\alpha u^\alpha\,. \tag{A.28}$$

At the level of jet bundles, this association $\sigma \mapsto \hat{\sigma}$ induces maps from the fibre $(J_q\pi)_x$ to every fibre $(J_q\hat{\pi})_\xi$ with $\pi^*(\xi) = x$. Namely, we map the equivalence class $[\sigma]^{(q)}_x$ to the class $[\hat{\sigma}]^{(q)}_\xi$. It is easy to see that this map is well-defined, i. e. independent of the section σ chosen as representative. Thus given a differential equation $\mathcal{R}_q \subseteq J_q\pi$, we may associate with it a set $\hat{\mathcal{R}}_q \subseteq J_q\hat{\pi}$ by applying these maps. Unfortunately, it does not seem possible to obtain easily a local representation of this set and we will consider a different equation as Drach transformation of $\mathcal{R}_q$.

We first note that all sections $\hat{\sigma} \in \Gamma_{loc}(\hat{\pi})$ obtained in the above described manner from sections $\sigma \in \Gamma_{loc}(\pi)$ satisfy

$$\frac{\partial^2 v}{\partial y_\alpha \partial y_\beta} = 0\,, \qquad 1 \le \alpha \le \beta \le m\,. \tag{A.29}$$

These equations form a differential equation $\hat{\mathcal{S}}_2 \subset J_2\hat{\pi}$. It is nothing but the image of $J_2\pi$ under the above maps and similarly its prolongations $\hat{\mathcal{S}}_q$ are the images of the higher-order jet bundles $J_q\pi$. Now we define as Drach transformation of $\mathcal{R}_q$ the differential equation $\hat{\mathcal{R}}_{q+1} = (\hat{\pi}^{q+1}_q)^{-1}(\hat{\mathcal{R}}_q) \cap \hat{\mathcal{S}}_{q+1}$.

By (A.28), we have for any multi index $\mu \in \mathbb{N}^n_0$ the identity

$$\frac{\partial^{|\mu|} u^\alpha}{\partial x^\mu} = \frac{\partial^{|\mu|+1} v}{\partial y_\alpha \partial x^\mu}\,. \tag{A.30}$$

Considering these identities as substitution rules, we may apply them to a local representation of $\mathcal{R}_q$ and obtain equations of order $q+1$ in $J_{q+1}\hat{\pi}$. Adding (A.29) plus its prolongations to order $q+1$ yields then a local representation of $\hat{\mathcal{R}}_{q+1}$.

Because of (A.29), any solution of $\hat{\mathcal{R}}_{q+1}$ is of the form

$$v(\mathbf{x}, \mathbf{y}) = y_\alpha u^\alpha(\mathbf{x}) + \Lambda(\mathbf{x}) \tag{A.31}$$

where $\mathbf{u}(\mathbf{x})$ is a solution of the original equation $\mathcal{R}_q$ and $\Lambda(\mathbf{x})$ is an arbitrary function. Obviously, this arbitrary function is of no interest for us; thus we may consider it as a "gauge symmetry" of $\hat{\mathcal{R}}_{q+1}$.

Proposition A.3.5. *Let $\hat{\alpha}_{q+1}^{(1)}, \dots, \hat{\alpha}_{q+1}^{(n+m)}$ be the Cartan characters of the transformed equation $\hat{\mathcal{R}}_{q+1}$ where we set $x^k = y_\alpha$ for $k = n + \alpha$ and $\alpha_q^{(1)}, \dots, \alpha_q^{(n)}$ the Cartan characters of the original equation $\mathcal{R}_q$. Then*

$$\hat{\alpha}_{q+1}^{(k)} = \begin{cases} 0 & \text{for } n < k \leq n+m\,, \\ \alpha_q^{(k)} + \binom{q+n-k}{q} & \text{for } 1 \leq k \leq n\,. \end{cases} \tag{A.32}$$

The differential equation $\hat{\mathcal{R}}_{q+1}$ is involutive, if and only if $\mathcal{R}_q$ is involutive.

Proof. Several ways exist to prove this proposition. One may proceed head-on and determine the indices of $\hat{\mathcal{R}}_{q+1}$; they are given by

$$\hat{\beta}_{q+1}^{(k)} = \begin{cases} \binom{q+n+m-k}{q} & \text{for } n < k \leq n+m\,, \\ \beta_q^{(k)} + \frac{m(m+1)}{2} \sum_{i=k}^{n+m} \binom{q+n+m-i-2}{q-2} & \text{for } 1 \leq k \leq n\,. \end{cases} \tag{A.33}$$

The binomial coefficients count the contribution of the equations in (A.29) and their prolongations. This yields (A.32) after some computations involving non-trivial identities of binomial coefficients.

A simpler and more direct way exploits (A.31). It shows that the solution of the equation $\hat{\mathcal{R}}_{q+1}$ does not contain any arbitrary functions of the new variables $\mathbf{y}$. Thus no parametric derivatives exist that are pure $\mathbf{y}$-derivatives and $\hat{\alpha}_{q+1}^{(k)} = 0$ for $k > n$. Concerning the $\mathbf{x}$-dependency we see that we recover all arbitrariness of $\mathcal{R}_q$ and have in addition the gauge symmetry. It contributes $\binom{q+n-k}{q}$ parametric derivatives of order $q+1$ and class k for $1 \leq k \leq n$. Hence $\hat{\alpha}_{q+1}^{(k)} = \alpha_q^{(k)} + \binom{q+n-k}{q}$ for $1 \leq k \leq n$.

The assertion about involution is trivial. (A.29) represents obviously an involutive equation; if it were not for the $\mathbf{x}$-dependency, it would be a finite type equation. Cross-derivatives between the original and these added equations vanish trivially. Thus the only source for possible integrability conditions or obstructions to involution is the original equation and there the $\mathbf{y}$-derivatives play no role at all. □

Using Hilbert polynomials, it is much easier to express (A.32) using the results of Section 8.3. Let $\hat{H}(s)$ be the Hilbert polynomial of $\hat{\mathcal{R}}_{q+1}$, $H(s)$ the one of $\mathcal{R}_q$ and $G(s)$ the one of the Lie pseudogroup defined by

$$\mathcal{G}_1 : \begin{cases} \bar{\mathbf{y}} = \mathbf{y}\,, \quad \bar{\mathbf{x}} = \mathbf{x}\,, \quad \dfrac{\partial \bar{v}}{\partial v} = 1\,, \\ \dfrac{\partial \bar{v}}{\partial y_\alpha} = 0\,, \qquad 1 \leq \alpha \leq m\,. \end{cases} \tag{A.34}$$

It describes the gauge symmetry $\bar{v} = v + \Lambda(\mathbf{x})$ of $\hat{\mathcal{R}}_{q+1}$. Then we get by Proposition 8.3.7 the relation

$$\hat{H}(s) = H(s-1) + \binom{q+n-1}{q}\,, \tag{A.35}$$

as a trivial calculation gives $G(s) = \binom{q+n-1}{q}$.

One may remove the gauge symmetry (and thus obtain an equivalence between the original and the transformed equation) by adding the gauge fixing condition

$$y_\alpha \frac{\partial v}{\partial y_\alpha} - v = 0\,. \tag{A.36}$$

One readily checks that this equation is compatible with those contained in the system for $\hat{\mathcal{R}}_{q+1}$ and thus yields only some trivial integrability conditions due to the fact that it is of lower order. Comparing with (A.31), we see that the sole effect of the condition (A.36) consists of enforcing that $\Lambda(\mathbf{x}) = 0$.

A.4 Modified Stirling Numbers

The modified Stirling numbers appear as combinatorial factors in some manipulations of the Cartan characters and the Hilbert polynomial. We introduce them in a fashion similar to the classical Stirling numbers of the first kind [178]. Let us denote the nth rising factorial powers (sometimes also called Pochhammer's symbol) by $x^{\overline{n}} = x(x+1)\cdots(x+n-1)$. Then we can define the Stirling numbers of the first kind[4] $S_n^{(k)}$ for $n \geq k$ by the identity

$$x^{\overline{n}} = \sum_{k=0}^{n} S_n^{(k)} x^k\,. \tag{A.37}$$

The modified Stirling numbers $s_k^{(n)}(q)$ arise, if we write $x = q+r+1$ for a non-negative integer q and then express the right hand side as a polynomial in r. Table A.1 on page 528 contains some explicit values.

Definition A.4.1. The *modified Stirling numbers* $s_k^{(n)}(q)$ are defined for all non-negative integers n, k, q with $n \geq k$ by the identity

$$(q+r+1)^{\overline{n}} = \sum_{k=0}^{n} s_{n-k}^{(n)}(q) r^k \tag{A.38}$$

(note the different position of the scripts compared with (A.37) and the use of $n-k$ instead of k).

We are mainly interested in the fact that the modified Stirling numbers appear naturally in the expansion of binomial coefficients as polynomials. Dividing (A.38) by $n!$ yields immediately the equality

$$\binom{q+r+n}{q+r} = \frac{1}{n!} \sum_{k=0}^{n} s_{n-k}^{(n)}(q)\, r^k\,. \tag{A.39}$$

[4] We follow here the notation of Abramowitz and Stegun [4] for the Stirling numbers of the first kind. Graham et al [178] propose the notation $\left[{n \atop k}\right]$ which we already use in a different context.

This observation also explains why we have used $q+r+1$ in (A.38) instead of the perhaps more natural $q+r$. Using the identity $S_{n+1}^{(k+1)} = \sum_{i=0}^{k} \binom{i}{k} S_n^{(i)}$ [178] one can show that $s_k^{(n)}(0) = S_{n+1}^{(n-k+1)}$.

With the help of the elementary symmetric polynomials one can easily derive a closed form expression for them. By definition, the elementary symmetric polynomial $\sigma_k^{(n)}$ is a polynomial of degree k in n variables $x^1,\dots,x^n$ that remains invariant under arbitrary permutations of the variables [98, Chapt. 7]. Thus $\sigma_1^{(n)} = x^1 + \cdots + x^n$, $\sigma_2^{(n)} = x^1x^2 + x^1x^3 + \cdots + x^2x^3 + \cdots + x^{n-1}x^n$ and so on until $\sigma_n^{(n)} = x^1x^2\cdots x^n$. Now we can write

$$s_k^{(n)}(q) = \begin{cases} 1 & \text{for } k=0\,, \\ \sigma_k^{(n)}(q+1, q+2, \dots, q+n) & \text{for } 0<k\leq n\,. \end{cases} \tag{A.40}$$

The modified Stirling numbers inherit from the elementary symmetric polynomials the following recursion relation:

$$s_k^{(n)}(q) = s_k^{(n-1)}(q) + (q+n)\,s_{k-1}^{(n-1)}(q)\,. \tag{A.41}$$

One can use this relation to construct explicit expressions for $s_k^{(n)}(q)$. They become fairly lengthy for larger values of k.

Lemma A.4.2.

$$s_n^{(n)}(q) = (q+n)!/q!\,, \tag{A.42a}$$

$$s_1^{(n)}(q) = \frac{1}{2}n(n+2q+1)\,, \tag{A.42b}$$

$$s_2^{(n)}(q) = \frac{1}{24}n(n-1)\left(3n^2 + 12qn + 5n + 12q^2 + 12q + 2\right). \tag{A.42c}$$

Proof. (A.42a) and (A.42b) follow immediately from Definition A.4.1. For (A.42c) we apply (A.41)

$$\begin{aligned} s_2^{(n)}(q) &= s_2^{(n-1)}(q) + (q+n)\,s_1^{(n-1)}(q) \\ &= \cdots = \sum_{l=2}^{n-1}(q+l+1)\,s_1^{(l)}(q) + s_2^{(2)}(q) \\ &= \frac{1}{2}\sum_{l=1}^{n-1}(q+l+1)\left[l^2 + (2q+1)l\right]. \end{aligned}$$

Evaluation of this sum yields the claim. □

One can also deduce recursion formulae relating the modified Stirling numbers for different values of q or express $s_k^{(n)}(q)$ directly by $s_l^{(m)}(0)$ with $0 \leq m \leq n$. Like most properties of these numbers these relations stem from a well-known formula for binomial coefficients.

Lemma A.4.3.

$$s_{n-k+1}^{(n)}(q) = n\, s_{n-k}^{(n-1)}(q) + s_{n-k+1}^{(n)}(q-1)\,. \tag{A.43}$$

Proof. The basis of this lemma is the following identity coming directly from the Pascal triangle

$$\binom{q+r+n}{q+r} = \sum_{k=0}^{n} \binom{q+r+k-1}{q+r-1}\,. \tag{A.44}$$

Plugging in (A.39) yields after a few simple index manipulations yet another recursion relation for the modified Stirling numbers

$$s_k^{(n)}(q) = \sum_{j=0}^{k} \frac{n!}{(n-j)!} s_{k-j}^{(n-j)}(q-1)\,. \tag{A.45}$$

Applying this relation in (A.43) proves at once the lemma. □

In order to express $s_k^{(n)}(q)$ through modified Stirling numbers with $q=0$, we must generalise (A.45). Each binomial coefficient $\binom{r+k}{k}$ represents by (A.39) a polynomial in r of degree k. Thus every polynomial in r of degree n can be written as a linear combination of $\binom{r}{0}$, $\binom{r+1}{1}$,..., $\binom{r+n}{n}$. Especially there exist numbers $d_k^{(n)}(q)$ such that

$$\binom{q+r+n}{q+r} = \sum_{k=0}^{n} d_k^{(n)}(q) \binom{r+k}{k}\,. \tag{A.46}$$

Applying repeatedly (A.44), one can express these coefficients as nested sums

$$d_k^{(n)}(q) = \sum_{k_1=k}^{n} \sum_{k_2=k_1}^{n} \cdots \sum_{k_{q-1}=k_{q-2}}^{n} 1\,. \tag{A.47}$$

For lower values of q these sums are easily evaluated. We omit, however, explicit expressions. Now we obtain similarly as above

$$s_k^{(n)}(q) = \sum_{j=0}^{k} \frac{n!}{(n-j)!} d_k^{(n)}(q) s_{k-j}^{(n-j)}(0)\,. \tag{A.48}$$

Finally, we list in Table A.1 some values of the modified Stirling numbers; the rows correspond to different values of n, the columns of k. We remark that many such integer sequences can be found in the encyclopedia [422] of which also an extended electronic version exists on the web.[5] It contains a number of generalisations of the classical Stirling numbers (with references).

[5] http://www.research.att.com/~njas/sequences

	q=0				q=1				q=2				q=3			
$n\backslash k$	1	2	3	4	1	2	3	4	1	2	3	4	1	2	3	4
1	1				2				3				4			
2	3	2			5	6			7	12			9	20		
3	6	11	6		9	26	24		12	47	60		15	74	120	
4	10	35	50	24	14	71	154	120	18	119	342	360	22	179	638	840

Table A.1 Some values of the modified Stirling numbers

Appendix B
Algebra

As the sun eclipses the stars by its brilliancy, so the man of knowledge will eclipse the fame of others in assemblies of the people if he proposes algebraic problems, and still more if he solves them.
Brahmagupta

A proper understanding of the meaning of the concept of involution in the form introduced in Chapter 3 requires certain ideas from (commutative) algebra collected in this chapter. Some classical references for all mentioned topics are [98, 99, 125]; older textbooks on commutative algebra are usually less constructive. The rather new books [185, 267, 268] put strong emphasis on computational issues (using the computer algebra systems SINGULAR and COCOA, respectively). Basic algebraic notions are introduced in [281, 415]; for the theory of non-commutative rings we mention [171, 276, 319].

The first section quickly reviews some basic algebraic structures like monoids, rings, algebras etc. Much of the material is elementary. Special emphasis is put on modules, as they are central for the algebraic analysis of the symbol of a differential equation. In particular, we introduce the Hilbert function and polynomial, as similar concepts appear for differential equations in Section 8.2. A simple method for the determination of the Hilbert function may be considered as one motivation for the introduction of involutive bases.

The second section reviews a few basic concepts from homological algebra. We introduce (co)homology modules for complexes of $\mathcal{R}$-modules and discuss resolutions. These notions are in particular used in Chapters 5 and 6. Section 10.5 requires in addition some knowledge about exact and derived functors. The third section concerns the more specialised topic of coalgebras and comodules; these concepts are only used in Chapter 6.

The fourth section gives an introduction into the theory of Gröbner bases of polynomial ideals and modules. These bases represent a fundamental computational tool in commutative algebra and algebraic geometry. Like most computations in a Euclidian vector space become significantly easier, if one chooses an orthonormal basis, almost any computation with ideals profits from the knowledge of Gröbner bases. Standard references are [5, 36, 98, 185]; most books on computer algebra also contain at least an introduction.

For us Gröbner bases are so important, as much of the theory of involutive bases is modelled on them and an involutive basis is a Gröbner basis with additional combinatorial properties. We recall the famous Buchberger algorithm for determining a

Gröbner basis in order to allow for a comparison with the completion algorithms derived in Chapter 4. However, deeper questions of the optimisation of the algorithm are not discussed.

Similarly, for lack of space applications of Gröbner bases are omitted; many can be found in the references above. They include in particular the problem of determining points on the variety defined by an ideal, in other words solving systems of polynomial equations. Other elementary applications are the union or intersection of ideals, computations with cosets, elimination problems, ...

B.1 Some Basic Algebraic Structures

Definition B.1.1. A set $\mathcal{S}$ with an operation $\cdot : \mathcal{S}\times\mathcal{S}\to\mathcal{S}$ is a *semigroup*, if $\cdot$ is associative, i. e. if $s_1\cdot(s_2\cdot s_3)=(s_1\cdot s_2)\cdot s_3$ for all $s_1,s_2,s_3\in\mathcal{S}$. The semigroup is *commutative* or *Abelian*, if $s_1\cdot s_2=s_2\cdot s_1$ for all $s_1,s_2\in\mathcal{S}$. A semigroup $\mathcal{S}$ is a *monoid*, if there exists a neutral element $e\in\mathcal{S}$ such that $e\cdot s=s\cdot e=s$ for all $s\in\mathcal{S}$. A monoid $\mathcal{S}$ is a *group*, if every $s\in\mathcal{S}$ has an inverse $s^{-1}\in\mathcal{S}$ such that $s\cdot s^{-1}=e$.

The for us most important example of an Abelian monoid is given by the multi indices $(\mathbb{N}_0^n,+)$ discussed in Appendix A.1. Its neutral element is $[0,\dots,0]$. The integers $\mathbb{Z}$ are both a monoid and a group: a monoid with respect to multiplication and a group with respect to addition; the respective neutral elements are 1 and 0. A non-commutative monoid $(\mathcal{C}^\infty(\mathbb{R}),\circ)$ is defined by the smooth functions $\mathbb{R}\to\mathbb{R}$ with the composition as operation and the identity as neutral element. The invertible functions form a group.

Let $(\mathcal{S},\cdot)$ be a monoid and $\mathcal{I}\subseteq\mathcal{S}$ a (nonempty) subset such that $\mathcal{S}\cdot\mathcal{I}\subseteq\mathcal{I}$, i. e. if we "multiply" an element of $\mathcal{I}$ by an arbitrary element of $\mathcal{S}$, then we always obtain again an element of the subset $\mathcal{I}$. We call such a subset $\mathcal{I}$ a *(left) monoid ideal*. In case that the monoid is not Abelian we may analogously define right monoid ideals by the condition $\mathcal{I}\cdot\mathcal{S}\subseteq\mathcal{I}$ or double sided monoid ideals by $\mathcal{S}\cdot\mathcal{I}\cdot\mathcal{S}\subseteq\mathcal{I}$. One should not confuse monoid ideals with *submonoids* which are (nonempty) subsets $\mathcal{T}\subseteq\mathcal{S}$ which are closed under the multiplication $\cdot$.

Definition B.1.2. A *(left) monoid module* over a monoid $\mathcal{S}$ with neutral element e is a set $\mathcal{M}$ together with an operation $\star:\mathcal{S}\times\mathcal{M}\to\mathcal{M}$ so that

(i) $\forall m\in\mathcal{M}:\ e\star m=m$;
(ii) $\forall s_1,s_2\in\mathcal{S},\forall m\in\mathcal{M}:\ (s_1\cdot s_2)\star m=s_1\star(s_2\star m)$.

A trivial example of a monoid module over $\mathcal{S}$ is a monoid ideal $\mathcal{I}\subseteq\mathcal{S}$ with the product $\cdot$ as operation. If $\mathcal{M}$ is a monoid module, a subset $\mathcal{G}\subset\mathcal{M}$ is a *generating set* of it, if $\mathcal{M}$ is the smallest monoid module containing $\mathcal{G}$. Any element of $\mathcal{M}$ may then be represented in the form $s\star g$ with some $s\in\mathcal{S}$ and $g\in\mathcal{G}$. We write $\mathcal{M}=\langle\mathcal{G}\rangle$. The monoid module $\mathcal{M}$ is *finitely generated*, if it possesses a finite generating set.

The monoid $(\mathcal{S},\cdot)$ is called *left Noetherian*, if every left monoid ideal $\mathcal{I}\subseteq\mathcal{S}$ is finitely generated. A right Noetherian monoid is defined analogously. We call $\mathcal{S}$

Noetherian, if it is both left and right Noetherian. It is well possible that a monoid is left but not right Noetherian or vice versa.

Two equivalent conditions are that either every ascending chain of monoid ideals $\mathcal{I}_1 \subseteq \mathcal{I}_2 \subseteq \mathcal{I}_3 \subseteq \cdots \subseteq \mathcal{S}$ becomes stationary, i. e. an index $N \geq 1$ exists such that $\mathcal{I}_N = \mathcal{I}_{N+1} = \mathcal{I}_{N+2} = \cdots$, or that any non-empty set of monoid ideals of $\mathcal{S}$ contains a maximal element with respect to the partial order given by set inclusion. It follows immediately from Dickson's Lemma A.1.2 that $(\mathbb{N}_0^n, +)$ is an example of a Noetherian monoid. This observation will be crucial for the theory of Gröbner bases in Appendix B.4.

Definition B.1.3. A *semiring* (sometimes also called *rng*) is a set $\mathcal{R}$ with two operations $+, \cdot : \mathcal{R} \times \mathcal{R} \to \mathcal{R}$ such that $(\mathcal{R}, +)$ is an Abelian group, the neutral element of which we denote by 0, and $(\mathcal{R}, \cdot)$ is a semigroup. Furthermore, the following distributive laws must hold:

(i) $\forall r_1, r_2, r_3 \in \mathcal{R}: \; r_1 \cdot (r_2 + r_3) = r_1 \cdot r_2 + r_1 \cdot r_3$;
(ii) $\forall r_1, r_2, r_3 \in \mathcal{R}: \; (r_1 + r_2) \cdot r_3 = r_1 \cdot r_3 + r_2 \cdot r_3$.

A semiring $\mathcal{R}$ is a *ring*, if $(\mathcal{R}, \cdot)$ is a monoid, the neutral element of which we denote by 1. If $r_1 \cdot r_2 = r_2 \cdot r_1$ for all ring elements $r_1, r_2 \in \mathcal{R}$, then we are dealing with a *commutative ring*. A *unit* is an element r of a ring $\mathcal{R}$ such that an inverse $r^{-1} \in \mathcal{R}$ with $rr^{-1} = 1$ exists. A *zero divisor* is an element $r \in \mathcal{R} \setminus \{0\}$ such that a further ring element $s \in \mathcal{R} \setminus \{0\}$ exists with either $r \cdot s = 0$ or $s \cdot r = 0$. An element of $\mathcal{R} \setminus \{0\}$ which is not a zero divisor is called *regular*. A ring without zero divisors is a *domain*.

The integers $\mathbb{Z}$ are a domain with neutral element 1. If $\mathcal{R}$ is a commutative ring (domain), the set $\mathcal{P} = \mathcal{R}[x^1, \dots, x^n]$ of all polynomials in the variables[1] $x^1, \dots, x^n$ with coefficients in $\mathcal{R}$ is again a ring (domain). Examples of non-commutative rings are rings of linear differential operators with variable coefficients or square matrices. The units of a ring $\mathcal{R}$ form a multiplicative group $(\mathcal{R}^\times, \cdot)$ with neutral element 1.

Definition B.1.4. A *skew field* $\Bbbk$ is a ring such that $\Bbbk^\times = \Bbbk \setminus \{0\}$, i. e. except 0 every element is invertible. If $\Bbbk$ is Abelian, it is a *field*. A field $\Bbbk$ is *algebraically closed*, if every non-constant univariate polynomial in $\Bbbk[x]$ has a zero in $\Bbbk$.

Skew fields are also known as *division rings*. Given any commutative domain $\mathcal{R}$, we may embed it in its *quotient field* $\operatorname{Quot}(\mathcal{R})$. Its elements are pairs $(r, s) \in \mathcal{R} \times \mathcal{R} \setminus \{0\}$ (usually written r/s) modulo the equivalence relation $r_1/s_1 \sim r_2/s_2$ if $r_1 s_2 = r_2 s_1$. The rational numbers $\mathbb{Q}$ are the quotient field of the integers $\mathbb{Z}$. Similarly, the rational functions $\mathcal{R}(x^1, \dots, x^n)$ are the quotient field of the polynomial ring $\mathcal{P} = \mathcal{R}[x^1, \dots, x^n]$. The analytic functions do not form a field but only a ring, as the quotient of two analytic functions is no longer analytic at the zeros of the denominator. But the *meromorphic functions*, i. e. the functions that are everywhere analytic except on a discrete set of points where they have a pole, form a field. We consider exclusively fields of *characteristic zero*; any such field contains the rational

[1] We follow here (as everywhere in this book) the convention in differential geometry and use superscripts to index variables.

numbers $\mathbb{Q}$ as a subfield. The classical example of an algebraically closed field is $\mathbb{C}$—this is the Fundamental Theorem of Algebra—whereas $\mathbb{R}$ is not algebraically closed (the polynomial x^2+1 has no zero in $\mathbb{R}$).

For a non-commutative domain $\mathcal{R}$ one must distinguish between its left and right quotient (skew) field and, in contrast to the commutative case, they do not always exist. $\mathcal{R}$ possesses a left (right) quotient field, if and only if it satisfies the *left (right) Ore condition* ensuring the existence of left (right) least common multiples [94, 276, 319, 343]: to each pair $r \in \mathcal{R}$, $s \in \mathcal{R}\setminus\{0\}$ there exists a pair $s' \in \mathcal{R}\setminus\{0\}$, $r' \in \mathcal{R}$ such that $s'r = r's$ (respectively, $rs' = sr'$). A domain satisfying this condition is called a *left (right) Ore domain*. Note that the left and the right Ore condition are independent: a domain $\mathcal{R}$ may satisfy one but not the other one.

If $\mathcal{R}$ satisfies the left Ore condition, then we may proceed as in the commutative case and introduce an equivalence relation on pairs $(r,s) \in \mathcal{R}\times\mathcal{R}\setminus\{0\}$ by saying $(r_1,s_1) \sim_l (r_2,s_2)$ if $u,v \in \mathcal{R}$ exist such $ur_1 = vr_2$ and $us_1 = vs_2 \in \mathcal{R}\setminus\{0\}$. We denote elements of $\mathrm{Quot}_l(\mathcal{R}) = (\mathcal{R}\times\mathcal{R}\setminus\{0\})/\sim_l$ by $s^{-1}r$. Let $s_1^{-1}r_1, s_2^{-1}r_2$ be two quotients; by the Ore condition elements $u,v \in \mathcal{R}$ exist with $ur_1 = vs_2$. The product on $\mathrm{Quot}_l(\mathcal{R})$ is then defined as $(s_1^{-1}r_1)(s_2^{-1}r_2) = (us_1)^{-1}(vr_2)$. For the addition on $\mathrm{Quot}_l(\mathcal{R})$, let $u_1,u_2 \in \mathcal{R}$ be such that $u_1s_1 = u_2s_2$, then $(s_1^{-1}r_1)+(s_2^{-1}r_2) = (u_1s_1)^{-1}(u_1r_1+u_2r_2)$. One easily checks that both definitions are independent of the choice of the representatives for the equivalence classes.

In a right Ore domain we proceed analogously and introduce the equivalence relation $(r_1,s_1) \sim_r (r_2,s_2)$ if two elements $u,v \in \mathcal{R}$ exist such that $r_1u = r_2v$ and $s_1u = s_2v \in \mathcal{R}\setminus\{0\}$. The elements of the arising right quotient field $\mathrm{Quot}_r(\mathcal{R}) = (\mathcal{R}\times\mathcal{R}\setminus\{0\})/\sim_r$ are denoted by rs^{-1} and their multiplication and addition is defined similar as above. If $\mathcal{R}$ satisfies both the left and the right Ore condition, then $\mathrm{Quot}_r(\mathcal{R}) = \mathrm{Quot}_l(\mathcal{R})$. Indeed, let $s^{-1}r$ be a left quotient; by the right Ore condition, a pair $s' \in \mathcal{R}\setminus\{0\}$, $r' \in \mathcal{R}$ with $rs' = sr'$ exists and thus $s^{-1}r = r'(s')^{-1}$.

Quotient fields are a special case of the more general concept of localisation. Let $\mathcal{R}$ be a ring and $\mathcal{S} \subseteq \mathcal{R}$ a multiplicatively closed subset, i. e. we have $1 \in \mathcal{S}$, $0 \notin \mathcal{S}$ and for any pair $r,s \in \mathcal{S}$ that also $rs \in \mathcal{S}$. Then the left *localisation* $\mathcal{S}^{-1}\mathcal{R}$ is defined like the left quotient field only with $\mathcal{R}\setminus\{0\}$ replaced by $\mathcal{S}$. Its existence is also decided by the left Ore condition—again with $\mathcal{R}\setminus\{0\}$ replaced by $\mathcal{S}$. The right localisation is defined correspondingly. Since we assume that $1 \in \mathcal{S}$, we can always embed $\mathcal{R}$ in $\mathcal{S}^{-1}\mathcal{R}$ by identifying $r \in \mathcal{R}$ with $1^{-1}r \in \mathcal{S}^{-1}\mathcal{R}$. An important special case is obtained by choosing for $\mathcal{S}$ the subset of all regular elements of $\mathcal{R}$; the localisation $\mathcal{S}^{-1}\mathcal{R}$ is then called the *total ring of left quotients* and correspondingly for right quotients. One easily shows that the localisation of a Noetherian ring is again Noetherian.

Definition B.1.5. Let $\mathcal{R}$ be a ring. An Abelian group $(\mathcal{M},+)$ is called a *(left) $\mathcal{R}$-module*, if a (left) $\mathcal{R}$-action $\cdot : \mathcal{R}\times\mathcal{M} \to \mathcal{M}$ exists such that:

(i) $\forall m \in \mathcal{M}:\ 1\cdot m = m$;
(ii) $\forall r_1,r_2 \in \mathcal{R}, \forall m \in \mathcal{M}:\ r_1\cdot(r_2\cdot m) = (r_1r_2)\cdot m$;
(iii) $\forall r \in \mathcal{R}, \forall m_1,m_2 \in \mathcal{M}:\ r\cdot(m_1+m_2) = r\cdot m_1 + r\cdot m_2$;
(iv) $\forall r_1,r_2 \in \mathcal{R}, \forall m \in \mathcal{M}:\ (r_1+r_2)\cdot m = r_1\cdot m + r_2\cdot m$.

This definition is an obvious extension of Definition B.1.2 of a monoid module. Large parts of the literature deal only with modules over *commutative* rings. We have not made this restriction, as we want to be able to consider modules over, say, rings of differential operators. One defines analogously *right* modules with a right action $\cdot : \mathcal{M} \times \mathcal{R} \to \mathcal{M}$ and obvious modifications of the axioms. If $\mathcal{R}$, $\mathcal{S}$ are two rings, then an *$\mathcal{R}$-$\mathcal{S}$-bimodule* is a left $\mathcal{R}$- and right $\mathcal{S}$-module. As we almost exclusively deal with left modules, we usually omit the "left" and simply speak of modules. Furthermore, for commutative rings we will usually not distinguish between left and right modules and consider any $\mathcal{R}$-module as an $\mathcal{R}$-$\mathcal{R}$-bimodule.

Let $\mathcal{G} \subset \mathcal{M}$ be a subset of $\mathcal{M}$. Then $\mathcal{G}$ is a *generating set* of $\mathcal{M}$, if any element $m \in \mathcal{M}$ can be written as a linear combination $m = \sum_{g\in\mathcal{G}} r_g g$ where only finitely many of the coefficients $r_g \in \mathcal{R}$ do not vanish. This representation is in general not unique; if it is, then we are dealing with a free module. We write $\mathcal{M} = \langle G \rangle$. A module is called *finitely generated*, if it possesses a finite generating set. A set $\mathcal{G}$ is called *linearly independent*, if $\sum_{g\in\mathcal{G}} r_g g = 0$ implies that all coefficients r_g vanish. A linearly independent generating set is a *basis* of the module $\mathcal{M}$; only in this case the above representation of the elements of $\mathcal{M}$ as linear combinations is unique.

If $\mathcal{G}$ is a *minimal generating set*, then it is not possible to remove an element of $\mathcal{G}$ without $\mathcal{G}$ loosing the property of being a generating set. Note that generally different minimal generating sets may have very different cardinalities; only if bases exist, they always possess the same number of elements. A special case arises for graded algebras (see below): as here the homogeneous components are vector spaces, any homogeneous generating set has the same cardinality.

Example B.1.6. The simplest examples of modules are *vector spaces*. Indeed, if the ring $\mathcal{R}$ is even a field, then an $\mathcal{R}$-module is nothing but a vector space over $\mathcal{R}$. Any Abelian group $(\mathcal{G},+)$ may be considered as a $\mathbb{Z}$-module with $n \cdot g = g + \cdots + g$ for any positive $n \in \mathbb{Z}$.

Notions like generating set or basis are of course familiar from linear algebra. One should, however, note that many facts which are true for vector spaces do not hold, in general, for modules, as in a ring we cannot divide by all elements. Assume, for example, that we are given a set of elements that are linearly dependent. In a vector space this implies that some of the elements can be expressed as linear combinations of the other ones. In a module this is not necessarily true. Take the free $\mathbb{Z}$-module $\mathbb{Z}^2$ and the two elements $m_1 = \binom{2}{4}$ and $m_2 = \binom{3}{6}$. Obviously, they are linearly dependent, as $3m_1 - 2m_2 = 0$. But over the ring $\mathbb{Z}$ none is a multiple of the other one, as the necessary divisions cannot be performed.

One consequence of these considerations is that—in contrast to vector spaces—not every module possesses a basis. A module with a basis is called *free*. A finitely generated $\mathcal{R}$-module is thus free, if and only if it is isomorphic to a module of the form $\mathcal{R}^m$ for some $m \in \mathbb{N}$, the *rank* of the module. For this reason we will usually identify elements of a free $\mathcal{R}$-module of rank m with m-dimensional vectors whose entries are elements of $\mathcal{R}$; addition and the action of $\mathcal{R}$ are then defined componentwise. ◁

Let $\mathcal{M}$ be an $\mathcal{R}$-module. We say that $m \in \mathcal{M}$ is a *torsion element*, if there exists a non zero divisor $r \in \mathcal{R}$ such that $rm = 0$. If $\mathcal{R}$ is a commutative ring (or satisfies an Ore condition), then the torsion elements form the *torsion submodule* $t(\mathcal{M})$; if $t(\mathcal{M}) = \mathcal{M}$, then we say that $\mathcal{M}$ is a *torsion module*. The module $\mathcal{M}$ is called *torsionfree*, if $t(\mathcal{M}) = 0$. A free module is always torsionfree.

Let $\mathcal{M}$ and $\mathcal{N}$ be two (left) $\mathcal{R}$-modules. A map $\phi : \mathcal{M} \to \mathcal{N}$ is called a *(left) module homomorphism*, if it respects the module structures of $\mathcal{M}$ and $\mathcal{N}$, i. e. if for all $m_1, m_2 \in \mathcal{M}$ and for all $r \in \mathcal{R}$ the equality $\phi(m_1 + r \cdot m_2) = \phi(m_1) + r \cdot \phi(m_2)$ holds. The case of right modules is treated analogously. We denote by $\mathrm{Hom}_{\mathcal{R}}(\mathcal{M}, \mathcal{N})$ the set of all such module homomorphisms. If $\mathcal{R}$ is commutative, then $\mathrm{Hom}_{\mathcal{R}}(\mathcal{M}, \mathcal{N})$ is again an $\mathcal{R}$-module with the obvious structure map $(r\phi)(m) = r\phi(m)$. Otherwise, $\mathrm{Hom}_{\mathcal{R}}(\mathcal{M}, \mathcal{N})$ is simply an Abelian group.

Remark B.1.7. An important special case is the *dual module* $\mathcal{M}^* = \mathrm{Hom}_{\mathcal{R}}(\mathcal{M}, \mathcal{R})$. If $\mathcal{M}$ is a left $\mathcal{R}$-module, then $\mathcal{M}^*$ is always a right $\mathcal{R}$-module (with structure map $(\phi r)(m) = \phi(m)r$ for any $r \in \mathcal{R}$, $\phi \in \mathcal{M}^*$), even if the ring $\mathcal{R}$ is not commutative. Given a homomorphism $\phi : \mathcal{M} \to \mathcal{N}$ between two left $\mathcal{R}$-module, the *dual map* is the homomorphism $\phi^* : \mathcal{N}^* \to \mathcal{M}^*$ of right $\mathcal{R}$-modules defined by $\phi^*\lambda = \lambda \circ \phi$.

Since $\mathcal{M}^*$ is a right $\mathcal{R}$-module, we can define the *bidual* $\mathcal{M}^{**} = \mathrm{Hom}_{\mathcal{R}}(\mathcal{M}^*, \mathcal{R})$ where of course right module homomorphisms are considered. Obviously, $\mathcal{M}^{**}$ is again a left $\mathcal{R}$-module and there is a natural homomorphism $\eta_{\mathcal{M}} : \mathcal{M} \to \mathcal{M}^{**}$ mapping $m \in \mathcal{M}$ to the homomorphism $\mu \in \mathcal{M}^{**}$ defined by $\mu(\phi) = \phi(m)$. In general $\eta_{\mathcal{M}}$ is neither injective nor surjective. A simple exception arises, if $\mathcal{M}$ is a free module, as then $\eta_{\mathcal{M}}$ is always an isomorphism. But if we dualise once more, then we may consider on one side the dual map $\eta^*_{\mathcal{M}} : \mathcal{M}^{***} \to \mathcal{M}^*$ and on the other side the natural morphism $\hat{\eta}_{\mathcal{M}} : \mathcal{M}^* \to \mathcal{M}^{***}$ defined in analogy to $\eta_{\mathcal{M}}$. It is not difficult to see that $\eta^*_{\mathcal{M}} \circ \hat{\eta}_{\mathcal{M}} = \mathrm{id}_{\mathcal{M}^*}$ and hence the map $\hat{\eta}_{\mathcal{M}}$ is always injective and η^* surjective.

If the map $\eta_{\mathcal{M}}$ is injective, i. e. if for every element $m \in \mathcal{M} \setminus \{0\}$ a homomorphism $\phi \in \mathcal{M}^*$ exists with $\phi(m) \neq 0$, then we call $\mathcal{M}$ *torsionless*. This terminology has its origin in the observation that if a module is torsionless, then it is also torsionfree. Indeed, if $m \in \mathcal{M} \setminus \{0\}$ is a torsion element, then there exists a non zero divisor $r \in \mathcal{R}$ such that $rm = 0$ and thus for any $\phi \in \mathcal{M}^*$ we have $\phi(rm) = r\phi(m) = 0$ implying that $\phi(m) = 0$. The converse is not true; for example $\mathbb{Q}$ is a torsionfree $\mathbb{Z}$-module but not torsionless (here even $\mathrm{Hom}_{\mathbb{Z}}(\mathbb{Q}, \mathbb{Z}) = 0$, as for any $\phi \in \mathrm{Hom}_{\mathbb{Z}}(\mathbb{Q}, \mathbb{Z})$ and any prime $p \in \mathbb{N}$ we have $p\phi(1/p) = \phi(1)$ and thus all primes divide $\phi(1)$ which is only possible if $\phi(1) = 0$). $\triangleleft$

Remark B.1.8. Assume that $\mathcal{R} \subseteq \mathcal{S}$ is an extension of commutative rings and that we are given an $\mathcal{R}$-module $\mathcal{M}$ and an $\mathcal{S}$-module $\mathcal{N}$ (obviously, the latter one may also be considered as an $\mathcal{R}$-module via the extension). Then we have a natural isomorphism between $\mathrm{Hom}_{\mathcal{R}}(\mathcal{M}, \mathcal{N})$ and $\mathrm{Hom}_{\mathcal{S}}(\mathcal{M}, \mathrm{Hom}_{\mathcal{R}}(\mathcal{S}, \mathcal{N}))$ mapping $\phi \in \mathrm{Hom}_{\mathcal{R}}(\mathcal{M}, \mathcal{N})$ to the homomorphism Ψ_ϕ defined by $\Psi_\phi(m)(s) = \phi(sm)$. Here we consider $\mathrm{Hom}_{\mathcal{R}}(\mathcal{S}, \mathcal{N})$ as an $\mathcal{S}$-module via the action $(s_1\phi)(s_2) = \phi(s_1 s_2)$. Indeed, it is trivial to see that the map $\phi \mapsto \Psi_\phi$ is injective and the preimage of any homomorphism $\Psi \in \mathrm{Hom}_{\mathcal{S}}(\mathcal{M}, \mathrm{Hom}_{\mathcal{R}}(\mathcal{S}, \mathcal{N}))$ is given by $\phi(m) = \Psi(m)(1)$. $\triangleleft$

Any finitely generated module is isomorphic to a factor module of a free module; one speaks of a *presentation by generators and relations*. Such a presentation is obtained by taking any finite generating set $\mathcal{G} = \{g_1, \dots, g_m\}$ of $\mathcal{M}$ and the free $\mathcal{R}$-module $\mathcal{R}^m$. We define the homomorphism

$$\phi : \begin{cases} \mathcal{R}^m & \longrightarrow & \mathcal{M} \\ (r_1, \dots, r_m) & \longmapsto & \sum_{i=1}^{m} r_i g_i \end{cases} \tag{B.1}$$

and set $\mathcal{N} = \ker\phi$. Then $\mathcal{M} \cong \mathcal{R}^m/\mathcal{N}$ (if $\mathcal{N} = 0$, then $\mathcal{M}$ is a free module) and we have an exact sequence (see Definition B.2.2)

$$0 \longrightarrow \mathcal{N} \xhookrightarrow{\iota} \mathcal{R}^m \xrightarrow{\phi} \mathcal{M} \longrightarrow 0 \tag{B.2}$$

where ι is the inclusion map. If the module $\mathcal{N}$ is again finitely generated, we say that $\mathcal{M}$ is *finitely presented*. In this case any finite generating set of $\mathcal{N}$ provides us with an analogous map $\psi : \mathcal{R}^n \to \mathcal{N}$ for some exponent $n \in \mathbb{N}$. Calling the concatenation $\iota \circ \psi$ again ψ, we obtain an exact sequence

$$\mathcal{R}^n \xrightarrow{\psi} \mathcal{R}^m \xrightarrow{\phi} \mathcal{M} \longrightarrow 0\,, \tag{B.3}$$

since $\ker\phi = \operatorname{im}\psi$. Hence there is an isomorphism $\mathcal{M} \cong \operatorname{coker}\psi = \mathcal{R}^m/\operatorname{im}\psi$ and we may identify any finitely presented $\mathcal{R}$-module with the cokernel of a map between two free $\mathcal{R}$-modules. The ring $\mathcal{R}$ is *left coherent*, if any finitely generated left $\mathcal{R}$-module is also finitely presented. Similarly, one defines a right coherent ring. We call $\mathcal{R}$ coherent, if it is both left and right coherent. Obviously, any (left/right) Noetherian ring is (left/right) coherent but the converse is not true.

In practice, it may be difficult to construct effectively a generating set $\mathcal{G}$ and its relations module $\mathcal{N}$ for a module $\mathcal{M}$. Fortunately, most of the modules we are dealing with automatically arise in a finite presentation by generators and relations, so that we do not bother about this problem.

Given a right $\mathcal{R}$-module $\mathcal{M}$ and a left $\mathcal{R}$-module $\mathcal{N}$, we may introduce their *tensor product* $\mathcal{M} \otimes_{\mathcal{R}} \mathcal{N}$. Consider the free Abelian group $\mathcal{F}$ of all formal sums $\sum_{j=1}^{r} c_j(m_j, n_j)$ with $c_j \in \mathbb{Z}$, $m_j \in \mathcal{M}$ and $n_j \in \mathcal{N}$. The group operation is the obvious addition. Let $\mathcal{G}$ be the subgroup generated by all elements of the form $(m+m', n) - (m, n) - (m', n)$ or $(m, n+n') - (m, n) - (m, n')$ or $(m \cdot r, n) - (m, r \cdot n)$ where $r \in \mathcal{R}$, $m, m' \in \mathcal{M}$ and $n, n' \in \mathcal{N}$. Then we define $\mathcal{M} \otimes_{\mathcal{R}} \mathcal{N} = \mathcal{F}/\mathcal{G}$ and $m \otimes n = \big[(m,n)\big] \in \mathcal{M} \otimes_{\mathcal{R}} \mathcal{N}$.

In general, $\mathcal{M} \otimes_{\mathcal{R}} \mathcal{N}$ is only an Abelian group, i. e. a $\mathbb{Z}$-module. If $\mathcal{M}$ is an $\mathcal{S}$-$\mathcal{R}$-bimodule and $\mathcal{N}$ an $\mathcal{R}$-$\mathcal{T}$-bimodule, then one easily verifies that the above construction yields an $\mathcal{S}$-$\mathcal{T}$-bimodule with $s(m \otimes n)t = (sm) \otimes (nt)$ for arbitrary elements $s \in \mathcal{S}$, $t \in \mathcal{T}$. Thus, in the for us most important case of a commutative ring $\mathcal{R}$ where we consider any $\mathcal{R}$-module as an $\mathcal{R}$-$\mathcal{R}$-bimodule, the tensor product is again an $\mathcal{R}$-module.

Remark B.1.9. It is very important to specify over which ring the tensor product is taken. As a simple example take $\mathcal{M} = \Bbbk$ with a field $\Bbbk$ and $\mathcal{N} = \mathcal{P} = \Bbbk[x]$. Then one possibility consists of considering both $\mathcal{M}$ and $\mathcal{N}$ as $\Bbbk$-linear spaces and we find that $\mathcal{M} \otimes_{\Bbbk} \mathcal{N} = \mathcal{P}$. But we may also treat $\mathcal{M}$ and $\mathcal{N}$ as $\mathcal{P}$-modules. Indeed, $\Bbbk$ is a $\mathcal{P}$-module with the action of a polynomial $f = \sum_{k=0}^{n} f_k x^k$ on a field element c given by $f \cdot c = f_0 c$. Now one readily checks that $\mathcal{M} \otimes_{\mathcal{P}} \mathcal{N} = \Bbbk$. In fact, one always has $\mathcal{M} \otimes_{\mathcal{R}} \mathcal{R} = \mathcal{R} \otimes_{\mathcal{R}} \mathcal{M} = \mathcal{M}$ for arbitrary $\mathcal{R}$-modules $\mathcal{M}$.

The case $\mathcal{P} \otimes_{\mathcal{P}} \Bbbk = \Bbbk$ shows that a tensor product may lead to a considerable "collapsing". An even more extreme example is obtained by considering for two coprime integers $m, n \in \mathbb{N}$ (i.e. $\gcd(m,n) = 1$) the factor spaces $\mathbb{Z}/m\mathbb{Z}$ and $\mathbb{Z}/n\mathbb{Z}$ as $\mathbb{Z}$-modules: $(\mathbb{Z}/m\mathbb{Z}) \otimes_{\mathbb{Z}} (\mathbb{Z}/n\mathbb{Z}) = 0$. ◁

If $\mathcal{M}$, $\bar{\mathcal{M}}$ are two right $\mathcal{R}$-modules and $\mathcal{N}$, $\bar{\mathcal{N}}$ two left $\mathcal{R}$-modules with homomorphism $f : \mathcal{M} \to \bar{\mathcal{M}}$ and $g : \mathcal{N} \to \bar{\mathcal{N}}$, then there exists a unique homomorphism $f \otimes g : \mathcal{M} \otimes_{\mathcal{R}} \mathcal{N} \to \bar{\mathcal{M}} \otimes_{\mathcal{R}} \bar{\mathcal{N}}$ with $(f \otimes g)(m \otimes n) = f(m) \otimes g(n)$ which is called the *tensor product* of f and g.

Definition B.1.10. A finite sequence $(r_1, \dots, r_k)$ of elements of the ring $\mathcal{R}$ is called *regular* for the $\mathcal{R}$-module $\mathcal{M}$ or shortly $\mathcal{M}$*-regular*, if r_1 is not a zero divisor of $\mathcal{M}$ and r_i is not a zero divisor for the factor module $\mathcal{M}/\langle r_1, \dots, r_{i-1} \rangle \mathcal{M}$ for $1 < i \leq k$. A regular sequence is *maximal*, if it is not possible to extend it without loosing the regularity. One can show that all maximal regular sequences for $\mathcal{M}$ have the same length. This length is called the *depth* of the module $\mathcal{M}$, written $\operatorname{depth} \mathcal{M}$.

The depth represents a measure for the size of a module. It can also be introduced via homological algebra (see Remark B.2.39 and Theorem 6.3.8) and in particular in older books it is therefore sometimes called the *cohomological dimension* of the module.

Definition B.1.11. Let $\mathcal{R}$ be a ring. An $\mathcal{R}$*-algebra* $\mathcal{A}$ is a (left) $\mathcal{R}$-module together with an $\mathcal{R}$-bilinear multiplication $\cdot : \mathcal{A} \times \mathcal{A} \to \mathcal{A}$.

The ring $\mathcal{P} = \mathcal{R}[x^1, \dots, x^n]$ of polynomials with coefficients in a commutative ring $\mathcal{R}$ is a simple example of an Abelian $\mathcal{R}$-algebra. More generally, let $\mathcal{M}$ be a finitely generated free $\mathcal{R}$-module. We introduce the *symmetric algebra* $S\mathcal{M}$ as follows.[2] Let $\{g_1, \dots, g_n\}$ be a basis of $\mathcal{M}$, then $S\mathcal{M}$ is the infinite-dimensional free $\mathcal{R}$-module generated by the basis $\{g^\mu = g_1^{\mu_1} \cdots g_n^{\mu_n} \mid \mu \in \mathbb{N}_0^n\}$. Thus $S\mathcal{M}$ is isomorphic to the polynomial ring $\mathcal{P}$ via the identification $x^i \leftrightarrow g_i$. However, it is important to note that the definition of $S\mathcal{M}$ is independent of the chosen basis of the free module $\mathcal{M}$, in contrast to this isomorphism which is obviously not canonical. Linear differential operators with variable coefficients form a non-commutative algebra over their coefficient ring.

The multiplication in an algebra is not necessarily associative; an associative algebra is always a ring. An important class of non-associative algebras are *Lie algebras* (see Definition C.5.5). For them the associativity condition is replaced by the Jacobi identity.

---

[2] A more satisfying definition of the symmetric algebra is as a quotient space of the full tensor algebra over $\mathcal{M}$ (see also Appendix B.3). However, for our purposes this "pedestrian" approach via a basis suffices.

Subgroups, subrings, submodules etc. are defined in the obvious way: a subset that is closed under the given operations (the addition respectively the multiplication of two elements of the subset yields again an element of the subset) and still satisfies the respective axioms. Thus in the case of a subring $\mathcal{U} \subseteq \mathcal{R}$ one requires for example that $(\mathcal{U}, +)$ is a subgroup of $(\mathcal{R}, +)$ and $\mathcal{U} \cdot \mathcal{U} \subseteq \mathcal{U}$. Furthermore, the neutral element $1 \in \mathcal{R}$ of the multiplication must lie in $\mathcal{U}$. A simple example of a subring is the *centre* of a ring $\mathcal{R}$ which is defined as the subset $\{s \in \mathcal{R} \mid \forall r \in \mathcal{R} : sr = rs\}$ of elements commuting with all other elements of $\mathcal{R}$.

Definition B.1.12. Let $\mathcal{R}$ be an arbitrary ring. An additive subgroup $\mathcal{I} \subseteq (\mathcal{R}, +)$ is called a *left ideal*, if $\mathcal{R} \cdot \mathcal{I} \subseteq \mathcal{I}$, a *right ideal*, if $\mathcal{I} \cdot \mathcal{R} \subseteq \mathcal{I}$, and a *two-sided ideal*, if it is both a left and a right ideal.

As we work almost exclusively with left ideals, we usually omit the "left" and simply speak of ideals. Note that a proper ideal is never a subring, since $1 \in \mathcal{I}$ would entail $\mathcal{I} = \mathcal{R}$. The trivial ideals are $0 = \{0\}$ and the whole ring $\mathcal{R}$. The sum, the product and the intersections of two ideals $\mathcal{I}, \mathcal{J} \subseteq \mathcal{R}$ are defined in the obvious way (note that trivially $\mathcal{I} \cdot \mathcal{J} \subseteq \mathcal{I} \cap \mathcal{J}$) and yields again an ideal; the quotient of $\mathcal{I}$ by $\mathcal{J}$ is the ideal

$$\mathcal{I} : \mathcal{J} = \{r \in \mathcal{R} \mid r \cdot \mathcal{J} \subseteq \mathcal{I}\} . \tag{B.4}$$

We list some special types of ideals for the case that $\mathcal{R}$ is a commutative ring. A *principal ideal* can be generated by a single element. A *maximal ideal* is a proper ideal $\mathfrak{m} \subsetneq \mathcal{R}$ such that no non-trivial ideal $\mathcal{I}$ exists with $\mathfrak{m} \subsetneq \mathcal{I} \subsetneq \mathcal{R}$. For a *prime ideal* $\mathfrak{p} \subseteq \mathcal{R}$ the fact $rs \in \mathfrak{p}$ implies $r \in \mathfrak{p}$ or $s \in \mathfrak{p}$. Trivial examples explaining the name are the ideals $\langle p \rangle \subset \mathbb{Z}$ generated by a prime number p. Furthermore, it is easy to show that any maximal ideal is prime.

Given an ideal $\mathcal{I} \subseteq \mathcal{R}$, we define its *radical* as the ideal

$$\sqrt{\mathcal{I}} = \{r \in \mathcal{R} \mid \exists k \in \mathbb{N} : r^k \in \mathcal{I}\} . \tag{B.5}$$

A *radical ideal* satisfies $\sqrt{\mathcal{I}} = \mathcal{I}$, i. e. $r^k \in \mathcal{I}$ for some $k \in \mathbb{N}$ implies for such an ideal $r \in \mathcal{I}$. Obviously, for any ideal $\mathcal{I} \subseteq \mathcal{R}$ the radical $\sqrt{\mathcal{I}}$ is a radical ideal and also any prime ideal is a radical ideal. A *primary ideal* $\mathfrak{q} \subseteq \mathcal{R}$ is an ideal such that $rs \in \mathfrak{q}$ and $r \notin \mathfrak{q}$ implies $s \in \sqrt{\mathfrak{q}}$. Obviously, the radical $\mathfrak{p} = \sqrt{\mathfrak{q}}$ of a primary ideal is prime and one then says that $\mathfrak{q}$ is $\mathfrak{p}$-primary. If $\mathcal{R}$ is a Noetherian ring, then one can easily show that there always exists an exponent $s \in \mathbb{N}$ such that $\mathfrak{p}^s \subseteq \mathfrak{q} \subseteq \mathfrak{p}$.

We are mainly interested in the special case of a polynomial ring $\mathcal{P} = \Bbbk[x^1, \dots, x^n]$ over a field $\Bbbk$. Given some polynomials $f_1, \dots, f_m \in \mathcal{P}$, the *variety* $\mathcal{V}(f_1, \dots, f_m) = \{\boldsymbol{\xi} \in \Bbbk^n \mid f_1(\boldsymbol{\xi}) = \cdots = f_m(\boldsymbol{\xi}) = 0\}$, i. e. the set of their common zeros, depends only on the ideal $\mathcal{I} = \langle f_1, \dots, f_m \rangle$ and therefore we usually write $\mathcal{V}(\mathcal{I})$. If $\mathcal{I}$, $\mathcal{J}$ are two ideals with $\mathcal{I} \subseteq \mathcal{J}$, then the corresponding varieties satisfy $\mathcal{V}(\mathcal{I}) \supseteq \mathcal{V}(\mathcal{J})$. One easily sees that always $\mathcal{V}(\mathcal{I}) = \mathcal{V}(\sqrt{\mathcal{I}})$ and Hilbert's famous Nullstellensatz asserts that over an algebraically closed field the varieties are in a one-to-one correspondence with radical ideals. For such a field the only maximal ideals in $\mathcal{P}$ are the ideals $\mathfrak{m}_{\boldsymbol{\zeta}} = \langle x^1 - \zeta^1, \dots, x^n - \zeta^n \rangle$ associated with the points $\boldsymbol{\zeta} \in \Bbbk^n$.

One can show that it is possible to define a topology, the *Zariski topology*, on $\Bbbk^n$ by saying that the varieties are the closed sets. Thus open sets in this topology are rather big: with the exception of the empty set, every open set consists of almost the whole space $\Bbbk^n$. If a property holds for all points on a Zariski open set, then one also says that it holds *generically*.

As for monoids, we call a ring $\mathcal{R}$ *Noetherian*, if any ideal $\mathcal{I} \subseteq \mathcal{R}$ is finitely generated or, equivalently, any ascending chain of ideals $\mathcal{I}_1 \subseteq \mathcal{I}_2 \subseteq \mathcal{I}_3 \subseteq \cdots \subseteq \mathcal{R}$ becomes eventually stationary, i. e. there exists an $N \in \mathbb{N}$ such that $\mathcal{I}_{N+k} = \mathcal{I}_N$ for all $k \geq 0$. In the case of a non-commutative product we must of course distinguish between a left and a right Noetherian ring (and these are independent properties—see Example 3.3.14). A trivial example of a Noetherian ring is a field $\Bbbk$, as its only ideals are 0 and $\Bbbk$ itself. Another important class of Noetherian rings is provided by the following result for which we present four different proofs in Section 3.3.

Theorem B.1.13 (Hilbert Basis Theorem). *For a Noetherian commutative ring $\mathcal{R}$ the polynomial ring $\mathcal{R}[x^1,\dots,x^n]$ in finitely many variables is Noetherian, too.*

Remark B.1.14. All known proofs of Hilbert's Basis Theorem require in some form the *Axiom of Choice* (cf. [36, Section 4.1] for a more detailed discussion). If one cannot accept this axiom, one will have great problems with the theory of Gröbner bases in Appendix B.4, as most of the algorithmic results there depend on it! $\triangleleft$

Reversion of the inclusions yields the descending chain condition and the notion of an *Artinian* ring $\mathcal{R}$. An equivalent condition is that any non-empty set of ideals of $\mathcal{R}$ contains a minimal element with respect to the partial order given by set inclusion.

Let $\mathcal{M}$ be a module over the commutative ring $\mathcal{R}$. Given an element $m \in \mathcal{M}$, we call the ideal $\operatorname{Ann}_{\mathcal{R}}(m) = \{r \in \mathcal{R} \mid r \cdot m = 0\} \subseteq \mathcal{R}$ the *annihilator* of m. A prime ideal $\mathfrak{p} \subseteq \mathcal{R}$ is *associated* to the module $\mathcal{M}$, if there exists an $m \in \mathcal{M}$ such that $\operatorname{Ann}_{\mathcal{R}}(m) = \mathfrak{p}$.[3] The set of all associated primes is denoted by $\operatorname{Ass}\mathcal{M}$. We have the following important result [125, Theorem 3.1].

Theorem B.1.15. *Let $\mathcal{R}$ be a Noetherian commutative ring and $\mathcal{M}$ a finitely generated $\mathcal{R}$-module. Then the set $\operatorname{Ass}\mathcal{M}$ of associated prime ideals is not empty and contains only finitely many elements.*

Remark B.1.16. Any ideal $\mathcal{I} \subseteq \mathcal{R}$ may be considered as an $\mathcal{R}$-module and thus we could apply the above definition of an associated prime ideal to it. However, in this case another definition is more common. Namely, one calls a prime ideal $\mathfrak{p} \subseteq \mathcal{R}$ associated to $\mathcal{I}$ (and writes $\mathfrak{p} \in \operatorname{Ass}\mathcal{I}$), if there exists a ring element $r \in \mathcal{R}$ such that $\mathfrak{p} = \mathcal{I} : \langle r \rangle = \{s \in \mathcal{R} \mid sr \in \mathcal{I}\}$. One easily verifies that this condition means nothing but that $\mathfrak{p}$ is an associated prime ideal in the above sense of the factor ring $\mathcal{R}/\mathcal{I}$ considered as an $\mathcal{R}$-module. $\triangleleft$

If $\mathcal{I} \subseteq \mathcal{R}$ is an arbitrary ideal, then a *primary decomposition* is a representation $\mathcal{I} = \mathfrak{q}_1 \cap \cdots \cap \mathfrak{q}_t$ where every factor $\mathfrak{q}_j$ is a primary ideal. A primary decomposition

[3] Not every annihilator ideal $\operatorname{Ann}_{\mathcal{R}}(m)$ is necessarily prime. But one can easily show that the *maximal* elements of the set of all annihilators are prime.

is *irredundant*, if all underlying prime ideals $\mathfrak{p}_j = \sqrt{\mathfrak{q}_j}$ are distinct and no primary ideal $\mathfrak{q}_j$ can be omitted in the decomposition. The existence of (irredundant) primary decompositions is guaranteed by the following result [125, Theorem 3.10] where we define $\operatorname{Ass}\mathcal{I}$ as in Remark B.1.16.

Theorem B.1.17. *Let $\mathcal{R}$ be a Noetherian commutative ring and $\mathcal{I} \subseteq \mathcal{R}$ an ideal. Then $\mathcal{I}$ possesses an irredundant primary decomposition $\mathcal{I} = \mathfrak{q}_1 \cap \cdots \cap \mathfrak{q}_t$ and $\operatorname{Ass}\mathcal{I} = \{\sqrt{\mathfrak{q}_1}, \dots, \sqrt{\mathfrak{q}_t}\}$.*

Note that even an irredundant primary decomposition is not unique, if embedded prime ideals exists: an associated prime ideal $\mathfrak{p} \in \operatorname{Ass}\mathcal{I}$ is called *embedded*, if there exists another associated prime ideal $\mathfrak{p}' \in \operatorname{Ass}\mathcal{I}$ with $\mathfrak{p}' \subset \mathfrak{p}$. For any embedded prime ideal $\mathfrak{p} \in \operatorname{Ass}\mathcal{I}$ we introduce the set $\operatorname{Ass}(\mathcal{I},\mathfrak{p}) = \{\mathfrak{p}' \in \operatorname{Ass}\mathcal{I} \mid \mathfrak{p}' \subseteq \mathfrak{p}\}$. Then one obtains only the following form of a uniqueness statement: the intersection of all the primary ideals $\mathfrak{q}'$ corresponding to the elements $\mathfrak{p}' \in \operatorname{Ass}(\mathcal{I},\mathfrak{p})$ is independent of the chosen primary decomposition.

Remark B.1.18. If $\mathcal{R} = \mathcal{P} = \Bbbk[x^1, \dots, x^n]$ is a polynomial ring, then a primary decomposition of any ideal $\mathcal{I} \subseteq \mathcal{P}$ can be effectively constructed with the help of Gröbner bases; Decker et al [103] discuss four different approaches. For *monomial* ideals this problem is almost trivial (two methods are given in [268, Tutorial 77]). Furthermore, in this case one can choose the primary ideals $\mathfrak{q}_j$ again monomial which implies that all associated prime ideals $\mathfrak{p}_j$ are monomial, too, and thus of the form $\mathfrak{p}_j = \langle x^{i_1}, \dots, x^{i_k}\rangle$ for some values $1 \leq i_1 < \cdots < i_k \leq n$. The primary ideal $\mathfrak{q}_j$ contains then for each index i_ℓ a term $(x^{i_\ell})^{q_\ell}$ and all minimal generators depend only on the variables $x^{i_1}, \dots, x^{i_k}$. ◁

Definition B.1.19. The ring $\mathcal{R}$ is *graded*, if it has as additive Abelian group $(\mathcal{R},+)$ a direct sum decomposition $\mathcal{R} = \bigoplus_{i\in\mathbb{Z}} \mathcal{R}_i$ such that $\mathcal{R}_i\mathcal{R}_j \subseteq \mathcal{R}_{i+j}$. A *graded $\mathcal{R}$-module* $\mathcal{M}$ is an $\mathcal{R}$-module possessing a direct sum decomposition $\mathcal{M} = \bigoplus_{i\in\mathbb{Z}} \mathcal{M}_i$ such that $\mathcal{R}_i \cdot \mathcal{M}_j \subseteq \mathcal{M}_{i+j}$. The elements of $\mathcal{R}_i$ or $\mathcal{M}_i$, respectively, are *homogeneous* of degree i. For a given degree $q \in \mathbb{Z}$ the *truncation* $\mathcal{M}_{\geq q}$ is defined as the module $\bigoplus_{p\geq q} \mathcal{M}_p$, i. e. we throw away all elements of degree less than q.

If $\mathcal{R}$ is a graded ring, then obviously $\mathcal{R}_0$ is a subring and each $\mathcal{R}_i$ an $\mathcal{R}_0$-module. We mainly deal with *non-negatively graded* rings and modules where all components $\mathcal{R}_i$, $\mathcal{M}_i$ with $i < 0$ are zero. A simple example of such a ring is the polynomial ring $\mathcal{P} = \Bbbk[x^1, \dots, x^n]$ where a natural grading is given by the total degree. Here $\mathcal{P}_0 = \Bbbk$ is even a field and the components $\mathcal{P}_i$ consisting of all homogeneous polynomials of degree i are $\Bbbk$-linear spaces. Other gradings are obtained by assigning a weight $w_k \in \mathbb{Z}$ to each variable x^k; now the component $\mathcal{P}_i$ is generated by all the terms x^μ with $\sum_{k=1}^n w_k\mu_k = i$.

If $\mathcal{M} = \bigoplus_{i\in\mathbb{Z}} \mathcal{M}_i$ is a graded module, then a submodule $\mathcal{N} \subseteq \mathcal{M}$ is graded, if it can be decomposed in the form $\mathcal{N} = \bigoplus_{i\in\mathbb{Z}}(\mathcal{N} \cap \mathcal{M}_i)$. A graded submodule can always be generated by homogeneous elements.

Definition B.1.20. A ring $\mathcal{R}$ is *filtered*, if it is equipped with a *filtration* Σ, i.e. an ascending chain $\Sigma_0 \subset \Sigma_1 \subset \Sigma_2 \subset \cdots$ of additive subgroups $\Sigma_i \subseteq \mathcal{R}$ such that $\bigcup_{i=0}^{\infty} \Sigma_i = \mathcal{R}$ and $\Sigma_i \cdot \Sigma_j \subseteq \Sigma_{i+j}$ for all $i, j \geq 0$. If $\mathcal{R}$ is a filtered ring, then a (left) $\mathcal{R}$-module $\mathcal{M}$ is *filtered*, if it is equipped with a *filtration* Γ consisting of an ascending chain $\Gamma_0 \subset \Gamma_1 \subset \Gamma_2 \subset \cdots$ of additive subgroups $\Gamma_i \subseteq \mathcal{M}$ such that $\bigcup_{i=0}^{\infty} \Gamma_i = \mathcal{M}$ and $\Sigma_i \cdot \Gamma_j \subseteq \Gamma_{i+j}$ for all $i, j \geq 0$.

If Σ is a filtration on $\mathcal{R}$, then we may introduce an *associated graded ring*

$$\mathrm{gr}_\Sigma \mathcal{R} = \Sigma_0 \oplus \Sigma_1/\Sigma_0 \oplus \Sigma_2/\Sigma_1 \oplus \cdots . \tag{B.6}$$

The multiplication on $\mathrm{gr}_\Sigma \mathcal{R}$ is defined as follows. Write $\bar{\Sigma}_i$ for the factor group Σ_i/Σ_{i-1} (we set $\bar{\Sigma}_0 = \Sigma_0$) and consider two arbitrary homogeneous elements $\bar{r} \in \bar{\Sigma}_i$ and $\bar{s} \in \bar{\Sigma}_j$. Let $r \in \Sigma_i$ and $s \in \Sigma_j$ be arbitrary representatives of these equivalence classes. Then we set $\bar{r} \cdot \bar{s} = \overline{r \cdot s} \in \bar{\Sigma}_{i+j}$. It is straightforward to verify that this definition is independent of the choice of the representatives r and s.

If furthermore Γ is a filtration on the left $\mathcal{R}$-module $\mathcal{M}$, then we may introduce an associated graded left $\mathrm{gr}_\Sigma \mathcal{R}$-module

$$\mathrm{gr}_\Gamma \mathcal{M} = \Gamma_0 \oplus \Gamma_1/\Gamma_0 \oplus \Gamma_2/\Gamma_1 \oplus \cdots . \tag{B.7}$$

The definition of the $\mathrm{gr}_\Sigma \mathcal{R}$-action on $\mathrm{gr}_\Gamma \mathcal{M}$ goes analogously to the definition of the product in $\mathrm{gr}_\Sigma \mathcal{R}$. We write again $\bar{\Gamma}_i = \Gamma_i/\Gamma_{i-1}$ (with $\bar{\Gamma}_0 = \Gamma_0$). For each non-vanishing element $m \in \mathcal{M}$ we denote by $\deg_\Gamma m$ the unique integer d such that $m \in \Gamma_d \setminus \Gamma_{d-1}$ and call the corresponding equivalence class $\sigma_\Gamma(m) \in \bar{\Gamma}_d$ its *Γ-symbol*.

Definition B.1.21. Let $\mathcal{R}$ be a graded ring with $\mathcal{R}_0$ a field and $\mathcal{M}$ a finitely generated graded $\mathcal{R}$-module. Then the homogeneous components $\mathcal{M}_r$ are finite-dimensional $\mathcal{R}_0$-linear spaces. The *Hilbert function* of $\mathcal{M}$ is the numerical function[4] defined by

$$h_\mathcal{M}(r) = \dim \mathcal{M}_r . \tag{B.8}$$

The Hilbert function measures the size of the graded module $\mathcal{M}$. As an $\mathcal{R}_0$-linear space, the whole module is generally infinite-dimensional, so that this dimension does not define a useful measure. Instead, we must use a more complicated object like the Hilbert function. Robbiano [384] (see also [268, Chapt. 5]) gives a nice introduction into the theory of Hilbert functions (and numerical functions in general).

Theorem B.1.22 (Hilbert). *Let the graded ring $\mathcal{R}$ be generated over $\mathcal{R}_0$ by n elements of degree 1 and $\mathcal{M}$ be a finitely generated, graded $\mathcal{R}$-module. Then there exists a value $r_0 \geq 0$ and a polynomial $H_\mathcal{M}(r)$ of degree less than n with rational coefficients such that $h_\mathcal{M}(r) = H_\mathcal{M}(r)$ for all $r \geq r_0$ (i. e. for sufficiently large degrees the Hilbert function becomes polynomial).*

---

[4] A *numerical function* $f(r)$ returns for every integer argument $r \in \mathbb{Z}$ an integer value, i. e. it defines a map $f : \mathbb{Z} \to \mathbb{Z}$.

The assumptions made on the ring $\mathcal{R}$ imply essentially that $\mathcal{R}$ is of the form $\mathcal{R} = \mathcal{R}_0[x^1,\dots,x^n]/\mathcal{I}$ with a homogeneous ideal $\mathcal{I} \subset \mathcal{R}_0[x^1,\dots,x^n]$, i. e. we are dealing with an affine algebra. For obvious reason, $H_{\mathcal{M}}(r)$ is called the *Hilbert polynomial* of $\mathcal{M}$. The number $D = 1 + \deg H_{\mathcal{M}}$ is the *(Krull) dimension* $\dim \mathcal{M}$. If the coefficient of r^{D-1} in $H_{\mathcal{M}}(r)$ is $a_{D-1}/(D-1)!$, then the *multiplicity* of $\mathcal{M}$ is defined as $\operatorname{mult} \mathcal{M} = a_{D-1}$.

Above we introduced $d = \operatorname{depth} \mathcal{M}$ as an alternative measure for the size of the module $\mathcal{M}$. One can show that always $D \geq d$ (see Section 5.2). Modules such that $D = d$ are called *Cohen–Macaulay*.

The *Hilbert series*[5] is the generating function of the Hilbert function $h_{\mathcal{M}}(r)$, i. e. the univariate series in some variable λ defined by

$$\mathcal{H}_{\mathcal{M}}(\lambda) = \sum_{r=0}^{\infty} h_{\mathcal{M}}(r)\lambda^r . \tag{B.9}$$

One can show (see Proposition 5.2.1 for the case of polynomial modules) that this formal series can always be summed in closed form yielding a rational function $\mathcal{H}_{\mathcal{M}}(\lambda) = f(\lambda)/(1-\lambda)^n$ where $f(\lambda)$ is a polynomial with integer coefficients. For a free module $\mathcal{M} \cong \mathcal{R}^m$ of rank m we get $\mathcal{H}_{\mathcal{M}}(\lambda) = m/(1-\lambda)^n$. The Krull dimension $D = \dim \mathcal{M}$ arises now as the order of the pole of $\mathcal{H}_{\mathcal{M}}$ at $\lambda = 1$. More precisely, by cancelling common factors we write $\mathcal{H}_{\mathcal{M}}(\lambda) = \bar{f}(\lambda)/(1-\lambda)^D$ with a polynomial $\bar{f}(\lambda)$. The degree r of this polynomial is precisely the degree from which on Hilbert function and polynomial coincide; this number is sometimes called the *Hilbert regularity* of the module $\mathcal{M}$. If $\bar{f}(\lambda) = \sum_{i=0}^{r} f_i \lambda^i$, then the Hilbert polynomial is given by [185, Cor. 5.1.5]

$$H_{\mathcal{M}}(r) = \sum_{i=0}^{r} f_i \binom{r+D-1-i}{D-1} . \tag{B.10}$$

Its degree is $D-1$ and its leading coefficient is $\bar{f}(1)/(D-1)!$.

These notions were originally introduced by Hilbert [213][6] for polynomial ideals $\mathcal{I} \subseteq \mathcal{P} = \Bbbk[x^1,\dots,x^n]$ (obviously, in this case $\mathcal{P}_0 = \Bbbk$ is indeed a field). Here one must distinguish between the *affine* and the *projective* Hilbert function. Let $\mathcal{I}$ be a homogeneous ideal in the polynomial ring $\mathcal{P}$, i. e. an ideal generated by homogeneous elements. The quotient $\mathcal{P}/\mathcal{I}$ is then a graded $\mathcal{P}$-module and we define the projective Hilbert function $h_{\mathcal{I}}$ as the Hilbert function of $\mathcal{P}/\mathcal{I}$. A particular case of a homogeneous ideal is a *monomial ideal*, i. e. an ideal with a basis consisting only of monomials. Then $h_{\mathcal{I}}(r)$ counts the terms of degree r that are *not* contained in $\mathcal{I}$. As we will see later, the general case can always be reduced to the monomial one with the help of Gröbner bases.

For non-homogeneous ideals $\mathcal{I}$, the quotient $\mathcal{P}/\mathcal{I}$ does not possess a natural grading. There exist two different ways to define nevertheless a Hilbert function. The first one consists of homogenising the ideal $\mathcal{I}$: we introduce an additional variable t

---

[5] The names *Poincaré* or *Hilbert-Poincaré series* are also frequently used.

[6] Hilbert used the term *characteristic function*.

and consider the ideal

$$\mathcal{I}^{(h)} = \langle f^{(h)} \mid f \in \mathcal{I} \rangle \subseteq \mathcal{P}[t] \tag{B.11}$$

where the homogeneous polynomial $f^{(h)} \in \mathcal{P}[t]$ is obtained from $f \in \mathcal{P}$ by multiplying each term x^μ contained in f by $t^{\deg f - |\mu|}$. Then we define the affine Hilbert function of $\mathcal{I}$ as the projective Hilbert function of $\mathcal{I}^{(h)}$: $h^a_{\mathcal{I}} = h_{\mathcal{I}^{(h)}}$.

In the second approach we exploit the total degree filtration of the polynomial ring $\mathcal{P}$. We denote by $\mathcal{P}_{\leq r}$ the vector space of all polynomials f with $\deg f \leq r$ and introduce the subspace $\mathcal{I}_{\leq r} = \mathcal{I} \cap \mathcal{P}_{\leq r}$. Now we define the affine Hilbert function by $h^a_I(r) = \dim \mathcal{P}_{\leq r} - \dim \mathcal{I}_{\leq r}$. It is not difficult to see that both approaches always yield the same result. If $\mathcal{I}$ is a homogeneous ideal, we have the obvious relation $h_{\mathcal{I}}(r) = h^a_{\mathcal{I}}(r) - h^a_{\mathcal{I}}(r-1)$.

The Hilbert polynomial $H_{\mathcal{I}}(r)$ is an example of a *numerical polynomial*, i.e. it defines a map $H_{\mathcal{I}} : \mathbb{Z} \to \mathbb{Z}$. These polynomials form a $\mathbb{Z}$-module and one can show—e.g. with a simple argument from interpolation theory—that any numerical polynomial of degree d can be written in the form

$$H_{\mathcal{I}}(r) = \sum_{i=0}^{d} a_i r^i = \sum_{i=0}^{d} b_i \binom{r}{i} . \tag{B.12}$$

The transformation between the coefficients a_i and b_i is a combinatorial operation leading to the modified Stirling numbers discussed in Appendix A.4.

Let $\mathcal{R} = \bigoplus_{k=0}^{\infty} \mathcal{R}_k$ be a non-negatively graded commutative ring. Then the subset $\mathcal{R}_+ = \bigoplus_{k=1}^{\infty} \mathcal{R}_k$ is trivially an ideal in $\mathcal{R}$; it is often called the *irrelevant ideal*. The *saturation* of a homogeneous ideal $\mathcal{I} \subseteq \mathcal{R}$ is the ideal[7]

$$\mathcal{I}^{\text{sat}} = \mathcal{I} : \mathcal{R}_+^{\infty} = \{ f \in \mathcal{R} \mid \exists k \in \mathbb{N} : f \cdot \mathcal{R}_k \subseteq \mathcal{I} \} \tag{B.13}$$

(the simple quotient $\mathcal{I} : \mathcal{R}_+$ is often called the *socle* of $\mathcal{I}$). One can show that for $q \gg 0$ we have $\mathcal{I}_q = \mathcal{I}_q^{\text{sat}}$ (see Section 5.5 for a proof). Hence all ideals with the same saturation have the same Hilbert polynomial and become identical for sufficiently large degrees; $\mathcal{I}^{\text{sat}}$ is the largest among all these ideals. The smallest degree s such that $\mathcal{I}_q = \mathcal{I}_q^{\text{sat}}$ for all $q \geq s$ is called the *satiety* $\operatorname{sat} \mathcal{I}$ of $\mathcal{I}$. An ideal $\mathcal{I}$ with $\mathcal{I} = \mathcal{I}^{\text{sat}}$ is called *saturated*.

Definition B.1.23. Let $\mathcal{I} \subseteq \mathcal{P} = \Bbbk[x^1, \dots, x^n]$ be a polynomial ideal. A *Noether normalisation* of the algebra $\mathcal{A} = \mathcal{P}/\mathcal{I}$ is a polynomial ring $\mathcal{R} = \Bbbk[z^1, \dots, z^D]$ together with an injective ring homomorphism $\mathcal{R} \to \mathcal{A}$ such that $\mathcal{A}$ is a finitely generated $\mathcal{R}$-module. If it is possible to take for the variables $z^1, \dots, z^D$ a subset of $\{x^1, \dots, x^n\}$, then the ideal $\mathcal{I}$ is said to be in *Noether position*.

Every affine algebra $\mathcal{A} = \mathcal{P}/\mathcal{I}$ possesses a Noether normalisation [125, Theorem 13.3], [185, Theorem 3.4.1]. If the field $\Bbbk$ is infinite (as we always assume), then one can choose the variables z^i as linear forms in $\mathcal{P}_1$; in other words, by a linear change of coordinates one can put every ideal in Noether position. A proof of

[7] More generally, the saturation of an ideal $\mathcal{I} \subseteq \mathcal{R}$ with respect to another ideal $\mathcal{J} \subseteq \mathcal{R}$ is defined as $\mathcal{I} : \mathcal{J}^{\infty} = \{ f \in \mathcal{R} \mid \exists k \in \mathbb{N} : f \cdot \mathcal{J}^k \subseteq \mathcal{I} \}$.

both claims using Pommaret bases is contained in Section 5.3. One can show that for any Noether normalisation the number D is the same and that D is nothing but the (Krull) dimension of $\mathcal{M}$. Assume for simplicity that already $\mathcal{R} = \Bbbk[x^1,\dots,x^D]$ leads to a Noether normalisation. Then one speaks of a *general Noether normalisation*, if for each $D < j \leq n$ a polynomial $f_j \in \mathcal{I}$ exists such that with respect to the degree lexicographic order $\mathrm{lt}_{\prec} f_j = x_j^{q_j}$

As we are mainly studying differential equations in this book, an interesting extension of the algebraic concepts introduced so far is *differential algebra*. Here one adds an additional operation satisfying the algebraic properties of a derivative. Some classical references are [251, 258, 383].

Definition B.1.24. Let $\mathcal{R}$ be a ring. A *derivation* on $\mathcal{R}$ is a map $\delta : \mathcal{R} \to \mathcal{R}$ that is linear with respect to the addition in $\mathcal{R}$, i.e. $\delta(r+s) = \delta(r) + \delta(s)$, and that satisfies the Leibniz rule $\delta(rs) = \delta(r)s + r\delta(s)$ for all $r,s \in \mathcal{R}$. A *differential ring* is a pair $(\mathcal{R},\delta)$ where δ is a derivation on $\mathcal{R}$. If $\mathcal{R}$ is even a field, then we have a *differential field*. A subset $\mathcal{I} \subseteq \mathcal{R}$ of a differential ring $(\mathcal{R},\delta)$ is a *differential ideal*, if $\mathcal{I}$ is an ideal of $\mathcal{R}$ and furthermore $\delta(\mathcal{I}) \subseteq \mathcal{I}$. Finally, a *differential module* over a differential ring $\mathcal{R}$ with derivation δ is a left $\mathcal{R}$-module $\mathcal{M}$ together with an additive map $\partial : \mathcal{M} \to \mathcal{M}$ such that $\partial(rm) = \delta(r)m + r\partial(m)$ for all ring elements $r \in \mathcal{R}$ and module elements $m \in \mathcal{M}$.

Elements r of a differential ring or field such that $\delta(r) = 0$ are called *constants*. One readily checks that they form again a ring or field, respectively. The definition above assumes that there is only one derivation. An obvious extension is a *partial differential ring* which is equipped with several commuting derivations.

A simple example of a differential ring is the ring of all smooth functions defined on some interval $\mathbb{I} \subseteq \mathbb{R}$ with the usual differentiation as derivation. In principle, every ring $\mathcal{R}$ may be considered as a differential ring, as the zero map trivially satisfies the conditions on a derivation. In the case of graded rings $\mathcal{R}$, one often requires that δ satisfies a graded Leibniz rule, i.e. depending on the grading of the elements $r,s \in \mathcal{R}$ we have $\delta(rs) = \delta(r)s \pm r\delta(s)$. Some examples of such graded derivations appear naturally in differential geometry (see Appendix C.2).

In analogy to the usual polynomial ring, one may introduce *differential polynomials*. Let $\mathcal{R}$ be a differential ring with derivation δ and u an indeterminate. Then we define the ring of differential polynomials in u as

$$\mathcal{R}\{u\} = \mathcal{R}[u,u',u'',\dots,u^{(k)},\dots]\,. \tag{B.14}$$

Thus $\mathcal{R}\{u\}$ is an ordinary polynomial ring in infinitely many unknowns $u^{(k)}$ with $u^{(0)} = u$. The derivation δ is extended from the ground ring $\mathcal{R}$ to $\mathcal{R}\{u\}$ by setting $\delta(u^{(k)}) = u^{(k+1)}$ so that we have again a differential ring. The generalisation to partial differential polynomials is straightforward. Similarly, we introduce *rational differential functions* as the quotient field $R\langle u\rangle$ of $\mathcal{R}\{u\}$.

A major complication in differential algebra is the fact that $\mathcal{R}\{u\}$ is no longer Noetherian. Indeed, because of the infinitely many unknowns we cannot apply Hilbert's Basis Theorem B.1.13 and it is not too difficult to find example of differential ideals without a finite basis [69].

B.2 Homological Algebra

At a few places we use notions from homological algebra. Thus we include some basic definitions for easier reference. For further details we refer to the textbooks [298, 346, 476] (most books on commutative algebra also contain the material we need). In the sequel, $\mathcal{R}$ will always be a ring with unit and if not explicitly stated otherwise all $\mathcal{R}$-modules are to be understood as left modules.

Many results in algebra (and other branches of mathematics) are formulated in the language of *commutative diagrams*. They represent a simple graphical way to express relations between maps. Typically, the maps are morphisms between spaces with some common structure, say groups, vector spaces, manifolds, bundles etc; our main interest will be in $\mathcal{R}$-modules. A randomly chosen example is

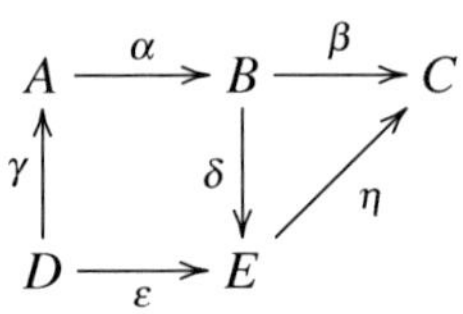

(B.15)

containing fives spaces with six maps between them. The statement that it commutes means only that if there are several ways to relate two spaces, then it does not matter which one we choose, as the outcome is always the same. Thus the diagram (B.15) encodes the relations $\eta \circ \delta = \beta$, $\delta \circ \alpha \circ \gamma = \varepsilon$ and $\eta \circ \varepsilon = \beta \circ \alpha \circ \gamma$.

Definition B.2.1. For each $i \in \mathbb{Z}$ let $\mathcal{M}^i$ be an $\mathcal{R}$-module and $d^i : \mathcal{M}^i \to \mathcal{M}^{i+1}$ a module homomorphism. Then

$$\cdots \longrightarrow \mathcal{M}^{i-1} \xrightarrow{d^{i-1}} \mathcal{M}^i \xrightarrow{d^i} \mathcal{M}^{i+1} \longrightarrow \cdots \tag{B.16}$$

is called a *(cochain) complex* with *differentials* d^i, if always $d^i \circ d^{i-1} = 0$ or, in other words, if $\operatorname{im} d^{i-1} \subseteq \ker d^i$. We will denote the whole complex by $(\mathcal{M}, d)$ or, if the maps d^i are obvious from the context, even shorter by $\mathcal{M}$.

It is well possible that all but finitely many of the modules $\mathcal{M}^i$ are the zero module with obvious maps between them. The zero modules are then usually omitted and we obtain a finite complex. A classical example of a finite complex is the *de Rham complex* of a manifold (see Example 10.5.6 and Remark C.2.9). In a *chain complex* the arrows in (B.16) are inverted (and one usually writes subscripts instead of superscripts), i. e. we have then $d_i : \mathcal{M}_i \to \mathcal{M}_{i-1}$. We will mostly work with cochain complexes and thus drop the "cochain". Most of the following material may be trivially adapted to chain complexes.

If $(\mathcal{M}, d)$ and $(\mathcal{N}, \delta)$ are two complexes of (right and left, respectively) $\mathcal{R}$-modules, then we introduce the *tensor product complex* $(\mathcal{L} = \mathcal{M} \otimes_{\mathcal{R}} \mathcal{N}, \partial)$ by setting $\mathcal{L}_k = \bigoplus_{i+j=k} (\mathcal{M}_i \otimes_{\mathcal{R}} \mathcal{N}_j)$ and $\partial = d \otimes \mathrm{id}_{\mathcal{N}} + \mathrm{id}_{\mathcal{M}} \otimes \delta$. A special case arises, if one factor is simply an $\mathcal{R}$-module. Then this factor is interpreted as a complex, where only the degree zero component does not vanish and where all differentials are the zero map.

Definition B.2.2. Let $(\mathcal{M},d)$ be a complex. It is *exact* at the module $\mathcal{M}^i$ for some index $i \in \mathbb{Z}$, if $\operatorname{im} d^{i-1} = \ker d^i$. We call the complex an *exact sequence*, if it is exact at $\mathcal{M}^i$ for all values $i \in \mathbb{Z}$.

Example B.2.3. The probably most important example of a finite exact complex is a *short exact sequence*. This is a finite complex of the form

$$0 \longrightarrow \mathcal{A} \xrightarrow{\alpha} \mathcal{B} \xrightarrow{\beta} \mathcal{C} \longrightarrow 0 \tag{B.17}$$

which is everywhere exact. The exactness at $\mathcal{A}$ implies that the map α is injective, as its kernel must be equal to $\{0_{\mathcal{A}}\}$, the image of the left zero module. Similarly, the exactness at $\mathcal{C}$ implies that the map β is surjective, as its image must be the whole module $\mathcal{C}$, the kernel of the final map. Finally, because of the exactness at $\mathcal{B}$ every solution $b \in \mathcal{B}$ of the equation $\beta(b) = 0_{\mathcal{B}}$ is of the form $b = \alpha(a)$ for some $a \in \mathcal{A}$.

The short exact sequence (B.17) *splits*, if a further morphism $\gamma : \mathcal{C} \to \mathcal{B}$ exists such that $\beta \circ \gamma = \mathrm{id}_{\mathcal{C}}$. In this case we have the isomorphism $\mathcal{B} \cong \mathcal{A} \oplus \mathcal{C}$. Indeed, as $\beta\bigl(\gamma(c)\bigr) = c$, the condition $\gamma(c) \in \ker\beta$ is equivalent to $c = 0$. Hence we find that $\operatorname{im}\gamma \cap \ker\beta = \{0_{\mathcal{B}}\}$ and γ is injective. Now any element $b \in \mathcal{B}$ may be decomposed as $b = \bigl[b - \gamma\bigl(\beta(b)\bigr)\bigr] + \gamma\bigl(\beta(b)\bigr)$ where obviously the first summand lies in $\ker\beta$ and the second in $\operatorname{im}\gamma$. Thus we can write $\mathcal{B} = \operatorname{im}\gamma \oplus \ker\beta$ and because of the injectivity of α and γ we have $\ker\beta = \operatorname{im}\alpha \cong \mathcal{A}$ and $\operatorname{im}\gamma \cong \mathcal{C}$.

An equivalent definition of a split exact sequence requires the existence of a morphism $\delta : \mathcal{B} \to \mathcal{A}$ such that $\delta \circ \alpha = \mathrm{id}_{\mathcal{A}}$. One shows similarly that then $\mathcal{B} = \operatorname{im}\alpha \oplus \ker\delta \cong \mathcal{A} \oplus \mathcal{C}$. Sequences of finite-dimensional vector spaces (or more generally free modules) always split. This fact is a simple consequence of the existence of bases and thus of complements. ◁

Example B.2.4. Let $\alpha : \mathcal{A} \to \mathcal{B}$ be an arbitrary module homomorphism between two $\mathcal{R}$-modules $\mathcal{A}$ and $\mathcal{B}$. It gives rise to the following exact sequence

$$0 \longrightarrow \ker\alpha \overset{\iota}{\hookrightarrow} \mathcal{A} \xrightarrow{\alpha} \mathcal{B} \xrightarrow{\pi} \operatorname{coker}\alpha \longrightarrow 0 \tag{B.18}$$

where the *cokernel* of α is defined as the quotient module $\operatorname{coker}\alpha = \mathcal{B}/\operatorname{im}\alpha$, ι is the inclusion map and π the canonical projection on the quotient space. ◁

Definition B.2.5. An $\mathcal{R}$-module $\mathcal{P}$ is called *projective*, if for any surjective morphism $\phi : \mathcal{B} \to \mathcal{C}$ between $\mathcal{R}$-modules and for any morphism $\gamma : \mathcal{P} \to \mathcal{C}$ a morphism $\beta : \mathcal{P} \to \mathcal{B}$ exists with $\phi \circ \beta = \gamma$. An $\mathcal{R}$-module $\mathcal{I}$ is *injective*, if for any injective morphism $\psi : \mathcal{A} \to \mathcal{B}$ between $\mathcal{R}$-modules and for any morphism $\alpha : \mathcal{A} \to \mathcal{I}$ a morphism $\beta : \mathcal{B} \to \mathcal{I}$ exists with $\beta \circ \psi = \alpha$. We call the ring $\mathcal{R}$ *left (right) self-injective*, if it is injective considered as a left (right) $\mathcal{R}$-module.

Pictorially, these definitions mean that in the two diagrams

(B.19)

fillers (the dotted arrows) exist such that the diagrams commute. One speaks of the *universal lifting property* of such modules. Note that the right diagram corresponds to the left one with all arrows reversed. Definition B.2.5 immediately implies the following result.

Proposition B.2.6. *A projective module $\mathcal{P}$ is a direct summand of any module $\mathcal{A}$ with a surjective map $\phi : \mathcal{A} \to \mathcal{P}$. An injective module $\mathcal{I}$ is a direct summand of any module $\mathcal{B}$ with an injective map $\psi : \mathcal{I} \to \mathcal{B}$.*

Most properties of projective modules can be transferred to injective ones by simple dualisation and vice versa. However, in some cases such a transfer is not possible. For example, for the following important test for injectivity no dual analogue for projective modules is known.

Proposition B.2.7 (Baer's Criterion). *An $\mathcal{R}$-module $\mathcal{M}$ is injective, if and only if for any ideal $\mathcal{I} \subseteq \mathcal{R}$ (considered as $\mathcal{R}$-module) and any morphism $\phi : \mathcal{I} \to \mathcal{M}$ of $\mathcal{R}$-modules an element $m \in \mathcal{M}$ exists such that $\phi(r) = rm$ (which is equivalent to the existence of an extension $\tilde{\phi} : \mathcal{R} \to \mathcal{M}$ of ϕ).*

Readers familiar with category theory will know the notion of a *functor* relating the objects and morphisms of two categories. We are mainly interested in functors relating $\mathcal{R}$-modules. Such a functor F maps a module $\mathcal{A}$ to another module $F(\mathcal{A})$. If F is a *covariant* functor, it furthermore maps any morphism $\alpha : \mathcal{A} \to \mathcal{B}$ to a morphism $F(\alpha) : F(\mathcal{A}) \to F(\mathcal{B})$. Here the monoid properties of maps must be preserved, i. e. $F(\mathrm{id}_{\mathcal{A}}) = \mathrm{id}_{F(\mathcal{A})}$ and $F(\beta \circ \alpha) = F(\beta) \circ F(\alpha)$, if $\beta : \mathcal{B} \to \mathcal{C}$ is a further morphism. By contrast, a *contravariant* functor induces a map $F(\alpha) : F(\mathcal{B}) \to F(\mathcal{A})$ satisfying similar rules as in the case of a covariant functor (this time the second condition reads $F(\beta \circ \alpha) = F(\alpha) \circ F(\beta)$, as the arrows are reversed).

We are mainly interested in four simple functors. For a given $\mathcal{R}$-module $\mathcal{A}$, we introduce the covariant functor $\mathcal{B} \mapsto \mathrm{Hom}_{\mathcal{R}}(\mathcal{A}, \mathcal{B})$ and the contravariant functor $\mathcal{B} \mapsto \mathrm{Hom}_{\mathcal{R}}(\mathcal{B}, \mathcal{A})$. In the first case any morphism $\beta : \mathcal{B} \to \mathcal{C}$ is mapped to the morphism $\beta_* : \mathrm{Hom}_{\mathcal{R}}(\mathcal{A}, \mathcal{B}) \to \mathrm{Hom}_{\mathcal{R}}(\mathcal{A}, \mathcal{C})$ defined by the concatenation $\beta_* \alpha = \beta \circ \alpha$ for any $\alpha : \mathcal{A} \to \mathcal{B}$ (this definition is an instance of a general construction in category theory and β_* is usually called the *push-forward* of β). In the contravariant case, β is mapped to the morphism $\beta^* : \mathrm{Hom}_{\mathcal{R}}(\mathcal{C}, \mathcal{A}) \to \mathrm{Hom}_{\mathcal{R}}(\mathcal{B}, \mathcal{A})$ defined by $\beta^* \gamma = \gamma \circ \beta$ for any $\gamma : \mathcal{C} \to \mathcal{A}$ (in the language of category theory, β^* is usually called the *pull-back* of β).

Two further functors are similarly defined with the help of the tensor product of $\mathcal{R}$-modules: for any $\mathcal{R}$-module $\mathcal{A}$ we introduce the covariant functors $\mathcal{A} \otimes \cdot$ and $\cdot \otimes \mathcal{A}$. We restrict here to the case that $\mathcal{R}$ is a commutative ring. Then, as discussed in the previous section, both functors map an $\mathcal{R}$-module again to an $\mathcal{R}$-module.

Applying a covariant functor F to a sequence of $\mathcal{R}$-modules of the form (B.17) yields the new sequence

$$0 \longrightarrow F(\mathcal{A}) \xrightarrow{F(\alpha)} F(\mathcal{B}) \xrightarrow{F(\beta)} F(\mathcal{C}) \longrightarrow 0\,. \tag{B.20}$$

By contrast, a contravariant functor F yields the "reversed" sequence

$$0 \longrightarrow F(\mathcal{C}) \xrightarrow{F(\beta)} F(\mathcal{B}) \xrightarrow{F(\alpha)} F(\mathcal{A}) \longrightarrow 0\,. \tag{B.21}$$

The functorial properties of F imply that both sequences are complexes, if (B.17) is a complex. Indeed in the covariant case we have $F(\beta)\circ F(\alpha) = F(\beta\circ\alpha) = 0$ and similarly for a contravariant functor. However, even if the original sequence (B.17) is exact, we cannot generally expect any of the above sequences to be again exact.

Definition B.2.8. The functor F is *exact*, if (B.20) (or (B.21), respectively) is exact whenever the sequence (B.17) is exact. Conversely, F is a *faithful* functor, if exactness of (B.20) (or (B.21), respectively) implies exactness of the sequence (B.17).

For an arbitrary $\mathcal{R}$-module $\mathcal{M}$ neither $\mathrm{Hom}_{\mathcal{R}}(\mathcal{M},\cdot)$ nor $\mathrm{Hom}_{\mathcal{R}}(\cdot,\mathcal{M})$ are exact; they are only *left exact*, i. e. the sequences (B.20) and (B.21), respectively, are everywhere exact except possibly at the right end.

Proposition B.2.9. *The functors* $\mathrm{Hom}_{\mathcal{R}}(\mathcal{M},\cdot)$ *and* $\mathrm{Hom}_{\mathcal{R}}(\cdot,\mathcal{M})$ *are left exact for arbitrary $\mathcal{R}$-modules $\mathcal{M}$. If $\mathcal{M}$ is a free $\mathcal{R}$-module, then* $\mathrm{Hom}_{\mathcal{R}}(\mathcal{M},\cdot)$ *is exact. If $\mathcal{R}$ is a (skew) field, then* $\mathrm{Hom}_{\mathcal{R}}(\cdot,\mathcal{M})$ *is exact, too.*

Proof. We give full details of the proof of the first assertion only for $\mathrm{Hom}_{\mathcal{R}}(\mathcal{M},\cdot)$; the contravariant case is left as an exercise for the reader. We consider again the short exact sequence (B.17). Assume that $\alpha_*\phi = \alpha\circ\phi = 0$ for some $\phi\in\mathrm{Hom}_{\mathcal{R}}(\mathcal{M},\mathcal{A})$. As α is injective, this can only be the case, if $\phi = 0$ and hence α_* is injective.

Assume that $\beta_*\psi = 0$ for some homomorphism $\psi\in\mathrm{Hom}_{\mathcal{R}}(\mathcal{M},\mathcal{B})$. This implies that $\mathrm{im}\,\psi\subseteq\ker\beta = \mathrm{im}\,\alpha$. As α is injective, the inverse α^{-1} is well-defined on $\mathrm{im}\,\psi\subseteq\mathrm{im}\,\alpha$ and we may introduce the map $\phi = \alpha^{-1}\circ\psi\in\mathrm{Hom}_{\mathcal{R}}(\mathcal{M},\mathcal{A})$. Obviously, we have $\alpha_*\phi = \psi$ and thus $\ker\beta_*\subseteq\mathrm{im}\,\alpha_*$. Together with the converse inclusion shown above, this observation implies the exactness at $\mathrm{Hom}_{\mathcal{R}}(\mathcal{M},\mathcal{B})$.

In the case that $\mathcal{M}$ is a free $\mathcal{R}$-module, i. e. $\mathcal{M}\cong\mathcal{R}^m$, we note that obviously $\mathrm{Hom}_{\mathcal{R}}(\mathcal{M},\mathcal{A})\cong\mathcal{A}^m$ via the isomorphism $\phi\leftrightarrow\big(\phi(\mathbf{e}_1),\dots,\phi(\mathbf{e}_m)\big)$ where $\{\mathbf{e}_1,\dots,\mathbf{e}_m\}$ denotes the standard basis of $\mathcal{R}^m$. Now the exactness of the functor $\mathrm{Hom}_{\mathcal{R}}(\mathcal{M},\cdot)$ is trivial.

If the ring $\mathcal{R}$ is a (skew) field, we are dealing with vector spaces and we may decompose $\mathcal{B} = \mathrm{im}\,\alpha\oplus\mathcal{T}$ for some linear subspace $\mathcal{T}\subseteq\mathcal{B}$. By the injectivity of α, any element of $\mathcal{B}$ possesses a representation of the form $b = \alpha(a)+t$ with uniquely determined elements $a\in\mathcal{A}$ and $t\in\mathcal{T}$. Given a morphism $\phi:\mathcal{A}\to\mathcal{R}$, we define $\psi:\mathcal{B}\to\mathcal{R}$ by setting $\psi(b) = \phi(a)$. This morphism ψ obviously satisfies $\alpha^*\psi = \phi$ and as ϕ was arbitrary, this fact proves the surjectivity of α^*. Thus $\mathrm{Hom}_{\mathcal{R}}(\cdot,\mathcal{R})$ is exact and the extension to arbitrary vector spaces goes as above. □

Note the slight asymmetry in this result. A sufficient condition for the exactness of the covariant functor $\operatorname{Hom}_{\mathcal{R}}(\mathcal{M},\cdot)$ is already the freeness of $\mathcal{M}$. For the contravariant functor $\operatorname{Hom}_{\mathcal{R}}(\cdot,\mathcal{M})$ we must impose the stronger condition that $\mathcal{R}$ is a (skew) field. Our proof in the latter case applies also for infinite-dimensional vector spaces, as it does not depend on bases but only on the existence of complements. Restricting to finite-dimensional spaces, we may represent morphisms by matrices. The pull-back is then represented by the transposed matrix and the exactness at $\operatorname{Hom}_{\mathcal{R}}(\mathcal{A},\mathcal{M})$ and $\operatorname{Hom}_{\mathcal{R}}(\mathcal{C},\mathcal{M})$ translates into elementary statements about full column and row rank, respectively, of matrices representing injective and surjective maps, respectively.

Example B.2.10. A simple example of a module for which the functor $\operatorname{Hom}_{\mathcal{R}}(\mathcal{M},\cdot)$ is not exact is the $\mathbb{Z}$-module $\mathcal{M}=\mathbb{Z}/2\mathbb{Z}$. Let us choose $\mathcal{A}=\mathcal{B}=\mathbb{Z}$ and $\mathcal{C}=\mathcal{M}$ together with the maps $\alpha(z)=2z$ and $\beta=\pi$, the canonical projection. Assume that $\phi:\mathcal{M}\to\mathcal{B}$ is a morphism. Obviously, this implies $\phi(0)=0$. But we also have $\phi(0)=\phi(2)=\phi(2\cdot 1)=2\phi(1)$ and hence $\phi(1)=0$, too. Thus the only morphism between $\mathcal{M}$ and $\mathcal{B}$ is the zero map and the identity map $\mathrm{id}_{\mathcal{M}}$ is not in $\operatorname{im}\pi_*$. With a slight modification one shows similarly that the functor $\operatorname{Hom}_{\mathbb{Z}}(\cdot,\mathcal{M})$ is also not exact for the module $\mathcal{M}=\mathbb{Z}/2\mathbb{Z}$. ◁

Proposition B.2.11. *An $\mathcal{R}$-module $\mathcal{P}$ is projective, if and only if the covariant functor* $\operatorname{Hom}_{\mathcal{R}}(\mathcal{P},\cdot)$ *is exact. An $\mathcal{R}$-module $\mathcal{I}$ is injective, if and only if the contravariant functor* $\operatorname{Hom}_{\mathcal{R}}(\cdot,\mathcal{I})$ *is exact.*

Proof. By the previous proposition, both functors are already left exact. The definition of a projective (or injective, respectively) module is trivially equivalent to the missing exactness at the right end. □

Remark B.2.12. As a corollary, we find that a module is projective, if and only if it is a direct summand of a free module. One direction follows easily from Proposition B.2.6. Let $\mathcal{P}$ be projective and choose a generating set $\mathcal{G}$ of it. Then we have an obvious surjective map from the free $\mathcal{R}$-module generated by $\mathcal{G}$ to $\mathcal{P}$. For the converse, let $\mathcal{F}=\mathcal{P}\oplus\mathcal{Q}$ be a direct sum decomposition of a free module $\mathcal{F}$. Then trivially $\operatorname{Hom}_{\mathcal{R}}(\mathcal{F},\cdot)\cong\operatorname{Hom}_{\mathcal{R}}(\mathcal{P},\cdot)\oplus\operatorname{Hom}_{\mathcal{R}}(\mathcal{Q},\cdot)$ and the exactness of the functor on the left hand side implies the exactness of the functors on the right hand side. Thus a free module is always projective (but not necessarily injective). ◁

In this book we are particularly interested in the special case that $\mathcal{R}$ is a polynomial ring $\Bbbk[x^1,\dots,x^n]$. Here one has the following surprising result known as *Serre's conjecture* or *Quillen–Suslin Theorem*, as it was first conjectured by Serre [413] and later proven independently by Quillen [366] and Suslin [444].

Theorem B.2.13 (Quillen–Suslin). *Any projective polynomial module is free.*

Above we introduced as further important functors the tensor products $\mathcal{A}\otimes_{\mathcal{R}}\cdot$ and $\cdot\otimes_{\mathcal{R}}\mathcal{A}$ with an $\mathcal{R}$-module $\mathcal{A}$. For them we encounter a similar situation as for $\operatorname{Hom}_{\mathcal{R}}(\mathcal{A},\cdot)$ and $\operatorname{Hom}_{\mathcal{R}}(\cdot,\mathcal{A})$ (which is not surprising, as in a certain sense these functors are dual to each other): they are exact only for a special class of modules. We skip the details and just present the results.

Proposition B.2.14. *The two functors $\mathcal{M} \otimes_{\mathcal{R}} \cdot$ and $\cdot \otimes_{\mathcal{R}} \mathcal{M}$ are right exact for arbitrary $\mathcal{R}$-modules $\mathcal{M}$.*

Definition B.2.15. An $\mathcal{R}$-module $\mathcal{F}$ is called *flat*, if the functor $\mathcal{F} \otimes_{\mathcal{R}} \cdot$ is exact.

Remark B.2.16. As one easily shows the existence of a canonical isomorphism $\mathcal{M} \otimes_{\mathcal{R}} (\mathcal{N}_1 \oplus \mathcal{N}_2) \cong (\mathcal{M} \otimes_{\mathcal{R}} \mathcal{N}_1) \oplus (\mathcal{M} \otimes_{\mathcal{R}} \mathcal{N}_2)$, a direct sum of modules is flat, if and only if each summand is flat. Now $\mathcal{R}$ considered as an $\mathcal{R}$-module is trivially flat, since $\mathcal{M} \otimes_{\mathcal{R}} \mathcal{R} \cong \mathcal{M}$ for any $\mathcal{R}$-module $\mathcal{M}$, and thus also any free $\mathcal{R}$-module is flat. By Remark B.2.12, any projective module is a direct summand of a free module and hence flat, too. $\triangleleft$

Example B.2.17. Assume that the ring $\mathcal{R}$ satisfies a left Ore condition for the subset $\mathcal{S}$ of its regular elements. We introduce the total ring of left quotients $\mathcal{Q} = \mathcal{S}^{-1}\mathcal{D}$ and claim that it is flat as a right $\mathcal{R}$-module. For a left $\mathcal{R}$-module $\mathcal{M}$ we can identify $\mathcal{Q} \otimes_{\mathcal{R}} \mathcal{M}$ with the localisation $\mathcal{S}^{-1}\mathcal{M}$. If $\phi : \mathcal{M} \to \mathcal{N}$ is a homomorphism between two left $\mathcal{R}$-modules, then the application of the functor $\mathcal{Q} \otimes_{\mathcal{R}} \cdot$ yields a homomorphism which we can identify with the map $\mathcal{S}^{-1}\phi : \mathcal{S}^{-1}\mathcal{M} \to \mathcal{S}^{-1}\mathcal{N}$ defined by $\mathcal{S}^{-1}\phi(s^{-1}m) = s^{-1}\phi(m)$. Since $\mathcal{Q} \otimes_{\mathcal{R}} \cdot$ is already right exact by Proposition B.2.14, if suffices to show that $\mathcal{S}^{-1}\phi$ is injective whenever ϕ is. But this fact follows immediately from the definition of $\mathcal{S}^{-1}\phi$.

Note that the same reasoning implies the following equalities for the transformed homomorphism: $\ker(\mathcal{Q} \otimes_{\mathcal{R}} \phi) = \mathcal{Q} \otimes_{\mathcal{R}} \ker\phi$ and $\operatorname{im}(\mathcal{Q} \otimes_{\mathcal{R}} \phi) = \mathcal{Q} \otimes_{\mathcal{R}} \operatorname{im}\phi$. $\triangleleft$

The following notion is not that common, but it is sometimes quite useful, in particular in connection with the question of faithfulness of a functor. We need it only in Section 10.5. More details can be found in [252] (which also contains the dual notion of a generator); our treatment follows [487].

Definition B.2.18. A left $\mathcal{R}$-module $\mathcal{M}$ is called a *cogenerator*, if it satisfies for all $\mathcal{R}$-modules $\mathcal{N}$ the equality

$$\bigcap_{\phi \in \operatorname{Hom}_{\mathcal{R}}(\mathcal{N},\mathcal{M})} \ker\phi = 0\,. \tag{B.22}$$

In other words, for every element $n \in \mathcal{N}$ of an $\mathcal{R}$-module $\mathcal{N}$ at least one homomorphism $\phi \in \operatorname{Hom}_{\mathcal{R}}(\mathcal{N},\mathcal{M})$ exists with $\phi(n) \neq 0$.

Lemma B.2.19. *An $\mathcal{R}$-module $\mathcal{M}$ is a cogenerator, if and only if for all homomorphisms $\psi \in \operatorname{Hom}_{\mathcal{R}}(\mathcal{N}_1,\mathcal{N}_2)$ with two arbitrary $\mathcal{R}$-modules $\mathcal{N}_1$, $\mathcal{N}_2$ the vanishing of $\psi^* = \operatorname{Hom}_{\mathcal{R}}(\psi,\mathcal{M})$ implies that ψ vanishes, too.*

Proof. Let us first assume that $\mathcal{M}$ is a cogenerator. If $\psi \in \operatorname{Hom}_{\mathcal{R}}(\mathcal{N}_1,\mathcal{N}_2)$ does not vanish, then an element $n_1 \in \mathcal{N}_1$ exists such that $\psi(n_1) \neq 0$. The assumption $\psi^* = 0$ implies for all $\phi \in \operatorname{Hom}_{\mathcal{R}}(\mathcal{N}_2,\mathcal{M})$ that $\psi^*\phi(n_1) = \phi\bigl(\psi(n_1)\bigr) = 0$. Thus the element $\psi(n_1) \neq 0$ is contained in $\ker\phi$ for all such ϕ and condition (B.22) is not satisfied for the module $\mathcal{N}_2$. As this observation contradicts our assumption that $\mathcal{M}$ is a cogenerator, $\psi^* = 0$ entails $\psi = 0$.

For the converse statement we use an indirect proof. Let $0 \neq n_2 \in \mathcal{N}_2$ be an element such that $\phi(n_2) = 0$ for all $\phi \in \mathrm{Hom}_{\mathcal{R}}(\mathcal{N}_2, \mathcal{M})$ (hence $\mathcal{M}$ cannot be a cogenerator). Now define $\psi \in \mathrm{Hom}_{\mathcal{R}}(\mathcal{R}, \mathcal{N}_2)$ by setting $\psi(1) = n_2$ (i. e. we choose $\mathcal{N}_1 = \mathcal{R}$). Then we obviously find that $\phi \circ \psi = 0$ for all $\phi \in \mathrm{Hom}_{\mathcal{R}}(\mathcal{N}_2, \mathcal{M})$ so that $\psi^* = 0$ although $\psi \neq 0$. □

Lemma B.2.20. *Let F be an exact contravariant functor mapping $\mathcal{R}$-modules into Abelian groups. Then the following two statements are equivalent.*

(i) *If $F(\mathcal{M}) = 0$ for an $\mathcal{R}$-module $\mathcal{M}$, then already $\mathcal{M} = 0$.*
(ii) *If $F(\phi) = 0$ for a module homomorphism ϕ, then already $\phi = 0$.*

Proof. Let us first show that (i) implies (ii). Any homomorphism $\phi : \mathcal{M} \to \mathcal{N}$ of $\mathcal{R}$-modules can be decomposed into two homomorphisms $\phi = \iota \circ \bar{\phi}$ with the canonical inclusion $\iota : \mathrm{im}\,\phi \hookrightarrow \mathcal{N}$ and the restriction $\bar{\phi} : \mathcal{M} \to \mathrm{im}\,\phi$ and then we have $F(\phi) = F(\bar{\phi}) \circ F(\iota)$. Since F is exact, the injectivity of ι implies that $F(\iota)$ is surjective and similarly the surjectivity of $\bar{\phi}$ makes $F(\bar{\phi})$ injective. Assume now that $F(\phi) = 0$. By the injectivity of $F(\bar{\phi})$, this is only possible if $F(\iota) = 0$. As the latter map is surjective, we find $\mathrm{im}\,F(\iota) = F(\mathrm{im}\,\phi) = 0$. By assumption, this fact implies $\mathrm{im}\,\phi = 0$ and hence $\phi = 0$.

For the converse direction, we note that $\mathcal{M} \neq 0$ implies $\mathrm{id}_{\mathcal{M}} \neq 0$ and thus, by assumption, $F(\mathrm{id}_{\mathcal{M}}) = \mathrm{id}_{F(\mathcal{M})} \neq 0$. Thus we must have $F(\mathcal{M}) \neq 0$. □

Proposition B.2.21. *Let F be an exact contravariant functor mapping $\mathcal{R}$-modules into Abelian groups. F is faithful, if and only if $F(\mathcal{M}) = 0$ implies $\mathcal{M} = 0$ for all $\mathcal{R}$-modules $\mathcal{M}$.*

Proof. One direction is very easy. If $F(\mathcal{M}) = 0$, then we have an exact sequence $0 \to F(\mathcal{M}) \to 0$. If F is faithful, we also have an exact sequence $0 \to \mathcal{M} \to 0$ which is only possible, if $\mathcal{M} = 0$.

For the converse, we use again an indirect proof. Assume that the sequence of $\mathcal{R}$-modules $\mathcal{M}_1 \xrightarrow{\phi} \mathcal{M} \xrightarrow{\psi} \mathcal{M}_2$ was *not* exact. There are two possible reasons for this. Firstly, the sequence could even fail to be a complex, i. e. $\psi \circ \phi \neq 0$. Then, by Lemma B.2.20, also $F(\phi) \circ F(\psi) \neq 0$ and the induced sequence could not be a complex either. Thus the functor F is faithful.

Alternatively, we could have $\psi \circ \phi = 0$ but $\mathrm{im}\,\phi \subsetneq \ker\psi$. Now Lemma B.2.20 asserts that $F(\psi \circ \phi) = F(\phi) \circ F(\psi) = 0$ and hence that the induced sequence is a complex, too. We introduce now the canonical inclusion $\iota : \ker\psi \hookrightarrow \mathcal{M}$ and the canonical projection $\pi : \mathcal{M} \to \mathcal{M}/\mathrm{im}\,\phi$. These maps lead to exact sequences $\ker\psi \to \mathcal{M} \to \mathcal{M}_2$ and $\mathcal{M}_1 \to \mathcal{M} \to \mathcal{M}/\mathrm{im}\,\phi$, respectively. As F is assumed to be an exact functor, applying it to these sequences yields again exact sequences. Hence we find that

$$\ker F(\iota) = \mathrm{im}\,F(\psi) \subseteq \ker F(\phi) = \mathrm{im}\,F(\pi)\,. \tag{B.23}$$

Since we assumed that $\mathrm{im}\,\phi \subsetneq \ker\psi$, it cannot be that $\pi \circ \iota = 0$ and hence we have $F(\iota) \circ F(\pi) \neq 0$ implying that $\ker F(\iota) \subsetneq \mathrm{im}\,F(\pi)$. Entering this observation into (B.23) shows that the induced sequence could not be exact and we again conclude that the functor F is faithful. □

Remark B.2.22. We are particularly interested in applying these results to the functor $F = \mathrm{Hom}_{\mathcal{R}}(\cdot, \mathcal{M})$. According to Proposition B.2.11, F is exact for an injective module $\mathcal{M}$, and, by Lemma B.2.19, the second statement in Lemma B.2.20 is now equivalent to $\mathcal{M}$ being a cogenerator. Thus Proposition B.2.21 asserts that for an injective module $\mathcal{M}$ the functor F is faithful, if and only if $\mathcal{M}$ is a cogenerator. ◁

In a (cochain) complex $(\mathcal{M}, d)$, an element $m^i \in \mathcal{M}^i$ such that $d^i(m^i) = 0$ is called a *cocycle* and if $m^i = d^{i-1}(m^{i-1})$ for some $m^{i-1} \in \mathcal{M}^{i-1}$ it is a *coboundary*. By definition of a complex, any coboundary is a cocycle; but in general the converse is not true. Only if $\mathcal{M}$ is an exact sequence, then any solution of the equation $d^i(m^i) = 0$ may be represented as $m^i = d^{i-1}(m^{i-1})$, i. e. any cocycle is a coboundary. Obviously, this representation of m^i is not unique: if $\bar{m}^{i-1} - m^{i-1} = d^{i-2}(m^{i-2})$ for some element $m^{i-2} \in \mathcal{M}^{i-2}$, then $d^{i-1}(\bar{m}^{i-1}) = d^{i-1}(m^{i-1}) = m^i$.

Definition B.2.23. Let $(\mathcal{M}, d)$ be a cochain complex of $\mathcal{R}$-modules. Its ith *cohomology module* is defined as the quotient

$$H^i(\mathcal{M}, d) = \ker d^i / \operatorname{im} d^{i-1} . \tag{B.24}$$

The direct sum of all cohomology modules $H^i(\mathcal{M}, d)$ defines a graded $\mathcal{R}$-module, the *cohomology* $H^\bullet(\mathcal{M}, d)$ of the complex $(\mathcal{M}, d)$.

As $\operatorname{im} d^{i-1} \subseteq \ker d^i$, the cohomology modules are well-defined. We may now characterise an exact sequence as a complex with vanishing cohomology. The non-zero elements of $H^i(\mathcal{M}, d)$ are sometimes called *obstructions*, as they obstruct the above mentioned idea of representing solutions of the equation $d^i(m^i) = 0$ in the form $m^i = d^{i-1}(m^{i-1})$. For a chain complex $(\mathcal{M}, d)$ one introduces similarly the notions of a *cycle* and a *boundary*. The quotient of the cycles by the boundaries defines now the *homology* $H_\bullet(\mathcal{M}, d)$.

For two cochain complexes $(\mathcal{M}, d)$ and $(\mathcal{N}, \delta)$ a family of maps $f^i : \mathcal{M}^i \to \mathcal{N}^i$ such that $\delta^i \circ f^i = f^{i+1} \circ d^i$ defines a *cochain map* $f : \mathcal{M} \to \mathcal{N}$. It induces maps $H^i(f) : H^i(\mathcal{M}, d) \to H^i(\mathcal{N}, \delta)$ by setting $H^i(f)\big([m^i]\big) = \big[f^i(m^i)\big]$. Two cochain maps $f, g : \mathcal{M} \to \mathcal{N}$ are *homotopic*, written $f \sim g$, if maps $s^i : \mathcal{M}^i \to \mathcal{N}^{i-1}$ exist such that $f^i - g^i = s^{i+1} \circ d^i + \delta^{i-1} \circ s^i$ (the map $s : \mathcal{M} \to \mathcal{N}$ is then called a *homotopy*). The following (non-commutative!) diagram shows all involved maps:

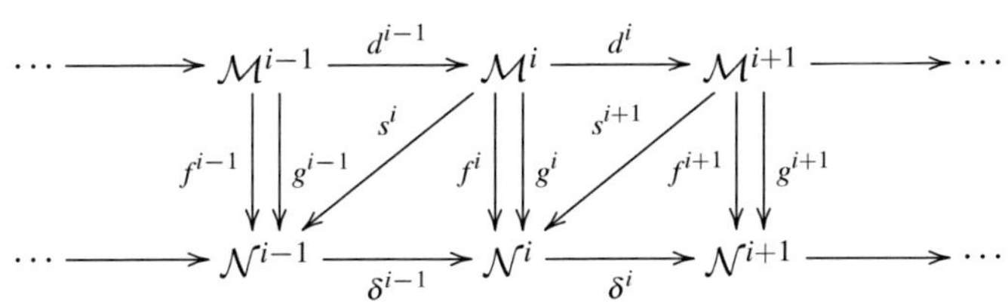

(B.25)

There is a similar notion of a chain map and homotopies between chain maps are then of the form $s_i : \mathcal{M}_i \to \mathcal{N}_{i+1}$.

Lemma B.2.24. *If the cochain maps $f, g : \mathcal{M} \to \mathcal{N}$ are homotopic, then their induced maps coincide: $H^\bullet(f) = H^\bullet(g)$.*

Proof. If $m_i \in \mathcal{M}_i$ is a cocycle, then $f^i(m^i) - g^i(m^i) = \delta^{i-1}\big(s^i(m^i)\big)$ and hence $H^i(f)\big([m_i]\big) = H^i(g)\big([m_i]\big)$. □

Assume now that $f : \mathcal{M} \to \mathcal{N}$ and $g : \mathcal{N} \to \mathcal{M}$ are two cochain maps satisfying $g \circ f \sim \mathrm{id}_{\mathcal{M}}$ and $f \circ g \sim \mathrm{id}_{\mathcal{N}}$. Then the two complexes $\mathcal{M}$, $\mathcal{N}$ are called *cochain equivalent*. The following result shows that such complexes behave identically with respect to cohomology.

Lemma B.2.25. *If the complexes $(\mathcal{M},d)$ and $(\mathcal{N},\delta)$ are cochain equivalent, then their cohomologies are isomorphic: $H^\bullet(\mathcal{M},d) \cong H^\bullet(\mathcal{N},\delta)$.*

Proof. By Lemma B.2.24, we find for the induced maps $H^\bullet(g \circ f) = \mathrm{id}_{H^\bullet(\mathcal{M},d)}$ and $H^\bullet(f \circ g) = \mathrm{id}_{H^\bullet(\mathcal{N},\delta)}$. One easily verifies that for arbitrary chain maps f, g the identity $H^\bullet(g \circ f) = H^\bullet(g) \circ H^\bullet(f)$ holds. Hence $H^\bullet(f)$ and $H^\bullet(g)$ are both isomorphisms (and inverse to each other). □

Remark B.2.26. An important special case arises, if for a complex $(\mathcal{M},d)$ the identity map is homotopic to the zero map, i. e. if a homotopy $s^i : \mathcal{M}^i \to \mathcal{M}^{i-1}$ exists such that $s^{i+1} \circ d^i + d^{i-1} \circ s^i = \mathrm{id}_{\mathcal{M}^i}$. By Lemma B.2.24, $\mathrm{id}_{H^i(\mathcal{M},d)}$ is then the zero map which is only possible, if $H^\bullet(\mathcal{M},d) = 0$. Hence, the complex $(\mathcal{M},d)$ must be exact and s is called a *contracting1 homotopy*. The construction of such contracting homotopies is a very useful tool for proving the exactness of sequences. ◁

Proposition B.2.27. *Given a short exact sequence of cochain complexes*

$$0 \longrightarrow \mathcal{A} \xrightarrow{\alpha} \mathcal{B} \xrightarrow{\beta} \mathcal{C} \longrightarrow 0 \tag{B.26}$$

with cochain maps α and β, there exists an exact sequence

$$\begin{gathered} \cdots \longrightarrow H^n(\mathcal{A},d_{\mathcal{A}}) \xrightarrow{H^n(\alpha)} H^n(\mathcal{B},d_{\mathcal{B}}) \xrightarrow{H^n(\beta)} \\ \xrightarrow{H^n(\beta)} H^n(\mathcal{C},d_{\mathcal{C}}) \xrightarrow{\delta^n} H^{n+1}(\mathcal{A},d_{\mathcal{A}}) \longrightarrow \cdots \end{gathered} \tag{B.27}$$

(B.27) is known as the *long exact homological sequence*. The main point in the proof of this result is to derive the *connecting homomorphisms* δ^n. We only sketch this derivation; the verification that the arising map is well-defined, independent of the made choices and satisfies the required properties consists of in principle straightforward but tedious diagram chases.

If we break the sequence (B.26) into its components, then the for the construction of δ^n relevant part is given by the commutative diagram

$$\begin{array}{ccccccccc} & & \vdots & & \vdots & & \vdots & & \\ & & \downarrow & & \downarrow & & \downarrow & & \\ 0 & \longrightarrow & \mathcal{A}^n & \xrightarrow{\alpha^n} & \mathcal{B}^n & \xrightarrow{\beta^n} & \mathcal{C}^n & \longrightarrow & 0 \\ & & \downarrow{\scriptstyle d^n_{\mathcal{A}}} & & \downarrow{\scriptstyle d^n_{\mathcal{B}}} & & \downarrow{\scriptstyle d^n_{\mathcal{C}}} & & \\ 0 & \longrightarrow & \mathcal{A}^{n+1} & \xrightarrow[\alpha^{n+1}]{} & \mathcal{B}^{n+1} & \xrightarrow[\beta^{n+1}]{} & \mathcal{C}^{n+1} & \longrightarrow & 0 \\ & & \downarrow & & \downarrow & & \downarrow & & \\ & & \vdots & & \vdots & & \vdots & & \end{array} \tag{B.28}$$

with exact rows. Any cohomology class in the module $H^n(\mathcal{C}, d_{\mathcal{C}})$ may be represented by an element $c \in \ker d^n_{\mathcal{C}}$. Since β^n is surjective, there exists an element $b \in \mathcal{B}_n$ such that $\beta^n(b) = c$. Because of the commutativity of (B.28), $d^n_{\mathcal{B}}(b) \in \ker \beta^{n+1}$ and hence, by the exactness of the rows, there finally exists an element $a \in \mathcal{A}^{n+1}$ such that $\alpha_{n+1}(a) = d^n_{\mathcal{B}}(b)$. Now we define $\delta^n\big([c]\big) = [a]$.

One might think that applying an exact functor to a complex leads to another complex with isomorphic (co)homology. The trivial example of the zero functor which always yields an exact sequence shows that this is not necessarily the case. However, in the important case of *dualisation* we obtain the expected result. Let

$$\cdots \longrightarrow \mathcal{V}^{i-1} \xrightarrow{d^{i-1}} \mathcal{V}^i \xrightarrow{d^i} \mathcal{V}^{i+1} \longrightarrow \cdots \tag{B.29}$$

be a cochain complex $(\mathcal{V}, d)$ of finite-dimensional vector spaces over a field $\Bbbk$. By definition, $\mathcal{V}^*_i = \mathrm{Hom}_{\Bbbk}(\mathcal{V}^i, \Bbbk)$ is the dual space of $\mathcal{V}^i$. The dual map $d^*_i : \mathcal{V}^*_{i+1} \to \mathcal{V}^*_i$ is the pull-back as introduced above: $d^*_i(\phi) = \phi \circ d^i$. Thus dualisation leads to the chain complex $(\mathcal{V}^*, d^*)$

$$\cdots \longleftarrow \mathcal{V}^*_{i-1} \xleftarrow{d^*_{i-1}} \mathcal{V}^*_i \xleftarrow{d^*_i} \mathcal{V}^*_{i+1} \longleftarrow \cdots \tag{B.30}$$

Of course, one expects that the cohomology of $(\mathcal{V}, d)$ and the homology of $(\mathcal{V}^*, d^*)$ are closely related.

Proposition B.2.28. *If $(\mathcal{V}, d)$ is a cochain complex of finite-dimensional vector spaces over a field $\Bbbk$, then $H_\bullet(\mathcal{V}^*, d^*) = \big(H^\bullet(\mathcal{V}, d)\big)^*$.*

Proof. Let $\phi \in \ker d^*_{i-1} \subseteq \mathcal{V}^*_i$ be a representative of a homology class in $H_i(\mathcal{V}^*, d^*)$. Then we have $d^*_{i-1}(\phi) = \phi \circ d^{i-1} = 0$ and hence $\phi\big|_{\mathrm{im}\, d^{i-1}} = 0$. This observation implies that the restriction $\phi_0 = \phi\big|_{\ker d^i}$ induces a well-defined map $\bar{\phi} \in \big(H^i(\mathcal{V}, d)\big)^*$. If $\psi = d^*_i(\rho) \in \mathrm{im}\, d^*_i$, then, by definition of the pull-back, $\psi\big|_{\ker d^i} = 0$ and hence $\phi_0 = (\phi + \psi)_0$ making our construction independent of the choice of the representative ϕ. Thus we may introduce a canonical vector space homomorphism $h_i : H_i(\mathcal{V}^*, d^*) \to \big(H^i(\mathcal{V}, d)\big)^*$ by setting $h_i\big([\phi]\big) = \bar{\phi}$.

There only remains to show that h_i is an isomorphism. Consider

$$0 \longrightarrow \ker d^i \hookrightarrow \mathcal{V}^i \xrightarrow{d^i} \operatorname{im} d^i \longrightarrow 0 \tag{B.31}$$

which is trivially an exact sequence. Since all involved spaces are finite-dimensional, the sequence splits. This implies the existence of a projection map $\pi : \mathcal{V}^i \to \ker d^i$ which is the identity on $\ker d^i$. This observation allows us to extend any map $\phi_0 : \ker d^i \to \mathbb{k}$ to a map $\phi = \phi_0 \circ \pi : \mathcal{V}^i \to \mathbb{k}$, i. e. to an element of $\mathcal{V}_i^*$.

Any element $\bar{\phi} \in \left(H^i(\mathcal{V},d)\right)^*$ can be represented by a functional $\phi_0 : \ker d^i \to \mathbb{k}$ such that $\phi_0\big|_{\operatorname{im} d^{i-1}} = 0$. The above described extension leads then to an element $\phi \in \mathcal{M}_i^*$ such that $\phi\big|_{\ker d^i} = \phi_0$. It follows trivially from our construction that it satisfies $h_i\big([\phi]\big) = \bar{\phi}$. Hence h_i is surjective and as a map between vector spaces of the same finite dimension thus bijective. □

More generally, the question what happens if a (left or right) exact functor is applied to a complex leads to the important concept of derived functors. They are usually defined via resolutions.

Definition B.2.29. Let $\mathcal{R}$ be a ring and $\mathcal{M}$ an $\mathcal{R}$-module. An exact sequence

$$\cdots \longrightarrow \mathcal{C}_j \xrightarrow{\psi_j} \mathcal{C}_{j-1} \longrightarrow \cdots \longrightarrow \mathcal{C}_1 \xrightarrow{\psi_1} \mathcal{C}_0 \xrightarrow{\pi} \mathcal{M} \longrightarrow 0 \tag{B.32}$$

of $\mathcal{R}$-modules is called a *resolution* of $\mathcal{M}$. Dually, a *coresolution* is an exact sequence of $\mathcal{R}$-modules of the form

$$0 \longrightarrow \mathcal{M} \xrightarrow{\iota} \mathcal{C}^0 \xrightarrow{\psi^0} \mathcal{C}^1 \longrightarrow \cdots \longrightarrow \mathcal{C}^j \xrightarrow{\psi^j} \mathcal{C}^{j+1} \longrightarrow \cdots . \tag{B.33}$$

The (co)resolution is *free*, *projective* or *injective*, if all modules $\mathcal{C}_j$ are free, projective or injective, respectively. If there exists an index ℓ such that $\mathcal{C}_\ell \neq 0$ but $\mathcal{C}_j = 0$ for all $j > \ell$, the (co)resolution is *finite* and its *length* is $\ell + 1$.

Remark B.2.30. Obviously, every resolution defines a chain complex $(\mathcal{C}, \psi)$ given by $\cdots \to \mathcal{C}_1 \to \mathcal{C}_0 \to 0$. Since (B.32) is assumed to be exact, all homology modules $H_j(\mathcal{C}, \psi)$ with $j > 0$ vanish and $H_0(\mathcal{C}, \psi) = \mathcal{C}_0 / \operatorname{im} \psi_1 = \mathcal{C}_0 / \ker \pi \cong \mathcal{M}$. We may furthermore consider $\mathcal{C} = \bigoplus_{k \geq 0} \mathcal{C}_k$ as a non-negatively graded module and ψ as a graded module homomorphism. Analogously, we may turn $\mathcal{M}$ into a non-negatively graded module which is trivial everywhere except in degree 0. Putting the zero differential on $\mathcal{M}$, we get $H_0(\mathcal{M}, 0) \cong \mathcal{M}$ and all other homology groups vanish. Extending now the map π to all of $\mathcal{C}$ by setting it zero in positive degree, we obtain a chain map $\mathcal{C} \to \mathcal{M} \to 0$ inducing an isomorphism on the homology.

Similarly, every coresolution defines a cochain complex $(\mathcal{C}, \psi)$, if we omit the module $\mathcal{M}$. Now all cohomology modules $H^j(\mathcal{C}, \psi)$ with $j > 0$ vanish because of the exactness of (B.33) and $H^0(\mathcal{C}, \psi) = \ker \psi^0 = \operatorname{im} \iota \cong \mathcal{M}$. Extending again ι trivially to the whole graded module $\mathcal{C}$, we obtain a cochain map $0 \to \mathcal{M} \to \mathcal{C}$ inducing an isomorphism on cohomology. ◁

Resolutions are not unique. A useful tool for analysing to what extent two projective resolutions of the same module $\mathcal{M}$ may differ is the following proposition. A similar dual result holds for injective coresolutions, but we leave its formulation as an exercise for the reader.

Proposition B.2.31. *Let $\mathcal{R}$ be a ring and*

$$\cdots \longrightarrow \mathcal{C}_j \xrightarrow{\phi_j} \mathcal{C}_{j-1} \longrightarrow \cdots \longrightarrow \mathcal{C}_1 \xrightarrow{\phi_1} \mathcal{C}_0 \tag{B.34}$$

a complex of projective $\mathcal{R}$-modules and set $\mathcal{M} = \operatorname{coker}\phi_1$. Let furthermore

$$\cdots \longrightarrow \mathcal{D}_j \xrightarrow{\psi_j} \mathcal{D}_{j-1} \longrightarrow \cdots \longrightarrow \mathcal{D}_1 \xrightarrow{\psi_1} \mathcal{D}_0 \xrightarrow{\pi_{\mathcal{N}}} \mathcal{N} \longrightarrow 0 \tag{B.35}$$

be a resolution of the $\mathcal{R}$-module $\mathcal{N}$. Then any homomorphism $f : \mathcal{M} \to \mathcal{N}$ comes from a chain map $g : \mathcal{C} \to \mathcal{D}$ which is unique up to homotopy.

If $\mathcal{C} \to \mathcal{M} \to 0$ and $\mathcal{D} \to \mathcal{M} \to 0$ are two different projective resolutions of the same module $\mathcal{M}$, then we can apply Proposition B.2.31 twice with $f = \mathrm{id}_{\mathcal{M}}$. It yields the existence of chain maps $g : \mathcal{C} \to \mathcal{D}$ and $h : \mathcal{D} \to \mathcal{C}$, respectively, such that $g \circ h \sim \mathrm{id}_{\mathcal{C}}$ and $h \circ g \sim \mathrm{id}_{\mathcal{D}}$. One says that any two projective resolutions of $\mathcal{M}$ are *homotopy equivalent*.

Over a Noetherian ring $\mathcal{R}$ it is straightforward to construct step by step a free resolution for any finitely generated $\mathcal{R}$-module $\mathcal{M}$. Let

$$0 \longrightarrow \mathcal{M}_0 \overset{\iota_0}{\hookrightarrow} \mathcal{F}_0 \xrightarrow{\phi_0} \mathcal{M} \longrightarrow 0 \tag{B.36}$$

be a finite presentation of $\mathcal{M}$ where $\mathcal{F}_0$ is a free module of rank s_0, ι_0 the inclusion map and ϕ_0 the map defined by (B.1). Since $\mathcal{R}$ is assumed to be Noetherian, $\mathcal{M}_0$ is again finitely generated and thus also possess a finite presentation with a free module $\mathcal{F}_1$ of rank s_1 leading to an analogous short exact sequence with maps ι_1 and ϕ_1. If we define $\psi_0 = \phi_0$ and $\psi_1 = \iota_0 \circ \phi_1$, we obtain the exact sequence

$$0 \longrightarrow \mathcal{M}_1 \overset{\iota_1}{\hookrightarrow} \mathcal{F}_1 \xrightarrow{\psi_1} \mathcal{F}_0 \xrightarrow{\psi_0} \mathcal{M} \longrightarrow 0\,. \tag{B.37}$$

We iterate this construction taking a finite presentation of $\mathcal{M}_1$ leading to a finitely generated module $\mathcal{M}_2$ and so on. Each finite presentation leads to a short exact sequence of the form (B.36) with associated maps ι_i and ϕ_i. Defining $\psi_i = \iota_{i-1} \circ \phi_i$ we obtain the following commutative diagram where the diagonals come from the various presentations and the central row defines a free resolution of $\mathcal{M}$:

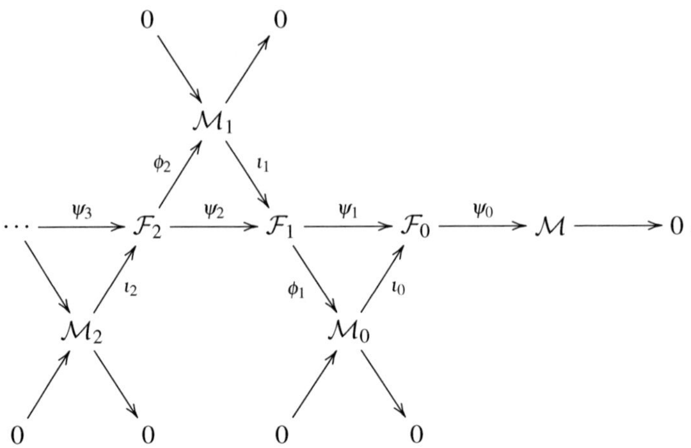

(B.38)

Assume that $\mathcal{R}$ is a non-negatively graded ring and write $\mathcal{R}_+ = \bigoplus_{d>0} \mathcal{R}_d$ for the irrelevant ideal. If the $\mathcal{R}$-module $\mathcal{M}$ is also graded, it is natural to consider graded resolutions. Among these, certain ones are distinguished.

Definition B.2.32. A free resolution of a graded $\mathcal{R}$-module $\mathcal{M}$ is called *minimal*, if the maps $\psi_j : \mathcal{R}^{s_j} \rightarrow \mathcal{R}^{s_{j-1}}$ satisfy $\operatorname{im} \psi_j \subseteq \mathcal{R}_+ \cdot \mathcal{R}^{s_{j-1}}$. In this case, the ranks s_j are the *Betti numbers* $\beta_j(\mathcal{M})$ of $\mathcal{M}$ and the length of the resolution is the *projective dimension* $\operatorname{proj}\dim \mathcal{M}$ of the module $\mathcal{M}$. The *global dimension* $\operatorname{gl}\dim \mathcal{R}$ of $\mathcal{R}$ is the supremum of the projective dimensions of all $\mathcal{R}$-modules.[8]

The adjective "minimal" refers to the fact that in a minimal resolution the free generators of any component $\mathcal{R}^{s_j}$ are mapped into a minimal generating set of its image in $\mathcal{R}^{s_{j-1}}$. Indeed, if their images do not form a minimal generating set, then at least one image can be expressed as an $\mathcal{R}$-linear combination of the other ones and we could not have $\operatorname{im} \psi_j \subseteq \mathcal{R}_+ \cdot \mathcal{R}^{s_{j-1}}$.

Minimal resolutions are unique up to isomorphism [125, Theorem 20.2], so that the Betti numbers and the projective dimension are indeed well-defined invariants of the module $\mathcal{M}$. It requires only some linear algebra in order to reduce any graded free resolution to the minimal one. In an abstract language, the main tool is the following technical result [104, Lemma 1.3.2].

Lemma B.2.33. *Let the diagram of $\mathcal{R}$-modules*

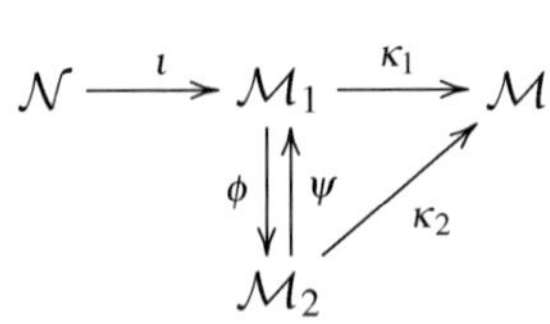

(B.39)

be commutative and its upper row exact. Then the sequence

[8] One can show that it suffices to take the supremum only over all finitely generated modules.

$$0 \longrightarrow \operatorname{im}(\mathrm{id}_{\mathcal{M}_1} - \psi\circ\phi) \xrightarrow{\rho_1} \operatorname{im}\iota \oplus \operatorname{im}(\mathrm{id}_{\mathcal{M}_2} - \phi\circ\psi) \xrightarrow{\rho_2} \mathcal{M}_2 \xrightarrow{\kappa_2} \mathcal{M} \tag{B.40}$$

with the maps $\rho_1(m_1) = \big(m_1, -\phi(m_1)\big)$ *and* $\rho_2(m_1,m_2) = \phi(m_1) + m_2$*, respectively, is well-defined and exact.*

Proof. We have $\kappa_1\big(m_1 - (\psi\circ\phi)(m_1)\big) = \kappa_1(m_1) - (\kappa_2\circ\phi)(m_1) = 0$ which implies by the exactness of the upper row of (B.39) that $\operatorname{im}(\mathrm{id} - \psi\circ\phi) \subseteq \operatorname{im}\iota$. It is also easy to see that $m_1 \in \operatorname{im}(\mathrm{id} - \psi\circ\phi)$ entails $\phi(m_1) \in \operatorname{im}(\mathrm{id} - \phi\circ\psi)$ and hence the sequence (B.40) is indeed well-defined. Its exactness at the first term and the fact that $\rho_2\circ\rho_1 = 0$ is trivial. If $\rho_2\big(\iota(n), (\mathrm{id} - \phi\circ\psi)(m_2)\big) = 0$, then we have with $m_1 = \iota(n)$ that $\phi(m_1) = -(\mathrm{id} - \phi\circ\psi)(m_2)$ and thus exactness at the second term.

$\kappa_2\big((\phi\circ\iota)(n) + (\mathrm{id} - \phi\circ\psi)(m_2)\big) = (\kappa_1\circ\iota)(n) + \kappa_2(m_2) - (\kappa_1\circ\psi)(m_2) = 0$ and hence $\operatorname{im}\rho_2 \subseteq \ker\kappa_2$. Finally, if $m_2 \in \ker\kappa_2$, then $(\kappa_1\circ\psi)(m_2) = 0$ and thus $\psi(m_2) \in \operatorname{im}\iota$. Hence we may write $m_2 = (\phi\circ\psi)(m_2) + (\mathrm{id} - \phi\circ\psi)(m_2) \in \operatorname{im}\rho_2$ proving the exactness at the last term. □

Let $E_0 = \{\mathbf{e}_1, \dots, \mathbf{e}_{s_0}\}$ be a basis of the first free module $\mathcal{F}_0$ appearing in the free resolution (B.32) and assume that its image $\psi_0(E_0)$ does not form a minimal generating set of the module $\mathcal{M}$. Then we may choose a minimal subset $E'_0 \subset E_0$ such that its image $\psi_0(E'_0)$ still generates $\mathcal{M}$. The linear span of E'_0 generates a free submodule $\mathcal{F}'_0 \subset \mathcal{F}_0$ and we can construct a diagram

$$\begin{array}{ccccccc} \cdots \xrightarrow{\psi_2} & \mathcal{F}_1 & \xrightarrow{\psi_1} & \mathcal{F}_0 & \xrightarrow{\psi_0} & \mathcal{M} & \longrightarrow 0 \\ & & & \pi\big\downarrow\big\uparrow\iota & \nearrow_{\psi'_0} & & \\ & & & \mathcal{F}'_0 & & & \end{array} \tag{B.41}$$

where ι denotes as usual the natural inclusion map and the map π is defined as follows: if $\mathbf{e}_i \in E'_0$, then $\pi(\mathbf{e}_i) = \mathbf{e}_i$; otherwise we know that coefficients $r_{\mathbf{e}} \in \mathcal{R}$ exist with $\psi_0(\mathbf{e}_i) = \sum_{\mathbf{e}\in E'_0} r_{\mathbf{e}}\psi_0(\mathbf{e})$ and set $\pi(\mathbf{e}_i) = \sum_{\mathbf{e}\in E'_0} r_{\mathbf{e}}\mathbf{e}$. If we finally define $\psi'_0 = \psi_0\circ\iota$, it is not difficult to verify that the diagram (B.41) commutes.

Thus Lemma B.2.33 leads to the exact sequence

$$0 \longrightarrow \operatorname{im}(\mathrm{id} - \iota\circ\pi) \hookrightarrow \operatorname{im}\psi_1 \xrightarrow{\pi} \mathcal{F}'_0 \xrightarrow{\psi'_0} \mathcal{M} \longrightarrow 0 \tag{B.42}$$

since $\pi\circ\iota = \mathrm{id}$. If $\mathbf{e}_i \in E'_0$, then obviously $(\mathrm{id} - \iota\circ\pi)(\mathbf{e}_i) = 0$. If $\mathbf{e}_i \in E_0 \setminus E'_0$, then $(\mathrm{id} - \iota\circ\pi)(\mathbf{e}_i) = \mathbf{e}_i - \sum_{\mathbf{e}\in E'_0} r_{\mathbf{e}}\mathbf{e}$ with some coefficients $r_{\mathbf{e}} \in \mathcal{R}$. Thus the images of the latter generators are linearly independent and $\operatorname{im}(\mathrm{id} - \iota\circ\pi)$ may be identified with a free submodule of $\operatorname{im}\psi_1$.

As next step we determine for each generator $\mathbf{e}_i \in E_0 \setminus E'_0$ an element $\mathbf{f}_i \in \mathcal{F}_1$ such that $\psi_1(\mathbf{f}_i) = (\mathrm{id} - \iota\circ\pi)(\mathbf{e}_i)$. The linear span of these elements $\mathbf{f}_i$ defines a free submodule $\iota_1 : \mathcal{F}'_1 \hookrightarrow \mathcal{F}_1$ and the restriction of ψ_1 to $\mathcal{F}'_1$ is injective. It is then straightforward to verify that

$$\cdots \longrightarrow \mathcal{F}_3 \xrightarrow{(\psi_3,0)} \mathcal{F}_2 \oplus \mathcal{F}_1' \xrightarrow{\psi_2 \oplus \iota_1} \mathcal{F}_1 \xrightarrow{\pi \circ \psi_1} \mathcal{F}_0' \xrightarrow{\psi_0'} \mathcal{M} \longrightarrow 0 \tag{B.43}$$

is again a free resolution of $\mathcal{M}$ but now with the last term minimal. If one of the other terms is not yet minimal, we may repeat the procedure there. In the case of a finite free resolution, we obtain after a finite number of steps a minimal resolution, since the whole process obviously stops if $\mathcal{F}_2 = 0$. For the special case of a graded module over a polynomial ring, we will return to the question of explicitly constructing a minimal resolution at the end of Appendix B.4.

Remark B.2.34. The minimal resolution of a graded module $\mathcal{M}$ is *pure*, if every module in the resolution has a generating set where all elements are of the same degree. A special case is a *linear* resolution where a value q exists such that the jth module in the resolution is generated in degree $q+j$. If $\mathcal{M}$ is a graded module, we denote by $\mathcal{M}_{\langle d\rangle} = \langle \mathcal{M}_d \rangle$ the graded module generated by the homogeneous component $\mathcal{M}_d$. Following Herzog and Hibi [210], we say that $\mathcal{M}$ is *componentwise linear*, if for every degree $d \geq 0$ the module $\mathcal{M}_{\langle d\rangle}$ has a linear resolution. ◁

Now we can finally define derived functors, a fundamental construction in homological algebra. They can be introduced for any *additive* functor, i. e. for any functor F satisfying $F(f+g) = F(f) + F(g)$ for arbitrary homomorphisms f, g.

Definition B.2.35. Let F be a covariant right exact additive functor and choose a projective resolution $\mathcal{P} \to \mathcal{M} \to 0$ of the $\mathcal{R}$-module $\mathcal{M}$ with differential ψ. Then the nth *left derived functor* $\mathcal{L}_n F$ maps $\mathcal{M}$ to the nth homology module $H_n\big(F(\mathcal{P}), F(\psi)\big)$. Similarly, if F is a covariant left exact additive functor and $0 \to \mathcal{I} \to \mathcal{M}$ an injective coresolution with differential ψ, then the nth *right derived functor* $\mathcal{R}^i F$ maps the module $\mathcal{M}$ to the nth cohomology module $H^n\big(F(\mathcal{I}), F(\psi)\big)$.

Obviously, this definition only makes sense, if the obtained functors are independent of the chosen (co)resolution. Thus let $\mathcal{P} \to \mathcal{M} \to 0$ and $\mathcal{P}' \to \mathcal{M} \to 0$ be two different projective resolutions of the same module $\mathcal{M}$. By Proposition B.2.31, a chain map $g : \mathcal{P} \to \mathcal{P}'$ exists satisfying $H_0(g) = \mathrm{id}_{\mathcal{M}}$. It induces another chain map $F(g) : F(\mathcal{P}) \to F(\mathcal{P}')$ which yields at the level of the homology a homomorphism $H\big(F(g)\big) : H_\bullet\big(F(\mathcal{P})\big) \to H_\bullet\big(F(\mathcal{P}')\big)$. If $g' : \mathcal{P} \to \mathcal{P}'$ is another chain map with the same property, then $g \sim g'$ and because of the additivity of F also $F(g) \sim F(g')$. By Lemma B.2.24, we have therefore $H_\bullet\big(F(g)\big) = H_\bullet\big(F(g')\big)$.

Reversing the role of the two resolutions $\mathcal{P}$ and $\mathcal{P}'$, we obtain a further chain map $h : \mathcal{P}' \to \mathcal{P}$ with $h \circ g \sim \mathrm{id}_{\mathcal{P}'}$ and $g \circ h \sim \mathrm{id}_{\mathcal{P}}$. But this fact implies that $H\big(F(g)\big) \circ H\big(F(h)\big)$ is the identity on $H_\bullet\big(F(\mathcal{P}')\big)$ and $H\big(F(h)\big) \circ H\big(F(g)\big)$ the identity on $H_\bullet\big(F(\mathcal{P})\big)$. Hence $H\big(F(g)\big)$ is a canonical isomorphism and the definition of a left derived functor is independent of the chosen resolution. A similar argument applies to injective coresolutions.

For contravariant functors we revert the role of the (co)resolutions: for a right exact functor we use an injective coresolution for defining its left derived functors and in the left exact case a projective resolution yields the right derived functors.

Definition B.2.36. Let $\mathcal{M}$, $\mathcal{N}$ be two $\mathcal{R}$-modules. The nth right derived functor for $\mathrm{Hom}_{\mathcal{R}}(\mathcal{M},\cdot)$ is denoted $\mathrm{Ext}^n_{\mathcal{R}}(\mathcal{M},\cdot)$ and for $\mathrm{Hom}_{\mathcal{R}}(\cdot,\mathcal{N})$ we write $\mathrm{Ext}^n_{\mathcal{R}}(\cdot,\mathcal{N})$. The Abelian groups $\mathrm{Ext}^n_{\mathcal{R}}(\mathcal{M},\mathcal{N})$ are called the *extension groups* for the two modules $\mathcal{M}$ and $\mathcal{N}$. The nth left derived functor for $\mathcal{M}\otimes_{\mathcal{R}}\cdot$ is denoted $\mathrm{Tor}^{\mathcal{R}}_n(\mathcal{M},\cdot)$ and similarly $\mathrm{Tor}^{\mathcal{R}}_n(\cdot,\mathcal{N})$ for $\cdot\otimes_{\mathcal{R}}\mathcal{N}$. The Abelian groups $\mathrm{Tor}^{\mathcal{R}}_n(\mathcal{M},\mathcal{N})$ are called the *torsion groups* for $\mathcal{M}$ and $\mathcal{N}$.

Of course, one must show now that these notations are meaningful, i. e. that it makes no difference whether we apply $\mathrm{Ext}^n_{\mathcal{R}}(\mathcal{M},\cdot)$ to $\mathcal{N}$ or $\mathrm{Ext}^n_{\mathcal{R}}(\cdot,\mathcal{N})$ to $\mathcal{M}$ (and similarly for $\mathrm{Tor}^{\mathcal{R}}_n(\mathcal{M},\cdot)$ and $\mathrm{Tor}^{\mathcal{R}}_n(\cdot,\mathcal{N})$, respectively). For the proof that this is indeed the case we refer to the literature. As the following result shows, these two functors measure to what extent modules are projective, injective or flat.

Proposition B.2.37. *Let $\mathcal{M}$, $\mathcal{N}$ be $\mathcal{R}$-modules.*

(i) $\mathrm{Ext}^0_{\mathcal{R}}(\mathcal{M},\mathcal{N})\cong\mathrm{Hom}_{\mathcal{R}}(\mathcal{M},\mathcal{N})$ *and* $\mathrm{Tor}^{\mathcal{R}}_0(\mathcal{M},\mathcal{N})\cong\mathcal{M}\otimes_{\mathcal{R}}\mathcal{N}$.
(ii) $\mathrm{Ext}^n_{\mathcal{R}}(\mathcal{M},\mathcal{N})=0$ *for all $n\geq 1$, if $\mathcal{M}$ is projective or $\mathcal{N}$ is injective.*
(iii) $\mathrm{Tor}^{\mathcal{R}}_n(\mathcal{M},\mathcal{N})=0$ *for all $n\geq 1$, if $\mathcal{M}$ is flat or $\mathcal{N}$ is projective.*

Remark B.2.38. For general $\mathcal{R}$-modules $\mathcal{M}$, $\mathcal{N}$ the extension groups $\mathrm{Ext}^n_{\mathcal{R}}(\mathcal{M},\mathcal{N})$ possess indeed only the structure of an Abelian group. However, for $\mathcal{N}=\mathcal{R}$ the functor $\mathrm{Hom}_{\mathcal{R}}(\cdot,\mathcal{R})$ corresponds to dualisation and as discussed in Remark B.1.7 for a left $\mathcal{R}$-module $\mathcal{M}$ the dual module is always a right $\mathcal{R}$-module. Hence in this case also all the extension groups $\mathrm{Ext}^n_{\mathcal{R}}(\mathcal{M},\mathcal{R})$ are right $\mathcal{R}$-modules. ◁

Remark B.2.39. Certain extension and torsion groups allow for an alternative characterisation of the depth of a module. For simplicity, we restrict to the polynomial ring $\mathcal{P}=\Bbbk[x^1,\dots,x^n]$ over a field $\Bbbk$. Given a $\mathcal{P}$-module $\mathcal{M}$, one can first show the existence of an isomorphism $\mathrm{Tor}^{\mathcal{P}}_i(\Bbbk,\mathcal{M})\cong\mathrm{Ext}^{n-i}_{\mathcal{P}}(\Bbbk,\mathcal{M})$ for any index $0\leq i\leq n$ (for all other values of i the groups always vanish). Then one can define $\operatorname{depth}\mathcal{M}$ as the smallest value i such that $\mathrm{Ext}^i_{\mathcal{P}}(\Bbbk,\mathcal{M})\neq 0$. ◁

B.3 Coalgebras and Comodules

Using an alternative approach to Definition B.1.11 of an algebra, we introduce the concept of a coalgebra. The polynomial and the exterior algebra, respectively, are examples of algebras that are simultaneously coalgebras; one speaks of *bialgebras*. As this structure is useful for a better description of the Spencer cohomology, we briefly recall here some basics. This material is only needed in Chapter 6; more details can be found in [59, 278, 232, 446].

Let $\mathcal{R}$ be a commutative ring and $\mathcal{A}$ an $\mathcal{R}$-module; we denote the action of $\mathcal{R}$ on $\mathcal{A}$ by $\rho:\mathcal{R}\otimes\mathcal{A}\to\mathcal{A}$. In order to make $\mathcal{A}$ to an associative $\mathcal{R}$-algebra, we need first a multiplication homomorphism $\mu:\mathcal{A}\otimes\mathcal{A}\to\mathcal{A}$ (all tensor products are over $\mathcal{R}$). Its associativity may be expressed by requiring that the diagram

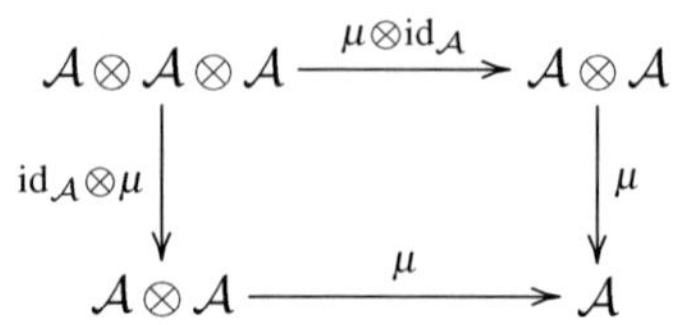

(B.44)

is commutative. The unit element $\mathbb{1}_{\mathcal{A}}$ of $\mathcal{A}$ may be defined with the help of an $\mathcal{R}$-linear map $\iota : \mathcal{R} \to \mathcal{A}$: we simply set $\mathbb{1}_{\mathcal{A}} = \iota(1)$. The usual properties of the unit are encoded in the commutativity of the diagram

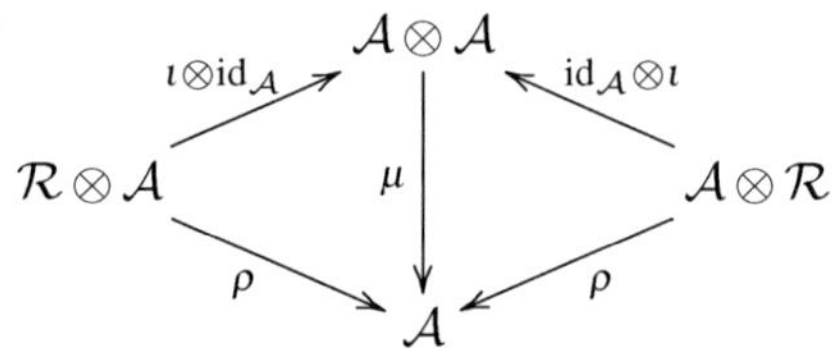

(B.45)

(we assume that the left and the right action of $\mathcal{R}$ on $\mathcal{A}$ are identical).

Definition B.3.1. A *(coassociative) coalgebra* over a commutative ring $\mathcal{R}$ is an $\mathcal{R}$-module $\mathcal{C}$ with a homomorphism $\Delta : \mathcal{C} \to \mathcal{C}\otimes\mathcal{C}$, the *coproduct*, and an $\mathcal{R}$-linear map $\varepsilon : \mathcal{C} \to \mathcal{R}$, the *counit*,[9] such that the diagrams

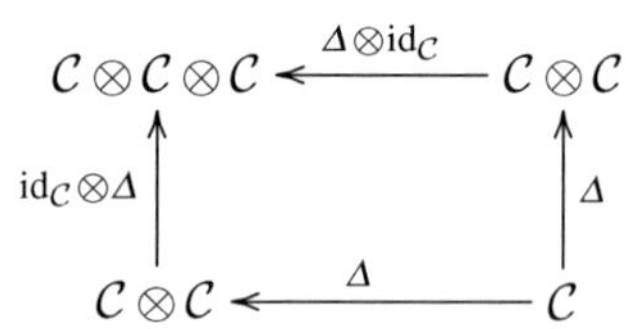

(B.46)

(expressing the coassociativity) and

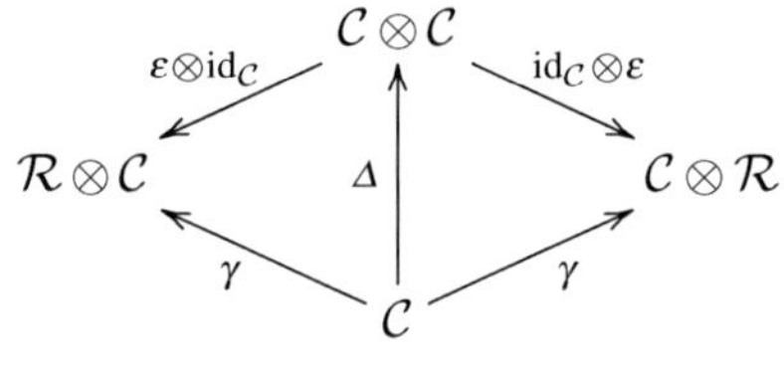

(B.47)

(here γ maps $c \in \mathcal{C}$ to $1\otimes c$ or $c\otimes 1$, respectively) commute.

The *Heyneman–Sweedler notation* is a handy short-form for explicit computations with coproducts. One writes $\Delta(c) = c_{(1)}\otimes c_{(2)}$. This must be understood as an abbreviation for a decomposition $\Delta(c) = \sum_{i=1}^{k} c_{1,i}\otimes c_{2,i}$. Thus the subscripts (1) and (2) should be read as indices which run in parallel and over which we sum. Note that the factors $c_{(1)}$ and $c_{(2)}$ are not uniquely defined, as for an element of $\mathcal{C}\otimes\mathcal{C}$ there are in general many different ways to decompose it into summands. Using this

[9] ε is sometimes also called *augmentation*.

notation, the coassociativity axiom (B.46) is equivalent to

$$c_{(1)(1)} \otimes c_{(1)(2)} \otimes c_{(2)} = c_{(1)} \otimes c_{(2)(1)} \otimes c_{(2)(2)} \tag{B.48a}$$

and—using the canonical isomorphism $\mathcal{R} \otimes \mathcal{C} \cong \mathcal{C}$—the counit relations (B.47) to

$$c_{(1)}\varepsilon(c_{(2)}) = c = \varepsilon(c_{(1)})c_{(2)} \ . \tag{B.48b}$$

Definition B.3.2. A *right comodule* over an $\mathcal{R}$-coalgebra $\mathcal{C}$ is a right $\mathcal{R}$-module $\mathcal{N}$ with a *coaction* $\rho : \mathcal{N} \to \mathcal{N} \otimes \mathcal{C}$ such that the two diagrams

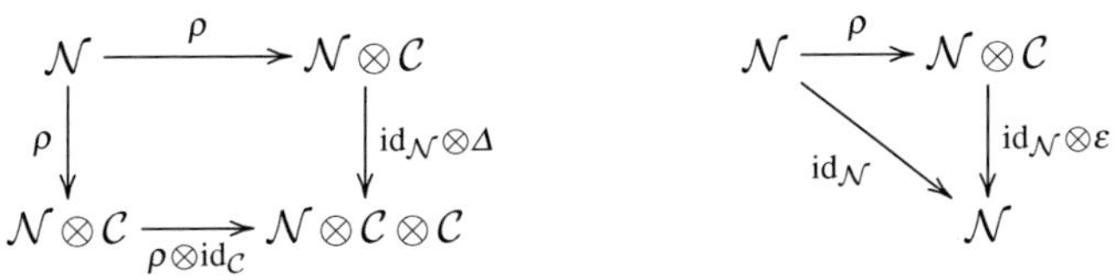

(B.49)

commute. The definition of left comodules goes analogously.

Using an extended Heyneman-Sweedler notation, we write the coaction concisely as $\rho(n) = n_{\langle 1 \rangle} \otimes n_{(2)}$. Then (B.49) may be expressed as

$$n_{\langle 1 \rangle} \otimes n_{(2)(1)} \otimes n_{(2)(2)} = n_{\langle 1 \rangle \langle 1 \rangle} \otimes n_{\langle 1 \rangle (2)} \otimes n_{(2)} \ , \tag{B.50a}$$

$$n_{\langle 1 \rangle}\varepsilon(n_{(2)}) = n \ . \tag{B.50b}$$

Any coalgebra $\mathcal{C}$ may be considered as a $\mathcal{C}$-comodule with the coproduct Δ as coaction. The *free comodule* $\mathcal{C}^m$ of rank m is defined by the coaction

$$\rho\left(\begin{pmatrix} c_1 \\ \vdots \\ c_m \end{pmatrix}\right) = \begin{pmatrix} (c_1)_{(1)} \\ \vdots \\ (c_m)_{(1)} \end{pmatrix} \otimes \Bigl((c_1)_{(2)} + \cdots + (c_m)_{(2)}\Bigr) \ . \tag{B.51}$$

Another special case is a *coideal* where the coaction is simply the coproduct Δ. Thus an additive subgroup $\mathcal{J} \subseteq \mathcal{C}$ is a (right) coideal, if $\Delta(\mathcal{J}) \subseteq \mathcal{J} \otimes \mathcal{C}$; the conditions (B.49) are trivially satisfied by the properties of a coproduct. More generally, a subset $\mathcal{L}$ of a comodule $\mathcal{N}$ is a right *subcomodule*, if $\rho(\mathcal{L}) \subseteq \mathcal{L} \otimes \mathcal{C}$.

The subcomodule $\mathcal{L} \subseteq \mathcal{N}$ *cogenerated* by a set $\mathcal{G} \subseteq \mathcal{N}$ is defined as the intersection of all subcomodules of $\mathcal{N}$ containing $\mathcal{G}$; we will write $\mathcal{L} = \rangle\mathcal{G}\langle$. If the underlying ring $\mathcal{R}$ is a field $\Bbbk$ and if we are dealing with finite-dimensional spaces, then we may describe the process of cogeneration as follows.

Let $\mathcal{V}$, $\mathcal{W}$ be two finite-dimensional $\Bbbk$-linear spaces with bases $\{v_1, \dots, v_k\}$ and $\{w_1, \dots, w_\ell\}$, respectively. Every element $u \in \mathcal{V} \otimes \mathcal{W}$ can be written in the form

$$u = \sum_{j=1}^{\ell} u_j \otimes w_j \ . \tag{B.52}$$

Here the vectors $u_j \in \mathcal{V}$ are uniquely determined and we introduce the vector subspace $L(u) = \langle u_1, \dots, u_\ell \rangle \subseteq \mathcal{V}$ (one could similarly introduce a linear subspace $R(u) \subseteq \mathcal{W}$ based on a decomposition $u = \sum_{i=1}^{k} v_i \otimes u_i$ with vectors $u_i \in \mathcal{W}$, but we will not need this). One easily sees that this construction is independent of the chosen bases. Given a whole subset $\mathcal{U} \subseteq \mathcal{V} \otimes \mathcal{W}$, we define $L(\mathcal{U})$ by summing $L(u)$ over all $u \in \mathcal{U}$.

Now let $\mathcal{N}$ be a comodule over $\mathcal{C}$ with coaction ρ. By assumption, $\mathcal{N}$ and $\mathcal{C}$ are also $\Bbbk$-linear spaces and we consider for any subset $\mathcal{U} \subseteq \mathcal{N}$ the space $\mathcal{L} = L\big(\rho(\mathcal{U})\big)$. It is not difficult to verify with the help of (B.50) that $\mathcal{L}$ is a subcomodule and in fact $\mathcal{L} = \rangle \mathcal{U} \langle$. In particular, if $\mathcal{U}$ is already a subcomodule, then $L\big(\rho(\mathcal{U})\big) = \mathcal{U}$.

An $\mathcal{R}$-linear map $f : \mathcal{N}_1 \to \mathcal{N}_2$ between two right $\mathcal{C}$-comodules $\mathcal{N}_1$, $\mathcal{N}_2$ with coactions ρ_1, ρ_2 is a *comodule morphism*, if the diagram

$$\begin{array}{ccc} \mathcal{N}_1 & \xrightarrow{\;f\;} & \mathcal{N}_2 \\ \big\downarrow{\scriptstyle \rho_1} & & \big\downarrow{\scriptstyle \rho_2} \\ \mathcal{N}_1 \otimes \mathcal{C} & \xrightarrow[f \otimes \mathrm{id}_{\mathcal{C}}]{} & \mathcal{N}_2 \otimes \mathcal{C} \end{array} \tag{B.53}$$

commutes, i. e. $f(n)_{\langle 1 \rangle} \otimes f(n)_{(2)} = f(n_{\langle 1 \rangle}) \otimes n_{(2)}$. The space $\mathrm{Hom}^{\mathcal{C}}(\mathcal{N}_1, \mathcal{N}_2)$ of all comodule morphisms between $\mathcal{N}_1$ and $\mathcal{N}_2$ has a natural $\mathcal{R}$-module structure via the exact sequence

$$0 \longrightarrow \mathrm{Hom}^{\mathcal{C}}(\mathcal{N}_1, \mathcal{N}_2) \longrightarrow \mathrm{Hom}_{\mathcal{R}}(\mathcal{N}_1, \mathcal{N}_2) \xrightarrow{\;\gamma\;} \mathrm{Hom}_{\mathcal{R}}(\mathcal{N}_1, \mathcal{N}_2 \otimes \mathcal{C}) \tag{B.54}$$

where $\gamma(\phi) = \rho_2 \circ \phi - (\phi \otimes \mathrm{id}_{\mathcal{C}}) \circ \rho_1$. Thus $\mathrm{Hom}^{\mathcal{C}}(\mathcal{N}, \cdot)$ (or $\mathrm{Hom}^{\mathcal{C}}(\cdot, \mathcal{N})$, respectively) defines a covariant (contravariant) functor from the category of right $\mathcal{C}$-modules into the category of $\mathcal{R}$-modules.

Projective and *injective* $\mathcal{C}$-comodules are defined as in the case of $\mathcal{R}$-modules by requiring the existence of fillers in the diagrams (B.19) (note that here no reversal of arrows occurs). Of course, all maps are now assumed to be comodule morphisms.

Lemma B.3.3. *Assume that the coalgebra $\mathcal{C}$ is flat as an $\mathcal{R}$-module. Then the functors* $\mathrm{Hom}^{\mathcal{C}}(\mathcal{N}, \cdot)$ *and* $\mathrm{Hom}^{\mathcal{C}}(\cdot, \mathcal{N})$ *are left exact.*

Proof. We treat only the case of $\mathrm{Hom}^{\mathcal{C}}(\cdot, \mathcal{N})$; the other case goes analogously. Let $\mathcal{N}_1 \to \mathcal{N}_2 \to \mathcal{N}_3 \to 0$ be an exact sequence of $\mathcal{C}$-comodules and consider the following diagram:

$$\begin{array}{ccccccc}
 & & 0 & & 0 & & 0 \\
 & & \downarrow & & \downarrow & & \downarrow \\
0 & \longrightarrow & \mathrm{Hom}^{\mathcal{C}}(\mathcal{N}_1,\mathcal{N}) & \longrightarrow & \mathrm{Hom}^{\mathcal{C}}(\mathcal{N}_2,\mathcal{N}) & \longrightarrow & \mathrm{Hom}^{\mathcal{C}}(\mathcal{N}_3,\mathcal{N}) \\
 & & \downarrow & & \downarrow & & \downarrow \\
0 & \longrightarrow & \mathrm{Hom}_{\mathcal{R}}(\mathcal{N}_1,\mathcal{N}) & \longrightarrow & \mathrm{Hom}_{\mathcal{R}}(\mathcal{N}_2,\mathcal{N}) & \longrightarrow & \mathrm{Hom}_{\mathcal{R}}(\mathcal{N}_3,\mathcal{N}) \\
 & & \downarrow & & \downarrow & & \downarrow \\
0 & \longrightarrow & \mathrm{Hom}_{\mathcal{R}}(\mathcal{N}_1,\mathcal{N}\otimes\mathcal{C}) & \longrightarrow & \mathrm{Hom}_{\mathcal{R}}(\mathcal{N}_2,\mathcal{N}\otimes\mathcal{C}) & \longrightarrow & \mathrm{Hom}_{\mathcal{R}}(\mathcal{N}_3,\mathcal{N}\otimes\mathcal{C})
\end{array} \tag{B.55}$$

All three columns are exact by (B.54) and the last two rows by Propositions B.2.9 and B.2.14. It is now a standard exercise in diagram chasing to show that the first row is exact, too. □

Corollary B.3.4. *Assume that the coalgebra $\mathcal{C}$ is flat as an $\mathcal{R}$-module. Then the $\mathcal{C}$-comodule $\mathcal{P}$ is projective, if and only if the functor* $\mathrm{Hom}^{\mathcal{C}}(\mathcal{P},\cdot)$ *is exact. Similarly, the $\mathcal{C}$-comodule $\mathcal{I}$ is injective, if and only if the functor* $\mathrm{Hom}^{\mathcal{C}}(\cdot,\mathcal{I})$ *is exact.*

Remark B.3.5. In the case of $\mathcal{R}$-modules we have seen that free modules are trivially projective but not necessarily injective (Remark B.2.12). For comodules it is just the other way round: if $\mathcal{R}$ is a field, then the free comodule $\mathcal{C}^m$ is injective (but not necessarily projective). For simplicity, we consider only $m = 1$; the extension to the general case is straightforward.

Let $\mathcal{N}$ be a right $\mathcal{C}$-comodule. Then any comodule morphism $f \in \mathrm{Hom}^{\mathcal{C}}(\mathcal{N},\mathcal{C})$ induces a module morphism $\tilde{f} \in \mathrm{Hom}_{\mathcal{R}}(\mathcal{N},\mathcal{R})$ by setting $\tilde{f} = \varepsilon \circ f$. Conversely, given a module morphism $\tilde{f} \in \mathrm{Hom}_{\mathcal{R}}(\mathcal{N},\mathcal{R})$, we may construct a comodule morphism $f \in \mathrm{Hom}^{\mathcal{C}}(\mathcal{N},\mathcal{C})$ by setting $f = (\tilde{f} \otimes \mathrm{id}_{\mathcal{C}}) \circ \Delta$. We claim that these two maps are inverse to each other. Indeed, we find by the $\mathcal{R}$-linearity of the involved maps and by (B.50b) that

$$\varepsilon\big(\tilde{f}(n_{\langle 1\rangle})n_{(2)}\big) = \tilde{f}(n_{\langle 1\rangle})\varepsilon(n_{(2)}) = \tilde{f}\big(n_{\langle 1\rangle}\varepsilon(n_{(2)})\big) = \tilde{f}(n)\,. \tag{B.56}$$

Similarly, it follows from the defining property of a comodule morphism and (B.48b) that

$$\varepsilon\big(f(n_{\langle 1\rangle})\big)n_{(2)} = \varepsilon\big(f(n)_{(1)}\big)f(n)_{(2)} = f(n)\,. \tag{B.57}$$

Thus we found a natural isomorphism between the functors $\mathrm{Hom}^{\mathcal{C}}(\cdot,\mathcal{C})$ and $\mathrm{Hom}_{\mathcal{R}}(\cdot,\mathcal{R})$. Since we know from Proposition B.2.9 that under our assumptions the latter one is exact, the former one must be exact, too, and we are done. ◁

The linear dual $\mathcal{C}^* = \mathrm{Hom}_{\mathcal{R}}(\mathcal{C},\mathcal{R})$ of a coalgebra $\mathcal{C}$ possesses a natural algebra structure via the *convolution product* $\star$. It is defined for arbitrary elements $\phi,\psi \in \mathcal{C}^*$ by the condition that the relation

$$\langle \phi \star \psi, c\rangle = \big\langle \phi \otimes \psi, \Delta(c)\big\rangle = \langle \phi, c_{(1)}\rangle\langle \psi, c_{(2)}\rangle \tag{B.58}$$

holds for all $c \in \mathcal{C}$. The unit element of $\mathcal{C}^*$ is simply the counit ε. If $\mathcal{N}$ is a (right) $\mathcal{C}$-comodule with coaction ρ, then its dual space $\mathcal{N}^*$ is naturally a right $\mathcal{C}^*$-module with the action $\rho^* : \mathcal{N}^* \otimes \mathcal{C}^* \to \mathcal{N}^*$ defined in similar manner by requiring that the relation

$$\big\langle \rho^*(\nu,\psi), n\big\rangle = \big\langle \nu \otimes \psi, \rho(n)\big\rangle = \langle \nu, n_{\langle 1\rangle}\rangle\langle \psi, n_{(2)}\rangle \tag{B.59}$$

holds for all $\nu \in \mathcal{N}^*$, $\psi \in \mathcal{C}^*$ and $n \in \mathcal{N}$.

For arbitrary subsets $\mathcal{L} \subseteq \mathcal{N}$ we define in the usual manner the *annihilator* $\mathcal{L}^0 = \{\nu \in \mathcal{N}^* \mid \nu(\ell) = 0 \forall \ell \in \mathcal{L}\} \subseteq \mathcal{N}^*$. Similarly, we introduce for any subset $\mathcal{L}^* \subseteq \mathcal{N}^*$ the annihilator $(\mathcal{L}^*)^0 = \{n \in \mathcal{N} \mid \lambda(n) = 0 \forall \lambda \in \mathcal{L}^*\} \subseteq \mathcal{N}$. The subset $\mathcal{L} \subseteq \mathcal{N}$ is *closed*, if $\mathcal{L} = (\mathcal{L}^0)^0$, and similarly for subsets $\mathcal{L}^* \subseteq \mathcal{N}^*$.

Proposition B.3.6. *Let $\mathcal{N}$ be a $\mathcal{C}$-comodule. If $\mathcal{L} \subseteq \mathcal{N}$ is a , then $\mathcal{L}^0 \subseteq \mathcal{N}^*$ is a submodule. Conversely, if the subset $\mathcal{L}^* \subseteq \mathcal{N}^*$ is a submodule, then $(\mathcal{L}^*)^0 \subseteq \mathcal{N}$ is a subcomodule. Thus let $\mathcal{L} \subseteq \mathcal{N}$ be a closed subset. Then $\mathcal{L}$ is a subcomodule, if and only if $\mathcal{L}^0$ is a submodule.*

Proof. We first note that by definition of the dual action

$$\big\langle \rho^*(\lambda,\psi), \lambda\big\rangle = \langle \lambda, \ell_{\langle 1\rangle}\rangle\langle \psi, \ell_{(2)}\rangle \tag{B.60}$$

for arbitrary $\ell \in \mathcal{N}$, $\lambda \in \mathcal{N}^*$ and $\psi \in \mathcal{C}^*$. Now consider this equation for $\lambda \in \mathcal{L}^0$ and $\ell \in \mathcal{L}$. If $\mathcal{L}$ is a subcomodule, then $\ell_{\langle 1\rangle} \in \mathcal{L}$ and we always obtain zero, as the right hand side vanishes. But this fact implies that $\rho^*(\lambda,\psi) \in \mathcal{L}^0$ for arbitrary ψ and hence $\mathcal{L}^0$ is a submodule.

Conversely, let $\mathcal{L}^*$ be a submodule and consider the above equation for $\lambda \in \mathcal{L}^*$ and $\ell \in (\mathcal{L}^*)^0$. We obtain again zero, as now the left hand side always vanishes. Without loss of generality, we may assume that the elements $\ell_{(2)}$ are linearly independent. Then by making appropriate choices for ψ, we find that $\langle \lambda, \ell_{\langle 1\rangle}\rangle = 0$ and hence $\ell_{\langle 1\rangle} \in (\mathcal{L}^*)^0$, so that we have a subcomodule. □

$\mathcal{C}$ is a *graded coalgebra*, if it is a graded $\mathcal{R}$-module $\mathcal{C} = \bigoplus_{k\in\mathbb{Z}} \mathcal{C}_k$ and the coproduct satisfies $\Delta(\mathcal{C}_k) \subseteq \sum_{r+s=k} \mathcal{C}_r \otimes \mathcal{C}_s$. Similarly, we speak of a *graded comodule* $\mathcal{N} = \bigoplus_{k\in\mathbb{Z}} \mathcal{N}_k$, if the coaction satisfies $\rho(\mathcal{N}_k) \subseteq \sum_{r+s=k} \mathcal{N}_r \otimes \mathcal{C}_s$. In such cases, we also consider the *graded dual* $\mathcal{N}^* = \bigoplus_{k\in\mathbb{Z}} \mathcal{N}_k^*$.

If $\mathcal{A}$ is an $\mathcal{R}$-algebra, we may define for a left $\mathcal{A}$-module $\mathcal{M}_L$ with left action η_L and a right $\mathcal{A}$-module $\mathcal{M}_R$ with right action η_R the *tensor product* over $\mathcal{A}$ by the exact sequence of Abelian groups

$$\mathcal{M}_R \otimes_{\mathcal{R}} \mathcal{A} \otimes_{\mathcal{R}} \mathcal{M}_L \xrightarrow{\ \sigma\ } \mathcal{M}_R \otimes_{\mathcal{R}} \mathcal{M}_L \longrightarrow \mathcal{M}_R \otimes_{\mathcal{A}} \mathcal{M}_L \longrightarrow 0 \tag{B.61}$$

where $\sigma = \eta_R \otimes \mathrm{id}_{\mathcal{M}_L} - \mathrm{id}_{\mathcal{M}_R} \otimes \eta_L$. As already noted in Appendix B.1, in general $\mathcal{M}_R \otimes_{\mathcal{A}} \mathcal{M}_L$ is only an Abelian group and no longer an $\mathcal{A}$-module.

Reverting the arrows in the above sequence (B.61) leads to the dual notion of the *cotensor product* over a coalgebra $\mathcal{C}$. Let $\mathcal{N}_L$ be a left $\mathcal{C}$-comodule with left coaction ρ_L and $\mathcal{N}_R$ a right $\mathcal{C}$-comodule with right coaction ρ_R. Then their cotensor product $\mathcal{N}_R \boxtimes_{\mathcal{C}} \mathcal{N}_L$ is defined by the exact sequence

$$\mathcal{N}_R \otimes_{\mathcal{R}} \mathcal{C} \otimes_{\mathcal{R}} \mathcal{N}_L \xleftarrow{\omega} \mathcal{N}_R \otimes_{\mathcal{R}} \mathcal{N}_L \longleftarrow \mathcal{N}_R \boxtimes_{\mathcal{C}} \mathcal{N}_L \longleftarrow 0 \tag{B.62}$$

where now $\omega = \rho_R \otimes \mathrm{id}_{\mathcal{N}_L} - \mathrm{id}_{\mathcal{N}_R} \otimes \rho_L$. Again we obtain in general only an Abelian group and not a $\mathcal{C}$-comodule with the trivial but important exceptions $\mathcal{N}_R \boxtimes_{\mathcal{C}} \mathcal{C} \cong \mathcal{N}_R$ and $\mathcal{C} \boxtimes_{\mathcal{C}} \mathcal{N}_L \cong \mathcal{N}_L$.

In close analogy to Definition B.2.36 of the torsion groups of two $\mathcal{R}$-modules, Eilenberg and Moore [123] introduced for two $\mathcal{C}$-comodules *cotorsion groups*. If $\mathcal{M}$, $\mathcal{N}$ are $\mathcal{C}$-comodules, then their cotorsion groups arise via the right derived functors of the functor $\cdot \boxtimes_{\mathcal{C}} \mathcal{N}$ (or $\mathcal{M} \boxtimes_{\mathcal{C}} \cdot$, respectively) defined by applying it to any injective coresolution of $\mathcal{M}$ and taking the cohomology. We use the notation $\mathrm{Cotor}^n_{\mathcal{C}}(\mathcal{M},\mathcal{N})$ for the nth cotorsion group.

If an algebra $\mathcal{A}$ may also be equipped with a coalgebra structure, then it is a *bialgebra* provided that Δ and ε are algebra morphisms (here $\mathcal{A} \otimes \mathcal{A}$ is given the tensor product structure $(a_1 \otimes b_1)(a_2 \otimes b_2) = a_1 a_2 \otimes b_1 b_2$). A particular class of bialgebras are *Hopf algebras*. They possess as additional structure an *antipode*: an endomorphism $S : \mathcal{A} \to \mathcal{A}$ such that $S(a_{(1)})a_{(2)} = \varepsilon(a) 1_{\mathcal{A}} = a_{(1)} S(a_{(2)})$.

Let $\mathcal{V}$ be a finite-dimensional vector space over a field $\Bbbk$ (all constructions remain valid for an $\mathcal{R}$-module $\mathcal{M}$, but for our purposes vector spaces are sufficient). Then we may introduce the *tensor algebra* $T\mathcal{V}$ by setting $T_0\mathcal{V} = \Bbbk$, $T_q\mathcal{V} = \mathcal{V} \otimes \cdots \otimes \mathcal{V}$ the q-fold product of $\mathcal{V}$ with itself and $T\mathcal{V} = \bigoplus_{q=0}^{\infty} T_q\mathcal{V}$. The product is given by

$$(v_1 \otimes \cdots \otimes v_q) \otimes (w_1 \otimes \cdots \otimes w_r) = v_1 \otimes \cdots \otimes v_q \otimes w_1 \otimes \cdots \otimes w_r \tag{B.63}$$

and makes $T\mathcal{V}$ to a graded algebra. Its unit is the $\Bbbk$-linear map $\iota : \Bbbk \to T\mathcal{V}$ defined by $\iota(1) = 1$.

$T\mathcal{V}$ is a Hopf algebra. Its coalgebra structure is given by the coproduct

$$\Delta(v_1 \otimes \cdots \otimes v_q) = \sum_{i=0}^{q} (v_1 \otimes \cdots \otimes v_i) \otimes (v_{i+1} \otimes \cdots \otimes v_q) \,. \tag{B.64}$$

and the counit $\varepsilon : T\mathcal{V} \to \Bbbk$ which is the identity on $T_0\mathcal{V}$ and zero everywhere else. In order to distinguish between these two structures, we will denote the tensor coalgebra by $\mathfrak{T}\mathcal{V}$. The antipode is defined by $S(v_1 \otimes \cdots \otimes v_q) = (-1)^q v_q \otimes \cdots \otimes v_1$

We introduce in the usual manner the *symmetric algebra* $S\mathcal{V}$ as the factor algebra of $T\mathcal{V}$ by the ideal generated by all differences $v \otimes w - w \otimes v$ with $v, w \in \mathcal{V}$. Thus it is commutative and the product is denoted by a dot or no symbol at all. The *exterior algebra* $E\mathcal{V}$ (we will also often write $\Lambda\mathcal{V}$ instead of $E\mathcal{V}$) arises similarly by factoring with respect to the ideal generated by all "squares" $v \otimes v$ with $v \in \mathcal{V}$. Its product is the wedge product $\wedge$ (see also Appendix C.2).

Both $S\mathcal{V}$ and $E\mathcal{V}$ inherit the Hopf algebra structure from $T\mathcal{V}$. The counit is always the augmentation ε mapping to the constant term. In both cases the coproduct is given by $\Delta(v) = v \otimes 1 + 1 \otimes v$ for $v \in \mathcal{V}$. For the *symmetric coalgebra* $\mathfrak{S}\mathcal{V}$ it is extended by the rule $\Delta(fg) = \Delta(f)\Delta(g)$ and similarly for the exterior coalgebra $\mathfrak{E}\mathcal{V}$ by $\Delta(f \wedge g) = \Delta(f) \wedge \Delta(g)$.

If $\{x^1,\dots,x^n\}$ is a basis of $\mathcal{V}$, then we may use as basis of the symmetric coalgebra $\mathfrak{S}\mathcal{V}$ all monomials x^μ with a multi index $\mu \in \mathbb{N}_0^n$ providing the well-known isomorphy of $S\mathcal{V}$ with the polynomial algebra $\Bbbk[x^1,\dots,x^n]$. A basis of the *exterior coalgebra* $\mathfrak{E}\mathcal{V}$ is given by all monomials x^I with a repeated index $I = (i_1,\dots,i_q)$ subject to the condition $i_1 < i_2 < \cdots < i_q$. In these bases the coproduct of $\mathfrak{S}\mathcal{V}$ is given by "Taylor expansion"

$$\Delta(f) = \sum_{\mu\in\mathbb{N}_0^n} \frac{1}{\mu!}\frac{\partial^{|\mu|}f}{\partial x^\mu} \otimes x^\mu \tag{B.65}$$

for any polynomial $f \in \Bbbk[x^1,\dots,x^n]$ and in the exterior coalgebra $\mathfrak{E}\mathcal{V}$ we find (see Appendix A.1 for the notations)

$$\Delta(x^I) = \sum_{J\cup K=I} \operatorname{sgn}(J\cup K)x^J \otimes x^K\,. \tag{B.66}$$

Consider the dual algebra $(\mathfrak{S}\mathcal{V})^*$. Denoting the dual basis to the terms x^μ by $\gamma_\mu(\mathbf{x})$, we obtain for the convolution product according to the remarks above

$$\gamma_\mu(\mathbf{x}) \star \gamma_\nu(\mathbf{x}) = \prod_{i=1}^n \binom{\mu_i+\nu_i}{\mu_i}\gamma_{\mu+\nu}(\mathbf{x})\,. \tag{B.67}$$

Introducing new "variables" $y^i = \gamma_{1_i}(\mathbf{x})$ (i. e. the dual basis to $\{x^1,\dots,x^n\}$ in $\mathcal{V}^*$), we find after a trivial computation $\gamma_\mu(\mathbf{x}) = \frac{y^\mu}{\mu!}$, i. e. the dual bases consists of *divided powers*, and hence $y^\mu \star y^\nu = y^{\mu+\nu}$. Thus the algebra $(\mathfrak{S}\mathcal{V})^*$ is isomorphic to $S(\mathcal{V}^*)$.

By definition, an additive subgroup $\mathcal{J} \subset \mathfrak{S}\mathcal{V}$ is a coideal, if and only if it satisfies $\Delta(\mathcal{J}) \subseteq \mathcal{J}\otimes\mathcal{C}$ which, by (B.65), is equivalent to the condition $\partial^{|\mu|}f/\partial x^\mu \in \mathcal{J}$ for all $f \in \mathcal{J}$. Similarly, a subset $\mathcal{N} \subseteq (\mathfrak{S}\mathcal{V})^m$ is a subcomodule, if and only if this condition holds in each component.

Let $\mathcal{F} \subset \mathfrak{S}_q\mathcal{V}$ be a finite set of homogeneous polynomials of degree q. We are interested in the homogeneous coideal $\mathcal{J} = \rangle\mathcal{F}\langle$ cogenerated by $\mathcal{F}$. Obviously, we must take for $\mathcal{J}_q$ the $\Bbbk$-linear span of $\mathcal{F}$. In a given basis $\{x^1,\dots,x^n\}$ of $\mathcal{V}$, we set for $0 < r \leq q$

$$\mathcal{J}_{q-r} = \left\{\frac{\partial^{|\mu|}f}{\partial x^\mu} \mid f \in \mathcal{J}_q,\ \mu \in \mathbb{N}_0^n,\ |\mu| = r\right\}. \tag{B.68}$$

It is easy to see that $\mathcal{J} = \bigoplus_{r=0}^q \mathcal{J}_r$ satisfies $\Delta(\mathcal{J}) \subseteq \mathcal{J}\otimes\mathcal{C}$ and that it is the smallest subset of $\mathfrak{S}\mathcal{V}$ containing $\mathcal{F}$ with this property. Note that, in contrast to the algebra case, we obtain components of lower degree and $\mathcal{J}$ is finite-dimensional as vector space. Again the extension to $(\mathfrak{S}\mathcal{V})^m$ is trivial using (B.68) componentwise.

B.4 Gröbner Bases for Polynomial Ideals and Modules

Throughout this section, we assume that $\mathcal{P} = \Bbbk[x^1,\dots,x^n]$ is a polynomial ring over a field $\Bbbk$ and develop the theory of Gröbner bases for ideals in $\mathcal{P}$. Gröbner bases can also be defined for polynomials over coefficient rings, but this generalisation leads to additional complications which we ignore here (see, however, Section 4.6).

Let $\mathcal{F} = \{f_1,\dots,f_r\}$ be a finite set of polynomials; it generates the following ideal in the ring $\mathcal{P}$:

$$\langle\mathcal{F}\rangle = \Bigl\{h \in \mathcal{P} \mid h = \sum_{i=1}^{r} g_i f_i,\ g_i \in \mathcal{P}\Bigr\}\,. \tag{B.69}$$

If $\mathcal{I} = \langle\mathcal{F}\rangle$ for some ideal $\mathcal{I} \subseteq \mathcal{P}$, then $\mathcal{F}$ is called a *basis*[10] of $\mathcal{I}$. By Hilbert's Basis Theorem B.1.13, any ideal in $\mathcal{P}$ has a finite basis (which is of course not unique), as any field is trivially a Noetherian ring. The theory of Gröbner bases is concerned with the choice of bases that are particularly useful because of their special properties not shared by arbitrary bases.

Recall that a monomial ideal is an ideal in $\mathcal{P}$ that can be generated by terms (or equivalently by monomials since we work over a field); the theory of such ideals is to a large extent of a combinatorial nature. An order $\prec$ (see Appendix A.1) associates with any ideal $\mathcal{I} \subseteq \mathcal{P}$ a monomial ideal, its *leading ideal* $\operatorname{lt}_\prec \mathcal{I}$ defined by

$$\operatorname{lt}_\prec \mathcal{I} = \langle \operatorname{lt}_\prec f \mid f \in \mathcal{I}\rangle \tag{B.70}$$

(note that, in contrast to Definition A.1.5 of $\operatorname{lt}_\prec \mathcal{F}$ for arbitrary sets $\mathcal{F} \subseteq \mathcal{P}$, we define $\operatorname{lt}_\prec \mathcal{I}$ not as the set $\{\operatorname{lt}_\prec f \mid f \in \mathcal{I}\}$ but as the ideal generated by it; this slight inconsistency should not lead to confusion).

Definition B.4.1. Let $\mathcal{I} \subseteq \mathcal{P}$ be a non-zero ideal. A finite set $\mathcal{G} \subset \mathcal{I}$ is a *Gröbner basis* of $\mathcal{I}$ for the term order $\prec$, if $\operatorname{lt}_\prec \mathcal{I} = \langle \operatorname{lt}_\prec \mathcal{G}\rangle$.

Remark B.4.2. Buchberger introduced the terminology "Gröbner basis" (using a different but equivalent definition) in his Ph.D. thesis [60] in order to honour his supervisor Gröbner. Some authors prefer to speak of *standard bases* (others reserve this term for Gröbner bases with respect to semigroup orders which are not monoid orders—see Section 4.5). Buchberger's thesis marks the official beginning of the history of Gröbner bases, but some of the basic ideas have appeared already much earlier. The allegedly first Gröbner basis can be found in an article by Gordan [172] dating from 1900. Term orders and leading ideals feature prominently in the work of Macaulay [299] on Hilbert functions. Around the same time, Janet [233] and Riquier [381] applied similar ideas in the context of differential equations (cf. the Notes at the end of Chapter 3); later Gröbner picked up this idea and used it repeatedly, in particular also for the analysis of differential equations [187]. Gröbner bases in a

[10] Obviously, every polynomial ideal is a module over the polynomial ring. One should note that strictly speaking these "ideal bases" are only generating sets and not bases, as only principal ideals are free modules. But this abuse of language has become standard in commutative algebra.

power series ring with respect to a semigroup order were applied both by Grauert [179] for the construction of versal deformation spaces and by Hironaka [218] in his famous work on the resolution of singularities of varieties (using the terminology "standard bases"). Both considered the corresponding division algorithm (see below) as a multivariate generalisation of the classical Weierstraß Division Theorem of complex analysis. ◁

Example B.4.3. Consider the set $\mathcal{F} = \{f_1 = \underline{xy} - y, f_2 = \underline{y^2} - x\} \subset \Bbbk[x,y]$ and let $\prec$ be an arbitrary degree compatible order. The underlining marks the leading terms of the elements of $\mathcal{F}$. If we set $\mathcal{I} = \langle \mathcal{F} \rangle$; then $\mathcal{F}$ is *not* a Gröbner basis of $\mathcal{I}$ with respect to $\prec$: the polynomial $f_3 = \underline{x^2} - x = y \cdot f_1 + (1-x) \cdot f_2$ is obviously an element of $\mathcal{I}$, but its leading term x^2 is not contained in the monomial ideal $\langle \mathrm{lt}_\prec \mathcal{F} \rangle = \langle xy, y^2 \rangle$ and thus $\langle \mathrm{lt}_\prec \mathcal{F} \rangle \subsetneq \mathrm{lt}_\prec \mathcal{I}$. We have not yet the tools to prove it, but the augmented set $\mathcal{G} = \{f_1, f_2, f_3\}$ is a Gröbner basis of $\mathcal{I}$ with respect to $\prec$. ◁

Note that Definition B.4.1 does not require $\mathcal{I} = \langle \mathcal{G} \rangle$; but we will later see that this is always the case. Obviously, a set $\mathcal{G}$ may well be a Gröbner basis of an ideal for one term order but not for another one; the chosen term order has a great influence on the basis. We sometimes simply say that a set $\mathcal{G}$ is a Gröbner basis meaning that it is a Gröbner basis of the ideal $\langle \mathcal{G} \rangle$ for some term order $\prec$ which should be obvious from the context.

Example B.4.4. Gröbner bases can be extremely sensitive to changes of the used term order $\prec$. We demonstrate this problem with the help of the simple ideal $\mathcal{I} = \langle z^5 + y^4 + x^3 - 1, z^3 + y^2 + x^2 - 1 \rangle \subset \mathbb{Q}[x,y,z]$. The (reduced) Gröbner basis of $\mathcal{I}$ with respect to the degree reverse lexicographic order consists of the following four polynomials:

$$\begin{aligned}
g_1 &= z^3 + y^2 + x^2 - 1\,, \\
g_2 &= z^2y^2 - y^4 + z^2x^2 - x^3 - z^2 + 1\,, \\
g_3 &= zy^4 + y^4 + 2y^2x^2 + zx^3 + x^4 - 2y^2 - 2x^2 - z + 1\,, \\
g_4 &= y^6 - y^4x^2 + z^2x^4 + 2zy^2x^2 + z^2x^3 + y^2x^3 + zx^4 - x^5 - 2z^2x^2 - \\
&\qquad 2y^2x^2 - zx^3 - x^4 - 2zy^2 - 2zx^2 + x^3 + y^2 + 3x^2 + 2z - 2\,.
\end{aligned} \tag{B.71}$$

Using the lexicographic order, we obtain for the same ideal a (reduced) basis comprising the following seven polynomials:

$$
\begin{aligned}
\bar{g}_1 &= y^{12} - y^{10} + 3y^8x^3 - 5y^8x^2 + 2y^8 - 10y^6x^4 + 20y^6x^2 - 10y^6 - \\
&\quad 7y^4x^6 + 30y^4x^4 - 6y^4x^3 - 30y^4x^2 + 13y^4 - 5y^2x^8 + 20y^2x^6 - \\
&\quad 30y^2x^4 + 20y^2x^2 - 5y^2 - x^{10} + x^9 + 5x^8 - 13x^6 + 10x^4 + 3x^3 - 5x^2\,, \\
\bar{g}_2 &= zx^{11} + 4zx^{10} + zx^9 - 10zx^8 - 4zx^7 + 8zx^6 - 3y^{10}x^3 - 2y^{10}x^2 + \\
&\quad 4y^{10}x + 4y^{10} + y^8x^5 - 2y^8x^4 - 2y^8x^3 - 6y^6x^6 + 10y^6x^5 + 20y^6x^4 - \\
&\quad 16y^6x^3 - 24y^6x^2 + 8y^6x + 8y^6 + 2y^4x^8 + 21y^4x^7 + 17y^4x^6 - \\
&\quad 78y^4x^5 - 64y^4x^4 + 90y^4x^3 + 76y^4x^2 - 32y^4x - 32y^4 + 17y^2x^9 + \\
&\quad 24y^2x^8 - 68y^2x^7 - 80y^2x^6 + 106y^2x^5 + 108y^2x^4 - 77y^2x^3 - \\
&\quad 70y^2x^2 + 20y^2x + 20y^2 + 7x^{11} + 7x^{10} - 31x^9 - 42x^8 + 55x^7 + \\
&\quad 77x^6 - 39x^5 - 62x^4 + 8x^3 + 20x^2\,, \\
\bar{g}_3 &= 24zy^2x - 24zy^2 - 3zx^{10} - 7zx^9 + 12zx^8 + 21zx^7 - 23zx^6 + 12zx^4 - \\
&\quad 12zx^3 - 24zx + 24z + 9y^{10}x^2 - 9y^{10}x - 7y^{10} - 3y^8x^4 + 11y^8x^3 - \\
&\quad 9y^8x^2 + 6y^8x - 6y^8 + 18y^6x^5 - 60y^6x^4 + 20y^6x^3 + 62y^6x^2 - \\
&\quad 24y^6x - 8y^6 - 6y^4x^7 - 53y^4x^6 + 44y^4x^5 + 226y^4x^4 - 165y^4x^3 - \\
&\quad 174y^4x^2 + 90y^4x + 38y^4 - 51y^2x^8 + 13y^2x^7 + 239y^2x^6 - 118y^2x^5 - \\
&\quad 302y^2x^4 + 170y^2x^3 + 129y^2x^2 - 87y^2x + 7y^2 - 21x^{10} + 14x^9 + \\
&\quad 93x^8 - 22x^7 - 206x^6 + 68x^5 + 127x^4 - 36x^3 - 17x^2 + 24x - 24\,, \\
\bar{g}_4 &= zy^4 + zx^3 - z + y^4 + 2y^2x^2 - 2y^2 + x^4 - 2x^2 + 1\,, \\
\bar{g}_5 &= z^2x^4 + z^2x^3 - 2z^2x^2 + zy^4 + 2zy^2x^2 - 2zy^2 + zx^4 - 2zx^2 + \\
&\quad z + y^6 - y^4x^2 + y^4 + y^2x^3 - y^2 - x^5 + x^3 + x^2 - 1\,, \\
\bar{g}_6 &= z^2y^2 + z^2x^2 - z^2 - y^4 - x^3 + 1\,, \\
\bar{g}_7 &= z^3 + y^2 + x^2 - 1\,.
\end{aligned}
\tag{B.72}
$$

Obviously, the second basis is much larger. It does not only contain almost the double number of generators; the generators are much more complicated (more terms and larger coefficients) and of higher degree (12 compared to 6). It is a general experience that Gröbner bases with respect to the lexicographic term order tend to be much larger than bases with respect to the degree reverse lexicographic order. In fact, the latter ones are usually the smallest and if one is free to choose an arbitrary order, one generally prefers it. ◁

While Definition B.4.1 is probably the simplest and most direct way to define a Gröbner basis, it does not make obvious why Gröbner bases are so important or how one can check whether a given set $\mathcal{G}$ is a Gröbner basis. In order to answer such questions we need a few additional concepts. The existence of Gröbner bases is a simple by-product of many proofs of Hilbert's Basis Theorem (see for example our proof of Theorem 3.3.13).

Definition B.4.5. For a given term order $\prec$, the polynomial $g \in \mathcal{P}$ is *reducible* with respect to the finite subset $\mathcal{F} \subset \mathcal{P}$, if there exists a term $t \in \operatorname{supp} g$ and a polynomial $f \in \mathcal{F}$ such that $\operatorname{lt}_{\prec} f \mid t$. A polynomial $g' \in \mathcal{P}$ that is not reducible with respect to $\mathcal{F}$ and such that $g' - g \in \langle \mathcal{F} \rangle$ is called a *normal form* of g with respect to $\mathcal{F}$.

Although it is important to note that in general a polynomial $g \in \mathcal{P}$ does not possess a unique normal form with respect to some set $\mathcal{F} \subset \mathcal{P}$, we will use the notation $g' = \mathrm{NF}_{\mathcal{F},\prec}(g)$. Algorithm B.1 allows us to determine a normal form. If g is not in normal form with respect to $\mathcal{F}$, we take the largest term $t \in \operatorname{supp} g$ divisible by the leading term of an element $f \in \mathcal{F}$ and its coefficient $c_t \in \Bbbk$. Then we compute the difference $g' = g - \big(c_t t / \operatorname{lm}_{\prec} f\big) f$; obviously, this operation eliminates the monomial $c_t t$ in g. If g' is still reducible, we iterate. As after each step the largest divisible term in $\operatorname{supp} g'$ becomes smaller and $\prec$ is a well-order by Lemma A.1.6, the algorithm terminates after a finite number of steps. Different normal forms may be obtained, if in Line /5/ several elements $f \in \mathcal{F}$ with $\operatorname{lt}_{\prec} f \mid t$ exist.

Algorithm B.1 Normal form

Input: finite set $\mathcal{F} \subset \mathcal{P}$, polynomial $g \in \mathcal{P}$, term order $\prec$
Output: a normal form $g' = \mathrm{NF}_{\mathcal{F},\prec}(g)$

```
1: S ← supp g ∩ ⟨lt≺ F⟩
2: g' ← g
3: while S ≠ ∅ do
4:     t ← max≺ S
5:     choose f ∈ F such that lt≺ f | t
6:     g' ← g' − (c_t t / lm≺ f) f
7:     S ← supp g' ∩ ⟨lt≺ F⟩
8: end while
9: return g'
```

Example B.4.6. We continue with the set $\mathcal{F} = \{f_1, f_2\}$ of Example B.4.3 and a degree compatible term order. If we reduce the polynomial $f_3 = \underline{xy^2} - x$ first with respect to f_2, we obtain the normal form $\underline{x^2} - x$. But if we start by reducing f_3 with respect to the first generator f_1, then the normal form 0 arises after one further reduction step. ◁

Remark B.4.7. Closely related to the normal form algorithm is the *division algorithm* used in many textbooks for the introduction of Gröbner bases. By a slight modification of Algorithm B.1, namely recording the individual reduction steps, one can determine a representation $g = \sum_{f \in \mathcal{F}} P_f f + g'$ with coefficients $P_f \in \mathcal{P}$. In Line /2/ we initialise $P_f \leftarrow 0$ for every $f \in \mathcal{F}$; in Line /6/ we compute in addition $P_f \leftarrow P_f + c_t t / \operatorname{lm}_{\prec} f$ for the f chosen in the previous line. This algorithm may be considered as a multivariate generalisation of the familiar univariate polynomial division and g' is sometimes called the *remainder* of the reduction. By construction, the "quotients" P_f satisfy $\operatorname{lt}_{\prec}(P_f f) \preceq \operatorname{lt}_{\prec} g$.

Although this chapter considers exclusively polynomials over a field $\Bbbk$, we briefly comment on the complications which arise, if the coefficients come from a ring $\mathcal{R}$ (this situation is studied in more detail in Section 4.6). In this case the simple fact that in Line /5/ the leading term $\mathrm{lt}_{\prec} f$ divides t does not suffice to guarantee that a reduction of g' is possible, as generally neither $\mathrm{lc}_{\prec} f$ divides the coefficient of t nor it is a unit in $\mathcal{R}$. One solution is to perform a *pseudo-division*: we simply multiply g' by some ring element $r' \in \mathcal{R}$ such that $r'g'$ can be reduced (the simplest choice would be $r' = \mathrm{lc}_{\prec} f$). Of course, the output g' is then only a *pseudo-remainder*:, i. e. we only know that $rg = \sum_{f\in\mathcal{F}} P_f f + g'$ for some $r \in \mathcal{R}$ where the "quotients" P_f satisfy the same estimate as above. ◁

Proposition B.4.8. *Let $\mathcal{I} \subseteq \mathcal{P}$ be a non-zero ideal and $\mathcal{G} \subset \mathcal{P}$ a finite set. Then the following statements are equivalent.*

(i) *The set $\mathcal{G}$ is a Gröbner basis of $\mathcal{I}$ for the term order $\prec$.*
(ii) *$\mathcal{I} = \langle \mathcal{G} \rangle$ and every polynomial $f \in \mathcal{P}$ possesses a unique normal form with respect to the set $\mathcal{G}$.*
(iii) *Every polynomial $f \in \mathcal{I}$ in the ideal can be written in the form*

$$f = \sum_{g\in\mathcal{G}} P_g g \tag{B.73}$$

where the polynomial coefficients $P_g \in \mathcal{P}$ satisfy $\mathrm{lt}_{\prec}(P_g g) \preceq \mathrm{lt}_{\prec} f$ for all generators $g \in \mathcal{G}$ such that $P_g \neq 0$.

Proof. Assume that $\mathcal{G}$ is a Gröbner basis of $\mathcal{I}$ with respect to the term order $\prec$. Let f_1 and f_2 be two normal forms of an arbitrary polynomial $f \in \mathcal{P}$; then obviously $f_1 - f_2 \in \mathcal{I}$. The leading term of $f_1 - f_2$ is an element of $\operatorname{supp} f_1 \cup \operatorname{supp} f_2$. By definition of a normal form this set cannot contain any term that is divisible by the leading term of an element of $\mathcal{G}$. On the other hand, by definition of a Gröbner basis, $\mathrm{lt}_{\prec}(f_1 - f_2)$ must be divisible by the leading term of an element of $\mathcal{G}$. This contradiction can only be resolved, if $f_1 - f_2 = 0$ and thus (i) implies (ii).

It is a bit more involved to show that (iii) follows from (ii); therefore we only sketch the proof. The key is to show that if f' is a normal form of $f \in \mathcal{P}$ with respect to $\mathcal{G}$, then it is also a normal form of any polynomial of the form $f - hg$ with $g \in \mathcal{G}$. This observation trivially implies that 0 is a normal form with respect to $\mathcal{G}$ whenever $f \in \mathcal{I}$. Since by assumption this normal form is unique, we may compute it with Algorithm B.1 in its extended form, i. e. the division algorithm mentioned in Remark B.4.7, which yields a representation of the form (B.73). As already remarked above, its coefficients P_g always satisfy the required condition.

Finally, we assume that (iii) holds. The condition on the coefficients obviously implies that for any $f \in \mathcal{I}$ the leading term $\mathrm{lt}_{\prec} f$ is divisible by $\mathrm{lt}_{\prec} g$ for some $g \in \mathcal{G}$. Thus $\mathrm{lt}_{\prec} \mathcal{I} = \langle \mathrm{lt}_{\prec} \mathcal{G} \rangle$ and consequently (i) is true. □

Example B.4.9. By the definition of an ideal basis, it is obvious that all elements of $\mathcal{I}$ can be written in the form (B.73)—compare with (B.69). The particular property of a Gröbner basis is the additional condition on the leading terms of the coefficients P_g; if it is satisfied, then the right hand side of (B.73) is called a *standard representation* of the polynomial $f \in \mathcal{P}$. It is important to note that it is *not* unique.

While the proposition above asserts that all possible normal form algorithms return the same result, they may achieve it in different ways and thus the associated division algorithms yield different standard representations. Consider again the Gröbner basis $\mathcal{G} = \{f_1, f_2, f_3\}$ of Example B.4.3. The polynomial $f = xy^2 - x \in \langle \mathcal{G} \rangle$ has two different standard representations: $f = yf_1 + f_2 = xf_2 + f_3$. ◁

Remark B.4.10. Proposition B.4.8 implies that any Gröbner basis $\mathcal{G}$ of an ideal $\mathcal{I}$ provides us with an effective method to choose unique representatives for the cosets in $\mathcal{P}/\mathcal{I}$, as all polynomials in a coset $[f]$ have the same normal form with respect to $\mathcal{G}$. More precisely: the normal form defines a vector space isomorphism between the algebra $\mathcal{P}/\mathcal{I}$ and the vector space generated by all terms *not* contained in the leading ideal $\mathrm{lt}_{\prec} \mathcal{I} = \langle \mathrm{lt}_{\prec} \mathcal{G} \rangle$. The latter ones are also called *standard terms*. This basic fact was already observed by Macaulay [299] (for this reason it is sometimes called *Macaulay's Theorem*). He concluded that every ideal $\mathcal{I}$ has the same Hilbert function as the *monomial* ideal $\mathrm{lt}_{\prec} \mathcal{I}$.[11]

An effective means to perform arithmetics with cosets was the original motivation for Buchberger to introduce Gröbner bases in his thesis [60]. Proposition B.4.8 also allows us to decide algorithmically the *ideal membership problem*: given a Gröbner basis $\mathcal{G}$, we have $f \in \mathcal{I}$, if and only if $\mathrm{NF}_{\mathcal{G},\prec}(f) = 0$. Using elimination orders, we can also effectively eliminate variables [459]. If $\prec$ is an elimination order for the variables $x^{k+1}, \dots, x^n$ and $\mathcal{G}$ is a Gröbner basis of the ideal $\mathcal{I}$, then $\mathcal{G} \cap \Bbbk[x^1, \dots, x^k]$ is a Gröbner basis of $\mathcal{I} \cap \Bbbk[x^1, \dots, x^k]$. ◁

Definition B.4.11. Let $\mathcal{G}$ be a Gröbner basis with respect to the term order $\prec$. The basis $\mathcal{G}$ is *minimal*, if it does not contain two distinct elements $g, g' \in \mathcal{G}$ such that $\mathrm{lt}_{\prec} g \mid \mathrm{lt}_{\prec} g'$. The basis $\mathcal{G}$ is called *reduced*, if every element $g \in \mathcal{G}$ is in normal form with respect to the set $\mathcal{G} \setminus \{g\}$ and if $\mathrm{lc}_{\prec} g = 1$ for all $g \in \mathcal{G}$.

Using a terminology introduced in Remark B.4.10, we may say that in a reduced Gröbner basis the support of any generator consists besides the leading term only of standard terms. Obviously, any reduced Gröbner basis is minimal. The terminology "minimal basis" refers to the following observation. A general ideal has many minimal bases, but every *monomial* ideal possesses a unique minimal *monomial* basis. If $\mathcal{G}$ is a minimal Gröbner basis of $\mathcal{I}$ with respect to the term order $\prec$, then the set $\mathrm{lt}_{\prec} \mathcal{G}$ is the unique minimal monomial basis of the leading ideal $\mathrm{lt}_{\prec} \mathcal{I}$.

Given an arbitrary Gröbner basis $\mathcal{G}$ of some ideal $\mathcal{I} \subseteq \mathcal{P}$, it is trivial to extract a minimal one from it: simply drop every generator $g' \in \mathcal{G}$ for which a different generator $g \in \mathcal{G}$ exists with $\mathrm{lt}_{\prec} g \mid \mathrm{lt}_{\prec} g'$. Obviously, the leading terms of the remaining elements of $\mathcal{G}$ still span $\mathrm{lt}_{\prec} \mathcal{I}$ so that we still have a Gröbner basis. In order to obtain a reduced Gröbner basis, we must perform a process known as *autoreduction* (see Algorithm B.2).

The correctness of Algorithm B.2 is trivial. For the termination we note that in each iteration of the `while` loop either a polynomial $h \in \mathcal{H}$ is eliminated or it is replaced by another polynomial $\bar{h}$ such that the maximal element of $\operatorname{supp} h \setminus \operatorname{supp} \bar{h}$ is

[11] In the case of the *affine* Hilbert function, i. e. for non-homogeneous ideals, this equality holds only for degree compatible term orders.

Algorithm B.2 Autoreduction of a set of polynomials

Input: a finite set $\mathcal{F} \subset \mathcal{P}$, a term order $\prec$
Output: an autoreduced set $\mathcal{H} \subset \mathcal{P}$ with $\langle \mathcal{H} \rangle = \langle \mathcal{F} \rangle$

$\mathcal{H} \leftarrow \mathcal{F}$
while $\exists h \in \mathcal{H} : \bar{h} = \texttt{NormalForm}\big(h, \mathcal{H} \setminus \{h\}\big) \neq h$ **do**
  $\mathcal{H} \leftarrow \mathcal{H} \setminus \{h\}$
  if $\bar{h} \neq 0$ **then**
    $\mathcal{H} \leftarrow \mathcal{H} \cup \{\bar{h}\}$
  end if
end while
return $\mathcal{H}$

greater than the maximal element of $\operatorname{supp} \bar{h} \setminus \operatorname{supp} h$ (more precisely, a term $t \in \operatorname{supp} h$ has been replaced by some terms which are all smaller than t). As any term order is a well-order by Lemma A.1.6, this observation implies that the loop can be iterated only a finite number of times.

It is now obvious that any polynomial ideal $\mathcal{I}$ has a reduced Gröbner basis: we just take some Gröbner basis of $\mathcal{I}$, apply Algorithm B.2 to it and finally divide every generator by its leading coefficient. In fact, the thus obtained basis is even unique (for the given term order).

Theorem B.4.12. *Let $\mathcal{I} \subseteq \mathcal{P}$ be an arbitrary non-zero ideal. Then $\mathcal{I}$ possesses a unique reduced Gröbner basis for any term order $\prec$.*

Proof. Let $\mathcal{G}$ and $\bar{\mathcal{G}}$ be two different reduced Gröbner bases of the ideal $\mathcal{I}$ for the same term order $\prec$. It follows from the remarks above that $\operatorname{lt}_\prec \mathcal{G} = \operatorname{lt}_\prec \bar{\mathcal{G}}$. If $\mathcal{G} \neq \bar{\mathcal{G}}$, there must exist two elements $g \in \mathcal{G}$ and $\bar{g} \in \bar{\mathcal{G}}$ such that $\operatorname{lt}_\prec g = \operatorname{lt}_\prec \bar{g}$ but $g \neq \bar{g}$. Obviously, $g - \bar{g} \in \mathcal{I}$ and thus its normal form with respect to both $\mathcal{G}$ and $\bar{\mathcal{G}}$ must vanish by Proposition B.4.8. But $\operatorname{supp}(g - g')$ consists only of standard terms and hence $g - \bar{g}$ is not reducible with respect to either basis implying that $g - \bar{g} = 0$ in contradiction to our assumptions. □

All the characterisations of Gröbner bases given so far are not very constructive, as they do not tell us how to find a Gröbner basis for a given ideal. This will change now with the help of the *S-polynomial*[12] $S_\prec(f,g)$ of two polynomials f,g. It is defined as the linear combination

$$S_\prec(f,g) = \frac{\operatorname{lcm}(\operatorname{lt}_\prec f, \operatorname{lt}_\prec g)}{\operatorname{lm}_\prec f} \cdot f - \frac{\operatorname{lcm}(\operatorname{lt}_\prec f, \operatorname{lt}_\prec g)}{\operatorname{lm}_\prec g} \cdot g \tag{B.74}$$

where lcm denotes as usual the least common multiple. Note that the factors are chosen such that the leading monomials cancel in the subtraction.

We saw above that for arbitrary sets $\mathcal{F}$ the normal form with respect to $\mathcal{F}$ is not unique. The S-polynomials account for this ambiguity. Assume that during the determination of a normal form of the polynomial h a monomial cx^μ is reducible with

[12] The S stands for *syzygy*; see below.

respect to two different elements $f_1, f_2 \in \mathcal{F}$ (i. e. the term $x^\nu = \operatorname{lcm}(\operatorname{lt}_\prec f_1, \operatorname{lt}_\prec f_2)$ divides x^μ). Then the reduction of h with respect to f_i yields $h_i = h - \frac{cx^\mu}{\operatorname{lm}_\prec f_i} f_i$ and for a suitably chosen constant $d \in \Bbbk$ the difference between the two reductions is $h_1 - h_2 = dx^{\mu-\nu} S_\prec(f_2, f_1)$. We can generalise this observation as follows.

Lemma B.4.13. *Let $f_1, \dots, f_r$ be polynomials such that $\operatorname{le}_\prec f_i = \mu$ for all $1 \leq i \leq r$. Let furthermore the coefficients $a_i \in \Bbbk$ be such that the polynomial $f = \sum_{i=1}^r a_i f_i$ satisfies $\operatorname{le}_\prec f \prec \mu$, i. e. the leading monomials cancel during the summation. Then coefficients $b_{ij} \in \Bbbk$ exist such that $f = \sum_{i<j} b_{ij} S_\prec(f_i, f_j)$.*

Proof. If $\operatorname{lc}_\prec f_i = c_i$, then our assumptions imply that $\sum_{i=1}^r a_i c_i = 0$. In addition, we have $S_\prec(f_i, f_j) = \frac{f_i}{c_i} - \frac{f_j}{c_j}$. This fact allows us to calculate

$$\begin{aligned} f &= a_1 c_1 \frac{f_1}{c_1} + \cdots + a_r c_r \frac{f_r}{c_r} \\ &= a_1 c_1 \left(\frac{f_1}{c_1} - \frac{f_2}{c_2} \right) + (a_1 c_1 + a_2 c_2) \left(\frac{f_2}{c_2} - \frac{f_3}{c_3} \right) + \cdots + \\ &\quad (a_1 c_1 + a_2 c_2 + \cdots + a_{r-1} c_{r-1}) \left(\frac{f_{r-1}}{c_{r-1}} - \frac{f_r}{c_r} \right) + \\ &\quad (a_1 c_1 + \cdots + a_r c_r) \frac{f_r}{c_r} . \end{aligned} \tag{B.75}$$

The last term cancels because of our choice of the coefficients a_i and in all other terms S-polynomials $S_\prec(f_i, f_j)$ appear. □

With the help of this technical result, we can derive the following finite characterisation of Gröbner bases which is the key to their actual computation and which represents one of the central results of Buchberger's thesis.

Theorem B.4.14 (Buchberger). *A finite set $\mathcal{G} \subset \mathcal{P}$ is a Gröbner basis of the ideal $\mathcal{I} = \langle \mathcal{G} \rangle$, if and only if for every pair $f, g \in \mathcal{G}$ the S-polynomial $S_\prec(f, g)$ reduces to zero with respect to $\mathcal{G}$, i. e. $\operatorname{NF}_{\mathcal{G},\prec}(S_\prec(f, g)) = 0$.*

Proof. One direction is trivial: if $\mathcal{G}$ is a Gröbner basis, then every S-polynomial reduces to zero by Part (ii) of Proposition B.4.8, as it is contained in $\mathcal{I}$.

For the converse, we use the characterisation of Gröbner bases via standard representations (Part (iii) of Proposition B.4.8). Let $\mathcal{G} = \{g_1, \dots, g_r\}$. Then every polynomial $f \in \langle \mathcal{G} \rangle$ can be written in the form $f = \sum_{i=1}^r h_i g_i$. This representation is not unique, but we choose one such that the term

$$x^\mu = \max_\prec \left\{ \operatorname{lt}_\prec (h_i g_i) \mid 1 \leq i \leq r \right\} \tag{B.76}$$

is minimal among all such representations (this conditions makes sense, as $\prec$ is a well-order). If $x^\mu = \operatorname{lt}_\prec f$, we found a standard representation for f and are done.

Otherwise $x^\mu \succ \operatorname{lt}_\prec f$ and we set $\mathcal{S} = \{ i \mid \operatorname{lt}_\prec (h_i g_i) = x^\mu \}$. If $\operatorname{lm}_\prec h_i = c_i x^{\nu^{(i)}}$, then we consider the polynomial $\bar{f} = \sum_{i \in \mathcal{S}} c_i x^{\nu^{(i)}} g_i$. By Lemma B.4.13, it possesses a representation

$$\bar{f} = \sum_{\substack{i,j \in \mathcal{S} \\ i<j}} b_{ij} S_{\prec}(x^{\nu^{(i)}} g_i, x^{\nu^{(j)}} g_j) \tag{B.77}$$

with coefficients $b_{ij} \in \Bbbk$. It is straightforward to verify that $S_{\prec}(x^{\nu^{(i)}} g_i, x^{\nu^{(j)}} g_j)$ is a multiple of $S_{\prec}(g_i, g_j)$. Hence, by assumption, all these S-polynomials reduce to zero modulo $\mathcal{G}$ and we obtain representations

$$S_{\prec}(x^{\nu^{(i)}} g_i, x^{\nu^{(j)}} g_j) = \sum_{k=1}^{r} h_{ijk} g_k \tag{B.78}$$

where $\max_{\prec}\{\mathrm{lt}_{\prec}(h_{ijk}g_k) \mid 1 \leq k \leq r\} = \mathrm{lt}_{\prec} S_{\prec}(x^{\nu^{(i)}} g_i, x^{\nu^{(j)}} g_j) \prec \mu$. Entering these representations into $\bar{f}$ and subsequently into f leads to a representation for f of the form $f = \sum_{i=1}^{r} \bar{h}_i g_i$ with $\max_{\prec}\{\mathrm{lt}_{\prec} \bar{h}_i g_i \mid 1 \leq i \leq r\} \prec x^{\mu}$ which contradicts the assumed minimality of x^{μ}. □

This theorem lies at the heart of the constructive theory of Gröbner bases and forms the foundation of the *Buchberger Algorithm B.3*. Given a finite set $\mathcal{G} \subset \mathcal{P}$, the algorithm first forms the set $\mathcal{S}$ of all pairs $\{g_1, g_2\}$ of different elements of $\mathcal{G}$. Then it takes one pair and computes the corresponding S-polynomial $S_{\prec}(g_1, g_2)$. If its normal form $\bar{g}$ with respect to $\mathcal{G}$ vanishes, the next pair is chosen. Otherwise $\bar{g}$ is added to $\mathcal{G}$ and we start anew. The algorithm stops with a Gröbner basis, when all S-polynomials reduce to zero.[13]

Algorithm B.3 Gröbner basis (Buchberger)

Input: finite set $\mathcal{F} \subset \mathcal{P}$, term order $\prec$
Output: Gröbner basis $\mathcal{G}$ of $\langle \mathcal{F} \rangle$

$\mathcal{G} \leftarrow \mathcal{F}$
$\mathcal{S} \leftarrow \{\{g_1, g_2\} \mid g_1, g_2 \in \mathcal{G}, g_1 \neq g_2\}$
while $\mathcal{S} \neq \emptyset$ **do**
  choose $\{g_1, g_2\} \in \mathcal{S}$
  $\mathcal{S} \leftarrow \mathcal{S} \setminus \{\{g_1, g_2\}\}$; $\bar{g} \leftarrow \mathrm{NF}_{\mathcal{G},\prec}(S_{\prec}(g_1, g_2))$
  if $\bar{g} \neq 0$ **then**
    $\mathcal{S} \leftarrow \mathcal{S} \cup \{\{\bar{g}, g\} \mid g \in \mathcal{G}\}$; $\mathcal{G} \leftarrow \mathcal{G} \cup \{\bar{g}\}$
  end if
end while
return $\mathcal{G}$

Theorem B.4.15. *Algorithm B.3 terminates for any finite set $\mathcal{F} \subset \mathcal{P}$ and any term order $\prec$ with a Gröbner basis of the ideal $\langle \mathcal{F} \rangle$.*

Proof. The main point is to show the termination of the algorithm. If it terminates, the outcome is clearly a Gröbner basis of the ideal $\langle \mathcal{F} \rangle$: we only add elements of

[13] Normal form computations may also be considered in the more abstract framework of term rewriting and critical pair completion. Buchberger's algorithm is then an instance of the famous *Knuth–Bendix algorithm* [255]. More on this relation can be found in [482].

this ideal so that the span does not change and the termination criterion is that all S-polynomials reduce to zero which by Theorem B.4.14 characterises Gröbner bases.

The termination follows from a Noetherian argument. The algorithm determines a sequence $\mathcal{G}_1 \subsetneq \mathcal{G}_2 \subsetneq \mathcal{G}_3 \subsetneq \cdots$ of finite sets. Consider the new element $\bar{g}$ added in Line /7/ to $\mathcal{G}_i$ in order to obtain $\mathcal{G}_{i+1}$. It results from a normal form computation with respect to $\mathcal{G}_i$ in Line /5/. This fact implies that $\mathrm{lt}_\prec \bar{g} \notin \langle \mathrm{lt}_\prec \mathcal{G}_i \rangle$ and we obtain a strictly ascending chain of monomial ideals: $\langle \mathrm{lt}_\prec \mathcal{G}_1 \rangle \subsetneq \langle \mathrm{lt}_\prec \mathcal{G}_2 \rangle \subsetneq \langle \mathrm{lt}_\prec \mathcal{G}_3 \rangle \subsetneq \cdots$. By Dickson's Lemma A.1.2, it is not possible that such a chain becomes infinite. Hence after a finite number of steps we must have $\mathcal{G}_i = \mathcal{G}_{i+1}$ and the algorithm terminates with a Gröbner basis of the ideal $\langle \mathcal{F} \rangle$. □

Example B.4.16. We apply Buchberger's Algorithm B.3 to the set $\mathcal{F} = \{f_1, f_2\}$ of Example B.4.3 for some degree compatible term order $\prec$. The S-polynomial of the two generators is $S_\prec(f_1, f_2) = x^2 - y^2$ and a normal form of it with respect to $\mathcal{F}$ is given by $f_3 = \underline{x^2} - x$. As one can easily check by explicitly computing the normal forms of the two new S-polynomials $S_\prec(f_1, f_3)$ and $S_\prec(f_2, f_3)$, the augmented set $\mathcal{G} = \{f_1, f_2, f_3\}$ is a Gröbner basis (even a reduced one) of the ideal $\langle \mathcal{F} \rangle$ for any degree compatible term order. ◁

Remark B.4.17. In general, Algorithm B.3 does not determine a reduced Gröbner basis. The simplest possibility to obtain at least a minimal basis is to eliminate in the output of Algorithm B.3 all elements $g \in \mathcal{G}$ for which another generator $g' \in \mathcal{G}$ exists with $\mathrm{lt}_\prec g' \mid \mathrm{lt}_\prec g$. As such eliminations do not affect the ideal $\langle \mathrm{lt}_\prec \mathcal{G} \rangle$, the result is still a Gröbner basis. Another possibility is to perform an autoreduction with Algorithm B.2 whenever a new generator is added to $\mathcal{G}$ (this modification may also be useful from an efficiency point of view). ◁

Note that it is *not* recommended to use Algorithm B.3 in the described form for actual computations! The above presentation is designed to make the simple underlying idea fairly transparent. However, it is not efficient and usable only for rather modest examples. Considerable optimisations are necessary to obtain an algorithm which is applicable in practice.

The choice of the pair $\{g_1, g_2\}$ in Line /4/ is important for the efficency. A number of strategies have been developed for it. The classical *normal selection strategy* chooses that pair for which the term $\mathrm{lcm}(\mathrm{lt}_\prec g_1, \mathrm{lt}_\prec g_2)$ is minimal with respect to the used term order $\prec$. It works very well for degree compatible term orders. Another popular approach is the "sugar cube" strategy [166] which is based on a simulated homogenisation of the input. In contrast to the normal selection strategy, this approach works also well for term orders like the purely lexicographic one.

Another crucial observation is that the algorithm spends most of its time computing normal forms. Buchberger found two criteria for predicting that certain S-polynomials reduce to zero. Their use often leads to drastic gains in efficiency, as the corresponding pairs $\{g_1, g_2\}$ can be thrown away without any computation.

Proposition B.4.18 (Buchberger's first criterion). *Let $g_1, g_2 \in \mathcal{P}$ be two polynomials with $\gcd(\mathrm{lt}_\prec g_1, \mathrm{lt}_\prec g_2) = 1$. Then $\mathrm{NF}_{\{g_1,g_2\},\prec}(S_\prec(g_1, g_2)) = 0$, i. e. the S-polynomial $S_\prec(g_1, g_2)$ reduces to zero modulo the set $\{g_1, g_2\}$.*

Proposition B.4.19 (Buchberger's second criterion). *Let $\mathcal{G} \subset \mathcal{P}$ be a finite set and $g_1, g_2, h \in \mathcal{G}$ such that $\operatorname{lt}_\prec h \mid \operatorname{lcm}(\operatorname{lt}_\prec g_1, \operatorname{lt}_\prec g_2)$. Then it suffices to treat in Algorithm B.3 the S-polynomials $S_\prec(g_1, h)$ and $S_\prec(g_2, h)$; the S-polynomial $S_\prec(g_1, g_2)$ may be ignored.*

Various ways exist to incorporate these criteria in Algorithm B.3, but we omit here a discussion. The proof of the first criterion is not very difficult; it follows from a direct computation. The second one is based on syzygy theory and we will briefly mention the underlying theory below.

Optimisation and implementation of the Buchberger algorithm (and the design of alternative algorithms for the construction of Gröbner bases) is still an active research topic in computer algebra and we must refer to the literature (cited in the above mentioned textbooks). The complexity of computing Gröbner bases and deciding the ideal membership problem was thoroughly analysed by Mayr and collaborators in a number of publications; a survey with many references is given in [317]. A by now classical result is that the ideal membership problem for polynomials with rational coefficients is in the worst case exponential in space [316, 318]. The same holds for determining a reduced Gröbner basis.

One should stress that this complexity result is less a question of designing clever algorithms but ideals exist where the *size* of the Gröbner basis is so huge, as Mayr and Meyer [318] demonstrated with an explicit example. Thus simply storing the basis requires already exponential space. At first sight, this observation seems to imply that it is hopeless to compute Gröbner bases. But one should always keep in mind that all these results concern the *worst case* complexity. There are some indications that most examples appearing in applications have a considerably lower complexity, as they are of a more geometric origin whereas all the known worst case examples are of a combinatorial nature. A more extensive discussion of this and related issues can be found in the survey article [33].

Remark B.4.20. Probably the simplest measure for estimating the complexity of a Gröbner basis computation is provided by the Castelnuovo–Mumford regularity $\operatorname{reg}\mathcal{I}$ of the ideal $\mathcal{I}$ defined by the basis (see Section 5.5 for its definition). Bayer and Stillman [34] showed that *generically* the Gröbner basis with respect to the degree reverse lexicographic term order has degree $\operatorname{reg}\mathcal{I}$. As this is typically the smallest Gröbner basis for a given ideal, we may consider generically $\operatorname{reg}\mathcal{I}$ as a lower bound for the degree of any Gröbner basis of $\mathcal{I}$. Of course, it is trivial to produce counterexamples to this observation: simply consider a monomial ideal; any monomial basis of it is a Gröbner basis, but in general its regularity is higher than the degree of its minimal basis. ◁

Gröbner bases may be defined for submodules of free polynomial modules in the same manner as for ideals; one only needs a bit more notation. We represent the elements of the free module $\mathcal{P}^m$ as vectors $\mathbf{f} = (f_1, \dots, f_m)$ with $f_\alpha \in \mathcal{P}$. The unit vectors $\{\mathbf{e}_1, \dots, \mathbf{e}_m\}$ with $e_{\alpha\beta} = \delta_{\alpha\beta}$ form the standard basis of $\mathcal{P}^m$; thus we may also write $\mathbf{f} = f_1\mathbf{e}_1 + \cdots + f_m\mathbf{e}_m$.

First, we must extend the concept of a term order. A term $\mathbf{t}$ is now a vector of the form $\mathbf{t} = t\mathbf{e}_\alpha$ for some α and with $t \in \mathbb{T}$ a term in $\mathcal{P}$. We denote the set of all terms by $\mathbb{T}^m$; it is a monoid module over $\mathbb{T}$. By a slight abuse of notation, we will sometimes "divide" terms: if $\mathbf{t} = r\mathbf{s}$, then we set $\mathbf{t}/\mathbf{s} = r\mathbf{e}_\alpha$; thus the result is again a vector. Terms belonging to different unit vectors cannot be divided. A *(module) term order* $\prec$ on $\mathbb{T}^m$ must satisfy the following two conditions: if $\mathbf{s}, \mathbf{t} \in \mathbb{T}^m$ are two arbitrary terms, then (i) $\mathbf{t} \preceq r\mathbf{t}$ and (ii) $\mathbf{s} \prec \mathbf{t}$ implies $r\mathbf{s} \prec r\mathbf{t}$ for all $r \in \mathbb{T}$. Obviously, these are simple generalisations of the conditions for a term order on $\mathbb{T}$ and assert that $\prec$ is compatible with the structure of $\mathbb{T}^m$ as a monoid module over $\mathbb{T}$.

In practice one uses on $\mathbb{T}^m$ mostly term orders derived from orders on $\mathbb{T}$. Two natural ways to "lift" an order are *TOP* and *POT*. These abbreviations stand for *Term Over Position* and *Position Over Term*, respectively. Let $\prec$ be a term order on $\mathbb{T}$. Then $s\mathbf{e}_\alpha \prec_{TOP} t\mathbf{e}_\beta$, if either $s \prec t$ or $s = t$ and $\alpha < \beta$, i.e. we rank the comparison of the terms s, t over the comparison of their position in the vector. We have $s\mathbf{e}_\alpha \prec_{POT} t\mathbf{e}_\beta$, if either $\alpha < \beta$ or $\alpha = \beta$ and $s \prec t$, i.e. the position in the vector is more important than the terms. It is easy to show that both $\prec_{TOP}$ and $\prec_{POT}$ are term orders on $\mathbb{T}^m$.

The notion of a class respecting term order is straightforwardly extended to orders on $\mathbb{T}^m$. In a similar manner to the proof of Lemma A.1.8 one shows that on terms of the same degree any class respecting term order is a *TOP* lift of the degree reverse lexicographic order for an arbitrary labelling of the unit vectors $\mathbf{e}_\alpha$.

Example B.4.21. A particular type of term orders for the module case will become important below. Let $\mathcal{F} = \{\mathbf{f}_1, \dots, \mathbf{f}_s\}$ be a subset of $\mathcal{P}^m$ and $\prec$ an arbitrary term order on $\mathbb{T}^m$. We define a new term order $\prec_\mathcal{F}$ on $\mathbb{T}^s$ by setting $s\mathbf{e}_\sigma \prec_\mathcal{F} t\mathbf{e}_\tau$ if either $\mathrm{lt}_\prec(s\mathbf{f}_\sigma) \prec \mathrm{lt}_\prec(t\mathbf{f}_\tau)$ or $\mathrm{lt}_\prec(s\mathbf{f}_\sigma) = \mathrm{lt}_\prec(t\mathbf{f}_\tau)$ and $\tau < \sigma$ (note that we reverse the order of the indices). One easily verifies that for any set $\mathcal{F}$ (not necessarily a Gröbner basis) $\prec_\mathcal{F}$ is a term order on $\mathbb{T}^s$.

If the underlying order $\prec$ is class respecting, then the induced order $\prec_\mathcal{F}$ is class respecting in a weighted sense: we must set $\deg(t\mathbf{e}_\tau) = \deg t + \deg \mathbf{f}_\tau$. If all elements $\mathbf{f}_\tau \in \mathcal{F}$ have the same degree, this observation implies obviously that we are dealing with a standard *TOP* lift of the degree reverse lexicographic order. $\triangleleft$

Given this definition of term orders on $\mathbb{T}^m$, we can straightforwardly extend all the notions introduced in the last section to submodules $\mathcal{M} \subseteq \mathcal{P}^m$. The leading term, coefficient, and monomial are defined as before; again we simply write $\mathrm{lt}_\prec \mathcal{M}$ for the monomial submodule generated by the leading terms of the elements of $\mathcal{M}$. Now we can easily generalise Definition B.4.1.

Definition B.4.22. Let $\mathcal{M}$ be a submodule of the module $\mathcal{P}^m$. A finite set $\mathcal{G} \subset \mathcal{M}$ is a *Gröbner basis* of $\mathcal{M}$ for a given term order $\prec$, if $\mathrm{lt}_\prec \mathcal{M} = \langle \mathrm{lt}_\prec \mathcal{G} \rangle$.

We may continue to use Algorithm B.1 for computing normal forms. In order to apply Buchberger's Algorithm B.3, we still have to generalise the S-polynomials. First, we introduce the least common multiple of two terms $\mathbf{s} = s\mathbf{e}_\alpha$ and $\mathbf{t} = t\mathbf{e}_\beta$ in $\mathbb{T}^m$: it is zero, if $\alpha \neq \beta$, and $\mathrm{lcm}(s,t)\mathbf{e}_\alpha$ otherwise. Then we define the S-"polynomial" of two elements $\mathbf{f}, \mathbf{g} \in \mathcal{P}^m$ by simply copying (B.74):

$$\mathbf{S}_\prec(\mathbf{f}, \mathbf{g}) = \frac{\mathrm{lcm}(\mathrm{lm}_\prec \mathbf{f}, \mathrm{lm}_\prec \mathbf{g})}{\mathrm{lm}_\prec \mathbf{f}} \cdot \mathbf{f} - \frac{\mathrm{lcm}(\mathrm{lm}_\prec \mathbf{f}, \mathrm{lm}_\prec \mathbf{g})}{\mathrm{lm}_\prec \mathbf{g}} \cdot \mathbf{g}\,. \tag{B.79}$$

Now all the results given in this section so far can be straightforwardly extended to the module case. In particular, it turns out that all presented propositions and theorems remain valid.

An important application of Gröbner bases is the construction of syzygies and syzygy resolutions. Like many other things in commutative algebra, the term *syzygy* goes back to Hilbert. It comes from the Greek word for yoke and indeed a syzygy defines a "yoke" connecting some module elements.

Definition B.4.23. Let $\mathcal{F} = \{\mathbf{f}_1, \dots, \mathbf{f}_s\}$ be a finite subset of the free module $\mathcal{P}^m$. A *syzygy* of $\mathcal{F}$ is an element $\mathbf{h} = (h_1, \dots, h_s) \in \mathcal{P}^s$ such that

$$\sum_{\tau=1}^{s} h_\tau \mathbf{f}_\tau = 0\,. \tag{B.80}$$

The syzygies of $\mathcal{F}$ form a submodule of $\mathcal{P}^s$, the *syzygy module* $\operatorname{Syz}(\mathcal{F})$.

Obviously, the syzygy module vanishes, $\operatorname{Syz}(\mathcal{F}) = 0$, if and only if the set $\mathcal{F}$ is the basis of a free module. The components of a syzygy may be understood as solutions of a linear system of equations over the ring $\mathcal{P}$: indeed, $\operatorname{Syz}(\mathcal{F})$ is just the kernel of the $\mathcal{P}$-linear map

$$\kappa_{\mathcal{F}} : \left\{ \begin{array}{ccc} \mathcal{P}^s & \longrightarrow & \mathcal{P}^m \\ (h_1, \dots, h_s)^t & \longmapsto & h_1\mathbf{f}_1 + \cdots + h_s\mathbf{f}_s \end{array} \right. . \tag{B.81}$$

With the help of syzygies many computational problems in commutative algebra like the intersection of submodules or their quotient can be solved in a fairly efficient way (see e. g. [5, §3.8]).

Remark B.4.24. By abuse of language, one sometimes speaks of the syzygy module $\operatorname{Syz}(\mathcal{M})$ of a submodule $\mathcal{M} \subseteq \mathcal{P}^m$. With this notation one actually means that one has chosen a generating set $\mathcal{F}$ of $\mathcal{M}$ and now considers $\operatorname{Syz}(\mathcal{F})$. Of course, a submodule $\mathcal{M}$ has many generating sets and these may contain a different number of elements so that the corresponding syzygy modules even live in free modules of different rank. However, if $\mathcal{F}_1$ is one generating set with s_1 elements and $\mathcal{F}_2$ another one with s_2 elements, then it is not difficult to show that $\operatorname{Syz}(\mathcal{F}_1) \oplus \mathcal{P}^{s_2} \cong \operatorname{Syz}(\mathcal{F}_2) \oplus \mathcal{P}^{s_1}$. Thus knowing the syzygy module of one generating set $\mathcal{F}_1$ suffices to compute it for any other generating set $\mathcal{F}_2$ (we will give the details of the construction below) and in a certain sense one may indeed speak of a module $\operatorname{Syz}(\mathcal{M})$. ◁

Remark B.4.25. If the elements of the set $\mathcal{F} = \{\mathbf{f}_1, \dots, \mathbf{f}_s\}$ are homogeneous, then the syzygy module $\operatorname{Syz}(\mathcal{F})$ may be considered as a graded submodule of $\mathcal{P}^s$ where the grading of the module $\mathcal{P}^s$ is defined as follows: we give each basis vector $\mathbf{e}_i$ the weight $\deg \mathbf{f}_i$, so that the degree of a term $x^\mu \mathbf{e}_i$ is given by $|\mu| + \deg \mathbf{f}_i$. ◁

A generating set of the syzygy module can be easily determined for a Gröbner basis. So let $\mathcal{G} = \{\mathbf{g}_1, \dots, \mathbf{g}_s\}$ be a Gröbner basis for an arbitrary term order $\prec$;

without loss of generality we may assume that $\mathcal{G}$ is reduced so that all leading coefficients are one. Furthermore, let $\mathbf{t}_\sigma = \operatorname{lt}_\prec \mathbf{g}_\sigma$ and $\mathbf{t}_{\sigma\tau} = \operatorname{lcm}(\mathbf{t}_\sigma, \mathbf{t}_\tau)$. It follows from Proposition B.4.8 (iii) that we can write

$$\mathbf{S}_\prec(\mathbf{g}_\rho, \mathbf{g}_\sigma) = \sum_{\tau=1}^{s} h_{\rho\sigma\tau}\mathbf{g}_\tau \tag{B.82}$$

where the leading terms of the polynomials $h_{\rho\sigma\tau} \in \mathcal{P}$ satisfy $\operatorname{lt}_\prec \mathbf{S}_\prec(\mathbf{g}_\rho, \mathbf{g}_\sigma) = \max_\prec \{\operatorname{lt}_\prec (h_{\rho\sigma\tau}\mathbf{g}_\tau) \mid 1 \leq \tau \leq s\}$. Now we define for $\rho \neq \sigma$

$$\mathbf{S}_{\rho\sigma} = \frac{\mathbf{t}_{\rho\sigma}}{\mathbf{t}_\rho}\mathbf{e}_\rho - \frac{\mathbf{t}_{\rho\sigma}}{\mathbf{t}_\sigma}\mathbf{e}_\sigma - \mathbf{h}_{\rho\sigma} \in \mathcal{P}^s \tag{B.83}$$

with $\mathbf{h}_{\rho\sigma} = \sum_{\tau=1}^{s} h_{\rho\sigma\tau}\mathbf{e}_\tau$. By the definition of the S-"polynomial" and (B.82), this is a syzygy in $\operatorname{Syz}(\mathcal{G})$. In fact, the syzygies obtained in this manner generate the syzygy module. We do not give a direct proof of the following theorem, as it follows immediately from the Schreyer Theorem B.4.27 below.

Theorem B.4.26. *Let $\mathcal{G}$ be a Gröbner basis. The syzygy module* $\operatorname{Syz}(\mathcal{G})$ *is generated by the set* $\mathcal{S}_\mathcal{G} = \{\mathbf{S}_{\rho\sigma} \mid 1 \leq \rho < \sigma \leq s\}$.

Thus we automatically obtain a generating set of $\operatorname{Syz}(\mathcal{G})$, if we compute the Gröbner basis $\mathcal{G}$ with Buchberger's algorithm using the extended form of Algorithm B.1, the division algorithm discussed in Remark B.4.7. It delivers us the coefficients $h_{\tau\sigma\rho}$ of the representation (B.82). An interesting result (given by Schreyer in his diploma thesis [395][14]) is that this generating set is even a Gröbner basis for the right term order, namely the one described in Example B.4.21.

Theorem B.4.27 (Schreyer). *The above defined set $\mathcal{S}_\mathcal{G}$ is a Gröbner basis of the syzygy module* $\operatorname{Syz}(\mathcal{G})$ *with respect to the term order $\prec_\mathcal{G}$.*

Proof. The key for proving this result lies in recognising the leading terms of the syzygies $\mathbf{S}_{\rho\sigma}$ with $\rho < \sigma$. We claim that $\operatorname{lt}_{\prec_\mathcal{G}} \mathbf{S}_{\rho\sigma} = \frac{\mathbf{t}_{\rho\sigma}}{\mathbf{t}_\rho}\mathbf{e}_\rho$. Because of the assumption $\rho < \sigma$, this term is indeed greater than $\frac{\mathbf{t}_{\rho\sigma}}{\mathbf{t}_\sigma}\mathbf{e}_\sigma$ with respect to $\prec_\mathcal{G}$ and, since $\operatorname{lt}_\prec (h_{\rho\sigma\tau}\mathbf{g}_\tau) \preceq \operatorname{lt}_\prec \mathbf{S}_\prec(\mathbf{g}_\rho, \mathbf{g}_\sigma) \prec \operatorname{lt}_\prec \left(\frac{\mathbf{t}_{\rho\sigma}}{\mathbf{t}_\rho}\mathbf{g}_\rho\right)$ in (B.82) for any $1 \leq \tau \leq s$, it is also greater than any term contained in $\operatorname{supp}\mathbf{h}_{\rho\sigma}$.

Consider an arbitrary syzygy $\mathbf{S} = \sum_{\rho=1}^{s} f_\rho \mathbf{e}_\rho \in \operatorname{Syz}(\mathcal{G})$. We define the monomial m_ρ by $\operatorname{lm}_{\prec_\mathcal{G}} (f_\rho\mathbf{e}_\rho) = m_\rho\mathbf{e}_\rho$. Then $\operatorname{lm}_{\prec_\mathcal{G}} \mathcal{S} = m_\tau\mathbf{e}_\tau$ for some value $1 \leq \tau \leq s$. We set $\tilde{\mathbf{S}} = \sum_\sigma m_\sigma\mathbf{e}_\sigma$ where the summation is over all σ such that $m_\sigma \operatorname{lm}_\prec \mathbf{g}_\sigma = m_\tau \operatorname{lm}_\prec \mathbf{g}_\tau$ (obviously, $\sigma \geq \tau$ since otherwise $m_\tau\mathbf{e}_\tau$ could not be the leading monomial of $\mathbf{S}$). We have $\tilde{\mathbf{S}} \in \operatorname{Syz}(\{\operatorname{lt}_\prec \mathbf{g}_\sigma \mid \sigma \geq \tau\})$ by construction. Since we are now dealing with the syzygies of a monomial set, it is easy to see that they are generated by the "binary" syzygies $\mathbf{s}_{\rho\sigma} = \frac{\mathbf{t}_{\rho\sigma}}{\mathbf{t}_\rho}\mathbf{e}_\rho - \frac{\mathbf{t}_{\rho\sigma}}{\mathbf{t}_\sigma}\mathbf{e}_\sigma$ with $\tau \leq \rho < \sigma$. By our considerations above, $\operatorname{lt}_{\prec_\mathcal{G}} \mathbf{S}_{\rho\sigma} = \operatorname{lt}_{\prec_\mathcal{G}} \mathbf{s}_{\rho\sigma}$ and hence there exists a syzygy $\mathbf{S}_{\rho\sigma} \in \mathcal{S}_\mathcal{G}$ such that $\operatorname{lt}_{\prec_\mathcal{G}} \mathbf{S}_{\rho\sigma} \mid \operatorname{lt}_{\prec_\mathcal{G}} \mathbf{S}$. But this fact implies that $\mathcal{S}_\mathcal{G}$ is a Gröbner basis as asserted. □

[14] Similar results were obtained around the same time by several authors, e. g. [380, 425, 484].

This result settles the question of determining effectively $\operatorname{Syz}(\mathcal{F})$, if $\mathcal{F}$ is a Gröbner basis. For arbitrary sets $\mathcal{F} = \{\mathbf{f}_1,\dots,\mathbf{f}_s\}$ the problem may be reduced to this case as follows. We compute a Gröbner basis $\mathcal{G} = \{\mathbf{g}_1,\dots,\mathbf{g}_t\}$ of $\langle\mathcal{F}\rangle$ with respect to an arbitrary term order $\prec$. Then there exist coefficients $m_{ij}, n_{ji} \in \mathcal{P}$ such that $\mathbf{f}_i = \sum_{j=1}^{t} m_{ij}\mathbf{g}_j$ and conversely $\mathbf{g}_j = \sum_{i=1}^{s} n_{ji}\mathbf{f}_i$ (the m_{ij} are easily obtained by a normal form computation, the n_{ji} arise automatically during the determination of $\mathcal{G}$). The matrix $M = (m_{ij})$ defines a homomorphism $\psi : \mathcal{P}^s \to \mathcal{P}^t$ by setting $\psi(\mathbf{h}) = \mathbf{h}\cdot M$. In exactly the same manner we define a homomorphism $\phi : \mathcal{P}^t \to \mathcal{P}^s$ by setting $\phi(\mathbf{k}) = \mathbf{k}\cdot N$ where $N = (n_{ji})$. One easily verifies that then the diagram

$$\begin{array}{ccccccc}
0 & \longrightarrow & \operatorname{Syz}(\mathcal{G}) & \longrightarrow & \mathcal{P}^t & \overset{\kappa_{\mathcal{G}}}{\searrow} & \\
 & & & & \phi\big\downarrow\big\uparrow\psi & & \mathcal{P}^m \\
0 & \longrightarrow & \operatorname{Syz}(\mathcal{F}) & \longrightarrow & \mathcal{P}^s & \underset{\kappa_{\mathcal{F}}}{\nearrow} &
\end{array} \tag{B.84}$$

is commutative with exact rows.

We may now apply Lemma B.2.33. In our case ι is the natural inclusion map and we identify $\operatorname{Syz}(\mathcal{G})$ and $\operatorname{Syz}(\mathcal{F})$ with their respective images in $\mathcal{P}^t$ and $\mathcal{P}^s$. As the lemma implies in particular that $\ker\kappa_{\mathcal{F}} = \operatorname{im}(\phi\circ\iota) + \operatorname{im}(\mathrm{id} - \phi\circ\psi)$, a generating set of $\operatorname{Syz}(\mathcal{F}) = \ker\kappa_{\mathcal{F}}$ consists of the elements $\mathbf{S}_{\rho\sigma}N$ with $\mathbf{S}_{\rho\sigma} \in \mathcal{S}_{\mathcal{G}}$ and the columns of the $(s\times s)$-matrix $\mathbb{1}_s - MN$.

Syzygy theory also underlies Buchberger's second criterion (Proposition B.4.19). Let $\mathcal{T} = \{t_1,\dots,t_s\} \subset \mathbb{T}$ be a finite set of terms. As above we set $t_{\rho\sigma} = \operatorname{lcm}(t_\rho, t_\sigma)$ and $\mathbf{S}_{\rho\sigma} = \frac{t_{\rho\sigma}}{t_\rho}\mathbf{e}_\rho - \frac{t_{\rho\sigma}}{t_\sigma}\mathbf{e}_\sigma$. As we are only dealing with terms, no polynomials $h_{\rho\sigma}$ arise. Setting $t_{\rho\sigma\tau} = \operatorname{lcm}(t_\rho, t_\sigma, t_\tau)$, it is a simple exercise to verify by direct computation that

$$\frac{t_{\rho\sigma\tau}}{t_{\rho\sigma}}\mathbf{S}_{\rho\sigma} + \frac{t_{\rho\sigma\tau}}{t_{\sigma\tau}}\mathbf{S}_{\sigma\tau} + \frac{t_{\rho\sigma\tau}}{t_{\tau\rho}}\mathbf{S}_{\tau\rho} = 0\,. \tag{B.85}$$

An obvious consequence is that if $t_\tau \mid t_{\rho\sigma}$, then $\mathbf{S}_{\rho\sigma} \in \langle\mathbf{S}_{\sigma\tau}, \mathbf{S}_{\tau\rho}\rangle$. More generally, we may formulate the following statement.

Lemma B.4.28. *Let $\mathcal{S} \subseteq \mathcal{S}_{\mathcal{T}} = \{\mathbf{S}_{\rho\sigma} \mid 1 \le \rho < \sigma \le s\}$ be a generating set of the syzygy module* $\operatorname{Syz}(\mathcal{T})$. *Assume that the three pairwise distinct indices ρ, σ, τ are such that*[15] *$\mathbf{S}_{\rho\sigma}, \mathbf{S}_{\sigma\tau}, \mathbf{S}_{\rho\tau} \in \mathcal{S}$ and $t_\tau \mid t_{\rho\sigma}$. Then the smaller set $\mathcal{S}\setminus\{\mathbf{S}_{\rho\sigma}\}$ still generates* $\operatorname{Syz}(\mathcal{T})$.

Proposition B.4.19 represents now a simple corollary of a generalisation of Theorem B.4.14. If $\mathcal{G} \subset \mathcal{P}$ is a finite set, we take $\mathcal{T} = \operatorname{lt}_\prec \mathcal{G}$. Given a subset $\mathcal{S} \subseteq \mathcal{S}_{\mathcal{T}}$ generating $\operatorname{Syz}(\mathcal{T})$, one can show that it suffices for $\mathcal{G}$ being a Gröbner basis that all S-polynomials $S_\prec(g_\rho, g_\sigma)$ with $\mathbf{S}_{\rho\sigma} \in \mathcal{S}$ reduce to zero modulo $\mathcal{G}$.

Recall the explicit construction of a free resolution for a finitely generated $\mathcal{R}$-module $\mathcal{M}$ via presentations described in Appendix B.2. Determining a presentation amounts to a syzygy computation once a generating set of $\mathcal{M}$ is known. Thus let

[15] If $\rho > \sigma$, then we understand that $\mathbf{S}_{\sigma\rho} \in \mathcal{S}$ etc.

the finite set $\mathcal{F}$ generate the $\mathcal{P}$-module $\mathcal{M}$ and take $\mathrm{Syz}(\mathcal{F})$ as $\mathcal{M}_0$. If the set $\mathcal{F}$ is a Gröbner basis of $\mathcal{M}$, the Schreyer Theorem B.4.27 provides us again with a Gröbner basis of $\mathrm{Syz}(\mathcal{F})$, so that we can iterate the whole construction leading to a *syzygy resolution* of $\mathcal{M}$.

As a Gröbner basis is rarely a minimal generating set, the obtained syzygy resolution will not be minimal for a graded module $\mathcal{M}$. One step of the minimisation process described in Appendix B.2 takes a very simple form. Let

$$\cdots \xrightarrow{\phi_{\ell+2}} \mathcal{F}_{\ell+1} \xrightarrow{\phi_{\ell+1}} \mathcal{F}_{\ell} \xrightarrow{\phi_{\ell}} \mathcal{F}_{\ell-1} \xrightarrow{\phi_{\ell-1}} \mathcal{F}_{\ell-2} \xrightarrow{\phi_{\ell-2}} \cdots \tag{B.86}$$

be a segment of a free resolution of $\mathcal{M}$. We denote by A_ℓ the matrix describing the map ϕ_ℓ with respect to the bases $\{\mathbf{e}_1,\dots,\mathbf{e}_{n_\ell}\}$ of $\mathcal{F}_\ell$ and $\{\bar{\mathbf{e}}_1,\dots,\bar{\mathbf{e}}_{n_{\ell-1}}\}$ of $\mathcal{F}_{\ell-1}$. Assume that at the position (i,j) we have in A_ℓ an entry $a_{ij} \in \Bbbk^\times \subset \mathcal{P}$ so that the resolution is not minimal. We introduce the two free submodules $\mathcal{F}'_\ell \subset \mathcal{F}_\ell$ and $\mathcal{F}'_{\ell-1} \subset \mathcal{F}_{\ell-1}$ as the linear spans of the bases obtained by removing the vectors $\mathbf{e}_j$ and $\bar{\mathbf{e}}_i$, respectively, and denote by π_ℓ the canonical projection $\mathcal{F}_\ell \to \mathcal{F}'_\ell$. Then consider the modified sequence

$$\cdots \xrightarrow{\phi_{\ell+2}} \mathcal{F}_{\ell+1} \xrightarrow{\psi_{\ell+1}} \mathcal{F}'_{\ell} \xrightarrow{\psi_{\ell}} \mathcal{F}'_{\ell-1} \xrightarrow{\psi_{\ell-1}} \mathcal{F}_{\ell-2} \xrightarrow{\phi_{\ell-2}} \cdots \tag{B.87}$$

where the three new maps are defined as follows: we set $\psi_{\ell+1} = \pi_\ell \circ \phi_{\ell+1}$ and $\psi_{\ell-1} = \phi_{\ell-1}|_{\mathcal{F}'_{\ell-1}}$; the map ψ_ℓ is given by $\psi_\ell(\mathbf{e}_k) = \phi_\ell(\mathbf{e}_k - \frac{a_{ik}}{a_{ij}}\mathbf{e}_j)$ for $k \neq j$. It is straightforward to verify that (B.87) is again a free resolution of $\mathcal{M}$. In matrix language, we subtracted $\frac{a_{ik}}{a_{ij}}$ times the jth column of A_ℓ from the kth column and then deleted the jth column and the ith row. As the ranks of the involved free modules strictly decrease, we obtain after a finite number of iterations a resolution where the ℓth map is minimal.

Hilbert's famous Syzygy Theorem states that for finitely generated polynomial modules always finite resolutions exist. A constructive proof of it may be obtained via Schreyer's Theorem B.4.27 and is contained in many textbooks on Gröbner bases (see e. g. [99, Chapt. 6, Thm. (2.1)]). A sharper version using Pommaret bases is given in Section 5.4.

Theorem B.4.29 (Hilbert). *Let $\mathcal{P} = \Bbbk[x^1,\dots,x^n]$. Then every finitely generated $\mathcal{P}$-module has a finite free resolution whose length is at most n.*

As it is not difficult to produce explicit examples of finitely generated $\mathcal{P}$-modules possessing a minimal free resolution of length n (the simplest example is given by the maximal ideal $\langle x^1,\dots,x^n\rangle$), an alternative formulation of Theorem B.4.29 is that the global dimension of $\mathcal{P}$ is n. More generally, one can show that for any commutative ring $\mathcal{R}$ the equality $\mathrm{gl\,dim}\big(\mathcal{R}[x]\big) = 1 + \mathrm{gl\,dim}\,\mathcal{R}$ holds.

Finally, we discuss the effect of a *specialisation* on the exactness of a sequence of polynomial modules.[16] Let $\mathcal{C} \to \mathcal{M} \to 0$ be a free resolution of some finitely

[16] The following considerations are taken from [273].

generated $\mathcal{P}$-module $\mathcal{M}$ where $\mathcal{C}$ is the exact sequence

$$\cdots \longrightarrow \mathcal{P}^{\ell_2} \xrightarrow{S} \mathcal{P}^{\ell_1} \xrightarrow{B} \mathcal{P}^{\ell_0} . \tag{B.88}$$

Using the standard bases of the free modules, we may consider S, B as matrices with polynomial entries. The exactness at $\mathcal{P}^{\ell_1}$ implies that not only $BS = 0$ but that in fact the columns of S (interpreted as elements of $\mathcal{P}^{\ell_1}$) generate the syzygies of the columns of B.

In a specialisation the variables x^i are substituted by field elements $\xi^i \in \Bbbk$. This operation transforms the polynomial matrices S, B into matrices $S(\boldsymbol{\xi})$, $B(\boldsymbol{\xi})$ over the field $\Bbbk$ and the exact sequence (B.88) into a complex of vector spaces of $\Bbbk$

$$\cdots \longrightarrow \Bbbk^{\ell_2} \xrightarrow{S(\boldsymbol{\xi})} \Bbbk^{\ell_1} \xrightarrow{B(\boldsymbol{\xi})} \Bbbk^{\ell_0} . \tag{B.89}$$

While it is trivial that (B.89) is indeed a complex, we cannot expect that for every vector $\boldsymbol{\xi} \in \Bbbk^n$ exactness is preserved (a trivial counterexample is usually given by the zero vector $\boldsymbol{\xi} = 0$).

In general, the rank of a matrix B over some ring $\mathcal{R}$ is defined via determinantal ideals [55, Chapt. 4]. Let $\mathcal{I}_k(B)$ denote the kth *Fitting ideal* of B generated by all $(k \times k)$-minors of B (one can show that the Fitting ideals depend only on the module $\mathcal{M}_0 = \operatorname{im} B \subseteq \mathcal{P}^{\ell_0}$). The *rank* $\operatorname{rank}_{\mathcal{R}} B$ of B is the largest value $r \in \mathbb{N}_0$ such that $\mathcal{I}_r(B) \neq 0$; we set $\mathcal{I}(B) = \mathcal{I}_r(B)$.

The polynomial ring $\mathcal{P} = \Bbbk[x^1, \dots, x^n]$ is a domain with the rational functions $\mathcal{Q} = \Bbbk(x^1, \dots, x^n)$ as field of fractions. Since $\mathcal{P} \subset \mathcal{Q}$ and since it does not matter whether we compute minors over $\mathcal{P}$ or over $\mathcal{Q}$, $\operatorname{rank}_{\mathcal{P}} B = \operatorname{rank}_{\mathcal{Q}} B$. The latter rank is the classical rank of linear algebra and may be determined by simple Gaussian elimination.

After a specialisation $\mathbf{x} \mapsto \boldsymbol{\xi}$, we can only say that

$$\operatorname{rank}_{\Bbbk} B(\boldsymbol{\xi}) \leq \operatorname{rank}_{\mathcal{P}} B , \tag{B.90}$$

although for generic vectors $\boldsymbol{\xi} \in \Bbbk^n$ equality holds. Those vectors $\boldsymbol{\xi} \in \Bbbk^n$ which lead to a smaller rank are called *characteristic* for B (they make denominators vanish which appear in the Gaussian elimination over $\mathcal{Q}$). More formally, they are defined by the zeros of $\mathcal{I}(B)$, i. e. they lie on the variety $\mathcal{V}\bigl(\mathcal{I}(B)\bigr)$. Since by Hilbert's Nullstellensatz varieties are in a one-to-one correspondence with radical ideals, it is natural to consider the radical $\sqrt{\mathcal{I}(B)}$.

Lemma B.4.30. *If the sequence* (B.88) *is exact at* $\mathcal{P}^{\ell_1}$*, then the Fitting ideals satisfy* $\sqrt{\mathcal{I}(B)} \subseteq \sqrt{\mathcal{I}(S)}$.

For a proof we refer to [125, p. 504]. By the considerations above, it implies that any vector $\boldsymbol{\xi} \in \Bbbk^n$ that is characteristic for S is also characteristic for B, since $\mathcal{V}\bigl(\mathcal{I}(S)\bigr) = \mathcal{V}\bigl(\sqrt{\mathcal{I}(S)}\bigr) \subseteq \mathcal{V}\bigl(\sqrt{\mathcal{I}(B)}\bigr) = \mathcal{V}\bigl(\mathcal{I}(B)\bigr)$. In fact, this argument immediately yields the following assertion.

Corollary B.4.31. *Let the entries of the matrix B be homogeneous polynomials and assume that* $\operatorname{rank}_{\mathcal{P}} B = \operatorname{rank}_{\Bbbk} B(\boldsymbol{\xi})$ *for all vectors* $\boldsymbol{\xi} \in \Bbbk^n \setminus \{0\}$. *Then the equality* $\operatorname{rank}_{\mathcal{P}} S = \operatorname{rank}_{\Bbbk} S(\boldsymbol{\xi})$ *holds for all vectors* $\boldsymbol{\xi} \in \Bbbk^n \setminus \{0\}$.

Lemma B.4.32. *Under the assumptions of Corollary B.4.31, the complex* (B.89) *is exact for all vectors* $\boldsymbol{\xi} \in \Bbbk^n \setminus \{0\}$.

Proof. Since the original sequence (B.88) is assumed to be exact, we have the rank equality $\ell_1 = \operatorname{rank}_{\mathcal{P}} B + \operatorname{rank}_{\mathcal{P}} S$ [125, p. 500]. Let $\boldsymbol{\xi} \in \Bbbk^n$ be an arbitrary non-zero vector. Using Corollary B.4.31, we get

$$\ell_1 = \operatorname{rank}_{\Bbbk} B(\boldsymbol{\xi}) + \operatorname{rank}_{\Bbbk} S(\boldsymbol{\xi}) = \dim \operatorname{im} B(\boldsymbol{\xi}) + \dim \operatorname{im} S(\boldsymbol{\xi}) . \tag{B.91}$$

Since $BS = 0$, we always have the inclusion $\operatorname{im} S(\boldsymbol{\xi}) \subseteq \ker B(\boldsymbol{\xi})$. Furthermore, the specialised matrix $B(\boldsymbol{\xi})$ trivially satisfies $\dim \operatorname{im} B(\boldsymbol{\xi}) = \ell_1 - \dim \ker B(\boldsymbol{\xi})$ implying $\dim \ker B(\boldsymbol{\xi}) = \dim \operatorname{im} S(\boldsymbol{\xi})$. Together with the inclusion above, this observation entails $\operatorname{im} S(\boldsymbol{\xi}) = \ker B(\boldsymbol{\xi})$ and hence the exactness at $\Bbbk^{\ell_1}$. Iteration yields the exactness of the whole complex. □

Applying the functor $\operatorname{Hom}_{\Bbbk}(\cdot, \Bbbk)$ to an exact sequence of vector spaces, i. e. dualising the sequence, leads again to an exact sequence by Proposition B.2.9. At the level of matrices this yields the following result on the transposed matrices.

Corollary B.4.33. *Under the assumptions of Corollary B.4.31, the complex*

$$\Bbbk^{\ell_0} \xrightarrow{B^t(\boldsymbol{\xi})} \Bbbk^{\ell_1} \xrightarrow{S^t(\boldsymbol{\xi})} \Bbbk^{\ell_2} \longrightarrow \cdots \tag{B.92}$$

is exact for all vectors $\boldsymbol{\xi} \in \Bbbk^n \setminus \{0\}$.

Appendix C
Differential Geometry

The doctrines of pure geometry often, and in many questions, give a simple and natural way to penetrate the origin of truths, to lay bare the mysterious chain which unites them, and to make them known individually, luminously and completely.

Michel Chasles

As in this book we are much concerned with a geometric approach to differential equations, we need some basic notions in differential geometry which are collected in this chapter. We start with a section introducing (differentiable) manifolds and some spaces associated with them: the tangent and the cotangent bundle. Special emphasis is given to fibred manifolds (bundles are only briefly introduced, as we hardly need their additional structures). The second section studies in some more detail vector fields and differential forms and the basic operations with them. Distributions of vector fields (or dual codistributions of one-forms, respectively) and the Frobenius theorem are the topic of the third section.

A fundamental object in differential geometry are connections. In our context, it should be noted that ordinary differential equations and certain partial differential equations correspond geometrically to connections (see Remark 2.3.6). The fourth section introduces connections on arbitrary fibred manifolds. At a few places we need some elementary results about Lie groups and algebras which are collected in the fifth section. The final section is concerned with (co)symplectic geometry as the basis of the theory of Hamiltonian systems.

Standard references on differential geometry are [3, 49, 81, 256, 282, 474], but they go much further than we need. For most of our purposes a working knowledge of some basic notions around manifolds is sufficient. Much of the material we need can also be found in a very accessible form in [342, Chapt. 1].

C.1 Manifolds

The jet bundle formalism is based on fibred manifolds, but before we study these we briefly recall the definition of an ordinary manifold. Loosely speaking, it is something that locally looks like an $\mathbb{R}^m$. When we speak of a manifold, we will always mean a finite-dimensional, smooth (C^∞) manifold.

Definition C.1.1. An m-dimensional *(differentiable) manifold* is a set $\mathcal{M}$ equipped with an *atlas* $\mathcal{A}$, that is a countable family of *coordinate charts* $(\mathcal{U} \subseteq \mathcal{M}, \phi_{\mathcal{U}})$ where $\phi_{\mathcal{U}} : \mathcal{U} \rightarrow \Omega_{\mathcal{U}}$ is a bijective map onto a connected open subset $\Omega_{\mathcal{U}} \subseteq \mathbb{R}^m$ such that:

(i) The charts in $\mathcal{A}$ cover the whole manifold: $\mathcal{M} = \bigcup_{(\mathcal{U},\phi_{\mathcal{U}})\in\mathcal{A}} \mathcal{U}$.
(ii) If two charts $(\mathcal{U},\phi_{\mathcal{U}}), (\mathcal{V},\phi_{\mathcal{V}}) \in \mathcal{A}$ overlap, the *transition function*

$$\psi_{\mathcal{U}\mathcal{V}} = \phi_{\mathcal{V}} \circ \phi_{\mathcal{U}}^{-1} : \phi_{\mathcal{U}}(\mathcal{U}\cap\mathcal{V}) \subseteq \mathbb{R}^m \longrightarrow \phi_{\mathcal{V}}(\mathcal{U}\cap\mathcal{V}) \subseteq \mathbb{R}^m \tag{C.1}$$

is a smooth function.
(iii) If $p_1 \in \mathcal{U}_1$ and $p_2 \in \mathcal{U}_2$ are two distinct points of the manifold $\mathcal{M}$, there exist open neighbourhoods $\Omega_1 \subseteq \Omega_{\mathcal{U}_1}$ of $\phi_{\mathcal{U}_1}(p_1)$ and $\Omega_2 \subseteq \Omega_{\mathcal{U}_2}$ of $\phi_{\mathcal{U}_2}(p_2)$ such that $\phi_{\mathcal{U}_1}^{-1}(\Omega_1) \cap \phi_{\mathcal{U}_2}^{-1}(\Omega_2) = \emptyset$.

The atlas $\mathcal{A}$ provides the set $\mathcal{M}$ with a *differentiable structure*. The charts endow the manifold $\mathcal{M}$ with a *topology*: the open sets are the inverse images $\phi_{\mathcal{U}}^{-1}(\Omega)$ of open sets $\Omega \subseteq \Omega_{\mathcal{U}}$ in $\mathbb{R}^m$. The third point in the definition above states that $\mathcal{M}$ with this topology is a Hausdorff space. These topological aspects are not so important for us, as we are mainly working locally, i. e. we stay within a chart $(\mathcal{U},\phi_{\mathcal{U}}) \in \mathcal{A}$ where we can map everything into $\mathbb{R}^m$ and then calculate in coordinates as usual. Typically, we do not explicitly mention the chart but simply say *"let* $\mathbf{x} = (x^1,\dots,x^m)$ *be local coordinates on* $\mathcal{M}$*"*. This means that we go into a chart $(\mathcal{U},\phi_{\mathcal{U}}) \in \mathcal{A}$ and identify each point $p \in \mathcal{U}$ with the vector $\mathbf{x} = \phi_{\mathcal{U}}(p) \in \mathbb{R}^m$.

One can also define analytic or C^k or topological manifolds; in this case one requires that the transition functions $\psi_{\mathcal{U}\mathcal{V}}$ describing changes of coordinates are of class C^ω or C^k or just homeomorphisms, respectively. Infinite-dimensional manifolds are usually obtained by modelling them on some Banach space instead of $\mathbb{R}^m$.

Example C.1.2. The simplest example of a manifold is provided by a connected open subset $\Omega \subseteq \mathbb{R}^m$. In this case $\mathcal{A}$ contains only one chart, namely $(\Omega, \mathrm{id}_\Omega)$.

A slightly more complicated example of an m-dimensional manifold is the surface $\mathcal{S}^m$ of the unit sphere in $\mathbb{R}^{m+1}$. Here it is not possible to cover the whole manifold with a single chart, as $\mathcal{S}^m$ is not homeomorphic to $\mathbb{R}^m$; one needs at least two charts using the stereographic projection. ◁

The space $\mathcal{C}^\infty(\mathcal{M},\mathbb{R})$ of smooth real-valued functions defined on the manifold $\mathcal{M}$ is denoted by $\mathcal{F}(\mathcal{M})$. It is a ring, as we can add and multiply such functions simply by adding and multiplying their values. A function $f \in \mathcal{F}(\mathcal{M})$ induces in a chart $(\mathcal{U},\phi_{\mathcal{U}})$ a function $f_{\mathcal{U}} = f \circ \phi_{\mathcal{U}}^{-1} : \Omega_{\mathcal{U}} \rightarrow \mathbb{R}$. We call f differentiable in $\mathcal{U}$, if $f_{\mathcal{U}}$ is differentiable. It is easy to see that differentiability at a point $p \in \mathcal{M}$ is independent of the used chart, as the transition functions are smooth.

Tangents to a curve or surface are familiar objects from elementary differential geometry. Their generalisation to an arbitrary manifold $\mathcal{M}$ leads to the tangent bundle $T\mathcal{M}$. We start by introducing the tangent space $T_p\mathcal{M}$ at an arbitrary point $p \in \mathcal{M}$. There exist several possibilities for its definition; we use a rather geometric one which also helps to understand the close relation between tangent and jet bundles, as jets are introduced in a fairly similar manner.

A curve on the manifold $\mathcal{M}$ is a differentiable map $\gamma : \mathbb{I} \subseteq \mathbb{R} \to \mathcal{M}$ from some real interval $\mathbb{I} \subseteq \mathbb{R}$ into $\mathcal{M}$. Without loss of generality, we assume that $0 \in \mathbb{I}$. For the construction of the tangent space $T_p\mathcal{M}$ we take all curves with $\gamma(0) = p$. Two such curves γ_1, γ_2 are *tangent* at p, if in some coordinate chart $(\mathcal{U}, \phi_\mathcal{U})$ with $p \in \mathcal{U}$ the tangent vectors $(\phi_\mathcal{U} \circ \gamma_1)'(0)$ and $(\phi_\mathcal{U} \circ \gamma_2)'(0)$ coincide (note that $\phi_\mathcal{U} \circ \gamma_i$ is a curve in $\mathbb{R}^m$ where we know tangent vectors from calculus). One can show that this definition is independent of the used chart.

Tangency induces an equivalence relation; let $[\gamma]_p$ be the equivalence class of all curves $\bar{\gamma}$ with $\bar{\gamma}(0) = p$ which are tangent at p to the curve γ. An element of $T_p\mathcal{M}$ is such an equivalence class of curves. It is not difficult to see that $T_p\mathcal{M}$ is a real vector space of the same dimension as the manifold $\mathcal{M}$, as a coordinate chart maps the elements of $T_p\mathcal{M}$ bijectively on vectors in $\mathbb{R}^m$.

Definition C.1.3. The *tangent bundle* $T\mathcal{M}$ of the manifold $\mathcal{M}$ is the union of all tangent spaces

$$T\mathcal{M} = \bigcup_{p \in \mathcal{M}} \{p\} \times T_p\mathcal{M} \,. \tag{C.2}$$

The *tangent bundle projection* is the canonical map $\tau_\mathcal{M} : T\mathcal{M} \to \mathcal{M}$ defined by $\tau_\mathcal{M}\bigl(\{p\} \times [\gamma]_p\bigr) = p$ for any $p \in \mathcal{M}$ and $[\gamma]_p \in T_p\mathcal{M}$.

Let $\mathcal{M}$ and $\mathcal{N}$ be two manifolds and $f : \mathcal{M} \to \mathcal{N}$ an arbitrary smooth map. It induces a map $Tf : T\mathcal{M} \to T\mathcal{N}$, the *tangent map*, which can be defined intrinsically as follows. Let $\gamma_\mathcal{M} : \mathbb{I} \subseteq \mathbb{R} \to \mathcal{M}$ be an arbitrary curve on $\mathcal{M}$. Together with f it defines a curve $\gamma_\mathcal{N} : \mathbb{I} \to \mathcal{N}$ on the manifold $\mathcal{N}$ by $\gamma_\mathcal{N} = f \circ \gamma_\mathcal{M}$. By definition, a point in $T_p\mathcal{M}$ is an equivalence class of such curves $\gamma_\mathcal{M}$ with $\gamma_\mathcal{M}(0) = p$. Obviously, all curves in this class are mapped on curves $\gamma_\mathcal{N}$ with $\gamma_\mathcal{N}(0) = f(p)$ and it is not difficult to see that in fact the class $[\gamma_\mathcal{M}]_p$ is mapped into the class $[\gamma_\mathcal{N}]_{f(p)}$. Thus we obtain an $\mathbb{R}$-*linear* map $T_pf : T_p\mathcal{M} \to T_{f(p)}\mathcal{N}$. Putting together these maps for all points p yields the tangent map Tf.

The tangent bundle $T\mathcal{M}$ is again a manifold, as the differentiable structure of $\mathcal{M}$ induces a differentiable structure on $T\mathcal{M}$. Namely, let $\mathcal{A}$ be an atlas of $\mathcal{M}$; then $T\mathcal{A} = \bigl\{\bigl(\tau_\mathcal{M}^{-1}(\mathcal{U}), T\phi_\mathcal{U}\bigr) \mid (\mathcal{U}, \phi_\mathcal{U}) \in \mathcal{A}\bigr\}$ is an atlas of $T\mathcal{M}$. We have $\dim T\mathcal{M} = 2\dim\mathcal{M}$. Let $(\mathcal{U}, \phi_\mathcal{U})$ be a coordinate chart with $p \in \mathcal{U}$ and $\phi_\mathcal{U}(p) = \mathbf{x} = (x^1, \dots, x^m)$. Consider for $1 \leq i \leq m$ the curves $\gamma_i : \mathbb{I} \to \mathcal{M}$ with $(\phi_\mathcal{U} \circ \gamma_i)(t) = \mathbf{x} + t \cdot \mathbf{e}_i$ where $\mathbf{e}_i$ is the ith unit vector in $\mathbb{R}^m$. We denote the vector $[\gamma_i]_p$ by $\partial_{x^i}|_p$. These vectors form a basis of $T_p\mathcal{M}$. One often denotes coordinates with respect to this basis by $\dot{\mathbf{x}} = (\dot{x}^1, \dots, \dot{x}^m)$, so that local coordinates on $T\mathcal{M}$ have the form $(\mathbf{x}, \dot{\mathbf{x}}) \in \mathbb{R}^m \times \mathbb{R}^m$.

Locally, the tangent map Tf corresponds to the linearisation of f via its Jacobian (sometimes denoted df). Let us introduce coordinates $(\mathbf{x}, \dot{\mathbf{x}})$ on $T\mathcal{M}$ (with some chart $(\mathcal{U}, \phi_\mathcal{U})$ on $\mathcal{M}$) and $(\mathbf{y}, \dot{\mathbf{y}})$ on $T\mathcal{N}$ (with some chart $(\mathcal{V}, \psi_\mathcal{V})$ on $\mathcal{N}$). We write $\mathbf{y} = \bar{f}(\mathbf{x})$, where $\bar{f} = \psi_\mathcal{V} \circ f \circ \phi_\mathcal{U}^{-1} : \mathcal{U} \to \mathcal{V}$ is the coordinate representation of f. Then Tf has the following form in these coordinates:

$$Tf : \begin{cases} T\mathcal{U} & \longrightarrow & T\mathcal{V} \\ (\mathbf{x}, \dot{\mathbf{x}}) & \longmapsto & \Bigl(\bar{f}(\mathbf{x}), \dfrac{\partial \bar{f}}{\partial \mathbf{x}}(\mathbf{x})\dot{\mathbf{x}}\Bigr) \end{cases} . \tag{C.3}$$

The *rank* of the map f at a point $p \in \mathcal{U}$ is defined as the rank of the Jacobian $\partial \bar{f}/\partial \mathbf{x}$ at the point $\phi_{\mathcal{U}}(p)$. One can show that this definition is independent of the chosen coordinate charts. If $\mathcal{K}$ is a further manifold and $g : \mathcal{N} \to \mathcal{K}$ a further map, the familiar *chain rule* can be expressed as $T(g \circ f) = Tg \circ Tf$.

Mostly, we are somewhat sloppy with our notations and do not distinguish between the map f and its coordinate representation $\bar{f}$. Thus we simply write $\mathbf{y} = f(\mathbf{x})$ omitting the coordinate maps $\phi_{\mathcal{U}}$ and $\psi_{\mathcal{V}}$.

Let $f : \mathcal{M} \to \mathcal{N}$ be a smooth map. We assume for simplicity that $\operatorname{rank} f$ is constant on the manifold $\mathcal{M}$. Maps with the maximal possible rank (which is obviously $\min\{\dim \mathcal{M}, \dim \mathcal{N}\}$) are of particular importance and have special names: if $\dim \mathcal{M} \leq \dim \mathcal{N}$, one speaks of an *immersion*; for $\dim \mathcal{M} \geq \dim \mathcal{N}$ of a *submersion*. Locally, immersions look like linear injections, submersions like projections (see [170, Section I.2] for a precise form of this statement). If f is a bijection and f^{-1} is also smooth, then f is called a *diffeomorphism* and the two manifolds $\mathcal{M}$ and $\mathcal{N}$ are *diffeomorphic*. Obviously, this requires that Tf is an isomorphism with $(Tf)^{-1} = T(f^{-1})$ and thus $\dim \mathcal{M} = \dim \mathcal{N}$. An *embedding* is an immersion $f : \mathcal{M} \to \mathcal{N}$ such that $\mathcal{M}$ is diffeomorphic to $\operatorname{im} f$; this requires in particular that $\operatorname{im} f$ is a manifold which is not necessarily the case for arbitrary immersions.

While it is intuitively fairly clear what a *submanifold* of an m-dimensional manifold $\mathcal{M}$ should be, providing a rigorous definition turns out to be surprisingly subtle. The simplest approach considers manifolds $\mathcal{N}$ for which an injective immersion $\iota : \mathcal{N} \to \mathcal{M}$ exists. Then the pair $(\mathcal{N}, \iota)$ is called an *immersed submanifold*. An alternative approach considers subsets $\mathcal{N} \subseteq \mathcal{M}$ such that for each point $p \in \mathcal{N}$ a chart $(\mathcal{U}, \phi)$ on $\mathcal{M}$ exists which maps $\mathcal{U} \cap \mathcal{N}$ into $\mathbb{R}^n \times \{0\} \subseteq \mathbb{R}^m$ for some $n \leq m$. In other words, in the corresponding local coordinates $(x^1, \dots, x^m)$ the set $\mathcal{U} \cap \mathcal{N}$ is mapped into the set $x^{n+1} = \cdots = x^m = 0$. Such a subset $\mathcal{N}$ is then called an n-dimensional *regular submanifold*.

It follows immediately that a regular submanifold $\mathcal{N}$ is an n-dimensional manifold, as it inherits the differentiable structure of $\mathcal{M}$. Furthermore, any regular submanifold is also an immersed submanifold, as we may simply take the natural inclusion map $\iota : \mathcal{N} \hookrightarrow \mathcal{M}$. The converse is, however, not true, as the map ι appearing in the definition of an immersed submanifold is not necessarily a homeomorphism onto its image in $\mathcal{M}$. In particular, it is not sufficient to consider only the image $\operatorname{im} \iota \subseteq \mathcal{M}$; one must also take the map ι into account, as the same image $\operatorname{im} \iota$ may be obtained in inequivalent ways.

We always assume that we are dealing with a regular submanifold, even if we describe it via an injective immersion, i. e. we always assume that ι is a homeomorphism onto its image and in fact identify $\mathcal{N}$ and $\operatorname{im} \iota$. One may contrast the two mentioned approaches to submanifolds as follows: in the immersed case, we are given a parametrisation of the subset $\operatorname{im} \iota$; in the regular case, the subset $\mathcal{N} \subseteq \mathcal{M}$ is described as solution set of some equations. The latter point of view may be formulated more precisely.

Proposition C.1.4. *Let $\mathcal{M}$ be a manifold and $F : \mathcal{M} \to \mathbb{R}^n$ a smooth map with $n \leq m = \dim \mathcal{M}$ such that the tangent map TF has maximal rank everywhere on* *the subset* $\mathcal{N} = \{p \in \mathcal{M} \mid F(p) = 0\}$. *Then* $\mathcal{N}$ *is a regular,* $(m-n)$*-dimensional submanifold of* $\mathcal{M}$.

This result provides us with a simple means to produce submanifolds. If the assumption on TF is satisfied, one says that 0 is a *regular value* of F. Locally, the converse of Proposition C.1.4 is also true by definition of a regular submanifold. For a submanifold generated this way, the following classical lemma describes in a simple way all smooth functions vanishing on it.

Lemma C.1.5 (Hadamard). *Let the regular submanifold* $\mathcal{N} \subseteq \mathcal{M}$ *be described as the zero set of a smooth map* $F : \mathcal{M} \to \mathbb{R}^n$ *whose tangent map* TF *has maximal rank on* $\mathcal{N}$. *Let* g *be a real-valued function defined in a neighbourhood* $\mathcal{U}$ *of* $\mathcal{N}$ *and vanishing on* $\mathcal{N}$. *Then smooth functions* $g_i \in \mathcal{F}(\mathcal{U})$ *exist such that* $g = \sum_{i=1}^{n} g_i F_i$ *where* F_i *is the ith component of* F.

Given two submanifolds $\mathcal{N}_1$ and $\mathcal{N}_2$ of a manifold $\mathcal{M}$, we say that they are *transversal* at a point $p \in \mathcal{N}_1 \cap \mathcal{N}_2$, if $T_p\mathcal{M} = T_p\mathcal{N}_1 + T_p\mathcal{N}_2$ (note that this requires that $\dim \mathcal{N}_1 + \dim \mathcal{N}_2 \geq \dim \mathcal{M}$; thus two curves in $\mathbb{R}^3$ can never be transversal). If this condition is satisfied at all points $p \in \mathcal{N}_1 \cap \mathcal{N}_2$, then $\mathcal{N}_1 \cap \mathcal{N}_2$ is again a submanifold and $T_p(\mathcal{N}_1 \cap \mathcal{N}_2) = T_p\mathcal{N}_1 \cap T_p\mathcal{N}_2$ (one speaks of a *clean intersection* whenever this last condition holds). For example, two curves $\mathcal{N}_1$ and $\mathcal{N}_2$ in a two-dimensional manifold $\mathcal{M}$ are transversal at a point $p \in \mathcal{N}_1 \cap \mathcal{N}_2$, if they do not have contact of higher order at p.

Definition C.1.6. The linear dual of the tangent space $T_p\mathcal{M}$, denoted by $T_p^*\mathcal{M}$, is called the *cotangent space* of the manifold $\mathcal{M}$ at the point $p \in \mathcal{M}$. Thus its elements are linear maps (functionals) $\lambda_p : T_p\mathcal{M} \to \mathbb{R}$. The *cotangent bundle* $T^*\mathcal{M}$ is the union of all cotangent spaces

$$T^*\mathcal{M} = \bigcup_{p \in \mathcal{M}} \{p\} \times T_p^*\mathcal{M} \,. \tag{C.4}$$

The *cotangent bundle projection* is the canonical map $\tau^*_{\mathcal{M}} : T^*\mathcal{M} \to \mathcal{M}$ defined by $\tau^*_{\mathcal{M}}\big(\{p\} \times \lambda_p\big) = p$ for any $\lambda_p \in T_p^*\mathcal{M}$.

A direct definition of the cotangent space $T_p^*\mathcal{M}$ goes as follows. We define on $\mathcal{F}(\mathcal{M})$ an equivalence relation by $f \sim g$, if their tangent maps at the point $p \in \mathcal{M}$ coincide: $T_pf = T_pg$. Then $T_p^*\mathcal{M}$ is the space of all equivalence classes $[f]$ for this relation. Indeed, since $T_r\mathbb{R} = \mathbb{R}$ for any $r \in \mathbb{R}$, we find that $T_pf : T_p\mathcal{M} \to \mathbb{R}$ is a linear map on $T_p\mathcal{M}$ and thus each such class $[f]$ may be considered as an element of the dual space. Conversely, it is trivial that we may find to any linear functional $\lambda : T_p\mathcal{M} \to \mathbb{R}$ a function $f \in \mathcal{F}(\mathcal{M})$ such that $\lambda = T_pf$.

Above we mentioned that in a local chart $(\mathcal{U}, \phi_{\mathcal{U}})$ the vectors $\partial_{x^i}|_p$ form a basis of $T_p\mathcal{M}$. The elements of the dual basis in $T_p^*\mathcal{M}$ are denoted by $\mathrm{d}x^i|_p$, i. e. we have the relations $\mathrm{d}x^i|_p(\partial_{x^j}|_p) = \delta^i_j$; they are the equivalence classes of the components $x^i : \mathcal{U} \to \mathbb{R}$ of the function $\phi_{\mathcal{U}} : \mathcal{U} \to \mathbb{R}^m$.

Fibred manifolds can be seen as a generalisation of the Cartesian product of two manifolds. As we will see, in the neighbourhood of a point a fibred manifold is in fact always of this form.

Definition C.1.7. A *fibred manifold* is a triple $(\mathcal{E},\pi,\mathcal{B})$ where $\mathcal{E}$ and $\mathcal{B}$ are manifolds and $\pi:\mathcal{E}\to\mathcal{B}$ is a surjective submersion. $\mathcal{E}$ is called the *total space* of the fibred manifold, $\mathcal{B}$ the *base space* and π the *projection*. The subset $\pi^{-1}(b)\subset\mathcal{E}$ is the *fibre* over the point $b\in\mathcal{B}$ and written $\mathcal{E}_b$.

We usually speak simply of the fibred manifold $\mathcal{E}$, if the base space $\mathcal{B}$ and the projection π are obvious from the context. If $\dim\mathcal{E}=n+m$ with $\dim\mathcal{B}=n$, then m is called the *fibre dimension* of the fibred manifold.

Example C.1.8. The simplest example of a fibred manifold is a *trivial fibred manifold* (we will see below that it is in fact even a *trivial bundle*). Its total space is of the form $\mathcal{E}=\mathcal{B}\times\mathcal{F}$ for some manifold $\mathcal{F}$ and the map π is the projection pr_1 on the first factor. In this case, each fibre can be identified with the manifold $\mathcal{F}$.

Given an arbitrary manifold $\mathcal{M}$, both the tangent bundle $(T\mathcal{M},\tau_{\mathcal{M}},\mathcal{M})$ and the cotangent bundle $T^*\mathcal{M},\tau^*_{\mathcal{M}},\mathcal{M})$ are fibred manifolds. In general, neither of them is a trivial fibred manifold. $\triangleleft$

Locally, any fibred manifold looks like a trivial one. With the help of the Implicit Function Theorem one can show that, because of the properties of a submersion, to each point $\xi\in\mathcal{E}$ a neighbourhood $\mathcal{U}_\xi\subset\mathcal{E}$, a manifold $\mathcal{F}_\xi$ and a diffeomorphism $\phi_\xi:\mathcal{U}_\xi\to\pi(\mathcal{U}_\xi)\times\mathcal{F}_\xi$ exist such that the diagram

$$\begin{array}{ccc} \mathcal{U}_\xi & \xrightarrow{\;\phi_\xi\;} & \pi(\mathcal{U}_\xi)\times\mathcal{F}_\xi \\ & {\scriptstyle \pi|_{\mathcal{U}_\xi}}\searrow \quad \swarrow{\scriptstyle \mathrm{pr}_1} & \\ & \pi(\mathcal{U}_\xi) & \end{array} \tag{C.5}$$

commutes. This property means that $\mathrm{pr}_1(\phi_\xi(\zeta))=\pi(\zeta)$ for any point $\zeta\in\mathcal{U}_\xi$. Thus any fibred manifold can be covered with subsets $\mathcal{U}_\xi$ which are diffeomorphic to products $\pi(\mathcal{U}_\xi)\times\mathcal{F}_\xi$. It is a natural thought to exploit this fact for the introduction of special coordinate charts on $\mathcal{E}$: *adapted coordinates*. Namely, we first map with ϕ_ξ on $\pi(\mathcal{U}_\xi)\times\mathcal{F}_\xi$ and then use charts on $\pi(\mathcal{U}_\xi)$ and $\mathcal{F}_\xi$, respectively. The arising coordinate maps are of the form $\mathcal{U}_\xi\to\mathbb{R}^n\times\mathbb{R}^m$ and if two points $\xi_1,\xi_2\in\mathcal{U}_\xi$ lie in the same fibre, i. e. $\pi(\xi_1)=\pi(\xi_2)$, then their first n coordinates are identical. We mostly denote adapted coordinates by $(\mathbf{x},\mathbf{u})=(x^1,\dots,x^n,u^1,\dots,u^m)$.

A submanifold $\mathcal{E}'\subseteq\mathcal{E}$ of the total space is a *fibred submanifold*, if the restriction $\pi|_{\mathcal{E}'}:\mathcal{E}'\to\mathcal{B}$ is still a surjective submersion. In other words, $(\mathcal{E}',\pi|_{\mathcal{E}'},\mathcal{B})$ must again be a fibred manifold.

Let $(\mathcal{E},\pi,\mathcal{B})$ and $(\mathcal{E}',\rho,\mathcal{B})$ be two fibred manifolds over the same base space. Their *fibred product* is the fibred manifold $(\mathcal{E}\underset{\mathcal{B}}{\times}\mathcal{E}',\pi\underset{\mathcal{B}}{\times}\rho,\mathcal{B})$ where the total space $\mathcal{E}\underset{\mathcal{B}}{\times}\mathcal{E}'$ consists of all pairs $(\xi,\zeta)\in\mathcal{E}\times\mathcal{E}'$ with $\pi(\xi)=\rho(\zeta)$ and the projection is defined by $(\pi \underset{\mathcal{B}}{\times} \rho)(\xi,\zeta) = \pi(\xi) = \rho(\zeta)$. It is straightforward to show that this construction indeed yields again a fibred manifold.

The total space $\mathcal{E} \underset{\mathcal{B}}{\times} \mathcal{E}'$ may also be considered as a fibred manifold over either $\mathcal{E}$ or $\mathcal{E}'$, respectively, with obvious projections on the first or second factor, respectively. In this case one speaks of a *pull-back* and uses the notation $\rho^*(\pi) : \rho^*(\mathcal{E}) \to \mathcal{E}'$ and $\pi^*(\rho) : \pi^*(\mathcal{E}') \to \mathcal{E}$, respectively.

Let $(\mathcal{E},\pi,\mathcal{B})$ and $(\mathcal{E}',\rho,\mathcal{B}')$ be two arbitrary fibred manifolds. A smooth map $F : \mathcal{E} \to \mathcal{E}'$ is a *fibred morphism*, if there exists a map $f : \mathcal{B} \to \mathcal{B}'$ (which is in fact uniquely determined and smooth) such that the diagram

$$\begin{array}{ccc} \mathcal{E} & \xrightarrow{F} & \mathcal{E}' \\ {\scriptstyle\pi}\downarrow & & \downarrow{\scriptstyle\rho} \\ \mathcal{B} & \xrightarrow{f} & \mathcal{B}' \end{array} \tag{C.6}$$

commutes, i. e. such that $\rho \circ F = f \circ \pi$. Thus F maps fibres into fibres: if $\xi_1,\xi_2 \in \mathcal{E}_b$, then $F(\xi_1),F(\xi_2) \in \mathcal{E}'_{f(b)}$. An example of a fibred morphism is the tangent map $Tf : T\mathcal{M} \to T\mathcal{N}$ of a map $f : \mathcal{M} \to \mathcal{N}$.

Fibre bundles are special fibred manifolds resembling even more a trivial one. We require that we may choose the subsets $\mathcal{U}_\xi$ in (C.5) in the form $\pi^{-1}(\mathcal{W}_b)$ where $\mathcal{W}_b$ is a neighbourhood in $\mathcal{B}$ of the point $b = \pi(\xi)$, i. e. U_ξ consists of complete fibres. A covering of $\mathcal{E}$ with such subsets is a *local trivialisation* of the bundle. One can show that to every local trivialisation a manifold $\mathcal{F}$ exists such that all the manifolds $\mathcal{F}_\xi$ are diffeomorphic to it. $\mathcal{F}$ is called the *typical fibre* of the bundle.

Special kinds of bundles arise, if the typical fibre $\mathcal{F}$ has additional structures. For us the most important ones are *vector bundles* where $\mathcal{F}$ is a vector space $\mathcal{V}$ and *affine bundles* where $\mathcal{F}$ is an affine space $\mathcal{A}$. Recall that every affine space $\mathcal{A}$ is modelled on a vector space $\mathcal{V}$: $\mathcal{A}$ is a set on which $\mathcal{V}$ acts free and transitively as an additive group (see Definition C.5.3); in other words, we identify the "difference" $\alpha - \beta$ of any two points $\alpha,\beta \in \mathcal{A}$ with a vector $v \in \mathcal{V}$. As we can do this over every point of the base space, a vector bundle is associated with every affine bundle. For a vector bundle $\pi : \mathcal{E} \to \mathcal{B}$ we may define its *dual bundle* $\pi^* : \mathcal{E}^* \to \mathcal{B}$. It has the same global structure but its typical fibre is the vector space $\mathcal{F}^*$ dual to the typical fibre $\mathcal{F}$ of $\mathcal{E}$.

Example C.1.9. Simple examples of vector bundles are both tangent and cotangent bundle of a manifold $\mathcal{M}$. By definition, they are dual to each other. Their typical fibre is the vector space $\mathbb{R}^m$ where $m = \dim \mathcal{M}$ (as $\mathbb{R}^m$ has a canonical scalar product, we may identify it with its dual space). The coordinates $(\mathbf{x},\dot{\mathbf{x}})$ on the tangent bundle $T\mathcal{M}$ are adapted. ◁

Remark C.1.10. The special form of the sets $\mathcal{U}_\xi$ in a fibre bundle seems to be a rather weak condition. In fact, it has a number of important consequences. For example, the differentiable structure of $\mathcal{E}$ is completely determined by those of $\mathcal{B}$ and $\mathcal{F}$.

Bundles also carry an additional structure, namely a group acting on the fibres. Indeed, if $\mathcal{W}_1$, $\mathcal{W}_2$ are two different neighbourhoods of a given point $b \in \mathcal{B}$ on the base manifold with maps $\phi_i : \pi^{-1}(\mathcal{W}_i) \to \mathcal{W}_i \times \mathcal{F}$, then by (C.5) the composition $\phi_1 \circ \phi_2^{-1} : (\mathcal{W}_1 \cap \mathcal{W}_2) \times \mathcal{F} \to (\mathcal{W}_1 \cap \mathcal{W}_2) \times \mathcal{F}$ is of the form $\mathrm{id} \times \phi_{12}$ with a diffeomorphism $\phi_{12} : \mathcal{F} \to \mathcal{F}$ called the transition map of the bundle. All the transition maps together form the *structure group*, a subgroup of the full diffeomorphism group of the typical fibre $\mathcal{F}$. As we do not need this additional structure, we refer to the literature for more details.

A very important special case are *principal fibre bundles* where the typical fibre is diffeomorphic to a Lie group $\mathcal{G}$ and the structure group is also $\mathcal{G}$ acting on itself by left multiplication. More precisely, a principal fibre bundle is a tuple $(\mathcal{P}, \pi, \mathcal{B}, \mathcal{G}, \Phi)$ with a fibred manifold $\pi : \mathcal{P} \to \mathcal{B}$ and a fibre-preserving action $\Phi : \mathcal{G} \times \mathcal{P} \to \mathcal{P}$ of the Lie group $\mathcal{G}$ on $\mathcal{P}$ such that the induced map $(\Phi, \mathrm{pr}_2) : \mathcal{G} \times \mathcal{P} \to \mathcal{P} \underset{\mathcal{B}}{\times} \mathcal{P}$ is a diffeomorphism. ◁

Definition C.1.11. A *(local) section* σ of the fibred manifold $\pi : \mathcal{E} \to \mathcal{B}$ is a smooth map $\sigma : \mathcal{U} \subseteq \mathcal{B} \to \mathcal{E}$ such that $\pi \circ \sigma = \mathrm{id}_{\mathcal{U}}$. The space of all local sections is denoted by $\Gamma_{loc}(\pi)$.

In adapted coordinates $(\mathbf{x}, \mathbf{u})$ on the fibred manifold $\mathcal{E}$, a section σ has the local form $\sigma(\mathbf{x}) = (\mathbf{x}, s(\mathbf{x}))$ where s is a smooth function, as the condition $\pi \circ \sigma = \mathrm{id}_{\mathcal{U}}$ implies that every point $b \in \mathcal{U}$ is mapped to a point in the fibre $\mathcal{E}_b$ above it. The notion of a section may be understood as a generalisation of the graph of a function. Indeed, if we are dealing with a trivial fibred manifold $\mathcal{E} = \mathcal{B} \times \mathcal{F}$, every smooth function $s : \mathcal{B} \to \mathcal{F}$ defines a section $\sigma(b) = (b, s(b))$ and the set $\mathrm{im}\,\sigma \subset \mathcal{B} \times \mathcal{F}$ is the graph of s.

A global section is defined on the whole base manifold $\mathcal{B}$. It is a very strong assumption that a fibred manifold $\pi : \mathcal{E} \to \mathcal{B}$ admits global sections. In fact, most fibred manifolds do not (one obvious exception are trivial ones). If such a global section σ exists, it defines an immersion of $\mathcal{B}$ into $\mathcal{E}$.

C.2 Vector Fields and Differential Forms

In this section we discuss in some more detail sections of the tangent and cotangent bundle of a manifold. They are fundamental objects in differential geometry.

Definition C.2.1. Sections of the bundle $T\mathcal{M}$ are called *vector fields*. We write $\mathfrak{X}(\mathcal{M})$ for $\Gamma_{loc}(\tau_{\mathcal{M}})$.

The set $\mathfrak{X}(\mathcal{M})$ possesses some algebraic structures (see Appendix B.1). In particular, it is an $\mathcal{F}(\mathcal{M})$-module: we can add vector fields and multiply them by functions contained in $\mathcal{F}(\mathcal{M})$. In local coordinates $(x^1, \dots, x^m)$, the module $\mathfrak{X}(\mathcal{M})$ is generated by the fields $\partial_{x^i} : p \mapsto \partial_{x^i}|_p$. Thus every vector field can be written locally as $X = v^i \partial_{x^i}$ with some functions $v^i \in \mathcal{F}(\mathcal{M})$.

The notation ∂_{x^i} already indicates that vector fields can also be interpreted as linear first-order differential operators acting on $\mathcal{F}(\mathcal{M})$. This is a more algebraic alternative to our geometric introduction of tangent vectors in the last section: a vector field is an $\mathbb{R}$-linear map $X : \mathcal{F}(\mathcal{M}) \to \mathcal{F}(\mathcal{M})$ that satisfies the Leibniz rule $X(fg) = X(f)g + fX(g)$ (such maps are also called *derivations*). The relation between the two approaches is provided by the tangent map. If $f \in \mathcal{F}(\mathcal{M})$, then $Tf : T\mathcal{M} \to T\mathbb{R} = \mathbb{R} \oplus \mathbb{R}$ and for any vector field $X \in \mathfrak{X}(\mathcal{M})$ we define the function $X(f) \in \mathcal{F}(\mathcal{M})$ by $X(f)(p) = T_p f(X_p)$ for any point $p \in \mathcal{M}$. In local coordinates, the vector field ∂_{x^i} has then the meaning one expects: $\partial_{x^i}(f) = \partial f / \partial x^i$.

With this interpretation, $\mathfrak{X}(\mathcal{M})$ can easily be given the structure of a real Lie algebra (see Definition C.5.5 below). For the multiplication we take the *Lie bracket*: $[X,Y] = X \circ Y - Y \circ X$ for any two fields $X, Y \in \mathfrak{X}(\mathcal{M})$. It is easy to verify that this bracket yields again a derivation on $\mathcal{F}(\mathcal{M})$, i. e. a vector field in $\mathfrak{X}(\mathcal{M})$. If the two vector fields are given in local coordinates by $X = v^i \partial_{x^i}$ and $Y = w^i \partial_{x^i}$, then their Lie bracket is the vector field

$$[X,Y] = \left(v^j \frac{\partial w^i}{\partial x^j} - w^j \frac{\partial v^i}{\partial x^j} \right) \partial_{x^i} \,. \tag{C.7}$$

Because of these two linear structures, module over the ring $\mathcal{F}(\mathcal{M})$ and Lie algebra over the field $\mathbb{R}$, one must be careful when speaking about linear combinations of vector fields. In order to avoid misunderstandings, one should always state where the coefficients live, if it is not evident from the context. The relation between the two structures is given by a kind of Leibniz rule: one easily verifies that for any smooth function $f \in \mathcal{F}(\mathcal{M})$ the identity $[X, fY] = f[X,Y] + X(f)Y$ holds.

Remark C.2.2. Consider an arbitrary smooth map $f : \mathcal{M} \to \mathcal{N}$ between two manifolds $\mathcal{M}$ and $\mathcal{N}$. Given a vector field $X \in \mathfrak{X}(\mathcal{M})$ on $\mathcal{M}$, we cannot expect that applying the tangent map Tf to X yields again a vector field $Y \in \mathfrak{X}(\mathcal{N})$ on $\mathcal{N}$. If the map f is not surjective, then Y will not be defined everywhere on $\mathcal{N}$ but only on $\operatorname{im} f$. If f is not injective, we encounter the following problem: given two points $p_1, p_2 \in \mathcal{M}$ such that $f(p_1) = f(p_2) = q \in \mathcal{N}$, there is no reason why the two vectors $T_{p_1} f(X_{p_1}), T_{p_2} f(X_{p_2}) \in T_q\mathcal{N}$ should coincide and hence we do not know how to define Y at q. We call two vector fields $X \in \mathfrak{X}(\mathcal{M})$ and $Y \in \mathfrak{X}(\mathcal{N})$ *f-related*, if $Tf \circ X = Y \circ f$; in other words: if the diagram

$$\begin{array}{ccc} T\mathcal{M} & \xrightarrow{Tf} & T\mathcal{N} \\ {\scriptstyle X}\uparrow & & \uparrow{\scriptstyle Y} \\ \mathcal{M} & \xrightarrow{f} & \mathcal{N} \end{array} \tag{C.8}$$

commutes. In this case $f(p_1) = f(p_2)$ implies that $T_{p_1} f(X_{p_1}) = T_{p_2} f(X_{p_2})$. Note that obviously only the values of the field Y on $\operatorname{im} f \subseteq \mathcal{N}$ matter. One easily verifies that if $X_1, X_2 \in \mathfrak{X}(\mathcal{M})$ are two vector fields on $\mathcal{M}$ which are f-related to $Y_1, Y_2 \in \mathfrak{X}(\mathcal{N})$, then their Lie bracket $[X_1, X_2]$ is f-related to the vector field $[Y_1, Y_2]$.

In the special case of a fibred manifold $\pi : \mathcal{E} \to \mathcal{B}$, a vector field $X \in \mathfrak{X}(\mathcal{E})$ such that $T_{p_1}\pi(X_{p_1}) = T_{p_2}\pi(X_{p_2})$ for all points p_1, p_2 with $\pi(p_1) = \pi(p_2)$ is called *projectable*. The unique vector field $Y \in \mathfrak{X}(\mathcal{B})$ on $\mathcal{B}$ which is π-related to X is then denoted by $T\pi(X)$.

If f is a diffeomorphism, then we can associate with every vector field $X \in \mathfrak{X}(\mathcal{M})$ the vector field $f_*X = Tf \circ X \circ f^{-1} \in \mathfrak{X}(\mathcal{N})$, called the *push-forward* of X. By a slight abuse of notation, we will also speak of a push-forward in the case of an injective map $f : \mathcal{M} \to \mathcal{N}$: we simply take the push-forward with respect to the diffeomorphism $f : \mathcal{M} \to \operatorname{im} f$ (but we must always keep in mind that now the vector field f_*X is defined only on $\operatorname{im} f \subseteq \mathcal{N}$). Conversely, if $Y \in \mathfrak{X}(\mathcal{N})$ is a vector field on $\mathcal{N}$, then we define its *pull-back* as the field $f^*Y = (f^{-1})_*Y = (Tf)^{-1} \circ Y \circ f$. According to the remark above, the push-forward and the Lie bracket commute:

$$[f_*X_1, f_*X_2] = f_*\big([X_1, X_2]\big) \tag{C.9}$$

for arbitrary vector fields $X_1, X_2 \in \mathfrak{X}(\mathcal{M})$. ◁

Vector fields are often used to model geometrically (autonomous) ordinary differential equations. Let X be a vector field on the manifold $\mathcal{M}$. An *integral curve* of X is a curve $\gamma : \mathbb{I} \subseteq \mathbb{R} \to \mathcal{M}$ such that $\dot\gamma(t) = X|_{\gamma(t)}$ for all $t \in \mathbb{I}$. In some local coordinates $(\mathbf{x})$, this definition means that the components $\gamma^i(t)$ satisfy the ordinary differential equation $\dot\gamma^i(t) = v^i\big(\gamma(t)\big)$ for $X = v^i \partial_{x^i}$. The classical existence and uniqueness theorem for ordinary differential equations implies that we have for every point $p \in \mathcal{M}$ exactly one integral curve γ with $\gamma(0) = p$.

The *flow* of the vector field X is the map $\phi : \mathcal{D} \subseteq \mathbb{R} \times \mathcal{M} \to \mathcal{M}$ defined by $\phi(t,p) = \gamma_p(t)$ where γ_p is the unique integral curve of X satisfying $\gamma_p(0) = p$. If the flow is defined on the whole space $\mathbb{R} \times \mathcal{M}$, then the vector field X is called *complete*. In general, the flow ϕ is defined for each $p \in \mathcal{M}$ only for values t in some open interval $\mathbb{I}_p \ni 0$. It is often convenient to consider ϕ as a function of p only for fixed t or conversely as a function of t only for fixed p; this yields a diffeomorphism $\phi_t : \mathcal{M} \to \mathcal{M}$ and a map $\phi_p : \mathbb{R} \to \mathcal{M}$, respectively. This notation should not be confused with partial derivatives of ϕ!

The computation of the flow for a given vector field X, i. e. the integration of the corresponding ordinary differential equation, is often referred to as *exponentiation* of X with the suggestive notation $\exp(tX)p = \phi(t,p)$. The definition of the flow implies then

$$\frac{d}{dt}\big[\exp(tX)p\big] = X_{\exp(tX)p}\,. \tag{C.10}$$

It follows again from the classical existence and uniqueness theorem for ordinary differential equations that

$$\exp(sX)\big(\exp(tX)p\big) = \exp\big((s+t)X\big)p \tag{C.11}$$

(which partially explains the terminology).

Lemma C.2.3 (Rectification Lemma). *Let X be a vector field that does not vanish at the point $p \in \mathcal{M}$. Then in a neighbourhood $\mathcal{U}$ of p local coordinates $(\mathbf{y})$ exist such that $X = \partial_{y^1}$ in $\mathcal{U}$ (the transformation to these coordinates is known as a* rectification *or a* straightening out *of the vector field X).*

Proof. Let $(\mathbf{x})$ be some local coordinates in a neighbourhood of the point p and $X = v^i(\mathbf{x})\partial_{x^i}$. Without loss of generality, we assume that $p = 0$ and that $X|_p = \partial_{x^1}$, i. e. $v^1(0) = 1$ and $v^k(0) = 0$ for $2 \leq k \leq m$. Because of the continuity of the components v^i, this assumption implies that $v^1(\mathbf{x}) > 0$ in some neighbourhood of p and thus the integral curves of X intersect there the hyperplane $x^1 = 0$ transversally. Hence, in a suitably chosen neighbourhood $\mathcal{U}$ of p every point $x \in \mathcal{U}$ with local coordinates $(\mathbf{x})$ may be considered as the image under the flow of a point y on this hyperplane with local coordinates $(0, y^2, \dots, y^m)$. Consequently, the flow of X induces in $\mathcal{U}$ a local diffeomorphism (i. e. a coordinate transformation) $(\mathbf{y}) \mapsto (\mathbf{x})$ by setting $\mathbf{x} = \exp(y^1 X)y$. In the new coordinates $\exp(tX)(y^1, \dots, y^m) = (y^1 + t, y^2, \dots, y^m)$ by (C.11), so that in $\mathcal{U}$ the flow corresponds to a simple translation in y^1-direction. But this observation immediately implies that $X = \partial_{y^1}$ in $\mathcal{U}$ by (C.10). □

The flow allows us an alternative definition of the Lie bracket. We define the *Lie derivative* of a vector field Y with respect to another field X by

$$\mathcal{L}_X Y = \frac{d}{dt}\left[\phi_t^* Y\right]_{t=0} \tag{C.12}$$

where ϕ_t^* is the pull-back of the constant time diffeomorphism for the flow ϕ of X. (C.12) may be interpreted as taking the limit of difference quotients along the flow lines of X. It is not difficult to show that $\mathcal{L}_X Y = [X, Y]$.

Recall that the cotangent bundle was defined as the dual of tangent bundle. Now we introduce as dual objects to vector fields differential forms.

Definition C.2.4. Sections of the bundle $T^*\mathcal{M}$ are called *(differential) one-forms*. We write $\Omega^1(\mathcal{M})$ for $\Gamma_{loc}(\tau^*_{\mathcal{M}})$.

Just like $\mathfrak{X}(\mathcal{M})$, the bundle $\Omega^1(\mathcal{M})$ possesses the structure of an $\mathcal{F}(\mathcal{M})$-module. In local coordinates $(\mathbf{x})$, a basis is given by the forms $dx^i : p \mapsto dx^i|_p$. Obviously, it is just the dual basis to $\{\partial_{x^i}\}$.

Given the name "one-forms," one expects of course that also two-forms and so on exist. Indeed, like over any finite-dimensional vector space, we can build at each point $p \in \mathcal{M}$ the *exterior algebra* $\Lambda(T_p^*\mathcal{M})$. It is a graded ring (see Appendix B.1) with $\Lambda^0(T_p^*\mathcal{M}) = \mathbb{R}$ and $\Lambda^1(T_p^*\mathcal{M}) = T_p^*\mathcal{M}$. The forms of higher degree are obtained with the *wedge product*: $\wedge : \Lambda^k(T_p^*\mathcal{M}) \times \Lambda^\ell(T_p^*\mathcal{M}) \to \Lambda^{k+\ell}(T_p^*\mathcal{M})$. It is associative and distributive but non-commutative. Instead we have for $\omega \in \Lambda^k(T_p^*\mathcal{M})$ and $\eta \in \Lambda^\ell(T_p^*\mathcal{M})$ that $\omega \wedge \eta = (-1)^{k\ell}\eta \wedge \omega$. The precise definition is obtained by anti-symmetrisation of the tensor product and requires the choice of a combinatorial factor. We will use the following convention

$$(\omega \wedge \eta)(v_1, \dots, v_{k+\ell}) = \\ \frac{1}{k!\ell!} \sum_{\pi \in S_{k+\ell}} \varepsilon(\pi)\omega(v_{\pi(1)}, \dots, v_{\pi(k)})\eta(v_{\pi(k+1)}, \dots, v_{\pi(k+\ell)}) \,. \tag{C.13}$$

Here $S_{k+\ell}$ is the symmetric group containing all permutations of the numbers $1,\dots,k+\ell$ and $\varepsilon(\pi)=\pm 1$ depending on whether we are dealing with an even or an odd permutation. Some authors use a factor $1/(k+\ell)!$, but we will stick to the convention above.

The union $\bigcup_{p\in\mathcal{M}}\{p\}\times\Lambda^k(T_p^*\mathcal{M})$ leads to a vector bundle. Its local sections are contained in $\Omega^k(\mathcal{M})$ and called *(differential) k-forms*. Zero-forms are (local) functions $\mathcal{M}\to\mathbb{R}$. A k-form $\omega\in\Omega^k(\mathcal{M})$ may be evaluated on k vector fields $X_i\in\mathfrak{X}(\mathcal{M})$ to a function in $\mathcal{F}(\mathcal{M})$:

$$\omega(X_1,\dots,X_k)(p)=\omega_p(X_{1,p},\dots,X_{k,p})\,. \tag{C.14}$$

In local coordinates $(\mathbf{x})$ on the manifold $\mathcal{M}$, a basis of $\Omega^k(\mathcal{M})$ is given by the forms $\mathrm{d}x^I=\mathrm{d}x^{i_1}\wedge\cdots\wedge\mathrm{d}x^{i_k}$ where $I=(i_1,\dots,i_k)$ is a sorted repeated index with $1\le i_1<i_2<\cdots<i_k\le n$ (see Appendix A.1 for our conventions on multi indices). However, typically one writes a k-form locally as $\omega=\omega_I\mathrm{d}x^I$ where the summation is over all repeated indices (i. e. including the unsorted ones) and where the smooth functions ω_I are assumed to be totally antisymmetric in the entries of I.

Because of the antisymmetry of the wedge product, it is easy to see that on an n-dimensional manifold any form of degree greater than n trivially vanishes. We write $\Omega(\mathcal{M})$ for the direct sum $\bigoplus_{k=0}^n\Omega^k(\mathcal{M})$. A nowhere vanishing n-form is called a *volume form*, as it permits the computation of volumes via integration. Note that such forms do not exist on every manifold but only on *orientable* ones. If ω_1 and ω_2 are two volume forms on a manifold $\mathcal{M}$, then a nowhere vanishing function $f\in\mathcal{F}(\mathcal{M})$ exists such that $\omega_1=f\omega_2$.

Remark C.2.5. Given a smooth map $f:\mathcal{N}\to\mathcal{M}$ between two manifolds $\mathcal{N}$, $\mathcal{M}$, we can *pull-back* forms. If $\omega\in\Omega^k(\mathcal{M})$ is a k-form on $\mathcal{M}$, then we define via the tangent map $Tf:T\mathcal{N}\to T\mathcal{M}$ a k-form $f^*\omega\in\Omega^k(\mathcal{N})$ on $\mathcal{N}$ by setting

$$(f^*\omega)_p(v_1,\dots,v_k)=\omega_{f(p)}\bigl(T_pf(v_1),\dots,T_pf(v_k)\bigr) \tag{C.15}$$

for arbitrary vectors $v_1,\dots,v_k\in T_p\mathcal{N}$. Note that in contrast to the pull-back of a vector field (cf. Remark C.2.2) it is here not necessary to require that f is a diffeomorphism. In particular, this construction can be applied to an inclusion map ι, i.e. when $\mathcal{N}$ is a submanifold of $\mathcal{M}$. In the case that f is a diffeomorphism, we can additionally *push-forward* forms: if now $\omega\in\Omega^k(\mathcal{N})$, then we set $f_*\omega=(f^{-1})^*\omega\in\Omega^k(\mathcal{M})$.

In local coordinates we get the following picture. Let $(x^1,\dots,x^n)$ be coordinates in some chart $\mathcal{U}\subseteq\mathcal{N}$ and $(y^1,\dots,y^m)$ coordinates in a chart $\mathcal{V}\subseteq\mathcal{M}$ where the charts are chosen such that $f(\mathcal{U})\subseteq\mathcal{V}$ and we can write $y^j=f^j(\mathbf{x})$. If $\omega=\omega_J\mathrm{d}y^J$ is a k-form defined on $\mathcal{V}$, then its pull-back takes on $\mathcal{U}$ the local form

$$f^*\omega=\frac{\partial f^J}{\partial x^I}\omega_J\mathrm{d}x^I\,. \tag{C.16}$$

Here $\partial f^J/\partial x^I$ is a short-hand for the product $(\partial f^{j_1}/\partial x^{i_1})\cdots(\partial f^{j_k}/\partial x^{i_k})$. ◁

There exist several important graded derivations on $\Omega(\mathcal{M})$ which thus acquires the structure of a differential ring. The grading of a derivation D is denoted by $|D|$ and means that $D : \Omega^k(\mathcal{M}) \to \Omega^{k+|D|}(\mathcal{M})$.

Definition C.2.6. The *interior derivative* with respect to a vector field $X \in \mathfrak{X}(\mathcal{M})$ is a map $\iota_X : \Omega^{k+1}(\mathcal{M}) \to \Omega^k(\mathcal{M})$ defined for $k > 0$ by

$$(\iota_X \omega)(X_1, \dots, X_k) = \omega(X, X_1, \dots, X_k) \tag{C.17}$$

for all forms $\omega \in \Omega^{k+1}(\mathcal{M})$ and vector fields $X_1, \dots, X_k \in \mathfrak{X}(\mathcal{M})$.

Thus the interior derivative reduces the form degree by one: $|\iota| = -1$. Its graded Leibniz rule is easily derived from the properties of the wedge product of forms: $\iota_X(\omega_1 \wedge \omega_2) = (\iota_X \omega_1) \wedge \omega_2 + (-1)^{k_1} \omega_1 \wedge (\iota_X \omega_2)$, if $\omega_i \in \Omega^{k_i}(\mathcal{M})$. In local coordinates, the interior derivative corresponds to what is classically called a "contraction" in tensor analysis: if $X = v^i \partial_{x^i}$ and $\omega = \omega_I \mathrm{d}x^I$, then

$$\iota_X \omega = \omega_{i_1 i_2 \cdots i_k} v^{i_1} \mathrm{d}x^{i_2} \wedge \cdots \wedge \mathrm{d}x^{i_k} \,. \tag{C.18}$$

The next derivation increases the form degree, but it is less easy to introduce. We take here an algebraic approach and define it via some characteristic properties. One can show that there exists one and only one map satisfying all these properties (see e. g. [81, Chapt. 3, Thm. 2.1]).

Definition C.2.7. The *exterior derivative* of differential forms is the unique map $\mathrm{d} : \Omega^k(\mathcal{M}) \to \Omega^{k+1}(\mathcal{M})$ satisfying the following properties.

(i) d is an $\mathbb{R}$-linear map: for all $a_1, a_2 \in \mathbb{R}$ and $\omega_1, \omega_2 \in \Omega^k(\mathcal{M})$

$$\mathrm{d}(a_1\omega_1 + a_2\omega_2) = a_1 \mathrm{d}\omega_1 + a_2 \mathrm{d}\omega_2 \,. \tag{C.19}$$

(ii) d satisfies a graded Leibniz rule: for all $\omega_1 \in \Omega^{k_1}(\mathcal{M})$ and $\omega_2 \in \Omega^{k_2}(\mathcal{M})$

$$\mathrm{d}(\omega_1 \wedge \omega_2) = (\mathrm{d}\omega_1) \wedge \omega_2 + (-1)^{k_1} \omega_1 \wedge \mathrm{d}\omega_2 \,. \tag{C.20}$$

(iii) d is nilpotent: $\mathrm{d} \circ \mathrm{d} = 0$.

(iv) If $f : \mathcal{M} \to \mathbb{R}$ is a zero-form, the one-form $\mathrm{d}f$ is locally given by

$$\mathrm{d}f(x) = \frac{\partial f}{\partial x^i}(x) \mathrm{d}x^i \,. \tag{C.21}$$

Given a k-form $\omega \in \Omega^k(\mathcal{M})$, an explicit coordinate-free expression for $\mathrm{d}\omega$ is

$$\begin{aligned} \mathrm{d}\omega(X_1, \dots, X_{k+1}) = &\sum_{i=1}^{k+1} (-1)^{i+1} X_i\big(\omega(X_1, \dots, \widehat{X_i}, \dots, X_{k+1})\big) \\ &+ \sum_{i<j} \omega\big([X_i, X_j], X_1, \dots, \widehat{X_i}, \dots, \widehat{X_j}, \dots, X_{k+1}\big) \end{aligned} \tag{C.22}$$

where the hat indicates that the corresponding vector is omitted. In particular, for a smooth function $f \in \mathcal{F}(\mathcal{M})$ we obtain $\mathrm{d}f(X) = X(f)$; thus the one-form $\mathrm{d}f$ corresponds to the fibre component of the tangent map Tf. An important property of the exterior derivative is that it commutes with pull-backs: if $f : \mathcal{M} \to \mathcal{N}$ is a smooth map, then $\mathrm{d} \circ f^* = f^* \circ \mathrm{d}$. In local coordinates, one obtains for $\omega = \omega_I \mathrm{d}x^I$ that

$$\mathrm{d}\omega = \frac{\partial \omega_I}{\partial x^i} \mathrm{d}x^i \wedge \mathrm{d}x^I \,. \tag{C.23}$$

Remark C.2.8. If $f : \mathcal{M} \to \mathbb{R}$ is a smooth function on $\mathcal{M}$ such that zero is a regular value of it, then by Proposition C.1.4 its zero set $\mathcal{N} = \{p \in \mathcal{M} \mid f(p) = 0\}$ is a regular submanifold of $\mathcal{M}$. Its tangent space $T_p\mathcal{N}$ at a point $p \in \mathcal{N}$ may be identified with a linear subspace of the full tangent space $T_p\mathcal{M}$. This subspace is easily determined with the help of the one-form $\mathrm{d}f \in \Omega^1(\mathcal{M})$: $T_p\mathcal{N} = \{v \in T_p\mathcal{M} \mid \mathrm{d}f_p(v) = 0\}$. As zero is a regular value, by definition the differential $\mathrm{d}f_p$ does not vanish for any point $p \in \mathcal{N}$ so that this condition is always non-trivial.

This observation follows directly from our geometric definition of tangent vectors as equivalence class of curves. If $\gamma : \mathbb{I} \to \mathcal{N}$ is a curve with $\gamma(0) = p$, then obviously $f\big(\gamma(t)\big) \equiv 0$. Differentiating this relation with respect to t yields $\mathrm{d}f\big(\dot{\gamma}(t)\big) \equiv 0$. Evaluation at $t = 0$ yields our result.

More generally, if the submanifold $\mathcal{N}$ is the common zero set of the functions $f^1, \dots, f^r \in \mathcal{F}(\mathcal{M})$, then any vector $v \in T_p\mathcal{N}$ satisfies $\mathrm{d}f^1_p(v) = \cdots = \mathrm{d}f^r_p(v) = 0$ and vice versa. The vector subspace $N^*_p\mathcal{N} \subset T^*_p\mathcal{N}$ spanned by $\mathrm{d}f^1_p, \dots, \mathrm{d}f^r_p$ is sometimes called the *conormal space* of $\mathcal{N}$ at the point p. $\lhd$

Remark C.2.9. We call a differential k-form $\omega \in \Omega^k(\mathcal{M})$ *exact*, if there exists a $(k-1)$-form $\eta \in \Omega^{k-1}(\mathcal{M})$ such that $\omega = \mathrm{d}\eta$. A form ω with $\mathrm{d}\omega = 0$ is *closed*. Obviously, any exact form is also closed, as $\mathrm{d}^2 = 0$. The *Poincaré Lemma* asserts that locally (more precisely, in star-shaped domains) the converse also holds: every closed form is exact. Globally, this is generally not true: the global closed forms that are not exact generate the *de Rham cohomology* of $\mathcal{M}$, a topological invariant. $\lhd$

The last derivation is again a Lie derivative along the flow lines of a vector field. It obviously preserves the form degree.

Definition C.2.10. The *Lie derivative* of differential k-forms with respect to a vector field $X \in \mathfrak{X}(\mathcal{M})$ is the map $\mathcal{L}_X : \Omega^k(\mathcal{M}) \to \Omega^k(\mathcal{M})$ defined by

$$\mathcal{L}_X \omega = \frac{d}{dt} \big[\phi_t^* \omega\big]_{t=0} \tag{C.24}$$

where ϕ denotes the flow of X.

For zero-forms, i. e. functions, this definition yields $\mathcal{L}_X f = Xf$. The Lie derivative $\mathcal{L}_X$ satisfies for arbitrary differential forms $\omega_1, \omega_2 \in \Omega(\mathcal{M})$ the following (ungraded!) Leibniz rule $\mathcal{L}_X(\omega_1 \wedge \omega_2) = \mathcal{L}_X(\omega_1) \wedge \omega_2 + \omega_1 \wedge \mathcal{L}_X(\omega_2)$. For a k-form ω we obtain as coordinate-free expression

$$\mathcal{L}_X\omega(X_1,\dots,X_k) = X\big(\omega(X_1,\dots,X_k)\big) + \\ + \sum_{i=1}^{k}\omega(X_1,\dots,[X,X_i],\dots,X_k)\,. \tag{C.25}$$

Locally, we find for $X = v^i\partial_{x^i}$ and $\omega = \omega_I\mathrm{d}x^I$ that

$$\mathcal{L}_X\omega = \Bigg(v^i\frac{\partial\omega_{i_1\cdots i_k}}{\partial x^i} + \\ \frac{\partial v^i}{\partial x^{i_1}}\omega_{ii_2\cdots i_k} + \cdots + \frac{\partial v^i}{\partial x^{i_k}}\omega_{i_1\cdots i_{k-1}i}\Bigg)\mathrm{d}x^{i_1}\wedge\cdots\wedge\mathrm{d}x^{i_k}\,. \tag{C.26}$$

If we introduce a graded commutator for derivations on a manifold $\mathcal{M}$, i.e. if we define $[D_1,D_2] = D_1\circ D_2 - (-1)^{|D_1||D_2|}D_2\circ D_1$, then the following equations prove that all three introduced derivations together form an infinite-dimensional Lie algebra with the graded commutator as Lie bracket:

$$[\iota_X,\mathrm{d}] = \mathcal{L}_X\,, \tag{C.27a}$$

$$[\iota_X,\iota_Y] = 0\,, \tag{C.27b}$$

$$[\mathcal{L}_X,\mathrm{d}] = 0\,, \tag{C.27c}$$

$$[\mathcal{L}_X,\iota_Y] = \iota_{[X,Y]}\,, \tag{C.27d}$$

$$[\mathcal{L}_X,\mathcal{L}_Y] = \mathcal{L}_{[X,Y]}\,. \tag{C.27e}$$

The first of these relations, explicitly written out $\mathcal{L}_X = \iota_X\circ\mathrm{d} + \mathrm{d}\circ\iota_X$ is very useful in computations and has therefore been nicknamed *Cartan's magic formula*.

Given a set of differential forms $\Theta \subseteq \Omega(\mathcal{M})$, one may consider the *algebraic ideal* $\langle\Theta\rangle_{\text{alg}} \subseteq \Omega(\mathcal{M})$ consisting of all finite linear combinations $\sum_{\theta\in\Theta}\omega_\theta\wedge\theta$ with coefficients $\omega_\theta \in \Omega(\mathcal{M})$ (although $\Omega(\mathcal{M})$ is a non-commutative ring, left and right ideals coincide here, as the wedge product is skew-symmetric). We may also think of $\Omega(\mathcal{M})$ as a differential ring with respect to the exterior derivative d and introduce the *differential ideal* $\langle\Theta\rangle_{\text{diff}}$ as the smallest algebraic ideal which contains Θ and is closed under d. Because of the nilpotency of d, we find that $\langle\Theta\rangle_{\text{diff}} = \langle\Theta\cup\mathrm{d}\Theta\rangle_{\text{alg}}$.

Remark C.2.11. Sometimes it is very useful to generalise to *vector valued forms*. Let $\nu : \mathcal{V} \to \mathcal{M}$ be an arbitrary vector bundle over $\mathcal{M}$ (in the simplest case the trivial bundle $\mathrm{pr}_1 : \mathcal{M}\times\mathcal{V} \to \mathcal{M}$ for some vector space $\mathcal{V}$). We define $\mathcal{V}$-valued k-forms as (local) sections of the bundle $\Lambda^k(T^*\mathcal{M}) \underset{\mathcal{M}}{\otimes} \mathcal{V}$ and use the notation $\Omega^k(\mathcal{M},\mathcal{V})$ for them. The evaluation of such a form on k vector fields $X_i \in \mathfrak{X}(\mathcal{M})$ yields a section in $\Gamma_{loc}(\nu)$. As an artifical example we may consider vector fields as $T\mathcal{M}$-valued zero-forms. i.e. $\mathfrak{X}(\mathcal{M}) = \Omega^0(\mathcal{M},T\mathcal{M})$.

Note that generally vector valued forms cannot be wedged with each other, as it is unclear how to multiply their vector part. An important exception occurs when the forms take their values in a Lie algebra $\mathfrak{g}$, as this allows for a natural modification of (C.13): if $\omega \in \Lambda^k(T_p^*\mathcal{M},\mathfrak{g})$ and $\eta \in \Lambda^\ell(T_p^*\mathcal{M},\mathfrak{g})$, then we define for arbitrary vectors $v_1,\dots,v_{k+\ell} \in T_p\mathcal{M}$

$$(\omega \wedge \eta)(v_1,\dots,v_{k+\ell}) = \frac{1}{k!\ell!}\sum_{\pi\in S_{k+\ell}} \varepsilon(\pi)\big[\omega(v_{\pi(1)},\dots,v_{\pi(k)}),\eta(v_{\pi(k+1)},\dots,v_{\pi(k+\ell)})\big] \tag{C.28}$$

where the square brackets denote the commutator in the Lie algebra $\mathfrak{g}$. Very often one finds the notation $[\omega,\eta]$ instead of $\omega\wedge\eta$.

With similar modifications also other expressions may be "rescued." Consider for example the coordinate-free expression (C.22) for the exterior derivative. If we interpret the terms in the first sum on the right hand side as Lie derivatives with respect to X_i, then the obvious extension of this expression to $T\mathcal{M}$-valued forms $\omega\in\Omega^k(\mathcal{M},T\mathcal{M})$ is given by

$$\begin{aligned} d\omega(X_1,\dots,X_{k+1}) &= \sum_{i=1}^{k+1}(-1)^{i+1}\big[X_i,\omega(X_1,\dots,\widehat{X}_i,\dots,X_{k+1})\big] \\ &\quad + \sum_{i<j}\omega\big([X_i,X_j],X_1,\dots,\widehat{X}_i,\dots,\widehat{X}_j,\dots,X_{k+1}\big)\,. \end{aligned} \tag{C.29}$$

The same holds for the expression (C.25) for the Lie derivative. ◁

C.3 Distributions and the Frobenius Theorem

A *distribution* $\mathcal{D}$ on a manifold $\mathcal{M}$ assigns to each point $p\in\mathcal{M}$ a vector subspace $\mathcal{D}_p\subseteq T_p\mathcal{M}$. We consider only *smooth* distributions which can be spanned by (local) vector fields. More precisely, consider the set $\mathfrak{X}_{\mathcal{D}}$ of all local vector fields X with the property $X_p\in\mathcal{D}_p$ for all points p in the domain of X. Then $\mathcal{D}$ is smooth, if conversely $\mathcal{D}_p=\langle\{X_p \mid X\in\mathfrak{X}_{\mathcal{D}}\}\rangle$. Note that our definition does not imply that all spaces $\mathcal{D}_p$ possess the same dimension, as it is often required. If this is the case, we speak of a distribution of *constant rank* otherwise of a *singular distribution*. The simplest example of a singular distribution is generated by a vector field with zeros.

Example C.3.1. Even if a distribution has constant rank r, we cannot automatically assume that it may be *globally* generated by r everywhere linearly independent vector fields. Figure C.1 shows a typical situation where a one-dimensional distribution cannot be generated by a smooth vector field. The distribution is defined on the manifold $\mathcal{M}=\mathbb{R}^2\setminus\{(1,0),(-1,0)\}$ as the tangent space of a one-parameter family of ellipses. In the limit the ellipses degenerate to the line segment connecting $(-1,0)$ and $(1,0)$. Obviously, everywhere outside this segment we easily find a smooth vector field by simply orienting the ellipses and choosing the corresponding unit tangent vector. However, no smooth continuation to the line segment is possible, as the limits from above and below point in opposite directions. *Locally*, we have no problem in defining the distribution by a single smooth vector field, as long as we avoid using charts around the singular points $(1,0)$ and $(-1,0)$.

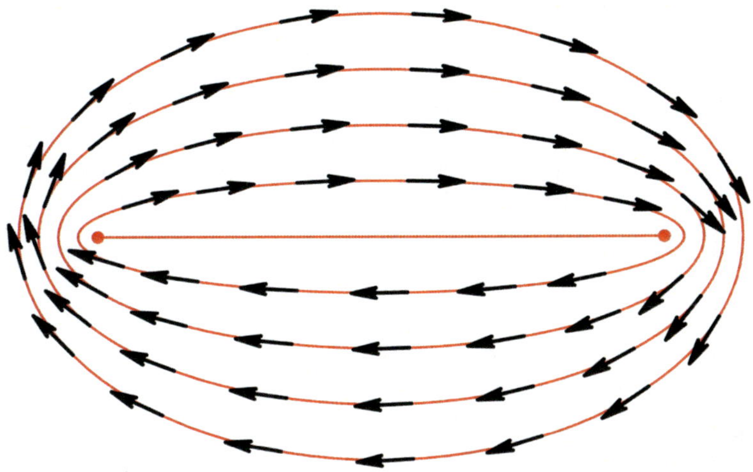

Fig. C.1 One-dimensional distribution not generated by a vector field

This is only a simple example. A more extensive discussion of the singularities of one-dimensional distributions is contained in [428, Chapt. 4, Add. 2] where the index of a distribution at a point on a manifold is introduced. In Figure C.1, the points $(1,0)$ and $(-1,0)$ are both of index $\frac{1}{2}$. ◁

A smooth distribution $\mathcal{D}$ is *involutive*, if it is closed under the Lie bracket: if X and Y are two vector fields contained in $\mathcal{D}$, then $[X,Y]$ lives in $\mathcal{D}$, too. An *integral manifold* of the distribution $\mathcal{D}$ is a submanifold $\mathcal{N} \subseteq \mathcal{M}$ such that $T_p\mathcal{N} \subseteq \mathcal{D}_p$ for all points $p \in \mathcal{N}$. Obviously, the simplest example of an integral manifold is the integral curve of a vector field (considered as a one-dimensional distribution).

The distribution is *integrable*, if every point $p \in \mathcal{M}$ lies on a integral manifold of dimension $\dim \mathcal{D}_p$. Many authors require that for an integral manifold always $T_p\mathcal{N} = \mathcal{D}_p$; then integrability is equivalent to the existence of integral manifolds. The (maximally extended) integral manifolds of maximal dimension are the leaves of a foliation of $\mathcal{M}$ (see e. g. [3, Theorem 4.4.7]).

Example C.3.2. The simplest example of an integrable and involutive distribution is $T\mathcal{M}$ itself: it is obviously closed under the Lie bracket and $\mathcal{M}$ is an integral manifold of maximal dimension. Another important example is the vertical bundle of a fibred manifold $\pi : \mathcal{E} \to \mathcal{B}$ (cf. Remark C.4.7) whose integral manifolds of maximal dimension are the fibres $\mathcal{E}_b$ with $b \in \mathcal{B}$. ◁

Theorem C.3.3 (Frobenius). *Let $\mathcal{D} \subseteq T\mathcal{M}$ be a distribution of constant rank on a manifold $\mathcal{M}$. Then $\mathcal{D}$ is integrable, if and only if it is involutive.*

Proof. One direction is easy. Assume that $\mathcal{D}$ is integrable and $\iota : \mathcal{N} \hookrightarrow \mathcal{M}$ is an integral manifold of maximal dimension; i. e. $\mathcal{D}|_{\mathcal{N}} = T\mathcal{N}$. Let X, Y be two vector fields contained in $\mathcal{D}$. Then their restrictions to $\mathcal{N}$ define vector fields in $\mathfrak{X}(\mathcal{N})$ and thus their Lie bracket is again contained in $T\mathcal{N}$. By (C.9), we have $\iota_*\big([X|_{\mathcal{N}}, Y|_{\mathcal{N}}]\big) = \big[\iota_*(X|_{\mathcal{N}}), \iota_*(Y|_{\mathcal{N}})\big] \in \mathcal{D}|_{\mathcal{N}}$ and as any point $p \in \mathcal{M}$ lies on some integral manifold we find that $[X,Y] \in \mathcal{D}$ everywhere on $\mathcal{M}$.

For the converse we use a local coordinate approach based on an induction over the Rectification Lemma C.2.3 for a single vector field. More precisely, we show that coordinates $(\mathbf{y})$ exists such that the vector fields $\partial_{y^1}, \dots, \partial_{y^r}$ (where $r = \operatorname{rank}\mathcal{D}$) locally generate the distribution $\mathcal{D}$. This fact implies our claim, as then obviously the manifolds locally described by $y^\ell = y_0^\ell = Const$ for $r < \ell \leq m$ are r-dimensional integral manifolds of $\mathcal{D}$.

Let X_1 be an arbitrary non-vanishing vector field in $\mathcal{D}$. By the Rectification Lemma C.2.3, local coordinates $(\bar{\mathbf{y}})$ exist such that $X_1 = \partial_{\bar{y}^1}$. For $r = 1$, this observation already suffices to prove our claim.

For $r > 1$ we choose $r-1$ further vector fields $X_k = \sum_{i=2}^m a_k^i(\bar{\mathbf{y}})\partial_{\bar{y}^i}$ with $2 \leq k \leq r$ which together with X_1 generate $\mathcal{D}$. Performing Gaussian elimination on the coefficient matrix (a_k^i), we may assume (after possibly renumbering the variables) that actually $X_k = \partial_{\bar{y}^k} + \sum_{i=r+1}^m b_k^i(\bar{\mathbf{y}})\partial_{\bar{y}^i}$ (in the terminology of Example 7.2.12 one says that we take a Jacobian system as basis). Then $[X_1, X_k] = \sum_{i=r+1}^m \frac{\partial b_k^i(\bar{\mathbf{y}})}{\partial \bar{y}^1}\partial_{\bar{y}^i}$. By assumption, the distribution $\mathcal{D}$ is involutive. Hence the vector fields arising this way must be contained in $\mathcal{D}$ and expressible as linear combinations of our basis. Obviously, because of the special form of our fields, this is possible, if and only if all coefficients b_k^i are independent of the variable $\bar{y}^1$.

Now we may reduce to the distribution of rank $r-1$ generated by $X_2, \dots, X_k$ and, by our induction assumption, there exists a change of coordinates $\bar{\mathbf{y}} \mapsto \mathbf{y}$ with $y^1 = \bar{y}^1$ and $y^i = \phi^i(\bar{y}^2, \dots, \bar{y}^m)$ for $2 \leq i \leq m$ such that $X_k = \partial_{y^k}$ for $2 \leq k \leq r$. This fact proves our claim. □

Theorem C.3.3 represents the modern geometric formulation of the Frobenius Theorem due to Chevalley [82] that is today found in almost any textbook on differential geometry. The original formulation by Frobenius [142] is concerned with the solvability of certain systems of partial differential equations and becomes better visible in Example 7.2.12. We considered here only the case of distributions of constant rank. Hermann [205, 206] discusses also singular distributions; further generalisations are studied by Sussman [445].

Remark C.3.4. If $\mathcal{D}$ is a distribution, then the *(first) derived distribution* is defined as $\mathcal{D}' = \mathcal{D} + [\mathcal{D}, \mathcal{D}]$, i. e. we add all Lie brackets of vector fields contained in $\mathcal{D}$. Thus for an involutive distribution $\mathcal{D} = \mathcal{D}'$, whereas for a general distribution $\mathcal{D} \subsetneq \mathcal{D}'$.

If the distribution $\mathcal{D}$ is generated by vector fields $X_1, \dots, X_r$ and if for the derived distribution we need additional generators $Y_1, \dots, Y_s$, then we may write

$$[X_i, X_j] = A_{ij}^k X_k + B_{ij}^a Y_a \tag{C.30}$$

with some smooth functions $A_{ij}^k, B_{ij}^a \in \mathcal{F}(\mathcal{M})$. (C.30) are often called the *structure equations* of the distribution $\mathcal{D}$. ◁

A *codistribution* Γ on a manifold $\mathcal{M}$ maps each point $p \in \mathcal{M}$ to a vector subspace $\Gamma_p \subseteq T_p^*\mathcal{M}$ of the cotangent bundle. The definitions of smoothness and constant rank is analogous to the ones for distributions. A codistribution Γ generated by some one-forms $\omega_1, \dots, \omega_r$ is classically called a *Pfaffian system*. Distributions and codistributions are dual to each other via the notion of an *annihilator*. Associated with each distribution $\mathcal{D} \subseteq T\mathcal{M}$ is its dual codistribution

$$\mathcal{D}^0 = \{\omega_p \in T_p^*\mathcal{M} \mid p \in \mathcal{M},\ \omega_p(\mathcal{D}_p) = 0\} \tag{C.31}$$

(of dimension $\dim \mathcal{M} - \dim \mathcal{D} = \operatorname{codim} \mathcal{D}$). Conversely, a codistribution $\Gamma \subseteq T^*\mathcal{M}$ induces the dual distribution

$$\Gamma^0 = \{X_p \in T_p\mathcal{M} \mid p \in \mathcal{M},\ \Gamma(X_p) = 0\}\ . \tag{C.32}$$

A codistribution Γ is *involutive*, if the exterior derivative of any form $\omega \in \Gamma$ vanishes modulo Γ, i. e. if $\mathrm{d}\omega \in \Gamma \wedge \Omega^1(\mathcal{M})$. In the case of a Pfaffian system this condition means that the algebraic and the differential ideal generated by the system coincide. As a simple test one must only check that $\mathrm{d}\omega_i \wedge \omega_1 \wedge \dots \wedge \omega_r = 0$ for $1 \leq i \leq r$. An *integral manifold* of a codistribution Γ is a submanifold $\mathcal{N} \subseteq \mathcal{M}$ such that the pull-back $\iota^*(\Gamma)$ with respect to the inclusion map $\iota : \mathcal{N} \hookrightarrow \mathcal{M}$ vanishes: $\iota^*\omega = 0$ for all $\omega \in \Gamma$. The codistribution Γ is *integrable*, if it possesses at any point $p \in \mathcal{M}$ an integral manifold of dimension $\operatorname{codim} \Gamma_p$.

Proposition C.3.5. *The distribution $\mathcal{D}$ is integrable (involutive), if and only if its annihilator $\mathcal{D}^0$ is integrable (involutive).*

Proof. We consider first the involutive case. Let $X, Y \in \mathcal{D}$ and $\omega \in \mathcal{D}^0$ be arbitrary elements. Then by (C.22)

$$\mathrm{d}\omega(X,Y) = X\big(\omega(Y)\big) - Y\big(\omega(X)\big) + \omega\big([X,Y]\big)\ . \tag{C.33}$$

Now, if $\mathcal{D}^0$ is involutive, then $\mathrm{d}\omega \in \mathcal{D}^0 \wedge \Omega^1(\mathcal{M})$ and thus $\mathrm{d}\omega(X,Y) = 0$. But this fact implies $[X,Y] \in (\mathcal{D}^0)^0 = \mathcal{D}$ and hence $\mathcal{D}$ is involutive, too. Conversely, if $\mathcal{D}$ is involutive, then $[X,Y] \in \mathcal{D}$ and $\omega\big([X,Y]\big) = 0$. This observation entails $\mathrm{d}\omega(X,Y) = 0$ and thus by some elementary linear algebra that $\mathrm{d}\omega \in \mathcal{D}^0 \wedge \Omega^1(\mathcal{M})$.

For the integrable case, let $\iota : \mathcal{N} \hookrightarrow \mathcal{M}$ be a submanifold. $\mathcal{N}$ is an integral manifold of $\mathcal{D}^0$, if and only if $\iota^*\omega = 0$ for all $\omega \in \mathcal{D}^0$. The latter statement is equivalent to $(\iota^*\omega)(X) = \omega(\iota_* X) = 0$ for all $X \in T\mathcal{N}$ and hence to $T\mathcal{N} \subseteq \mathcal{D}|_{\mathcal{N}}$, i. e. to $\mathcal{N}$ being an integral manifold of $\mathcal{D}$. □

As a trivial corollary we obtain a dual version of the Frobenius Theorem: a codistribution of constant rank is integrable, if and only if it is involutive. As in our proof of Theorem C.3.3, the local coordinate formulation says that for an involutive codistribution Γ coordinates $(\mathbf{y})$ exist such that Γ is generated by $\{\mathrm{d}y^1, \dots, \mathrm{d}y^r\}$ where $r = \operatorname{rank} \Gamma$. The integral manifolds of Γ are then locally described by the equations $y^1 = Const, \dots, y^r = Const$.

Remark C.3.6. In Remark C.3.4 we introduced for a distribution $\mathcal{D}$ its derived distribution $\mathcal{D}'$. Similarly, one can define for any codistribution Γ a *derived codistribution* Γ'. One simple approach is via duality: $\Gamma' = \big((\Gamma^0)'\big)^0$. Using (C.22), it is not difficult to see that alternatively we may set

$$\Gamma' = \{\omega \in \Gamma \mid d\omega \equiv 0 \mod \Gamma\}\,. \tag{C.34}$$

Both definitions show that generally $\Gamma' \subseteq \Gamma$ with equality holding, if and only if Γ is involutive. ◁

Remark C.3.7. Higher forms may also be used to measure to what extent a distribution $\mathcal{D}$ is involutive. We consider the "normal" bundle[1] $\mathcal{K} = T\mathcal{M}/\mathcal{D}$ and define the *curvature* of $\mathcal{D}$ as the vector valued two-form $\Omega_{\mathcal{D}} \in \Lambda^2\mathcal{D} \otimes \mathcal{K}$ given by $\Omega_{\mathcal{D}}(X,Y) = \overline{[X,Y]}$ (the bar denotes the equivalence class in $\mathcal{K}$) for arbitrary vector fields $X, Y \in \mathcal{D}$. Obviously, the curvature $\Omega_{\mathcal{D}}$ vanishes, if and only if the considered distribution $\mathcal{D}$ is involutive. ◁

C.4 Connections

Let $\pi : \mathcal{E} \to \mathcal{B}$ be a fibred manifold. The projection π induces in a natural way the notion of a vertical vector which is tangential to the fibres. These vectors can be easily characterised via the tangent map $T\pi : T\mathcal{E} \to T\mathcal{B}$. In adapted coordinates we have simply $\pi(\mathbf{x},\mathbf{u}) = \mathbf{x}$. Thus it follows from (C.3) that $T\pi(\mathbf{x},\mathbf{u};\dot{\mathbf{x}},\dot{\mathbf{u}}) = (\mathbf{x},\dot{\mathbf{x}})$.

Definition C.4.1. The *vertical bundle* of the fibred manifold $\pi : \mathcal{E} \to \mathcal{B}$ is the subbundle $V\pi = \ker T\pi \subset T\mathcal{E}$.

In adapted coordinates, vertical vectors at a given point $\xi \in \mathcal{E}$ are of the form $v = \dot{u}^\alpha \partial_{u^\alpha}|_\xi$, i. e. all their components $\dot{x}^i$ in the directions $\partial_{x^i}|_\xi$ corresponding to the coordinates in the base space $\mathcal{B}$ vanish. This immediately implies that at any point $\xi \in \mathcal{E}$ the dimension of the vertical space $V_\xi\pi$ is just the fibre dimension of $\mathcal{E}$.

Example C.4.2. In the case of the trivial fibred manifold $\mathcal{E} = \mathcal{B} \times \mathcal{F}$ we have the decomposition $T\mathcal{E} = T\mathcal{B} \times T\mathcal{F}$ and we can identify $V\pi = \mathcal{E} \times T\mathcal{F}$. ◁

Lemma C.4.3. *Let $\pi : \mathcal{A} \to \mathcal{B}$ be an arbitrary affine bundle modelled on the vector bundle $\bar{\pi} : \mathcal{V} \to \mathcal{B}$. Then there exists a canonical isomorphism $V_{(b,a)}\pi \cong \mathcal{V}_b$ for each point $(b,a) \in \mathcal{A}$.*

Proof. By definition of an affine bundle, the fibre $\mathcal{A}_b$ at a point $b \in \mathcal{B}$ carries an action of the Abelian group $\mathcal{V}_b$ which we write simply as an addition $a + v$. Choosing a point $a \in \mathcal{A}_b$ and a vector $v \in \mathcal{V}_b$, we may consider in $\mathcal{A}_b$ the curve $\gamma_{a,v} : \mathbb{I} \to \mathcal{A}_b$ given by $t \mapsto a + tv$. Here $\mathbb{I}$ is as usual a real interval containing zero. Since the curve

[1] Alternatively, we could use the derived distribution $\mathcal{D}'$ and set $\mathcal{K} = \mathcal{D}'/\mathcal{D}$.

lies completely in the fibre $\mathcal{A}_b$, the tangent vector $\dot{\gamma}_{a,v}(0)$ defines an element of the vertical space $V_{(b,a)}\pi$ and it is trivial to see that the map $v \mapsto \dot{\gamma}_{a,v}(0)$ represents an isomorphism $\mathcal{V}_b \to V_{(b,a)}\mathcal{A}$. □

Remark C.4.4. Obviously, we may apply the same construction to any vector bundle $\bar{\pi} : \mathcal{V} \to \mathcal{B}$ by considering the addition in the fibre $\mathcal{V}_b$ at an arbitrary point $b \in \mathcal{B}$ as an action of the Abelian group $\mathcal{V}_b$ on itself. Thus we also have an analogous canonical isomorphism $V_{(b,v)}\pi \cong \mathcal{V}_b$ for every point $(b,v) \in \mathcal{V}$. This isomorphism is usually exploited in form of the *vertical lift* $\lambda : \mathcal{V} \underset{\mathcal{B}}{\times} \mathcal{V} \to V\bar{\pi}$. If (b,v_1,v_2) is a point in $\mathcal{V} \underset{\mathcal{B}}{\times} \mathcal{V}$, then we consider the curve $\gamma_{v_1,v_2} : t \mapsto v_1 + tv_2$ which lies completely in the fibre $V_{(b,v_1)}$ and define $\lambda(b,v_1,v_2) = \dot{\gamma}_{v_1,v_2}(0) = (b,v_1;0,v_2)$. ◁

If vertical vectors exist, one expects also horizontal vectors. However, their introduction requires an additional structure on the fibred manifold $\mathcal{E}$, namely a *connection*, as it is not possible to give an intrinsic definition of a horizontal vector for arbitrary fibred manifolds. The simple idea that in adapted coordinates such vectors should have the form $v = \dot{x}^i\partial_{x^i}|_\xi$ does not work, as this form does not remain invariant under changes to other adapted coordinates.

Connections represent a fundamental geometric structure on fibred manifolds and they appear in many places. In particular, some differential equations may be interpreted as connections (see Remark 2.3.6). We consider here a very general concept of connections on arbitrary fibred manifolds which is due to Ehresmann [119]. Special forms have been around already since the middle of the 19th century, in particular for the case that $\mathcal{E} = T\mathcal{B}$.[2] Some references are [183, 207, 256, 257, 261] (note that most textbooks on differential geometry consider exclusively the case of principal connections on principal fibre bundles where one imposes additional symmetry conditions).

Connections are so important and versatile, because many different points of view exist for them. We will use the simplest one via horizontal vectors, i. e. via a distribution, for our definition. But there exists a number of alternative characterisations via certain maps or forms that are all equivalent.

Definition C.4.5. A *connection* on a fibred manifold $\pi : \mathcal{E} \to \mathcal{B}$ is defined by a smooth distribution $H\pi$, the *horizontal bundle*, such that the tangent bundle decomposes as $T\mathcal{E} = H\pi \oplus V\pi$.

Obviously, we have $\dim H_\xi\pi = \dim\mathcal{B}$ at each point $\xi \in \mathcal{E}$ (i. e. the horizontal distribution is of constant rank) and each vector $v_\xi \in T_\xi\mathcal{E}$ may now be decomposed uniquely: $v_\xi = v_\xi^H + v_\xi^V$ with $v_\xi^H \in H_\xi\pi$ and $v_\xi^V \in V_\xi\pi$. This direct decomposition of the tangent bundle $T\mathcal{E}$ induces two natural bundle morphisms: the *horizontal projector* $P^H : T\mathcal{E} \to H\pi$ with $P^H(\xi,v_\xi) = (\xi,v_\xi^H)$ and the complementary *vertical projector* $P^V : T\mathcal{E} \to V\pi$ with $P^V(\xi,v_\xi) = (\xi,v_\xi^V)$. Conversely, each of these projectors uniquely specifies the connection.

[2] Note that if $\mathcal{B}$ has further structures, it may even be possible to define canonically a unique connection. This is for example the case in Riemannian geometry where compatibility with the metric singles out a particular connection, namely the Levi–Civita connection.

Remark C.4.6. We may consider both the vertical and the horizontal projector of a connection as vector valued forms: $P^V \in \Omega^1(\mathcal{E}, V\pi)$ and $P^H \in \Omega^1(\mathcal{E}, H\pi)$. One often calls the vertical projector P^V *connection form*. In the special case of a principal fibre bundle all fibres are diffeomorphic to a Lie group $\mathcal{G}$. Hence each vertical vector may be identified with an element of the corresponding Lie algebra $\mathfrak{g}$ and we may consider P^V as living in $\Omega^1(\mathcal{E}, \mathfrak{g})$. ◁

Alternatively, one may define a connection via its *horizontal lift*. This is a map $A : \mathcal{E} \underset{\mathcal{B}}{\times} T\mathcal{B} \to T\mathcal{E}$ which is linear in the second factor and a right inverse to the map $\rho = (\tau_{\mathcal{E}}, T\pi) : T\mathcal{E} \to \mathcal{E} \underset{\mathcal{B}}{\times} T\mathcal{B}$, i.e. the map A defines a splitting of the following exact sequence of vector bundles

$$0 \longrightarrow V\pi \xrightarrow{\ \iota\ } T\mathcal{E} \underset{A}{\overset{\rho}{\rightleftarrows}} \mathcal{E} \underset{\mathcal{B}}{\times} T\mathcal{B} \longrightarrow 0\,. \tag{C.35}$$

Hence we have $\rho \circ A = \mathrm{id}_{\mathcal{E} \underset{\mathcal{B}}{\times} T\mathcal{B}}$ and one easily verifies that $A \circ \rho = P^H$.

In adapted coordinates ρ maps a vector $\big((\mathbf{x}, \mathbf{u}); (\dot{\mathbf{x}}, \dot{\mathbf{u}})\big) \in T\mathcal{E}$ into the point $\big((\mathbf{x}, \mathbf{u}); (\mathbf{x}, \dot{\mathbf{x}})\big) \in \mathcal{E} \underset{\mathcal{B}}{\times} T\mathcal{B}$. This observation implies that the horizontal lift A is locally of the form $\big((\mathbf{x}, \mathbf{u}); (\mathbf{x}, \dot{\mathbf{x}})\big) \mapsto \big((\mathbf{x}, \mathbf{u}); (\dot{\mathbf{x}}, \bar{A}(\mathbf{x}, \mathbf{u})\dot{\mathbf{x}})\big)$ where $\bar{A}(\mathbf{x}, \mathbf{u})$ denotes an $m \times n$ matrix. The horizontal spaces of the connection defined by the map A are now given by $H_\xi \pi = A(\xi, T_{\pi(\xi)}\mathcal{B})$. This fact explains the name "horizontal lift:" $A(\xi, \cdot)$ takes an arbitrary vector in $T_{\pi(\xi)}\mathcal{B}$ and lifts it to a horizontal vector in $T_\xi \mathcal{E}$. Each connection uniquely determines such a horizontal lift and conversely each horizontal lift defines uniquely a connection.

Figure C.2 sketches this approach to a connection. It shows a fibred manifold $\pi : \mathcal{E} \to \mathcal{B}$ and the fibre $\mathcal{E}_b$ over some point $b \in \mathcal{B}$. At two points $\xi_1, \xi_2 \in \mathcal{E}_b$ the vertical spaces are shown as the tangent spaces of the fibre. Furthermore, one can see at these two points the horizontal spaces of some connection; obviously they are complements to the vertical spaces. Finally, the horizontal lifts of a vector $v \in T_b\mathcal{B}$ to vectors in $T_{\xi_1}\mathcal{E}$ and $T_{\xi_2}\mathcal{E}$, respectively, are shown.

A connection always yields a *covariant derivative*. It is defined as the map $\nabla : \mathfrak{X}(\mathcal{B}) \times \Gamma_{loc}(\pi) \to \mathfrak{X}(\mathcal{E})$ given by $\nabla(X, \sigma) = P^V(\sigma_* X) = \sigma_* X - A \circ (\sigma, X)$. One often writes $\nabla_X \sigma$ instead of $\nabla(X, \sigma)$ and speaks of a covariant derivative with respect to X. Obviously, the vector field $\nabla_X \sigma$ is everywhere vertical.

Usually, one expects that applying a derivative to a function (or a section in our formalism) yields again a function which is clearly not the case for the covariant derivative ∇_X. However, if $\pi : \mathcal{E} \to \mathcal{B}$ is a vector bundle, we may recover this point of view: via the vertical lift λ introduced in Remark C.4.4, we can transform the vertical vector field $\nabla_X \sigma$ into the section $\mathrm{pr}_2 \circ \lambda^{-1}(\nabla_X \sigma) \in \Gamma_{loc}(\pi)$ and thus may consider ∇_X as a map $\Gamma_{loc}(\pi) \to \Gamma_{loc}(\pi)$.[3]

---

[3] In some applications it is customary to consider instead the covariant derivative ∇ as a map $\Gamma_{loc}(\pi) \to \Omega^1(\mathcal{B}, \mathcal{E})$ and to call this map connection.

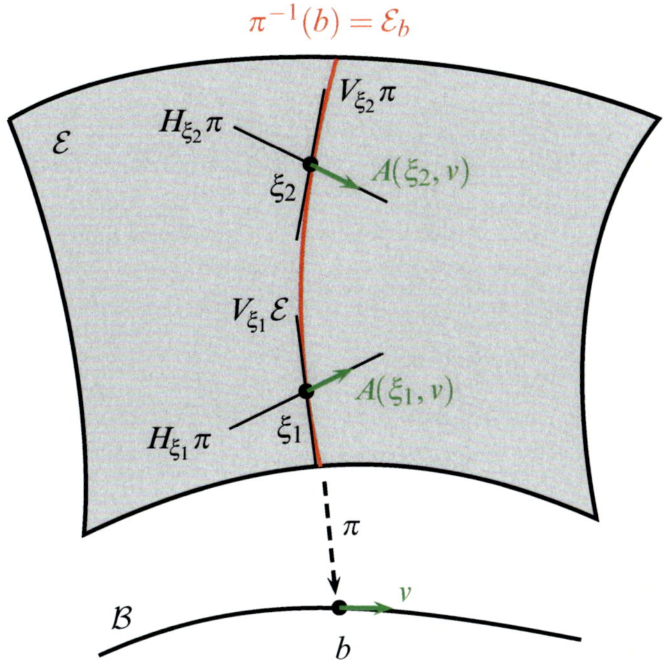

Fig. C.2 Fibred manifold with connection

For arbitrary connections, ∇_X does not possess the usual properties associated with a derivative, as generally $\nabla_X(\sigma_1 + \sigma_2) \neq \nabla_X\sigma_1 + \nabla_X\sigma_2$. Equality holds only for *linear connections* where the map $\bar{A}(\mathbf{x},\mathbf{u})$ introduced in the above discussion of the horizontal lift is linear in its second argument.

Finally, we consider all these constructions in adapted coordinates $(\mathbf{x},\mathbf{u})$ on the fibred manifold $\pi : \mathcal{E} \to \mathcal{B}$ with $\dim\mathcal{B} = n$ and $\dim\mathcal{E} = m+n$. By definition, the horizontal bundle is locally generated by n vector fields of the form $\partial_{x^i} + \Gamma_i^{\alpha}(\mathbf{x},\mathbf{u})\partial_{u^\alpha}$. Thus a connection is specified by mn smooth functions $\Gamma_i^{\alpha} \in \mathcal{F}(\mathcal{E})$. For the special case of a linear connection we have $\Gamma_i^{\alpha}(\mathbf{x},\mathbf{u}) = \Gamma_{i\beta}^{\alpha}(\mathbf{x})u^{\beta}$ and the coefficients $\Gamma_{i\beta}^{\alpha} \in \mathcal{F}(\mathcal{B})$ are known as *Christoffel symbols*. Written as one-form, the horizontal projector is thus obviously $P^H = dx^i \otimes (\partial_{x^i} + \Gamma_i^{\alpha}\partial_{u^\alpha})$ and correspondingly we obtain for the horizontal lift $A\bigl((\mathbf{x},\mathbf{u}), \dot{x}^i\partial_{x^i}\bigr) = \dot{x}^i(\partial_{x^i} + \Gamma_i^{\alpha}(\mathbf{x},\mathbf{u})\partial_{u^\alpha})$. Similarly, the vertical projector (and thus the connection form) is given by $P^V = (du^{\alpha} - \Gamma_i^{\alpha}dx^i) \otimes \partial_{u^\alpha}$. If the section $\sigma \in \Gamma_{loc}(\pi)$ is locally defined by $\sigma(\mathbf{x}) = \bigl(\mathbf{x},\mathbf{s}(\mathbf{x})\bigr)$, then its covariant derivative in the ith coordinate direction is

$$\nabla_{\partial_{x^i}}\sigma(\mathbf{x}) = \left(\frac{\partial s^{\alpha}}{\partial x^i}(\mathbf{x}) - \Gamma_i^{\alpha}\bigl(\mathbf{x},\mathbf{s}(\mathbf{x})\bigr)\right)\partial_{u^\alpha}\,. \tag{C.36}$$

Remark C.4.7. The vertical bundle $V\pi$ is a trivial example of an integrable distribution: the fibres are integral manifolds of maximal dimension. By the Frobenius Theorem C.3.3 it is thus also involutive (this fact can also be easily shown directly, since (C.9) implies $T\pi\big([X,Y]\big) = \big[T\pi(X), T\pi(Y)\big] = 0$). In contrast to the vertical bundle, the horizontal bundle is in general not involutive. Connections with an involutive horizontal distribution are called *flat*.

The *curvature* of a connection is defined as the curvature of its horizontal bundle as introduced in Remark C.3.7. In this special case, one commonly identifies the "normal" bundle K with the vertical bundle $V\pi$. Then the curvature two-form may be defined for arbitrary vector fields on $\mathcal{E}$ via the projectors associated with the connection: $\Omega_{H\mathcal{E}}(X,Y) = P^V\big([P^H X, P^H Y]\big)$. It is now easy to see that the curvature contains the obstructions to the involution of the horizontal bundle and hence the curvature vanishes for flat connections.

In the special case of a connection on a principal fibre bundle where P^V may be considered as a $\mathfrak{g}$-valued form for some Lie algebra $\mathfrak{g} \cong V\pi$, a straightforward computation using (C.28) and (C.29) yields that

$$\Omega = \mathrm{d}P^V + \frac{1}{2}P^V \wedge P^V \,, \tag{C.37}$$

a relation known as the *Maurer–Cartan formula*. In this situation, one may also introduce an *exterior covariant derivative* by setting $\mathrm{D}\omega = \mathrm{d}\omega + P^V \wedge \omega$ for all $\mathfrak{g}$-valued forms $\omega \in \Omega(\mathcal{E}, \mathfrak{g})$. ◁

C.5 Lie Groups and Algebras

Lie groups combine algebra and geometry. This combination leads to many structures and makes Lie groups a powerful tool for many different purposes. As we will see below, any Lie group has automatically associated with it a Lie algebra. An in depth treatment of both Lie groups and algebras is given in [467]; for our purposes the material contained in [342, Chapt. 1] is sufficient.

Definition C.5.1. A *Lie group* is a group $\mathcal{G}$ which is simultaneously a manifold such that both the group multiplication $(g,h) \mapsto g \cdot h$ and the inversion $g \mapsto g^{-1}$ are smooth maps between manifolds.

Trivial examples of Lie groups are $\mathbb{R}^k$ with the addition or the group $GL(n,\mathbb{R})$ of invertible $n \times n$ matrices with real entries. Most of the time we will restrict to *connected* groups. *Lie subgroups* are submanifolds $\mathcal{H} \subset \mathcal{G}$ which are again groups.

In many cases we are not interested in the full Lie group but only in a neighbourhood of the identity element. This leads to the concept of a *local Lie group*: a manifold $\mathcal{G}$ with two operations, multiplication and inversion, which are, however, only defined in a neighbourhood of the identity element. The usual group axioms like $f \cdot (g \cdot h) = (f \cdot g) \cdot h$ are required to hold only, if all appearing products and inversions are defined.

Example C.5.2. Let $\mathcal{G} =]-1,1[\subset \mathbb{R}$ and define

$$x \cdot y = \frac{2xy - x - y}{xy - 1}, \qquad x^{-1} = \frac{x}{2x-1}. \tag{C.38}$$

One easily verifies that all group axioms are satisfied with $0 \in \mathcal{G}$ as neutral element, but obviously the inversion is defined only for the neighbourhood $|x| < \frac{1}{2}$. Hence $\mathcal{G}$ is a one-dimensional local Lie group. ◁

We are mainly interested in Lie groups $\mathcal{G}$ acting on some manifold $\mathcal{M}$. Then a diffeomorphism on $\mathcal{M}$ is associated with each group element. For this reason one often speaks of a *transformation group*.

Definition C.5.3. An *action* of a Lie group $\mathcal{G}$ on a manifold $\mathcal{M}$ is defined by a map $\Phi : \mathcal{G} \times \mathcal{M} \to \mathcal{M}$ such that $\Phi(e,p) = p$ and $\Phi\big(g, \Phi(h,p)\big) = \Phi(g \cdot h, p)$ for all points $p \in \mathcal{M}$ and all group elements $g, h \in \mathcal{G}$. The action Φ is *free*, if $\Phi(g,p) = p$ for some point $p \in \mathcal{M}$ implies that $g = e$, the neutral element. The action Φ is *transitive*, if to any two points $p, q \in \mathcal{M}$ a group element $g \in \mathcal{G}$ exists such that $\Phi(g,p) = q$. The set $\mathcal{O}_p = \{\Phi(g,p) \mid g \in \mathcal{G}\}$ is the *orbit* of $\mathcal{G}$ through the point p.

In the case of a local action of a local Lie group, the map Φ is only defined for group elements sufficiently close to the identity element e and the second condition is only required to hold, if both sides are defined.

Example C.5.4. Every Lie group operates on itself via left multiplication: we simply define $\Phi(g,h) = L_g(h) = g \cdot h$. The flow of a complete vector field $X \in \mathfrak{X}(\mathcal{M})$ on a manifold $\mathcal{M}$ represents an action of the Lie group $(\mathbb{R}, +)$ on $\mathcal{M}$. If X is not complete, we obtain only a local group action. ◁

As already mentioned, any Lie group has associated with it a Lie algebra. Historically, Lie algebras were introduced this way. Today, they represent a mathematical structure of its own right and with its own theory.

Definition C.5.5. A *Lie algebra* $\mathfrak{A}$ (over a field $\Bbbk$) is a vector space together with a product $[\cdot,\cdot] : \mathfrak{A} \times \mathfrak{A} \to \mathfrak{A}$ (usually called *bracket*) satisfying three conditions.

(i) The bracket is bilinear over the field $\Bbbk$: for all $\mathbf{a}, \mathbf{b}, \mathbf{c} \in \mathfrak{A}$, $\lambda \in \Bbbk$ we have $[\mathbf{a} + \lambda\mathbf{b}, \mathbf{c}] = [\mathbf{a}, \mathbf{c}] + \lambda[\mathbf{b}, \mathbf{c}]$ and correspondingly for the second argument.

(ii) The bracket is skew-symmetric: $[\mathbf{a}, \mathbf{b}] = -[\mathbf{b}, \mathbf{a}]$ for all $\mathbf{a}, \mathbf{b} \in \mathfrak{A}$.

(iii) The *Jacobi identity* holds:

$$\forall \mathbf{a}, \mathbf{b}, \mathbf{c} \in \mathfrak{A} : \big[\mathbf{a}, [\mathbf{b}, \mathbf{c}]\big] + \big[\mathbf{b}, [\mathbf{c}, \mathbf{a}]\big] + \big[\mathbf{c}, [\mathbf{a}, \mathbf{b}]\big] = 0. \tag{C.39}$$

If $\mathfrak{A}$ is a finite-dimensional Lie algebra, we may choose a basis $\{\mathbf{e}_1, \dots, \mathbf{e}_r\}$ of $\mathfrak{A}$ as a $\Bbbk$-linear space. The bracket is then uniquely defined, as soon as its values on the basis vectors are known. This fact leads to the *structure constants*: a set of r^3 scalars $C_{ij}^k \in \Bbbk$ such that $[\mathbf{e}_i, \mathbf{e}_j] = C_{ij}^k \mathbf{e}_k$. The defining properties of a Lie algebra imply that $C_{ij}^k = -C_{ji}^k$ and $C_{ij}^k C_{kl}^m + C_{jl}^k C_{ki}^m + C_{li}^k C_{kj}^m = 0$. Conversely, any set of r^3 scalars C_{ij}^k satisfying these properties defines an abstract r-dimensional Lie algebra.

We have already seen that the vector fields on a manifold $\mathcal{M}$ form a real Lie algebra $\mathfrak{X}(\mathcal{M})$ of the same dimension as $\mathcal{M}$ with the Lie bracket as multiplication. Given a Lie group $\mathcal{G}$, we introduce now its associated Lie algebra $\mathfrak{g}$. It is easy to find an appropriate vector space, namely the tangent space $T_e\mathcal{G}$ at the identity element. Less trivial is the construction of a suitable bracket. Therefore we extend each vector in $T_e\mathcal{G}$ uniquely to a vector field on $\mathcal{G}$ and then use the Lie brackets of these fields.

For each element $g \in \mathcal{G}$, we consider the right[4] multiplication $R_g : \mathcal{G} \to \mathcal{G}$ defined by $R_g(h) = h \cdot g$. Obviously, R_g is a diffeomorphism with inverse $(R_g)^{-1} = R_{g^{-1}}$. A vector field $X \in \mathfrak{X}(\mathcal{G})$ is called *right invariant*, if it is invariant under the tangent map of R_g: $T_hR_g(X_h) = X_{g \cdot h}$. It is not difficult to verify that the right invariant vector fields form a vector space isomorphic to $T_e\mathcal{G}$. Indeed, if X is right invariant, then it is completely fixed by its value $X_e \in T_e\mathcal{G}$ since $X_h = T_eR_h(X_e)$.

Definition C.5.6. The *Lie algebra* of the Lie group $\mathcal{G}$ is the vector space $\mathfrak{g} \cong T_e\mathcal{G}$ of right invariant vector fields together with the Lie bracket.

This definition makes sense, as one easily verifies with the help of (C.9) that the Lie bracket of two right invariant vector fields is again right invariant. Depending on the context, we will consider elements of $\mathfrak{g}$ sometimes as right invariant vector fields and sometimes as vectors in $T_e\mathcal{G}$; both points of view have their advantages.

Example C.5.7. The Lie algebra $\mathfrak{gl}(n,\mathbb{R})$ of the Lie group $GL(n,\mathbb{R})$ is the vector space of all $n \times n$ matrices and its bracket is the matrix commutator $[A,B] = AB - BA$. One can show that any finite-dimensional Lie algebra is isomorphic to a subalgebra of $\mathfrak{gl}(n,\mathbb{R})$ (Ado's Theorem). Thus any finite-dimensional Lie algebra may be considered as a matrix algebra (note that this is *not* true for Lie groups). $\triangleleft$

Lie considered only local transformation groups. The abstract definition of a Lie group via manifolds is due to Cartan [73] (in the same article he also introduced the modern definition of a manifold via charts). The results about the relation between a Lie group and its Lie algebra are often referred to as *Lie's Fundamental Theorems*. The first Fundamental Theorem describes the Lie algebra via right (or left) invariant vector fields on the Lie group. The second one concerns the construction of a local Lie group to a given Lie algebra and the third one says that a Lie algebra is uniquely determined by its structure constants.

C.6 Symplectic Geometry and Generalisations

Symplectic geometry forms the foundation of Hamiltonian mechanics. As Hamiltonian systems appear at several places as examples, we briefly recapitulate a few

[4] We could alternatively use the left multiplication; in fact, this choice is more common in the literature. We would then obtain a Lie algebra whose bracket is the negative of the bracket we will derive. Our choice is more convenient for applications in Lie symmetry theory as discussed by Olver [342], Exercise 1.33.

basic facts. More details can be found in most text books on geometric mechanics, see e. g. [2, 26, 28, 309]. The more general case of Poisson manifolds is extensively discussed in [43, 466].

Definition C.6.1. A *symplectic manifold* is a manifold $\mathcal{M}$ equipped with a non-degenerated, closed two-form $\omega \in \Omega^2(\mathcal{M})$, the *symplectic two-form.*

The conditions on ω mean that $d\omega = 0$ and that for any non-zero vector field $X \in \mathfrak{X}(\mathcal{M})$ the one-form $\iota_X \omega$ does not vanish. By the Poincaré Lemma, we may locally write $\omega = d\theta$ for some one-form θ. If this is even globally possible, one speaks of an *exact* symplectic manifold. The non-degeneracy trivially implies that a symplectic manifold must be even-dimensional.

In local coordinates $(\mathbf{x})$ on $\mathcal{M}$, the symplectic two-form has a representation $\omega = \frac{1}{2}B_{ij}(\mathbf{x})dx^i \wedge dx^j$ with an antisymmetric matrix B. The non-degeneracy implies in particular that B must be regular.

Symplectic geometry has certain similarities with Riemannian geometry. In both cases we are given at each point of a manifold a bilinear non-degenerate form. In Riemannian geometry the form is symmetric, namely the metric, whereas here we are dealing with a antisymmetric form. Nevertheless, many operations familiar from Riemannian geometry can be performed similarly in symplectic geometry. For example, a Riemannian metric induces a notion of orthogonality in the tangent bundle. Similarly, if $\mathcal{D} \subseteq T\mathcal{M}$ is a distribution on a symplectic manifold, then we define its *symplectic complement* as the distribution whose value at a point $p \in \mathcal{M}$ is given by

$$\mathcal{D}_p^{\perp} = \{v \in T_p\mathcal{M} \mid \omega_p(v,w) = 0 \;\forall w \in \mathcal{D}_p\}\;. \tag{C.40}$$

Example C.6.2. The prototype of a symplectic manifold is the cotangent bundle $\mathcal{M} = T^*\mathcal{Q}$ of some manifold $\mathcal{Q}$ (the *configuration space*). It is canonically equipped with an exact symplectic structure. Consider the one-form $\theta \in \Omega^1(\mathcal{M})$ defined at a point $\alpha \in \mathcal{M}$ by $\theta_\alpha(v) = \alpha\big(T\tau_{\mathcal{Q}}^*(v)\big)$. Thus for evaluating θ on some vector $v \in T\mathcal{M}$, we project v with the help of the tangent map of the cotangent bundle projection $\tau_{\mathcal{Q}}^* : \mathcal{M} = T^*\mathcal{Q} \to \mathcal{Q}$ to a vector in $T\mathcal{Q}$ and evaluate the one-form $\alpha \in T^*\mathcal{Q}$ on this vector. The symplectic two-form of $\mathcal{M}$ is then obtained as $\omega = -d\theta$ and hence is trivially closed.

If we take adapted coordinates $(\mathbf{q},\mathbf{p})$ on the cotangent bundle $\mathcal{M} = T^*\mathcal{Q}$, then we find $\theta = p_i dq^i$ and $\omega = dq^i \wedge dp_i$. Indeed, let the one-form $\alpha \in \mathcal{M}$ lie in the fibre over the point $q \in \mathcal{Q}$ with local coordinates $\mathbf{q}$. Then its fibre component may be written as the form $\alpha_q = p_i dq^i \in T_q^*\mathcal{Q}$. By definition, θ is an element of $\Omega^1(\mathcal{M})$. Let us denote adapted coordinates on $T^*\mathcal{M}$ by $(\mathbf{q},\mathbf{p};\mathbf{a},\mathbf{b})$. Thus θ_α is of the form $a_i dq^i + b^i dp_i$. Let $v = v^i\partial_{q^i} + w_i\partial_{p_i}$ be a vector in $T_\alpha\mathcal{M}$; its projection $T_\alpha\tau_{\mathcal{Q}}^*(v)$ is simply given by $v^i\partial_{q^i} \in T_q\mathcal{Q}$. Evaluating α_q on this projected vector yields $p_i v^i$. Hence we find $a_i = p_i$ and $b^i = 0$.

The local form of ω is a trivial consequence of the definition of the exterior derivative d and immediately implies its non-degeneracy, as we find for any vector field $X = v^i\partial_{q^i} + w_i\partial_{p_i}$ that $\iota_X\omega = v^i dp_i - w_i dq^i$ vanishes only for $X = 0$.

By contrast, the tangent bundle $T\mathcal{Q}$ does *not* carry a canonical symplectic structure. Without additional ingredients (like e. g. a metric), it is not possible to pull back the symplectic structure on $T^*\mathcal{Q}$. ◁

A fundamental result is the *Darboux Theorem* which asserts that locally all symplectic manifolds look alike, i. e. as long as we restrict to local questions we may always assume that we are dealing with a cotangent bundle. More precisely, this theorem says that around every point on a symplectic manifold $\mathcal{M}$ a chart $\mathcal{U}$ with local coordinates $(\mathbf{q},\mathbf{p})$ exists such that the symplectic two-form has everywhere[5] in $\mathcal{U}$ the form $\omega = dq^i \wedge dp_i$. Such coordinates are called *Darboux coordinates*.

Because of the non-degeneracy, the symplectic structure induces two bundle isomorphisms $\flat : T\mathcal{M} \to T^*\mathcal{M}$ and $\sharp : T^*\mathcal{M} \to T\mathcal{M}$, called the *musical isomorphisms*. They are defined by $\flat(X) = \iota_X\omega$ and $\sharp = \flat^{-1}$. In Riemannian geometry similar isomorphisms exist; they represent the proper mathematical formulation of the "lowering" and "raising" of indices in tensor analysis.

Definition C.6.3. Let $(\mathcal{M},\omega)$ be a symplectic manifold and $H : \mathcal{M} \to \mathbb{R}$ a smooth function. The *Hamiltonian vector field* for H is $X_H = \sharp(dH)$, i. e. X_H is the (unique) solution of the equation $\iota_{X_H}\omega = dH$.

In Darboux coordinates, one easily computes that

$$X_H = \frac{\partial H}{\partial p_i}\partial_{q^i} - \frac{\partial H}{\partial q_i}\partial_{p^i}\,. \tag{C.41}$$

The field X_H is sometimes called the *symplectic gradient* of the function H, as its construction is in analogy to the definition of a gradient vector field on a Riemannian manifold. The ordinary differential equation corresponding to X_H has the form familiar from classical mechanics textbooks:

$$\dot{q}^i = \frac{\partial H}{\partial p_i}\,, \qquad \dot{p}_i = -\frac{\partial H}{\partial q_i}\,. \tag{C.42}$$

The function H is called the *Hamiltonian* of the vector field X_H. A trivial property of it is its preservation by the flow of X_H. Indeed, $\mathcal{L}_{X_H}H = X_H(H) = 0$, as one can easily see from (C.41).

A characteristic property of the flow of a Hamiltonian vector field is that it preserves the symplectic structure. Indeed, we obtain immediately from (C.27a) that

$$\mathcal{L}_{X_H}\omega = d\iota_{X_H}\omega + \iota_{X_H}d\omega = d^2H = 0\,. \tag{C.43}$$

Here we have exploited the definition of a Hamiltonian vector field X_H and that the symplectic two-form ω is closed. Because of the Poincaré Lemma, the inverse holds at least locally: to every vector field X whose flow preserves the symplectic structure there exists locally a Hamiltonian H such that $X_H = X$.

[5] Note the striking difference to Riemannian geometry: there locally geodesic coordinates lead to a special form of the connection form only at one point and thus Riemannian manifolds may look very differently even locally. Here we get a normal form of ω in a whole neighbourhood.

Traditionally, Hamiltonian mechanics is often formulated via *Poisson brackets* operating on the smooth functions. For non-degenerated brackets this approach is equivalent to the introduction of a symplectic structure.

Definition C.6.4. A *Poisson manifold* is a manifold $\mathcal{M}$ equipped with a bracket $\{\cdot,\cdot\} : \mathcal{F}(\mathcal{M}) \times \mathcal{F}(\mathcal{M}) \to \mathcal{F}(\mathcal{M})$ satisfying the following properties.

(i) The bracket $\{\cdot,\cdot\}$ is bilinear with respect to the addition in $\mathcal{F}(\mathcal{M})$ and the multiplication by constants.

(ii) The bracket is antisymmetric: $\forall F, G \in \mathcal{F}(\mathcal{M}) : \{F,G\} = -\{G,F\}$.

(iii) The bracket behaves like a derivation with respect to the multiplication in $\mathcal{F}(\mathcal{M})$, i. e. we have the Leibniz rule

$$\{FG,H\} = F\{G,H\} + \{F,H\}G \qquad \forall F,G,H \in \mathcal{F}(\mathcal{M}) . \tag{C.44}$$

(iv) The Jacobi identity holds:

$$\{\{F,G\},H\} + \{\{H,F\},G\} + \{\{G,H\},F\} = 0 \qquad \forall F,G,H \in \mathcal{F}(\mathcal{M}) . \tag{C.45}$$

A non-constant function $C \in \mathcal{F}(\mathcal{M})$ is called a *Casimir function*, if any Poisson bracket with it vanishes: $\{C,F\} = 0$ for all $F \in \mathcal{F}(\mathcal{M})$. A Poisson manifold is *non-degenerate*, if no Casimir functions exist on it.

Any symplectic manifold $(\mathcal{M}, \omega)$ is also a Poisson manifold with the Poisson bracket defined by

$$\{F,G\} = \omega(X_F, X_G) = \mathrm{d}F(X_G) = X_G(F) . \tag{C.46}$$

The non-degeneracy of the symplectic two-form ω immediately implies the non-degeneracy of this Poisson bracket. The converse is not true: a Poisson manifold is symplectic, if and only if it is non-degenerate.

One can show that in local coordinates $(\mathbf{x})$ on $\mathcal{M}$ any Poisson bracket is of the form $\{F,G\} = (\nabla F)^t J(\mathbf{x}) \nabla G$ with a *Poisson matrix* J. The Poisson manifold is non-degenerate, if this matrix is regular. In this case the corresponding symplectic two-form ω is locally described by the matrix $B = J^{-1}$. Darboux coordinates are characterised as those coordinates in which J is the symplectic matrix

$$J = \begin{pmatrix} 0 & \mathbb{1} \\ -\mathbb{1} & 0 \end{pmatrix} . \tag{C.47}$$

This condition leads to the familiar coordinate form

$$\{F,G\} = \frac{\partial F}{\partial q^i}\frac{\partial G}{\partial p_i} - \frac{\partial F}{\partial p_i}\frac{\partial G}{\partial q^i} \tag{C.48}$$

of the Poisson bracket (C.46).

Example C.6.5. Let $\mathfrak{A}$ be an n-dimensional real Lie algebra and $\{A_1,\dots,A_n\}$ a basis of it with corresponding structure constants C_{ij}^k. As the dual space $\mathfrak{A}^*$ is again an n-dimensional real vector space, it is trivially an n-dimensional manifold and we have $T\mathfrak{A}^* = \mathfrak{A}^* \times \mathfrak{A}^*$. We denote the dual basis by $\{A^1,\dots,A^n\}$, i.e. $A^i(A_j) = \delta^i_j$.

The manifold $\mathfrak{A}^*$ carries a natural Poisson structure, namely a *Lie-Poisson structure*. Let $\phi, \psi \in \mathcal{F}(\mathfrak{A}^*)$ be two smooth functions on $\mathfrak{A}^*$; we define

$$\{\phi,\psi\}(\lambda) = \lambda\big([T_\lambda\phi, T_\lambda\psi]\big)\,. \tag{C.49}$$

Here λ is an arbitrary element of $\mathfrak{A}^*$ and we use the following identification. The tangent map $T_\lambda\phi$ is a linear map $T_\lambda\mathfrak{A}^* \to \mathbb{R}$. As $T_\lambda\mathfrak{A}^* \cong \mathfrak{A}^*$, we may consider it as an element of the bidual space $\mathfrak{A}^{**}$ which may be canonically identified with $\mathfrak{A}$. Thus it makes sense to compute the Lie bracket of $T_\lambda\phi$ and $T_\lambda\psi$ and to evaluate λ on the result. The Jacobi identity for the Lie bracket implies the Jacobi identity for this Poisson bracket and similarly for the skew-symmetry.

In coordinates, this structure looks as follows. Let $\lambda = \lambda_i A^i$. We identify $T_\lambda\phi$ with $\frac{\partial\phi}{\partial\lambda_i}A_i \in \mathfrak{A}$. Then one readily computes that

$$\{\phi,\psi\}(\lambda) = C_{ij}^k\lambda_k\frac{\partial\phi}{\partial\lambda_i}\frac{\partial\psi}{\partial\lambda_j}\,. \tag{C.50}$$

Thus the Poisson matrix is given by $J_{ij}(\lambda) = C_{ij}^k\lambda_k$ and its entries are the linear functions of the coordinates λ_i. Reversing the arguments above, one can show that in fact any Poisson structure where the entries of the Poisson matrix are linear functions stems from a Lie algebra. ◁

Hamiltonian vector fields may now be written in the form of differential operators as follows: $X_F(G) = \{F,G\}$. One can show that the Hamiltonian vector fields are closed under the Lie bracket and thus form a Lie algebra. More precisely, we have the identity $[X_F,X_G] = -X_{\{F,G\}}$. As it follows immediately from Definition C.6.4 that for any Poisson manifold $\big(\mathcal{F}(\mathcal{M}),\{\cdot,\cdot\}\big)$ is a Lie algebra, too, we may say that the map $F \mapsto X_F$ defines a Lie algebra anti-homomorphism from $\mathcal{F}(\mathcal{M})$ into the algebra of Hamiltonian vector fields.

Given some local coordinates $\mathbf{x}$ on $\mathcal{M}$, we may now express the equations of motion in the form $\dot{x}^i = \{x^i,H\}$, as the right hand side represents the components of X_H in the given coordinates. One easily checks that in Darboux coordinates this yields (C.42). More generally, let $F \in \mathcal{F}(\mathcal{M})$ be any smooth function. Its variation along integral curves of a Hamiltonian vector field X_H is then given by

$$\frac{\mathrm{d}}{\mathrm{d}t}F = \frac{\partial F}{\partial x^i}\dot{x}^i = \frac{\partial F}{\partial x^i}\{x^i,H\} = \{F,H\}\,. \tag{C.51}$$

Symplectic geometry allows us only the treatment of autonomous systems. There is no intrinsic notion of time; it appears at most as a curve parameter for the integral curves of a Hamiltonian vector field. If we want to treat explicitly time dependent systems, we need *cosymplectic geometry* [11, 66, 291].

Definition C.6.6. A *cosymplectic manifold* is a $(2n+1)$-dimensional manifold $\mathcal{M}$ equipped with a closed two-form $\omega \in \Omega^2(\mathcal{M})$ and a closed one-form $\eta \in \Omega^1(\mathcal{M})$ such that $\omega^n \wedge \eta$ is a volume form on $\mathcal{M}$.

The condition that $\omega^n \wedge \eta$ is a volume form generalises the non-degeneracy condition on the symplectic two-form on a symplectic manifold. It implies that any vector field X such that $\iota_X \omega = 0$ must satisfy $\iota_X \eta \neq 0$. For dimensional reasons, this condition implies the existence of a unique vector field $R \in \mathfrak{X}(\mathcal{M})$, the *Reeb vector field*, such that $\iota_R \omega = 0$ and $\iota_R \eta = 1$.

Example C.6.7. Every symplectic manifold can trivially be extended to a cosymplectic manifold as follows. Let $(\mathcal{M}, \omega)$ be a symplectic manifold and $\mathrm{d}t$ the canonical one-form on $\mathbb{R}$. Then the product manifold $\mathcal{M} \times \mathbb{R}$ together with the forms $\mathrm{pr}_1^* \omega$ and $\mathrm{pr}_2^*(\mathrm{d}t)$ (here pr_i denotes as usual the projection on the ith factor) is a cosymplectic manifold. $\lhd$

The cosymplectic structure induces canonically a vector bundle isomorphism $\chi : T\mathcal{M} \to T^*\mathcal{M}$ generalising the musical isomorphism $\flat$ on a symplectic manifold. It is defined by $\chi(X) = \iota_X \omega + (\iota_X \eta)\eta$. The Reeb vector field is now given by $R = \chi^{-1}(\eta)$. The Darboux theorem for cosymplectic manifolds asserts that there always exists a local coordinate system $(\mathbf{q}, \mathbf{p}, t)$ such that $\omega = \mathrm{d}q^i \wedge \mathrm{d}p_i$ and $\eta = \mathrm{d}t$ (implying $R = \partial_t$).

The isomorphism χ allows us to introduce in analogy to the symplectic complement (C.40) of a distribution $\mathcal{D} \subseteq T\mathcal{M}$ its *cosymplectic complement*

$$\mathcal{D}_p^{\perp} = \{ v \in T_p\mathcal{M} \mid \chi(v)(w) = 0 \; \forall w \in \mathcal{D}_p \} . \tag{C.52}$$

On a cosymplectic manifold $\mathcal{M}$ we can associate three different types of "gradient fields" with a given function $H : \mathcal{M} \to \mathbb{R}$.

(i) *Cosymplectic gradient*: $\nabla H = \chi^{-1}(\mathrm{d}H) = H_t \partial_t + H_{p_i} \partial_{q^i} - H_{q^i} \partial_{p_i}$.
(ii) *Hamiltonian vector field*: $X_H = \chi^{-1}\big(\mathrm{d}H - R(H)\eta\big) = H_{p_i} \partial_{q^i} - H_{q^i} \partial_{p_i}$.
(iii) *Evolution vector field*: $E_H = R + X_H = \partial_t + H_{p_i} \partial_{q^i} - H_{q^i} \partial_{p_i}$.

Note that in Darboux coordinates these three vector fields differ only in their component in t-direction. One can show that the Hamiltonian vector fields form an involutive distribution.

In physical applications, the cosymplectic manifold $\mathcal{M}$ is often called the *extended phase space* of the studied system. The Reeb vector field induces an intrinsic clock describing the flow of time. The trajectories of the system described by the Hamiltonian H are the integral curves of the evolution vector field E_H. Thus our equations of motion are now in Darboux coordinates:

$$\dot{q}^i = \frac{\partial H}{\partial p_i} , \qquad \dot{p}_i = -\frac{\partial H}{\partial q^i} , \qquad \dot{t} = 1 . \tag{C.53}$$

Obviously, they are almost identical with the usual Hamiltonian equations (C.42); the only difference is that now an explicit time dependency of the Hamiltonian H is permitted, as time is part of the extended phase space.

An interesting special case arises, if the cosymplectic manifold $\mathcal{M}$ carries in addition the structure of a fibred manifold $\pi : \mathcal{M} \to \mathcal{B}$ such that $\dim \mathcal{B} = 1$ and the vector field $T\pi(R) \in \mathfrak{X}(\mathcal{B})$ does nowhere vanish, i. e. the Reeb vector field R is everywhere transversal to the fibration. In this case the Hamiltonian vector fields span the vertical bundle $V\pi$; in other words the fibres are the integral manifolds of the distribution defined by them. Furthermore, each fibre is a symplectic manifold with the symplectic two-form given by the pull-back of ω on the fibre.

Because of its transversality, the Reeb vector field R defines the horizontal bundle of a canonical connection on $\mathcal{M}$. The corresponding vertical projector $P^V[R] : T\mathcal{M} \to V\pi$ is given by $P^V[R](X_\xi) = X_\xi - \mu R_\xi$ where the factor $\mu \in \mathbb{R}$ is determined by the relation $T_\xi\pi(X_\xi) = \mu\, T_\xi\pi(R_\xi)$. Similarly, the horizontal lift is given by $A[R](\xi, v) = \mu R_\xi$ for $v = \mu\, T_\xi\pi(R_\xi)$.

References

The numbers in parantheses at the end of each reference correspond to the pages on which it is cited in the text.

1. Ablowitz M, Segur H (1981) Solitons and the Inverse Scattering Transform. SIAM Studies in Applied Mathematics, SIAM, Philadelphia *(344)*
2. Abraham R, Marsden J (1978) Foundations of Mechanics, 2nd edn. Benjamin-Cummings, Reading *(611)*
3. Abraham R, Marsden J, Ratiu T (1988) Manifolds, Tensor Analysis, and Applications. Applied Mathematical Sciences 75, Springer-Verlag, New York *(585, 601)*
4. Abramowitz M, Stegun I (eds) (1965) Handbook of Mathematical Functions. Dover, New York *(525)*
5. Adams W, Loustaunau P (1994) An Introduction to Gröbner Bases. Graduate Studies in Mathematics 3, American Mathematical Society, Providence *(165, 529, 579)*
6. Agmon S (1965) Lectures on Elliptic Boundary Value Problems. Van Nostrand Mathematical Studies 2, Van Nostrand, New York *(504)*
7. Agmon S, Douglis A, Nirenberg L (1959) Estimates near the boundary for solutions of elliptic partial differential equations satisfying general boundary conditions. I. Comm Pure Appl Math 12:623–727 *(504)*
8. Agmon S, Douglis A, Nirenberg L (1964) Estimates near the boundary for solutions of elliptic partial differential equations satisfying general boundary conditions. II. Comm Pure Appl Math 17:35–92 *(504)*
9. Agranovich M (1997) Elliptic boundary problems. In: Agranovich M, Egorov Y, Shubin M (eds) Partial Differential Equations IX, Encyclopaedia of Mathematical Sciences 79, Springer-Verlag, Berlin, pp 1–144 *(443, 504)*
10. Ahmad S, Anwar I (2008) An upper bound for the regularity of ideals of Borel type. Commun Alg 36:670–673 *(233)*
11. Albert C (1989) Le théorème de réduction de Marsden-Weinstein en géométrie cosymplectique et de contact. J Geom Phys 6:627–649 *(614)*
12. Alekseevskij D, Vinogradov A, Lychagin V (1991) Basic ideas and concepts of differential geometry. In: Gamkrelidze R (ed) Geometry I, Encyclopaedia of Mathematical Sciences 28, Springer-Verlag, Berlin, pp 1–264 *(61)*
13. Amann H (1990) Ordinary Differential Equations. Walter de Gruyter, Berlin *(359, 360)*
14. Amasaki M (1990) Application of the generalized Weierstrass preparation theorem to the study of homogeneous ideals. Trans Amer Math Soc 317:1–43 *(102)*
15. Amasaki M (2000) Generic Gröbner bases and Weierstrass bases of homogeneous submodules of graded free modules. J Pure Appl Algebra 152:3–16 *(102)*

16. Ampère A (1815) Considérations Générales sur les Intégrales des Équations aux Différentielles Partielles. J École Polytéchnique 10:549–611 *(353)*
17. Anderson I, Kamran N, Olver P (1993) Internal, external, and generalized symmetries. Adv Math 100:53–100 *(409)*
18. Apel J (1988) Gröbnerbasen in Nichtkommutativen Algebren und ihre Anwendung. PhD thesis, Universität Leipzig *(102)*
19. Apel J (1998) The computation of Gröbner bases using an alternative algorithm. In: Bronstein M, Grabmeier J, Weispfenning V (eds) Symbolic Rewriting Techniques, Progress in Computer Science and Applied Logic 15, Birkhäuser, Basel, pp 35–45 *(164)*
20. Apel J (1998) Theory of involutive divisions and an application to Hilbert function computations. J Symb Comp 25:683–704 *(101, 161, 162)*
21. Apel J (2003) On a conjecture of R.P. Stanley. Part I – monomial ideals. J Algebr Comb 17:39–56 *(229)*
22. Apel J (2003) On a conjecture of R.P. Stanley. Part II – quotients modulo monomial ideals. J Algebr Comb 17:57–74 *(229)*
23. Apel J, Hemmecke R (2005) Detecting unnecessary reductions in an involutive basis computation. J Symb Comp 40:1131–1149 *(162)*
24. Arajs E, Shapeev V, Yanenko N (1974) Realization of Cartan's method of exterior forms on an electronic computer. Sov Math Dokl 15:203–205 *(327)*
25. Arnold E (2003) Modular algorithms for computing Gröbner bases. J Symb Comp 35: 403–419 *(163)*
26. Arnold V (1978) Mathematical Methods of Classical Mechanics. Graduate Texts in Mathematics 60, Springer-Verlag, New York *(50, 611)*
27. Arnold V (1988) Geometrical Methods in the Theory of Ordinary Differential Equations, 2nd edn. Grundlehren der mathematischen Wissenschaften 250, Springer-Verlag, New York *(363, 409, 426)*
28. Arnold V, Kozlov V, Neishtadt A (1988) Mathematical aspects of classical and celestial mechanics. In: Arnold V (ed) Dynamical Systems III, Springer-Verlag, Berlin *(611)*
29. Aschenbrenner M, Leykin A (2009) Degree bounds for Gröbner bases in algebras of solvable type. J Pure Appl Alg 213:1578–1605 *(103)*
30. Auslander M, Bridger M (1969) Stable Module Theory. Mem. Amer. Math. Soc. 94, American Mathematical Society, Providence (RI) *(505)*
31. Bäcklund A (1880) Zur Theorie der partiellen Differentialgleichungen erster Ordnung. Math Ann 17:285–328 *(344)*
32. Baclawski K, Garsia A (1981) Combinatorial decompositions of rings. Adv Math 39: 155–184 *(228)*
33. Bayer D, Mumford D (1993) What can be computed in algebraic geometry? In: Eisenbud D, Robbiano L (eds) Computational Algebraic Geometry and Commutative Algebra, Symposia Mathematica 34, Cambridge University Press, Cambridge, pp 1–48 *(229, 232, 234, 577)*
34. Bayer D, Stillman M (1987) A criterion for detecting m-regularity. Invent Math 87:1–11 *(216, 232, 233, 234, 262, 577)*
35. Bayer D, Charalambous H, Popescu S (1999) Extremal Betti numbers and applications to monomial ideals. J Alg 221:497–512 *(232)*
36. Becker T, Weispfenning V (1993) Gröbner Bases. Graduate Texts in Mathematics 141, Springer-Verlag, New York *(229, 511, 529, 538)*
37. Belanger J, Hausdorf M, Seiler W (2001) A *MuPAD* library for differential equations. In: [162], pp 25–42 *(515)*
38. Bell A, Goodearl K (1988) Uniform rank over differential operator rings and Poincaré-Birkhoff-Witt extensions. Pacific J Math 131:13–37 *(81)*
39. Berger R (1992) The quantum Poincaré-Birkhoff-Witt theorem. Comm Math Phys 143:215–234 *(83)*
40. Bergman G (1978) The diamond lemma for ring theory. Adv Math 29:178–218 *(103)*
41. Bermejo I, Gimenez P (2006) Saturation and Castelnuovo-Mumford regularity. J Alg 303:592–617 *(230, 232, 233)*

42. Bernstein I, Rosenfeld B (1973) Homogeneous spaces of infinite Lie algebras and characteristic classes of foliations. Uspekhi Mat Nauk 28:103–138 *(60, 61)*
43. Bhaskara K, Viswanath K (1988) Poisson Algebras and Poisson Manifolds. Pitman Research Notes in Mathematics 174, Longman Scientific & Technical, Harlow *(611)*
44. Billera L, Cushman R, Sanders J (1988) The Stanley decomposition of the harmonic oscillator. Indagat Math 50:375–394 *(228)*
45. Björk J (1979) Rings of Differential Operators. North-Holland Mathematical Library 21, North-Holland, Amsterdam *(86, 103, 506)*
46. Björk J (1993) Analytical D-Modules and Applications. Mathematics and Its Applications 247, Kluwer, Dordrecht *(82)*
47. Blinkov Y (2001) Method of separative monomials for involutive divisions. Prog Comp Software 27:139–141 *(101)*
48. Bluman G, Kumei S (1989) Symmetries and Differential Equations. Applied Mathematical Sciences 81, Springer-Verlag, New York *(326, 407)*
49. Boothby W (1986) An Introduction to Differentiable Manifolds and Riemannian Geometry. Pure and Applied Mathematics 120, Academic Press, Orlando *(585)*
50. Borel A (ed) (1987) Algebraic D-Modules. Perspectives in Mathematics 2, Academic Press, Boston *(82)*
51. Boulier F, Lazard D, Ollivier F, Petitot M (1995) Representation for the radical of a finitely generated differential ideal. In: Levelt A (ed) Proc. ISSAC '95, ACM Press, New York, pp 158–166 *(327)*
52. Brandt F (1994) Bäcklund transformations and zero-curvature representations of systems of partial differential equations. J Math Phys 35:2463–2484 *(344)*
53. Brenan K, Campbell S, Petzold L (1996) Numerical Solution of Initial-Value Problems in Differential-Algebraic Equations. Classics in Applied Mathematics 14, SIAM, Philadelphia *(6)*
54. Bronstein M, Petkovšek M (1996) An introduction to pseudo-linear algebra. Theor Comp Sci 157:3–33 *(79)*
55. Brown W (1993) Matrices Over Commutative Rings. Pure and Applied Mathematics, Marcel Dekker, New York *(583)*
56. Bruns W, Conca A (2003) Gröbner bases, initial ideals and initial algebras. Preprint math.AC/0308102 *(234)*
57. Bruns W, Vetter U (1988) Determinantal Rings. Lecture Notes in Mathematics 1327, Springer-Verlag, Berlin *(505)*
58. Bryant R, Chern S, Gardner R, Goldschmidt H, Griffiths P (1991) Exterior Differential Systems. Mathematical Sciences Research Institute Publications 18, Springer-Verlag, New York *(60, 260, 323, 327)*
59. Brzezinski T, Wisbauer R (2003) Corings and Comodules. London Mathematical Society Lecture Notes Series 309, Cambridge University Press, Cambridge *(559)*
60. Buchberger B (1965) Ein Algorithmus zum Auffinden der Basiselemente des Restklassenringes nach einem nulldimensionalen Polynomideal. PhD thesis, Universität Innsbruck, (Engl. translation: J. Symb. Comput. 41 (2006) 475–511) *(567, 572)*
61. Bueso J, Gómez-Torrecillas J, Lobillo F, Castro-Jiménez F (1998) An introduction to effective calculus in quantum groups. In: Caenepeel S, Verschoren A (eds) Rings, Hopf Algebras, and Brauer Groups, Lecture Notes in Pure and Applied Mathematics 197, Marcel Dekker, New York, pp 55–83 *(102, 234)*
62. Bueso J, Gómez-Torrecillas J, Lobillo F (2001) Homological computations in PBW modules. Alg Represent Theo 4:201–218 *(102)*
63. Bueso J, Gómez-Torrecillas J, Verschoren A (2003) Algorithmic Methods in Non-Commutative Algebra. Mathematical Modelling: Theory and Applications 17, Kluwer, Dordrecht *(102, 165)*
64. Burman R (1977) On the strengths of field equations. Czech J Phys B 27:113–116 *(354)*
65. Calmet J, Seiler W, Tucker R (eds) (2006) Global Integrability of Field Theories. Universitätsverlag Karlsruhe, Karlsruhe *(622, 625)*

66. Cantrijn F, de León M, Lacomba E (1992) Gradient vector fields on cosymplectic manifolds. J Phys A: Math Gen 25:175–188 *(614)*
67. Cantrijn F, Ibort A, de León M (1996) Hamiltonian structures on multisymplectic manifolds. Rend Sem Mat Univ Pol Torino 54:225–236 *(61)*
68. Carathéodory C (1956) Variationsrechnung und Partielle Differentialgleichungen erster Ordnung, Band I. Teubner, Leipzig *(38, 292)*
69. Carrà Ferro G (1987) Gröbner bases and differential algebra. In: Huguet L, Poli A (eds) Proc. AAECC-5, Lecture Notes in Computer Science 350, Springer-Verlag, Berlin, pp 129–140 *(326, 543)*
70. Carrà Ferro G (2007) A survey on differential Gröbner bases. In: Rosenkranz M, Wang D (eds) Gröbner Bases in Symbolic Analysis, Radon Series on Computation and Applied Mathematics 2, Walter de Gruyter, Berlin, pp 43–73 *(326)*
71. Cartan E (1901) Sur l'intégration des systèmes d'équations aux différentielles totales. Ann École Normale 18:241–311 *(426)*
72. Cartan E (1915) Sur l'intégration de certains systèmes indéterminés d'équations différentielles. J Reine Angew Math 145:86–91, (also: Œuvres Complètes d'Élie Cartan, Partie II, Vol. 2, pp. 1169–1174) *(339)*
73. Cartan E (1930) La Théorie des Groupes Finis et Continus et l'Analysis Situs. Mém. Sci. Math. 42, Gauthier-Villars, Paris, (also: Œuvres Complètes d'Élie Cartan, Vol. 1, pp. 1165–1225)) *(610)*
74. Cartan E (1945) Les Systèmes Différentielles Extérieurs et leurs Applications Géométriques. Hermann, Paris *(60, 323, 327)*
75. Cartan E, Einstein A (1979) Lettres sur la Parallélisme Absolue 1929–1932, edited by R. Debever. Palais des Académies, Bruxelles *(354)*
76. Cauchy A (1842) Mémoire sur l'application du calcul des limites à l'intégration d'un système d' équations aux dérivées partielles. Compt Rend Acad Sci 15:85, (also: Œuvres Complètes d'Augustin Cauchy, Série I, Tome VII, pp. 33–49) *(427)*
77. Cauchy A (1842) Mémoire sur l'emploi du calcul des limites dans l'intégration des équations aux dérivées partielles. Compt Rend Acad Sci 15:44, (also: Œuvres Complètes d'Augustin Cauchy, Série I, Tome VII, pp. 17–33) *(427)*
78. Cauchy A (1842) Mémoire sur les systèmes d'équations aux dérivées partielles d'ordres quelconques, et sur leur réduction à des systèmes d'équations linéaires du premier ordre. Compt Rend Acad Sci 15:131, (also: Œuvres Complètes d'Augustin Cauchy, Série I, Tome VII, pp. 52–58) *(427)*
79. Caviglia G, Sbarra E (2005) Characteristic-free bounds for the Castelnuovo-Mumford regularity. Compos Math 141:1365–1373 *(230, 233)*
80. Chen Y, Gao X (2001) Involutive directions and new involutive divisions. Comp Math Appl 41:945–956 *(162)*
81. Chern S, Chen W, Lam K (1999) Lectures on Differential Geometry. Series on University Mathematics 1, World Scientific, Singapore *(585, 597)*
82. Chevalley C (1946) Theory of Lie Groups I. Princeton University Press, Princeton *(602)*
83. Chinea D, de León M, Marrero J (1994) The constraint algorithm for time-dependent Lagrangians. J Math Phys 35:3410–3447 *(304)*
84. Chrastina J (1998) The Formal Theory of Differential Equations. Folia Mathematica 6, Masaryk University, Brno *(61, 260)*
85. Chua L (1969) Introduction to Nonlinear Network Theory. McGraw-Hill, New York *(426)*
86. Chua L, Deng A (1989) Impasse points. Part I: Numerical aspects. Int J Circuit Theory Appl 17:213–235 *(368, 426)*
87. Chua L, Deng A (1989) Impasse points. Part II: Analytical aspects. Int J Circuit Theory Appl 17:271–282 *(368, 426)*
88. Chyzak F, Quadrat A, Robertz D (2005) Effective algorithms for parametrizing linear control systems over Ore algebras. Appl Alg Eng Comm Comp 16:319–376 *(505, 506)*
89. Cimpoeaş M (2006) Monomial ideals with linear upper bound regularity. Preprint math.AC/0611064 *(233)*

90. Cimpoeaş M (2008) A stable property of Borel type ideals. Commun Alg 36:674–677 *(233)*
91. Cimpoeaş M (2009) Some remarks on Borel type ideals. Commun Alg 37:724–727 *(233)*
92. Clairaut A (1734) Solution de plusieurs problèmes où il s'agit de trouver des courbes dont la propriété consiste dans une certaine relation entre leurs branches, exprimée par une équation donnée. Mémoires Math Phys Acad Roy Sciences pp 196–215 *(330, 354)*
93. Clebsch A (1866) Über die simultane Integration linearer partieller Differentialgleichungen. J Reine Angew Math 65:257–268 *(294)*
94. Cohn P (1977) Algebra II. John Wiley, London *(532)*
95. Copson E (1975) Partial Differential Equations. Cambridge University Press, Cambridge *(373)*
96. Cosner C (1991) On the definition of ellipticity for systems of partial differential equations. J Math Anal Appl 158:80–93 *(504)*
97. Coutinho S (1995) A Primer of Algebraic D-Modules. London Mathematical Society Student Texts 33, Cambridge University Press *(82)*
98. Cox D, Little J, O'Shea D (1992) Ideals, Varieties, and Algorithms. Undergraduate Texts in Mathematics, Springer-Verlag, New York *(228, 526, 529)*
99. Cox D, Little J, O'Shea D (1998) Using Algebraic Geometry. Graduate Texts in Mathematics 185, Springer-Verlag, New York *(165, 529, 582)*
100. Damiano A, Sabadini I, Struppa D (2007) Computational methods for the construction of a class of Noetherian operators. Exp Math 16:41–53 *(506)*
101. Damiano A, Struppa D, Vajiac A, Vajiac M (2009) Hartogs phenomena and antisyzygies for systems of differential equations. J Geom Anal 19:288–300 *(505)*
102. Dautray R, Lions J (1988) Mathematical Analysis and Numerical Methods for Science and Technology 2: Functional and Variational Methods. Springer-Verlag, Heidelberg *(440)*
103. Decker W, Greuel G, Pfister G (1999) Primary decomposition: Algorithms and comparisons. In: Matzat B, Greuel G, Hiss G (eds) Algorithmic Algebra and Number Theory, Springer-Verlag, Heidelberg, pp 187–220 *(539)*
104. Derksen H, Kemper G (2002) Computational Invariant Theory. Encyclopedia of Mathematical Sciences 130, Springer-Verlag, Berlin *(556)*
105. Deser S (1991) Gravity and gauge theories in three dimensions. In: Nath P, Reucroft S (eds) Particles, Strings and Cosmology, World Scientific, River Edge, pp 560–572 *(53)*
106. Dickson L (1913) Finiteness of the odd perfect and primitive abundant numbers with n distinct prime numbers. Amer J Math 35:413–426 *(512)*
107. Dirac P (1950) Generalized Hamiltonian dynamics. Can J Math 2:129–148 *(302)*
108. Dirac P (1958) Generalized Hamiltonian dynamics. Proc Roy Soc A 246:326–332 *(302)*
109. Dirac P (1964) Lectures on Quantum Mechanics. Belfer Graduate School Monograph Series 3, Yeshiva University, New York *(302)*
110. Dito G, Sternheimer D (2002) Deformation quantization: Genesis, developments and metamorphoses. In: Halbout G (ed) Deformation Quantization, IRMA Lectures in Mathematical and Theoretical Physics 1, Walter de Gruyter, Berlin, pp 9–54 *(82)*
111. Douglis A, Nirenberg L (1955) Interior estimates for elliptic systems of partial differential equations. Comm Pure Appl Math 8:503–538 *(442, 504)*
112. Drach J (1897) Sur les systèmes complètement orthogonaux dans l'espace à n dimensions et sur la réduction des systèmes différentielles les plus généraux. Compt Rend Acad Sci 125:598–601 *(522)*
113. Drinfeld V (1985) Hopf algebras and the quantum Yang-Baxter equations. Sov Math Dokl 32:254–258 *(83)*
114. Dubé T (1990) The structure of polynomial ideals and Gröbner bases. SIAM J Comp 19: 750–773 *(228)*
115. Dubois-Violette M (1984) The theory of overdetermined linear systems and its applications to non-linear field equations. J Geom Phys 1:139–172 *(260)*
116. Dudnikov P, Samborski S (1996) Linear overdetermined systems of partial differential equations. Initial and initial-boundary value problems. In: Shubin M (ed) Partial Differen-

tial Equations VIII, Encyclopaedia of Mathematical Sciences 65, Springer-Verlag, Berlin, pp 1–86 *(504)*

117. Ehrenpreis L (1961) A fundamental principle for systems of linear differential equations with constant coefficients and some of its applications. In: Proc. Intl. Symp. Linear Spaces, Jerusalem Academic Press, Jerusalem, pp 161–174 *(505)*
118. Ehrenpreis L (1970) Fourier Analysis in Several Complex Variables. Pure and Applied Mathematics 17, Wiley, New York *(466, 506)*
119. Ehresmann C (1950) Les connections infinitésimales dans un espace fibré différentiable. In: Colloque de Topologie (Espaces Fibrés), Thone, Liège, pp 29–55 *(605)*
120. Ehresmann C (1951) Les prolongements d'une variété différentiable. Comp Rend Acad Sci 233:598–600, 777–779, 1081–1083 *(58)*
121. Ehresmann C (1951) Les prolongements d'une variété différentiable. Comp Rend Acad Sci 234:1028–1030, 1424–1425 *(58)*
122. Ehresmann C (1953) Introduction à la théorie des structures infinitésimals es des pseudo-groupes de Lie. In: Géométrie Différentielle, Colloq. Intern. Centre Nat. Rech. Sci. 52, CNRS, Paris, pp 97–110 *(58)*
123. Eilenberg S, Moore J (1966) Homology and fibrations I: Coalgebras, cotensor product and its derived functors. Comment Math Helv 40:199–236 *(565)*
124. Einstein A (1956) The Meaning of Relativity, 5th edn. Princeton University Press, Princeton *(352, 354, 355)*
125. Eisenbud D (1995) Commutative Algebra with a View Toward Algebraic Geometry. Graduate Texts in Mathematics 150, Springer-Verlag, New York *(212, 232, 261, 529, 538, 539, 542, 556, 583, 584)*
126. Eisenbud D, Goto S (1984) Linear free resolutions and minimal multiplicity. J Alg 88:89–133 *(232)*
127. Eisenbud D, Reeves A, Totaro B (1994) Initial ideals, Veronese subrings and rates of algebras. Adv Math 109:168–187 *(232, 233)*
128. Eliahou S, Kervaire M (1990) Minimal resolutions of some monomial ideals. J Alg 129:1–25 *(230)*
129. Evans G (2005) Noncommutative involutive bases. PhD thesis, University of Wales, Bangor *(103)*
130. Fackerell E (1985) Isovectors and prolongation structures by Vessiot's vector field formulation of partial differential equations. In: Martini R (ed) Geometric Aspects of the Einstein Equations and Integrable Systems, Lecture Notes in Physics 239, Springer-Verlag, Berlin, pp 303–321 *(426, 427)*
131. Fesser D (2008) On Vessiot's theory of partial differential equations. PhD thesis, Fachbereich Mathematik, Universität Kassel *(290, 422, 423, 426)*
132. Fesser D, Seiler W (2006) Vessiot connections of partial differential equations. In: [65], pp 111–134 *(426)*
133. Fesser D, Seiler W (2009) Existence and construction of Vessiot connections. SIGMA 5:092 *(426)*
134. Fesser D, Saller D, Seiler W (2002) A differential equations approach to Hamiltonian systems. Rep Math Phys 49:345–359 *(56, 61, 324, 325)*
135. Finzi A (1947) Sur les Systèmes d'Équations aux Derivées Partielles qui, comme les Systèmes Normaux, Comportent Autant d'Équations que de Fonctions Inconnues. Proc Kon Neder Akad Wetenschappen 50:136–142, 143–150, 288–297, 351–356 *(295)*
136. Fordy A (1990) Prolongation structures of nonlinear evolution equations. In: Fordy A (ed) Soliton Theory: A Survey of Results, Manchester University Press, Manchester, chap 15, pp 403–426 *(427)*
137. Forsyth A (1890) Theory of Differential Equations I: Exact Equations and Pfaff's Problem. Cambridge University Press, Cambridge *(294)*
138. Forsyth A (1900) Theory of Differential Equations II: Ordinary Equations, Not Linear. Cambridge University Press, Cambridge *(353)*

139. Forsyth A (1900) Theory of Differential Equations III: Ordinary Equations, Not Linear. Cambridge University Press, Cambridge *(333)*
140. Forsyth A (1906) Theory of Differential Equations V: Partial Differential Equations. Cambridge University Press, Cambridge *(353)*
141. Forsyth A (1906) Theory of Differential Equations VI: Partial Differential Equations. Cambridge University Press, Cambridge *(353)*
142. Frobenius G (1877) Über das Pfaffsche Problem. J Reine Angew Math 82:230–315 *(602)*
143. Fröberg R (1979) Some complex constructions with applications to Poincaré series. In: Séminaire d'Algèbre Paul Dubreil, 31ème année (Paris, 1977–1978), Lecture Notes in Mathematics 740, Springer-Verlag, pp 272–284 *(231, 232)*
144. Fröhler S, Oberst U (1998) Continuous time-varying linear systems. Syst Control Lett 35: 97–110 *(505)*
145. Fuhrmann P (2002) A study of behaviors. Lin Alg Appl 351/352:303–380 *(505)*
146. Galligo A (1974) A propos du théorème de préparation de Weierstrass. In: Norguet F (ed) Fonctions de Plusieurs Variables Complexes, Lecture Notes in Mathematics 409, Springer-Verlag, Berlin, pp 543–579 *(212)*
147. Gantmacher F (1998) The Theory of Matrices II. AMS Chelsea, Providence *(425)*
148. García-Román M, García-Román S (2005) Gröbner bases and syzygies on bimodules over PBW algebras. J Symb Comp 40:1039–1052 *(165)*
149. Gardner R, Shadwick W (1987) A simple characterisation of the contact system on $J^k(E)$. Rocky Mountain J Math 17:19–21 *(27)*
150. Garey M, Johnson D (1977) The rectilinear Steiner problem is NP-complete. SIAM J Appl Math 32:826–834 *(515)*
151. Gatermann K (2000) Computer Algebra Methods for Equivariant Dynamical Systems. Lecture Notes in Mathematics 1728, Springer-Verlag, Berlin *(228)*
152. Gelfand I, Zelevinsky A, Kapranov M (1989) Hypergeometric functions and toral manifolds. Funct Anal Appl 23:94–106 *(147)*
153. Gerdt V (1999) Completion of linear differential systems to involution. In: [161], pp 115–137 *(229)*
154. Gerdt V (2000) On the relation between Pommaret and Janet bases. In: Ghanza V, Mayr E, Vorozhtsov E (eds) Computer Algebra in Scientific Computing — CASC 2000, Springer-Verlag, Berlin, pp 167–182 *(164)*
155. Gerdt V (2002) On an algorithmic optimization in computation of involutive bases. Prog Comp Softw 28:62–65 *(162)*
156. Gerdt V, Blinkov Y (1998) Involutive bases of polynomial ideals. Math Comp Simul 45: 519–542 *(100, 101, 102, 106, 161, 162)*
157. Gerdt V, Blinkov Y (1998) Minimal involutive bases. Math Comp Simul 45:543–560 *(162)*
158. Gerdt V, Gogilidze S (1999) Constrained Hamiltonian systems and Gröbner bases. In: [161], pp 139–146 *(301)*
159. Gerdt V, Blinkov Y, Yanovich D (2001) Construction of Janet bases I: Monomial bases. In: [162], pp 233–247 *(101)*
160. Gerdt V, Blinkov Y, Yanovich D (2001) Construction of Janet bases II: Polynomial bases. In: [162], pp 249–263 *(162)*
161. Ghanza V, Mayr E, Vorozhtsov E (eds) (1999) Computer Algebra in Scientific Computing — CASC '99. Springer-Verlag, Berlin *(623, 633)*
162. Ghanza V, Mayr E, Vorozhtsov E (eds) (2001) Computer Algebra in Scientific Computing — CASC 2001. Springer-Verlag, Berlin *(618, 623)*
163. Giachetta G, Mangiarotti L, Sardanashvily G (1997) New Lagrangian and Hamiltonian Methods in Field Theory. World Scientific, Singapore *(59)*
164. Gianni P, Trager B, Zacharias G (1988) Gröbner bases and primary decomposition of polynomial ideals. J Symb Comp 6:149–167 *(165)*
165. Giesbrecht M, Reid G, Zhang Y (2002) Non-commutative Gröbner bases in Poincaré-Birkhoff-Witt extensions. In: Ghanza V, Mayr E, Vorozhtsov E (eds) Computer Algebra

in Scientific Computing — CASC 2002, Fakultät für Informatik, Technische Universität München *(102, 165)*

166. Giovani A, Mora T, Niesi G, Robbiano L, Traverso C (1991) "One sugar cube, please" or selection strategies in the Buchberger algorithm. In: Watt S (ed) Proc. ISSAC '91, ACM Press, New York, pp 49–54 *(576)*
167. Goldschmidt H (1965) Existence theorems for analytic linear partial differential equations. Ann Math 82:246–270 *(260)*
168. Goldschmidt H (1969) Integrability criteria for systems of non-linear partial differential equations. J Diff Geom 1:269–307 *(260, 323, 324)*
169. Goldschmidt H, Sternberg S (1973) The Hamilton-Cartan formalism in the calculus of variations. Ann Inst Fourier 23:203–267 *(28)*
170. Golubitsky M, Guillemin V (1973) Stable Mappings and Their Singularities. Graduate Texts in Mathematics 14, Springer-Verlag, New York *(58, 588)*
171. Goodearl K, RB Warfield J (2004) An Introduction to Noncommutative Noetherian Rings, 2nd edn. London Mathematical Society Student Texts 61, Cambridge University Press, Cambridge *(529)*
172. Gordan P (1900) Les Invariants des Formes Binaires. J Math Pure Appl 6:141–156 *(104, 567)*
173. Gotay M, Nester J, Hinds G (1978) Presymplectic manifolds and the Dirac-Bergmann theory of constraints. J Math Phys 19:2388–2399 *(302, 303)*
174. Gotay M, Isenberg J, Marsden J (1998) Momentum maps and classical relativistic fields I: Covariant field theory. Preprint physics/9801019 *(59, 61)*
175. Goursat E (1942) Cours d'Analyse Mathématique, vol II. Gauthier-Villars, Paris *(427)*
176. Gräbe HG (1994) The tangent cone algorithm and homogenization. J Pure Appl Alg 97:303–312 *(165)*
177. Gräbe HG (1995) Algorithms in local algebra. J Symb Comp 19:545–557 *(165)*
178. Graham R, Knuth D, Patashnik O (1989) Concrete Mathematics. Addison-Wesley, Reading *(525, 526)*
179. Grauert H (1972) Über die Deformation isolierter Singularitäten analytischer Mengen. Invent Math 15:171–198 *(568)*
180. Green E (1994) An introduction to noncommutative Gröbner bases. In: Fischer K, Loustaunau P, Shapiro J, Green E, Farkas D (eds) Computational Algebra, Lecture Notes in Pure and Applied Algebra 151, Marcel Dekker, pp 167–190 *(103)*
181. Green M (1984) Koszul cohomology and the geometry of projective varieties. J Diff Geom 19:125–171 *(261)*
182. Green M (1998) Generic initial ideals. In: Elias J, Giral J, Miró-Roig R, Zarzuela S (eds) Six Lectures in Commutative Algebra, Progress in Mathematics 166, Birkhäuser, Basel, pp 119–186 *(230)*
183. Greub W, Halperin S, Vanstone R (1973) Connections, Curvature and Cohomology II. Pure and Applied Mathematics 47-II, Academic Press, New York *(605)*
184. Greuel GM, Pfister G (1996) Advances and improvements in the theory of standard bases and syzygies. Arch Math 66:163–176 *(165)*
185. Greuel GM, Pfister G (2002) A SINGULAR Introduction to Commutative Algebra. Springer-Verlag, Berlin *(165, 188, 230, 529, 541, 542)*
186. Gröbner W (1938) Über eine neue idealtheoretische Grundlegung der Algebraischen Geometrie. Math Ann 115:333–358 *(229)*
187. Gröbner W (1939) Über die algebraischen Eigenschaften der Integrale von linearen Differentialgleichungen mit konstanten Koeffizienten. Monatsh Math Phys 47:247–284 *(507, 567)*
188. Gröbner W (1949) Moderne Algebraische Geometrie. Springer-Verlag, Wien *(229)*
189. Guillemin V, Sternberg S (1964) An algebraic model of transitive differential geometry. Bull Amer Math Soc 70:16–47 *(232, 262)*
190. Hadamard J (1952) Lectures on Cauchy's Problem in Linear Partial Differential Equations. Dover, New York *(375)*

191. Haggmann B, Bryant P (1984) Solutions of singular constrained differential equations: A generalization of circuits containing capacitor-only loops and inductor-only cutsets. IEEE Trans Circ Syst 31:1015–1029 *(426)*
192. Hartley D (1997) EDS: A REDUCE package for exterior differential systems. Comp Phys Comm 100:177 *(327)*
193. Hartley D (1997) Involution analysis for non-linear exterior differential systems. Math Comp Model 25:51–62 *(327)*
194. Hartley D, Tucker R (1991) A constructive implementation of the Cartan-Kähler theory of exterior differential systems. J Symb Comp 12:655–667 *(164, 327)*
195. Hashemi A (2007) Polynomial-time algorithm for Hilbert series of Borel type ideals. Albanian J Math 1:145–155 *(233)*
196. Hashemi A (2008) Efficient algorithm for computing Noether normalization. In: Kapur D (ed) Computer Mathematics (ASCM 2007), Lecture Notes in Computer Science 5081, Springer-Verlag, Berlin, pp 97–107 *(230)*
197. Hausdorf M (2000) Geometrisch-Algebraische Vervollständigung allgemeiner Systeme von Differentialgleichungen. Diploma Thesis, Fakultät für Informatik, Universität Karlsruhe *(507)*
198. Hausdorf M, Seiler W (2002) An efficient algebraic algorithm for the geometric completion to involution. Appl Alg Eng Comm Comp 13:163–207 *(164, 507)*
199. Hausdorf M, Seiler W, Steinwandt R (2002) Involutive bases in the Weyl algebra. J Symb Comp 34:181–198 *(165)*
200. Hausdorf M, Sahbi M, Seiler W (2006) δ- and quasi-regularity for polynomial ideals. In: [65], pp 179–200 *(230)*
201. Henneaux M, Teitelboim C (1992) Quantization of Gauge Systems. Princeton University Press *(325)*
202. Hereman W (1994) Review of symbolic software for the computation of Lie symmetries of differential equations. Euromath Bull 2:45–82 *(326)*
203. Hereman W (1995) Symbolic software for Lie symmetry analysis. In: Ibragimov N (ed) CRC Handbook of Lie Group Analysis of Differential Equations, Volume 3: New Trends in Theoretical Development and Computational Methods, CRC Press, Boca Raton, Florida, chap 13 *(326)*
204. Hermann G (1926) Die Frage der endlich vielen Schritte in der Theorie der Polynomideale. Math Ann 95:736–788 *(233)*
205. Hermann R (1962) The differential geometry of foliations II. J Math Mech 11:303–315 *(602)*
206. Hermann R (1964) Cartan connections and the equivalence problem for geometric structures. Contrib Diff Eq 3:199–248 *(602)*
207. Hermann R (1975) Gauge Fields and Cartan-Ehresmann Connections, Part A. Interdisciplinary Mathematics X, Math Sci Press, Brookline *(605)*
208. Hermann R (1976) The Geometry of Non-Linear Differential Equations, Bäcklund Transformations, and Solitons, Part A. Interdisciplinary Mathematics XII, Math Sci Press, Brookline *(333, 353)*
209. Hermann R (1976) Pseudopotentials of Estabrook and Wahlquist, the geometry of solitons, and the theory of connections. Phys Rev Lett 36:835–836 *(427)*
210. Herzog J, Hibi T (1999) Componentwise linear ideals. Nagoya Math J 153:141–153 *(231, 558)*
211. Herzog J, Popescu D, Vladoiu M (2003) On the Ext-modules of ideals of Borel type. In: Commutative Algebra, Contemp. Math. 331, Amer. Math. Soc., Providence, pp 171–186 *(230)*
212. Herzog J, Soleyman Jahan A, Yassemi S (2008) Stanley decompositions and partionable simplicial complexes. J Algebr Comb 27:113–125 *(229)*
213. Hilbert D (1890) Über die Theorie der algebraischen Formen. Math Ann 36:473–534, (also: Gesammelte Abhandlungen, Band II, pp. 199–257) *(104, 541)*
214. Hilbert D (1893) Über die vollen Invariantensysteme. Math Ann 42:313–373, (also: Gesammelte Abhandlungen, Band II, pp. 287–344) *(104)*

215. Hilbert D (1912) Über den Begriff der Klasse von Differentialgleichungen. Math Ann 73:95–108, (also: Gesammelte Abhandlungen, Band III, pp. 81–93) *(339)*
216. Hile G, Protter M (1977) Properties of overdetermined first order elliptic systems. Arch Rat Mech Anal 66:267–293 *(504)*
217. Hillion P (1997) Beware of Maxwell's divergence equations. J Comp Phys 132:154–155 *(458)*
218. Hironaka H (1964) Resolution of singularities of an algebraic variety over a field of characteristic zero. Ann Math 79:109–326 *(102, 568)*
219. Hoşten S, Smith G (2002) Monomial ideals. In: Eisenbud D, Grayson D, Stillman M, Sturmfels B (eds) Computations in Algebraic Geometry with Macaulay 2, Algorithms and Computation in Mathematics 8, Springer-Verlag, Berlin, pp 73–100 *(229)*
220. Hoenselaers C (1977) The strength of a system of differential equations. Prog Theor Phys 58:1185–1190 *(354)*
221. Holmgren E (1901) Über Systeme von linearen, partiellen Differentialgleichungen. Ofversigt af kongl Vetenskapsakad Förh 58:91–103 *(504)*
222. Hörmander L (1966) An Introduction to Complex Analysis in Several Variables. North-Holland, Amsterdam *(506)*
223. Hörmander L (1969) Linear Partial Differential Operators. Grundlehren der mathematischen Wissenschaften 116, Springer-Verlag, Berlin *(372)*
224. Hörmander L (1983) The Analysis of Linear Partial Differential Operators I. Grundlehren der mathematischen Wissenschaften 256, Springer-Verlag, Berlin *(440, 504)*
225. Hörmander L (1983) The Analysis of Linear Partial Differential Operators II. Grundlehren der mathematischen Wissenschaften 257, Springer-Verlag, Berlin *(440, 494)*
226. Hubert E (1996) The general solution of an ordinary differential equation. In: Lakshman Y (ed) Proc. ISSAC '96, ACM Press, New York, pp 189–195 *(353)*
227. Hubert E (1997) Detecting degenerate behaviors in first order algebraic differential equations. Theor Comp Sci 187:7–25 *(353)*
228. Hubert E (2003) Notes on triangular sets and triangulation-decomposition algorithms. II: Differential systems. In: Winkler F, Langer U (eds) Symbolic and Numerical Scientific Computation, Lecture Notes in Computer Science 2630, Springer-Verlag, Berlin, pp 40–87 *(327)*
229. Hyman J, Shashkov M (1999) Mimetic discretizations for Maxwell's equations. J Comp Phys 151:881–909 *(458)*
230. Ince E (1956) Ordinary Differential Equations. Dover, New York *(353)*
231. Ivey T, Landsberg J (2003) Cartan for Beginners: Differential Geometry via Moving Frames and Exterior Differential Systems. Graduate Studies in Mathematics 61, American Mathematical Society, Providence *(60, 323, 409)*
232. Jacobson N (1989) Basic Algebra II, 2nd edn. Freeman, New York *(559)*
233. Janet M (1920) Sur les systèmes d'équations aux dérivées partielles. J Math Pure Appl 3:65–151 *(101, 228, 567)*
234. Janet M (1924) Les modules de formes algébriques et la théorie générale des systèmes différentiels. Ann École Norm Sup 41:27–65 *(101, 262)*
235. Janet M (1929) Leçons sur les Systèmes d'Équations aux Dérivées Partielles. Cahiers Scientifiques, Fascicule IV, Gauthier-Villars, Paris *(100, 228, 229, 307, 428)*
236. Jiang B, Wu J, Povelli L (1996) The origin of spurious solutions in computational electrodynamics. J Comp Phys 125:104–123 *(458)*
237. Jimbo M (1985) A q-difference analogue of $U(\mathfrak{g})$ and the Yang-Baxter equations. Lett Math Phys 10:63–69 *(83)*
238. John F (1982) Partial Differential Equations. Applied Mathematical Sciences 1, Springer-Verlag, New York *(372, 409, 427, 504)*
239. Johnson H (1962) Classical differential invariants and applications to partial differential equations. Math Ann 148:308–329 *(59)*
240. Johnson H (1964) Bracket and exponential for a new type of vector field. Trans Amer Math Soc 15:432–437 *(15)*

241. Johnson H (1964) A new type of vector field and invariant differential systems. Trans Amer Math Soc 15:675–678 *(15)*
242. Johnson H (1965) Algebraic aspects of prolongations. Proc Amer Math Soc 16:123–125 *(261)*
243. Johnson H (1966) An algebraic approach to exterior differential systems. Pacific J Math 17:423–434 *(261)*
244. Johnson H (1967) The non-invariance of hyperbolicity in partial differential equations. Pacific J Math 22:419–430 *(504)*
245. Johnson H (1969) Determination of hyperbolicity by partial prolongation. Pacific J Math 30:679–695 *(504)*
246. Johnson J (1971) On Spencer's cohomology theory for linear partial differential equations. Trans Amer Math Soc 154:137–149 *(260)*
247. Johnston M, Rogers C, Schief W, Seiler W (1995) On moving pseudo-spherical surfaces: A generalized Weingarten system and its formal analysis. Lie Groups Appl 1:124–136 *(321)*
248. Kähler E (1934) Einführung in die Theorie der Systeme von Differentialgleichungen. Teubner, Leipzig *(60, 323, 327)*
249. Kandry-Rody A, Weispfenning V (1990) Non-commutative Gröbner bases in algebras of solvable type. J Symb Comp 9:1–26 *(81, 102, 165)*
250. Kaneko A (1988) Introduction to Hyperfunctions. Mathematics and its Applications, Kluwer, Dordrecht *(460)*
251. Kaplansky I (1957) An Introduction to Differential Algebra. Hermann, Paris *(543)*
252. Kasch F (1977) Moduln und Ringe. Teubner, Stuttgart *(549)*
253. Kashiwara M, Kawai T, Kimura T (1986) Foundations of Algebraic Analysis. Princeton University Press, Princeton *(460)*
254. Kato G, Struppa D (1999) Fundamentals of Algebraic Microlocal Analysis. Pure and Applied Mathematics 217, Marcel Dekker, New York *(460)*
255. Knuth D, Bendix P (1970) Simple word problems in universal algebras. In: Leech J (ed) Computational Problems in Abstract Algebra, Pergamon, Oxford, pp 263–297 *(575)*
256. Kobayashi S, Nomizu K (1963) Foundations of Differential Geometry I. Interscience Publisher, New York *(585, 605)*
257. Kolář I, Michor P, Slovák J (1993) Natural Operations in Differential Geometry. Springer-Verlag, Berlin *(605)*
258. Kolchin E (1973) Differential Algebra and Algebraic Groups. Academic Press, New York *(326, 337, 342, 543)*
259. Kondratieva M, Levin A, Mikhalev A, Pankratiev E (1999) Differential and Difference Dimension Polynomials. Mathematics and its Applications 461, Kluwer, Dordrecht *(337, 342)*
260. König D (1936) Theorie der endlichen und unendlichen Graphen. Akademische Verlagsgesellschaft, Leipzig *(512)*
261. Koszul J (1960) Lectures on Fibre Bundles and Differential Geometry. Springer-Verlag, Berlin *(605)*
262. Kowalevsky S (1875) Zur Theorie der partiellen Differentialgleichungen. J Reine Angew Math 80:1–32 *(372, 373, 427)*
263. Krasilshchik I, Lychagin V, Vinogradov A (1986) Geometry of Jet Spaces and Nonlinear Partial Differential Equations. Gordon & Breach, New York *(61, 260, 427)*
264. Kredel H (1993) Solvable Polynomial Rings. Verlag Shaker, Aachen *(78, 102, 103, 138, 165)*
265. Kredel H, Weispfenning V (1988) Computing dimension and independent sets for polynomial ideals. J Symb Comp 6:231–247 *(229)*
266. Kreiss HO, Lorenz J (1989) Initial-Boundary Value Problems and the Navier-Stokes Equations. Pure and Applied Mathematics 136, Academic Press, Boston *(451, 456)*
267. Kreuzer M, Robbiano L (2000) Computational Commutative Algebra 1. Springer-Verlag, Berlin *(131, 529)*
268. Kreuzer M, Robbiano L (2005) Computational Commutative Algebra 2. Springer-Verlag, Berlin *(529, 539, 540)*

269. Kruglikov B, Lychagin V (2002) Mayer brackets and solvability of PDEs I. Diff Geom Appl 17:251–272 *(38)*
270. Kruglikov B, Lychagin V (2006) Mayer brackets and solvability of PDEs II. Trans Amer Math Soc 358:1077–1103 *(38, 429)*
271. Kruglikov B, Lychagin V (2008) Geometry of differential equations. In: Krupka D, Saunders D (eds) Handbook on Global Analysis, Elsevier, Amsterdam, pp 725–771 *(427)*
272. Krupchyk K, Tuomela J (2006) The Shapiro-Lopatinskij condition for elliptic boundary value problems. LMS J Comput Math 9:287–329 *(504)*
273. Krupchyk K, Seiler W, Tuomela J (2006) Overdetermined elliptic systems. Found Comp Math 6:309–351 *(504, 582)*
274. Kuranishi M (1957) On E. Cartan's prolongation theorem of exterior differential systems. Amer J Math 79:1–47 *(325)*
275. Kuranishi M (1967) Lectures on Involutive Systems of Partial Differential Equations. Publicações da Sociedade de Mathemática de Sã Paulo, São Paolo *(58)*
276. Lam T (1999) Lectures on Modules and Rings. Graduate Texts in Mathematics 189, Springer-Verlag, New York *(529, 532)*
277. Lam T (2000) On the equality of row rank and column rank. Expos Math 18:161–163 *(473)*
278. Lambe L, Radford D (1997) Introduction to the Quantum Yang-Baxter Equation and Quantum Groups: An Algebraic Approach. Kluwer, Dordrecht *(559)*
279. Lambe L, Seiler W (2002) Differential equations, Spencer cohomology, and computing resolutions. Georg Math J 9:723–772 *(260)*
280. Landsberg J, Manivel L (2003) On the projective geometry of rational homogeneous varieties. Comment Math Helv 78:65–100 *(261)*
281. Lang S (1984) Algebra. Addison-Wesley, Reading *(529)*
282. Lang S (1999) Fundamentals of Differential Geometry. Graduate Texts in Mathematics 191, Springer-Verlag, New York *(585)*
283. Lazard D (1983) Gröbner bases, Gaussian elimination and resolution of systems of algebraic equations. In: van Hulzen J (ed) Proc. EUROCAL '83, Lecture Notes in Computer Science 162, Springer-Verlag, Berlin, pp 146–156 *(164)*
284. Lemaire F (2003) An orderly linear PDE with analytic initial conditions with a non-analytic solution. J Symb Comp 35:487–498 *(428)*
285. Levandovskyy V (2002) On Gröbner bases for non-commutative G-algebras. In: Calmet J, Hausdorf M, Seiler W (eds) Proc. Under- and Overdetermined Systems of Algebraic or Differential Equations, Fakultät für Informatik, Universität Karlsruhe, pp 99–118 *(78, 208)*
286. Levandovskyy V (2005) Non-commutative computer algebra for polynomial algebras: Gröbner bases, applications and implementation. PhD thesis, Fachbereich Mathematik, Universität Kaiserslautern *(78, 102, 103, 165, 208)*
287. Levin A (2007) Computation of the strength of systems of difference equations with constant coefficients using Gröbner bases. In: [387], pp 43–73 *(355)*
288. Lewy H (1957) An example of a smooth linear partial differential equation without solution. Ann Math 66:155–158 *(372)*
289. Leykin A (2004) Algorithmic proofs of two theorems of Stafford. J Symb Comp 38:1535–1550 *(86)*
290. Li H (2002) Noncommutative Gröbner Bases and Filtered-Graded Transfer. Lecture Notes in Mathematics 1795, Springer-Verlag, Berlin *(103)*
291. Libermann P, Marle C (1987) Symplectic Geometry and Analytical Mechanics. Reidel, Dordrecht *(614)*
292. Lie S, Engel F (1888) Theorie der Transformationsgruppen, vol 1. Teubner, Leipzig *(44)*
293. Logar A (1989) A computational proof of the Noether normalization lemma. In: Mora T (ed) Proc. AAECC-6, Lecture Notes in Computer Science 357, Springer-Verlag, Berlin, pp 259–273 *(230)*
294. Lomadze V, Zerz E (2000) Partial differential equations of Krull dimension zero. Proc. Math. Theory of Networks and Systems (MTNS 2000) *(506)*
295. Lusztig G (1990) Quantum groups at roots of 1. Geom Dedi 35:89–113 *(83)*

296. Lychagin V (1995) Homogeneous geometric structures and homogeneous differential equations. In: Lychagin V (ed) The Interplay between Differential Geometry and Differential Equations, Amer. Math. Soc. Transl. 167, Amer. Math. Soc., Providence, pp 143–164 *(58, 393, 409, 427)*
297. Lyubeznik G (1988) A new explicit finite resolution of ideals generated by monomials in an R-sequence. J Pure Appl Alg 51:193–195 *(232)*
298. Mac Lane S (1995) Homology. Classics in Mathematics, Springer-Verlag, New York *(544)*
299. Macaulay F (1927) Some properties of enumeration in the theory of modular systems. Proc London Math Soc 26:531–555 *(567, 572)*
300. Maclagan D, Smith G (2005) Uniform bounds on multigraded regularity. J Alg Geom 14:137–164 *(228, 232)*
301. Malgrange B (1964) Systèmes différentiels à coefficients constants. In: Semin. Bourbaki 15 (1962/63), 246, Paris *(505)*
302. Malgrange B (1966) Cohomologie de Spencer (d'après Quillen). Publications du Séminaire Mathématique, Orsay *(260)*
303. Malgrange B (2003) Cartan involutiveness = Mumford regularity. Contemp Math 331: 193–205 *(262)*
304. Malgrange B (2005) Systèmes Différentiels Involutifs. Panoramas et Synthèses 19, Societé Mathématique de France, Paris *(260, 327)*
305. Mall D (1998) On the relation between Gröbner and Pommaret bases. Appl Alg Eng Comm Comp 9:117–123 *(230)*
306. Mansfield E (1991) Differential Gröbner bases. PhD thesis, Macquarie University, Sydney *(261, 326)*
307. Mansfield E (1996) A simple criterion for involutivity. J London Math Soc 54:323–345 *(260)*
308. Mariwalla K (1974) Applications of the concept of strength of a system of partial differential equations. J Math Phys 15:468–473 *(339, 354)*
309. Marsden J, Ratiu T (1994) Introduction to Mechanics and Symmetry. Texts in Applied Mathematics 17, Springer-Verlag, New York *(611)*
310. Matsumoto T, Chua L, Kawakami H, Ichiraku S (1981) Geometric properties of dynamic nonlinear networks: Transversality, local-solvability and eventual passivity. IEEE Trans Circ Syst 28:406–428 *(426)*
311. Matsushima Y (1953) On a theorem concerning the prolongation of a differential system. Nagoya Math J 6:1–16 *(262)*
312. Matsushima Y (1954–55) Sur les algèbres de Lie linéaires semi-involutives. In: Colloque de Topologie de Strasbourg, Université de Strasbourg, p 17 *(262)*
313. Matthews N (1987) On the strength of Maxwell's equations. J Math Phys 28:810–814 *(354)*
314. Matthews N (1992) The strength of Einstein's equations. Gen Rel Grav 24:17–33 *(354)*
315. Mayer A (1872) Ueber unbeschränkt integrable Systeme von linearen totalen Differentialgleichungen und die simultane Integration linearer partieller Differentialgleichungen. Math Ann 5:448–470 *(294)*
316. Mayr E (1989) Membership in polynomial ideals over $\mathbb{Q}$ is exponential space complete. In: Monien B, Cori R (eds) Theoretical Aspects of Computer Science — STACS '89, Lecture Notes in Computer Science 349, Springer-Verlag, pp 400–406 *(577)*
317. Mayr E (1997) Some complexity results for polynomial ideals. J Compl 13:303–325 *(577)*
318. Mayr E, Meyer A (1982) The complexity of the word problems for commutative semigroups and polynomial ideals. Adv Math 46:305–329 *(233, 577)*
319. McConnell J, Robson J (1987) Non-commutative Noetherian Rings. Wiley *(81, 82, 84, 91, 103, 234, 475, 529, 532)*
320. Mendella G, Marmo G, Tulczyjew W (1995) Integrability of implicit differential equations. J Phys A: Math Gen 28:149–163 *(324)*
321. Méray C, Riquier C (1890) Sur la convergence des développements des intégrales ordinaires d'un système d'équations différentielles partielles. Ann Sci Ec Norm Sup 7:23–88 *(228)*
322. Miller E, Sturmfels B (2005) Combinatorial Commutative Algebra. Graduate Texts in Mathematics 227, Springer-Verlag, New York *(70, 191, 232)*

323. Mishra B (1993) Algorithmic Algebra. Texts and Monographs in Computer Science, Springer-Verlag, New York *(326)*
324. Mizohata S (1981) On the Cauchy-Kowalewski theorem. In: Nachbin L (ed) Mathematical Analysis and Applications, Part B, Advances in Mathematics Supplementary Studies 7B, Academic Press, New York, pp 617–652 *(374)*
325. Modugno M (1999) Covariant quantum mechanics, unpublished manuscript, Dept. of Mathematics, University of Florence *(59)*
326. Molino P (1982) Exterior differential systems and partial differential equations. Lecture notes, Mathematisch Instituut, Universiteit van Amsterdam *(60, 327, 427)*
327. Molino P (1983) Connexions adaptées à un système différentiel extérieur et prolongements d'Estabrook-Wahlquist. Astérisque 107/108:229–241, (Proc. IIIeme Rencontre de Géométrie du Schnepfenried) *(427)*
328. Mora T (1982) An algorithm to compute the equations of tangent cones. In: Calmet J (ed) Proc. EUROCAM '82, Lecture Notes in Computer Science 144, Springer-Verlag, Berlin, pp 158–165 *(164)*
329. Mora T (1986) Gröbner bases for non-commutative polynomial rings. In: Calmet J (ed) Proc. AAECC-3, Lecture Notes in Computer Science 223, Springer-Verlag, Berlin, pp 353–362 *(103)*
330. Mora T (1994) An introduction to commutative and noncommutative Gröbner bases. Theor Comp Sci 134:131–173 *(103)*
331. Morimoto M (1993) An Introduction to Sato's Hyperfunctions. Transl. Math. Monogr. 129, American Mathematical Society, Providence *(460)*
332. Muñoz-Lecanda M, Román-Roy N (1999) Implicit quasilinear differential systems: A geometrical approach. Electr J Diff Eqs Paper No. 10 *(324)*
333. Munz C, Omnes P, Schneider R, Sonnendrücker E, Voß U (2000) Divergence correction techniques for Maxwell solvers based on a hyperbolic model. J Comp Phys 161:484–511 *(458)*
334. Murai S, Hibi T (2008) Gotzmann ideals of the polynomial ring. Math Z 260:629–646 *(231)*
335. Murphy G (1979) On the strength of Einstein's unified field equations. Int J Theor Phys 18:323–327 *(355)*
336. Noether E, Schmeidler W (1920) Moduln in nichtkommutativen Bereichen, insbesondere aus Differential- und Differenzausdrücken. Math Zeit 8:1–35 *(79)*
337. Oberst U (1990) Multidimensional constant linear systems. Acta Appl Math 20:1–175 *(466, 505, 506)*
338. Oberst U (1996) Finite dimensional systems of partial differential or difference equations. Adv Appl Math 17:337–356 *(506)*
339. Oberst U (1999) The construction of Noetherian operators. J Alg 222:595–620 *(506)*
340. Oberst U, Pauer F (2001) The constructive solution of linear systems of partial difference and differential equations with constant coefficients. Multidim Syst Signal Proc 12:253–308 *(229, 428)*
341. Ollivier F (1990) Standard bases of differential ideals. In: Sakata S (ed) Applied Algebra, Algebraic Algorithms and Error-Correcting Codes (AAECC-8), Lecture Notes in Computer Science 508, Springer-Verlag, Berlin, pp 304–321 *(326)*
342. Olver P (1986) Applications of Lie Groups to Differential Equations. Graduate Texts in Mathematics 107, Springer-Verlag, New York *(55, 58, 59, 295, 326, 327, 407, 585, 608, 610)*
343. Ore O (1931) Linear equations in non-commutative fields. Ann Math 32:463–477 *(532)*
344. Ore O (1933) Theory of non-commutative polynomials. Ann Math 34:480–508 *(79)*
345. Ortega J, Rheinboldt W (2000) Iterative Solution of Nonlinear Equations in Several Variables. Classics in Applied Mathematics 30, SIAM, Philadelphia *(368)*
346. Osborne M (2000) Basic Homological Algebra. Graduate Texts in Mathematics 196, Springer-Verlag, New York *(544)*
347. Palais R (1965) Seminar on the Atiyah-Singer Index Theorem. Annals of Mathematical Studies 57, Princeton University Press, Princeton *(58)*

348. Palamodov V (1970) Linear Differential Operators with Constant Coefficients. Grundlehren der mathematischen Wissenschaften 168, Springer-Verlag, Berlin *(466, 505, 506)*
349. Pazy A (1983) Semigroups of Linear Operators and Applications to Partial Differential Equations. Applied Mathematical Sciences 44, Springer-Verlag, New York *(387)*
350. Penney R (1965) On the dimensionality of the real world. J Math Phys 6:1607–1611 *(355)*
351. Peradzynski Z (1989) On overdetermined hyperbolic systems. Notes Numer Fluid Mech 24:482–488 *(504)*
352. Pillai H, Shankar S (1999) A behavioral approach to control of distributed systems. SIAM J Control Optim 37:388–408 *(505)*
353. Pirani F, Robinson D, Shadwick W (1979) Local Jet Bundle Formulation of Bäcklund Transformations. Mathematical Physics Studies 1, Reidel, Dordrecht *(343)*
354. Plesken W, Robertz D (2005) Janet's approach to presentations and resolutions for polynomials and linear PDEs. Arch Math 84:22–37 *(162, 231)*
355. Polderman J, Willems J (1998) Introduction to Mathematical Systems Theory. Texts in Applied Mathematics 26, Springer-Verlag, New York *(505)*
356. Pommaret J (1978) Systems of Partial Differential Equations and Lie Pseudogroups. Gordon & Breach, London *(164, 260, 324, 327, 387)*
357. Pommaret J (1994) Partial Differential Equations and Group Theory. Kluwer, Dordrecht *(506)*
358. Pommaret J (1995) Einstein equations do not admit a generic potential. In: Janyska J, Kolár I, Slovák J (eds) Proc. 6th International Conference on Differential Geometry and Applications, Dept. of Mathematics, Masaryk University, Brno, Czech Republic, (electronically available at www.emis.de/proceedings/6ICDGA) *(470, 471)*
359. Pommaret J (2001) Partial Differential Control Theory II: Control Systems. Mathematics and Its Applications 530, Kluwer, Dordrecht *(506)*
360. Pommaret J, Quadrat A (1998) Generalized Bezout identity. Appl Alg Eng Comm Comp 9:91–116 *(506)*
361. Pommaret J, Quadrat A (1999) Algebraic analysis of linear multidimensional control systems. IMA J Math Control Inf 16:275–297 *(475)*
362. Pritchard F (2003) On implicit systems of differential equations. J Diff Equat 194:328–363 *(324)*
363. Protter M (1988) Overdetermined first order elliptic systems. In: Schäfer P (ed) Proc. Maximum Principles and Eigenvalue Problems in Partial Differential Equations, Pitman Research Notes in Mathematics 175, Longman Scientific & Technical, Harlow, pp 68–81 *(504)*
364. van der Put M, Singer M (2003) Galois Theory of Linear Differential Equations. Grundlehren der mathematischen Wissenschaften 328, Springer-Verlag, Berlin *(363)*
365. Quillen D (1964) Formal properties of over-determined systems of linear partial differential equations. PhD thesis, Harvard University, Cambridge *(260, 261, 262)*
366. Quillen D (1976) Projective modules over polynomial rings. Invent Math 36:161–171 *(548)*
367. Rabier P (1989) Implicit differential equations near a singular point. J Math Anal Appl 144:425–449 *(363, 426)*
368. Rabier P, Rheinboldt W (1994) A geometric treatment of implicit differential algebraic equations. J Diff Eq 109:110–146 *(324)*
369. Rabier P, Rheinboldt W (1994) On impasse points of quasilinear differential algebraic equations. J Math Anal Appl 181:429–454 *(368, 426)*
370. Rabier P, Rheinboldt W (1994) On the computation of impasse points of quasilinear differential algebraic equations. Math Comp 62:133–154 *(368, 426)*
371. Rees D (1956) A basis theorem for polynomial modules. Proc Cambridge Phil Soc 52:12–16 *(228)*
372. Reich S (1991) On an existence and uniqueness theory for nonlinear differential-algebraic equations. Circ Sys Sig Proc 10:343–359 *(324)*
373. Reid G (1991) Algorithms for reducing a system of PDEs to standard form, determining the dimension of its solution space and calculating its Taylor series solution. Eur J Appl Math 2:293–318 *(326, 428)*

374. Reid G, Wittkopf A, Boulton A (1996) Reduction of systems of nonlinear partial differential equations to simplified involutive forms. Eur J Appl Math 7:635–666 *(326)*
375. Reißig G (1995) Semi-implicit differential-algebraic equations constitute a normal form. IEEE Trans Circ Syst I 42:399–402 *(426)*
376. Reißig G (1996) Differential-algebraic equations and impasse points. IEEE Trans Circ Syst I 43:122–133 *(368, 426)*
377. Reißig G (1998) Beiträge zu Theorie und Anwendungen impliziter Differentialgleichungen. PhD thesis, Fakultät für Elektrotechnik, Universität Dresden *(426)*
378. Reißig G, Boche H (2003) On singularities of autonomous implicit ordinary differential equations. IEEE Trans Circ Syst I 50:922–931 *(426)*
379. Renardy M, Rogers R (1993) An Introduction to Partial Differential Equations. Texts in Applied Mathematics 13, Springer-Verlag, New York *(330, 387, 427, 504, 518)*
380. Richman F (1974) Constructive aspects of Noetherian rings. Proc Amer Math Soc 44: 436–441 *(580)*
381. Riquier C (1910) Les Systèmes d'Équations aux Derivées Partielles. Gauthier-Villars, Paris *(100, 228, 229, 427, 428, 567)*
382. Ritt J (1932) Differential Equations from the Algebraic Standpoint. Amer. Math. Soc., Providence (RI) *(428)*
383. Ritt J (1966) Differential Algebra. Dover, New York, (Original: AMS Colloquium Publications, Vol. XXXIII, 1950) *(326, 353, 543)*
384. Robbiano L (1990) Introduction to the theory of Hilbert functions. Queen's Papers Pure Appl Math 85:B1–B26, queen's Curves Seminar Vol. VII *(540)*
385. Robertz D (2009) Noether normalization guided by monomial cone decompositions. J Symb Comp 44:1359–1373 *(228, 230)*
386. Rogers C, Shadwick W (1982) Bäcklund Transformations and Their Applications. Mathematics in Science and Engineering 161, Academic Press, New York *(343)*
387. Rosenkranz M, Wang D (eds) (2007) Gröbner Bases in Symbolic Analysis. Radon Series on Computation and Applied Mathematics 2, Walter de Gruyter, Berlin *(628, 633)*
388. Roy Chowdhury A (2000) Lie Algebraic Methods in Integrable Systems. Research Notes in Mathematics 415, Chapman&Hall/CRC, Boca Raton *(427)*
389. Ruiz C (1975) L'identification fondamentale en théorie de jets. Compt Rend Acad Sci Ser A 280:1625–1627 *(58)*
390. Ruiz C (1975) Propriétés de dualité du prolongement formel des systèmes différentiels extérieur. CR Acad Sci Ser A 280:1625–1627 *(261)*
391. Saito M, Sturmfels B, Takayama N (2000) Gröbner Deformations of Hypergeometric Differential Equations. Algorithms and Computation in Mathematics 6, Springer-Verlag, Berlin *(82, 142, 148)*
392. Sarges H (1976) Ein Beweis des Hilbertschen Basissatzes. J Reine Angew Math 283/284:436–437 *(103)*
393. Saunders D (1987) Jet fields, connections and second-order differential equations. J Phys A: Math Gen 20:3261–3270 *(20)*
394. Saunders D (1989) The Geometry of Jet Bundles. London Mathematical Society Lecture Notes Series 142, Cambridge University Press, Cambridge *(58)*
395. Schreyer F (1980) Die Berechnung von Syzygien mit dem verallgemeinerten Weierstraßschen Divisionssatz. Master's thesis, Fakultät für Mathematik, Universität Hamburg *(580)*
396. Schü J, Seiler W, Calmet J (1993) Algorithmic methods for Lie pseudogroups. In: Ibragimov N, Torrisi M, Valenti A (eds) Proc. Modern Group Analysis: Advanced Analytical and Computational Methods in Mathematical Physics, Kluwer, Dordrecht, pp 337–344 *(507)*
397. Schutz B (1975) On the strength of a system of partial differential equations. J Math Phys 16:855–856 *(355)*
398. Seiler W (1994) Analysis and application of the formal theory of partial differential equations. PhD thesis, School of Physics and Materials, Lancaster University *(504)*
399. Seiler W (1994) On the arbitrariness of the general solution of an involutive partial differential equation. J Math Phys 35:486–498 *(342, 354, 355)*

400. Seiler W (1995) Applying AXIOM to partial differential equations. Internal Report 95–17, Universität Karlsruhe, Fakultät für Informatik *(507)*
401. Seiler W (1995) Arbitrariness of the general solution and symmetries. Acta Appl Math 41:311–322 *(354)*
402. Seiler W (1999) Indices and solvability for general systems of differential equations. In: [161], pp 365–385 *(324)*
403. Seiler W (1999) Numerical integration of constrained Hamiltonian systems using Dirac brackets. Math Comp 68:661–681 *(324)*
404. Seiler W (2000) Involution and constrained dynamics III: Intrinsic degrees of freedom count. Tech Mech 20:137–146 *(354)*
405. Seiler W (2002) Completion to involution and semi-discretisations. Appl Num Math 42: 437–451 *(504)*
406. Seiler W (2002) Involution — the formal theory of differential equations and its applications in computer algebra and numerical analysis. Habilitation thesis, Dept. of Mathematics and Computer Science, Universität Mannheim *(XVI, 262)*
407. Seiler W (2002) Taylor and Lyubeznik resolutions via Gröbner bases. J Symb Comp 34: 597–608 *(232)*
408. Seiler W (2007) Spencer cohomology, differential equations, and Pommaret bases. In: [387], pp 171–219 *(262)*
409. Seiler W (2009) A combinatorial approach to involution and δ-regularity I: Involutive bases in polynomial algebras of solvable type. Appl Alg Eng Comm Comp 20:207–259 *(102, 165)*
410. Seiler W (2009) A combinatorial approach to involution and δ-regularity II: Structure analysis of polynomial modules with Pommaret bases. Appl Alg Eng Comm Comp 20:261–338 *(164, 228, 230, 231, 232, 233)*
411. Seiler W, Tucker R (1995) Involution and constrained dynamics I: The Dirac approach. J Phys A: Math Gen 28:4431–4451 *(325)*
412. Seiler W, Weber A (2003) Deciding ellipticity by quantifier elimination. In: Ghanza V, Mayr E, Vorozhtsov E (eds) Computer Algebra in Scientific Computing — CASC 2003, TU München, pp 347–355 *(504)*
413. Serre JP (1955) Faisceaux algébriques cohérents. Ann Math 61:191–278 *(548)*
414. Serre JP (2000) Local Algebra. Springer-Verlag, Berlin *(262)*
415. Shafarevich I (1997) Basic Notions of Algebra. Springer-Verlag, Berlin *(529)*
416. Shinbrot M, Welland R (1976) The Cauchy-Kowalewskaya theorem. J Math Anal Appl 55:757–772 *(427)*
417. Sidman J, Sullivant S (2009) Prolongations and computational algebra. Can J Math 61:930–949 *(261)*
418. Siklos S (1996) Counting solutions of Einstein's equations. Class Quant Grav 13:1931–1948 *(355)*
419. Singer I, Sternberg S (1965) The infinite groups of Lie and Cartan I: The transitive groups. J Anal Math 15:1–114 *(261)*
420. Sit W (1992) An algorithm for solving parametric linear systems. J Symb Comp 13:353–394 *(326)*
421. Sit W (2002) The Ritt-Kolchin theory for differential polynomials. In: Guo L, Keigher W, Cassedy P, Sit W (eds) Differential Algebra and Related Topics, World Scientific, pp 1–70 *(327)*
422. Sloane N, Plouffe S (1995) The Encyclopedia of Integer Sequences. Academic Press, San Diego *(527)*
423. Smale S (1972) On the mathematical foundations of electrical circuit theory. J Diff Geom 7:193–210 *(426)*
424. Śniatycki J (1974) Dirac brackets in geometric dynamics. Ann Inst Henri Poincaré A 20: 365–372 *(324)*
425. Spear D (1977) A constructive approach to commutative ring theory. In: Proc. MACSYMA Users' Conference, NASA CP-2012, NASA, pp 369–376 *(580)*

426. Spencer D (1962) Deformation of structures on manifolds defined by transitive, continuous pseudogroups: I–II. Ann Math 76:306–445 *(260)*
427. Spencer D (1969) Overdetermined systems of linear partial differential equations. Bull Amer Math Soc 75:179–239 *(260)*
428. Spivak M (1979) A Comprehensive Introduction to Differential Geometry, vol III, 2nd edn. Publish or Perish, Berkeley *(601)*
429. Stafford J (1977) Weyl algebras are stably free. J Alg 48:297–304 *(86)*
430. Stafford J (1978) Module structure of Weyl algebras. J London Math Soc, II Ser 18:429–442 *(86)*
431. Stanley R (1978) Hilbert functions of graded algebras. Adv Math 28:57–83 *(228)*
432. Stanley R (1982) Linear diophantine equations and local cohomology. Invent Math 68: 175–193 *(228, 229)*
433. Stanley R (1983) Combinatorics and Commutative Algebra. Birkhäuser *(189)*
434. Stetter H (2004) Numerical Polynomial Algebra. SIAM, Philadelphia *(494)*
435. Stormark O (2000) Lie's Structural Approach to PDE Systems. Encyclopedia of Mathematics and its Applications 80, Cambridge University Press, Cambridge *(280, 426)*
436. Sturmfels B (1993) Algorithms in Invariant Theory. Texts and Monographs in Symbolic Computation, Springer-Verlag, Wien *(228, 230)*
437. Sturmfels B (1996) Gröbner Bases and Convex Polytopes. University Lecture Series 8, American Mathematical Society, Providence *(147)*
438. Sturmfels B (2002) Solving Polynomial Equations. CBMS, American Mathematical Society, Philadelphia *(506)*
439. Sturmfels B, White N (1991) Computing combinatorial decompositions of rings. Combinatorica 11:275–293 *(114, 164, 178, 228)*
440. Sturmfels B, Trung N, Vogel W (1995) Bounds on degrees of projective schemes. Math Ann 302:417–432 *(229)*
441. Sudarshan E, Mukunda N (1974) Classical Dynamics: A Modern Perspective. John Wiley *(325)*
442. Sué M (1991) Involutive systems of differential equations: Einstein's strength versus Cartan's degré d'arbitraire. J Math Phys 32:392–399 *(355)*
443. Sundermeyer K (1982) Constrained Dynamics. Lecture Notes in Physics 169, Springer-Verlag, New York *(325)*
444. Suslin A (1976) Projective modules over a polynomial ring are free. Soviet Math Dokl 17:1160–1164 *(548)*
445. Sussman H (1973) Orbits of families of vector fields and integrability of distributions. Trans Amer Math Soc 180:171–188 *(602)*
446. Sweedler M (1969) Hopf Algebras. Mathematics Lecture Note Series, Benjamin, New York *(559)*
447. Sweeney J (1968) The D-Neumann problem. Acta Math 120:223–277 *(261)*
448. Szatkowski A (1992) Geometric characterization of singular differential algebraic equations. Int J Sys Sci 23:167–186 *(324)*
449. Takens F (1976) Constrained equations: A study of implicit differential equations and their discontinous solutions. In: Hilton P (ed) Structural Stability, the Theory of Catastrophes, and Applications in the Sciences, Lecture Notes in Mathematics 525, Springer-Verlag, Berlin, pp 143–234 *(426)*
450. Tarkhanov N (1995) Complexes of Differential Operators. Mathematics and its Applications 340, Kluwer, Dordrecht *(504)*
451. Taylor B (1715) Methodus Incrementorum Directa et Inversa. Innys, London *(353)*
452. Taylor D (1960) Ideals generated by monomials in an *R*-sequence. PhD thesis, University of Chicago *(231)*
453. Taylor M (1996) Partial Differential Equations. Texts in Applied Mathematics 23, Springer-Verlag, New York *(322, 439, 451)*
454. Thomas J (1937) Differential Systems. Colloquium Publications XXI, American Mathematical Society, New York *(101, 228)*

455. Tikhonov A (1935) Théorèmes d'unicité pour l'équation de la chaleur. Rec Math Moscou 42:199–215 *(373, 374)*
456. Topunov V (1989) Reducing systems of linear differential equations to a passive form. Acta Appl Math 16:191–206 *(326)*
457. Traverso C (1996) Hilbert functions and the Buchberger algorithm. J Symb Comput 22: 355–376 *(162)*
458. Tresse A (1894) Sur les invariants différentiels des groupes continus de transformations. Acta Math 18:1–88 *(228)*
459. Trinks W (1978) Über B. Buchbergers Verfahren, Systeme algebraischer Gleichungen zu lösen. J Num Th 10:475–488 *(165, 572)*
460. Trung N (2002) Evaluations of initial ideals and Castelnuovo–Mumford regularity. Proc Amer Math Soc 130:1265–1274 *(232)*
461. Tsujishita T (1990) Formal geometry of systems of differential equations. Sugaku Exp 3: 25–73 *(61)*
462. Tuomela J (1997) On singular points of quasilinear differential and differential-algebraic equations. BIT 37:968–977 *(368, 426)*
463. Tuomela J (1998) On the resolution of singularities of ordinary differential equations. Num Algo 19:247–259 *(368, 426)*
464. Tuomela J, Arponen T (2000) On the numerical solution of involutive ordinary differential systems. IMA J Num Anal 20:561–599 *(426)*
465. Ufnarovski V (1998) Introduction to noncommutative Gröbner basis theory. In: Buchberger B, Winkler F (eds) Gröbner Bases and Applications, London Mathematical Society Lecture Notes Series 251, Cambridge University Press, Cambridge, pp 259–280 *(103)*
466. Vaisman I (1994) Lectures on the Geometry of Poisson Manifolds. Progress in Mathematics 118, Birkhäuser, Basel *(611)*
467. Varadarajan V (1984) Lie Groups, Lie Algebras, and Their Representations. Graduate Texts in Mathematics 102, Springer-Verlag, New York *(80, 81, 608)*
468. Vasconcelos W (1997) Computational Methods in Commutative Algebra and Algebraic Geometry. Algorithms and Computations in Mathematics 2, Springer-Verlag, Berlin *(169, 229)*
469. Vassiliou P (2001) Vessiot structure for manifolds of (p,q)-hyperbolic type: Darboux integrability and symmetry. Trans Amer Math Soc 353:1705–1739 *(426)*
470. Vessiot E (1924) Sur une théorie nouvelle des problèmes généraux d'intégration. Bull Soc Math Fr 52:336–395 *(392, 426)*
471. Wahlquist H, Estabrook F (1975) Prolongation structures of nonlinear evolution equations. J Math Phys 16:1–7 *(427)*
472. Waldmann S (2003) Deformation quantization: Observable algebras, states and representation theory. In: Dragovich B (ed) Proc. Summer School Modern Mathematical Physics, Sveske Fizičkih Nauka A; 16,1, Institute of Physics, Belgrade, pp 83–107 *(82)*
473. Walter W (1985) An elementary proof of the Cauchy-Kowalevsky theorem. Amer Math Monthly 92:115–126 *(427)*
474. Warner F (1971) Foundations of Differentiable Manifolds and Lie Groups. Scott, Foreman, Glenview, London *(585)*
475. von Weber E (1900) Partielle Differentialgleichungen. In: Enzyklopädie der mathematischen Wissenschaften, Vol. II, Part 1.1, Teubner, Leipzig, chap A5, pp 294–399 *(XVI)*
476. Weibel C (1994) An Introduction to Homological Algebra. Cambridge Studies in Advanced Mathematics 38, Cambridge University Press, Cambridge *(544)*
477. Weispfenning V (1992) Comprehensive Gröbner bases. J Symb Comp 14:1–29 *(327)*
478. Willems J (1986) From time series to linear systems I: Finite dimensional time invariant systems. Automatica 22:561–580 *(505)*
479. Willems J (1986) From time series to linear systems II: Exact modelling. Automatica 22: 675–694 *(505)*
480. Willems J (1987) From time series to linear systems III: Approximate modelling. Automatica 23:87–115 *(505)*

481. Willems J (1991) Paradigms and puzzles in the theory of dynamical systems. IEEE Trans Autom Control 36:256–294 *(505)*
482. Winkler F (1989) Knuth-Bendix procedure and Buchberger algorithm — a synthesis. In: Gonnet G (ed) Proc. ISSAC '89, ACM Press, New York, pp 55–67 *(575)*
483. Wu W (1991) On the construction of Gröbner basis of a polynomial ideal based on Riquier-Janet theory. Syst Sci Math Sci 4:194–207 *(101)*
484. Zacharias G (1978) Generalized Gröbner bases in commutative polynomial rings. Bachelor thesis, Computer Science Lab., MIT *(580)*
485. Zerz E (2000) Topics in Multidimensional Linear Systems Theory. Lecture Notes in Control and Information Sciences 256, Springer-Verlag, London *(475, 506)*
486. Zerz E (2001) Extension modules in behavioural linear systems theory. Multidim Syst Signal Proc 12:309–327 *(505)*
487. Zerz E (2006) Algebraic systems theory. Lecture Notes, Lehrstuhl D für Mathematik, RWTH Aachen *(475, 549)*
488. Zerz E, Seiler W, Hausdorf M (to appear) On the inverse syzygy problem. Comm Alg *(505, 506)*
489. Zharkov A, Blinkov Y (1993) Involution approach to solving systems of algebraic equations. In: Jacob G, Oussous N, Steinberg S (eds) Proc. Int. IMACS Symp. Symbolic Computation, Lille, pp 11–17 *(101, 162)*

Glossary

Index

Page numbers in italics indicate where a term is defined or first introduced. If several numbers are marked, then the term appears either with different meanings or in different contexts.